ec-, ex-, ecto   out, outside, away from. Example: excrete, to remove materials from the body.

en-, em-   in, inside. Example: encysted, enclosed in a cyst.

entero-   intestine. Example: *Enterobacter,* a bacterium found in the intestine.

eo-   dawn, early. Example: *Eobacterium,* a 3.4-billion-year-old fossilized bacterium.

epi-   upon, over. Example: epidemic, number of cases of a disease over the normally expected number.

erythro-   red. Example: erythema, redness of the skin.

eu-   well, proper. Example: eucaryote, a proper cell.

exo-   outside, outer layer. Example: exogenous, from outside the body.

extra-   outside, beyond. Example: extracellular, outside the cells of an organism.

firmi-   strong. Example: *Bacillus firmus* forms resistant endospores.

flagell-   a whip. Example: flagellum, a projection from a cell; in eucaryotic cells, it pulls cells in a whiplike fashion.

flav-   yellow. Example: *Flavobacterium* cells produce yellow pigment.

fruct-   fruit. Example: fructose, fruit sugar.

-fy   to make. Example: magnify, to make larger.

galacto-   milk. Example: galactose, monosaccharide from milk sugar.

gamet-   to marry. Example: gamete, a reproductive cell.

gastr-   stomach. Example: gastritis, inflammation of the stomach.

gel-   to stiffen. Example: gel, a solidified colloid.

-gen   an agent that initiates. Example: pathogen, any agent that produces disease.

-genesis   formation. Example: pathogenesis, production of disease.

germ, germin-   bud. Example: germ, part of an organism capable of developing.

-gony   reproduction. Example: schizogony, multiple fission producing many new cells.

gracili-   thin. Example: *Aquaspirillum gracile,* a thin cell.

halo-   salt. Example: halophile, an organism that can live in high salt concentrations.

haplo-   one, single. Example: haploid, half the number of chromosomes or one set.

hema-, hemato-, hemo-   blood. Example: *Hemophilus,* a bacterium that requires nutrients from red blood cells.

hepat-   liver. Example: hepatitis, inflammation of the liver.

herpes   creeping. Example: herpes, or shingles, lesions appear to creep along the skin.

hetero-   different, other. Example: heterotroph, obtains organic nutrients from other organisms; other feeder.

hist-   tissue. Example: histology, the study of tissues.

hom-, homo-   same. Example: homofermenter, an organism that produces only lactic acid from fermentation of a carbohydrate.

hydr-, hydro-   water. Example: dehydration, loss of body water.

hyper-   excess. Example: hypertonic, having a greater osmotic pressure in comparison with another.

hypo-   below, deficient. Example: hypotonic, having a lesser osmotic pressure in comparison with another.

im-   not, in. Example: impermeable, not permitting passage.

inter-   between. Example: intercellular, between the cells.

intra-   within, inside. Example: intracellular, inside the cell.

io-   violet. Example: iodine, a chemical element that produces a violet vapor.

iso-   equal, same. Example: isotonic, having the same osmotic pressure when compared with another.

-itis   inflammation of. Example: colitis, inflammation of the large intestine.

kin-   movement. Example: streptokinase, an enzyme that lyses or moves fibrin.

lacti-   milk. Example: lactose, the sugar in milk.

lepis-   scaly. Example: leprosy, disease characterized by skin lesions.

lepto-   thin. Example: *Leptospira,* thin spirochete.

leuko-   whiteness. Example: leukocyte, a white blood cell.

lip-, lipo-   fat, lipid. Example: lipase, an enzyme that breaks down fats.

-logy   the study of. Example: pathology, the study of changes in structure and function brought on by disease.

lopho-   tuft. Example: lophotrichous, having a group of flagella on one side of a cell.

luc-, luci-   light. Example: luciferin, a substance in certain organisms that emits light when acted upon by the enzyme luciferase.

lute-, luteo-   yellow. Example: *Micrococcus luteus,* yellow colonies.

-lysis   loosening, to break down. Example: hydrolysis, chemical decomposition of a compound into other compounds as a result of taking up water.

macro-   largeness. Example: macromolecules, large molecules.

mendosi-   faulty. Example: mendosicutes, archaeobacteria lacking peptidoglycan.

meningo-   membrane. Example: meningitis, inflammation of the membranes of the brain.

meso-   middle. Example: mesophile, an organism whose optimum temperature is in the middle range.

meta-   beyond, between, transition. Example: metabolism, chemical changes occurring within a living organism.

micro-   smallness. Example: microscope, an instrument used to make small objects appear larger.

Microbiology

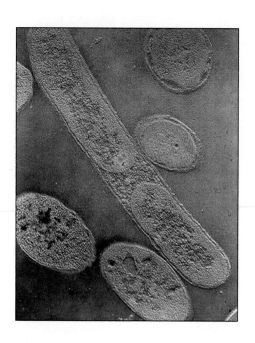

Microbiology
AN INTRODUCTION

Fifth Edition

Gerard J. Tortora
BERGEN COMMUNITY COLLEGE

Berdell R. Funke
NORTH DAKOTA STATE UNIVERSITY

Christine L. Case
SKYLINE COLLEGE

THE BENJAMIN/CUMMINGS PUBLISHING COMPANY, INC.

Redwood City, California • Menlo Park, California
Reading, Massachusetts • New York • Don Mills, Ontario
Wokingham, U.K. • Amsterdam • Bonn • Sydney
Singapore • Tokyo • Madrid • San Juan

Microbiology Editor: Anne Scanlan-Rohrer
Senior Developmental Editor: Suzanne Olivier
Assistant Editor: Leslie With
Editorial Assistants: Karmen Butterer and Hilair Chism
Production Editor: Donna Linden
Art Supervisor: Donna Kalal
Photo Editor: Kelli d'Angona-West
Photo Assistant: Lisa Lougee
Design Manager: Michele Carter
Composition and Film Manager: Lillian Hom
Composition and Film Assistant: Vivian McDougal
Manufacturing Supervisor: Merry Free Osborn
Marketing Manager: David Harris
Copy Editor: Alan Titche
Proofreaders: Margaret Moore and Rosemary Sheffield
Text Designer: Paul Uhl
Illustrations: Precision Graphics
Photo Researcher: Linda Sykes/Photo Search
Page Layout: Detta Penna
Composition/Digital Prepress: Thompson Type
Film Preparation: H&S Graphics
Printing: Rand McNally & Company
Cover Designer: Yvo Riezebos
Cover Photographs: background: © CNRI/Science Photo Library/Photo Researchers; inset: © Suzanne DeChillo/NYT Pictures

Credits appear following the Glossary.

Library of Congress Cataloging-in-Publication Data
Tortora, Gerard J.
   Microbiology : an introduction / Gerard J. Tortora, Berdell R. Funke, Christine L. Case. —5th ed.
     p. cm.
   This text has been revised.
   Includes bibliographical references and index.
   ISBN 0-8053-8496-0
   1. Microbiology. I. Funke, Berdell R. II. Case, Christine L., 1948– . III. Title.
QR41.2.T67 1994
576—dc20   94-17346

1 2 3 4 5 6 7 8 9 10—RNV—98 97 96 95 94

The Benjamin/Cummings Publishing Company, Inc.
390 Bridge Parkway

About the Authors

GERARD J. TORTORA

Jerry Tortora is a professor of biology and teaches microbiology and human anatomy and physiology at Bergen Community College in Paramus, New Jersey, where he has been a faculty member for the past 27 years. He received his M.A. in biology from Montclair State College in 1965. He belongs to numerous biology/microbiology organizations, such as the American Society of Microbiology (ASM), Human Anatomy and Physiology Society (HAPS), American Association for the Advancement of Science (AAAS), National Education Association (NEA), New Jersey Education Association (NJEA), and the Metropolitan Association of College and University Biologists (MACUB). Jerry is the author of a number of biological science textbooks.

BERDELL R. FUNKE

Bert Funke received his Ph.D., M.S., and B.S. in microbiology from Kansas State University. He has spent his professional years as a professor of microbiology at North Dakota State University. He is currently teaching introductory microbiology, including laboratory sections, general microbiology, food microbiology, soil microbiology, and clinical parasitology. He has also taught pathogenic microbiology. As a research scientist in the Experiment Station at North Dakota State, he has published numerous papers in soil microbiology and food microbiology.

CHRISTINE L. CASE

Chris Case is a registered microbiologist and a professor of microbiology at Skyline College in San Bruno, California, where she has taught for the past 24 years. She received her Ed.D. in curriculum and instruction from Nova University and her M.A. in microbiology from San Francisco State University. Presently she is an active member of the ASM, the Northern California Association of Microbiologists (NCASM), Northern California Chapter of the Society for Industrial Microbiology (NCA-SIM), and the Scientific Research Society of North America (Sigma Xi). In addition to teaching, Chris contributes regularly to the professional literature, develops innovative educational methodologies, and maintains a personal and professional commitment to conservation and the importance of science in society. Chris is also an avid photographer, and many of her photographs appear in this volume.

Preface

Microbiology: An Introduction, Fifth edition, is a comprehensive text for students in a wide variety of programs, including allied-health sciences, biological sciences, environmental studies, animal science, forestry, agriculture, home economics, and liberal arts. It is a beginning text, assuming no previous study of biology or chemistry.

During the twelve years since the publication of the first edition, this book has been used at more than 900 colleges and universities by over 535,000 students. We have been gratified to hear from instructors and students alike that the book has become a favorite among their textbooks—a learning tool that is both effective and enjoyable.

Overall Features of the Revision

Our primary goal for the fifth edition of *Microbiology* was to update the book throughout, making certain it reflects the important new discoveries of the past few years. We also wanted to give our book a fresh new look with full-color illustrations, new micrographs and photographs, and a more inviting design. At the same time, we wanted to maintain a manageable length and the emphasis on the clear coverage of fundamental principles of microbiology. Every page of the book was scrutinized with these goals in mind. The following list highlights the resulting major changes in this edition:

- One of the most noticeable changes is the use of full color in illustrations *throughout* the book to focus attention on the key elements of the discussion. All of the illustrations have been redrawn, and most of them have been rethought as well, creating an updated illustration program that fully supports and enhances the text.

- One of the features of the new art program is the more consistent use of symbols and colors. For example, molecules such as phosphate groups (P) and ATP (ATP) are the same color and shape throughout the book, allowing students to progress from the familiar parts of illustrated processes to the unfamiliar ones with confidence.

- Orientation diagrams, which are miniature versions of overview illustrations with appropriate parts highlighted, have been added to the figures in several chapters to help students relate the various steps of complex processes to the process as a whole. Each part of the process is represented by a consistent shape and color in these diagrams to help students follow pathways and see mechanisms. In addition, more overview and summary diagrams have been added.

- Most figures showing processes now include numbered steps (①, ②, ③) and descriptions next to each step in the process. These allow students to connect the explanation with the process more easily and clearly.

- The life cycle illustrations have been reworked to further clarify the scale of the organisms involved. Many life cycles now also include photographs of these organisms.

- This edition includes more figures that incorporate both illustrations and photographs, allowing students to connect concepts to the real world.

- New illustrations or photographs have been added to many tables throughout the book. These illustrated tables help students to better visualize key concepts and compare related information.

- New micrograph icons used throughout the book denote types of micrographs. As in earlier editions, included are state-of-the-art scanning electron micrographs (SEM) and transmission electron micrographs (TEM) that dramatically show microbial structures, and light micrographs (LM) that more closely resemble what is usually seen in a microbiology laboratory. Red SEM and TEM icons indicate the use of colorization in electron micrographs.

- Another notable feature of this edition is the placement of Learning Objectives throughout each chapter rather than at the beginning of each chapter. The objectives are placed at each major heading, where they will be most useful for students as they encounter new concepts.

- Critical Thinking questions that require students to apply their analytical skills to their new knowledge have replaced the Challenge questions.

- New to this edition are Clinical Applications questions, actual case histories that provide students with data to solve clinical problems. These questions require critical thinking and, in some cases, quantitative analysis.

- New Applications of Microbiology boxes appear in Chapters 1–6, 9, 18, and 28 to present students with practical applications of the concepts they are studying.

- Word Roots Used in Microbiology are now on the endpapers of the textbook, making them easily accessible. As in earlier editions, this list provides basic rules of pronunciation and some phonetic pronunciations for genus and species names used in the text.

Overall Features Retained from Earlier Editions

We have retained in this new edition the features that made the previous editions so popular. These include the following:

- **An appropriate balance between microbiological fundamentals and applications and between medical applications and other applied areas.** As in previous editions, basic principles are given greater emphasis than applications, and health-related applications are emphasized. Applications are integrated throughout the text, and considerable attention is devoted to microorganisms in habitats outside the human body. We hope that all students will gain an appreciation for the fascinating diversity of microbial life, the central roles of microorganisms in nature, and the importance of microorganisms in our daily lives.

- **Straightforward presentation of complex topics.** Each section of the text has been revised with the student in mind to maintain the clarity of explanation for which our book has been known.

- **Phonetic pronunciations.** Throughout the text, phonetic pronunciations are provided in parentheses for the majority of genus and species names for microorganisms. The pronunciations are typically provided where the terms are introduced. A comprehensive, consolidated list of all the pronunciations cited in the book follows the appendices.

- **Appendices, further readings, and glossary at the end of the book heighten its usefulness.** Appendix A is the Classification of Bacteria According to *Bergey's Manual*. Appendix B is a Most Probable Numbers (MPN) Table, and Appendix C describes Methods for Taking Clinical Samples. Appendix D, Biochemical Pathways, provides more detailed illustrations of key biochemical concepts. Appendix E provides background on Exponents, Exponential Notation, Logarithms, and Generation Time. Following the appendices are a list of Further Readings and a Glossary that provides definitions of all important terms used in the text.

Highlights of Changes in Content and Organization

- All data in the text, tables, and figures have been updated, through 1993 where possible.

Part One: Fundamentals of Microbiology

- Chapter 1, The Microbial World and You, includes new sections on the emergence of both a strain of *Escherichia coli* in undercooked meat and unpasteurized milk as a cause of diarrhea and a drug-resistant strain of *Mycobacterium tuberculosis*. There is also a new section on gene therapy, and an increased emphasis on the contributions of women in the field of microbiology.

- Chapter 8, Microbial Genetics, has been carefully revised to include updated information on molecular biology. It also includes new art featuring the use of orientation diagrams.

- Chapter 9, Recombinant DNA and Biotechnology, has been carefully revised and updated, including an expanded discussion of bioethics. Also, methods for making recombinant cells, cloning genes, and for making products are covered in the order of their use in the industry.

Part Two: Survey of the Microbial World

- Chapter 11, Bacteria, includes a new *Chlamydia* life cycle illustration, and a discussion and micrograph of the recently discovered "huge" bacterium *Epulopiscium fishelsoni*.

- Viral names adopted by the International Committee on Taxonomy of Viruses (ICTV) are introduced in Chapter 13, Viruses, and are used in conjunction with the common names throughout the book. In addition, the ICTV definition of a viral species is given. The section on oncogenic viruses has been updated to reflect new information on viruses in the genera *Hepadnavirus* and *Lymphocryptovirus*.

Part Three: Interaction Between Microbe and Host

- In Chapter 15, Microbial Mechanisms of Pathogenicity, there is an expanded section on viral mechanisms for evading host defenses. There is also new material dealing with the *Limulus* Amoebocyte Lysate (LAL) assay, and how protozoa and helminths each evade host defenses.

- Chapter 17, Specific Defenses of the Host: The Immune Response, has been reorganized to emphasize the basic concepts being presented.

- Chapter 18, Practical Applications of Immunology, includes updated information on diagnostic tests. In addition, coverage of autoimmune diseases has been incorporated into one section for ease of comparison.

- In Chapter 19, Disorders Associated with the Immune System, the discussion of AIDS has been rewritten in response to current research findings and applications. The new diagnostic classification system is introduced, and new art shows the details of HIV–T cell interactions.

Part Four: Microorganisms and Human Disease

- In Chapter 24, Microbial Diseases of the Respiratory System, a new discussion of *Hantavirus* Pulmonary Syndrome has been added. In addition, a full page of new art that illustrates the progression of tuberculosis accompanies the rewritten and updated text discussion. The chapter also includes a new illustration of the life cycle of *Pneumocystis carinii*.

Part Five: Environmental and Applied Microbiology

- Chapter 27, Soil and Water Microbiology, has new photographs of an ozone generator and the MUG test for coliforms.

- Chapter 28, Applied and Industrial Microbiology, has new art and photographs of a commercial irradiation facility, aseptic packaging machinery, and methane production from a landfill. It also includes a new discussion of biosensors.

Course Sequences

We have organized the book in what we think is a useful fashion, but we recognize that the material might be effectively presented in a number of other sequences. For those who wish to follow another sequence, we have made each chapter as independent as possible and have included numerous cross-references. Thus, Survey of the Microbial World, Part Two, could be studied at the beginning of a course, immediately after Chapter 1. Environmental and Applied Microbiology, Part Five, could follow Parts One and Two. Or, Chapter 8, Microbial Genetics, and Chapter 9, Recombinant DNA and Biotechnology, can be covered with Part Five. Because Chapters 7 and 20 both deal with the control of microbial growth, they could be covered together. The material on Microorganisms and Human Disease, Part Four, readily lends itself to rearrangement of selective coverage. The various diseases are organized into chapters according to the host organ system most affected. The *Instructor's Guide* provides detailed guidelines for organizing the disease material in several other ways.

Supplementary Materials

Study Guide

The *Study Guide*, by Berdell Funke, will help students master and review major concepts and facts from the text. Each chapter of the *Study Guide* begins with learning objectives and a chapter summary organized by the text headings. Important terms are printed in boldface type and are defined, and important figures and tables from the text are included, some with labels replaced by blanks for the students to fill in. Following the summary is an extensive self-testing section containing matching questions, fill-in questions, critical thinking questions, and diagrams. The answers to the *Study Guide* questions for each chapter appear in the back of the guide.

Lab Manual

Laboratory Experiments in Microbiology: Brief Edition (Fourth edition, 1994), by Ted Johnson and Christine Case, has been updated throughout with new exercises. This revised edition includes critical thinking questions and new color photos. The manual is organized so that lab report forms immediately follow each exercise. A detailed *Lab Manual Instructor's Guide* facilitates preparation for laboratory sessions.

Instructor's Guide

The *Instructor's Guide*, by Christine Case, includes many practical suggestions for using the text in a course. Suggested course outlines and corresponding text pages provide maximum flexibility in organizing your course. Important fifth edition changes are highlighted for each chapter to facilitate the transition from the fourth to the fifth edition. Also included are answers to study questions from the text. Special case studies of clinical issues ask students to be "medical detectives" by solving the problem presented.

MicroShow: The Videodisc

This is a double-sided, twelve-inch laserdisc containing stepped, still, and animated text art, original animations, and film sequences to accompany *Microbiology*. MicroShow is available to qualified adopters and is accompanied by a complete barcode correlation manual.

MMWR Update

The *MMWR Update* newsletter, written by Christine Case, updates the popular MMWR feature of the text. Each newsletter includes topics of interest to microbiology instructors and students from the MMWR presented in a pedagogical format that uses critical thinking, case studies, and clinical applications. The newsletter is distributed three times a year to adopters.

Answer Pamphlet

This pamphlet contains all answers to the text end-of-chapter questions in a format that can be slipped easily into the back of the instructor's text. It is also available to students with instructor approval.

Software
Benjamin/Cummings Testing Software is available for IBM PC and Macintosh personal computers.

Transparencies
Acetate overhead transparencies of 200 full-color line drawings from the text are available from the publisher to qualified adopters.

Transparency Masters
Transparency masters for all the line drawings in the text are available in a reproducible format from the publisher to qualified adopters.

Slides
Two slide sets to complement instructors' lectures are available in one package to qualified adopters. The topics of the slide sets are (1) Microbial Agents of Disease and (2) Human Immunodeficiency Virus.

Acknowledgments

In the preparation of this textbook, we have benefited from the guidance and advice of a large number of microbiology instructors across the country. David Bourgaize, of Colby College, Maine, was largely responsible for the revisions of Chapters 8 and 9 on microbial genetics and biotechnology. Many dedicated teachers participated in focus groups to help plan the revisions. Reviewers offered constructive criticism and valuable suggestions at various stages of manuscript preparation. Consultants, focus group participants, and reviewers are listed on the following pages. We gratefully acknowledge our debt to these individuals.

We offer special thanks to the staff of Benjamin/Cummings for their dedication to excellence. Anne Scanlan-Rohrer, our sponsoring editor, successfully kept us all focused on where we wanted this revision to go. Suzanne Olivier's careful attention to detail in her developmental edit of both the text and art served to keep concepts and information clear. The scientific and editorial contributions of Jane Reece to Chapter 17 were invaluable. Editorial assistants Hilair Chism and Karmen Butterer provided the help necessary to making those deadlines. Donna Linden and Donna Kalal expertly guided the text and art, respectively, through the production process. Kelli d'Angona-West, the photo editor, made sure we had clear and striking images throughout the book, while Lisa Lougee searched tirelessly for fantastic cover images. Michele Carter provided guidance in overall art and design, while Paul Uhl and Yvo Riezebos gave us beautiful and functional book and cover designs. Don Kesner at Precision Graphics brought invaluable insight and expertise to the rendering of the art. Lillian Hom and Vivian McDougal coordinated composition and film production, and Merry Free Osborn guided the book surely through the manufacturing process. In addition, Leslie With, Karmen Butterer, and Larry Olsen were responsible for the editorial management and production of the ancillaries.

Finally, we have an enduring appreciation for our students, whose comments and suggestions both provide insight and remind us of their needs. This text is for them.

GERARD J. TORTORA

BERDELL R. FUNKE

CHRISTINE L. CASE

Consultants

Wallis L. Jones, DeKalb Community College
Diana Kaftan, Diablo Valley College
Ken Keudell, Western Illinois University
Juhee Kim, California State University, Long Beach
Alan Konopka, Purdue University
Walter Koostra, University of Michigan
Robert I. Krasner, Providence College
John Lammert, Gustavus Adolphus College
John Lewis, San Bernardino Valley College
Peter Ludovici, University of Arizona
Thomas Maier, University of Colorado, Denver
Eleanor K. Marr, Dutchess Community College
William Matthai, Tarrant County Junior College
J.R. Milam, University of Florida, Gainesville
Robert Mitchell, Community College of Philadelphia
Frank Mittermeyer, Elmhurst College
Carolyn Mohr, St. Mary's College of California
Remo Morlli, San Francisco State University
Roger Nichols, Weber State College
Elinor O'Brien, Boston College
J. Dennis O'Malley, University of Arkansas
Dennis Opheim, Texas A & M University
Joel Ostroff, Brevard Community College
Helen Oujesky, University of Texas, San Antonio
Robert Pengra, South Dakota State University
Jane Phillips, University of Wisconsin, Madison
Christine Pootjes, Pennsylvania State University, University Park
Joseph R. Powell, Florida Junior College

Marian Price, Onondaga Community College
Ralph Rascati, Kennesaw College
Luis A. Rodriguez, San Antonio Community College
Robert Satterfield, College of Du Page
Violet Schirone, Suffolk County College
Robert Schoenhoff, Jefferson Community College
Michael H. Scholla, E.I. DuPont de Nemours & Company
Louis Shainberg, Mount San Antonio College
Gregory L. Shipley, Texas A & M University
Jeffrey J. Sich, Youngstown State University
Josephine Smith, Montgomery County Community College
Cynthia V. Sommer, University of Wisconsin, Milwaukee
Helen Sowers, California State University, Hayward
Lisa Steiner, Massachusetts Institute of Technology
Bernice Stewart, Prince Georges Community College
Teresa Ann Thomas, Southwestern College
Mary Lou Tortorello, Cornell University
Dan Trubovitz, Pasadena City College
Robert Twarog, University of North Carolina, Chapel Hill
James E. Urban, Kansas State University
Pat Vary, Northern Illinois University
Roberta Wald, Portland Community College
James White, Prairie State College
Mary K. Wicksten, Texas A & M University
Fred Williams, Iowa State University
Brian J. Wilkinson, Illinois State University, Normal
Hideo Yonenaka, San Francisco State University
Shanna Yonenaka, San Francisco State University

A Student's Guide to *Microbiology*

CHAPTER 5

Microbial Metabolism

N͏ow that you are familiar with the structure of procaryotic cells, we can discuss the activities that allow these microbes to thrive. The life support activity of even the most structurally simple organism involves a large number of complex biochemical reactions. Most, although not all, of the biochemical processes of bacteria also occur in eucaryotic microbes and in the cells of multicellular organisms, including humans. However, the reactions that are unique to bacteria are fascinating because they allow microorganisms to do things we cannot do. For example, some bacteria (the chemoautotrophs) can grow on diets of such inorganic substances as carbon dioxide, iron, sulfur, hydrogen gas, and ammonia.

This chapter examines some representative chemical reactions that either produce energy (the catabolic reactions) or use energy (the anabolic reactions) in microorganisms. We will also look at how these various reactions are integrated within the cell.

CATABOLIC AND ANABOLIC REACTIONS

LEARNING OBJECTIVE

- *Define metabolism, and describe the fundamental differences between anabolism and catabolism.*

We use the term **metabolism** to refer to the sum of all chemical reactions within a living organism. Because chemical reactions either release or require energy, metabolism can be viewed as an energy-balancing act. Accordingly, metabolism can be divided into two classes of chemical reactions: those that release energy and those that require energy. In living cells, the chemical reactions that release energy are generally the ones involved in **catabolism**, the brea[k]...pounds into simpler o[...]**bolic**, or *degradative,* [...] requiring reactions are [...] building of complex or [...] These reactions are ca[...] tions. Anabolic proces[...] thesis reactions (reacti[...] energy to form new ch[...] processes are the forma[...] nucleic acids from nuc[...] simple sugars. These b[...] materials for cell growt[...] hydrolytic reactions [...] which chemical bonds [...] energy; when they a[...] leased. An example of [...] down sugars into carbo[...]

Catabolic reactions [...] anabolic reactions. Thi[...] energy-releasing react[...]

Learning Objectives structure each chapter

Learning Objectives highlight the major concepts in each chapter. Introduced at major headings in each chapter, the Objectives direct you to important points you will need to master. As you can see in this example from Chapter 5 on Microbial Metabolism. The Learning Objectives tell you the most important concepts to keep in mind as you read this section on Catabolic and Anabolic Reactions.

Text and illustrations are fully integrated

Text and art have been carefully integrated to help you visualize and understand important concepts as they are presented. Taking this integration one step further, many tables, such as this one on the Effects of Oxygen on Growth of Various Types of Bacteria, now include illustrations that provide overviews of important concepts and help connect those concepts to organisms, experiments, and applications.

TABLE 6.1 Effect of Oxygen on Growth of Various Types of Bacteria

	a. OBLIGATE AEROBES	b. FACULTATIVE ANAEROBES	c. OBLIGATE ANAEROBES	d. AEROTOLERANT ANAEROBES	e. MICROAEROPHILES
Effect of oxygen on growth	Only aerobic growth; oxygen required	Both aerobic and anaerobic growth; greater growth in presence of oxygen	Only anaerobic growth; ceases in presence of oxygen	Only anaerobic growth; continues in presence of oxygen	Only aerobic growth; oxygen required in low concentrations
Bacterial growth in tube of solid growth medium					
Explanation of growth patterns	Growth occurs only where high concentrations of oxygen have diffused into the medium.	Growth is best where most oxygen is present, but occurs throughout tube.	Growth occurs only where there is no oxygen.	Growth occurs evenly; oxygen has no effect.	Growth occurs only where low concentrations of oxygen have diffused into medium.
Explanation of oxygen's effects	Presence of enzymes catalase and SOD allows toxic forms of oxygen to be neutralized; can use oxygen.	Presence of enzymes catalase and SOD allows toxic forms of oxygen to be neutralized; can use oxygen.	Lacks enzymes to neutralize harmful forms of oxygen; cannot tolerate oxygen.	Presence of one enzyme, SOD, allows harmful forms of oxygen to be partially neutralized; tolerates oxygen.	Catalase and SOD are absent or ineffective. Pathogens use hosts' enzymes.

Oxygen

We are accustomed to thinking of molecular oxygen (O_2) as a necessity of life, but it is actually a poisonous gas. Molecular oxygen did not exist in the atmosphere during most of our planet's history—in fact, it is possible that life could not have arisen had oxygen been present. However, many current forms of life have metabolic systems that require oxygen for aerobic respiration. As we have seen, hydrogen atoms that have been stripped from organic compounds combine with oxygen to form water. This process yields a great deal of energy while neutralizing a potentially toxic gas—a very neat solution, all in all.

Microbes that use molecular oxygen, called **aerobes**, produce more energy from nutrients than do microbes that do not use oxygen. Organisms that require oxygen to live are called **obligate aerobes** (Table 6.1a).

Obligate aerobes are at a disadvantage, in that oxygen is poorly soluble in the water of their environment. Therefore, many of the aerobic bacteria have developed, or retained, the ability to continue growing in the absence of oxygen. Such organisms, including some that exist in your intestinal tract, are called **facultative anaerobes** (Table 6.1b). In other words, facultative anaerobes can use oxygen when it is present but are able to continue growth by using fermentation or anaerobic respiration when oxygen is not available. However, their efficiency in producing energy decreases in the absence of oxygen. Examples of facultative anaerobes include the familiar *E. coli* and many yeasts.

Recall from Chapter 5 that many microbes are able to substitute other electron acceptors, such as nitrate ions, for oxygen in anaerobic respiration.

Obligate anaerobes (Table 6.1c) are bacteria that are unable to use molecular oxygen for energy-yielding reactions. In fact, most are harmed by it. The genus *Clostridium*, which contains the species that cause tetanus and botulism, is the most familiar example. These bacteria do use oxygen atoms in cellular materials; the atoms are usually obtained from water.

Understanding how organisms can be harmed by oxygen requires a brief discussion of the toxic forms of oxygen:

1. **Singlet oxygen** is normal molecular oxygen (O_2) that has been boosted into a higher-energy state and is extremely reactive. It is present in phagocytic cells, specialized animal cells that ingest and eliminate small foreign bodies such as bacteria. It plays a role in the destruction of such foreign bodies after they are ingested by the phagocytes.

2. **Superoxide free radicals** ($O_2^-\cdot$) are formed in small amounts during the normal respiration of organisms that use oxygen as a final electron acceptor, forming water (see Figure 5.17). In the presence of oxygen, obligate anaerobes also appear to form some superoxide

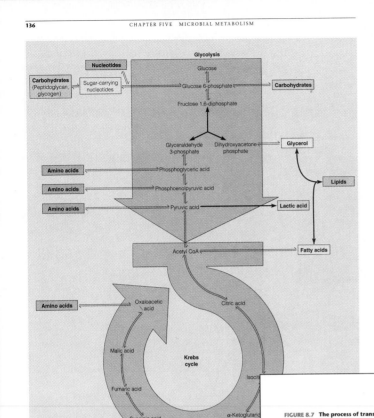

FIGURE 5.29 Integration of metabolism. Key intermediates are shown. Although not indicated in the figure, amino acids and ribose are used in the synthesis of purine and pyrimidine nucleotides (see Figure 5.28). The double arrows indicate amphibolic pathways.

Full color illustrations make learning easier

In all illustrations, colors are used consistently to represent the same concepts. This idea, called color-coding, makes it easier to recognize structures and connect concepts. Many commonly used concepts are consistent in both color and shape throughout the text—for example, ATP ( ) and phosphate groups ( ). This figure from Chapter 5 on the integration of metabolism shows another example of both color-coding and consistent symbols (i.e., the Krebs cycle is always this pink circular shape).

Step-by-step illustrations with orientation diagrams guide you through processes

The illustration program includes many step-by-step diagrams to walk you through important processes. These diagrams almost always include numbered steps and step descriptions, as illustrated in the figure from Chapter 8 showing the process of transcription. Many of these step-by-step figures also provide orientation diagrams such as the one at the top of this figure. These are miniature versions of previous overview illustrations with parts highlighted to show you where you are in the process. For example, transcription is part of the overall process in the flow of genetic information. Such diagrams keep the various steps of each process connected to central concepts.

FIGURE 8.7   The process of transcription. The inset diagram indicates the relationship of transcription to the overall flow of genetic information within a cell.

FIGURE 8.8   RNA processing in eucaryotic cells.

FIGURE 3.8 Transmission and scanning electron microscopy. The illustrations show the pathways of electron beams used to create images of the specimens. The photographs show lymphocytes, a kind of white blood cell, viewed through both of these types of electron microscopes.

Electron beam
Electromagnetic condenser lens
Specimen
Electromagnetic objective lens
Electromagnetic projector lens
Fluorescent screen or photographic plate
Viewing eyepiece

Electron gun
Primary electron beam
Electromagnetic lenses
Viewing screen
Electron collector
Secondary electrons
Specimen
Amplifier

TEM   1 μm   SEM   1 μm

(a) Transmission. (Top) In a transmission electron microscope, electrons pass through the specimen and are scattered. Magnetic lenses focus the image onto a fluorescent screen or photographic plate. (Bottom) This colorized transmission electron micrograph (TEM) shows a thin slice of lymphocyte. In this type of microscopy, the internal structures present in the slice can be seen.

(b) Scanning. (Top) In a scanning electron microscope, primary electrons sweep across the specimen and knock electrons from its surface. These secondary electrons are picked up by a collector, amplified, and transmitted onto a viewing screen or photographic plate. (Bottom) In this colorized scanning electron micrograph (SEM), the surface structures of a lymphocyte can be seen. Note the three-dimensional appearance of this cell, in contrast to the...

New micrographs clarify the world of microbiology

The many micrographs in this edition allow you to more clearly visualize the microorganisms you are studying. New icons below the micrographs identify how a micrograph was taken (whether it is a light micrograph [LM] or an electron micrograph). For the electron micrographs, red icons indicate the use of colorization. Shown here are two colorized electron micrographs, one scanning (SEM) and one transmission (TEM). Art and photos are juxtaposed in this figure so you can better connect concepts with reality.

---

useful for the study of the very smallest specimens, such as virus particles, bacterial flagella, and protein molecules.

Transmission electron microscopy has high resolution and is extremely valuable for examining different layers of specimens. However, it does have certain disadvantages. Because electrons have limited penetrating power, only a very thin section of a specimen (about 100 nm) can be studied effectively. Thus, the specimen has no three-dimensional aspect. In addition, specimens must be fixed, dehydrated, and viewed under a high vacuum. These treatments not only kill the specimen but also cause some shrinkage and distortion, sometimes to

the extent that there...
tures in a prepared ce...
of the method of pr...
Chapter 4).

Scanning Electron...
The **scanning electro...**
lem of sectioning asso...
microscope. A scann...
striking three-dimen...
3.8b). In scanning el...
produces a finely foc...

Boxes provide more in-depth information

An innovative box program supports the text with interesting stories and clinical examples of microbiology in action. The three types of boxes are Applications of Microbiology, Microbiology in the News, and the Morbidity and Mortality Weekly Report (MMWR). This Applications box from Chapter 28 gives you information on biosensors, a new application of microbiology and genetic engineering. Many boxes also include full-color photographs or illustrations.

Applications of Microbiology

BIOSENSORS— BACTERIA THAT DETECT TOXIC POLLUTANTS

EACH YEAR IN THE U.S.A., industrial plants generate 265 million metric tons of hazardous waste, 80% of which make their way into landfills. Burying these chemicals doesn't remove them from the ecosystem, however; it just moves them to other places, where they may still find their way into bodies of water. Traditional chemical analyses to locate these chemicals are expensive and can't distinguish between chemicals that affect biological systems from those that lie inert in the environment.

In response to this problem, Olga Selifonova and her colleagues at the Center for Environmental Diagnostics and Bioremediation in Florida are among a group of scientists developing *biosensors*—bacteria that can locate biologically active pollutants. Biosensors do not require costly chemicals or equipment, and they work quickly—within minutes.

In order to work, these bacterial biosensors require both a receptor that is activated in the presence of pollutants and a reporter that will make such a change apparent. Biosensors use the *lux* operon from *Vibrio* or *Photobacterium* as a reporter. This operon contains inducer and structural genes for the enzyme luciferase. In the presence of a coenzyme called $FMNH_2$, luciferase reacts with the molecule in such a way that the enzyme–substrate complex emits blue-green light, which then oxidizes the $FMNH_2$ to produce FMN. Therefore, a bacterium containing the *lux* gene will emit visible light (see photos) when the receptor is activated.

The *lux* operon is readily transferred to many bacteria. It has already been transferred to the plant pathogen *Xanthomonas* in order to monitor the progression of its infection in plants. Because the altered bacteria emit light, their presence can be detected by photographic film. *Bradyrhizobium* has also been engineered to contain the *lux* genes in an effort to understand root nodule formation in soybean plants.

In another application, *Lactococcus* bacteria containing the *lux* operon are

being used to detect the presence of antibiotics in milk that is to be used for cheese production. (If milk contains antibiotics, the cheese starter cultures will not grow.) Because the emission of light requires a living cell, the presence of antibiotics is measured as a decrease in light output by such recombinant *Lactococcus* bacteria.

The Microtox System developed in Great Britain uses the marine bacterium *Photobacterium* directly to detect toxic pollutants. These bacteria cannot emit light if they are killed by pollutants.

Other environmental biosensors use recombinant bacteria such as *Escherichia coli*. Some *E. coli* bacteria have been genetically engineered to contain both a *mer* resistance operon, which acts as a receptor, and the *lux* operon. The *mer* acts as the promotor for the *lux* structural genes. When such an *E. coli* cell is placed in a soil sample containing mercury, the mercury acts as an inducer, activating the production of luciferase in the bacteria.

Detecting harmful pollutants in soil and water is necessary to protect humans and animals. However, after detection, bioremediation processes are still required to remove the pollutants.

(a)

(b)

Vibrio fischeri emits light when energy is released by the transport of electrons to luciferase. These photos show colonies of *V. fischeri* photographed **(a)** in daylight and **(b)** in the dark, illuminated by their own light.

End-of-chapter material reinforces key ideas

Extensive material at the end of each chapter supports your understanding of important concepts. This material includes a conceptually organized Study Outline and several types of Study Questions: Review, Critical Thinking, and Clinical Applications. These questions may include illustrations or photos, or perhaps a table for you to complete.

STUDY QUESTIONS   **187**

STUDY QUESTIONS

Review

1. Name the cause of cell death resulting from damage to each of the following:

   (a) Cell wall   (c) Proteins
   (b) Plasma membrane (d) Nucleic acids

2. The thermal death time for a suspension of *B. subtilis* endospores is 30 minutes in dry heat and less than 10 minutes in an autoclave. Which type of heat is more effective? Why?

3. If pasteurization does not achieve sterilization, why is food treated by pasteurization?

4. Thermal death point is not considered an accurate measure of the effectiveness of heat sterilization. List three factors that can alter thermal death point.

5. The antimicrobial effect of gamma radiation is due to _____ . The antimicrobial effect of ultraviolet radiation is due to _____ .

6. A bacterial culture was in log phase in the following figure. At time *x*, an antibacterial compound was added to the culture. Which line indicates addition of a bactericidal compound? A bacteriostatic compound? How can you tell? Explain why the viable count does not immediately drop to zero in line *a*.

7. Fill in the following table:

METHOD OF STERILI-ZATION	TEMP.	TIME	TYPE OF HEAT	PREF'D USE	TYPE OF ACTION
Auto-claving					
Hot air					
Pasteur-ization					

8. How do the examples in question 7 illustrate the concept of equivalent treatments?

9. Heat-labile solutions such as glucose–minimal salts broth can be sterilized by _____ .

10. How do salts and sugars preserve foods? Why are these considered physical rather than chemical methods of control? Name one food that is preserved with sugar and one preserved with salt. How do you account for the occasional growth of *Penicillium* mold in jelly, which is 50% sucrose?

11. List five factors to consider before selecting a disinfectant.

12. Give the method of action and at least one standard use of each of the following types of disinfectants:

    (a) Phenolics   (e) Heavy metals
    (b) Iodine   (f) Aldehydes
    (c) Chlorine   (g) Ethylene oxide
    (d) Alcohol   (h) Oxidizing agents

13. The use-dilution values for two disinfectants tested under the same conditions are Disinfectant A—1:2; Disinfectant B—1:10,000. If both disinfectants are designed for the same purpose, which would you select?

14. A large hospital washes burn patients in a stainless steel tub. After each patient, the tub is cleaned with a quat. It was noticed that 14 of 20 burn patients acquired *Pseudomonas* infections after being bathed. Provide an explanation for this high rate of infection.

Critical Thinking

1. The filter paper method was used to evaluate three disinfectants. The results were as follows:

DISINFECTANT	ZONE OF INHIBITION
X	0 mm
Y	
Z	

   (a) Which disin...
       the organism...
   (b) Can you det...
       tericidal or b...

2. Why is each of th...
   disinfectants?
   (a) *Mycobacteriu...*
   (b) *Pseudomonas...*
   (c) *Bacillus*

---

188   CHAPTER SEVEN   CONTROL OF MICROBIAL GROWTH

3. The use-dilution test was used to evaluate two disinfectants against *Salmonella choleraesuis*. The results were as follows:

TIME OF EXPOSURE (MIN)	BACTERIAL GROWTH AFTER EXPOSURES		
	DISINFEC-TANT A	DISINFECTANT B DILUTED WITH DISTILLED WATER	DISINFECTANT B DILUTED WITH TAP WATER
10	+	–	+
20	+	–	–
30	–	–	–

   (a) Which disinfectant was the most effective?
   (b) Which disinfectant should be used against *Staphylococcus?*

4. To determine the lethal action of microwave radiation, two 10^5 suspensions of *E. coli* were prepared. One cell suspension was exposed to microwave radiation while wet, whereas the other was lyophilized (freeze-dried) and then exposed to radiation. The results are shown in the following figure. Dashed lines indicate the temperature of the samples. What is the most likely method of lethal action of microwave radiation? How do you suppose these data might differ for *Clostridium*?

Clinical Applications

1. *Entamoeba histolytica* and *Giardia lamblia* were isolated from the stool sample of a 45-year-old man, and *Shigella sonnei* was isolated from the stool sample of an 18-year-old woman. Both patients experienced diarrhea and severe abdominal cramps, and, prior to onset of digestive symptoms, both had been treated by the same chiropractor.

   The chiropractor had administered colonic irrigations (enemas) to these patients. The device used for this treatment was a gravity-dependent apparatus with 12 liters of tap water. There were no check valves to prevent backflow, so all parts of the apparatus could have become contaminated with feces during each colonic treatment. The chiropractor provided colonic treatment to four or five patients per day. Between patients, the adaptor piece that is inserted into the rectum was placed in a "hot-water sterilizer."

   What two errors were made by the chiropractor?

2. Between March 9 and April 12, five chronic peritoneal dialysis patients at one hospital became infected with *Pseudomonas aeruginosa*. Four patients developed peritonitis (inflammation of the abdominal cavity), and one developed a skin infection at the catheter insertion site. All patients with peritonitis had low-grade fever, cloudy peritoneal fluid, and abdominal pain. All patients had permanent indwelling peritoneal catheters, which the nurse wiped with gauze soaked with an iodophor (Prepodyne®) solution each time the catheter was connected to or disconnected from the machine tubing. Aliquots of the iodophor were transferred from stock bottles to small in-use bottles.

   Cultures from the dialysate concentrate and the internal areas of the dialysis machines were negative; iodophor from a small in-use plastic container yielded a pure culture of *P. aeruginosa*.

   What improper technique led to this infection?

3. Eleven patients received injections of methylprednisolone and lidocaine to relieve the pain and inflammation of arthritis at the same orthopedic surgery office. All the patients developed septic arthritis caused by *Serratia marcescens*. Unopened bottles of methylprednisolone from the same lot numbers tested sterile; the methylprednisolone was preserved with a quat. Cotton balls were used to wipe multiple-use injection vials before the medication was drawn into a disposable syringe. The site of injection on each patient was also wiped with a cotton ball. The cotton balls were soaked in benzalkonium chloride, and fresh cotton balls were added as the jar was emptied. Opened methylprednisolone and the jar of cotton balls contained *S. marcescens*.

   How was the infection transmitted? What part of the routine procedure caused the contamination?

You will find additional study aids and information in the five appendices at the end of the textbook, as well as a list of Further Readings, Pronunciation of Scientific Names, and a Glossary and Index. In addition, you will find a list of Word Roots Used in Microbiology in the endpapers (the very beginning and ending pages) of the text.

Brief Contents

Contents

CHAPTER 3 **Observing Microorganisms Through a Microscope** 52

CHAPTER 4 **Functional Anatomy of Procaryotic and Eucaryotic Cells** 70

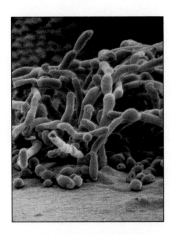

CHAPTER 7   Control of Microbial Growth   167

CHAPTER 8   Microbial Genetics   189

PART TWO
Survey of the Microbial World 249

CHAPTER 11 Bacteria 273

CHAPTER 12 Fungi, Algae, Protozoa, and Multicellular Parasites 295

CHAPTER 13 Viruses 332

PART THREE
Interaction Between Microbe and Host 365

CHAPTER 14 Principles of Disease and Epidemiology 366

CHAPTER 15 Microbial Mechanisms of Pathogenicity 390

CHAPTER 16 Nonspecific Defenses of the Host 406

CHAPTER 17 Specific Defenses of the Host: The Immune Response 425

CHAPTER 18 Practical Applications of Immunology 448

CHAPTER 19 Disorders Associated with the Immune System 466

CHAPTER 20 Antimicrobial Drugs 491

PART FOUR
Microorganisms and Human Disease 515

CHAPTER 21 Microbial Diseases of the Skin and Eyes 516

CHAPTER 22   Microbial Diseases of the Nervous System   537

CHAPTER 23   Microbial Diseases of the Cardiovascular and Lymphatic Systems   558

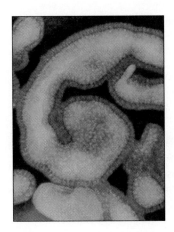

CHAPTER 24   Microbial Diseases of the Respiratory System   586

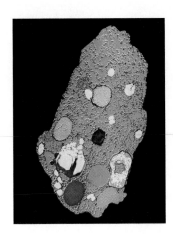

CHAPTER 25   Microbial Diseases of the Digestive System   612

List of Boxes

Fundamentals of Microbiology

At a ski area in Lake Tahoe, California, a technician monitors the making of artificial snow. As at some 300 ski areas worldwide, this snow is made with Snowmax®, a protein produced by genetically engineered bacteria that causes water to freeze at higher than normal temperatures.

The Microbial World and You

THE OVERALL THEME of this textbook concerns the relationship of microbes to our lives. This relationship involves not only their familiar harmful effects, such as causing disease and spoiling food, but also their many beneficial effects. In this chapter we will introduce the many ways in which microbes affect our lives. First, they have been fruitful subjects of study for many years, as you will see in our short history of microbiology. Then we introduce the incredible diversity of microorganisms. The chapter next discusses the ecological importance of microbes, which help maintain balance in the environment by recycling chemical elements such as oxygen and nitrogen between the soil and the atmosphere. We will also see how microbes are used in commercial and industrial applications to produce foods, chemicals, and drugs (such as penicillin) and to treat sewage, control pests, and clean up pollutants. Finally, we will discuss microbes as the cause of diseases such as AIDS, toxic shock syndrome, and Lyme disease.

In the summer of 1985, newspapers announced that a well-known actor was suffering from AIDS (acquired immunodeficiency syndrome). The American public had heard about the disease, but until then AIDS had mistakenly not been perceived as much of a health threat by the majority of Americans. While the actor sought treatment in hospitals in France, Americans became increasingly aware of the illness that would end the life of one of their screen idols, as well as the lives of over 8000 other Americans that same year.

The general public was just beginning to worry about what would become the most frightening epidemic of the century. Yet medical researchers had been gathering information about this devastating disease for several years. The same concepts of microbiology that had helped them identify other diseases, their means of transmission, and treatments were being applied to the study of AIDS.

AIDS first came to public attention in 1981 with reports from Los Angeles that a few young homosexual men had died of a previously rare type of pneumonia known as *Pneumocystis* (nü-mō-sis'tis) pneumonia. These men had experienced a severe weakening of the immune system, which normally fights infectious diseases. Soon these cases were correlated with an unusual number of occurrences of a rare form of cancer, Kaposi's sarcoma, among young homosexual men. Similar increases in such rare diseases were found among hemophiliacs and intravenous drug users.

By the end of 1993, nearly 350,000 people in the U.S.A. had been diagnosed as having AIDS, and more than 60% of them had died as a result of the disease. A great many more people had tested positive for the presence of the AIDS virus in their blood. As of 1992, health officials estimated that more than 1.5 million Americans carried the virus. In 1993, the World Health Organization estimated that over 13 million people worldwide had become infected with the AIDS virus. And this number will increase to 40 million by the year 2000. Scientists studying AIDS are still uncertain whether all individuals carrying the

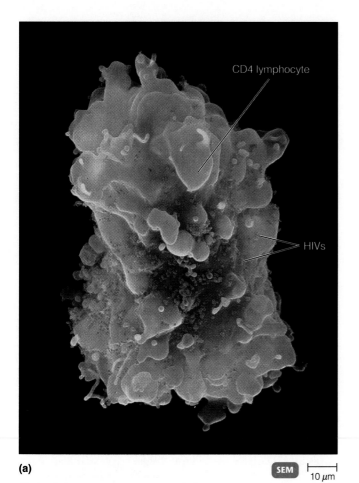

CD4 lymphocyte

HIVs

(a)

SEM | 10 μm |

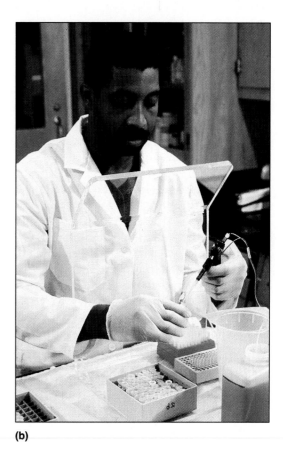

(b)

FIGURE 1.1 Human immunodeficiency virus (HIV). (a) This micrograph shows HIVs (small blue spheres), the virus that causes AIDS, budding from one of its target cells, a white blood cell called a CD4 lymphocyte. **(b)** Laboratory technician performing the standard enzyme-linked immunosorbent assay (ELISA) that is used to indirectly detect the presence of HIV in the body.

virus will eventually develop the disease or whether some may be able to live normally without any symptoms.

Researchers quickly discovered that the cause of AIDS was a previously unknown virus (Figure 1.1). The virus, now called **human immunodeficiency virus (HIV)**, destroys certain cells of the immune system, one of the body's defense systems. Sickness and death result from microorganisms or cancerous cells that might otherwise have been defeated by the body's natural defenses. So far, the disease has been inevitably fatal once symptoms develop.

By studying disease patterns, medical researchers found that HIV could be spread through sexual intercourse, by contaminated needles, from infected mothers to their fetuses before birth, and by blood transfusions—in short, by transmission of body fluids from one person to an-

other. Since 1985, blood used for transfusions has been carefully checked for the presence of HIV, and it is now quite unlikely that the virus can be spread by this route.

The majority of individuals with AIDS are in the sexually active age group, and because heterosexual partners of AIDS sufferers are at high risk of infection, public health officials are concerned that even more teenagers will contract AIDS. Forty-six percent of U.S. teenage girls with AIDS became infected through heterosexual intercourse, as compared with 30% of all U.S. women with AIDS. In several U.S. cities, AIDS is now the major cause of death of young men and women.

The first drug used to treat AIDS (although it does not provide a cure) was zidovudine (AZT). Other drugs include dideoxyinosine (ddI) and dideoxycytidine (ddC). Still other drugs are in clinical trials. In addition, considerable effort is being expended in developing a vaccine against AIDS. Public health officials have also focused on prevention through education.

In the months and years to come, microbiological techniques will be applied intensively to help scientists learn more about the structure of this deadly virus, how it is transmitted, how it grows in cells and causes disease, how drugs can be directed against it, and whether an effective vaccine can be developed.

AIDS poses one of this century's most formidable health threats, but it is not the first serious epidemic of a sexually transmitted disease. Syphilis was also once a fatal epidemic disease. As recently as 1941, syphilis caused an estimated 14,000 deaths in the United States. With few drugs available for treatment and no vaccines to prevent it, efforts to control the disease focused mainly on restraining sexual behavior and on the use of condoms. The eventual development of drugs to treat syphilis contributed significantly to preventing the spread of the disease. Reported cases of syphilis dropped from a record high of 575,000 in 1943 to 34,179 in 1992.

Just as microbiological techniques helped researchers in the fight against syphilis, in 1978 they helped scientists discover the cause of toxic shock syndrome (TSS).

TSS caught the attention of the public for the first time when a number of otherwise healthy young women suddenly died. Microbiologists were able to show that the cause of death was **toxins** (poisonous substances) produced by the bacterium *Staphylococcus aureus* (staf-i-lō-kok′kus ô′rē-us), a microorganism present on human skin (see Figure 11.11). In their research on how such a common bacterium could become so toxic, scientists noticed that all the victims used tampons during menstruation. In 1985, a Harvard research team found that certain tampon fibers absorb magnesium that is normally present in the vagina. When the amount of magnesium in the vagina is reduced, *S. aureus* produces large amounts of toxin. The association of TSS with a particular brand of tampon resulted in the recall of that brand and the publication of cautions about the use of tampons in general. (It was subsequently learned that TSS also occurs in males and nonmenstruating females, for example, following postsurgical infections.)

In 1975, a cluster of disease cases first diagnosed as rheumatoid arthritis was reported in young people living near the city of Lyme, Connecticut. Today this disease is known as **Lyme disease.** By 1982, microbiologists determined that the disease is caused by a bacterium (*Borrelia burgdorferi* [bôr-rel′ē-ä burg-dôr′fėr-ē]) transmitted to humans by ticks that live on field mice and deer. These ticks are so small that their bites go unnoticed. Within a few weeks of the tick bite, a rash may appear at the site, although some people never develop one. The typical rash resembles a bull's-eye, but there are many variations. The rash often is accompanied by joint stiffness, fever and chills, headache, stiff neck, nausea, and low back pain. In the next stage, some patients may develop cardiac abnor-

malities or neurological problems. In the third stage, arthritis, usually affecting the larger joints, is the principal complication. Lyme disease responds well to antibiotics, especially if they are given early in the course of the disease.

In 1993, illness attributed to the ubiquitous bacterium *Escherichia coli* (esh-ėr-i′kē-ä kō′lī or kō′lē) was reported on the front pages of major newspapers. *E. coli* is a normal inhabitant of the large intestine of vertebrates, including humans, and its presence is beneficial because it helps produce certain vitamins and breaks down otherwise undigestible foodstuffs (see Chapter 25). However, a strain of *E. coli* called *E. coli* O157:H7 causes bloody diarrhea when it grows in the intestines. This strain was first recognized in 1982 and since then has emerged as a public health problem. It is now one of the leading causes of diarrhea worldwide. The recent outbreaks of *E. coli* O157:H7 in the United States, associated with contamination of undercooked meat and unpasteurized beverages, have made public health officials aware that new methods of testing for bacteria in food must be developed.

All the diseases we have mentioned (AIDS, TSS, Lyme disease, and diarrhea) are caused by viruses or bacteria, two types of microorganisms. This book introduces you to the enormous variety of microscopic organisms. It shows you how microbiologists use specific techniques and procedures to study the microbes that cause such diseases as AIDS, TSS, and Lyme disease. You will also learn about the body's responses to microbial infection and the ways certain drugs combat microbial diseases. Finally, you will learn about the many beneficial roles that microbes play in the world around us.

MICROBES IN OUR LIVES

LEARNING OBJECTIVE

• *List several ways in which microbes affect our lives.*

For many people, the words *germ, microbe,* and *microorganism* bring to mind a group of tiny creatures that do not quite fit any of the categories in that old question, "Is it animal, vegetable, or mineral?" Microorganisms are minute living things that individually are too small to be seen with the naked eye. The group includes bacteria (see Chapter 11), fungi (yeasts and molds), protozoa, and microscopic algae (Chapter 12). It also includes viruses, those noncellular entities sometimes regarded as being at the border of life and nonlife (see Chapter 13). Figure 1.6 on page 16 gives you an idea of the diversity of the microorganisms that you will learn more about in later chapters.

We tend to associate these small organisms only with uncomfortable infections, major diseases such as AIDS, or such common inconveniences as spoiled food. However, the majority of microorganisms make crucial contributions to the welfare of the world's inhabitants by helping to maintain the balance of living organisms and chemicals in our environment. Marine and freshwater microorganisms form the basis of the food chain in oceans, lakes, and rivers. Soil microbes help to break down wastes and incorporate nitrogen gas from the air into organic compounds, thereby recycling chemical elements in the soil, water, and air. Certain bacteria and algae play important roles in photosynthesis, a food- and oxygen-generated process that is critical to life on Earth. (Photosynthesis is the light-driven synthesis of carbohydrates from carbon dioxide.) Humans and many other animals depend on the bacteria in their intestines for digestion and synthesis of some vitamins that their bodies require, including some B vitamins for metabolism and vitamin K for blood clotting.

Microorganisms also have many commercial applications. They are used in the synthesis of such chemical products as acetone, organic acids, enzymes, alcohols, and many drugs. The process for producing acetone and butanol by bacteria was discovered in 1914 by **Chaim Weizmann**, a Polish-born chemist working in England for Winston Churchill. When World War I broke out in August of that year, the production of acetone was very important for making cordite (gunpowder) for use in the manufacture of munitions. Weizmann's discovery played a significant role in determining the outcome of the war. After the war, Weizmann refused all honors proposed by the British government. However, he did use his influence to encourage the British to help establish a Jewish homeland in Palestine. In 1949, Weizmann was elected the first president of Israel.

The food industry also uses microbes in producing vinegar, sauerkraut, pickles, alcoholic beverages, green olives, soy sauce, buttermilk, cheese, yogurt, and bread (see the box on p. 6). In addition, bacteria and other microbes can now be manipulated to produce substances that they normally do not synthesize. Through this technology, called *genetic engineering,* bacteria can produce important therapeutic substances such as insulin, human growth hormone, and interferon.

Though only a minority of microorganisms are **pathogenic** (disease-producing), practical knowledge of microbes is necessary for medicine and the related health sciences. For example, hospital workers must be able to protect patients from common bacteria that are normally harmless but pose a threat to the sick and injured.

Today we understand that microorganisms are found almost everywhere. Yet not long ago, before the invention of the microscope, microbes were unknown to scientists. Thousands of people died in devastating epidemics, the causes of which were not understood. Food spoilage often could not be controlled, and entire families died because vaccinations and antibiotics were not available to fight infections.

We can get an idea of how our current concepts of microbiology developed by looking at a few of the historic milestones in microbiology that have changed our lives.

A BRIEF HISTORY OF MICROBIOLOGY

LEARNING OBJECTIVES

- *Explain the importance of observations made by Hooke and van Leeuwenhoek.*
- *Compare the theories of spontaneous generation and biogenesis.*
- *Identify the contributions to microbiology made by Needham, Spallanzani, Virchow, and Pasteur.*
- *Identify the importance of Koch's Postulates.*
- *Explain how Pasteur's work influenced Lister and Koch.*
- *Identify the contributions to microbiology made by Ehrlich, Fleming, and Dubos.*
- *Define immunology, virology, and microbial genetics.*

Now that we have generally examined how microbes relate to our everyday lives, we will take a look at how the field of microbiology developed from its beginning hundreds of years ago to its current high-technology state.

The First Observations

One of the most important discoveries in the history of biology occurred in 1665 with the help of a relatively crude microscope. An Englishman, **Robert Hooke,** reported to the world that life's smallest structural units were "little boxes," or "cells," as he called them. Using his improved version of a compound microscope (one that uses two sets of lenses), Hooke was able to see individual cells. Hooke's discovery marked the beginning of the cell theory—the theory that *all living things are composed of cells.* Subsequent investigations into the structure and functions of cells were based on this theory.

Though Hooke's microscope was capable of showing protozoa and probably bacteria, he lacked the staining techniques (see Chapter 3) that would have allowed him to see such microbes clearly. The Dutch merchant and amateur scientist **Anton van Leeuwenhoek** was probably the first to actually observe live microorganisms through magnifying lenses. Between 1673 and 1723, he wrote a series of letters to the Royal Society of London describing the "animalcules" he saw through his simple, single-lens microscope. Van Leeuwenhoek's detailed drawings of "animalcules" in rain water, in liquid in which peppercorns had soaked, and in material scraped from his teeth have since been identified as representations of bacteria and protozoa (Figure 1.2).

The Debate over Spontaneous Generation

After van Leeuwenhoek discovered the previously "invisible" world of microorganisms, the scientific community

Applications of Microbiology

WHAT MAKES SOURDOUGH BREAD DIFFERENT?

IMAGINE BEING A miner during the California gold rush. You've just made bread dough from your last supplies of flour and salt when someone yells, "Gold!" Temporarily forgetting your hunger, you run off to the gold fields. Many hours later you return. The dough has been rising longer than usual, but you are too cold, tired, and hungry to start a new batch. Later you find that your bread tastes different from previous batches; it is slightly sour. During the gold rush, miners baked so many sour loaves that they were nicknamed "sourdoughs."

Conventional bread is made from flour, water, sugar, salt, shortening, and a living microbe, yeast. The yeast belongs to the Kingdom Fungi and is named *Saccharomyces cerevisiae.* When flour is mixed with water, an enzyme in the flour breaks its starch into two sugars, maltose and glucose. After the ingredients for the bread are mixed, the yeast metabolizes the sugars and produces alcohol (ethanol) and carbon dioxide as waste products. This metabolic process is called fermentation. The dough rises as carbon dioxide bubbles get trapped in the sticky matrix. The alcohol evaporates during baking. It and the carbon dioxide gas form spaces that remain in the bread.

Originally, breads were leavened by wild yeast from the air, which had been trapped in the dough. Later,

bakers kept a starter culture of yeast—dough from the last batch of bread—to leaven each new batch of dough. Sourdough bread is made with a special sourdough starter culture that is added to flour, water, and salt. Perhaps the most famous sourdough bread made today comes from San Francisco, where a handful of bakeries have continuously cultivated their starters for more than 100 years and have meticulously maintained the starters to keep out unwanted microbes that can produce different and undesirable flavors. After bakeries in other areas made several unsuccessful attempts to match the unique flavor of San Francisco sourdough, rumors attributed the taste to a unique local climate or contamination from bakery walls. T. F. Sugihara and L. Kline from the United States Department of Agriculture (U.S.D.A.) set out to debunk these rumors and to determine the microbiological reason for the bread's different taste so that it could be made in other areas.

The U.S.D.A. workers found that sourdough is eight to ten times more

acidic than conventional bread because of the presence of lactic and acetic acids. These acids account for the sour flavor of the bread. The workers isolated and identified the yeast in the starter as *Saccharomyces exiguus,* a unique yeast that does not ferment maltose and thrives in the acidic environment of this dough. The sourdough question was not answered, however, because the yeast did not produce the acids and did not use maltose. Sugihara and Kline searched the starter for a second agent capable of fermenting maltose and producing the acids (see the figure). The bacterium they isolated, so carefully guarded all those years, was classified into the genus *Lactobacillus.* Many members of this genus are used in dairy fermentations and are part of the normal microbiota of humans and other mammals. Analyses of cell structure and genetic composition showed that the sourdough bacterium is genetically different from other previously characterized lactobacilli. It has been given the name *Lactobacillus sanfrancisco.*

---

of the time became interested in the origins of these tiny living things. Until the second half of the nineteenth century, many scientists and philosophers believed that some forms of life could arise spontaneously from nonliving matter; they called this hypothetical process **spontaneous** **generation.** Not much more than a hundred years ago, people commonly believed that toads, snakes, and mice could be born of moist soil; that flies could emerge from manure; and that maggots, the larvae of flies, could arise from decaying corpses.

FIGURE 1.2 Anton van Leeuwenhoek's microscopic observations. (a) A replica of a simple microscope made by Anton van Leeuwenhoek to observe living organisms too small to be seen with the unaided eye. The specimen was placed on the tip of the adjustable point and viewed from the other side through the tiny, nearly spherical lens. The highest magnification possible with his microscopes was about 300× (times). **(b)** Some of van Leeuwenhoek's drawings of bacteria, made in 1683. He was the first person to see the microorganisms we now call bacteria and protozoa.

Location of specimen

Lens

(a) Microscope replica

(b) Drawings of bacteria

Evidence Pro and Con

A strong opponent of spontaneous generation, the Italian physician **Francesco Redi**, set out in 1668 (even before van Leeuwenhoek's discovery of microscopic life) to demonstrate that maggots do not arise spontaneously from decaying meat. Redi filled three jars with decaying meat and sealed them tightly. Then he arranged three other jars similarly but left them open. Maggots appeared in the open vessels after flies entered the jars and laid their eggs, but the sealed containers showed no signs of maggots. Still, Redi's antagonists were not convinced; they claimed that fresh air was needed for spontaneous generation. So Redi set up a second experiment, in which three jars were covered with a fine net instead of being sealed. No larvae appeared in the gauze-covered jars, even though air was present. Maggots appeared only if flies were allowed to leave their eggs on the meat.

Redi's results were a serious blow to the long-held belief that large forms of life could arise from nonlife. However, many scientists still believed that small organisms, such as van Leeuwenhoek's "animalcules," were simple enough to be generated from nonliving materials.

The case for the spontaneous generation of microorganisms seemed to be strengthened in 1745 when **John Needham**, an Englishman, found that even after he heated nutrient fluids (chicken broth and corn broth) before pouring them into covered flasks, the cooled solutions were soon teeming with microorganisms. Needham claimed that the microbes developed spontaneously from the fluids. Twenty years later, **Lazzaro Spallanzani**, an Italian scientist, suggested that microorganisms from the air probably had entered Needham's solutions after they were boiled. Spallanzani showed that nutrient fluids heated *after* being sealed in a flask did not develop microbial growth. Needham responded by claiming that the "vital force" necessary for spontaneous generation had been destroyed by the heat and was kept out of the flasks by the seals.

This intangible "vital force" was given all the more credence shortly after Spallanzani's experiment, when **Laurent Lavoisier** showed the importance of oxygen to life. Spallanzani's observations were criticized on the grounds that there was not enough oxygen in the sealed flasks to support microbial life.

The Theory of Biogenesis

The issue was still unresolved in 1858, when the German scientist **Rudolf Virchow** challenged spontaneous generation with the concept of **biogenesis**, the claim that living cells can arise only from preexisting living cells. Arguments about spontaneous generation continued until 1861, when the issue was resolved by the French scientist **Louis Pasteur.**

With a series of ingenious and persuasive experiments, Pasteur demonstrated that microorganisms are indeed present in the air and can contaminate seemingly sterile solutions, but air itself does not create microbes. He began by filling several short-necked flasks with beef broth and then boiling their contents. Some were then left open and allowed to cool. In a few days, these flasks were found to be contaminated with microbes. The other flasks, sealed after boiling, remained free of microorganisms. From these results, Pasteur reasoned that microbes in the air were the agents responsible for contaminating nonliving matter such as the broths in Needham's flasks.

Pasteur next placed broth in open-ended long-necked flasks and bent the necks into S-shaped curves (Figure 1.3). The contents of these flasks were then boiled and cooled. The broth in the flasks did not decay and showed no signs of life, even after months. Pasteur's unique design allowed

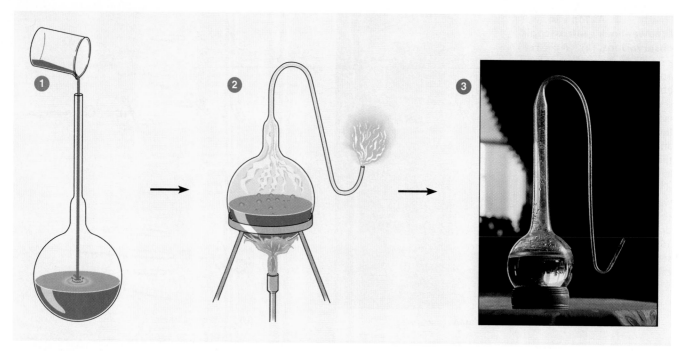

FIGURE 1.3   Pasteur's experiment disproving the theory of spontaneous generation. ①Pasteur first poured beef broth into a long-necked flask. ②Next he heated the neck of the flask and bent it into an S-shaped curve; then he boiled the broth for several minutes. ③ Microorganisms did not appear in the cooled solution, even after long periods, as you can see in this recent photograph of an actual flask used by Pasteur in a similar experiment.

air to pass into the flask, but the curved neck trapped any airborne microorganisms that might contaminate the broth. (Some of these original vessels, which were later sealed, are on display at the Pasteur Institute in Paris. Like the flask shown in Figure 1.3, they still show no sign of contamination more than 100 years later.)

Pasteur showed that microorganisms can be present in nonliving matter—on solids, in liquids, and in the air. Furthermore, he demonstrated conclusively that microbial life can be destroyed by heat and that methods can be devised to block the access of airborne microorganisms to nutrient environments. These discoveries form the basis of the **aseptic techniques** (techniques to prevent contamination by unwanted microorganisms) that are now the standard practice in laboratory and many medical procedures. Modern aseptic techniques are among the first and most important things that a beginning microbiologist learns.

Pasteur's work provided evidence that microorganisms cannot originate from mystical forces present in nonliving materials. Rather, any appearance of "spontaneous" life in nonliving solutions can be attributed to microorganisms that were already present in the air or in the fluids themselves. Scientists now believe that a form of spontaneous generation probably did occur on the primitive Earth when life first began, but they agree that this does not happen under today's environmental conditions.

The Golden Age of Microbiology

For about 60 years, beginning with the work of Pasteur, there was an explosion of discoveries in microbiology. The period from 1857 to 1914 has been appropriately named the **Golden Age of Microbiology**. During this period, rapid advances, spearheaded mainly by Pasteur and Robert Koch, led to the establishment of microbiology as a science. Discoveries during these years included both the agents of many diseases and the role of immunity in the prevention and cure of disease. During this productive period, microbiologists studied the chemical activities of microorganisms, improved the techniques for performing microscopy and culturing microorganisms, and developed vaccines and surgical techniques. Some of the major events that occurred during the Golden Age of Microbiology are listed in Figure 1.4.

Fermentation and Pasteurization

One of the key steps that established the relationship between microorganisms and disease occurred when a group of French merchants asked Pasteur to find out why wine and beer soured. They hoped to develop a method that would prevent spoilage when those beverages were shipped long distances. At the time, many scientists believed that air converted the sugars in these fluids into alcohol. Pasteur found instead that microorganisms called

1665   Hooke—First observation of cells
1673   van Leeuwenhoek—First observation of live microorganisms
1735   Linnaeus—Nomenclature for organisms
1798   Jenner—First vaccine
1835   Bassi—Silkworm fungus
1840   Semmelweis—Childbirth fever
1853 DeBary—Fungal plant disease

1857   Pasteur—Fermentation
1861   Pasteur—Disproved spontaneous generation
1864   Pasteur—Pasteurization
1867   Lister—Aseptic surgery
1870   Abbé—Abbé condenser and oil immersion objective
1876   *Koch—Germ theory of disease
1879   Neisser—*Neisseria gonorrhoeae*
1880   Pasteur—Immunization techniques
1881   *Koch—Pure cultures
1882   *Koch—*Mycobacterium tuberculosis*
        Hess—Agar (solid) media

**GOLDEN
AGE OF**   1883   *Koch—*Vibrio cholerae*
MICROBIOLOGY   1884   *Metchnikoff—Phagocytosis
        Gram—Gram-staining procedure
        Escherich—*Escherichia coli*
1887   Petri—Petri dish
1890   *von Bering—Diphtheria antitoxin
        *Ehrlich—Theory of immunity
1892   Winogradsky—Sulfur cycle
1898   Shiga—*Shigella dysenteriae*
1910   *Ehrlich—Syphilis

1928   *Fleming, Chain, Florey—Penicillin
        Griffith—Transformation in bacteria
1934   Lancefield—Streptococcal antigens
1935   *Stanley, Northrup, Sumner—Crystallized virus
1941   Beadle and Tatum—Relationship between genes and enzymes
1943   *Delbrück and Luria—Viral infection of bacteria
1944   Avery, MacLeod, McMarty—Genetic material is DNA
1946   Lederberg and Tatum—Conjugation
1953   *Watson and Crick—DNA structure
1957   *Jacob and Monod—Protein synthesis regulation
1959   Stewart—Viral cause of cancer
1962   *Edelman and Porter—Antibodies
1964   Epstein, Achong, Barr—Epstein-Barr virus as cause of human cancer
1969   Whittaker—Five-kingdom classification system
1971   *Nathans, Smith, Arber—Restriction enzymes (used for genetic engineering)
1973   Berg, Boyer, Cohen—Genetic engineering
1975   Dulbecco, Temin, Baltimore—Reverse transcriptase
1978   *Arber, Smith, Nathans—Restriction endonucleases
        *Mitchell—Chemiosmotic mechanism
1981   Margulis—Origin of eucaryotic cells
1982   *Klug—Structure of tobacco mosaic virus
1983 *McClintock—Transposons

1988   *Deisenhofer, Huber, Michel—Bacterial photosynthesis pigments
1989   *Bishop and Varmus—Oncogenes
1990   Murray and Thomas—Organ transplants

Louis Pasteur (1822–1895)

Robert Koch (1843–1910)

Rebecca C. Lancefield (1895–1981)

*Nobel laureate

FIGURE 1.4   Milestones in microbiology, highlighting those that occurred during the Golden Age of Microbiology.

yeasts convert the sugars to alcohol in the absence of air. This process, called **fermentation** (see Chapter 5), is used to make wine and beer. Souring and spoilage are caused by different microorganisms called bacteria. In the presence of air, bacteria change the alcohol in the beverage into vinegar (acetic acid).

Pasteur's solution to the spoilage problem was to heat the beer and wine just enough to kill most of the bacteria; the process does not greatly affect the flavor of the beverage. This process, called **pasteurization**, is now commonly used to kill potentially harmful bacteria in milk as well as in some alcoholic drinks. Showing the connection between spoilage of food and microorganisms was a major step toward establishing the relationship between disease and microbes.

The Germ Theory of Disease

As we have seen, the fact that many kinds of diseases are related to microorganisms was unknown until relatively recently. Before the time of Pasteur, effective treatments for many diseases were discovered by trial and error, but the causes of the diseases were unknown.

The realization that yeasts play a crucial role in fermentation was the first link between the activity of a microorganism and physical and chemical changes in organic materials. This discovery alerted scientists to the possibility that microorganisms might have similar relationships with plants and animals—specifically, that microorganisms might cause disease. This idea was called the **germ theory of disease.**

The germ theory was a difficult concept for many people to accept because for centuries disease was believed to be punishment for an individual's crimes or misdeeds. When the inhabitants of an entire village became ill, people often blamed the disease on foul odors from sewage or on poisonous vapors from swamps. Most people born in Pasteur's time found it inconceivable that "invisible" microbes could travel through the air to infect plants and animals or remain on clothing and bedding to be transmitted from one person to another. But gradually scientists accumulated the information needed to support the new germ theory.

In 1865, Pasteur was called upon to help fight silkworm disease, which was ruining the silk industry throughout Europe. Years earlier, in 1835, **Agostino Bassi,** an amateur microscopist, had proved that another silkworm disease was caused by a fungus. Using data provided by Bassi, Pasteur found that the more recent infection was caused by a protozoan, and he developed a method for recognizing afflicted silkworm moths.

In the 1860s, **Joseph Lister,** an English surgeon, applied the germ theory to medical procedures. Lister was aware that in the 1840s, the Hungarian physician **Ignaz Semmelweis** had demonstrated that physicians who went from one obstetrical patient to another without disinfecting their hands transmitted puerperal (childbirth) fever.

Lister had also heard of Pasteur's work connecting microbes to animal diseases. Disinfectants were unknown at the time, but Lister knew that phenol (carbolic acid) kills bacteria, so he began treating surgical wounds with a solution of it. The practice so reduced the incidence of infections and deaths that other surgeons quickly adopted it. Lister's technique was one of the earliest medical attempts to control infections caused by microorganisms. In fact, his findings proved that microorganisms cause surgical wound infections.

The first proof that bacteria actually cause disease came from **Robert Koch** in 1876. Koch, a brilliant German physician, was Pasteur's young rival in the race to discover the cause of anthrax, a disease that was destroying cattle and sheep in Europe. Koch discovered rod-shaped bacteria now known as *Bacillus anthracis* (bä-sil′lus an-thrā′sis) in the blood of cattle that had died of anthrax. He cultured the bacteria on nutrients and then injected samples of the culture into healthy animals. When these animals became sick and died, Koch isolated the bacteria in their blood and compared them with the bacteria originally isolated. He found that the two sets of blood cultures contained the same bacteria.

Koch thus established a sequence of experimental steps for directly relating a specific microbe to a specific disease. These steps are known today as **Koch's postulates** (see Figure 14.3). During the past 100 years, these same criteria have been invaluable in investigations proving that specific microorganisms cause many diseases. Koch's postulates, their limitations, and their application to disease will be discussed in greater detail in Chapter 14.

Vaccination

Often a treatment or preventive procedure is developed before scientists know why it works. The smallpox vaccine is an example of this. In 1798, almost 70 years before Koch established that a specific microorganism causes anthrax, **Edward Jenner,** a young British physician, embarked on an experiment to find a way to protect people from smallpox.

Smallpox epidemics were greatly feared. The disease periodically swept through Europe, killing thousands, and it wiped out 90% of the Native Americans on the East Coast when European settlers first brought the infection to the New World.

When a young milkmaid told Jenner that she couldn't get smallpox because she already had been sick from cowpox—a much milder disease—he decided to put the girl's story to a test. First Jenner collected scrapings from cowpox blisters. Then he inoculated a healthy volunteer with the cowpox material by scratching the person's arm with a pox-contaminated needle. The scratch turned into a raised bump. In a few days, the volunteer became mildly sick but recovered and never again contracted either cowpox or smallpox. The process was called *vaccination,* from the Latin word *vacca,* for cow. The protection from disease provided by vaccination (or by recovery from the disease itself) is called **immunity.** We will discuss the mechanisms of immunity in Chapter 17.

Years after Jenner's experiment, in about 1880, Pasteur discovered why vaccinations work. He found that the bacterium that causes fowl cholera lost its ability to cause disease (lost its *virulence,* or became *avirulent*) after it was grown in the laboratory for long periods. However, it—and other microorganisms with decreased virulence—was able to induce immunity against subsequent infections by its virulent counterparts. The discovery of this phenomenon provided a clue to Jenner's successful experiment with cowpox. Both cowpox and smallpox are caused by viruses. Even though cowpox virus is not a laboratory-produced derivative of smallpox virus, it is so closely related to the smallpox virus that it can induce immunity to both viruses. Pasteur used the term *vaccine* for cultures of avirulent microorganisms used for preventive inoculation.

Jenner's experiment was the first time in a Western culture that a living viral agent—the cowpox virus—was used to produce immunity. Physicians in ancient China had immunized patients by removing scales from drying pustules of a person suffering from a mild case of smallpox, grinding the scales to a fine powder, and inserting the powder into the nose of the person to be protected.

Some vaccines are still produced from avirulent microbial strains that stimulate immunity to the related virulent strain. Other vaccines are made from killed virulent microbes, from isolated components of virulent microorganisms, or by genetic engineering techniques.

launched a search for such a bullet. In 1910, after testing hundreds of substances, he found a chemotherapeutic agent called *salvarsan,* an arsenic derivative effective against syphilis. Before this discovery, the only known chemical in Europe's medical arsenal was an extract from the bark of a South American tree, *quinine,* which had been used by Spanish conquistadors to treat malaria.

By the late 1930s, researchers had developed several other synthetic drugs that could destroy microorganisms. Most of these drugs were derivatives of dyes. In addition, *sulfa drugs* were discovered at about the same time.

A Fortunate Accident—Antibiotics

In contrast to the sulfa drugs, which were deliberately developed from a series of industrial chemicals, the first antibiotic was discovered by accident. **Alexander Fleming,** a Scottish physician and bacteriologist, almost tossed out some culture plates that had been contaminated by mold. Fortunately, he took a second look at the curious pattern of growth on the contaminated plates. There was a clear area around the mold where the bacterial culture had stopped growing (Figure 1.5). Fleming was looking at a mold that could inhibit the growth of a bacterium. The mold was later identified as *Penicillium notatum* (pen-i-sil'lē-um nō-tä'tum), and in 1928 Fleming named the mold's active inhibitor *penicillin.* Thus, penicillin is an antibiotic produced by a fungus. The enormous usefulness of penicillin was not apparent until the 1940s, when it was finally mass-produced and tested clinically.

The Birth of Modern Chemotherapy: Dreams of a "Magic Bullet"

After the relationship between microorganisms and disease was established, medical microbiologists next focused on the search for substances that could destroy pathogenic microorganisms without damaging the infected animal or human. Treatment of disease by using chemical substances is called **chemotherapy.** (The term also commonly refers to chemical treatment of noninfectious diseases, such as cancer.) The chemotherapeutic agents prepared from chemicals in the laboratory are called **synthetic drugs.** Chemicals produced naturally by bacteria and fungi to act against other microorganisms are called **antibiotics.** The success of chemotherapy is based on the fact that some chemicals are more poisonous to microorganisms than to the hosts infected by the microbes.

The First Synthetic Drugs
Paul Ehrlich, a German physician, was the imaginative thinker who fired the first shot in the chemotherapy revolution. As a medical student, Ehrlich speculated about a "magic bullet" that could hunt down and destroy a pathogen without harming the infected host. Ehrlich

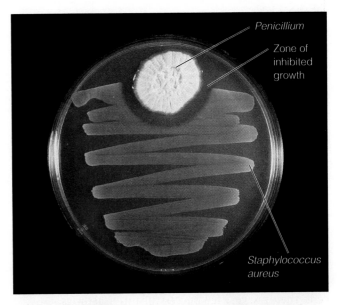

FIGURE 1.5 Inhibition of bacterial growth around the mold *Penicillium chrysogenum.* Colonies of the bacterium *Staphylococcus aureus* do not grow in the vicinity of the contaminating *Penicillium* colony near the top of the photo. The reason is that the mold secretes penicillin, which kills the bacteria.

TABLE 1.1   Selected Nobel Prizes Awarded for Research in Microbiology

NOBEL LAUREATES	YEAR OF PRESENTATION	COUNTRY OF BIRTH	CONTRIBUTION
Emil A. von Behring	1901	Germany	Developed a diphtheria antitoxin
Robert Koch	1905	Germany	Cultured tuberculosis bacteria
Paul Ehrlich	1908	Germany	Developed theories on immunity
Elie Metchnikoff	1908	Russia	Described phagocytosis, the intake of solid materials by cells
Alexander Fleming, Ernst Chain, and Howard Florey	1945	Scotland England	Discovered penicillin
Selman A. Waksman	1952	Ukraine	Discovered streptomycin
Hans A. Krebs	1953	Germany	Discovered chemical steps of the Krebs cycle in carbohydrate metabolism
John F. Enders, Thomas H. Weller, and Frederick C. Robbins	1954	United States	Cultured poliovirus in cell cultures
Joshua Lederberg, George Beadle, and Edward Tatum	1958	United States	Described genetic control of biochemical reactions in the fungus *Neurospora*
James D. Watson, Frances H.C. Crick, and Maurice A. F. Wilkins	1962	United States England New Zealand	Identified the physical structure of DNA
François Jacob, Jacques Monod, and André Lwoff	1965	France	Described how protein synthesis is regulated in bacteria
Robert Holley, Har Gobind Khorana, and Marshall W. Nirenberg	1968	United States India United States	Discovered the genetic code for amino acids
Max Delbrück, Alfred D. Hershey, and Salvador E. Luria	1969	Germany United States Italy	Described the mechanism of viral infection of bacterial cells
Gerald M. Edelman and Rodney R. Porter	1972	United States England	Described the nature and structure of antibodies

Interest in penicillin came after 1939, when **René Dubos,** a French microbiologist, discovered two antibiotics called *gramicidin* and *tyrocidine.* Both were produced by a bacterium, *Bacillus brevis,* cultured from soil.

Since the early discoveries of antibiotics, many others have been developed. Unfortunately, antibiotics and other chemotherapeutic drugs are not without problems. Many antimicrobial chemicals are too toxic to humans for practical use; they kill the pathogenic microbes, but they also damage the infected host. For reasons we discuss later, toxicity to humans is a particular problem in the development of drugs for treating viral diseases. Because viral growth is so closely linked with the life processes of normal host cells, there are very few successful antiviral drugs.

Another major problem associated with antimicrobial drugs is the emergence and spread of new varieties of microorganisms that are resistant to antibiotics. Over the years, more and more microbes have developed resistance to antibiotics that at one time were very effective against them. Drug resistance is due to an adaptive response of microbes that allows them to tolerate a certain amount of an antibiotic that would normally inhibit them (see the box in Chapter 26). These changes might include the production by microbes of chemicals (enzymes) that inactivate antibiotics, changes in the surface of a microbe that prevent an antibiotic from attaching to it, and prevention of an antibiotic from entering the microbe.

The recent increase in the number of tuberculosis cases has brought a drug-resistant strain of the microbe *Mycobacterium tuberculosis* to public attention. However, this increase is the result of more than drug resistance alone. Individuals infected with HIV are much more likely to develop tuberculosis because their immune systems are impaired. Factors such as increased homelessness, drug abuse, and crowded living conditions have also contributed to this rise in tuberculosis cases.

The quest to solve drug resistance and other problems requires sophisticated research techniques and correlated studies that were never dreamed of in the days of Koch and Pasteur.

TABLE 1.1 Continued

NOBEL LAUREATES	YEAR OF PRESENTATION	COUNTRY OF BIRTH	CONTRIBUTION
Renato Dulbecco, Howard Temin, and David Baltimore	1975	United States	Discovered reverse transcriptase and described how RNA viruses could cause cancer
Daniel Nathans, Hamilton Smith, and Werner Arber	1978	United States United States Switzerland	Described the action of restriction enzymes (now used in genetic engineering)
Peter Mitchell	1978	England	Described the chemiosmotic mechanism for ATP synthesis
Paul Berg	1980	United States	Performed experiments in gene splicing (genetic engineering)
Walter Gilbert	1980	United States	Discovered method of DNA sequencing
Aaron Klug	1982	South Africa	Described the structure of tobacco mosaic virus (TMV)
Barbara McClintock	1983	United States	Discovered transposons
César Milstein, Georges J.F. Köhler, and Niels Kai Jerne	1984	Argentina Germany Denmark	Developed technique for producing monoclonal antibodies
Susumu Tonegawa	1987	Japan	Described genetics of antibody production
Johann Deisenhofer, Robert Huber, and Hartmut Michel	1988	Germany	Described the structure of bacterial photosynthesis pigments
J. Michael Bishop and Harold E. Varmus	1989	United States	Discovered cancer-causing genes called oncogenes
Joseph E. Murray and E. Donnall Thomas	1990	United States	Performed first successful transplants by using immunosuppressive agents
Edmond H. Fisher and Edwin G. Krebs	1992	United States	Discovered protein kinases, enzymes that regulate cell growth
Richard J. Roberts and Phillip A. Sharp	1993	Great Britain United States	Discovered genes can be separated on different segments of DNA
Kary B. Mullis	1993	United States	Discovered polymerase chain reaction
Michael Smith	1993	Canada	Discovered site-directed mutagenesis

Modern Developments in Microbiology

The groundwork laid during the Golden Age of Microbiology provided the basis for several monumental achievements during the twentieth century (Table 1.1). New branches of microbiology were developed, including immunology and virology. Most recently, the development of a set of new methods called recombinant DNA technology has revolutionized research and practical applications in all areas of microbiology. DNA stands for deoxyribonucleic acid, a molecule that contains the genetic code (genes).

Immunology

Immunology, the study of immunity, actually dates back in Western culture to Jenner's first vaccine in 1798. Since then, knowledge about the immune system accumulated steadily and has expanded rapidly in the twentieth century. Vaccines are now available for numerous diseases, including measles, mumps, polio, rubella, and hepatitis B.

The smallpox vaccine is so effective that not a single naturally acquired case of the disease has been reported worldwide since 1977. In 1960, interferon, a substance generated by the body's own immune system, was discovered. Interferon inhibits replication of viruses and has triggered considerable research related to treatment of viral diseases and cancer. One of today's biggest challenges for immunologists is learning how the immune system might be stimulated to ward off HIV, the virus responsible for AIDS.

A major advance in immunology occurred in 1933, when **Rebecca Lancefield** proposed a classification system for streptococci based on certain components in the cell walls of the bacteria. By studying how these foreign components reacted with components of the human immune system (see Chapter 10), Lancefield was able to classify streptococci into serotypes (variants within a species). She classified the various serotypes using an alphabetical system (A through O). Streptococci are responsible for a variety of diseases, such as sore throat (strep throat), scarlet fever, and septicemia (blood poisoning).

Virology

The study of viruses, **virology**, actually originated during the Golden Age of Microbiology. In 1892, **Dmitri Iwanowski** reported that the organism that caused mosaic disease of tobacco was so small that it passed through filters fine enough to stop all known bacteria. At the time, Iwanowski was not aware that the organism in question was a virus in the sense that we now understand the term. In 1935, **Wendell Stanley** demonstrated that the organism, called tobacco mosaic virus (TMV), was fundamentally different from other microbes and so simple and homogeneous that it could be crystallized like a chemical compound. Stanley's work facilitated the study of viral structure and chemistry. Since the development of the electron microscope in the 1940s, microbiologists have been able to observe the structure of viruses in detail, and today much is known about their structure and activity.

Recombinant DNA Technology

The set of methods called **recombinant DNA technology** had its origins in two related fields. The first, **microbial genetics**, studies the mechanisms by which microorganisms inherit traits. The second, **molecular biology**, specifically studies how genetic information is carried in molecules of DNA and how DNA directs the synthesis of proteins.

Although molecular biology encompasses all organisms, much of our knowledge of how genes determine specific traits has been revealed through experiments with bacteria. Until the 1930s, all genetic research was based on the study of plant and animal cells. But in the 1940s, scientists turned to unicellular organisms, primarily bacteria, which have several advantages for genetic and biochemical research. For one thing, bacteria are less complex than plants and animals. For another, the life cycles of many bacteria require less than an hour, so scientists can cultivate very large numbers of individuals for study in a relatively short time.

Once science turned to the study of unicellular life, progress in genetics began to appear rapidly. In 1941, **George W. Beadle** and **Edward L. Tatum** demonstrated the relationship between genes and enzymes. DNA was established as the hereditary material in 1944 by **Oswald Avery, Colin MacLeod,** and **Maclyn McCarty.** In 1946, **Joshua Lederberg** and **Edward L. Tatum** discovered that genetic material could be transferred from one bacterium to another by a process called conjugation. Then, in 1953, **James Watson** and **Francis Crick** proposed a model for the structure and replication of DNA. The early 1960s witnessed a further explosion of discoveries relating to the way DNA controls protein synthesis. In 1961 **François Jacob** and **Jacques Monod** discovered messenger RNA (ribonucleic acid), a chemical involved in protein synthesis, and later they made the first major discoveries about regulation of gene function in bacteria. During the same period, scientists were able to break the genetic code

and thus understand the biochemical signals that DNA transmits.

Because of the knowledge gained in such research, microorganisms can now be genetically engineered to manufacture large amounts of human hormones and other urgently needed medical substances. In the late 1960s, **Paul Berg** showed that fragments of human or animal DNA that code for important proteins (genes) can be attached to bacterial DNA. The resulting hybrid was the first example of **recombinant DNA.** When recombinant DNA is inserted into bacteria, the bacteria can be used to make large quantities of the desired protein.

NAMING AND CLASSIFYING MICROORGANISMS

LEARNING OBJECTIVES

- *Recognize the system of scientific nomenclature that uses genus and specific epithet names.*
- *List the five kingdoms of living organisms and the major characteristics of each kingdom.*

The system of nomenclature (naming) for organisms in use today was established in 1735 by **Carolus Linnaeus.** Scientific names are latinized because Latin was the language traditionally used by scholars. Scientific nomenclature assigns each organism two names—the **genus** is the first name and is always capitalized; the **specific epithet** (**species** name) follows and is not capitalized. The organism is referred to by both the genus and the specific epithet, and both names are underlined or italicized. By custom, after a scientific name has been mentioned once, it can be abbreviated with the initial of the genus followed by the specific epithet.

Scientific names can, among other things, describe the organism, honor a researcher, or identify the habitat of the species. For example, consider *Staphylococcus aureus,* a bacterium that is commonly found on the skin of humans and, as we saw earlier, is involved in TSS. *Staphylo-* describes the clustered arrangement of the cells; *coccus* indicates that they are shaped like spheres. The specific epithet, *aureus,* is Latin for golden, the color of many colonies of this bacterium. The genus of the bacterium *Escherichia coli* is named for a scientist, Theodor Escherich, whereas its specific epithet, *coli,* reminds us that *E. coli* live in the colon, or large intestine.

Before the existence of microbes was known, all organisms were grouped into either the animal kingdom or the plant kingdom. When microscopic organisms with characteristics of animals or plants were discovered late in the seventeenth century, a new system of classification was needed. Still, biologists could not agree on the criteria for classifying the new organisms they were seeing until the late 1960s.

In 1969, **Robert Whittaker** of Cornell University devised a five-kingdom classification system that is based on the cellular organization and nutritional patterns of organisms. The five-kingdom system is widely accepted. It groups all organisms into the following kingdoms:

- Procaryotae or Monera (eubacteria and archaeobacteria)*

- Protista (slime molds, protozoa, and some algae)

- Fungi (unicellular yeasts, multicellular molds, and mushrooms)

- Plantae (some algae and all mosses, ferns, conifers, and flowering plants)

- Animalia (including, among others, sponges, worms, insects, and vertebrates)

Classification will be discussed in more detail in Part Two. For now, however, let us simply introduce the major groups of microorganisms described in this text.

THE DIVERSITY OF MICROORGANISMS

LEARNING OBJECTIVE

- *Differentiate among the major groups of organisms studied in microbiology.*

Bacteria

Bacteria are very small, relatively simple, single-celled (unicellular) organisms whose genetic material is not enclosed in a special nuclear membrane (Figure 1.6a). For this reason, bacteria are called **procaryotes**, from Greek words meaning prenucleus. Bacteria make up the kingdom that Whittaker calls Monera; others, including microbiologists, call it Procaryotae. (Criteria for classifying bacteria will be discussed in Part Two.)

Bacterial cells generally appear in one of several shapes. **Bacillus** (rodlike), **coccus** (spherical or ovoid), and *spiral* (corkscrew or curved) are the most common shapes, but some bacteria are star-shaped or square (see Figures 4.1 through 4.4). Individual bacteria may form pairs, chains, clusters, or other groupings; such formations are usually characteristic of a particular genus or species.

Bacteria are enclosed in cell walls that are largely composed of a substance called *peptidoglycan*. (By contrast, cellulose is the main substance of plant and algal cell walls.)

---

*Debate over the proper spelling of the term *archaeobacteria* currently exists. *Archaeobacteria* is used in *Bergey's Manual of Systematic Bacteriology;* other references use *archaebacteria.*

Bacteria generally reproduce by dividing into two equal daughter cells; this process is called *binary fission*. For nutrition, most bacteria use organic chemicals, which in nature can be derived from either dead or living organisms. Some bacteria can manufacture their own food by photosynthesis, and some can derive nutrition from inorganic substances. Many bacteria can "swim" by using moving appendages called *flagella*.

Fungi

Fungi are **eucaryotes**—organisms whose cells have a distinct nucleus that contains the cell's genetic material (DNA) and is surrounded by a special envelope called the nuclear membrane. Organisms in the Kingdom Fungi (see Figure 10.1) may be unicellular or multicellular. Large multicellular fungi, such as mushrooms, may look somewhat like plants, but they cannot carry out photosynthesis, as most plants can. True fungi have cell walls composed primarily of a substance called *chitin*. The unicellular forms of fungi, *yeasts,* are oval microorganisms that are larger than bacteria. The most typical fungi are *molds* (Figure 1.6b). Molds form visible masses called *mycelia,* which are composed of long filaments (*hyphae*) that branch and intertwine. The cottony growths sometimes found on bread and fruit are mold mycelia. Fungi can reproduce sexually or asexually. They obtain nourishment by absorbing solutions of organic material from their environment—whether soil, seawater, freshwater, or an animal or plant host.

Protozoa

Protozoa are unicellular, eucaryotic microbes. They belong to the Kingdom Protista. Protozoa are classified according to their means of locomotion. Amoebas move by using extensions of their cytoplasm called *pseudopods* (false feet) (Figure 1.6c). Other protozoa have long flagella or numerous shorter appendages for locomotion called *cilia*. Protozoa have a variety of shapes and live either as free entities or as parasites (organisms that derive nutrients from living hosts) that absorb or ingest organic compounds from their environment. Protozoa can reproduce sexually or asexually.

Algae

Algae are photosynthetic eucaryotes with a wide variety of shapes and both sexual and asexual reproductive forms (Figure 1.6d). The algae of interest to microbiologists are members of the Kingdom Protista and are usually unicellular. The cell walls of many algae, like those of plants, are composed of cellulose. Algae are abundant in fresh and

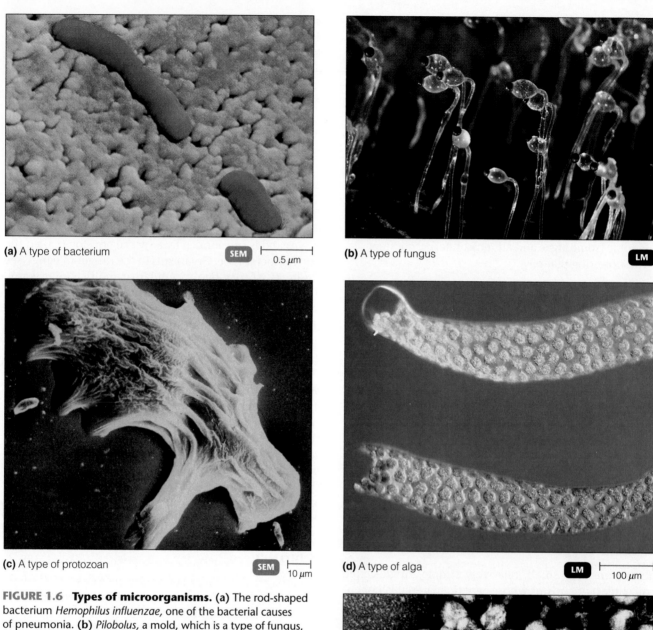

(a) A type of bacterium   `SEM`   0.5 μm

(b) A type of fungus   `LM`

(c) A type of protozoan   `SEM`   10 μm

(d) A type of alga   `LM`   100 μm

FIGURE 1.6   Types of microorganisms. (a) The rod-shaped bacterium *Hemophilus influenzae,* one of the bacterial causes of pneumonia. **(b)** *Pilobolus,* a mold, which is a type of fungus, aiming its sporangia toward the light. When released, the sporangia will stick to grasses and be eaten by grazing animals. **(c)** An amoeba, a protozoan, approaching food. **(d)** The alga *Characiosiphon.* **(e)** Several T4 bacteriophages (viruses that attack bacteria) attached to *E. coli.*

(e) A type of virus   `TEM`   10 μm

salt water, in soil, and in association with plants. As photosynthesizers, algae need light and air for food production and growth, but they do not generally require organic compounds from the environment. As a result of photosynthesis, algae produce oxygen and carbohydrates, which are utilized by other organisms, including animals. Thus they play an important role in the balance of nature.

Viruses

Viruses (Figure 1.6e) are very different from the other microbial groups mentioned here. They are so small that most can be seen only with an electron microscope, and they are acellular (not cellular). Structurally very simple, a virus particle contains a core made of only one type of nucleic acid, either DNA or RNA. This core is surrounded by a protein coat. Sometimes the coat is encased by an additional layer, a lipid membrane called an envelope. All living cells have RNA *and* DNA, can carry out chemical reactions, and can reproduce as self-sufficient units. Viruses can reproduce only by using the cellular machinery of other organisms. Thus, all viruses are parasites of other forms of life.

Multicellular Animal Parasites

Although multicellular animal parasites are not strictly microorganisms, they are of medical importance and so will be discussed in this text. The two major groups of parasitic worms are the flatworms and the roundworms, collectively called **helminths** (see Chapter 12). During some stages of their life cycle, helminths are microscopic in size. Laboratory identification of these organisms includes many of the same techniques used for the identification of microbes.

MICROBES AND HUMAN WELFARE

LEARNING OBJECTIVES

- *List at least four beneficial activities of microorganisms.*
- *List two examples of biotechnology that use genetic engineering and two examples that do not.*

As mentioned earlier, only a minority of all microorganisms are pathogenic. Microbes that cause food spoilage, such as soft spots on fruits and vegetables, decomposition of meats, rancidity of fats and oils, and ropiness of bread, are also a minority. The vast majority of microbes benefit humans as well as other animals and plants in numerous

ways. The following sections outline some of these beneficial activities. In later chapters we will discuss these activities in greater detail.

Recycling Vital Elements

Discoveries made by two microbiologists in the 1880s have formed the basis for today's understanding of the biochemical cycles that support life on Earth. **Martinus Beijerinck** and **Sergei Winogradsky** were the first to show how bacteria help recycle vital elements between the soil and the atmosphere. **Microbial ecology**, the study of the relationship between microorganisms and their environment, originated with the work of Beijerinck and Winogradsky. Today, microbial ecology has branched out and includes the study of how microbial populations interact in various environments with plants and animals. Among the concerns of microbial ecologists are water pollution and toxic chemicals in the environment.

The chemical elements nitrogen, carbon, oxygen, sulfur, and phosphorus are essential for life and are available in limited, though large, amounts. Nitrogen, oxygen, and carbon (as carbon dioxide) exist as gases in the atmosphere. Sulfur and phosphorus are stored in the Earth's crust. Microorganisms are mostly responsible for converting these elements into forms that can be used by plants and animals.

The **nitrogen cycle** is one example of how microorganisms recycle an element (see Chapter 27). Nitrogen is a major constituent of all living cells and is present in proteins and in nucleic acids (DNA and RNA). Although 79% of the atmosphere is gaseous nitrogen, animals and plants cannot use it in that form. Certain soil bacteria and cyanobacteria (Monera) in water combine gaseous nitrogen with other elements (a process called *nitrogen fixation*) to form compounds that other organisms can use. When other bacteria help to decompose dead plants and animals, nitrogen-containing chemicals are released into the soil; still other microbes then convert the soil nitrogen back into nitrogen gas.

In the **carbon cycle**, green plants and algae remove carbon dioxide from the air and use sunlight to convert the carbon dioxide into food by the process of photosynthesis (see Chapter 27). When this food is consumed by other organisms, including humans, carbon dioxide is produced in the process called *respiration*. Microorganisms, mostly bacteria, play another key role in the carbon cycle by returning carbon dioxide to the atmosphere when they decompose organic wastes and dead plants and animals. Algae, higher plants, and certain bacteria (cyanobacteria) also participate in the **oxygen cycle** by recycling oxygen to the air during photosynthesis. The nitrogen, carbon, and oxygen cycles, as well as those involving sulfur and phosphorus, will be discussed in greater detail in Chapters 5 and 27.

Sewage Treatment: Using Microbes to Recycle Water

With our growing awareness of the need to preserve the environment, we are conscious of our responsibility to recycle precious water and prevent the pollution of rivers and oceans. One major pollutant is sewage, which consists of human excrement, waste water, industrial wastes, and surface run-off. Sewage is about 99.9% water with a few hundredths of one percent suspended solids. The remainder is a variety of dissolved materials.

Sewage plant treatments remove the undesirable materials and harmful microorganisms. Treatments combine various physical and chemical processes with the action of beneficial microbes. Large solids such as paper, wood, glass, gravel, and plastic are removed from sewage; left behind are liquid and organic materials that bacteria convert into such by-products as carbon dioxide, nitrates, phosphates, sulfates, ammonia, hydrogen sulfide, and methane. (We shall discuss sewage treatment in detail in Chapter 28.)

Bioremediation: Using Microbes to Clean Up Pollutants

In 1988, scientists began using microbes to clean up pollutants and toxic wastes produced by various industrial processes. For example, some bacteria can actually turn pollutants into energy sources that they consume; others produce enzymes that break down toxins into less harmful substances. By using bacteria in these ways—a process known as **bioremediation**—toxins can be removed from underground wells, chemical spills, toxic waste sites, and oil spills, such as the *Exxon Valdez* disaster of 1989 (see the box in Chapter 2, p. 31). In some cases, microorganisms indigenous to the environment are used; in others, genetically engineered microbes are used. Among the most commonly used bioremedial microbes are certain species of bacteria of the genera *Pseudomonas* and *Bacillus*.

Insect Pest Control by Microorganisms

Besides spreading diseases, insects can cause devastating crop damage. Insect pest control is therefore important both for agriculture and for the prevention of human disease.

The bacterium *Bacillus thuringiensis* (bä-sil′lus thür-in-jē-en′sis) has been used extensively in the United States to control such pests as alfalfa caterpillars, bollworms, corn borers, cabbageworms, tobacco budworms, and fruit tree leaf rollers. It is incorporated into a dusting powder that is applied to the crops these insects eat. The bacteria produce protein crystals that are toxic to the digestive systems of the insects.

By using microbial rather than chemical insect control, farmers can avoid harming the environment. Many chemical insecticides, such as DDT, remain in the soil as toxic pollutants and are eventually incorporated into the food chain.

Modern Biotechnology and Genetic Engineering

Earlier in this chapter we touched on the commercial use of microorganisms to produce some common foods and chemicals. Such practical applications of microbiology are called **biotechnology.** Although biotechnology has been used in some form for centuries, techniques have become much more sophisticated in the past few decades. And in the last 15 years biotechnology has undergone a revolution through the advent of **genetic engineering,** the use of recombinant DNA technology to expand the potential of bacteria as miniature biochemical factories.

The applications of genetic engineering are increasing with each passing year. Recombinant DNA techniques have been used thus far to produce a number of natural proteins that are otherwise expensive and difficult to isolate and purify. Such proteins have great potential for medical use. Among them are insulin, a hormone used to lower blood sugar levels in diabetics; human growth hormone, required for growth during childhood; interferon, an antiviral (and possibly anticancer) substance; factor VIII, a clotting substance missing in the blood of most hemophiliacs; tissue plasminogen activator, a substance used to dissolve blood clots in coronary arteries; erythropoietin, a hormone used to stimulate red blood cell production; beta-endorphin, a substance that suppresses pain; and monoclonal antibodies, which are used to diagnose and treat cancer and assist in AIDS research.

Several vaccines have been developed through recombinant DNA techniques. These include vaccines against hepatitis B, malaria, and influenza. Scientists are using recombinant DNA techniques in attempts to develop vaccines against several other microorganisms, including those that cause AIDS and herpes. Recombinant DNA techniques have also been used in the diagnosis and treatment of cancer in its early stages.

A very exciting and important outcome of recombinant DNA techniques is **gene therapy**—inserting a missing gene or replacing a defective one in human cells. This technique uses a harmless virus to carry the missing or new gene into certain host cells, where the gene is picked up and inserted into the appropriate chromosome. Since 1990, gene therapy has been used to treat patients with ADA (adenosine deaminase) deficiency, a cause of severe combined immunodeficiency disease (SCID); Duchenne's muscular dystrophy; and cystic fibrosis. Results are still being evaluated. Certain genetic diseases such as hemophilia, diabetes, sickle-cell anemia, phenylketonuria

(PKU), one type of hypercholesterolemia, and Huntington's disease may also be treatable by gene therapy in the future.

Beyond medical applications, recombinant DNA techniques have also been applied to agriculture. For example, genetically altered strains of bacteria have been developed to protect fruit against frost damage, and bacteria are being engineered to control insects that damage crops. Bacteria have also been used to improve the appearance, flavor, and shelf life of fruits and vegetables. Potential agricultural uses of genetic engineering include drought resistance, resistance to insects and microbial diseases, and increased temperature tolerance in crops.

SEM $\vdash\!\!\dashv$ 0.5 μm

FIGURE 1.7 **Several types of bacteria found as members of the normal microbiota inside the human mouth.**

MICROBES AND HUMAN DISEASE

LEARNING OBJECTIVE

- *Define normal microbiota.*

We all live in a world filled with microbes, from birth until death, and we all have a variety of microorganisms on and inside our bodies. These microorganisms make up our **normal microbiota** or *flora* (Figure 1.7). Although the normal microbiota do us no harm, and in some cases actually benefit us, under some circumstances they can make us sick or infect people we contact.

When is a microbe a welcome part of a healthy human, and when is it a harbinger of disease? The distinction between health and disease is in large part a balance between the natural defenses of the body and the disease-producing properties of microorganisms. Whether our bodies overcome the offensive tactics of a particular microbe depends on our **resistance.** Important natural resistance is provided by the barrier of the skin, mucous membranes, cilia, stomach acid, and antimicrobial chemicals such as interferon. Microbes can be destroyed by white blood cells, the inflammatory response, fever, and our powerful immune systems. Sometimes, when our natural defenses are not strong enough to overcome an invader, they have to be supplemented by antibiotics or other drugs.

In the chapters that follow you will become acquainted with the principles and tools that microbiologists use to understand the many interactions between microbes and their hosts.

STUDY OUTLINE

MICROBES IN OUR LIVES (pp. 4–5)

1. Living things too small to be seen with the unaided eye are called microorganisms.
2. Microorganisms are important in the maintenance of an ecological balance on Earth.
3. Some microorganisms live in humans and other animals and are needed to maintain the animal's health.
4. Some microorganisms are used to produce foods and chemicals.
5. Some microorganisms cause disease.

A BRIEF HISTORY OF MICROBIOLOGY

(pp. 5–14)

The First Observations (p. 5)

1. Robert Hooke observed that plant material was composed of "little boxes"; he introduced the term *cell* (1665).
2. Hooke's observations were the groundwork for development of the cell theory, the concept that all living things are composed of cells.
3. Anton van Leeuwenhoek, using a simple microscope, was the first to observe microorganisms (1673).

The Debate over Spontaneous Generation
(pp. 5–8)

1. Until the mid-1880s, many people believed in spontaneous generation, the idea that living organisms could arise from nonliving matter.
2. Francesco Redi demonstrated that maggots appear on decaying meat only when flies are able to lay eggs on the meat (1668).
3. John Needham claimed that microorganisms could arise spontaneously from heated nutrient broth (1745).
4. Lazzaro Spallanzani repeated Needham's experiments and suggested that Needham's results were due to microorganisms in the air entering his broth (1765).
5. Rudolf Virchow introduced the concept of biogenesis: Living cells can arise only from preexisting cells (1858).
6. Louis Pasteur demonstrated that microorganisms are in the air everywhere and offered proof of biogenesis (1861).
7. Pasteur's discoveries led to the development of aseptic techniques used in laboratory and medical procedures to prevent contamination by microorganisms that are in the air.

The Golden Age of Microbiology (pp. 8–11)

1. Rapid advances in the science of microbiology were made between 1857 and 1914.

Fermentation and Pasteurization (pp. 8–10)

1. Pasteur found that yeast ferments sugars to alcohol and that bacteria can oxidize the alcohol to acetic acid.
2. A heating process called pasteurization is used to kill bacteria in some alcoholic beverages and milk.

The Germ Theory of Disease (p. 10)

1. Agostino Bassi (1835) and Pasteur (1865) showed a causal relationship between microorganisms and disease.
2. Joseph Lister introduced the use of a disinfectant to clean surgical dressings in order to control infections in humans (1860s).
3. Robert Koch proved that microorganisms cause disease. He used a sequence of procedures called Koch's postulates (1876), which are used today to prove that a particular microorganism causes a particular disease.

Vaccination (pp. 10–11)

1. In a vaccination, immunity (resistance to a particular disease) is conferred by inoculation with a vaccine.
2. In 1798, Edward Jenner demonstrated that inoculation with cowpox material provides humans with immunity from smallpox.
3. About 1880, Pasteur discovered that avirulent bacteria could be used as a vaccine for fowl cholera; he coined the word *vaccine.*
4. Modern vaccines are prepared from living avirulent microorganisms or killed pathogens, from isolated components of pathogens, and by recombinant DNA techniques.

The Birth of Modern Chemotherapy: Dreams of a "Magic Bullet" (pp. 11–12)

1. Chemotherapy is the chemical treatment of a disease.
2. Two types of chemotherapeutic agents are synthetic drugs (chemically prepared in the laboratory) and antibiotics (substances produced naturally by bacteria and fungi to inhibit the growth of other microorganisms).
3. Paul Ehrlich introduced an arsenic-containing chemical called salvarsan to treat syphilis (1910).
4. Alexander Fleming observed that the mold (fungus) *Penicillium* inhibited the growth of a bacterial culture. He named the active ingredient penicillin (1928).
5. Penicillin has been used clinically as an antibiotic since the 1940s.
6. In 1939, René Dubos discovered two antibiotics produced by the bacterium *Bacillus.*
7. Researchers are tackling the problem of drug-resistant microbes.

Modern Developments in Microbiology (pp. 13–14)

1. The study of AIDS, analysis of interferon action, and the development of new vaccines are among the current research interests in immunology.
2. New techniques in molecular biology and electron microscopy have provided tools for advancement of our knowledge of virology.
3. The development of recombinant DNA technology has helped advance all areas of microbiology.

NAMING AND CLASSIFYING MICROORGANISMS (pp. 14–15)

1. In a nomenclature system designed by Carolus Linnaeus (1735), each living organism is assigned two names.
2. The two names consist of a genus and specific epithet, both of which must be underlined or italicized.
3. In the five-kingdom system, all organisms are classified into Procaryotae (or Monera), Protista, Fungi, Plantae, or Animalia.

THE DIVERSITY OF MICROORGANISMS
(pp. 15–17)

Bacteria (p. 15)

1. Bacteria are unicellular organisms. Because they have no nucleus, the cells are described as procaryotic.
2. The three major basic shapes of bacteria are bacillus, coccus, and spiral.
3. Most bacteria have a peptidoglycan cell wall; they divide by binary fission; and they may possess flagella.
4. Bacteria can use a wide range of chemical substances for their nutrition.

Fungi (p. 15)

1. Fungi (mushrooms, molds, and yeasts) have eucaryotic cells (with a true nucleus). Most fungi are multicellular.
2. Fungi obtain nutrients by absorbing organic material from their environment.

Protozoa (p. 15)

1. Protozoa are unicellular eucaryotes and are classified according to their means of locomotion.
2. Protozoa obtain nourishment by absorption or ingestion through specialized structures.

Algae (pp. 15–17)

1. Algae are unicellular or multicellular eucaryotes that obtain nourishment by photosynthesis.
2. Algae produce oxygen and carbohydrates that are used by other organisms.

Viruses (p. 17)

1. Viruses are noncellular entities that are parasites of cells.
2. Viruses consist of a nucleic acid core (DNA or RNA) surrounded by a protein coat. An envelope may surround the coat.

Multicellular Animal Parasites (p. 17)

1. The principal groups of multicellular animal parasites are flatworms and roundworms, collectively called helminths.
2. The microscopic stages in the life cycle of helminths are identified by traditional microbiologic procedures.

MICROBES AND HUMAN WELFARE (pp. 17–19)

1. Microorganisms degrade dead plants and animals and recycle chemical elements to be used by living plants and animals.
2. Bacteria are used to decompose organic matter in sewage.
3. Bioremediation processes use bacteria to clean up toxic wastes.
4. Bacteria that cause diseases in insects are being used as biological controls of insect pests. Biological controls are specific for the pest and do not harm the environment.
5. Using microbes to make products such as foods and chemicals is called biotechnology.
6. Using recombinant DNA, bacteria can produce important human proteins, such as insulin, beta-endorphin, and hepatitis B vaccine.
7. In gene therapy, viruses are used to carry replacements for defective or missing genes into human cells.
8. Genetically engineered bacteria are used in agriculture to protect plants from frost and insects and to improve the shelf life of produce.

MICROBES AND HUMAN DISEASE (p. 19)

1. Everyone has microorganisms in and on the body; these make up the normal microbiota or flora.
2. The disease-producing properties of a species of microbe and the host's resistance are important factors in determining whether a person will contract a disease.

STUDY QUESTIONS

Review

1. How did the idea of spontaneous generation come about?
2. Some proponents of spontaneous generation believed that air is necessary for life. They thought that Spallanzani did not really disprove spontaneous generation because he hermetically sealed his flasks to keep air out. How did Pasteur's experiments address the air question without allowing the microbes in the air to ruin his experiment?
3. Briefly state the role played by microorganisms in each of the following:
   (a) Biological control of pests
   (b) Recycling of elements
   (c) Normal microbiota
   (d) Sewage treatment
   (e) Human insulin production
   (f) Vaccine production

4. Into which field of microbiology would the following scientists best fit?

RESEARCHER WHO	FIELD
____ Studies biodegradation of toxic wastes	(a) Immunology
____ Studies the causative agent of AIDS	(b) Microbial ecology
____ Studies the production of human proteins by bacteria	(c) Microbial genetics
____ Studies the symptoms of AIDS	(d) Microbial physiology
____ Studies the production of TSS toxin by *S. aureus*	(e) Molecular biology
____ Studies the life cycle of *B. burgdorferi*	(f) Virology
____ Develops gene therapy for a disease	

5. Match the following people to their contribution toward the advancement of microbiology.

____	Avery, MacLeod, and McCarty	(a)	Developed vaccine against smallpox
____	Beadle and Tatum	(b)	Discovered antibiotics produced by bacteria
____	Berg	(c)	Discovered how DNA controls protein synthesis in a cell
____	Dubos		
____	Ehrlich		
____	Fleming	(d)	Discovered penicillin
____	Hooke	(e)	Discovered that DNA can be transferred from one bacterium to another
____	Iwanowski		
____	Jacob and Monod		
____	Jenner	(f)	Disproved spontaneous generation
____	Koch	(g)	First to characterize a virus
____	Lancefield		
____	Lederberg and Tatum	(h)	First to use disinfectants in surgical procedures
____	Lister	(i)	First to observe bacteria
____	Pasteur	(j)	First to observe cells in plant material and name them
____	Stanley		
____	van Leeuwenhoek		
____	Virchow	(k)	Observed that viruses are filterable
____	Weizmann	(l)	Proved that DNA is the hereditary material

(m) Proved that microorganisms can cause disease

(n) Said living cells arise from preexisting living cells

(o) Showed that genes code for enzymes

(p) Spliced animal DNA to bacterial DNA

(q) Used bacteria to produce acetone

(r) Used the first synthetic chemotherapeutic agent

(s) Proposed a classification system for streptococci based on antigens in their cell walls

6. The genus name of a bacterium is "erwinia" and the specific epithet is "carotovora." Write the scientific name of this organism correctly. Using this name as an example, explain how scientific names are chosen.

7. Match the following microorganisms to their descriptions.

____	Algae	(a)	Not composed of cells
____	Bacteria	(b)	Cell wall made of chitin
____	Fungi	(c)	Cell wall made of peptidoglycan
____	Protozoa	(d)	Cell wall made of cellulose; photosynthetic
____	Viruses	(e)	Complex cell structure lacking a cell wall

Critical Thinking

1. How did the theory of biogenesis lead the way for the germ theory of disease?

2. Even though the germ theory of disease was not demonstrated until 1876, why did Semmelweis (1840) and Lister (1867) argue for the use of aseptic techniques?

3. Find at least three supermarket products made by microorganisms. (*Hint:* The label will state the scientific name of the organism or include the word *culture, fermented,* or *brewed.*)

Clinical Applications

1. The prevalence of arthritis in the U.S.A. is 1 in 100,000 children. However, 1 in 10 children in Lyme, Connecticut, developed arthritis between June and September in 1973. Allen Steere, a rheumatologist at Yale University, investigated the cases in Lyme and found that 25% of the patients remembered having a skin rash during their arthritic episode and that the disease was treatable with penicillin. Steere concluded that this was a new infectious disease and did not have an environmental, genetic, or immunologic cause.
   (a) What was the factor that caused Steere to reach his conclusion?
   (b) What is the disease?
   (c) Why was the disease more prevalent between June and September?

2. In 1864, Lister observed that patients recovered completely from simple fractures, but compound fractures had "disastrous consequences." He knew that the application of phenol (carbolic acid) to fields in the town of Carlisle prevented cattle disease. In 1864, Lister treated compound fractures with phenol, and his patients recovered without complications. How was Lister influenced by Pasteur's work? Why was Koch's work still needed?

Chemical Principles

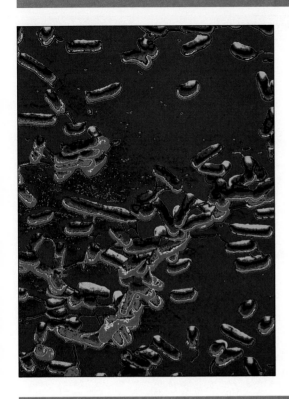

W E CAN SEE a tree rot and smell milk going sour, but we might not realize what is happening on a microscopic level. In both cases, microbes are conducting chemical operations. The tree rots when microorganisms induce the hydrolysis of wood. (Hydrolysis is a chemical reaction that involves the addition and subsequent splitting of water.) Milk turns sour from the production of lactic acid by bacteria. Most of the activities of microorganisms are the result of a series of chemical reactions.

Like all organisms, microorganisms must use nutrients to make chemical building blocks that are used for growth and for all the other functions essential to life. For most microorganisms, synthesizing these building blocks requires them to break down nutrient substances and use the energy released to assemble the resulting molecular fragments into new substances. These chemical reactions take place minute by minute in countless microenvironments that are, in many cases, smaller than the head of a pin.

All matter—whether air, rock, or a living organism—is made up of small units called **atoms.** Atoms interact with each other in certain combinations to form **molecules.** Living cells are made up of molecules, some of which are very complex. The science of the interaction of atoms and molecules is called **chemistry.**

The chemistry of microbes is among the microbiologist's concerns. To understand the changes that occur in microorganisms and the changes microbes make in the world around us, we need to know how molecules are formed and how they interact.

STRUCTURE OF ATOMS

LEARNING OBJECTIVE

• *Discuss the structure of an atom and its relation to the chemical properties of elements.*

Atoms are the smallest units of matter that enter into chemical reactions. Each atom (Figure 2.1) has a centrally located **nucleus** and particles called **electrons** that move around the nucleus in patterns called electronic configurations. The nuclei of most atoms are stable—that is, they do not change spontaneously—and nuclei do not participate in chemical reactions. The nucleus is made up of positively (+) charged particles called **protons** and uncharged (neutral) particles called **neutrons.** The nucleus, therefore, bears a net positive charge. Neutrons and protons have approximately the same weight, which is about 1840 times that of an electron. The charge on electrons is negative (−), and in all atoms the number of protons is equal to the number of electrons. Because the total positive charge of the nucleus equals the total negative charge of the electrons, each atom is electrically neutral.

The number of protons in an atomic nucleus ranges from one (in a hydrogen atom) to more than 100 (in the largest atoms known). Atoms are often listed by their **atomic number,** the number of protons in the nucleus. The total number of protons and neutrons in an atom is its approximate **atomic weight.**

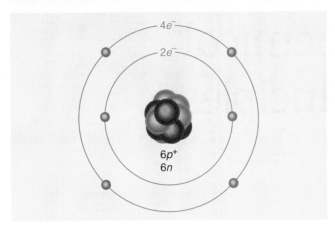

FIGURE 2.1   Atomic structure. Illustrated here is a very simplified version of a carbon atom (C). It contains six protons (p^+, gray) and six neutrons (n, purple) in its centrally located nucleus. Six electrons (e^-, magenta) orbit the nucleus.

Chemical Elements

All atoms with the same atomic number behave the same way chemically and are classified as the same **chemical element.** Each element has its own name and a one- or two-letter symbol, usually derived from the English or Latin name for the element. For example, the symbol for the element hydrogen is H, and the symbol for carbon is C. The symbol for sodium is Na—the first two letters of its Latin name, *natrium*—to distinguish it from nitrogen, N, and from sulfur, S. There are 92 naturally occurring elements. However, only about 26 elements are commonly found in living things; these are listed in Table 2.1, including their atomic numbers and weights. The elements most abundant in living matter are hydrogen, carbon, nitrogen, and oxygen.

Most elements have several **isotopes**—atoms with different numbers of neutrons in their nuclei. All isotopes of an element have the same number of protons in their nuclei, but their atomic weights differ because of the difference in the number of neutrons. For example, in a natural sample of oxygen, all the atoms will contain eight protons. However, 99.76% of the atoms will have eight neutrons, 0.04% will contain nine neutrons, and the remaining 0.2% will contain ten neutrons. Therefore, the three isotopes composing a natural sample of oxygen will have atomic weights of 16, 17, and 18, although all will have the atomic number 8. Atomic numbers are written as a subscript to the left of an element's chemical symbol. Atomic weights are written as a superscript above the atomic number. Thus, natural oxygen isotopes are designated as $^{16}_{8}O$, $^{17}_{8}O$, and $^{18}_{8}O$. Rare isotopes of certain elements are extremely useful in biological research, medical diagnosis, and treatment of some disorders.

TABLE 2.1   The Elements of Life*

ELEMENT	SYMBOL	ATOMIC NUMBER	APPROXIMATE ATOMIC WEIGHT
Hydrogen	H	1	1
Carbon	C	6	12
Nitrogen	N	7	14
Oxygen	O	8	16
Fluorine	F	9	19
Sodium	Na	11	23
Magnesium	Mg	12	24
Silicon	Si	14	28
Phosphorus	P	15	31
Sulfur	S	16	32
Chlorine	Cl	17	35
Potassium	K	19	39
Calcium	Ca	20	40
Vanadium	V	23	51
Chromium	Cr	24	52
Manganese	Mn	25	55
Iron	Fe	26	56
Cobalt	Co	27	59
Nickel	Ni	28	59
Copper	Cu	29	64
Zinc	Zn	30	65
Arsenic	As	33	75
Selenium	Se	34	79
Molybdenum	Mo	42	96
Tin	Sn	50	119
Iodine	I	53	127

*Hydrogen, carbon, nitrogen, and oxygen are the most abundant chemical elements in living organisms.

Electronic Configurations

In an atom, electrons are arranged in **electron shells,** which are regions corresponding to different **energy levels.** The arrangement is called an **electronic configuration.** Shells are layered outward from the nucleus, and each shell can hold a characteristic maximum number of electrons—two electrons in the innermost shell (lowest energy level), eight electrons in the second shell, and eight electrons in the third shell, if it is the atom's outermost (valence) shell. The fourth, fifth, and sixth electron shells can each accommodate 18 electrons, although there are some exceptions to this generalization. Table 2.2 shows

TABLE 2.2 Electronic Configurations for the Atoms of Some Elements Found in Living Organisms

ELEMENT	FIRST ELECTRON SHELL	SECOND ELECTRON SHELL	THIRD ELECTRON SHELL	DIAGRAM	NUMBER OF VALENCE (OUTERMOST) SHELL ELECTRONS	NUMBER OF UNFILLED SPACES	MAXIMUM NUMBER OF BONDS FORMED
Hydrogen	1	—	—	H	1	1	1 (by either losing or gaining 1 electron)
Carbon	2	4	—	C	4	4	4 (by either losing or gaining 4 electrons)
Nitrogen	2	5	—	N	5	3	3 (by gaining or losing 3 electrons)
Oxygen	2	6	—	O	6	2	2 (by gaining 2 electrons)
Magnesium	2	8	2	Mg	2	6	2 (by losing 2 electrons)
Phosphorus	2	8	5	P	5	3	5 (by losing 5 electrons)
Sulfur	2	8	6	S	6	2	2 (by gaining 2 electrons)

Key: ● electron
◌ unfilled space
◯ atomic nucleus

the electronic configurations for atoms of some elements found in living organisms.

There is a tendency for the outermost shell to be filled with the maximum number of electrons. An atom can give up, accept, or share electrons with other atoms to fill this shell. The chemical properties of atoms are largely a function of the number of electrons in the outermost electron shell. When its outer shell is filled, the atom is chemically stable, or inert: it does not tend to react with other atoms. Helium (atomic number 2) and neon (atomic number 10) are examples of atoms of inert gases that have filled outer shells. Helium has two electrons in the first and only shell, and neon has two electrons in the first shell and eight electrons in the second shell, its outermost shell.

When an atom's outer electron shell is only partially filled, the atom is chemically unstable. Such an atom reacts with other atoms, and this reaction depends, in part, on the degree to which the outer energy levels are filled. Note the number of electrons in the outer energy levels of the atoms in Table 2.2. We will see later how the number correlates with the chemical reactivity of the elements.

HOW ATOMS FORM MOLECULES: CHEMICAL BONDS

LEARNING OBJECTIVE

- *Define ionic bond, covalent bond, hydrogen bond, molecular weight, and mole.*

When the outermost energy level of an atom is not completely filled by electrons, you can think of it as having either unfilled spaces or extra electrons in that energy level, depending on whether it is easier for the atom to gain or lose electrons. For example, an atom of oxygen, with two electrons in the first energy level and six in the second, has two unfilled spaces in the second electron shell; an atom of magnesium has two extra electrons in its outermost shell. The most chemically stable configuration for any atom is to have its outermost shell filled, as do the inert gases. Therefore, for these two atoms to attain that state, oxygen must gain two electrons, and magnesium must lose two electrons. All atoms tend to combine so that the extra electrons in the outermost shell of one atom fill the spaces of the outermost shell of the other atom; for example, oxygen and magnesium combine so that the outermost shell of each atom has the full complement of eight electrons.

The **valence,** or combining capacity, of an atom is the number of extra or missing electrons in its outermost electron shell. For example, hydrogen has a valence of 1 (one unfilled space, or one extra electron), oxygen has a valence of 2 (two unfilled spaces), carbon has a valence of 4 (four unfilled spaces, or four extra electrons), and magnesium has a valence of 2 (two extra electrons).

Basically, atoms achieve the full complement of electrons in their outermost energy shells by combining to form molecules, which are made up of atoms of one or more elements. A molecule that contains at least two different kinds of atoms, such as H_2O, the water molecule, is called a **compound.** In H_2O, the subscript 2 indicates that there are two atoms of hydrogen; the lack of another subscript indicates that there is only one atom of oxygen. Molecules hold together because the valence electrons of the combining atoms form attractive forces, called **chemical bonds,** between the atomic nuclei. Therefore, valence may also be viewed as the bonding capacity of an element. Because energy is required for chemical bond formation, each chemical bond possesses a certain amount of potential chemical energy. Atoms of a given element can form only a specific number of chemical bonds, based on their electron configurations. Hydrogen can form one chemical bond with another atom, oxygen can form two chemical bonds with other atoms, carbon can form four chemical bonds, and magnesium can form two chemical bonds.

In general, atoms form bonds in one of two ways: by gaining or losing electrons from their outer electron shell or by sharing outer electrons. When atoms have gained or lost outer electrons, the chemical bond is called an ionic bond. When outer electrons are shared, the bond is called a covalent bond. Although we will discuss ionic and covalent bonds separately, the kinds of bonds actually found in molecules do not belong entirely to either category. Instead, bonds range from the highly ionic to the highly covalent.

Ionic Bonds

Atoms are electrically neutral when the number of positive charges (protons) equals the number of negative charges (electrons). But when an isolated atom gains or loses electrons, this balance is upset. If the atom gains electrons, it acquires an overall negative charge; if the atom loses electrons, it acquires an overall positive charge. Such a negatively or positively charged atom (or group of atoms) is called an **ion.**

Consider the following examples (Figure 2.2). Sodium (Na) has 11 protons and 11 electrons, with one electron in its outer electron shell. Sodium tends to lose the single outer electron; it is an *electron donor* (Figure 2.2a). When sodium donates an electron to another atom, it is left with 11 protons and only 10 electrons and so has an overall charge of +1. This positively charged sodium atom is called a sodium ion and is written as Na^+. Chlorine (Cl) has a total of 17 electrons, 7 of them in the outer electron shell. Because this outer shell can hold 8 electrons, chlorine tends to pick up an electron that has been lost by another atom; it is an *electron acceptor* (see Figure 2.2a). By accepting an electron, chlorine totals 18 electrons. However, it still has only 17 protons in its nucleus. The chloride ion therefore has a charge of -1 and is written as Cl^-.

FIGURE 2.2 Ionic bond formation. (a) A sodium atom (Na), left, loses one electron to an electron acceptor and forms a sodium ion (Na⁺). A chlorine atom (Cl), right, accepts one electron from an electron donor to become a chloride ion (Cl⁻). **(b)** The sodium and chloride ions are attracted because of their opposite charges and are held together by an ionic bond to form a molecule of sodium chloride.

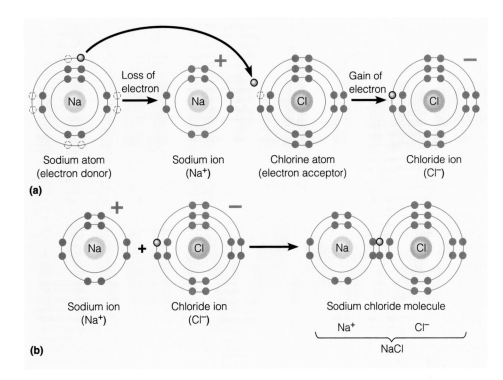

The opposite charges of the sodium ion (Na⁺) and chloride ion (Cl⁻) attract each other. The attraction, an ionic bond, holds the two atoms together, and a molecule is formed (Figure 2.2b). The formation of this molecule, called sodium chloride (NaCl) or table salt, is a common example of ionic bonding. Thus, an **ionic bond** is an attraction between ions of opposite charge that holds them together to form a stable molecule. Put another way, an ionic bond is an attraction between atoms in which one atom loses electrons and another atom gains electrons. Strong ionic bonds, such as those that hold Na⁺ and Cl⁻ together in salt crystals, have only limited importance in living cells. But the weaker ionic bonds formed in aqueous (water) solutions are important in biochemical reactions of microbes and other organisms. For example, weaker ionic bonds assume a role in certain antigen-antibody reactions—that is, reactions in which molecules produced by the immune system (antibodies) combine with foreign substances (antigens) in order to combat infection.

In general, an atom whose outer electron shell is less than half filled will lose electrons and form positively charged ions, called **cations**. Examples of cations are the potassium ion (K⁺), calcium ion (Ca²⁺), and sodium ion (Na⁺). When an atom's outer electron shell is more than half filled, the atom will gain electrons and form negatively charged ions, called **anions**. Examples are the iodide ion (I⁻), chloride ion (Cl⁻), and sulfide ion (S²⁻).

Hydrogen is an example of an atom whose outer level is exactly half filled. The first energy level of a hydrogen atom can hold two electrons, but it contains only one electron. Hydrogen can lose its electron and become a positive

ion (H⁺), which is precisely what has happened when a hydrogen ion combines with a chloride ion to form hydrochloric acid (HCl). Hydrogen can gain an electron and become a negative ion (H⁻), which can form ionic bonds with positive ions, but H⁻ is not very significant in living systems. Hydrogen atoms can also form a more important and altogether different kind of bond, a covalent bond.

Covalent Bonds

A **covalent bond** is a chemical bond formed by two atoms sharing one or more pairs of electrons. Covalent bonds are stronger and far more common in organisms than are true ionic bonds. In the hydrogen molecule, H₂, two hydrogen atoms share a pair of electrons. Each hydrogen atom has its own electron plus one electron from the other atom (Figure 2.3a). The shared pair of electrons actually orbits the nuclei of both atoms. Therefore, the outer electron shells of both atoms are filled. When only one pair of electrons is shared between atoms, a *single covalent bond* is formed. For simplicity, a single covalent bond is expressed as a single line between the atoms (H—H). When two pairs of electrons are shared between atoms, a *double covalent bond* is formed, expressed as two single lines (=) (Figure 2.3b). A *triple covalent bond*, expressed as three single lines (≡), occurs when three pairs of electrons are shared (Figure 2.3c).

The principles of covalent bonding that apply to atoms of the same element also apply to atoms of different elements. Methane (CH₄) is an example of covalent bonding

(a) Single covalent bond between two hydrogen atoms

(b) Double covalent bond between two oxygen atoms

(c) Triple covalent bond between two nitrogen atoms

(d) Single covalent bonds between four hydrogen atoms and a carbon atom, forming the methane molecule

FIGURE 2.3   Covalent bond formation.

between atoms of different elements (Figure 2.3d). The outer electron shell of the carbon atom can hold eight electrons but has only four; each hydrogen atom can hold two electrons but has only one. Consequently, in the methane molecule, the carbon atom gains four hydrogen electrons to complete its outer shell, and each hydrogen atom completes its pair by sharing one electron from the carbon atom. Each outer electron of the carbon atom orbits both the carbon nucleus and a hydrogen nucleus. Each hydrogen electron orbits both its own nucleus and the carbon nucleus.

Elements such as hydrogen and carbon, whose outer electron shells are half-filled, form covalent bonds quite easily. In fact, in living organisms, carbon almost always forms covalent bonds; it almost never becomes an ion. However, many atoms whose outer electron shells are almost full will also form covalent bonds. An example is oxygen. For now it is not necessary to know why some atoms tend to form covalent bonds rather than ionic bonds. But it is important to understand the basic principles of bond formation because chemical reactions are nothing more than the making or breaking of bonds between atoms. *Remember:* Covalent bonds are formed by the *sharing* of electrons between atoms. Ionic bonds are formed by *attractions* between atoms that have lost or gained electrons and are therefore positively or negatively charged.

Hydrogen Bonds

Another chemical bond of special importance to all organisms is the **hydrogen bond**, in which a hydrogen atom that is covalently bonded to one oxygen or nitrogen atom is attracted to another oxygen or nitrogen atom. Such bonds are weak and do not bind atoms into molecules. However, they do serve as bridges between different molecules or between various portions of the same molecule.

When hydrogen combines with atoms of oxygen or nitrogen, the relatively large positive nucleus of these larger atoms attracts the hydrogen electron more strongly than does the small hydrogen nucleus. Thus, in a molecule of water (H_2O), all the electrons tend to be closer to the oxygen nucleus than to the hydrogen nuclei. The oxygen portion of the molecule thus has a slightly negative charge, and the hydrogen portion of the molecule has a slightly positive charge (Figure 2.4a). When the positively charged end of one molecule is attracted to the negatively charged end of another molecule, a hydrogen bond is formed (Figure 2.4b). This attraction can also occur between hydrogen and other atoms of the same molecule, especially in large molecules. Because nitrogen and oxygen have unshared pairs of electrons, they are the elements most frequently involved in hydrogen bonding.

Hydrogen bonds are considerably weaker than either ionic or covalent bonds; they have only about 5% of the strength of covalent bonds. Consequently, hydrogen bonds are formed and broken relatively easily. This property accounts for the temporary bonding that occurs between certain atoms of large and complex molecules, such as proteins and nucleic acids. Even though hydrogen bonds are relatively weak, large molecules containing several hundred of these bonds have considerable strength and stability.

FIGURE 2.4 Hydrogen bond formation in water. (a) In a water molecule, the electrons of the hydrogen atoms are strongly attracted to the oxygen atom. Therefore, the part of the water molecule containing the oxygen atom has a slightly negative charge, and the part containing hydrogen atoms has a slightly positive charge. **(b)** In a hydrogen bond between water molecules, the hydrogen of one water molecule is attracted to the oxygen of another water molecule. Many water molecules may be attracted to each other by hydrogen bonds (black dots).

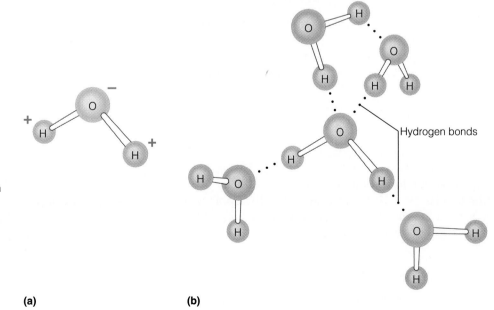

(a)   **(b)**

Molecular Weight and Moles

You have seen that bond formation results in the creation of molecules. Molecules are often discussed in terms of units of measure called molecular weight and moles. The **molecular weight** of a molecule is the sum of the atomic weights of all its atoms. To relate the molecular level to the laboratory level, we use a unit called the mole. One **mole** of a substance is its molecular weight expressed in grams. For example, one mole of water weighs 18 grams because the molecular weight of H_2O is 18 [$(2 \times 1) + 16$]. The word *mole* can also be applied to atoms or even to ions. Thus, one mole of hydrogen ions is equal to 1 gram.

To determine the number of moles in a certain quantity of a substance, the mass of the substance (in grams) is divided by the molecular weight of the substance. For example, to calculate the number of moles in 500 grams of glucose ($C_6H_{12}O_6$), first calculate the number of grams in one mole [$(6 \times 12) + (12 \times 1) + (6 \times 16) = 180$]. Then divide 500 grams by 180 grams per mole to get 2.78 moles.

CHEMICAL REACTIONS

LEARNING OBJECTIVES

- *Diagram three basic types of chemical reactions.*
- *Identify the role of enzymes in chemical reactions.*

As we said earlier, **chemical reactions** involve the making or breaking of bonds between atoms. After a chemical reaction, the total number of atoms remains the same, but there are new molecules with new properties because the atoms have been rearranged.

Energy of Chemical Reactions

Some change of energy occurs whenever bonds between atoms are formed or broken during chemical reactions. This energy is called **chemical energy.** When a chemical bond is formed, energy is required. Such a chemical reaction that requires energy is called an **endergonic reaction,** meaning that energy is directed inward. When a bond is broken, energy is released. A chemical reaction that releases energy is an **exergonic reaction,** meaning that energy is directed outward.

In this section we will look at three basic types of chemical reactions common to all living cells. By becoming familiar with these reactions, you will be able to understand the specific chemical reactions we will discuss later, particularly in Chapter 5.

Synthesis Reactions

When two or more atoms, ions, or molecules combine to form new and larger molecules, the reaction is called a **synthesis reaction.** To synthesize means to put together,

and a synthesis reaction *forms new bonds.* Synthesis reactions can be expressed in the following way:

A   +   B $\longrightarrow$ AB
Atom, ion,   Atom, ion,   Combine   New molecule
or molecule A   or molecule B   to form   AB

The combining substances, A and B, are called the *reactants;* the substance formed by the combination, AB, is the *product.* The arrow indicates the direction in which the reaction proceeds.

Pathways of synthesis reactions in living organisms are collectively called anabolic reactions, or simply **anabolism.** The combining of sugar molecules to form starch, and of amino acids to form proteins, are two examples of anabolism.

Decomposition Reactions

The reverse of a synthesis reaction is a **decomposition reaction.** To decompose means to break down into smaller parts, and in a decomposition reaction *bonds are broken.* Typically, decomposition reactions split large molecules into smaller molecules, ions, or atoms. A decomposition reaction occurs in the following way:

AB $\longrightarrow$ A   +   B
Molecule AB   Breaks   Atom, ion,   Atom, ion,
            down into   or molecule A   or molecule B

Decomposition reactions that occur in living organisms are collectively called catabolic reactions, or simply **catabolism.** An example of catabolism is the breakdown of sucrose (table sugar) into simpler sugars, glucose and fructose, during digestion. Bacterial decomposition of petroleum is discussed in the box on the facing page.

Exchange Reactions

All chemical reactions are based on synthesis and decomposition. Many reactions, such as **exchange reactions,** are actually part synthesis and part decomposition. An exchange reaction works in the following way:

AB + CD $\longrightarrow$ AD + BC
            Recombine
            to form

First, the bonds between A and B and between C and D are broken in a decomposition process. New bonds are then formed between A and D and between B and C in a synthesis process. For example, an exchange reaction occurs when sodium hydroxide (NaOH) and hydrochloric acid (HCl) react to form table salt (NaCl) and water (H_2O), as follows:

NaOH + HCl $\longrightarrow$ NaCl + H_2O

BIOREMEDIATION—BACTERIAL BANQUETERS ATTEND OIL SPILL

READERS OF SCIENCE fiction have long realized that beings from another planet might have quite a different chemical makeup from earthlings and might be able to eat, drink, and breathe the substances we find deadly. Such aliens could be invaluable in helping clean up pollutants such as oil and mercury, which harm plants, animals, and humans alike. Fortunately, however, we need not wait for a visit from outer space to find creatures whose unusual chemistry can be harnessed for environmental cleanup. Although many bacteria have dietary requirements similar to ours—that's why they cause food spoilage—others metabolize (or process chemically) the substances we might expect at a banquet of extraterrestrials—heavy metals, sulfur, nitrogen gas, petroleum, and even polychlorinated biphenyls (PCBs) and mercury.

Bacteria have several advantages as pollution fighters. They can extract pollutants that have combined with soil and water and hence cannot be simply shoveled away. In addition, they may chemically alter a harmful substance so that it becomes harmless or even beneficial. Bacteria that are able to degrade many pollutants are naturally present in soil and water; their use to degrade pollutants is called bioremediation. However, their small numbers make them inefficient

to deal with large-scale contamination. Scientists are now working to improve the efficiency of natural pollution fighters and, in some cases, are altering organisms by genetic engineering to give them exactly the right chemical appetites.

One of the most promising successes for bioremediation occurred on an Alaskan beach following the *Exxon Valdez* oil spill. Several naturally occurring bacteria in the genus *Pseudomonas* are able to degrade oil for their carbon and energy requirements. In the presence of air, they remove two carbons at a time from a large petroleum molecule (see the figures).

The bacteria degrade the oil too slowly to be helpful in cleaning up an oil spill. However, scientists hit upon a very simple way to speed up the process—with no need for genetic engineering. They simply dumped ordinary nitrogen and phosphorus plant fertilizers onto the beach (a process called bioaugmentation). The number of oil-degrading bacteria increased compared with that on unfertilized control beaches, and the test beach is now free of oil.

Another group of bacteria is being investigated for its ability to clean up mercury contamination. Mercury is present in such common substances as discarded leftover paint and can leak into soil and water from garbage dumps. One species of bacteria that is common in the environment, *Desulfovibrio desulfuricans,* actually makes the mercury more dangerous by adding a methyl group, converting it into the highly toxic substance methyl mercury. Methyl mercury in ponds or marshes sticks to small organisms such as plankton, which are eaten by

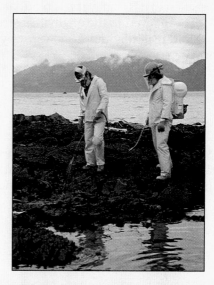

larger organisms, which in turn are eaten by fish. Fish and human poisonings have been attributed to the ingestion of methyl mercury.

However, other bacteria, such as species of *Pseudomonas,* may offer the solution. To avoid mercury poisoning, these bacteria first convert methyl mercury to mercuric ion:

$$CH_3Hg \longrightarrow CH_4 + Hg^{2+}$$

Methyl mercury   Methane   Mercuric ion

They then convert the positively charged mercuric ion to the relatively harmless elemental form by adding electrons, which they take from hydrogen atoms:

$$Hg^{2+} + 2H \xrightarrow{2e^-} Hg + 2H^+$$

Mercuric ion   Hydrogen atoms   Elemental mercury   Hydrogen ions

These bacteria work too slowly in nature for cleaning up human-made toxic spills, but scientists are experimenting with bioaugmentation and other techniques to increase their effectiveness. Unlike some forms of environmental cleanup, in which dangerous substances are removed from one place only to be dumped in another, bacterial cleanup eliminates the toxic substance and often returns a harmless or useful substance to the environment.

Typical saturated hydrocarbon found in petroleum   Two-carbon unit can be metabolized in cell

The Reversibility of Chemical Reactions

All chemical reactions are, in theory, reversible; that is, they can occur in either direction. In practice, however, some reactions do this more easily than others. A chemical reaction that is readily reversible (that is, when the end product can revert to the original molecules) is termed a **reversible reaction** and is indicated by two arrows, as shown here:

$$A + B \xrightleftharpoons[\text{Breaks down into}]{\text{Combines to form}} AB$$

Some reversible reactions occur because neither the reactants nor the end products are very stable. Other reactions will reverse only under special conditions:

$$A + B \xrightleftharpoons[\text{Water}]{\text{Heat}} AB$$

Whatever is written above or below the arrows indicates the special condition under which the reaction in that direction occurs. In this case, A and B react to produce AB only when heat is applied, and AB breaks down into A and B only in the presence of water.

How Chemical Reactions Occur

The **collision theory** explains how chemical reactions occur and how certain factors affect the rates of those reactions. The basis of the collision theory is that all atoms, ions, and molecules are continuously moving and are thus continuously colliding with one another. The energy transferred by the particles in the collision can disrupt their electron structures enough so that chemical bonds are broken or new bonds are formed.

Factors Affecting Chemical Reactions

Several factors determine whether a collision will cause a chemical reaction: the velocities of the colliding particles, their energy, and their specific chemical configurations. Up to a point, the higher the particles' velocities, the greater the probability that their collision will cause a reaction. Also, each chemical reaction requires a specific level of energy. But even if colliding particles possess the minimum energy needed for reaction, no reaction will take place unless the particles are properly oriented toward each other.

Let us assume that molecules of substance AB (the reactant) are to be converted to molecules of substances A and B (the products). In a given population of molecules of substance AB, at a specific temperature, some molecules will possess relatively little energy; the majority of the population will possess an average amount of energy; and a small portion of the population will have high energy. If only the high-energy AB molecules are able to react and be converted to A and B molecules, then only relatively few molecules at any one time possess enough energy to react in a collision. The collision energy required for a chemical reaction is its **activation energy**, which is the amount of energy needed to disrupt the stable electronic configuration of any specific molecule so that the electrons can be rearranged.

The **reaction rate**—the frequency of collisions containing sufficient energy to bring about a reaction—depends on the number of reactant molecules at or above the activation energy level. One way to increase the reaction rate of a substance is to raise its temperature. By causing the molecules to move faster, heat increases both the frequency of collisions and the number of molecules that attain activation energy. The number of collisions also increases when pressure is increased or when the reactants are more concentrated (because the distance between molecules is thereby decreased).

Enzymes and Chemical Reactions

Living systems contain large protein molecules called **enzymes** that act as catalysts. A **catalyst** is a substance that changes the rate of a chemical reaction, usually by increasing it, without being altered itself. As catalysts, enzymes typically accelerate chemical reactions. The three-dimensional enzyme molecule has an *active site,* a region that will interact with a specific chemical substance (see Figure 5.3). The chemical substance (reactant) an enzyme acts on is referred to as the enzyme's **substrate.**

The enzyme orients the substrate into a position that increases the probability of a reaction. The **enzyme-substrate complex** formed by the temporary binding of enzyme and reactants enables the collisions to be more effective and lowers the activation energy of the reaction (Figure 2.5). The enzyme therefore speeds up the reaction by increasing the number of AB molecules that attain sufficient activation energy to react.

An enzyme is able to accelerate a reaction without the need for an increase in temperature. This ability is crucial to living systems because a significant temperature increase would destroy cellular proteins. The crucial function of enzymes, therefore, is to speed up biochemical reactions at a temperature that is compatible with the normal functioning of the cell.

IMPORTANT BIOLOGICAL MOLECULES

Biologists and chemists divide compounds into two principal classes, inorganic and organic. **Inorganic compounds** are defined as molecules, usually small, that typ-

FIGURE 2.5 Energy requirements of a chemical reaction. This graph shows the progress of the reaction AB → A + B both without (black line) and with (magenta line) an enzyme. The presence of an enzyme lowers the activation energy of the reaction (see arrows). Thus, more molecules of reactant AB are converted to products A and B because more molecules of reactant AB possess the activation energy needed for the reaction.

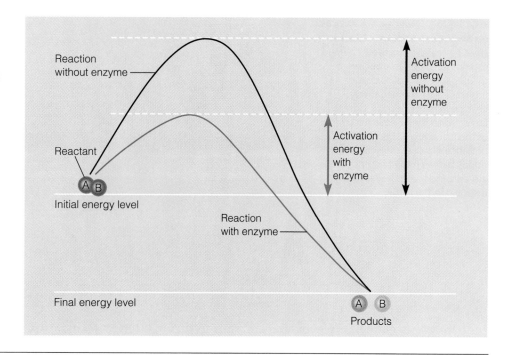

ically lack carbon and in which ionic bonds may play an important role. Inorganic compounds include water, oxygen, carbon dioxide, and many salts, acids, and bases.

Organic compounds always contain carbon and hydrogen. Carbon is a unique element in the chemistry of life because it has four electrons in its outer shell and four unfilled spaces. It can combine with a variety of atoms, including other carbon atoms, to form straight or branched chains and rings (see Figure 2.9). Carbon chains form the basis of many organic compounds in living cells, including sugars, amino acids, and vitamins. Organic compounds are held together mostly or entirely by covalent bonds. Some organic molecules, such as polysaccharides, proteins, and nucleic acids, are very large and usually contain thousands of atoms. Such giant molecules are called *macromolecules*. In the following section we will discuss inorganic and organic compounds that are essential for cells.

INORGANIC COMPOUNDS

LEARNING OBJECTIVES

- *List several properties of water that are important to living systems.*
- *Define acid, base, salt, and pH.*

Water

All living organisms require a wide variety of inorganic compounds for growth, repair, maintenance, and reproduction. Water is one of the most important, as well as

one of the most abundant, of these compounds, and it is particularly vital to microorganisms. Outside the cell, nutrients are dissolved in water, which facilitates their passage through cell membranes. And inside the cell, water is the medium for most chemical reactions. In fact, water is by far the most abundant component of almost all living cells. Water makes up 5% to 95% or more of each cell, the average being between 65% and 75%. Simply stated, no organism can survive without water.

Water has structural and chemical properties that make it particularly suitable for its role in living cells. As we discussed, the total charge on the water molecule is neutral, but the oxygen region of the molecule has a slightly negative charge and the hydrogen region has a slightly positive charge (see Figure 2.4a). Any molecule having such an unequal distribution of charges is called a **polar molecule.** The polar nature of water gives it four characteristics that make it a useful medium for living cells.

First, every water molecule is capable of forming four hydrogen bonds with nearby water molecules (see Figure 2.4b). This property results in a strong attraction between water molecules. Because of this strong attraction, a great deal of heat is required to separate water molecules from each other to form water vapor; thus, water has a relatively high boiling point (100°C). Because water has such a high boiling point, it exists in the liquid state on most of the Earth's surface. Furthermore, the hydrogen bonding between water molecules affects the density of water, depending on whether it occurs as ice or a liquid. For example, the hydrogen bonds in the crystalline structure of water (ice) make ice take up more space. As a result, ice has fewer molecules than an equal volume of liquid water. This makes its crystalline structure less dense than liquid water. For this reason ice floats and can serve as an

FIGURE 2.6 How water acts as a solvent for sodium chloride (NaCl). **(a)** The positively charged sodium ion (Na^+) is attracted to the negative part of the water molecule. **(b)** The negatively charged chloride ion (Cl^-) is attracted to the positive part of the water molecule. In the presence of water molecules, the bonds between the Na^+ and Cl^- are disrupted, and the NaCl dissolves in the water.

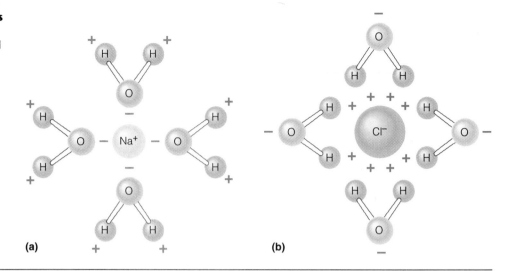

(a)   (b)

insulating layer on the surfaces of lakes and streams that harbor living organisms.

Second, the polarity of water makes it an excellent dissolving medium or **solvent.** Many polar substances undergo **dissociation** (separation) into individual molecules in water—that is, they dissolve—because the negative part of the water molecule is attracted to the positive part of the molecules in the **solute** (dissolving substance), and the positive part of the water molecules is attracted to the negative part of the solute molecules. Substances (such as salts) that are composed of atoms (or groups of atoms) held together by ionic bonds tend to dissociate into separate cations and anions in water. Thus, the polarity of water allows molecules of many different substances to separate and become surrounded by water molecules (Figure 2.6).

Third, polarity accounts for water's characteristic role as a reactant or product in many chemical reactions. Its polarity facilitates the splitting and rejoining of hydrogen ions (H^+) and hydroxide ions (OH^-). Water is a key reactant in the digestive processes of organisms, whereby larger molecules are broken down into smaller ones. Water molecules are also involved in synthetic reactions—water is an important source of the hydrogen and oxygen that are incorporated into numerous organic compounds in living cells.

Finally, the relatively strong hydrogen bonding between water molecules (see Figure 2.4b) makes water an excellent temperature buffer. A given quantity of water requires a great gain of heat to increase its temperature and a great loss of heat to decrease its temperature, as compared with many other substances. Normally, heat absorption by molecules increases their kinetic energy and thus increases their rate of motion and their reactivity. In water, however, heat absorption first breaks hydrogen bonds rather than increasing the rate of motion. Therefore, much more heat must be applied to raise the temperature of water than to raise the temperature of a non-hydrogen-bonded liquid. The reverse is true as water cools.

Thus, water more easily maintains a constant temperature than other solvents and tends to protect a cell from fluctuations in environmental temperatures.

Acids, Bases, and Salts

As we saw in Figure 2.6, when inorganic salts such as sodium chloride (NaCl) are dissolved in water, they undergo *dissociation,* or **ionization;** that is, they break apart into ions. Substances called acids and bases show similar behavior.

An **acid** can be defined as a substance that dissociates into one or more hydrogen ions (H^+) and one or more negative ions (anions). Thus, an acid can also be defined as a proton (H^+) donor. A **base** dissociates into one or more positive ions (cations) plus one or more negatively charged hydroxide ions (OH^-) that can accept, or combine with, protons. Thus, sodium hydroxide (NaOH) is a base because it dissociates to release hydroxide ions, which have a strong attraction for protons and are among the most important proton acceptors. A **salt** is a substance that dissociates in water into cations and anions, neither of which is H^+ or OH^-. Figure 2.7 shows common examples of each type of compound and how they dissociate in water.

Acid–Base Balance

An organism must maintain a fairly constant balance of acids and bases to remain healthy. In the aqueous environment within organisms, acids dissociate into hydrogen ions (H^+) and anions. Bases, on the other hand, dissociate into hydroxide ions (OH^-) and cations. The more hydrogen ions that are free in a solution, the more acidic the solution is. Conversely, the more hydroxide ions that are free in a solution, the more basic, or alkaline, it is.

Biochemical reactions—that is, chemical reactions in

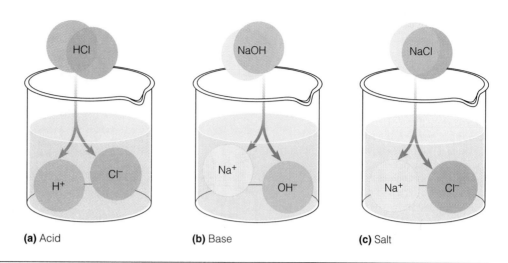

FIGURE 2.7 **Acids, bases, and salts.** **(a)** In water, hydrochloric acid (HCl) dissociates into H⁺ and Cl⁻. **(b)** Sodium hydroxide (NaOH), a base, dissociates into OH⁻ and Na⁺ in water. **(c)** In water, table salt (NaCl) dissociates into positive ions (Na⁺) and negative ions (Cl⁻), neither of which are H⁺ or OH⁻.

(a) Acid   **(b)** Base   **(c)** Salt

living systems—are extremely sensitive to even small changes in the acidity or alkalinity of the environments in which they occur. In fact, H⁺ and OH⁻ are involved in almost all biochemical processes, and the functions of a cell are modified greatly by any deviation from its narrow band of normal H⁺ and OH⁻ concentrations. For this reason, the acids and bases that are continually formed in an organism must be kept in balance.

It is convenient to express the amount of H⁺ in a solution by a logarithmic pH scale, which ranges from 0 to 14 (Figure 2.8). The **pH** of a solution is calculated by $-\log_{10}[H^+]$, the negative logarithm to the base 10 of the hydrogen ion concentration, determined in moles per liter [H⁺]. For example, if the H⁺ concentration of a solution is 1.0×10^{-4} moles/liter, or 10^{-4}, its pH equals $-\log_{10}10^{-4} = -(-4) = 4$; this is about the pH of wine. The pH values of some human body fluids and other common substances are also shown in Figure 2.8. In the laboratory, you will usually measure the pH of a solution with a pH meter or with chemical test papers.

Acidic solutions contain more H⁺ than OH⁻ and have a pH lower than 7. If a solution has more OH⁻ than H⁺, it is a basic, or alkaline, solution. In pure water, a small percentage of the molecules are dissociated into H⁺ and OH⁻, so it has a pH of 7. Because the concentrations of H⁺ and OH⁻ are equal, this pH is said to be the pH of a neutral solution.

The pH scale is logarithmic, so a change of one whole number represents a tenfold change from the previous concentration. Thus, a solution of pH 1 has ten times more H⁺ ions than a solution of pH 2 and has 100 times more H⁺ ions than a solution of pH 3.

Keep in mind that the pH of a solution can be changed. We can increase its acidity by adding substances that will increase the concentration of hydrogen ions. As a living organism takes up nutrients, carries out chemical reactions, and excretes wastes, its balance of acids and bases tends to change, and the pH fluctuates. Fortunately, organisms possess natural pH **buffers**, compounds that help

keep the pH from changing drastically. But the pH in our environment's water and soil can be altered by waste products from organisms, pollutants from industry, or fertilizers used in agricultural fields or gardens. When bacteria are grown in a laboratory medium, they excrete waste products such as acids that can alter the pH of the medium. If this effect were to continue, the medium would become acidic enough to inhibit bacterial enzymes and cause the death of the bacteria. To prevent this problem, pH buffers are added to the culture medium. One very effective pH buffer for some culture media uses a mixture of K_2HPO_4 and KH_2PO_4 (see Table 6.4).

Different microbes function best within different pH ranges, but most organisms grow best in environments with a pH between 6.5 and 8.5. Among microbes, fungi are best able to tolerate acidic conditions, whereas the procaryotes called cyanobacteria tend to do well in alkaline habitats. *Propionibacterium acnes* (prō-pē-on-ē-bak-ti′rē-um ak′nēz), a bacterium that contributes to acne, has as its natural environment human skin, which tends to be slightly acidic, with a pH of about 4. *Thiobacillus ferrooxidans* (thī-ō-bä-sil′lus fer-rō-oks′i-danz) is a bacterium that grows on elemental sulfur and produces sulfuric acid (H_2SO_4). Its pH range for optimum growth is from 1 to 3.5. The sulfuric acid produced by this bacterium in mine water is important in dissolving uranium and copper from low-grade ore. (See the box in Chapter 28.)

ORGANIC COMPOUNDS

LEARNING OBJECTIVES

- *Distinguish between organic and inorganic compounds.*
- *Identify the building blocks of carbohydrates, simple lipids, phospholipids, proteins, and nucleic acids.*
- *Identify the role of ATP in cellular activities.*

FIGURE 2.8   The pH scale. The concentrations of H^+ and OH^- are equal at pH 7, which is the neutral point. As pH values decrease from 14 to 0, the H^+ concentration increases. Thus, the lower the pH, the more acidic the solution; the higher the pH, the more basic the solution. If the pH value of a solution is below 7, the solution is acidic; if the pH is above 7, the solution is basic (alkaline). The approximate pH values of some human body fluids and common substances are shown next to the pH scale.

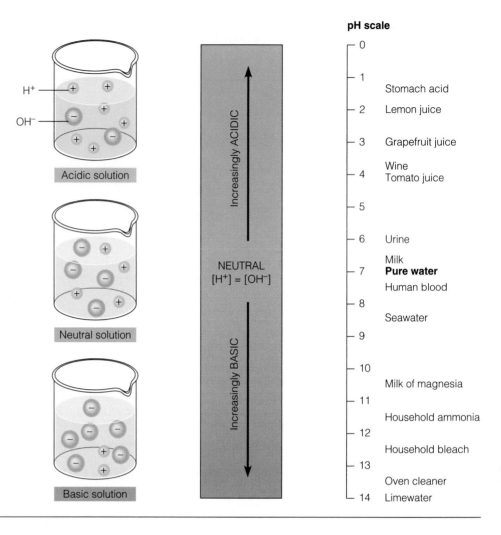

Inorganic compounds, excluding water, constitute about 1% to 1.5% of living cells. These relatively simple components, whose molecules have only a few atoms, cannot be used by cells to perform complicated biological functions. Organic molecules, whose carbon atoms can combine in an enormous variety of ways with other carbon atoms and atoms of other elements, are relatively complex and thus are capable of more complicated biological functions.

Structure and Chemistry

In the formation of organic molecules, carbon's four outer electrons can participate in up to four covalent bonds, and carbon atoms can bond to each other to form straight-chain, branched-chain, or ring structures (Figure 2.9).

When single bonds hold a carbon atom to four other atoms, those four atoms form a skeleton of a three-dimensional shape called a **tetrahedron** (Figure 2.10). Carbon atoms can also bond via double or triple bonds; the molecular shape then is not tetrahedral.

In addition to carbon, the most common elements in organic compounds are hydrogen (which can form one bond), oxygen (two bonds), and nitrogen (three bonds).

Sulfur (two bonds) and phosphorus (five bonds) appear less often. Other elements are found, but only in a relatively few organic compounds. The elements that are most abundant in living organisms (see footnote, Table 2.1) are the same as those that are most abundant in organic compounds.

The chain of carbon atoms in an organic molecule is called the **carbon skeleton**; there is a huge number of combinations possible for carbon skeletons. Most of these carbons are bonded to hydrogen atoms. The bonding of other elements with carbon and hydrogen forms characteristic **functional groups** (Table 2.3). Functional groups are specific groups of atoms that are responsible for most of the characteristic chemical properties and many of the physical properties of a particular organic compound. Thus, functional groups help us to classify organic compounds. For example, the —OH group is present in each of the following molecules:

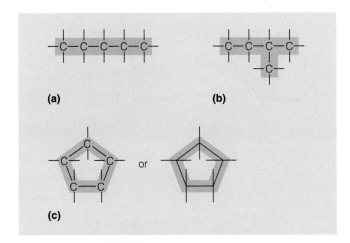

FIGURE 2.9 Various bonding patterns of carbon.
(a) Straight-chain structures. (b) Branched-chain structures.
(c) Ring. In ring structures, the carbon atoms at the corners are often not shown.

FIGURE 2.10 Methane, a molecule in the shape of a tetrahedron. This figure compares the structural formula of methane using two types of models that represent its three-dimensional shape. A ball-and-stick model emphasizes the bond angles of the molecule. In methane, carbon's four covalent bonds angle toward the corners of an imaginary tetrahedron, outlined here with a dotted magenta line. A space-filling model portrays a molecule's shape more accurately.

Isopropanol

Toluol

Because the characteristic reactivity of the molecules is based on the —OH group, they are grouped together in a class called alcohols. The —OH group is called the *hydroxyl group* and is not to be confused with the *hydroxide ion* (OH⁻) of bases. The hydroxyl group of alcohols does not ionize at neutral pH; it is covalently bonded to a carbon atom.

When a class of compounds is characterized by a certain functional group, the letter *R* can be used to stand for the remainder of the molecule. For example, alcohols in general may be written R—OH.

Frequently, more than one functional group is found in a single molecule. For example, an amino acid molecule contains both amino and carboxyl groups. The amino acid glycine has the following structure:

Amino group

Carboxyl group

Most of the organic compounds found in living organisms are quite complex; a large number of carbon atoms form the skeleton, and many functional groups are at-

TABLE 2.3 Representative Functional Groups and the Compounds in Which They Are Found

FUNCTIONAL GROUP	NAME OF GROUP	CLASS OF COMPOUNDS
R—O—H	Hydroxyl	Alcohol
R—C(=O)H	Carbonyl (terminal)*	Aldehyde
R—C(=O)—R	Carbonyl (internal)*	Ketone
R—C(H)(H)—NH_2	Amino	Amine
R—C(=O)—O—R'	Ester	Ester
R—C(H)(H)—O—C(H)(H)—R'	Ether	Ether
R—C(H)(H)—SH	Sulfhydryl	Sulfhydryl
R—C(=O)—OH	Carboxyl	Organic acid

*A "terminal" carbonyl group is at an end of a molecule. In contrast, an "internal" carbonyl is at least one C atom removed from either end.

FIGURE 2.11 Dehydration synthesis and hydrolysis. (a) In dehydration synthesis (left to right), the monosaccharides glucose and fructose combine to form a molecule of the disaccharide sucrose. A molecule of water is lost in the reaction. **(b)** In hydrolysis (right to left), the sucrose molecule breaks down into the smaller molecules glucose and fructose. For the hydrolysis reaction to proceed, water must be added to the sucrose.

tached. In organic molecules, it is important that each of the four bonds of carbon is satisfied (attached to another atom) and that each of the attaching atoms has its characteristic number of bonds satisfied. Because of this, such molecules are chemically stable.

Small organic molecules can be combined into very large molecules called **macromolecules.** Macromolecules are usually **polymers,** large molecules formed by covalent bonding of many repeating small molecules called **monomers.** When two monomers join together, the reaction usually involves the elimination of a hydrogen atom from one monomer and a hydroxyl group from the other; the hydrogen atom and the hydroxyl group combine to produce water:

$$X - OH + H - Y \longrightarrow X - Y + H_2O$$

This type of exchange reaction is called **dehydration synthesis** or a **condensation reaction** because a molecule of water is released (Figure 2.11). Such macromolecules as carbohydrates, lipids, proteins, and nucleic acids are assembled in the cell, essentially by dehydration synthesis. However, other molecules must also participate to provide energy for bond formation. ATP, the cell's chief energy provider, is discussed at the end of this chapter.

Carbohydrates

The **carbohydrates** are a large and diverse group of organic compounds that includes sugars and starches. Carbohydrates perform a number of major functions in living systems. For instance, one type of sugar (deoxyribose) is a building block of deoxyribonucleic acid (DNA), the molecule that carries hereditary information. Other sugars help to form the cell walls of bacterial cells. Simple carbohydrates are used in the synthesis of amino acids and fats or fatlike substances, which are used to build structures and provide an emergency source of energy. Macromolecular carbohydrates function as food reserves. The principal

function of carbohydrates, however, is to fuel cell activities with a ready source of energy.

Carbohydrates are made up of carbon, hydrogen, and oxygen atoms. The ratio of hydrogen to oxygen atoms is always 2:1 in simple carbohydrates. This ratio can be seen in the formulas for the carbohydrates ribose ($C_5H_{10}O_5$), glucose ($C_6H_{12}O_6$), and sucrose ($C_{12}H_{22}O_{11}$). Although there are exceptions, the general formula for carbohydrates is $(CH_2O)_n$, where n indicates that there are three or more CH_2O units. Carbohydrates can be divided into three major groups on the basis of size: monosaccharides, disaccharides, and polysaccharides.

Monosaccharides

Simple sugars are called **monosaccharides;** each molecule contains from three to seven carbon atoms. The number of carbon atoms in the molecule of a simple sugar is indicated by the prefix in its name. For example, simple sugars with three carbons are called trioses. There are also tetroses (four-carbon sugars), pentoses (five-carbon sugars), hexoses (six-carbon sugars), and heptoses (seven-carbon sugars). Pentoses and hexoses are extremely important to living organisms. Deoxyribose is a pentose found in DNA. Glucose, a common hexose, is the main energy-supplying molecule of living cells.

Disaccharides

Disaccharides are formed when two monosaccharides bond in a dehydration synthesis reaction. For example, molecules of two monosaccharides, glucose and fructose, combine to form a molecule of the disaccharide sucrose (table sugar) and a molecule of water (see Figure 2.11). Similarly, the dehydration synthesis of the monosaccharides glucose and galactose forms the disaccharide lactose (milk sugar).

It may seem odd that glucose and fructose should have the same chemical formula (see Figure 2.11), even though they are different monosaccharides. The positions of the oxygens and carbons vary in the two different molecules,

FIGURE 2.12 Structural formulas of simple lipids.
(a) Glycerol. (b) Lauric acid, a fatty acid. In the condensed formula for lauric acid, COOH(CH$_2$)$_{10}$CH$_3$, the subscript 10 means that the molecule has ten CH$_2$ units. (c) The chemical combination of a molecule of glycerol and three fatty acid molecules, in this case lauric acid, forms (d) one molecule of fat (triglyceride) and three molecules of water in a dehydration synthesis reaction. The bond between glycerol and each fatty acid is called an ester linkage. The addition of three water molecules to a fat forms glycerol and three fatty acid molecules in a hydrolysis reaction.

(a) Glycerol

(b) Lauric acid, COOH(CH$_2$)$_{10}$CH$_3$

(c) Glycerol + 3 fatty acid molecules

(d) Fat (triglyceride)

and consequently the molecules have different physical and chemical properties. Two molecules with the same chemical formula but different structures and properties are called **isomers.**

Disaccharides can be broken down into smaller, simpler molecules when water is added. This chemical reaction, the reverse of dehydration synthesis, is called **hydrolysis,** which means to split water. A molecule of sucrose, for example, may be hydrolyzed (digested) into its components of glucose and fructose by reacting with the H$^+$ and OH$^-$ of water. This reaction is also represented in Figure 2.11.

As you will see in Chapter 4, the cell walls of bacterial cells are composed of disaccharides and proteins that together form a substance called peptidoglycan.

Polysaccharides

Carbohydrates in the third major group, the **polysaccharides,** consist of eight or more monosaccharides joined through dehydration synthesis. Polysaccharides often have side chains branching off the main structure and are classified as macromolecules. Like disaccharides, polysaccharides can be split apart into their constituent sugars through hydrolysis. Unlike monosaccharides and disaccharides, however, they usually lack the characteristic sweetness of sugars such as fructose and sucrose and usually are not soluble in water.

One important polysaccharide is *glycogen,* which is composed of glucose subunits and is synthesized as a storage material by animals and some bacteria. *Cellulose,* another important glucose polymer, is the main component of the cell walls of plants and most algae. The polysaccharide *dextran,* which is produced as a sugary slime by certain bacteria, is used in a blood plasma substitute. *Starch* is a polymer of glucose produced by plants and used as food

by humans. In addition, polysaccharides and/or proteins make up the capsule of some bacteria.

Lipids

If lipids were suddenly to disappear from the Earth, all living cells would collapse in a pool of fluid, because lipids are essential to the structure and function of membranes that separate living cells from their environment. **Lipids** are a second major group of organic compounds found in living matter. Like carbohydrates, they are composed of atoms of carbon, hydrogen, and oxygen, but lipids lack the 2:1 ratio between hydrogen and oxygen atoms. Even though lipids are a very diverse group of compounds, they share one common characteristic: that they are very *nonpolar* molecules; unlike water, they do not have a positive and a negative end (pole). Hence, most are insoluble in water but dissolve readily in nonpolar solvents, such as ether and chloroform. Lipids function in energy storage and provide part of the structure of membranes and cell walls.

Simple Lipids

Simple lipids, called *fats* or *triglycerides,* contain an alcohol called *glycerol* and a group of compounds known as *fatty acids.* Glycerol molecules have three carbon atoms to which are attached three hydroxyl (—OH) groups (Figure 2.12a). Fatty acids consist of long hydrocarbon chains (composed only of carbon and hydrogen atoms) ending in a carboxyl (—COOH, organic acid) group (Figure 2.12b). Most common fatty acids contain an even number of carbon atoms.

A fat molecule is formed when a molecule of glycerol

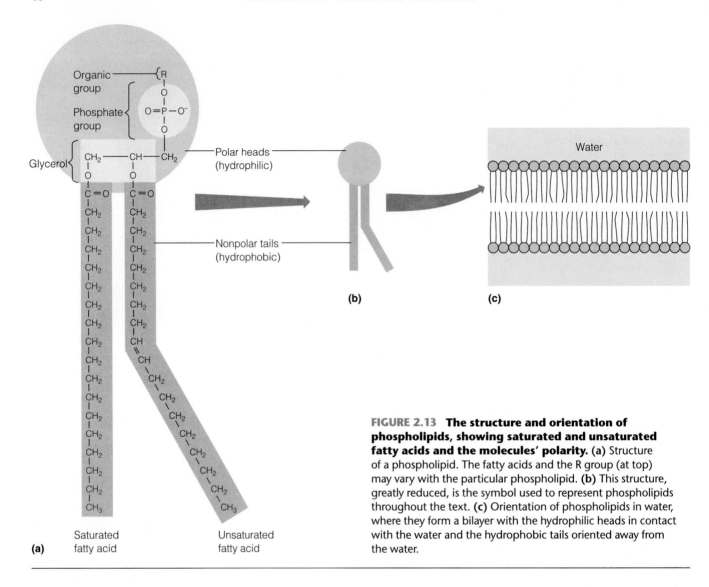

FIGURE 2.13 The structure and orientation of phospholipids, showing saturated and unsaturated fatty acids and the molecules' polarity. (a) Structure of a phospholipid. The fatty acids and the R group (at top) may vary with the particular phospholipid. (b) This structure, greatly reduced, is the symbol used to represent phospholipids throughout the text. (c) Orientation of phospholipids in water, where they form a bilayer with the hydrophilic heads in contact with the water and the hydrophobic tails oriented away from the water.

combines with one to three fatty acid molecules (Figure 2.12c) to form a monoglyceride, diglyceride, or triglyceride (Figure 2.12d). In the reaction, one to three molecules of water are formed (dehydration), depending on the number of fatty acid molecules reacting. The chemical bond formed where the water molecule is removed is an *ester linkage.* In the reverse reaction, hydrolysis, a fat molecule is broken down into its component fatty acid and glycerol molecules.

Because the fatty acids that form lipids have different structures, there is a wide variety of lipids. For example, three molecules of fatty acid A might combine with a glycerol molecule, or one molecule each of fatty acids A, B, and C might unite with a glycerol molecule. Simple lipids serve as energy storage materials.

The terms *saturated fat* and *unsaturated fat* refer to the structure of the hydrocarbon chains of the fatty acids in a fat molecule. If there are no double bonds between carbon atoms in the hydrocarbon chain, then the carbon skeleton contains the maximum number of hydrogen atoms. Such

a fat is said to be *saturated,* meaning that it is saturated with hydrogen atoms (Figure 2.13a, at left). Most animal fats are saturated fats. If there is one or more double bonds, which reduces the number of hydrogen atoms that can be attached to the carbon skeleton, then the fat molecule is referred to as *unsaturated.* Wherever a double bond occurs in a hydrocarbon chain of an unsaturated fat, the fatty acid has a kink (bend) in its structure. Plant fats are usually unsaturated.

Complex Lipids

Complex lipids contain such elements as phosphorus, nitrogen, and sulfur in addition to the carbon, hydrogen, and oxygen found in simple lipids. The complex lipids called *phospholipids* are made up of glycerol, two fatty acids, and, in place of a third fatty acid, a phosphate group bonded to one of several organic groups (Figure 2.13a). Phospholipids are the lipids that build membranes; they are thus essential to a cell's survival. Phospholipids have polar as well as nonpolar regions (see Figures 2.13a, b and

4.13). When placed in water, phospholipid molecules twist and orient themselves in such a way that all polar (hydrophilic) portions will orient themselves toward the polar water molecules (Figure 2.13c), with which they then form hydrogen bonds. (*Hydrophilic* means water-loving.) Polar portions consist of a phosphate group and glycerol. In contrast to the polar regions, all nonpolar (hydrophobic) parts of the phospholipid make contact only with the nonpolar portions of neighboring molecules. (*Hydrophobic* means water-fearing.) Nonpolar portions consist of fatty acids. This characteristic behavior makes phospholipids particularly suitable to being a major component of the membranes that enclose cells. Phospholipids enable the membrane to act as a barrier that separates the contents of the cell from the water-based environment in which it lives.

Some complex lipids are useful in identifying certain bacteria. For example, the cell wall of *Mycobacterium tuberculosis* (mī-kō-bak-ti′rē-um tü-bėr-kū-lō′sis), the bacterium that causes tuberculosis, is distinguished by its lipid-rich content. The cell wall contains complex lipids such as waxes and glycolipids (lipids with carbohydrates attached) that give the bacterium distinctive staining characteristics. Cell walls rich in such complex lipids are characteristic of all members of the genus *Mycobacterium*.

Steroids

Steroids are structurally very different from the lipids described previously. Figure 2.14 shows the structure of the steroid cholesterol, with the characteristic ring structure of steroids. The —OH group in cholesterol makes it a *sterol*. Sterols are important constituents of the plasma membranes of animal cells and of one group of bacteria (mycoplasmas), and they are also found in fungi and plants. Animals synthesize steroid hormones and vitamin D from sterols.

Proteins

Proteins are organic molecules that contain carbon, hydrogen, oxygen, and nitrogen. Some also contain sulfur. If you were to separate and weigh all the groups of organic compounds in a living cell, the proteins would tip the scale. Hundreds of different proteins can be found in any single cell, and together they make up 50% or more of a cell's dry weight.

Proteins are essential ingredients in all aspects of cell structure and function. We have already mentioned enzymes, the proteins that catalyze biochemical reactions. But proteins have other vital functions as well. *Carrier proteins* help transport certain chemicals into and out of cells. Other proteins, such as the *bacteriocins* produced by many bacteria, kill other bacteria. Toxins produced by certain disease-causing microorganisms are also proteins. Some proteins play a key role in contraction of animal muscle

FIGURE 2.14 Cholesterol, a steroid. Note the four "fused" carbon rings, which are characteristic of steroid molecules. The hydrogen atoms attached to the carbons at the corners of the rings have been omitted.

cells and movement of microbial and other types of cells. Other proteins are integral parts of cell structures such as walls, membranes, and cytoplasmic components. Still others, such as the hormones of certain organisms, have regulatory functions. As we will see in Chapter 17, proteins called antibodies play an important role in vertebrate immune systems. Normally, antibodies recognize components of the body as "self" and antigens (foreign matter) as "nonself." Upon exposure to any antigen, the body produces antibodies that recognize, bind to, and destroy that antigen.

Amino Acids

Just as the monosaccharides are the building blocks of larger carbohydrate molecules and fatty acids and glycerol are the building blocks of fats, **amino acids** are the building blocks of proteins. Amino acids contain at least one carboxyl (—COOH) group and one amino (—NH_2) group attached to the same carbon atom, called an alpha-carbon (written C_α) (Figure 2.15). Such amino acids are called *alpha-amino acids*. Also attached to the alpha-carbon is a side group (R group), which is the amino acid's distinguishing factor. The side group can be a hydrogen atom, an unbranched or branched chain of atoms, or a ring structure that is cyclic (all carbon) or heterocyclic (when an atom other than carbon is included in the ring). Figure 2.15b illustrates the structural formula of an amino acid—in this case, valine—that has a branched-chain side group. The side group can contain functional groups, such as the sulfhydryl group (—SH), the hydroxyl group (—OH), or additional carboxyl or amino groups. These side groups and the carboxyl and alpha-amino groups affect the total structure of a protein, as will be described later. The structures and standard abbreviations of the 20 amino acids found in proteins are shown in Table 2.4.

Amino acids exist in either of two configurations called *stereoisomers,* designated by D and L. These two configurations are mirror images, corresponding to "right-handed"

FIGURE 2.15 Structure of amino acids. (a) General structural formula for an amino acid. The alpha-carbon (C_α) is shown in the center. The letter R can stand for any of a number of groups of atoms. Different amino acids have different R groups, also called side groups. **(b)** Structural formula for the amino acid valine. Note the branched-chain side group.

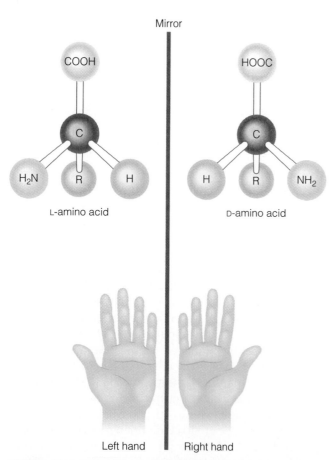

FIGURE 2.16 The L- and D-isomers of an amino acid, shown with ball-and-stick models. The two isomers, like left and right hands, are mirror images of each other and cannot be superimposed on one another. (Try it!)

(D) and "left-handed" (L) three-dimensional shapes (Figure 2.16). The amino acids found in proteins are always the L-isomers (except for glycine, the simplest amino acid, which does not have stereoisomers). However, D-amino acids occasionally occur in nature—for example, in certain bacterial cell walls and antibiotics. (Many other kinds of organic molecules also can exist in D and L forms. One example is the sugar glucose, which occurs in nature as D-glucose.)

Although only 20 different amino acids occur naturally in proteins, a single protein molecule can contain from 50 to hundreds of amino acid molecules, which can be arranged in an almost infinite number of ways to make proteins of different lengths, compositions, and structures. The number of proteins is practically endless, and every living cell produces many different proteins.

Peptide Bonds

Amino acids bond between the carbon atom of the carboxyl (—COOH) group of one amino acid and the nitrogen atom of the amino (—NH_2) group of another (Figure 2.17). The bonds between amino acids are called **peptide bonds.** For the formation of every peptide bond between two amino acids, one water molecule is released; thus peptide bond formation occurs by dehydration synthesis. In Figure 2.17 the resulting compound is called a *dipeptide* because it consists of two amino acids joined by a peptide bond. Adding another amino acid to a dipeptide would form a *tripeptide*. Further additions of amino acids would produce a long, chainlike molecule called a *polypeptide,* a protein molecule.

Levels of Protein Structure

Proteins vary tremendously in structure. Different proteins have different architectures and different three-dimensional shapes. This variation in structure is directly related to their diverse functions.

When a cell makes a protein, the polypeptide chain folds spontaneously to assume a certain shape. One reason for folding of the polypeptide is that some parts of a protein are attracted to water and other parts are repelled by it. In practically every case, the function of a protein depends on its ability to recognize and bind to some other molecule. For example, an enzyme binds specifically with its substrate. A hormonal protein binds to a receptor on a cell whose function it will alter. An antibody binds to an antigen (foreign substance) that has invaded the body. The unique shape of each protein permits it to interact with specific other molecules in order to carry out specific functions.

Proteins are described in terms of four levels of organization: primary, secondary, tertiary, and quaternary. The *primary structure* is the unique order (the sequence) in which the amino acids are linked together (Figure 2.18a). This sequence is genetically determined. Alterations in sequence can have profound metabolic effects. For example, a single incorrect amino acid in a blood protein can produce the deformed hemoglobin molecule characteristic of

TABLE 2.4 The 20 Amino Acids Found in Proteins*

GLYCINE (Gly)

Hydrogen
atom

ALANINE (Ala)

Unbranched
chain

VALINE (Val)

Branched
chain

LEUCINE (Leu)

Branched
chain

ISOLEUCINE (Ile)

Branched
chain

SERINE (Ser)

Hydroxyl
(—OH) group

THREONINE (Thr)

Hydroxyl
(—OH) group

CYSTEINE (Cys)

Sulfhydryl
(—SH) group

METHIONINE (Met)

Sulfhydryl
(—SH) group

GLUTAMIC ACID (Glu)

Additional
carboxyl
(—COOH)
group, acidic

ASPARTIC ACID (Asp)

Additional
carboxyl
(—COOH)
group, acidic

LYSINE (Lys)

Additional amino
(—NH$_2$)
group, basic

ARGININE (Arg)

Additional amino
(—NH$_2$)
group, basic

ASPARAGINE (Asn)

Additional amino
(—NH$_2$)
group, basic

GLUTAMINE (Gln)

Additional amino
(—NH$_2$)
group, basic

PHENYLALANINE (Phe)

Cyclic

TYROSINE (Tyr)

Cyclic

HISTIDINE (His)

Heterocyclic

TRYPTOPHAN (Trp)

Heterocyclic

PROLINE (Pro)

Heterocyclic

*Shown are the proteins' names, including the three-letter abbreviations in parentheses (above), their structural
formulas (middle), and characteristic R group (below).

FIGURE 2.17 Peptide bond formation by dehydration synthesis. The amino acids glycine and alanine combine to form a dipeptide. The newly formed bond between the nitrogen atom of glycine and the carbon atom of alanine is called a peptide bond.

sickle-cell anemia. But proteins do not exist as long, straight chains. Each polypeptide chain folds and coils in specific ways into a relatively compact structure with a characteristic three-dimensional shape.

A protein's *secondary structure* is the localized, repetitious twisting or folding of the polypeptide chain. This aspect of a protein's shape results from hydrogen bonds joining the atoms of peptide bonds at different locations along the polypeptide chain. The two types of secondary protein structures are clockwise spirals called helices (singular: helix) and pleated sheets, which form from roughly parallel portions of the chain (Figure 2.18b). Both structures are held together by hydrogen bonds between oxygen or nitrogen atoms that are part of the polypeptide's backbone.

Tertiary structure refers to the overall three-dimensional structure of a polypeptide chain (Figure 2.18c). The folding is not repetitive or predictable, as in secondary structure. Whereas secondary structure involves hydrogen bonding between atoms of the amino and carbonyl groups involved in the peptide bonds, tertiary structure involves interactions between various amino acid side groups in the polypeptide chain. Thus, the tertiary structure of a protein is elaborately irregular. Hydrogen bonds and other relatively weak interactions play an important role in tertiary structure. In addition, sulfhydryl groups (—SH) on two amino acid subunits can form a covalent, disulfide link (—S—S—) by removal of the hydrogen atoms. Such a reaction, in which electrons (in this case belonging to hydrogen atoms) are removed from a molecule, is called an oxidation reaction.

Some proteins have a *quaternary structure,* which consists of an aggregation of two or more individual polypeptide chains (subunits) that operate as a single functional unit. Figure 2.18d shows a hypothetical protein consisting of two identical polypeptide chains. More commonly, proteins have two or more kinds of polypeptide subunits. The bonds that hold a quaternary structure together are basically the same as those that maintain tertiary structure. The overall shape of a protein may be globular (compact and roughly spherical, as in the figure) or fibrous (threadlike).

If a protein encounters a hostile environment in terms of temperature, pH, or salt concentrations, it may unravel and lose its characteristic shape. This process is called **denaturation** (see Figure 5.5). As a result of denaturation, the protein is no longer functional. This is discussed in more detail in Chapter 5 with regard to denaturation of enzymes.

The proteins we have been discussing are *simple proteins,* which contain only amino acids. *Conjugated proteins* are combinations of amino acids with other organic or inorganic components. Conjugated proteins are named by their non-amino-acid component. Thus, glycoproteins contain sugars, nucleoproteins contain nucleic acids, metalloproteins contain metal atoms, lipoproteins contain lipids, and phosphoproteins contain phosphate groups. An example of a microbial phosphoprotein is phospholipase C. Synthesized by the bacterium *Clostridium perfringens* (klôs-tri′dē-um pér-frin′jens), this enzyme breaks down red blood cells and induces some of the symptoms of gas gangrene.

Nucleic Acids

In 1944, three American microbiologists, Oswald Avery, Colin MacLeod, and Maclyn McCarty, discovered that a substance called **deoxyribonucleic acid (DNA)** is the substance of which genes are made. Nine years later, James Watson and Francis Crick, working with molecular models and X-ray information supplied by Maurice Wilkins and Rosalind Franklin, identified the physical structure of DNA. In addition, Crick suggested a mechanism for DNA replication and how it works as the hereditary material. DNA, and another substance called **ribonucleic acid (RNA)**, are together referred to as **nucleic acids** because they were first discovered in the nuclei of cells. Just as amino acids are the structural units of proteins, so nucleotides are the structural units of nucleic acids.

Each **nucleotide** is composed of three parts: a nitrogen-containing base, a pentose (five-carbon) sugar called **deoxyribose** or **ribose**, and a phosphate group (phosphoric acid). The nitrogen-containing bases are cyclic com-

FIGURE 2.18 Protein structure. In this figure, polypeptide chains are shown as ribbons. **(a)** Primary structure, the amino acid sequence. Part of the polypeptide chain has been expanded to illustrate this. Each amino acid is designated by its three-letter abbreviation, as given in Table 2.4. **(b)** Secondary structures: helix and pleated sheet. One amino acid is highlighted in green. **(c)** Tertiary structure, the overall three-dimensional folding of a polypeptide chain. **(d)** Quaternary structure, the relationship between several polypeptide chains that make up a protein. Shown here is the quaternary structure of a hypothetical protein composed of two identical polypeptide chains.

(a) Primary structure

— Ala — Ser — Gly — Leu — His --

Hydrogen bonds

Helix

Pleated sheet

(b) Secondary structure

(c) Tertiary structure

Disulfide link

(d) Quaternary structure

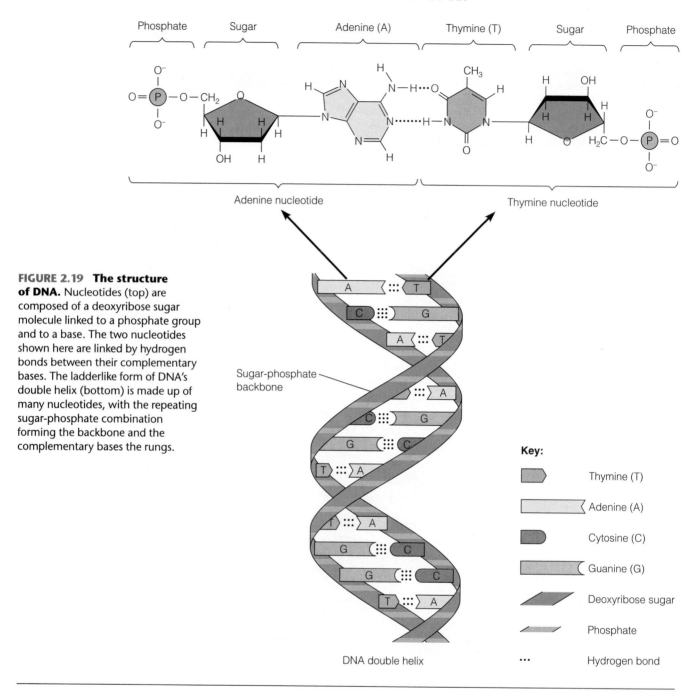

FIGURE 2.19 The structure of DNA. Nucleotides (top) are composed of a deoxyribose sugar molecule linked to a phosphate group and to a base. The two nucleotides shown here are linked by hydrogen bonds between their complementary bases. The ladderlike form of DNA's double helix (bottom) is made up of many nucleotides, with the repeating sugar-phosphate combination forming the backbone and the complementary bases the rungs.

pounds made up of carbon, hydrogen, oxygen, and nitrogen atoms. The bases are named adenine (A), thymine (T), cytosine (C), guanine (G), and uracil (U). A and G are double-ring structures called **purines,** whereas T, C, and U are single-ring structures referred to as **pyrimidines.**

Nucleotides are named according to their nitrogenous base. Thus, a nucleotide containing thymine is called a *thymine nucleotide,* one containing adenine is called an *adenine nucleotide,* and so on. The term **nucleoside** refers to the combination of a purine or pyrimidine plus a pentose sugar; it does not contain a phosphate group.

DNA

According to the model proposed by Watson and Crick, a DNA molecule consists of two long strands wrapped around each other to form a **double helix** (Figure 2.19). The double helix looks like a twisted ladder. Each strand of the DNA double helix is composed of many nucleotides.

Each strand of DNA composing the double helix has a "backbone" consisting of alternating deoxyribose and phosphate groups. The deoxyribose of one nucleotide is joined to the phosphate group of the next. (Refer to Figure 8.3 to see how nucleotides are bonded.) The nitrogen-containing bases make up the rungs of the ladder. Note that the purine A is always paired with the pyrimidine T and that the purine G is always paired with the pyrimidine C. The bases are held together by hydrogen bonds; A—T is held by two hydrogen bonds, and G—C is held by three.

The order in which the nitrogen-base pairs occur along the backbone is extremely specific and in fact contains the genetic instructions for the organism. A certain segment of a nucleotide chain constitutes a gene, and a single DNA molecule may contain thousands of genes. Genes determine all hereditary traits, and they control all the activities that take place within cells.

A very important consequence of nitrogen-containing base pairing is that if the sequence of bases of one strand is known, then the sequence of the other strand is also known. For example, if one strand has the sequence . . . ATGC . . . , then the other strand has the sequence . . . TACG . . . . Because the sequence of bases of one strand is determined by the sequence of bases of the other, the bases are said to be *complementary*. The actual transfer of information becomes possible because of DNA's unique structure and will be discussed further in Chapter 8.

RNA

RNA, the second principal kind of nucleic acid, differs from DNA in several respects. Whereas DNA is double-stranded, RNA is usually single-stranded. The five-carbon sugar in the RNA nucleotide is **ribose**, which has one more oxygen atom than deoxyribose. Also, one of RNA's bases is uracil (U) instead of thymine (Figure 2.20). The other three bases (A, G, C) are the same as DNA. Three major kinds of RNA have been identified in cells. These are referred to as **messenger RNA**, **ribosomal RNA**, and **transfer RNA**. As we shall see in Chapter 8, each type of RNA has a specific role in protein synthesis.

Adenosine Triphosphate (ATP)

Adenosine triphosphate (ATP) is the principal energy-carrying molecule of all cells and is indispensable to the life of the cell. It stores the chemical energy released by some chemical reactions, and it provides the energy for reactions that require energy. ATP consists of an adenosine unit, composed of adenine and ribose, with three phosphate groups (abbreviated Ⓟ) attached (Figure 2.21). In other words, it is an adenine nucleotide (also called adenosine monophosphate, or AMP) with two extra phosphate groups. ATP is called a high-energy molecule because it releases a large amount of usable energy when it loses its terminal phosphate group to become **adenosine diphosphate (ADP)**. This reaction can be represented as follows:

$$\text{ATP} \rightleftharpoons \text{ADP} + \text{Ⓟ} + \text{Energy}$$

Adenosine triphosphate	Adenosine diphosphate	Inorganic phosphate

FIGURE 2.20 **A uracil nucleotide of RNA.**

FIGURE 2.21 **Structure of ATP.** High-energy phosphate bonds are indicated by wavy lines. When ATP breaks down to ADP and inorganic phosphate, a large amount of chemical energy is released for use in other chemical reactions.

A cell's supply of ATP at any particular time is limited. Whenever the supply needs replenishing, the reaction goes in the reverse direction—the addition of a phosphate group to ADP and the input of energy produces more ATP. The energy required to attach the terminal phosphate group to ADP is supplied by the cell's various decomposition reactions, particularly the decomposition of glucose. ATP can be stored in every cell, where its potential energy is not released until needed.

INTRODUCTION (p. 23)

1. The science of the interaction of atoms and molecules is called chemistry.
2. The metabolic activities of microorganisms involve complex chemical reactions.
3. Nutrients are broken down by microbes to obtain energy and to make new cells.

STRUCTURE OF ATOMS (pp. 23–26)

1. Atoms are the smallest units of chemical elements that enter into chemical reactions.
2. Atoms consist of a nucleus, which contains protons and neutrons, and electrons that move around the nucleus.
3. The atomic number is the number of protons in the nucleus; the total number of protons and neutrons is the atomic weight.

Chemical Elements (p. 24)

1. Atoms with the same atomic number and same chemical behavior are classified as the same chemical element.
2. Chemical elements are designated by abbreviations called chemical symbols.
3. About 26 elements are commonly found in living cells.
4. Atoms that have the same atomic number (are of the same element) but different atomic weights are called isotopes.

Electronic Configurations (pp. 24–26)

1. In an atom, electrons are arranged around the nucleus in electron shells.
2. Each shell can hold a characteristic maximum number of electrons.
3. Chemical properties of an atom are largely due to the number of electrons in its outermost shell.

HOW ATOMS FORM MOLECULES: CHEMICAL BONDS (pp. 26–30)

1. Molecules are made up of two or more atoms; molecules consisting of at least two different kinds of atoms are called compounds.
2. Atoms form molecules in order to fill their outermost electron shells.
3. Attractive forces that bind the atomic nuclei of two atoms together are called chemical bonds.
4. The combining capacity of an atom—the number of chemical bonds the atom can form with other atoms—is its valence.

Ionic Bonds (pp. 26–27)

1. A positively or negatively charged atom or group of atoms is called an ion.
2. A chemical attraction between ions of opposite charge is called an ionic bond.
3. To form an ionic bond, one ion is an electron donor; the other ion is an electron acceptor.

Covalent Bonds (pp. 27–29)

1. In a covalent bond, atoms share pairs of electrons.
2. Covalent bonds are stronger than ionic bonds and are far more common in organisms.

Hydrogen Bonds (p. 29)

1. A hydrogen bond exists when a hydrogen atom covalently bonded to one oxygen or nitrogen atom is attracted to another oxygen or nitrogen atom.
2. Hydrogen bonds form weak links between different molecules or between parts of the same large molecule.

Molecular Weight and Moles (p. 30)

1. The molecular weight is the sum of the atomic weights of all the atoms in a molecule.
2. A mole of an atom, ion, or molecule is equal to its atomic or molecular weight expressed in grams.
3. The number of moles of a substance equals its mass in grams divided by its molecular weight.

CHEMICAL REACTIONS (pp. 30–32)

1. Chemical reactions are the making or breaking of chemical bonds between atoms.

Energy of Chemical Reactions (p. 30)

1. A change of energy occurs during chemical reactions.
2. Endergonic reactions require energy; exergonic reactions release energy.
3. In a synthesis reaction, atoms, ions, or molecules are combined to form a larger molecule.
4. In a decomposition reaction, a larger molecule is broken down into its component molecules, ions, or atoms.
5. In an exchange reaction, two molecules are decomposed, and their subunits are used to synthesize two new molecules.
6. The products of reversible reactions can readily revert back to form the original reactants.

How Chemical Reactions Occur (p. 32)

1. For a chemical reaction to take place, the reactants must collide with each other.

2. The minimum collision energy that can produce a chemical reaction is called its activation energy.
3. Specialized proteins called enzymes accelerate chemical reactions in living systems by lowering the activation energy.

IMPORTANT BIOLOGICAL MOLECULES
(pp. 32–47)

INORGANIC COMPOUNDS (pp. 33–36)

1. Inorganic compounds are usually small, ionically bonded molecules.
2. Water and many common acids, bases, and salts are examples of inorganic compounds.

Water (pp. 33–34)
1. Water is the most abundant substance in cells.
2. Because water is a polar molecule, it is an excellent solvent.
3. Water is a reactant in many of the decomposition reactions of digestion.
4. Water is an excellent temperature buffer.

Acids, Bases, and Salts (p. 34)
1. An acid dissociates into H^+ and anions.
2. A base dissociates into OH^- and cations.
3. A salt dissociates into negative and positive ions, neither of which is H^+ or OH^-.

Acid–Base Balance (pp. 34–35)
1. The term pH refers to the concentration of H^+ in a solution.
2. A solution with a pH of 7 is neutral; a pH below 7 indicates acidity; a pH above 7 indicates alkalinity.
3. A pH buffer, which stabilizes the pH inside a cell, can be used in culture media.

ORGANIC COMPOUNDS (pp. 35–47)

1. Organic compounds always contain carbon and hydrogen.
2. Carbon atoms form up to four bonds with other atoms.
3. Organic compounds are mostly or entirely covalently bonded, and many of them are large molecules.

Structure and Chemistry (pp. 36–38)
1. A chain of carbon atoms forms a carbon skeleton.
2. The letter *R* may be used to denote a particular functional group of atoms within an organic molecule.
3. Functional groups of atoms are responsible for most of the properties of organic molecules.
4. Frequently encountered classes of molecules are R—OH (alcohols), R—COOH (organic acids), H_2N—R—COOH (amino acids).

5. Small organic molecules may combine into very large molecules called macromolecules.
6. Monomers usually bond together by dehydration synthesis or condensation reactions that form water and a polymer.

Carbohydrates (pp. 38–39)
1. Carbohydrates are compounds consisting of atoms of carbon, hydrogen, and oxygen, with hydrogen and oxygen in a 2:1 ratio.
2. Carbohydrates include sugars and starches.
3. Carbohydrates can be divided into three types: monosaccharides, disaccharides, and polysaccharides.
4. Monosaccharides contain from three to seven carbon atoms.
5. Isomers are two molecules with the same chemical formula but different structures and properties—for example, glucose ($C_6H_{12}O_6$) and fructose ($C_6H_{12}O_6$).
6. Monosaccharides may form disaccharides and polysaccharides by dehydration synthesis.
7. Polysaccharides and disaccharides may be broken down by hydrolysis, a reaction involving the splitting of water molecules.

Lipids (pp. 39–41)
1. Lipids are a diverse group of compounds distinguished by their insolubility in water.
2. Simple lipids (fats) consist of a molecule of glycerol and three molecules of fatty acids.
3. A saturated fat has no double bonds between carbon atoms in the fatty acids; an unsaturated fat has one or more double bonds.
4. Phospholipids are complex lipids consisting of glycerol, two fatty acids, and a phosphate group.
5. Steroids have carbon ring systems with functional hydroxyl and carbonyl groups.

Proteins (pp. 41–44)
1. Amino acids are the building blocks of proteins.
2. Amino acids consist of carbon, hydrogen, oxygen, nitrogen, and sometimes sulfur.
3. Twenty amino acids occur naturally.
4. By linking amino acids, peptide bonds (formed by dehydration synthesis) allow the formation of polypeptide chains.
5. Proteins have four levels of structure: primary (sequence of amino acids), secondary (regular coils or pleats), tertiary (overall three-dimensional structure of a polypeptide), and quaternary (two or more polypeptide chains).
6. Conjugated proteins consist of amino acids combined with other organic or inorganic compounds.

Nucleic Acids (pp. 44–47)
1. Nucleic acids—DNA and RNA—are macromolecules consisting of repeating nucleotides.

2. A nucleotide is composed of a pentose, a phosphate group, and a nitrogen-containing base. A nucleoside is composed of a pentose and a nitrogen-containing base.

3. A DNA nucleotide consists of deoxyribose (a pentose) and one of the following nitrogen-containing bases: thymine or cytosine (pyrimidines) or adenine or guanine (purines).

4. DNA consists of two strands of nucleotides wound in a double helix. The strands are held together by hydrogen bonds between purine and pyrimidine nucleotides: A—T and G—C.

5. An RNA nucleotide consists of ribose (a pentose) and one of the following nitrogenous bases: cytosine, guanine, adenine, or uracil.

Adenosine Triphosphate (ATP) (p. 47)

1. ATP stores chemical energy for various cellular activities.

2. When the bond to ATP's terminal phosphate group is broken, energy is released.

3. The energy from decomposition reactions is used to regenerate ATP from ADP and phosphate.

STUDY QUESTIONS

Review

1. What is a chemical element?

2. Diagram the electronic configuration of a carbon atom.

3. How does $^{14}_{6}C$ differ from $^{12}_{6}C$?

4. What type of bonding exists between water molecules?

5. What type of bonds holds the following atoms together?
   (a) Li^+ and Cl^- in LiCl
   (b) Carbon and oxygen atoms in CO_2
   (c) Oxygen atoms in O_2
   (d) A carbon atom from glutamic acid and a nitrogen atom from lysine in:

Glutamic acid   Lysine

   (e) A hydrogen atom of one nucleotide to a nitrogen or oxygen atom of another nucleotide in:

Deoxyribose-phosphate   Deoxyribose-phosphate

Guanine   Cytosine

6. Classify the following inorganic molecules as an acid, a base, or a salt. Their dissociation products are shown to help you.
   (a) $HNO_3 \rightarrow H^+ + NO_3^-$
   (b) $H_2SO_4 \rightarrow 2H^+ + SO_4^{2-}$
   (c) $NaOH \rightarrow Na^+ + OH^-$
   (d) $MgSO_4 \rightarrow Mg^{2+} + SO_4^{2-}$

7. Vinegar, pH 3, is how many times more acidic than pure water, pH 7?

8. Calculate the molecular weight of $C_6H_{12}O_6$.

9. How many moles are in 360 grams of glucose?

10. Classify the following types of chemical reactions.
    (a) Glucose + fructose $\rightarrow$ sucrose + H_2O
    (b) Lactose $\rightarrow$ glucose + galactose
    (c) $NH_4Cl + H_2O \rightarrow NH_4OH + HCl$
    (d) ATP $\rightarrow$ ADP + Ⓟ

11. Bacteria use the enzyme urease to obtain nitrogen in a form they can use from urea in the following reaction:

$$CO(NH_2)_2 + H_2O \longrightarrow 2NH_4^+ + CO_2$$

Urea   Ammonium ion   Carbon dioxide

    What purpose does the enzyme serve in this reaction? What type of reaction is this?

12. Classify the following as subunits of either a carbohydrate, lipid, protein, or nucleic acid.
    (a) $CH_3-(CH_2)_7-CH=CH-(CH_2)_7-COOH$
          Oleic acid

    (b)
          Serine

(c) $C_6H_{12}O_6$
(d) Thymine nucleotide

13. Add the appropriate functional group(s) to this ethyl group to produce each of the following compounds: ethanol, acetic acid, acetaldehyde, ethanolamine, diethyl ether.

$$H-\underset{\underset{H}{|}}{\overset{\overset{H}{|}}{C}}-\underset{\underset{H}{|}}{\overset{\overset{H}{|}}{C}}-$$

14. Water plays an important role in the following reversible reaction:

$$HO-\underset{\underset{O}{\|}}{\overset{\overset{H}{|}}{C}}-\underset{\underset{H}{|}}{\overset{\overset{H}{|}}{C}}-N-\underset{\underset{O}{\|}}{\overset{\overset{H}{|}}{C}}-\underset{\underset{H}{|}}{\overset{\overset{H}{|}}{C}}-N-H + H_2O \rightleftharpoons$$

$$HO-\underset{\underset{O}{\|}}{\overset{\overset{H}{|}}{C}}-\underset{\underset{H}{|}}{\overset{\overset{H}{|}}{C}}-N-H + HO-\underset{\underset{O}{\|}}{\overset{\overset{H}{|}}{C}}-\underset{\underset{H}{|}}{\overset{\overset{H}{|}}{C}}-N-H$$

(a) What direction is the hydrolysis reaction (left to right or right to left)?
(b) What direction is the dehydration synthesis reaction?
(c) Circle the atoms involved in the formation of water.
(d) Identify the peptide bond.

15. The energy-carrying property of the ATP molecule is due to energy dynamics that favor the breaking of bonds between _____ . What type of bonds are these?

16. The following diagram shows a protein molecule. Indicate the regions of primary, secondary, and tertiary structure. Does this protein have quaternary structure?

17. Draw a simple lipid, and show how it could be modified to a phospholipid.

18. ATP is an energy-storage compound. From where does it get this energy?

Critical Thinking

1. When you blow bubbles into a glass of water, the following reactions take place:

$$H_2O + CO_2 \xrightarrow{A} H_2CO_3 \xrightarrow{B} H^+ + HCO_3^-$$

(a) What type of reaction is A?
(b) What does reaction B tell you about the type of molecule H_2CO_3 is?

2. What are the common structural characteristics of ATP and DNA molecules?

3. What happens to the relative amount of unsaturated lipids when *E. coli* grown at 25°C are then grown at 37°C?

Applications

1. Laundry detergents and drain cleaners contain enzymes from *Bacillus subtilis*. What is the value of these enzymes in clothes washing and drain cleaning?

2. *Thiobacillus ferrooxidans* was responsible for destroying buildings in the Midwest by causing changes in the earth. The original rock, which contained lime ($CaCO_3$) and pyrite (FeS_2), expanded as bacterial metabolism caused gypsum ($CaSO_4$) crystals to form. How did *T. ferrooxidans* bring about the change from lime to gypsum?

3. When growing in an animal, *Bacillus anthracis* produces a capsule that is resistant to phagocytosis. The capsule is composed of D-glutamic acid. Why is this capsule resistant to digestion by the host's phagocytes? (Phagocytes are white blood cells that engulf bacteria.)

Observing Microorganisms Through a Microscope

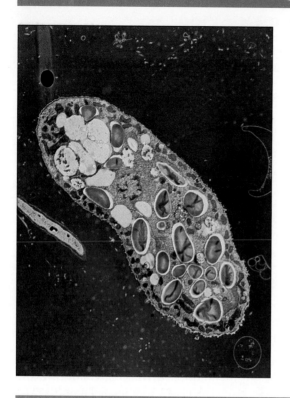

MICROORGANISMS ARE MUCH TOO small to be seen with the unaided eye. In order to make them visible, they must be observed with a microscope. The word microscope is derived from the terms *micro,* which means small, and *skopein,* which means to see. In Anton van Leeuwenhoek's day, looking through a microscope meant something completely different than it does today. Back in his day, the observer would look through rainbow rings and shadows and multiple images. Modern microbiologists, however, have access to microscopes that produce, with great clarity, magnifications that range anywhere from ten to thousands of times greater than those of van Leeuwenhoek's single lens (see Figure 1.2a). Part of this chapter will describe how different types of microscopes function and why one type of microscope might be used in preference to another.

Some microbes are more readily visible than others because of their larger size or more easily observable features. Many microbes, however, must undergo several staining procedures before their cell walls, membranes, and other structures lose their opaque or colorless natural state. The last part of this chapter will explain some of the more commonly used methods of preparing specimens for examination through a light microscope.

You may be wondering how we are going to sort out, measure, and count the specimens we will be studying. This chapter opens with a discussion of how to use the metric system for measuring microbes.

UNITS OF MEASUREMENT

LEARNING OBJECTIVE

- *List the metric units of measurement used for microorganisms, and know their equivalents.*

Because microorganisms and their component parts are so very small, they are measured in units that are unfamiliar to many of us in everyday life. When measuring microorganisms, we use the metric system. The standard unit of length in the metric system is the meter (m). A major advantage of the metric system is that the units are related to each other by factors of 10. Thus, 1 m equals 10 decimeters (dm) or 100 centimeters (cm) or 1000 millimeters (mm). Units in the U.S. system of measure do not have the advantage of conversion by a factor of 10. For example, we use 3 ft or 36 in. to equal 1 yd.

Microorganisms and their structural components are measured in even smaller units, such as micrometers, nanometers, and angstroms. A **micrometer (μm)**, formerly known as a micron (μ), is equal to 0.000001 m (10^{-6} m). The prefix *micro* indicates that the unit following it should be divided by one million, or 10^6 (see Exponential Notation in Appendix E). A **nanometer (nm)**, formerly known as a millimicron (mμ), is equal to 0.000000001 m (10^{-9} m). The prefix *nano* tells us that the

TABLE 3.1 Metric Units of Length and U.S. Equivalents

METRIC UNIT	MEANING OF PREFIX	METRIC EQUIVALENT	U.S. EQUIVALENT
1 kilometer (km)	*kilo* = 1000	1000 m = 10^3 m	3280.84 ft or 0.62 mi; 1 mi = 1.61 km
1 meter (m)		Standard unit of length	39.37 in. or 3.28 ft or 1.09 yd
1 decimeter (dm)	*deci* = 1/10	0.1 m = 10^{-1} m	3.94 in.
1 centimeter (cm)	*centi* = 1/100	0.01 m = 10^{-2} m	0.394 in.; 1 in. = 2.54 cm
1 millimeter (mm)	*milli* = 1/1000	0.001 m = 10^{-3} m	
1 micrometer (μm)	*micro* = 1/1,000,000	0.000001 m = 10^{-6} m	
1 nanometer (nm)	*nano* = 1/1,000,000,000	0.000000001 m = 10^{-9} m	

unit after it should be divided by one billion (10^9). An **angstrom (Å)** is equal to 0.0000000001 m (10^{-10} m). The angstrom is no longer an official unit of measure; however, because of its widespread presence in scientific literature, you should be familiar with it. Its accepted equivalent is 0.1 nm. Table 3.1 presents the basic metric units of length and some of their U.S. equivalents. In Table 3.1, you can compare the microscopic units of measurement with the commonly known macroscopic units of measurement, such as centimeters, meters, and kilometers. If you look ahead to Figure 3.2 on page 55, you will see the relative sizes of various organisms on the metric scale.

MICROSCOPY: THE INSTRUMENTS

LEARNING OBJECTIVES

- *Diagram the path of light through a compound microscope.*

- *Define total magnification and resolution.*

- *Identify a use for darkfield, phase-contrast, DIC, and fluorescence microscopy, and compare each with brightfield illumination.*

- *Explain how electron microscopy differs from light microscopy.*

The simple microscope used by Anton van Leeuwenhoek in the seventeenth century had only one lens and was similar to a magnifying glass. Contemporaries of van Leeuwenhoek, such as Robert Hooke, built microscopes with multiple lenses, called compound microscopes. In fact, a Dutch spectacle-maker, Zacharias Janssen, is credited with making the first compound microscope around 1600. However, these early compound microscopes were of poor quality. It was not until about 1830 that a significantly better microscope was developed by Joseph Jackson Lister (the father of Joseph Lister). Various improvements on Lister's microscope were made that resulted in the development of the modern compound microscope, the

kind used in microbiology laboratories today. Microscopic studies of live specimens have revealed dramatic interactions between microbes (see the box on p. 57).

Compound Light Microscopy

A modern **compound light microscope** has a series of lenses and uses visible light as its source of illumination (Figure 3.1a). With a compound light microscope, we can examine very small specimens as well as some of their fine detail, or **ultrastructure.** A series of finely ground lenses (Figure 3.1b) forms a clearly focused image that is many times larger than the specimen itself. This magnification is achieved when light rays from an **illuminator,** the light source, are passed through a **condenser,** which has lenses that direct the light rays through the specimen. From here, light rays pass into the **objective lenses,** the lenses closest to the specimen. The image of the specimen is magnified again by the **ocular lens,** or **eyepiece.**

We can calculate the **total magnification** of a specimen by multiplying the objective lens magnification (power) by the ocular lens magnification (power). Most microscopes used in microbiology have several objective lenses, including 10× (low power), 40× (high power), and 100× (oil immersion). Most oculars magnify specimens by a factor of 10. Multiplying the magnification of a specific objective lens with that of the ocular, we see that the total magnifications would be 100× for low power, 400× for high power, and 1000× for oil immersion. Some compound light microscopes can achieve a magnification of 2000× with the oil immersion lens.

Resolution (also called *resolving power*) is the ability of the lenses to distinguish fine detail and structure. Specifically, it refers to the ability of the lenses to distinguish between two points a specified distance apart. For example, if a microscope has a resolving power of 0.4 nm, it can distinguish two points as separate objects if they are at least 0.4 nm apart. A general principle of microscopy is that the shorter the wavelength of light used in the instrument, the greater the resolution. The white light used in a com-

FIGURE 3.1 The compound light microscope.

Ocular (eyepiece) Remagnifies the image formed by the objective lens

Body tube Transmits the image from the objective lens to the ocular

Arm

Objective lenses Primary lenses that magnify the specimen

Stage Holds the microscope slide in position

Condenser Focuses light through specimen

Diaphragm Controls the amount of light entering the condenser

Coarse focusing knob

Illuminator Light source

Base

Fine focusing knob

(a) Principal parts and functions

Ocular lens

Line of vision

Path of light

Prism

Body tube

Objective lenses

Specimen

Condenser lenses

Illuminator

Base with source of illumination

(b) The path of light (bottom to top)

pound light microscope has a relatively long wavelength and cannot resolve structures smaller than 0.2 μm. This fact and other practical considerations limit the magnification achieved by even the best compound light microscopes to about 2000$\times$.

Figure 3.2 shows various specimens that can be resolved by the human eye, light microscope, and electron microscope.

To obtain a clear, finely detailed image under a compound light microscope, specimens must be made to contrast sharply with their medium. To attain such contrast, we must change the refractive index of specimens from that of their medium. The **refractive index** is a measure of the relative velocity at which light passes through a material. We change the refractive index of specimens by staining them, a procedure we will discuss shortly. Light rays move in a straight line through a single medium. After staining, when light rays pass through the two materials (the specimen and its medium) with different refractive indexes, the rays change direction from a straight path by bending or changing an angle (refracting) at the boundary between the materials and increase the image's contrast between the specimen and the medium. As the light rays travel away from the specimen, they spread out to resolve points that are close together in the image.

These light rays enter the objective lens, and the image is magnified.

To achieve high magnification (100$\times$) with good resolution, the lens must be small. Although we want light traveling through the specimen and medium to refract differently, we do not want to lose light rays after they have passed through the stained specimen. To preserve the direction of light rays at the highest magnification, immersion oil is placed between the glass slide and the oil immersion objective lens (Figure 3.3). The immersion oil has the same refractive index as glass, so the oil becomes part of the optics of the glass of the microscope. Unless immersion oil is used, light rays are refracted as they enter the air from the slide, and the objective lens would have to be increased in diameter to capture them. The oil has the same effect as increasing the objective diameter; therefore, it improves the resolving power of the lenses. If oil is not used with an oil immersion objective lens, the image becomes fuzzy, with poor resolution.

Under usual operating conditions, the field of vision in a compound light microscope is brightly illuminated. By focusing the light, the condenser produces a **brightfield** illumination (Figure 3.4a).

It is not always desirable to stain a specimen. However, an unstained cell has little contrast with its surroundings

FIGURE 3.2 Relationships between the sizes of various specimens and the resolution of the human eye, light microscope, and electron microscope. The area in yellow shows the size range of the organisms we will be studying in this book.

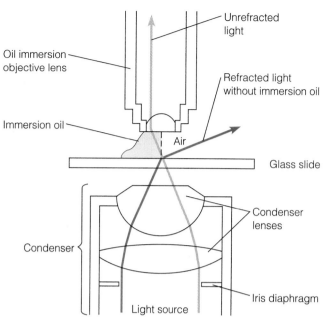

FIGURE 3.3 Refraction in the compound microscope using an oil immersion objective lens. Because the refractive indexes of the glass microscope slide and immersion oil are the same, the light rays do not refract when passing from one to the other when the oil immersion objective lens is used. This method produces images with better resolution at magnifications greater than 100×.

microscope, cannot be stained by standard methods, or are so distorted by staining that their characteristics then cannot be identified. Instead of the normal condenser, a darkfield microscope uses a darkfield condenser that contains an opaque disc. The disc blocks light that would enter the objective directly. Only light that is reflected off (turned back from) the specimen enters the objective lens (Figure 3.4b). Because there is no direct background light, the specimen appears light against a black background—the dark field. This technique is frequently used to examine unstained microorganisms suspended in liquid. One use for darkfield microscopy is the examination of very thin spirochetes, such as *Treponema pallidum* (tre-pō-nē′mä pal′li-dum), the causative agent of syphilis.

Phase-Contrast Microscopy

Another way to observe microorganisms is with a **phase-contrast microscope.** Phase-contrast microscopy is especially useful because it permits detailed examination of internal structures in *living* microorganisms. In addition, it is not necessary to fix (attach the microbes to the microscope slide) or stain the specimen—procedures that could harm or kill the microorganisms.

The principle of phase-contrast microscopy is based on slight variations in refractive index. As rays pass from the

and is therefore difficult to see. Unstained cells are more easily observed with the modified compound microscopes described in the next section.

Darkfield Microscopy

A **darkfield microscope** is used for examining live microorganisms that either are invisible in the ordinary light

light source through the specimen, their velocity may be altered by differences in the thickness and physical properties of various portions of the specimen. Light rays passing through the specimen are diffracted (bent) differently and travel different pathways (out of phase with one another) to reach the eye of the viewer (Figure 3.4c). These phase differences are seen through the microscope as different degrees of brightness. Details of the internal structures of the specimen also become more sharply defined

in phase-contrast microscopy. The internal details of a cell appear as degrees of brightness against a dark background.

A phase-contrast microscope uses a special condenser that contains an annular (ring-shaped) diaphragm. The diaphragm allows a ring of light to pass through the condenser, focusing light on the specimen and a ring-shaped diffraction (phase) plate in the objective lens. The diffracted and undiffracted rays are then brought into phase with each other to produce the image that meets the eye.

FIGURE 3.4 **Brightfield, darkfield, and phase-contrast microscopy.** The illustrations show the contrasting light pathways of each of these types of microscopy. The photographs compare the same *Paramecium* specimen using these different microscopy techniques.

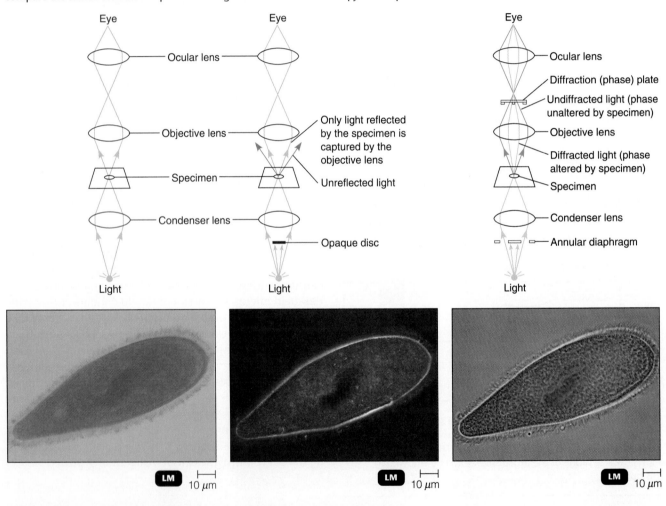

(a) Brightfield. (Top) The path of light in brightfield microscopy, the type of illumination produced by regular compound light microscopes. (Bottom) Brightfield illumination shows internal structures and the outline of the transparent sheath (external covering).

(b) Darkfield. (Top) The darkfield microscope uses a special condenser with an opaque disc that eliminates all light in the center of the beam. The only light that reaches the specimen comes in at an angle; thus, only light reflected by the specimen (gold rays) reaches the objective lens. (Bottom) Against the black background seen with darkfield microscopy, edges of the cell are bright, some internal structures seem to sparkle, and the sheath is almost visible.

(c) Phase-contrast. (Top) In phase-contrast microscopy, light rays are diffracted differently and travel different pathways to reach the eye of the viewer. Diffracted light rays are indicated in gold; undiffracted rays are shown in red. (Bottom) Phase-contrast microscopy shows greater differentiation among internal structures and also shows the sheath wall. The wide light band around the cell is an artifact resulting from this type of microscopy—a "phase halo."

CAN A BACTERIAL PREDATOR MAKE EGGS SAFER?

EVEN THROUGH A microscope, appearances can be deceptive, and first impressions can be quite misleading. For example, at first glance, the bacterium *Bdellovibrio* looks like a perfectly ordinary 1-μm-long curved, gram-negative rod. *Bdellovibrio* cells exhibit the normal runs and tumbles of flagellated bacteria, although they move about ten times faster than other bacteria of the same size. However, more patient observation or use of a microscope equipped with a video camera reveals that *Bdellovibrio*'s behavior is unlike that of any other known bacteria.

A unique microbial drama occurs when *Bdellovibrio* cells are placed in a suspension with other bacteria, such as *E. coli*. Apparently drawn by chemical attractants, *Bdellovibrio* speeds toward and rams the larger, slower *E. coli*. The two cells become attached and rotate rapidly. As the two bacteria whirl about, the *Bdellovibrio* paralyzes the *E. coli* by an unknown mechanism that most probably involves one of *Bdellovibrio*'s protein products. *Bdellovibrio* kills *E. coli* without breaking down its contents.

Next, in preparation for moving into its newly acquired home, *Bdellovibrio* adds structural components to the outer membrane of *E. coli*. This activity, which stabilizes the outer membrane, is another ability unique to *Bdellovibrio*. After it penetrates *E. coli*'s cell wall and detaches its flagellum, *Bdellovibrio* snuggles into the periplasmic space (that is, the space between the plasma membrane and the cell wall) and prepares for a feast.

Macromolecules leak from *E. coli*'s cytoplasm into the periplasmic space. *Bdellovibrio* digests the macromole-cules and grows into a long, coiled filament. In about two or three hours, the cytoplasmic contents of the host (including its DNA) are digested, and the *Bdellovibrio* filament fragments into as many as 20 individual curved rods. Each rod develops a flagellum and is released when the host cell is broken open by a *Bdellovibrio* enzyme. The new *Bdellovibrio* cells scurry away to feed on other bacteria.

Bdellovibrio was once regarded as simply a parasite. However, thorough studies have shown that *Bdellovibrio* does not merely siphon off its host's resources and energy; it actually digests the cell's contents as food, so *Bdellovibrio* is now considered to be a bacterial predator. The ecological importance of *Bdellovibrio* is that it, along with protozoa and bacteriophages, helps to control the population growth of soil bacteria.

In addition, now that its habits are known, *Bdellovibrio* may have a new career as a bacterial exterminator. Researchers led by Richard Whiting at the U.S. Department of Agriculture are investigating the use of *Bdellovibrio* to control *Salmonella* (a gramnegative bacterium) in chickens and eggs, where it is one of the most common causes of food poisoning in the United States. *Salmonella* can infect the bird's ovaries and enter the egg before it is laid, or it can penetrate the eggshell from bird droppings. Although *Salmonella* is destroyed by cooking, foods that contain raw eggs (e.g., mayonnaise) are a problem. Of the 91,678 cases of foodborne infection reported in the United States from 1983 through 1987, roughly half were caused by *Salmonella* associated with eggs. For example, one outbreak was traced to a blender used to make scrambled eggs. *Salmonella* left in the blender contaminated other foods throughout the day and caused numerous cases of food poisoning.

Because *Bdellovibrio* cells prey on gram-negative bacteria, the U.S.D.A. researchers hope that if *Bdellovibrio* can be introduced into chickens via food or water, it will be able to survive and kill salmonellae. As harmful as they are to other bacteria, *Bdellovibrio* cells are safe for humans. If the U.S.D.A. project succeeds, picnickers need not even be aware of the violent microbiological drama acted out in defense of their chicken salad.

E. coli plasma membrane

Bdellovibrio

E. coli cell wall

TEM

1 μm

Bdellovibrio. A *Bdellovibrio* bacterium has penetrated the cell wall of an *E. coli* cell. The *Bdellovibrio*, having killed the prey cell, uses the *E. coli* cell's contents for its elongation and multiplication.

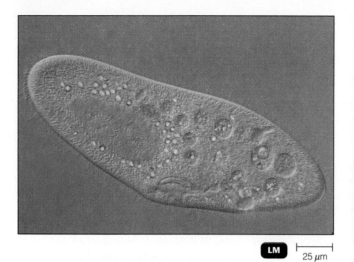

LM | 25 μm

FIGURE 3.5 Differential interference contrast (DIC) microscopy. Like phase-contrast, DIC uses differences in refractive indexes to produce an image, in this case of the protozoan *Paramecium*. The colors in the image are produced by prisms that split the two light beams used in this process.

LM | 10 μm

FIGURE 3.6 Fluorescence microscopy. Fluorescence microscopy has been used to visualize this *Mycobacterium tuberculosis* cell after it absorbed the fluorochrome dye auramine O.

Differential Interference Contrast (DIC) Microscopy

Differential interference contrast (DIC) microscopy is similar to phase-contrast microscopy in that it uses differences in refractive indexes. However, a DIC microscope uses two beams of light instead of one. In addition, prisms split each light beam, adding contrasting colors to the specimen. Therefore, the resolution of a DIC microscope is higher than that of a standard phase-contrast microscope. In addition, the image is brightly colored and appears nearly three-dimensional (Figure 3.5).

Fluorescence Microscopy

Fluorescence microscopy takes advantage of the fluorescence of substances. Fluorescent substances absorb short wavelengths of light (ultraviolet) and give off light at a longer wavelength that can be seen by the use of special light filters. Some organisms fluoresce naturally under special lighting; if the specimen to be viewed does not naturally fluoresce, it is stained with one of a group of fluorescent dyes called *fluorochromes*. When microorganisms stained with a fluorochrome are examined under a fluorescence microscope with an ultraviolet or near-ultraviolet light source, they appear as luminescent, bright objects against a dark background.

Fluorochromes have special attractions for different microorganisms. For example, the fluorochrome auramine O, which glows yellow when exposed to ultraviolet light, is strongly absorbed by *Mycobacterium tuberculosis* (mī-kō-bak-ti′rē-um tü-bėr-kū-lō′sis), the bacterium

that causes tuberculosis. When the dye is applied to a sample of material suspected of containing the bacterium, the bacterium can be detected by the appearance of bright yellow organisms against a dark background (Figure 3.6). *Bacillus anthracis,* the causative agent of anthrax, appears apple green when stained with another fluorochrome, fluorescein isothiocyanate (FITC).

The principal use of fluorescence microscopy is a diagnostic technique called the **fluorescent-antibody technique,** or **immunofluorescence. Antibodies** are natural defense molecules that are produced by humans and many animals in reaction to a foreign substance, or **antigen.** Fluorescent antibodies for a particular antigen are obtained as follows: An animal is injected with a specific antigen, such as a bacterium; the animal then begins to produce specific antibodies against that antigen. After a sufficient time, the antibodies are removed from the serum of the animal. Next, as you can see in Figure 3.7a, a fluorochrome is chemically combined with the antibodies. These fluorescent antibodies are then added to a microscope slide containing an unknown bacterium. If this unknown bacterium is the same bacterium that was injected into the animal, the fluorescent antibodies bind to antigens on the surface of the bacterium, causing it to fluoresce.

This technique can detect bacteria or other disease-producing microorganisms (Figure 3.7b) even within cells, tissues, or other clinical specimens. It is especially useful in diagnosing syphilis and rabies. We will say more about antigen-antibody reactions and immunofluorescence in Chapter 18.

Electron Microscopy

Objects smaller than 0.2 μm, such as viruses or the internal structures of cells, must be examined with an **electron microscope**. In electron microscopy, a beam of electrons is used instead of light. Free electrons travel in waves. The resolving power of the electron microscope is far greater than that of the other microscopes described here so far. The better resolution of electron microscopes is due to the shorter wavelengths of electrons; the wavelengths of electrons are about 100,000 times smaller than the wavelengths of visible light. Thus, electron microscopes are used to examine structures too small to be resolved with light microscopes.

Instead of using glass lenses, an electron microscope uses electromagnetic lenses to focus onto a specimen a beam of electrons traveling through an evacuated tube. There are two types of electron microscopes: the transmission electron microscope and the scanning electron microscope.

Transmission Electron Microscopy

In the **transmission electron microscope**, a finely focused beam of electrons from an electron gun passes through a specially prepared, ultrathin section of the specimen (Figure 3.8a). The beam is focused on a small area of the specimen by an electromagnetic condenser lens that performs roughly the same function as the condenser of a light microscope—to direct the beam of electrons in a straight line to illuminate the specimen.

Electron microscopes use electromagnetic lenses rather than glass lenses, as in a light microscope. Electromagnetic lenses control illumination, focus, and magnification. Instead of being placed on a glass slide, as in light microscopes, the specimen is usually placed on a copper mesh grid. The beam of electrons passes through the specimen and then through an electromagnetic objective lens, which magnifies the image. Finally, the electrons are focused by an electromagnetic projector lens, rather than by an ocular lens as in a light microscope, onto a fluorescent screen or photographic plate. The final image, called a **transmission electron micrograph**, appears as many light and dark areas, depending on the number of electrons absorbed by different areas of the specimen.

In practice, the transmission electron microscope can resolve objects as close together as 2.5 nm, and objects are generally magnified 10,000× to 100,000×. Because most microscopic specimens are so thin, the contrast between their ultrastructures and the background is weak. Contrast can be greatly enhanced by use of a "stain" that absorbs electrons and produces a darker image in the stained region. Salts of various heavy metals, such as lead, osmium, tungsten, and uranium, are commonly used as stains. These metals can be fixed onto the specimen (positive staining) or used to increase the electron opacity of the surrounding field (negative staining). Negative staining is

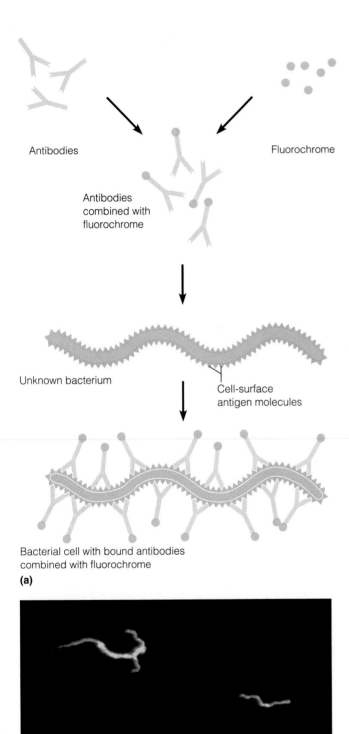

Antibodies

Fluorochrome

Antibodies combined with fluorochrome

Unknown bacterium

Cell-surface antigen molecules

Bacterial cell with bound antibodies combined with fluorochrome

(a)

(b)   **LM** ⊢───┤ 10 μm

FIGURE 3.7 The principle of immunofluorescence.
(a) A type of fluorochrome is combined with antibodies against a specific type of bacterium. When the preparation is added to bacterial cells on a microscope slide, the antibodies attach to the bacterial cells and the cells fluoresce when illuminated with ultraviolet light. **(b)** In the fluorescent treponemal antibody absorption (FTA-ABS) test for syphilis shown here, *Treponema pallidum* shows up as light regions against a darker background.

FIGURE 3.8 Transmission and scanning electron microscopy. The illustrations show the pathways of electron beams used to create images of the specimens. The photographs show lymphocytes, a kind of white blood cell, viewed through both of these types of electron microscopes.

(a) Transmission. (Top) In a transmission electron microscope, electrons pass through the specimen and are scattered. Magnetic lenses focus the image onto a fluorescent screen or photographic plate. (Bottom) This colorized trasmission electron micrograph (TEM) shows a thin slice of lymphocyte. In this type of microscopy, the internal structures present in the slice can be seen.

(b) Scanning. (Top) In a scanning electron microscope, primary electrons sweep across the specimen and knock electrons from its surface. These secondary electrons are picked up by a collector, amplified, and transmitted onto a viewing screen or photographic plate. (Bottom) In this colorized scanning electron micrograph (SEM), the surface structures of a lymphocyte can be seen. Note the three-dimensional appearance of this cell, in contrast to that of the cell in part (a).

---

useful for the study of the very smallest specimens, such as virus particles, bacterial flagella, and protein molecules.

Transmission electron microscopy has high resolution and is extremely valuable for examining different layers of specimens. However, it does have certain disadvantages. Because electrons have limited penetrating power, only a very thin section of a specimen (about 100 nm) can be studied effectively. Thus, the specimen has no three-dimensional aspect. In addition, specimens must be fixed, dehydrated, and viewed under a high vacuum. These treatments not only kill the specimen but also cause some shrinkage and distortion, sometimes to

the extent that there may appear to be additional structures in a prepared cell. Structures that appear as a result of the method of preparation are called *artifacts* (see Chapter 4).

Scanning Electron Microscopy

The **scanning electron microscope** overcomes the problem of sectioning associated with a transmission electron microscope. A scanning electron microscope provides striking three-dimensional views of specimens (Figure 3.8b). In scanning electron microscopy, an electron gun produces a finely focused beam of electrons called the primary electron beam. These electrons pass through electromagnetic lenses and are directed over the surface of the specimen. The primary electron beam knocks electrons out of the surface of the specimen, and the secondary electrons thus produced are transmitted to an electron collector, amplified, and used to produce an image on a viewing screen or photographic plate. The scanning electron microscope is especially useful in studying the surface structures of intact cells and viruses. In practice, it can resolve objects as close together as 20 nm, and objects are generally magnified $1000\times$ to $10,000\times$.

The various types of microscopy just described are summarized in Table 3.2 (pp. 64–65).

PREPARATION OF SPECIMENS FOR LIGHT MICROSCOPY

LEARNING OBJECTIVES

- *Differentiate between an acidic dye and a basic dye.*
- *Compare simple, differential, and special stains.*
- *List the steps in preparing a Gram stain, and describe the appearance of gram-positive and gram-negative cells after each step.*
- *Explain why each of the following is used: capsule stain, endospore stain, flagellar stain.*

Because most microorganisms appear almost colorless when viewed through a standard light microscope, we often must prepare them for observation. One of the ways this can be done is by staining (coloring). Next we will discuss several different staining procedures.

Preparing Smears for Staining

Most studies of the shapes and cellular arrangements of microorganisms are made with stained preparations. **Staining** simply means coloring the microorganisms with a dye that emphasizes certain structures. Before the microorganisms can be stained, however, they must be attached (or fixed) to the microscope slide; without **fixing**, the stain might wash them off the slide.

When a specimen is to be fixed, a thin film of material containing the microorganisms is spread over the surface of the slide. This film, called a **smear**, is allowed to air dry. In most staining procedures the slide is then slowly passed through the flame of a Bunsen burner several times, smear side up. Air drying and flaming fix the microorganisms to the slide and usually kill them. Stain is applied and then washed off with water; then the slide is blotted with absorbent paper. The stained microorganisms are now ready for microscopic examination.

Stains are salts composed of a positive and a negative ion, one of which is colored and is known as the chro-

mophore. The color of so-called **basic dyes** is in the positive ion; in **acidic dyes**, it is in the negative ion. Bacteria are slightly negatively charged at pH 7.0. Thus, the colored positive ion in a basic dye is attracted to the negatively charged bacterial cell. Basic dyes, which include crystal violet, methylene blue, and safranin, are more commonly used than acidic dyes. Acidic dyes are not attracted to most types of bacteria because the dye's negative ions are repelled by the negatively charged bacterial surface, so the stain colors the background instead. This preparation of colorless bacteria against a colored background is called **negative staining.** It is valuable in the observation of overall cell shapes, sizes, and capsules because the cells are made highly visible against a contrasting dark background (see Figure 3.11a). Distortions of cell size and shape are minimized because heat fixing is not necessary and the cells do not pick up the stain. Examples of acidic dyes are eosin, nigrosin, and India ink.

To apply acidic or basic dyes, microbiologists use three kinds of staining techniques: simple, differential, and special.

Simple Stains

A **simple stain** is an aqueous or alcohol solution of a single basic dye. Although different dyes bind specifically to different parts of cells, the primary purpose of a simple stain is to highlight the entire microorganism so that cellular shapes and basic structures are visible. The stain is applied to the fixed smear for a certain length of time and then washed off, and the slide is dried and examined. Occasionally, a chemical is added to the solution to intensify the stain; such an additive is called a **mordant**. One function of a mordant is to increase the affinity of a stain for a biological specimen; another is to coat a structure (such as a flagellum) to make it thicker and easier to see after it is stained with a dye. Some of the simple stains commonly used in the laboratory are methylene blue, carbolfuchsin, crystal violet, and safranin.

Differential Stains

Unlike simple stains, **differential stains** react differently with different kinds of bacteria and thus can be used to distinguish among them. The differential stains most frequently used for bacteria are the Gram stain and the acid-fast stain.

Gram Stain

The **Gram stain** was developed in 1884 by the Danish bacteriologist Hans Christian Gram. It is one of the most useful staining procedures because it divides bacteria into two large groups, gram-positive and gram-negative.

In this procedure (Figure 3.9), ① a heat-fixed smear is covered with a basic purple dye, usually crystal violet. Be-

Crystal violet
Iodine
Alcohol
Safranin

1 Application of crystal violet (purple dye)

2 Application of iodine (mordant)

3 Alcohol wash (decolorization)

4 Application of safranin (counterstain)

(a)

(b)

LM   1 μm

FIGURE 3.9 Gram-staining procedure. (a) ① A heat-fixed smear of cocci and rods is first covered with a basic purple dye (primary stain) such as crystal violet, and then the dye is washed off. ② Then the smear is covered with iodine (a mordant) and washed off. At this time, both gram-positive and gram-negative bacteria are purple. ③ The slide is washed with ethanol or an alcohol–acetone solution (a decolorizer) and then washed with water. Now gram-positive cells are purple and gram-negative cells are colorless. ④ In the final step, safranin is added as a counterstain, and the slide is washed, dried, and examined microscopically. Gram-positive bacteria retain the purple dye, even through the alcohol wash. Gram-negative bacteria appear pink because they pick up the safranin counterstain. **(b)** Photomicrograph of gram-stained bacteria. *Staphylococcus aureus* (purple) are gram-positive, and *Escherichia coli* (pink) are gram-negative.

cause the purple stain imparts its color to all cells, it is referred to as a **primary stain.** ② After a short time, the purple dye is washed off, and the smear is covered with iodine, a mordant. When the iodine is washed off, both gram-positive and gram-negative bacteria appear dark violet or purple. ③ Next, the slide is washed with ethanol or an ethanol–acetone solution. This solution is a **decolorizing agent,** which removes the purple from the cells of some species but not from others. ④ The alcohol is rinsed off, and the slide is then stained with safranin, a basic red dye. The smear is washed again, blotted dry, and examined microscopically.

The purple dye and the iodine combine with each bacterium and color it dark violet or purple. Bacteria that re-

tain this color after the alcohol has attempted to decolorize them are classified as **gram-positive;** bacteria that lose the dark violet or purple color after decolorization are classified as **gram-negative** (Figure 3.9b). Because gram-negative bacteria are colorless after the alcohol wash, they are no longer visible. This is why the basic dye safranin is applied; it turns the gram-negative bacteria pink. Stains such as safranin that have a contrasting color to the primary stain are called **counterstains.** Because gram-positive bacteria retain the original purple stain, they are not affected by the safranin counterstain.

As you will see in the next chapter, different kinds of bacteria react differently to the Gram stain, probably because structural differences in their cell walls affect the re-

tention or escape of a combination of crystal violet and iodine, called the crystal violet–iodine (CV–I) complex.
Among other differences, gram-positive bacteria have a
thicker peptidoglycan cell wall than gram-negative bacteria (see Chapter 4). In addition, gram-negative bacteria
contain a layer of lipopolysaccharide as part of their cell
wall. When applied to both gram-positive and gram-
negative cells, crystal violet and then iodine readily enter
the cells. Inside the cells, the crystal violet and iodine
combine to form CV–I. This complex is larger than the
crystal violet molecule that entered the cells, and, because
of its size, it cannot be washed out of the intact peptidoglycan layer of gram-positive cells by alcohol. Consequently, gram-positive cells retain the color of the crystal
violet dye. In gram-negative cells, however, the alcohol
wash disrupts the outer lipopolysaccharide layer, and the
CV–I complex is washed out through the thin layer of peptidoglycan. As a result, gram-negative cells are rendered
colorless until counterstained with safranin, after which
they are pink. In summary, the gram-positive cells retain
the dye and remain purple. The gram-negative cells do not
retain the dye; they are colorless until counterstained with
a red dye, after which they appear pink.

The Gram method is one of the most important staining techniques in medical microbiology. But Gram staining results are not universally applicable because some
bacterial cells stain poorly or not at all. The Gram reaction
is most consistent when it is used on young, growing
bacteria.

In many cases, the Gram reaction of a bacterium provides valuable information for the treatment of disease.
For example, gram-positive bacteria tend to be killed easily by penicillin and sulfonamide drugs. Gram-negative
bacteria resist these drugs, but they are much more susceptible to such drugs as streptomycin, chloramphenicol,
and tetracycline. Thus, Gram-stain identification of a bacterium can help determine which drug will be most effective against a disease.

Acid-Fast Stain

Another important differential stain (one that divides
bacteria into distinctive groups) is the **acid-fast stain**,
which binds strongly only to bacteria that have a waxy
material in their cell walls. Microbiologists use this stain
to identify all bacteria in the genus *Mycobacterium*, including the two important disease producers *Mycobacterium tuberculosis* and *Mycobacterium leprae* (lep′rī). This
stain is also used to identify the disease-producing strains
of the genus *Nocardia* (nō-kär′dē-ä).

In the acid-fast staining procedure, the red dye carbolfuchsin is applied to a fixed smear, and the slide is gently
heated for several minutes. (Heating enhances penetration and retention of the dye.) Then the slide is cooled and
washed with water. The smear is next treated with acid-
alcohol, a decolorizer, which removes the red stain from
bacteria that are not acid-fast. The acid-fast microorganisms retain the red color because the carbolfuchsin is more

FIGURE 3.10 Acid-fast bacteria. The *Mycobacterium leprae*
cells that have infected this tissue have been stained red with the
acid-fast stain.

soluble in the cell wall waxes than in the acid-alcohol (Figure 3.10). In non-acid-fast bacteria, whose cell walls lack
the waxy components, the carbolfuchsin is rapidly removed during decolorization, leaving the cells colorless.
The smear is then stained with a methylene blue counterstain. Non-acid-fast cells appear blue after application of
the counterstain.

Special Stains

Special stains are used to color and isolate specific parts
of microorganisms, such as endospores and flagella, and
to reveal the presence of capsules.

Negative Staining for Capsules

Many microorganisms contain a gelatinous covering
called a **capsule**, which we will discuss in our examination of the procaryotic cell in Chapter 4. In medical microbiology, demonstrating the presence of a capsule is a
means of determining the organism's **virulence**, the degree to which a pathogen can cause disease. Capsule staining is more difficult than other types of staining procedures because capsular materials are soluble in water and
may be dislodged or removed during rigorous washing. To
demonstrate the presence of capsules, a microbiologist can
mix the bacteria in a solution containing a fine colloidal
suspension of colored particles (usually India ink or nigrosin) to provide a dark background and then can stain
the bacteria with a simple stain, such as safranin (Figure
3.11a). Because of their chemical composition, capsules do
not accept most biological dyes, such as safranin, and thus
appear as halos surrounding each stained bacterial cell.
The use of India ink illustrates a negative-staining technique; negative stains do not penetrate the capsule, and
thus they provide a contrast between the capsule and the
surrounding dark medium.

TABLE 3.2 Summary of Various Types of Microscopes

MICROSCOPE TYPE	DISTINGUISHING FEATURES	TYPICAL IMAGE	PRINCIPAL USES
Light			
Brightfield	Uses visible light as a source of illumination; cannot resolve structures smaller than about 0.2 μm; specimen appears against a bright background.	LM 10 μm	Commonly used to observe various stained (killed) specimens and to count microbes; does not resolve very small specimens, such as viruses. Instrument is inexpensive and simple to use.
Darkfield	Uses a special condenser with an opaque disc that blocks light from entering the objective directly; light reflected by specimen enters the objective, and the specimen appears light against a black background.	LM 10 μm	Commonly used to examine living microorganisms that are invisible in brightfield microscopy, do not stain easily, or are distorted by staining; frequently used to detect *Treponema pallidum* in the diagnosis of syphilus.
Phase-contrast	Uses a special condenser and diffraction plate to diffract light rays so that they are out of phase with one another; the specimen appears as different degrees of brightness and contrast.	LM 10 μm	Commonly used to provide detailed examination of the internal structures of living specimens; no staining is required.
Differential interference contrast	Like phase-contrast, uses differences in refractive indexes to produce images. Uses two beams of light separated by prisms; the specimen appears colored due to the prism effect.	LM 25 μm	Used to provide three-dimensional image; no staining is required.
Fluorescence	Uses an ultraviolet or near-ultraviolet source of illumination that causes fluorescent compounds in a specimen to emit light.	LM 10 μm	Principal use is for fluorescent-antibody techniques (immunofluorescence) to identify and rapidly detect microorganisms in tissues or clinical specimens.

TABLE 3.2   Continued

MICROSCOPE TYPE	DISTINGUISHING FEATURES	TYPICAL IMAGE	PRINCIPAL USES
Electron			
Transmission	Uses a beam of electrons instead of light; because of the shorter wavelength of electrons, structures smaller than 0.2 μm can be resolved. The image produced is two-dimensional.	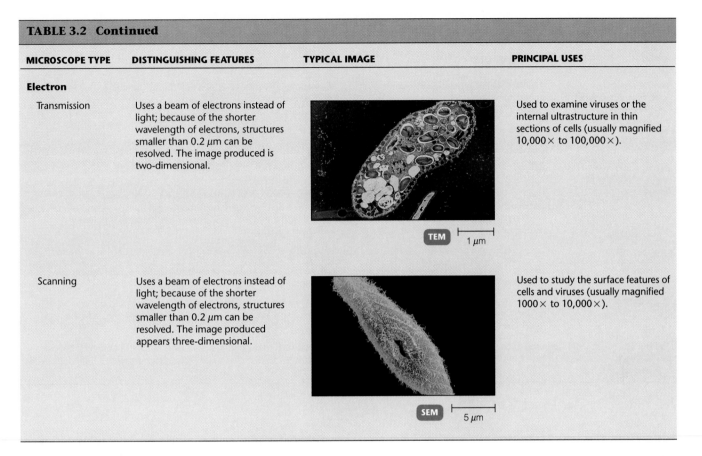 TEM ⊢ 1 μm	Used to examine viruses or the internal ultrastructure in thin sections of cells (usually magnified 10,000$\times$ to 100,000$\times$).
Scanning	Uses a beam of electrons instead of light; because of the shorter wavelength of electrons, structures smaller than 0.2 μm can be resolved. The image produced appears three-dimensional.	SEM ⊢ 5 μm	Used to study the surface features of cells and viruses (usually magnified 1000$\times$ to 10,000$\times$).

Capsule

(a) Negative staining of capsules   LM ⊢ 10 μm

Endospore

(b) Staining of endospores   LM ⊢ 10 μm

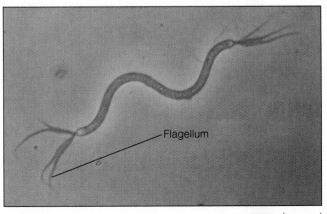

Flagellum

FIGURE 3.11   Special staining. (a) India ink provides a dark background, so the capsules of these bacteria, *Klebsiella pneumoniae,* show up as light areas surrounding the stained cells. **(b)** Endospores are seen as blue ovals in these rod-shaped cells of the bacterium *Bacillus cereus,* using the Schaeffer-Fulton endospore stain. **(c)** Flagella are shown as wavy extensions from the ends of these cells of the bacterium *Spirillum volutans.* In relation to the bodies of the cells, the flagella are much thicker than normal because layers of the stain have accumulated from treatment of the specimen with a mordant.

(c) Staining of flagella   LM ⊢ 10 μm

TABLE 3.3 Summary of Various Stains and Their Uses

STAIN	PRINCIPAL USES
Simple (methylene blue, carbolfuchsin, crystal violet, safranin)	Aqueous or alcohol solution of a single basic dye. (Sometimes a mordant is added to intensify the stain.) Used to highlight microorganisms to determine cellular shapes and arrangements.
Differential	React differently with different kinds of bacteria in order to distinguish among them.
Gram	Divides bacteria into two large groups, gram-positive and gram-negative. Gram-positive bacteria retain the crystal violet stain and appear purple. Gram-negative bacteria do not retain the crystal violet stain and remain colorless until counterstained with safranin and then appear pink.
Acid-fast	Used to distinguish *Mycobacterium* species and some species of *Nocardia*. Acid-fast bacteria, once stained with carbolfuchsin and treated with acid-alcohol, remain red because they retain the carbolfuchsin stain. Non-acid-fast bacteria, when stained and treated the same way and then stained with methylene blue, appear blue because they lose the carbolfuchsin stain and are then able to accept the methylene blue stain.
Special	Used to color and isolate various structures, such as capsules, endospores, and flagella; sometimes used as a diagnostic aid.
Negative	Used to demonstrate the presence of capsules. Because capsules do not accept most stains, the capsules appear as unstained halos around bacterial cells and stand out against a dark background.
Endospore	Used to detect the presence of endospores in bacteria. When malachite green is applied to a heat-fixed smear of bacterial cells, the stain penetrates the endospores and stains them green. When safranin (red) is then applied, it stains the remainder of the cells red or pink.
Flagella	Used to demonstrate the presence of flagella. A mordant is used to build up the diameters of flagella until they become visible microscopically when stained with carbolfuchsin.

Endospore (Spore) Staining

An **endospore** is a special resistant, dormant structure formed within a cell that protects a microorganism from adverse environmental conditions. Although endospores are relatively uncommon in bacterial cells, they can be formed by seven genera of bacteria. Endospores cannot be stained by ordinary methods, such as simple staining and Gram staining, because the dyes do not penetrate the wall of the endospore.

The most commonly used endospore stain is the *Schaeffer-Fulton endospore stain*. Malachite green, the primary stain, is applied to a heat-fixed smear and heated to steaming for about five minutes. The heat helps the stain to penetrate the endospore wall. Then the preparation is washed for about 30 seconds with water to remove the malachite green from all of the cells' parts except the endospores. Next, safranin, a counterstain, is applied to the smear to stain portions of the cell other than endospores. In a properly prepared smear, the endospores appear green within red or pink cells (Figure 3.11b). Because endospores are highly refractive, they can be detected under the light microscope when unstained, but they cannot be differentiated from inclusions of stored material without a special stain.

Flagella Staining

Bacterial **flagella** are structures of locomotion too small to be seen with light microscopes without staining. A tedious and delicate staining procedure uses a mordant and the stain carbolfuchsin to build up the diameters of the flagella until they become visible under the light microscope. Microbiologists use the number and arrangement of flagella as diagnostic aids. Figure 3.11c depicts stained flagella.

A summary of stains is presented in Table 3.3. In the next chapter we will take a closer look at the structure of microbes and how they protect, nourish, and reproduce themselves.

UNITS OF MEASUREMENT (pp. 52–53)

1. The standard unit of length is the meter (m).
2. Microorganisms are measured in micrometers, μm $(10^{-6}$ m); nanometers, nm $(10^{-9}$ m); and angstroms, Å $(10^{-10}$ m).

MICROSCOPY: THE INSTRUMENTS (pp. 53–61)

Compound Light Microscopy (pp. 53–55)

1. The most common microscope used in microbiology is the compound light microscope, which uses two sets of lenses: ocular and objective.
2. We calculate the total magnification of an object by multiplying the magnification of the objective lens by the magnification of the ocular lens.
3. The compound light microscope uses visible light.
4. The maximum resolving power (ability to distinguish two points) of a compound light microscope is 0.2 μm; maximum magnification is 2000×.
5. Specimens are stained to increase the difference between the refractive indexes of the specimen and the medium.
6. Immersion oil is used with the oil immersion lens to reduce light loss between the slide and the lens.
7. Brightfield illumination is used for stained smears.
8. Unstained cells are more productively observed using darkfield, phase-contrast, or fluorescence microscopy. These types of microscopy use modified compound microscopes.

Darkfield Microscopy (p. 55)

1. The darkfield microscope shows a light silhouette of an organism against a dark background.
2. It is most useful to detect the presence of extremely small organisms.

Phase-Contrast Microscopy (pp. 55–56)

1. A phase-contrast microscope uses a special condenser to enhance differences in the refractive indexes of the cell's parts and its surroundings.
2. It allows the detailed observation of living organisms.

Differential Interference Contrast (DIC) Microscopy (p. 58)

1. The DIC microscope provides a colored, three-dimensional image of the object being observed.
2. It allows detailed observations of living cells.

Fluorescence Microscopy (p. 58)

1. In fluorescence microscopy, specimens are first stained with fluorochromes and then viewed through a compound microscope by using an ultraviolet (or near-ultraviolet) light source.
2. The microorganisms appear as bright objects against a dark background.
3. Fluorescence microscopy is used primarily in a diagnostic procedure called fluorescent-antibody technique.

Electron Microscopy (pp. 59–61)

1. A beam of electrons, instead of light, is used with an electron microscope.
2. Electromagnets, instead of glass lenses, control focus, illumination, and magnification.
3. Thin sections of organisms can be seen in an electron micrograph produced using a transmission electron microscope. Magnification: 10,000× to 100,000×. Resolving power: 2.5 nm.
4. Three-dimensional views of the surfaces of whole microorganisms can be obtained with a scanning electron microscope. Magnification: 1000× to 10,000×. Resolving power: 20 nm.

PREPARATION OF SPECIMENS FOR LIGHT MICROSCOPY (pp. 61–66)

Preparing Smears for Staining (p. 61)

1. Staining means coloring a microorganism with a dye to make some structures more visible.
2. Fixing uses air and heat to attach microorganisms to a slide.
3. A smear is a thin film of material used for microscopic examination.
4. Bacteria are negatively charged, and the colored positive ion of a basic dye will stain bacterial cells.
5. The colored negative ion of an acidic dye will stain the background of a bacterial smear; a negative stain is produced.

Simple Stains (p. 61)

1. A simple stain is an aqueous or alcohol solution of a single basic dye.
2. It is used to make cellular shapes and arrangements visible.
3. A mordant may be used to improve bonding between the stain and the specimen.

Differential Stains (pp. 61–63)

1. Differential stains, such as the Gram stain and acid-fast stain, divide bacteria into groups according to their reactions to the stains.

2. The Gram-stain procedure uses a purple stain (crystal violet), iodine as a mordant, an alcohol decolorizer, and a red counterstain.

3. Gram-positive bacteria retain the purple stain after the decolorization step; gram-negative bacteria do not and thus appear pink from the counterstain.

4. Acid-fast bacteria, such as members of the genera *Mycobacterium* and *Nocardia,* retain carbolfuchsin after acid-alcohol decolorization and appear red; non-acid-fast bacteria take up the methylene blue counterstain and appear blue.

Special Stains (pp. 63–66)

1. Stains such as the endospore stain and flagella stain color only certain parts of microbes.

2. Negative staining is used to make microbial capsules visible.

STUDY QUESTIONS

Review

1. Fill in the following blanks.

   $1 \ \mu m =$ _____ m

   1 _____ $= 10^{-9}$ m

   $1 \ \text{Å} =$ _____ m

   $1 \ \mu m =$ _____ nm $=$ _____ Å

2. Label the parts of the compound light microscope in the following figure:

(a) _____

(b) _____

(c) _____

(d) _____

(e) _____

3. Calculate the total magnification of the nucleus of a cell being observed through a compound light microscope with a 10× ocular lens and the oil immersion lens.

4. Which type of microscope would be best to use to observe each of the following?

   (a) A stained bacterial smear

   (b) Unstained bacterial cells when the cells are small and no detail is needed

   (c) Unstained live tissue when it is desirable to see some intracellular detail

   (d) A sample that emits light when illuminated with ultraviolet light

   (e) Intracellular detail of a cell that is 1 μm long

   (f) Unstained live cells in which intracellular structures are shown in color

5. An electron microscope differs from a light microscope in that _____ focused by _____ is used instead of light, and the image is viewed not through the ocular lenses but on _____ .

6. The maximum magnification of a compound microscope is _____ ; that of an electron microscope, _____ . The maximum resolution of a compound microscope is _____ ; that of an electron microscope, _____ .

7. One advantage of a scanning electron microscope over a transmission electron microscope is _____ .

8. Acidic dyes stain the (cells/background) in a smear and are used for (negative/simple) stains.

9. Basic dyes stain the (cells/background) in a smear and are used for (negative/simple) stains.

10. Why do basic dyes stain bacterial cells? Why don't acidic dyes stain bacterial cells?

11. When is it most appropriate to use:

    (a) A simple stain?   (c) A negative stain?

    (b) A differential stain?   (d) A flagella stain?

12. Why is a mordant used in the Gram stain? In the flagella stain?

13. What is the purpose of a counterstain in the acid-fast stain?

14. What is the purpose of a decolorizer in the Gram stain? In the acid-fast stain?

15. Choose from the following terms to fill in the blanks: counterstain, decolorizer, mordant, primary stain. In the endospore stain, safranin is the _____ . In the Gram stain, safranin is the _____ .

16. Fill in the following table regarding the Gram stain.

| STEPS | APPEARANCE AFTER THIS STEP OF | |
	GRAM-POSITIVE CELLS	GRAM-NEGATIVE CELLS
Crystal violet		
Iodine		
Alcohol–acetone		
Safranin		

Critical Thinking

1. In a Gram stain, one step could be omitted and still allow differentiation between gram-positive and gram-negative cells. What is that one step?

2. Using a good compound light microscope with a resolving power of 0.3 μm, a 10× ocular lens, and a 100× oil immersion lens, would you be able to discern two objects separated by 3 μm? 0.3 μm? 300 nm? 3000 Å?

3. Why isn't the Gram stain used on acid-fast bacteria? If you did Gram stain acid-fast bacteria, what would their Gram reaction be? What is the Gram reaction of non-acid-fast bacteria?

Clinical Applications

1. In 1882, Paul Erhlich described a method for staining *Mycobacterium* and noted, "It may be that all disinfecting agents which are acidic will be without effect on this [tubercle] bacillus, and one will have to be limited to alkaline agents." How did he reach this conclusion without testing disinfectants?

2. Laboratory diagnosis of *Neisseria gonorrhoeae* infection is based on microscopic examination of Gram-stained pus. Identify the bacteria in this figure. What is the disease?

LM   10 μm

Functional Anatomy of Procaryotic and Eucaryotic Cells

ESPITE THEIR COMPLEXITY and variety, all living cells can be divided into two groups, procaryotes and eucaryotes, based on their ultrastructure as seen with the electron microscope. Plants and animals are entirely composed of eucaryotic cells. In the microbial world, bacteria and cyanobacteria (blue-green algae) are procaryotes. Other cellular microbes—fungi (yeasts and molds), protozoa, and algae—are eucaryotes.

Viruses, as noncellular elements with some cell-like properties, do not fit into any organizational scheme of living cells. They are genetic particles that replicate but are unable to perform the usual chemical activities of living cells. Viral structure and activity will be discussed in Chapter 13. In this chapter, we will concentrate on describing procaryotic and eucaryotic cells.

COMPARISON OF PROCARYOTIC AND EUCARYOTIC CELLS: AN OVERVIEW

LEARNING OBJECTIVE

- *Compare and contrast the overall cell structure of procaryotes and eucaryotes.*

Procaryotes and eucaryotes are chemically similar, in the sense that they both contain nucleic acids, proteins, lipids, and carbohydrates. They use the same kinds of chemical reactions to metabolize food, build proteins, and store energy. It is primarily the structure of cell walls and membranes, and the presence or absence of *organelles* (specialized cellular structures that have specific functions), that distinguish procaryotes from eucaryotes.

The chief distinguishing characteristics of **procaryotes** (from the Greek for prenucleus) are as follows:

1. Their genetic material (DNA) is not enclosed within a membrane.

2. They lack other membrane-bounded organelles.

3. Their DNA is not associated with histone proteins (special chromosomal proteins found in eucaryotes).

4. Their cell walls almost always contain the complex polysaccharide peptidoglycan.

5. They usually divide by **binary fission.** During this process, the DNA is copied and the cell splits into two cells. Binary fission involves fewer structures and processes than eucaryotic cell division.

Eucaryotes (from the Greek for true nucleus) have linear structures of DNA called chromosomes; these are found in the cell's nucleus, which is separated from the cytoplasm by a nuclear membrane. The DNA of eucaryotic chromosomes is consistently associated with chromosomal proteins called histones and with nonhistones. Eucaryotes also have a mitotic apparatus (various cellular structures that participate in a type of nuclear division called mitosis) and a number of organelles, including mitochondria, endoplasmic reticulum, and sometimes chloroplasts. We will examine the particular characteristics and functions of these organelles in more detail later in this chapter.

THE PROCARYOTIC CELL

The members of the procaryotic world make up a vast heterogeneous group of very small unicellular organisms. Procaryotes include eubacteria and archaeobacteria. The majority of bacteria, including the photosynthesizing cyanobacteria, are included in the eubacteria. Although eubacteria and archaeobacteria look similar, they are different in chemical composition, as will be described later. The thousands of species of bacteria are differentiated by many factors, including morphology (shape), chemical composition (often detected by staining reactions), nutritional requirements, biochemical activities, and source of energy (sunlight or chemicals).

SIZE, SHAPE, AND ARRANGEMENT OF BACTERIAL CELLS

LEARNING OBJECTIVE

- *Identify the three basic shapes of bacteria.*

There are a great many sizes and shapes among bacteria. Most bacteria range from 0.20 to 2.0 μm in diameter and from 2 to 8 μm in length. They have a few basic shapes: spherical **coccus** (plural, *cocci,* meaning berries), rod-shaped **bacillus** (plural, *bacilli,* meaning little staffs), and **spiral.**

Cocci are usually round but can be oval, elongated, or flattened on one side. When cocci divide to reproduce, the cells can remain attached to one another. Cocci that remain in pairs after dividing are called **diplococci;** those that divide and remain attached in chainlike patterns are called **streptococci** (Figure 4.1a). Those that divide in two planes and remain in groups of four are known as **tetrads** (Figure 4.1b). Those that divide in three planes and remain attached in cubelike groups of eight are called **sarcinae** (Figure 4.1c). Those that divide in multiple planes and form grapelike clusters or broad sheets are called **staphylococci** (Figure 4.1d). These group characteristics are frequently helpful in the identification of certain cocci.

Bacilli divide only across their short axis, so there are fewer groupings of bacilli than of cocci. Most bacilli appear as single rods (Figure 4.2a). **Diplobacilli** appear in pairs after division (Figure 4.2b), and **streptobacilli** occur in chains (Figure 4.2c). Some bacilli look like straws. Others have tapered ends, like cigars. Still others are oval and

look so much like cocci that they are called **coccobacilli** (Figure 4.2d).

The term *bacillus* has two meanings in microbiology. As we have just used it, bacillus refers to a bacterial shape. When capitalized and italicized, it refers to a specific genus. For example, the bacterium *Bacillus anthracis* is the causative agent of anthrax.

Spiral bacteria have one or more twists; they are never straight. Bacteria that look like curved rods are called **vibrios** (Figure 4.3a). Others, called **spirilla,** have a helical shape, like a corkscrew, and fairly rigid bodies (Figure 4.3b). Yet another group of spirals are helical and flexible; they are called **spirochetes** (Figure 4.3c). Unlike the spirilla, which use whiplike external appendages called flagella to move, spirochetes move by means of an axial filament, which resembles a flagellum but is contained under an external flexible sheath.

In addition to the three basic shapes, there are star-shaped cells (genus *Stella*), recently discovered square, flat cells (halophilic archaeobacteria) of the genus *Haloarcula* (Figure 4.4), and triangular cells.

The shape of a bacterium is determined by heredity. Genetically, most bacteria are **monomorphic;** that is, they maintain a single shape. However, a number of environmental conditions can alter that shape. If the shape is altered, identification becomes difficult. Moreover, some bacteria, such as *Rhizobium* (rī-zō′bē-um) and *Corynebacterium* (kô-rī-nē-bak-ti′rē-um), are genetically **pleomorphic,** which means they can have many shapes, not just one.

The structure of a typical procaryotic (bacterial) cell is shown in Figure 4.5 (p. 74). We will discuss its components according to the following organization: (1) structures external to the cell wall, (2) the cell wall itself, and (3) structures internal to the cell wall.

STRUCTURES EXTERNAL TO THE CELL WALL

LEARNING OBJECTIVE

- *Describe the structure and function of the glycocalyx, flagella, axial filaments, fimbriae, and pili.*

Among the structures external to the procaryotic cell wall are the glycocalyx, flagella, axial filaments, fimbriae, and pili. The glycocalyx can take the form of a capsule or a slime layer.

Glycocalyx

Glycocalyx is the general term used for substances that surround cells. The bacterial glycocalyx is a viscous (sticky), gelatinous polymer that is external to the cell wall

FIGURE 4.1 Arrangements of cocci. The number of planes in which the cell divides determines the arrangement of cells. Shown are diagrams (left) and corresponding photos (right). **(a)** Division in one plane produces diplococci and streptococci. **(b)** Division in two planes produces tetrads. **(c)** Division in three planes produces sarcinae, and **(d)** division in multiple planes produces staphylococci.

FIGURE 4.2 Bacilli. Shown are diagrams (left) and corresponding photos (right). **(a)** Single bacilli. **(b)** Diplobacilli. In the uppermost photo, a few joined pairs of bacilli serve as examples of diplobacilli. **(c)** Streptobacilli. **(d)** Coccobacilli.

FIGURE 4.3 Spiral bacteria. Diagrams (left) and corresponding photos (right) of **(a)** vibrios, **(b)** spirilla, and **(c)** spirochetes.

(a) Star-shaped bacteria

(b) Square bacteria

FIGURE 4.4 Star-shaped and square bacteria. **(a)** *Stella* (star-shaped). **(b)** *Haloarcula*, a type of halophilic archaeobacteria (square cells).

FIGURE 4.5   Structure of a typical procaryotic (bacterial) cell. Both the diagram and photograph show a cell cut lengthwise to reveal the internal structures.

and is composed of polysaccharide, polypeptide, or both. Its chemical composition varies widely with the species. For the most part, it is made inside the cell and excreted to the cell surface. If the substance is organized and is firmly attached to the cell wall, the glycocalyx is described as a **capsule.**

The presence of a capsule can be determined by using negative staining, such as the India ink method mentioned in Chapter 3 (see Figure 3.11a). If the substance is unorganized and only loosely attached to the cell wall, the glycocalyx is described as a **slime layer.**

In certain species, capsules are important in contributing to bacterial virulence (the degree to which a pathogen causes disease). Capsules often protect pathogenic bacteria from phagocytosis by the cells of the host. Phagocytosis, which is discussed in Chapter 16, is a process by which certain white blood cells engulf and destroy microbes (see Figure 16.8). *Streptococcus pneumoniae* (strep-tō-kok′kus nü-mō′nē-ī) causes pneumonia when the cells are protected by a polysaccharide capsule. Unencapsulated *S. pneumoniae* cells are readily phagocytized and cannot cause pneumonia. *Bacillus anthracis* produces a capsule of D-glutamic acid. Recall from Chapter 2 that the D forms of amino acids are unusual. Because only encapsulated *B. anthracis* causes anthrax, the capsule may prevent its being destroyed by phagocytosis.

Another function of the sticky glycocalyx is to allow a bacterium to attach to various surfaces in its natural en-

vironment in order to survive. Through attachment, bacteria can grow on diverse surfaces such as rocks in fast-moving streams, plant roots, human teeth and tissues, and even other bacteria. *Streptococcus mutans* (mū′tans), an important cause of dental caries, attaches itself to the surface of teeth by a glycocalyx. The capsule of *Klebsiella pneumoniae* (kleb-sē-el′lä) prevents phagocytosis and allows this bacterium to adhere to and colonize the respiratory tract.

S. mutans may use its capsule as a source of nutrition by breaking it down and utilizing the sugars when energy stores are low. A glycocalyx can protect a cell against dehydration. Also, its viscosity may inhibit the movement of nutrients from the cell.

Flagella

Some procaryotic cells carry **flagella** (singular, *flagellum*, meaning whip), which are long filamentous appendages that propel bacteria (see Figure 4.5).

Bacterial cells have four arrangements of flagella (Figure 4.6): **monotrichous** (single polar flagellum), **amphitrichous** (single flagellum at each end of the cell), **lophotrichous** (two or more flagella at one or both poles of the cell), and **peritrichous** (flagella distributed over the entire cell).

(a) Monotrichous   LM   10 μm

(b) Amphitrichous   LM   1 μm

(c) Lophotrichous   LM   10 μm

(d) Peritrichous   TEM   1 μm

FIGURE 4.6   Arrangement of bacterial flagella. Four basic types of flagellar arrangements.

A flagellum has three basic parts (Figure 4.7). The long outermost region, the *filament,* is constant in diameter and contains the globular (roughly spherical) protein *flagellin* arranged in several chains that intertwine and form a helix around a hollow core. Flagellar proteins serve to identify certain pathogenic bacteria. In most bacteria, filaments are not covered by a membrane or sheath, as in eucaryotic cells. The filament is attached to a slightly wider *hook,* which consists of a different protein. The third portion of a flagellum is the *basal body,* which anchors the flagellum to the cell wall and plasma membrane.

The basal body is composed of a small central rod inserted into a series of rings. Gram-negative bacteria contain two pairs of rings; the outer pair of rings is anchored to various portions of the cell wall, and the inner pair of rings is anchored to the plasma membrane. In gram-positive bacteria, only the inner pair is present. As you will see later in the chapter, the flagella (and cilia) of eucaryotic cells are more complex than those of procaryotic cells.

Bacteria with flagella are motile; that is, they have the ability to move on their own. Each procaryotic flagellum is a semirigid, helical rotor that moves the cell by rotating from the basal body. The rotation of a flagellum is either clockwise or counterclockwise around its long axis. (Eucaryotic flagella, by contrast, undulate in a wavelike motion.) The movement of a procaryotic flagellum results from rotation of its basal body and is similar to the movement of the shaft of an electric motor. As the flagella ro-

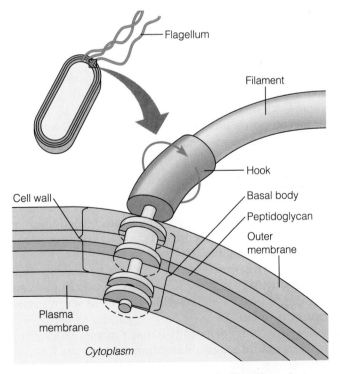

Flagellum

Filament

Hook

Cell wall

Basal body

Peptidoglycan

Outer membrane

Plasma membrane

Cytoplasm

FIGURE 4.7   Structure of a procaryotic flagellum. Parts and attachment of a flagellum of a gram-negative bacterium.

FIGURE 4.8   One type of bacterial motility. This diagram shows a bacterium "running" and "tumbling." Note that the direction of flagellar rotation determines which of these movements occurs.

tate, they form a bundle that pushes against the surrounding liquid and propels the bacterium. Although the exact mechanochemical basis for this biological "motor" is not completely understood, we know that it depends on the cell's continuous generation of energy.

Bacterial cells can alter the speed and direction of rotation of flagella and thus are capable of various patterns of motility (Figure 4.8). When a bacterium moves in one direction for a length of time, the movement is called a "run" or "swim." "Runs" are interrupted by periodic, abrupt, random changes in direction called "tumbles." Then, a "run" resumes. "Tumbles" are caused by a reversal of flagellar rotation. Some species of bacteria endowed with many flagella (*Proteus* [prō′tē-us], for example) can "swarm," or show rapid wavelike movement across a solid culture medium.

One advantage of motility is that it enables a bacterium to move toward a favorable environment or away from an adverse one. The movement of a bacterium toward or away from a particular stimulus is called **taxis.** Such stimuli include chemicals (**chemotaxis**) and light (**phototaxis**). Motile bacteria contain receptors in various locations, such as in or just under the cell wall. These receptors pick up chemical stimuli, such as oxygen, ribose, and galactose. In response to the stimuli, information is passed to the flagella. If the chemotactic signal is positive, called an *attractant,* the bacteria move toward the stimulus with many runs and few tumbles. If the chemotactic signal is negative, called a *repellent,* the frequency of tumbles increases as the bacteria move away from the stimulus (see the box on p. 80).

Axial Filaments

Spirochetes are a group of bacteria that have unique structure and motility. One of the best-known spirochetes is *Treponema pallidum,* the causative agent of syphilis. Another spirochete is *Borrelia burgdorferi,* the causative agent of Lyme disease. Spirochetes move by means of **axial filaments,** bundles of fibrils that arise at the ends of the cell beneath the outer sheath and spiral around the cell (Figure 4.9). Axial filaments, which are anchored at one end of the spirochete, have a structure similar to that of flagella. The rotation of the filaments produces a movement of the outer sheath that propels the spirochetes in a spiral motion. This type of movement is similar to the way a corkscrew moves through a cork. This corkscrew motion probably allows a bacterium such as *T. pallidum* to move effectively through body tissues.

Fimbriae and Pili

Many gram-negative bacteria contain hairlike appendages that are shorter, straighter, and thinner than flagella and are used for attachment rather than for motility. These structures, which consist of a protein called *pilin* arranged helically around a central core, are divided into two types, *fimbriae* and *pili,* and have very different functions. (Some microbiologists use the two terms interchangeably to refer to all such structures, but we distinguish between them.)

Fimbriae can occur at the poles of the bacterial cell, or they can be evenly distributed over the entire surface of the cell. They can number anywhere from a few to several hundred per cell (Figure 4.10). Like the glycocalyx, fimbriae allow a cell to adhere to surfaces, including the surfaces of other cells. For example, fimbriae attached to the bacterium *Neisseria gonorrhoeae* (nī-se′rē-ä go-nôr-rē′ī), the causative agent of gonorrhea, help the microbe to colonize mucous membranes. Once colonization occurs, the bacteria can cause disease. When fimbriae are absent (because of genetic mutation), colonization cannot happen, and no disease ensues.

Pili are usually longer than fimbriae and number only one or two per cell. Pili join bacterial cells in preparation for the transfer of DNA from one cell to another. For this reason, they are sometimes also called **sex pili** (see Chapter 8).

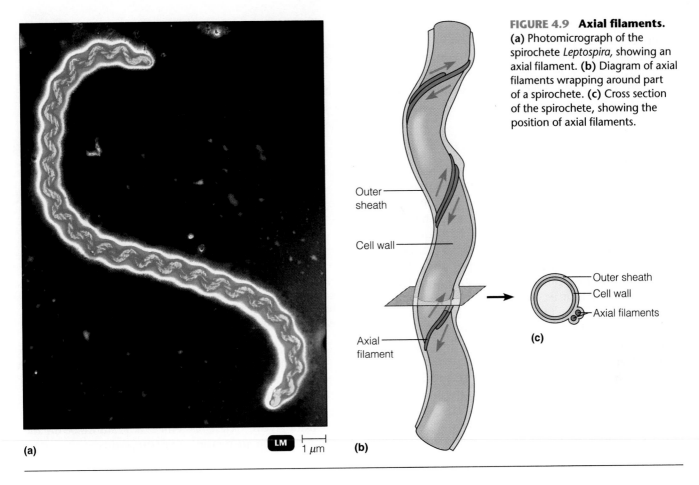

FIGURE 4.9 Axial filaments.
(a) Photomicrograph of the spirochete *Leptospira,* showing an axial filament. **(b)** Diagram of axial filaments wrapping around part of a spirochete. **(c)** Cross section of the spirochete, showing the position of axial filaments.

THE CELL WALL

LEARNING OBJECTIVE

• *Compare and contrast the cell walls of gram-positive bacteria, gram-negative bacteria, archaeobacteria, and mycoplasmas.*

The **cell wall** of the bacterial cell is a complex, semirigid structure that is responsible for the characteristic shape of the cell. The cell wall surrounds the underlying, fragile plasma (cytoplasmic) membrane and protects it and the internal parts of the cell from adverse changes in the surrounding environment (see Figure 4.5). Almost all procaryotes have cell walls.

The major function of the cell wall is to prevent bacterial cells from rupturing when the osmotic pressure inside the cell is greater than that outside the cell. It also helps maintain the shape of a bacterium and serves as a point of anchorage for flagella. As the volume of a bacterial cell increases, its plasma membrane and cell wall extend as needed. Clinically, the cell wall is important because it contributes to the ability of some species to cause disease and is the site of action of some antibiotics. In addition,

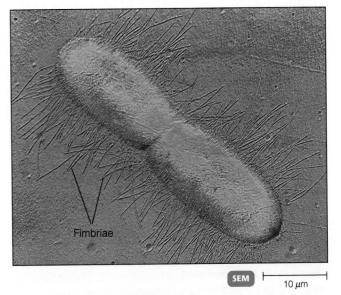

FIGURE 4.10 Fimbriae. The fimbriae seem to bristle from this *E. coli* cell, which is beginning to divide.

the chemical composition of the cell wall is used to differentiate major types of bacteria.

Although the cells of some eucaryotes, including plants, algae, and fungi, have cell walls, their walls differ chemically from those of procaryotes, are simpler in structure, and are less rigid.

Composition and Characteristics

The bacterial cell wall is composed of a macromolecular network called **peptidoglycan (murein)**, which is present either alone or in combination with other substances. Peptidoglycan consists of a repeating disaccharide attached by polypeptides to form a lattice that surrounds and protects the entire cell. The disaccharide portion is made up of monosaccharides called N-acetylglucosamine (NAG) and N-acetylmuramic acid (NAM) (from *murus,* meaning wall), which are related to glucose. The structural formulas for NAG and NAM are shown in Figure 4.11.

The various components of peptidoglycan are assembled in the cell wall (Figure 4.12a). Alternating NAM and NAG molecules are linked in rows of 10 to 65 sugars to form a carbohydrate "backbone" (the glycan portion of peptidoglycan). Adjacent rows are linked by polypeptides (the peptide portion of peptidoglycan). Although the structure of the polypeptide link varies, it always includes *tetrapeptide side chains,* which consist of four amino acids attached to NAMs in the backbone. The amino acids occur in an alternating pattern of D and L forms (see Figure 2.16). This is unique because the amino acids found in other proteins are L forms. Parallel tetrapeptide side chains may be directly bonded to each other or linked by a *peptide cross bridge,* consisting of a short chain of amino acids.

Gram-Positive Cell Walls
In most gram-positive bacteria, the cell wall consists of many layers of peptidoglycan, forming a thick, rigid structure (Figure 4.12b). (By contrast, gram-negative cell walls contain only a thin layer of peptidoglycan.) In addition, the cell walls of gram-positive bacteria contain *teichoic acids,* which consist primarily of an alcohol (such as glycerol or ribitol) and phosphate. There are two classes of teichoic acids: *lipoteichoic acid,* which spans the peptidoglycan layer and is linked to the plasma membrane, and *wall teichoic acid,* which is linked to the peptidoglycan layer. Because of their negative charge (from the phosphate groups), teichoic acids may bind and regulate the movement of cations (positive ions) into and out of the cell. They may also assume a role in cell growth, preventing extensive wall breakdown and possible *cell lysis* (rupturing). Finally, teichoic acids provide much of the wall's antigenic specificity and thus make it possible to identify bacteria by serological means (see Chapter 10).

The cell walls of gram-positive streptococci are covered with various polysaccharides that allow them to be grouped into medically significant types. The cell walls of

FIGURE 4.11 N-acetylglucosamine (NAG) and N-acetylmuramic acid (NAM) joined as in peptidoglycan. The light purple areas show the differences between the two molecules. The linkage between them is called a β-1,4 linkage.

acid-fast bacteria, such as *Mycobacterium,* consist of as much as 60% mycolic acid, a waxy lipid, whereas the rest is peptidoglycan. These bacteria can be stained with the Gram stain and are considered gram-positive.

Gram-Negative Cell Walls
The cell walls of gram-negative bacteria consist of one or a very few layers of peptidoglycan and an outer membrane (Figure 4.12c). The peptidoglycan is bonded to lipoproteins (lipids covalently linked to proteins) in the outer membrane and is in the *periplasmic space,* a space between the outer membrane and the plasma membrane. The periplasmic space contains a high concentration of degradative enzymes and transport proteins. Gram-negative cell walls do not contain teichoic acids. Because the cell walls of gram-negative bacteria contain only a small amount of peptidoglycan, they are more susceptible to mechanical breakage.

The *outer membrane* of the gram-negative cell consists of lipoproteins, lipopolysaccharides (LPS), and phospholipids (see Figure 4.12c). The outer membrane has several specialized functions. Its strong negative charge is an important factor in evading phagocytosis and the action of complement, two components of the defenses of the host (discussed in detail in Chapter 16). The outer membrane also provides a barrier to certain antibiotics (for example, penicillin), digestive enzymes such as lysozyme, detergents, heavy metals, bile salts, and certain dyes.

However, the outer membrane does not provide a barrier to all substances in the environment because nutrients must pass through to sustain the metabolism of the cell. Part of the permeability of the outer membrane is due to proteins in the membrane, called **porins**, that form channels. Porins permit the passage of molecules such as nu-

FIGURE 4.12 Bacterial cell walls. (a) Structure of peptidoglycan in gram-positive bacteria. Together the carbohydrate backbone (glycan portion) and tetrapeptide side chains (peptide portion) make up peptidoglycan. The frequency of peptide cross bridges and the number of amino acids in these bridges vary with the species of bacterium. (b) Gram-positive cell wall. (c) Gram-negative cell wall.

N-acetylglucosamine (NAG)
N-acetylmuramic acid (NAM)
Side chain amino acid
Cross bridge amino acid

Tetrapeptide side chain
Peptide cross bridge
Carbohydrate "backbone"

(a) Structure of peptidoglycan in gram-positive bacteria

Wall teichoic acid
Lipoteichoic acid
Peptidoglycan
Cell wall
Plasma membrane
Protein

(b) Gram-positive cell wall

Porin protein
Lipopolysaccharide { O polysaccharide / Lipid A }
Phospholipid
Lipoprotein
Enzymes and other active substances
Cell wall { Outer membrane / Peptidoglycan }
Plasma membrane
Periplasmic space
Protein

(c) Gram-negative cell wall

cleotides, disaccharides, peptides, amino acids, vitamin B_{12}, and iron. However, they can also make bacteria vulnerable to attack by providing attachment sites for viruses and harmful substances.

The lipopolysaccharide component of the outer membrane provides two important characteristics of gram-negative bacteria. First, the polysaccharide portion is composed of sugars, called *O polysaccharides*, that function as antigens and are useful for distinguishing species of gram-negative bacteria. For example, the over 2000 *Salmonella* (sal-mōn-el'lä) serovars (variations within a species) can be distinguished by serological means. This role is comparable to that of teichoic acids in gram-positive cells. Second, the lipid portion of the lipopolysaccharide, called *lipid A,* is referred to as *endotoxin* and is toxic when in the host's bloodstream or gastrointestinal tract. It causes fever and

Applications of Microbiology

FROM MICROBES TO MAGNETIC TAPE?

FOR NEARLY 100 years, biologists have suspected that birds can use the Earth's magnetic field for orientation during flight. During the 1970s, magnetotaxis (response to a magnetic field) was demonstrated in some birds, as well as in honeybees and some bacteria. Around 1980, magnetic material was found in these birds and honeybees, as well as in some dolphins.

In 1975, Richard P. Blakemore reported that sulfide-rich mud contained highly motile bacteria that ". . . swam in the same geographic direction even when the microscope was turned around, moved to another location, or covered with a pasteboard box. To my astonishment, when a magnet was brought near the microscope, the hundreds of swimming cells instantly turned and rushed away from the end of the magnet! They were always attracted by the end that also attracted the north-seeking end of a compass needle and they were repelled by its opposite end. Their swimming speed was very fast, on the order of 100 μm per second, and the entire population swerved in unison as the magnet was moved about nearby."*

Magnetotactic bacteria synthesize magnetite—iron oxide (Fe_3O_4)—and store it in inclusions called magnetosomes, as shown in the photo. The magnetosomes act like magnets and cause the bacterial cells to align on the Earth's geomagnetic field. All bacteria with magnetosomes are gram-negative and have fimbriae or a polysaccharide glycocalyx to provide attachment. These bacteria are found in sediments throughout the world. In the Northern Hemisphere, north-seeking bacteria predominate; in the Southern Hemisphere, south-seeking bacteria predominate.

Aquaspirillum magnetotacticum (see photo) metabolizes organic molecules for carbon and energy and requires a microaerophilic environment. It was found that when *A. magnetotacticum* cells are separated from an attachment site, they swim along what appear to be geomagnetic lines. They move downward, toward either the north or south pole, until reaching a suitable attachment site.

TEM |————————| 1 μm

Aquaspirillum magnetotacticum, showing chain of magnetosomes. Outer membrane of gram-negative wall is also visible.

How these bacteria synthesize magnetite is currently being studied, and the function of magnetite in the cell's metabolism is not yet clear. The formation of ferric iron (Fe^{3+}) from ferrous iron (Fe^{2+}) could provide adenosine triphosphate (ATP) for the cell, but the estimated amount of ATP is so small that the magnetite must serve another function. In vitro, the magnetosomes can decompose hydrogen peroxide, which forms in cells in the presence of oxygen and is usually degraded by the enzyme catalase before it reaches toxic levels. Researchers speculate that the magnetosomes protect the cell against hydrogen peroxide accumulation in vivo.

MORE THAN A CURIOSITY

The synthesis and function of magnetosomes are of interest to microbiologists both to further our understanding of cells and perhaps to provide a model to explain the formation of similar iron oxides in birds, insects, and other animals. The ecology of magnetic microbes may also provide a tool for scientists studying the Earth's magnetic field. Evidence suggests that the Earth's magnetic field may have reversed or shifted several times. Examination of the orientation of fossils of magnetotactic bacteria preserved in rock or sediment may shed light on the movement of the continents and magnetic poles in relation to one another.

In 1987, researchers discovered other bacteria that live in the bottom sediments of lakes and bays and that produce magnetite. The Tokyo Institute of Agriculture and TDK Corporation are developing culture methods to obtain large quantities of magnetite from bacteria. They hope to use the magnetic particles in the production of magnetic tapes for recording sound and data.

*Reproduced, with permission, from the *Annual Review of Microbiology,* Volume 36, © 1982 by Annual Reviews Inc. and Richard P. Blakemore. (Richard P. Blakemore, "Magnetotactic bacteria." *Annual Review of Microbiology* 36:217–238, 1982.)

TABLE 4.1 Some Comparative Characteristics of Gram-Positive and Gram-Negative Bacteria

CHARACTERISTIC	GRAM-POSITIVE	GRAM-NEGATIVE
	LM 10 μm	LM 10 μm
Gram reaction	Retain crystal violet dye and stain dark violet or purple	Can be decolorized to accept counterstain (safranin) and stain red
Peptidoglycan layer	Thick (multilayered)	Thin (single-layered)
Teichoic acids	Present in many	Absent
Periplasmic space	Absent	Present
Outer membrane	Absent	Present
Lipopolysaccharide (LPS) content	Virtually none	High
Lipid and lipoprotein content	Low (acid-fast bacteria have lipids linked to peptidoglycan)	High (due to presence of outer membrane)
Flagellar structure	2 rings in basal body	4 rings in basal body
Toxins produced	Primarily exotoxins	Primarily endotoxins
Resistance to physical disruption	High	Low
Cell wall disruption by lysozyme	High	Low (requires pretreatment to destabilize the outer membrane)
Susceptibility to penicillin and sulfonamide	High	Low
Susceptibility to streptomycin, chloramphenicol, and tetracycline	Low	High
Inhibition by basic dyes	High	Low
Susceptibility to anionic detergents	High	Low
Resistance to sodium azide	High	Low
Resistance to drying	High	Low

shock. The nature and importance of endotoxins and other bacterial toxins will be discussed in Chapter 15.

A comparison of some characteristics of gram-positive and gram-negative bacteria is presented in Table 4.1.

Atypical Cell Walls

Among procaryotes, there are cells that naturally have no walls or have very little wall material. These include members of the genus *Mycoplasma* (mī-kō-plaz'mä) and related organisms. Mycoplasmas are the smallest known bacteria that can grow and reproduce outside living host cells. Because of their size and because they have no cell walls, they pass through most bacterial filters and were first mistaken for viruses. Their plasma membranes are unique among bacteria in having lipids called *sterols,* which are thought to help protect them from osmotic lysis.

Archaeobacteria may lack walls or may have unusual walls composed of polysaccharides and proteins but not peptidoglycan. These walls do, however, contain a substance similar to peptidoglycan called *pseudomurein.* Pseudomurein contains N-acetyltalosaminuronic acid instead of NAM and lacks the D amino acids found in bacterial cell walls. An archaeobacterium is described in the box in Chapter 5 (p. 131).

Recall from Chapter 3 that the mechanism for the Gram stain is related to the structure and chemical composition of the cell wall. You might want to review that mechanism before continuing to the next section.

Damage to the Cell Wall

Chemicals that damage bacterial cell walls, or interfere with their synthesis, often do not harm the cells of an animal host because the bacterial cell wall is made of chemicals unlike those in eucaryotic cells. Thus, cell wall synthesis is the target for some antimicrobial drugs. One way that the cell wall can be damaged is by exposure to the digestive enzyme *lysozyme*. This enzyme occurs naturally in some eucaryotic cells and is a constituent of tears, mucus, and saliva. Lysozyme is particularly active on the major cell wall components of most gram-positive bacteria, making them vulnerable to rupture, or lysis. Lysozyme catalyzes the hydrolysis of the bonds between the sugars in the repeating disaccharide "backbone" of peptidoglycan. This act is analogous to cutting the steel supports of a bridge with a cutting torch: the gram-positive cell wall is almost completely destroyed by lysozyme. The cellular contents that remain surrounded by the plasma membrane may remain intact if osmotic lysis does not occur; this wall-less cell is termed a **protoplast**. Typically, a protoplast is spherical and is still capable of carrying on metabolism.

When lysozyme is applied to gram-negative cells, usually the wall is not destroyed to the same extent as in gram-positive cells; some of the outer membrane also remains. In this case, the cellular contents, plasma membrane, and remaining outer wall layer are called a **spheroplast**, also a spherical structure. For lysozyme to exert its effect on gram-negative cells, the cells are first treated with ethylenediaminetetraacetic acid (EDTA). EDTA weakens ionic bonds in the outer membrane and so damages it, giving the lysozyme access to the peptidoglycan layer.

Protoplasts and spheroplasts burst in pure water or very dilute salt or sugar solutions because the water molecules from the surrounding fluid rapidly move into and enlarge the cell, which has a much lower internal concentration of water. This rupturing is called **osmotic lysis** and will be discussed in detail shortly.

Certain antibiotics, such as penicillin, destroy bacteria by interfering with the formation of the peptide cross bridges of peptidoglycan, thus preventing the formation of a functional cell wall. Most gram-negative bacteria are not as susceptible to penicillin as gram-positive bacteria are because the outer membrane of gram-negative bacteria forms a barrier that inhibits the entry of this and other substances, and gram-negative bacteria have fewer peptide cross bridges. However, gram-negative bacteria are quite susceptible to some β-lactam antibiotics that penetrate the outer membrane better than penicillin. Antibiotics will be discussed in more detail in Chapter 20.

STRUCTURES INTERNAL TO THE CELL WALL

LEARNING OBJECTIVES

- *Describe the structure, chemistry, and functions of the procaryotic plasma membrane.*
- *Define simple diffusion, osmosis, facilitated diffusion, active transport, and group translocation.*
- *Identify the functions of the nuclear area, ribosomes, and inclusions.*
- *Describe the functions of endospores, sporulation, and endospore germination.*

Thus far, we have discussed the procaryotic cell wall and structures external to it. We will now look inside the procaryotic cell and discuss the structures and functions of the plasma membrane and other components within the cytoplasm of the cell.

Plasma (Cytoplasmic) Membrane

The **plasma (cytoplasmic) membrane** (or *inner membrane*) is a thin structure lying inside the cell wall and enclosing the cytoplasm of the cell (see Figure 4.5). The plasma membrane of procaryotes consists primarily of phospholipids (see Figure 2.13), which are the most abundant chemicals in the membrane, and proteins. Eucaryotic plasma membranes also contain carbohydrates and sterols, such as cholesterol. Because they lack sterols, procaryotic plasma membranes are less rigid than eucaryotic membranes. One exception to this is the wall-less procaryote called *Mycoplasma*, which contains membrane sterols.

Structure

In electron micrographs, procaryotic and eucaryotic plasma membranes (and the outer membranes of gram-negative bacteria) look like two-layered structures; there are two dark lines with a light space between the lines (Figure 4.13a, right). The phospholipid molecules are arranged in two parallel rows, called a *phospholipid bilayer* (Figure 4.13b). Each phospholipid molecule contains a polar head, composed of a phosphate group and glycerol that is hydrophilic ("water-loving") and soluble in water, and nonpolar tails, composed of fatty acids that are hydrophobic ("water-fearing") and are insoluble in water (Figure 4.13c). The polar heads are on the two surfaces of the phospholipid bilayer, and the nonpolar tails are in the interior of the bilayer.

The protein molecules in the membrane can be arranged in a variety of ways. Some, called *peripheral proteins,* are easily removed from the membrane by mild treatments and are thought to lie at the inner or outer sur-

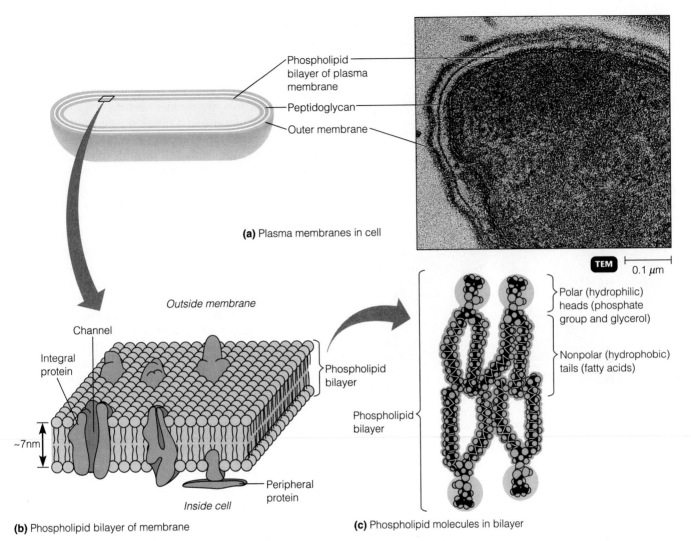

(a) Plasma membranes in cell

Phospholipid bilayer of plasma membrane

Peptidoglycan

Outer membrane

TEM   0.1 μm

Outside membrane

Channel

Integral protein

~7nm

Phospholipid bilayer

Peripheral protein

Inside cell

(b) Phospholipid bilayer of membrane

Polar (hydrophilic) heads (phosphate group and glycerol)

Nonpolar (hydrophobic) tails (fatty acids)

Phospholipid bilayer

(c) Phospholipid molecules in bilayer

FIGURE 4.13 Plasma membrane. (a) Micrograph and corresponding diagram showing the phospholipid bilayer forming the inner plasma membrane of the gram-negative bacterium *Bacillus brevis*. Layers of the cell wall, including the outer plasma membrane, can be seen outside this inner membrane. **(b)** Drawing of a portion of the inner membrane showing the phospholipid bilayer and proteins. The outer membrane of gram-negative bacteria is also a phospholipid bilayer. **(c)** Space-filling models of several molecules as they are arranged in the phospholipid bilayer.

face of the membrane. They may function as enzymes that catalyze chemical reactions, as a "scaffold" for support, and as mediators of changes in membrane shape during movement. Other proteins, called *integral proteins,* can be removed from the membrane only after disrupting the bilayer (by using detergents, for example). Some integral proteins are believed to penetrate the membrane completely. Some of these proteins contain channels through which substances enter and exit the cell.

Studies have demonstrated that the phospholipid and protein molecules in membranes are not static but move quite freely within the membrane surface. This movement is most probably associated with the many functions performed by the plasma membrane. Because the fatty acid tails cling together, phospholipids in the presence of water form a self-sealing bilayer, with the result that breaks and

tears in the membrane will heal themselves. The membrane must be about as viscous as olive oil to allow membrane proteins to move freely enough to perform their functions without destroying the structure of the membrane. This dynamic arrangement of phospholipid and protein is referred to as the *fluid mosaic model.*

Functions

The most important function of the plasma membrane is to serve as a selective barrier through which materials enter and exit the cell. In this function, plasma membranes have **selective permeability** (sometimes called *semipermeability*). This term indicates that certain molecules and ions pass through the membrane, but others are prevented from passing through it. The permeability of the membrane depends on several factors. Large mole-

TEM |————| 1 μm

FIGURE 4.14 Chromatophores. Membranous structures that are sites of photosynthesis can be seen in this micrograph of *Rhodospirillum rubrum,* a purple (nonsulfur) bacterium.

cules (such as proteins) cannot pass through the plasma membrane, possibly because these molecules are larger than the channels in integral proteins. But smaller molecules (such as water, oxygen, carbon dioxide, and some simple sugars) usually pass through easily. Ions penetrate the membrane very slowly. Substances that dissolve easily in lipids (such as oxygen, carbon dioxide, and nonpolar organic molecules) enter and exit more easily than other substances because the membrane consists mostly of phospholipids. The movement of materials across plasma membranes also depends on carrier molecules, which will be described shortly.

Plasma membranes are also important to the breakdown of nutrients and the production of energy. The plasma membranes of bacteria contain enzymes capable of catalyzing the chemical reactions that break down nutrients and produce ATP. In some bacteria, pigments and enzymes involved in photosynthesis (the conversion of light energy into chemical energy) are found in infoldings of the plasma membrane that extend into the cytoplasm. These membranous structures are called **chromatophores** or **thylakoids** (Figure 4.14).

When viewed with an electron microscope, bacterial plasma membranes often appear to contain one or more large, irregular folds called **mesosomes.** Many functions have been proposed for mesosomes. However, recent evidence indicates that they are artifacts (see Chapter 3), not true cell structures. Mesosomes are believed to be folds in the plasma membrane that develop when cells are dehydrated before electron microscopy. When a new electron microscopy technique made it possible to view cells without dehydration, no mesosomes were seen.

Destruction of the Plasma Membrane by Antimicrobial Agents

Because the plasma membrane is vital to the bacterial cell, it is not surprising that several antimicrobial agents exert their effects at this site. In addition to the chemicals that damage the cell wall and hence indirectly expose the membrane to injury, many compounds specifically damage plasma membranes. These compounds include certain alcohols and quaternary ammonium compounds, which are used as disinfectants. By disrupting the membrane's phospholipids, a group of antibiotics known as the *polymyxins* cause leakage of intracellular contents and subsequent cell death. This mechanism is discussed in Chapter 20.

Movement of Materials Across Membranes

When the concentration of a substance is higher on one side of a membrane than on another, a *concentration gradient* (difference) exists. If the substance can cross the membrane, it will move to the more dilute side until the concentrations are equal, or until other forces stop its movement.

Materials move across plasma membranes of both procaryotic and eucaryotic cells by two kinds of processes: passive and active. In *passive processes,* substances cross the membrane from an area of high concentration to an area of low concentration (move with the concentration gradient), without any expenditure of energy (ATP) by the cell. In *active processes,* the cell must use energy (ATP) to move substances from areas of low concentration to areas of high concentration (against the concentration gradient).

Passive Processes

Passive processes include simple diffusion, facilitated diffusion, and osmosis.

Simple Diffusion. **Simple diffusion** is the net (overall) movement of molecules or ions from an area of high concentration to an area of low concentration (Figure 4.15). The movement from areas of high to low concentration continues until the molecules or ions are evenly distributed. The point of even distribution is called *equilibrium.* Cells rely on simple diffusion to transport certain small molecules, such as oxygen and carbon dioxide, across their cell membranes.

FIGURE 4.15   The principle of simple diffusion. (a) After a dye pellet is put into a beaker of water, the molecules of dye in the pellet diffuse into the water from an area of high dye concentration to areas of low dye concentration. **(b)** The dye potassium permanganate in the process of diffusing.

(a)

(b)

Facilitated Diffusion. When **facilitated diffusion** occurs, the substance (glucose, for example) to be transported combines with a *carrier protein* in the plasma membrane. Such carriers are sometimes called *permeases*. The carrier can transport a substance across the membrane from an area of high concentration to one of low concentration. Carrier proteins bind to the substance they transport on the outer surface of the plasma membrane and, by a poorly understood mechanism, transport the substance through the membrane to the inner surface and release it into the cytoplasm. One proposed mechanism for facilitated diffusion is that the carrier protein remains in place in the membrane but undergoes a change in shape that transports substances from one side of the membrane to the other (Figure 4.16). Facilitated diffusion is similar to simple diffusion in that the cell does not need to expend energy because the substance moves from a high to a low concentration. The process differs from simple diffusion in its use of carriers.

In some cases, molecules that bacteria need are too large to be transported into the cells by the methods just described. Most bacteria, however, produce enzymes that can break down large molecules into simpler ones (such as proteins into amino acids, or polysaccharides into simple sugars). Such enzymes, which are released by the bacteria into the surrounding medium, are appropriately called *extracellular enzymes*. Once the enzymes degrade the large molecules, the subunits are transported by permeases into the cell. For example, specific carriers retrieve DNA bases, such as the purine guanine, from extracellular media and bring them into the cell's cytoplasm.

Osmosis. Osmosis is the net movement of solvent molecules across a selectively permeable membrane from an area with a high concentration of solvent molecules to an area of low concentration. In living systems, the chief solvent is water.

Osmosis may be demonstrated with the apparatus shown in Figure 4.17a and b. A sack constructed from cel-

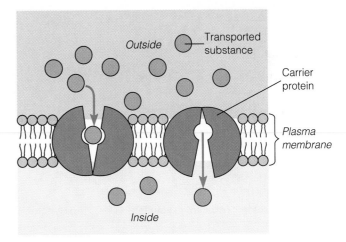

FIGURE 4.16   Facilitated diffusion. Carrier proteins in the plasma membrane transport molecules across the membrane from an area of high concentration to one of low concentration (with the concentration gradient). The carrier molecule probably undergoes a change in shape to transport the substance. The process does not require ATP.

lophane, which is a selectively permeable membrane, is filled with a solution of 20% sucrose (table sugar). The opening of the cellophane sack is plugged with a rubber stopper, through which a glass tube is fitted. The cellophane sack is placed into a beaker containing distilled water. Initially, the concentrations of water on either side of the membrane are different. Because of the sucrose molecules, the concentration of water is lower inside the cellophane sack than outside. Therefore, water moves from the beaker (where its concentration is higher) into the cellophane sack (where its concentration is lower).

There is no movement of sugar out of the cellophane sack into the beaker, however, because the cellophane is impermeable to molecules of sugar—the sugar molecules are too large to go through the pores of the membrane. As

FIGURE 4.17 The principle of osmosis. (a) Setup at the beginning of an osmotic pressure experiment. Water molecules start to move from the beaker into the sack along the concentration gradient. **(b)** Setup at equilibrium. The osmotic pressure exerted by the solution in the sack pushes water molecules from the sack back into the beaker to balance the rate of water entry into the sack. The height of solution in the glass tube at equilibrium is a measure of the osmotic pressure. **(c)–(e)** Effects of various solutions on bacterial cells.

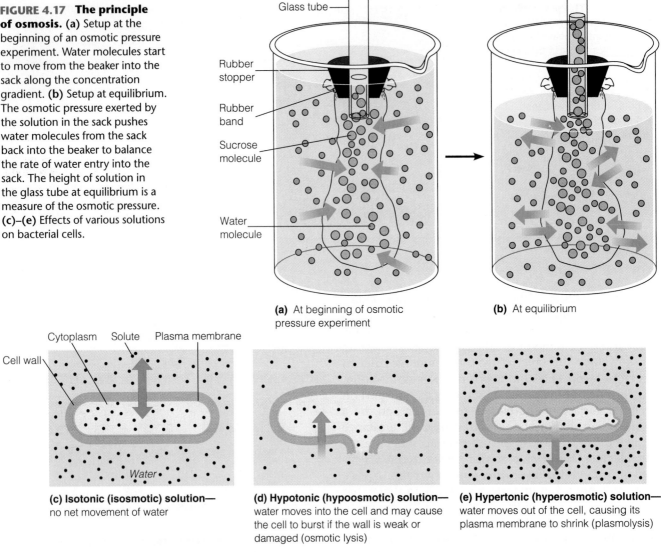

(a) At beginning of osmotic pressure experiment

(b) At equilibrium

(c) Isotonic (isosmotic) solution— no net movement of water

(d) Hypotonic (hypoosmotic) solution— water moves into the cell and may cause the cell to burst if the wall is weak or damaged (osmotic lysis)

(e) Hypertonic (hyperosmotic) solution— water moves out of the cell, causing its plasma membrane to shrink (plasmolysis)

water moves into the cellophane sack, the sugar solution becomes increasingly dilute, and, because the cellophane sack has expanded to its limit as a result of an increased volume of water, water begins to move up the glass tube. In time, the water that has accumulated in the cellophane sack and the glass tube exerts a downward pressure that forces water molecules out of the cellophane sack and back into the beaker. This movement of water through a selectively permeable membrane produces a pressure called osmotic pressure. **Osmotic pressure** is the pressure required to prevent the movement of pure water (water with no solutes) into a solution containing some solutes. In other words, osmotic pressure is the pressure needed to stop the flow of water across the selectively permeable membrane (cellophane). When water molecules leave and enter the cellophane sack at the same rate, equilibrium is reached.

A bacterial cell may be subjected to three kinds of osmotic solutions: isotonic, hypotonic, or hypertonic. An **isotonic (isosmotic) solution** (*iso* means equal) is one in

which the overall concentrations of solutes are the same on both sides of the membrane. Water leaves and enters the cell at the same rate (no net change); the cell's contents are in equilibrium with the solution outside the cell wall (Figure 4.17c).

Earlier we mentioned that lysozyme and certain antibiotics damage bacterial cell walls, causing the cells to rupture, or lyse. Such rupturing occurs because bacterial cytoplasm usually contains such a high concentration of solutes that, when the wall is weakened or removed, additional water enters the cell by osmosis. The damaged (or removed) cell wall cannot constrain the swelling of the cytoplasmic membrane, and the membrane bursts. This is an example of osmotic lysis caused by immersion in a hypotonic solution. A **hypotonic (hypoosmotic) solution** (*hypo* means under or less) outside the cell is a medium whose concentration of solutes is lower than that inside the cell. Most bacteria live in hypotonic solutions, and swelling is contained by the cell wall. Cells with weak cell walls, such as gram-negative bacteria, may burst or undergo osmotic lysis as a result of excessive water intake (Figure 4.17d).

A **hypertonic (hyperosmotic) solution** (*hyper* means above) is a medium having a higher concentration of solutes than the cell has. Most bacterial cells placed in a hypertonic solution shrink and collapse or plasmolyze because water leaves the cells by osmosis (Figure 4.17e). Keep in mind that the terms *isotonic, hypotonic,* and *hypertonic* describe the concentration of solutions outside the cell *relative* to the concentration inside the cell.

Active Processes

Active processes include active transport and group translocation.

Active Transport. In performing **active transport**, the cell *uses energy* in the form of adenosine triphosphate (ATP) to move substances across the plasma membrane. The movement of a substance in active transport is usually from outside to inside, even though the concentration might be much higher inside the cell. Like facilitated diffusion, active transport depends on carrier proteins in the plasma membrane (see Figure 4.16). There appears to be a different carrier for each transported substance or group of closely related transported substances.

Group Translocation. In active transport, the substance that crosses the membrane is not altered by transport across the membrane. In **group translocation,** a special form of active transport that occurs exclusively in procaryotes, the substance is chemically altered during transport across the membrane. Once the substance is altered and inside the cell, the plasma membrane is impermeable to it, so it remains inside the cell. This important mechanism allows a cell to accumulate various substances even though they may be in low concentrations outside the cell. Group translocation requires energy supplied by high-energy phosphate compounds, such as phosphoenolpyruvic acid (PEP).

One example of group translocation is the transport of the sugar glucose, which is often used in growth media for bacteria. While a specific carrier protein is transporting the glucose molecule across the membrane, a phosphate group is added to the sugar. This phosphorylated form of glucose, which cannot be transported out, can then be used in the cell's metabolic pathways.

Simple diffusion and facilitated diffusion are useful mechanisms for transporting substances into cells when the concentrations of the substances are greater outside the cell. However, when a bacterial cell is in an environment in which nutrients are in low concentration, the cell must use active transport and group translocation to accumulate the needed substances.

Some eucaryotic cells (those without cell walls) can use two additional active transport processes called phagocytosis and pinocytosis. These processes, which do not occur in bacteria, are described in later sections.

Cytoplasm

For a procaryotic cell, the term **cytoplasm** refers to the internal matrix of the cell contained inside the plasma membrane (see Figure 4.5). Cytoplasm is about 80% water and contains primarily proteins (enzymes), carbohydrates, lipids, inorganic ions, and many low–molecular-weight compounds. Inorganic ions are present in much higher concentrations in cytoplasm than in most media. Cytoplasm is thick, aqueous, semitransparent, and elastic. The major structures in the cytoplasm are DNA, particles called ribosomes, and reserve deposits called inclusions.

Procaryotic cytoplasm lacks certain features of eucaryotic cytoplasm, such as a cytoskeleton and cytoplasmic streaming. These features will be described later.

Nuclear Area

The nuclear area, or **nucleoid,** of a bacterial cell (see Figure 4.5) contains a single long circular molecule of double-stranded DNA, the **bacterial chromosome** (see Figure 8.1a). This is the cell's genetic information, which carries all the information required for the cell's structures and functions. Unlike the chromosomes of eucaryotic cells, bacterial chromosomes do not include histones and are not surrounded by a nuclear envelope (membrane). The nuclear area can be spherical, elongated, or dumbbell-shaped. In actively growing bacteria, as much as 20% of the cell volume is occupied by DNA because such cells presynthesize nuclear material for future cells. The chromosome is attached to the plasma membrane. Proteins in the plasma membrane are believed to be responsible for replication of the DNA and segregation of the new chromosomes to daughter cells in cell division (see Figure 6.9).

In addition to the bacterial chromosome, bacteria often contain small circular, double-stranded DNA molecules called **plasmids** (see the F factor in Figure 8.26a). These molecules are extrachromosomal genetic elements; that is, they are not connected to the main bacterial chromosome, and they replicate independently of chromosomal DNA. Research indicates that plasmids are associated with plasma membrane proteins. Plasmids usually contain from five to 100 genes that are generally not crucial for the survival of the bacterium under normal environmental conditions; plasmids may be gained or lost without harming the cell. Under certain conditions, however, plasmids are an advantage to cells. Plasmids may carry genes for such activities as antibiotic resistance, tolerance to toxic metals, production of toxins, and synthesis of enzymes. Plasmids can be transferred from one bacterium to another. In fact, plasmid DNA is used for gene manipulation in biotechnology.

Ribosomes

All eucaryotic and procaryotic cells contain **ribosomes**, which function as the sites of protein synthesis. Cells that have high rates of protein synthesis, such as those that are actively growing, have a great number of ribosomes. The cytoplasm of a procaryotic cell contains tens of thousands of these very small structures, which give the cytoplasm a granular appearance (see Figure 4.5).

Ribosomes are composed of two subunits, each subunit being composed of protein and a type of RNA called ribosomal RNA (rRNA). Procaryotic ribosomes differ from eucaryotic ribosomes in the number of proteins and rRNA molecules they contain; they are also somewhat smaller and less dense than ribosomes of eucaryotic cells. Accordingly, procaryotic ribosomes are called 70S ribosomes (Figure 4.18), and those of eucaryotic cells are known as 80S ribosomes. The letter *S* refers to Svedberg units, which indicate the relative rate of sedimentation during ultra–high-speed centrifugation. Sedimentation rate is a function of the size, weight, and shape of a particle. The subunits of a 70S ribosome are a small 30S subunit containing one molecule of rRNA and a larger 50S subunit containing two molecules of rRNA.

Several antibiotics, such as streptomycin, neomycin, and tetracyclines, work by inhibiting protein synthesis on the ribosomes. Because of differences in procaryotic and eucaryotic ribosomes, the microbial cell can be killed by the antibiotic while the eucaryotic host cell remains unaffected.

Inclusions

Within the cytoplasm of procaryotic (and eucaryotic) cells are several kinds of reserve deposits, known as **inclusions**. Some inclusions are common to a wide variety of bacteria, whereas others are limited to a small number of species and therefore serve as a basis for identification. Some of the more prominent bacterial inclusions are briefly described in the following paragraphs.

Metachromatic Granules

These large inclusions, which take their name from the fact that they sometimes stain red with certain blue dyes such as methylene blue, are collectively known as **volutin**. Volutin represents a reserve of inorganic phosphate (polyphosphate) that can be used in the synthesis of ATP. It is generally formed by cells that grow in phosphate-rich environments. Metachromatic granules are found in algae, fungi, and protozoa, as well as in bacteria. These granules are characteristic of *Corynebacterium diphtheriae* (kô-rī-nē-bak-ti′rē-um dif-thi′rē-ī), the causative agent of diphtheria; thus, they have diagnostic significance.

Polysaccharide Granules

These inclusions typically consist of glycogen and starch, and their presence can be demonstrated when iodine is

FIGURE 4.18 Procaryotic ribosome. A small 30S subunit **(a)** and a large 50S subunit **(b)** make up the complete 70S procaryotic ribosome **(c)**.

applied to the cells. In the presence of iodine, glycogen granules appear reddish brown and starch granules appear blue.

Lipid Inclusions

Lipid inclusions appear in various species of *Mycobacterium*, *Bacillus*, *Azotobacter* (ä-zō-tō-bak′tėr), *Spirillum* (spī-ril′lum), and other genera. A common lipid-storage material, one unique to bacteria, is the polymer *poly-β-hydroxybutyric acid*. Lipid inclusions are revealed by use of fat-soluble dyes, such as Sudan dyes.

Sulfur Granules

Certain bacteria—for example, the "sulfur bacteria" that belong to the genus *Thiobacillus*—derive energy by oxidizing sulfur and sulfur-containing compounds. These bacteria may deposit sulfur granules in the cell, where they serve as an energy reserve.

Carboxysomes

These are polyhedral and hexagonal inclusions that contain the enzyme ribulose 1,5-diphosphate carboxylase. Bacteria that use carbon dioxide as their sole source of carbon require this enzyme for carbon dioxide fixation during photosynthesis. Among the bacteria containing carboxysomes are nitrifying bacteria, cyanobacteria, and thiobacilli.

Gas Vacuoles

These are hollow cavities found in many aquatic procaryotes, including cyanobacteria, anoxygenic photosynthetic bacteria, and halobacteria. Each vacuole consists of rows of several individual *gas vesicles*, which are hollow cylinders covered by protein. Gas vacuoles maintain buoyancy so that the cells can remain at the depth in the water appropriate for them to receive sufficient amounts of oxygen, light, and nutrients.

For a discussion of inclusions called magnetosomes in magnetotactic bacteria, see the box on page 80.

Endospores

When essential nutrients are depleted or when water is unavailable, certain gram-positive bacteria, such as those

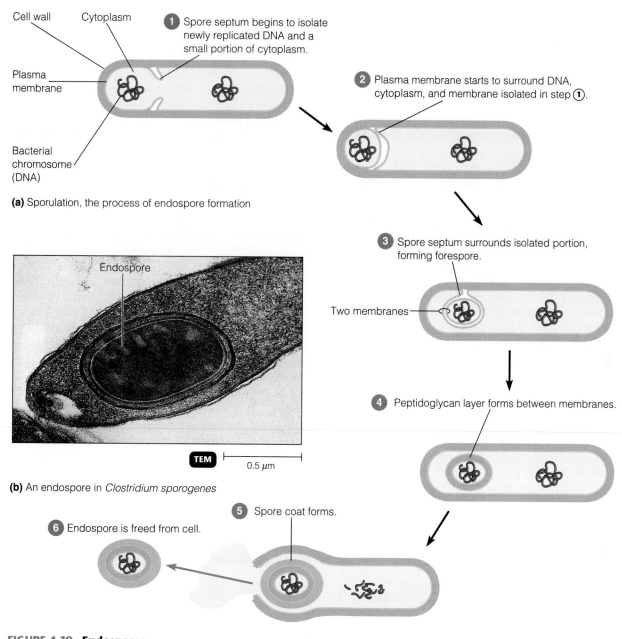

(a) Sporulation, the process of endospore formation

Cell wall   Cytoplasm

Plasma
membrane

Bacterial
chromosome
(DNA)

1 Spore septum begins to isolate newly replicated DNA and a small portion of cytoplasm.

2 Plasma membrane starts to surround DNA, cytoplasm, and membrane isolated in step **1**.

3 Spore septum surrounds isolated portion, forming forespore.

Two membranes

4 Peptidoglycan layer forms between membranes.

5 Spore coat forms.

6 Endospore is freed from cell.

Endospore

TEM   0.5 μm

(b) An endospore in *Clostridium sporogenes*

FIGURE 4.19   Endospores.

of the genera *Clostridium* and *Bacillus,* form specialized "resting" cells called **endospores.** Unique to bacteria, endospores are highly durable dehydrated cells with thick walls and additional layers. They are formed internal to the bacterial cell membrane. When released into the environment, they can survive extreme heat, lack of water, and exposure to many toxic chemicals and radiation. For example, 7500-year-old endospores of *Thermoactinomyces vulgaris* (thėr-mō-ak-tin-ō-mī′sēs vul-ga′ris) from the freezing muds of Elk Lake in Minnesota have germinated when rewarmed and placed in a nutrient medium. Although true endospores are found in gram-positive bacteria, one

gram-negative species, *Coxiella burnetii* (käks-ē-el′lä bėr-ne′tē-ē), the cause of Q fever, forms endosporelike structures that resist heat and chemicals and can be stained with endospore stains.

The process of endospore formation within a vegetative (parent) cell is known as **sporulation** or **sporogenesis** (Figure 4.19a). It is not clear what biochemical events trigger this process. In the first observable stage of sporulation, a newly replicated bacterial chromosome and a small portion of cytoplasm are isolated by an ingrowth of the plasma membrane called a *spore septum.* The spore septum becomes a double-layered membrane that surrounds the chromosome and cytoplasm. This structure, entirely enclosed within the original cell, is called a *forespore*. Thick layers of peptidoglycan are laid down between the two membrane layers. Then a thick *spore coat* of protein forms around the outside membrane (Figure 4.19b). This coat is responsible for the resistance of endospores to many harsh chemicals.

The diameter of the endospore may be the same as, smaller than, or larger than the diameter of the vegetative cell. Depending on the species, the endospore might be located *terminally* (at one end), *subterminally* (near one end), or *centrally* inside the vegetative cell. When the endospore matures, the vegetative cell wall dissolves (lyses), killing the cell, and the endospore is freed (see Figure 4.19a).

Most of the water present in the forespore cytoplasm is eliminated by the time sporulation is complete, and endospores do not carry out metabolic reactions. The highly dehydrated endospore core contains only DNA, small amounts of RNA, ribosomes, enzymes, and a few important small molecules. The latter include a strikingly large amount of an organic acid called *dipicolinic acid* (found in the cytoplasm), which is accompanied by a large number of calcium ions. These cellular components are essential for resuming metabolism later.

Endospores can remain dormant for thousands of years. An endospore returns to its vegetative state by a process called **germination.** Germination is triggered by physical or chemical damage to the endospore's coat. The endospore's enzymes then break down the extra layers surrounding the endospore, water enters, and metabolism resumes. Because one vegetative cell forms a single endospore, which, after germination, remains one cell, sporulation in bacteria is *not* a means of reproduction. This process does not increase the number of cells.

Endospores are important from a clinical viewpoint and in the food industry because they are resistant to processes that normally kill vegetative cells. Such processes include heating, freezing, desiccation, use of chemicals, and radiation. Whereas most vegetative cells are killed by temperatures above 70°C, endospores can survive in boiling water for several hours or more. Endospores of thermophilic (heat-loving) bacteria can survive in boiling water for 19 hours. Endospore-forming bacteria are a problem in the food industry because they are likely to survive underprocessing, and if conditions for growth occur, some species produce toxins and disease. Special methods used to control organisms that produce endospores are discussed in Chapter 7.

As noted in Chapter 3, endospores are difficult to stain for detection. Thus, a specially prepared stain must be used along with heat. (The Schaeffer–Fulton endospore stain is commonly used.)

Having examined the functional anatomy of the procaryotic cell, we will now look at the functional anatomy of the eucaryotic cell.

THE EUCARYOTIC CELL

As mentioned earlier, eucaryotic organisms include algae, protozoa, fungi, higher plants, and animals. The eucaryotic cell (Figure 4.20) is typically larger and structurally more complex than the procaryotic cell. By comparing the structure of the procaryotic cell in Figure 4.5 with that of the eucaryotic cell, the differences between the two types of cells become apparent.

To review briefly, eucaryotic cells are much larger than procaryotic cells, typically 10–100 μm in diameter as compared to 0.20–2.0 μm. Also, eucaryotic cells contain membrane-bounded organelles; procaryotic cells do not. **Organelles** are specialized structures that perform specific functions. Although both types of cells carry on the same basic functions, procaryotic cells generally do not localize these functions in specific organelles. The genetic material (DNA) of procaryotic cells is not membrane-bounded or associated in a regular way with protein. The genetic material of eucaryotic cells *is* membrane-bounded, organized into chromosomes, and closely associated with histones and other proteins.

The following discussion of eucaryotic cells will parallel our discussion of procaryotic cells by starting with structures that extend to the outside of the cell. The principal differences between procaryotic and eucaryotic cells are summarized in Table 4.2 on page 98.

FLAGELLA AND CILIA

LEARNING OBJECTIVE

• *Differentiate between procaryotic and eucaryotic flagella.*

Many types of eucaryotic cells have projections that are used for cellular locomotion or for moving substances along the surface of the cell. These projections contain cytoplasm and are enclosed by the plasma membrane. If the projections are few and are long in relation to the size of the cell, they are called **flagella.** If the projections are numerous and short, resembling hairs, they are called **cilia.**

Euglenoid algae use a flagellum for locomotion, whereas protozoa, such as *Tetrahymena* (tet-rä-hī′me-nä), use cilia for locomotion (Figure 4.21a and b). Both flagella and cilia consist of nine pairs of microtubules arranged in a ring, plus two single microtubules in the center of the ring (Figure 4.21c). A procaryotic flagellum rotates, but a eucaryotic flagellum moves in a wavelike manner (Figure 4.21d). To help keep foreign material out of the lungs, ciliated cells of the human respiratory system move the material along the surface of the cells in the bronchial tubes and trachea toward the throat and mouth (see Figure 16.4).

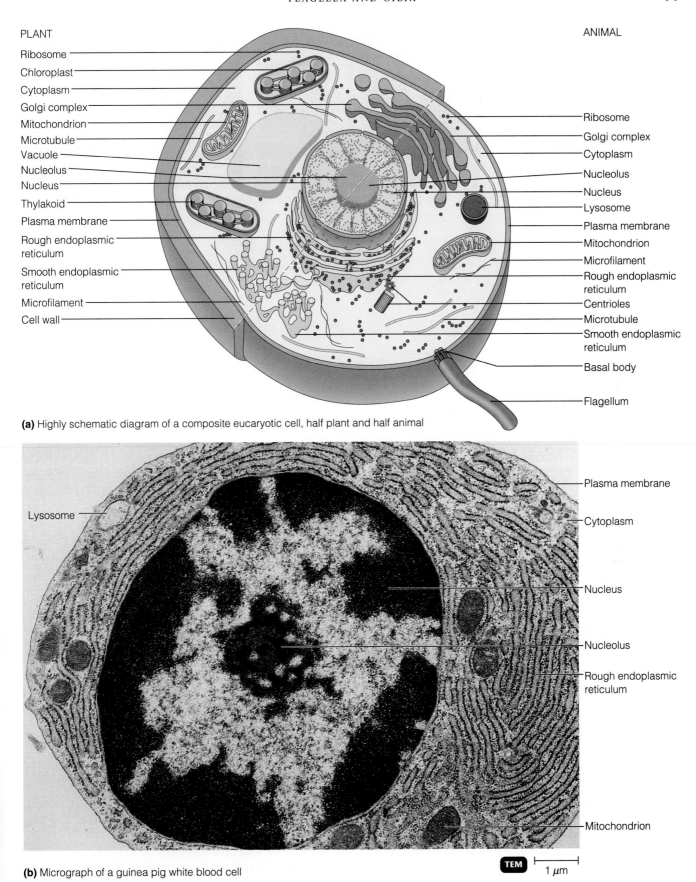

PLANT

Ribosome
Chloroplast
Cytoplasm
Golgi complex
Mitochondrion
Microtubule
Vacuole
Nucleolus
Nucleus
Thylakoid
Plasma membrane
Rough endoplasmic reticulum
Smooth endoplasmic reticulum
Microfilament
Cell wall

ANIMAL

Ribosome
Golgi complex
Cytoplasm
Nucleolus
Nucleus
Lysosome
Plasma membrane
Mitochondrion
Microfilament
Rough endoplasmic reticulum
Centrioles
Microtubule
Smooth endoplasmic reticulum
Basal body
Flagellum

(a) Highly schematic diagram of a composite eucaryotic cell, half plant and half animal

Lysosome

Plasma membrane
Cytoplasm
Nucleus
Nucleolus
Rough endoplasmic reticulum
Mitochondrion

TEM 1 μm

(b) Micrograph of a guinea pig white blood cell

FIGURE 4.20   Structure of eucaryotic cells.

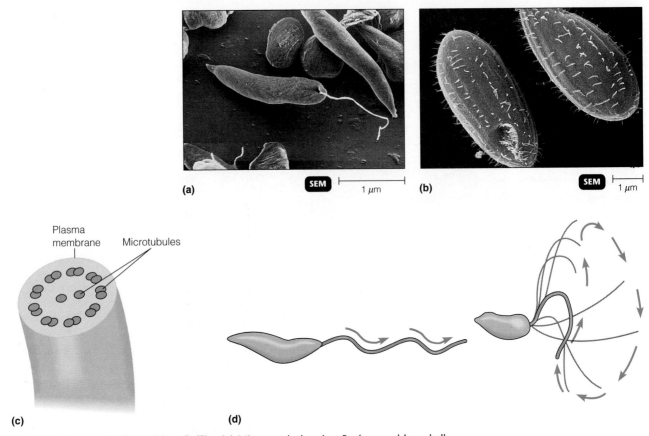

FIGURE 4.21   Eucaryotic flagella and cilia. (a) Micrograph showing *Euglena,* a chlorophyll-containing flagellate, with its flagellum. **(b)** Micrograph showing two *Tetrahymena,* a common freshwater ciliate, with their cilia. **(c)** Internal structure of a flagellum (or cilium) showing the arrangement of microtubules. **(d)** Movement of a eucaryotic flagellum.

THE CELL WALL AND GLYCOCALYX

LEARNING OBJECTIVE

- *Compare and contrast procaryotic and eucaryotic cell walls and glycocalyxes.*

Some eucaryotic cells have cell walls, although these walls are generally much simpler than those of procaryotic cells. Most algae have cell walls consisting of the polysaccharide *cellulose* (as do all plants). Cell walls of some fungi also contain cellulose, but in most fungi the principal structural component of the cell wall is the polysaccharide *chitin,* a polymer of N-acetylglucosamine (NAG) units. (Chitin is also the main structural component of the exoskeleton of crustaceans and insects.) The cell walls of yeasts contain the polysaccharides *glucan* and *mannan.* In eucaryotes that lack a cell wall, the plasma membrane

may be the outer covering; however, cells that have direct contact with the environment may have coatings outside the plasma membrane. Protozoa do not have a typical cell wall; instead, they have a flexible outer covering called a *pellicle.*

In other eucaryotic cells, including animal cells, the plasma membrane is covered by a **glycocalyx** ("sugar coat"), a layer of material containing substantial amounts of sticky carbohydrates. Some of these carbohydrates are covalently bonded to proteins and lipids in the plasma membrane, forming glycoproteins and glycolipids that anchor the glycocalyx to the cell. The glycocalyx strengthens the cell surface, helps attach cells together, and may contribute to cell-to-cell recognition.

Eucaryotic cells do not contain peptidoglycan, the framework of the procaryotic cell wall. This is significant medically because antibiotics, such as penicillins and cephalosporins, act against peptidoglycan and therefore do not affect human eucaryotic cells.

THE PLASMA (CYTOPLASMIC) MEMBRANE

LEARNING OBJECTIVE

- *Compare and contrast procaryotic and eucaryotic plasma membranes.*

In eucaryotic cells that lack a cell wall, the **plasma membrane** is the external covering of the cell. In function and basic structure, the eucaryotic and procaryotic plasma membranes are very similar. There are, however, differences in the types of proteins found in their membranes. Eucaryotic membranes also contain carbohydrates, which serve as receptor sites that assume a role in such functions as cell-to-cell recognition. These carbohydrates also provide attachment sites for bacteria. Eucaryotic plasma membranes also contain *sterols,* complex lipids not found in procaryotic plasma membranes (with the exception of the mycoplasmas). Sterols seem to be associated with the ability of the membranes to resist lysis due to increased osmotic pressure.

Substances can cross eucaryotic and procaryotic plasma membranes by simple diffusion, facilitated diffusion, osmosis, or active transport. Group translocation does not occur in eucaryotic cells. However, eucaryotic cells can use a mechanism called **endocytosis.** This occurs when a segment of the plasma membrane surrounds a particle or large molecule, encloses it, and brings it into the cell. Endocytosis is one of the ways that viruses can enter animal cells.

Two very important types of endocytosis are *phagocytosis* and *pinocytosis*. During phagocytosis, cellular projections called pseudopods engulf particles and bring them into the cell. Phagocytosis is used by white blood cells to destroy bacteria and foreign substances (see Figure 16.8 and further discussion in Chapter 16). In pinocytosis, the plasma membrane folds inward, bringing extracellular fluid into the cell along with whatever substances are dissolved in the fluid.

CYTOPLASM

LEARNING OBJECTIVE

- *Compare and contrast procaryotic and eucaryotic cytoplasms.*

The **cytoplasm** of eucaryotic cells encompasses the matrix inside the plasma membrane and outside the nucleus (see Figure 4.20). The cytoplasm is the substance in which various cellular components are found. A major difference between eucaryotic and procaryotic cytoplasm is that eucaryotic cytoplasm has a complex internal structure, consisting of exceedingly small rods called *microfilaments* and *intermediate filaments* and cylinders called *microtubules*. Together, they form the **cytoskeleton.** The cytoskeleton provides support and shape and assists in transporting substances through the cell (and even in moving the entire cell, as in phagocytosis). The movement of eucaryotic cytoplasm from one part of the cell to another, which helps to distribute nutrients and move the cell over a surface, is called *cytoplasmic streaming.* Another difference between procaryotic and eucaryotic cytoplasm is that many of the important enzymes found in the cytoplasmic fluid of procaryotes are sequestered in the organelles of eucaryotes.

ORGANELLES

LEARNING OBJECTIVES

- *Define organelle.*
- *Describe the functions of the nucleus, endoplasmic reticulum, ribosomes, Golgi complex, mitochondria, chloroplasts, lysosomes, and centrioles.*

Organelles are specialized structures characteristic of eucaryotic cells. They include the nucleus, endoplasmic reticulum, Golgi complex, mitochondria, chloroplasts, lysosomes, and centrioles. In addition, eucaryotic cells contain ribosomes that are larger and denser than procaryotic ribosomes.

Nucleus

The most characteristic eucaryotic organelle is the nucleus (see Figure 4.20). The **nucleus** is usually spherical or oval (Figure 4.22), is frequently the largest structure in the cell, and contains almost all of the cell's hereditary information (DNA). Some DNA is also found in mitochondria and in chloroplasts of photosynthetic organisms.

The nucleus is separated from the cytoplasm by a double membrane called the **nuclear envelope.** Each of the two membranes resembles the plasma membrane in structure. Minute **nuclear pores** in the membrane allow the nucleus to communicate with the membranous network in the cytoplasm, called the endoplasmic reticulum (see Figure 4.20a). Substances entering and exiting the nucleus are believed to pass through the tiny pores. Within the nuclear envelope is a gelatinous fluid called **nucleoplasm.** One or more spherical bodies called **nucleoli** (singular: *nucleolus)* are also present. Nucleoli are actually condensed regions of chromosomes where ribosomal RNA is being synthesized. Ribosomal RNA is an essential component of ribosomes (described shortly).

Finally, there is the DNA, which is combined with a number of proteins, including several basic proteins called

FIGURE 4.22   Eucaryotic nucleus. The inset indicates the relationship of the nucleus to the whole cell. **(a)** Micrograph of a nucleus. **(b)** Diagram of a nucleus.

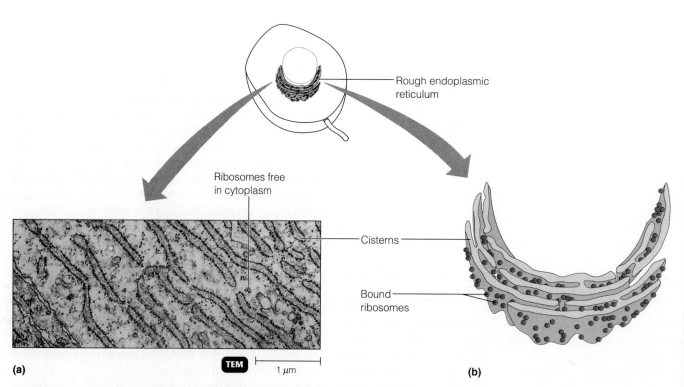

FIGURE 4.23   Rough endoplasmic reticulum and ribosomes. (a) Micrograph of rough endoplasmic reticulum and ribosomes in cross section. **(b)** Three-dimensional drawing of rough endoplasmic reticulum.

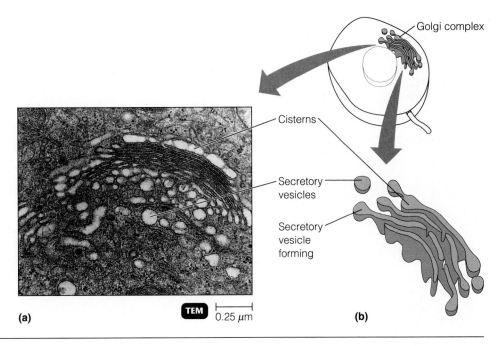

Golgi complex

Cisterns

Secretory vesicles

Secretory vesicle forming

TEM |——| 0.25 μm

(a)   **(b)**

histones and nonhistones. The combination of about 165 base pairs of DNA and 9 molecules of histones is referred to as a *nucleosome.* When the cell is not reproducing, the DNA and its associated proteins appear as a threadlike mass called **chromatin.** Before nuclear division, the chromatin coils into shorter and thicker rodlike bodies called **chromosomes.** Nucleosomes are the basic structural units of chromosomes. Procaryotic chromosomes do not undergo this process, do not have histones, are not formed of nucleosomes, and are not enclosed in a nuclear envelope.

Eucaryotic cells divide by two elaborate mechanisms called mitosis and meiosis (see Chapter 8). Neither occurs in procaryotic cells.

Endoplasmic Reticulum

Within the cytoplasm of eucaryotic cells there is an extensive network of flattened membranous sacs (**cisterns**) called the **endoplasmic reticulum,** or **ER** (Figure 4.23). The ER network is continuous with the nuclear envelope (see Figure 4.20). The ER synthesizes and stores lipids and proteins, and it provides a pathway for transport of those molecules to other parts of the cell. The ends of the ER sacs can pinch off into small spheres called **secretory vesicles,** which then transport substances within the cell. Many lipids and proteins produced by the ER are transported to the Golgi complex (described shortly) in this manner as well.

Two types of ER are found in most eucaryotic cells. The first has numerous ribosomes bound to it and thus looks rough through a microscope; this is called **rough ER.** The bound ribosomes help rough ER function in protein syn-

thesis. The second type has no ribosomes attached to it and thus is called **smooth ER.** It has diverse metabolic functions, including lipid synthesis.

Ribosomes

Attached to the outer surface of rough ER are ribosomes, which are also found free in the cytoplasm (see Figure 4.23a). As in procaryotes, ribosomes are the sites of protein synthesis in the cell.

The ribosomes of the eucaryotic ER and cytoplasm are somewhat larger and denser than those of procaryotic cells. These eucaryotic ribosomes are 80S ribosomes, each of which consists of a large 60S subunit containing three molecules of rRNA and a smaller 40S subunit with one molecule of rRNA. Chloroplasts and mitochondria contain 70S ribosomes, which may indicate their evolution from procaryotes. This hypothesis is discussed later in this chapter. The role of ribosomes in protein synthesis will be discussed in more detail in Chapter 8.

Golgi Complex

The **Golgi complex** is generally near the nucleus. Similar in form to the ER, it consists of four to six flattened sacs (cisterns) that are stacked on each other like a pile of dishes with expanded areas at their ends (Figure 4.24). Also like the ER, these expanded areas detach from the cisterns to form secretory vesicles that can export substances from the cell as well as transporting them within it.

The Golgi complex receives newly synthesized proteins and lipids from the ER and then sorts, packages, and de-

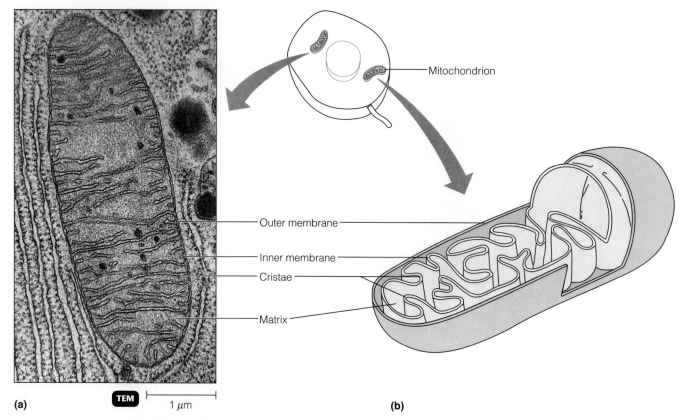

(a)   TEM |———| 1 μm   **(b)**

FIGURE 4.25 Mitochondria. (a) Micrograph of a mitochondrion (in longitudinal section) from a rat pancreas cell. **(b)** Three-dimensional drawing of a mitochondrion.

livers them by secretory vesicles. Substances it transports within the cell include lipids that become part of the plasma membrane and enzymes that are carried in lysosomes (described shortly). Proteins destined for export from the cell pass into secretory vesicles and are then released to the exterior of the cell when the vesicles fuse with the cell's plasma membrane.

Mitochondria

Spherical or rod-shaped organelles called **mitochondria** appear throughout the cytoplasm of most eucaryotic cells (see Figure 4.20). A mitochondrion consists of a double membrane similar in structure to the plasma membrane (Figure 4.25). The outer mitochondrial membrane is smooth, but the inner mitochondrial membrane is arranged in a series of folds called **cristae**. The center of the mitochondrion is a semifluid substance called the *matrix*. Because of the nature and arrangement of the cristae, the inner membrane provides an enormous surface area on which chemical reactions can occur. Some proteins that function in cellular respiration, including the enzyme that makes ATP, are located on the cristae of the inner mitochondrial membrane, and many of the metabolic steps involved in cellular respiration are concen-

trated in the matrix (see Chapter 5). Mitochondria are frequently called the "powerhouses of the cell" because of their central role in the production of ATP.

Mitochondria contain 70S ribosomes and some DNA of their own, as well as the machinery necessary to replicate, transcribe, and translate the information encoded by their DNA. In addition, mitochondria can reproduce more or less on their own by growing and dividing in two.

Chloroplasts

Algae and green plants contain a unique organelle called a **chloroplast** (Figure 4.26a). A chloroplast is a membrane-bounded structure that contains both the pigment chlorophyll and the enzymes required for the light-gathering phases of photosynthesis (see Chapter 5). The chlorophyll is contained in flattened membrane sacs called *thylakoids;* stacks of thylakoids are called *grana* (the singular is *granum*) (Figure 4.26b). Like mitochondria, chloroplasts contain 70S ribosomes, DNA, and enzymes involved in protein synthesis. They are capable of multiplying on their own within the cell. Interestingly, the way both chloroplasts and mitochondria multiply—by increasing in size and then dividing in two—is strikingly reminiscent of bacterial multiplication.

Chloroplast

Chloroplast

Thylakoids

Granum

(a)

TEM | 0.5 μm

(b)

FIGURE 4.26 **Chloroplasts.** Photosynthesis occurs in chloroplasts; the light-trapping pigments are located on the thylakoids. **(a)** Micrograph of chloroplasts in a cell. **(b)** Three-dimensional drawing of a chloroplast.

Lysosomes

As viewed under the electron microscope, **lysosomes**, which are formed from Golgi complexes, appear to be membrane-enclosed spheres free in the eucaryotic cytoplasm. Unlike mitochondria, lysosomes have only a single membrane and lack detailed structure (see Figure 4.20). But they contain powerful digestive enzymes capable of breaking down many kinds of molecules. Moreover, these enzymes can also digest bacteria that enter the cell. Human white blood cells, which use phagocytosis to ingest bacteria, contain large numbers of lysosomes.

Vacuoles

A **vacuole** is a space or cavity in the cytoplasm of a cell that is enclosed by a membrane called a tonoplast. In plant cells, vacuoles may occupy from 5% to 90% of the cell volume, depending on the type of cell. Vacuoles are derived from the Golgi complex and have several diverse functions. Some vacuoles serve as storage organelles for substances such as proteins, sugars, organic acids, and inorganic ions. Other vacuoles function as lysosomes that store hydrolytic enzymes that can digest stored macromolecules. Many plant cells also store metabolic wastes and poisons that would otherwise be injurious if they accumulated in the cytoplasm. Finally, vacuoles may take up water to enable plant cells to increase in size and to also provide rigidity to leaves and stems.

Centrioles

Located near the nucleus are a pair of cylindrical structures, the **centrioles** (see Figure 4.20). Each centriole is a ring of nine evenly spaced bundles, each of which consists of three microtubules. The two centrioles are situated so that the long axis of one is at right angles to the long axis of the other. Centrioles play roles in eucaryotic cell division and in the formation of cilia and flagella (as basal bodies).

The principal differences between procaryotic and eucaryotic cells are presented in Table 4.2. Most of the differences have been discussed in this chapter; a few will be treated in greater detail in subsequent chapters.

EVOLUTION OF EUCARYOTES

LEARNING OBJECTIVE

- *Discuss evidence that supports the endosymbiotic hypothesis of eucaryotic evolution.*

TABLE 4.2 Principal Differences Between Procaryotic and Eucaryotic Cells

CHARACTERISTIC	PROCARYOTIC	EUCARYOTIC
Size of cell	Typically 0.20–2.0 μm in diameter	Typically 10–100 μm in diameter
Nucleus	No nuclear membrane or nucleoli	True nucleus, consisting of nuclear membrane and nucleoli
Membrane-bounded organelles	Absent	Present; examples include lysosomes, Golgi complex, endoplasmic reticulum, mitochondria, and chloroplasts
Flagella	Consist of two protein building blocks	Complex; consist of multiple microtubules
Glycocalyx	Present as a capsule or slime layer	Present in some cells that lack a cell wall
Cell wall	Usually present; chemically complex (typical bacterial cell wall includes peptidoglycan)	When present, chemically simple
Plasma membrane	No carbohydrates and generally lacks sterols	Sterols and carbohydrates that serve as receptors present
Cytoplasm	No cytoskeleton or cytoplasmic streaming	Cytoskeleton; cytoplasmic streaming
Ribosomes	Smaller size (70S)	Larger size (80S); small size (70S) in organelles
Chromosome (DNA) arrangement	Single circular chromosome; lacks histones	Multiple linear chromosomes with histones
Cell division	Binary fission	Mitosis
Sexual reproduction	No meiosis; transfer of DNA fragments only	Involves meiosis

At present, there are two major views on the origin of eucaryotes. According to the **autogenous hypothesis**, specialized internal membranes that derived from a procaryotic plasma membrane evolved into organelles characteristic of eucaryotic cells. According to this view, the nuclear envelope, endoplasmic reticulum, Golgi complex, and organelles bounded by single membranes (such as lysosomes) evolved this way. These organelles are sometimes referred to as the *endomembrane system.*

The other view—the **endosymbiotic hypothesis**—proposes that the forerunners of eucaryotic cells were associations of small, symbiotic (living together) procaryotic cells living within larger procaryotic cells. This hypothesis focuses primarily on the origin of mitochondria and chloroplasts (which multiply in a way that is similar to that of bacteria). Good evidence indicates that the ancestors of mitochondria were probably oxygen-requiring heterotrophic bacteria. (Heterotrophs obtain food molecules by eating other organisms or their by-products.) Chloroplasts are believed to be descendants of photosynthetic procaryotes. (Photosynthetic organisms use light energy to produce food molecules.) It has been suggested that oxygen-requiring heterotrophs and photosynthetic procaryotes both gained entry into the large procaryotic cell as undigested prey or internal parasites. The development of eucaryotes from procaryotes may have occurred by a combination of the two mechanisms. The nuclear envelope and other cellular membranes may have evolved as outlined in the autogenous hypothesis, whereas mitochondria and chloroplasts may have evolved by the endosymbiotic mechanism.

•   •   •

Our next concern is to examine microbial metabolism. In Chapter 5 you will learn about the importance of enzymes to microbes and the ways microbes produce and use energy.

STUDY OUTLINE

COMPARISON OF PROCARYOTIC AND EUCARYOTIC CELLS: AN OVERVIEW (pp. 70–71)

1. Procaryotic and eucaryotic cells are similar in their chemical composition and chemical reactions.
2. Procaryotic cells lack membrane-bounded organelles (including a nucleus).
3. Peptidoglycan is found in procaryotic cell walls but not in eucaryotic cell walls.

THE PROCARYOTIC CELL (pp. 71–90)

1. Bacteria are unicellular, and most of them multiply by binary fission.
2. Bacterial species are differentiated by morphology, chemical composition, nutritional requirements, biochemical activities, and source of energy.

SIZE, SHAPE, AND ARRANGEMENT OF BACTERIAL CELLS (p. 71)

1. Most bacteria are from 0.20 to 2.0 μm in diameter and from 2 to 8 μm in length.
2. The three basic bacterial shapes are coccus (spheres), bacillus (rods), and spiral (twisted).
3. Pleomorphic bacteria can assume several shapes.

STRUCTURES EXTERNAL TO THE CELL WALL (pp. 71–76)

Glycocalyx (pp. 71–74)

1. The glycocalyx (capsule or slime layer) is a gelatinous polysaccharide and/or polypeptide covering.
2. Capsules may protect pathogens from phagocytosis.
3. Capsules enable adherence to surfaces, prevent desiccation, and may provide nutrients.

Flagella (pp. 74–76)

1. Flagella are relatively long filamentous appendages consisting of a filament, hook, and basal body.
2. Procaryotic flagella rotate to push the cell.
3. Motile bacteria exhibit taxis; positive taxis is movement toward an attractant, and negative taxis is movement away from a repellent.

Axial Filaments (p. 76)

1. Spiral cells that move by means of an axial filament are called spirochetes.
2. Axial filaments are similar to flagella, except that they wrap around the cell.

Fimbriae and Pili (p. 76)

1. Fimbriae and pili are short, thin appendages.
2. Fimbriae help cells adhere to surfaces.
3. Pili join cells for the transfer of DNA from one cell to another.

THE CELL WALL (pp. 77–82)

Composition and Characteristics (pp. 78–81)

1. The cell wall surrounds the plasma membrane and protects the cell from changes in osmotic pressure.
2. The bacterial cell wall consists of peptidoglycan, a polymer consisting of NAG and NAM and short chains of amino acids.
3. Gram-positive cell walls consist of many layers of peptidoglycan and also contain teichoic acids.
4. Gram-negative bacteria have a lipoprotein–lipopolysaccharide–phospholipid outer membrane surrounding a thin peptidoglycan layer.
5. The outer membrane protects the cell from phagocytosis and from penicillin, lysozyme, and other chemicals.
6. Porins are proteins that permit small molecules to pass through the outer membrane; specific channel proteins allow other molecules to move through the outer membrane.

7. The lipopolysaccharide component of the outer membrane consists of sugars that function as antigens and lipid A, which is an endotoxin.

Atypical Cell Walls (pp. 81–82)

1. *Mycoplasma* is a bacterial genus that naturally lacks cell walls.
2. Archaeobacteria have pseudomurein; they lack peptidoglycan.

Damage to the Cell Wall (p. 82)

1. In the presence of lysozyme, gram-positive cell walls are destroyed, and the remaining cellular contents are referred to as a protoplast.
2. In the presence of lysozyme, gram-negative cell walls are not completely destroyed, and the remaining cellular contents are referred to as a spheroplast.
3. Protoplasts and spheroplasts are subject to osmotic lysis.
4. Antibiotics such as penicillin interfere with cell wall synthesis.

STRUCTURES INTERNAL TO THE CELL WALL (pp. 82–90)

Plasma (Cytoplasmic) Membrane (pp. 82–84)

1. The plasma membrane encloses the cytoplasm and is a phospholipid bilayer with protein (fluid mosaic).
2. The plasma membrane is selectively permeable (has selective permeability).
3. Plasma membranes carry enzymes for metabolic reactions, such as nutrient breakdown, energy production, and photosynthesis.
4. Mesosomes—irregular infoldings of the plasma membrane—are now considered artifacts.
5. Plasma membranes can be destroyed by alcohols and polymyxins.

Movement of Materials Across Membranes (pp. 84–87)

1. Movement across the membrane may be by passive processes, in which materials move from higher to lower concentration and no energy is expended by the cell.
2. In simple diffusion, molecules and ions move until equilibrium is reached.
3. In facilitated diffusion, substances are transported by permeases across membranes from high to low concentration.
4. Osmosis is the movement of water from high to low concentration across a selectively semipermeable membrane until equilibrium is reached.
5. In active transport, materials move from low to high concentrations by permeases, and the cell must expend energy.
6. In group translocation, energy is expended to modify chemicals and transport them across the membrane.

Cytoplasm (p. 87)

1. Cytoplasm is the fluid component inside the plasma membrane.
2. The cytoplasm is mostly water, with inorganic and organic molecules, DNA, ribosomes, and inclusions.

Nuclear Area (p. 87)

1. The nuclear area contains the DNA of the bacterial chromosome.
2. Bacteria can also contain plasmids, which are extrachromosomal DNA circles.

Ribosomes (p. 88)

1. The cytoplasm of a procaryote contains numerous 70S ribosomes; ribosomes consist of rRNA and protein.
2. Protein synthesis occurs at ribosomes; it can be inhibited by certain antibiotics.

Inclusions (p. 88)

1. Inclusions are reserve deposits found in procaryotic and eucaryotic cells.
2. Among the inclusions found in bacteria are metachromatic granules (inorganic phosphate), polysaccharide granules (usually glycogen or starch), lipid inclusions, sulfur granules, carboxysomes (ribulose 1,5-diphosphate carboxylase), and gas vacuoles.

Endospores (pp. 88–90)

1. Endospores are resting structures formed by some bacteria for survival during adverse environmental conditions.
2. The process of endospore formation is called sporulation; the return of an endospore to its vegetative state is called germination.

THE EUCARYOTIC CELL (pp. 90–98)

FLAGELLA AND CILIA (p. 90)

1. Flagella are few and long in relation to cell size; cilia are numerous and short.
2. Flagella and cilia are used for motility, and cilia also move substances along the surface of the cells.
3. Both flagella and cilia contain a "nine pairs + two single" arrangement and microtubules.

THE CELL WALL AND GLYCOCALYX (p. 92)

1. The cell walls of most algae and some fungi consist of cellulose.
2. The main material of fungal cell walls is chitin.
3. Yeast cell walls consist of glucan and mannan.
4. Animal cells are surrounded by a glycocalyx; this makes the cell stronger and provides a means of attachment to other cells.

THE PLASMA (CYTOPLASMIC) MEMBRANE (p. 93)

1. Like the procaryotic plasma membrane, the eucaryotic plasma membrane is a phospholipid bilayer containing proteins.
2. Eucaryotic plasma membranes contain carbohydrates attached to the proteins and sterols not found in procaryotic cells (except *Mycoplasma* cells).
3. Eucaryotic cells can move materials across the plasma membrane by the passive processes used by procaryotes, in addition to active transport and endocytosis (phagocytosis and pinocytosis).

CYTOPLASM (p. 93)

1. The cytoplasm of eucaryotic cells includes everything inside the plasma membrane and external to the nucleus.
2. The chemical characteristics of the cytoplasm of eucaryotic cells resemble those of the cytoplasm of procaryotic cells.
3. Eucaryotic cytoplasm has a cytoskeleton and exhibits cytoplasmic streaming.

ORGANELLES (pp. 93–97)

1. Organelles are specialized membrane-bounded structures in the cytoplasm.
2. Organelles are characteristic of eucaryotic cells.
3. The nucleus, which contains DNA in the form of chromosomes, is the most characteristic eucaryotic organelle.
4. The nuclear envelope is connected to a system of parallel membranes in the cytoplasm called the endoplasmic reticulum (ER).
5. The ER provides a surface for chemical reactions, serves as a transporting network, and stores synthesized molecules. Protein synthesis and transport occur on rough ER; lipid synthesis occurs on smooth ER.
6. 80S ribosomes are found in the cytoplasm or attached to the rough endoplasmic reticulum.
7. The Golgi complex consists of cisterns. It functions in membrane formation and protein secretion.
8. Mitochondria are the primary sites of ATP production. They contain small 70S ribosomes and DNA, and they multiply by fission.
9. Chloroplasts contain chlorophyll and enzymes for photosynthesis. Like mitochondria, they contain 70S ribosomes and DNA and multiply by fission.
10. Lysosomes are formed from Golgi complexes. They store powerful digestive enzymes.
11. Vacuoles are membrane-enclosed cavities derived from the Golgi complex usually found in plant cells that store various substances, serve as lysosomes, increase cell size, and provide rigidity to leaves and stems.
12. A pair of cylindrical structures called centrioles, which are involved in cell division, are near the nucleus.

EVOLUTION OF EUCARYOTES (pp. 97–98)

1. Two hypotheses offer explanations of the origin of eucaryotes from procaryotic ancestors.
2. The autogenous hypothesis states that organelles (for example, the nucleus and Golgi complex) evolved from internal membranes derived from the plasma membrane.
3. According to the endosymbiotic hypothesis, organelles (for example, mitochondria and chloroplasts) evolved from symbiotic procaryotes living inside other procaryotic cells.

STUDY QUESTIONS

Review

1. Diagram each of the following flagellar arrangements:
   (a) Lophotrichous
   (b) Monotrichous
   (c) Peritrichous

2. Endospore formation is called _____ . It is initiated by _____ . Formation of a new cell from an endospore is called _____ . This process is triggered by _____ .

3. Draw the following bacterial shapes (listed in a, b, and c). Show how d, e, and f are special conditions of a, b, and c, respectively.

   (a) Spiral   (d) Spirochetes
   (b) Bacillus   (e) Streptobacilli
   (c) Coccus   (f) Staphylococci

4. List three differences between procaryotic and eucaryotic cells.

5. Match the structures to their functions.

   ____ Cell wall   (a) Attachment to surfaces
   ____ Endospore   (b) Cell wall formation
   ____ Fimbriae   (c) Motility
   ____ Flagella   (d) Protection from osmotic lysis
   ____ Glycocalyx   (e) Protection from phagocytes
   ____ Pili   (f) Resting
   ____ Plasma   (g) Protein synthesis
        membrane   (h) Selective permeability
   ____ Ribosomes   (i) Transfer of genetic material

6. Of what value to the cell is each of the following?
   (a) Metachromatic granules
   (b) Polysaccharide granules
   (c) Lipid inclusions
   (d) Sulfur granules
   (e) Carboxysomes
   (f) Gas vacuoles

7. Why is an endospore called a resting structure? Of what advantage is an endospore to a bacterial cell?

8. Explain what would happen in the following circumstances.
   (a) A suspension of bacteria is placed in distilled water.
   (b) A suspension of bacteria is placed in distilled water with lysozyme.
   (c) A suspension of bacteria is placed in an aqueous solution of lysozyme and 10% sucrose.
   (d) A suspension of gram-negative bacteria is placed in distilled water with penicillin.

9. Compare and contrast the following:
   (a) Simple diffusion and facilitated diffusion
   (b) Active transport and facilitated diffusion
   (c) Active transport and group translocation

10. Why are mycoplasmas resistant to antibiotics that interfere with cell wall synthesis?

11. Answer the following questions using the diagrams provided, which represent cross sections of bacterial cell walls.
    (a) Which diagram represents a gram-positive bacterium? How can you tell?
    (b) Explain how the Gram stain works to distinguish between these two types of cell walls.
    (c) Why does penicillin have no effect on most gram-negative cells?
    (d) How do essential molecules enter cells through each wall?

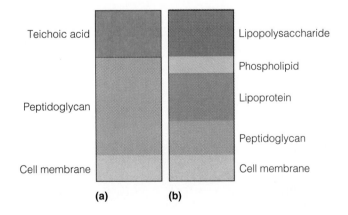

12. Starch is readily metabolized by many cells, but a starch molecule is too large to cross the plasma membrane. How does a cell obtain the glucose molecules from a starch polymer? How does the cell transport these glucose molecules across the plasma membrane?

13. Match the following characteristics of eucaryotic cells with their functions.

   ____ Chloroplasts   (a) Protein synthesis
   ____ Rough endoplasmic   (b) Photosynthesis
        reticulum   (c) ATP production
   ____ Golgi complex   (d) Digestive enzyme
   ____ Lysosomes   storage
   ____ Mitochondria   (e) Secretion

14. Eucaryotic cells might have evolved from early procaryotic cells living in close association. What do you know about eucaryotic organelles that would support this theory?

15. What process would a eucaryotic cell use to ingest a procaryotic cell? To ingest a virus?

Critical Thinking

1. Why can procaryotic cells be smaller than eucaryotic cells and still carry on all the functions of life?

2. The smallest eucaryotic cell is the motile alga *Micromonas*. What is the minimum number of organelles this alga must have?

3. Two types of procaryotic cells have been distinguished: eubacterial and archaeobacterial. How do these cells differ from each other? How are they similar?

4. In 1985, a 0.5-mm cell was discovered in surgeonfish and named *Epulopiscium fishelsoni* (see Figure 11.22). It was presumed to be a protozoan. In 1993, researchers determined that *Epulopiscium* was actually a gram-positive bacterium. What do you suppose caused the initial identification of this organism as a protozoan? What evidence would change the identification to eubacterium?

5. Recently, researchers found that when *E. coli* cells are exposed to a hyperosmotic solution, the bacteria produce a permease that can move K^+ into the cell. Of what value is the active transport of K^+, which requires ATP?

Clinical Applications

1. A woman with a blood-borne *Enterobacter* infection was treated with cephalosporin, the preferred antibiotic for *Enterobacter* infections. After treatment, *Enterobacter* could not be cultured from her blood, indicating that the bacteria were killed. However, her symptoms became worse. Annually, nearly half of similar patients die. Explain why antibiotic treatment made her symptoms increase.

2. *Clostridium botulinum* is a strict anaerobe; that is, it is killed by the molecular oxygen (O_2) present in air. Humans can die of botulism from eating foods in which *C. botulinum* is growing. How does this bacterium survive on plants picked for human consumption? Why are home-canned foods most often the source of botulism?

3. Within a three-day period at a large hospital, five patients undergoing hemodialysis developed fever and chills. *Pseudomonas aeruginosa* and *Klebsiella pneumoniae* were isolated from three of the patients. *P. aeruginosa*, *K. pneumoniae*, and *Enterobacter agglomerans* were isolated from the dialysis system. Why do all three bacteria cause similar symptoms?

Microbial Metabolism

Now that you are familiar with the structure of procaryotic cells, we can discuss the activities that allow these microbes to thrive. The life support activity of even the most structurally simple organism involves a large number of complex biochemical reactions. Most, although not all, of the biochemical processes of bacteria also occur in eucaryotic microbes and in the cells of multicellular organisms, including humans. However, the reactions that are unique to bacteria are fascinating because they allow microorganisms to do things we cannot do. For example, some bacteria (the chemoautotrophs) can grow on diets of such inorganic substances as carbon dioxide, iron, sulfur, hydrogen gas, and ammonia.

This chapter examines some representative chemical reactions that either produce energy (the catabolic reactions) or use energy (the anabolic reactions) in microorganisms. We will also look at how these various reactions are integrated within the cell.

CATABOLIC AND ANABOLIC REACTIONS

LEARNING OBJECTIVE

- *Define metabolism, and describe the fundamental differences between anabolism and catabolism.*

We use the term **metabolism** to refer to the sum of all chemical reactions within a living organism. Because chemical reactions either release or require energy, metabolism can be viewed as an energy-balancing act. Accordingly, metabolism can be divided into two classes of chemical reactions: those that release energy and those that require energy. In living cells, the chemical reactions that release energy are generally the ones involved in **catabolism**, the breakdown of complex organic compounds into simpler ones. These reactions are called **catabolic**, or *degradative*, reactions. However, the energy-requiring reactions are mostly involved in **anabolism**, the building of complex organic molecules from simpler ones. These reactions are called **anabolic**, or *biosynthetic*, reactions. Anabolic processes often involve dehydration synthesis reactions (reactions that release water) and require energy to form new chemical bonds. Examples of anabolic processes are the formation of proteins from amino acids, nucleic acids from nucleotides, and polysaccharides from simple sugars. These biosynthetic reactions generate the materials for cell growth. Catabolic reactions are generally hydrolytic reactions (reactions that use water and in which chemical bonds are broken). Chemical bonds store energy; when they are broken, chemical energy is released. An example of catabolism occurs when cells break down sugars into carbon dioxide and water.

Catabolic reactions furnish the energy needed to drive anabolic reactions. This coupling of energy-requiring and energy-releasing reactions is made possible through the molecule adenosine triphosphate (ATP). (You can review its structure in Figure 2.20.) ATP stores energy derived from catabolic reactions and releases it later to drive anabolic reactions and perform other cellular work. Recall from Chapter 2 that a molecule of ATP consists of an adenine, a ribose, and three phosphate groups. When the terminal phosphate group is split from ATP, adenosine diphosphate (ADP) is formed, and energy is released to drive anabolic reactions. Using Ⓟ to represent a phosphate group, we can write this reaction as:

$$ATP \longrightarrow ADP + Ⓟ + energy$$

Then, the energy from catabolic reactions is used to combine ADP and a Ⓟ to resynthesize ATP:

$$ADP + Ⓟ + energy \longrightarrow ATP$$

Thus, anabolic reactions are coupled to ATP breakdown, and catabolic reactions are coupled to ATP synthesis. This concept of coupled reactions is very important; we will see why by the end of this chapter. For now, you should know that the chemical composition of a living cell is constantly changing: some molecules are being broken down while others are being synthesized. This balanced flow of chemicals and energy maintains the life of a cell.

The role of ATP in coupling anabolic and catabolic reactions is shown in Figure 5.1. Only part of the energy released in catabolism is actually available for cellular functions, for part of the energy is lost to the environment as heat. Because the cell must use energy to maintain life, it has a continuous need for new external sources of energy.

Before we discuss how cells produce energy, let us first consider the principal properties of a group of proteins involved in almost all biologically important chemical reactions. These proteins, the enzymes, were described briefly in Chapter 2. It is important to understand that a cell's metabolic pathways are determined by its enzymes, which are in turn determined by the cell's genetic makeup.

ENZYMES

LEARNING OBJECTIVES

- *Describe the mechanism of enzymatic action.*
- *List the factors that influence enzymatic activity.*

We indicated in Chapter 2 that chemical reactions occur when chemical bonds are formed or broken. In order for reactions to take place, atoms, ions, or molecules must collide. Whether a collision produces a reaction depends on the speed of the particles, the amount of energy required to trigger the reaction (called **activation energy**), and the specific configuration of the particles. Paradoxically, the

FIGURE 5.1 The relationship between anabolism, catabolism, and ATP.

inherent physiological temperature and pressure of organisms are too low for chemical reactions to occur quickly enough to maintain the life of the organism. Raising the temperature and pressure and the number of reacting molecules would increase the frequency of collisions and the rate of chemical reactions. However, such changes could also damage or kill the organism. The solution to this problem is a class of proteins called **enzymes.** Enzymes can speed up chemical reactions in several ways. For example, an enzyme may bring two reactant molecules close together and may properly orient them to react. Whatever the method, the result is that the enzyme lowers the activation energy for the reaction without increasing the temperature or pressure inside the cell.

Substances that can speed up a chemical reaction without being altered themselves are called **catalysts.** In living cells, enzymes serve as biological catalysts. As catalysts, enzymes are specific. Each acts on a specific substance, called the enzyme's **substrate** (or substrates, when there are two or more reactants), and each catalyzes only one reaction. For example, sucrose (table sugar) is the substrate of the enzyme sucrase, which catalyzes the hydrolysis of sucrose to glucose and fructose. The specificity of enzymes is made possible by their structures. Enzymes are generally large globular proteins that range in molecular weight from about 10,000 to several million. Each of the thousands of known enzymes has a characteristic three-dimensional shape with a specific surface configuration as a result of its primary, secondary, and tertiary structures (see Figure 2.18). The unique configuration of each enzyme enables it to "find" the correct substrate from among the large number of diverse molecules in the cell.

Enzymes are extremely efficient. Under optimum conditions, they can catalyze reactions at rates 10^8 to 10^{10} times (up to 10 billion times) higher than those of comparable reactions without enzymes. The **turnover number** (maximum number of substrate molecules an enzyme molecule converts to product each second) is generally between 1 and 10,000 and can be as high as 500,000. For example, the enzyme DNA polymerase I, which participates

TABLE 5.1 Enzyme Classification Based on Type of Chemical Reaction Catalyzed

CLASS	TYPE OF CHEMICAL REACTION CATALYZED	EXAMPLES
Oxidoreductase	Oxidation–reduction in which oxygen and hydrogen are gained or lost	Cytochrome oxidase, lactate dehydrogenase
Transferase	Transfer of functional groups, such as an amino group, acetyl group, or phosphate group	Acetate kinase, alanine deaminase
Hydrolase	Hydrolysis (addition of water)	Lipase, sucrase
Lyase	Removal of groups of atoms without hydrolysis	Oxalate decarboxylase, isocitrate lyase
Isomerase	Rearrangement of atoms within a molecule	Glucose-phosphate isomerase, alanine racemase
Ligase	Joining of two molecules (using energy usually derived from breakdown of ATP)	Acetyl-CoA synthetase, DNA ligase

FIGURE 5.2 Components of a holoenzyme. Many enzymes require both an apoenzyme (protein portion) and a cofactor (nonprotein portion) to become active. The cofactor can be a metal ion or an organic molecule called a coenzyme (as shown here). The apoenzyme and cofactor together make up the holoenzyme, or whole enzyme. The substrate is the reactant acted upon by the enzyme.

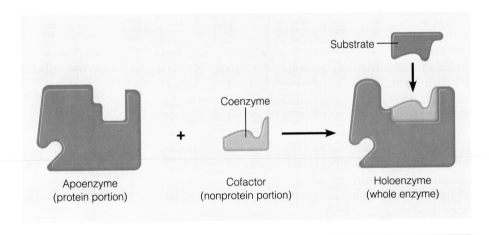

Apoenzyme (protein portion) + Cofactor (nonprotein portion) → Holoenzyme (whole enzyme)

Coenzyme

Substrate

in the synthesis of DNA, has a turnover number of 15, whereas the enzyme lactate dehydrogenase, which removes hydrogen atoms from lactic acid, has a turnover number of 1000.

Enzymes are subject to various cellular controls. Two main types are control of enzyme *synthesis* (see Chapter 8) and control of enzyme *activity* (*how much* enzyme is present vs. *how active* it is). Many enzymes exist in the cell in both active and inactive forms. The rate at which enzymes switch between these two forms is determined by the cellular environment.

Naming Enzymes

The names of enzymes usually end in *-ase*. All enzymes can be grouped into six classes, according to the type of chemical reaction they catalyze (Table 5.1). Enzymes within each of the major classes are named according to the more specific types of reactions they assist. For example, the class called oxidoreductases is involved with oxidation–reduction reactions (described shortly). During some oxidation reactions, a substrate loses hydrogen; in others, a substrate gains oxygen. Enzymes in the oxidore-ductase class that remove hydrogen are called *dehydrogenases;* those that add oxygen are called *oxidases*. As you will see later, dehydrogenase and oxidase enzymes have even more specific names, such as lactate dehydrogenase and cytochrome oxidase, depending on the specific substrates on which they act.

Enzyme Components

Although some enzymes consist entirely of proteins, most consist of both a protein portion called an **apoenzyme** and a nonprotein component called a **cofactor.** Together, the apoenzyme and cofactor form a **holoenzyme,** or whole enzyme (Figure 5.2). If the cofactor is removed, the apoenzyme will not function. The cofactor can be a metal ion or a complex organic molecule called a **coenzyme.**

Coenzymes may assist the enzyme by accepting atoms removed from the substrate or by donating atoms required by the substrate. Some coenzymes act as electron carriers, removing electrons from the substrate and donating them to other molecules in subsequent reactions. Many coenzymes are derived from vitamins (Table 5.2). Two of the most important coenzymes in cellular metabolism are

TABLE 5.2   Selected Vitamins and Their Coenzymatic Functions

VITAMIN	FUNCTION
Vitamin B_1 (thiamine)	Part of coenzyme cocarboxylase; has many functions, including the metabolism of pyruvic acid
Vitamin B_2 (riboflavin)	Coenzyme in flavoproteins; active in electron transfers
Niacin (nicotinic acid)	Part of NAD molecule; active in electron transfers
Vitamin B_6 (pyridoxine)	Coenzyme in amino acid metabolism
Vitamin B_{12} (cyanocobalamin)	Coenzyme (methyl cyanocobalamide) involved in transfer of methyl groups; active in amino acid metabolism
Pantothenic acid	Part of coenzyme A molecule; involved in metabolism of pyruvic acid and lipids
Biotin	Involved in carbon dioxide fixation reactions, fatty acid synthesis
Folic acid	Coenzyme used in synthesis of purines and pyrimidines
Vitamin E	Needed for cellular and macromolecular syntheses
Vitamin K	Coenzyme used in electron transport (naphthoquinones and quinones)

nicotinamide adenine dinucleotide (NAD^+) and **nicotinamide adenine dinucleotide phosphate** ($NADP^+$). Both compounds contain derivatives of the B vitamin nicotinic acid (niacin), and both function as electron carriers. Whereas NAD^+ is primarily involved in catabolic (energy-yielding) reactions, $NADP^+$ is primarily involved in anabolic (energy-requiring) reactions. The flavin coenzymes, such as **flavin mononucleotide (FMN)** and **flavin adenine dinucleotide (FAD)**, contain derivatives of the B vitamin riboflavin and are also electron carriers. Another important coenzyme, **coenzyme A (CoA)**, contains a derivative of pantothenic acid, another B vitamin. This coenzyme plays an important role in the synthesis and breakdown of fats and in a series of oxidizing reactions called the Krebs cycle. We will come across all of these coenzymes in our discussion of metabolism.

As noted earlier, some cofactors are metal ions, including iron, copper, magnesium, manganese, zinc, calcium, and cobalt. Such cofactors may help catalyze a reaction by forming a bridge between the enzyme and a substrate. For example, magnesium (Mg^{2+}) is required by many phosphorylating enzymes (enzymes that transfer a phosphate group from ATP to another substrate). The Mg^{2+} can form a link between the enzyme and the ATP molecule. Most trace elements required by living cells are probably used in some such way to activate cellular enzymes.

Mechanism of Enzymatic Action

Although scientists do not completely understand how enzymes lower the activation energy of chemical reactions, the general sequence of events in enzyme action is as follows (Figure 5.3a):

1. The surface of the substrate contacts a specific region of the surface of the enzyme molecule, called the **active site.**

2. A temporary intermediate compound forms, called an **enzyme–substrate complex.**

3. The substrate molecule is transformed by the rearrangement of existing atoms, the breakdown of the substrate molecule, or combination with another substrate molecule.

4. The transformed substrate molecules—the products of the reaction—are released from the enzyme molecule because they no longer fit in the active site of the enzyme.

5. The unchanged enzyme is now free to react with other substrate molecules.

As a result of these events, an enzyme speeds up a chemical reaction.

As mentioned earlier, enzymes have *specificity* for particular substrates. For example, a specific enzyme may be able to hydrolyze a peptide bond only between two specific amino acids. Other enzymes can hydrolyze starch but not cellulose; even though both starch and cellulose are polysaccharides composed of glucose subunits, the orientations of the subunits in the two polysaccharides differ. Enzymes have this specificity because the three-dimensional shape of the active site fits the substrate somewhat as a lock fits with its key (Figure 5.3b). However, the active site and substrate are flexible, and they change shape somewhat as they meet to fit together more tightly. The substrate is usually much smaller than the enzyme, and relatively few of the enzyme's amino acids make up the active site.

A certain compound can be a substrate for a number of different enzymes that catalyze different reactions, so the fate of a compound depends on the enzyme that acts upon it. Glucose 6-phosphate, a molecule that is important in cell metabolism, can be acted upon by at least four different enzymes, and each reaction will yield a different product.

FIGURE 5.3 Mechanism of enzymatic action. (a) ① The substrate contacts the active site on the enzyme to form ② an enzyme–substrate complex. ③ The substrate is then transformed into products, ④ the products are released, and ⑤ the enzyme is recovered unchanged. In the example shown, the transformation into products involves a breakdown of the substrate into two products. Other transformations, however, may occur. **(b)** (Left) Molecular model of the enzyme in step ① of part (a). The active site of the enzyme can be seen here as a groove on the surface of the protein. (Right) As the enzyme and substrate meet in step ② of part (a), they change shape slightly to fit together more tightly.

Factors Influencing Enzymatic Activity

Several factors influence the activity of an enzyme. Among the more important are temperature, pH, substrate concentration, and presence or absence of inhibitors.

Temperature

The rate of most chemical reactions increases as the temperature increases. Molecules move more slowly at lower temperatures than at higher temperatures and so may not have enough energy to cause a chemical reaction. For enzymatic reactions, however, elevation beyond a certain temperature drastically reduces the rate of reaction (Figure 5.4a). This decrease is due to the enzyme's **denaturation,** the loss of its characteristic three-dimensional structure (tertiary configuration) (Figure 5.5). Denaturation of a protein involves breakage of hydrogen bonds and other noncovalent bonds.

A common example of denaturation is the transformation of uncooked egg white (a protein called albumin) to a hardened state by heat. As might be expected, denaturation of an enzyme changes the arrangement of the amino acids in the active site, altering its shape and causing the enzyme to lose its catalytic ability. In some cases, denaturation is partially or fully reversible. However, if denaturation continues until the enzyme has lost its solubility and coagulates (as with cooked albumin), the enzyme cannot regain its original properties. Enzymes can also be denatured by concentrated acids, bases, heavy-metal ions (such as lead, arsenic, or mercury), alcohol, and ultraviolet radiation.

pH

Most enzymes have an optimum pH at which their activity is characteristically maximal. Above or below this pH value, enzyme activity, and therefore the reaction rate, de-

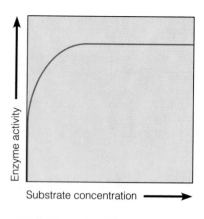

(a) Temperature. The enzymatic activity (rate of reaction catalyzed by the enzyme) increases with increasing temperature until the enzyme, a protein, is denatured by heat and inactivated. At this point, the reaction rate falls steeply.

(b) pH. The enzyme illustrated is most active at about pH 5.5.

(c) Substrate concentration. With increasing concentration of substrate molecules, the rate of reaction increases until the active sites on all the enzyme molecules are filled, at which point the maximum rate of reaction is reached.

FIGURE 5.4 Factors that influence enzymatic activity, plotted for a hypothetical enzyme.

clines (Figure 5.4b). When the H⁺ concentration (pH) in the medium is changed, many of the enzyme's amino acids are affected and the protein's three-dimensional structure is altered. Extreme changes in pH can cause denaturation.

Substrate Concentration

There is a maximum rate at which a certain amount of enzyme can catalyze a specific reaction. Only when the concentration of substrate(s) is extremely high can this maximum rate be attained. Under conditions of high substrate concentration, the enzyme is said to be in **saturation;** that is, its active site is always occupied by substrate or product molecules. In this condition, a further increase in substrate concentration will not affect the reaction rate because all active sites are already in use (Figure 5.4c). If a substrate's concentration exceeds a cell's saturation level for a particular enzyme, the rate of reaction can be increased only if the cell produces additional enzyme molecules. However, under normal cellular conditions, enzymes are not saturated with substrate(s). At any given time, many of the enzyme molecules are inactive for lack of substrate; thus, the rate of reaction is likely to be influenced by the substrate concentration.

Inhibitors

As you will see in later chapters, an effective way to control the growth of bacteria is to control their enzymes. Certain poisons, such as cyanide, arsenic, and mercury, combine with enzymes and prevent them from functioning. As a result, the cells stop functioning and die.

Enzyme inhibitors are classified according to their mechanism of action as either competitive or noncom-

petitive inhibitors. **Competitive inhibitors** fill the active site of an enzyme and compete with the normal substrate for the active site. A competitive inhibitor is able to do this because its shape and chemical structure are similar to those of the normal substrate (Figure 5.6b). However, unlike the substrate, it does not undergo any reaction to form products. Some competitive inhibitors bind irreversibly to amino acids in the active site, preventing any further interactions with the substrate. Others bind reversibly, alternately occupying and leaving the active site; these slow the enzyme's interaction with the substrate. Reversible competitive inhibition can be overcome by increasing the substrate concentration. As active sites become available, more substrate molecules than competitive inhibitor molecules are available to attach to the active sites of enzymes. One good example of a competitive inhibitor is sulfanilamide (a sulfa drug), which inhibits the enzyme whose normal substrate is para-aminobenzoic acid (PABA):

Sulfanilamide   PABA

PABA is an essential nutrient used by many bacteria in the synthesis of folic acid, a vitamin that functions as a coenzyme. When sulfanilamide is administered to bacteria, the enzyme that normally converts PABA to folic acid combines instead with the sulfanilamide. Folic acid is not synthesized, and the bacteria cannot grow. Because human

FIGURE 5.5 Denaturation of a protein. Breakage of the noncovalent bonds (such as hydrogen bonds) that hold the active protein in its three-dimensional shape renders the denatured protein nonfunctional.

Active (functional) protein   Denatured protein

FIGURE 5.6 Enzyme inhibitors. (a) Uninhibited enzyme and its normal substrate. **(b)** Competitive inhibitor. **(c)** One type of noncompetitive inhibitor, causing allosteric inhibition.

NORMAL BINDING OF SUBSTRATE

Substrate
Active site
Enzyme

(a)

ACTION OF ENZYME INHIBITORS

Competitive inhibitor
Altered active site
Noncompetitive inhibitor

(b)   **(c)**

cells do not use PABA to make their folic acid, sulfanilamide kills bacteria but does not harm human cells.

Noncompetitive inhibitors do not compete with the substrate for the enzyme's active site; instead, they interact with another part of the enzyme (Figure 5.6c). In this process, called **allosteric** ("other space") **inhibition**, the inhibitor binds to a site on the enzyme other than the substrate's binding site. This binding causes the active site to change its shape, making it nonfunctional. As a result, the enzyme's activity is reduced. This effect can be either reversible or irreversible, depending on whether or not the active site can return to its original shape. In some cases, allosteric interactions can activate an enzyme rather than inhibit it. Another type of noncompetitive inhibition can operate on enzymes that require metal ions for their activity. Certain chemicals can bind or tie up the metal ion activators and thus prevent an enzymatic reaction. Cyanide can bind the iron in iron-containing enzymes, and fluoride can bind calcium or magnesium. Substances such as cyanide and fluoride are sometimes called *enzyme poisons* because they permanently inactivate enzymes.

Feedback Inhibition

Allosteric inhibitors play a role in a kind of biochemical control called **feedback inhibition** (or **end-product inhibition**). This control mechanism stops the cell from wasting chemical resources by making more of a substance than it needs. In some metabolic reactions, several steps are required for the synthesis of a particular chemical compound, called the *end-product*. The process is similar to an assembly line, with each step catalyzed by a separate enzyme (Figure 5.7). In many biosynthetic pathways, the final product can allosterically inhibit the activity of one of the enzymes earlier in the pathway. This phenomenon is feedback inhibition.

Feedback inhibition generally acts on the first enzyme in a metabolic pathway (similar to shutting down an assembly line by stopping the first worker). Because the enzyme is inhibited, the product of the first enzymatic reaction in the pathway is not synthesized. Because that unsynthesized product would normally be the substrate for the second enzyme in the pathway, the second reaction stops immediately as well. Thus, even though only the first enzyme in the pathway is inhibited, the entire pathway shuts down and no new end-product is formed. By inhibiting the first enzyme in the pathway, the cell also keeps metabolic intermediates from accumulating. As the existing end-product is used up by the cell, the first enzyme's allosteric site will more often remain unbound, and the pathway will resume activity.

The bacterium *Escherichia coli* can be used to demonstrate feedback inhibition in the synthesis of the amino acid isoleucine, which is required for the cell's growth. In this metabolic pathway, five steps are taken to enzymati-

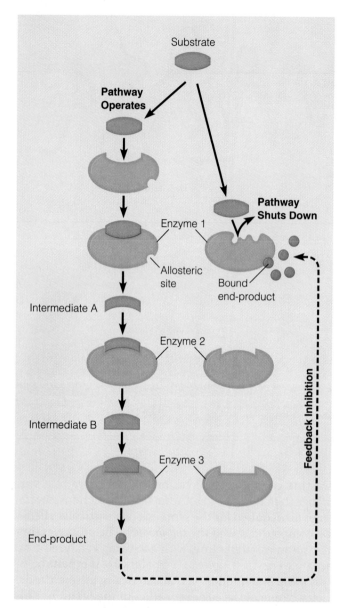

FIGURE 5.7   Feedback inhibition. A series of enzymes makes an end-product that inhibits the first enzyme in the series, thus shutting down the entire pathway when sufficient end-product has been made.

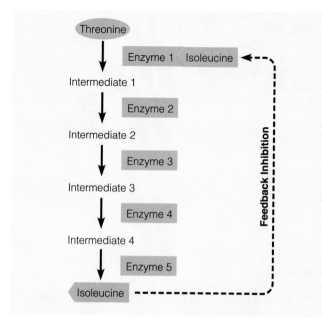

FIGURE 5.8   Feedback inhibition in *E. coli.* Shown here is a sequence of reactions in which threonine is converted to isoleucine. If isoleucine is added to the culture medium, *E. coli* no longer synthesizes isoleucine because the isoleucine combines with the enzyme that catalyzes the first step of the synthesis. Thus, isoleucine inhibits the synthesis of more isoleucine.

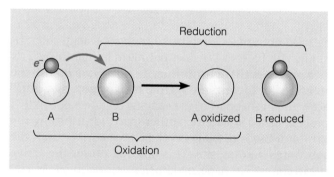

FIGURE 5.9   Oxidation–reduction. An electron is transferred from molecule A to molecule B. In the process, molecule A is oxidized and molecule B is reduced.

cally convert the amino acid threonine to isoleucine (Figure 5.8). If isoleucine is added to the growth medium for *E. coli*, it inhibits the first enzyme in the pathway and the bacteria stop synthesizing isoleucine. This condition is maintained until the supply of isoleucine is depleted. This type of feedback inhibition is also involved in regulating the cells' production of other amino acids, as well as vitamins, purines, and pyrimidines. In the following section on energy production methods, you will see the key roles enzymes play in both catabolic and anabolic reactions.

ENERGY PRODUCTION

LEARNING OBJECTIVES

- *Explain what is meant by oxidation–reduction.*
- *List and provide examples of three types of phosphorylation reactions that generate ATP.*

Nutrient molecules, like all molecules, have energy associated with the electrons that form bonds between their atoms. When this energy is spread throughout the mole-

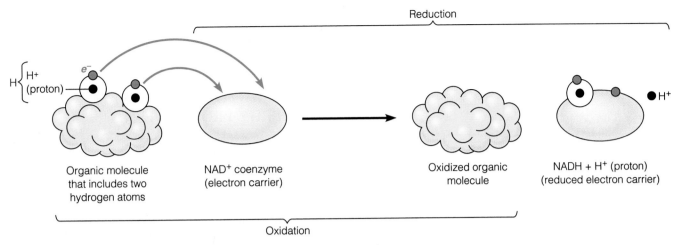

FIGURE 5.10 **Representative biological oxidation.** Two electrons and two protons (altogether equivalent to two hydrogen atoms) are transferred from an organic substrate molecule to a coenzyme, NAD^+. NAD^+ actually receives one hydrogen atom and one electron, and one proton is released into the medium. NAD^+ is reduced to NADH, which is a more energy-rich molecule.

cule, it is difficult for the cell to use. Various reactions in catabolic pathways, however, concentrate the energy into the bonds of ATP, which serves as a convenient energy carrier. ATP is generally referred to as having "high-energy" bonds. Actually, a better term is probably unstable bonds. Although the amount of energy in these bonds is not exceptionally large, it can be released quickly and easily. In a sense, ATP is similar to a highly flammable liquid such as kerosene. Although a large log might eventually burn to produce more heat than a cup of kerosene, the kerosene is easier to ignite and provides heat more quickly and conveniently. In a similar way, the "high-energy" unstable bonds of ATP provide the cell with readily available energy for anabolic reactions.

Before discussing the catabolic pathways, we will consider two general aspects of energy production: the concept of oxidation–reduction and the mechanisms of ATP generation.

Oxidation–Reduction

Oxidation is the removal of electrons (e^-) from an atom or molecule, a reaction that often produces energy. Figure 5.9 shows an example of an oxidation in which molecule A loses an electron to molecule B. Molecule A is oxidized (meaning that it has lost one or more electrons), whereas molecule B has undergone **reduction** (meaning that it has gained one or more electrons). Oxidation and reduction reactions are always coupled; in other words, each time one substance is oxidized, another is simultaneously reduced. The pairing of these reactions is called **oxidation–reduction.**

In many cellular oxidations, electrons and protons (hydrogen ions, H^+) are removed at the same time; this is equivalent to the removal of hydrogen atoms, because a hydrogen atom is made up of one proton and one electron (see Table 2.2). Because most biological oxidations involve the loss of hydrogen atoms, they are also called **dehydrogenation** reactions. Figure 5.10 shows an example of a biological oxidation. An organic molecule is oxidized by the loss of two hydrogen atoms, and a molecule of NAD^+ is reduced. As you may recall from our earlier discussion of coenzymes, NAD^+ assists enzymes by accepting hydrogen atoms removed from the substrate, in this case the organic molecule. As shown in Figure 5.10, NAD^+ accepts two electrons and one proton. One proton (H^+) is left over and is released into the surrounding medium. The reduced coenzyme, NADH, contains more energy than NAD^+. This energy can be used to generate ATP in later reactions.

An important point to remember about biological oxidation–reduction reactions (also called *redox* reactions) is that cells use them in catabolism to extract energy from nutrient molecules. Cells take nutrients, some of which serve as energy sources, and degrade them from highly reduced compounds (with many hydrogen atoms) to highly oxidized compounds. For example, when a cell oxidizes a molecule of glucose ($C_6H_{12}O_6$) to CO_2 and H_2O, the energy in the glucose molecule is removed in a stepwise manner and ultimately is trapped by ATP, which can then serve as an energy source for energy-requiring reactions. Compounds such as glucose that have many hydrogen atoms are highly reduced compounds, containing a large amount of potential energy. Thus, glucose is a valuable nutrient for organisms.

Generation of ATP

Much of the energy released during oxidation–reduction reactions is trapped within the cell by the formation of ATP. Specifically, a phosphate group, $\circledP$, is added to ADP with the input of energy to form ATP:

$$\overbrace{\text{Adenosine}—\circledP \sim \circledP}^{\text{ADP}} + \text{Energy} + \circledP \longrightarrow$$
$$\underbrace{\text{Adenosine}—\circledP \sim \circledP \sim \circledP}_{\text{ATP}}$$

The symbol ~ designates a "high-energy" bond—that is, one that can readily be broken to release usable energy. The high-energy bond that attaches the third $\circledP$ in a sense contains the energy stored in this reaction. When this $\circledP$ is removed, usable energy is released. The addition of $\circledP$ to a chemical compound is called **phosphorylation**. Organisms use three mechanisms of phosphorylation to generate ATP from ADP.

Substrate-Level Phosphorylation

In **substrate-level phosphorylation**, ATP is generated when a high-energy $\circledP$ is directly transferred from a phosphorylated compound (a substrate) to ADP. Generally, the $\circledP$ has acquired its energy during an earlier reaction in which the substrate itself was oxidized. The following example shows only the carbon skeleton and the $\circledP$ of a typical substrate:

$$\text{C—C—C} \sim \circledP + \text{ADP} \longrightarrow \text{C—C—C} + \text{ATP}$$

Oxidative Phosphorylation

When **oxidative phosphorylation** occurs, electrons transferred from organic compounds to electron carriers (usually to NAD^+) are passed through a series of different electron carriers to molecules of oxygen (O_2) or other inorganic molecules. This process occurs in the plasma membrane of procaryotes and in the inner mitochondrial membrane of eucaryotes. The series of electron carriers used in oxidative phosphorylation is called an **electron transport chain** (see Figure 5.14). The transfer of electrons from one electron carrier to the next releases energy, some of which is used to generate ATP from ADP through a process called *chemiosmosis*, to be described shortly.

Photophosphorylation

The third mechanism of phosphorylation, **photophosphorylation**, occurs only in photosynthetic cells, which contain light-trapping pigments such as chlorophylls. In photosynthesis, organic molecules, especially sugars, are synthesized with the energy of light from the energy-poor building blocks carbon dioxide and water. Photophosphorylation starts this process by converting light energy to the chemical energy of ATP and NADPH, which, in turn, are used to synthesize organic molecules. As in ox-idative phosphorylation, an electron transport chain is involved.

BIOCHEMICAL PATHWAYS OF ENERGY PRODUCTION

LEARNING OBJECTIVE

- *Explain the overall function of biochemical pathways.*

Organisms release and store energy from organic molecules by a series of controlled reactions rather than in a single burst. If the energy were released all at once, as a large amount of heat, it could not be readily used to drive chemical reactions and would, in fact, damage the cell. To extract energy from organic compounds and store it in chemical form, organisms pass electrons from one compound to another through a series of oxidation–reduction reactions.

A sequence of enzymatically catalyzed chemical reactions occurring in a cell is called a **biochemical pathway**. Let us write out a hypothetical pathway that converts starting material A to end-product F in a series of five steps:

The first step is the conversion of molecule A to molecule B. The curved arrow indicates that the reduction of coenzyme NAD^+ to NADH is coupled to that reaction; the electrons and protons come from molecule A. Similarly, the two arrows in the third step show a coupling of two reactions. As C is converted to D, ADP is converted to ATP; the energy needed comes from C as it transforms into D. The reaction converting D to E is readily reversible, as indicated by the double arrow. In the fifth step, the curved arrow leading from O_2 indicates that O_2 is a reactant in the reaction. The curved arrows leading to CO_2 and H_2O indicate that these substances are secondary products produced in the reaction, in addition to F, the end-product that (presumably) interests us the most. Secondary products such as CO_2 and H_2O shown here are sometimes called "by-products" or "waste products." Keep in mind that almost every reaction in a biochemical pathway is catalyzed by a specific enzyme; sometimes the name of the enzyme is printed near the arrow.

FIGURE 5.11 Overview of respiration and fermentation. Glycolysis produces ATP and reduces NAD⁺ to NADH while oxidizing glucose to pyruvic acid. In respiration, the pyruvic acid is converted into the first reactant in the Krebs cycle, which produces ATP and reduces NAD^+ (and another electron carrier called $FADH_2$) while giving off CO_2. The NADH from both processes carries electrons to the electron transport chain, in which their energy is used to produce a great deal of ATP. In fermentation, the pyruvic acid and the electrons carried by NADH from glycolysis are incorporated into fermentation end-products. A small version of this figure will be included in figures throughout the chapter to indicate the relationships of different reactions to the overall processes.

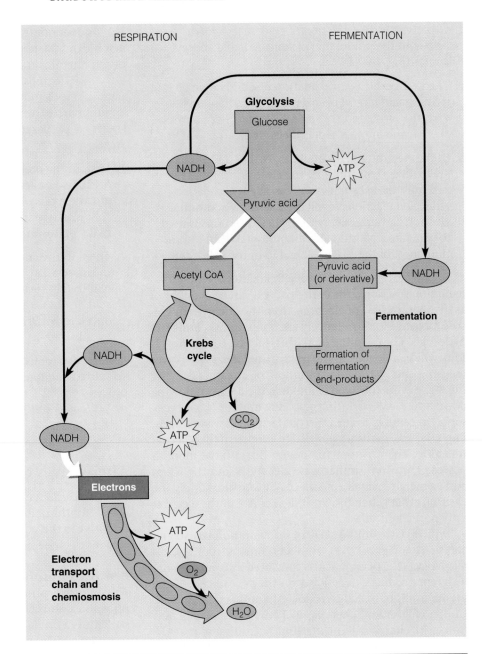

CARBOHYDRATE CATABOLISM

LEARNING OBJECTIVES

- *Describe the chemical reactions of glycolysis.*
- *Explain the products of the Krebs cycle.*
- *Describe the chemiosmotic model for ATP generation.*
- *Compare and contrast aerobic and anaerobic respiration.*
- *Describe the chemical reactions of and list some products of fermentation.*
- *Describe how lipids and proteins are prepared for glycolysis.*

Most microorganisms oxidize carbohydrates as their primary source of cellular energy. Carbohydrate catabolism, the breakdown of carbohydrate molecules to produce energy, is therefore of great importance in cell metabolism. Glucose is the most common carbohydrate energy source used by cells. Microorganisms can also catabolize various lipids and proteins for energy production, as you will see later.

To produce energy from glucose, microorganisms use two general processes: respiration and fermentation. Both processes usually start with the same first step, glycolysis, but follow different subsequent pathways (Figure 5.11). Before examining the details of respiration and fermentation, we will first look at a general overview of the processes.

As shown in Figure 5.11, the respiration of glucose typically occurs in three principal stages: glycolysis, the Krebs cycle, and the electron transport chain.

1. Glycolysis is the oxidation of glucose to pyruvic acid with the production of some ATP and energy-containing NADH.

2. The Krebs cycle is the oxidation of a derivative of pyruvic acid (acetyl coenzyme A) to carbon dioxide with the production of some ATP, energy-containing NADH, and another reduced electron carrier, $FADH_2$.

3. In the electron transport chain, NADH and $FADH_2$ are oxidized, contributing the electrons they have carried from the substrates to a "cascade" of oxidation–reduction reactions involving a series of additional electron carriers. Energy from these reactions is used to generate a considerable amount of ATP. In respiration, most of the ATP is generated in the third step.

Because respiration involves a long series of oxidation–reduction reactions, the entire process can be thought of as involving a flow of electrons from the energy-rich glucose molecule to the relatively energy-poor CO_2 and H_2O molecules. The coupling of ATP production to this flow is somewhat analogous to the production of electric power by using energy provided by a flowing stream. Carrying the analogy further, you could imagine a stream flowing down a gentle slope during glycolysis and the Krebs cycle, supplying energy to turn two old-fashioned waterwheels. Then the stream rushes down a steep slope in the electron transport chain, supplying energy for a large modern power plant. In a similar way, glycolysis and the Krebs cycle generate a small amount of ATP and also supply the electrons that generate a great deal of ATP at the electron transport chain stage.

Typically, the initial stage of fermentation is also glycolysis (see Figure 5.11). However, once glycolysis has taken place, the pyruvic acid is converted into one or more different products, depending on the type of cell. These products might include alcohol (ethanol) and lactic acid. Unlike respiration, there is no Krebs cycle or electron transport chain in fermentation. Accordingly, the ATP yield, which comes only from glycolysis, is much lower.

Glycolysis

Glycolysis, the oxidation of glucose to pyruvic acid, is usually the first stage in carbohydrate catabolism. Most microorganisms use this pathway; in fact, it occurs in most living cells.

Glycolysis is also called the **Embden–Meyerhof pathway.** The word *glycolysis* means splitting of sugar, and this is exactly what happens. The enzymes of glycolysis catalyze the splitting of glucose, a six-carbon sugar, into two three-carbon sugars. These sugars are then oxidized, releasing energy, and their atoms are rearranged to form two molecules of pyruvic acid. During glycolysis NAD^+ is reduced to NADH, and there is a net production of two ATP molecules by substrate-level phosphorylation. Gly-colysis does not require oxygen; it can occur under either aerobic or anaerobic conditions. This pathway is a series of ten chemical reactions, each catalyzed by a different enzyme. The steps are outlined in Figure 5.12; see also Appendix D for a more detailed representation of glycolysis. To summarize the process, glycolysis consists of two basic stages, a preparatory stage and an energy-conserving stage.

1. First, in the preparatory stage (steps ①–⑤ in Figure 5.12), two molecules of ATP are used as a six-carbon glucose molecule is phosphorylated, restructured, and split into two three-carbon compounds: glyceraldehyde 3-phosphate and dihydroxyacetone phosphate.

2. Then, in the energy-conserving stage (steps ⑥–⑩ in Figure 5.12), the two three-carbon molecules are oxidized in several steps to two molecules of pyruvic acid. In these reactions, two molecules of NAD^+ are reduced to NADH and four molecules of ATP are formed by substrate-level phosphorylation.

Because two molecules of ATP were needed to get glycolysis started and four molecules of ATP are generated by the process, *there is a net gain of two molecules of ATP for each molecule of glucose that is oxidized.*

Alternatives to Glycolysis

Many bacteria have another pathway in addition to glycolysis for the oxidation of glucose. The most common alternative is the pentose phosphate pathway; another alternative is the Entner–Doudoroff pathway.

Pentose Phosphate Pathway

The **pentose phosphate pathway** (or **hexose monophosphate shunt**) operates simultaneously with glycolysis. This is a cyclic pathway that provides a means for the breakdown of five-carbon sugars (pentoses) as well as glucose (see the detailed figure in Appendix D). A key feature of this pathway is that it produces important intermediate pentoses that act as precursors in the synthesis of (1) nucleic acids, (2) glucose from carbon dioxide in photosynthesis, and (3) certain amino acids. The pathway is an important producer of the reduced coenzyme NADPH from NADP. Like NAD^+, NADP carries an electron and a hydrogen atom. Twelve molecules of NADPH can be produced from each molecule of glucose if the glucose is completely oxidized. NADPH is then used in various biosynthetic reactions in the cell. Unlike glycolysis, the pentose phosphate pathway yields a net gain of only one molecule of ATP for each molecule of glucose oxidized. Bacteria that use the pentose phosphate pathway include *Bacillus subtilis* (su'til-us), *E. coli*, *Leuconostoc mesenteroides* (lü-kō-nos'tok mes-en-ter-oi'dēz), and *Enterococcus faecalis* (fē-kāl'is).

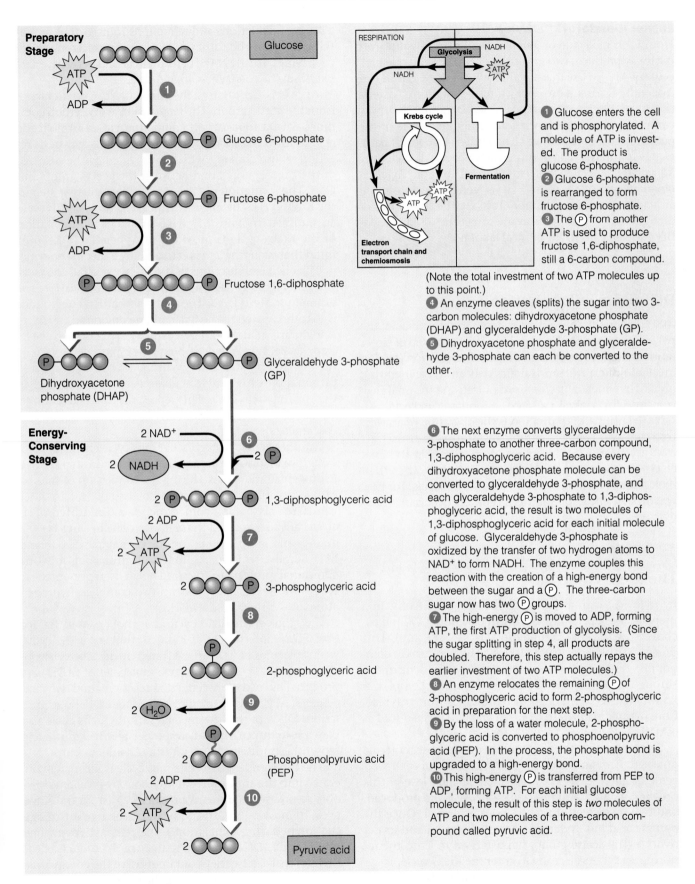

Preparatory Stage

Glucose

RESPIRATION

Glycolysis NADH

NADH ATP

Krebs cycle

Fermentation

Electron transport chain and chemiosmosis

ATP ADP

Glucose 6-phosphate

Fructose 6-phosphate

ATP ADP

Fructose 1,6-diphosphate

Dihydroxyacetone phosphate (DHAP)

Glyceraldehyde 3-phosphate (GP)

Energy-Conserving Stage

2 NAD⁺

2 NADH

2 P

2 P

1,3-diphosphoglyceric acid

2 ADP

2 ATP

2 3-phosphoglyceric acid

2 2-phosphoglyceric acid

2 H₂O

2 Phosphoenolpyruvic acid (PEP)

2 ADP

2 ATP

2 Pyruvic acid

1 Glucose enters the cell and is phosphorylated. A molecule of ATP is invested. The product is glucose 6-phosphate.

2 Glucose 6-phosphate is rearranged to form fructose 6-phosphate.

3 The Ⓟ from another ATP is used to produce fructose 1,6-diphosphate, still a 6-carbon compound. (Note the total investment of two ATP molecules up to this point.)

4 An enzyme cleaves (splits) the sugar into two 3-carbon molecules: dihydroxyacetone phosphate (DHAP) and glyceraldehyde 3-phosphate (GP).

5 Dihydroxyacetone phosphate and glyceraldehyde 3-phosphate can each be converted to the other.

6 The next enzyme converts glyceraldehyde 3-phosphate to another three-carbon compound, 1,3-diphosphoglyceric acid. Because every dihydroxyacetone phosphate molecule can be converted to glyceraldehyde 3-phosphate, and each glyceraldehyde 3-phosphate to 1,3-diphosphoglyceric acid, the result is two molecules of 1,3-diphosphoglyceric acid for each initial molecule of glucose. Glyceraldehyde 3-phosphate is oxidized by the transfer of two hydrogen atoms to NAD⁺ to form NADH. The enzyme couples this reaction with the creation of a high-energy bond between the sugar and a Ⓟ. The three-carbon sugar now has two Ⓟ groups.

7 The high-energy Ⓟ is moved to ADP, forming ATP, the first ATP production of glycolysis. (Since the sugar splitting in step 4, all products are doubled. Therefore, this step actually repays the earlier investment of two ATP molecules.)

8 An enzyme relocates the remaining Ⓟ of 3-phosphoglyceric acid to form 2-phosphoglyceric acid in preparation for the next step.

9 By the loss of a water molecule, 2-phosphoglyceric acid is converted to phosphoenolpyruvic acid (PEP). In the process, the phosphate bond is upgraded to a high-energy bond.

10 This high-energy Ⓟ is transferred from PEP to ADP, forming ATP. For each initial glucose molecule, the result of this step is *two* molecules of ATP and two molecules of a three-carbon compound called pyruvic acid.

FIGURE 5.12 Outline of the reactions of glycolysis. The inset indicates the relationship of glycolysis to the overall processes of respiration and fermentation. A more detailed version of glycolysis is presented in Appendix D.

Entner–Doudoroff Pathway

From each molecule of glucose, the **Entner–Doudoroff pathway** produces two molecules of NADPH and one molecule of ATP for use in cellular biosynthetic reactions (see Appendix D for a more detailed representation). Bacteria that have the enzymes for the Entner–Doudoroff pathway can metabolize glucose without either glycolysis or the pentose phosphate pathway. The Entner–Doudoroff pathway is found in some gram-negative bacteria, including *Rhizobium, Pseudomonas* (sū-dō-mō'nas), and *Agrobacterium* (ag-rō-bak-ti'rē-um); it is generally not found among gram-positive bacteria. Tests for the ability to oxidize glucose by this pathway are sometimes used to identify *Pseudomonas* in the clinical laboratory.

Definition of Respiration

After glucose has been broken down to pyruvic acid, the pyruvic acid can be channeled into the next step of either fermentation (to be described later) or respiration (see Figure 15.11). **Respiration** is defined as an ATP-generating process in which molecules are oxidized and the final electron acceptor is (almost always) an inorganic molecule. An essential feature of respiration is the operation of an electron transport chain. In **aerobic respiration** the final electron acceptor is O_2; in **anaerobic respiration** the final electron acceptor is an inorganic molecule other than molecular oxygen or, rarely, an organic molecule. For now we will describe respiration as it typically occurs in an aerobic cell.

The Krebs Cycle

Pyruvic acid cannot enter the Krebs cycle directly. In a preparatory step, it must lose one molecule of CO_2 and become a two-carbon compound (Figure 5.13, top). This process is called **decarboxylation**. The two-carbon compound, called an *acetyl group,* attaches to coenzyme A through a high-energy bond; the resulting complex is known as *acetyl coenzyme A (acetyl CoA).* During this reaction, pyruvic acid is also oxidized and NAD+ is reduced to NADH.

Remember that the oxidation of one glucose molecule produces two molecules of pyruvic acid, so for each molecule of glucose, two molecules of CO_2 are lost in this preparatory step, two molecules of NADH are produced, and two molecules of acetyl CoA are formed. Once the pyruvic acid has undergone decarboxylation and its derivative (the acetyl group) has attached to CoA, the resulting acetyl CoA is ready to enter the Krebs cycle.

The **Krebs cycle,** also called the *tricarboxylic acid (TCA) cycle* or *citric acid cycle,* is a series of biochemical reactions in which the large amount of potential chemical energy stored in acetyl CoA is released step by step. In this cycle, a series of oxidations and reductions transfer that potential energy, in the form of electrons, to electron carrier coenzymes, chiefly NAD+. The pyruvic acid derivatives are oxidized; the coenzymes are reduced.

As acetyl CoA enters the Krebs cycle, CoA detaches from the acetyl group. The two-carbon acetyl group combines with a four-carbon compound called oxaloacetic acid to form the six-carbon citric acid. This synthesis reaction requires energy, which is provided by the cleavage of the high-energy bond between the acetyl group and CoA. The formation of citric acid is thus the first step in the Krebs cycle. The major chemical reactions of this cycle are outlined in Figure 5.13; a more detailed representation of the Krebs cycle is provided in Appendix D. Keep in mind that each reaction is catalyzed by a specific enzyme.

The chemical reactions of the Krebs cycle fall into several general categories; one of these is decarboxylation. For example, in step ③ isocitric acid, a six-carbon compound, is decarboxylated to the five-carbon compound called α-ketoglutaric acid. Another decarboxylation takes place in step ④. Because one decarboxylation has taken place in the preparatory step and two in the Krebs cycle, all three carbon atoms in pyruvic acid are eventually released as CO_2 by the Krebs cycle. This represents the conversion to CO_2 of all six carbon atoms contained in the original glucose molecule.

Another general category of Krebs cycle chemical reactions is oxidation–reduction. For example, in step ③, two hydrogen atoms are lost during the conversion of the six-carbon isocitric acid to a five-carbon compound. In other words, the six-carbon compound is oxidized. Hydrogen atoms are also released in the Krebs cycle in steps ④, ⑥, and ⑧ and are picked up by the coenzymes NAD+ and FAD. Because NAD+ picks up two electrons but only one additional proton, its reduced form is represented as NADH; however, FAD picks up two complete hydrogen atoms and is reduced to $FADH_2$.

If we look at the Krebs cycle as a whole, we see that for every two molecules of acetyl CoA that enter the cycle, four molecules of CO_2 are liberated by decarboxylation, six molecules of NADH and two molecules of $FADH_2$ are produced by oxidation–reduction reactions, and two molecules of ATP are generated by substrate-level phosphorylation. Many of the intermediates in the Krebs cycle also play a role in other pathways, especially in amino acid biosynthesis (discussed later in this chapter).

The CO_2 produced in the Krebs cycle is ultimately liberated into the atmosphere as a gaseous by-product of aerobic respiration. (Humans produce CO_2 from the Krebs cycle in most cells of the body and discharge it through the lungs during exhalation.) The reduced coenzymes NADH and $FADH_2$ are the most important products of the Krebs cycle because they contain most of the energy originally stored in glucose. During the next phase of respiration, a series of reductions indirectly transfers the energy stored in those coenzymes to ATP. These reactions are collectively called the electron transport chain.

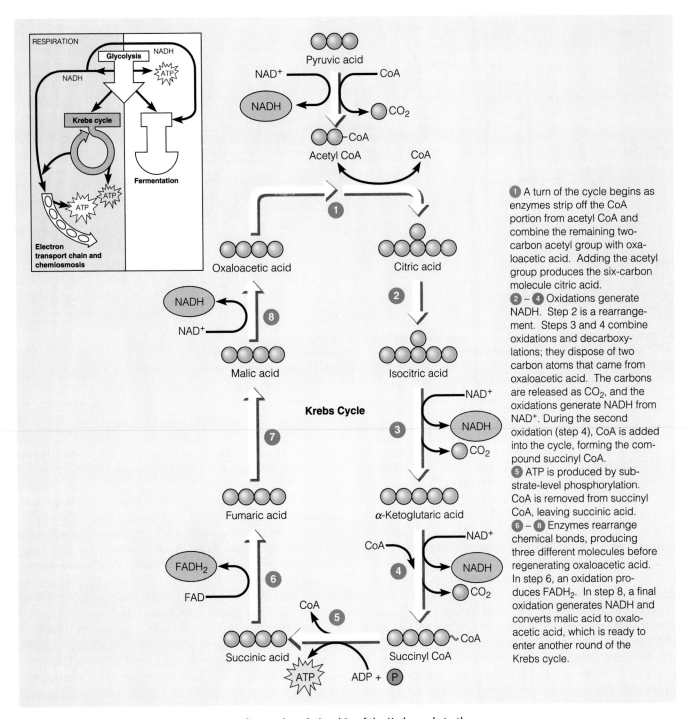

① A turn of the cycle begins as enzymes strip off the CoA portion from acetyl CoA and combine the remaining two-carbon acetyl group with oxaloacetic acid. Adding the acetyl group produces the six-carbon molecule citric acid.

② – ④ Oxidations generate NADH. Step 2 is a rearrangement. Steps 3 and 4 combine oxidations and decarboxylations; they dispose of two carbon atoms that came from oxaloacetic acid. The carbons are released as CO_2, and the oxidations generate NADH from NAD^+. During the second oxidation (step 4), CoA is added into the cycle, forming the compound succinyl CoA.

⑤ ATP is produced by substrate-level phosphorylation. CoA is removed from succinyl CoA, leaving succinic acid.

⑥ – ⑧ Enzymes rearrange chemical bonds, producing three different molecules before regenerating oxaloacetic acid. In step 6, an oxidation produces $FADH_2$. In step 8, a final oxidation generates NADH and converts malic acid to oxaloacetic acid, which is ready to enter another round of the Krebs cycle.

FIGURE 5.13 The Krebs cycle. The inset indicates the relationship of the Krebs cycle to the overall process of respiration.

Electron Transport Chain

An **electron transport chain** consists of a sequence of carrier molecules that are capable of oxidation and reduction. As electrons are passed through the chain, there is a stepwise release of energy, which is used to drive the chemiosmotic generation of ATP, to be described shortly. The final oxidation is irreversible. In eucaryotic cells, the electron transport chain is contained in the inner membrane of mitochondria; in procaryotic cells it is found in the plasma membrane.

There are three classes of carrier molecules in electron transport chains. The first are **flavoproteins.** These proteins contain flavin, a coenzyme derived from riboflavin (vitamin B_2), and are capable of performing alternating oxidations and reductions. One important flavin coenzyme is flavin mononucleotide (FMN). The second class of carrier molecules are **cytochromes**, proteins with an iron-

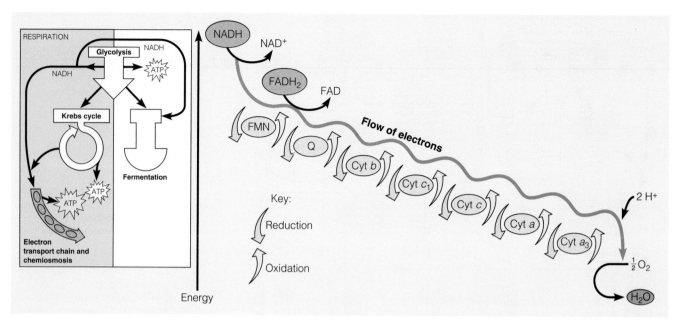

FIGURE 5.14   **An electron transport chain.** The inset indicates the relationship of the electron transport chain to the overall process of respiration. In the mitochondrial electron transport chain, the electrons pass along the chain in a gradual and stepwise fashion, so energy is released in manageable quantities. See Figure 5.16 to see where ATP is formed.

containing group (heme) capable of existing alternately as a reduced form (Fe^{2+}) and an oxidized form (Fe^{3+}). The cytochromes involved in electron transport chains include cytochrome b (cyt b), cytochrome c_1 (cyt c_1), cytochrome c (cyt c), cytochrome a (cyt a), and cytochrome a_3 (cyt a_3). The third class is known as **ubiquinones** (or *coenzyme* Q), symbolized Q; these are small nonprotein carriers.

The electron transport chains of bacteria are somewhat diverse, in that the particular carriers used by a bacterium and the order in which they act may differ from those of other bacteria and from those of eucaryotic mitochondrial systems. Even a single bacterium may have several types of electron transport chains. However, keep in mind that all electron transport chains achieve the same basic goal, that of releasing energy as electrons are transferred from higher-energy compounds to lower-energy compounds. Much is known about the electron transport chain in the mitochondria of eucaryotic cells, so this is the chain we will describe.

The first step in the mitochondrial electron transport chain involves the transfer of high-energy electrons from NADH to FMN, the first carrier in the chain (Figure 5.14). This transfer actually involves the passage of a hydrogen atom with two electrons to FMN, which then picks up an additional H^+ from the surrounding aqueous medium. As a result of the first transfer, NADH is oxidized to NAD^+, and FMN is reduced to $FMNH_2$. In the second step in the electron transport chain, $FMNH_2$ passes $2H^+$ to the other side of the mitochondrial membrane (see Figure 5.16) and passes two electrons to Q. As a result, $FMNH_2$ is oxidized

to FMN. Q also picks up an additional $2H^+$ from the surrounding aqueous medium and releases it on the other side of the membrane.

The next part of the electron transport chain involves the cytochromes. Electrons are passed successively from Q to cyt b, cyt c_1, cyt c, cyt a, and cyt a_3. Each cytochrome in the chain is reduced as it picks up electrons and is oxidized as it gives up electrons. The last cytochrome, cyt a_3, passes its electrons to molecular oxygen (O_2), which becomes negatively charged and then picks up protons from the surrounding medium to form H_2O.

Note that Figure 5.14 shows $FADH_2$, which is derived from the Krebs cycle, as another source of electrons. However, $FADH_2$ adds its electrons to the electron transport chain at a lower level than NADH. Because of this, the electron transport chain produces about one-third less energy for ATP generation when $FADH_2$ donates electrons than when NADH is involved.

An important feature of the electron transport chain is the presence of some carriers, such as FMN and Q, that accept and release protons as well as electrons, and other carriers, such as cytochromes, that transfer electrons only. Electron flow down the chain is accompanied at several points by the active transport (pumping) of protons from the matrix side of the inner mitochondrial membrane to the opposite side of the membrane. The result is a buildup of protons on one side of the membrane. Just as water behind a dam stores energy that can be used to generate electricity, this buildup of protons provides energy for the generation of ATP by the chemiosmotic mechanism.

FIGURE 5.15 Chemiosmosis.
An overview of the mechanism of chemiosmosis in the mitochondrial membrane. The numbered steps are described in the text.

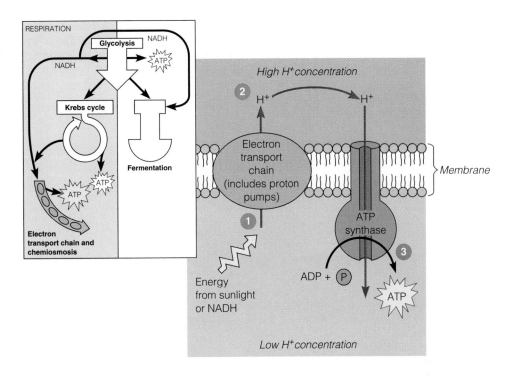

Chemiosmotic Mechanism of ATP Generation

The chemiosmotic mechanism of ATP synthesis—**chemiosmosis**—was first proposed by British biochemist Peter Mitchell in 1961. Although it took a number of years to gain full acceptance, its description is now recognized as a major milestone in the history of biochemistry.

What is chemiosmosis, and how does it drive the energy-requiring synthesis of ATP from ADP? To understand the answers to these questions, we need to recall several concepts that were introduced in Chapter 4 as part of the section Movement of Materials Across Membranes (pp. 84–87). Recall that substances diffuse passively across membranes from regions of high concentration to regions of low concentration; this diffusion yields energy. Recall also that the movement of substances *against* such a concentration gradient *requires* energy and that, in such an active transport of molecules or ions across biological membranes, the required energy is usually provided by ATP. In chemiosmosis, the energy released when a substance moves along a gradient is used to *synthesize* ATP. The "substance" in this case refers to protons. In cellular respiration, chemiosmosis is responsible for most of the ATP that is generated. The steps of chemiosmosis are as follows (Figure 5.15):

1. As energetic electrons from NADH (or sunlight) pass down the electron transport chain, some of the carriers in the chain pump (actively transport) protons across the membrane. Such carrier molecules are called *proton pumps*.

2. The phospholipid membrane is normally impermeable to protons, so this one-directional pumping establishes a proton gradient (a difference in the concentrations of protons on the two sides of the membrane). In addition to a concentration gradient, there is an electrical charge gradient. The excess H^+ on one side of the membrane makes that side positively charged compared with the other side. The resulting electrochemical gradient has potential energy, called the *proton motive force*.

3. The protons on the side of the membrane with the higher proton concentration can diffuse across the membrane only through special protein channels that contain an enzyme called *adenosine triphosphatase (ATP synthase)*. When this flow occurs, energy is released and is used by the enzyme to synthesize ATP from ADP and ℗.

Figure 5.16 shows in detail how the electron transport chain operates in eucaryotes to drive the chemiosmotic mechanism. Within the mitochondrial membrane, the carriers of the electron transport chain are organized into three complexes, with Q transporting electrons between the first and second complex, and cyt *c* transporting them between the second and third complex. Three components of the system pump protons: the first and second complexes and Q. At the end of the chain, electrons join with protons and oxygen (O_2) in the matrix fluid to form water (H_2O). Thus, O_2 is the final electron acceptor.

Both procaryotic and eucaryotic cells use the chemiosmotic mechanism to generate energy for ATP production.

FIGURE 5.16   Electron transport and the chemiosmotic generation of ATP. In the mitochondrial membrane, electron carriers are organized into three complexes, and protons (H^+) are pumped across the membrane at three points. In a eucaryotic cell, they are pumped from the matrix side of the mitochondrial membrane to the opposite side. The flow of electrons is shown in magenta.

However, in eucaryotic cells, the inner mitochondrial membrane contains the electron transport carriers and ATP synthase, whereas in most procaryotes, the plasma membrane does so. An electron transport chain also operates in photophosphorylation and is located in the thylakoid membrane of cyanobacteria and of eucaryotic chloroplasts.

Summary of Aerobic Respiration

The various electron transfers in the electron transport chain generate about 34 molecules of ATP from each molecule of glucose oxidized: approximately three from each of the ten molecules of NADH (a total of 30) and approximately two from each of the two molecules of $FADH_2$ (a total of 4). To arrive at the total number of ATP molecules generated for each molecule of glucose, the 34 from chemiosmosis are added to those generated by oxidation in glycolysis and the Krebs cycle. In aerobic respiration among procaryotes, a total of 38 molecules of ATP can be generated from one molecule of glucose. Note that four of

those ATPs come from substrate-level phosphorylation in glycolysis and the Krebs cycle. Table 5.3 provides a detailed accounting of the ATP yield during procaryotic aerobic respiration.

Aerobic respiration among eucaryotes produces a total of only 36 molecules of ATP. There are fewer ATPs than in procaryotes because some energy is lost when electrons are shuttled across the mitochondrial membranes that separate glycolysis (in the cytoplasm) from the electron transport chain. No such separation exists in procaryotes. We can now summarize the overall reaction for aerobic respiration in procaryotes as follows:

$$C_6H_{12}O_6 + 6O_2 + 38ADP + 38\textcircled{P} \longrightarrow$$
Glucose   Oxygen

$$6CO_2 + 6H_2O + 38ATP$$
Carbon   Water
dioxide

A summary of the various stages of aerobic respiration in procaryotes is presented in Figure 5.17.

TABLE 5.3   Summary of ATP Produced During Procaryotic Aerobic Respiration of One Glucose Molecule

SOURCE		ATP YIELD (METHOD)
Glycolysis		
1. Oxidation of glucose to pyruvic acid		2 ATP (substrate-level phosphorylation)
2. Production of 2NADH		6 ATP (oxidation phosphorylation in electron transport chain)
Preparatory Step		
1. Formation of acetyl CoA produces 2NADH		6 ATP (oxidative phosphorylation in electron transport chain)
Krebs Cycle		
1. Oxidation of succinyl CoA to succinic acid		2 GTP (equivalent of ATP; substrate-level phosphorylation)
2. Production of 6NADH		18 ATP (oxidative phosphorylation in electron transport chain)
3. Production of 2FADH		4 ATP (oxidative phosphorylation in electron transport chain)
		Total: 38 ATP

Electron transport chain and chemiosmosis

FIGURE 5.17 Summary of aerobic respiration in procaryotes. Glucose is broken down completely to carbon dioxide and water, and ATP is generated. This process has three major phases: glycolysis, the Krebs cycle, and the electron transport chain. The key event in the process is that electrons are picked up from intermediates of glycolysis and the Krebs cycle by NAD$^+$ or FAD and are carried by NADH or FADH$_2$ to the electron transport chain. NADH is also produced in the conversion of pyruvic acid to acetyl CoA. Most of the ATP generated by aerobic respiration is made by the chemiosmotic mechanism during the electron transport chain phase; this is called oxidative phosphorylation.

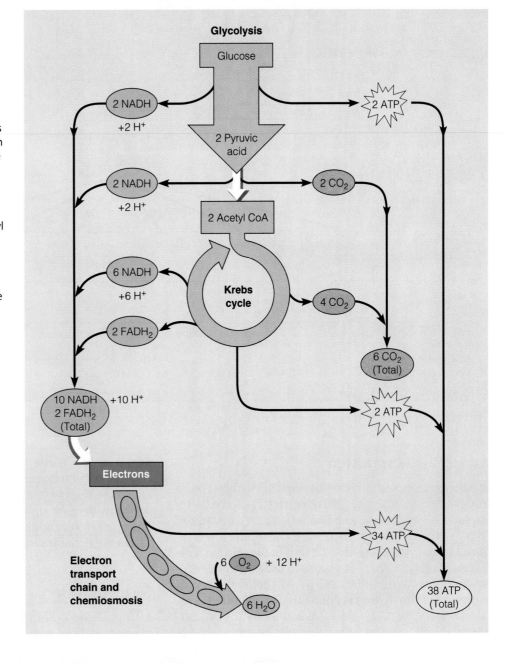

FIGURE 5.18 Fermentation.
The inset indicates the relationship of fermentation to the overall energy-producing processes. **(a)** Overview of fermentation. The first step is glycolysis, the conversion of glucose to pyruvic acid. In the second step, the reduced coenzymes from glycolysis or its alternatives (NADH, NADPH) donate their electrons and hydrogen ions to pyruvic acid or a derivative to form a fermentation end-product. **(b)** End-products of various microbial fermentations.

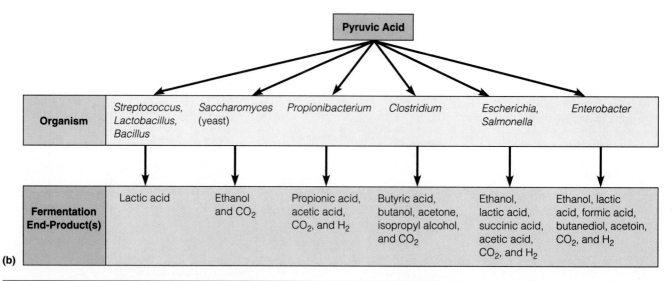

(b)

Anaerobic Respiration

In anaerobic respiration, the final electron acceptor is an inorganic substance other than oxygen (O_2). Some bacteria, such as *Pseudomonas* and *Bacillus,* can use a nitrate ion (NO_3^-) as a final electron acceptor; the nitrate ion is reduced to a nitrite ion (NO_2^-), nitrous oxide (N_2O), or nitrogen gas (N_2). Other bacteria, such as *Desulfovibrio* (dē-sul-fō-vib′rē-ō), use sulfate (SO_4^{2-}) as the final electron acceptor to form hydrogen sulfide (H_2S). Still other bacteria use carbonate (CO_3^{2-}) to form methane (CH_4). Anaerobic respiration by bacteria using nitrate and sulfate as

final acceptors is essential for the nitrogen and sulfur cycles that occur in nature. The amount of ATP generated in anaerobic respiration varies with the organism and the pathway. Because only part of the Krebs cycle operates under anaerobic conditions, the ATP yield is never as high as in aerobic respiration, but it can be nearly as high.

Fermentation

After glucose has been broken down into pyruvic acid, the pyruvic acid can be completely broken down in respiration, as previously described, or can be converted to an organic product in fermentation (see Figure 5.11). **Fermentation** can be defined in several ways (see the box on p. 124), but we will define it here as a process that:

1. Releases energy from sugars or other organic molecules, such as amino acids, organic acids, purines, and pyrimidines.

2. Does not require oxygen (but sometimes can occur in its presence).

3. Does not require use of the Krebs cycle or an electron transport chain.

4. Uses an organic molecule as the final electron acceptor.

5. Produces only small amounts of ATP (only one or two ATP molecules for each molecule of starting material) because much of the original glucose energy remains in the chemical bonds of the organic end-products, such as lactic acid or ethanol.

During fermentation, electrons are transferred (along with protons) from reduced coenzymes (NADH, NADPH) to pyruvic acid or its derivatives (Figure 5.18a). Those final electron acceptors are reduced to the end-products shown in Figure 5.18b. In the process, NAD⁺ and NADP⁺ are regenerated and can enter another round of glycolysis. An essential function of the second stage of fermentation is to ensure a steady supply of NAD^+ and $NADP^+$ so that glycolysis can continue. In fermentation, ATP is generated only during glycolysis.

Various microorganisms can ferment various substrates; the end-products depend on the particular microorganism, the substrate, and the enzymes that are present and active. Chemical analyses of these end-products are useful in identifying microorganisms. We next consider two of the more important processes: lactic acid fermentation and alcohol fermentation.

Lactic Acid Fermentation

During glycolysis, which is the first phase of **lactic acid fermentation**, a molecule of glucose is oxidized to two molecules of pyruvic acid (Figure 5.19; see also Figure 5.12). This oxidation generates the energy that is used to form the two molecules of ATP. In the next step, the two molecules of pyruvic acid are reduced by two molecules of NADH to form two molecules of lactic acid (Figure 5.19a). Because lactic acid is the end-product of the reaction, it undergoes no further oxidation, and most of the energy produced by the reaction remains stored in the lactic acid. Thus, this fermentation yields only a small amount of energy. The overall reaction is:

Glucose + 2ADP + 2℗ ⟶ 2 Lactic acid + 2ATP

Two important genera of lactic acid bacteria are *Streptococcus* and *Lactobacillus* (lak-tō-bä-sil′lus). Because these microbes produce only lactic acid, they are referred to as

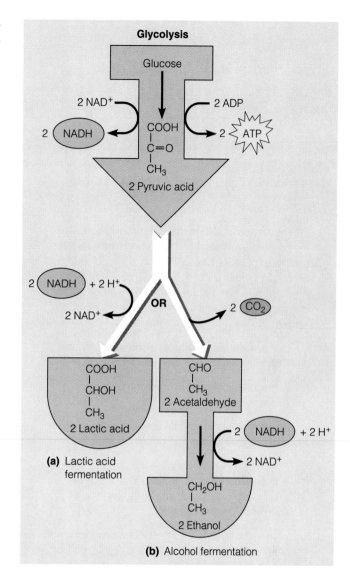

FIGURE 5.19 Types of fermentation.

homolactic (or *homofermentative*). Lactic acid fermentation can result in food spoilage. However, the process can also produce yogurt from milk, sauerkraut from fresh cabbage, and pickles from cucumbers.

Alcohol Fermentation

Alcohol fermentation also begins with the glycolysis of a molecule of glucose to yield two molecules of pyruvic acid and two molecules of ATP. In the next reaction, the two molecules of pyruvic acid are converted to two molecules of acetaldehyde and two molecules of CO_2 (Figure 5.19b). The two molecules of acetaldehyde are next reduced by two molecules of NADH to form two molecules of ethanol. Again, alcohol fermentation is a low–energy-yield process because most of the energy contained in the initial glucose molecule remains in the ethanol, the end-product.

Alcohol fermentation is carried out by a number of bacteria and yeasts. The ethanol and carbon dioxide produced by the yeast *Saccharomyces* (sak-ä-rō-mī′sēs) are waste products for yeast cells but are useful to humans.

Applications of Microbiology

WHAT IS FERMENTATION?

TO MANY PEOPLE, fermentation simply means the production of alcohol: grains and fruits are fermented to produce beer and wine. If a food soured, you might say it was "off" or fermented. Here are some definitions of fermentation. They range from informal, general usage to more scientific definitions. Fermentation is:

1. Any process that produces alcoholic beverages or acidic dairy products (general use).

2. Any spoilage of food by microorganisms (general use).

3. Any large-scale microbial process occurring with or without air (common definition used in industry).

4. Any energy-releasing metabolic process that takes place only under anaerobic conditions (becoming more scientific).

5. All metabolic processes that release energy from a sugar or other organic molecule, do not require oxygen or an electron transport system, and use an organic molecule as the final electron acceptor. (This is the definition we will use in this book.)

Ethanol made by yeasts is the alcohol in alcoholic beverages, and carbon dioxide made by yeasts causes bread dough to rise (see the box in Chapter 1, p. 6).

Organisms that produce lactic acid as well as other acids or alcohols are known as **heterolactic** (or *heterofermentative*) and often use the pentose phosphate pathway. The overall reaction for a typical heterolactic fermentation is:

Glucose + ADP + $\textcircled{P}$ $\longrightarrow$

Lactic acid + Ethanol + CO_2 + ATP

Table 5.4 lists some of the various microbial fermentations used by industry to convert inexpensive raw materials into useful end-products. A summary of aerobic respiration, anaerobic respiration, and fermentation is given in Table 5.5 (p. 126).

Lipid Catabolism

Our discussion of energy production has emphasized the oxidation of glucose, the main energy-supplying carbohydrate. However, microbes also oxidize lipids and proteins, and the oxidations of all these nutrients are related.

Fats, as you recall, are lipids consisting of fatty acids and glycerol. Microbes produce extracellular enzymes called **lipases** that break fats down into their fatty acid and glycerol components. Each component is then metabolized separately.

Many types of microbes can convert glycerol into dihydroxyacetone phosphate, one of the three-carbon intermediates formed during glycolysis (see Figure 5.12). The dihydroxyacetone phosphate is then metabolized by glycolysis and the Krebs cycle, as shown in Figure 5.20.

Fatty acids are catabolized somewhat differently. The mechanism of fatty acid oxidation is called **beta oxidation**. In this process, carbon fragments from a long chain of fatty acid are removed two at a time and acetyl CoA is formed from them. As the molecules of acetyl CoA form, they enter the Krebs cycle, as do the acetyl CoA molecules formed by the oxidation of pyruvic acid (see Figure 5.13). Thus, the Krebs cycle functions in the oxidation of glycerol, fatty acids, and glucose and is the stage at which all three pathways converge. Beta oxidation of petroleum is discussed in the box in Chapter 2, page 31.

Protein Catabolism

Proteins are too large to pass unaided through plasma membranes. Microbes produce extracellular *proteases* and *peptidases,* which break down proteins into their component amino acids, which can cross the membrane. However, before amino acids can be catabolized, they must be converted to substances that can enter the Krebs cycle. In one such conversion, called **deamination**, the amino group of an amino acid is removed and converted to an ammonium ion (NH_4^+), which can be excreted from the cell. The remaining organic acid can enter the Krebs cycle. Other conversions involve **decarboxylation** (removal of —COOH) and **dehydrogenation**. These conversions are complex; just remember that amino acids can be made ready to enter the Krebs cycle by various methods.

A summary of the interrelation of carbohydrate, lipid, and protein catabolism is shown in Figure 5.21.

PHOTOSYNTHESIS

LEARNING OBJECTIVES

- *Compare and contrast cyclic and noncyclic photophosphorylation.*
- *Compare and contrast the light and dark reactions of photosynthesis.*
- *Compare and contrast oxidative phosphorylation and photophosphorylation.*

In all of the metabolic pathways just discussed, organisms obtain energy for cellular work by oxidizing organic compounds. But where do organisms obtain these organic compounds? Some, including animals and many microbes, feed on matter produced by other organisms. For example, bacteria may catabolize compounds from dead plants and animals or may obtain nourishment from a living host.

Other organisms synthesize complex organic compounds from simple inorganic substances. The major mechanism for such synthesis is a process called **photosynthesis**, which is used by plants and many microbes. Essentially, photosynthesis is the conversion of light energy from the sun into chemical energy. The chemical energy is then used to convert CO_2 from the atmosphere to more reduced carbon compounds, primarily sugars. The word *photosynthesis* summarizes the process: *photo* means light, and *synthesis* refers to the assembly of organic compounds. This synthesis of sugars by using carbons from CO_2 gas is also called **carbon fixation**. Continuation of life as we know it on Earth depends on the recycling of carbon in this way. Cyanobacteria, algae, and green plants all contribute to this vital recycling with photosynthesis.

Photosynthesis can be summarized as follows:

$$6CO_2 + 12H_2O + \text{Light energy} \longrightarrow$$

$$C_6H_{12}O_6 + 6O_2 + 6H_2O$$

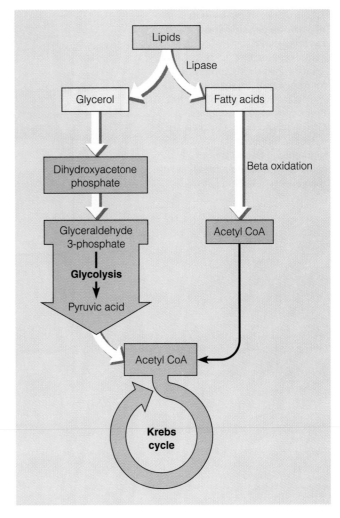

FIGURE 5.20 Lipid catabolism.

TABLE 5.4 Some Industrial Uses for Different Types of Fermentations

FERMENTATION END-PRODUCT(S)	INDUSTRIAL OR COMMERCIAL USE	STARTING MATERIAL	MICROORGANISM
Ethanol	Beer	Malt extract	*Saccharomyces cerevisiae* (yeast, a fungus)
	Wine	Grape or other fruit juices	*Saccharomyces ellipsoideus* (yeast, a fungus)
	Fuel	Agricultural wastes	*Saccharomyces cerevisiae*
Acetic acid	Vinegar	Ethanol	*Acetobacter* (bacterium)
Lactic acid	Cheese, yogurt	Milk	*Lactobacillus, Streptococcus* (bacteria)
	Rye bread	Grain, sugar	*Lactobacillus bulgaricus* (bacterium)
	Sauerkraut	Cabbage	*Lactobacillus plantarum* (bacterium)
	Summer sausage	Meat	*Pediococcus* (bacterium)
Propionic acid and carbon dioxide	Swiss cheese	Milk	*Propionibacterium freudenreichii* (bacterium)
Acetone and butanol	Pharmaceutical, industrial uses	Molasses	*Clostridium acetobutylicum* (bacterium)
Glycerol	Pharmaceutical, industrial uses	Molasses	*Saccharomyces cerevisiae*
Citric acid	Flavoring	Molasses	*Aspergillus* (fungus)
Methane	Fuel	Acetic acid	*Methanosarcina* (bacterium)
Sorbose	Vitamin C (ascorbic acid)	Sorbitol	*Acetobacter* (bacterium)

FIGURE 5.21 Catabolism of various organic food molecules. Proteins, carbohydrates, and lipids can all be sources of electrons and protons for respiration. These food molecules enter glycolysis or the Krebs cycle at various points. Glycolysis and the Krebs cycle are catabolic funnels through which high-energy electrons from all kinds of organic molecules flow on their energy-releasing pathways.

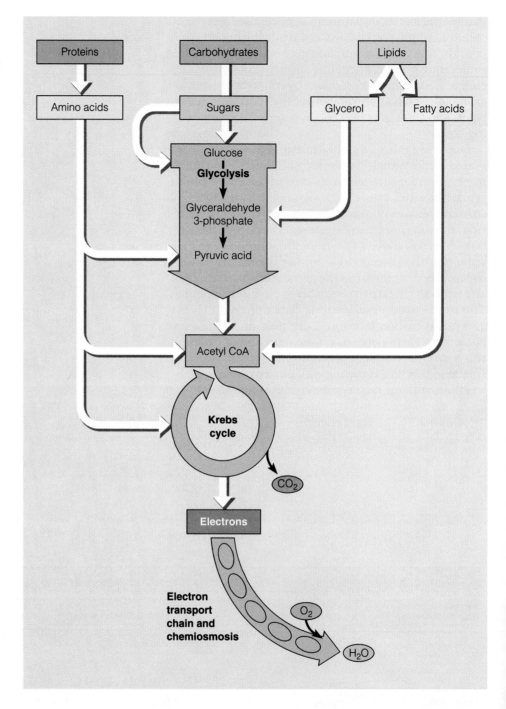

TABLE 5.5   Comparison of Aerobic Respiration, Anaerobic Respiration, and Fermentation

ENERGY-PRODUCING PROCESS	GROWTH CONDITIONS	FINAL HYDROGEN (ELECTRON) ACCEPTOR	TYPE OF PHOSPHORYLATION USED TO GENERATE ATP	ATP MOLECULES PRODUCED PER GLUCOSE MOLECULE
Aerobic respiration	Aerobic	Molecular oxygen (O_2)	Substrate-level and oxidative	36 or 38*
Anaerobic respiration	Anaerobic	Usually an inorganic substance (such as NO_3^-, SO_4^{2-}, or CO_3^{2-}), but not molecular oxygen (O_2)	Substrate-level and oxidative	Variable (fewer than 38 but more than 2)
Fermentation	Aerobic or anaerobic	An organic molecule	Substrate-level	2

*In procaryotic aerobic respiration, 38 ATP molecules are produced; in eucaryotic aerobic respiration, 36 ATP molecules are produced.

FIGURE 5.22
Photophosphorylation. (a) In cyclic photophosphorylation, electrons released from chlorophyll by light return to chlorophyll after passage along the electron transport chain. The energy from electron transfer is converted to ATP. **(b)** In noncyclic photophosphorylation, electrons released from chlorophyll are replaced by electrons from water. The chlorophyll electrons are passed along the electron transport chain to the electron acceptor NADP⁺. NADP⁺ combines with electrons and with hydrogen ions from water, forming NADPH.

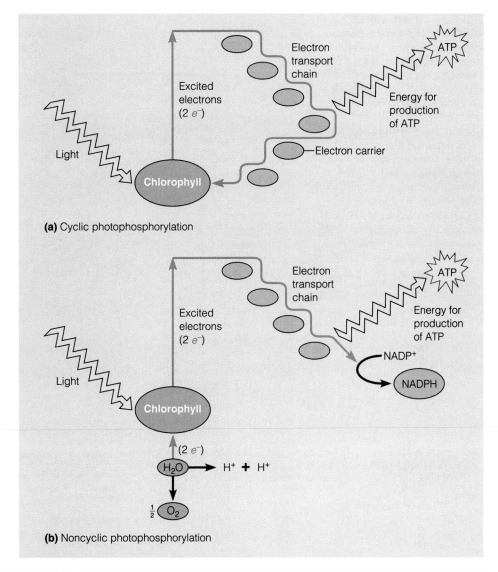

(a) Cyclic photophosphorylation

(b) Noncyclic photophosphorylation

The Light Reactions: Photophosphorylation

Photophosphorylation was introduced earlier in this chapter as one of the three ways that ATP is formed. Recall that

In the course of photosynthesis, electrons are taken from the hydrogen atoms of water, an energy-poor molecule, and incorporated into sugar, an energy-rich molecule. The energy boost is supplied by light energy, although indirectly.

Photosynthesis takes place in two stages. In the first stage, called the **light reactions**, light energy is used in a process that converts ADP and Ⓟ to ATP. In addition, in the predominant form of the light reactions, the electron carrier NADP is reduced to NADPH. The coenzyme NADPH, like NADH, is an energy-rich carrier of electrons. In the second stage, called the **dark (light-independent) reactions**, these electrons are used along with energy from ATP to reduce CO_2 to sugar.

photophosphorylation occurs only in photosynthetic cells. In this mechanism, light energy is absorbed by chlorophyll molecules in the photosynthetic cell, exciting some of the molecules' electrons. The chlorophyll principally used by green plants, algae, and cyanobacteria is *chlorophyll a*. It is located in the membranous thylakoids of chloroplasts in algae and green plants (see Figure 4.26) and in the thylakoids found in the photosynthetic structures of cyanobacteria. Other bacteria use *bacteriochlorophylls*.

The excited electrons jump from the chlorophyll to the first of a series of carrier molecules, an electron transport chain similar to that used in respiration. As electrons are passed along the series of carriers, protons are pumped across the membrane and ADP is converted to ATP by chemiosmosis. In **cyclic photophosphorylation** the electron eventually returns to chlorophyll (Figure 5.22a). In **noncyclic photophosphorylation**, which is the more common process, the electrons released from chlorophyll do not return to chlorophyll but become incorporated into NADPH (Figure 5.22b). The electrons lost from chlorophyll are replaced by electrons from water (H_2O) or

FIGURE 5.23 Simplified version of the Calvin–Benson cycle. This diagram shows three turns of the cycle, in which three molecules of CO_2 are fixed and one molecule of glyceraldehyde phosphate is produced and leaves the cycle. Two molecules of glyceraldehyde 3-phosphate are needed to make one molecule of glucose. Therefore, the cycle must turn six times for each glucose molecule produced, requiring a total investment of 6 molecules of CO_2, 18 molecules of ATP, and 12 molecules of NADPH. A more detailed version of this cycle is presented in Appendix D.

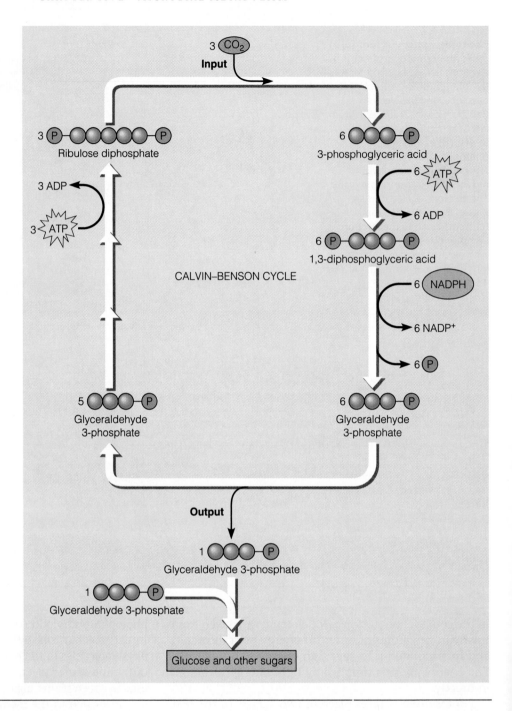

another oxidizable compound, such as hydrogen sulfide (H_2S). To summarize: the products of noncyclic photophosphorylation are ATP (formed by chemiosmosis using energy released in an electron transport chain), O_2 (from water molecules), and NADPH (in which the hydrogen electrons and protons were derived ultimately from water).

The Dark Reactions: Calvin–Benson Cycle

The dark reactions are so called because no light is directly required for them to occur. They include a complex cyclic pathway called the **Calvin–Benson cycle**, in which CO_2 is "fixed"—that is, used to synthesize sugars (Figure 5.23; see also Appendix D). For the net synthesis of one molecule of a three-carbon sugar called glyceraldehyde 3-phosphate, the cycle of reactions must occur three times, fixing three molecules of CO_2 (note that this sugar is also a key intermediate in glycolysis). These three turns of the cycle require the input of nine molecules of ATP and six molecules of NADPH. To make glucose, a six-carbon sugar, three more turns of the cycle are required. Thus, to make one molecule of glucose, the cycle must turn six times and requires an investment of $6CO_2$, 18ATP, and 12NADPH.

SUMMARY OF ENERGY PRODUCTION MECHANISMS

LEARNING OBJECTIVE

• *Categorize the various nutritional patterns among organisms according to carbon source and mechanisms of carbohydrate catabolism and ATP generation.*

In the living world, energy passes from the sun to photosynthetic organisms and then to other organisms in the form of the potential energy contained in the bonds of chemical compounds. To obtain energy in a usable form, a cell must have an electron (or hydrogen) donor, which serves as an initial energy source within the cell. Electron donors can be as diverse as photosynthetic pigments, glucose or other organic compounds, elemental sulfur, ammonia, or hydrogen gas (Figure 5.24a). Next, electrons removed from the chemical energy sources are transferred to electron carriers, such as the coenzymes NAD^+, $NADP^+$, and FAD (Figure 5.24b). This transfer is an oxidation–reduction reaction; the initial energy source is oxidized as this first electron carrier is reduced. During this phase, some ATP is produced. In the third stage, electrons are transferred from electron carriers to their final electron acceptors in further oxidation–reduction reactions (Figure 5.24c), producing more ATP.

In aerobic respiration, oxygen (O_2) serves as the final electron acceptor. In anaerobic respiration, inorganic substances other than oxygen, such as nitrate ions (NO_3^-) or sulfate ions (SO_4^{2-}), serve as the final electron acceptors. In fermentation, organic compounds serve as the final electron acceptors. In aerobic and anaerobic respiration, a series of electron carriers called an electron transport chain releases energy that is used by the mechanism of chemiosmosis to synthesize ATP. Regardless of their energy sources, all organisms use similar oxidation–reduction reactions to transfer electrons and similar mechanisms to use the energy released to produce ATP.

Nutritional Patterns Among Organisms

We have looked in detail at some of the energy-generating metabolic pathways that are used by animals and plants, as well as by many microbes. Microbes are distinguished by their great metabolic diversity, however, and some can sustain themselves on inorganic substances by using pathways that are unavailable to either plants or animals. All organisms, including microbes, can be classified metabolically according to their *nutritional pattern*—that is, their source of energy and their source of carbon.

First considering the energy source, we can generally classify organisms as phototrophs or chemotrophs. **Phototrophs** use light as their primary energy source, whereas **chemotrophs** depend on oxidation–reduction reactions of inorganic or organic compounds for energy. For their

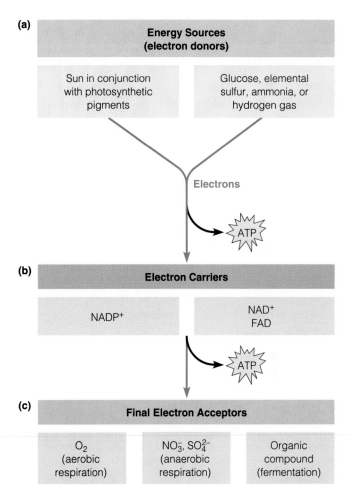

(a) **Energy Sources (electron donors)**

Sun in conjunction with photosynthetic pigments

Glucose, elemental sulfur, ammonia, or hydrogen gas

Electrons

ATP

(b) **Electron Carriers**

$NADP^+$

NAD^+
FAD

ATP

(c) **Final Electron Acceptors**

O_2 (aerobic respiration)

NO_3^-, SO_4^{2-} (anaerobic respiration)

Organic compound (fermentation)

FIGURE 5.24 ATP production. The production of ATP requires **(a)** an energy source (electron donor), **(b)** the transfer of electrons to an electron carrier during an oxidation–reduction reaction, and **(c)** the transfer of electrons to a final electron acceptor.

principal carbon source, **autotrophs** ("self-feeders") use carbon dioxide, and **heterotrophs** (feeders on others) require an organic carbon source. Autotrophs are also referred to as *lithotrophs* (rock eating), and heterotrophs are referred to as *organotrophs*.

If we combine the energy and carbon sources, we derive the following nutritional classifications for organisms: *photoautotrophs, photoheterotrophs, chemoautotrophs,* and *chemoheterotrophs*. This nutritional classification of microbes is summarized in Table 5.6. Almost all of the medically important microorganisms discussed in this book are chemoheterotrophs. Typically, infectious organisms catabolize substances obtained from the host.

Photoautotrophs

Photoautotrophs use light as a source of energy and carbon dioxide as their chief source of carbon. They include photosynthetic bacteria (green sulfur and purple sulfur

TABLE 5.6 Nutritional Classification of Organisms

NUTRITIONAL TYPE	ENERGY SOURCE	CARBON SOURCE	EXAMPLES
*Photoautotroph	Light	Carbon dioxide (CO_2)	Photosynthetic bacteria (green sulfur and purple sulfur bacteria), cyanobacteria, algae, plants (some green and purple sulfur bacteria may grow chemotrophically)
*Photoheterotroph	Light	Organic compounds	Purple nonsulfur and green nonsulfur bacteria (some purple nonsulfur bacteria may grow photoautotrophically and chemotrophically)
Chemoautotroph	Usually electrons from inorganic compounds	Carbon dioxide (CO_2)	Hydrogen, sulfur, iron, and nitrifying bacteria
Chemoheterotroph	Usually electrons from organic compounds	Organic compounds	Most bacteria and all fungi, protozoa, and animals

*The electron source for noncyclic photophosphorylation for photosynthetic bacteria is H_2, H_2S, or S; for algae and plants the source is H_2O; for purple nonsulfur and green nonsulfur bacteria, it is organic compounds; for cyanobacteria the source is usually H_2O.

TABLE 5.7 Comparison of Eucaryotic and Selected Procaryotic Photosynthesis

CHARACTERISTIC	EUCARYOTES	PROCARYOTES		
	ALGAE, PLANTS	CYANOBACTERIA	GREEN SULFUR BACTERIA	PURPLE SULFUR BACTERIA
Substance that reduces CO_2	H atoms of H_2O	H atoms of H_2O	Sulfur, sulfur compounds, H_2 gas	Sulfur, sulfur compounds, H_2 gas
Oxygen production	Oxygenic	Oxygenic (and anoxygenic)	Anoxygenic	Anoxygenic
Light-trapping pigment	Chlorophyll *a*	Chlorophyll *a*	Bacteriochlorophyll *a*	Bacteriochlorophyll *a* or *b*
Site of photosynthesis	Chloroplasts with thylakoids	Thylakoids	Chlorosomes	Intracytoplasmic membrane
Environment	Aerobic	Aerobic (and anaerobic)	Anaerobic	Anaerobic

bacteria and cyanobacteria), algae, and green plants. In the photosynthetic reactions of cyanobacteria, algae, and green plants, the hydrogen atoms of water are used to reduce carbon dioxide, and oxygen gas is given off. Because this photosynthetic process produces O_2, it is sometimes called **oxygenic.**

In addition to the cyanobacteria (see Figure 11.20), there are several other families of photosynthetic procaryotes. Each is classified according to the way it reduces CO_2. These bacteria cannot use H_2O to reduce CO_2 and cannot carry on photosynthesis when oxygen is present (they must have an *anaerobic* environment). Consequently, their photosynthetic process does not produce O_2 and is called **anoxygenic.** Two of the families are photoautotrophs: the green sulfur and purple sulfur bacteria. The **green sulfur bacteria,** such as *Chlorobium* (klô-rō'bē-um) use sulfur (S), sulfur compounds (such as hydrogen sulfide, H_2S), or hydrogen gas (H_2) to reduce carbon dioxide and form organic compounds. Applying the energy from light and the appropriate enzymes, these bacteria ox-

idize sulfide (S^{2-}) or sulfur (S) to sulfate (SO_4^{2-}), or hydrogen gas to water (H_2O). The **purple sulfur bacteria,** such as *Chromatium* (krō-mā'tē-um), also use sulfur, sulfur compounds, or hydrogen gas to reduce carbon dioxide. They are distinguished from the green sulfur bacteria by their type of chlorophyll and location of stored sulfur.

The chlorophylls used by these photosynthetic bacteria are called *bacteriochlorophylls,* and they absorb light at longer wavelengths than that absorbed by chlorophyll *a.* Bacteriochlorophylls of green sulfur bacteria are found in vesicles called *chlorosomes* (or *chlorobium vesicles*) underlying and attached to the plasma membrane. In the purple sulfur bacteria, the bacteriochlorophylls are located in invaginations of the plasma membrane (*intracytoplasmic membranes*).

Several characteristics that distinguish eucaryotic photosynthesis from procaryotic photosynthesis are presented in Table 5.7. See the box opposite for a discussion of an exceptional photosynthetic system that exists in *Halobacterium*. The system does not use chlorophyll.

BACTERIA MAKE A FASTER, SMARTER COMPUTER

THE INTERESTING ARCHAEOBACTERIUM *Halobacterium* lives where very little else can grow. This bacterium is found in salt lakes, in salt licks on ranches, in salt flats, or in any other environment with a concentration of salt that is from five to seven times greater than that of the ocean. Halobacteria are easy to detect because they turn their environment purple (see photo). These bacteria cannot ferment carbohydrates and do not contain chlorophyll, so it was assumed that all their energy comes from oxidative phosphorylation. The exciting discovery of a new system of photophosphorylation arose through the study of the plasma membrane of *Halobacterium halobium*.

Researchers found that the plasma membrane of *H. halobium* fragments into two fractions (red and purple) when the cell is broken down and its components are sorted. The red fraction, which comprises most of the membrane, contains cytochromes, flavoproteins, and other parts of the electron transport chain that carries out oxidative phosphorylation. The purple fraction is more interesting. This purple membrane occurs in distinct patches of hexagonal lattices within the plasma membrane. The purple color comes from a protein that makes up 75% of the purple membrane. This protein is similar to the retinal pigment in the rod cells of the human eye, rhodopsin, so the protein was named bacteriorhodopsin. At the time it was discovered, its function was not known.

During the 1970s, further study of *H. halobium* suggested some startling explanations. When starved cells in anaerobic, dark environments were exposed to either light or oxygen, ATP synthesis increased. *H. halobium* can grow in the presence of either light or oxygen but cannot grow when neither is present.

This unexpected result suggested that *Halobacterium* can obtain energy by using either of two systems, one that operates in the presence of oxygen (oxidative phosphorylation) and one that operates in the presence of light (some kind of photophosphorylation). The rate of ATP synthesis by *H. halobium* was found to be highest when the cells receive light that is between 550 and 600 nm in wavelength; this range exactly corresponds to the absorption spectrum of bacteriorhodopsin.

Researchers hypothesized that bacteriorhodopsin, like chlorophyll-containing systems, acts as a proton pump to create a proton gradient across a cell membrane; in this case, the gradient is created across the purple membrane. The proton gradient can do cellular work—can drive the synthesis of ATP or transport solutes.

HALOBACTERIUM REPLACES SEMICONDUCTORS

At Syracuse University's Center of Molecular Electronics, Robert Birge grows *Halobacterium* for five days in 5-liter batches and then extracts bacteriorhodopsin from the cells for a novel use. Birge has developed a computer chip made of a thin layer of bacteriorhodopsin.

Conventional computers store information on thin wafers of silicon. Computers process information by "reading" a series of zeros and ones produced as electrons flow through switches etched in the silicon. Electrons passing through a switch represents a one; a switch that stops the electron flow represents a zero. However, silicon can't hold enough information or process information fast enough for such applications as artificial intelligence or robot vision.

In contrast, the bacteriorhodopsin chip will be able to store more information than a silicon chip and process the information faster, more like a human brain. The bacteriorhodopsin chip works with light, which of course moves at the speed of light, much faster than the flow of electrons. Green light causes the protein to fold; a folded protein is read as a one, whereas an unfolded protein represents a zero. Laser light is used to "see" the configuration of the protein.

At present, the protein chip needs to be stored at −4°C to maintain its structure, but Birge and his coworkers are hopeful they will solve this problem. The military is already apparently using the protein chips in their combat planes; if such a plane crashes, the cooling system will go off, thus destroying the chip and keeping classified information from being stolen. Eventually, these smaller, faster, and higher-capacity chips will probably make it possible to develop computers that perform functions closer to human intelligence, such as acting as eyes for blind people.

Halobacterium (purple color) grows in the high-salt concentration of solar evaporation ponds used for manufacturing salt around San Francisco Bay.

Photoheterotrophs

Photoheterotrophs use light as a source of energy but cannot convert carbon dioxide to sugar; rather, they use organic compounds, such as alcohols, fatty acids, other organic acids, and carbohydrates, as sources of carbon. They are anoxygenic. Among the photoheterotrophs are the **green nonsulfur bacteria**, such as *Chloroflexus* (klô-rō-flex′us), and **purple nonsulfur bacteria**, such as *Rhodopseudomonas* (rō-dō-sū-dō-mō′nas).

Chemoautotrophs

Chemoautotrophs use the electrons from reduced inorganic compounds as a source of energy, and they use CO_2 as their principal source of carbon (see Figure 27.1, the carbon cycle). Inorganic sources of energy for these organisms include hydrogen sulfide (H_2S) for *Beggiatoa* (bej-jē-ä-tō′ä); elemental sulfur (S) for *Thiobacillus thiooxidans;* ammonia (NH_3) for *Nitrosomonas* (nī-trō-sō-mō′näs); nitrite ions (NO_2^-) for *Nitrobacter* (nī-trō-bak′tèr); hydrogen gas (H_2) for *Hydrogenomonas* (hī-drō-je-nō-mō′näs); and ferrous iron (Fe^{2+}) for *Thiobacillus ferrooxidans*. The energy derived from the oxidation of these inorganic compounds is eventually stored in ATP, which is produced by oxidative phosphorylation.

Chemoheterotrophs

When we discuss photoautotrophs, photoheterotrophs, and chemoautotrophs, it is easy to categorize the energy source and carbon source because they occur as separate entities. However, in chemoheterotrophs, the distinction is not so clear because the energy source and carbon source are usually the same organic compound—glucose, for example. **Chemoheterotrophs** specifically use the electrons from hydrogen atoms in organic compounds as their energy source.

Heterotrophs are further classified according to their source of organic molecules: **saprophytes** live on dead organic matter, and **parasites** derive nutrients from a living host. Most bacteria, and all fungi, protozoa, and animals, are chemoheterotrophs.

We will next consider how cells use ATP pathways for the synthesis of organic compounds such as carbohydrates, lipids, proteins, and nucleic acids.

BIOCHEMICAL PATHWAYS OF ENERGY USE (ANABOLISM)

LEARNING OBJECTIVE

- *Describe the major types of anabolism and their relationship to catabolism.*

Up to now we have been considering energy production. Through the oxidation of organic molecules, organisms produce energy by aerobic respiration, anaerobic respiration, and fermentation. Much of this energy is given off as heat. The complete metabolic oxidation of glucose to carbon dioxide and water is considered a very efficient process, but about 45% of the energy of glucose is lost as heat. Cells use the remaining energy, which is trapped in the bonds of ATP, in a variety of ways. Microbes use ATP to provide energy for the transport of substances across plasma membranes—the process called active transport that we discussed in Chapter 4. Microbes also use some of their energy for flagellar motion (also discussed in Chapter 4). Most of the ATP, however, is used in the production of new cellular components. This production is a continuous process in cells, and, in general, is faster in procaryotic cells than in eucaryotic cells.

Autotrophs build their organic compounds by fixing carbon dioxide in the Calvin–Benson cycle (see Figure 5.23). This requires both energy (ATP) and electrons (from the oxidation of NADPH). Heterotrophs, by contrast, must have a ready source of organic compounds for biosynthesis—the production of needed cellular components, usually from simpler molecules. The cells use these compounds as both the carbon source and the energy source. We will next consider the biosynthesis of a few representative classes of biological molecules: carbohydrates, lipids, amino acids, purines, and pyrimidines. As we do so, keep in mind that synthesis reactions require a net input of energy.

Biosynthesis of Polysaccharides

Microorganisms synthesize sugars and polysaccharides. The carbon atoms required to synthesize glucose are derived from the intermediates produced during processes such as glycolysis and the Krebs cycle and from lipids or amino acids. After synthesizing glucose (or other simple sugars), bacteria may assemble it into the more complex polysaccharides. One such example is glycogen. For bacteria to build glucose into glycogen, glucose units must be phosphorylated and linked. The product of glucose phosphorylation is glucose 6-phosphate. Such a process involves the expenditure of energy, usually in the form of ATP, and dehydration (loss of water). The key intermediate in the synthesis of glycogen in bacteria is called *adenosine diphosphoglucose (ADPG)*, which is formed when a molecule of ATP is added to glucose 6-phosphate (Figure 5.25). Once the building block is synthesized, it is linked with similar units through dehydration synthesis to form glycogen.

Using a nucleotide called uridine triphosphate (UTP) as a source of energy, animals synthesize glycogen and many other carbohydrates from *uridine diphosphoglucose, UDPG* (see Figure 5.25). A compound related to UDPG, called *UDP-N-acetylglucosamine (UDPNAc)*, is a key starting material in the biosynthesis of peptidoglycan, the substance that forms bacterial cell walls. UDPNAc is formed from fructose 6-phosphate, and the reaction also uses UTP.

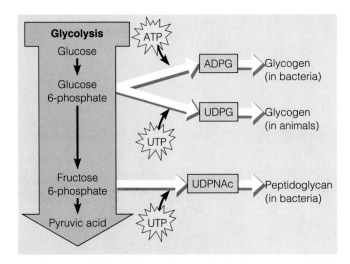

FIGURE 5.25   Biosynthesis of polysaccharides.

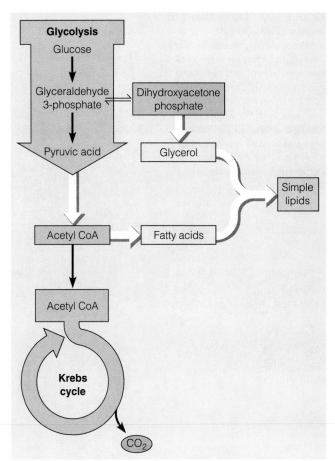

FIGURE 5.26   Biosynthesis of simple lipids.

Biosynthesis of Lipids

Because lipids vary in chemical composition, they are synthesized by a variety of routes. Cells synthesize fats by joining glycerol and fatty acids. The glycerol portion of the fat is derived from dihydroxyacetone phosphate, an intermediate formed during glycolysis. Fatty acids, which are long-chain hydrocarbons (hydrogen linked to carbon), are built up when two-carbon fragments of acetyl CoA are successively added to each other (Figure 5.26). As with polysaccharide synthesis, the building units of fats and other lipids are linked via dehydration reactions that require energy, not always in the form of ATP. In the synthesis of phospholipids, for example, the energy is derived from the hydrolysis of a nucleotide called cytidine triphosphate (CTP).

As mentioned previously, the most important role of lipids is as structural components of biological membranes, and most membrane lipids are phospholipids. A lipid of a very different structure, cholesterol, is also found in plasma membranes of eucaryotic cells. Waxes are lipids that are important components of the cell wall of acid-fast bacteria. Other lipids, such as carotenoids, provide the red, orange, and yellow pigments of some microorganisms. Some lipids form portions of chlorophyll molecules. Lipids also function in energy storage. Recall that the breakdown products of lipids after biological oxidation feed into the Krebs cycle.

Biosynthesis of Amino Acids and Proteins

Amino acids are required for protein biosynthesis. Some microbes, such as *E. coli,* contain the enzymes necessary to use starting materials, such as glucose and inorganic salts, for the synthesis of all the amino acids they need. Organisms with the necessary enzymes can synthesize all amino acids directly or indirectly from intermediates of carbohydrate metabolism (Figure 5.27a). Other microbes require that the environment provide some preformed amino acids.

One important source of the *precursors* (intermediates) used in amino acid synthesis is the Krebs cycle. Adding an amine group to pyruvic acid or to an appropriate organic acid of the Krebs cycle converts the acid into an amino acid. This process is called *amination*. If the amine group comes from a preexisting amino acid, the process is called *transamination* (Figure 5.27b). Other precursors for amino acid biosynthesis are derived from the pentose phosphate pathway and from the Entner–Doudoroff pathway.

Most amino acids within cells are destined to be building blocks for protein synthesis. Proteins play major roles in the cell as enzymes, structural components, and toxins, to name just a few uses. The joining of amino acids to form proteins involves dehydration synthesis and requires energy in the form of ATP. The mechanism of protein synthesis involves genes and is discussed in Chapter 8.

FIGURE 5.27 Biosynthesis of amino acids. (a) Pathways of amino acid biosynthesis through amination or transamination of intermediates of carbohydrate metabolism. **(b)** Transamination, a process by which new amino acids are made with the amine groups from old amino acids. Glutamic acid and aspartic acid are both amino acids; the other two compounds are intermediates in the Krebs cycle.

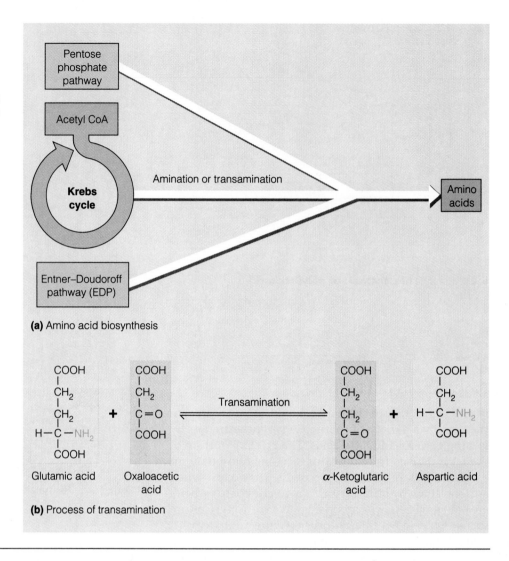

(a) Amino acid biosynthesis

(b) Process of transamination

Biosynthesis of Purines and Pyrimidines

As you may recall from Chapter 2, the informational molecules DNA and RNA consist of repeating units called *nucleotides,* each of which consists of a purine or pyrimidine, a pentose (five-carbon sugar), and a phosphate group. The five-carbon sugars of nucleotides are derived from either the pentose phosphate pathway or the Entner–Doudoroff pathway. Certain amino acids—aspartic acid, glycine, and glutamine—made from intermediates produced during glycolysis and in the Krebs cycle participate in the biosyntheses of purines and pyrimidines (Figure 5.28). The carbon and nitrogen atoms derived from these amino acids form the purine and pyrimidine rings, and the energy for synthesis is provided by ATP. DNA contains all the information necessary to determine the specific structures and functions of cells. Both RNA and DNA are required for pro-

tein synthesis. In addition, such nucleotides as ATP, NAD[+], and NADP[+] assume roles in stimulating and inhibiting the rate of cellular metabolism. The synthesis of DNA and RNA from nucleotides is discussed in Chapter 8.

INTEGRATION OF METABOLISM

LEARNING OBJECTIVE

• *Define amphibolic pathways.*

We have seen thus far that the metabolic processes of microbes produce energy from light, inorganic compounds, and organic compounds. Reactions also occur in which

FIGURE 5.28 Biosynthesis of purine and pyrimidine nucleotides.

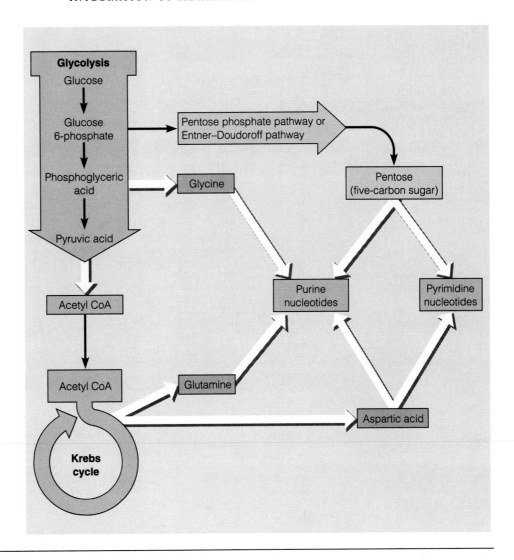

energy is used for biosynthesis. With such a variety of activity, you might imagine that anabolic and catabolic reactions occur independently of each other in space and time. Actually, anabolic and catabolic reactions are joined through a group of common intermediates (identified as key intermediates in Figure 5.29). Both anabolic and catabolic reactions also share some biochemical pathways, such as the Krebs cycle. For example, reactions in the Krebs cycle not only participate in the oxidation of glucose but also produce intermediates that can be converted to amino acids. Biochemical pathways that function in both anabolism and catabolism are called **amphibolic pathways**, meaning that they are dual-purpose.

Amphibolic pathways bridge the reactions that lead to the breakdown and synthesis of carbohydrates, lipids, proteins, and nucleotides. Such pathways enable simultaneous reactions to occur in which the breakdown product formed in one reaction is used in another reaction to synthesize a different compound, and vice versa. Because various intermediates are common to both anabolic and catabolic reactions, mechanisms exist that regulate synthesis and breakdown pathways and allow these reactions to occur simultaneously. One such mechanism involves the use of different coenzymes for opposite pathways. For example, NAD^+ is involved in catabolic reactions, whereas $NADP^+$ is involved in anabolic reactions. Enzymes can also coordinate anabolic and catabolic reactions by accelerating or inhibiting the rates of biochemical reactions. The energy stores of a cell can also affect the rates of biochemical reactions. For example, if ATP begins to accumulate, an enzyme shuts down glycolysis; this control helps to synchronize the rates of glycolysis and the Krebs cycle. Thus, if citric acid consumption increases, either because of a demand for more ATP or because anabolic pathways are draining off intermediates of the citric acid cycle, glycolysis accelerates and meets the demand.

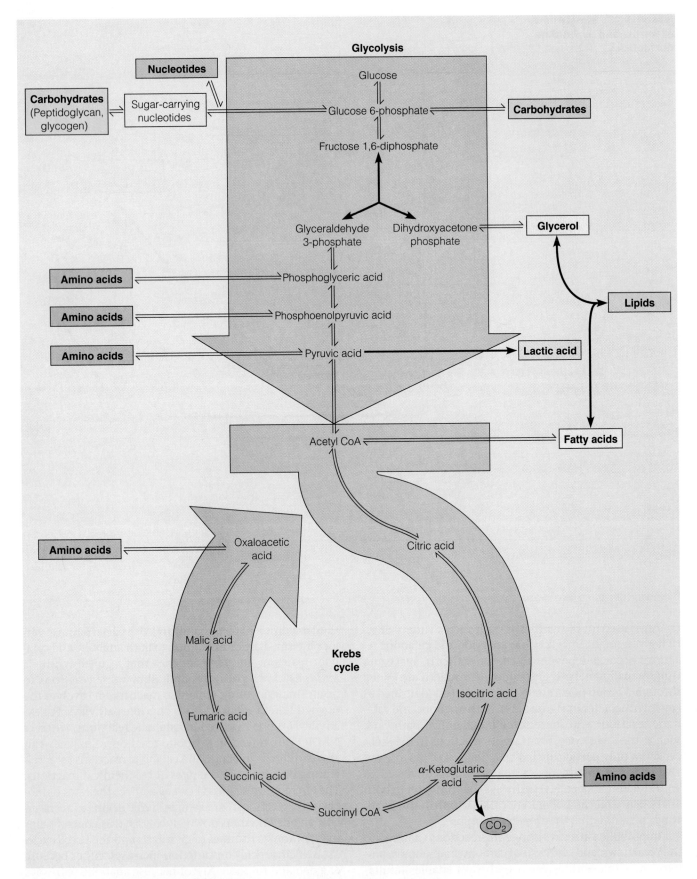

FIGURE 5.29 Integration of metabolism. Key intermediates are shown. Although not indicated in the figure, amino acids and ribose are used in the synthesis of purine and pyrimidine nucleotides (see Figure 5.28). The double arrows indicate amphibolic pathways.

CATABOLIC AND ANABOLIC REACTIONS
(pp. 103–104)

1. The sum of all chemical reactions within a living organism is known as metabolism.
2. Catabolism refers to chemical reactions that result in the breakdown of more complex organic molecules into simpler substances. Catabolic reactions usually release energy.
3. Anabolism refers to chemical reactions in which simpler substances are combined to form more complex molecules. Anabolic reactions usually require energy.
4. The energy of catabolic reactions is used to drive anabolic reactions.
5. The energy for chemical reactions is stored in ATP.

ENZYMES (pp. 104–110)

1. Enzymes are proteins, produced by living cells, that catalyze chemical reactions.
2. Enzymes are generally globular proteins with characteristic three-dimensional shapes.
3. Enzymes catalyze chemical reactions by lowering the activation energy.
4. Enzymes are efficient, are able to operate at relatively low temperatures, and are subject to various cellular controls.

Naming Enzymes (p. 105)

1. Enzymes' names usually end in -ase.
2. The six classes of enzymes are defined on the basis of the types of reactions they catalyze.

Enzyme Components (pp. 105–106)

1. Most enzymes are holoenzymes, consisting of a protein portion (apoenzyme) and a nonprotein portion (cofactor).
2. The cofactor can be a metal ion (iron, copper, magnesium, manganese, zinc, calcium, or cobalt) or a complex organic molecule known as a coenzyme (NAD, NADP, FMN, FAD, and coenzyme A).

Mechanism of Enzymatic Action (p. 106)

1. When an enzyme and substrate combine, the substrate is transformed, and the enzyme is recovered.
2. Enzymes are characterized by specificity, which is a function of their active sites.

Factors Influencing Enzymatic Activity (pp. 107–109)

1. At high temperatures, enzymes undergo denaturation and lose their catalytic properties; at low temperatures, the reaction rate decreases.
2. The pH at which enzymatic activity is maximal is known as the optimum pH.

3. Within limits, enzymatic activity increases as substrate concentration increases.
4. Competitive inhibitors compete with the normal substrate for the active site of the enzyme. Noncompetitive inhibitors act on other parts of the apoenzyme or on the cofactor and decrease the enzyme's ability to combine with the normal substrate.

Feedback Inhibition (pp. 109–110)

1. Feedback inhibition occurs when the end-product of a pathway inhibits an enzyme's activity in the pathway.

ENERGY PRODUCTION (pp. 110–112)

Oxidation–Reduction (p. 111)

1. Oxidation is the removal of one or more electrons from a substrate. Protons (H^+) are often removed with the electrons.
2. Reduction of a substrate refers to its gain of one or more electrons.
3. Each time a substance is oxidized, another is simultaneously reduced.
4. NAD^+ is the oxidized form; NADH is the reduced form.
5. Glucose is a reduced molecule; energy is released during a cell's oxidation of glucose.

Generation of ATP (p. 112)

1. Energy released during certain metabolic reactions can be trapped to form ATP from ADP and Ⓟ (phosphate). Addition of a Ⓟ to a molecule is called phosphorylation.
2. During substrate-level phosphorylation, a high-energy Ⓟ from an intermediate in catabolism is added to ADP.
3. During oxidative phosphorylation, energy released as electrons are passed to a series of electron acceptors (an electron transport chain) and finally to O_2 or another inorganic compound.
4. During photophosphorylation, energy from light is trapped by chlorophyll, and electrons are passed through a series of electron acceptors. The electron transfer releases energy used for the synthesis of ATP.

BIOCHEMICAL PATHWAYS OF ENERGY PRODUCTION (p. 112)

1. A series of enzymatically catalyzed chemical reactions called biochemical pathways store energy in and release energy from organic molecules.

CARBOHYDRATE CATABOLISM (pp. 113–124)

1. Most of a cell's energy is produced from the oxidation of carbohydrates.

2. Glucose is the most commonly used carbohydrate.
3. The two major types of glucose catabolism are respiration, in which glucose is completely broken down, and fermentation, in which it is partially broken down.

Glycolysis (p. 114)

1. The most common pathway for the oxidation of glucose is glycolysis. Pyruvic acid is the end-product.
2. Two ATP and two NADH molecules are produced from one glucose molecule.

Alternatives to Glycolysis (pp. 114–116)

1. The pentose phosphate pathway is used to metabolize five-carbon sugars; one ATP and 12 NADPH molecules are produced from one glucose molecule.
2. The Entner–Doudoroff pathway yields one ATP and two NADPH molecules from one glucose molecule.

Definition of Respiration (p. 116)

1. During respiration, organic molecules are oxidized. Energy is generated from the electron transport chain.
2. In aerobic respiration, O_2 functions as the final electron acceptor.
3. In anaerobic respiration, the final electron acceptor is an inorganic molecule other than O_2.

The Krebs Cycle (p. 116)

1. Decarboxylation of pyruvic acid produces one CO_2 molecule and one acetyl group.
2. Two-carbon acetyl groups are oxidized in the Krebs cycle. Electrons are picked up by NAD^+ and FAD for the electron transport chain.
3. From one molecule of glucose, oxidation produces six molecules of NADH, two molecules of $FADH_2$, and two molecules of ATP.
4. Decarboxylation produces six molecules of CO_2.

Electron Transport Chain (pp. 117–118)

1. Electrons are brought to the electron transport chain by NADH.
2. The electron transport chain consists of carriers, including flavoproteins, cytochromes, and ubiquinones.

Chemiosmotic Mechanism of ATP Generation
(pp. 119–120)

1. Protons being pumped across the membrane generate a proton motive force as electrons move through a series of acceptors or carriers.
2. Energy produced from movement of the protons back across the membrane is used by ATP synthase to make ATP from ADP and Ⓟ.
3. In eucaryotes, electron carriers are located on the inner mitochondrial membrane; in procaryotes, electron carriers are in the plasma membrane.

Summary of Aerobic Respiration (p. 120)

1. In aerobic procaryotes, 38 ATP molecules can be produced from complete oxidation of a glucose molecule in glycolysis, the Krebs cycle, and the electron transport chain.
2. In eucaryotes, 36 ATP molecules are produced from complete oxidation of a glucose molecule.

Anaerobic Respiration (p. 122)

1. The final electron acceptors in anaerobic respiration include NO_3^-, SO_4^{2-}, and CO_3^{2-}.
2. The total ATP yield is less than aerobic respiration because only part of the Krebs cycle operates under anaerobic conditions.

Fermentation (pp. 122–124)

1. Fermentation releases energy from sugars or other organic molecules by oxidation.
2. O_2 is not required in fermentation.
3. Two ATP molecules are produced by substrate-level phosphorylation.
4. Electrons removed from the substrate reduce NAD^+.
5. The final electron acceptor is an organic molecule.
6. In lactic acid fermentation, pyruvic acid is reduced by NADH to lactic acid.
7. In alcohol fermentation, acetaldehyde is reduced by NADH to produce ethanol.
8. Heterolactic fermenters can use the pentose phosphate pathway to produce lactic acid and ethanol.

Lipid Catabolism (p. 124)

1. Lipases hydrolyze lipids into glycerol and fatty acids.
2. Glycerol is catabolized by conversion to dihydroxyacetone phosphate, and fatty acids are catabolized by beta oxidation.
3. Catabolic products can be further broken down in glycolysis and the Krebs cycle.

Protein Catabolism (p. 124)

1. Before amino acids can be catabolized, they must be converted to various substances that enter the Krebs cycle.
2. Transamination, decarboxylation, and dehydrogenation reactions convert the amino acids to be catabolized.

PHOTOSYNTHESIS (pp. 124–128)

1. Photosynthesis is the conversion of light energy from the sun into chemical energy; the chemical energy is used for carbon fixation.

The Light Reactions: Photophosphorylation
(pp. 127–128)

1. Chlorophyll *a* is used by green plants, algae, and cyanobacteria; it is found in thylakoid membranes.
2. Electrons from chlorophyll pass through an electron transport chain, from which ATP is produced by chemiosmosis.
3. In cyclic photophosphorylation, the electrons return to the chlorophyll.

4. In noncyclic photophosphorylation, the electrons are used to reduce NADP, and electrons are returned to chlorophyll from H_2O or H_2S.
5. When H_2O is oxidized by green plants, algae, and cyanobacteria, O_2 is produced.

The Dark Reactions: Calvin–Benson Cycle
(p. 128)
1. CO_2 is used to synthesize sugars.
2. Synthesis of glyceraldehyde 3-phosphate requires three molecules of CO_2, nine molecules of ATP, and six molecules of NADPH.

SUMMARY OF ENERGY PRODUCTION MECHANISMS (pp. 129–132)
1. Sunlight is converted to chemical energy in oxidation–reduction reactions carried on by phototrophs. Chemotrophs can use this chemical energy.
2. In oxidation–reduction reactions, energy is derived from the transfer of electrons.
3. To produce energy, a cell needs an electron donor (organic or inorganic), a system of electron carriers, and a final electron acceptor (organic or inorganic).

Nutritional Patterns Among Organisms (pp. 129–132)
1. Photoautotrophs obtain energy by photophosphorylation and fix carbon from CO_2 via the Calvin–Benson cycle to synthesize organic compounds.
2. Cyanobacteria are oxygenic phototrophs. Green sulfur bacteria and purple sulfur bacteria are anoxygenic phototrophs.
3. Photoheterotrophs use light as an energy source and an organic compound for their carbon source or electron donor.
4. Chemoautotrophs use inorganic compounds as their energy source and carbon dioxide as their carbon source.
5. Chemoheterotrophs use complex organic molecules as their carbon and energy sources.

BIOCHEMICAL PATHWAYS OF ENERGY USE (ANABOLISM) (pp. 132–134)

Biosynthesis of Polysaccharides (p. 132)
1. Glycogen is formed from ADPG.
2. UDPNAc is the starting material for the biosynthesis of peptidoglycan.

Biosynthesis of Lipids (p. 133)
1. Lipids are synthesized from fatty acids and glycerol.
2. Glycerol is derived from dihydroxyacetone phosphate, and fatty acids are built from acetyl CoA.

Biosynthesis of Amino Acids and Proteins (p. 133)
1. Amino acids are required for protein biosynthesis.
2. All amino acids can be synthesized either directly or indirectly from intermediates of carbohydrate metabolism, particularly from the Krebs cycle.

Biosynthesis of Purines and Pyrimidines (p. 134)
1. The sugars composing nucleotides are derived from either the pentose phosphate pathway or the Entner–Doudoroff pathway.
2. Carbon and nitrogen atoms from certain amino acids form the backbones of the purines and pyrimidines.

INTEGRATION OF METABOLISM (pp. 134–135)
1. Anabolic and catabolic reactions are integrated through a group of common intermediates.
2. Such integrated pathways are referred to as amphibolic pathways.

STUDY QUESTIONS

Review

1. Define metabolism.
2. Distinguish between catabolism and anabolism. How are these processes related?
3. Using the diagrams opposite, show
   (a) Where the substrate will bind.
   (b) Where the competitive inhibitor will bind.
   (c) Where the noncompetitive inhibitor will bind.
   (d) Which of the four elements could be the inhibitor in feedback inhibition.

4. What will the effect of the reactions in Question 3 be?

Enzyme   Substrate   Competitive inhibitor   Noncompetitive inhibitor

5. Why are most enzymes active at one particular temperature? Why are enzymes less active below this temperature? What happens above this temperature?

6. Which substances in each of the following reactions are being oxidized? Which are being reduced?

$$
\begin{array}{c}
\text{H} \\
| \\
\text{C}{=}\text{O} + \text{NADH} + \text{H}^+ \longrightarrow \text{H}{-}\overset{\displaystyle |}{\underset{\displaystyle |}{\text{C}}}{-}\text{OH} + \text{NAD}^+ \\
| \\
\text{CH}_3 \qquad\qquad\qquad \text{CH}_3
\end{array}
$$

(a) Acetaldehyde   Ethanol

$$
\begin{array}{c}
\text{COOH} \qquad\qquad \text{COOH} \\
| \qquad\qquad\qquad | \\
\text{H}{-}\text{C}{-}\text{H} \qquad\qquad \text{C}{-}\text{H} \\
\qquad\qquad + \text{FAD} \longrightarrow \qquad\quad + \text{FADH}_2 \\
\text{H}{-}\text{C}{-}\text{H} \qquad\qquad \text{C}{-}\text{H} \\
| \qquad\qquad\qquad | \\
\text{COOH} \qquad\qquad \text{COOH}
\end{array}
$$

(b) Succinic acid   Fumaric acid

7. What is the fate of pyruvic acid in an organism that uses respiration? Fermentation?

8. List four compounds that can be made from pyruvic acid by an organism that uses fermentation only.

9. Fill in the table below with the carbon source and energy source of each type of organism.

ORGANISM	CARBON SOURCE	ENERGY SOURCE
Photoautotroph		
Photoheterotroph		
Chemoautotroph		
Chemoheterotroph		

10. There are three mechanisms for the phosphorylation of ADP to produce ATP. Write the name of the mechanism that describes each of the reactions in the following table.

ATP GENERATED BY	REACTION
	An electron, liberated from chlorophyll by light, is passed down an electron transport chain.
	Cytochrome c passes two electrons to cytochrome a.
	$\begin{array}{cc} \text{CH}_2 & \text{CH}_3 \\ \| & \| \\ \text{C}{-}\text{O} {\sim} \text{P} \longrightarrow \text{C}{=}\text{O} \\ \| & \| \\ \text{COOH} & \text{COOH} \\ \text{Phosphoenolpyruvic acid} & \text{Pyruvic acid} \end{array}$

11. Define *oxidation–reduction,* and differentiate between the following terms:
    (a) Aerobic and anaerobic respiration
    (b) Respiration and fermentation
    (c) Cyclic and noncyclic photophosphorylation

12. The pentose phosphate pathway produces only one ATP. List four advantages of this pathway for the cell.

Use the following diagrams for Questions 13 through 21.

(a)

(b)

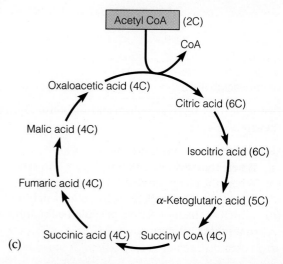

(c)

13. Name the pathways diagrammed in a, b, and c.

14. Show where glycerol is catabolized and where fatty acids are catabolized.

15. Show where the amino acid glutamic acid is catabolized:

$$HOOC-CH_2-CH_2-\overset{\overset{\displaystyle H}{|}}{\underset{\underset{\displaystyle NH_2}{|}}{C}}-COOH$$

16. Show how these pathways are related.

17. Where is ATP required in pathways a and b?

18. Where is CO_2 released in pathways a and b?

19. Show where a long-chain hydrocarbon such as petroleum is catabolized.

20. Where is NADH (or $FADH_2$ or NADPH) used and produced in these pathways?

21. Identify four places where anabolic and catabolic pathways are integrated.

22. Explain how ATP is a key intermediate in metabolism.

23. An enzyme and substrate are combined. The rate of reaction begins as shown in the following graph. To complete the graph, show the effect of increasing substrate concentration on a constant enzyme concentration. Show the effect of increasing temperature.

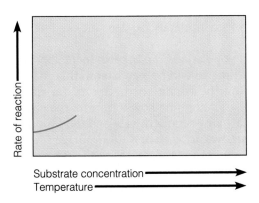

Critical Thinking

1. Write your own definition of the chemiosmotic mechanism of ATP generation. On Figure 5.16, mark the following using the appropriate letter:
   (a) The acidic side of the membrane
   (b) The side with a positive electrical charge
   (c) Potential energy
   (d) Kinetic energy

2. Explain why, even under ideal conditions, *Streptococcus* grows slowly.

3. Why must NADH be reoxidized? How does this happen in an organism that uses respiration? Fermentation?

4. The following graph shows the normal rate of reaction of an enzyme and its substrate (black) and the rate when an excess of competitive inhibitor is present (red). Explain why the graph appears as it does.

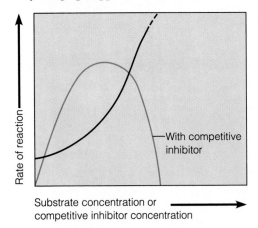

5. Compare and contrast carbohydrate catabolism and energy production in the following bacteria:
   (a) *Pseudomonas*, an aerobic chemoheterotroph
   (b) *Spirulina*, an oxygenic photoautotroph
   (c) *Ectothiorhodospira*, an anoxygenic photoautotroph

6. How much ATP could be obtained from the complete oxidation of one molecule of glucose? From one molecule of butterfat containing one glycerol and three 12-carbon chains?

7. The chemoautotroph *Thiobacillus* can obtain energy from oxidation of arsenic ($As^{3+} \rightarrow As^{5+}$). How does this reaction provide energy? How can this bacterium be put to use by humans?

Clinical Applications

1. *Hemophilus influenzae* requires hemin (X factor) to synthesize cytochromes and NAD^+ (V factor) from other cells. For what does it use these two growth factors? What diseases does *H. influenzae* cause?

2. The drug HIVID®, also called ddC, inhibits DNA synthesis. It is used to treat HIV infection and AIDS. Compare the following illustration to Figure 2.19. How does this drug work?

3. The bacterial enzyme streptokinase is used to digest fibrin (blood clots) in patients with atherosclerosis. Why doesn't injection of streptokinase cause a streptococcal infection? How do we know the streptokinase will digest fibrin only and not good tissues?

Microbial Growth

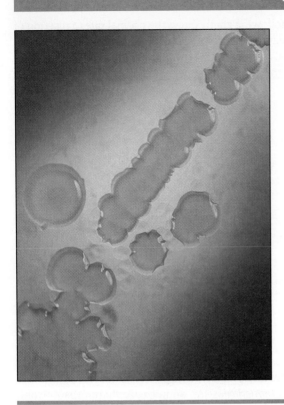

WHEN WE TALK about microbial growth, we are really referring to the *number* of cells, not the *size* of the cells. Microbes that are "growing" are increasing in number, accumulating into *clumps* of hundreds, *colonies* (accumulations of cells large enough to be seen without a microscope) of hundreds of thousands, or *populations* of billions. For the most part, we are not concerned with the growth of individual cells. Although individual cells approximately double in size during their lifetime, this change is not very significant compared with the size increases observed during the lifetime of plants and animals.

REQUIREMENTS FOR GROWTH

LEARNING OBJECTIVES

- *Classify microbes into five groups on the basis of preferred temperature range.*
- *Identify how and why the pH of culture media is controlled.*
- *Explain the importance of osmotic pressure to microbial growth.*
- *Provide a use for each of the four elements (carbon, nitrogen, sulfur, and phosphorus) needed in large amounts for microbial growth.*
- *Explain how microbes are classified on the basis of oxygen requirements.*
- *Identify ways in which aerobes avoid damage by toxic forms of oxygen.*

The requirements for microbial growth can be divided into two main categories, physical and chemical. Physical aspects include temperature, pH, and osmotic pressure. Chemical requirements include water, sources of carbon and nitrogen, minerals, oxygen, and organic growth factors.

Microbial populations can become incredibly large in a very short time, as will be seen later in this chapter. By understanding the conditions necessary for microbial growth, we can determine how to control the growth of microbes that cause serious disease and food spoilage. At the same time, we can learn how to encourage the growth of helpful microbes and those we wish to study. The manager of a sewage treatment plant and the manager of a brewery may not seem to have much in common, but both would be interested in promoting rapid microbial activity.

We will examine in this chapter the physical and chemical requirements for microbial growth, the various kinds of culture media, bacterial division, the phases of microbial growth, and the methods of measuring microbial growth. In later chapters we will see how unwanted microbial growth can be controlled.

Physical Requirements

Temperature

Most microorganisms grow well at the temperatures favored by humans. However, certain bacteria are capable of growing in extreme cold or extreme heat, temperatures

FIGURE 6.1 Typical growth rate of different types of microorganisms in response to temperature. The optimum growth (fastest reproduction) is represented by the peak of the curve. Note that the reproductive rate drops off very quickly at temperatures only a little above the optimum. At either extreme of the temperature range, the reproductive rate is much lower than the rate at the optimum temperature.

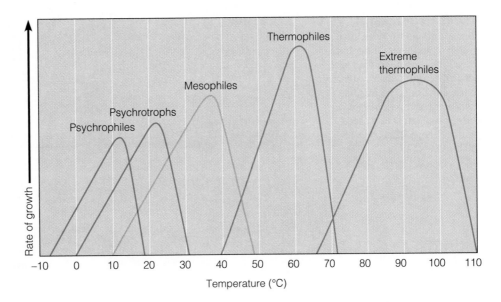

that would certainly hinder the survival of most eucaryotic organisms.

Microorganisms are divided into three primary groups on the basis of their preferred range of temperature: **psychrophiles** (cold-loving microbes), **mesophiles** (moderate-temperature–loving microbes), and **thermophiles** (heat-loving microbes). As we will see, some of these primary groups can be subdivided. Most bacteria grow only within a limited range of temperatures, and their maximum and minimum growth temperatures are about 30°C apart. They grow poorly at some temperatures within their range and well at others. Each bacterial species grows at particular minimum, optimum, and maximum temperatures. The **minimum growth temperature** is the lowest temperature at which the species will grow. The **optimum growth temperature** is the temperature at which the species grows best. The **maximum growth temperature** is the highest temperature at which growth is possible. By graphing the growth response over a temperature range, we can see that the optimum growth temperature is usually near the top of the range; above that temperature, the rate of growth drops off rapidly (Figure 6.1). This presumably happens because the high temperature has inactivated necessary enzymatic systems of the cell.

The ranges and maximum growth temperatures that define bacteria as psychrophiles, mesophiles, or thermophiles are not rigidly defined. Psychrophiles, for example, were originally considered simply to be organisms capable of growing at 0°C. However, there seem to be two fairly distinct groups of organisms capable of growth at that temperature. One group, composed of psychrophiles in the strictest sense, can grow at 0°C but has an optimum growth temperature of about 15°C. Most of these organisms are so sensitive to higher temperatures that they will not even grow in a reasonably warm room (20°C). Such organisms are found mostly in the oceans' depths or in cer-

tain Arctic regions and seldom cause problems in food preservation. The other group that can grow at 0°C has higher optimum temperatures, usually between 20° and 30°C. Organisms of this type are much more common than psychrophiles and are the most likely to be encountered in low-temperature food spoilage because they grow fairly well at refrigerator temperatures. We will use the term **psychrotrophs** for the group of microorganisms responsible for such spoilage.

Refrigeration is the most common method of preserving household food supplies. It is based on the principle that microbial reproductive rates decrease at low temperatures. Although microbes usually survive even subfreezing temperatures (they might become entirely dormant), they gradually decline in number. Some species decline faster than others. Psychrotrophs actually do not grow well at low temperatures, except in comparison with other organisms, but given time they are able to slowly degrade food. Such spoilage might take the form of mold mycelium, slime on food surfaces, or off tastes or colors in foods. The temperature inside a properly set refrigerator will greatly slow the growth of most spoilage organisms and will entirely prevent the growth of all but a few pathogenic bacteria.

Mesophiles, with an optimum growth temperature of between 25° and 40°C, are the most common type of microbe. Organisms that have adapted to live in the bodies of animals usually have an optimum temperature close to that of their host. The optimum temperature for many pathogenic bacteria is about 37°C, and incubators for clinical cultures are usually set at about this temperature. The mesophiles include most of our common spoilage and disease organisms.

Thermophiles are microorganisms capable of growth at high temperatures. Many of these organisms have an optimum growth temperature between 50°C and 60°C, about

FIGURE 6.2 Plasmolysis. If the concentration of solutes, such as sodium chloride, is higher in the surrounding medium than in the cell (hypertonic), then water tends to leave the cell. The plasma membrane shrinks away from the cell wall (an action called plasmolysis), and cell growth is inhibited.

Normal cell in isotonic solution   Plasmolyzed cell in hypertonic solution

the temperature of water from a hot-water tap. Such temperatures can also be reached in sunlit soil and in thermal waters such as hot springs. Remarkably, many thermophiles cannot grow at temperatures below about 45°C. Endospores formed by thermophilic bacteria are unusually heat-resistant and may survive the usual heat treatment given canned goods. Although elevated storage temperatures may cause surviving endospores to germinate and grow, thereby spoiling the food, these thermophilic bacteria have not been found to be a public health problem. Thermophiles are important in organic compost piles, in which the temperature can rise rapidly to 50° or 60°C.

Some microbes, members of the archaeobacteria, have an optimum growth temperature of 80°C or higher. These bacteria are called **extreme thermophiles**, or sometimes **hyperthermophiles.** Most of these organisms live in hot springs associated with volcanic activity; sulfur is usually important in their metabolic activity. The known record for bacterial growth at high temperatures is about 110°C near deep-ocean hydrothermal vents. The immense pressure in the ocean depths prevents water from boiling even at temperatures well above 100°C.

pH

As you will recall from Chapter 2, pH refers to the acidity or alkalinity of a solution. Most bacteria grow best in a narrow range of pH near neutrality, between pH 6.5 and 7.5. Very few bacteria grow at an acidic pH below about 4.0. This is why a number of foods, such as sauerkraut, pickles, and many cheeses, are preserved from spoilage by acids produced by bacterial fermentation. Nonetheless, some bacteria, **acidophiles**, are remarkably tolerant of acidity. One type of chemoautotrophic bacteria, which is found in the drainage water from coal mines and which oxidizes sulfur to form sulfuric acid, can survive at a pH of

1. Molds and yeasts will grow over a greater pH range than bacteria will, but the optimum pH of molds and yeasts is generally below that of bacteria, usually about pH 5 to 6. Alkalinity also inhibits microbial growth but is rarely used to preserve foods.

When bacteria are cultured in the laboratory, they often produce acids that eventually interfere with their own growth. To neutralize the acids and maintain the proper pH, chemicals called **buffers** are included in the growth medium. The peptones and amino acids in some media act as buffers, and many media also contain phosphate salts. Phosphate salts have the advantage of exhibiting their buffering effect in the pH growth range of most bacteria. They are also nontoxic; in fact, they provide phosphorus, an essential nutrient.

Osmotic Pressure

Microbes obtain almost all of their nutrients in solution from the surrounding water. They therefore require water for growth and are actually about 80% to 90% water. High osmotic pressures have the effect of removing necessary water from a cell. When a microbial cell is in a solution that has a higher concentration of solutes than in the cell (hypertonic), the cellular water passes out through the plasma membrane to the high salt concentration. (See the discussion of osmosis in Chapter 4, and review Figure 4.17 for the three types of solution a cell may encounter.) This osmotic loss of water causes **plasmolysis**, or shrinkage of the cell's plasma (cytoplasmic) membrane (Figure 6.2). What is so important about this phenomenon is that the growth of the cell is inhibited as the plasma membrane pulls away from the cell wall. Thus, the addition of salts (or other solutes) to a solution, and the resulting increase in osmotic pressure, can be used to preserve foods. Salted fish, honey, and sweetened condensed milk are preserved largely by this mechanism; the high salt or sugar concentrations draw water out of any microbial cells that are present and thus prevent their growth. These effects of osmotic pressure are roughly related to the number of molecules in a volume of solution. A low–molecular-weight compound such as sodium chloride has a greater antimicrobial effect per gram than does, for example, sucrose, which has a higher molecular weight.

Some bacteria, **extreme halophiles**, have adapted so well to high salt concentrations that they actually require them for growth. Bacteria from such saline waters as the Dead Sea often require nearly 30% salt, and the inoculating loop (a device for handling bacteria in the laboratory) used to transfer them must first be dipped into a saturated salt solution. More common are **facultative halophiles**, which do not require high salt concentrations but are able to grow at salt concentrations up to 2%, a concentration that inhibits the growth of many other bacteria. A few species of facultative halophiles can even tolerate 15% salt. A halophile is described in the box in Chapter 5, page 131.

Most microbes, however, must be grown in a medium that is nearly all water. For example, the concentration of agar (a complex polysaccharide isolated from algae) used to solidify microbial growth media is usually about 1.5%. If markedly higher concentrations are used, the growth of some bacteria can be inhibited by the increased osmotic pressure.

If the osmotic pressure is unusually low (hypotonic), such as in distilled water, for example, water tends to enter the cell rather than leave it. Some microbes that have a relatively weak cell wall may by lysed by such treatment.

Chemical Requirements

Carbon

Besides water, one of the most important requirements for microbial growth is a source of carbon. Carbon is the structural backbone of living matter; it is needed for all the organic compounds that make up a living cell. As we saw in Chapter 2, its four valences allow it to be used in constructing extended and complicated organic molecules (see Figures 2.9 and 2.10). One-half of the dry weight of a typical bacterial cell is carbon.

Chemoheterotrophs get most of their carbon from the source of their energy, namely, organic materials such as proteins, carbohydrates, and lipids. Chemoautotrophs and photoautotrophs derive their carbon from carbon dioxide.

Because the utilization of carbon is such a basic trait of life on Earth, NASA has tested for life on Mars by looking for a sign that carbon was being metabolized. If carbon compounds were being formed or broken down, living organisms might be present.

Nitrogen, Sulfur, and Phosphorus

In addition to carbon, other elements are needed by microbes for the synthesis of cellular material. For example, protein synthesis requires considerable amounts of nitrogen as well as some sulfur. The syntheses of DNA and RNA also require nitrogen and some phosphorus, as does the synthesis of ATP, the molecule so important for storage and transfer of chemical energy within the cell. Nitrogen makes up about 12% to 15% of the dry weight of a bacterial cell, and sulfur and phosphorus together constitute another 3%.

Organisms use nitrogen primarily to form the amino group of the amino acids of proteins. Many bacteria meet this requirement by decomposing protein-containing material and reincorporating the amino acids into newly synthesized proteins and other nitrogen-containing compounds. Other bacteria use nitrogen from ammonium ions (NH_4^+), which are already in the reduced form and are usually found in organic cellular material. Still other bacteria are able to derive nitrogen from nitrates (compounds that dissociate to give the nitrate ion, NO_3^-, in solution).

Some important bacteria, including the cyanobacteria, use gaseous nitrogen (N_2) directly from the atmosphere. The process by which they take up N_2 is called **nitrogen fixation.** Some organisms that can use this method are free-living, mostly in the soil, but others live in *symbiosis* (close association) with the roots of certain plants. The most important symbiotic nitrogen-fixing bacteria belong to the genera *Rhizobium* and *Bradyrhizobium* (brad-ē-rī-zō'bē-um) and are associated with such legumes as clover, soybeans, alfalfa, beans, and peas. The nitrogen fixed in the symbiosis is used by both the plant and the bacterium (see Chapter 27). This reaction greatly increases soil fertility and is therefore very important in agriculture.

Sulfur is used to synthesize sulfur-containing amino acids and vitamins such as thiamine and biotin. The box on page 149 describes an unusual ecosystem that is based on a supply of hydrogen sulfide (H_2S) that would be toxic to most organisms. Important natural sources of sulfur include the sulfate ion (SO_4^{2-}), hydrogen sulfide, and the sulfur-containing amino acids. Phosphorus is essential for synthesis of nucleic acids and the phospholipids of cell membranes. Among other places, it is also found in the energy bonds of ATP. An important source of phosphorus is the phosphate ion (PO_4^{3-}). Potassium, magnesium, and calcium are also elements that microorganisms require, often as cofactors for enzymes (see Chapter 5).

Trace Elements

Microbes require very small amounts of other mineral elements, such as iron, copper, molybdenum, and zinc; these are referred to as **trace elements.** Most are essential for activity of certain enzymes, usually as cofactors. Although these elements are sometimes added to a laboratory medium, they are usually assumed to be naturally present in tap water and other components of media. Even most distilled waters contain adequate amounts, but tap water is sometimes specified to ensure that these trace minerals will be present in culture media.

TABLE 6.1   Effect of Oxygen on Growth of Various Types of Bacteria

	a. OBLIGATE AEROBES	b. FACULTATIVE ANAEROBES	c. OBLIGATE ANAEROBES	d. AEROTOLERANT ANAEROBES	e. MICROAEROPHILES
Effect of oxygen on growth	Only aerobic growth; oxygen required	Both aerobic and anaerobic growth; greater growth in presence of oxygen	Only anaerobic growth; ceases in presence of oxygen	Only anaerobic growth; continues in presence of oxygen	Only aerobic growth; oxygen required in low concentrations
Bacterial growth in tube of solid growth medium					
Explanation of growth patterns	Growth occurs only where high concentrations of oxygen have diffused into the medium.	Growth is best where most oxygen is present, but occurs throughout tube.	Growth occurs only where there is no oxygen.	Growth occurs evenly; oxygen has no effect.	Growth occurs only where low concentrations of oxygen have diffused into medium.
Explanation of oxygen's effects	Presence of enzymes catalase and SOD allows toxic forms of oxygen to be neutralized; can use oxygen.	Presence of enzymes catalase and SOD allows toxic forms of oxygen to be neutralized; can use oxygen.	Lacks enzymes to neutralize harmful forms of oxygen; cannot tolerate oxygen.	Presence of one enzyme, SOD, allows harmful forms of oxygen to be partially neutralized; tolerates oxygen.	Catalase and SOD are absent or ineffective. Pathogens use hosts' enzymes.

Oxygen

We are accustomed to thinking of molecular oxygen (O_2) as a necessity of life, but it is actually a poisonous gas. Molecular oxygen did not exist in the atmosphere during most of our planet's history—in fact, it is possible that life could not have arisen had oxygen been present. However, many current forms of life have metabolic systems that require oxygen for aerobic respiration. As we have seen, hydrogen atoms that have been stripped from organic compounds combine with oxygen to form water. This process yields a great deal of energy while neutralizing a potentially toxic gas—a very neat solution, all in all.

Microbes that use molecular oxygen, called **aerobes**, produce more energy from nutrients than do microbes that do not use oxygen. Organisms that require oxygen to live are called **obligate aerobes** (Table 6.1a).

Obligate aerobes are at a disadvantage, in that oxygen is poorly soluble in the water of their environment. Therefore, many of the aerobic bacteria have developed, or retained, the ability to continue growing in the absence of oxygen. Such organisms, including some that exist in your intestinal tract, are called **facultative anaerobes** (Table 6.1b). In other words, facultative anaerobes can use oxygen when it is present but are able to continue growth by using fermentation or anaerobic respiration when oxygen is not available. However, their efficiency in producing energy decreases in the absence of oxygen. Examples

of facultative anaerobes include the familiar *E. coli* and many yeasts.

Recall from Chapter 5 that many microbes are able to substitute other electron acceptors, such as nitrate ions, for oxygen in anaerobic respiration.

Obligate anaerobes (Table 6.1c) are bacteria that are unable to use molecular oxygen for energy-yielding reactions. In fact, most are harmed by it. The genus *Clostridium*, which contains the species that cause tetanus and botulism, is the most familiar example. These bacteria do use oxygen atoms in cellular materials; the atoms are usually obtained from water.

Understanding how organisms can be harmed by oxygen requires a brief discussion of the toxic forms of oxygen:

1. **Singlet oxygen** is normal molecular oxygen (O_2) that has been boosted into a higher-energy state and is extremely reactive. It is present in phagocytic cells, specialized animal cells that ingest and eliminate small foreign bodies such as bacteria. It plays a role in the destruction of such foreign bodies after they are ingested by the phagocytes.

2. **Superoxide free radicals** ($O_2^-\cdot$) are formed in small amounts during the normal respiration of organisms that use oxygen as a final electron acceptor, forming water (see Figure 5.17). In the presence of oxygen, obligate anaerobes also appear to form some superoxide

free radicals. These free radicals are so toxic to cellular components that all organisms attempting to grow in atmospheric oxygen must produce an enzyme, **superoxide dismutase (SOD)**, to neutralize them. Their toxicity is caused by their great instability, which leads them to avidly steal an electron from a neighboring molecule, which in turn becomes a radical and steals an electron, and so on. Unless this reaction series is quickly neutralized, extreme damage to the cell results. Aerobic bacteria, facultative anaerobes growing aerobically, and aerotolerant anaerobes (discussed later) produce this enzyme. Using SOD, they convert the superoxide free radical into molecular oxygen (O_2) and hydrogen peroxide (H_2O_2):

$$O_2^- \cdot + O_2^- \cdot + 2H^+ \longrightarrow H_2O_2 + O_2$$

3. The hydrogen peroxide produced in this reaction contains the **peroxide anion** O_2^{2-} and is also toxic. In the following chapter we will encounter it as the active principle in the antimicrobial agents hydrogen peroxide and benzoyl peroxide. Because the hydrogen peroxide produced during normal aerobic respiration is toxic, microbes have developed enzymes to neutralize it. The most familiar of these is **catalase**, which converts it into water and oxygen:

$$2H_2O_2 \longrightarrow 2H_2O + O_2$$

Catalase is easily detected by its action on hydrogen peroxide. When a drop of hydrogen peroxide is added to a colony of bacterial cells producing catalase, oxygen bubbles are released. Anyone who has put hydrogen peroxide on a wound will recognize that cells in human tissue also contain catalase. The other enzyme that breaks down hydrogen peroxide is **peroxidase**; it differs from catalase in that its reaction does not produce oxygen:

$$H_2O_2 + 2H^+ \longrightarrow 2H_2O$$

4. The **hydroxyl radical** (OH ·) is another intermediate form of oxygen and probably the most reactive. It is formed in the cellular cytoplasm by ionizing radiation. It is also produced by a reaction between superoxide free radicals and peroxide, so most aerobic respiration produces some hydroxyl radicals.

Obligate anaerobes usually produce neither superoxide dismutase nor catalase. Because aerobic conditions probably lead to an accumulation of superoxide free radicals in their cytoplasm, they are extremely sensitive to oxygen.

Aerotolerant anaerobes (Table 6.1d) cannot use oxygen for growth, but they tolerate it fairly well. On the surface of a solid medium, they will grow without the special techniques (discussed later) required for obligate anaerobes. Many of the aerotolerant bacteria characteristically ferment carbohydrates to lactic acid. As lactic acid accumulates, it inhibits the growth of aerobic competitors and establishes an ecological niche for the lactic acid producers. A common example of lactic-acid–producing aerotolerant anaerobes is the lactobacilli used in the production of many acidic fermented foods, such as pickles and cheese. In the laboratory, they are handled and grown much like any other bacteria, but they make no use of the oxygen in the air. These bacteria can tolerate oxygen because they possess superoxide dismutase (see the bottom of Table 6.1d) or an equivalent system that neutralizes the toxic forms of oxygen previously discussed.

A few bacteria are **microaerophiles** (Table 6.1e). They are aerobic; they do require oxygen. However, they grow only in oxygen concentrations lower than those in air. In a test tube of solid nutrient medium, they grow only at a depth where small amounts of oxygen have diffused into the medium; they do not grow at the oxygen-rich surface or below the narrow zone of adequate oxygen. This limited tolerance is probably due to their sensitivity to superoxide free radicals and peroxides, which they produce in lethal concentrations under oxygen-rich conditions.

Organic Growth Factors

Organic growth factors are essential organic compounds that an organism is unable to synthesize; they must be directly obtained from the environment. One group of organic growth factors for humans is vitamins. Most vitamins function as coenzymes, the accessories required by certain enzymes in order to function. Many bacteria can synthesize all their own vitamins and are not dependent on outside sources. However, some bacteria lack the enzymes needed for the synthesis of certain vitamins, and for them those vitamins are organic growth factors. Other organic growth factors required by some bacteria are amino acids, purines, and pyrimidines.

CULTURE MEDIA

LEARNING OBJECTIVES

- *Distinguish between chemically defined and complex media.*
- *Justify the use of each of the following: anaerobic techniques, living host cells, candle jars, selective and differential media, enrichment media.*
- *Define colony and clone.*
- *Describe how pure cultures can be isolated by using the streak plate method.*

A nutrient material prepared for the growth of microorganisms in a laboratory is called a **culture medium.** Some bacteria can grow well on just about any culture medium; others require special media, and still others cannot grow on any nonliving medium yet developed. The microbes that grow and multiply in or on a culture medium are referred to as a **culture.**

Suppose we want to grow a culture of a particular microorganism, perhaps the microbes from a particular clinical specimen. What criteria must the culture medium meet? First, it must contain the right nutrients for the particular microorganism we want to grow. It should also contain sufficient moisture, a properly adjusted pH, and a suitable level of oxygen, perhaps none at all. The medium must initially be **sterile**—that is, it must initially contain no living microorganisms—so that the culture will contain only the microorganisms (and their offspring) we add to the medium. Finally, the growing culture should be incubated at the proper temperature.

A wide variety of media are available for the growth of microorganisms in the laboratory. Most of these media, which are available from commercial sources, have premixed components and require only the addition of water and sterilization. Media are constantly being developed or revised for use in the isolation and identification of bacteria that are of interest to researchers in such fields as food, water, and clinical microbiology.

When it is desirable to grow bacteria on a solid medium, a solidifying agent such as agar is added to the medium. **Agar** is a complex polysaccharide derived from a marine alga, and it has long been used as a thickener in foods such as jellies, soups, and ice cream.

Agar has some very important properties that make it valuable to microbiology, and no really satisfactory substitute has ever been found. Few microbes can degrade agar, so it remains solid. Also, agar melts at about the boiling point of water but remains liquid until the temperature drops to about 40°C. For laboratory use, agar is held in water baths at about 50°C. At this temperature, it does not injure most bacteria when it is poured over a bacterial inoculum. Once the agar has solidified, it can be incubated at temperatures approaching 100°C before it liquefies; this property is particularly useful when thermophilic bacteria are being grown.

Agar media are usually contained in test tubes or Petri plates. The test tubes are called *slants* when they are allowed to solidify with the tube held at an angle so that a large surface for growth is available. When the agar solidifies in a vertical tube, it is called a *deep*. Petri plates, named for their inventor, are shallow dishes with a lid that nests over the bottom to prevent contamination.

Chemically Defined Media

To be appropriately prepared for microbial growth, a medium must provide an energy source as well as sources of carbon, nitrogen, sulfur, phosphorus, and any necessary growth factors that the organism is unable to synthesize. A **chemically defined medium** is one whose exact chemical composition is known. Table 6.2 shows a chemically defined medium used for the growth of a

TABLE 6.2 Chemically Defined Medium for the Growth of a Chemoautotrophic Bacterium Capable of Using Ammonium Ions for Energy

CONSTITUENT*	AMOUNT
Ammonium sulfate [$(NH_4)_2SO_4$]	0.5 g
Potassium phosphate, monobasic (KH_2PO_4)	0.2 g
Calcium chloride ($CaCl_2$)	0.04 g
Magnesium sulfate ($MgSO_4$)	0.04 g
Ferric citrate ($FeC_6H_5O_7$)	0.0005 g
Water	1 liter

*Carbon source is naturally dissolved CO_2.
SOURCE: R. M. Atlas, *Handbook of Microbiological Media*, Ann Arbor, MI: CRC Press, 1993.

TABLE 6.3 Chemically Defined Medium for the Growth of a Typical Chemoheterotroph, Such As *E. coli*

CONSTITUENT	AMOUNT
Glucose	5.0 g
Ammonium phosphate, monobasic ($NH_4H_2PO_4$)	1.0 g
Sodium chloride (NaCl)	5.0 g
Magnesium sulfate ($MgSO_4 \cdot 7H_2O$)	0.2 g
Potassium phosphate, dibasic (K_2HPO_4)	1.0 g
Water	1 liter

chemoautotrophic organism capable of extracting energy from the oxidation of ammonium ions to nitrite ions. For a chemoheterotroph, the chemically defined medium must contain organic growth factors that serve as a source of carbon and energy. For example, as shown in Table 6.3, glucose is included in the medium for growing the chemoheterotroph *E. coli*.

As Table 6.4 (p. 150) demonstrates, many organic growth factors must be provided in the chemically defined medium used to cultivate a species of *Neisseria*. Organisms that require many growth factors are described as *fastidious*. Organisms of this type, such as *Lactobacillus*, are sometimes used in tests that determine the concentration of a particular vitamin in a substance. To do such a *microbiological assay* (test), a growth medium is prepared that contains all the growth requirements of the bacterium except the vitamin being assayed. Then the medium, test substance, and bacterium are combined, and the growth of bacteria is measured. This bacterial growth, which is reflected by the amount of lactic acid produced, will be proportional to the amount of vitamin in the test substance. The more lactic acid, the more the *Lactobacillus* cells have been able to grow, so the more vitamin is present.

STUDYING HYDROTHERMAL BACTERIA

....

UNTIL HUMANS EXPLORED the ocean floor, scientists believed that only a few forms of life could survive in that high-pressure, completely dark, oxygen-poor environment. Then, in 1977, the first manned vehicle able to penetrate to the bottom of the deepest oceans carried two scientists 2600 meters below the surface at the Galápagos Rift, about 350 km northeast of the Galápagos Islands. There, amid the vast expanse of barren basalt pillows, the scientists found unexpectedly rich oases of life, including mollusks, crustaceans, and worms (see photo). How do such creatures survive in these harsh conditions?

Because of the extreme temperatures and pressures in the environments of deep-sea organisms, most experiments are conducted *in situ*— that is, directly on the sea floor. Tubes of liquid media are prepared and lowered to the sea floor, where they are inoculated and incubated. When the tubes are retrieved, microbial growth can be determined indirectly by measuring metabolic activity.

A few bacterial samples have been taken to the surface and studied. Researchers have been studying the metabolism of these organisms ever since their discovery to determine how life can be sustained without solar energy.

ECOSYSTEM OF THE HYDROTHERMAL VENTS

Life at the surface of the world's oceans depends on photosynthetic organisms, such as bacteria and plankton, to harness the Sun's energy and make carbohydrates. At the ocean floor, where no light penetrates, photosynthesis is not possible. The scientists found that the so-called primary producers at the ocean floor are chemoautotrophic bacteria. Using such inorganic compounds as hydrogen sulfide (H_2S) as a source of energy and carbon dioxide (CO_2) as a source of carbon, the chemoautotrophs create an environment that supports higher life forms. Hydrothermal vents in the sea floor supply the H_2S and CO_2. As superheated water from within the Earth rises through fractures in the Earth's crust called vents, it reacts with surrounding rock and dissolves metal ions, sulfides, and CO_2. The ecosystems of these vents depend on an abundance of sulfur compounds in the hot water. The concentration of sulfide (S^{2-}) is three times greater than the concentration of molecular oxygen in and around the vents. Such high sulfide concentrations are toxic to many organisms, but not to the creatures that inhabit this exotic environment.

Mats of bacteria grow along the sides of the vents, where temperatures exceed 100°C. This is the highest temperature that any organism is known to tolerate. Above the vent, where temperatures are about 30°C, the concentration of bacteria is about four times greater than that in water farther from the vents, and the growth rate of bacteria is equal to that found in productive, sunlit coastal waters. The bacteria in and around the vents create an environment in which bacteria consumers, such as clams, mussels, and worms, and decomposers can exist.

AN UNEXPECTED BIOTECHNOLOGICAL BENEFIT

DNA polymerases isolated from archaeobacteria living near deep-sea vents are being used in polymerase chain reactions (PCRs). PCR is a technique that allows many copies of DNA to be made. In PCR, single-stranded DNA is made by heating a chromosome fragment to 98°C and cooling it so that DNA polymerase can copy each strand. Heating and cooling are repeated every few minutes, producing billions of copies of DNA in a few hours. The original PCR procedure used DNA polymerases from mesophiles, which are denatured at 98°C and so had to be repeatedly added to the reaction mixture. However, DNA polymerase from *Thermococcus litoralis*, called Vent $_R$, and from *Pyrococcus*, called Deep Vent$_R$, are not denatured at 98°C. Therefore, automatic thermocyclers can now be used to repeat the heating and cooling cycles, allowing many copies of DNA to be made easily and quickly. Vent$_R$ and Deep Vent$_R$ add bases to DNA at a rate of about 1000 bases per minute.

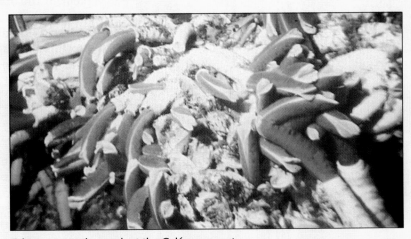

Tubeworms and mussels at the Galápagos vent.

TABLE 6.4 Chemically Defined Medium for the Growth of a Fastidious Chemoheterotrophic Bacterium, Such as *Neisseria gonorrhoeae*

CONSTITUENT	AMOUNT
Carbon and energy sources	
Glucose	9.1 g
Starch	9.1 g
Sodium acetate	1.8 g
Sodium citrate	1.4 g
Oxaloacetate	0.3 g
Salts	
Potassium phosphate, dibasic (K_2HPO_4)	12.7 g
Sodium chloride (NaCl)	6.4 g
Potassium phosphate, monobasic (KH_2PO_4)	5.5 g
Sodium bicarbonate ($NaHCO_3$)	1.2 g
Potassium sulfate (K_2SO_4)	1.1 g
Sodium sulfate (Na_2SO_4)	0.9 g
Magnesium chloride ($MgCl_2$)	0.5 g
Ammonium chloride (NH_4Cl)	0.4 g
Potassium chloride (KCl)	0.4 g
Calcium chloride ($CaCl_2$)	0.006 g
Ferric nitrate [$Fe(NO_3)_3$]	0.006 g
Amino acids	
Cysteine	1.5 g
Arginine, proline (each)	0.3 g
Glutamic acid, methionine (each)	0.2 g
Asparagine, isoleucine, serine (each)	0.2 g
Cystine	0.06 g
Organic growth factors	
Calcium pantothenate	0.02 g
Thiamine	0.02 g
Nicotinamide adenine dinucleotide	0.01 g
Uracil	0.006 g
Biotin	0.005 g
Hypoxanthine	0.003 g
Reducing agent	
Sodium thioglycolate	0.00003 g
Water	1 liter

SOURCE: R. M. Atlas, *Handbook of Microbiological Media*, Ann Arbor, MI: CRC Press, 1993.

TABLE 6.5 Composition of Nutrient Agar, a Complex Medium for the Growth of Heterotrophic Bacteria

CONSTITUENT	AMOUNT
Peptone (partially digested protein)	5.0 g
Beef extract	3.0 g
Sodium chloride	8.0 g
Agar	15.0 g
Water	1 liter

Complex Media

Chemically defined media are usually reserved for laboratory experimental work or for the growth of autotrophic bacteria. Most heterotrophic bacteria and fungi are routinely grown on **complex media**, those media whose exact chemical composition varies slightly from batch to batch. These complex media are made up of nutrients such as extracts from yeasts, meat, or plants, or digests of proteins from these and other sources. Table 6.5 gives one widely used recipe.

In complex media such as this one, the energy, carbon, nitrogen, and sulfur requirements of the growing microorganisms are met largely by protein. Protein is a large, relatively insoluble molecule that few microorganisms can utilize directly, but a partial digestion by acids or enzymes reduces protein to shorter chains of amino acids called *peptones*. These small, soluble fragments can be digested by the bacteria.

The vitamins and other organic growth factors are provided by meat extracts or yeast extracts. The soluble vitamins and minerals from the meats or yeasts are dissolved in the extracting water, which is then evaporated so that these factors are concentrated. (These extracts also supplement the organic nitrogen and carbon compounds.) Yeast extracts are particularly rich in the B vitamins. If this type of medium is in liquid form, it is called **nutrient broth.** When agar is added, it is called **nutrient agar.** (This terminology can be confusing; just remember that agar itself is not a nutrient.)

Anaerobic Growth Media and Methods

The cultivation of anaerobic bacteria poses a special problem. Because anaerobes might be killed by exposure to oxygen, special media called **reducing media** must be used. These media contain ingredients, such as sodium thioglycolate, that chemically combine with dissolved oxygen and deplete the oxygen in the culture medium. To routinely grow and maintain pure cultures of obligate anaerobes, microbiologists use reducing media stored in

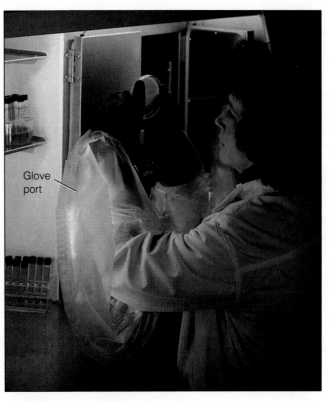

FIGURE 6.3 Anaerobic container used to cultivate anaerobic bacteria on Petri plates. When water is mixed with the chemical packet containing sodium bicarbonate and sodium borohydride, hydrogen and carbon dioxide are generated. Reacting on the surface of a palladium catalyst in a screened reaction chamber, the hydrogen and the atmospheric oxygen in the jar combine to form water. The oxygen is thus removed. An anaerobic indicator is also in the container. The indicator contains methylene blue, which is blue when oxidized (as shown here) and turns colorless when the oxygen is removed from the container.

FIGURE 6.4 Anaerobic chamber. The technician is holding a Petri dish that is inside an anaerobic chamber. Such chambers are filled with an inert, oxygen-free gas. Organisms and materials enter and leave through an air lock, and equipment is manipulated by means of the glove ports visible in the photo.

ordinary, tightly capped test tubes. These media are heated shortly before use, so that any absorbed oxygen is driven off.

When the culture must be grown in Petri plates so that individual colonies can be observed, special jars are used that hold several Petri plates in an oxygen-free atmosphere (Figure 6.3). The oxygen is removed by the following process: A packet of chemicals (sodium bicarbonate and sodium borohydride) in the jar is moistened with a few milliliters of water, and the jar is sealed. Hydrogen and carbon dioxide are produced by the reaction of the chemicals with the water. A palladium catalyst in the jar combines the oxygen in the jar with the hydrogen produced by the chemical reaction, and water is formed. As a result, the oxygen disappears in a short time. Moreover, the carbon dioxide that is produced aids the growth of many anaerobic bacteria.

Researchers working with anaerobes on a regular basis use anaerobic glove boxes, which are transparent chambers equipped with air locks and filled with inert gases (Figure 6.4). The technicians are able to manipulate the

equipment by inserting their hands into airtight rubber gloves fitted to the wall of the chamber.

Colonies of anaerobes can also be grown in deep test tubes instead of Petri plates. The natural atmosphere in the test tube is replaced with an inert gas, such as nitrogen. A bacterial inoculum can be mixed with melted nutrient medium in the tube, and the tube, tightly capped, is rolled on horizontal rollers. The nutrient medium solidifies against the interior walls of the test tube (usually referred to as **roll tubes**), and colonies that appear there can be counted and picked, as on a shallow layer of agar in a Petri plate.

Special Culture Techniques

Some bacteria have never been successfully grown on artificial laboratory media. *Mycobacterium leprae*, the leprosy bacillus, is now usually grown in armadillos, which have a relatively low body temperature that matches the requirements of the microbe. Another example is the

Lid

Glass jar

Tubes with liquid media

Candle

Petri plates with solid media (inverted)

(a) Candle jar

Petri plate with bacterial culture

Gas generator

(b) CO_2-generating packet

FIGURE 6.5 Equipment used to produce CO_2-rich environments. (a) Candle jar. Plates and tubes inoculated with, for example, *Neisseria meningitidis* are placed in a jar with a lighted candle, and the jar is sealed. The candle will go out when the atmosphere contains approximately 10% carbon dioxide. **(b)** CO_2-generating packet. The packet consists of a bag containing a Petri plate and a CO_2 gas generator. The gas generator is crushed to mix the chemicals it contains and start the reaction that produces CO_2. This gas reduces the oxygen concentration in the bag.

syphilis spirochete, although certain specialized strains of this microbe have been grown on laboratory media. With few exceptions, the obligate intracellular bacteria, such as the rickettsias and the chlamydias, do not grow on artificial media. They, like viruses, can reproduce only in a living host cell.

Many clinical laboratories have special *carbon dioxide incubators* in which to grow aerobic bacteria that require concentrations of CO_2 higher or lower than that found in the atmosphere. Desired carbon dioxide levels are maintained by electronic controls. High levels of carbon dioxide are also obtained with simple *candle jars* or CO_2-generating packets (Figure 6.5a). Cultures are placed in a large sealed jar containing a lighted candle, which con-

sumes oxygen. The candle stops burning when the atmosphere in the jar has a low concentration of oxygen (but one that is still adequate for the growth of aerobic bacteria) and has reached the high CO_2 concentration required by certain clinically important bacteria, such as those causing brucellosis, gonorrhea, and meningococcal meningitis. Candle jars are still used occasionally, but now commercially available chemical packets are used more often to generate carbon dioxide atmospheres in containers (Figure 6.5b). When only one or two Petri plates of cultures are to be incubated, clinical laboratory investigators often use small plastic bags with self-contained chemical gas generators that are activated by crushing the packet or moistening it with a few milliliters of water. These packets are sometimes specially designed to provide precise carbon dioxide and oxygen concentrations for culture of organisms such as the microaerophilic *Campylobacter* bacteria.

Selective and Differential Media

In clinical and public health microbiology, it is frequently necessary to detect the presence of specific microorganisms associated with disease or poor sanitation. For this task, selective and differential media are used. **Selective media** are designed to suppress the growth of unwanted bacteria and encourage the growth of the desired microbes. For example, bismuth sulfite agar is one medium used to isolate the typhoid bacterium, the gram-negative *Salmonella typhi* (tī′fē), from feces. Bismuth sulfite inhibits gram-positive bacteria and most gram-negative intestinal bacteria (other than *S. typhi,*) as well. Sabouraud's dextrose agar, which has a pH of 5.6, is used to isolate fungi that outgrow most bacteria at this pH. Dyes such as brilliant green selectively inhibit gram-positive bacteria, and this dye is the basis of a medium called Brilliant Green agar that is used to isolate the gram-negative *Salmonella.*

Differential media make it easier to distinguish colonies of the desired organism from other colonies growing on the same plate. Similarly, pure cultures of microorganisms have identifiable reactions with differential media in tubes or plates. Blood agar (which contains red blood cells) is a dark, reddish-brown medium that microbiologists often use to identify bacterial species that destroy red blood cells. These species, such as *Streptococcus pyogenes* (pī-äj′en-ēz), the bacterium that causes strep throat, show a clear ring around their colonies where they have lysed the surrounding blood cells (Figure 6.6).

Sometimes, selective and differential media are used together. Suppose we want to isolate the common bacterium *Staphylococcus aureus,* found in the nasal passages. One of the characteristics of this organism is its tolerance for high concentrations of sodium chloride; another characteristic is its ability to ferment the carbohydrate mannitol to form acid. There is an isolation medium called mannitol salt agar medium that combines features of both selective and differential media. Mannitol salt agar contains 7.5% sodium chloride, which will discourage the growth of competing organisms and thus *select for* (favor the growth of) *S. aureus*. This salty medium also contains a pH indicator that changes color if the mannitol in the medium is fermented to acid; the mannitol-fermenting colonies of *S. aureus* are thus differentiated from colonies of bacteria that do not ferment mannitol. Bacteria that grow at the high salt concentration *and* ferment mannitol to acid can be readily identified by the color change. These are probably colonies of *S. aureus,* and their identification can be confirmed by additional tests.

Another medium that is both selective and differential is MacConkey agar. This medium contains bile salts and crystal violet, which inhibit the growth of gram-positive bacteria. Because this medium also contains lactose, gram-negative bacteria that can grow on lactose can be differentiated from similar bacteria that cannot. The ability to distinguish between lactose fermenters (red or pink colonies) and nonfermenters (colorless colonies) is useful in distinguishing between the pathogenic *Salmonella* bacteria and other related bacteria. Figure 6.7 shows the appearance of bacterial colonies on several differential media.

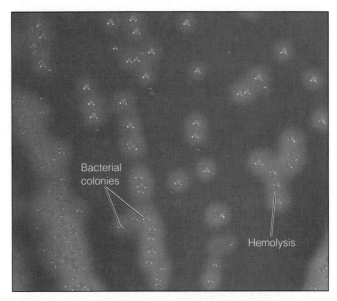

FIGURE 6.6 **Blood agar, a differential medium containing red blood cells.** The bacteria have lysed the red blood cells (hemolysis), so clear areas have formed around the colonies.

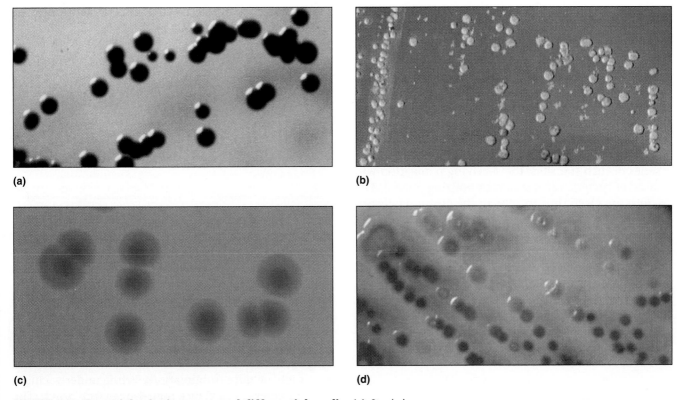

(a)

(b)

(c)

(d)

FIGURE 6.7 **Bacterial colonies on several differential media.** (a) *Staphylococcus aureus* on tellurite-glycine medium. (b) *Escherichia coli* on eosin methylene blue (EMB) medium. (c) *Enterobacter aerogenes* on EMB medium showing characteristic dark-centered colonies. (d) On Pseudomonas agar P (PSP) medium, *Pseudomonas aeruginosa* produces a blue-green water-soluble pigment.

TABLE 6.6 Culture Media	
TYPE OF MEDIA	**PURPOSE**
Chemically defined	Grow chemoautotrophs and photoautotrophs, and for microbiological assays
Complex	Grow most chemoheterotrophic organisms
Reducing	Grow obligate anaerobes
Selective	Suppress unwanted microbes, encourage desired microbes
Differential	Distinguish colonies of desired microbes from others
Enrichment	Similar to selective media but designed to increase numbers of desired microbes to detectable levels

Enrichment Culture

Because bacteria present in small numbers can be missed, especially if other bacteria are present in much larger numbers, it is sometimes necessary to use an **enrichment culture.** This is often the case for soil or fecal samples. The medium for an enrichment culture is usually liquid and provides nutrients and environmental conditions that favor the growth of a particular microbe but are not suitable for the growth of other types of microbes. In this sense, it is also a selective medium, but it is designed to increase very small numbers of the desired type of microbe to detectable levels.

Suppose we want to isolate from a soil sample a microbe that can grow on phenol and is present in much smaller numbers than other species. If the soil sample is placed in a liquid enrichment medium in which phenol is the only source of carbon and energy, microbes unable to metabolize phenol will not grow. The culture medium is allowed to incubate for a few days, and then a small amount of it is transferred into another flask of the same medium. After a series of such transfers, the surviving population will consist of bacteria capable of metabolizing phenol. The bacteria are given time to grow in the medium between transfers; this is the enrichment stage. Any nutrients in the original inoculum are rapidly diluted out with the successive transfers. When the last dilution is streaked onto a solid medium of the same composition, only those colonies of organisms that are capable of using phenol should grow. A remarkable aspect of this particular technique is that phenol is normally lethal to most bacteria.

Table 6.6 summarizes the purposes of the main types of culture media.

Obtaining Pure Cultures

Most infectious materials, such as pus, sputum, and urine, contain several kinds of bacteria; so do samples of soil, water, or foods. If these materials are plated out onto the surface of a solid medium, colonies will form that are exact copies of the same organism. A visible **colony** theoretically arises from a single spore or vegetative cell or from a group of the same microorganisms (in other words, a **clone** of cells) attached to one another in clumps or chains. Microbial colonies often have a distinctive appearance that distinguishes one microbe from another (see Figure 6.7). The bacteria must be distributed widely enough so that the colonies are visibly separated from each other.

Most bacteriological work requires pure cultures, or clones, of bacteria. The isolation method most commonly used to get pure cultures is the **streak plate method** (Figure 6.8). A sterile inoculating loop is dipped into a mixed culture that contains more than one type of microbe and is streaked in a pattern (that shown in the figure is only one example of such a pattern) over the surface of the nutrient medium. As the pattern is traced, bacteria are rubbed off the loop onto the medium in paths of fewer and fewer cells. The last cells to be rubbed off the loop are far enough apart to grow into isolated colonies. These colonies can be picked up with an inoculating loop and transferred to a test tube of nutrient medium to form a pure culture containing only one type of bacterium.

The streak plate method works well when the organism to be isolated is present in large numbers relative to the population. However, when the microbe to be isolated is present only in very small numbers, its numbers must be greatly increased by selective enrichment before it can be isolated with the streak plate method.

PRESERVING BACTERIAL CULTURES

LEARNING OBJECTIVE

- *Explain how microbes are preserved by deep-freezing and freeze-drying.*

Refrigeration can be used for short-term storage of bacterial cultures. Two common methods of preserving microbial cultures for long periods are deep-freezing and lyophilization. **Deep-freezing** is a process in which a pure culture of microbes is placed in a suspending liquid and quick-frozen at temperatures ranging from $-50°$ to $-95°C$. The culture can usually be thawed and used up to several years later. During **lyophilization (freeze-drying),** a suspension of microbes is quickly frozen at temperatures ranging from $-54°$ to $-72°C$, and the water is removed by a high vacuum (sublimation). While under vacuum, the container is sealed by a high-temperature torch. The remaining powderlike residue that contains the surviving microbes can be stored for years. The microbes can be revived at any time by hydration with a suitable liquid nutrient medium.

(a)

(b)

FIGURE 6.8 **Streak plate method for isolation of pure cultures of bacteria.** **(a)** The direction of streaking is indicated by arrows. Streak series 1 is made from the original bacterial mixture. The inoculating loop is sterilized following each streak series. In series 2 and 3, the loop picks up bacteria from the previous series, diluting the number of cells each time. There are numerous variants of such patterns. **(b)** In series 3 of this example, note that well-isolated colonies of two different types of bacteria have been obtained.

GROWTH OF BACTERIAL CULTURES

LEARNING OBJECTIVES

- *Define bacterial growth, including binary fission.*
- *Compare the phases of microbial growth and describe their relation to generation time.*
- *Differentiate between direct and indirect methods of measuring cell growth.*
- *Explain four direct methods of measuring cell growth.*
- *Explain three indirect methods of measuring cell growth.*

Being able to graphically represent the enormous populations resulting from the growth of bacterial cultures is an essential part of microbiology. It is also necessary to be able to determine microbial numbers, either directly, by counting, or indirectly, by measuring their metabolic activity.

Bacterial Division

As we mentioned at the beginning of this chapter, bacterial growth refers to an increase in bacterial numbers, not an increase in the size of the individual cells. Bacteria normally reproduce by binary fission (Figure 6.9). The first step in division is cell elongation and duplication of the chromosomal DNA. The cell wall and cell membrane then begin to grow inward from all sides at a point between the two regions of the chromosomal DNA. Eventually, the ingrowing cell walls meet, forming a cross wall, and two individual cells are formed, each of which is essentially identical to the parent cell.

A few bacterial species **bud**; that is, they form a small initial outgrowth that enlarges until its size approaches that of the parent cell, and then it separates. Some filamentous bacteria (certain actinomycetes) reproduce by producing chains of spores carried externally at the tips of the filaments. A few filamentous species simply fragment, and the fragments initiate the growth of new cells.

Generation Time

For purposes of calculating the generation time of bacteria, we will consider only reproduction by binary fission, which is by far the most common method. As you can see in Figure 6.10, one cell's division produces two cells, two cells' division produces four cells, and so on. When the arithmetic number of cells in each generation is expressed as a power of 2, the exponent tells the number of doublings (generations) that have occurred. (You might at this point wish to refer to the discussion of exponential notation in Appendix E.)

The time required for a cell to divide (and its population to double) is called the **generation time.** It varies considerably among organisms and with environmental

FIGURE 6.9 Binary fission in bacteria. (a) Diagram of sequence of cell division. **(b)** A thin section of a cell of *Bacillus licheniformis* starting to divide.

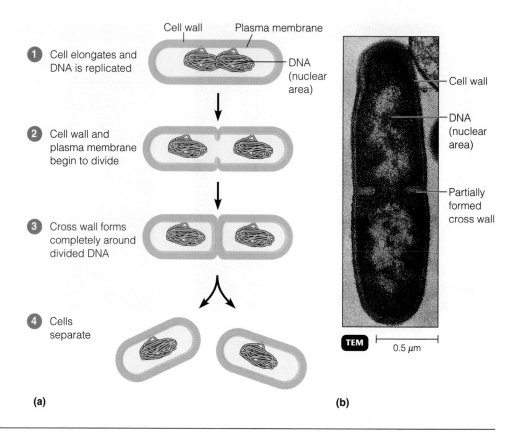

① Cell elongates and DNA is replicated

② Cell wall and plasma membrane begin to divide

③ Cross wall forms completely around divided DNA

④ Cells separate

Cell wall　　Plasma membrane

DNA (nuclear area)

(a)

Cell wall

DNA (nuclear area)

Partially formed cross wall

TEM　　0.5 μm

(b)

conditions such as temperature. Most bacteria have a generation time of from 1 to 3 hours; others require more than 24 hours per generation. If binary fission continues unchecked, an enormous number of cells will be produced. If a doubling occurred every 20 minutes—which is the case for *E. coli* under favorable conditions—after 20 generations a single initial cell would increase to over 1 million cells. This would require a little less than 7 hours. In 30 generations, or 10 hours, the population would be 1 billion, and in 24 hours it would be a number trailed by 21 zeroes. It is difficult to graph population changes of such enormous magnitude by using arithmetic numbers. This is why logarithmic scales rather than arithmetic scales are generally used to graph bacterial growth. Understanding logarithmic representations of bacterial populations requires some use of mathematics and is necessary for anyone studying microbiology.

Logarithmic Representation of Bacterial Populations

To illustrate the difference between logarithmic and arithmetic graphing of bacterial populations, let us express 20 bacterial generations both logarithmically and arithmetically. In five generations (2^5), there would be 32 cells; in ten generations (2^{10}), there would be 1024 cells, and so on. (If your calculator has a y^x key and a *log* key, you can duplicate the calculations in the third column.)

GENERATION NUMBER	ARITHMETIC NUMBER OF CELLS	LOG_{10} OF ARITHMETIC NUMBER OF CELLS
0	1	0
5 (2^5) =	32	1.51
10 (2^{10}) =	1,024	3.01
15 (2^{15}) =	32,768	4.52
16 (2^{16}) =	65,536	4.82
17 (2^{17}) =	131,072	5.12
18 (2^{18}) =	262,144	5.42
19 (2^{19}) =	524,288	5.72
20 (2^{20}) =	1,048,576	6.02

In Figure 6.11, note that the arithmetically plotted line (solid) does not clearly show the population changes in the early stages of the growth curve at this scale. In fact, the first ten generations do not appear to leave the baseline. Furthermore, another one or two arithmetic generations graphed to the same scale would greatly increase the height of the graph and take the line off the page.

The dashed line in Figure 6.11 shows how these plotting problems can be avoided by graphing the log_{10} of the population numbers. The log_{10} of the population is plotted at 5, 10, 15, and 20 generations. Note that a straight line is formed and that a thousand times this

FIGURE 6.10 Cell division. When the arithmetic number of cells in each generation is expressed as a power of 2, the exponent tells the number of doublings (generations) that have occurred.

Arithmetic numbers of cells	Numbers expressed as a power of 2	Visual representation of numbers
1	2^0	
2	2^1	
4	2^2	
8	2^3	
16	2^4	
32	2^5	

population (1,000,000,000, or $\log_{10}$ 9.0) could be accommodated in relatively little extra space. However, this advantage is obtained at the cost of distorting our "common sense" perception of the actual situation. We are not accustomed to thinking in logarithmic relationships, but it is necessary for a proper understanding of graphs of microbial populations.

Phases of Growth

When a few bacteria are inoculated into a liquid growth medium and the population is counted at intervals, it is possible to plot a **bacterial growth curve** that shows the growth of cells over time (Figure 6.12). There are four basic phases of growth.

Lag Phase

For a while, the number of cells changes very little because the cells do not immediately reproduce in a new medium. This period of little or no cell division is called the **lag phase**, and it can last for an hour or several days. During this time, however, the cells are not dormant. The microbial population is undergoing a period of intense metabolic activity involving, in particular, DNA and enzyme synthesis. (The situation is analogous to a factory newly equipped to produce automobiles—there is considerable initial activity but no immediate increase in the automobile population.)

Log Phase

Eventually, the cells begin to divide and enter a period of growth, or logarithmic increase, called the **log, or exponential growth, phase.** Cellular reproduction is most active during this period, and generation time reaches a constant minimum. Because the generation time is constant, a logarithmic plot of growth during the log phase is a straight line. The log phase is the time when cells are most active metabolically.

However, during their log phase of growth, microorganisms are particularly sensitive to adverse conditions. Radiation and many antimicrobial drugs—for example, the antibiotic penicillin—exert their effect by interfering with some important step in the growth process and are therefore most harmful to cells during this phase.

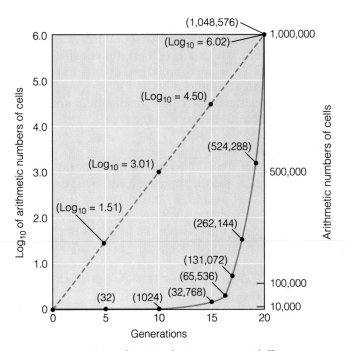

FIGURE 6.11 Growth curve for an exponentially increasing population, plotted logarithmically (dashed line) and arithmetically (solid line).

FIGURE 6.12 Bacterial growth curve, showing the four typical phases of growth.

Stationary Phase

If exponential growth continues unchecked, startlingly large numbers of cells could arise. For example, a single bacterium (at a weight of 9.5×10^{-13} g per cell) dividing every 20 minutes for only 25.5 hours can theoretically produce a population equivalent in weight to that of an 80,000-ton aircraft carrier. In reality, this does not happen. Eventually, the growth rate slows, the number of microbial deaths balances the number of new cells, and the population stabilizes. The metabolic activities of individual surviving cells also slow at this stage. This period of equilibrium is called the **stationary phase.**

What causes exponential growth to stop is not always clear. In addition to the exhaustion of nutrients and accumulation of waste products, harmful changes in pH may also play a role. In a specialized apparatus called a *chemostat,* a population can be kept in the exponential growth phase indefinitely by continually draining off spent medium and adding fresh medium.

Death Phase

Usually, the number of deaths soon exceeds the number of new cells formed, and the population enters the **death, or logarithmic decline, phase.** This phase continues until the population is diminished to a tiny fraction of the number of cells in the previous phase, or the population might die out entirely. Some species pass through the entire series of phases in only a few days; others retain some surviving cells almost indefinitely. Microbial death will be discussed further in the next chapter.

Direct Measurement of Microbial Growth

The growth of microbial populations can be measured in a number of ways. Some methods measure cell numbers; other methods measure the population's total mass, which is often directly proportional to cell numbers. Population numbers are usually recorded as the number of cells in a milliliter of liquid or in a gram of solid material. Because bacterial populations are usually very large, most methods of counting them are based on direct or indirect counts of very small samples; calculations then determine the size of the total population. Assume, for example, that a millionth of a milliliter (10^{-6} ml) of sour milk is found to contain 70 bacterial cells. Then there must be 70 times 1 million, or 70 million, cells per milliliter.

However, it is not practical to measure out a millionth of a milliliter of liquid or a millionth of a gram of food. Therefore, the procedure is done indirectly, in a series of dilutions. For example, if we add 1 milliliter of milk to 99 milliliters of water, each milliliter of this dilution now has one-hundredth as many bacteria as each milliliter of the original sample had. By making a series of such dilutions, we can readily estimate the number of bacteria in our original sample. To count microbial populations in foods (such as hamburger), a homogenate of one part food

to nine parts water is finely ground in a food blender. Samples can then be transferred with a pipette for further dilutions or cell counts.

Plate Counts

The most frequently used method for the measurement of bacterial populations is the **plate count.** An important advantage of this method is that it measures the number of viable cells. One disadvantage may be that it takes some time, usually 24 hours or more, for visible colonies to form. This can be a serious problem in some applications, such as quality control of milk, when it is not possible to hold a particular lot for this amount of time.

The plate count is based on three assumptions: that each bacterium grows and divides to produce a single colony, that the original inoculum is homogeneous, and that no aggregates of cells are present.

When a plate count is performed, it is important that only a limited number of colonies develop in the plate. When too many colonies are present, some cells are overcrowded and do not develop; these conditions cause inaccuracies in the count. Generally, only plates with from 25 to 250 colonies are counted. To ensure that some colony counts will be within this range, the original inoculum is diluted several times in a process called **serial dilution** (Figure 6.13).

Serial Dilutions. Let us say, for example, that a milk sample has 10,000 bacteria per milliliter. If 1 ml of this sample were plated out, there would theoretically be 10,000 colonies formed in the Petri plate of medium. Obviously, this would not produce a countable plate. If 1 ml of this sample were transferred to a tube containing 9 ml of sterile water, each milliliter of fluid in this tube would now contain 1000 bacteria. If 1 ml of this sample were inoculated into a Petri plate, there would still be too many potential colonies to count on a plate. Therefore, another serial dilution could be made. One milliliter containing 1000 bacteria would be transferred to a second tube of 9 ml of water. Each milliliter of this tube would now contain only 100 bacteria, and if 1 ml of the contents of this tube were plated out, potentially 100 colonies would be formed—an easily countable number.

Pour Plates and Spread Plates. A plate count is done by either the pour plate or the spread plate method. The **pour plate method** follows the procedure shown in Figure 6.14a. Either 1.0 ml or 0.1 ml of dilutions of the bacterial suspension is introduced into a Petri dish. The nutrient medium, in which the agar is kept liquid by holding it in a water bath at about 50°C, is poured over the sample, which is then mixed into the medium by gentle agitation of the plate. When the agar solidifies, the plate is incubated. With the pour plate technique, colonies will grow within the nutrient agar (from cells suspended in the nutrient medium as the agar solidifies) as well as on the surface of the agar plate. This technique has some drawbacks

Calculation: Number of colonies on plate × reciprocal of dilution of sample = number of bacteria/ml
(For example, if 32 colonies were on a plate of $^1/_{1000}$ dilution, then the count is 32 × 1000 = 32,000/ml in sample.)

FIGURE 6.13 **Plate counts and serial dilutions.** In serial dilutions, the original inoculum is diluted in a series of dilution tubes. In our example, each succeeding dilution tube will have only one-tenth the number of microbial cells as the preceding tube. Then samples of the dilution are used to inoculate Petri plates, on which colonies grow and can be counted. This count is then used to estimate the number of bacteria in the original sample.

because some relatively heat-sensitive microorganisms may be damaged by the melted agar and will therefore be unable to form colonies. Also, when certain differential media are used, the distinctive appearance of the colony on the surface is essential for diagnostic purposes. Colonies that form beneath the surface of a pour plate are not satisfactory for such tests. To avoid these problems, the **spread plate method** is frequently used instead (Figure 6.14b). A 0.1 ml inoculum is added to the surface of a pre-poured, solidified agar medium. The inoculum is then spread uniformly over the surface of the medium with a specially shaped sterilized glass rod. This method positions all the colonies on the surface and avoids contact of the cells with melted agar.

Filtration

When the quantity of bacteria is very small, as in lakes or relatively pure streams, bacteria can be counted by **filtration methods** (Figure 6.15). One hundred milliliters or more of water are passed through a thin membrane filter whose pores are too small to allow bacteria to pass. Thus, the bacteria are sieved out and retained on the surface of

the filter. This filter is then transferred to a Petri dish containing a pad soaked in liquid nutrient medium, where colonies arise from the bacteria on the filter's surface. This method is applied frequently to coliform bacteria, which are indicators of fecal pollution of food or water (see Chapter 27). The colonies formed by these bacteria are distinctive when a differential nutrient medium is used.

Most Probable Number Method

Another method for the determination of the number of bacteria in a sample is the **most probable number (MPN) method.** Figure 6.16 illustrates the MPN technique. This statistical estimating method is based on the fact that the greater the number of bacteria in a sample, the more dilution is needed to reduce the density to the point at which no bacteria are left to grow in the tubes in a dilution series. The MPN method is most useful when the microbes being counted will not grow on solid media (such as the chemoautotrophic nitrifying bacteria). It is also useful when growth of bacteria in a liquid differential medium is used to identify the microbes (such as coliform bacteria in water testing). The MPN is only a statement that there

FIGURE 6.14 Methods for preparation of plates for plate counts. (a) Pour plate. (b) Spread plate.

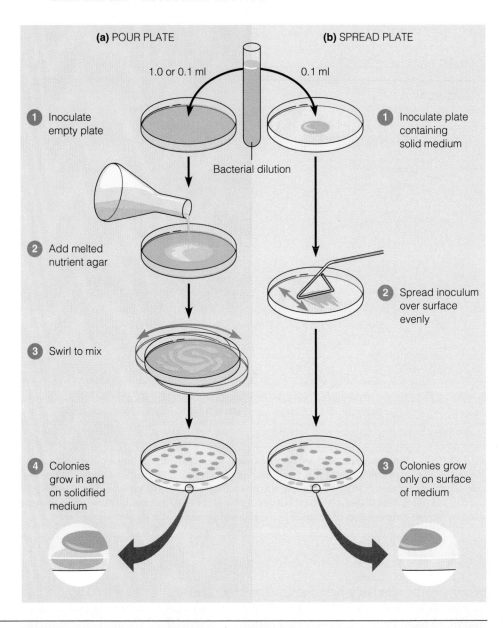

(a) POUR PLATE

1.0 or 0.1 ml

1 Inoculate empty plate

Bacterial dilution

2 Add melted nutrient agar

3 Swirl to mix

4 Colonies grow in and on solidified medium

(b) SPREAD PLATE

0.1 ml

1 Inoculate plate containing solid medium

2 Spread inoculum over surface evenly

3 Colonies grow only on surface of medium

FIGURE 6.15 Counting bacteria by filtration. (a) The bacteria in 100 ml of water were sieved out onto the surface of a membrane filter. (b) Such a filter, with the bacteria much more widely spaced, was placed on a pad saturated with liquid nutrient medium, and the individual bacteria grew into visible colonies. Twenty-seven colonies are visible, so we would record 27 bacteria per 100 ml of water sample.

(a)   **SEM** |———| 10 μm   (b)

is a 95% chance that the bacterial population falls within a certain range and that the MPN is statistically the most probable number.

Direct Microscopic Count

In the method known as the **direct microscope count**, a measured volume of a bacterial suspension is placed inside a defined area on a microscope slide. In the *Breed count method* for milk bacteria, for example, a 0.01-ml sample is spread over a marked square centimeter of slide, stain is added so that the bacteria can be seen, and the sample is viewed under the oil immersion objective. The area of the viewing field of this objective can be determined. Once the number of bacteria has been counted in several different fields, the average number of bacteria per viewing field can be calculated. From these data, the number of bacteria in the square centimeter over which the sample was spread can also be calculated. Because this area on the slide contained 0.01 ml of sample, the number of bacteria in each milliliter of the suspension is the number of bacteria in the sample times 100.

A specially designed slide called a *Petroff-Hausser counter* is also used in direct microscopic counts (Figure 6.17). A shallow well of known volume is indented into the surface of a microscope slide inscribed with squares of known areas and covered with a thin cover glass. The well is filled with the microbial suspension. The average number of bacteria in each of a series of these squares is calculated and then multiplied by the factor that produces the count per milliliter. Motile bacteria are difficult to count by this method, and, as happens with other microscopic methods, dead cells are about as likely to be counted as live ones. In addition to these disadvantages, a rather high concentration of cells is required—about 10 million bacteria per milliliter. The chief advantage of microscopic counts is that no incubation time is required, and they are usually reserved for applications in which time is the primary consideration. This advantage also holds for *electronic cell counters*, which automatically count the number of cells in a measured volume of liquid. These instruments are used in some research laboratories and hospitals.

Estimations of Bacterial Numbers by Indirect Methods

It is not always necessary to count microbial cells to estimate their numbers. In science and industry, microbial numbers and activity are determined by some of the following indirect means as well.

Turbidity

For some types of experimental work, estimating **turbidity** is a practical way of monitoring bacterial growth. As bacteria multiply in a liquid medium, the medium becomes *turbid*, or cloudy with cells.

Volume of inoculum for each set of five tubes	Tubes of nutrient medium (sets of five tubes)	Number of positive tubes in set
10 ml		5
1 ml		3
0.1 ml		1

(a) Example

Combination of positives	MPN index/ 100 ml	95% confidence limits	
		Lower	Upper
4-2-0	22	9	56
4-2-1	26	12	65
4-3-0	27	12	67
4-3-1	33	15	77
4-4-0	34	16	80
5-0-0	23	9	86
5-0-1	30	10	110
5-0-2	40	20	140
5-1-0	30	10	120
5-1-1	50	20	150
5-1-2	60	30	180
5-2-0	50	20	170
5-2-1	70	30	210
5-2-2	90	40	250
5-3-0	80	30	250
5-3-1	110	40	300
5-3-2	140	60	360

(b) MPN table

FIGURE 6.16 Most probable number (MPN) determination. (a) In this example, there are three sets of tubes and five tubes in each set. Each tube in the first set of five tubes receives 10 ml of the inoculum, such as a sample of water. Each tube in the second set of five tubes receives 1 ml of the sample, and the third set, 0.1 ml each. In our example, there were enough bacteria in the sample so that all five tubes in the first set showed bacterial growth and were recorded as positive. In the second set of tubes, which received only one-tenth as much inoculum, only three tubes were positive. In the third set of tubes, which received one-hundredth as much inoculum, only one tube was positive. **(b)** Tables of MPN allow us to calculate for a sample the microbial numbers that are statistically likely to lead to such a result. The number of positive tubes is recorded for each set: 5, 3, and 1. If we look up this combination in an MPN table (see Appendix B for complete MPN table), we find that the MPN index per 100 ml is 110. Statistically, this means that 95% of the water samples that give this result contain between 40 and 300 bacteria, with 110 being the most frequent number.

FIGURE 6.17 **Direct microscopic count of bacteria with a Petroff-Hausser cell counter.** The average number of cells within a large square multiplied by a factor of 1,250,000 gives the number of bacteria per milliliter.

Grid with 25 large squares

Cover glass

Slide

1 Bacterial suspension is added here and fills the shallow well over the squares by capillary action.

Bacterial suspension

Cover glass

Slide

Location of squares

2 Cross section of a cell counter. The depth under the cover glass is known, and the area of the squares is known, so the volume of the bacterial suspension over the squares can be calculated (depth × area).

3 Microscopic count—all cells in several large squares are counted, and the numbers are averaged. The large square shown here has 14 bacterial cells.

4 The volume of fluid over the large square is 1/1,250,000th of a milliliter. If it contains 14 cells, as shown here, then there are 14 times 1,250,000 (17,500,000) cells in a milliliter.

FIGURE 6.18 **Turbidity estimation of bacterial numbers.** The amount of light striking the light-sensitive detector on the spectrophotometer is inversely proportional to the number of bacteria under standardized conditions. The less light transmitted, the more bacteria in the sample.

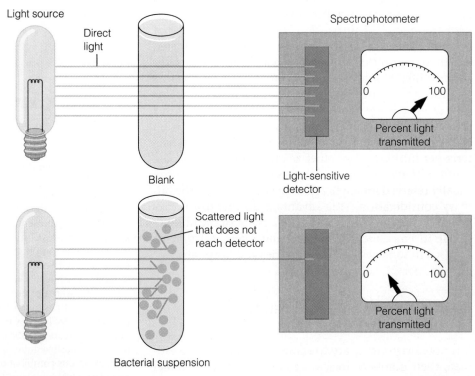

Light source

Direct light

Spectrophotometer

0   100

Percent light transmitted

Blank

Light-sensitive detector

Scattered light that does not reach detector

0   100

Percent light transmitted

Bacterial suspension

The instrument used to measure turbidity is a spectrophotometer, or colorimeter. In the spectrophotometer, a beam of light is transmitted through a bacterial suspension to a photoelectric cell (Figure 6.18). As bacterial numbers increase, less light will reach the photoelectric cell. This change of light will register on the instrument's scale as the *percentage of transmission*. Also printed on the instrument's scale is a logarithmic expression called the *absorbance* (sometimes called *optical density,* or *OD*), a value derived from the percentage of transmission. The absorbance is used to plot bacterial growth. When the bacteria are in logarithmic growth or decline, a graph of absorbance versus time will form an approximately straight line. If absorbance readings are matched with plate counts of the same culture, this correlation can be used in future estimations of bacterial numbers obtained by measuring turbidity.

More than a million cells per milliliter must be present for the first traces of turbidity to be visible. About 10 million to 100 million cells per milliliter are needed to make a suspension turbid enough to be read on a spectrophotometer. Therefore, turbidity is not a useful measure of contamination of liquids by relatively small numbers of bacteria.

Metabolic Activity

Another indirect way to estimate bacterial numbers is to measure the *metabolic activity* of the population. This method assumes that the amount of a certain metabolic product, such as acid or CO_2, is in direct proportion to the number of bacteria present. Bacterial numbers can also be estimated by a reduction test, which measures oxygen uptake directly or indirectly. In this test, a dye that changes color in the presence or absence of oxygen, such as methylene blue, is added to a medium such as milk, and the filled tube is tightly capped. The bacteria then use the oxygen as they metabolize the milk. Because methylene blue is blue in the presence of oxygen and colorless in its absence, the faster the dye (and thus the milk) loses color, the faster the oxygen is being depleted and the more bacteria are presumed to be present in the milk. Reduction tests are frequently used in microbiology teaching laboratories, but they lack accuracy and are seldom used in commercial applications.

Dry Weight

For filamentous organisms, such as molds, the usual measuring methods are less satisfactory. A plate count would not measure this increase in filamentous biomass. In plate counts of molds, the number of asexual spores is counted instead, but this is often not a good measure of growth. One of the better ways to measure the growth of filamentous organisms is by *dry weight*. In this procedure, the fungus is removed from the growth medium, filtered to remove extraneous material, placed in a weighing bottle, and dried in a desiccator. For bacteria, the same basic procedure is followed. The bacteria are usually removed from the culture medium by centrifugation.

• • •

You now have a basic understanding of the requirements for, and measurements of, microbial growth. In Chapter 7, we will look at how this growth is controlled in laboratories, hospitals, industry, and our homes.

STUDY OUTLINE

REQUIREMENTS FOR GROWTH (pp. 142–147)

1. The growth of a population is an increase in the number of cells or in mass.
2. Requirements for microbial growth are divided into physical and chemical needs.

Physical Requirements (pp. 142–145)

1. On the basis of growth range of temperature, microbes are classified as psychrophiles (cold-loving), mesophiles (moderate-temperature–loving), and thermophiles (heat-loving).
2. The minimum growth temperature is the lowest temperature at which a species will grow; the optimum growth temperature is the temperature at which it grows best; and the maximum growth temperature is the highest temperature at which growth is possible.
3. Most bacteria grow best at a pH between 6.5 and 7.5.
4. In a hypertonic solution, most microbes undergo plasmolysis; halophiles can tolerate high salt concentrations.

Chemical Requirements (pp. 145–147)

1. All organisms require a carbon source; chemoheterotrophs use an organic molecule, and autotrophs typically use carbon dioxide.
2. Nitrogen is needed for protein and nucleic acid synthesis. Nitrogen can be obtained from decomposition of proteins or from NH_4^+ or NO_3^-; a few bacteria are able to fix N_2.
3. On the basis of oxygen requirements, organisms are classified as obligate aerobes, facultative anaerobes, obligate anaerobes, aerotolerant anaerobes, and microaerophiles.
4. Aerobes, facultative anaerobes, and aerotolerant anaerobes must have the enzymes superoxide dismutase ($2O_2^- + 2H^+ \rightarrow O_2 + H_2O_2$) and either catalase ($2H_2O_2 \rightarrow 2H_2O + O_2$) or peroxidase ($H_2O_2 + 2H^+ \rightarrow 2H_2O$).
5. Other chemicals required for microbial growth include sulfur, phosphorus, trace elements, and, for some microorganisms, organic growth factors.

CULTURE MEDIA (pp. 147–154)

1. A culture medium is any material prepared for the growth of bacteria in a laboratory.
2. Microbes that grow and multiply in or on a culture medium are known as a culture.
3. Agar is a common solidifying agent for a culture medium.

Chemically Defined Media (p. 148)

1. A chemically defined medium is one in which the exact chemical composition is known.

Complex Media (p. 150)

1. A complex medium is one in which the exact chemical composition is not known.

Anaerobic Growth Media and Methods (pp. 150–151)

1. Reducing media chemically remove molecular oxygen (O_2) that might interfere with the growth of anaerobes.
2. Petri plates can be incubated in an anaerobic jar or anaerobic glove box.

Special Culture Techniques (pp. 151–152)

1. Some parasitic and fastidious bacteria must be cultured in living animals or in cell cultures.
2. CO_2 incubators or candle jars are used to grow bacteria requiring an increased CO_2 concentration.

Selective and Differential Media (pp. 152–153)

1. By inhibiting unwanted organisms with salts, dyes, or other chemicals, selective media allow growth of only the desired microbes.
2. Differential media are used to distinguish among different organisms.

Enrichment Culture (p. 154)

1. An enrichment culture is used to encourage the growth of a particular microorganism in a mixed culture.

Obtaining Pure Cultures (p. 154)

1. A clone is a population of cells arising from a single cell.
2. Colonies are visible clones that have grown on solid media.
3. Pure cultures are usually obtained by the streak plate method.

PRESERVING BACTERIAL CULTURES (p. 154)

1. Microbes can be preserved for long periods of time by deep-freezing or lyophilization (freeze-drying).

GROWTH OF BACTERIAL CULTURES

(pp. 155–163)

Bacterial Division (p. 155)

1. The normal reproductive method of bacteria is binary fission, in which a single cell divides into two identical cells.
2. Some bacteria reproduce by budding, aerial spore formation, or fragmentation.

Generation Time (pp. 155–156)

1. The time required for a cell to divide or a population to double is known as the generation time.

Logarithmic Representation of Bacterial Populations (pp. 156–157)

1. Bacterial division occurs according to a logarithmic progression (2 cells, 4 cells, 8 cells, etc.).

Phases of Growth (pp. 157–158)

1. During the lag phase, there is little or no change in the number of cells, but metabolic activity is high.
2. During the log phase, the bacteria multiply at the fastest rate possible under the conditions provided.
3. During the stationary phase, there is an equilibrium between cell division and death.
4. During the death phase, the number of deaths exceeds the number of new cells formed.

Direct Measurement of Microbial Growth

(pp. 158–161)

1. A standard plate count reflects the number of viable microbes and assumes that each bacterium grows into a single colony.
2. A plate count may be done by either the pour plate method or spread plate method.
3. In filtration, bacteria are retained on the surface of a membrane filter and then are transferred to a culture medium to grow and subsequently be counted.
4. The most probable number (MPN) method can be used for microbes that will grow in a liquid medium; it is a statistical estimation.
5. In a direct microscopic count, the microbes in a measured volume of a bacterial suspension are counted with the use of a specially designed slide.

Estimations of Bacterial Numbers by Indirect Methods (pp. 161–163)

1. A spectrophotometer is used to determine turbidity by measuring the amount of light that passes through a suspension of cells.
2. An indirect way of estimating bacterial numbers is measurement of the metabolic activity of the population, for example, acid production or oxygen consumption.
3. For filamentous organisms such as fungi, measuring dry weight is a convenient method of growth measurement.

STUDY QUESTIONS

Review

1. Describe binary fission.

2. What physical and chemical properties of agar make it useful in culture media?

3. Draw a typical bacterial growth curve. Label and define each of the four phases.

4. Macronutrients (needed in relatively large amounts) are often listed as CHONPS. What does each of these letters indicate, and why are they needed by the cell?

5. Most bacteria grow best at pH _____ .

6. Why can high concentrations of salt or sugar be used to preserve food?

7. Explain five bacterial categories based on organisms' requirements for oxygen.

8. Define the following terms and explain the importance of each enzyme:

   (a) Catalase   (d) Superoxide free radical
   (b) Hydrogen peroxide   (e) Superoxide dismutase
   (c) Peroxidase

9. Which of the following curves best depicts the log phase of:
   (a) A thermophile incubated at room temperature?
   (b) *Listeria monocytogenes* growing in a human?
   (c) A psychrophile incubated at 9°C?

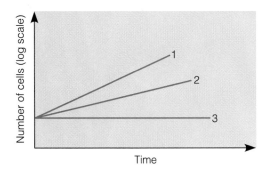

10. *Clostridium* can be cultured in an anaerobic incubator or in the presence of atmospheric oxygen if thioglycolate is added to the nutrient broth. Compare these two techniques. Using terms from Question 8, explain why elaborate culture techniques are used for *Clostridium*.

11. Seven methods of measuring microbial growth were explained in this chapter. Categorize each as either a direct or an indirect method.

12. By deep-freezing, bacteria can be stored without harm for extended periods. Why do refrigeration and freezing preserve foods?

13. A pastry chef accidentally inoculated a cream pie with six *S. aureus* cells. If *S. aureus* has a generation time of 60 minutes, how many cells would be in the cream pie after seven hours?

14. Nitrogen and phosphorus added to beaches following an oil spill encourage growth of natural oil-degrading bacteria. Explain why the bacteria do not grow if nitrogen and phosphorus are not added.

15. A normal fecal sample has a high concentration of coliform bacteria. A patient with typhoid fever will excrete *Salmonella* along with normal bacteria. Feces from a suspected case of typhoid fever are inoculated on an S-S agar plate. After proper incubation, coliform bacteria form red colonies, and *Salmonella* bacteria form colorless colonies with a black center. Is S-S agar selective? Differential? How can you tell?

16. Differentiate between complex and chemically defined media.

17. Draw the following growth curves for *E. coli*, starting with 100 cells with a generation time of 30 min at 35°C.
    (a) The cells are incubated for 5 hours at 35°C.
    (b) After 5 hours, the temperature is changed to 20°C for 2 hours.
    (c) After 5 hours at 35°C, the temperature is changed to 5°C for 2 hours followed by 35°C for 5 hours.

Critical Thinking

1. *E. coli* was incubated with aeration in a nutrient medium containing two carbon sources, and the following growth curve was made from this culture.

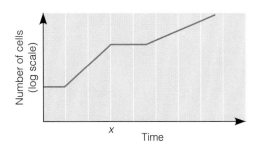

   (a) Explain what happened at the time marked *x*.
   (b) Which substrate provided "better" growth conditions for the bacteria? How can you tell?

2. *Clostridium* and *Streptococcus* are both catalase-negative. *Streptococcus* grows by fermentation. Why is *Clostridium* killed by oxygen, whereas *Streptococcus* is not?

3. Design an enrichment medium and a procedure for growing an endospore-forming bacterium that fixes nitrogen and uses cellulose for its carbon source.

4. Most laboratory media contain a fermentable carbohydrate and peptone because the majority of bacteria require carbon, nitrogen, and energy sources in these forms. How are these three needs met by glucose–minimal salts medium? (See Table 6.4.)

5. Flask A contains yeast cells in glucose–minimal salts broth incubated at 30°C with aeration. Flask B contains yeast cells in glucose–minimal salt broth incubated at 30°C in an anaerobic jar. The yeasts are facultative anaerobes.
   (a) Which culture produced more ATP?
   (b) Which culture produced more alcohol?
   (c) Which culture had the shorter generation time?
   (d) Which culture had the greater cell mass?
   (e) Which culture had the higher absorbance?

Clinical Applications

1. Assume that, after washing your hands, you leave ten bacterial cells on a new bar of soap. You then decide to do a plate count of the soap after it was left in the soap dish for 24 hours. You dilute 1 g of the soap 1:10^6 and plate it on standard plate count agar. After 24 hours of incubation, there are 168 colonies. How many bacteria were on the soap? How did they get there?

2. Heat lamps are commonly used to maintain foods at about 50°C for as long as 12 hours in cafeteria serving lines. The following experiment was conducted to determine whether this practice poses a potential health hazard.

   Beef cubes were surface-inoculated with 500,000 bacterial cells and incubated at 43–53°C to establish temperature limits for bacterial growth. The following results were obtained from standard plate counts performed on beef cubes at 6 and 12 hours after inoculation:

	TEMP. (°C)	BACTERIA/G OF BEEF	
		6 HR	12 HR
S. aureus	43	140,000,000	740,000,000
	51	810,000	59,000
	53	650	300
S. typhimurium	43	3,200,000	10,000,000
	51	950,000	83,000
	53	1,200	300
C. perfringens	43	1,200,000	3,600,000
	51	120,000	3,800
	53	300	300

Draw the growth curves for each organism. What holding temperature would you recommend? Assuming that cooking kills bacteria in foods, how could these bacteria contaminate the cooked foods? What disease does each organism cause? (*Hint:* Refer to Chapter 25.)

Control of Microbial Growth

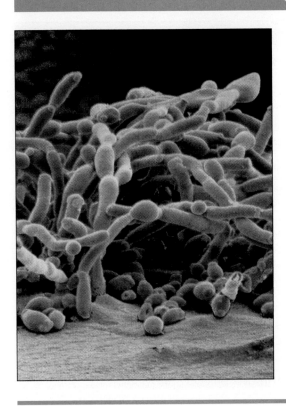

THE SCIENTIFIC CONTROL of microbial growth began only about 100 years ago. Prior to that time, epidemics periodically killed thousands of people. In some hospitals, 25% of delivering mothers died of infections carried by the hands and instruments of attending nurses and physicians. Ignorance of microbes was such that, during the American Civil War, a surgeon might have cleaned his scalpel on his bootsole between incisions.

THE HISTORY AND CONCEPT OF MICROBIAL CONTROL

LEARNING OBJECTIVES

- *Identify the contributions made by Semmelweis and Lister to the control of microbes.*
- *Define the following key terms related to microbial control: sterilization, disinfection, antisepsis, germicide, bacteriostasis, asepsis, degerming, and sanitation.*

Ignatz Semmelweis (1816–1865), a Hungarian physician working in Vienna, and Joseph Lister (1827–1912), an English physician, first introduced the concept of microbial control. At the obstetrics ward in the Vienna General Hospital, Semmelweis required all personnel to wash their hands in chlorinated lime; that procedure significantly lowered the infection rate. Lister, meanwhile, read about

Pasteur's work with microbes and concluded that the number of infected surgical wounds (sepsis) could be decreased through procedures that prevented the access of microbes to the wound. At that time, mortality rates of about 10% were common in hospitals' surgical wards; contemporary estimates were that a person had a better chance of surviving battle than hospitalization. Although Lister did not fully understand wound contamination by hands and instruments, he had considerable success treating surgical wounds with solutions of phenol (carbolic acid); he even operated in a carbolic acid aerosol. The use of this chemical was suggested, in part, by its use (in the crude form of creosote) to preserve railroad ties. Today, the methods that prevent microbial contamination of surgical wounds are known as **aseptic surgery.**

We have come a long way in controlling microbial growth since the time of Semmelweis and Lister. This chapter will discuss how microbial growth can be controlled by physical methods and chemical agents. Physical methods include the use of heat, low temperatures, desiccation, osmotic pressure, filtration, and radiation. Chemical agents include several groups of substances that destroy microbes or limit microbial growth on body surfaces and on inanimate objects. In Chapter 20, we will discuss methods for the control of microbes once infection has occurred, mainly antibiotic chemotherapy.

Before we begin our survey of physical and chemical methods used to control microbial growth, we need to define the terms we will use to discuss this control. Table 7.1 contains a list of terms related to the control of microbial

TABLE 7.1   Terminology Related to the Control of Microbial Growth

TERM	DEFINITION	COMMENTS
Terms Related to Destruction of Organisms		
Sterilization	The process of destroying all forms of microbial life on an object or in a material. This includes the destruction of endospores—the most resistant form of microbial life. *Sterilization* is absolute; there are no degrees of sterilization.	Typical temperatures required: moist heat, 121°C for 15 minutes; dry heat, 170°C for 120 minutes. Other methods are ionizing radiation and gases such as ethylene oxide.
Disinfection	The process of destroying vegetative pathogens but not necessarily endospores or viruses. Usually, a *disinfectant* is a chemical applied to an object or a material. Disinfectants tend to reduce or inhibit growth; they usually do not sterilize.	Term usually applied to use of liquid chemical solutions on surfaces or to elimination of pathogens in water—for example, by chlorination.
Antisepsis	Chemical disinfection of the skin, mucous membranes, or other living tissues.	Term especially applied to treatment of wounds. *Antisepsis* is a specific kind of *disinfection*.
Germicide (*cide,* kill)	A chemical agent that rapidly kills microbes but not necessarily their endospores.	A *bactericide* kills bacteria, a *sporicide* kills endospores, a *fungicide* kills fungi, a *virucide* kills viruses, and an *amoebicide* kills amoebas.
Terms Related to Suppression of Organisms		
Bacteriostasis (*stasis,* halt)	A condition in which bacterial growth and multiplication are inhibited, but the bacteria are not killed. If the bacteriostatic agent is removed, bacterial growth and multiplication may resume. *Fungistasis* refers to the inhibition of fungal growth.	Refrigeration is bacteriostatic; many chemicals such as dyes are bacteriostatic rather than bactericidal.
Asepsis (*asepsis,* without infection)	The absence of pathogens from an object or area. Aseptic techniques are designed to prevent the entry of pathogens into the body. Whereas *surgical asepsis* is designed to exclude all microbes, *medical asepsis* is designed to exclude microbes associated with communicable diseases.	Air filtration; ultraviolet lights; personnel masks, gloves, and gowns; and instrument sterilization are all factors in achieving *asepsis*.
Degerming	The removal of transient microbes from the skin by mechanical cleansing or by the use of an antiseptic.	For routine injections, alcohol swabs are often used; before surgery, iodine-containing products are often used.
Sanitization	The reduction of pathogens on eating utensils to safe public health levels by mechanical cleansing or chemicals.	Any chemicals used must be compatible with safety and palatability of foods.

growth. Note that "microbial control" does not always mean killing the microbes; some situations require that they only be inhibited or removed. The term *sterilization,* however, refers to those situations in which all forms of life must be destroyed (something very difficult to do with chemicals).

CONDITIONS INFLUENCING MICROBIAL CONTROL

LEARNING OBJECTIVE

- *Explain how microbial growth is affected by the type of microbes, the microbe's physiological state, and the ambient environmental conditions.*

Usually, if we want to sterilize an object, such as a laboratory flask, we do not bother to identify the species of mi-

crobes on it. Instead, we launch a general attack calculated to be strong enough to kill the most resistant microbial forms that could possibly be present—endospores. The time it takes to reduce or eliminate microbial growth on any object depends, in part, on the number and species of microbes on the object. It also depends on the following factors.

Temperature

Everyone is familiar with the use of ice or mechanical refrigeration to control microbial growth. Like all chemical reactions, the biochemical reactions required for growth are slowed considerably by low temperatures. However, chemical disinfectants are also often inhibited by low temperatures. Because their activity is due to temperature-dependent chemical reactions, disinfectants tend to work somewhat better in warm solutions. You might have noticed that directions on disinfectant containers frequently specify that a warm solution be used.

Type of Microbe

Many disinfectants and antiseptics tend to have a greater effect on gram-positive bacteria, as a group, than on gram-negative bacteria, although this difference is not as clear-cut as it is with antibiotic activity. A certain group of gram-negative bacteria, the pseudomonads (genus *Pseudomonas*), is of special interest (see Chapter 11). Pseudomonads, which are common in the environment, are unusually resistant to chemical activity and will even grow actively *in* some disinfectants and antiseptics (see box on p. 170). These bacteria are also resistant to many antibiotics. This resistance to antimicrobials is probably related to the characteristics of the *porins* (structural openings in the wall) of pseudomonads (see Chapter 4). Because of these factors, pseudomonads are very troublesome in a hospital setting.

The mycobacteria are another group of non–endospore-forming organisms that exhibit greater than normal resistance to disinfectants. This group includes *Mycobacterium tuberculosis,* the microbe that causes tuberculosis, and a number of mycobacteria that are pathogenic in people with depressed immune systems (opportunistically pathogenic). Special tuberculocidal tests have been developed to evaluate the effectiveness of chemical antimicrobials acting on the mycobacteria.

The endospores of bacteria and cysts of protozoa are also highly resistant to chemical agents, as are some viruses. The resistance of these organisms to chlorine disinfection in water treatment is a particularly important factor in public health efforts.

Physiological State of the Microbe

Microorganisms that are actively growing tend to be more susceptible to chemical agents than are older cells. This might be because an actively metabolizing and reproducing cell has more points of vulnerability than does an older, possibly dormant cell. When microorganisms have formed endospores, the endospores are generally more resistant than the vegetative cells. The endospores of the resistant strains of *Clostridium botulinum* (bo-tū-lī′num), for example, can withstand boiling for several hours.

Environment

Organic matter frequently interferes with the action of chemical agents. In hospitals, the presence of organic matter in vomit and feces influences the selection of disinfectants because some disinfectants are more effective than others under these conditions. Even bacteria found in food tend to be protected by the organic matter of the food. For example, bacteria in cream are protected by pro-teins and fats and can survive more heat than bacteria in skim milk. Heat is much more effective in killing bacteria under acidic conditions than at a neutral pH. Some disinfectants also work best at an acidic pH.

ACTIONS OF MICROBIAL CONTROL AGENTS

LEARNING OBJECTIVE

- *Describe the effects of microbial control agents on cellular structures.*

In this section we will examine the mechanisms of action of control agents—that is, the way various agents actually kill or inhibit microbes.

Alteration of Membrane Permeability

The plasma membrane, located just inside the cell wall, is the target of many microbial control agents. This membrane actively controls the passage of nutrients into the cell and the elimination of wastes from the cell. Damage to the lipids or proteins of the plasma membrane by antimicrobial agents, such as quaternary ammonium compounds, typically causes cellular contents to leak into the surrounding medium and interferes with the growth of the cell. Several types of chemical agents and antibiotics work at least partially in this manner.

Damage to Proteins and Nucleic Acids

Bacteria are sometimes thought of as "little bags of enzymes." Enzymes, which are primarily protein, are vital to all cellular activities. You might recall that the functional properties of proteins are due to their three-dimensional shape. This shape is maintained by chemical bonds that link adjoining portions of the amino acid chain as it folds back and forth upon itself. Some of those bonds are hydrogen bonds, which are susceptible to breakage by heat or certain chemicals; breakage results in denaturation of the protein. Covalent bonds, which are stronger, are also subject to attack. For example, disulfide bridges, which play an important role in protein structure by joining amino acids with exposed sulfhydryl (—SH) groups, can be broken by certain chemicals or sufficient heat.

Because DNA and RNA carry the genetic message, damage to these nucleic acids by radiation or chemicals is frequently lethal to the cell; this damage prevents both replication and normal functioning.

M M W R

Morbidity and Mortality Weekly Report

HOSPITAL-ACQUIRED INFECTION FROM CONTAMINATED ENDOSCOPES— WISCONSIN

FLEXIBLE FIBER-OPTIC ENDOSCOPES are widely used in the U.S.A. for diagnostic and therapeutic procedures in which the interior of a hollow organ (such as the rectum or urethra) needs to be visualized. Between uses, endoscopes should be cleaned and disinfected either manually or by using automated machines. However, outbreaks have occurred among patients exposed to contaminated endoscopes or to inadequately disinfected bronchoscopes. This report summarizes one outbreak of hospital-acquired infection linked to endoscopes contaminated during cleaning and disinfection by automated reprocessing machines.

Pseudomonas aeruginosa colonization or infection of the biliary tract, respiratory tract, or bloodstream occurred in 115 of 1349 patients undergoing gastrointestinal (GI) endoscopic procedures. The endoscopes were routinely washed using an automated reprocessing machine that flushed with a detergent solution, disinfected with one of two liquid chemical germicides (2% glutaraldehyde; 2% glutaraldehyde/7.05% phenol/1.2% sodium phenate diluted 1:16 in tap water), and rinsed with tap water.

An investigation performed by the hospital indicated that a thick biofilm of *P. aeruginosa* had formed in the detergent holding tank, inlet water hose, and air vents of the automated machine. Attempts to disinfect the machine by the manufacturer's instructions using commercial preparations of glutaraldehyde were unsuccessful.

P. aeruginosa serotype 10 was the predominant serotype recovered from the automated machine, from available isolates from patients with post-endoscopy colonization or infection, and from endoscopes that had been sampled after disinfection in the machine. Molecular subtyping (by electrophoresis of endonuclease-digested DNA) confirmed that the *P. aeruginosa* serotype 10 isolates recovered from the reprocessing machine and from infected patients were identical.

Hospital personnel began manually rinsing machine-washed endoscope channels and external surfaces with 70% isopropyl alcohol and drying the channels and surfaces with forced air for 10–20 minutes. Infection rates were lower (28 of 996 patients), although the automated machine remained colonized with *P. aeruginosa*. No additional cases have occurred.

Editorial Note: In the U.S.A., use of automated endoscope reprocessing machines has increased—in part because of the complex and time-consuming process of manually cleaning and disinfecting endoscopes. Centers for Disease Control and Prevention has previously recommended that diagnostic and therapeutic de-

vices that contact mucous membranes during use (e.g., endoscopes and bronchoscopes) should be either sterilized or subjected to high-level disinfection. However, the findings in this report emphasize the potential for some automated endoscope reprocessing machines to become colonized with heterotrophic organisms such as *P. aeruginosa.*

In the hospital described in this report, the automated reprocessing machine was identified as the source of contamination. This contamination was present in multiple sites, including the detergent and water holding tanks, water hoses, and air vents. *P. aeruginosa,* an organism commonly found in water, may colonize water distribution pipes, tubing, and reservoirs. Once affixed to the inner surface of pipes, *P. aeruginosa* could be protected from the bactericidal action of glutaraldehyde, probably by a glycocalyx film. Bacteria that come free may end up infecting patients.

New types of plastic and metals are being sought that prevent this bacterial biofilm from developing. Bacteria that adhere to plastic and metal are also a source of infection for people with implants such as heart valves and pacemakers. New materials that prevent bacterial adhesion would be beneficial to these people as well.

SOURCE: Adapted from *MMWR* 40(39):675–678 (10/4/91).

RATE OF MICROBIAL DEATH

LEARNING OBJECTIVE

- *Describe the patterns of microbial death caused by treatments with microbial control agents.*

When bacterial populations are heated or treated with antimicrobial chemicals, they usually die at a constant rate. For example, suppose a population of 1 million microbes has been treated for 1 minute and 90% of the population has died. We are now left with 100,000 microbes. If the population is treated for another minute, 90% of *those* microbes die, and we are left with 10,000 survivors. In other words, for each minute that the treatment is applied, 90% of the remaining population is killed (Table 7.2). If the death curve is plotted logarithmically, the death rate is seen to be constant, as shown by the straight line in Figure 7.1. Obviously, the more microbes there are at the beginning, the longer it will take to kill them all.

Now that you have an understanding of the basic principles of microbial control, we will examine some representative physical and chemical methods.

PHYSICAL METHODS OF MICROBIAL CONTROL

LEARNING OBJECTIVES

- *Compare the effectiveness of moist heat (boiling, autoclaving, pasteurization) and dry heat.*
- *Describe how filtration, cold, desiccation, and osmotic pressure suppress microbial growth.*
- *Explain how radiation kills cells.*

As early as the Stone Age, it is likely that humans were already using some physical methods of microbial control to preserve foods. Drying (desiccation) and salting (osmotic pressure) were probably among the earliest. We next examine the physical means of controlling microbes.

Heat

Probably the most common method by which microbes are now killed is heat. A visit to any supermarket will demonstrate that canned goods—in which microorganisms have been killed by moist heat—outnumber the foods preserved by other methods. Laboratory media and glassware and hospital instruments are also usually sterilized by heat.

Heat is not only the most widely applicable and effective agent for sterilization but also the most economical and easily controlled. Heat appears to kill microbes by de-

TABLE 7.2 Example of Microbial Death Rate

TIME (MINUTES)	DEATHS PER MINUTE	NUMBER OF SURVIVORS
0	0	1,000,000
1	900,000	100,000
2	90,000	10,000
3	9,000	1,000
4	900	100
5	90	10
6	9	1

SOURCE: O. Rahn, *Physiology of Bacteria.* New York: McGraw-Hill, 1932.

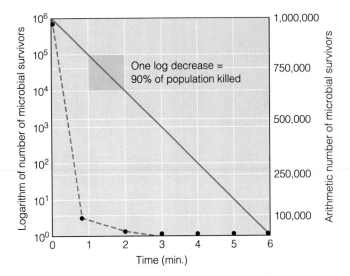

FIGURE 7.1 Microbial death curve. The curve is plotted logarithmically (solid line) and arithmetically (broken line). In this case, the cells are dying at a rate of 90% each minute.

naturing their enzymes. To sterilize with heat, the degree of heat resistance of the bacteria must be considered. Heat resistance varies among different microbes; these differences can be expressed through the concept of thermal death point. **Thermal death point (TDP)** is the lowest *temperature* at which all of the microorganisms in a liquid suspension will be killed in 10 minutes.

Another factor to be considered in sterilization is the length of time required for the material to be rendered sterile. This is expressed as **thermal death time (TDT)**, the minimal length of *time* in which all bacteria in a liquid culture will be killed at a given temperature. Both TDP and TDT are useful guidelines that indicate the severity of treatment required to kill a given population of bacteria.

Decimal reduction time (DRT or D value) is a third concept related to the bacteria's degree of heat resistance. It is the time, in minutes, in which 90% of the population

FIGURE 7.2 Autoclave. The entering steam forces the air out of the bottom (blue arrows). The automatic ejector valve remains open as long as an air-steam mixture is passing out of the waste line. When all the air has been ejected, the higher temperature of the pure steam closes the valve, and the pressure in the chamber increases.

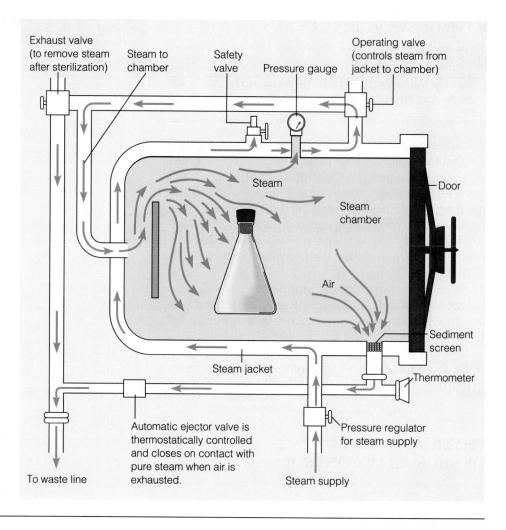

of bacteria at a given temperature will be killed (in Figure 7.1, DRT is 1 minute). DRT is especially useful in the canning industry.

The heat used in sterilization can be applied in the form of moist heat or dry heat. We consider moist heat first.

Moist Heat

Moist heat kills microorganisms primarily by the coagulation of proteins, which is caused by breakage of the hydrogen bonds that hold the proteins in their three-dimensional structure. This coagulation process is familiar to anyone who has watched an egg white frying. Protein coagulation, or denaturation, occurs more quickly in the presence of water.

One type of moist heat sterilization is boiling (100°C at sea level), which kills vegetative forms of bacterial pathogens, many viruses, and fungi and their spores within about 10 minutes. Free-flowing (unpressurized) steam is equivalent in temperature to boiling water. Endospores and some viruses, however, are not destroyed so quickly. One type of hepatitis virus, for example, can sur-

vive up to 30 minutes of boiling, and some bacterial endospores have resisted boiling for more than 20 hours. Boiling is therefore not always a reliable sterilization procedure. However, boiling for a few minutes will kill most pathogens and generally will make food or water safe to eat or drink.

Reliable sterilization with moist heat requires temperatures above that of boiling water. These high temperatures are most commonly achieved by steam under pressure in an **autoclave** (Figure 7.2). As the preferred method of sterilization, autoclaving is used unless the material to be sterilized can be damaged by heat or moisture.

The higher the pressure in the autoclave, the higher the temperature. For example, when free-flowing steam at a temperature of 100°C is placed under a pressure of 1 atmosphere above sea level pressure—that is, about 15 pounds of pressure per square inch (psi)—the temperature rises to 121°C. Increasing the pressure to 20 psi raises the temperature to 126°C. The relationship between temperature and pressure is shown in Table 7.3. Sterilization in an autoclave is most effective when the organisms either contact the steam directly or are contained in a small vol-

TABLE 7.3 Relationship Between Pressure and Temperature of Steam at Sea Level*

PRESSURE (PSI IN EXCESS OF ATMOSPHERIC PRESSURE)	TEMPERATURE (°C)
0 psi	100
5 psi	110
10 psi	116
15 psi	121
20 psi	126
30 psi	135

*At higher altitudes, the pressure indicated on the pressure gauge would be higher for a given temperature.

TABLE 7.4 Effect of Container Size on Autoclave Sterilization Times for Liquid Solutions*

CONTAINER SIZE	LIQUID VOLUME	STERILIZATION TIME (MINUTES)
Test tube: 18 × 150 mm	10 ml	15
Erlenmeyer flask: 125 ml	95 ml	15
Erlenmeyer flask: 2000 ml	1500 ml	30
Fermentation bottle: 9000 ml	6750 ml	70

*Sterilization times in the autoclave include the time needed for the contents of the containers to reach sterilization temperatures. For smaller containers this is only 5 minutes or less, but for a 9000-ml bottle it might be as much as 70 minutes. A container is usually not filled past 75% of its capacity.

ume of aqueous (primarily water) liquid. Under these conditions, steam at a pressure of about 15 psi (121°C) will kill *all* organisms and their endospores in about 15 minutes.

Autoclaving is used to sterilize culture media, instruments, dressings, intravenous equipment, applicators, solutions, syringes, transfusion equipment, and numerous other items that can withstand high temperatures and pressures. Large industrial autoclaves are called *retorts,* but the same principle is used for the common household pressure cooker in the home canning of foods.

Heat requires extra time to reach the center of solid materials, such as canned meats, because such materials do not develop the efficient convection currents that occur in liquids. Heating large containers also requires extra time. Table 7.4 shows the different time requirements for sterilizing liquids in various sizes of containers. Unlike sterilizing aqueous solutions, in order to sterilize the surface of a solid, steam must actually contact it. To sterilize dry glassware, bandages, and the like, care must be taken to ensure that steam contacts all surfaces. For example, aluminum foil is impervious to steam and should not be used to wrap materials that are to be sterilized; paper should be used instead. Care should also be taken to avoid trapping air in the bottom of a dry container because trapped air will not be replaced by steam, which is lighter than air. The trapped air is the equivalent of a small hot-air oven, which, as we shall see shortly, requires a higher temperature and longer time to sterilize materials. Containers that can trap air should be placed in a tipped position so that the steam will force out the air. Products that do not permit penetration by moisture, such as mineral oil or petroleum jelly, are not sterilized by the same treatments that would sterilize aqueous solutions.

Steam under pressure fails to sterilize when the air is not completely exhausted. This is usually due to the premature closing of the autoclave's automatic ejector valve (see Figure 7.2). Anyone familiar with home canning knows that the steam must flow vigorously out of the valve in the lid for several minutes to carry with it all the air before the pressure cooker is sealed. If the air is not completely exhausted, the container will not reach the temperature expected for a given pressure.

Pasteurization

You may recall from Chapter 1 that Louis Pasteur, in the early days of microbiology, found a practical method of preventing the spoilage of beer and wine. Pasteur used mild heating, which was sufficient to kill the organisms that caused the particular spoilage problem without seriously damaging the taste of the product. The same principle was later applied to milk to produce what we now call pasteurized milk. Milk was first pasteurized to eliminate the tuberculosis bacterium. Many relatively heat-resistant (*thermoduric*) organisms survive pasteurization, but these are unlikely to cause disease or cause refrigerated milk to spoil in a short time.

In the classic pasteurization treatment of milk, the milk was exposed to a temperature of about 63°C for 30 minutes. Most milk pasteurization today uses higher temperatures, at least 72°C, but for only 15 seconds. This treatment, known as **high-temperature short-time (HTST) pasteurization,** is applied as the milk flows continuously past a heat exchanger. In addition to killing pathogens, HTST pasteurization lowers total bacterial counts, so the milk keeps well under refrigeration.

Milk can also be sterilized—something quite different from pasteurization—by **ultra-high-temperature (UHT) treatments** so that it can be stored without refrigeration. This is more useful in parts of the world where refrigeration facilities are not always available. In the United States, sterilization is sometimes used on the small containers of coffee creamers found in restaurants. To avoid giving the milk a cooked taste, a UHT system was devised in which the liquid milk never touches a surface hotter than the

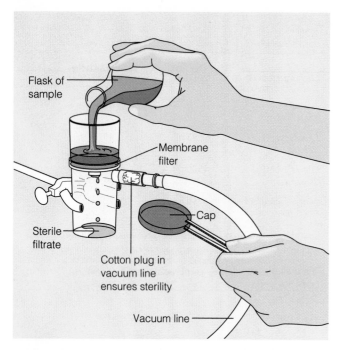

Flask of sample

Membrane filter

Cap

Sterile filtrate

Cotton plug in vacuum line ensures sterility

Vacuum line

FIGURE 7.3   Filter sterilization with a disposable, presterilized plastic unit. The sample is placed into the upper chamber and forced through the membrane filter by a vacuum in the lower chamber. Pores in the membrane filter are smaller than the bacteria, so bacteria are retained on the filter. The sterilized sample can then be decanted from the lower chamber. Similar equipment with removable filter disks is used to count bacteria in samples (see Figure 6.15).

milk itself while being heated by steam. The milk falls in a thin film through a chamber of superheated steam and reaches 140°C in less than a second. It is held for 3 seconds in a holding tube and then cooled in a vacuum chamber, where the steam flashes off. With this process, in less than 5 seconds the milk temperature rises from 74°C to 140°C and drops back to 74°C.

The heat treatments we have just discussed illustrate the concept of **equivalent treatments**—as the temperature is increased, much less time is needed to kill the same number of microbes. For example, the destruction of highly resistant endospores might take 70 minutes at 115°C, whereas only 7 minutes would be needed at 125°C. Both treatments yield the same result. The concept of equivalent treatments also explains why classic pasteurization at 63°C for 30 minutes, HTST treatment at 72°C for 15 seconds, and UHT treatment at 140°C for less than a second can have similar effects.

Dry Heat Sterilization

Dry heat kills by oxidation effects. A simple analogy is the slow charring of paper in a heated oven, even when the temperature remains below the ignition point of paper. One of the simplest methods of dry heat sterilization is *direct flaming.* You will use this procedure many times in the microbiology laboratory when you sterilize inoculating

loops. To sterilize the inoculating loop, you heat the wire to a red glow, which is 100% effective. A similar principle is used in *incineration,* an effective way to sterilize and dispose of contaminated paper cups, bags, and dressings.

Another form of dry heat sterilization is **hot-air sterilization.** Items to be sterilized by this procedure are placed in an oven. Generally, a temperature of about 170°C maintained for nearly 2 hours ensures sterilization. The longer period and higher temperature (relative to moist heat) are required because the heat in water is more readily transferred to a cool body than is the heat in air. For example, imagine the different effects of immersing your hand in boiling water at 100°C (212°F) and of holding it in a hot-air oven at the same temperature for the same time.

Filtration

Filtration is the passage of a liquid or gas through a screen-like material with pores small enough to retain microorganisms (this is often the same apparatus used for counting; see Figure 6.15). A vacuum that is created in the receiving flask helps gravity pull the liquid through the filter. Filtration is used to sterilize heat-sensitive materials, such as some culture media, enzymes, vaccines, and antibiotic solutions.

Some operating theaters and rooms occupied by burn patients receive filtered air to lower the numbers of airborne microbes. **High-efficiency particulate air (HEPA) filters** remove almost all microorganisms larger than about 0.3 μm in diameter.

In the early days of microbiology, hollow candle-shaped filters of unglazed porcelain were used to filter liquids. The long and indirect passageways through the walls of the filter adsorbed the bacteria. Unseen pathogens that passed through such filters (causing such diseases as rabies) were called *filterable viruses.*

In recent years, **membrane filters**, composed of such substances as cellulose esters or plastic polymers, have become popular for industrial and laboratory use (Figure 7.3; see also Figure 6.15). These filters are only 0.1 mm thick. The pores of a membrane filter are uniform in size. In some brands, plastic film is irradiated so that very uniform holes, where the radiation particles have passed, are etched in the plastic. The pores of membrane filters include 0.22- and 0.45-μm sizes, intended for bacteria. Some very flexible bacteria, such as spirochetes, or the wall-less mycoplasma, will sometimes pass through such filters, however. Filters are available with pores as small as 0.01 μm, a size that will retain viruses and even some large protein molecules.

Cold

The effect of low temperatures on microbes depends on the particular microbe and the intensity of the application. For example, at temperatures of ordinary refrigera-

FIGURE 7.4 Radiant energy spectrum. Visible light and other forms of radiant energy radiate through space as waves of various lengths. Ionizing radiation, such as gamma rays and X rays, has a wavelength shorter than 1 nm. Nonionizing radiation, such as ultraviolet (UV) light, has a wavelength between 1 nm and about 380 nm, where the visible spectrum begins.

tors (0° to 7°C), the metabolic rate of most microbes is so reduced that they cannot reproduce or synthesize toxins. In other words, ordinary refrigeration has a bacteriostatic effect. Yet, psychrotrophs do grow slowly at refrigerator temperatures and will alter the appearance and taste of foods after a time.

Surprisingly, some bacteria can grow at temperatures several degrees below freezing (see Chapter 6). Rapidly attained subfreezing temperatures tend to render microbes dormant but do not necessarily kill them. Slow freezing is most harmful to bacteria; the ice crystals that form and grow disrupt the cellular and molecular structure of the bacteria. One-third of the population of some vegetative bacteria might survive a year of freezing, whereas other species might have very few survivors.

Desiccation

To grow and multiply, microbes require water. In the absence of water—a condition known as **desiccation**—microbes cannot grow or reproduce but can remain viable for years. Then, when water is made available to them, they can resume their growth and division. This ability is used in the laboratory when microbes are preserved by lyophilization, a process described in Chapter 6.

The resistance of vegetative cells to desiccation varies with the species and the organism's environment. For example, the gonorrhea bacterium can withstand dryness for only about an hour, but the tuberculosis bacterium can remain viable for months. A normally susceptible bacterium is much more resistant if it is embedded in mucus, pus, or feces. Viruses are generally resistant to desiccation, but they are not as resistant as bacterial endospores. This ability of certain dried microbes and endospores to remain viable is important in a hospital setting. Dust, clothing, bedding, and dressings might contain infectious microbes in dried mucus, urine, pus, and feces.

Osmotic Pressure

The use of high concentrations of salts and sugars to preserve food is based on the effects of *osmotic pressure*. High concentrations of these substances create a hypertonic environment that causes water to leave the microbial cell (see Chapters 4 and 6). This process resembles preservation by desiccation, in that both methods deny the cell the moisture it needs for growth. As water leaves the microbial cell, the plasma membrane shrinks away from the cell wall (plasmolysis), and the cell stops growing, although it might not immediately die. The principle of osmotic pressure is used in the preservation of foods. For example, concentrated salt solutions are used to cure meats, and thick sugar solutions are used to preserve fruits.

As a general rule, molds and yeasts are much more capable than bacteria of growing in materials with low moisture or high osmotic pressures. This property of molds, sometimes combined with their ability to grow under acidic conditions, is the reason fruits and grains are spoiled by molds rather than by bacteria. It is also part of the reason why molds are able to form "mildew" on a damp wall or a shower curtain.

Radiation

Radiation has various effects on cells, depending on its wavelength, intensity, and duration. There are two types of radiation that will kill microorganisms, or sterilizing radiation: ionizing and nonionizing.

Ionizing radiation, such as *gamma rays, X rays,* or *high-energy electron beams,* has a wavelength shorter than that of nonionizing radiation, less than about 1 nm. Therefore, it carries much more energy (Figure 7.4). Gamma rays are emitted by radioactive cobalt, and electron beams are produced by accelerating electrons to high energies in special machines. X rays, which are produced by machines in a manner similar to the production of electron beams, are similar in nature to gamma rays. Gamma rays penetrate deeply but may require hours for sterilization of large masses; high-energy electron beams have much lower penetrating power but usually require only a few seconds of exposure. The principal effect of ionizing radiation is the ionization of water, which forms highly reactive hydroxyl radicals (see the discussion of toxic forms of oxygen on pp. 146–147). These radicals react with cellular organic components, especially DNA, and kill the cell.

The food industry has recently renewed its interest in the use of radiation for food preservation. (This subject will be discussed more fully in Chapter 28.) Low-level ionizing radiation, used for years in many countries, has been approved in the U.S.A. for certain purposes, such as decontaminating spices.

Ionizing radiation, especially high-energy electron beams, is increasingly used for the sterilization of pharmaceuticals and disposable dental and medical supplies, such as plastic syringes, surgical gloves, suturing materials, and catheters. Radiation has almost entirely replaced gases for sterilizing these items. Sterilizing gases will be discussed later in this chapter.

Nonionizing radiation has a wavelength longer than that of ionizing radiation, usually greater than about 1 nm. The best example of nonionizing radiation is ultraviolet (UV) light. UV light damages the DNA of exposed cells by causing bonds to form between adjacent thymines in DNA chains (see Figure 8.19). These *thymine dimers* inhibit correct replication of the DNA during reproduction of the cell. The UV wavelengths most effective for killing microorganisms are about 260 nm; these wavelengths are specifically absorbed by cellular DNA. UV radiation is also used to control microbes in the air. A UV or "germicidal" lamp is commonly found in hospital rooms, nurseries, operating rooms, and cafeterias. UV light is also used to sterilize vaccines, serum, and toxins, and occasionally municipal waste waters and drinking waters. A major disadvantage of UV light as a sterilizer is that the radiation is not very penetrating, so the organisms to be killed must be directly exposed to the rays. Organisms protected by solids and such coverings as paper, glass, and textiles are not affected. Another potential problem is that UV light can damage human eyes, and prolonged exposure can cause burns and skin cancer in humans.

Sunlight contains some UV radiation, but the shorter wavelengths—those most effective against bacteria—are screened out by the ozone layer of the atmosphere. The antimicrobial effect of sunlight is due almost entirely to the formation of singlet oxygen in the cytoplasm (see Chapter 6). Many pigments produced by bacteria provide protection from sunlight.

Microwaves do not have much direct effect on microorganisms but kill them indirectly by heating the medium, such as food.

Table 7.5 summarizes the physical methods of microbial control.

CHEMICAL METHODS OF MICROBIAL CONTROL

LEARNING OBJECTIVES

- *List the factors related to effective disinfection.*
- *Interpret results of use-dilution and filter paper tests.*
- *Identify the methods of action and preferred uses of chemical disinfectants.*
- *Differentiate between halogens used as antiseptics and as disinfectants.*
- *Identify the appropriate uses for surface-active agents.*
- *List the advantages of glutaraldehyde over other chemical disinfectants.*
- *Identify the method of sterilizing plastic labwares.*

Chemical agents are used to control the growth of microbes on both living tissue and inanimate objects. Unfortunately, few chemical agents achieve sterility; most of them merely reduce microbial populations to safe levels or remove vegetative forms of pathogens from objects. A common problem in disinfection is the selection of an agent. Just as there is no single physical method of microbial control that can be used in every situation, there is no one disinfectant that will be appropriate for all circumstances.

Principles of Effective Disinfection

To select a disinfectant for a particular job, we must first understand its action. For example, what are the properties of the disinfectant? What is it designed to do? Simply by reading the label, we can learn a great deal about a disinfectant's properties. We should also remember that the concentration of a disinfectant affects its action. A disinfectant should always be diluted exactly as specified by the manufacturer. Solutions that are too weak may be ineffective or be bacteriostatic instead of bactericidal. Solutions that are too strong can be dangerous to humans who come in contact with them.

Also to be considered is the nature of the material being disinfected. For example, are organic materials present that might interfere with the action of the disinfectant? Similarly, the pH of the medium often has a great effect on a disinfectant's activity.

Another very important consideration is whether the disinfectant will easily make contact with the microbes. An area might need to be scrubbed and rinsed before the disinfectant is applied. In general, disinfection is a gradual process. Thus, to be effective, a disinfectant might need to be left on a surface for several hours.

A final factor to keep in mind is that the higher the temperature at which the disinfectant is applied, the more effective it usually is.

TABLE 7.5 Summary of Physical Methods Used to Control Microbial Growth

METHOD	MECHANISM OF ACTION	COMMENT	PREFERRED USE
Heat			
1. Moist heat			
a. Boiling or flowing steam	Denaturation	Kills vegetative bacterial and fungal pathogens and many viruses within 10 minutes; less effective on endospores.	Dishes, basins, pitchers, various equipment.
b. Autoclaving	Denaturation	Very effective method of sterilization; at about 15 psi of pressure (121°C), all vegetative cells and their endospores are killed in about 15 minutes.	Microbiological media, solutions, linens, utensils, dressings, equipment, and other items that can withstand temperature and pressure.
2. Pasteurization	Denaturation	Heat treatment for milk (72°C for about 15 seconds) that kills all pathogens and some nonpathogens.	Milk, cream, and certain alcoholic beverages (beer and wine).
3. Dry heat			
a. Direct flaming	Burning contaminants to ashes	Very effective method of sterilization.	Inoculating loops.
b. Incineration	Burning to ashes	Very effective method of sterilization.	Paper cups, contaminated dressings, animal carcasses, bags, and wipes.
c. Hot-air sterilization	Oxidation	Very effective method of sterilization, but requires temperature of 170°C for about 2 hours.	Empty glassware, instruments, needles, and glass syringes.
Filtration	Separation of bacteria from suspending liquid	Passage of a liquid or gas through a screenlike material that traps microbes; most filters in use consist of cellulose acetate or nitrocellulose.	Useful for sterilizing liquids (toxins, enzymes, vaccines) that are destroyed by heat.
Cold			
1. Refrigeration	Decreased chemical reactions and possible changes in proteins	Has a bacteriostatic effect.	Food, drug, and culture preservation.
2. Deep-freezing (see Chapter 6)	Decreased chemical reactions and possible changes in proteins	An effective method for preserving microbial cultures, in which cultures are quick-frozen between −50° and −95°C.	Food, drug, and culture preservation.
3. Lyophilization (see Chapter 6)	Decreased chemical reactions and possible changes in proteins	Most effective method for long-term preservation of microbial cultures; water removed by high vacuum at low temperature.	Food, drug, and culture preservation.
Desiccation	Disruption of metabolism	Involves removing water from microbes; primarily bacteriostatic.	Food preservation.
Osmotic Pressure	Plasmolysis	Results in loss of water from microbial cells.	Food preservation.
Radiation			
1. Ionizing	Destruction of DNA by gamma rays and high-energy electron beams	Not widespread in routine sterilization.	Used for sterilizing pharmaceuticals and medical and dental supplies.
2. Nonionizing	Damage to DNA by UV light	Radiation not very penetrating.	Control of closed environment with UV (germicidal) lamp.

Evaluating a Disinfectant

Use-Dilution Tests

There is an obvious need to evaluate the effectiveness of disinfectants and antiseptics. For many years the standard test was the phenol coefficient test, which compared the activity of a given disinfectant with that of phenol. However, the standard currently in use is the **American Offi**-cial Analytical Chemist's use-dilution method. For most purposes, three strains of bacteria are used in this test: *Salmonella choleraesuis, Staphylococcus aureus,* and *Pseudomonas aeruginosa.* To perform a use-dilution test, metal carrier rings are dipped into standardized cultures of the test bacteria grown in liquid media, removed, and dried at 37°C for a short time. The dried cultures are then placed into a solution of the disinfectant at the concen-

Zone of inhibition

Chlorine

Hexachlorophene   O-phenylphenol

Quat

Staphylococcus aureus (+)

Chlorine

Hexachlorophene   O-phenylphenol

Quat

Escherichia coli (–)

Chlorine

Hexachlorophene   O-phenylphenol

Quat

Pseudomonas aeruginosa (–)

FIGURE 7.5 Evaluation of disinfectants by the filter paper method. In this familiar teaching-laboratory experiment, paper disks are soaked in a solution of the disinfectant and placed on the surface of a nutrient medium on which a culture of test bacteria has been spread to produce uniform growth.

At the top of each plate the tests show that chlorine (as sodium hypochlorite) was effective against all of the test bacteria but was more effective against gram-positive bacteria. At the bottom row of each plate the tests show that the quaternary ammonium compound ("quat") was also more effective

against the gram-positive bacteria, but it did not affect the pseudomonads at all. At the left side of each plate, the tests show that hexachlorophene was effective against gram-positive bacteria only. At the right sides, O-phenylphenol was ineffective against pseudomonads but was almost equally effective against the gram-positive bacteria and the gram-negative bacteria used in the test. Notice that all four test chemicals worked against the gram-positive test bacteria, but only one of the four chemicals affected pseudomonads.

tration recommended by the manufacturer and left there for 10 minutes at 20°C. Following this exposure, the carrier rings are transferred to a medium that will permit the growth of any surviving bacteria. The effectiveness of the disinfectant can then be determined by the number of resulting cultures.

Variations of this method are used for testing the effectiveness of antimicrobial agents against endospores, mycobacteria that cause tuberculosis, and fungi, because they are difficult to control with chemicals. Also, tests of antimicrobials intended for special purposes, such as dairy utensil disinfection, may substitute other test bacteria. Virucidal chemicals are usually tested against cultures of Newcastle disease virus (a disease of birds and domestic fowl). After exposure to the chemical, the cultures are injected into embryonated chicken eggs; if any viruses survive, they will kill the embryos.

Filter Paper Method

The *filter paper method* is often used in teaching laboratories to evaluate the efficacy of a chemical agent. A disk of filter paper is soaked with a chemical and placed on the surface of an agar plate that has been previously inoculated and incubated with the test organism. After incubation, if the chemical agent is effective, a clear zone representing inhibition of growth can be observed around the disk (Figure 7.5).

Types of Disinfectants

Phenol and Phenolics

Phenol (carbolic acid), the substance first used by Joseph Lister in his operating room, is now rarely used as an antiseptic or disinfectant because it irritates the skin and has a disagreeable odor. It is often used in throat lozenges for its local anesthetic effect but has little antimicrobial effect at the low concentrations used. At concentrations above 1% (such as in some throat sprays), however, phenol has a significant antibacterial effect. The structure of a phenol molecule is shown in Figure 7.6a.

Derivatives of phenol, called **phenolics**, contain a molecule of phenol that has been chemically altered to reduce its irritating qualities or increase its antibacterial activity in combination with a soap or detergent. Phenolics exert antimicrobial activity by injuring plasma membranes, inactivating enzymes, and denaturing proteins. Phenolics are frequently used as disinfectants because they remain active in the presence of organic compounds, they are stable, and they persist for long periods after application. For these reasons, phenolics are suitable agents for disinfecting pus, saliva, and feces.

One of the most frequently used phenolics is derived from coal tar, a group of chemicals called *cresols*. A very important cresol is *o-phenylphenol* (Figures 7.5 and 7.6b), the main ingredient in most formulations of Lysol®. Cresols are very good surface disinfectants.

FIGURE 7.6 Structure of phenolics and chlorhexidine.
(a) Phenol. **(b)** O-phenylphenol. **(c)** Hexachlorophene.
(d) Chlorhexidine.

Another phenolic, used in the past much more widely than now, is *hexachlorophene* (Figure 7.6c). This phenolic was originally an ingredient in soaps and lotions (such as pHisoHex®) used for surgical scrubs and hospital microbial control procedures, and in cosmetic soaps, deodorants, feminine hygiene sprays, and even toothpaste. As a bacteriostatic agent, hexachlorophene is especially effective against gram-positive staphylococci and streptococci (see Figure 7.5), which can cause infections of the skin. Therefore, it was once heavily used to control staphylococcal and streptococcal infections in hospital nurseries. However, in 1972 it was found that *excessive* use of hexachlorophene, such as bathing infants with it several times a day, could lead to neurological damage. Hexachlorophene is still used in hospitals to control nosocomial (hospital-acquired) infections and as a scrub for hospital personnel. However, it is no longer casually used by the general public. At present in the U.S.A., a prescription is required for the purchase of a 3% or stronger solution of hexachlorophene.

Chlorhexidine

Chlorhexidine, which is not a phenol—although its structure (Figure 7.6d) and application are similar to those of hexachlorophene—is frequently used for disinfection of skin and mucous membranes. Combined with a detergent or alcohol, it is also used for surgical hand scrubs and preoperative skin preparation in patients. In such applications, its strong affinity for binding to the skin or mucous membranes is an advantage, as is its low toxicity. Its killing effect is related to the damage it causes to the plasma membrane. It is effective against most vegetative bacteria but is not sporicidal. The only viruses affected are certain enveloped (lipophilic) types (see Chapter 13).

Halogens

The *halogen elements,* particularly iodine and chlorine, are effective antimicrobial agents, both alone (as I_2 or Cl_2 in solution) and as constituents of inorganic or organic compounds. *Iodine* (I_2) is one of the oldest and most effective antiseptics. It is effective against all kinds of bacteria, many endospores, various fungi, and some viruses. One proposed mechanism for the activity of iodine is that it combines with the amino acid tyrosine, a common component of many enzymes and other cellular proteins (Figure 7.7). As a result, microbial protein function is inhibited. More generally, iodine also oxidizes the —SH groups of certain amino acids that are important for maintaining the structure of proteins.

Iodine is available as a **tincture**—that is, in solution in aqueous alcohol—and as an iodophor. An **iodophor** is a combination of iodine and an organic molecule, from which the iodine is released slowly. Iodophors have the antimicrobial activity of iodine, but they do not stain and are less irritating. The most common commercial preparations are Betadine® and Isodine®. These are *povidone-iodines;* povidone is a surface-active iodophor that improves the wetting action. Iodines are used mainly for skin disinfection and wound treatment. Note that *Pseudomonas* can survive for long periods in iodophores.

Chlorine (Cl_2), as a gas or in combination with other chemicals, is another widely used disinfectant. Its germicidal action is caused by the hypochlorous acid (HOCl) that forms when chlorine is added to water:

(1) Cl_2 + H_2O ⇌ H^+ + Cl^- + HOCl

Chlorine Water   Hydrogen Chloride Hypochlorous
                       ion   ion   acid

(2) HOCl ⇌ H^+ + OCl^-

Hypochlorous   Hydrogen Hypochlorite
acid   ion   ion

Exactly how hypochlorous acid exerts its killing power is not known. It is a strong oxidizing agent that prevents much of the cellular enzyme system from functioning. Hypochlorous acid is the most effective form of chlorine because it is neutral in electrical charge and diffuses as rapidly as water through the cell wall. Because of its negative charge, the hypochlorite ion (OCl^-) cannot enter the cell freely.

A liquid form of compressed chlorine gas is used extensively for disinfecting municipal drinking water, water in swimming pools, and sewage. Several compounds of chlorine are also effective disinfectants. For example, solutions of *calcium hypochlorite* [$Ca(OCl)_2$] are used to disinfect dairy equipment and restaurant eating utensils. Another chlorine compound, *sodium hypochlorite* (NaOCl; see Figure 7.5), is used as a household disinfectant and a bleach (Clorox®) and as a disinfectant in dairies, food-processing establishments, and hemodialysis systems. When the quality of drinking water is in question, household

FIGURE 7.7 A proposed explanation of the mechanism for the antimicrobial activity of iodine. Iodine reacts with tyrosine, an amino acid found in protein. The diiodotyrosine formed inhibits the function of the protein in the cell. Iodine probably also reacts with other cellular constituents.

Tyrosine → Diiodotyrosine

TABLE 7.6 Germicidal Action of Various Concentrations of Ethanol in Aqueous Solution Against *Streptococcus pyogenes*

CONCENTRATION OF ETHANOL	SECONDS				
	10	20	30	40	50
100%	−*	−	−	−	−
95%	+†	+	+	+	+
90%	+	+	+	+	+
80%	+	+	+	+	+
70%	+	+	+	+	+
60%	+	+	+	+	+
50%	−	−	+	+	+
40%	−	−	−	−	−

*− = no germicidal action (bacterial growth)
†+ = germicidal action (no bacterial growth)
Note: The highlighted area of the table indicates bacteria killed by germicidal action.
SOURCE: Adapted from S. S. Block, *Disinfection, Sterilization, and Preservation,* 4th ed., Philadelphia: Lea & Febiger, 1951, p. 194.

bleach can provide a rough equivalent of municipal chlorination. After two drops of bleach are added to a liter of water (four drops if the water is cloudy) and the mixture has sat for 30 minutes, the water is considered safe for drinking under emergency conditions.

Another group of chlorine compounds, the *chloramines,* consist of chlorine and ammonia. They are used as disinfectants, antiseptics, or sanitizing agents. Chloramines are very stable compounds that release chlorine over long periods. They are relatively effective in organic matter, but they have the disadvantages of acting more slowly and being less effective purifiers than many other chlorine compounds. Chloramines are used to sanitize glassware and eating utensils and to treat dairy and food-manufacturing equipment. Ammonia is usually mixed with chlorine in municipal water treatment systems to form chloramines. The chloramines control taste and odor problems caused by the reaction of chlorine with other nitrogenous compounds in the water. Because chloramines are less effective as germicides, sufficient chlorine must be added to ensure a residual of chlorine in the

form of HOCl. (Chloramines are toxic to aquarium fish, but pet shops sell chemicals to neutralize them.)

Alcohols

Alcohols effectively kill bacteria and fungi but not endospores and nonenveloped viruses. The mechanism of action of alcohol is usually protein denaturation, but alcohol can also disrupt membranes and dissolve many lipids, including the lipid component of enveloped viruses. Alcohols have the advantage of acting and then evaporating rapidly and leaving no residue. When the skin is swabbed (degermed) before an injection, most of the microbial control activity comes from simply *wiping away* dirt and microorganisms. However, alcohols are unsatisfactory antiseptics when applied to wounds. They cause coagulation of a layer of protein under which bacteria continue to grow.

Two of the most commonly used alcohols are ethanol and isopropanol. The recommended optimum concentration of *ethanol* is 70%, but concentrations between 60% and 95% seem to kill as fast (Table 7.6). Pure ethanol is less effective than aqueous solutions (ethanol mixed with water) because denaturation requires water. *Isopropanol,* often sold as rubbing alcohol, is slightly superior to ethanol as an antiseptic and disinfectant. Moreover, it is less volatile, less expensive, and more easily obtained than ethanol.

Ethanol and isopropanol are often used to enhance the effectiveness of other chemical agents. For example, an aqueous solution of Zephiran™ (described shortly) kills about 40% of the population of a test organism in two minutes, whereas a tincture of Zephiran™ kills about 85% in the same period. You will be able to compare the effectiveness of tinctures and aqueous solutions in Figure 7.10.

Heavy Metals and Their Compounds

Several heavy metals, such as silver, mercury, and copper, can be germicidal or antiseptic. The ability of very small amounts of heavy metals, especially silver and copper, to exert antimicrobial activity is referred to as **oligodynamic action** (*oligo* means few). This action can be seen when we place a coin or other clean piece of metal containing silver or copper on an inoculated Petri plate. Extremely small amounts of metal diffuse from the coin and inhibit the growth of bacteria for some distance around the coin (Figure 7.8). This effect is produced by the action of heavy metal ions on microbes. When the metal ions combine with the —SH groups on cellular proteins, denaturation results.

Silver is used as an antiseptic in a 1% *silver nitrate* solution. At one time, many states required that the eyes of newborns be treated with a few drops of silver nitrate to guard against an infection of the eyes called gonorrheal ophthalmia neonatorum, which the infants might have contracted as they passed through the birth canal. In recent years, antibiotics have replaced silver nitrate for this purpose.

Inorganic mercury compounds, such as *mercuric chloride,* probably have the longest history of use as disinfectants. They have a very broad spectrum of activity; their effect is primarily bacteriostatic. However, their use is now limited because of their toxicity, corrosiveness, and ineffectiveness in organic matter. At present, the primary use of mercurials is to control mildew in paints. In addition, Mercurochrome antiseptic, an organic mercury compound, is often found in home medicine chests.

Copper in the form of *copper sulfate* is used chiefly to destroy green algae (algicide) that grow in reservoirs, swimming pools, and fish tanks. If the water does not contain excessive organic matter, copper sulfate is effective in concentrations of one part per million (ppm) of water. To prevent mildew, copper compounds such as copper 8-hydroxyquinoline are sometimes included in paint.

Another metal used as an antimicrobial is zinc. *Zinc chloride* is a common ingredient in mouthwashes, and *zinc oxide* is probably the most widely used antifungal in paints, mainly because it is often part of the pigment formulation.

Surface-Active Agents

Surface-active agents, or **surfactants,** can decrease surface tension among molecules of a liquid. Such agents include soaps and detergents. Soap has little value as an antiseptic, but it does have an important function in the mechanical removal of microbes through scrubbing. The skin normally contains dead cells, dust, dried sweat, microbes, and oily secretions from oil glands. Soap breaks the oily film into tiny droplets, a process called **emulsification,** and the water and soap together lift up the emulsified oil and debris and float them away as the lather is washed off. In this sense, soaps are good degerming agents. Many so-called deodorant soaps contain compounds such as *triclocarban* that strongly inhibit gram-positive bacteria.

Acid-anionic surface-active sanitizers are very important in the cleaning of dairy utensils and equipment. Their sanitizing ability is related to the negatively charged portion (anion) of the molecule, which reacts with the plasma membrane. These sanitizers act on a wide spectrum of microbes, including troublesome thermoduric bacteria. They are nontoxic, noncorrosive, and fast acting.

Quaternary Ammonium Compounds (Quats)

The most widely used surface-active agents are the cationic detergents, especially the **quaternary ammonium compounds (quats).** Their cleansing ability is related to

FIGURE 7.8 Oligodynamic action of heavy metals. Clear zones where bacterial growth has been inhibited are seen around the sombrero charm (pushed aside), the dime, and the penny. The charm and the dime contain silver; the penny contains copper.

FIGURE 7.9 The ammonium ion and a quaternary ammonium compound, benzalkonium chloride (Zephiran™). Note how other groups replace the hydrogens of the ammonium ion.

Ammonium ion   Benzalkonium chloride

the positively charged portion—the cation—of the molecule. Their name is derived from the fact that they are modifications of the four-valence ammonium ion, NH_4^+ (Figure 7.9). Quaternary ammonium compounds are strongly bactericidal against gram-positive bacteria and somewhat less active against gram-negative bacteria.

Quats are also fungicidal, amoebicidal, and virucidal against enveloped viruses. They do not kill endospores or tuberculosis bacteria. Their chemical mode of action is unknown, but they probably affect the plasma membrane. They change the cell's permeability and cause the loss of essential cytoplasmic constituents, such as potassium.

Two popular quats are Zephiran™, a brand name of benzalkonium chloride (see Figure 7.9), and Cepacol®, a brand name of cetylpyridinium chloride. They are strongly antimicrobial, colorless, odorless, tasteless, stable, easily diluted, and nontoxic, except at high concentrations. If your mouthwash bottle fills with foam when shaken, the mouthwash probably contains a quat. However, organic matter interferes with their activity, and they are rapidly neutralized by soaps and anionic detergents.

Anyone associated with medical applications of quats should remember that certain bacteria, such as some species of *Pseudomonas,* not only survive in quaternary ammonium compounds but actively grow in them. This resistance occurs not only to the disinfectant solution but also to the moistened gauze and bandages, whose fibers tend to neutralize the quats. This resistance is another example of the peculiar and troublesome characteristics of the pseudomonads. Their ability to metabolize unusual substrates for growth permits pseudomonads to grow in unlikely places, such as in simple saline solutions and even distilled water. They are using as nutrients such things as traces of soap residues and cap-liner adhesives.

Before we move on to the next group of chemical agents, refer to Figure 7.10, which compares the effectiveness of some of the antimicrobials we have discussed so far.

Organic Acids and Derivatives

A number of organic acids are used as preservatives to control mold growth. *Sorbic acid* (often as the salt *potassium sorbate*) is used to inhibit mold growth in acidic foods, such as cheese. *Benzoic acid* (or its salt *sodium benzoate*) is an antifungal agent that is effective at low pH values and is used widely in soft drinks and other acidic foods. The *parabens,* such as *methylparaben* and *propylparaben,* are often used to inhibit mold growth in liquid cosmetics and shampoos. The parabens are derivatives of benzoic acid but work at a neutral pH. *Calcium propionate* prevents mold growth in bread.

The activity of these organic acids is not related to their acidity but to their ability to inhibit enzymatic and metabolic activity. Organic acids are listed on the labels of many foods and a few cosmetic preparations. By and large, the body metabolizes these organic acids readily, so their use is quite safe.

Aldehydes

Aldehydes are among the most effective antimicrobials. Two examples are formaldehyde and glutaraldehyde. They inactivate proteins by forming covalent crosslinks with a number of organic functional groups on proteins ($-NH_2$, $-OH$, $-COOH$, and $-SH$). *Formaldehyde gas* is an excellent disinfectant. However, it is more commonly available as *formalin,* a 37% aqueous solution of formaldehyde gas. Formalin was once used extensively to preserve biological specimens and inactivate bacteria and viruses in vaccines.

Glutaraldehyde is a chemical relative of formaldehyde that is less irritating and more effective than formaldehyde. Glutaraldehyde is used to sterilize hospital instruments, including respiratory-therapy equipment. When

FIGURE 7.10   Comparison of the effectiveness of various antiseptics. The steeper the downward slope of the killing curve of the antiseptic, the more effective it is. A 1% iodine in 70% ethanol solution is the most effective; soap and water are the least effective. Note that a tincture of Zephiran™ is more effective than an aqueous solution of the same antiseptic.

used in a 2% solution (Cidex™), it is bactericidal, tuberculocidal, and virucidal in 10 minutes and sporicidal in 3 to 10 hours. It is probably the only liquid chemical disinfectant that can be considered a possible sterilizing agent (sterilant). However, such a long exposure time is required for sporicidal activity that chemical agents cannot be relied on as sterilants.

Both glutaraldehyde and formalin are used by morticians for embalming.

Gaseous Chemosterilizers

Gaseous chemosterilizers are chemicals that sterilize in a closed chamber (similar to an autoclave). A gas suitable for this method is *ethylene oxide:*

$$H_2C\!\!-\!\!CH_2$$
$$\diagdown\!\!O\!\!\diagup$$

Its activity depends on the denaturation of proteins: The proteins' labile hydrogens, such as $-SH$, $-COOH$, or $-OH$, are replaced by alkyl groups (*alkylation*), such as $-CH_2CH_2OH$. Ethylene oxide kills all microbes and endospores but requires a lengthy exposure period of 4 to 18 hours. It is toxic and explosive in its pure form, so it is usually mixed with a nonflammable gas, such as carbon dioxide or nitrogen. One of its principal advantages is that it is highly penetrating, so much so that ethylene oxide was chosen to sterilize spacecraft sent to land on the Moon and some of the planets. Using heat to sterilize the electronic gear on these vehicles was not practical.

Because of their ability to sterilize without heat, gases like ethylene oxide are also widely used on medical supplies and equipment. Examples include disposable sterile plasticware (such as syringes and Petri plates), textiles, sutures, lensed instruments, artificial heart valves, heartlung machines, and mattresses. Many large hospitals have ethylene oxide chambers, some large enough to sterilize mattresses, as part of their sterilizing equipment. Propylene oxide and beta-propiolactone are also used for gaseous sterilization:

$$H_3C-CH-CH_2 \qquad H_2C-CH_2$$
$$| \qquad\qquad | \quad |$$
$$O \qquad\qquad O-C=O$$

Propylene oxide   Beta-propiolactone

The principal disadvantage of all these gases is that they are suspected carcinogens, especially beta-propiolactone. For this reason, there has been concern about exposure of hospital workers to ethylene oxide from such sterilizers; certainly, it is wise to minimize exposure.

Oxidizing Agents

Oxidizing agents exert antimicrobial activity by oxidizing cellular components of the treated microbes. Examples of oxidizing agents are ozone and hydrogen peroxide. *Ozone* (O_3) is a highly reactive form of oxygen that is generated by passing oxygen through high-voltage electrical discharges. It is responsible for the air's rather fresh odor after a lightning storm, in the vicinity of electric sparking, or around an ultraviolet light. Ozone is often used to supplement chlorine in the disinfection of water, because it helps neutralize tastes and odors. Although ozone is a more effective killing agent, its residual activity is difficult to maintain in water, and it is more expensive than chlorine.

Hydrogen peroxide is an antiseptic found in many household medicine cabinets and in hospital supply rooms. It is not a good antiseptic for open wounds because it is quickly broken down to water and gaseous oxygen by the action of the enzyme catalase, which is present in human cells (see Chapter 6). However, hydrogen peroxide does effectively disinfect inanimate objects, an application in which it is even sporicidal, especially at elevated temperatures. On a nonliving surface, the normally protective enzymes of aerobic bacteria and facultative anaerobes are overwhelmed by the high concentrations of peroxide used. Because of these factors, the food industry is increasing its use of hydrogen peroxide for aseptic packaging (see Chapter 28). The packaging materials pass through a hot solution of the chemical before being assembled into a container. In addition, many wearers of contact lenses are familiar with disinfection by hydrogen peroxide. After disinfection, a platinum catalyst in the lens-disinfecting kit destroys residual hydrogen peroxide so it does not persist on the lens, where it might be an irritant.

Oxidizing agents are useful in irrigation of deep wounds, where the oxygen released makes an environment that inhibits the growth of anaerobic bacteria.

Benzoyl peroxide is another compound useful for treating wounds infected by anaerobic pathogens, but it is probably more familiar as the main ingredient in over-the-counter medications for acne, which is caused by a type of anaerobic bacterium infecting hair follicles.

• • •

A summary of chemical agents used to control microbial growth is presented in Table 7.7.

The compounds discussed in this chapter are not generally useful in the treatment of diseases. Because antibiotics are used in chemotherapy, antibiotics and the pathogens against which they are active will be discussed together, in Chapter 20.

TABLE 7.7 Summary of Chemical Agents Used to Control Microbial Growth

CHEMICAL AGENT	MECHANISM OF ACTION	PREFERRED USE	COMMENT
Phenol and Phenolics			
1. Phenol	Disruption of plasma membrane, denaturation, inactivation of enzymes.	Rarely used	Seldom used as a disinfectant or antiseptic because of its irritating qualities and disagreeable odor.
2. Phenolics	Disruption of plasma membrane, denaturation, inactivation of enzymes.	Environmental surfaces, instruments, skin surfaces, and mucous membranes.	Derivatives of phenol that are reactive even in the presence of organic material; examples include o-phenylphenol and hexachlorophene.
Chlorhexidine	Disruption of plasma membrane.	Skin degerming, especially for surgical scrubs.	Bactericidal to gram-positives and gram-negatives. Nontoxic, persistent.
Halogens	Iodine inhibits protein function and is a strong oxidizing agent; chlorine forms the strong oxidizing agent hypochlorous acid, which alters cellular components.	Effective antiseptic available as a tincture and an iodophor; chlorine gas is used to disinfect water; chlorine compounds are used to disinfect dairy equipment, eating utensils, household items, and glassware.	Iodine and chlorine may act alone or as components of inorganic and organic compounds.

continued

TABLE 7.7 Continued

CHEMICAL AGENT	MECHANISM OF ACTION	PREFERRED USE	COMMENT
Alcohols	Denaturation and lipid dissolution.	Thermometers and other instruments; in swabbing the skin with alcohol before an injection, most of the disinfecting action probably comes from a simple wiping away (cleansing) of dirt and some microbes.	Bactericidal and fungicidal, but not effective against endospores or nonenveloped viruses; commonly used alcohols are ethanol and isopropanol.
Heavy Metals and Their Compounds	Denaturation of enzymes and other essential proteins.	Silver nitrate may be used to prevent gonococcal eye infections; Mercurochrome disinfects skin and mucous membranes; copper sulfate is an algicide.	Heavy metals such as silver and mercury are germicidal or antiseptic.
Surface-Active Agents			
1. Soaps and acid-anionic detergents	Mechanical removal of microbes through scrubbing.	Skin degerming and removal of debris.	Many deodorant soaps contain antimicrobials such as triclocarban.
2. Acid-anionic detergents	Not certain; may involve enzyme inactivation or disruption.	Sanitizers in dairy and food-processing industries.	Wide spectrum of activity, nontoxic, noncorrosive, fast acting.
3. Cationic detergents (quaternary ammonium compounds)	Enzyme inhibition, protein denaturation, and disruption of plasma membranes.	Antiseptic for skin, instruments, utensils, rubber goods.	Bactericidal, bacteriostatic, fungicidal, and virucidal against enveloped viruses; examples of quats are Zephiran™ and Cepacol®.
Organic Acids	Metabolic inhibition, mostly affecting molds; action not related to their acidity.	Sorbic acid and benzoic acid effective at low pH; parabens much used in cosmetics, shampoos; calcium propionate used in bread; all are mainly antifungals.	Widely used to control molds and some bacteria in foods and cosmetics.
Aldehydes	Protein inactivation	Glutaraldehyde (Cidex™) is less irritating than formaldehyde and is used for sterilization of medical equipment.	Very effective antimicrobials; glutaraldehyde is considered a liquid sterilant.
Gaseous Sterilants	Denaturation	Excellent sterilizing agent, especially for objects that would be damaged by heat.	Ethylene oxide is most commonly used.
Oxidizing Agents	Oxidation	Contaminated surfaces; some deep wounds, in which they are very effective against oxygen-sensitive anaerobes.	Ozone is gaining attention as a potential replacement for chlorination; hydrogen peroxide is a poor antiseptic but a good disinfectant.

STUDY OUTLINE

THE HISTORY AND CONCEPT OF MICROBIAL CONTROL (pp. 167–168)

1. Ignatz Semmelweis introduced handwashing with chlorinated lime by hospital personnel to reduce infections.
2. Joseph Lister introduced aseptic surgery, including heat sterilization of instruments and disinfection of wounds.
3. Microbial growth can be controlled by physical and chemical methods.
4. Control of microbial growth can prevent infections and food spoilage.

5. Sterilization is the process of destroying all microbial life on an object.
6. Disinfection is the process of reducing or inhibiting microbial growth.

CONDITIONS INFLUENCING MICROBIAL CONTROL (pp. 168–169)

1. A general rule of disinfection is that the method used should be strong enough to kill the most resistant microbes that can be found on the object to be disinfected.

2. The number and species of microorganisms present will affect the rate of disinfection.

Temperature (p. 168)

1. Biochemical reactions occur more rapidly at warm temperatures.
2. Disinfectant activity is enhanced by warm temperatures.

Type of Microbe (p. 169)

1. Gram-positive bacteria are generally more susceptible to disinfectants than gram-negative bacteria.
2. Pseudomonads can even grow in some disinfectants and antiseptics.
3. *M. tuberculosis* is resistant to many disinfectants.

Physiological State of the Microbe (p. 169)

1. An actively growing microorganism tends to be less resistant to chemical agents than an older microorganism.
2. Endospores are more resistant to chemical agents and physical methods.

Environment (p. 169)

1. Organic matter and pH level frequently affect the actions of chemical control agents.
2. Examples of such organic matter include vomit, feces, pus, and food.

ACTIONS OF MICROBIAL CONTROL AGENTS (p. 169)

Alteration of Membrane Permeability (p. 169)

1. The susceptibility of the plasma membrane is due to its lipid and protein components.
2. Certain chemical control agents damage the plasma membrane by altering its permeability.

Damage to Proteins and Nucleic Acids (p. 169)

1. Some microbial control agents damage cellular proteins by breaking hydrogen and covalent bonds.
2. Other agents interfere with DNA and RNA replication and protein synthesis.

RATE OF MICROBIAL DEATH (p. 171)

1. Bacterial populations subjected to heat or antimicrobial chemicals usually die at a constant rate.
2. Such a death curve, when plotted logarithmically, shows this constant death rate as a straight line.

PHYSICAL METHODS OF MICROBIAL CONTROL (pp. 171–176)

Heat (pp. 171–174)

1. Heat is frequently used to eliminate microorganisms; it is economical and easily controlled.
2. Moist heat kills microbes by denaturing enzymes.

3. Thermal death point (TDP) is the lowest temperature at which all the bacteria in a liquid culture will be killed in 10 minutes.
4. Thermal death time (TDT) is the length of time required to kill all bacteria in a liquid culture at a given temperature.
5. Decimal reduction time (DRT) is the length of time in which 90% of a bacterial population will be killed at a given temperature.
6. Boiling (100°C) kills many vegetative cells and viruses within 10 minutes.
7. Autoclaving (steam under pressure) is the most effective method of moist heat sterilization. The steam must directly contact the material to be sterilized.
8. In pasteurization, a high temperature is used for a short time (72°C for 15 seconds) to destroy pathogens without altering the flavor of the food.
9. Methods of dry heat sterilization include direct flaming, incineration, and hot-air sterilization. Dry heat kills by oxidation.
10. Different methods that produce the same effect (reduction in microbial growth) are called equivalent treatments.

Filtration (p. 174)

1. Filtration is the passage of a liquid or gas through a filter with pores small enough to retain microbes.
2. Microbes can be removed from air by high-efficiency particulate air filters.
3. Membrane filters composed of nitrocellulose or cellulose acetate are commonly used to filter out bacteria, viruses, and even large proteins.

Cold (pp. 174–175)

1. The effectiveness of low temperatures depends on the particular microorganism and the intensity of the application.
2. Most microorganisms do not reproduce at ordinary refrigerator temperatures (0° to 7°C).
3. Many microbes survive (but do not grow) at the subzero temperatures used to store foods.

Desiccation (p. 175)

1. In the absence of water, microorganisms cannot grow but can remain viable.
2. Viruses and endospores can resist desiccation.

Osmotic Pressure (p. 175)

1. Microorganisms in high concentrations of salts and sugars undergo plasmolysis.
2. Molds and yeasts are more capable of growing in materials with low moisture or high osmotic pressure than bacteria are.

Radiation (pp. 175–176)

1. The effects of radiation depend on its wavelength, intensity, and duration.

2. Ionizing radiation (gamma rays, X rays, and high-energy electron beams) has a high degree of penetration and exerts its effect primarily by ionizing water and forming highly reactive hydroxyl radicals.

3. Ultraviolet radiation, a form of nonionizing radiation, has a low degree of penetration and causes cell damage by making thymine dimers in DNA that interfere with the DNA replication; the most effective germicidal wavelength is 260 nm.

4. Microwaves can kill microbes indirectly as materials get hot.

CHEMICAL METHODS OF MICROBIAL CONTROL (pp. 176–184)

1. Chemical agents are used on living tissue (as antiseptics) and on inanimate objects (as disinfectants).
2. Few chemical agents achieve sterility.

Principles of Effective Disinfection (p. 177)

1. Careful attention should be paid to the properties and concentration of the disinfectant to be used.
2. The presence of organic matter, degree of contact with microorganisms, and temperature should also be considered.

Evaluating a Disinfectant (pp. 177–178)

1. In the use-dilution test, bacterial (*S. choleraesuis, S. aureus,* and *P. aeruginosa*) survival in the manufacturer's recommended dilution of a disinfectant is determined.
2. Viruses, endospore-forming bacteria, mycobacteria, and fungi can also be used in the use-dilution test.
3. In the filter paper method, a disk of filter paper is soaked with a chemical and placed on an inoculated agar plate; a clear zone of inhibition indicates effectiveness.

Types of Disinfectants (pp. 178–183)

Phenol and Phenolics (pp. 178–179)

1. Phenolics exert their action by injuring plasma membranes, inactivating enzymes, and denaturing proteins.
2. Common phenolics are cresols and hexachlorophene.

Chlorhexidine (p. 179)

1. Chlorhexidine damages plasma membranes of vegetative cells.

Halogens (pp. 179–180)

1. Some halogens (iodine and chlorine) are used alone or as components of inorganic or organic solutions.
2. To inactivate enzymes and other cellular proteins, iodine combines with the amino acid tyrosine.
3. Iodine is available as a tincture (in solution with alcohol) or as an iodophor (combined with an organic molecule).
4. The germicidal action of chlorine is based on the formation of hypochlorous acid when chlorine is added to water.

5. Chlorine is used as a disinfectant in gaseous form (Cl_2) or in the form of a compound, such as calcium hypochlorite, sodium hypochlorite, and chloramines.

Alcohols (p. 180)

1. Alcohols exert their action by denaturing proteins and dissolving lipids.
2. In tinctures, they enhance the effectiveness of other antimicrobial chemicals.
3. Aqueous ethanol (60% to 90%) and isopropanol are used as disinfectants.

Heavy Metals and Their Compounds (pp. 180–181)

1. Silver, mercury, copper, and zinc are used as germicidals.
2. They exert their antimicrobial action through oligodynamic action. When heavy metal ions combine with —SH groups, proteins are denatured.

Surface-Active Agents (p. 181)

1. Surface-active agents decrease the tension between molecules that lie on the surface of a liquid; soaps and detergents are examples.
2. Soaps have limited germicidal action but assist in the removal of microorganisms through scrubbing.
3. Acid-anionic detergents are used to clean dairy equipment.

Quaternary Ammonium Compounds (Quats) (pp. 181–182)

1. Quats are cationic detergents attached to NH_4^+.
2. By disrupting plasma membranes, they allow cytoplasmic constituents to leak out of the cell.
3. They are most effective against gram-positive bacteria.

Organic Acids and Derivatives (p. 182)

1. Sorbic acid, benzoic acid, and propionic acid inhibit fungal metabolism. They are used as food preservatives.

Aldehydes (p. 182)

1. Aldehydes such as formaldehyde and glutaraldehyde exert their antimicrobial effect by inactivating proteins.
2. They are among the most effective chemical disinfectants.

Gaseous Chemosterilizers (pp. 182–183)

1. Ethylene oxide is the gas most frequently used for sterilization.
2. It penetrates most materials and kills all microorganisms by protein denaturation.

Oxidizing Agents (p. 183)

1. Ozone and peroxide are used as antimicrobial agents.
2. They exert their effect by oxidizing molecules inside cells.

STUDY QUESTIONS

Review

1. Name the cause of cell death resulting from damage to each of the following:

   (a) Cell wall   (c) Proteins
   (b) Plasma membrane   (d) Nucleic acids

2. The thermal death time for a suspension of *B. subtilis* endospores is 30 minutes in dry heat and less than 10 minutes in an autoclave. Which type of heat is more effective? Why?

3. If pasteurization does not achieve sterilization, why is food treated by pasteurization?

4. Thermal death point is not considered an accurate measure of the effectiveness of heat sterilization. List three factors that can alter thermal death point.

5. The antimicrobial effect of gamma radiation is due to _____ . The antimicrobial effect of ultraviolet radiation is due to _____ .

6. A bacterial culture was in log phase in the following figure. At time *x*, an antibacterial compound was added to the culture. Which line indicates addition of a bactericidal compound? A bacteriostatic compound? How can you tell? Explain why the viable count does not immediately drop to zero in line *a*.

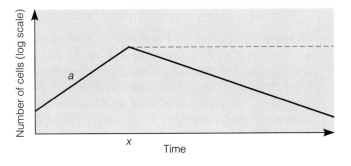

7. Fill in the following table:

METHOD OF STERILIZATION	TEMP.	TIME	TYPE OF HEAT	PREF'D USE	TYPE OF ACTION
Autoclaving					
Hot air					
Pasteurization					

8. How do the examples in question 7 illustrate the concept of equivalent treatments?

9. Heat-labile solutions such as glucose–minimal salts broth can be sterilized by _____ .

10. How do salts and sugars preserve foods? Why are these considered physical rather than chemical methods of control? Name one food that is preserved with sugar and one preserved with salt. How do you account for the occasional growth of *Penicillium* mold in jelly, which is 50% sucrose?

11. List five factors to consider before selecting a disinfectant.

12. Give the method of action and at least one standard use of each of the following types of disinfectants:

    (a) Phenolics   (e) Heavy metals
    (b) Iodine   (f) Aldehydes
    (c) Chlorine   (g) Ethylene oxide
    (d) Alcohol   (h) Oxidizing agents

13. The use-dilution values for two disinfectants tested under the same conditions are Disinfectant A—1:2; Disinfectant B—1:10,000. If both disinfectants are designed for the same purpose, which would you select?

14. A large hospital washes burn patients in a stainless steel tub. After each patient, the tub is cleaned with a quat. It was noticed that 14 of 20 burn patients acquired *Pseudomonas* infections after being bathed. Provide an explanation for this high rate of infection.

Critical Thinking

1. The filter paper method was used to evaluate three disinfectants. The results were as follows:

DISINFECTANT	ZONE OF INHIBITION
X	0 mm
Y	5 mm
Z	10 mm

   (a) Which disinfectant was the most effective against the organism?
   (b) Can you determine whether compound Y was bactericidal or bacteriostatic?

2. Why is each of the following bacteria often resistant to disinfectants?
   (a) *Mycobacterium*
   (b) *Pseudomonas*
   (c) *Bacillus*

3. The use-dilution test was used to evaluate two disinfectants against *Salmonella choleraesuis*. The results were as follows:

TIME OF EXPOSURE (MIN)	BACTERIAL GROWTH AFTER EXPOSURES		
	DISINFECTANT A	DISINFECTANT B DILUTED WITH DISTILLED WATER	DISINFECTANT B DILUTED WITH TAP WATER
10	+	−	+
20	+	−	−
30	−	−	−

(a) Which disinfectant was the most effective?

(b) Which disinfectant should be used against *Staphylococcus?*

4. To determine the lethal action of microwave radiation, two 10^5 suspensions of *E. coli* were prepared. One cell suspension was exposed to microwave radiation while wet, whereas the other was lyophilized (freeze-dried) and then exposed to radiation. The results are shown in the following figure. Dashed lines indicate the temperature of the samples. What is the most likely method of lethal action of microwave radiation? How do you suppose these data might differ for *Clostridium?*

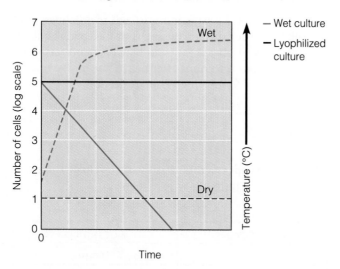

Clinical Applications

1. *Entamoeba histolytica* and *Giardia lamblia* were isolated from the stool sample of a 45-year-old man, and *Shigella sonnei* was isolated from the stool sample of an 18-year-old woman. Both patients experienced diarrhea and severe abdominal cramps, and, prior to onset of digestive symptoms, both had been treated by the same chiropractor.

The chiropractor had administered colonic irrigations (enemas) to these patients. The device used for this treatment was a gravity-dependent apparatus using 12 liters of tap water. There were no check valves to prevent backflow, so all parts of the apparatus could have become contaminated with feces during each colonic treatment. The chiropractor provided colonic treatment to four or five patients per day. Between patients, the adaptor piece that is inserted into the rectum was placed in a "hot-water sterilizer."

What two errors were made by the chiropractor?

2. Between March 9 and April 12, five chronic peritoneal dialysis patients at one hospital became infected with *Pseudomonas aeruginosa*. Four patients developed peritonitis (inflammation of the abdominal cavity), and one developed a skin infection at the catheter insertion site. All patients with peritonitis had low-grade fever, cloudy peritoneal fluid, and abdominal pain. All patients had permanent indwelling peritoneal catheters, which the nurse wiped with gauze soaked with an iodophor (Prepodyne®) solution each time the catheter was connected to or disconnected from the machine tubing. Aliquots of the iodophor were transferred from stock bottles to small in-use bottles.

Cultures from the dialysate concentrate and the internal areas of the dialysis machines were negative; iodophor from a small in-use plastic container yielded a pure culture of *P. aeruginosa*.

What improper technique led to this infection?

3. Eleven patients received injections of methylprednisolone and lidocaine to relieve the pain and inflammation of arthritis at the same orthopedic surgery office. All the patients developed septic arthritis caused by *Serratia marcescens*. Unopened bottles of methylprednisolone from the same lot numbers tested sterile; the methylprednisolone was preserved with a quat. Cotton balls were used to wipe multiple-use injection vials before the medication was drawn into a disposable syringe. The site of injection on each patient was also wiped with a cotton ball. The cotton balls were soaked in benzalkonium chloride, and fresh cotton balls were added as the jar was emptied. Opened methylprednisolone and the jar of cotton balls contained *S. marcescens*.

How was the infection transmitted? What part of the routine procedure caused the contamination?

Microbial Genetics

VIRTUALLY ALL THE microbial traits you have read about in earlier chapters are controlled or influenced by heredity. The inherited traits of microbes include their shape and structural features (morphology), their biochemical reactions (metabolism), their ability to move or behave in various ways, and their ability to interact with other organisms—perhaps causing disease. Individual organisms transmit these characteristics to their offspring through genes, which are the units of the hereditary material (DNA) that contain the information that determines these characteristics.

All living organisms find it advantageous to share in a gene pool. If an environmental change is encountered, the organisms that have genes that are advantageous under these new conditions (presumably acquired by random mutations) will have a better chance of reproducing, eventually increasing their relative numbers in the population. Eucaryotic organisms, including plants and animals, use sexual reproduction—with its improved survival value—to share in this general gene pool. Procaryotic cells, bacteria, do not have the capacity for sexual reproduction as such. Instead, they have acquired other mechanisms to avoid the genetic uniformity that might be lethal to the species if certain conditions, such as development of an antibiotic, arose.

Chromosomes are cellular structures that physically carry hereditary information; the chromosomes contain the genes. **Genetics** is the science of heredity; it includes the study of what genes are, how they carry information, how they are replicated and passed to subsequent generations of cells or passed between organisms, and how the

expression of their information within an organism determines the particular characteristics of that organism.

STRUCTURE AND FUNCTION OF THE GENETIC MATERIAL

LEARNING OBJECTIVES

- *Define genetics, chromosome, gene, genetic code, genotype, and phenotype.*
- *Describe how DNA serves as genetic information.*
- *Describe DNA replication.*
- *Describe protein synthesis, including transcription, RNA processing, and translation.*

Genes are made of DNA (except in some viruses, in which they are made of RNA). We saw in Chapter 2 that DNA is a macromolecule composed of repeating units called *nucleotides*. You may recall that each nucleotide consists of a nitrogenous base—adenine (A), thymine (T), cytosine (C), or guanine (G)—deoxyribose (a pentose sugar), and a phosphate group (see Figure 2.19). The DNA within a cell exists as long strands of nucleotides twisted together in pairs to form a double helix. Each strand has a string of alternating sugar and phosphate groups—its *sugar-phosphate backbone*—and a nitrogenous base is attached to each sugar in the backbone. The two strands are held together by hydrogen bonds between their nitrogenous bases. The bases are paired in a specific way: adenine always pairs with thymine, and cytosine always pairs with guanine. Because of this specific base pairing, the base sequence of one DNA strand determines the sequence of the other strand. The two strands of DNA are thus *complementary*. You can think of these complementary DNA sequences as being like a positive photograph and its negative. As we will see, the complementary structure of DNA helps explain how DNA stores and transmits genetic information.

A **gene** can be defined as a segment of DNA (a sequence of nucleotides in DNA) that codes for a functional product. This product is usually a messenger RNA (mRNA) molecule, which ultimately results in the formation of a protein. Alternatively, it can be a ribosomal RNA (rRNA) or a transfer RNA (tRNA). As we will see, all of these types of RNA are involved in the process of protein synthesis.

Genetic information is encoded by the sequence of bases along a strand of DNA, in much the same way as our written language uses a linear sequence of letters to form words and sentences. The genetic language, however, uses an alphabet with only four letters—the four kinds of nitrogenous bases in DNA (or RNA). But 1000 of these four bases, the number contained in an average gene, can be arranged in 4^{1000} different ways. This astronomically large number explains how genes can be varied enough to provide all the information a cell needs to grow and perform its functions. The genetic code, which determines how a nucleotide sequence is converted into the amino acid sequence of a protein, is discussed in more detail later in this chapter.

Much of cellular metabolism is concerned with translating the genetic message of genes into specific proteins. The DNA sequence is *transcribed* (copied) to produce a specific molecule of RNA called messenger RNA (mRNA). The information encoded in the mRNA is then *translated* into a specific sequence of amino acids that forms a protein.

When the ultimate molecule for which a gene codes (a protein, for example) has been produced, we say that the gene has been *expressed*. Thus, gene expression involves the transcription of DNA to produce specific molecules of RNA. If the RNA is mRNA, the information encoded in it is then translated into protein. This process can be symbolized as follows:

$$\text{DNA} \xrightarrow{\text{Transcription}} \text{mRNA} \xrightarrow{\text{Translation}} \text{Protein}$$

To express genetic information, the structure of DNA must be disrupted. Although an individual hydrogen bond is weak, the numerous hydrogen bonds along even a short stretch of DNA provide sufficient bonding energy to make the double helix a relatively stable structure. However, genetic information can generally be extracted from the DNA only when the strands are separated. Thus, many proteins function to separate these strands at appropriate times. Along with gene expression, this also allows the two separated strands of DNA to serve as templates for replication, which leads to the production of two daughter DNA molecules.

The specific sequence of nucleotides contained in a cell's DNA is usually duplicated accurately by enzymes before the cell divides. This duplication allows DNA to carry genetic information from cell to cell and from generation to generation. The information in the sequence of nucleotides specifies the complete collection of proteins to be found in the cell, which in turn determines the characteristics of the cell and transfers these characteristics to subsequent generations of cells. Later in this chapter we will describe several important experiments that clearly identified DNA as the genetic material of cells.

DNA can change, or *mutate*. Although many genetic mutations cause harm or kill the cell, others create new characteristics that give the offspring inheriting the mutation a better chance of survival in new environments. Thus, mutation is—in the long run—an advantage that contributes to the successful evolution of species. We will discuss various types of mutation later in this chapter.

Genotype and Phenotype

The **genotype** of an organism is its genetic makeup, the information that codes for all the particular characteristics of the organism. The genotype represents *potential* properties, but not the properties themselves. **Phenotype** refers to *actual, expressed* properties, such as the organism's ability to perform a particular chemical reaction. Phenotype, then, is the manifestation of genotype.

In molecular terms, an organism's genotype is its collection of genes, its entire DNA. What constitutes the organism's phenotype in molecular terms? In a sense, an organism's phenotype is its collection of proteins. Most of a cell's properties derive from the structures and functions of its proteins. In microbes, most proteins are either enzymatic (catalyze particular reactions) or structural (participate in large functional complexes such as membranes or ribosomes). Even phenotypes that depend on structural macromolecules other than protein (such as lipids or polysaccharides) rely indirectly on proteins. For instance, the structure of a complex lipid or polysaccharide molecule results from the catalytic activities of enzymes that synthesize, process, and degrade those structures. Thus, although it is not completely accurate to say that phenotypes are due only to proteins, it is a useful simplification.

DNA and Chromosomes

Bacteria typically have a single circular chromosome consisting of a single circular molecule of DNA (Figure 8.1a) with associated proteins. The chromosome is looped and folded and attached at one or several points to the plasma membrane. The DNA of *E. coli*, the most-studied bacterial

(a) A procaryotic chromosome   TEM   1 μm

(b) A replicated eucaryotic   SEM   1 μm
chromosome

FIGURE 8.1 Chromosomes. (a) A procaryotic chromosome.
The tangled mass and looping strands of DNA emerging from
this disrupted *E. coli* cell are part of its single chromosome.
(b) A replicated eucaryotic chromosome. Two copies of one
of the 46 chromosomes found in a normal human cell are shown.
The individual fibers are strands of DNA with attached proteins.

species, has about 4 million base pairs and is about 1 mm
long—1000 times longer than the entire cell. However,
DNA is very thin and is tightly packed inside the cell, so
that this twisted, coiled macromolecule takes up only
about 10% of the cell's volume.

The DNA in eucaryotic chromosomes (Figure 8.1b) is
even more tightly coiled (condensed) than is procaryotic
DNA. Eucaryotic chromosomes also contain much more
protein than do procaryotic chromosomes; this mixture
of eucaryotic DNA and protein is called **chromatin.** In eucaryotic cells, a group of proteins known as histones form
complexes around which DNA is wound. The detailed
structure of the eucaryotic chromosome (and that of the
procaryotic chromosome as well) and the precise arrangement of DNA with proteins are still under investigation.
Researchers believe that a detailed understanding of chromosomal structure will help reveal how the cell turns
genes on and off to produce crucial proteins when needed.
This regulation of gene expression governs both the differentiation of eucaryotic cells into the various types of
cells found in multicellular organisms and an individual
cell's ongoing activities.

We will next examine more closely how DNA functions as the genetic material in replication and protein
synthesis.

DNA Replication

In DNA replication, one "parental" double-stranded DNA
molecule is converted to two identical "daughter" molecules. The complementary structure of the nitrogenous
base sequences in the DNA molecule is the key to understanding DNA replication. Because the bases along the two
strands of double-helical DNA are complementary, one
strand can act as a template for the production of the
other strand.

DNA replication requires the presence of complex cellular proteins that direct a particular sequence of events.
When replication begins, the two strands of parental
DNA are unwound and separated from each other in one
small DNA segment after another. Free nucleotides present in the cytoplasm of the cell are matched up to the exposed bases of the single-stranded parental DNA. Where
thymine is present on the original strand, only adenine
can fit into place on the new strand; where guanine is
present on the original strand, only cytosine can fit into
place, and so on. Any bases that are improperly base-
paired are removed and replaced by replication enzymes.
Once aligned, the newly added nucleotide is joined to the
growing DNA strand by an enzyme called DNA polymerase. Then the parental DNA is unwound a bit further

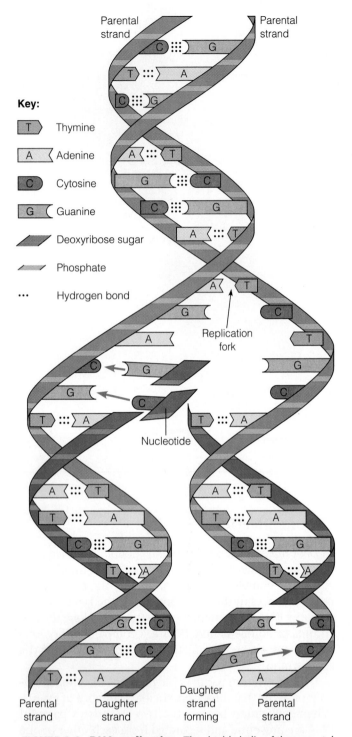

Key:

- T   Thymine
- A   Adenine
- C   Cytosine
- G   Guanine
-   Deoxyribose sugar
-   Phosphate
- ···   Hydrogen bond

FIGURE 8.2   DNA replication. The double helix of the parental DNA separates as weak hydrogen bonds between the nucleotides on opposite strands break in response to the action of replication enzymes. Next, hydrogen bonds form between new complementary nucleotides and each strand of the parental template to form new base pairs. Enzymes catalyze the formation of sugar-phosphate bonds between sequential nucleotides on each resulting daughter strand.

to allow the addition of the next nucleotides. The point at which replication occurs is called the **replication fork** (Figure 8.2).

As the replication fork moves along the parental DNA, each of the unwound single strands combines with new nucleotides. The original strand and this newly synthesized daughter strand then rewind. Because each new double-stranded DNA molecule contains one original, conserved strand and one new strand, the process of replication is referred to as **semiconservative replication.**

Before discussing DNA replication in more detail, let's take a closer look at the structure of DNA (Figure 8.3) in order to understand the important concept that the paired DNA strands are oriented in opposite directions to each other. Note in the detail inset to Figure 8.3 that the carbon atoms of the sugar component of each nucleotide are numbered 1′ (pronounced "one prime") to 5′. Although the two strands of DNA are complementary, their backbones have different chemical senses of direction in terms of the orientation of the sugar groups. Each phosphate group in the sugar-phosphate backbone of DNA attaches the 5′ carbon of one sugar to the 3′ carbon of the next sugar. (The nitrogenous base attaches to the 1′ carbon.) If the sugar-phosphate backbone is broken by the action of an enzyme or if the DNA is a linear molecule, a hydroxyl group is found at one end of each DNA strand. The end of the DNA strand that has the hydroxyl attached to the 3′ carbon of the sugar is called the 3′ end of the DNA strand; the end having a phosphate group attached to the 5′ carbon of the sugar is called the 5′ end. The way in which the two strands of DNA fit together dictates that if one strand has the chemical sense of direction 5′ ⟶ 3′, then the other must have the reverse 3′ ⟶ 5′ direction. The two strands of the DNA double helix are therefore said to be *antiparallel.*

The antiparallel structure of DNA affects the replication process. DNA polymerases can join new nucleotides to the growing daughter strand only by adding them to its 3′ end; that is, a DNA strand grows only in the 5′ ⟶ 3′ direction. Therefore, as the site of the replication fork moves along the parental DNA, the two new DNA strands must grow in different directions. One new DNA strand, called the *leading strand,* is synthesized continuously in the 5′ ⟶ 3′ direction (from a template parental strand running 3′ ⟶ 5′). In contrast, the *lagging strand* of new DNA is synthesized *discontinuously* in fragments of about 1000 nucleotides, which must later be joined to make one continuous strand (Figure 8.4). Thus the *overall* growth of the lagging strand in the 3′ ⟶ 5′ direction is achieved by assembling short stretches that have each been synthesized in the usual 5′ ⟶ 3′ direction.

The process of DNA synthesis requires many proteins and enzymes. For the rest of this discussion we will describe the process of DNA replication in procaryotes, for it is simpler and better understood than the corresponding process in eucaryotes.

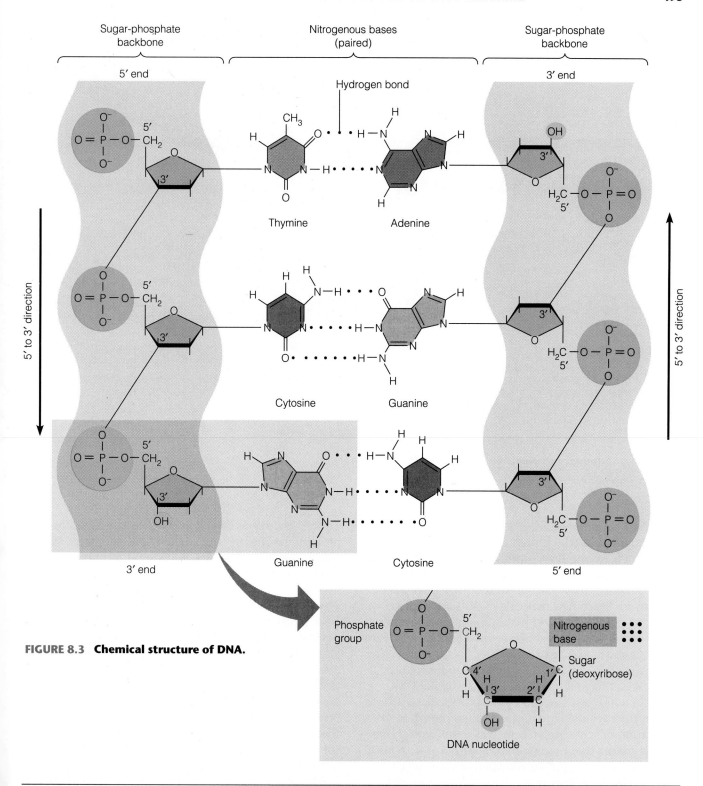

FIGURE 8.3   Chemical structure of DNA.

At the site of the replication fork, enzymes unwind the parental double helix. Stabilizing proteins bind to the single-stranded regions and prevent them from rewinding. DNA polymerase cannot initiate synthesis of a DNA strand entirely on its own, but it can attach additional nucleotides to the free 3′ end of an existing DNA or RNA strand to elongate the strand. While the leading strand of DNA is being synthesized by DNA polymerase, an RNA polymerase starts the synthesis of each DNA fragment of the lagging strand with a short stretch of RNA nucleotides, called an *RNA primer*. DNA synthesis catalyzed by DNA polymerase can then proceed in the 5′ ⟶ 3′ direction

FIGURE 8.4   Summary of events at the DNA replication fork.

from the RNA primers. The RNA is later digested away by DNA polymerase, which then synthesizes DNA to fill in the gap left by the RNA. Another enzyme, *DNA ligase,* joins the discontinuous fragments with covalent bonds to make a continuous lagging strand. DNA polymerase "proofreads" each new molecule of DNA and replaces incorrectly paired bases before adding a new base. Each newly synthesized DNA strand forms a double helix with its parental strand. The two daughter DNA double helices that are the final result of the process are indistinguishable from each other and from the original parental DNA.

In bacteria, the replication process begins at a unique site on the bacterial chromosome called the *origin.* DNA replication of some bacteria, such as *E. coli,* goes *bidirectionally* around the chromosome (Figure 8.5). Two replication forks move in opposite directions away from the origin of replication. Because the bacterial chromosome is a closed loop, the replication forks eventually meet when the replication of the chromosome is completed. Much evidence shows an association between the bacterial cell membrane and the site on the DNA where replication begins. The DNA–membrane attachment in procaryotes might ensure that in cell division each daughter cell receives one copy of the DNA molecule—that is, one complete chromosome.

In eucaryotes, replication is made more complex by the greater amount of DNA and the presence of many more proteins bound to this DNA. There are many sites at which replication begins, allowing it to finish in a reasonable time.

The Rate of DNA Replication

DNA synthesis is a surprisingly fast process: about 1000 nucleotides per second in *E. coli* growing at 37°C. At first glance, this speed seems improbable, considering that nucleotide substrates must be synthesized and then must diffuse to the replication fork. Furthermore, several attempts are probably made by wrong nucleotides to pair at each position before the correct bases pair up. Nevertheless, the speed and specificity of DNA replication are governed by the same principles that guide all chemical reactions.

Under some conditions—namely, log-phase growth in a rich nutrient medium (see Figure 6.12)—*E. coli* cells can grow faster than the two replication forks can complete the circular chromosome. Under these conditions, the cell initiates *multiple* replication forks at the origin on the chromosome; a new pair of forks begins before the last pair has finished. In this way, the overall rate of DNA synthesis matches the rate at which the cell divides. Similarly, when the cell's growth greatly slows, the initiation of DNA synthesis at the origin of replication may be delayed. The rate at which each replication fork moves is generally con-

(a)

0.25 μm

(b)

—Origin of replication

Replication fork

Parental strand

Daughter strand

Replication fork

Termination of replication

FIGURE 8.5 Replication of bacterial DNA.
(a) Autoradiograph of the replication of an *E. coli* chromosome. (In the corresponding diagram at right, the arrows point to the two replication forks.) The chromosome is about one-third replicated. Note that one of the new helices is crossed over the other one. **(b)** Diagrammatic representation of the bidirectional replication of a circular bacterial DNA molecule. The new strands are shown in blue.

stant (at a given temperature); however, by regulating how often replication is initiated, the cell controls its overall rate of DNA synthesis to match its rate of growth and cell division.

The Flow of Genetic Information

DNA replication makes possible the flow of genetic information from one generation to the next. As shown in Figure 8.6, the DNA of a cell replicates before cell division, so that each daughter cell receives a chromosome that is identical to the parent's. Within each metabolizing cell, the genetic information contained in DNA also flows in another way: it is transcribed into mRNA and then translated into protein. We describe the processes of transcription and translation in the next section.

RNA and Protein Synthesis

How is the information in DNA used to make the proteins that control cell activities? In a process called *transcription*, genetic information in DNA is copied, or transcribed, into a complementary base sequence of RNA. The cell then uses the information encoded in this RNA to synthesize specific proteins through the process of *translation*. We will now take a closer look at these two processes as they occur in a bacterial cell.

Transcription

Transcription is the synthesis of a complementary strand of RNA from a DNA template. As we mentioned earlier, there are three kinds of RNA in bacterial cells: messenger

FIGURE 8.6 Two ways genetic information can flow. Genetic information can be transferred between generations of cells, through replication of DNA. Genetic information is also used within a cell to produce the proteins the cell needs to function. This information is transferred through the processes of transcription and translation. The cell represented in this diagram is a bacterium that has a single circular chromosome.

RNA, ribosomal RNA, and transfer RNA. Ribosomal RNA forms an integral part of ribosomes, the cellular machinery for protein synthesis. Transfer RNA is also involved in protein synthesis, as we will see later. **Messenger RNA (mRNA)** carries the coded information for making specific proteins from DNA to ribosomes, where proteins are synthesized.

During transcription, a strand of mRNA is synthesized using a specific gene—a portion of the cell's DNA—as a template (Figure 8.7). In other words, the genetic information stored in the sequence of nitrogenous bases of DNA is rewritten so that the same information appears in the base sequence of mRNA. As in DNA replication, a G in the DNA template dictates a C in the mRNA being made; a C in the DNA template dictates a G in the mRNA; and a T in the DNA template dictates an A in the mRNA. However, an A in the DNA template dictates a uracil (U) in the mRNA because RNA contains U instead of T. (U has a chemical structure slightly different from T [see Figure 2.20], but it base-pairs in the same way.) If, for example, the template portion of DNA has the base sequence ATG-CAT, then the newly synthesized mRNA strand will have the complementary base sequence UACGUA.

The process of transcription requires both an enzyme called *RNA polymerase* and a supply of RNA nucleotides. Transcription begins after RNA polymerase binds to the DNA at a site called **promoter.** Only one of the two DNA strands serves as the template for RNA synthesis for a given gene. Like DNA, RNA is synthesized in the 5′ ⟶ 3′ direction. The endpoint for transcription of the gene is signaled by a **terminator** region in the DNA. At this re-

gion, RNA polymerase and newly formed, single-stranded mRNA are released from the DNA.

In eucaryotic organisms, transcription takes place in the nucleus. The mRNA must be completely synthesized and moved through the nuclear membrane to the cytoplasm before translation can begin. In addition, the RNA undergoes further processing before it leaves the nucleus. In eucaryotic cells the regions of genes that code for proteins are often interrupted by noncoding DNA. Thus, eucaryotic genes are composed of **exons,** the regions of DNA *expressed,* and **introns,** the *intervening* regions of DNA that do not code for protein (Figure 8.8). In the nucleus of a eucaryotic cell, RNA polymerase synthesizes from the entire gene a long, continuous RNA product (including all exons and introns) called an *RNA transcript.* This long RNA is then processed by other enzymes, which remove the intron-derived RNA and splice together the exon-derived RNA, yielding an mRNA capable of directing protein synthesis. The resulting RNA leaves the nucleus and becomes the mRNA of the cytoplasm. Some introns are actually cut out by the RNA itself, which acts as a splicing enzyme on itself. The discovery of such RNA enzymes, which are called **ribozymes,** was the first identified instance of a nonprotein enzyme. Since then, other RNA molecules that act as enzymes, including some not involved in splicing, have been found.

In summary, in both procaryotes and eucaryotes the genetic information for protein synthesis is stored in DNA and is passed to mRNA during transcription. The mRNA then acts as the source of information for protein synthesis.

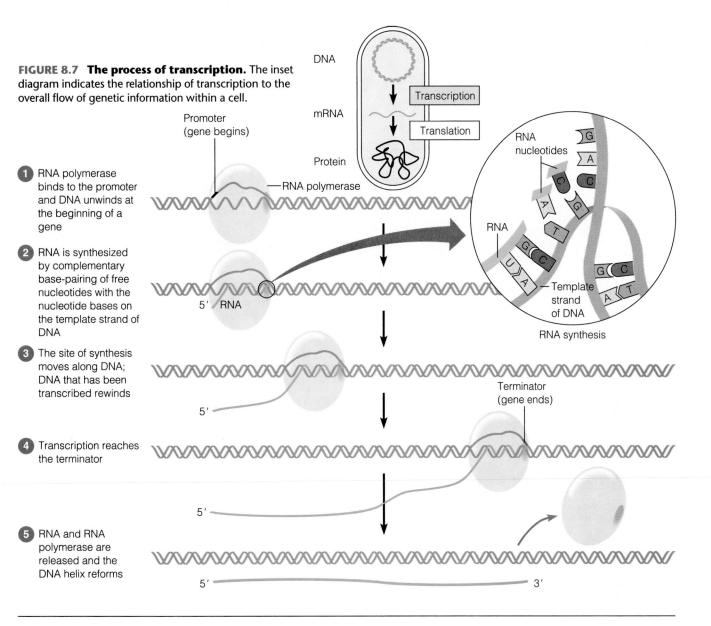

FIGURE 8.7 The process of transcription. The inset diagram indicates the relationship of transcription to the overall flow of genetic information within a cell.

DNA

Transcription

mRNA

Translation

Protein

RNA polymerase

Promoter (gene begins)

RNA nucleotides

RNA

Template strand of DNA

RNA synthesis

1. RNA polymerase binds to the promoter and DNA unwinds at the beginning of a gene

2. RNA is synthesized by complementary base-pairing of free nucleotides with the nucleotide bases on the template strand of DNA

5′ RNA

3. The site of synthesis moves along DNA; DNA that has been transcribed rewinds

5′

Terminator (gene ends)

4. Transcription reaches the terminator

5′

5. RNA and RNA polymerase are released and the DNA helix reforms

5′   3′

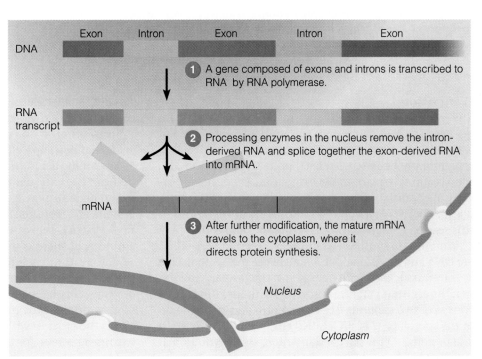

FIGURE 8.8 RNA processing in eucaryotic cells.

Exon | Intron | Exon | Intron | Exon

DNA

1. A gene composed of exons and introns is transcribed to RNA by RNA polymerase.

RNA transcript

2. Processing enzymes in the nucleus remove the intron-derived RNA and splice together the exon-derived RNA into mRNA.

mRNA

3. After further modification, the mature mRNA travels to the cytoplasm, where it directs protein synthesis.

Nucleus

Cytoplasm

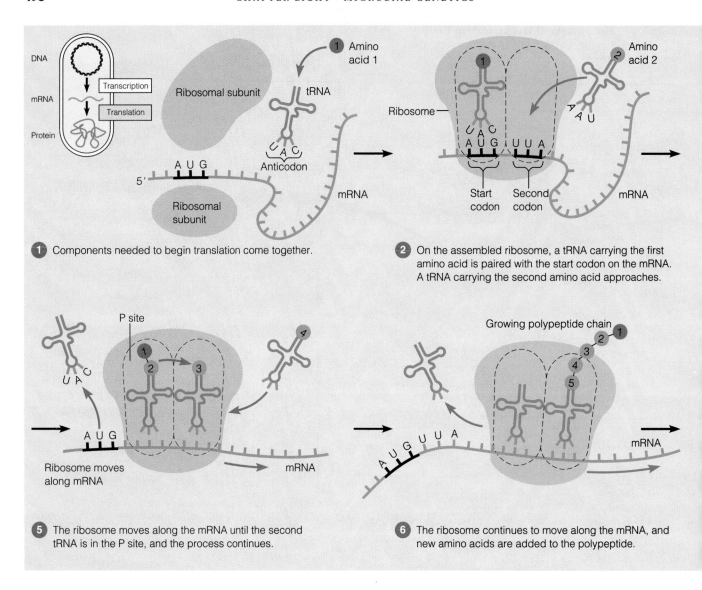

FIGURE 8.9 **The process of translation.** The inset indicates the relationship of translation to the overall flow of genetic information within a cell.

Translation

The process through which the nitrogenous-base sequence of mRNA determines the amino acid sequence of a protein is called **translation.** In bacteria, which lack a membrane-enclosed nucleus, both transcription and translation take place in the cytoplasm. In eucaryotic organisms, translation begins when mRNA enters the cytoplasm. The events involved in translation are described in this section and are shown in Figures 8.9 and 8.10.

Proteins are synthesized in the 5′ ⟶ 3′ direction, as are DNA and RNA. First, the 5′ end of the mRNA molecule becomes associated with a ribosome, the primary cellular machinery that catalyzes protein synthesis. Ribosomes consist of two subunits (see Figure 4.18) and are made up of the special type of RNA called **ribosomal RNA (rRNA)** and proteins. At the start of translation, the two ribosomal

subunits come together with the mRNA and several other components of the process.

In solution in the cytoplasm are 20 different amino acids that participate in protein synthesis (see Table 2.4). (These amino acids can be synthesized by the cell or taken up from the external medium.) However, before the appropriate amino acids can be joined together to form a protein, they must be *activated* by attachment to **transfer RNA (tRNA),** shown in Figure 8.10a. For each different amino acid there is a specific tRNA, and during amino acid activation each amino acid attaches to its specific tRNA. This attachment is accomplished by an *amino-acid activating enzyme* and with energy from ATP (Figure 8.10b).

Now we are ready to see how the mRNA actually determines the order in which amino acids are linked together to form a protein. Each set of three nucleotides of mRNA,

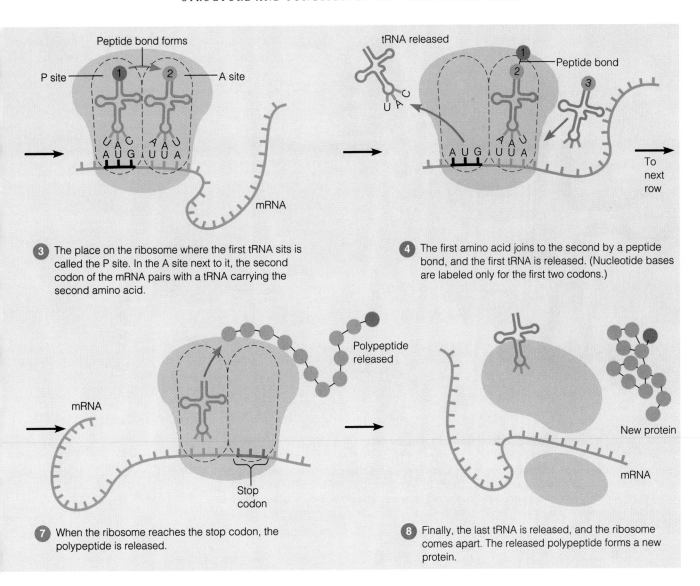

3 The place on the ribosome where the first tRNA sits is called the P site. In the A site next to it, the second codon of the mRNA pairs with a tRNA carrying the second amino acid.

4 The first amino acid joins to the second by a peptide bond, and the first tRNA is released. (Nucleotide bases are labeled only for the first two codons.)

7 When the ribosome reaches the stop codon, the polypeptide is released.

8 Finally, the last tRNA is released, and the ribosome comes apart. The released polypeptide forms a new protein.

called a **codon**, specifies (codes for) a single amino acid (Figure 8.10c). For example, the sequence

AUGCCAGGCAAA

contains four codons specifying the amino acids methionine (AUG), proline (CCA), glycine (GGC), and lysine (AAA). If the bases are grouped differently, the same sequence could code for other amino acids. For example,

A UGC CAG GCA AA

would encode cysteine (UGC), glutamine (CAG), and alanine (GCA). Similarly,

AU GCC AGG CAA A

would encode alanine (GCC), arginine (AGG), and glutamine (CAA). These groupings are called *reading frames*.

Which of these three possible groupings is chosen during protein synthesis? As we shall see, a reading frame is determined by the position of the very first codon of the gene.

The tRNA molecules "read" the coded message on the mRNA. On one special part of each tRNA molecule is a set of three nucleotides, called an **anticodon**, that is complementary to the codon for the amino acid carried by the tRNA (Figure 8.10c). During translation, the anticodon of a molecule of tRNA hydrogen-bonds to its complementary codon on mRNA. For example, a tRNA with the anticodon CGA pairs with the mRNA codon GCU (anticodons and codons are always read in the 5′ ⟶ 3′ direction). The pairing of codon and anticodon can occur only at two sites de-

(b) Activation of an amino acid

(a) Structure of tRNA

(c) Base-pairing of tRNA anticodon to complementary mRNA codon

FIGURE 8.10   **The structure and function of transfer RNA. (a)** Structure of tRNA is shown in two-dimensional form. Each "box" represents a nucleotide. Note the regions of hydrogen bonding between base pairs and the loops of unpaired bases, an arrangement typical of RNA molecules. **(b)** Each amino acid is activated by attachment to tRNA. **(c)** The anticodon of tRNA pairs with its complementary codon on an mRNA strand. The tRNA shown here carries the amino acid alanine. Anticodons are always read in the 5′ ⟶ 3′ direction; hence, the anticodon for alanine is read C-G-A.

fined by the ribosome: the *A, or aminoacyl, site* and the *P, or peptidyl, site*.

Now let's return to Figure 8.9 and follow the process of translation. ① and ② After the first tRNA (with its amino acid) base-pairs with mRNA at the *start codon,* ③ a second tRNA molecule (with its amino acid) moves into position on the second codon. ④ The two amino acids are then joined by a peptide bond, the first tRNA molecule leaves the ribosome, and ⑤ the ribosome moves along the mRNA to the next codon. The detached tRNA can now pick up another molecule of its specific amino acid. Each tRNA can be used over and over again. ⑥ As the proper amino acids are brought into line one by one, peptide bonds form between the amino acids, and a polypeptide chain is formed.

⑦ Any one of three special *stop codons* in the mRNA signals the end of a polypeptide chain, at which time the chain is released from the ribosome. ⑧ The ribosome then comes apart into its two subunits ready to begin the cycle again. In procaryotic cells, the translation of mRNA into protein can begin even before transcription is complete.

Before the ribosome moving along the mRNA completes translation of that message, another ribosome, and then another, can attach and begin translation of the same mRNA molecule; a number of ribosomes can be attached at different positions to the same mRNA molecule (Figure 8.11). In this way, a single mRNA strand can be translated simultaneously into several identical protein molecules. The combination of an mRNA strand with several attached ribosomes is called a **polyribosome** (or *polysome*).

In summary, the synthesis of a protein requires that the genetic information in DNA first be transferred to a molecule of mRNA by a process called transcription. Then, in a process called translation, the mRNA directs the assembly of amino acids into a polypeptide: the mRNA attaches to a ribosome, tRNA delivers the amino acids to mRNA according to the complementary base-pairings of codons and anticodons, and the amino acids are then linked to form a polypeptide. Recall that proteins can be composed of one or more polypeptide chains, each of which may range in length from 50 to several hundred amino acids.

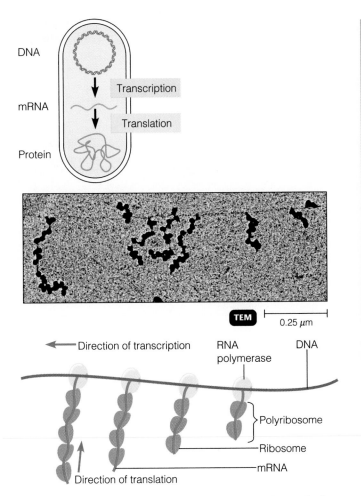

FIGURE 8.11 Simultaneous transcription and translation in bacteria. The micrograph and diagram show these processes in a single bacterial gene. Many molecules of mRNA are being synthesized simultaneously. The longest mRNA molecules were the first to be transcribed at the promoter. Note the ribosomes attached to the newly forming mRNA. The newly synthesized polypeptides are not shown.

FIGURE 8.12 The genetic code. The three nucleotides in an mRNA codon are designated, respectively, as the first position, second position, and third position of the codon, going in the $5' \longrightarrow 3'$ direction on the mRNA. Each set of three nucleotides specifies a particular amino acid, represented by a three-letter abbreviation (see Table 2.4). The codon AUG, which specifies the amino acid methionine, is also the start of protein synthesis. The word "Stop" identifies the nonsense codons that signal the termination of protein synthesis.

Because a typical polypeptide has about 300 amino acids, the typical mRNA must have about 300 codons and thus be at least 900 nucleotides long. The DNA of *E. coli* is long enough to contain about 4000 genes and can therefore specify about 4000 different kinds of proteins. At any one time a cell contains only a few copies of some proteins but several thousand copies of other proteins. Table 8.1 lists the main components of transcription and translation.

Protein synthesis in eucaryotes is fundamentally the same, although eucaryotic ribosomes are larger and many more accessory factors are involved.

The Genetic Code

The **genetic code** is the set of rules that relates the nitrogenous base sequence of DNA, the corresponding codons of mRNA, and the amino acids for which the codons stand (Figure 8.12). Codons are written in terms of their base sequence in mRNA. Note that there are 64 possible codons but only 20 amino acids. This means that most amino acids are signaled by several alternative codons; this situation is referred to as the *degeneracy* of the code. For example, leucine has six codons and alanine has four codons. Degeneracy allows for a certain amount of change, or mutation, in the DNA without affecting the protein ultimately produced.

Of the 64 codons, 61 are sense codons and 3 are nonsense codons. **Sense codons** code for amino acids, and **nonsense codons** (also called *stop codons*) do not. Rather, the nonsense codons—UAA, UAG, and UGA—signal the end of the protein molecule's synthesis. The start codon that initiates the synthesis of the protein molecule is AUG, which is also the codon for methionine. The initiating methionine is often removed later, so not all proteins begin with methionine.

TABLE 8.1 Major Components of Protein Synthesis

	TRANSCRIPTION	TRANSLATION
Messenger RNA (mRNA)	Synthesized as copy of information in DNA	Acts as coded message specifying amino acid sequence of proteins
Monomer building blocks (subunits)	RNA nucleotides (4)	Amino acids (20)
Proteins		
Enzymes	RNA polymerase	Amino-acid activating enzymes; ribosomal catalysis of peptide bond formation; initiation, elongation, and termination factors
Structural proteins	None	Ribosomal proteins
Other RNAs	tRNA and rRNA are themselves made by transcription from DNA	
Transfer RNA (tRNA)		tRNAs with amino acids attached
Ribosomal RNA (rRNA)		rRNAs, structural components of ribosomes
Chemical energy	From RNA nucleoside triphosphates (ATP, GTP, CTP, UTP)	From ATP and GTP

REGULATION OF GENE EXPRESSION IN BACTERIA

LEARNING OBJECTIVES

- *Explain the regulation of gene expression in bacteria by induction, repression, and catabolic repression.*
- *Relate gene regulation to bacterial survival.*

In Chapter 5 we learned that the bacterial cell carries out an enormous number of metabolic reactions. Some of these reactions are concerned with biosynthesis and are known as anabolic reactions; others are concerned with degradation and are known as catabolic reactions. The common feature of all metabolic reactions is that they are catalyzed by enzymes.

We have seen that genes, through transcription and translation, direct the synthesis of proteins, many of which serve as enzymes—the very enzymes used for cellular metabolism. Therefore, a cell's genetic machinery and its metabolic machinery are integrated and interdependent. We will look next at how gene expression is regulated. Because protein (enzyme) synthesis requires a tremendous expenditure of energy (ATP), the regulation of protein synthesis is important to the cell's energy economy. The cell conserves energy by making only those proteins needed at a particular time.

Many genes, perhaps from 60% to 80%, are not regulated but are instead *constitutive*, meaning that their products are constantly produced at a fixed rate. Usually these genes, which are effectively turned on all the time, code for enzymes that the cell needs in fairly large amounts for its major life processes; the enzymes of glycolysis are examples. The production of other enzymes is regulated so that they are present only when needed. In bacteria, the regulation of gene expression occurs primarily during transcription.

Repression and Induction

Mechanisms called *repression* and *induction* regulate the transcription of mRNAs and consequently the synthesis of enzymes from these mRNAs. These two genetic control mechanisms regulate the formation and amounts of enzymes in the cell, not the activities of the enzymes.

Repression

One regulatory mechanism that inhibits gene expression and decreases the synthesis of enzymes is called **repression**. Repression is usually a response to the overabundance of an end-product of a metabolic pathway; it causes a decrease in the rate of synthesis of the enzymes leading to the formation of that product. Repression is mediated by regulatory proteins called *repressors*, which block the ability of RNA polymerase to initiate transcription from the repressed genes.

Induction

Induction is the process that turns on the transcription of a gene or genes. A substance that acts to induce transcription of a gene is called an *inducer*, and enzymes that are synthesized in the presence of inducers are *inducible enzymes*. The genes required for lactose metabolism in *E. coli* are a well-known example of an inducible system. One of these genes codes for the enzyme β-galactosidase, which splits the substrate lactose into two simple sugars, glucose and galactose. (β is the Greek letter "beta"; here it refers to the type of linkage that joins the glucose and galactose.) If *E. coli* is placed into a medium into which no lactose is present, the organisms contain almost no β-galactosidase;

(a) Structure of the operon. This segment of the DNA molecule of *E. coli* shows the genes involved in lactose catabolism. In addition to the promoter (*P*) and operator (*O*) sites, the lactose operon contains three structural genes, *Z*, *Y*, and *A*, which code for the proteins β-galactosidase, permease, and transacetylase, respectively. For this operon the regulatory gene (*I*) is located next to the promoter (more often it is some distance away from the operon).

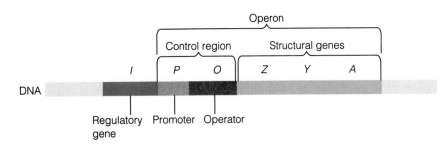

(b) Control of the operon in the absence of lactose (repression). The repressor protein binds with the operator, preventing transcription from the operon.

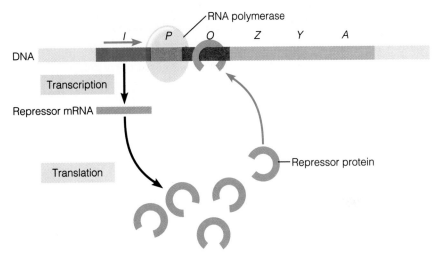

(c) Control of the operon in the presence of lactose (induction). When the inducer allolactose binds to the repressor protein, the inactivated repressor can no longer block transcription. The structural genes are transcribed, ultimately resulting in the production of the three enzymes needed for lactose catabolism.

FIGURE 8.13 The *lac* operon.

however, when lactose is added to the medium, the bacterial cells produce a large quantity of the enzyme. Lactose is converted in the cell to the related compound allolactose, which is the inducer for these genes; the presence of lactose thus indirectly induces the cells to synthesize more enzyme. This response, which is under genetic control, is termed **enzyme induction.**

Operon Model of Gene Expression

Details of the control of gene expression by induction and repression are described by the operon model. François Jacob and Jacques Monod formulated this general model

in 1961 to account for the regulation of protein synthesis. They based their model on studies of the induction of the enzymes of lactose catabolism in *E. coli*. In addition to β-galactosidase, these enzymes include permease, which is involved in the transport of lactose into the cell, and transacetylase, which metabolizes certain disaccharides other than lactose.

The genes for the three enzymes involved in lactose uptake and utilization are next to each other on the bacterial chromosome and are regulated together (Figure 8.13a). These genes, which determine the structures of proteins, are called **structural genes** to distinguish them from an adjoining control region on the DNA. When lactose is introduced into the culture medium, the *lac* structural genes

FIGURE 8.14 Growth rate of *E. coli* on glucose and lactose. The steeper the straight line, the faster the growth. **(a)** Bacteria growing on glucose as the sole carbon source grow faster than on lactose. **(b)** Bacteria growing in a medium containing glucose and lactose first consume the glucose and then, after a short lag time, the lactose. During the lag time, intracellular cyclic AMP increases, the *lac* operon is transcribed, more lactose is transported into the cell, and β–galactosidase is synthesized to break down lactose.

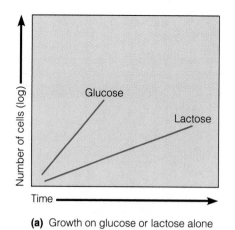

(a) Growth on glucose or lactose alone

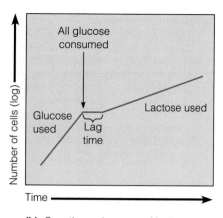

(b) Growth on glucose and lactose combined

are all transcribed and translated rapidly and simultaneously. We will now see how this regulation occurs.

In the control region of the *lac* operon are two relatively short segments of DNA. One, the *promoter,* is the region of DNA where RNA polymerase initiates transcription. The other is the *operator,* which is like a traffic light that acts as a go or stop signal for transcription of the structural genes. A set of operator and promoter sites, and the structural genes they control, are what define an **operon;** thus, the combination of the three *lac* structural genes and the adjoining control regions is called the *lac* operon.

Near the *lac* operon on the bacterial DNA is a regulatory gene called the *I* gene, which codes for a *repressor* protein. When lactose is absent (Figure 8.13b), the repressor protein binds tightly to the operator site. This binding prevents RNA polymerase from transcribing the adjacent structural genes; consequently, no mRNA is made and no enzymes are synthesized. But when lactose is present (Figure 8.13c), some of it is transported into the cells and is converted into the inducer allolactose. The inducer binds to the repressor protein and alters it so that it cannot bind to the operator site. In the absence of an operator-bound repressor protein, RNA polymerase can transcribe the structural genes into mRNA, which is then translated into enzymes. This is why, in the presence of lactose, enzymes are produced. Lactose is said to induce enzyme synthesis, and the *lac* operon is called an inducible operon.

In repressible operons, the repressor requires the presence of a small molecule called a **corepressor,** which helps the repressor bind to the operator gene. Without the corepressor, the repressor is unable to bind. The enzymes involved in several biosynthetic pathways are regulated in this manner.

Regulation of the lactose operon also depends on the level of glucose in the medium, which in turn controls the intracellular level of the small molecule **cyclic AMP,** a substance that is derived from ATP and serves as a cellular

alarm signal. Enzymes that metabolize glucose are constitutive, and cells grow at their maximal rate with glucose as their carbon source because they can use it most efficiently (Figure 8.14a). When glucose is no longer available, the cell responds by producing cyclic AMP. The cyclic AMP binds to a protein called *catabolite activator protein* (CAP). This complex, which binds to the *lac* promoter, is required for the transcription of the *lac* operon, as is the presence of lactose. When both of these requirements are met, transcription of the structural genes in the lactose operon can begin. As a result, the cell can grow on lactose (Figure 8.14b). Thus, transcription of the *lac* operon requires both the presence of lactose and the absence of glucose.

Cyclic AMP is an example of an *alarmone,* a chemical alarm signal the cell uses to respond to environmental or nutritional stress. (In this case, the stress is the lack of glucose.) The same mechanism involving cyclic AMP allows the cell to grow on other sugars. Inhibition of the metabolism of alternative carbon sources by glucose is termed **catabolic repression** (or the *glucose effect*). When glucose is available, the level of cAMP in the cell is low, and consequently CAP is not bound.

Table 8.2 summarizes the components involved in regulation of the lactose operon.

Global Regulation

In response to environmental stress, bacterial cells can turn on or off large sets of genes that are all involved in the same metabolic pathway. This mechanism is called **global regulation.** For example, when bacteria are subjected to increased temperatures, they respond by rapidly synthesizing a set of about 20 proteins (called heat shock proteins) that help them survive the higher temperatures. Similar responses are generated following oxygen, nitrogen, carbon,

TABLE 8.2 Components Involved in Regulation of the Lactose Operon

	COMPONENT	FUNCTION
Proteins	RNA polymerase	Binds to the promoter site on DNA and catalyzes the synthesis of mRNA
	lac repressor	Binds the inducer allolactose to the operator site on DNA
	CAP (catabolite activator protein)	Binds cyclic AMP, then binds to the promoter site; required for efficient transcription from certain promoters
Small molecules	Lactose	Forms the inducer allolactose inside the cell
	Cyclic AMP (cAMP)	Increases when glucose is lacking, then binds to CAP
Sites in DNA	Promoter site	Attachment site for RNA polymerase and CAP
	Operator site	Attachment site for repressor protein (when not bound by inducer)

or phosphate starvation, or during the formation of endospores in certain gram-positive bacteria. All of the genes of these pathways must be similarly regulated.

MUTATION: CHANGE IN THE GENETIC MATERIAL

LEARNING OBJECTIVES

- *Define mutagen.*
- *Classify mutations by type, and describe how mutations are prevented or repaired.*
- *Outline methods of direct and indirect selection of mutants.*
- *Identify the purpose of and outline the procedure for the Ames test.*

A **mutation** is a change in the base sequence of DNA. We can reasonably expect that such a change in the base sequence of a gene will sometimes cause a change in the product encoded by that gene. For example, when the gene for an enzyme mutates, the enzyme encoded by the gene may become inactive or less active because its amino acid sequence has changed. Such a change in genotype may be disadvantageous, or even lethal, if the cell loses a phenotypic trait it needs. However, a mutation can be beneficial if, for instance, the altered enzyme encoded by the mutant gene has a new or enhanced activity that benefits the cell.

Many simple mutations are silent (neutral); the change in DNA base sequence causes no change in the activity of the product encoded by the gene. Silent mutations commonly occur when one nucleotide is substituted for another in the DNA, especially at a location corresponding to the third position of the mRNA codon. Because of the degeneracy of the genetic code, the resulting new codon might still code for the same amino acid. Even if the amino acid is changed, the function of the protein may not change if the amino acid is in a nonvital portion of the protein or is chemically very similar to the original amino acid.

Types of Mutations

The most common type of mutation involving single base pairs is **base substitution** or **point mutation**, in which a single base at one point in the DNA is replaced with a different base. Then, when the DNA replicates, the result is a substituted base pair (Figure 8.15). For example, A-T might be substituted for G-C, or C-G for G-C. If a base substitution occurs within a gene that codes for a protein, then the mRNA transcribed from the gene will carry an incorrect base at that position. When the mRNA is translated into protein, the incorrect base may cause the insertion of an incorrect amino acid in the protein. If the base substitution results in an amino acid substitution in the synthesized protein, this change in the DNA is known as a **missense mutation** (Figure 8.16a and b).

By creating a stop (nonsense) codon in the middle of an mRNA molecule, some base substitutions effectively prevent the synthesis of a functional protein. Instead, only a fragment of the protein is synthesized. A base substitution resulting in a nonsense codon is thus called a **nonsense mutation** (Figure 8.16c).

Besides base-pair mutations, there are also changes in DNA called **frameshift mutations**, in which one or a few nucleotide pairs are deleted or inserted in the DNA (Figure 8.16d). This mutation can shift the "translational reading frame"—that is, the three-by-three grouping of nucleotides recognized as codons by the tRNAs during translation. For example, deleting one nucleotide pair in the middle of a gene causes changes in many amino acids downstream from the site of the original mutation.

FIGURE 8.15
Base substitution. This mutation leads to an altered protein in a granddaughter cell.

① During DNA replication, a thymine is incorporated opposite guanine by mistake.

Parental DNA

Replication

Daughter DNA

Daughter DNA

Daughter DNA

② In the next round of replication, adenine pairs with the new thymine, yielding an A-T pair in place of the original G-C pair.

Replication

Granddaughter DNA

③ When mRNA is transcribed from the DNA containing this substitution, a codon is produced that, during translation, codes for a different amino acid, tyrosine instead of cysteine.

Transcription

Translation

mRNA   U G U   U A U   U G U   U G U

Amino acids   Cysteine   Tyrosine   Cysteine   Cysteine

Frameshift mutations almost always result in a long stretch of missense and the production of an inactive protein from the mutated gene. In most cases, a nonsense codon will eventually be encountered and thereby terminate translation.

Base substitutions and frameshift mutations may occur spontaneously because of occasional mistakes made during DNA replication. These **spontaneous mutations** occur without known intervention of mutation-causing agents. Agents in the environment, such as certain chemicals and radiation, that directly or indirectly bring about mutations are called **mutagens.** Next we discuss two types of mutagens: chemicals and radiation.

Mutagens
Chemical Mutagens

One of the many chemicals known to be mutagenic is nitrous acid. Figure 8.17 shows how exposure of DNA to nitrous acid can convert the base adenine (A) to a form that no longer pairs with thymine (T) but instead pairs with cytosine (C). When DNA containing such modified adenines replicates, one daughter DNA molecule will have a base-pair sequence different from that of the parent DNA. Eventually, some A-T base pairs of the parent will have been changed to G-C base pairs in a granddaughter cell. Nitrous acid makes a specific base-pair change in DNA. Like all mutagens, it alters DNA at random locations.

Another type of chemical mutagen is the **base analog.** These molecules are structurally similar to normal nitrogenous bases, but they have slightly altered base-pairing properties. Examples are 2-aminopurine and 5-bromouracil. The 2-aminopurine molecule (Figure 8.18a) is incorporated into DNA in place of adenine (A) but can sometimes pair with cytosine (C). The 5-bromouracil molecule (Figure 8.18b) is incorporated into DNA in place of thymine (T) but often pairs with guanine (G). When base analogs are given to growing cells, the analogs are randomly incorporated into cellular DNA in place of the normal bases. Then, during DNA replication, the analogs cause mistakes in base-pairing. The incorrectly paired bases will be copied during subsequent replication of the DNA, resulting in base-pair substitutions in the progeny cells. Some antiviral and antitumor drugs are base analogs.

Still other chemical mutagens cause small deletions or insertions, which can result in frameshifts. For instance, under certain conditions, benzpyrene, which is present in smoke and soot, is an effective *frameshift mutagen.* Likewise, aflatoxin—produced by *Aspergillus flavus* (a-spèr-jil'lus flā'vus), a mold that grows on peanuts and grain—is a frameshift mutagen, as are the acridine dyes used experimentally against herpesvirus infections. Frameshift mutagens usually have the right size and chemical properties to slip between the stacked base pairs of the DNA double helix. Frameshift mutagens may work by slightly offsetting the two strands of DNA, leaving a gap or bulge in one strand or the other. When the staggered DNA strands are copied during DNA synthesis, one or more base pairs can be inserted or deleted in the new double-stranded DNA. Interestingly, frameshift mutagens are often potent carcinogens.

Radiation

X rays and gamma rays are forms of radiation that are potent mutagens because of their ability to *ionize* atoms and molecules. The penetrating rays of ionizing radiation cause electrons to pop out of their usual shells. The affected electrons bombard other molecules and cause more damage, and many of the resulting ions and free radicals (molecular fragments with unpaired electrons) are very reactive. Some of these ions can combine with bases in DNA, resulting in errors in DNA replication and repair that produce mutations. An even more serious outcome is the breakage of covalent bonds in the sugar-phosphate backbone of DNA, which causes physical breaks in chromosomes.

Another form of mutagenic radiation is ultraviolet light (UV), a nonionizing component of ordinary sunlight. However, the most mutagenic component of UV light (260 nm) is screened out by the ozone layer. The most important effect of direct UV light on DNA is the formation of harmful covalent bonds between certain bases. Adjacent thymines in a DNA strand can cross-link to form thymine dimers. Such dimers, unless repaired, may cause

(a) Normal DNA molecule

(b) Missense mutation

(c) Nonsense mutation

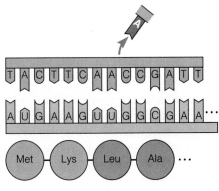

(d) Frameshift mutation

FIGURE 8.16 Types of mutations and their effects on the amino acid sequences of proteins.

FIGURE 8.17   Mutagenesis by nitrous acid (HNO₂). The nitrous acid alters an adenine in such a way that it pairs with cytosine instead of thymine.

FIGURE 8.18   Base analogs and the nitrogenous bases they replace. (a) Adenine and 2-aminopurine. (b) Thymine and 5-bromouracil.

serious damage or death to the cell because it cannot properly transcribe or replicate such DNA.

Bacteria and other organisms have enzymes that can repair radiation damage. In the presence of visible light, photoreactivating enzymes can perform light excision repair to fix damage done by UV light. These enzymes bind to thymine dimers and, using light energy, split the bonds forming the dimer, thereby reconstituting the thymine nucleotides. Dark excision repair is shown in Figure 8.19. Enzymes cut out the distorted cross-linked thymines by

opening a wide gap. They then fill in the gap with newly synthesized DNA that is complementary to the undamaged strand. By this means, the original base-pair sequence is restored. The last step is the covalent sealing of the DNA backbone by the enzyme DNA ligase. Occasionally, such a repair process makes an error, and the original base-pair sequence is not properly restored. The result of this error is a mutation. Exposure to UV light in humans (for example, through heavy tanning) results in a large number of thymine dimers in skin cells. The repair of these dimers results in an accumulation of other mutations, contributing to the incidence of skin cancer.

Frequency of Mutation

The **mutation rate** is the probability that a gene will mutate when a cell divides. The rate is usually stated as a power of 10, and because mutations are very rare, the exponent is always a negative number. For example, if there is one chance in 10,000 that a gene will mutate when the cell divides, the mutation rate is 1/10,000, which is expressed as 10^{-4}. Spontaneous mistakes in DNA replication occur at a very low rate—perhaps only once in 10^9 replicated base pairs (a mutation rate of 10^{-9}). Because the average gene has about 10^3 base pairs, the spontaneous rate of mutation is about once in 10^6 (a million) replicated genes.

Mutations usually occur more or less randomly along a chromosome. The occurrence of random mutations at low frequency is an essential aspect of the adaptation of species to their environments, for evolution requires that genetic diversity be generated randomly and at a low rate. For example, in a bacterial population of significant size—say, greater than 10^7 cells—a few new mutant cells will always be produced in every generation. Most mutations are either harmful and likely to be removed from the gene pool when the individual cell dies, or are neutral. However, a few may be beneficial. For example, a mutation that confers antibiotic resistance is beneficial if a population of bacteria is regularly exposed to antibiotics. Once such a trait has appeared through mutation, cells carrying the mutated gene are more likely than other cells to survive and reproduce. Soon most of the cells in the population will have the gene; an evolutionary change will have occurred, although on a small scale.

A mutagen usually increases the spontaneous rate of mutation, which is about once in 10^{-6} replicated genes, by a factor of 10 to 1000 times. That is, in the presence of a mutagen, the normal rate of 10^{-6} mutations per replicated gene becomes a rate of 10^{-5} to 10^{-3} per replicated gene. Mutagens are used experimentally to enhance the production of mutant cells for research on the genetic properties of microorganisms and for commercial use.

Identifying Mutants

Mutants can be detected by selecting or testing for an altered phenotype. Whether or not a mutagen is used, mutant cells with specific mutations are always rare compared with other cells in the population. The problem is detecting such a rare event.

Experiments are performed more easily with bacteria than with other organisms because bacteria reproduce rapidly, and thus large numbers of organisms (more than 10^9 per milliliter of broth) can easily be used. Furthermore, because bacteria generally have only one copy of each gene per cell, the effects of a mutated gene are not masked by the presence of a normal version of the gene, as in many eucaryotic organisms.

Positive (direct) selection involves the detection of mutant cells by rejection of the unmutated parent cells. For example, suppose we were trying to find mutant bacteria that are resistant to penicillin. When the bacterial cells are plated on a medium containing penicillin, the mutant can be identified directly. The few cells in the population that are resistant (mutants) will grow and form colonies, whereas the normal, penicillin-sensitive parental cells cannot grow.

To identify mutations in other kinds of genes, **negative (indirect) selection** can be used. This process selects a cell that cannot perform a certain function. This is done by the **replica-plating** technique. For example, suppose we wanted to use replica plating to identify a bacterial cell that has lost the ability to synthesize histidine (Figure 8.20). First, about 100 bacterial cells are inoculated onto an agar plate. This plate, called the master plate, contains a medium with histidine on which all cells will grow. After several hours of incubation, each cell reproduces to form a colony. Then a pad of sterile velvet is pressed over the master plate, and some of the cells from each colony adhere to the velvet. Next, the velvet is pressed down onto two (or more) sterile plates. One plate contains a medium

Ultraviolet light

Thymine dimer

New DNA

1 Exposure to ultraviolet light causes adjacent thymines to become cross-linked, forming a thymine dimer and disrupting their normal base pairing.

2 An enzyme cuts out and removes the damaged DNA.

3 DNA polymerase fills the gap by synthesizing new DNA, using the intact strand as a template.

4 DNA ligase seals the remaining gap by joining the old and new DNA.

FIGURE 8.19 Creation and repair of a thymine dimer caused by ultraviolet light. After exposure to UV light, adjacent thymines can become cross-linked, forming a thymine dimer. In the absence of visible light, this mechanism is used in a cell to repair the damage.

with histidine, and one contains a medium without histidine on which the original, nonmutant bacteria can grow. Any colony that grows on the medium with histidine on the master plate but that cannot synthesize its own histidine will not be able to grow on the medium without histidine. The mutant colony can then be identified on the master plate. Of course, because mutants are so rare (even

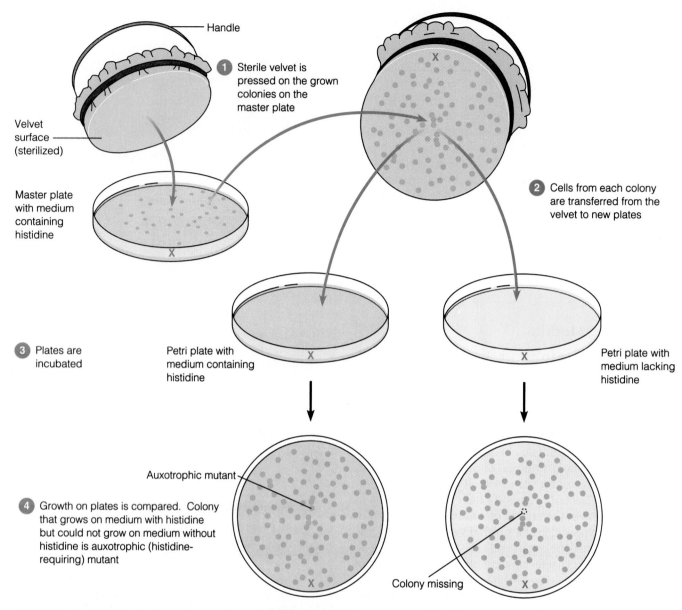

Handle

1 Sterile velvet is pressed on the grown colonies on the master plate

Velvet surface (sterilized)

Master plate with medium containing histidine

2 Cells from each colony are transferred from the velvet to new plates

3 Plates are incubated

Petri plate with medium containing histidine

Petri plate with medium lacking histidine

Auxotrophic mutant

4 Growth on plates is compared. Colony that grows on medium with histidine but could not grow on medium without histidine is auxotrophic (histidine-requiring) mutant

Colony missing

FIGURE 8.20 Replica plating. This procedure identifies auxotrophic mutants—bacteria that have lost the ability to synthesize an essential nutrient. In this example, the auxotrophic mutant cannot synthesize histidine. The plates must be carefully marked (with an X here) to maintain orientation so that colony positions are known in relation to the original master plate.

those induced by mutagens), many plates must be screened with this technique to isolate a specific mutant.

The replica-plating technique is a very effective means of isolating mutants that require one or more new growth factors. Any mutant microorganism possessing a nutritional requirement not possessed by the parent is known as an **auxotroph**. For example, an auxotroph may lack an enzyme needed to synthesize a particular amino acid and will therefore require that amino acid as a growth factor in its nutrient medium.

Identifying Chemical Carcinogens

Many known mutagens have been found to be **carcinogens**—substances that cause cancer in animals, including humans. In recent years, chemicals in the environment, the workplace, and the diet have been implicated as causes of cancer in humans. The usual subjects of tests to determine potential carcinogens are animals, and the testing procedures are both time-consuming and expensive. Now there are faster and less expensive procedures for the

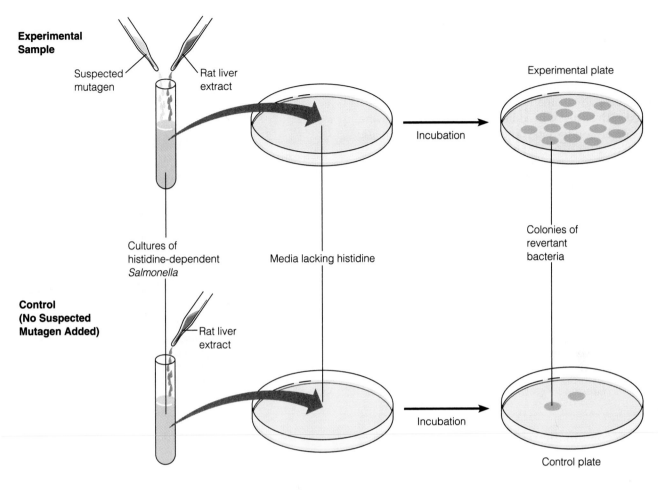

Experimental Sample

Suspected mutagen

Rat liver extract

Experimental plate

Incubation

Colonies of revertant bacteria

Cultures of histidine-dependent *Salmonella*

Media lacking histidine

Control (No Suspected Mutagen Added)

Rat liver extract

Incubation

Control plate

① Two cultures are prepared of *Salmonella* bacteria that have lost the ability to synthesize histidine (histidine-dependent).

② The suspected mutagen is added to the experimental sample only; rat liver extract (an activator) is added to both samples.

③ Each sample is poured onto a plate of medium lacking histidine. The plates are then incubated at 37°C for two days. Only bacteria whose histidine-dependent phenotype has mutated back (reverted) to histidine-synthesizing will grow into colonies.

④ The numbers of colonies on the experimental and control plates are compared. The control plate may show a few spontaneous histidine-synthesizing revertants. The test plates will show an increase in the number of histidine-synthesizing revertants if the test chemical is indeed a mutagen and potential carcinogen. The higher the concentration of mutagen used, the more revertant colonies will result.

FIGURE 8.21   The Ames test. This test is used to screen mutagens suspected of causing cancer in humans.

preliminary screening of potential carcinogens. One of these, called the **Ames test,** uses bacteria as carcinogen indicators.

The Ames test is based on the observation that exposure of mutant bacteria to mutagenic substances may cause new mutations that reverse the effect (the change in phenotype) of the original mutation. These are called back-mutations or reversions. Specifically, the test measures the reversion of histidine auxotrophs of *Salmonella* (His⁻ cells—mutants that have lost the ability to synthesize the amino acid histidine) to histidine-synthesizing

cells (His⁺) after treatment with a mutagen (Figure 8.21). Bacteria are incubated in both the presence and absence of the substance being tested. Because many chemicals must be activated (transformed chemically into forms that are chemically reactive) by animal enzymes for mutagenic or carcinogenic activity to appear, the chemical to be tested and the mutant bacteria are incubated together with rat liver extract, a rich source of activation enzymes. If the substance being tested is mutagenic, it will cause reversion of His⁻ bacteria to His⁺ bacteria at a rate higher than the spontaneous reversion rate. The number of observed revertants provides an indication of the degree to which a substance is mutagenic and therefore possibly carcinogenic.

The test can be used in many ways. Several potential mutagens can be qualitatively tested by spotting the individual chemicals on small paper disks on a single plate inoculated with bacteria. In addition, mixtures such as wine, blood, smoke condensates, and extracts of foods can also be tested to see if they contain mutagenic substances.

About 90% of the substances found by the Ames test to be mutagenic have also been shown to be carcinogenic in animals. Furthermore, the more mutagenic substances have generally been found to be more carcinogenic.

GENETIC TRANSFER AND RECOMBINATION

LEARNING OBJECTIVES

- *Compare mechanisms of genetic recombination in bacteria.*
- *Define plasmid and transposon, and discuss their functions.*
- *Relate mechanisms for genetic change to microbial evolution.*

Genetic recombination refers to the exchange of genes between two DNA molecules to form new combinations of genes on a chromosome. Figure 8.22 shows one type of genetic recombination occurring between two pieces of DNA, which we will call A and B and regard as chromosomes for the sake of simplicity. If these two chromosomes break and rejoin as shown (a process called **crossing over**), some of the genes carried by these chromosomes are shuffled. The original chromosomes have recombined, so that each now carries a portion of the other's genes.

If A and B represent DNA from different individuals, how are they brought close together enough to recombine? In eucaryotes, genetic recombination is an ordered process that usually occurs as part of the sexual cycle of the organism. Recombination generally takes place during the formation of reproductive cells, such that these cells contain recombinant DNA. In bacteria, genetic recombination can happen in a number of ways, which we will discuss in the following sections.

Like mutation, genetic recombination contributes to a population's genetic diversity, which is the source of variation in evolution. In highly evolved organisms such as present-day microorganisms, recombination is more likely than mutation to be beneficial because recombination is less likely to destroy the function of a gene and may bring together combinations of genes that enable the organism to carry out a valuable new function.

Genetic material can be transferred between bacteria in several ways. In all of the mechanisms, the transfer involves a **donor cell** that gives a portion of its total DNA to a **recipient cell**. Once transferred, part of the donor's DNA

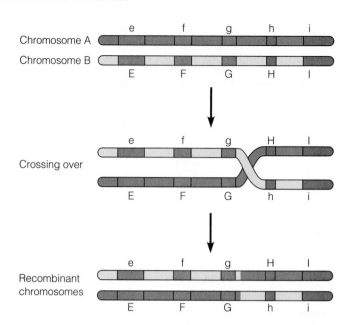

FIGURE 8.22 Genetic recombination, by crossing over between two related chromosomes. Chromosomes A and B each carry one version of genes E through I. The chromosomes cross over by breaking and rejoining, sometimes in more than one location. The result is two recombinant chromosomes, each of which carries genes originating from both chromosomes.

is usually incorporated into the recipient's DNA; the remainder is degraded by cellular enzymes. The recipient cell that incorporates donor DNA into its own DNA is called a *recombinant*. The transfer of genetic material between bacteria is by no means a frequent event; it may occur in only 1% or less of an entire population. Let us now examine in detail the specific types of genetic transfer.

Transformation in Bacteria

During the process of **transformation**, genes are transferred from one bacterium to another as "naked" DNA in solution. This process was first demonstrated more than 65 years ago, although it was not understood at that time. Not only did transformation show that genetic material could be transferred from one bacterial cell to another, but study of this phenomenon eventually led to the conclusion that DNA is the genetic material. The initial experiment on transformation was performed by Frederick Griffith in England in 1928 while he was working with two strains of *Streptococcus pneumoniae*. One, a virulent strain (meaning it will cause disease), has a polysaccharide capsule and causes pneumonia; the other, an avirulent strain, lacks the capsule and does not cause disease.

Griffith was interested in determining whether injections of heat-killed bacteria of the encapsulated strain could be used to vaccinate mice against pneumonia. As he expected, injections of the killed bacteria did not make mice ill; neither did injections of live nonencapsulated

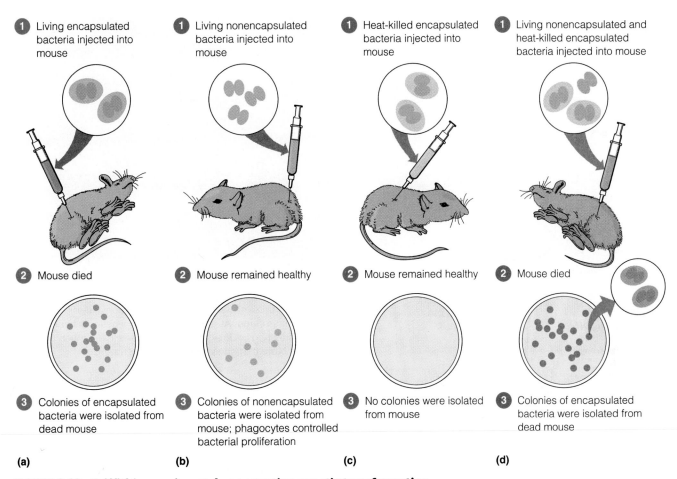

1 Living encapsulated bacteria injected into mouse

2 Mouse died

3 Colonies of encapsulated bacteria were isolated from dead mouse

(a)

1 Living nonencapsulated bacteria injected into mouse

2 Mouse remained healthy

3 Colonies of nonencapsulated bacteria were isolated from mouse; phagocytes controlled bacterial proliferation

(b)

1 Heat-killed encapsulated bacteria injected into mouse

2 Mouse remained healthy

3 No colonies were isolated from mouse

(c)

1 Living nonencapsulated and heat-killed encapsulated bacteria injected into mouse

2 Mouse died

3 Colonies of encapsulated bacteria were isolated from dead mouse

(d)

FIGURE 8.23   Griffith's experiment demonstrating genetic transformation.
(a) Living encapsulated bacteria caused disease and death when injected into a mouse.
(b) Living nonencapsulated bacteria are readily destroyed by the phagocytic defenses of the host, so the mouse remained healthy after injection. **(c)** After being killed by heat, encapsulated bacteria lost the ability to cause disease. **(d)** However, the combination of living nonencapsulated bacteria and heat-killed encapsulated bacteria (neither of which alone cause disease) did cause disease. Somehow, the live nonencapsulated bacteria were transformed by the dead encapsulated bacteria so that they acquired the ability to form a capsule and so cause disease. Subsequent experiments proved the transforming factor to be DNA.

bacteria. However, when the dead encapsulated bacteria were mixed with live nonencapsulated bacteria and injected into the mice, many of the mice died. In the blood of the dead mice Griffith found living, encapsulated bacteria. Hereditary material (genes) from the dead bacteria had entered the live cells and changed them genetically so that their offspring were encapsulated and therefore virulent (Figure 8.23).

Subsequent investigations based on Griffith's research revealed that bacterial transformation could be carried out without mice. A broth (liquid culture medium) was inoculated with live nonencapsulated bacteria. Dead encapsulated bacteria were then added to the broth. After incubation, the culture was found to contain living bacteria that were encapsulated and virulent. The nonencapsulated bacteria had been transformed; they had acquired a new hereditary trait by incorporating genes from the killed encapsulated bacteria.

The next step was to extract various chemical components from the killed cells to determine which component caused the transformation. These crucial experiments were performed by Oswald T. Avery and his associates Colin M. MacLeod and Maclyn McCarty in the United States. After years of research, they announced in 1944 that the component responsible for transforming harmless *S. pneumoniae* into virulent strains was DNA. Their results provided one of the conclusive indications that DNA was indeed the carrier of genetic information.

Since the time of Griffith's experiment, considerable information has been gathered about transformation. In nature, some bacteria, perhaps after death and cell lysis, release their DNA into the environment. Other bacteria can then encounter the DNA and, depending on the particular species and growth conditions, take up fragments of DNA and integrate them into their own chromosomes by recombination. A recipient cell with this new combina-

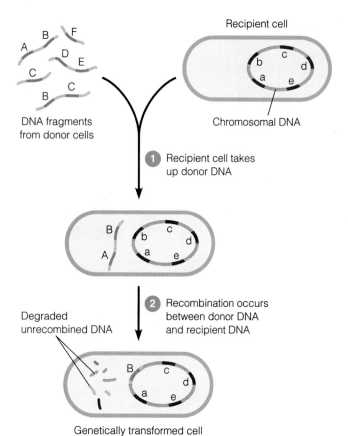

FIGURE 8.24 **Mechanism of genetic transformation in bacteria.**

Recipient cell

DNA fragments from donor cells

Chromosomal DNA

1 Recipient cell takes up donor DNA

2 Recombination occurs between donor DNA and recipient DNA

Degraded unrecombined DNA

Genetically transformed cell

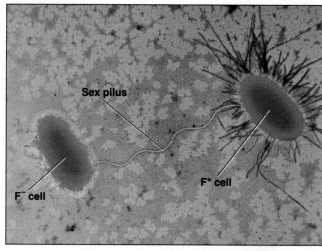

Sex pilus

F⁻ cell

F⁺ cell

TEM   1 μm

FIGURE 8.25 **Bacterial conjugation.** The sex pilus connecting these cells undergoing conjugation allows transfer of genetic information. At the actual time of genetic exchange the cells' connecting bridge contracts and the cells are much closer together. Note that one cell has numerous fimbriae.

tion of genes is a kind of hybrid, or recombinant cell (Figure 8.24). All the descendants of such a recombinant cell will be identical to it. Transformation occurs naturally among very few genera of bacteria. These genera include *Bacillus, Hemophilus* (hē-mä′fi-lus), *Neisseria, Acinetobacter* (a-si-ne′tō-bak-tėr), and certain strains of the genera *Streptococcus* and *Staphylococcus*.

Transformation works best when the donor and recipient cells are very closely related and when the recipient cells are in the late log phase of growth. Even though only a small portion of a cell's DNA is transferred to the recipient, the molecule that must pass through the recipient cell wall and membrane is still very large. When a recipient cell is in a physiological state in which it can take up the donor DNA, it is said to be competent. **Competence** results from alterations in the cell wall that make it permeable to the large DNA molecules.

The well-understood and widely used bacterium *E. coli* is not naturally competent for transformation. However, a simple laboratory treatment enables *E. coli* to readily take up DNA. The discovery of this treatment has enabled *E. coli* to be used for genetic engineering, as will be discussed in Chapter 9.

Conjugation in Bacteria

Conjugation is another mechanism by which genetic material is transferred from one bacterium to another. Conjugation is mediated by one kind of *plasmid*, which is a circular piece of DNA that replicates independently from the cell's chromosome. However, plasmids differ from bacterial chromosomes in that the genes they carry are usually not essential for the growth of the cell under normal conditions. The plasmids responsible for conjugation are transmissible between cells during conjugation.

Conjugation differs from transformation in two major ways. First, conjugation requires direct cell-to-cell contact. Second, the conjugating cells must generally be of opposite "mating type"; donor cells must carry the plasmid, and recipient cells usually do not. In gram-negative bacteria, the plasmid carries genes that code for the synthesis of *sex pili* (singular: *pilus*), projections from the donor's cell surface that contact the recipient and help to bring the two cells into direct contact. Gram-positive bacterial cells produce sticky surface molecules that cause cells to come into direct contact with each other. In the process of conjugation, the plasmid is replicated during transfer of a single-stranded copy of the plasmid DNA to the recipient, where the complementary strand is synthesized.

Because most experimental work on conjugation has been done with *E. coli*, we will describe the process in this organism. In *E. coli*, the *F factor* was the first plasmid observed to be transferred between cells during conjugation (Figure 8.25). Donors carrying F factors (F⁺ cells) transfer the plasmid to recipients (F⁻ cells), which become F⁺ cells as a result (Figure 8.26a). In some cells carrying F factors, the factor integrates into the chromosome, converting the

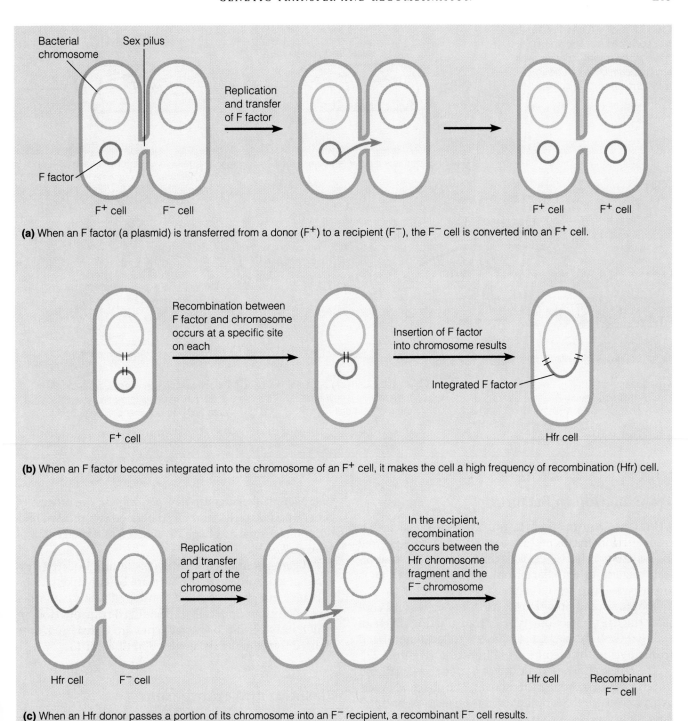

(a) When an F factor (a plasmid) is transferred from a donor (F⁺) to a recipient (F⁻), the F⁻ cell is converted into an F⁺ cell.

(b) When an F factor becomes integrated into the chromosome of an F⁺ cell, it makes the cell a high frequency of recombination (Hfr) cell.

(c) When an Hfr donor passes a portion of its chromosome into an F⁻ recipient, a recombinant F⁻ cell results.

FIGURE 8.26 Conjugation in *E. coli.*

F⁺ cell to an *Hfr (high frequency of recombination) cell* (Figure 8.26b). When conjugation occurs between an Hfr and an F⁻ cell, the Hfr cell's chromosome (with its integrated F factor) replicates, and a parental strand of the chromosome is transferred to the recipient cell (Figure 8.26c). Replication of the Hfr chromosome begins in the middle of the integrated F factor, and a small piece of the F factor leads the chromosomal genes into the F⁻ cell. Usually, the chromosome breaks before it is completely transferred. Once within the recipient cell, donor DNA can recombine with the recipient's DNA. (Donor DNA that is not integrated is degraded.) Therefore, by conjugation with an Hfr cell, an F⁻ cell may acquire new versions of chromosomal genes (just as in transformation). However, it remains an F⁻ cell because it did not receive a complete F factor during conjugation.

FIGURE 8.27 Transduction by bacteriophage. Shown here is generalized transduction, in which any bacterial DNA can be transferred from one cell to another.

1 A phage infects the donor bacterial cell.

2 Phage DNA and proteins are made, and the bacterial chromosome is broken down into pieces.

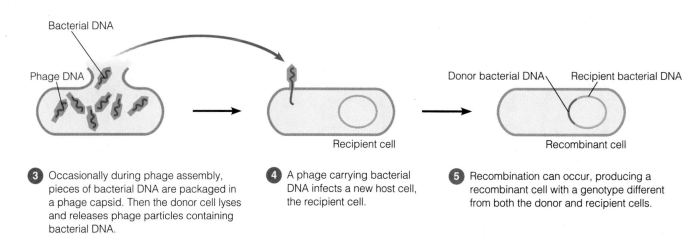

3 Occasionally during phage assembly, pieces of bacterial DNA are packaged in a phage capsid. Then the donor cell lyses and releases phage particles containing bacterial DNA.

4 A phage carrying bacterial DNA infects a new host cell, the recipient cell.

5 Recombination can occur, producing a recombinant cell with a genotype different from both the donor and recipient cells.

Transduction in Bacteria

A third mechanism of genetic transfer between bacteria is **transduction**. In this process, bacterial DNA is transferred from a donor cell to a recipient cell inside a virus that infects bacteria, called a **bacteriophage** (or **phage**). (Phages are discussed further in Chapter 13.)

To understand how transduction works, we will consider the life cycle of one type of transducting phage of *E. coli;* this phage carries out a process called **generalized transduction**. In the process of infection (Figure 8.27), ① the phage attaches to the donor bacterial cell wall and injects its DNA into the bacterium. ② The phage DNA acts as a template for the synthesis of new phage DNA and also directs the synthesis of phage protein coats. During phage development inside the infected cell, the bacterial chromosome is broken apart by phage enzymes, and ③ at least some pieces of bacterial DNA are mistakenly packaged inside phage protein coats. (Even plasmid DNA or DNA of another virus that is inside the cell can be given phage protein coats.) The resulting phage particles then carry bacterial DNA instead of phage DNA. ④ When the released phage particles later infect a new population of bacteria, bacterial genes will be transferred to the newly infected recipient cells at low frequency. ⑤ Transduction of cellular DNA by a virus can lead to recombination between the DNA of the donor host cell and the DNA of the recipient host cell. Transduction is thus another way, in addition to mutation, transformation, and conjugation,

that bacteria can acquire new genotypes. The process of generalized transduction just described is typical of bacteriophages such as phage P1 of *E. coli* and phage P22 of *Salmonella.*

All genes contained within a bacterium infected by a generalized transducting phage are equally likely to be packaged in a phage coat and transferred. There is another type of transduction, called **specialized transduction**, in which only certain bacterial genes are transferred. Specialized transduction is discussed in Chapter 13.

Plasmids and Transposons

Plasmids and transposons are genetic elements that provide additional mechanisms for genetic change. They occur in both procaryotic and eucaryotic organisms, but this discussion focuses on their role in genetic change in procaryotes.

Plasmids

Plasmids are self-replicating, gene-containing circular pieces of DNA about 1% to 5% the size of the bacterial chromosome. They are found mainly in bacteria but also in some eucaryotic microorganisms, such as *Saccharomyces cerevisiae* (se-ri-vis'ē-ī). Recall that the F factor is a plasmid that carries genes for sex pili and for transfer of the plasmid to another cell. Although plasmids are usually dispensable, under certain conditions genes carried by plasmids can be crucial to the survival and growth of the cell. For example, **dissimilation plasmids** code for enzymes that catalyze the catabolism of certain unusual sugars and hydrocarbons. Some species of *Pseudomonas* can actually use such exotic substances as toluene, camphor, and hydrocarbons of petroleum as primary carbon and energy sources because they have catabolic enzymes encoded by genes carried on plasmids. Other plasmids code for proteins that enhance the pathogenicity of a bacterium. The strain of *E. coli* that causes infant diarrhea and traveler's diarrhea carries plasmids that code for toxin production and for bacterial attachment to intestinal cells. Without these plasmids, *E. coli* is a harmless resident of the large intestine; with them, it is pathogenic. Still other plasmids contain genes for the synthesis of bacteriocins, toxic proteins that kill other bacteria. These plasmids have been found in many bacterial genera, and they are useful markers for the identification of certain bacteria in clinical laboratories. The box on page 220 describes a commercial application of bacteriocins.

Resistance factors (R factors) are plasmids of great medical importance. R factors carry genes that confer upon their host cell resistance to antibiotics, heavy metals, or cellular toxins. Many R factors contain two groups of genes. One group is called the **resistance transfer factor (RTF)** and includes genes for plasmid replication and conjugation. The other group, the **r-determinant**, has the resistance genes; it codes for the production of enzymes that inactivate certain drugs or toxic substances (Figure 8.28). Different R factors, when present in the same cell, can recombine to produce R factors with new combinations of genes in their r-determinants.

R factors present very serious problems for the treatment of infectious diseases with antibiotics. The widespread use of antibiotics in medicine and agriculture (many animal feeds contain antibiotics) has led to the preferential survival (selection) of bacteria that have R factors, so populations of resistant bacteria grow larger and larger. The transfer of resistance between bacterial cells of a population, and even between bacteria of different genera, also contributes to the problem. (See the box in Chapter 26.)

Plasmids are an important tool for genetic engineering, which is discussed in Chapter 9.

Transposons

Transposons (also called *transposable genetic elements*) are small segments of DNA that can move (be "transposed") from one region of the DNA molecule to another. In the popular press, they are often called "jumping genes." These pieces of DNA measure from 700 to 40,000 base pairs long.

Barbara McClintock of the United States discovered transposons in corn, but transposons occur in all organisms and have been studied most thoroughly in microorganisms. They may move from one site on a chromosome to another site on the same chromosome or to another chromosome or plasmid. As you might imagine, frequent

(a)

SEM | 1 μm

r-determinant: Genes for resistance to various antibiotics

Resistance transfer factor (RTF): Genes for replication and conjugation

(b)

FIGURE 8.28 R factors, a type of plasmid. (a) Micrograph showing two different resistance (R) factors. The larger loop of DNA is a plasmid isolated from the bacterium *Bacteroides fragilis* and encodes resistance to the antibiotic clindamycin. The smaller loop of DNA is a plasmid from *E. coli* and encodes resistance to tetracycline. (b) Diagram of an R factor, which has two parts. One part, the resistance transfer factor (RTF), contains the genes needed for plasmid replication and transfer of the plasmid by conjugation. The other part, the r-determinant, carries genes for resistance to five different antibiotics.

movement of transposons could wreak havoc within a cell. For example, as transposons move about on chromosomes, they may insert themselves *within* genes, inactivating them. Fortunately, transposition occurs relatively rarely. The frequency of transposition is comparable to the spontaneous mutation rate that occurs in bacteria—that is, from 10^{-5} to 10^{-7} per generation.

All transposons contain the information for their own transposition. As shown in Figure 8.29a, the simplest transposons, also called **insertion sequences**, contain only a gene that codes for an enzyme (transposase, which catalyzes the cutting and resealing of DNA that occurs in transposition) and recognition sites. Recognition sites are short inverted repeat sequences of DNA that the enzyme recognizes as recombination sites between the transposon and the chromosome.

Complex transposons also carry other genes not connected with the transposition process. For example, bacterial transposons may contain genes for toxins or for antibiotic resistance (Figure 8.29b). R factors are frequently made up of a collection of transposons (Figure 8.29c).

Transposons with antibiotic resistance genes are of particular practical interest, but there is no limitation on the kinds of genes that transposons can have. Thus, transposons provide a powerful natural mechanism for the movement of genes from one chromosome to another. Furthermore, because they may be carried between cells on plasmids or viruses, they can also spread from one organism—or even species—to another. Transposons are thus a potentially powerful mediator of evolution in organisms.

Recombination in Eucaryotes

In transformation, conjugation, and transduction in bacteria, only a small segment of the DNA of the donor cell enters and recombines with the DNA of the recipient cell. In eucaryotes, genetic recombination occurs during sexual reproduction. In contrast to the processes in bacteria, sexual reproduction is a regular event for most eucaryotes and is necessary for the survival of many species.

Each species of eucaryotic organism has a characteristic **chromosome number**—that is, the number of chromosomes in each cell. For example, the human chromosome number is 46, meaning that each ordinary body cell *(somatic cell)* has 46 chromosomes. These chromosomes fall into two sets of 23 chromosomes. Each set is derived from one parent and carries genes for virtually all the activities of human cells.

Sexual reproduction in eucaryotes involves the fusion of the nuclei of two parent cells that are haploid (Figure 8.30). A **haploid cell** is one that contains only one of each chromosome; the number of chromosomes in such a cell is called the haploid number, symbolized N. In humans, the haploid number of chromosomes is 23. In more complex organisms, the haploid cells that fuse are called **gametes.** (In animals and plants, the male gametes are sperm or pollen, respectively, and the female gametes are ova.) Fusion of the gametes produces a fertilized ovum or **zygote,** which has received half its chromosomes from one parent and half from the other (Figure 8.30a). The zygote is a **diploid** cell because it contains two sets of chromosomes. The diploid number ($2N$) is twice the haploid number (N). The two sets of chromosomes are said to be *homologous;* that is, for each chromosome in one set, there is, in the other set, a corresponding chromosome that carries genes for the same traits in the same order (the exception is the chromosomes that determine sex; in humans, the X and Y chromosomes). Subsequent growth and development of the zygote produces a mature multicellular organism capable of sexual reproduction. Most cells in this mature organism are produced by **mitosis,** a eucary-

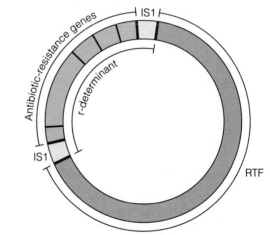

(a) Insertion sequence "IS1"

(b) Complex transposon "Tn1681"

(c) R plasmid "R1" with r-determinant transposon

FIGURE 8.29   Transposons and insertion sequences.
(a) The simplest type of transposon is the insertion sequence (IS)—a segment of DNA that contains a gene for transposase (the enzyme that catalyzes transposition). The transposase gene is bounded at each end by inverted repeat (IR) sequences that function as recognition sites for the transposon. IS1 is one example of an insertion sequence, shown here with simplified inverted repeat sequences. **(b)** Complex transposons carry other genetic material in addition to transposase genes. The example shown here, Tn1681, carries the gene for a bacterial toxin and has complete copies of the insertion sequence IS1 at each end. **(c)** Plasmids often include at least one transposon in their genetic makeup. The presence of an insertion sequence at each end of the group of antibiotic-resistance genes here makes the r-determinant of this R1 plasmid a transposon.

otic cell division process in which the chromosomes are duplicated and then one copy of each is distributed to each of two daughter cells.

Many eucaryotic microorganisms also form haploid cells of opposite mating types that can fuse to form a zygote. However, eucaryotic microorganisms are often capable of asexual reproduction as well. For example, consider yeasts, which are unicellular fungi. In some yeasts, such as *S. cerevisiae*, asexual reproduction, in which diploid and haploid cells simply divide by mitosis, can go on for many generations. In yeasts, periods of asexual reproduction alternate with sexual reproduction, in which haploid cells join to form a zygote (Figure 8.30b).

FIGURE 8.30 **Reproduction in eucaryotes.** (a) Human reproductive cycle, which is an example of sexual reproduction. (b) Reproductive cycle of the yeast *Saccharomyces cerevisiae,* which can involve both sexual and asexual reproduction. Plus and minus signs are used to distinguish opposite mating types. (Meiosis actually produces four haploid cells, not just the two shown, because the chromosomes are duplicated during the process.)

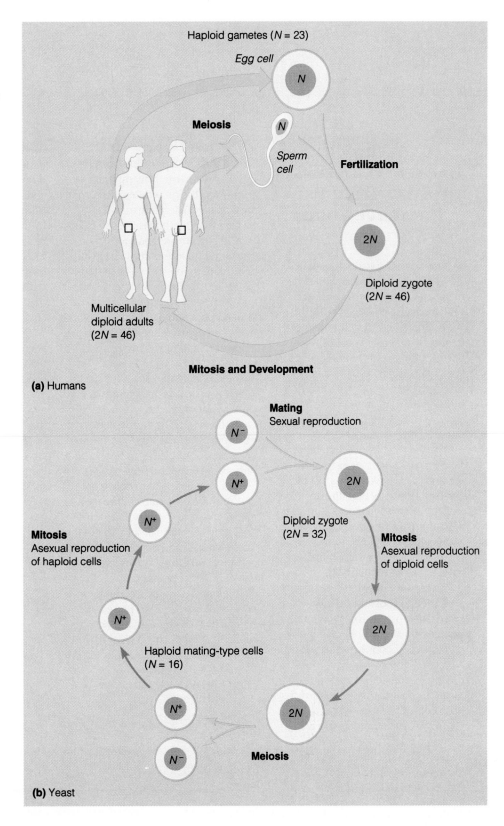

Microbiology in the News

BACTERIOCINS: BACTERIAL DEFENSES THAT HELP US PRESERVE FOODS

FOODBORNE INFECTIONS BY gram-negative bacteria may strike as many as 4 million Americans each year and cause at least 500 deaths. For example, *Salmonella* is the most frequently reported cause of foodborne outbreaks of gastroenteritis in the United States, and the number of cases has been increasing since 1976. In 1990, 8591 *Salmonella enteritidis* cases were reported. Food containing poultry, eggs, or dairy products are most often the vehicles for foodborne salmonellosis. Another example is *E. coli* O157:H7, which was first recognized as a cause of bloody diarrhea affecting humans when two outbreaks of hemorrhagic colitis occurred in Michigan and Oregon. Since then, illnesses caused by this strain of *E. coli* have increased dramatically.

These foodborne infections are attributed to substandard food-preparation practices such as improper food storage or holding temperatures and poor hygiene by food handlers. Food contaminated with pathogenic bacteria may not be detected by standard procedures such as plate counts because the bacteria are injured in cooking or freezing and cannot grow on regular culture media. These bacteria are able to grow, however, in the warm, nutritive environment of food left on a warming table.

Lowered salt and nitrate levels are popular trends in processed food; however, these compounds were used to inhibit the growth of pathogenic bacteria such as *Clostridium botulinum*. *Clostridium* endospores will germinate

in foods stored at temperatures above 10°C if the endospores are not killed by heat treatment. New methods of controlling the growth of *C. botulinum* are currently being researched.

One such method is the use of lactic acid bacteria in food production. The fermentations of these bacteria contribute to the flavor and preservation of foods. This antimicrobial action was assumed to be caused by the bacteria's production of organic acids and hydrogen peroxide until 1944, when a powerful antimicrobial agent was isolated from *Lactococcus lactis*. The agent was identified as a bacteriocin, which is a peptide that inhibits or kills closely related species.

In the 1990s, the attention of food microbiologists has turned to bacteriocins produced by the lactic acid bacteria as possible food preservatives. The advantages of bacteriocins are that they are heat-stable, they are readily digested in the human intestine, they are nontoxic and do not induce allergies, and they are already consumed in foods such as fermented dairy and meat products.

The only bacteriocin currently approved for use as a food preservative is nisin, which is produced by *Lactococcus lactis*. Nisin is bactericidal against gram-positive bacteria such as *Clostridium*, it inhibits endospore germination, and it has recently been

shown to kill the gram-negative bacterium *Salmonella*. Nisin is used in processed cheese to prevent growth and gas production by *C. sporogenes*. Nisin and other bacteriocins produced by the lactic acid bacteria work by attaching themselves to the plasma membrane of target cells. This alters the membrane's permeability and decreases the proton motive force required to make ATP. The operon that codes for bacteriocin production also carries a gene for resistance to the bacteriocin so that the cell producing the bacteriocin does not kill itself. Food microbiologists are screening natural populations of lactic acid bacteria looking for other bacteriocins that will kill pathogens transmitted by food. Additionally, genetic engineering techniques might be used to increase the spectrum of activity, solubility, and stability of known bacteriocins. Bacteriocins that show promise against some important pathogens are listed in the accompanying table.

BACTERIOCIN	PRODUCED BY	BACTERICIDAL AGAINST
Lactocin F	*Lactobacillus acidophilus*	*Enterococcus faecalis*
Lactocin S	*L. sake*	Gram-positive bacteria
Plantacin	*L. plantarum*	Inhibits endospore germination
Mesenterocin 5	*Leuconostoc mesenteroides*	*Listeria monocytogenes*
Pediocin	*Pediococcus acidilactici*	Gram-positive bacteria; inhibits endospore germination
Hiraecin S	*Enterococcus hirae*	*Listeria monocytogenes*
Propionicin	*Propionibacterium thoenii*	Gram-negative bacteria

Eventually, for sexual reproduction to recur in either unicellular or multicellular organisms, some of the diploid cells must give rise to haploid cells. Diploid cells produce haploid cells by a special cell division process called **meiosis,** in which the homologous chromosomes are separated and a single chromosome from each pair goes to each daughter cell. In the early stages of meiosis the homologous chromosomes come together in pairs (a process called synapsis), and portions of homologous chromosomes can be exchanged by crossing over (see Figure 8.22). Crossing over is in fact genetic recombination, and in eucaryotic cells it can often be observed microscopically.

In eucaryotic microorganisms, meiosis may be followed by a period of asexual reproduction in which generations of haploid organisms are produced by mitosis. Alternatively, another round of sexual reproduction can be initiated when two haploid mating-type cells fuse and their nuclei unite to form a new zygote, as happens in more complex eucaryotic organisms. Because these two haploid cells are usually from different individuals, new genetic combinations result. The regular events of crossing over in meiosis, segregation of chromosomes into haploid cells, and formation of diploid zygotes generate enormous diversity in every generation.

GENES AND EVOLUTION

LEARNING OBJECTIVE

- *Discuss how genetic mutation and recombination provide material for natural selection to act on.*

We have now seen how gene activity can be controlled by the cell's internal regulatory mechanisms and how genes themselves can be altered or rearranged by mutation, transposition, and recombination. All these processes provide diversity in the descendants of cells, and diversity is the driving force of evolution. The different kinds of microorganisms that exist today are the result of a long history of evolution. Microorganisms have continually changed by altering their genetic properties and adapting to many different habitats.

STUDY OUTLINE

STRUCTURE AND FUNCTION OF THE GENETIC MATERIAL (pp. 189–201)

1. Genetics is the study of what genes are, how they carry information, how their information is expressed, and how they are replicated and passed to subsequent generations or other organisms.
2. DNA in cells exists as a double-stranded helix; the two strands are held together by hydrogen bonds between specific nitrogenous base pairs: A-T and C-G.
3. A gene is a segment of DNA, a sequence of nucleotides, that codes for a functional product, usually a protein.
4. When a gene is expressed, DNA is transcribed to produce RNA; mRNA is then translated into proteins.
5. The DNA in a cell is duplicated before the cell divides, so each daughter cell receives the same genetic information.

Genotype and Phenotype (p. 190)

1. Genotype is the genetic composition of an organism— its entire DNA.
2. Phenotype is the expression of the genes—the proteins of the cell and the properties they confer on the organism.

DNA and Chromosomes (pp. 190–191)

1. The DNA in a chromosome exists as one long double helix associated with various proteins that regulate genetic activity.
2. Bacterial DNA is circular; the chromosome of *E. coli,* for example, contains about 4 million base pairs and is approximately 1000 times longer than the cell.
3. The DNA of eucaryotic chromosomes is more condensed and is associated with much protein, including histones. This mixture of DNA and proteins is called chromatin.

DNA Replication (pp. 191–195)

1. During DNA replication, the two strands of the double helix separate at the replication fork, and each strand is used as a template by DNA polymerases to synthesize two new strands of DNA according to the rules of nitrogenous base-pairing.
2. The result of DNA replication is two new strands of DNA, each having a base sequence complementary to one of the original strands.
3. Because each double-stranded DNA molecule contains one original and one new strand, the replication process is called semiconservative.

4. DNA is synthesized in one chemical direction called 5′ ⟶ 3′. At the replication fork, one daughter strand (the leading strand) is synthesized continuously, the other (the lagging strand), discontinuously.

5. DNA synthesis of new fragments on the lagging strand moves in the opposite direction from the movement of the replication fork. Completion of the lagging strand requires RNA primers, RNA polymerase, DNA ligase, and DNA polymerase.

6. DNA polymerase proofreads new molecules of DNA and removes mismatched bases before continuing DNA synthesis.

7. DNA can be synthesized at 1000 nucleotides per second in *E. coli* grown at 37°C.

8. Because multiple replication forks occur in log phase, the generation time for bacteria can be shorter than the time it takes for the bacterial chromosome to replicate.

9. Each daughter bacterium receives a chromosome identical to the parent's.

10. Information contained in the DNA is transcribed into RNA and translated into proteins.

RNA and Protein Synthesis (pp. 195–201)

1. During transcription, the enzyme RNA polymerase synthesizes a strand of RNA from one strand of double-stranded DNA, which serves as a template.

2. RNA is synthesized from nucleotides containing the bases A, C, G, and U, which pair with the bases of the DNA sense strand.

3. The starting point for transcription, where RNA polymerase binds to DNA, is the promoter site; the region of DNA that is the end point of transcription is the terminator site.

4. RNA is synthesized in the 5′ ⟶ 3′ direction.

5. In procaryotes, transcription occurs in the cytoplasm; in eucaryotes, transcription occurs in the nucleus.

6. After transcription in eucaryotes, RNA and proteins remove intron RNA from the RNA transcript and join the exon RNA together to form mRNA.

7. The enzymes that cut and splice the RNA transcript are composed of RNA and are called ribozymes.

8. In procaryotes, translation can begin before transcription is completed.

9. In eucaryotes, translation begins after the mRNA enters the cytoplasm.

10. Translation is the process in which the information in the nucleotide-base sequence of mRNA is used to dictate the amino acid sequence of a protein.

11. The mRNA associates with ribosomes, which consist of rRNA and protein.

12. Three-base segments of mRNA that specify amino acids are called codons. mRNA is read in the 5′ ⟶ 3′ direction.

13. Specific amino acids are attached to molecules of tRNA. Another portion of the tRNA has a base triplet called an anticodon.

14. The base-pairing of codon and anticodon at the ribosome results in specific amino acids being brought to the site of protein synthesis.

15. The ribosome moves along the mRNA strand as amino acids are joined to form a growing polypeptide.

16. Translation ends when the ribosome reaches a stop codon on the mRNA.

17. A polyribosome is an mRNA with several ribosomes attached to it.

The Genetic Code (p. 201)

1. The genetic code refers to the relationship among the nucleotide-base sequence of DNA, the corresponding codons of mRNA, and the amino acids for which the codons code.

2. The genetic code is degenerate; that is, most amino acids are coded for by more than one codon.

3. Of the 64 codons, 61 are sense codons (which code for amino acids), and 3 are nonsense codons (which do not code for amino acids and are stop signals for translation).

4. The start codon, AUG, codes for methionine.

REGULATION OF GENE EXPRESSION IN BACTERIA (pp. 202–205)

1. Regulating protein synthesis at the gene level is energy-efficient because proteins are synthesized only as they are needed.

2. Constitutive enzymes produce products at a fixed rate. Examples are genes for the enzymes in glycolysis.

3. For these gene regulatory mechanisms, the control is aimed at mRNA synthesis.

4. Repression controls the synthesis of one or several (repressible) enzymes.

5. When cells are exposed to a particular end-product, the synthesis of enzymes related to that product decreases.

6. In the presence of certain chemicals (inducers), cells synthesize more enzymes. This process is called induction.

7. An example of induction is the production of β-galactosidase by *E. coli* in the presence of lactose, so lactose can be metabolized.

8. The formation of enzymes is determined by structural genes.

9. In bacteria, a group of coordinately regulated structural genes with related metabolic functions and the promoter and operator sites that control their transcription are called an operon.

10. In the operon model for an inducible system, a regulatory gene codes for the repressor protein.

11. When the inducer is absent, the repressor binds to the operator and no mRNA is synthesized.

12. When the inducer is present, it binds to the repressor so that it cannot bind to the operator; thus, mRNA is made and enzyme synthesis is induced.

13. In repressible systems, the repressor requires a corepressor in order to bind to the operator site; thus, the corepressor controls enzyme synthesis.
14. Transcription of structural genes for catabolic enzymes (such as β-galactosidase) is induced by the absence of glucose. cAMP and catabolic activator protein must bind to a promoter in the presence of an alternative carbohydrate (such as lactose).
15. The presence of glucose inhibits metabolism of alternative carbon sources by catabolic repression (the glucose effect).
16. Global regulation refers to the control of large sets of genes.
17. Some bacteria make special proteins following an environmental stress such as heat or deprivation of nutrients.

MUTATION: CHANGE IN THE GENETIC MATERIAL (pp. 205–212)

1. A mutation is a change in the nitrogenous-base sequence of DNA; that change causes a change in the product coded for by the mutated gene.
2. Many mutations are neutral, some are disadvantageous, and others are beneficial.

Types of Mutations (pp. 205–206)
1. A base substitution occurs when one base pair in DNA is replaced with a different base pair.
2. Alterations in DNA can result in missense mutations (which cause amino acid substitutions) or nonsense mutations (which create stop codons).
3. In a frameshift mutation, one or a few base pairs are deleted or added to DNA.
4. Mutagens are agents in the environment that cause permanent changes in DNA.
5. Some mutations occur without the presence of a mutagen; these are called spontaneous mutations.

Mutagens (pp. 206–208)
1. Chemical mutagens include base-pair mutagens (for example, nitrous acid), base analogs (for example, 2-aminopurine and 5-bromouracil), and frameshift mutagens (for example, benzpyrene).
2. Ionizing radiation causes the formation of ions and free radicals that react with DNA; base substitutions or breakage of the sugar-phosphate backbone results.
3. Ultraviolet radiation is nonionizing; it causes bonding between adjacent thymines.
4. Damage to DNA caused by ultraviolet radiation can be repaired by enzymes that cut out and replace the damaged portion of DNA.
5. Photoreactivation enzymes repair thymine dimers in the presence of visible light.

Frequency of Mutation (pp. 208–209)
1. Mutation rate is the probability that a gene will mutate when a cell divides; the rate is expressed as 10 to a negative power.
2. Mutations usually occur randomly along a chromosome.
3. A low rate of spontaneous mutations is beneficial in providing the genetic diversity needed for evolution.

Identifying Mutants (pp. 209–210)
1. Mutants can be detected by selecting or testing for an altered phenotype.
2. Positive selection involves the selection of mutant cells and rejection of nonmutated cells.
3. Replica plating is used for negative selection—to detect, for example, auxotrophs that have nutritional requirements not possessed by the parent (nonmutated) cell.

Identifying Chemical Carcinogens (pp. 210–212)
1. The Ames test is a relatively inexpensive and rapid test for identifying possible chemical carcinogens.
2. The test assumes that a mutant cell can revert to a normal cell in the presence of a mutagen and that many mutagens are carcinogens.
3. Histidine auxotrophs of *Salmonella* are exposed to an enzymatically treated potential carcinogen, and reversions to the nonmutant state are selected.

GENETIC TRANSFER AND RECOMBINATION (pp. 212–221)

1. Genetic recombination, which refers to the rearrangement of genes from separate groups of genes, usually involves DNA from different organisms; it contributes to genetic diversity.
2. In crossing over, genes from two chromosomes are recombined into one chromosome containing some genes from each original chromosome.
3. Mechanisms for genetic transfer in bacteria involve a portion of the cell's DNA being transferred from donor to recipient.
4. When some of the donor's DNA has been integrated into the recipient's DNA, the resultant cell is called a recombinant.

Transformation in Bacteria (pp. 212–214)
1. During this process, genes are transferred from one bacterium to another as "naked" DNA in solution.
2. This process was first demonstrated in *Streptococcus pneumoniae*.
3. It occurs naturally among a few genera of bacteria.

Conjugation in Bacteria (pp. 214–215)

1. This process requires contact between living cells.
2. One type of genetic donor cell is an F+; recipient cells are F−. F cells contain plasmids called F factors; these are transferred to the F− cells during conjugation.
3. When the plasmid becomes incorporated into the chromosome, the cell is called an Hfr (high-frequency recombinant).
4. During conjugation, an Hfr can transfer chromosomal DNA to an F− cell. Usually, the Hfr chromosome breaks before it is fully transferred.

Transduction in Bacteria (p. 216)

1. In this process, DNA is passed from one bacterium to another in a bacteriophage and is then incorporated into the recipient's DNA.
2. In generalized transduction, any bacterial genes can be transferred.

Plasmids and Transposons (pp. 216–218)

1. Plasmids are self-replicating circular molecules of DNA carrying genes that are not usually essential for survival of the cell.
2. There are several types of plasmids, including the F factor, dissimilation plasmids, plasmids carrying genes for toxins or bacteriocins, and resistance factors.
3. Transposons are small segments of DNA that can move from one region of a chromosome to another region of the same chromosome or to a different chromosome or a plasmid.

4. Transposons are found in the main chromosomes of organisms, in plasmids, and in the genetic material of viruses. They vary from simple (insertion sequences) to complex.
5. Complex transposons can carry any type of gene, including antibiotic-resistance genes, and are thus a natural mechanism for moving genes from one chromosome to another.

Recombination in Eucaryotes (pp. 218–221)

1. In eucaryotes, genetic recombination is associated with sexual reproduction.
2. Sexual reproduction involves the fusion of haploid gametes to form a diploid zygote.
3. Haploid gametes are produced through meiosis, in which portions of chromosomes can be exchanged in the recombination process of crossing over.
4. In eucaryotic microorganisms, sexual reproduction often alternates with asexual reproduction.

GENES AND EVOLUTION (p. 221)

1. Diversity is the precondition of evolution.
2. Genetic mutation and recombination provide a diversity of organisms, and the process of natural selection allows the growth of those best adapted for a given environment.

STUDY QUESTIONS

Review

1. Briefly describe the components of DNA, and explain its functional relationship to RNA and protein.

2. Draw a diagram showing a portion of a chromosome undergoing replication.
   (a) Identify the replication fork.
   (b) What is the role of DNA polymerase? Of RNA polymerase?
   (c) What is the role of DNA ligase?
   (d) Why is the lagging strand synthesized discontinuously?
   (e) How does this process represent semiconservative replication?

3. The following is a code for a strand of DNA.

	(3′)																		(5′)
DNA	A	T	A	T	_	_	_	T	T	T	_	_	_	_	_	_	_	_	_
	1	2	3	4	5	6	7	8	9	10	11	12	13	14	15	16	17	18	19
mRNA											C	G	U				U	G	A
tRNA												U	G	G					
Amino acid			Met																

ATAT = Promoter sequence

   (a) Using the genetic code provided in Figure 8.12, fill in the blanks to complete the segment of DNA shown.
   (b) Fill in the blanks to complete the sequence of amino acids coded for by this strand of DNA.
   (c) Write the code for the complementary strand of DNA completed in part (a).

Use the sense strand for questions (d) through (i).

(d) What would be the effect if C was substituted for T at base 10?

(e) What would be the effect if A was substituted for G at base 11?

(f) What would be the effect if G was substituted for T at base 14?

(g) What would be the effect if C was inserted between bases 9 and 10?

(h) How would ultraviolet radiation affect this strand of DNA?

(i) Identify a nonsense sequence in this strand of DNA.

4. Describe translation, and be sure to include the following terms: ribosome, rRNA, amino acid activation, tRNA, anticodon, and codon.

5. Contrast the structures of a bacterial chromosome and a eucaryotic chromosome.

6. Explain how you would find an antibiotic-resistant mutant by direct selection and how you would find an antibiotic-sensitive mutant by indirect selection.

7. Match the following examples of mutagens.

____ A mutagen that is incorporated into DNA in place of a normal base

____ A mutagen that causes the formation of highly reactive ions

____ A mutagen that alters adenine so that it base-pairs with cytosine

____ A mutagen that causes insertions

____ A mutagen that causes the formation of pyrimidine dimers

(a) Frameshift mutagen
(b) Base analog
(c) Base-pair mutagen
(d) Ionizing radiation
(e) Nonionizing radiation

8. Describe the principle of the Ames test for identifying chemical carcinogens.

9. Differentiate between transformation and transduction.

10. Define *plasmids,* and explain the relationship between F plasmids and conjugation.

11. Use the following metabolic pathway to answer the questions that follow it.

Substrate A $\xrightarrow{\text{enzyme } a}$ Intermediate B $\xrightarrow{\text{enzyme } b}$ End-product C

(a) If enzyme a is inducible and is not being synthesized at present, a _____ protein must be bound tightly to the _____ site. When the inducer is present, it will bind to the _____ so that _____ can occur.

(b) If enzyme a is repressible, end-product C, called a _____ , causes the _____ to bind to the _____ . What causes derepression?

(c) If enzyme a is constitutive, what effect, if any, will the presence of A or C have on it?

12. Identify three ways of preventing mistakes in DNA.

13. How does a decrease in glucose result in transcription of new catabolic genes?

14. Define the following terms:
(a) Genotype
(b) Phenotype
(c) Recombination

15. Which sequence is the best target for damage by UV radiation: AGGCAA, CTTTGA, or GUAAAU? Why aren't all bacteria killed when they are exposed to sunlight?

16. You are provided with cultures with the following characteristics:

Culture 1: F^+, genotype $A^+ B^+ C^+$
Culture 2: F^-, genotype $A^- B^- C^-$

(a) Indicate the possible genotypes of a recombinant cell resulting from the conjugation of cultures 1 and 2.

(b) Indicate the possible genotypes of a recombinant cell resulting from conjugation of the two cultures after the F^+ has become an Hfr.

17. Why are semiconservative replication and degeneracy of the genetic code advantageous to the survival of species?

18. Why are mutation and recombination important in the process of natural selection and the evolution of organisms?

Critical Thinking

1. Base analogs and ionizing radiation are used in the treatment of cancer. These mutagens can cause cancer, so how do you suppose they are used to treat the disease?

2. Replication of the *E. coli* chromosome takes from 40 to 45 minutes, but the organism has a generation time of 26 minutes. How does the cell have time to make complete chromosomes for each daughter cell? For each granddaughter cell?

3. *Pseudomonas* has a plasmid containing the *mer* operon, which includes the gene for mercuric reductase. This enzyme catalyzes the reduction of the mercuric ion Hg^{2+} to the uncharged form of mercury, Hg^0. Hg^{2+} is quite toxic to cells; Hg^0 is not.

(a) What do you suppose is the inducer for this operon?

(b) The protein encoded by one of the *mer* genes binds Hg^{2+} in the periplasmic gel and brings it into the cell. Why would a cell bring in a toxin?

(c) What is the value of the *mer* operon to *Pseudomonas?*

Clinical Applications

1. Chloroquine and erythromycin are used to treat microbial infections. Chloroquine acts by fitting between base pairs in the DNA molecule. Erythromycin binds in front of the A site on the 50S subunit of a ribosome.

   (a) What steps in protein synthesis are inhibited by each drug?

   (b) Which drug is more effective against bacteria? Why?

   (c) Which drug will have effects on the host's cells? Why?

   (d) Use the index to identify the disease for which chloroquine is primarily used. Why is it more effective than erythromycin for treating this disease?

2. Once inside a host, pathogenic bacteria produce altered cell walls, anti-oxidizing agents, and fimbriae. Of what value is this global regulation to the bacteria?

3. A dentist showed symptoms of HIV infection in late 1986 and was diagnosed with AIDS in 1987. He closed his practice in 1989. All of the dentist's employees, including the dental hygienists, tested negative for HIV antibodies. The dentist wrote an open letter to his former patients, which prompted 591 people to be tested for HIV antibodies. There were seven seropositive individuals. HIV, the virus that causes AIDS, was isolated from these seven individuals, and the amino acid sequences for the viral coat were determined. Of the three amino acid sequences shown below, which one was not contracted from the dentist? Why can these amino acid sequences be used to identify the source of a virus?

PATIENT	VIRAL AMINO ACID SEQUENCE
A	Asn Gln Thr Ala Ala Ser Lys Asn Ile Asp Ala Glu Leu
B	Asn Leu His Ser Asp Lys Ile Asn Ile Ile Leu Gln Leu
C	Asn Gln Thr Ala Asp Ser Ile Val Ile Asp Ala Cys Leu

Recombinant DNA and Biotechnology

DURING THE NINETEENTH CENTURY scientists showed that microorganisms were responsible for the formation of many products, including bread, beer, wine, and cheese. As the science of microbiology developed, microbiologists learned how to isolate and grow microorganisms in pure culture. Once they had obtained pure strains, they could select the ones that were most efficient in brewing beer, for example; this was a logical extension of the reasoning long used in selecting desirable breeds of animals or strains of plants. Much more recently, scientists working with antibiotic-producing microbes discovered that they could create new strains by exposing the microbes to mutagenic radiation or chemicals. Some of these variant strains produced more, or better, antibiotics. This mutational approach was responsible for many advances in the industrial production of antibiotics following World War II.

After the war, scientists discovered recombination of DNA in microbes. Recombination occurs naturally in microorganisms, generally between the DNAs of closely related strains (see Figure 8.22). In bacteria, the DNA of a donor strain may be brought into contact with the DNA of a recipient strain by transformation, conjugation, or transduction. Recombination is more likely to be a beneficial form of genetic change than are mutations, most of which are harmful to the bacterium and are not commercially useful.

THE ADVENT OF RECOMBINANT-DNA TECHNOLOGY

LEARNING OBJECTIVE

- *Compare and contrast genetic engineering, recombinant DNA, and biotechnology.*

During the 1970s and 1980s, the practical application of microorganisms expanded almost beyond imagination with the development of new, artificial techniques for making recombinant DNA. Although natural recombination makes it possible for closely related organisms to exchange genes, the new techniques make it possible to transfer genes between completely unrelated species. These techniques are so powerful that the term **recombinant DNA** is now widely understood to mean any artificial manipulation of genes, whether within a particular species or between different species.

A gene from a vertebrate animal, including a human, can be inserted into the DNA of a bacterium, or a gene from a virus into a yeast. In many cases, the recipient can then be made to express the gene, which may code for a commercially useful product. Thus, bacteria with genes for human insulin are now being used to produce insulin for treating diabetes, and a vaccine for hepatitis B is being made by yeast carrying a gene for part of the hepatitis virus (the yeast produces a viral protein). Scientists hope that such an approach may prove useful in producing vaccines against other infectious agents, eliminating the need to use whole viruses as in conventional vaccines.

The new recombinant-DNA techniques can also be used to make thousands of copies of the same DNA molecule—to *amplify* DNA, thus generating sufficient DNA for various kinds of experimentation and analysis. This technique has found practical application in the field of criminology. DNA from a few human blood cells or from semen found at a crime scene can be amplified and then compared with the DNA of suspects. Because each person's DNA is unique (except for identical twins), the procedure has great potential usefulness to criminologists.

Artificial gene manipulation is popularly known as **genetic engineering.** In fact, the term **biotechnology,** which generally had been defined to include all industrial applications of biological systems or processes, has increasingly become identified in the public mind as the industrial application of genetic engineering.

Genetic engineering was made possible by the discovery and development of a number of tools and techniques. Perhaps most important was the discovery of **restriction enzymes,** bacterial enzymes that can be used to cut DNA from different sources into pieces that are easy to recombine in vitro (*in vitro* means "in glass"—that is, in a test tube rather than inside a living organism). Genetic manipulation required the development of methods for inserting recombinant DNA molecules into cells by using so-called vectors. If a mosquito carrying the virus for yellow fever bites and infects a human, the mosquito is considered a "disease vector" because it can transmit the virus from one host to another. The term **vector,** or *cloning vector,* has been adopted to describe a self-replicating DNA molecule that is used as a carrier to transmit a gene from one organism into another.

OVERVIEW OF RECOMBINANT-DNA PROCEDURES

LEARNING OBJECTIVE

- *Identify the roles of a clone and a vector in genetic engineering.*

Figure 9.1 presents an overview of some of the procedures typically used in genetic engineering and some promising applications. ① and ② After several preparatory steps, ③ the gene of interest is inserted into the vector DNA in vitro. In our example, the vector is a plasmid. The DNA molecule chosen as a vector must be a self-replicating type, such as a plasmid or a viral genome. ④ Next, this recombinant vector DNA is taken up by a cell such as a bacterium, where it can multiply. ⑤ The cell containing the recombinant vector is then grown in culture to form a **clone** of many genetically identical cells (see Chapter 6, p. 154), each of which carries a copy of the vector. This cell clone therefore contains many copies of the gene of interest. This is why DNA vectors are often called gene-cloning vectors, or simply cloning vectors. (In addition to referring to a culture of identical cells, the word *clone* is also routinely used as a verb, to describe the entire process, as in "to clone a gene.")

The final step can go in alternative directions, depending upon whether the gene itself or the product of the gene is of interest. ⑥Ⓐ From the cell clone, the genetic engineer may isolate ("harvest") large quantities of the gene of interest, which may then be used for a variety of purposes. The gene may even be inserted into another vector for introduction into another kind of cell (such as a plant or animal cell). Alternatively, ⑥Ⓑ if the gene of interest is expressed (transcribed and translated) in the cell clone, ⑦ its protein product can be harvested and used for a variety of purposes.

The advantages of genetic engineering for obtaining such proteins is illustrated by one of its early successes, somatotropin, which is also known as human growth hormone. Some individuals do not produce adequate amounts of somatotropin and so are stunted in growth. In the past, growth hormone needed to correct this deficiency had to be obtained from human pituitary glands at autopsy. (Growth hormone from other animals is not effective in humans.) This practice was not only expensive but also dangerous because on several occasions neurological diseases were transmitted with the hormone. Human growth hormone produced by genetically engineered *E. coli* is therefore an important product of modern biotechnology.

BIOTECHNOLOGY TOOLS AND TECHNIQUES

LEARNING OBJECTIVES

- *Define restriction enzymes, and outline how they are used to make recombinant DNA.*
- *Describe the use of plasmid and viral vectors.*
- *Describe five ways of getting DNA into a cell.*

The science of genetic engineering has advanced to such a degree that many routine cloning procedures are performed using prepackaged materials and procedures that are very much like cookbook recipes. Genetic engineers have a grab bag of methods from which to choose, depending upon the ultimate application of their experiments. We will describe next some of the most important tools and techniques, and later we will consider some specific applications that use these tools.

FIGURE 9.1 Typical genetic engineering procedures with examples of applications.

FIGURE 9.2 The role of a restriction enzyme in making recombinant DNA.

Recognition sites

1 Restriction enzyme cuts (magenta arrows) double-stranded DNA at its particular recognition sites, which are shown in blue.

DNA   Cut ↓   Cut ↓
G A A T T C   G A A T T C
C T T A A G   C T T A A G
↑ Cut   ↑ Cut

2 These cuts produce a DNA fragment with two "sticky" ends.

A A T T C   A A T T C
G   G
G   G
C T T A A   C T T A A
"Sticky" end

A A T T C   G
G   C T T A A
DNA from another source, perhaps a plasmid

3 When two such fragments of DNA cut by the same restriction enzyme come together, they can join by base-pairing.

A A T T C   G A A T T C   G
G   C T T A A G   C T T A A

Restriction Enzymes

Genetic engineering has its technical roots in the discovery of a special class of DNA-cutting enzymes that exist in many bacteria. These enzymes, called *restriction enzymes*, were first isolated in 1970. However, their role in nature had actually been observed earlier, when certain bacteriophages were found to have a restricted host range. If these phages were used to infect bacteria other than their usual hosts, "restriction" enzymes in the new host destroyed almost all the phage DNA. The purified forms of these bacterial enzymes are used by today's genetic engineers.

What is important for genetic engineering is that a **restriction enzyme** recognizes and cuts, or digests, only one particular sequence of nucleotide bases in DNA and that it cuts this sequence in the same way each time. Typical restriction enzymes used in cloning experiments recognize four-, six-, or eight-base sequences. Many restriction enzymes make *staggered* cuts in the two strands of a DNA molecule—that is, cuts that are not directly opposite each other (Figure 9.2). ① Note in this figure that whereas the blue base sequences on the two strands are the same, they run in opposite directions. ② Staggered cuts leave stretches of single-stranded DNA at the ends of the DNA fragments. These are called *sticky ends* because they can "stick" to complementary stretches of single-stranded DNA by base-pairing. Hundreds of restriction enzymes are known, each producing DNA fragments with characteristic ends.

③ If two fragments of DNA from different sources have been produced by the action of the same restriction en-

4 The joined fragments will usually form either a linear molecule or a circular one, as shown here for a plasmid. Other combinations of fragments can also occur, however.

5 The enzyme DNA ligase is used to unite the backbones of the two DNA fragments, producing a molecule of recombinant DNA.

Recombinant DNA

zyme, the two pieces will have identical sets of sticky ends and can be spliced (recombined) in vitro. ④ The sticky ends first join spontaneously by hydrogen bonding (base-pairing) in either a linear or a circular form. ⑤ Then the enzyme DNA ligase is used to covalently link the backbones of the DNA pieces, producing a recombinant-DNA molecule.

Vectors

A great variety of different types of DNA molecules can serve as vectors. The most important property they must have is that of self-replication; once in a cell, a vector must be capable of replicating. Any DNA that is cloned in the vector will be replicated in the process. Thus, vectors serve as vehicles for the replication of desired DNA sequences.

Plasmids are one of the primary vectors in use, particularly variants of R factor plasmids. Plasmid DNA can be cut with the same restriction enzymes as the DNA to be cloned, so that all pieces of the DNA will have the same sticky ends. When the pieces are mixed, the DNA to be cloned will become inserted into the plasmid (see Figure 9.2). Note that other possible combinations of fragments can occur as well, including the plasmid re-forming a circle with no DNA inserted.

Some plasmids are capable of existing in several different species. Such **shuttle vectors** can be used to move cloned DNA sequences among organisms, such as among bacterial, yeast, and mammalian cells, or among bacterial, fungal, and plant cells. Shuttle vectors can be very useful in the process of genetically modifying multicellular organisms—for example, by trying to insert herbicide resistance genes into plants.

A different kind of vector is virus DNA. This type of vector can usually accept much larger pieces of foreign DNA than can plasmids. After the DNA has been inserted into the virus vector, it can then be cloned in the virus's host cells. The choice of a suitable vector depends upon many factors, including the organism that will receive the new gene and the size of the DNA to be cloned.

Methods for Inserting Foreign DNA into Cells

In nature, plasmids are usually transferred between closely related microbes by cell-to-cell contact, such as in conjugation. In genetic engineering, a plasmid must be inserted into a cell by *transformation*. Many cell types, including *E. coli*, yeast, and mammalian cells, do not naturally transform; however, simple chemical treatments can make all of these cell types *competent*, or able to take up external DNA. For *E. coli*, the procedure for making cells competent is to soak them in a solution of calcium chloride for a brief period. Following this treatment, the now-competent cells are mixed with the cloned DNA. Many of these cells will take up the DNA.

There are a variety of other means of transferring DNA to cells. Several of these processes require the enzymatic removal of cell walls, producing *protoplasts*. The process called **electroporation** begins with keeping the protoplasts in a solution with an osmotic pressure high enough to keep the cells intact. An electric current is then used to form microscopic pores in the cell membranes of the protoplasts; the DNA then enters the cells through the pores.

Electroporation is generally applicable to all cells; those with cell walls require conversion to protoplasts first.

The process of **protoplast fusion** also takes advantage of the properties of protoplasts. Protoplasts in solution will fuse at a low but significant rate; the addition of polyethylene glycol increases the frequency of fusion (Figure 9.3a). In the new hybrid cell, the DNA derived from the two "parent" cells may undergo natural recombination. This method is especially valuable in the genetic manipulation of plant cells (Figure 9.3b).

A remarkable way of introducing foreign DNA into plant cells is to literally shoot it directly through the thick cellulose walls using a "gene gun." Microscopic particles of tungsten or gold are coated with DNA and propelled by a burst of helium through the plant cell walls. Some of the cells express the introduced DNA as if it were their own.

DNA can also be introduced directly into an animal cell by **microinjection.** This technique requires the use of a glass micropipette with a diameter that is much smaller than the cell. It can then be used to puncture the cell membrane, and DNA can be injected through it (Figure 9.4).

Thus, there are a great variety of different restriction enzymes, vectors, and methods of inserting DNA into cells. Whatever the means by which foreign DNA is inserted into a cell, such DNA will survive only if it either is present on a self-replicating vector or is incorporated into one of the cell's chromosomes by recombination.

SOURCES OF DNA

LEARNING OBJECTIVES

- *Describe how a gene library is made.*
- *Differentiate between cDNA and synthetic DNA.*

We have seen how genes can be cloned into vectors by using restriction enzymes and can be transformed or transferred into a variety of cell types. Let's backtrack now to ask how genetic engineers get the genes they are interested in. There are three main sources of genes: gene libraries containing natural copies of genes, gene libraries containing "complementary DNA" copies of genes made from mRNA, and artificial synthesis of DNA.

Gene Libraries

Isolating specific genes as individual pieces of DNA after an entire genome has been digested by restriction enzymes is seldom practical. Therefore, researchers interested in genes from a particular organism usually start by cutting up its genome with restriction enzymes, splicing as many of the genome fragments as possible into copies

(a) Process of protoplast fusion

(b) Leaf cell protoplasts fusing   **LM**   10 μm

FIGURE 9.3   Protoplast fusion. **(a)** Diagram of protoplast fusion with bacterial cells. **(b)** Two protoplast leaf cells from a tobacco plant are shown fusing. Removal of the cellulose wall left only the delicate plasma membrane to bind the cell contents together, allowing exchange of DNA.

LM   10 μm

FIGURE 9.4   Injection of foreign DNA into a fertilized mouse egg. The egg is first immobilized by applying mild suction to the large, blunt holding pipette (right). Several hundred copies of the gene of interest are then injected into the nucleus of the cell through the sharp end of the micropipette (left).

of a plasmid or phage vector, and then introducing the recombinant vectors into bacterial cells. The goal is to make a collection of clones large enough to ensure that at least one clone exists for every gene in the organism. This collection of clones containing different DNA fragments is called a **gene library;** each "book" is a bacterial or phage strain that contains a fragment of the genome (Figure 9.5). Such libraries are essential for maintaining and retrieving DNA clones; they can even be purchased commercially.

Recall that the genes of eucaryotic cells generally contain both exons (stretches of DNA that code for protein) and introns (intervening stretches of DNA that do not code for protein). When the RNA transcript of such a gene

is converted to mRNA, the introns are removed. In cloning genes of eucaryotic cells, it is desirable to use a version of the gene that lacks introns because a gene that includes introns may be too large to work with easily. In addition, if such a gene is put into a bacterial cell, the bacterium will not have the machinery to remove the introns from the

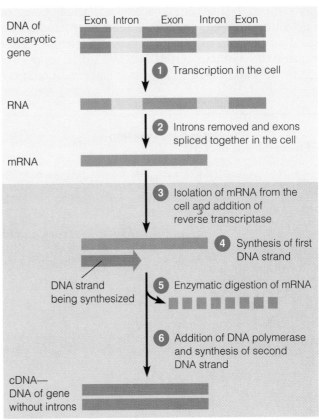

FIGURE 9.5 Gene libraries. A gene library is a collection of a large number of DNA fragments from a genome. Each fragment, containing about one gene, is carried by a vector, either a plasmid carried within a bacterial cell or a phage.

FIGURE 9.6 Making complementary DNA (cDNA) for a eucaryotic gene. Reverse transcriptase catalyzes the synthesis of double-stranded DNA from an RNA template.

RNA transcript, and therefore it will not be able to make the correct protein product. However, an artificial gene that contains only exons can be produced by using an enzyme called *reverse transcriptase* to synthesize **complementary DNA,** or **cDNA,** from an mRNA template (Figure 9.6). This synthesis is the reverse of the normal DNA-to-RNA transcription process. A DNA copy of mRNA is produced by reverse transcriptase. Following this, the mRNA is enzymatically digested away. DNA polymerase then synthesizes a complementary strand of DNA, creating a double-stranded piece of DNA containing the information from the mRNA. cDNAs produced from a mixture of all the mRNAs in a cell can then be cloned to form a cDNA library.

The cDNA method is the most common method of obtaining eucaryotic genes. A difficulty with this method is that long molecules of mRNA may not get completely reverse-transcribed into DNA: the reverse transcription often aborts, forming only parts of the desired gene.

Synthetic DNA

Under certain circumstances, genes can be made in vitro with the help of DNA synthesis machines (Figure 9.7). A

FIGURE 9.7 DNA synthesis machine. Short sequences of DNA can be synthesized by instruments such as this one.

keyboard on the machine is used to enter the sequence of nucleotides desired, much as letters are entered into a word processor to compose a sentence. A microprocessor controls the synthesis of the DNA from stored supplies of nucleotides and the other necessary reagents. A chain of

over 120 nucleotides can be synthesized by this method. Unless the gene is very small, at least several chains must be synthesized separately and linked together to form an entire gene.

The difficulty of this approach, of course, is that the se-

FIGURE 9.8   One method for selection of recombinant bacteria.

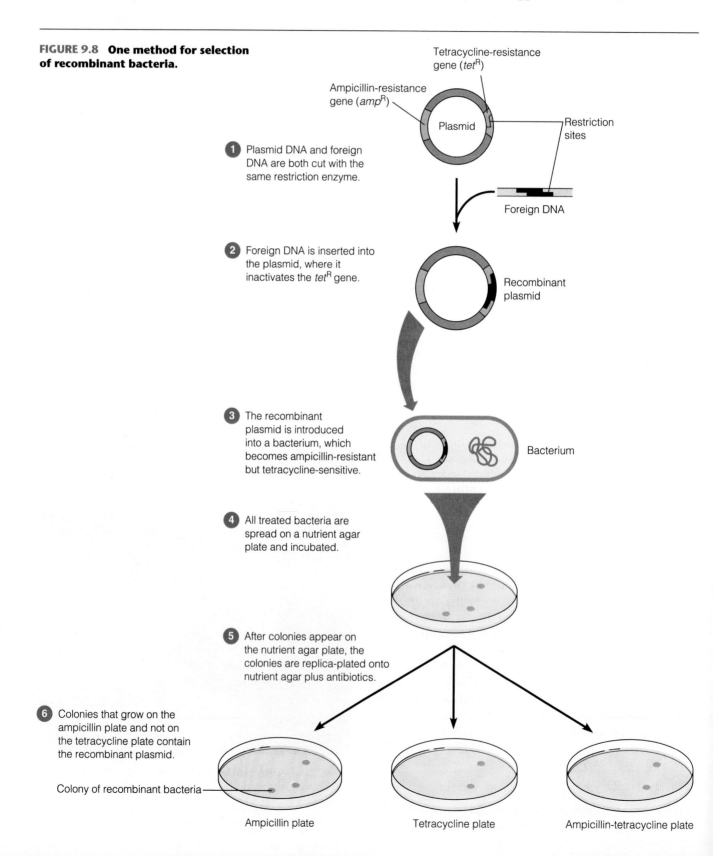

quence of the gene must be known before it can be synthesized. If the gene has not already been isolated, then the only way to predict the DNA sequence is by knowing the amino acid sequence of the protein product of the gene. If this amino acid sequence is known, then in principle one can work backward through the genetic code to obtain the DNA sequence. Unfortunately, the degeneracy of the code prevents an unambiguous determination; thus, if the protein contains a leucine, for example, which of the six codons for leucine is the one in the gene?

For these reasons, it is rare to clone a gene by synthesizing it directly, although some commercial products, such as insulin and interferon, are produced from synthesis genes. Both of these products are encoded by small genes, allowing the synthetic approach to be used. Synthetic DNA plays a much more useful role in selection procedures, as we will see.

SELECTING A CLONE

LEARNING OBJECTIVE

- *Explain how each of the following is used to locate a clone: antibiotic-resistance genes, DNA probes, antibodies.*

Cells containing new genes can be identified and selected by a variety of methods. Figure 9.8 illustrates a common trick for identifying bacterial clones that have acquired recombinant plasmids. The R plasmid used as a vector carries two antibiotic-resistance genes, one conferring resistance to ampicillin (amp^R) and the other to tetracycline (tet^R). Bacteria carrying the nonrecombinant version of the plasmid will be resistant to both antibiotics. However, bacteria carrying recombinant plasmids will be resistant only to ampicillin because the plasmid's restriction site is in the middle of the tetracycline-resistance gene. When the foreign DNA inserts there, it inactivates the tet^R gene. After treatment to get the vectors into the bacteria, the cells are spread onto agar plates that contain both appropriate nutrients and either tetracycline or ampicillin. Only cells that have been transformed will grow on media containing ampicillin. Cells containing recombinant molecules will not be able to grow on media containing tetracycline.

Another method for identifying cells carrying a specific cloned gene is *colony hybridization* (Figure 9.9). Hybridization is also the basis of Southern blots and Western blots. In colony hybridization, short segments (usually 15–20 bases) of single-stranded DNA, consisting of a sequence of nucleotides unique to the gene being sought, are synthesized as previously described. These small DNA molecules, called **probes**, are radioactively labeled so that they can be located later. Meanwhile, bacteria from the library that carry different fragments of foreign DNA are grown into colonies on a plate of nutrient medium. After incubation,

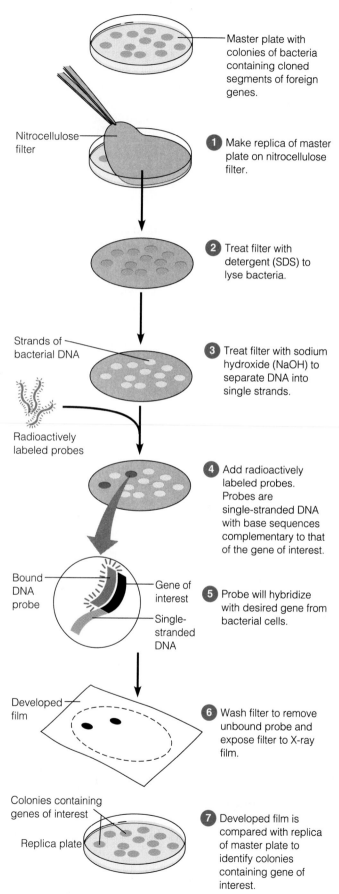

1. Make replica of master plate on nitrocellulose filter.

Master plate with colonies of bacteria containing cloned segments of foreign genes.

Nitrocellulose filter

2. Treat filter with detergent (SDS) to lyse bacteria.

3. Treat filter with sodium hydroxide (NaOH) to separate DNA into single strands.

Strands of bacterial DNA

Radioactively labeled probes

4. Add radioactively labeled probes. Probes are single-stranded DNA with base sequences complementary to that of the gene of interest.

5. Probe will hybridize with desired gene from bacterial cells.

Bound DNA probe

Gene of interest

Single-stranded DNA

6. Wash filter to remove unbound probe and expose filter to X-ray film.

Developed film

Colonies containing genes of interest

Replica plate

7. Developed film is compared with replica of master plate to identify colonies containing gene of interest.

FIGURE 9.9   Colony hybridization: use of a DNA probe to identify a cloned gene of interest.

① the colonies are transferred onto a membrane support, such as a nitrocellulose filter or a piece of nylon. ② The filter is then treated with detergent to gently break open the bacterial cells. ③ The cells' DNA is separated into single strands by further chemical treatments. ④ Then the radioactively labeled probe is added to the filter. ⑤ Any bacterial DNA that base-pairs with the probe can be identified by exposure to photographic film ⑥ and ⑦. Living cells from the same clone can then be obtained from the untreated replica plate.

A useful method of selection is based on this hybridization procedure. If the amino acid sequence of a protein is known, then it is usually possible to find a short 5–7 amino acid segment that results in an unambiguous (or nearly so) DNA sequence. A probe is then synthesized based on this sequence, and this probe is used to identify the complete gene in the library through hybridization. This method is valuable because very little information about the protein product—only a short amino acid sequence—is required to isolate the entire gene. Armed with the capability of producing any desired probe by synthetic DNA methods, genetic engineers can imagine cloning virtually any gene.

An approach analogous to hybridization, but based on protein rather than DNA, is called *recognition*. Recognition can be used to identify colonies that have made the protein product of the foreign gene. Here the "probe" consists of radioactively labeled antibodies that specifically attach to molecules of the protein product (antibodies are discussed in Chapter 17). Colonies producing the desired protein can be identified by the radioactive tag.

MAKING A GENE PRODUCT

LEARNING OBJECTIVE

- *List one advantage of engineering each of the following:* E. coli, Saccharomyces cerevisiae, *vaccinia virus, mammalian cells.*

We have just seen that one way of identifying cells carrying a particular gene is by assaying for the gene product. Such products are themselves, of course, a frequent objective of genetic engineering. Most of the earliest work in genetic engineering used *E. coli* to synthesize the gene products. Besides being easily grown, *E. coli* has the advantage that researchers are very familiar with this organism and its genetics. For example, some inducible promoters, such as that of the *lac* operon, have been cloned, and cloned genes can be attached to such promoters. The synthesis of great amounts of the cloned gene product can then be directed by the addition of an inducer. Such a method has been used to produce gamma interferon in *E. coli* (Figure 9.10). However, *E. coli* also has several disadvantages. Like other gram-negative bacteria, it pro-

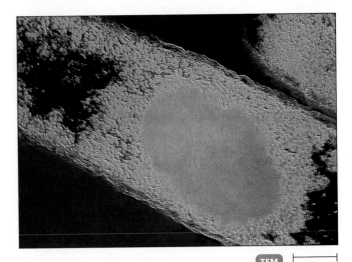

TEM | 0.25 μm

FIGURE 9.10 *E. coli* **genetically engineered to produce gamma interferon, which has antiviral and other medical applications.** The product, visible here as an orange mass, can be released by lysis of the cell.

duces endotoxins as part of the outer layer of its cell wall. Because endotoxins cause fever and shock in animals, their accidental presence in products intended for human use would be a serious problem.

Another disadvantage of *E. coli* is that it does not usually secrete protein products. To obtain a product, cells must usually be broken open and the product purified from the resulting "soup" of cell components. Recovering the product from such a mixture is expensive when done on an industrial scale. It is more economical to have an organism secrete the product so that it can be recovered continuously from the growth medium. One approach has been to link the product to a natural *E. coli* protein that the bacterium does secrete. However, gram-positive bacteria, such as *Bacillus subtilis,* are more likely to secrete their products and are often preferred industrially for that reason.

Another microbe being used as a vehicle for the expression of genetically engineered genes is baker's yeast, *Saccharomyces cerevisiae.* Its genome is only about four times larger than that of *E. coli* and is probably the best understood eucaryotic genome. Yeasts may carry plasmids, and the plasmids are easily transferred into yeast cells after their cell walls have been removed. As eucaryotic cells, yeasts may be more successful in expressing foreign eucaryotic genes than bacteria. Furthermore, yeasts are likely to continuously secrete the product. Because of all these factors, yeasts have become the eucaryotic workhorse of biotechnology.

Mammalian cells in culture, even human cells, can be used much like bacteria to produce genetically engineered products. Scientists have developed effective methods of growing certain mammalian cells in culture as hosts for growing viruses (see Chapter 13). In genetic engineering, mammalian cells are often the best suited to make protein products for medical use; these products include hormones, cytokines (which regulate cells of the immune system), and interferon (a natural antiviral substance that is also used to treat some cancers). Using mammalian cells to make foreign gene products on an industrial scale often requires a preliminary step of cloning the gene in bacteria. Consider the example of colony-stimulating factor (CSF). CSF, a protein produced naturally in tiny amounts by white blood cells, is valuable because it stimulates the growth of certain cells that protect against infection. To produce huge amounts of CSF industrially, the gene is first inserted into a plasmid, and bacteria are used to make multiple copies of the plasmid (as in Figure 9.1). The recombinant plasmids are then inserted into mammalian cells that are grown in huge tanks.

Plant cells can also be grown in culture, altered by recombinant-DNA techniques, and then used to generate genetically engineered plants. Such plants may prove useful as sources of valuable products, such as plant alkaloids (the painkiller codeine, for example), the isoprenoids that are the basis of synthetic rubber, and melanin (the animal skin pigment) for use in sunscreens. We will return to the topic of genetic engineering of plants later in this chapter.

APPLICATIONS OF GENETIC ENGINEERING

LEARNING OBJECTIVES

- *List at least five applications of genetic engineering.*
- *Diagram the Southern blotting procedure and provide an example of its use.*
- *Diagram DNA fingerprinting and provide an example of its use.*
- *Outline the PCR and provide an example of its use.*
- *Outline genetic engineering with* Agrobacterium.

We have now described the entire sequence of events in cloning a gene. As indicated earlier, such cloned genes can be applied in a variety of ways. One is to produce important substances, such as insulin, more efficiently and less expensively. Another is to obtain information from the cloned DNA that is useful for either basic research or medical applications. A third is to use cloned genes to alter the characteristics of cells or organisms, as when a gene for herbicide resistance is cloned into plants. The box on page 240 describes how researchers develop a product. We will now consider some examples of each of these.

Genetically Engineered Products for Medical Therapy

An extremely valuable pharmaceutical product is the hormone insulin, a small protein produced by the pancreas that controls the body's uptake of glucose from blood. For many years insulin-deficient diabetics have controlled their disease by the injection of insulin obtained from the pancreases of slaughtered animals. Obtaining this insulin is an expensive process, and the insulin from animals is not as effective as human insulin.

Because of the value of human insulin and the small size of the protein, the production of human insulin by recombinant-DNA techniques was an early goal for the pharmaceutical industry. To produce the hormone, synthetic genes were first constructed for each of the two short polypeptide chains that make up the insulin molecule. The small size of these chains—only 21 and 30 amino acids long—made it possible to use synthetic genes. Following the strategy discussed earlier, each of the two synthetic genes was inserted into a plasmid vector and linked to the end of a gene coding for the bacterial enzyme β-galactosidase, so that the insulin polypeptide was coproduced with the enzyme. Two different *E. coli* bacterial cultures were used, one to produce each of the insulin polypeptide chains. The polypeptides were then recovered from the bacteria, separated from the β-galactosidase, and chemically joined to make human insulin. This accomplishment was one of the early commercial successes of genetic engineering and illustrates a number of the principles and procedures we have discussed in this chapter.

Another human hormone that is now being produced commercially by genetic engineering is somatostatin. This hormone, which is involved in limiting growth, is now being produced in *E. coli* for research purposes at the rate of 10,000 somatostatin molecules per cell. At one time 500,000 sheep brains were needed to produce 5 mg of animal somatostatin for experimental purposes. By contrast, only 8 liters of a genetically engineered bacterial culture are now required to obtain the equivalent amount of the human hormone.

There has been much excitement over the life-saving potential of a medicine, tissue-plasminogen activator (t-PA), which helps dissolve blood clots. Consequently, it speeds recoveries from heart attacks and also helps prevent recurrences. If this medicine is administered very soon after an initial attack, patients have a much lowered chance of having a recurrence. Tissue-plasminogen activator, a product of genetic engineering, is produced by mammalian cell culture; streptokinase, an older product, is a natural product of bacteria. Although both are effective, t-PA is much more expensive to produce than is streptokinase.

Earlier we mentioned a recombinant-DNA approach to vaccines that uses vaccinia virus engineered to carry genes for surface proteins of a pathogen. A related approach already in use is to harvest such proteins from engineered yeast cells and then to use the purified protein as a vaccine against hepatitis B. The term for a vaccine consisting only of such a limited protein portion of a pathogen is *subunit vaccine*. Table 9.1 lists some other important genetically engineered products used in medical therapy.

TABLE 9.1   Some Pharmaceutical Products of Genetic Engineering

PRODUCT	COMMENTS
Tissue plasminogen activator (t-PA)	Dissolves the fibrin of blood clots; therapy for heart attacks; produced by mammalian cell culture
Erythropoietin (EPO)	Treatment of anemia; produced by mammalian cell culture
Human insulin	Therapy for diabetics; better tolerated than insulin extracted from animals; produced by *E. coli*
Interleukin-2 (IL-2)	Possible treatment for cancer; stimulates the immune system; produced by *E. coli*
Alpha- and gamma-interferon	Possible cancer and virus-disease therapy; produced by *E. coli, S. cerevisiae*
Tumor necrosis factor (TNF)	Causes disintegration of tumor cells; produced by *E. coli*
Human growth hormone (somatotropin)	Corrects growth deficiencies in children; produced by *E. coli*
Epidermal growth factor (EGF)	Heals wounds, burns, ulcers; produced by *E. coli*
Prourokinase	Anticoagulant; therapy for heart attacks; produced by *E. coli* and yeast
Factor VIII	Treatment of hemophilia; improves clotting; produced by mammalian cell culture
Colony-stimulating factor (CSF)	Counteracts effects of chemotherapy; improves resistance to infectious disease such as AIDS; treatment of leukemia; produced by *E. coli, S. cerevisiae*
Superoxide dismutase	Minimizes damage caused by oxygen free radicals during reperfusion of oxygen-deprived tissues; produced by *S. cerevisiae* and *Pichia pastoris* (yeast).
Monoclonal antibodies	Possible therapy for cancer and transplant rejection; used in diagnostic tests; produced by mammalian cell culture (from fusion of cancer cell and antibody-producing cell)
Hepatitis B vaccine	Produced by *S. cerevisiae* that carries hepatitis-virus gene on a plasmid
Bone morphogenic proteins	Induce new bone formation; useful in healing fractures and reconstructive surgery; produced by mammalian cell culture
Taxol	Plant product used for treatment for ovarian cancer; produced in *E. coli*

Obtaining Information from DNA for Basic Research and Medical Applications

Recombinant-DNA technology can be used to make products, but this is not its only important application. Because of its ability to produce many copies of DNA, it can serve as a sort of DNA "printing press." Once a large amount of a particular piece of DNA is available, various analytic techniques, which we will discuss in this section, can be used to "read" the information contained in the DNA.

What kind of information can be obtained from cloned DNA? One kind is provided by the process of **DNA sequencing**, which is the determination of the exact sequence of nucleotide bases in DNA. Recent automation of this process enables a researcher to determine the sequence of over 1000 bases per day. The sequences of entire virus genomes are now relatively easy to obtain. Larger projects are also in progress: the genome of *E. coli* is almost entirely sequenced, and a massive international project to sequence all the DNA of a human, called the Human Genome Project, has begun.

An example of the use of human DNA sequencing is the recent identification and cloning of the mutant gene that causes cystic fibrosis (CF). CF is characterized by over-secretion of mucus leading to blocked respiratory passageways. The sequence of the mutated gene can be used as a diagnostic tool in a hybridization technique called **Southern blotting** (Figure 9.11).

In this technique, ① human DNA is first digested with a restriction enzyme, yielding thousands of fragments of all different sizes. ② The different-sized fragments are then separated by *gel electrophoresis* (see Chapter 10). ③ and ④ The separated fragments are transferred onto a filter by blotting. ⑤ The fragments on the filter are then exposed to a radioactive probe made from the cloned gene of interest, in this case the CF gene. The probe will hybridize to this mutant gene but not to the normal gene. ⑥ Fragments to which the probe bound are identified by exposing the filter to X-ray film. With this method, any person's DNA can be tested for the presence of the mutated gene.

This process, which is called **genetic screening**, can now be used to screen for several hundred genetic diseases. Such screening procedures can be done for adults planning to have children and also on fetal tissue.

Gene therapy may eventually provide cures for some genetic diseases. It is possible to imagine removing some cells from a person and transforming them with a normal gene to replace a defective or mutated gene. When these cells are returned to the person, they should function normally. For example, a rare and deadly mutation can result in adenosine deaminase deficiency, a disease that leaves the individual without a functional immune system. Re-

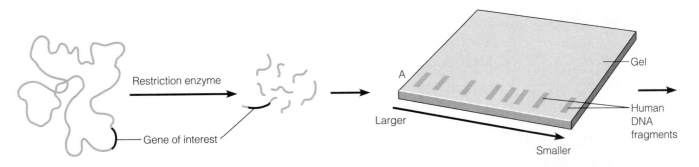

1. DNA containing the gene of interest is extracted from human cells and cut into fragments by restriction enzymes.

2. The fragments are separated according to size by gel electrophoresis. Each band consists of many copies of a particular DNA fragment. The bands are invisible but can be made visible by a dye that glows under ultraviolet light.

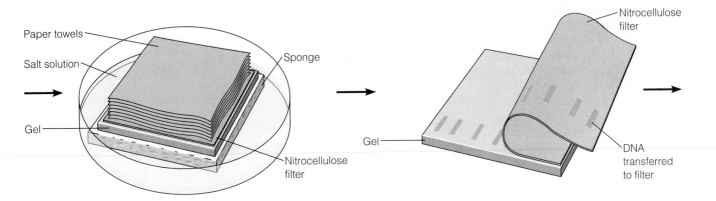

3. The DNA bands are transferred to a nitrocellulose filter by blotting. The solution passes through the gel and filter to the paper towels.

4. This produces a nitrocellulose filter with DNA fragments positioned exactly as on the gel.

5. The filter is exposed to radioactively labeled probe for a specific gene. The probe will base-pair (hybridize) with a short sequence present on the gene.

6. The filter is then exposed to X-ray film. The fragment containing the gene of interest is identified by a band on the developed film.

FIGURE 9.11 Southern blotting.

Applications of Microbiology

FROM PLANT DISEASE TO SHAMPOO AND SALAD DRESSING

XANTHOMONAS CAMPESTRIS is a gram-negative rod that causes a disease called black rot in plants. After gaining access to a plant's vascular tissues, the bacteria use the glucose transported in those tissues to produce a sticky, gumlike substance. This substance builds up to form gumlike masses, which eventually block the plant's transport of nutrients. The gum that makes up these masses, xanthan, is composed of a high-molecular-weight polymer of mannose (see photo).

Unlike its effects in plants, xanthan is not digested by humans and has no adverse effects when ingested by humans. Consequently, xanthan can be used as a thickener in foods such as dairy products and salad dressings, and in cosmetics such as cold creams and shampoos.

So when researchers at the U.S. Department of Agriculture (USDA) wanted to find some useful product that could be made out of whey, a liquid waste produced in abundance by the dairy industry, they thought of turning it into xanthan. However, because whey is mostly water and lactose, researchers had to figure out how to get X. campestris to produce xanthan using lactose rather than glucose.

One research team decided to engineer X. campestris to hydrolyze lactose more efficiently. They used an F⁺ strain of E. coli that contains a plasmid with a lac transposon. Organisms of this strain were incubated with X. campestris, and, as proven by later restriction enzyme digests, the plasmids containing the lac gene were transferred by conjugation. However, deletion mutations subsequently occurred in the X. campestris DNA that reduced their β-galactosidase (lactose-utilizing) ability.

Another research team working with the USDA at Stauffer Chemical Company used a simpler approach, which was based on satisfying only two requirements: that the bacteria grow on whey and make xanthan. First, they inoculated a whey medium with X. campestris and incubated it for 24 hours. Then they transferred an inoculum of this culture to a flask of lactose broth, to select a lactose-utilizing cell. The strain did not have to make xanthan from this broth; it only had to grow and use lactose.

A lactose-utilizing strain was isolated through serial transfers, selecting for the strain with the best ability to grow. After incubation for 10 days, an inoculum was transferred to another flask of lactose broth, and the procedure was repeated two more times. When transferred to a flask of whey medium, the final lactose-utilizing bacteria grew in the whey and the culture became extremely viscous—xanthan was being produced.

The process then had to be fine-tuned. The researchers addressed and solved problems such as how to sterilize the whey without destroying necessary ingredients, and how to handle the extremely viscous fermentations that resulted from the procedure. The final result was a process in which 40 g/L of whey powder is converted into 30 g/L of xanthan gum. A quick survey of labels in your neighborhood supermarket will demonstrate just how successful this project was.

Xanthomonas campestris producing gooey xanthan.   **SEM**

cent gene therapy experiments concerning this disease have had positive results. The number of such gene therapy trials will certainly increase as technical improvements are made and initial attempts are successful. However, there is a great deal of preliminary work to do, and cures may not be possible for all genetic diseases.

Southern blotting is also used in a method of identifi-

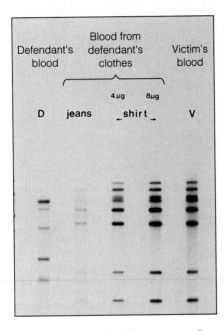

FIGURE 9.12 DNA fingerprints from a murder case. This is the developed film from a Southern blotting test done on blood taken from the defendant, the defendant's clothes, and the victim. Analysis of this film reveals that DNA from bloodstains on the defendant's clothes matches the DNA fingerprint of the victim but differs from the DNA fingerprint of the defendant. This is evidence that the blood on the defendant's clothes came from the victim.

cation known as **DNA fingerprinting.** The DNA of every organism is unique; DNA fingerprinting can be used to identify bacterial or viral pathogens (see Chapter 10), but its primary application is in forensic medicine. Probes constructed from variable regions in human DNA can be used in Southern blotting to compare DNA from hair, blood, or semen left at the scene of a crime with that of suspects. In a recent application of this new forensic technique, DNA fingerprints were used as evidence that the blood found on a murder suspect's clothes came from the murder victim (Figure 9.12). However, such evidence is often deemed inadmissible in court because these scientific techniques are still being developed.

Southern blotting requires substantial amounts of DNA, much more than is available from a crime scene sample. Small samples of DNA can be quickly *amplified*—increased to quantities large enough for analysis—with a method called the **polymerase chain reaction** (PCR). This procedure was originally developed by Kary Mullis in 1986; he received the Nobel Prize for this work in 1993.

Starting with just one gene-sized piece of DNA, PCR can be used to make literally billions of copies in only a few hours (Figure 9.13). In the PCR technique, ① a solution containing the piece of DNA to be amplified is first heated to 98°C in a test tube to separate the two strands of DNA, which will serve as the initial templates for DNA synthesis. ② To this DNA is added a supply of the four nucleotides (for assembly into new DNA) and the enzyme for catalyzing the synthesis, DNA polymerase (see Chapter 8). Short pieces of nucleic acid called primers are also added to help start the reaction. ③ Then, during a period of incubation at 60°C, the primers hybridize to the ends of the fragments to be amplified, and the polymerase synthesizes new complementary strands. ④ After each cycle of synthesis, the DNA is heated to convert all the new DNA into single strands. Each newly synthesized DNA strand serves

FIGURE 9.13 Polymerase chain reaction.

① Separate target DNA stands by heating at 98°C for 5 minutes.

② Add primers, nucleotides (ATP, CTP, GTP, and TTP), and DNA polymerase.

③ Cool to 60°C and incubate for a few minutes; during this time, primers attach to single-stranded DNA, and DNA polymerase synthesizes complementary strands.

④ Repeat cycle of heating and cooling until enough copies of target DNA have been produced.

in turn as a template for more new DNA. As a result, the process proceeds exponentially. PCR is made possible by the use of DNA polymerase taken from a thermophilic bacterium such as *Thermus aquaticus*; the enzyme from such organisms can survive the heating phase without being destroyed. Thirty cycles, completed in just a few hours, will increase the amount of target DNA by more than a billion times.

PCR can be applied to any situation that requires the amplification of DNA. Especially noteworthy are diagnostic tests that use PCR to detect the presence of infectious agents, such as the AIDS virus, in situations in which they would otherwise be undetectable. In one dramatic case, PCR was used to amplify DNA of the AIDS virus in tissue samples from a British sailor who died in 1959 of a then-mysterious cause. Thus, PCR was used over 30 years after his death to establish that this sailor had the earliest documented case of AIDS. Because of the usefulness of the PCR technique, machines that perform it automatically are becoming common fixtures in research laboratories.

DNA probes such as the ones used to screen gene libraries are promising tools for rapid identification of microorganisms. For use in medical diagnosis, these probes are derived from the DNA of a pathogenic microbe and are labeled (with a radioactive tag, for example). The probe then serves a diagnostic function by combining with the DNA of the pathogen to reveal its location in body tissue (or perhaps its presence in food). Probes are also being used in nonmedical aspects of microbiology—for example, for locating and identifying specific microbes in soil. We will discuss probes further in Chapter 10.

Agricultural Applications of Recombinant-DNA Technology

The process of selecting for genetically desirable plants has always been a time-consuming one. Performing conventional plant crosses is laborious and involves waiting for the planted seed to germinate and for the plant to mature. Plant breeding has been revolutionized by the use of plant cells grown in culture. Clones of plant cells, including cells that have been genetically altered by recombinant-DNA techniques, can be grown in large numbers. These cells can then be induced to regenerate whole plants, from which seeds can be harvested.

Recombinant DNA can be introduced into plant cells in several ways. Previously we mentioned protoplast fusion and the use of DNA-coated "bullets." The most elegant method, however, makes use of a plasmid called the **Ti plasmid** (*Ti* stands for *tumor inducing*), which occurs naturally in the bacterium *Agrobacterium tumefaciens*. This bacterium infects certain plants, in which the Ti plasmid causes the formation of a tumorlike growth called a crown gall (Figure 9.14). A part of the Ti plasmid, called T-DNA, integrates into the genome of the infected plant. The T-DNA stimulates local cellular growth (the crown gall),

FIGURE 9.14   **Crown gall disease in a pecan tree.** The tumorlike growth is stimulated by a gene on the Ti plasmid carried by the bacterium *Agrobacterium tumefaciens,* which has infected the tree.

and at the same time it causes production of certain products that are used by the bacteria as a source of nutritional carbon and nitrogen.

For plant scientists, the attraction of the Ti plasmid is that it provides a vehicle for the introduction of genetically engineered DNA into a plant (Figure 9.15). A scientist can insert foreign genes into the T-DNA, put the recombinant plasmid back into the *Agrobacterium* cell, and use the bacterium to insert the recombinant Ti plasmid into a plant cell. The plant cell with the foreign gene can then be used to generate a new plant. With luck, the new plant will express the foreign gene. Unfortunately, *Agrobacterium* does not infect grasses, so it cannot be used to improve grains such as wheat, rice, or corn.

A noteworthy accomplishment of this approach has been the introduction into plants of resistance to the herbicide glyphosate. Normally, this herbicide kills both weeds and useful plants by inhibiting an enzyme necessary for making certain essential amino acids. *Salmonella* bacteria happen to have this enzyme, and some salmonellae have a mutant enzyme that is resistant to the herbicide. When the DNA for this enzyme is introduced into a crop plant, the crop becomes resistant to the herbicide, which then kills only the weeds. There are now a variety of plants in which different herbicide and pesticide resistances have been engineered.

Perhaps the most exciting potential use of genetic engineering in plants concerns nitrogen fixation, the ability to convert the nitrogen gas in the air to compounds that living cells can use (see Chapter 27). The availability of

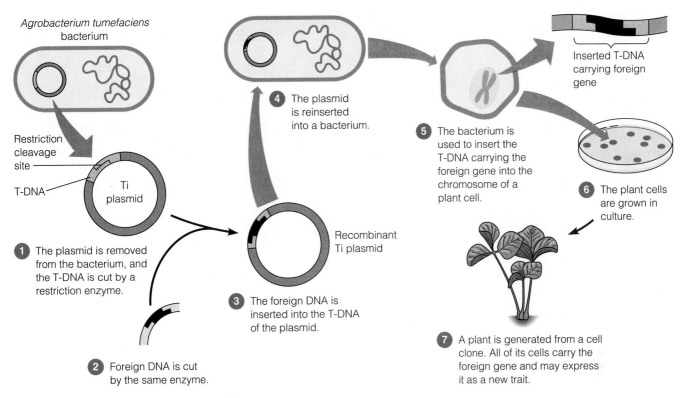

Agrobacterium tumefaciens bacterium

Restriction cleavage site

T-DNA

Ti plasmid

1 The plasmid is removed from the bacterium, and the T-DNA is cut by a restriction enzyme.

2 Foreign DNA is cut by the same enzyme.

3 The foreign DNA is inserted into the T-DNA of the plasmid.

Recombinant Ti plasmid

4 The plasmid is reinserted into a bacterium.

5 The bacterium is used to insert the T-DNA carrying the foreign gene into the chromosome of a plant cell.

Inserted T-DNA carrying foreign gene

6 The plant cells are grown in culture.

7 A plant is generated from a cell clone. All of its cells carry the foreign gene and may express it as a new trait.

FIGURE 9.15 Using the Ti plasmid as a vector for genetic engineering in plants.

such nitrogen-containing nutrients is usually the main factor limiting crop growth. But in nature, only certain bacteria have genes for carrying out this process. In addition, some plants, such as alfalfa, benefit from a symbiotic relationship with these microbes. Some species of the symbiotic bacterium *Rhizobium* have already been genetically engineered for enhanced nitrogen fixation. In the future, *Rhizobium* strains may be designed that can colonize such crop plants as corn and wheat, perhaps eliminating their requirement for nitrogen fertilizer. The ultimate goal would be to introduce functioning nitrogen-fixation genes directly into the plants. Although this goal cannot be achieved with our current knowledge, work toward it will continue because of the potential for dramatically increasing the world's food supply.

An example of a genetically engineered bacterium now in agricultural use is *Pseudomonas fluorescens* that has been engineered to produce a toxin normally produced by *Bacillus thuringiensis.* This toxin kills certain plant pathogens, such as the European corn borer. The genetically altered *Pseudomonas,* which produces much more toxin than *B. thuringiensis,* can be added to plant seeds and in time will enter the vascular system of the growing plant. Its toxin is ingested by the feeding borer larvae and kills them (but it is harmless to humans and other warm-blooded animals).

Animal husbandry has also benefited from genetic engineering. We have seen how one of the early commercial products of genetic engineering was human growth hormone. By similar methods, it is possible to manufacture bovine growth hormone (BGH). When BGH is injected into beef cattle, it increases their weight gain; in dairy cows it also causes a 10% increase in milk production. Such procedures have met with resistance from consumers, especially in Europe, primarily as a result of as yet unsubstantiated fears that some of the bovine growth hormone would be present in the milk or meat of these cattle and might be harmful to humans.

Table 9.2 lists these and several other genetically engineered products used in agriculture and animal husbandry.

SAFETY ISSUES AND THE ETHICS OF GENETIC ENGINEERING

LEARNING OBJECTIVE

- *List the advantages of and problems associated with the use of genetic engineering techniques.*

There will always be concern about the safety of any new technology, and genetic engineering and biotechnology are certainly no exceptions. One reason for this concern is that it is nearly impossible to prove that something is entirely safe under all conceivable conditions. People worry that the same techniques that can alter a microbe or plant to make them useful to humans could also inadvertently make them pathogenic to humans or otherwise dangerous to living organisms, or could create an eco-

TABLE 9.2   Some Agriculturally Important Products of Genetic Engineering

PRODUCT	COMMENTS
Agricultural Products	
Pseudomonas syringae, ice-minus bacterium	Lacks normal protein product that initiates undesirable ice formation on plants
Pseudomonas fluorescens bacterium	Has toxin-producing gene from insect pathogen *Bacillus thuringiensis;* toxin kills root-eating insects that ingest bacteria
Rhizobium meliloti bacterium	Modified for enhanced nitrogen fixation
Animal Husbandry Products	
Porcine growth hormone (PGH)	Improves weight gain in swine; produced by *E. coli*
Bovine growth hormone (BGH)	Improves weight gain and milk production in cattle; produced by *E. coli*
Other Food Production Products	
Rennin	Causes formation of milk curds for dairy products; produced by *Aspergillus niger*
Cellulase	Enzymes that degrade cellulose to make animal feedstocks; produced by *E. coli*

logical nightmare. Therefore, laboratories engaged in recombinant-DNA research must meet rigorous standards of control to avoid either accidentally releasing genetically engineered organisms into the environment or exposing humans to any risk of infection. To reduce risk further, microbiologists engaged in genetic engineering often delete from the microbes' genomes certain genes that are essential for growth in environments outside the laboratory. Finally, recombinant-DNA–carrying microbes intended for use in the environment (in agriculture, for example) may be engineered to contain "suicide genes"—genes that eventually turn on to produce a toxin that kills the microbes, thus ensuring that the microbes do not survive in the environment for very long after they have accomplished their task.

A variety of moral and ethical issues are also raised by these developing technologies. Is it ethical to tamper with the genes of an individual? Thus far, gene therapy is considered appropriate only for patients with diseases for which there is no other available treatment. However, if initial tests are successful, there likely will be pressure to develop more relaxed guidelines for such procedures.

If genetic screening for diseases becomes routine, who should have access to this information? Should employers or insurance companies have the right to know the results of such tests? Restricting access to such information will be very difficult, which raises questions concerning the right to privacy. How can we be assured that such information will not be used to discriminate against certain groups?

Applications of genetic screening techniques are not limited to adults. The ability to diagnose a genetic disease in a fetus adds even more controversy to the abortion debate. Genetic counseling, which provides advice and counseling to prospective parents with family histories of genetic disease, is becoming more important in considerations about whether or not to have children. As more is learned about genetic causes for various diseases, such as cancer or Huntington's disease, reproductive choices might become more difficult for some families.

What extra burdens does genetic engineering place upon our already overtaxed health care system? Genetic screening and gene therapy are expensive procedures, and we need to consider how they will be delivered to the public as the technology develops. Will a sufficient number of genetic counselors be available to work with people? Will expensive medical cures and treatments be available only to those who can afford them?

There are probably just as many harmful applications of a new technology as there are helpful ones. It is particularly easy to imagine genetic engineering being used to develop new and powerful biological weapons. In addition, because such research efforts are performed under top-secret conditions, it is virtually impossible for the general public to learn of them.

Perhaps more than most new technologies, genetic engineering holds the promise of affecting human life in previously unimaginable ways. It is important that society and individuals be given every opportunity to understand the potential impacts of these new developments.

THE FUTURE OF GENETIC ENGINEERING

LEARNING OBJECTIVE

- *Discuss some possible results of sequencing the human genome.*

Like the invention of the microscope, the development of recombinant-DNA techniques is causing profound changes in science, agriculture, and human health care. With this technology only a bit more than 20 years old, it is difficult to predict exactly what changes will occur. However, it is likely that within another 20 years, many of the treatments and diagnostic methods discussed in this book will have been replaced by far more powerful techniques based on the unprecedented ability to manipulate DNA precisely. One monumental project involving much of the new technology is the previously mentioned *Human Genome Project.* The goal of this project is to map the 100,000 or so genes in human DNA and to sequence the entire genome, approximately 3 billion nucleotide pairs.

The technical aspects of the project have been compared to reconstructing the contents of several sets of shredded encyclopedias. Reaching the mapping goal and interpreting the data will probably require cooperation among thousands of workers all over the world over several decades. Even before it is completed, however, the project is of immense value to our understanding of biology. It will also eventually be of great medical benefit, especially for the diagnosis of, and possibly for the repair of, genetic diseases. James Watson, codiscoverer of the structure of DNA, has said, "A more important set of instruction books will never be found by human beings. When finally interpreted, the genetic messages encoded within our DNA molecules will provide the ultimate answers to the chemical underpinnings of human existence."

STUDY OUTLINE

THE ADVENT OF RECOMBINANT-DNA TECHNOLOGY (pp. 227–228)

1. Closely related organisms can exchange genes in natural recombination.
2. Genes can be transferred between unrelated species via laboratory manipulations called genetic engineering.
3. Recombinant DNA refers to DNA that has been artificially manipulated to combine genes from two different sources.
4. Biotechnology includes all industrial applications of microorganisms as well as industrial uses of genetically engineered cells.
5. A DNA molecule used to carry a desired gene from one organism to another is called a vector.

OVERVIEW OF RECOMBINANT-DNA PROCEDURES (p. 228)

1. A desired gene is inserted into a DNA vector such as a plasmid or viral genome.
2. The vector inserts the DNA into a new cell, which is grown to form a clone.
3. Large quantities of the gene or the gene product can be harvested from the clone.

BIOTECHNOLOGY TOOLS AND TECHNIQUES (pp. 228–231)

1. Prepackaged kits are available for many genetic engineering techniques.

Restriction Enzymes (p. 230)
1. A restriction enzyme recognizes and cuts only one particular nucleotide sequence in DNA.

2. Some restriction enzymes produce sticky ends—short stretches of single-stranded DNA at the ends of the DNA fragments.
3. Fragments of DNA produced by the same restriction enzyme will spontaneously join by hydrogen-bonding. DNA ligase can covalently link the DNA backbones.

Vectors (p. 231)
1. A plasmid containing a new gene can be inserted into a cell by transformation.
2. Shuttle vectors are plasmids that can exist in several different species.
3. A viral vector containing a new gene can be inserted into a cell by transduction.

Methods for Inserting Foreign DNA into Cells (p. 231)
1. Cells can take up naked DNA by transformation. Chemical treatments are used to make cells that do not naturally transform competent to take up DNA.
2. Pores made in protoplasts and animal cells by electric current in the process of electroporation can provide entrance for new pieces of DNA.
3. Protoplast fusion is the joining of cells whose cell walls have been removed.
4. Foreign DNA can be introduced into plant cells by shooting DNA-coated particles into the cells.
5. Foreign DNA can be injected into animal cells by using a fine glass micropipette.

SOURCES OF DNA (pp. 231–235)

1. Gene libraries can be made by cutting up an entire genome with restriction enzymes and inserting the fragments into bacterial plasmids or phages.

2. cDNA made from mRNA by reverse transcription can be cloned in gene libraries.
3. Synthetic DNA can be made in vitro by a DNA synthesis machine.

SELECTING A CLONE (pp. 235–236)

1. Antibiotic-resistance markers on plasmid vectors are used to identify cells containing the engineered vector by direct selection.
2. In hybridization, cells carrying the desired gene are grown on a plate of nutrient medium. After replica plating, one plate is treated to lyse the cells and separate the DNA into single strands.
3. A short piece of radioactively labeled DNA called a probe is added to the plate.
4. Bacterial colonies containing DNA that base-pairs with the probe are identified by the radioactive label.
5. A radioactively labeled antibody against a protein product can be used to identify colonies carrying the gene encoding that protein.
6. The desired colony is selected from the replica plate.

MAKING A GENE PRODUCT (pp. 236–237)

1. *E. coli* is used to produce proteins by genetic engineering because it is easily grown and its genetics are well understood.
2. Efforts must be made to ensure that *E. coli*'s endotoxin does not contaminate a product intended for human use.
3. To recover the product, *E. coli* must be lysed or the gene must be linked to a gene that produces a naturally secreted protein.
4. Yeasts can be genetically engineered and are likely to continuously secrete the gene product.
5. Animal viruses can be engineered to carry a gene for a pathogen's surface protein. When the virus is used as a vaccine, the host develops an immunity to the pathogen.
6. Mammalian cells can be engineered to produce proteins such as hormones for medical use.
7. Plant cells can be engineered and used to produce plants with new properties.

APPLICATIONS OF GENETIC ENGINEERING (pp. 237–243)

1. Cloned DNA is used to produce products, study the cloned DNA, and alter the phenotype of an organism.

Genetically Engineered Products for Medical Therapy (p. 237)

1. Synthetic genes linked to the β-galactosidase gene in a plasmid vector were inserted into *E. coli*, allowing *E. coli* to produce and secrete the two polypeptides used to make human insulin.
2. Cells can be engineered to produce a pathogen's surface protein, which can be used as a subunit vaccine.

Obtaining Information from DNA for Basic Research and Medical Applications (pp. 238–242)

1. Recombinant-DNA techniques can be used to increase understanding of DNA, for genetic fingerprinting, and for gene therapy.
2. DNA sequencing machines are used to determine the nucleotide base sequence in a gene.
3. Southern blotting can be used to locate a gene in a cell.
4. Genetic screening uses Southern blotting to look for mutations responsible for inherited disease in humans.
5. Gene therapy can be used to cure genetic diseases by replacing the defective or missing gene.
6. Southern blotting is used in DNA fingerprinting to compare DNA recovered from a crime scene with that of a suspect.
7. The polymerase chain reaction (PCR) is used to make multiple copies of a desired piece of DNA enzymatically.
8. PCR can be used to increase the amounts of DNA in samples to detectable levels. This may allow sequencing of genes, diagnosis of genetic diseases, or detection of viruses.
9. DNA probes can be used to quickly identify a pathogen in body tissue or food.

Agricultural Applications of Recombinant-DNA Technology (pp. 242–243)

1. Cells from plants with desirable characteristics can be cloned to produce many identical cells. These cells can then be used to produce whole plants from which seeds can be harvested.
2. Plant cells can be engineered by using the Ti plasmid vector.
3. The tumor-producing T genes are replaced with desired genes, and the recombinant DNA is inserted into *Agrobacterium*. The bacterium naturally transforms its plant hosts.
4. *Rhizobium* has been engineered for enhanced nitrogen fixation.
5. *Pseudomonas* has been engineered to produce *Bacillus thuringiensis* toxin against insects.
6. Bovine growth hormone is being produced by *E. coli*.

SAFETY ISSUES AND THE ETHICS OF GENETIC ENGINEERING (pp. 243–244)

1. Strict safety standards are used to avoid accidental release of genetically engineered microorganisms.
2. Some microbes used in genetic engineering have been altered so that they cannot survive outside of the laboratory.
3. Microorganisms that are intended for use in the environment may be engineered to contain suicide genes so that the organisms do not persist in the environment.
4. Genetic technology raises ethical questions such as (a) should employers and insurance companies have access to a person's genetic records, (b) will some people be targeted for either breeding or sterilization, and (c) will genetic counseling be available to everyone?

THE FUTURE OF GENETIC ENGINEERING

(pp. 244–245)

1. Genetic engineering techniques may provide new treatments for disease and new diagnostic tools.

2. Genetic engineering techniques are being used to map the human genome through the Human Genome Project.

3. This will provide tools for diagnosis and possibly repair of genetic diseases.

STUDY QUESTIONS

Review

1. Differentiate between recombinant DNA and genetic engineering.

2. Compare and contrast the following terms:
   (a) cDNA and gene
   (b) Restriction fragment and gene
   (c) DNA probe and gene
   (d) DNA polymerase and DNA ligase
   (e) Recombinant DNA and cDNA

3. How is each of the following used in genetic engineering?
   (a) Plasmid
   (b) Viral genome
   (c) Antibiotic-resistance genes
   (d) Restriction enzyme

4. How is each of the following used to select a clone?
   (a) Radioactively labeled DNA probe
   (b) Radioactively labeled antibody

5. Differentiate between a gene library and synthetic DNA.

6. Differentiate among the following terms. Which one is "hit and miss"—that is, does not add a specific gene to a cell?
   (a) Protoplast fusion
   (b) Gene gun
   (c) Microinjection
   (d) Electroporation

7. Some commonly used restriction enzymes are listed in the following table. The cutting site is indicated by ↓. Indicate which enzymes produce sticky ends. Of what value are sticky ends in making recombinant DNA?

ENZYME	BACTERIAL SOURCE	RECOGNITION SEQUENCE
*Bam*H I	*Bacillus amyloliquefaciens*	G↓G A T C C C C T A G↑G
*Eco*R I	*Escherichia coli*	G↓A A T T C C T T A A↑G
Hae III	*Hemophilus aegyptius*	G G↓C C C C↑G G
Hind III	*Hemophilus influenzae*	A↓A G C T T T T C G A↑A

8. Suppose you want multiple copies of a gene you have synthesized. How would you obtain the necessary copies by cloning? By PCR?

9. Describe a genetic engineering experiment in two or three sentences. Use the following terms: intron, exon, DNA, mRNA, cDNA, RNA polymerase, reverse transcriptase.

10. List at least two examples of the use of genetic engineering in medicine and in agriculture.

11. You are attempting to insert a gene for salt-water tolerance into a plant by using the Ti plasmid. In addition to the desired gene, you add a gene for tetracycline resistance (tet^R) to the plasmid. What is the purpose of the tet^R gene?

Critical Thinking

1. Using the following map of plasmid pMICRO, give the number of restriction fragments that would result from digesting pMICRO with *Eco*R I, *Hin*d III, and both enzymes together. What is the smallest fragment that would contain the tetracycline-resistance gene?

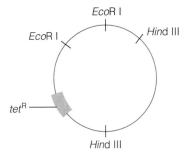

2. Design an experiment using vaccinia virus to make a vaccine against the AIDS virus (HIV).

3. The polymerase chain reaction involves heating (to 90°C for 10 seconds) to break the hydrogen bonds in DNA, cooling to allow a primer to hydrogen-bond to the single strands, and incubation to allow DNA polymerase to build complementary strands of DNA. This procedure is repeated 30 times an hour to get large quantities of the

desired DNA. Why did the use of DNA polymerase from the bacterium *Thermus aquaticus* allow researchers to add the necessary reagents to tubes in a preprogrammed heating block?

Clinical Applications

1. PCR has been used to examine oysters for the presence of *Vibrio cholerae*. Oysters from different areas were homogenized, and DNA was extracted from the homogenates. The DNA was digested by the restriction enzyme *Hinc* II. A primer for the hemolysin gene of *V. cholerae* was used for the PCR reaction. After PCR, each sample was electrophoresed and stained with a probe for the hemolysin gene. Which of the oyster samples were (was) positive for *V. cholerae*? How can you tell? Why look for *V. cholerae* in oysters? What is the advantage of PCR over conventional biochemical tests to identify the bacteria?

2. Suppose your biotechnology company is trying to make a subunit vaccine consisting of a surface protein from hepatitis C virus (HCV). Place the following steps for this experiment in the correct order. (Which one of the items listed is not appropriate?)

(a) Culture *E. coli* on nutrient agar containing tetracycline.

(b) Determine the chemical structure of the HCV surface protein.

(c) Determine the nucleotide sequence that codes for the HCV surface protein.

(d) Digest HCV genome with restriction enzyme *Eco*R I.

(e) Digest plasmid DNA with *Eco*R I.

(f) Digest plasmid DNA with *Pst* I.

(g) Separate HCV genome fragments by electrophoresis.

(h) Isolate a plasmid containing a tetracycline-resistance gene.

(i) Locate surface protein gene by using a radioactively labeled DNA probe.

(j) Mix plasmid and HCV surface protein gene with DNA ligase.

(k) Remove surface protein from gel.

(l) Synthesize short chain of DNA that is complementary to surface protein gene.

(m) Test colonies with radioactively labeled antibodies against HCV surface protein.

(n) Transform *E. coli* with recombinant plasmid.

3. Using the restriction enzyme *Eco*R I, the following gel electrophoresis patterns were obtained from digests of various DNA molecules from a transformation experiment. Can you conclude from these data that the transformation occurred? Explain why or why not.

After grapes have been crushed to make wine, what's left goes to recycling companies to be converted into fertilizer. Here, a California recycler takes the temperature of grape compost being reduced to fertilizer by bacteria and fungi. The action of these microbes creates the heat that makes the compost give off steam.

Classification of Microorganisms

THE SCIENCE OF CLASSIFICATION, especially the classification of living forms, is called **taxonomy**, from the Greek words for law and order. The objective of taxonomy is to classify living organisms—that is, to establish the relationships between one group of organisms and another and to differentiate between them. A taxonomic system enables us to classify organisms that have not yet been studied in detail; that is, a previously unknown organism can be identified and then grouped or classified with other organisms that have similar characteristics. For example, the causative agent of cat scratch disease has never been identified, but recent data suggest that the cause may be a previously unknown eubacterium. This bacterium is evidently most closely related to the rickettsia *Rochalimaea quintana* (rō-chä-lē′mā-ä kin′tän-ä).

Taxonomy also provides a common reference for identifying organisms already classified. For example, when a bacterium suspected of causing a specific disease is isolated from a patient, characteristics of that isolate are matched to lists of characteristics of previously classified bacteria to identify the isolate. After the bacterium has been identified, drugs can be selected that affect that bacterium.

Finally, taxonomy is a basic and necessary tool for scientists, as it provides a universal language of communication. Modern taxonomy is an exciting and dynamic field; new techniques in molecular biology and genetics are providing new insights into classification and evolution.

In this chapter you will learn the various classification systems, the different criteria used for classification, and tests that are used to identify bacteria that have already been classified.

THE STUDY OF PHYLOGENETIC RELATIONSHIPS

LEARNING OBJECTIVES

- *Define taxonomy.*
- *Discuss the limitations of a two-kingdom classification system.*
- *List the characteristics of the Kingdom Procaryotae that differentiate it from other kingdoms.*
- *List the major characteristics used to differentiate among kingdoms in the five-kingdom system and in the proposed three-kingdom system.*

In 1859, **Charles Darwin** proposed a mechanism (natural selection) for the theory of evolution, which explains that the many similarities among organisms are a result of their descent from common ancestors. The arrangement of organisms into taxonomic categories, collectively called *taxa* (singular: **taxon**), reflects degrees of relatedness among organisms; that is, the hierarchy of taxa shows evolutionary or **phylogenetic** (from a common ancestor) relationships.

TABLE 10.1 A Comparison of Procaryotic Cells and Eucaryotic Cells and Organelles

	PROCARYOTIC CELLS	EUCARYOTIC CELLS	EUCARYOTIC ORGANELLES (MITOCHONDRIA AND CHLOROPLASTS)
DNA	Circular	Linear	Circular
Histones	Histone-like proteins	Yes	No
Ribosomes	70S	80S	70S
Growth	Binary fission	Mitosis	Binary fission

Since Aristotle's time, living organisms had been categorized as either plants or animals. In 1857, **Carl von Nägeli**, a contemporary of Pasteur, proposed that bacteria be placed in the plant kingdom. In 1866, **Ernst Haeckel** proposed the Kingdom Protista to include both organisms lacking a nucleus (bacteria) and simple nucleated organisms. Because of disagreements over the definition of protists, for the next 100 years biologists continued to follow von Nägeli's placement of bacteria in the plant kingdom. However, when biologists began to use electron microscopes, the physical differences between these types of cells became apparent. The term *procaryote* was introduced in 1937 by **Edward Chatton** to distinguish cells with no nucleus from the nucleated cells of plants and animals. Nearly 30 years later, **Roger Stanier** provided the current definition of procaryotes: cells in which the nuclear material (nucleoplasm) is not surrounded by a nuclear membrane. In 1968, the Kingdom Procaryotae, proposed by **R.G.E. Murray**, was accepted by biologists.

We now know that the most fundamental taxonomic or phylogenetic division is that between procaryotes and eucaryotes. One widely held theory (discussed on p. 98) is that eucaryotic cells evolved from procaryotic cells living inside one another (as endosymbionts). In fact, the similarities between procaryotic cells and eucaryotic organelles (Table 10.1) provide striking evidence for this endosymbiotic relationship. The oldest known fossils are the remains of organisms that lived more than 3.5 billion years ago; they are procaryotes. Eucaryotic cells evolved more recently, about 1.4 billion years ago.

A Five-Kingdom System

In 1969, **Robert H. Whittaker** founded the **five-kingdom system** of biological classification, which considers procaryotes to be the ancestors to all eucaryotes (Figure 10.1).

In this system, all procaryotes are included in the Kingdom **Procaryotae** (or Monera). The four eucaryotic kingdoms are distinguished according to nutritional requirements, patterns of development, tissue differentiation, and possession of "9 + 2" flagella. (Recall from Chapter 4 that eucaryotic flagella and cilia are composed of nine outer pairs plus two inner microtubules, whereas procaryotic flagella are composed of repeating flagellin subunits.)

Simple eucaryotic organisms, mostly unicellular, are grouped as the Kingdom **Protista.** Protists generally have flagella at some time during their life cycle. The multicellular protists lack tissue organization. The Kingdom Protista includes funguslike water molds, slime molds, protozoa, and the primitive eucaryotic algae.

Fungi, plants, and animals make up the three kingdoms of more complex eucaryotic organisms, most of which are multicellular.

The Kingdom **Fungi** includes the unicellular yeasts, multicellular molds, and macroscopic species such as mushrooms. To obtain raw materials for vital functions, a fungus absorbs dissolved organic matter through its plasma membrane. The cells of a multicellular fungus are commonly joined to form thin tubes called hyphae. The hyphae are divided into multinucleated units by cross walls that have holes so that cytoplasm can flow between the cell-like units. Most fungi lack flagella. Fungi develop from spores or from fragments of hyphae.

The Kingdom **Plantae** (plants) includes some algae and all mosses, ferns, conifers, and flowering plants. All members of this kingdom are multicellular. To obtain energy, a plant uses photosynthesis; this process converts carbon dioxide and water into organic molecules used by the cell.

The kingdom of multicellular organisms called **Animalia** (animals) includes sponges, various worms, insects, and animals with backbones (vertebrates). Animals obtain nutrients and energy by ingesting organic matter through a mouth of some kind.

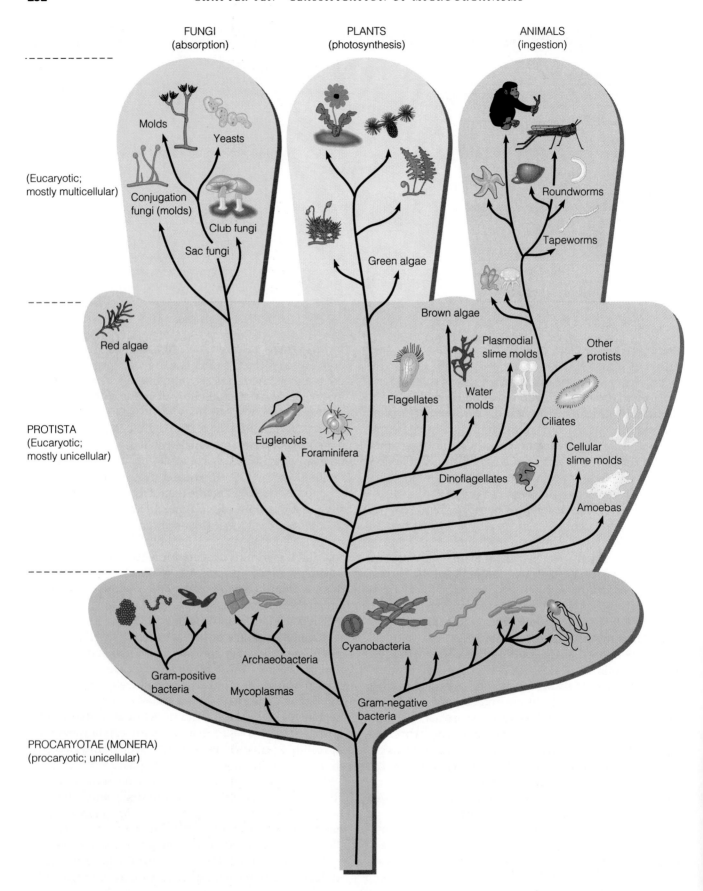

FIGURE 10.1 Five-kingdom system. This is a commonly accepted system of classification.

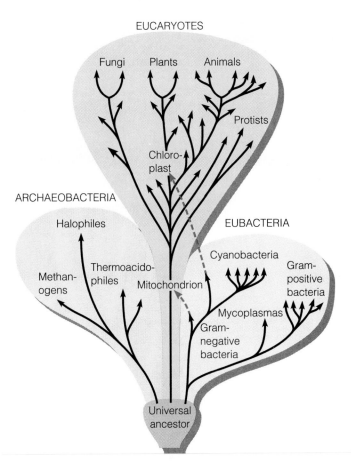

EUCARYOTES

Fungi   Plants   Animals

Protists

Chloro-
plast

ARCHAEOBACTERIA

Halophiles

EUBACTERIA

Cyanobacteria

Gram-
positive
bacteria

Methan-
ogens

Thermoacido-
philes   Mitochondrion

Mycoplasmas

Gram-
negative
bacteria

Universal
ancestor

FIGURE 10.2 Proposed three-kingdom system. This system of classification can be used as an alternative to the five-kingdom system or as three domains above those five kingdoms.

A Proposed Three-Kingdom System

In 1978, **Carl R. Woese** proposed a three-kingdom system of classification (Figure 10.2). The grouping of all bacteria into Procaryotae in the five-kingdom system had been based on microscopic observations; Woese's proposal was based on modern techniques in molecular biology and biochemistry that revealed that there are actually two types of procaryotic cells. This discovery was based on the observation that ribosomes are not the same in all cells (as you will recall from Chapter 4). In particular, comparing the sequences of nucleotides in ribosomal RNA (rRNA) from different kinds of cells shows that there are three distinctly different cell groups: the eucaryotes and two different types of procaryotes—the **eubacteria,** or "true bacteria," and the **archaeobacteria** (from *archaios,* meaning ancient).*

Woese believed, therefore, that the archaeobacteria and the eubacteria (the Procaryotae in the five-kingdom system), although similar in appearance, should form their own separate branches on the evolutionary tree (see Figure 10.2). All eucaryotes (divided into Protista, Fungi,

Plantae, and Animalia in Whittaker's system), he believed, should be grouped into a third kingdom. Other workers have suggested raising Woese's three kingdoms to the status of *domains* or *superkingdoms.* In this scheme, animals, plants, fungi, and protists would then be kingdoms in the eucaryote domain. Differences in membrane lipid structure, transfer RNA molecules, and sensitivity to antibiotics also support classification of the three cell types into separate kingdoms (Table 10.2).

Archaeobacteria differ from eubacteria in a number of other ways. For example, their cell walls never contain peptidoglycan, they often live in extreme environments, and they carry out unusual metabolic processes. Archaeobacteria include three groups:

1. The methanogens, strict anaerobes that produce methane (CH_4) from carbon dioxide and hydrogen.

2. Extreme halophiles, which require high concentrations of salt for survival.

3. Thermoacidophiles, which normally grow in hot, acidic environments.

The evolutionary relationship of the three kingdoms is the subject of current research by Woese and others. Originally, archaeobacteria were thought to be the most primitive group, whereas eubacteria were assumed to be more closely related to eucaryotes. However, studies of rRNA indicate that a universal ancestor split into three lineages. That split led to the archaeobacteria, the eubacteria, and what eventually became the nucleoplasm of the eucaryotes. This original nucleoplasmic cell was procaryotic. However, infoldings in its plasma membrane may have surrounded the nuclear region to produce a true nucleus. That cell provided the original host in which endosymbiotic eubacteria developed into organelles. An example of a modern procaryote living in a eucaryotic cell is shown in Figure 10.3. The cyanobacterium-like cell and the protistan host require each other for survival.

Acceptance of the three-kingdom system depends, at least in part, on answering the question about the origin of eucaryotes. At present, the five-kingdom system is accepted by most biologists as the most useful taxonomic tool.

Taxonomists are searching for a natural classification system that depicts phylogenetic relationships. As you can see, taxonomy has developed into a tool for clarifying the evolution of organisms as well as their relationships to each other.

CLASSIFICATION OF ORGANISMS

LEARNING OBJECTIVES

- *Explain why scientific names are used.*
- *List the major taxa.*
- *Define bacterial species.*

---

*Debate over the proper spelling of the term *archaeobacteria* currently exists. *Archaeobacteria* is used in *Bergey's Manual of Determinative Bacteriology,* 9th edition; other references use *archaebacteria.*

TABLE 10.2 Some Characteristics of Archaeobacteria, Eubacteria, and Eucaryotes

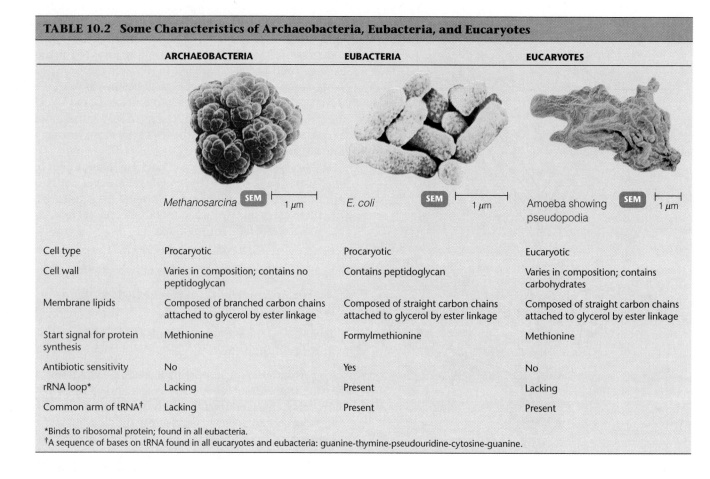

	ARCHAEOBACTERIA	EUBACTERIA	EUCARYOTES
	Methanosarcina SEM ⊢ 1 μm	*E. coli* SEM ⊢ 1 μm	Amoeba showing pseudopodia SEM ⊢ 1 μm
Cell type	Procaryotic	Procaryotic	Eucaryotic
Cell wall	Varies in composition; contains no peptidoglycan	Contains peptidoglycan	Varies in composition; contains carbohydrates
Membrane lipids	Composed of branched carbon chains attached to glycerol by ester linkage	Composed of straight carbon chains attached to glycerol by ester linkage	Composed of straight carbon chains attached to glycerol by ester linkage
Start signal for protein synthesis	Methionine	Formylmethionine	Methionine
Antibiotic sensitivity	No	Yes	No
rRNA loop*	Lacking	Present	Lacking
Common arm of tRNA†	Lacking	Present	Present

*Binds to ribosomal protein; found in all eubacteria.
†A sequence of bases on tRNA found in all eucaryotes and eubacteria: guanine-thymine-pseudouridine-cytosine-guanine.

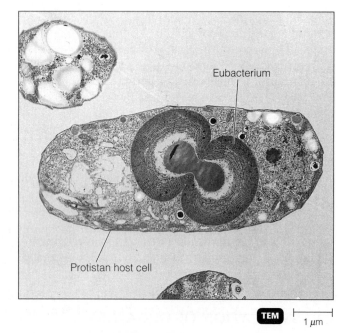

Eubacterium

Protistan host cell

TEM ⊢ 1 μm

FIGURE 10.3 *Cyanophora paradoxa.* This organism, in which the protistan host and the eubacterium require each other for survival, provides a modern example of how eucaryotic cells might have evolved.

Living organisms are grouped according to similar characteristics (classification), and each organism is assigned a unique scientific name. The rules for classifying and naming, used by biologists worldwide, are discussed next.

Scientific Nomenclature

In a world inhabited by millions of living organisms, biologists must be sure they know exactly which organism is being discussed. We cannot use common names because the same name is often used for many different organisms in different locales. For example, there are two different plants with the common name Spanish moss—and neither one is a moss. Three different animals are referred to as a gopher. Because common names are rarely specific and can often be misleading, a system of scientific names, referred to as scientific nomenclature, was developed in the eighteenth century by **Carolus Linnaeus.** Linnaeus established the use of Latin names because scholars wrote in Latin at that time.

This system gives every organism two names. These names are the **genus** name and **specific epithet (species)**, and both names are printed underlined or italicized. The genus name is always capitalized and is always a noun. The species name is written in lowercase and is usually an adjective. Because this system gives two names to each organism, the system is called **binomial nomenclature.**

Let us consider a few examples. Our own genus and specific epithet are *Homo sapiens* (hō′mō sā′pē-ens); the noun, or genus, means *man,* and the adjective, or specific epithet, means *wise.* A mold that contaminates bread is called *Rhizopus nigricans* (rī′zō-pŭs nī′gri-kans). *Rhizo-* (root) describes rootlike structures on the fungus; *nigr-* (black) identifies the color of its spore sacs. One bacterium that causes pneumonia is called *Klebsiella pneumoniae;* the genus honors Edwin Klebs, a colleague of Robert Koch, and the specific epithet describes a diseased condition. The genus and specific epithet are spelled out the first time they are used (*Klebsiella pneumoniae*), but thereafter the genus name may be abbreviated (*K. pneumoniae*).

Rules for assigning names to newly classified bacteria and for assigning bacteria to taxa are established by the International Committee on Systematic Bacteriology and are published in the *Bacteriological Code.* Descriptions of bacteria and evidence for their classifications are published in the *International Journal of Systematic Bacteriology* before being incorporated into a reference called *Bergey's Manual.* According to the *Bacteriological Code,* scientific names are to be taken from Latin (a genus name can be taken from Greek) or latinized by the addition of the appropriate suffix. Suffixes for order and family are *-ales* and *-aceae,* respectively (Figure 10.4).

As new laboratory techniques make more detailed characterizations of bacteria possible, two *genera* (the plural of genus) may be reclassified as a single genus, or a genus may be divided into two or more genera. For example, the genera "Diplococcus" and *Streptococcus* were combined in 1974; the only diplococcal species is now called *Streptococcus pneumoniae.* (The *Bacteriological Code* states that the older genus name should be retained in such cases.) In 1984, DNA hybridization studies indicated that "Streptococcus faecalis" and "Streptococcus faecium" were only distantly related to the other streptococcal species; consequently, a new genus called *Enterococcus* was created, and these species were renamed *E. faecalis* and *E. faecium* because the rules require that the original specific epithets be retained.

The Taxonomic Hierarchy

All organisms can be grouped into a series of subdivisions that make up the taxonomic hierarchy. For eucaryotic organisms, a species is a group of closely related organisms that breed among themselves. (Bacterial species will be discussed shortly.) A genus consists of species that differ from each other in certain ways but are related by descent. For example, *Quercus* (kwer′kus), the genus name for oak, consists of all types of oak trees (white oak, red oak, bur oak, velvet oak, and so on). Even though each species of oak differs from every other species, they are all related genetically. Just as a number of species make up a genus, related genera make up a **family.** A group of similar families constitutes an **order,** and a group of similar orders makes up a **class.** Related classes, in turn, make up a **division.** (In zoology, the comparable term **phylum** is used.) Thus, a particular organism (or species) has a genus name and specific epithet and belongs to a family, order, class, and phylum or division. All phyla or divisions that are related to each other make up a **kingdom** (see Figure 10.4).

A Phylogenetic Hierarchy

In a phylogenetic hierarchy, grouping organisms according to common properties implies that a group of organisms evolved from a common ancestor; each species retains some of the characteristics of the ancestor. Some of the information used to classify and to determine evolutionary relationships in higher organisms comes from fossils. Bones, shells, or stems that contain mineral matter or have left imprints in rock that was once mud are examples of fossils.

The structures of most microorganisms are not readily fossilized. Some notable exceptions are the following:

1. A marine protist whose fossilized colonies form the White Cliffs of Dover, England.

2. Stromatolites, the fossilized remains of microbial communities that flourished between 0.5 and 2 billion years ago.

3. The fossilized cyanobacteria found in 3- to 3.5-billion-year-old rocks in Western Australia. These are the oldest known fossils. Some fossils of procaryotes are shown in Figure 10.5.

Because fossil evidence is not available for the majority of procaryotes, their phylogeny must be based on other kinds of evidence. Conclusions from DNA hybridization studies (discussed shortly) of selected orders and families of eucaryotes are in agreement with the fossil records. This has encouraged workers to use DNA and RNA hybridization and RNA sequencing to gain an understanding of the evolutionary relationships among procaryotic groups.

Classification of Bacteria

The taxonomic classification scheme for bacteria may be found in *Bergey's Manual of Systematic Bacteriology.* In *Bergey's Manual,* bacteria are divided into four divisions: three divisions consist of eubacterial cells, and the fourth division consists of the archaeobacteria. Each division is divided into classes; in all, there are seven classes of bacteria. (Descriptions of the divisions and classes are listed in Table 10.3.) Classes are divided into orders; orders, into families; families, into genera; and genera, into species.

A bacterial species is defined somewhat differently than a **eucaryotic species,** which is a group of closely related

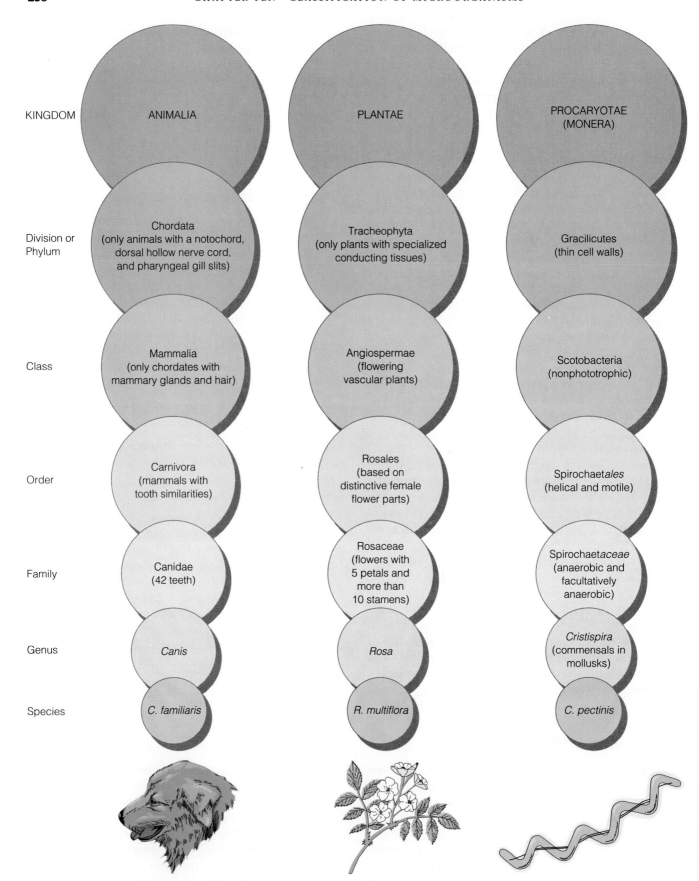

FIGURE 10.4   The taxonomic hierarchy.

(a) Coccoid cyanobacteria `TEM` ⊢————⊣ 10 μm

(b) Filamentous procaryotes `LM` ⊢————⊣ 10 μm

FIGURE 10.5 Fossilized procaryotes. (a) Coccoid cyanobacteria from the Late Precambrian (ca. 850 million years ago) of central Australia. **(b)** Filamentous procaryotes from the Early Precambrian (ca. 3.5 billion years ago) of Western Australia.

organisms that can interbreed. Unlike reproduction in eucaryotic organisms, cell division in bacteria is not directly tied to sexual conjugation, which is infrequent and does not always need to be species-specific. A **bacterial species,** therefore, is defined simply as a population of cells with similar characteristics. (The types of characteristics will be discussed later in this chapter.) The members of a bacterial species are essentially indistinguishable from each other but are distinguishable from members of other species, usually on the basis of several features. In some cases, pure cultures of the same species are not identical in all ways. Each such group is called a **strain,** which is a group of cells derived from a single cell. Strains are identified by numbers, letters, or names that follow the specific epithet.

Bergey's Manual provides a reference for identifying bacteria in the laboratory, as well as a classification scheme for bacteria. Suggested evolutionary relationships are given for most groups. Information used to build (and modify) the phylogenetic models comes from analyses of nucleotide sequences in DNA and RNA, from DNA hybridizations, and from chemical analyses of cellular components. (All these techniques will be described later in this chapter.) One scheme for the evolutionary relationships of bacteria is shown in Figure 10.6.

Not all bacteria have been surveyed, and those needing further investigation are temporarily placed in descriptive categories or sections, such as "Endosymbionts" and "Dissimilatory Sulfate- or Sulfur-Reducing Bacteria."

Classification of Viruses

Viruses are not classified as part of any of the five kingdoms discussed in this chapter. Viruses are not composed of cells, and they use the anabolic machinery within liv-

TABLE 10.3 Descriptions of the Divisions and Classes in the Kingdom Procaryotae (Monera)

DIVISIONS	CLASSES
Thin, gram-negative cell walls	Nonphotosynthetic bacteria Anaerobic photosynthetic bacteria Cyanobacteria
Thick, gram-positive cell walls	Rods and cocci Actinomycetes and related organisms
Wall-less eubacteria	Mycoplasmas
Unusual cell walls	Archaeobacteria

ing host cells to multiply. A viral genome (genetic material) can direct biosynthesis inside a host cell, and some viral genomes can become incorporated into the host genome. The niche of a virus is its specific host cell, so viruses may be more closely related to their hosts than to other viruses. In 1991, the International Committee on Taxonomy of Viruses defined a **viral species** as a population of viruses with similar characteristics (including morphology, genes, and enzymes) that occupies a particular ecological niche.

Viruses are obligatory intracellular parasites, so they must have evolved after a suitable host cell had evolved. There are two hypotheses on the origin of viruses: (1) they arose from independently replicating strands of nucleic acids (such as plasmids), and (2) they developed from degenerative cells that, through many generations, gradually lost the ability to survive independently but could survive when associated with another cell. Viruses will be discussed in Chapter 13.

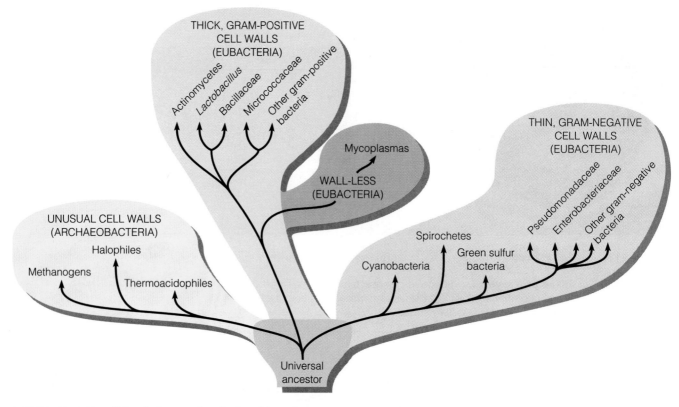

FIGURE 10.6 Possible phylogenetic relationships of procaryotes. Arrows indicate possible major lines of descent of bacterial groups. The DNA base composition of the mycoplasmas is more similar to that of the gram-positive bacteria than to that of the gram-negative division.

CRITERIA FOR CLASSIFICATION AND IDENTIFICATION OF MICROORGANISMS

LEARNING OBJECTIVES

- *Compare and contrast classification and identification.*
- *Explain the purpose of* Bergey's Manual.
- *Describe how staining and biochemical tests are used to identify bacteria.*
- *Explain how serological tests and phage typing can be used to identify an unknown bacterium.*
- *Describe how a new bacterium can be classified by the following molecular methods: amino acid sequencing, protein profiles, base composition, DNA fingerprinting, PCR, and hybridization.*
- *Differentiate between Western blotting and Southern blotting.*
- *Provide the rationale for the use of numerical taxonomy.*

A classification scheme provides a list of characteristics and a means for comparison to aid in the identification of an organism. Once an organism is identified, it can be placed into a previously devised classification scheme. Microorganisms are *identified* for practical purposes—for example, to determine an appropriate treatment for an infection. They are not necessarily identified by the same

techniques by which they are classified. Most identification procedures are easily performed in a laboratory and use as few procedures or tests as possible.

Bergey's Manual of Determinative Bacteriology has been a widely used reference since the first edition was published in 1923. *Bergey's Manual of Determinative Bacteriology* (9th ed., 1994) does not classify bacteria according to evolutionary relatedness but provides identification (determinative) schemes based on such criteria as cell wall composition, morphology, differential staining, oxygen requirements, and biochemical testing.

In 1984, the first edition of *Bergey's Manual of Systematic Bacteriology* was published. In recent years, microbiologists have been acquiring information on relationships among bacteria and are constructing identification schemes that reflect evolutionary or *systematic* relationships used in *Bergey's Manual of Systematic Bacteriology.*

Bergey's Manual of Systematic Bacteriology is divided into four volumes. Volume 1 includes wall-less eubacteria and some gram-negative eubacteria. Volume 2 describes gram-positive eubacteria. Volume 3 includes some gram-negative eubacteria (photosynthetic bacteria; chemolithotrophic bacteria; budding, appendaged, and sheathed bacteria; gliding and fruiting bacteria); and the archaeobacteria. Volume 4 describes the actinomycetes.

Medical microbiology (the branch of microbiology dealing with human pathogens) has dominated interest

FIGURE 10.7 Use of metabolic characteristics to identify selected genera of enteric bacteria.

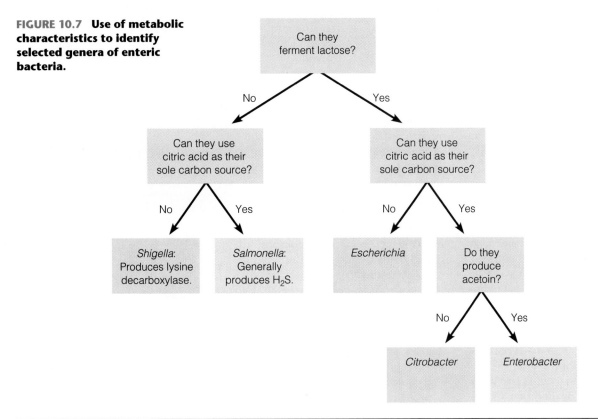

Morphological Characteristics

Higher organisms are frequently classified according to observed anatomical detail. However, many microorganisms look too similar to be classified by their structures. Through a microscope, organisms that might differ in metabolic or physiological properties may look alike. Literally hundreds of bacterial species are small rods or small cocci.

Cell morphology tells us little about phylogenetic relationships. However, morphological characteristics are still useful in identifying bacteria. For example, differences in such structures as endospores or flagella can be helpful.

Differential Staining

One of the first steps in the identification of bacteria is differential staining. As we have seen from previous chapters, most bacteria are either gram-positive or gram-negative. Other differential stains, such as the acid-fast stain, can be useful for a more limited group of microorganisms. Recall that these stains are based on the chemical composition of cell walls and therefore are not useful in identifying either the wall-less bacteria or the archaeobacteria with unusual walls.

Biochemical Tests

Enzymatic activities are widely used to differentiate among bacteria. Even closely related bacteria can usually be separated into distinct species by subjecting them to biochemical tests, such as their ability to ferment an assortment of selected carbohydrates. For one example of the use of biochemical tests to identify bacteria (in this instance, in marine mammals), see the box on page 260.

Enteric, gram-negative bacteria are a large heterogeneous group of microbes whose natural habitat is the intestinal tract of humans and animals. All members of the family Enterobacteriaceae are oxidase-negative. Among the enteric bacteria are members of the genera *Escherichia, Enterobacter* (en-te-rō-bak'tėr), *Shigella* (shi-gel'lä), *Citrobacter* (sit'rō-bak-tėr), and *Salmonella. Escherichia, Enterobacter,* and *Citrobacter,* which ferment lactose to produce acid and gas, can be distinguished from *Salmonella* and *Shigella,* which do not. Further biochemical testing, as represented in Figure 10.7, can differentiate among the genera.

The time needed to identify bacteria can be reduced considerably by the use of selective and differential media

MASS DEATHS OF DOLPHINS SPUR VETERINARY MICROBIOLOGY

DURING AN AVERAGE YEAR, ten or twelve dead dolphins are found washed ashore on beaches between New Jersey and Virginia. However, in one recent year the death toll climbed to more than 200, and scientists believe that hundreds more may have died offshore. Because this mortality represents a significant part of the Atlantic dolphin population, some experts are worried that this population may ultimately be destroyed. Moreover, alarmed swimmers have asked whether they are also in danger from whatever is killing the dolphins. The U.S. Office of Naval Research is continuing to fund marine mammal research at colleges and universities to find the cause of the dolphins' death.

Large numbers of opportunistic pathogens, including *Edwardsiella tarda* and 55 species of *Vibrio,* were found in the dead dolphins. These bacteria are a part of the normal microbiota of dolphins and of coastal waters and can cause disease only if the animals' immune systems, their normal defense against infection, have been weakened. To find the ultimate cause of the deaths, scientists must find out what has weakened the animals' immune systems.

One possibility is a chemical spill. High levels of polychlorinated biphenyls (PCBs), which are chemical pollutants, have been found in the 53 dolphins tested. Daniel Martineau of Cornell University says that PCBs are strong immunosuppressants. Another possibility is a viral infection that affects the immune system. The cause of thousands of seal deaths in northern Europe in April 1988 is now known to be a previously undiscovered virus. Could a new virus be infecting dolphins as well?

INFORMATION IS SCARCE

Such questions are the concern of veterinary microbiology, which until recently has been a neglected branch of medical microbiology. Although the diseases of such animals as cattle, chickens, and mink have been studied—partly because of their availability to researchers—the microbiology of wild animals, especially marine mammals, is a relatively newly emerging field. Gathering samples of animals that live in the open ocean and performing bacteriological analyses on them are very difficult. Currently, the animals being studied are those that live in captivity (see photo) and those that come onto the shore to breed, such as the northern fur sea lion.

The scientists are identifying bacteria in marine mammals by using conventional test batteries (see illustration) and rapid-identification test kits developed for human microbiology. The bacteria are compared with species described in *Bergey's Manual* to assign names or to identify them, although they are not always a perfect match. Perhaps new species of bacteria will be found in the marine mammals.

Marine mammal researchers examine a Pacific bottlenosed dolphin.

Veterinary microbiologists hope that increased study of the microbiology of wild animals, including marine mammals, not only will promote improved wildlife management but also will provide models for the study of human diseases.

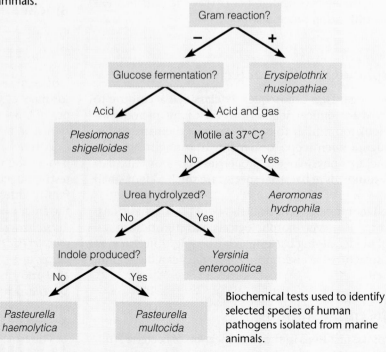

Gram reaction?

− +

Glucose fermentation? — *Erysipelothrix rhusiopathiae*

Acid — Acid and gas

Plesiomonas shigelloides — Motile at 37°C?

No — Yes

Urea hydrolyzed? — *Aeromonas hydrophila*

No — Yes

Indole produced? — *Yersinia enterocolitica*

No — Yes

Pasteurella haemolytica — *Pasteurella multocida*

Biochemical tests used to identify selected species of human pathogens isolated from marine animals.

FIGURE 10.8 One type of rapid identification method for bacteria: Enterotube® II from Roche Diagnostics. ① One tube containing media for 15 tests is inoculated. ② After incubation, the tube is observed for results. ③ The value for each positive test is circled, and the numbers from each group of tests are added to give each portion of the I.D. value. ④ Comparing the resultant I.D. value with a computerized listing shows that the organism in the tube is *Klebsiella pneumoniae*. This example shows results for a typical strain of *K. pneumoniae;* however, other strains may produce different test results, which are listed in the Atypical Test Results column.

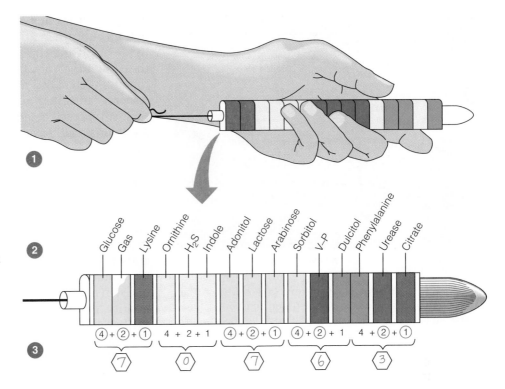

ID Value	Organism	Atypical Test Results
70742	*K. pneumoniae*	V–P⁻, Citrate⁻
70743	*K. pneumoniae*	V–P⁻
70762	*K. pneumoniae*	Citrate⁻
70763	*K. pneumoniae*	None

or by rapid identification methods. Recall from Chapter 6 (p. 152) that selective media contain ingredients that suppress the growth of competing organisms and encourage the growth of desired ones, and that differential media allow the desired organism to form a colony that is somehow distinctive.

Rapid identification tools are manufactured for groups of medically important bacteria, such as the enterics. Such tools are designed to perform several biochemical tests simultaneously and can identify bacteria within 4 to 24 hours. The results of each test are assigned a number (which varies with the tool being used) based on the relative reliability and importance of each test.

In the example shown in Figure 10.8, ① an unknown enteric bacterium is inoculated into a tube that is designed to perform 15 biochemical tests. ② After incubation, results in each compartment are observed. ③ Positive results are marked on the scoring form. Note that each test is assigned a value; the number derived from scoring all the tests is called the I.D. value. Fermentation of glucose is important, and a positive reaction is valued at 4, as compared

with the production of acetoin (V–P test), for which a positive reaction is assigned the value of 2. ④ A computerized interpretation of the simultaneous test results is essential and is provided by the manufacturer. In this example, the test results indicate that the bacterium is *Klebsiella pneumoniae*, with the typical V–P-positive and citrate-positive test results.

Bergey's Manual does not evaluate the relative importance of each biochemical test and does not always describe strains. In clinical diagnosis of a disease, a particular species and even a particular strain must be identified in order to proceed with proper treatment. To this end, specific series of biochemical tests have been developed for fast identification in hospital laboratories.

Serology

Serology is the science that deals with blood serum and particularly with immune responses evident in serum (see Chapter 18). Microorganisms are antigenic; that is,

(a) Positive test   **(b)** Negative test

FIGURE 10.9   Slide agglutination test. (a) Positive test. The grainy appearance is due to the clumping (agglutination) of the bacteria. Agglutination results when the bacteria are mixed with antibodies that were produced in response to the same strain. **(b)** Negative test. The bacteria are still evenly distributed in the saline and antiserum.

(a)

(b)

FIGURE 10.10   Use of a computer scanner to read the results of an ELISA test. (a) A technician uses a micropipette to add samples to a microplate for an ELISA. **(b)** ELISA results are then read by the computer scanner.

microorganisms that enter an animal's body stimulate it to form antibodies. Antibodies are proteins that circulate in the blood and combine in a highly specific way with the bacteria that caused their production. For example, the immune system of a rabbit injected with killed typhoid bacteria (antigens) responds by producing antibodies against typhoid bacteria. *Antisera* (singular: **antiserum**) are commercially available solutions of such antibodies used in the identification of many medically important microorganisms. If an unknown bacterium is isolated from a patient, it can be tested against known antisera and often identified quickly.

In a procedure called a **slide agglutination test**, samples of an unknown bacterium are placed in a drop of saline on each of several slides. Then a different known antiserum is added to each sample. The bacteria agglutinate (clump) when mixed with antibodies that were produced in response to that species or strain of bacterium; a positive test is indicated by the presence of agglutination. Positive and negative slide agglutination tests are shown in Figure 10.9.

Serological techniques can differentiate not only among microbial species but also among strains within species. Rebecca Lancefield (Chapter 1, p. 13) was able to classify serotypes of streptococci by studying serological reactions. She found that the different antigens in the cell walls of various serotypes of streptococci stimulate the formation of different antibodies. In contrast, because closely related bacteria also produce some of the same antigens, serological testing can be used to screen bacterial isolates for possible similarities. If an antiserum reacts with proteins from different bacterial species or strains, these bacteria can be tested further for relatedness. An application of serotyping is discussed in the box in Chapter 23.

Enzyme-linked immunosorbent assays (ELISAs) (Figure 18.15) are now widely used because they are fast and can be read by a computer scanner (Figure 10.10). In a direct ELISA, known antibodies are placed in the wells of a microplate, and an unknown type of bacteria is added to each well. A reaction between the known antibodies and the bacteria provides identification of the bacteria. An ELISA is used in AIDS testing to detect the presence of antibodies against Human Immunodeficiency Virus (HIV), the virus that causes AIDS (see Figure 1.1b).

Another serological test, the **Western blot**, is used to identify bacterial antigens in a patient's serum (Figure 10.11). For example, Lyme disease, caused by *Borrelia burgdorferi*, is often diagnosed by the Western blot. ① Proteins (including bacterial proteins) in the patient's serum are separated by a process called electrophoresis (described shortly). ② The proteins are then transferred to a filter by blotting. ③ Next, antibodies tagged with a dye are washed over the filter. If the specific antigen (in this case, *Borrelia* proteins) is present in the serum, the antibodies will combine with it and will be visible as a colored band on the filter. The Western blot is also used, like the ELISA, to diagnose HIV infection.

1 If Lyme disease is suspected in a patient, a sample of the patient's serum is taken. Electrophoresis is used to separate proteins in the serum sample. Each band consists of many molecules of a particular protein (antigen). The bands are not visible at this point.

Serum samples

B

Proteins from B

Polyacrylamide gel

A

Proteins from A

Larger

Smaller

2 The bands are transferred to a nitrocellulose filter by blotting.

Paper towels

Salt solution

Sponge

Gel

Nitrocellulose filter

3 The filter with proteins (antigens) is positioned exactly as it was on the gel. The filter is then washed with antibodies against a particular antigen. The antibodies are tagged with a dye so that they are visible when they combine with their specific antigen (shown here in purple).

B

A

4 The test is read. If the tagged antibodies stick to the filter, evidence of the presence of the microorganism in question—in this case, *Borrelia burgdorferi*—has been found in the patient's serum.

FIGURE 10.11 The Western blot. Proteins separated by electrophoresis can be detected by their reactions with antibodies.

Phage Typing

Phage typing, like serological testing, looks for similarities among bacteria. Both techniques are useful in tracing the origin and course of a disease outbreak. **Phage typing** is a test for determining which phages a bacterium is susceptible to. Recall from Chapter 8 that bacteriophages (phages) are bacterial viruses and that they usually cause lysis of the bacterial cells they infect. They are highly specialized, in that they usually infect only members of a particular species, or even particular strains within a species. One bacterial strain might be susceptible to two different phages, whereas another strain of the same species might be susceptible to those two phages plus a third phage. Bacteriophages will be discussed further in Chapter 13.

FIGURE 10.12 Phage typing of a strain of *Salmonella typhi*. The tested strain was grown over the entire plate. Plaques, or areas of lysis, were produced by bacteriophages, indicating that the strain was sensitive to infection by these phages. Bacteria with this pattern of phage sensitivity are designated as *Salmonella typhi* phage type A.

The sources of food-associated infections can be traced by phage typing (see the box in Chapter 25). One version of this procedure starts with a plate totally covered with bacteria growing on agar. A drop of each different phage type to be used in the test is then placed on the bacteria. Wherever the phages are able to infect and lyse the bacterial cells, clearings in the bacterial growth (called plaques) appear (Figure 10.12). Such a test might show, for instance, that bacteria isolated from a surgical wound have the same pattern of phage sensitivity as those isolated from the operating surgeon or surgical nurses. This establishes that the surgeon or a nurse is the source of infection.

Amino Acid Sequencing

A great deal of information can be derived from the study of the amino acid sequences of proteins. The sequence of amino acids in a protein directly reflects the base sequence of the DNA that encodes the protein. The longer the time between the evolution of two organisms, the more changes in the DNA sequence (mutations) encoding the same protein in the two organisms. Thus, the comparison of amino acid sequences from proteins of two different organisms can help determine the similarity in the DNA sequence and therefore the evolutionary relatedness of the species. The more similar the proteins, the more closely related are the organisms.

The sequencing of amino acids in several important

proteins, such as cytochrome *c*, has been useful in the study of phylogenetic relationships among different organisms (Figure 10.13). Amino acid sequencing is limited to use with groups of organisms that share a common protein. Also, it cannot be assumed that all organisms that share a protein are related. Although unlikely, it is not impossible for two unrelated organisms to evolve the same solution to a problem—for example, the production of a certain enzyme to transport a needed nutrient.

Protein Analysis

Amino acid sequencing focuses on the sequence of one or a few known proteins in different organisms. Protein analysis compares the pattern of all proteins found in the cells of different organisms. Because a cell's proteins are products of its genes, each species is expected to produce a unique set of proteins. This concept is the basis for making profiles of the proteins in a cell.

Protein profiles are obtained by **polyacrylamide gel electrophoresis (PAGE)**. In this process, proteins from lysed cells are dissolved in a detergent, and a sample of the resulting solution is placed on a thin slab of polyacrylamide gel. In a process called **electrophoresis**, an electric current is passed through the gel. While the charge is applied, different proteins migrate through the gel at different rates, which depend on the size and charge of each protein. After the gel is stained, it shows a characteristic pattern of the cells' proteins (Figure 10.14).

We assume that identical organisms have identical protein profiles, that the profiles of closely related organisms show only minor differences, and that the profiles of distantly related organisms show major differences. As procedures and materials are developed further, this method may provide a useful tool for identification as well as classification. By comparing the protein profile of an unknown cell with a set of standard profiles, a laboratory technician might be able to identify an organism within 24 hours of its isolation.

PAGE analysis is now being used to compare one group of proteins, the cytochromes, in different organisms. Although the presence of certain cytochromes is affected by the environment, analysis of bacterial cytochromes has disclosed some patterns that may be useful for classifying organisms. The following table shows the cytochromes (listed in the order in which they are arranged in the electron transport system) that are common to three metabolic categories of organisms. The letters correspond to the cytochromes that are present in these cells.

ORGANISMS	CYTOCHROMES
Gram-positive chemoheterotrophs	*b, c, a, a$_3$, o*
Gram-negative chemoheterotrophs	*b, c, d, o, a*
Gram-negative chemoautotrophs	*b, c, a, a$_3$, o, a*

FIGURE 10.13 Amino acid sequences for a fragment of cytochrome c from tuna and from two species of bacteria. Three-letter abbreviations indicate the amino acids. Amino acids of the same chemical type are shown in the same shade of blue (for example, Glu and Asp are both "acidic amino acids," having an extra —COOH group). The similarity among the three sequences is striking, although the two bacteria are clearly more closely related to each other than to tuna.

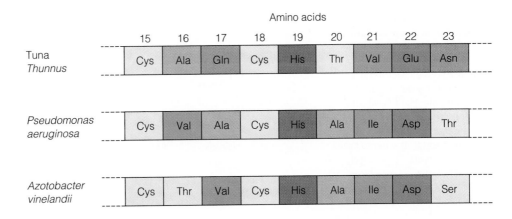

A limitation of the PAGE process is that different proteins may not be resolved from one another because they migrate to the same position. However, **Patrick O'Farrell** developed a more sensitive version called **two-dimensional PAGE**, shown in Figure 10.15a. This process differs from the PAGE process because the protein solution is exposed to electrophoresis twice, once to separate proteins by charge and again to separate them by size. ① Proteins from lysed cells are mixed with a special pH buffer, and a sample of the mixture is placed on a polyacrylamide gel. ② When exposed to the electric field, the buffer molecules arrange themselves in a pH gradient (basic to acidic), and each protein migrates through the gel until each reaches its **isoelectric point**—the pH at which the protein is no longer charged and therefore no longer migrates. This process is called **isoelectric focusing.** ③ The gel is then placed on a larger polyacrylamide slab gel and exposed to an electric field in a detergent buffer at a 90° angle to the electric field in the isoelectric focusing gel. This process separates the proteins at each pH by size. The gel is stained to make the proteins visible. Because two proteins are unlikely to have both the same isoelectric point and the same size, comparison of protein patterns from two bacteria shows differences between the two species (Figures 10.15b and 10.15c).

Base Composition of Nucleic Acids

A classification technique that has come into wide use among taxonomists because it has a good possibility of at least suggesting evolutionary relationships is the determination of the nitrogenous-base composition of DNA. This base composition is usually expressed as percentage of guanine plus cytosine (G + C). The base composition of a single species is theoretically a fixed property; thus, a comparison of the G + C content in different species can reveal the degree of species relatedness. As we saw in Chapter 8, each guanine (G) in DNA has a complementary cytosine (C). Similarly, each adenine (A) in the DNA has a complementary thymine (T). Therefore, the percentage of DNA bases that are G-C pairs also tells us the percentage that are A-T pairs (G-C + A-T = 100%). Two organisms that are closely related and hence have many identical or similar genes will have similar amounts of the various bases in their DNA. However, if there is a difference of more than 10% in their percentage of G-C pairs (for example, if one bacterium's DNA contains 40% G-C and another bacterium has 60% G-C), then these two organisms are probably not related. Of course, two organisms that have the same percentage of G-C are not necessarily closely related; other supporting data are needed to draw conclusions about evolutionary relationships.

Actually determining the entire *sequence* of bases in an

FIGURE 10.14 PAGE patterns. Twelve strains of *Bifidobacterium* were isolated from humans and other mammals. All 12 strains produced identical results from biochemical tests. The PAGE patterns, however, reveal that there are two species. Lanes 1–8 are a new species, *B. globosum*; lanes 9–12 are *B. pseudolongum*.

FIGURE 10.15 Two-dimensional polyacrylamide gel electrophoresis. (a) The production of a 2D PAGE. ① A sample of cell extract is placed at one end of a gel strip. ② During electrophoresis, the buffer establishes a pH gradient in the gel, and charged proteins migrate through the gel to their isoelectric points. ③ The gel strip is then placed on a larger gel slab and submitted to a second electrophoresis at right angles to the first. In this case, a detergent is used to eliminate the differences in charge, so the migration rates of the proteins depend only on their sizes. The results: **(b)** *Vibrio* and **(c)** *Enterobacter* protein patterns. Proteins shown in boxes are not found in both species. (Note that proteins are present in different amounts, as shown by the sizes of spots.)

(a) Production of a 2D PAGE

(b) 2D PAGE of *Vibrio* proteins

(c) 2D PAGE of *Enterobacter* proteins

FIGURE 10.16 DNA fingerprints. Plasmids from five different bacteria were digested with the same restriction enzyme. Each digest was put in a different well (origin) in the agarose gel. (Wells 1 and 7 are empty.) An electric current was then applied to the gel to separate the fragments by size and electric charge. The DNA was made visible by staining with ethidium bromide, which fluoresces under ultraviolet light. Comparison of the lanes shows that none of the DNA (and therefore none of the bacteria) are identical.

organism's DNA is now possible with modern biochemical methods, but this is currently impractical for all but the smallest microorganisms (viruses) because of the great amount of time required. However, the use of restriction enzymes allows the comparison of base sequences of different organisms. Restriction enzymes cut a molecule of DNA everywhere a specific base sequence occurs, producing restriction fragments (as discussed in Chapter 9). For example, the enzyme *Eco*R I cuts DNA at the arrows in the sequence

$$...\text{G}^{\downarrow}\text{A T T C}...$$
$$...\text{C T T A A}_{\uparrow}\text{G}...$$

In this technique, the DNA from two microorganisms is treated with the same restriction enzyme, and the restriction fragments produced are separated by electrophoresis on a thin layer of agar. A comparison of the number and sizes of restriction fragments that are produced from different organisms provides information about their genetic similarities and differences; the more similar the patterns (called **DNA fingerprints**), the more closely related the organisms are expected to be (Figure 10.16).

DNA fingerprinting is used to determine the source of hospital-acquired infections. In 1990, patients undergoing

FIGURE 10.17 DNA–DNA hybridization. The greater the amount of pairing between DNA strands from different organisms (hybridization), the more closely the organisms are related.

Organism *A* DNA

Organism *B* DNA

1 Heat to separate strands

2 Combine single strands of DNA

3 Cool to allow renaturation of double-stranded DNA

4 Determine degree of hybridization

Complete hybridization: Organisms identical

Partial hybridization: Organisms related

No hybridization: Organisms unrelated

coronary-bypass surgery at one hospital developed infections caused by *Rhodococcus bronchialis* (rō-dō-kok′kus bron-kē′al-is). The DNA fingerprints of the patients' bacteria and the bacteria of one nurse were identical. The hospital was thus able to break the chain of transmission of this infection by encouraging this nurse to use aseptic technique. See page 170 for another example.

In a situation in which a microbe cannot be cultured by conventional methods, the causative agent of an infectious disease might not be recognized. However, the **polymerase chain reaction (PCR)** can be used to increase the amount of microbial DNA to levels that can be tested by gel electrophoresis. (PCR is described in Chapter 9, pp. 241–242.) In 1992, this technique was used to identify as the causative agent of Whipple's Disease a previously unknown bacterium, now named *Tropheryma whippelii* (trō-fėr-ē-mä whip-pėl′ē-ē). Whipple's Disease was first described in 1907 by George Whipple as a gastrointestinal and nervous system disorder caused by an unknown bacillus. No one had been able to identify the bacterium, and thus there was no reliable method to diagnose the disease.

An indirect determination of the degree of similarity between the DNA nucleotide sequences of two organisms can be made with the method of nucleic acid hybridization, our next topic.

Nucleic Acid Hybridization

If a double-stranded molecule of DNA is subjected to heat, the complementary strands will separate as the hydrogen bonds between the bases break. If the single strands are then cooled slowly, they will reunite to form a double-stranded molecule identical to the original double strand. (This reunion occurs because the single strands have complementary sequences.) When this technique is applied to separated DNA strands from two different organisms, it is possible to determine the extent of similarity between the base sequences of the two organisms. This method is known as **nucleic acid hybridization**. The procedure assumes that if two species are similar or related, a major portion of their nucleic acid sequences will also be similar. The procedure measures the ability of DNA strands from one organism to hybridize (bind through complementary basepairing) with the DNA strands of another organism (Figure 10.17). The greater the degree of hybridization, the greater the degree of relatedness.

Recall from Chapter 8 that RNA is single-stranded and is transcribed from one strand of DNA; a particular strand of RNA, therefore, is complementary to the strand of DNA from which it was transcribed and will hybridize with that separated strand of DNA. DNA–RNA hybridization can thus be used to determine relatedness between DNA from one organism and RNA from another organism in the same way that DNA–DNA hybridization is used.

Nucleic acid hybridization can be used to identify unknown microorganisms by **Southern blotting** (see Figure 9.11). In addition, rapid identification methods using tools called **DNA probes** are being developed. One such method (Figure 10.18) involves breaking DNA extracted from *Salmonella* into fragments with a restriction enzyme, and then selecting a specific fragment as the probe for *Sal-*

FIGURE 10.18 DNA probe used to identify bacteria.
① and ② Method of producing *Salmonella* DNA probes. ③ to ⑤ Preparation of unknown bacterial DNA from a sample of food. ⑥ and ⑦ DNA probes are used to identify *Salmonella* DNA from sample.

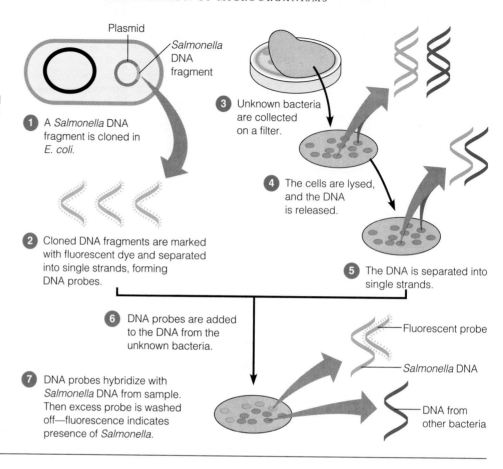

Plasmid

Salmonella DNA fragment

1 A *Salmonella* DNA fragment is cloned in *E. coli.*

2 Cloned DNA fragments are marked with fluorescent dye and separated into single strands, forming DNA probes.

3 Unknown bacteria are collected on a filter.

4 The cells are lysed, and the DNA is released.

5 The DNA is separated into single strands.

6 DNA probes are added to the DNA from the unknown bacteria.

7 DNA probes hybridize with *Salmonella* DNA from sample. Then excess probe is washed off—fluorescence indicates presence of *Salmonella.*

Fluorescent probe

Salmonella DNA

DNA from other bacteria

monella. This fragment must be able to hybridize with the DNA of all *Salmonella* strains but not with the DNA of closely related enteric bacteria. ① The chosen DNA fragment is cloned in a plasmid in *E. coli*, producing hundreds of specific *Salmonella* DNA fragments. ② These fragments are tagged with radioactive isotopes or a fluorescent dye and are separated into single strands of DNA. ③ to ⑥ The resulting DNA probes can then be mixed with single-stranded DNA prepared from a food sample suspected of containing *Salmonella.* ⑦ If *Salmonella* is present, the DNA probes will hybridize with the DNA of the *Salmonella,* and hybridization can be detected by the radioactivity or fluorescence of the probes.

Flow Cytometry

Flow cytometry can be used to identify bacteria in a sample without culturing the bacteria. In a *flow cytometer,* a moving fluid containing bacteria is forced through a small opening (see Figure 18.14). The simplest method detects the presence of bacteria by detecting the difference in electrical conductivity between cells and the surrounding medium. If the fluid passing through the opening is illuminated by a laser, the scattering of light provides information about the cell size, shape, density, and surface,

which is analyzed by a computer (see Figure 18.14). Fluorescence can be used to detect naturally fluorescent cells, such as *Pseudomonas,* or cells tagged with fluorescent dyes.

Milk can be a vehicle for transmission of disease. A proposed test that uses flow cytometry (see Chapter 18) to detect *Listeria* in milk could save time because the bacteria need not be cultured for identification. Antibodies against *Listeria* can be labeled with a fluorescent dye and added to the milk to be tested. The milk is passed through the flow cytometer, which records the light scattering of the antibody-labeled cells.

Numerical Taxonomy

Numerical taxonomy (Figure 10.19) has been used to infer evolutionary relationships among organisms. Many microbial characteristics are listed, such as the ability to use a certain carbohydrate, to form a pigment, or to produce a certain acid, and the presence or absence of each characteristic is scored for each organism. Amino acid sequences, the percentage of G-C pairs, and the percentage of hybridization observed between two nucleic acid strands can be listed. Tests must be selected carefully because each test is weighted equally; that is, a positive reaction is assigned a value of 1, and a negative result 0, un-

(a) Initial matrix

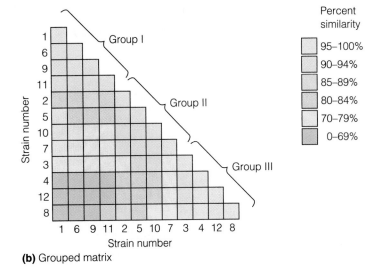

(b) Grouped matrix

Percent similarity

	95–100%
	90–94%
	85–89%
	80–84%
	70–79%
	0–69%

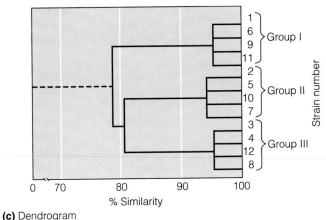

(c) Dendrogram

FIGURE 10.19 Numerical taxonomy. A similarity matrix (array) can be constructed with a computer to show relationships among different organisms. After as many as 100 different tests, each organism tested is given a numerical score. The closer the scores of two organisms, the more similar they are. **(a)** The initial matrix for 12 organisms. **(b)** Then, organisms with similar test results are grouped together in another matrix. Groups I, II, and III represent three species, each containing four strains. **(c)** A dendrogram constructed from the similarity values shows that the species in Group I are closely related to each other but not to the other organisms tested. The strains in Groups II and III form another family that is divided into two genera, each consisting of four species.

like the rapid identification methods, in which a relative weighting can be assigned to each test.

A computer then matches the characteristics of each organism against those of other organisms. The greater the number of characteristics shared by two organisms, the closer the taxonomic relationship is assumed to be. A match of 90% similarity or higher usually indicates a single taxonomic unit, or species. The computer can also generate a dendrogram (Figure 10.19c) that illustrates phylogenetic relationships derived from similarities from hybridization and other genetic experiments.

• • •

Morphological characteristics, differential staining, and biochemical testing were the only identification tools available just a few years ago. Technological advancements are making it possible to use nucleic acid and protein analysis techniques, once reserved for classification, for routine identification. (Practical examples are described in the boxes in Chapters 14 and 24.) A summary of taxonomic criteria and their uses is provided in Table 10.4.

TABLE 10.4 A Summary of Taxonomic Criteria Used for Classification and Identification of Bacteria

	USED FOR	
CRITERION	**CLASSIFICATION**	**IDENTIFICATION**
Morphological characteristics	No (yes for cyanobacteria)	Yes
Differential staining	Yes (for cell wall type)	Yes
Biochemical testing	No	Yes
Serology	Yes	Yes
Phage typing	No	Yes
Amino acid sequencing	Yes	No
Protein analysis	Yes	No
Base composition of nucleic acids	Yes	No
Nucleic acid hybridization	Yes	Yes (DNA probes)
Flow cytometry	No	Yes
Numerical taxonomy	Yes	No

INTRODUCTION (p. 250)

1. Taxonomy is the science of the classification of organisms, with the goal of showing relationships among organisms.
2. Taxonomy provides a means to identify organisms.

THE STUDY OF PHYLOGENETIC RELATIONSHIPS (pp. 250–253)

1. Phylogeny refers to evolutionary history.
2. The taxonomic hierarchy shows evolutionary or phylogenetic relationships among organisms.
3. Living organisms were once divided into just plants and animals; bacteria were classified with plants.
4. Bacteria were separated into the Kingdom Procaryotae in 1968.

A Five-Kingdom System (pp. 251–252)

1. Living organisms can be classified into five kingdoms.
2. Procaryotic organisms are placed in the Kingdom Procaryotae (Monera).
3. Eucaryotic organisms may be classified into the Kingdoms Protista, Fungi, Plantae, or Animalia.
4. Protista are mostly unicellular organisms.
5. Fungi are absorptive heterotrophs that develop from spores.
6. Multicellular photoautotrophs are placed in the Kingdom Plantae.
7. Multicellular ingestive heterotrophs are classified as Animalia.

A Proposed Three-Kingdom System (p. 253)

1. Another proposal divides living organisms into three kingdoms.
2. In this system, plants, animals, fungi, and protists belong to a single kingdom.
3. Eubacteria (with peptidoglycan) form a second kingdom.
4. Archaeobacteria (with unusual cell walls) form a third kingdom.

CLASSIFICATION OF ORGANISMS (pp. 253–257)

Scientific Nomenclature (pp. 254–255)

1. According to scientific nomenclature, each organism is assigned two names (binomial nomenclature): a genus and a specific epithet, or species.
2. Rules for the assignment of names to bacteria are published in the *Bacteriological Code*.

The Taxonomic Hierarchy (p. 255)

1. A eucaryotic species is a group of organisms that interbreeds but does not breed with individuals of another species.
2. Similar species are grouped into a genus; similar genera are grouped into a family; families, into an order; orders, into a class; classes, into a division or phylum; and phyla, into a kingdom.

A Phylogenetic Hierarchy (p. 255)

1. Organisms are grouped into taxa according to phylogenetic (from a common ancestor) relationships.
2. Some of the information for eucaryotic relationships is obtained from the fossil record.
3. Procaryotic relationships are determined by molecular biology techniques.

Classification of Bacteria (pp. 255–257)

1. *Bergey's Manual of Systematic Bacteriology* is the standard reference on bacterial classification.
2. In *Bergey's Manual,* bacteria are divided into four divisions.
3. A division is divided into classes; classes are composed of orders; orders consist of families; a family is divided into genera; and a genus is composed of species.
4. A group of bacteria derived from a single cell is called a strain.
5. Closely related strains constitute a species.

Classification of Viruses (p. 257)

1. Viruses are not placed in a kingdom. They are not composed of cells and cannot grow without a host cell.
2. A viral species is a population of viruses with similar characteristics that occupies a particular ecological niche.

CRITERIA FOR CLASSIFICATION AND IDENTIFICATION OF MICROORGANISMS (pp. 258–269)

1. *Bergey's Manual of Systematic Bacteriology* and *Bergey's Manual of Determinative Bacteriology* are references for laboratory identification of bacteria.
2. Morphological characteristics are useful in the identification of microorganisms, especially when aided by differential staining techniques.
3. The possession of various enzymes as determined by biochemical tests is used in the identification of microorganisms.
4. Serological tests, involving the reactions of microorganisms with specific antibodies, are useful in determining the identity of strains and species as well as relationships

among organisms. ELISA and Western blot are examples of serological tests.

5. Phage typing is the identification of bacterial species and strains by the determination of their susceptibility to various phages.

6. The sequences of amino acids in proteins of related organisms are similar.

7. Related organisms have similar proteins; this characteristic can be ascertained by PAGE "fingerprints."

8. The percentage of G-C pairs in the nucleic acid of cells can be used in the classification of organisms.

9. The number and sizes of DNA fragments produced by restriction enzymes are used to determine genetic similarities.

10. PCR can be used to detect small amounts of microbial DNA in a sample.

11. Single strands of DNA, or of DNA and RNA, from related organisms will hydrogen-bond to form a double-stranded molecule; this bonding is called nucleic acid hybridization.

12. Southern blotting and DNA probes are examples of hybridization techniques.

13. Flow cytometry measures physical and chemical characteristics of cells.

14. Numerical taxonomy uses many characteristics to show relatedness among organisms.

STUDY QUESTIONS

Review

1. What is taxonomy?

2. Explain why a two-kingdom system is no longer acceptable for classification.

3. List and define the five kingdoms used in the five-kingdom system of classification.

4. What characterizes the Kingdom Procaryotae, and what organisms are placed in this kingdom?

5. Discuss the evidence that supports a three-kingdom system of classification.

6. What is binomial nomenclature?

7. Why is binomial nomenclature preferable to the use of common names?

8. Using *Escherichia coli* and *Entamoeba coli* as examples, explain why the genus name must always be written out the first time you use it in a report.

9. Put the following terms in the correct sequence from the most general to the most specific: order, class, genus, kingdom, species, division, family.

10. Define species.

11. List the ten bases discussed in this chapter for the classification of microorganisms. Separate your list into those tests used primarily for taxonomic classification and those used primarily for identification of microorganisms already classified.

12. Higher organisms are arranged into taxonomic groups on the basis of evolutionary relationships. Why is this type of classification only now being developed for bacteria?

13. Describe the use of a DNA probe and PCR for:
    (a) Rapid identification of an unknown bacterium.
    (b) Determining which of a group of bacteria are most closely related.

14. Can you tell which of the following organisms are most closely related? Are any two the same species? On what did you base your answer?

CHARACTERISTIC	A	B	C	D
Morphology	Rod	Coccus	Coccus	Rod
Gram reaction	+	−	−	+
Glucose utilization	Fermentative	Oxidative	Fermentative	Fermentative
Cytochrome oxidase	Present	Present	Absent	Absent
G-C moles %	48–52	23–40	30–40	49–53

Critical Thinking

1. Here is some additional information on the organisms in Question 14:

ORGANISMS	% DNA HYBRIDIZATION
A and B	5–15
A and C	5–15
A and D	70–90
B and C	10–20
B and D	2–5

Which of these organisms are most closely related? Compare this answer with your response to Question 14.

2. In *Bergey's Manual*, the gram-positive cocci include the genera *Micrococcus* and *Staphylococcus*. The reported G + C content of *Micrococcus* is 66–75 moles %, and of *Staphylococcus*, 30–40 moles %. According to this information, would you conclude that these two genera are closely related?

3. Compare and contrast *Bergey's Manual of Systematic Bacteriology* and *Bergey's Manual of Determinative Bacteriology*.

4. SF medium is a selective medium, developed in the 1940s, to test for fecal contamination of milk and water. Only certain gram-positive cocci can grow in this medium. Why is it named "SF"? Using this medium, which genus will you culture? (Hint: Refer to Chapter 11.)

Clinical Applications

1. Using *Bergey's Manual*, identify the organism with the following characteristics: gram-negative rod; facultative anaerobe; motile; oxidase-negative; fermentative; acid and gas from lactose; indole produced; methyl red positive; V–P negative; citrate negative; H_2S not produced; does not grow in KCN broth.

2. A 55-year-old veterinarian was admitted to a hospital with a two-day history of fever, chest pain, and cough. Gram-positive cocci were detected in his sputum, and he was treated for lobar pneumonia with penicillin. The next day, another Gram stain of his sputum revealed gram-negative rods, and he was switched to ampicillin and gentamicin. A sputum culture showed biochemically inactive gram-negative rods identified as *Enterobacter agglomerans*. After fluorescent-antibody staining and phage typing, *Yersinia pestis* was identified in the patient's sputum and blood, and chloramphenicol and tetracycline were administered. The patient died three days after admission to the hospital. Tetracycline was given to his 220 contacts (hospital personnel, family, and coworkers). What disease did the patient have? Discuss what went wrong in the diagnosis and how his death might have been prevented. Why were the 220 other people treated?

Bacteria

M OST OF US have been conditioned to think of bacteria as invisible, potentially harmful little creatures. Actually, relatively few species of bacteria cause disease in humans, animals, plants, or any other organisms. Indeed, once you have completed a course in microbiology you will realize that without bacteria, much of life as we know it on Earth would not be possible. In fact, all organisms made up of eucaryotic cells probably evolved from bacterialike organisms, which were some of the earliest forms of life on Earth. Because bacteria are structurally simple and readily cultured and manipulated in the laboratory, they have been essential tools in the study of basic life processes, including physiology and genetics. They are also an essential element in the making of important foods such as cheese.

Many bacteria that may look similar under a microscope are, in fact, vastly different one from another; certainly they are not readily grouped by physical appearance, as we distinguish insects or birds. In 1988, a physician wrote to the *Journal of the American Medical Association (JAMA)* to inquire why ". . . we spoke of *Neisseria catarrhalis;* now I read about *Branhamella catarrhalis*." The editors of *JAMA* replied that things are even worse than we thought: studies of DNA composition have entirely eliminated *Branhamella* as a genus—it is now a subgenus of the genus *Moraxella.* As our understanding of bacterial genetics improves, many of the present bacterial groupings will be impacted. In this chapter you will gain some understanding of how bacterial groups are differentiated from each other and how important bacteria are in the world of microbiology.

BACTERIAL GROUPS

LEARNING OBJECTIVES

- *List at least six characteristics used to classify and identify bacteria according to* Bergey's Manual of Systematic Bacteriology.
- *List two major characteristics of each section of* Bergey's Manual *described in this chapter.*
- *Identify the sections in* Bergey's Manual *that contain species of medical importance.*

This chapter will introduce you to many of the more common bacteria. All bacteria share the property of procaryotic cellular organization, but it is difficult to correlate structural and functional characteristics to show evolutionary relationships. As we mentioned in the previous chapter, the most widely accepted taxonomic classification for bacteria is *Bergey's Manual of Systematic Bacteriology. Bergey's Manual* divides bacteria into four divisions (or phyla) according to the characteristics of cell walls (see Table 10.3). Each division is subdivided into *sections* according to such characteristics as Gram stain reaction, cell shape, cell arrangements, oxygen requirements, motility, and nutritional and metabolic properties. Sections are named according to these characteristics—for example, *spirochetes, gram-negative aerobic rods and cocci, anoxygenic photosynthetic bacteria,* and *endospore-forming rods and cocci.* Each section consists of a number of genera. In some sections, genera are grouped into families and orders; in other sections, they are not (see Table 11.1 and Appendix A).

TABLE 11.1   Summary of Selected Characteristics of Bacterial Groups, from *Bergey's Manual of Systematic Bacteriology*

NAME OF SECTION	IMPORTANT GENERA	HABITAT	GRAM REACTION	SPECIAL FEATURES
Spirochetes	*Treponema, Borrelia, Leptospira*	Aquatic; animal parasites	–	Helical morphology; motility by axial filaments; several important pathogens
Aerobic/microaerophilic, motile, helical/vibrioid, gram-negative bacteria	*Spirillum, Azospirillum, Campylobacter, Bdellovibrio*	Soil and aquatic environments; human intestinal tract and oral cavity	–	Helical morphology; motility by flagella, not axial filaments; vibrioids do not have a complete turn. Includes nitrogen-fixing bacteria and some pathogens
Nonmotile (or rarely motile), gram-negative, curved bacteria*	*Spirosoma, Meniscus*	Aquatic and sedimentary environments	–	Uncommon; mostly aquatic, not pathogenic; form S-shapes, C-shapes, rings
Gram-negative; aerobic rods and cocci	*Pseudomonas, Legionella, Neisseria, Brucella, Bordetella, Francisella, Rhizobium, Agrobacterium*	Soil; water; animal parasites	–	Organisms of medical, industrial, and environmental importance
Facultatively anaerobic, gram-negative rods	*Escherichia, Salmonella, Shigella, Klebsiella, Yersinia, Vibrio, Enterobacter, Hemophilus, Gardnerella, Pasteurella*	Soil; plants; animal respiratory and intestinal tracts	–	Many important pathogens
Anaerobic, gram-negative straight, curved, or helical rods	*Bacteroides, Fusobacterium*	Animals and insects	–	Obligate anaerobes, mostly of intestinal tract; some common in mouth and genital tract
Dissimilatory sulfate-reducing or sulfur-reducing bacteria	*Desulfovibrio*	Anaerobic sediments	–	Reduce oxidized forms of sulfur to H_2S
Anaerobic gram-negative cocci	*Veillonella*	Mostly animal intestinal tracts	–	Nonmotile anaerobes
Rickettsias and chlamydias	*Rickettsia, Coxiella, Chlamydia*	Parasites of arthropods and animals	–	Obligate intracellular bacteria; many important pathogens
Mycoplasmas	*Mycoplasma*	Parasites of animals, plants, insects	–	Pleomorphic; lack cell walls; some pathogens
Endosymbionts*	*Holospora, Blattabacterium*	Intracellular symbionts of protozoa, insects, helminths, plants, etc.	–	Assorted bacteria that live symbiotically in protozoa, insects, and fungi
Gram-positive cocci	*Staphylococcus, Streptococcus, Lactococcus, Enterococcus*	Soils; skin and mucous membranes of animals	+	Some important pathogens and industrial organisms

This chapter briefly describes the principal bacterial groups according to the basic organization of *Bergey's Manual.* Our discussion emphasizes bacteria considered to be of practical importance, those important in medicine, or those that illustrate biologically unusual or interesting principles.

Spirochetes

Spirochetes are found in contaminated water, in sewage, soil, and decaying organic matter, and within the bodies of humans and animals. One of the first microorganisms described by Anton van Leeuwenhoek in the 1600s was a large spirochete taken from saliva and tooth scrapings. These bacteria are typically coiled, like a metal spring. Some are tightly coiled; others resemble a spring that is stretched.

All spirochetes are actively motile and achieve their motility by means of two or more *axial filaments*. The axial filaments are enclosed in the space between an outer sheath and the body of the cell.

One end of each axial filament is attached near a pole

TABLE 11.1 Continued

NAME OF SECTION	IMPORTANT GENERA	HABITAT	GRAM REACTION	SPECIAL FEATURES
Endospore-forming rods and cocci	Bacillus, Clostridium	Soil; animal intestinal tract	+	Bacillus: aerobic or facultative anaerobes; Clostridium: anaerobic
Regular, nonsporing, gram-positive rods	Lactobacillus, Listeria	Dairy products; genital and oral cavities, animal feces	+	Lactobacillus forms lactic acid from carbohydrates, important industrially; Listeria: an animal pathogen
Irregular, nonsporing, gram-positive rods	Gardnerella, Corynebacterium, Propionibacterium, Actinomyces	Human pathogens; soil organisms	+	Pleomorphic morphology; several important pathogens
Mycobacteria	Mycobacterium	Soil; plants; animals	+	Mycobacterium is an important pathogen; acid-fast
Nocardioforms	Nocardia	Soil and animals	+	Form branched filaments; reproduce by fragmentation; often acid-fast; some pathogens
Budding and/or appendaged bacteria	Hyphomicrobium, Caulobacter	Mostly aquatic, some soil	−	Possess prosthecae; Hyphomicrobium reproduces by budding; Caulobacter is stalked
Sheathed bacteria	Sphaerotilus	Aquatic cause of sewage treatment problems	−	Cells encased in hollow sheath
Nonphotosynthetic, nonfruiting, gliding bacteria	Cytophaga, Beggiatoa	Aquatic	−	Cytophaga degrades cellulose; Beggiatoa oxidizes H_2S
Gliding, fruiting bacteria	Myxococcus	Dung, soil	−	Cells aggregate to form a fruiting body
Aerobic chemoautotrophic bacteria	Nitrosomonas, Nitrobacter, Thiobacillus	Soil	−	Nitrifying and sulfur-oxidizing bacteria; agriculturally and environmentally important
Archaeobacteria	Methanobacterium, Halobacterium, Sulfolobus	Anaerobic sediments; environments of extreme temperature and osmotic pressure	Varies	Not related to other bacterial groups; no peptidoglycan in cell walls; methane producers useful in sewage treatment
Anoxygenic photosynthetic bacteria	Chromatium, Rhodospirillum, Chlorobium	Anaerobic sediments	−	Includes the green and purple sulfur and nonsulfur bacteria; green and purple sulfur bacteria use H_2S as an electron donor and release sulfur
Oxygenic photosynthetic bacteria (cyanobacteria)	Chroococcus, Anabaena	Aquatic	−	Produce oxygen during photosynthesis; many species fix atmospheric nitrogen
Actinomycetes	Streptomyces, Frankia, Micromonospora	Soil; some aquatic	+	Common in soil; branching filaments with reproductive conidiospores; Frankia involved in nitrogen-fixing symbiosis with plants

*Not discussed in this chapter. See Appendix A.

of the cell (see Figure 4.9). By rotating its axial filament, the cell rotates in the opposite direction to move through liquids like a corkscrew. The rotating flagella of flagellated bacteria also conform to this motion (see Chapter 4), which is very efficient in moving bacteria through liquids. To a bacterium, water is as viscous as molasses is to an animal. However, a bacterium can typically move about 100 times its body length in a second (or about 50 μm/sec), whereas a large fish such as a tuna can move only about 10 times its body length in this time.

The spirochetes include a number of important patho-genic bacteria. The best known is the genus *Treponema*, which includes *Treponema pallidum*, the cause of syphilis (Figure 11.1). Members of the genus *Borrelia* cause relaps-ing fever and Lyme disease, serious diseases that are usu-ally transmitted by ticks or lice.

Leptospirosis is a disease usually spread to humans by water contaminated by *Leptospira* (lep-tō-spī'rä) species. The bacteria are excreted in the urine of such animals as dogs, rats, and swine, so domestic dogs and cats are rou-tinely immunized against it. The cells of *Leptospira* typi-cally have hooked ends (see Figure 26.4).

FIGURE 11.1 *Treponema pallidum.* Spirochetes such as *T. pallidum,* the causative agent of syphilis, are helical and have axial filaments under their outer sheath that allow them to move by a corkscrewlike rotation.

Aerobic/Microaerophilic, Motile, Helical/Vibrioid Gram-Negative Bacteria

A distinctly helical shape does not serve as the universal criterion for classification of a bacterium as a spirochete. Some helical bacteria are not included with the spirochetes because they lack axial filaments; they use flagella instead. These bacteria possess a single flagellum at one or both poles or sometimes tufts of flagella at these locations (Figure 11.2a). Helical or spiral bacteria, unlike spirochetes, are rigid helices or curved rods. These bacteria are frequently microaerophilic.

Most spiral bacteria are harmless aquatic organisms, such as *Spirillum volutans* (vō'lū-tans). This is an unusually large bacterium, up to 60 μm long. The box in Chapter 4, page 80, describes a member of the genus *Aquaspirillum* that is able to respond to magnetic fields.

Agricultural microbiologists have been interested in members of the genus *Azospirillum* (ā-zō-spī-ril'lum), a soil bacterium that grows in close association with the roots of many plants, especially tropical grasses. It uses nutrients excreted by the plants and in return fixes nitrogen from the atmosphere. This form of nitrogen fixation is most significant in some tropical grasses and in sugar cane, although the organism can be isolated from the root system of many temperate-climate plants, such as corn.

The spiral bacteria also include pathogenic organisms, One species, *Campylobacter fetus* (kam-pī-lō-bak'tėr fē'tus), causes abortion in domestic animals. Another species, *C. jejuni* (jē-jū'nē), causes outbreaks of foodborne intestinal disease (see Chapter 25). *Helicobacter* (hē'lik-ō-bak-tėr) causes ulcers in humans (see Figure 11.2b).

Sometimes, helical bacteria do not make a complete turn or twist; these are **vibrioids** (see the box in Chapter 3, p. 57). A particularly interesting example is the comma-shaped bacterium *Bdellovibrio*, which attacks other bacteria. It reproduces within the periplasmic space of the gram-negative bacteria upon which it preys. The

cell elongates into a tight spiral, which then fragments almost simultaneously into a number of individual flagellated cells. The host cell then lyses, releasing them.

Gram-Negative Aerobic Rods and Cocci

This group contains many microbial species of medical, industrial, and environmental importance. Some of the more interesting are those of the genus *Pseudomonas* (Figure 11.3). The common name for these bacteria is *pseudomonads*. These organisms are rod-shaped and have polar flagella. Many species excrete extracellular, water-soluble pigments that diffuse into their media. One species, *Pseudomonas aeruginosa*, produces a soluble, blue-green pigmentation. Under certain conditions, particularly in weakened hosts, this organism can infect the urinary tract, burns, and wounds and can cause septicemia, abscesses, and meningitis. Other pseudomonads produce soluble fluorescent pigments that glow when illuminated by ultraviolet light.

Pseudomonads are very common in soil and other natural environments and are generally no threat to a healthy individual. These bacteria are less efficient than some other heterotrophic bacteria in utilizing many of the common nutrients, but they have compensating characteristics. For example, many grow at refrigerator temperatures; these impart off tastes and colors to foods. Pseudomonads are also capable of synthesizing an unusually large number of enzymes and probably contribute significantly to the decomposition of chemicals, such as pesticides, that are added to soils.

In hospitals and other places where pharmaceuticals are prepared, the ability of pseudomonads to grow on minute traces of unusual carbon sources, such as soap residues or cap liner adhesives found in a solution, has been unexpectedly troublesome (see the box in Chapter 7, p. 170). They are even capable of growth in some antiseptics, such as quaternary ammonium compounds. Their resistance to most antibiotics has also been a source of medical concern. This resistance is probably related to the characteristics of the cell wall porins (see Chapter 4). Some of these considerations will be discussed in more detail later. Many pseudomonad genes that code for unusual metabolic characteristics, including resistance to antibiotics, are carried on plasmids (see Chapter 8).

Although pseudomonads are classified as aerobic, like many other bacteria they are capable of substituting nitrate for oxygen as a terminal electron acceptor. This process, anaerobic respiration, yields almost as much energy as aerobic respiration yields (see Chapter 5). In this way, pseudomonads cause important losses of valuable nitrogen in fertilizer and soil. Nitrate is the form of fertilizer nitrogen most easily used by plants. Under anaerobic conditions, as in waterlogged soil, pseudomonads eventually convert nitrate into nitrogen gas (N_2), which is lost to the atmosphere (see Chapter 27).

FIGURE 11.2 Helical/vibrioid bacteria. (a) Spiral cells of *Aquaspirillum sinosum*. This freshwater bacterium is motile by means of tufts of polar flagella. **(b)** *Helicobacter pylori,* an example of a helical bacterium that does not make a complete twist, resulting in a comma shape called vibrioid. This bacterium moves by means of multiple polar flagella.

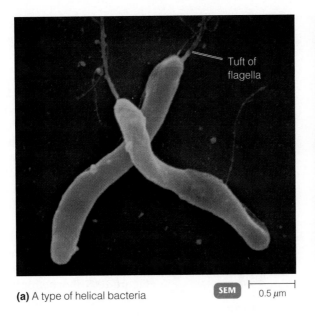

Tuft of flagella

(a) A type of helical bacteria

SEM · 0.5 μm

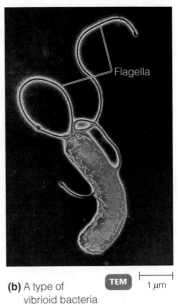

Flagella

(b) A type of vibrioid bacteria

TEM · 1 μm

Several genera of gram-negative aerobic rods and cocci are of medical importance.

Legionella is a relatively recently discovered genus, which now contains numerous species. Originally isolated during a search for the cause of a pneumonia now known as legionellosis, these bacteria did not grow on usual laboratory isolation media or stain with the usual histological staining techniques. After intensive effort, they were isolated through tests of guinea pigs inoculated with infected tissue taken from patients. A specially buffered charcoal–yeast extract agar has since been developed for use in their isolation and growth. This organism is now known to be relatively common in streams, and it colonizes such habitats as warm-water supply lines in hospitals and water in cooling towers of air conditioning systems.

Neisseria is a genus of non-endospore-forming diplococci. These bacteria are aerobic or facultatively anaerobic. Frequently found on human mucous membranes, they are able to grow well only at temperatures near body temperature. Pathogenic *Neisseria* species include the gonococcus bacterium *Neisseria gonorrhoeae*, the causative agent of gonorrhea (Figure 11.4; see also the box in Chapter 26), and *N. meningitidis* (me-nin-ji'ti-dis), the agent of meningococcal meningitis (see Chapter 22).

Members of the genus *Moraxella* (mô-raks-el'lä) are strictly aerobic coccobacilli. The coccobacilli are egg-shaped, a shape intermediate between cocci and rods. *Moraxella lacunata* (la-kü-nä'tä) is implicated in conjunctivitis (pink eye), an inflammation of the conjunctiva, the membrane that covers the eye and lines the eyelids.

Brucella (brü-sel'lä) is a small nonmotile coccobacillus. All species of *Brucella* are obligate parasites of mammals and cause the disease brucellosis. Of medical interest is the ability of *Brucella* to survive phagocytosis, an important element of our defense against bacteria (see Chapter 16).

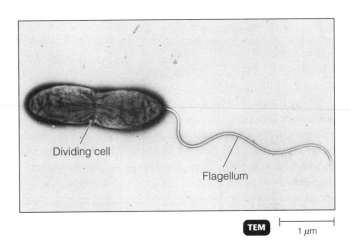

Dividing cell

Flagellum

TEM · 1 μm

FIGURE 11.3 *Pseudomonas andropogonis.* Note the single polar flagellum.

Bordetella (bôr-de-tel'lä) is a nonmotile rod. Virulent forms have capsules. *Bordetella pertussis* (pėr-tus'sis) is the cause of whooping cough (pertussis).

Francisella (fran-sis-el'lä) is a genus of small, pleomorphic bacteria that grow only on complex media enriched with blood or tissue extracts. *Francisella tularensis* (tü-lä-ren'sis) causes the disease tularemia.

Thus far in our discussion of this group of bacteria, most genera have been of medical importance. There are also several genera of industrial and environmental interest.

Rhizobium and *Bradyrhizobium* are especially important to agriculture. These organisms cause formation of nodules on the roots of leguminous plants, such as beans or clover. They live within these nodules in a symbiotic relationship with the plants and fix nitrogen from the air (see Chapter 27).

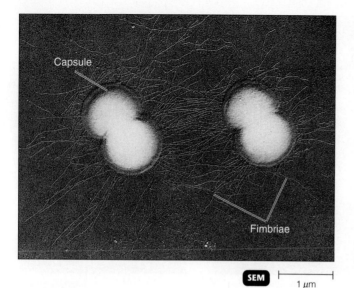

Capsule

Fimbriae

SEM   ⊢——⊣
            1 μm

FIGURE 11.4   The gram-negative coccus *Neisseria gonorrhoeae*. Note the paired arrangement (diplococcus). The fimbriae enable the organism to attach to mucous membranes and thus contribute to its pathogenicity; *N. gonorrhoeae* causes gonorrhea.

Some nitrogen-fixing bacteria, such as *Azotobacter* and *Azomonas* (ā-zō-mō′nas), are free-living in the soil. They are frequently used in laboratory demonstrations of nitrogen fixation. However, to fix agriculturally significant amounts of nitrogen, they would require energy sources, such as carbohydrates, that are in limited supply in soil.

Agrobacterium tumefaciens (tü-me-fā′shenz) is a plant pathogen that causes a disease of plants called crown gall. The tumorlike gall grows when *A. tumefaciens* inserts a plasmid containing bacterial genetic information into the plant's chromosomal DNA (see Figure 9.14). For this reason, microbial geneticists are deeply interested in this organism. They hope that eventually it may be a vehicle for inserting useful genetic information into plants—perhaps even the ability to fix nitrogen into plants such as wheat and corn.

Acetobacter (ä-sē-tō-bak′tėr) and *Gluconobacter* (glü-kon-ō-bak′tėr) are industrially important aerobic organisms that convert ethanol to vinegar. *Zoogloea* (zō-ō-glē′ä) will be discussed in Chapter 27 in the context of aerobic sewage treatment processes, such as the activated sludge system. As they grow, *Zoogloea* organisms form unusual fluffy, slimy masses that are essential to the proper operation of such systems.

Facultatively Anaerobic Gram-Negative Rods

From a medical viewpoint, facultatively anaerobic gram-negative rods are a very important group of bacteria. Many of them cause diseases of the gastrointestinal tract and other organs. Discussed here are three families: Enterobacteriaceae, Vibrionaceae, and Pasteurellaceae.

Enterobacteriaceae (Enterics)

The family Enterobacteriaceae, or **enterics**, as they are commonly called, includes a group of bacteria that inhabit the intestinal tracts of humans and other animals. Some species are permanent residents, others are found in only a fraction of the population, and still others are present only as agents of disease conditions. Most enterics are active fermenters of glucose and other carbohydrates.

Because of the clinical importance of enterics, there are many techniques for their isolation and identification. An identification key to some enterics is shown in Figure 10.7 and a modern tool using 15 biochemical tests is shown in Figure 10.8. Biochemical tests are especially important in clinical laboratory work and in food and water microbiology. Enterics can also be distinguished from each other according to antigens present on their surfaces—that is, by serology (see Chapter 18).

Enterics include motile as well as nonmotile species; those that are motile have peritrichous flagella. Many enterics have fimbriae (see Figure 4.10) that help them adhere to surfaces of mucous membranes. Specialized sex pili aid in the exchange of genetic information between cells, which often includes antibiotic resistance (see Figure 8.25).

Enterics, like most bacteria, produce proteins called *bacteriocins* that cause the lysis of closely related species of bacteria. Bacteriocins might help maintain the ecological balance of various enterics in the intestine.

Among the important genera included as enterics are *Escherichia, Salmonella, Shigella, Klebsiella, Serratia* (ser-rä′tē-ä), *Proteus* (prō′tē-us), *Yersinia* (yėr-sin′ē-ä), *Erwinia* (ėr-wi′nē-ä), and *Enterobacter*.

Escherichia.   The facultative anaerobe *E. coli* is one of the most common inhabitants of the intestinal tract and is probably the most familiar organism in microbiology. As you may remember from previous chapters, a great deal is known about the biochemistry and genetics of *E. coli,* and it continues to be an important tool for basic biological research. Its presence in water or food is also important as an indication of fecal contamination (see Chapter 27). *E. coli* is not usually considered to be pathogenic. However, it can be a common cause of urinary tract infections, and certain strains produce enterotoxins that commonly cause traveler's diarrhea and occasionally cause very serious foodborne disease (see Chapter 25).

Salmonella.   Almost all members of this genus are potentially pathogenic. Accordingly, there are extensive biochemical and serological tests to clinically isolate and identify salmonellae. They are common inhabitants of the intestinal tracts of many animals, especially poultry and cattle. Under unsanitary conditions, they can contaminate food.

Typhoid fever, caused by *Salmonella typhi,* is the most severe illness caused by any member of the genus *Salmonella.* A less severe gastrointestinal disease caused by other salmonellae is called *salmonellosis.* Salmonellosis is one of the most common forms of foodborne illness.

Although many members of the genus *Salmonella* have specieslike names, such as *S. typhimurium* (tī-fi-mûr′ē-um) and *S. dublin* (dub′lin), no individual species are recognized. Instead, this genus is taxonomically divided into about 2000 **serovars** (serotypes); that is, they are differentiated by serological means. Serovars can be further differentiated by special biochemical or physiological properties into **biovars,** or biotypes.

When the salmonellae are injected into appropriate animals, their flagella, capsules, and cell walls cause the animal to form antibodies that are specific for each of these structures. (This topic is more fully discussed in Chapter 17.) These specific antibodies, which are available commercially, can be used to differentiate *Salmonella* serovars by a system known as the Kauffmann–White scheme. The Kauffmann–White scheme designates an organism by numbers and letters that correspond to specific antigens on the organism's capsule, body, and flagella. For example, the antigenic formula of the bacterium *S. typhimurium* is 1,4,[5],12,i,1,2. Some salmonellae are named only by their antigenic formulas.

Shigella. Species of *Shigella* are responsible for a disease called bacillary dysentery or shigellosis. Unlike salmonellae, they are found only in humans. These organisms are second only to *E. coli* as a cause of traveler's diarrhea. Some strains of *Shigella* can cause a life-threatening dysentery.

Klebsiella. *Klebsiella pneumoniae* is a major cause of septicemia in hospital pediatric wards. It also causes one form of pneumonia, which is contracted especially by people with compromised immune systems (see Chapter 17), such as those suffering from chronic alcoholism.

Serratia. *Serratia marcescens* (mär-ses′sens) is distinguished by its production of red pigment. *S. marcescens* is important as a cause of hospital-acquired (nosocomial) infections. The organism causes urinary and respiratory tract infections and has been found on catheters, in saline irrigation solutions, and in other supposedly sterile solutions.

Proteus. *Proteus* organisms are very actively motile. They are implicated in infections of the urinary tract and wounds and in infant diarrhea (Figure 11.5).

Yersinia. *Yersinia pestis* (pes′tis) causes plague, the Black Death of medieval Europe. Urban rats in some parts of the world, and ground squirrels in the American Southwest, carry these bacteria. Fleas usually transmit the organisms among animals and to humans, although contact with an-

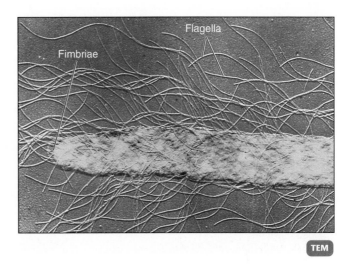

TEM

FIGURE 11.5 *Proteus mirabilis,* **a gram-negative rod.** Note the flagella distributed about the body, and the short fimbriae.

imals and respiratory droplets from infected people can be involved in transmission.

Erwinia. *Erwinia* species are primarily plant pathogens; some cause plant soft-rot diseases. These species produce enzymes that hydrolyze the pectin between individual plant cells and cause the plants to rot.

Enterobacter. Two *Enterobacter* species, *E. cloacae* (klō-ā′kī) and *E. aerogenes* (ā-rä′jen-ēz), can cause urinary tract infections and nosocomial infections. They are widely distributed in humans and animals, as well as in water, sewage, and soil (see Chapter 27).

Vibrionaceae

This is a family of gram-negative, facultatively anaerobic rods, many of which are slightly curved. They are found mostly in aquatic habitats. *Vibrio* (vib′rē-ō) is the most important genus in the family.

Vibrio. Members of the genus *Vibrio* are rods that are frequently slightly curved (Figure 11.6). One important pathogen is *Vibrio cholerae* (kol′ér-ī), the causative agent of cholera. The disease is characterized by a profuse and watery diarrhea. *V. parahaemolyticus* (pa-rä-hē-mō-li′ti-kus) causes a less serious form of gastroenteritis. Usually inhabiting coastal salt waters, it is transmitted to humans mostly by raw or undercooked shellfish.

Pasteurellaceae

The third medically important family among the gram-negative facultatively anaerobic rods includes *Pasteurella* (pas-tyér-el′lä) and *Hemophilus.*

Pasteurella. This genus is primarily known as a pathogen of domestic animals. It causes septicemia in cattle, fowl

TEM ⊢——⊣
1 µm

FIGURE 11.6 *Vibrio cholerae.* Note the characteristic slight curvature of the rods.

cholera in chickens and other fowl, and pneumonias in several types of animals. The best-known species is *Pasteurella multocida* (mul-tō′si-dä), which can be transmitted to humans by dog and cat bites.

Hemophilus. *Hemophilus* (also spelled *Haemophilus*) is a very important genus of pathogenic bacteria. These organisms commonly inhabit the mucous membranes of the upper respiratory tract, mouth, vagina, and intestinal tract. The best-known species that affects humans is *Hemophilus influenzae* (in-flü-en′zī), named long ago based on the erroneous conclusion that it was responsible for influenza.

The name *Hemophilus* is derived from the bacteria's requirement for blood in their culture medium. They are unable to synthesize important parts of the cytochrome system needed for respiration, and they obtain these substances from the heme fraction, known as the X factor, of blood hemoglobin. In some formulations, such as chocolate agar, the blood is heated, which causes it to turn a dark brown. The culture medium must also supply the cofactor nicotinamide adenine dinucleotide (NAD⁺ or NADP⁺), which is known as V factor. Clinical laboratories use tests for the requirement of X and V factors to identify isolates as *Hemophilus* species.

Hemophilus is responsible for several important diseases. It is the most common cause of meningitis in young children and is a frequent cause of earaches. Other clinical conditions caused by *H. influenzae* include epiglottitis (a life-threatening condition in which the epiglottis becomes infected), septic arthritis in children, bronchitis, and pneumonia.

Gardnerella

Gardnerella vaginalis (gärd-nė-rel′lä va-jin-al′is), a bacterium that causes one of the most common forms of vaginitis, is not presently assigned to any family. In 1955, researchers named it *Hemophilus vaginalis,* and it was placed with *Hemophilus* in *Bergey's Manual of Determinative Bacteriology,* 8th edition, because it did not seem to fit anywhere else. Its pleomorphic (highly irregular) form and occasional gram-positive reaction have caused it to be classified at times with the genus *Corynebacterium,* which will be discussed later in this chapter. The organism has a cell wall that is structurally similar to that of a gram-positive cell; however, the wall is quite thin, which leads to a gram-negative staining reaction. The current edition of *Bergey's Manual* places this genus in the group of gram-negative rods that we are now discussing. However, *Bergey's Manual* also includes *Gardnerella* in a group of irregular, non-spore-forming gram-positive rods, which we will discuss later in this chapter.

At least five genera in *Bergey's Manual* have a Gram-staining reaction that does not match their cell wall anatomy, and they are similarly grouped in two different taxonomic sections. Such dual classification is unheard of in botany or zoology today, and illustrates the difficulty of applying taxonomy to bacteria. The types of evolutionary criteria used to classify plants and animals are not so evident among bacteria. For example, although a bat and a bird superficially seem closely related, it can be determined that the bat is a mammal and that the bird occupies another evolutionary branch more closely related to reptiles. Such relationships among bacteria may eventually become clearer as we learn to detect similarities and differences in ribosomal RNA and chromosomal DNA of bacteria. A footnote in Appendix A lists the other bacterial genera that, for one reason or another, have been placed in two different sections.

Anaerobic, Gram-Negative, Straight, Curved, and Helical Rods

Among the anaerobic gram-negative bacteria, the genus *Bacteroides* (bak-tė-roi′dēz) is a large group of microbes that live in the human intestinal tract (Figure 11.7a). Some *Bacteroides* species also reside in the oral cavity, genital tract, and upper respiratory tract. *Bacteroides* organisms are non-endospore-forming and nonmotile. Infections due to *Bacteroides* often result from puncture wounds or surgery and are a frequent cause of peritonitis.

Another genus of gram-negative anaerobic bacteria is *Fusobacterium* (fü-sō-bak-ti′rē-um). These microbes are long and slender, with pointed rather than blunt ends (Figure 11.7b). In humans, they are found most often in the gums and may be responsible for many dental abscesses.

FIGURE 11.7 Gram-negative anaerobes. (a) Chains of *Bacteroides hypermegas.* **(b)** *Fusobacterium.* Note the slender shape with pointed ends.

(a) *Bacteroides* LM 5 μm

(b) *Fusobacterium* LM 5 μm

FIGURE 11.8 Rickettsias. (a) The triangular body is a chicken-embryo cell infected with intracellular rickettsias. Note the scattered rickettsias in the cytoplasm and the two distinct compact masses of rickettsia in the oval nuclear area of the cell. **(b)** Micrograph of *Rickettsia prowazekii,* which causes epidemic typhus.

Scattered rickettsias

Chicken embryo cell

Nucleus

Masses of rickettsia in nucleus

(a) Rickettsias in chicken embryo cell LM 50 μm

(b) Rickettsia TEM 0.25 μm

Dissimilatory Sulfate-Reducing or Sulfur-Reducing Bacteria

The sulfur-reducing bacteria are obligately anaerobic organisms that use oxidized forms of sulfur, such as sulfates (SO_4^{2-}) or elemental sulfur (S^0), rather than oxygen as electron acceptors. The product of this reduction is hydrogen sulfide (H_2S). The activity of these bacteria releases millions of tons of H_2S into the atmosphere every year. These reactions are part of the sulfur cycle and are discussed in Chapter 27.

These bacteria are found in anaerobic muds and sediments and also in the intestinal tracts of humans and animals. *Desulfovibrio* is the best-known genus. Still other sulfur bacteria, which we will discuss later in this chapter, are able to use H_2S either as part of photosynthesis or as an autotrophic energy source. Altogether, the sulfur bacteria of different types are some of the most interesting microorganisms known; their means of energy production are physiologically unique and ecologically crucial.

Anaerobic Gram-Negative Cocci

The cells of anaerobic gram-negative cocci typically occur in pairs, but they can occur singly, in clusters, or in chains. They are nonmotile and non-endospore-forming. Bacteria of the genus *Veillonella* (vi-lo-nel'lä) are found as part of the normal microbiota of the mouth and are components of dental plaque.

Rickettsias and Chlamydias

Both rickettsias and chlamydias are obligate intracellular parasites, which means that they can reproduce only within a host cell. In this respect, they are similar to viruses. However, in morphological and in limited biochemical aspects, they resemble bacteria and are therefore classified as such. A comparison of rickettsias, chlamydias, and viruses appears in Table 13.1, page 333.

Rickettsias (Figure 11.8) are rod-shaped bacteria or coccobacilli (0.8 to 2.0 μm in length) that have a high degree of pleomorphism. They are gram-negative and divide by binary fission. One distinguishing feature of most rickettsias is that they are transmitted to humans by insects and ticks. The one exception is *Coxiella burnetii*, which causes Q fever. Although cattle ticks harbor the organism, it is most commonly transmitted by aerosols or contaminated milk. A sporogenic cycle has been reported for *C. burnetii* (see Figure 24.11). This might explain the bacterium's relatively high resistance to pasteurization temperatures and antimicrobial chemicals.

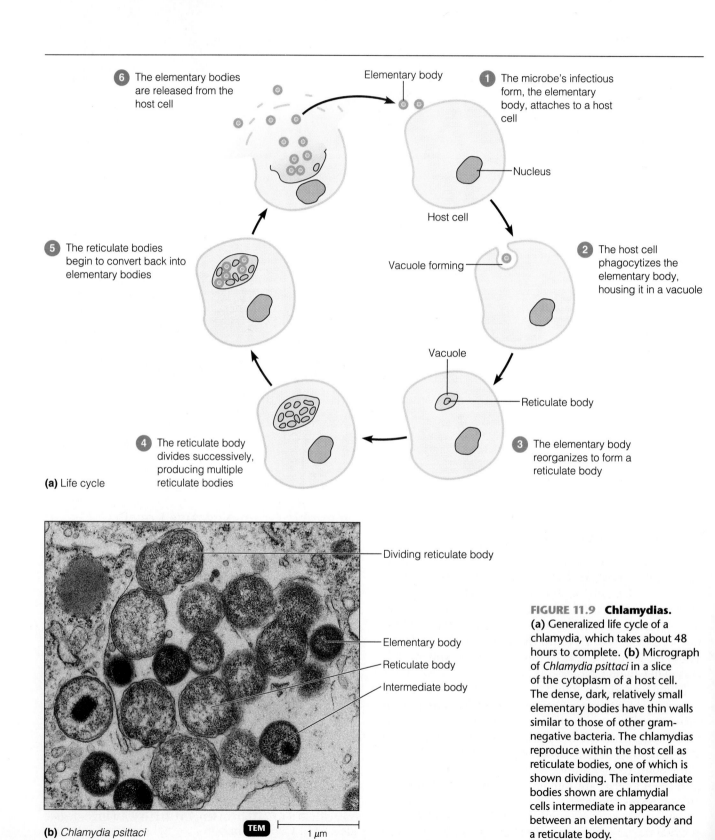

6 The elementary bodies are released from the host cell

Elementary body

1 The microbe's infectious form, the elementary body, attaches to a host cell

Nucleus

Host cell

5 The reticulate bodies begin to convert back into elementary bodies

Vacuole forming

2 The host cell phagocytizes the elementary body, housing it in a vacuole

Vacuole

Reticulate body

4 The reticulate body divides successively, producing multiple reticulate bodies

3 The elementary body reorganizes to form a reticulate body

(a) Life cycle

Dividing reticulate body

Elementary body

Reticulate body

Intermediate body

(b) *Chlamydia psittaci*

TEM

1 μm

FIGURE 11.9 Chlamydias.
(a) Generalized life cycle of a chlamydia, which takes about 48 hours to complete. **(b)** Micrograph of *Chlamydia psittaci* in a slice of the cytoplasm of a host cell. The dense, dark, relatively small elementary bodies have thin walls similar to those of other gram-negative bacteria. The chlamydias reproduce within the host cell as reticulate bodies, one of which is shown dividing. The intermediate bodies shown are chlamydial cells intermediate in appearance between an elementary body and a reticulate body.

Diseases caused by rickettsias include epidemic typhus, caused by *Rickettsia prowazekii* (ri-ket′sē-ä prou-wä-ze′kē-ē) and transmitted by lice; endemic murine typhus (not to be confused with typhoid fever), caused by *R. typhi* (tī′fē) and transmitted by rat fleas; and Rocky Mountain spotted fever, caused by *R. rickettsii* (ri-ket′sē-ē) and transmitted by ticks. In humans, rickettsial infections damage the permeability of capillaries; severe infections can cause the cardiovascular system to collapse. Rickettsias are usually cultivated in the yolk sac of chicken embryos.

Chlamydias have a developmental cycle that is perhaps their most distinguishing characteristic (Figure 11.9a). They are gram-negative coccoid bacteria that range in size from 0.2 to 1.5 μm (Figure 11.9b). Unlike most rickettsias, chlamydias do not require insects or ticks for transmission; they are transmitted to humans by interpersonal contact or by airborne respiratory routes.

There are only three species of chlamydias. *Chlamydia trachomatis* (kla-mi′dē-ä trä-kō′mä-tis) is the causative agent of trachoma, a common cause of blindness in humans. *C. trachomatis* seems to be the primary causative agent of nongonococcal urethritis, which might now be the most common sexually transmitted disease in the United States, and lymphogranuloma venereum, another sexually transmitted disease. *C. psittaci* (sit′tä-sē) is the causative agent of psittacosis (ornithosis). A third species, *C. pneumoniae,* has recently been identified as the cause of a mild pneumonia. Chlamydias can be cultivated in laboratory animals, cell cultures, or the yolk sac of chicken embryos.

FIGURE 11.10 *Mycoplasma pneumoniae.* Bacteria such as *M. pneumoniae* have no cell walls, and their morphology is irregular (pleomorphic). This micrograph shows filaments with a series of bulges; the organism reproduces by fragmentation of the filaments at these sites.

FIGURE 11.11 *Staphylococcus aureus.* Note the grapelike clusters of these gram-positive cocci.

Mycoplasmas

Mycoplasmas are bacteria that do not form cell walls. The majority of *Mycoplasma* species are aerobes or facultative anaerobes. Because they lack cell walls, they are highly pleomorphic (Figure 11.10). They can produce filaments that resemble fungi; hence their name (*myco* means fungus). *Mycoplasma* cells are very small, ranging in size from 0.1 to 0.25 μm. The most significant human pathogen among mycoplasmas is *M. pneumoniae,* the causative agent of primary atypical pneumonia (commonly called "walking" pneumonia).

Also included in this group of bacteria are *Spiroplasma* (spī-ro-plaz′mä) species, cells with a tight corkscrew morphology that are serious plant pathogens and are common parasites of plant-feeding insects. *Ureaplasma* (ū-rē-ä-plaz′mä) species, so called because they can enzymatically split the urea in urine, are occasionally associated with urinary tract infections. *Thermoplasma* species, which were originally isolated from spontaneously heating coal refuse piles, are harmless organisms occasionally found in household hot water systems. This genus is taxonomically grouped with the archaeobacteria as well as with the mycoplasmas.

Mycoplasmas can be grown on artificial media that provide them with sterols (if necessary) and other special nutritional or physical requirements. Colonies are less than 1 mm in diameter and have a characteristic "fried-egg" appearance when viewed under magnification. Because of the small colony size on artificial media, cell culture methods are often more satisfactory.

Gram-Positive Cocci

Most gram-positive cocci of medical importance are members of the genera *Staphylococcus* and *Streptococcus.*

Staphylococcus

Staphylococci typically occur in grapelike clusters (Figure 11.11). The most important staphylococcal species is

Staphylococcus aureus, named for its yellow-pigmented colonies (*aureus* means golden). Members of this species are aerobes or facultative anaerobes.

Understanding the characteristics of the staphylococci may help in understanding the reasons for their pathogenicity, which takes many forms. They grow comparatively well under conditions of high osmotic pressure and low moisture, which partially explains why they can grow and survive in nasal secretions (many of us carry the bacteria in our noses) and on the skin. This ability also explains how *S. aureus* can grow in certain foods with high osmotic pressures (such as ham and other cured meats) or in low-moisture foods that tend to inhibit the growth of other organisms. Furthermore, the yellow pigment probably confers some protection from the antimicrobial effects of sunlight.

S. aureus produces many toxins that contribute to the bacterium's pathogenicity by increasing its ability to invade the body or damage tissue. The infection of surgical wounds by *S. aureus* is a common problem in hospitals, as we will see in the box in Chapter 14. The bacterium's ability to develop resistance quickly to such antibiotics as penicillin contributes to its danger in hospitals. *S. aureus* is the agent of toxic shock syndrome, a severe infection causing high fever and vomiting and sometimes death. *S. aureus* also produces an *enterotoxin* that causes vomiting and nausea when ingested and is one of the most common causes of food poisoning.

Streptococcus

Members of the genus *Streptococcus* are spherical grampositive bacteria. They are probably responsible for more illnesses and cause a greater variety of diseases than any other group of bacteria. Among the diseases caused by streptococci are scarlet fever, pharyngitis (sore throat), and pneumococcal pneumonia.

Streptococci typically appear in chains that can contain as few as four to six cocci or as many as 50 or more (Figure 11.12). One species, *Streptococcus pneumoniae,* is usually found only in pairs (see Figure 24.10). Streptococci do not use oxygen, although most are aerotolerant. A few are obligately anaerobic.

One basis for the classification of streptococci is their action on blood agar. *Alpha-hemolytic* species produce a substance called alpha-hemolysin that reduces hemoglobin (red) to methemoglobin (green); this reduction causes a greenish zone to surround the colony. *Beta-hemolytic* species produce a hemolysin that forms a clear zone of hemolysis on blood agar (see Figure 6.6). Some species have no apparent effect on red blood cells; these are referred to as *nonhemolytic* or sometimes, less logically, as gammahemolytic.

Like staphylococci, pathogenic streptococci produce a number of extracellular substances that contribute to disease. Among them are products that destroy phagocytic cells (which are essential to the body's defenses), enzymes

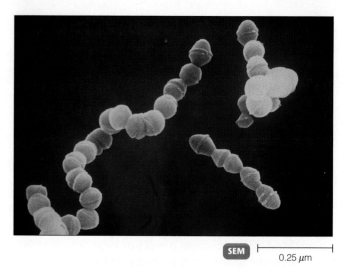

SEM　　　　0.25 μm

FIGURE 11.12　*Streptococcus.* Note the chains of cells characteristic of most streptococci; many of the spherical cells are dividing and are somewhat oval in appearance.

that digest connective tissue of the host and spread the infection, and enzymes that lyse fibrin. Fibrin is a fibrous protein that is deposited when blood clots and that limits movement of pathogens in infected areas.

Streptococci that are primarily pathogenic are grouped in the genus *Streptococcus.* The genus *Enterococcus* contains streptococci normally inhabiting the intestinal tract; they are occasional opportunistic pathogens. Streptococci of importance to the dairy industry for their production of lactic acid are grouped in the genus *Lactococcus.*

Endospore-Forming Gram-Positive Rods and Cocci

The formation of endospores by bacteria is important both to medicine and to the food industry because of the endospores' resistance to heat and many chemicals. The majority of endospore-forming rods and cocci are grampositive. With respect to oxygen requirements, they can be strict aerobes, facultative anaerobes, obligate anaerobes, or microaerophiles. Cocci of the genus *Sporosarcina* are saprophytic soil bacteria. The two important rod-shaped genera are *Bacillus* and *Clostridium* (Figure 11.13).

Bacillus anthracis causes anthrax, a disease of cattle, sheep, and horses that can be transmitted to humans. The anthrax bacillus is a nonmotile facultative anaerobe ranging in length from 4 to 8 μm. It is one of the largest bacterial pathogens. Smears of *B. anthracis* from growth in tissue show a capsule. The endospores are centrally located.

Bacillus thuringiensis is probably the best-known microbial insect pathogen (Figure 11.13b). It produces intracellular crystals of toxic glycoproteins when it sporulates. Commercial preparations containing endospores and

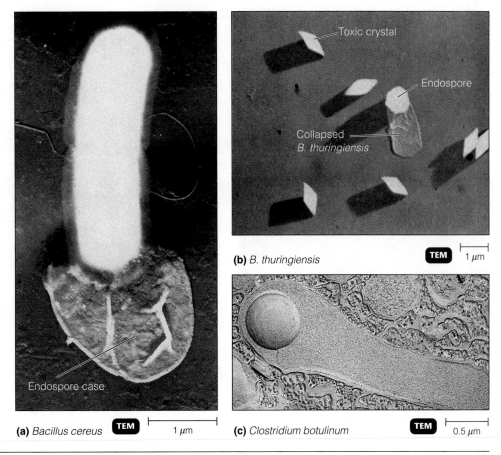

FIGURE 11.13 Representative gram-positive, endospore-forming rods. (a) A germinating cell of *Bacillus cereus.* (b) *Bacillus thuringiensis;* the diamond-shaped crystal shown next to the endospore is toxic to insects that ingest it. (c) *Clostridium botulinum,* showing the terminal location of the endospore.

(a) *Bacillus cereus*   **TEM**   1 μm

Endospore case

(b) *B. thuringiensis*   **TEM**   1 μm

Toxic crystal

Endospore

Collapsed
B. thuringiensis

(c) *Clostridium botulinum*   **TEM**   0.5 μm

crystalline toxin of this bacterium are sold in gardening supply shops to be sprayed on plants. When ingested by an insect, the toxin quickly causes paralysis of the insect's gut, and the insect ceases to feed. In some insects the ingested endospores germinate, and bacterial growth is a factor in the death of the insect. Certain other bacteria closely associated with plant roots have been genetically engineered to contain this toxin and kill insects feeding on the roots.

Members of the genus *Clostridium* are obligate anaerobes. They vary in length from 3 to 8 μm, and in most species the cells containing endospores appear swollen (Figure 11.13c). Diseases associated with clostridia include tetanus, caused by *C. tetani* (te'tan-ē); botulism, caused by *C. botulinum;* and gas gangrene, caused by *C. perfringens* (pėr-frin'jens) and other clostridia. *C. perfringens* is also the cause of a common form of foodborne diarrhea. However, beneficial uses are also being found for *C. botulinum* toxin (see the box on p. 288).

Regular Nonsporing Gram-Positive Rods

The genus *Lactobacillus* is representative of nonsporing gram-positive rods. Lactobacilli lack a cytochrome system and are unable to use oxygen as an electron acceptor. Instead, these aerotolerant bacteria produce lactic acid from simple carbohydrates. The acidity inhibits competing bacteria, creating an ecological niche in which the lactobacilli have an advantage. In humans, lactobacilli are located in the vagina, intestinal tract, and oral cavity. Common industrial uses of lactobacilli are in the production of sauerkraut, pickles, buttermilk, and yogurt. Typically, a succession of lactobacilli, each more acid-tolerant than the other, participate in these lactic acid fermentations.

A pathogen in this group, *Listeria monocytogenes* (lis-te'rē-ä mo-nō-sī-tô'je-nēz), can contaminate food, especially dairy products and processed delicatessen meats. *L. monocytogenes* survives within phagocytic cells and is capable of growth at refrigeration temperatures. If it infects a pregnant woman, the organism poses the threat of stillbirth or serious damage to the fetus.

Irregular Nonsporing Gram-Positive Rods

The organisms in this group are often grouped under the general term *corynebacteria* (*coryne* means club-shaped). They tend to be pleomorphic (highly irregular in morphology); their morphology often varies with the age of

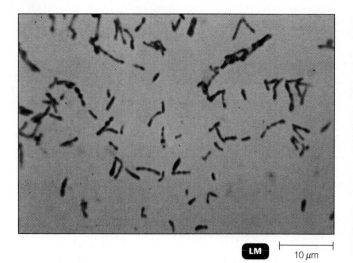

LM |— 10 μm

FIGURE 11.14 *Corynebacterium diphtheriae.* Note the
irregular, club-shaped morphology.

SEM |— 0.25 μm

FIGURE 11.15 *Actinomyces sp.* Note the branched filamentous
morphology.

the cells. They may be aerobic, anaerobic, or microaerophilic. The best-known and most widely studied
species is *Corynebacterium diphtheriae,* the causative agent
of diphtheria (Figure 11.14). One related species, *Propionibacterium acnes,* is commonly found on human skin and is
implicated in acne.

The genus *Actinomyces* (ak-tin-ō-mī′sēs) consists of anaerobes that are found in the mouth and throat of humans and animals. They occasionally form filaments
(Figure 11.15) that can fragment into coryneform cells.
One species, *Actinomyces israelii* (isrā′lē-ē), causes actino-
mycosis, a tissue-destroying disease usually affecting the
head, neck, or lungs.

Mycobacteria

Mycobacteria are aerobic, non-endospore-forming, nonmotile, rod-shaped organisms. Their name (*myco* means
fungus) was suggested by their occasional exhibition of filamentous growth. Most of the pathogenic species are acid-
fast. A number of species of *Mycobacterium* are found in
the soil. Others are important pathogens: *Mycobacterium
tuberculosis* causes tuberculosis, and *Mycobacterium leprae*
causes leprosy.

Nocardioforms

The best-known genus in this group is *Nocardia.* These organisms morphologically resemble *Actinomyces;* however,
they are aerobic. To reproduce, they form rudimentary filaments, which fragment into short rods. The structure of
their cell wall resembles that of the mycobacteria; therefore, they are often acid-fast. *Nocardia* species are common

in soil. Some species, such as *Nocardia asteroides* (as-tėr-
oi′dēz), occasionally cause chronic, difficult-to-treat pulmonary nocardiosis. *N. asteroides* is also one of the causative
agents of mycetoma, a localized destructive infection of
the feet or hands. A hospital-acquired infection caused by
Rhodococcus, another member of this group, is described
in the box in Chapter 16.

Gliding, Sheathed, and Budding and/or Appendaged Bacteria

Appendaged Bacteria

This is an unusual group of bacteria that is linked taxonomically by the presence of **prosthecae** (singular: *prostheca*). Prosthecae are protrusions, such as stalks and buds.
Most of these organisms are found in low-nutrient aquatic
environments such as lakes. One of the best studied is the
genus *Caulobacter* (kô-lō-bak′tėr), which has stalks that
anchor the organisms to surfaces (Figure 11.16a). This
arrangement tends to increase their nutrient uptake because they are exposed to a continuously changing flow
of water and because the stalk increases the surface-to-
volume ratio of the cell. Also, if the surface to which they
anchor is a living host, these bacteria can use the host's
excretions as nutrients. When the nutrient concentration
is exceptionally low, the size of the stalk increases, evidently to provide an even greater surface area for nutrient
absorption.

(a) A type of appendaged bacteria   **TEM** ⊢——————⊣ 1 μm

FIGURE 11.16 Gliding or appendaged bacteria.
(a) *Caulobacter.* The stalked cell is dividing, and a flagellated swarmer cell is being formed. **(b)** Slime trails of the gliding, fruiting myxobacterium *Myxococcus fulvus.* Under appropriate conditions, these bacteria aggregate and form a stalk bearing a fruiting body similar to that shown in c. **(c)** *Stigmatella aurantiaca,* another example of a gliding, fruiting bacterium. Numerous gliding vegetative cells have aggregated to form the stalk shown here, on which four spores (fruiting bodies) have formed. When growth conditions improve, these spores can germinate into new vegetative cells with gliding motility.

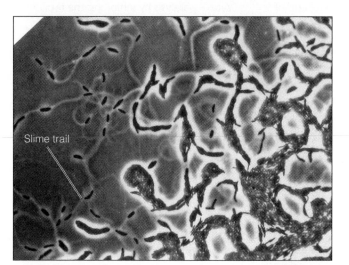

Slime trail

(b) The gliding form of a type of gliding,   **LM** ⊢—⊣ 10 μm
fruiting bacteria

(c) The fruiting form of a type of   ⊢—⊣ 1 μm
gliding, fruiting bacteria

Another unusual characteristic of *Caulobacter* is that reproduction by fission does not result in two essentially identical cells. When a stalked *Caulobacter* cell divides, the result is one stalked cell and one flagellated swarmer cell (see Figure 11.16a). The swarmer cell is highly motile and confers the advantage of movement to new locations. After a period of motility, the swarmer cell loses its flagellum and becomes another stalked cell. Because of this unusual reproductive cycle, biologists studying the mechanics and genetics of cellular differentiation have found *Caulobacter* to be a useful research tool.

Gliding, Nonfruiting Bacteria
Gliding bacteria are classified as such on the basis of their method of motility. Embedded in a layer of slime, they glide over surfaces, often leaving a visible trail. Members of the genus *Cytophaga* (sī-täf'äg-ä) are important cellulose

degraders in the soil. Another gliding bacterium is *Beggiatoa* (bej-jē-ä-tō'ä); this organism morphologically resembles certain cyanobacteria, but it is not photosynthetic. Nutritionally, it uses hydrogen sulfide (H_2S) as an autotrophic energy source and accumulates granules of sulfur within the cytoplasm. Its affinity for H_2S allows it to have a symbiotic relationship with rice. Rice grows in muds in which bacteria such as *Desulfovibrio* produce H_2S, to which the plant is sensitive. The *Beggiatoa* cells lower the amounts of toxic H_2S and, in return, benefit from low levels of oxygen that the plant transports to the root area.

Gliding, Fruiting Bacteria
This group of bacteria, the *myxobacteria* (from *myxa;* Gr, nasal mucus), have a remarkable life cycle. Vegetative cells move by gliding and leave behind a slime trail (Figure 11.16b), and large numbers of these gram-negative or-

Microbiology in the News

DEADLY BOTULINUM TOXIN PROVIDES CURE

T HE WORD *BOTULISM* is frightening; when it is mentioned in association with a commercial product, sales of that product plummet. This fatal disease was identified in 1870, when it was connected to eating sausage (*botulus* means sausage). Since then, botulism has been associated with a variety of improperly processed meats and vegetables. We now know that botulism is caused by ingesting food contaminated with the neurotoxin produced by *Clostridium botulinum;* this toxin is one of the most potent toxins known. However, the toxin can be inactivated by boiling food for one minute before it is eaten. Many food preparation techniques are geared toward preventing the growth of *C. botulinum.*

Botulism is a paralytic illness; patients typically develop cranial nerve palsies, followed by descending paralysis that can lead to respiratory failure and death. Treatment for botulism usually requires the early administration of antitoxin, intensive care, and the use of a respirator to assist breathing. Fortunately, the disease is not common; in 1991, there were 114 cases of botulism reported in the U.S.A., many fewer than the 48,000 reported cases of salmonellosis, for example.

Since the 1970s, two types of botulism have been identified: adult (foodborne) botulism and infant botulism. In infant botulism, *C. botulinum* bacteria themselves are ingested in contaminated foods, particularly honey. These bacteria colonize the infant's intestine and produce botulinum toxin as they grow. In 1993, 21 cases of foodborne botulism and 55 cases of infant botulism were reported in the U.S.A.

In 1990, however, the Food and Drug Administration approved a beneficial use for the usually deadly botulinum toxin. People who experience either painful muscle spasms (associated with diseases or conditions such as cerebral palsy, stroke, or spinal cord injury) or involuntary tremors (associated with diseases such as Parkinson's) are being treated with sublethal doses of botulinum toxin. Because the toxin causes paralysis by inhibiting the release of acetylcholine from nerves to muscle, when given in very small doses it paralyzes the muscles involved in these unwanted movements. To date, botulinum toxin has been used to treat over 3700 people with crossed-eye disorder. Although this treatment relieves the spasms, it is not permanent because new nerve endings will eventually grow to replace the toxin-damaged nerve endings.

ganisms aggregate. An important source of nutrition is the lysis of bacteria they encounter along the way. Where the moving cells aggregate, they differentiate and form a stalked fruiting body that contains resting cells called *myxospores* (Figure 11.16c). Under proper conditions, the myxospores germinate and form new vegetative gliding cells. This is probably the most complex life cycle among procaryotic cells.

Budding Bacteria
Budding bacteria do not divide by fission into nearly identical cells; members of the genus *Hyphomicrobium* (hī-fō-mī-krō'bē-um) and a few other genera instead form buds (Figure 11.17a). The process in principle resembles the asexual reproductive processes of many yeasts. The parent cell retains its identity while the bud increases in size until it separates as a complete new cell.

Sheathed Bacteria
Sheathed bacteria, such as *Sphaerotilus natans* (sfe-rä'ti-lus nā'tans), are found in fresh water and in sewage. These gram-negative bacteria with polar flagella form a hollow, filamentous sheath in which to live (Figure 11.17b). They probably contribute to bulking, an important problem in sewage treatment (see Chapter 27).

Chemoautotrophic Bacteria

These organisms are of great importance to the environment and to agriculture. They are autotrophs capable of using inorganic chemicals as energy sources and carbon dioxide as the only source of carbon. They possess great synthesizing ability. Especially important for agriculture are the nitrifying bacteria. The energy sources of the gen-

Bud

(a) A type of budding bacteria   SEM   1 μm

Sheath

Bacterial cell

(b) A type of sheathed bacteria   TEM   1 μm

FIGURE 11.17 Budding or sheathed bacteria. (a) The budding bacterium *Hyphomicrobium* sp. **(b)** The sheathed bacterium *Sphaerotilus natans*. This organism, which is found in sewage and aquatic environments, forms elongated sheaths in which the bacteria live. The bacteria have flagella (not visible here) and can eventually swim free of the sheath.

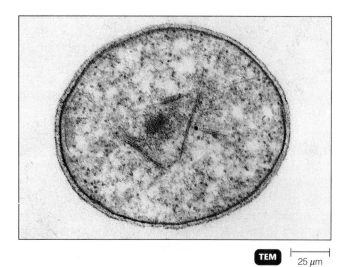

TEM   25 μm

FIGURE 11.18 Archaeobacteria. Members of this bacterial group are capable of growth under extreme conditions. For example, this *Pyrococcus* is an extremely thermophilic, heterotrophic, anaerobic archaeobacterium that was isolated from hydrothermal deep sea vent rock samples. It grows in a temperature range of 66°–110°C—higher than the boiling point of water.

Archaeobacteria

This exceptionally interesting group of bacteria includes microbes that are highly unusual in morphology and in their environmental niches. Prominent among them are bacteria that are extreme halophiles, surviving in very high concentrations of salt. Others tolerate extremes of heat (Figure 11.18) and acidity. The obligately anaerobic, methane-producing bacteria (methanogens) are of considerable economic importance. The archaeobacteria are grouped together because of their related ribosomal RNA sequences. Also, their cell walls lack the peptidoglycan common to most other bacterial groups.

Examples of extreme halophiles include *Halobacterium* (hal-ō-bak-ti′-rē-um) (see the box in Chapter 5, p. 131) and *Halococcus* (hal-ō-kok′kus), which live in high concentrations of sodium chloride (NaCl) and actually require such environmental conditions for growth. Other archaeobacteria thrive in acidic, sulfur-rich hot springs. Such an organism is *Sulfolobus* (sul-fō-lō′bus), which has a pH optimum of about 2 and a temperature optimum of more than 70°C. An example of thermophiles is described in the box in Chapter 6, page 149.

The methane-producing bacteria are used in sewage treatment processes (see Chapter 27). These archaeobacteria derive energy from combining hydrogen (H_2) with CO_2 to form methane (CH_4). An essential part of the treatment of sewage sludge is encouraging the growth of these organisms in anaerobic digestion tanks to convert sewage sludge into CH_4.

era *Nitrobacter* and *Nitrosomonas* are reduced nitrogenous compounds, such as ammonium (NH_4^+) and nitrite (NO_2^-); the bacteria convert these compounds into nitrates (NO_3^-). Nitrate is a nitrogen form that is mobile in soil and is therefore likely to be encountered and used by plants. *Thiobacillus* and other sulfur-oxidizing bacteria are parts of the important sulfur cycle. These bacteria are capable of obtaining energy by oxidizing the reduced forms of sulfur, such as hydrogen sulfide (H_2S) or elemental sulfur (S^0), into sulfates (SO_4^{2-}).

Phototrophic Bacteria

Three groups of procaryotes use light as an energy source. These **phototrophs** are the purple bacteria, the green bacteria, and the cyanobacteria (blue-green algae). The energetics of phototrophic bacteria were discussed in Chapter 5.

Purple and Green Phototrophic Bacteria

The so-called purple and green bacteria (which are not necessarily these colors) are generally anaerobic (Figure 11.19). Their habitat is usually the deep sediments of lakes and ponds. Like plants, algae, and the cyanobacteria, purple and green bacteria carry out photosynthesis. However, unlike plantlike photosynthesis, the photosynthesis of purple or green phototrophic bacteria is **anoxygenic**—it does not produce oxygen. Plants, algae, and the cyanobacteria produce oxygen (O_2) from water (H_2O) as they carry out photosynthesis, as follows:

$$(1) \quad 2H_2O + CO_2 \xrightarrow{\text{light}} (CH_2O) + H_2O + O_2$$

The *purple sulfur* and *green sulfur bacteria* use reduced sulfur compounds, such as hydrogen sulfide (H_2S), instead of water, and they produce granules of sulfur (S) rather than oxygen, as follows:

$$(2) \quad 2H_2S + CO_2 \xrightarrow{\text{light}} (CH_2O) + H_2O + 2S^0$$

At one time, an important question in biology concerned the source of the oxygen produced by plant photosynthesis; was it from CO_2 or from H_2O? Until the advent of radioisotope tracers, which traced the oxygen in water and carbon dioxide and finally settled the question, the comparison of equations (1) and (2) was the best evidence that the oxygen source was indeed H_2O.

Other phototrophs, the *purple nonsulfur* and *green nonsulfur bacteria,* use organic compounds, such as acids and carbohydrates, for the photosynthetic reduction of carbon dioxide. Morphologically, the photosynthetic bacteria are very diverse, with spirals, rods, cocci, and even budding forms.

Cyanobacteria

The **cyanobacteria** are essentially aerobic. They carry out oxygen-producing photosynthesis, much as plants and the eucaryotic algae do (see Chapter 12). Some, however, can use reduced sulfur compounds for anoxygenic photosynthesis. A considerable number of cyanobacteria are capable of fixing nitrogen. Specialized cells called *heterocysts* (Figure 11.20a) contain enzymes that fix nitrogen gas (N_2) into ammonium (NH_4^+) for use by the growing cell. *Gas vacuoles* are found in many species that grow in water. The gas vacuole is a series of chambers or gas vesicles surrounded by a protein wall that is permeable to air but not to water. Gas vacuoles provide buoyancy that helps the

FIGURE 11.19   Purple sulfur bacteria. This phase-contrast micrograph of cells of the genus *Chromatium* shows the intracellular sulfur granules as multicolored objects.

cell to move to favorable environments. Cyanobacteria that are motile move about by gliding.

The cyanobacteria are morphologically varied. They have unicellular forms that divide by simple fission (Figure 11.20b), colonial forms that divide by multiple fission, and filamentous forms (Figure 11.20c) that reproduce by fragmentation of the filaments. The filamentous forms usually exhibit some differentiation of cells, which are often bound together within an envelope or sheath.

The cyanobacteria, especially those that fix nitrogen, are extremely important to the environment. They occupy environmental niches similar to those occupied by the eucaryotic algae described in Chapter 12 (see Roles of Algae in Nature, p. 309), but the ability of cyanobacteria to fix nitrogen makes them even more adaptable. The environmental role of the cyanobacteria is presented more fully in Chapter 27, in the discussion of eutrophication.

Actinomycetes

Actinomycetes are filamentous bacteria. Superficially, their morphology resembles that of the filamentous fungi; however, the filaments of actinomycetes consist of procaryotic cells with diameters much smaller than those of molds. Some actinomycetes further resemble molds by using externally carried asexual spores for reproduction.

Actinomycetes are very common inhabitants of soil, where a filamentous habit of growth has advantages. The organism can bridge water-free gaps between soil particles to move to a new nutritional site. This morphology also gives the organism a much higher surface-area-to-volume ratio and improves its nutritional efficiency in the highly competitive soil environment.

One genus, *Frankia* (frank'ē-ä), causes nitrogen-fixing nodules to form in alder tree roots, much as *Rhizobium*

(a) Filamentous   **LM**   10 μm

Heterocyst

(b) Nonfilamentous   **LM**   5 μm

(c) Branching filamentous   **LM**   10 μm

FIGURE 11.20 Cyanobacteria. (a) A filamentous cyanobacterium, *Anabaena spiroides*. Note the heterocysts, in which nitrogen-fixing activity is located. The terminal cell of a filament rounds up and no longer divides. **(b)** A nonfilamentous cyanobacterium, *Gleocapsa*, as seen with a phase-contrast microscope. Groups of two, three, or four cells are held together by the glycocalyx. **(c)** *Anabaena spiroides*, showing helical chains of rod-shaped cells.

causes nodules on the roots of legumes (see Chapter 27). The best-known genus of actinomycetes is *Streptomyces* (strep-tō-mī'sēs), which is one of the bacteria most commonly isolated from soil (Figure 11.21). The reproductive asexual spores of *Streptomyces*, called conidiospores, are formed at the ends of aerial filaments. If each conidiospore lands on a suitable substrate, it is capable of germinating into a new colony. These organisms are strict aerobes. They often produce extracellular enzymes that enable them to utilize proteins, polysaccharides such as starch or cellulose, and many other organic materials found in soil. *Streptomyces* produces a gaseous compound called *geosmin*, which gives fresh soil its typical musty odor. Species of *Streptomyces* are valuable because they produce most of our commercial antibiotics (see Chapter 20).

A Giant Procaryote

In 1991, a dramatic exception to the general rule that procaryotes are smaller than eucaryotes was discovered. In the process of examining the gut contents of a tropical surgeonfish, a group of Australian scientists discovered a highly unusual organism. Because it was more than half a millimeter in length and visible to the unaided eye (Figure 11.22), it was originally considered to be a protozoan. However, further analysis has proved that this organism, which lacks a nucleus, is indeed a giant bacterium. This new type of bacterium, which feeds on the algae eaten by its surgeonfish host, has been assigned the genus name *Epulopiscium*, which means "guest at the banquet of a fish."

In the next chapter we will turn our attention to algae, fungi, protozoa, and helminths.

(a)

(b)

FIGURE 11.21 *Streptomyces.* **(a)** Drawing of a typical *Streptomyces* showing filamentous, branching growth with asexual reproductive spores at filament tips. **(b)** A tangled mass of the filamentous bacterium *Streptomyces lividans* can be seen. A short chain of reproductive conidiospores, carried at the ends of the filaments, can be seen near the center of the photo.

FIGURE 11.22   A giant procaryote, *Epulopiscium fishelsoni.* Several of these bacteria, each roughly a million times the size of *Escherichia coli,* are shown on the head of a pin.

STUDY OUTLINE

BACTERIAL GROUPS (pp. 273–292)

1. *Bergey's Manual* divides bacteria into sections based on Gram-stain reaction, cellular morphology, oxygen requirements, and nutritional properties.
2. In some cases, the sections include families and orders, and some bacteria are included as genera of uncertain affiliation.

3. Spirochetes are long, thin, helical cells that move by means of an axial filament.
4. Aerobic/microaerophilic, motile, helical/vibrioid gram-negative bacteria move by means of one or more polar flagella.
5. Gram-negative aerobic rods and cocci have polar flagella (if flagellated) and can utilize a wide variety of organic compounds.

6. Facultatively anaerobic gram-negative rods have flagella (if motile) and include the enterics, Vibrionaceae, Pasteurellaceae, and the genus *Gardnerella*.
7. Members of the anaerobic gram-negative straight, curved, and helical rods can be found in humans.
8. Dissimilatory sulfate-reducing or sulfur-reducing bacteria are anaerobes that are important in the sulfur cycle.
9. Anaerobic gram-negative cocci are normal microbiota of the human mouth.
10. Rickettsias and chlamydias are obligate intracellular parasites.
11. Mycoplasmas are bacteria that lack cell walls.
12. Gram-positive cocci include the catalase-positive *Staphylococcus* and catalase-negative *Streptococcus, Enterococcus,* and *Lactococcus*.
13. Endospore-forming gram-positive rods and cocci may be aerobic, facultatively anaerobic, or anaerobic.
14. The diverse group of regular, nonsporing, gram-positive rods includes *Lactobacillus* and *Listeria*.
15. Irregular, nonsporing, gram-positive rods include the irregular-staining corynebacteria.

16. Pathogenic species of mycobacteria are acid-fast.
17. Nocardioform bacteria may be acid-fast; they form short filaments.
18. Bacteria with unusual morphologies are discussed in this book in terms of the following groups: budding and/or appendaged; gliding, nonfruiting; gliding, fruiting; budding; and sheathed.
19. The chemoautotrophic bacteria play important roles in the cycles of elements in the environment.
20. Extreme halophiles, acidophiles, thermophiles, and methane-producing bacteria are included in the archaeobacteria.
21. Photosynthetic purple and green bacteria are included in the group of anoxygenic phototrophic bacteria; they do not produce molecular oxygen.
22. Cyanobacteria produce molecular oxygen during photosynthesis; they are oxygenic phototrophs.
23. Actinomycetes produce mycelia and reproduce by external spores.
24. *Epulopiscium fishelsoni* is a bacterial giant, visible to the unaided eye.

STUDY QUESTIONS

Review

1. The following outline is a key that can be used to identify the medically important groups of bacteria. Fill in the name of the group in the space provided.

NAME OF GROUP AND
REPRESENTATIVE GENUS

I. Gram-positive
  A. Endospore-forming   _____
  B. Nonsporing
    1. Cocci   _____
    2. Rods
      a. Regular   _____
      b. Irregular   _____
      c. Acid-fast   _____
  C. Mycelium produced
    1. Acid-fast   _____
    2. Produce chains
      of conidia   _____

II. Gram-negative
  A. Cells are helical or curved
    1. Axial filament   _____
    2. No axial filament
      a. Aerobic   _____
      b. Anaerobic   _____
  B. Cells are rods or cocci
    1. Aerobic, nonfermenting   _____
    2. Facultatively anaerobic   _____
  C. Intracellular parasites   _____
III. Lacking cell wall   _____

2. Compare and contrast each of the following:
  (a) Cyanobacteria and algae
  (b) Actinomycetes and fungi
  (c) *Bacillus* and *Lactobacillus*

(d) *Pseudomonas* and *Escherichia*

(e) *Leptospira* and *Spirillum*

(f) *Veillonella* and *Bacteroides*

(g) *Rickettsia* and *Chlamydia*

(h) *Thermoplasma* and *Mycoplasma*

3. Matching:

I. Gram-positive	(a) Archaeobacteria
A. Nitrogen-fixing _____	(b) Cyanobacteria
II. Gram-negative	(c) *Cytophaga*
A. Phototrophic	(d) *Desulfovibrio*
1. Anoxygenic _____	(e) *Frankia*
2. Oxygenic _____	(f) *Hyphomicrobium*
B. Chemoautotrophic	(g) Methanogenic
1. Oxidize inorganics,	bacteria
such as NO_2^- _____	(h) Myxobacteria
2. Reduce CO_2 to	(i) *Nitrobacter*
CH_4 _____	(j) Purple bacteria
C. Chemoheterotrophic	(k) *Sphaerotilus*
1. Move via a	(l) *Sulfolobus*
slime layer _____	
2. Form	
myxospores _____	
3. Reduce sulfate to H_2S	
a. Anaerobic _____	
b. Thermophilic _____	
4. Long filaments,	
found in sewage _____	
5. Form projections	
from the cell _____	
III. Unusual cell wall	
(lacking peptidoglycan) _____	

Critical Thinking

1. Place each section listed in Table 11.1 in the appropriate division:

(a) Typical gram-positive

(b) Typical gram-negative

(c) Wall-less

(d) Unusual walls

2. Where is each of the following classified in *Bergey's Manual?* Why are they described more than once?

(a) *Gardnerella*

(b) *Desulfotomaculum*

(c) *Gallionella*

3. Identify the genus that best fits each of the following descriptions:

(a) This gram-negative genus is well suited to degrade hydrocarbons in an oil spill.

(b) This gram-positive genus presents the greatest source of bacterial damage to the beekeeping industry.

(c) This gram-negative genus is most probably able to oxidize arsenic and asbestos compounds and to clean up polluted water.

(d) This group can produce a fuel used for home heating and for generating electricity.

4. Identify the industrial product associated with each of the following:

(a) *Acetobacter*

(b) *Streptomyces*

(c) *Lactobacillus*

5. Explain why (a) a farmer would buy a bag of *Rhizobium leguminosarum* to spread in the field; and (b) a gardener would buy a box of *Bacillus thuringiensis* to spread on the lawn.

Clinical Applications

1. After contact with a patient's spinal fluid, a lab technician developed fever, nausea, and purple lesions on her neck and extremities. A throat culture grew gram-negative diplococci. What is the genus of the bacteria?

2. Between April 1 and May 15 of one year, 22 children in three states developed diarrhea, fever, and vomiting. The children had each received pet ducklings. Gram-negative, facultatively anaerobic bacteria were isolated from both the patients' and the ducks' feces; the bacteria were identified as serovar C2. What is the genus of these bacteria?

3. A woman complaining of lower abdominal pain with a temperature of 39°C gave birth soon after to a baby with respiratory difficulties. Blood cultures from the infant revealed gram-positive rods. The woman had a history of eating unheated hot dogs during her pregnancy. Which organism is most likely involved?

4. Four members of a family developed blurred vision, facial drooping, and dry mouth after eating. One man experienced respiratory failure. The patients had no apparent infections; however, gram-positive, endospore-forming rods were found in the moloha (salted, uneviscerated fish) they had eaten. What organism caused the patients' symptoms?

Fungi, Algae, Protozoa, and Multicellular Parasites

IN THIS CHAPTER we will examine the eucaryotic microorganisms: fungi, algae, protozoa, and parasitic helminths, or worms. (See Table 12.1 for a comparison of their characteristics.) Although most adult helminths are not microscopic, they do have microscopic stages in their development, and some of these stages cause disease; therefore, microbiologists study helminths. Microbiologists also study arthropods (jointed-legged animals) because some of them transmit microbial diseases.

FUNGI

LEARNING OBJECTIVES

- *List the defining characteristics of fungi.*
- *List the defining characteristics of the four phyla of fungi described in this chapter.*
- *Differentiate between asexual and sexual reproduction, and describe each of these processes in fungi.*

The study of fungi is called **mycology.** The fungi include yeasts, molds, and fleshy fungi. Yeasts are unicellular organisms. Molds are multicellular filamentous organisms, such as mildews, rusts, and smuts. Fleshy fungi are multicellular, filamentous organisms that produce a thick (fleshy) reproductive body. The fleshy fungi include mushrooms, puffballs, and coral fungi.

All fungi are chemoheterotrophs, requiring organic compounds for energy and carbon. Fungi are aerobic or facultatively anaerobic; only a few anaerobic fungi are known. Most fungi are saprophytes in soil and water; there they primarily decompose plant material. Like bacteria, fungi contribute significantly to the decomposition of matter and recycling of nutrients. Through the use of extracellular enzymes, such as cellulases and pectinases, fungi are the primary decomposers of the hard parts of plants, which cannot be digested by most animals. Table 12.2 lists the basic differences between fungi and eubacteria. Of the more than 100,000 species of fungi, only about 100 are pathogenic for humans and other animals. However, thousands of fungi are pathogenic to plants. Virtually every economically important plant is attacked by one or more of the fungi.

Characteristics of Fungi

Yeast identification, like bacterial identification, uses biochemical tests. However, multicellular fungi are identified on the basis of physical appearance, including colony characteristics and reproductive spores.

Vegetative Structures
Fungal colonies are described as **vegetative** structures because they are composed of the cells involved in catabolism and growth.

Molds and Fleshy Fungi. The **thallus** (body) of a mold or fleshy fungus consists of long filaments of cells joined together; these filaments are called **hyphae** (singular: *hypha*). In most molds, the hyphae contain crosswalls called **septa** (singular: *septum*), which divide the hyphae into distinct, uninucleate (one-nucleus) cell-like units. These hyphae are called **septate hyphae** (Figure 12.1a). In a few classes of fungi, however, the hyphae contain no septa and appear as long, continuous cells with many nuclei. These hyphae are called **coenocytic hyphae** (Figure 12.1b). Even in fungi with septate hyphae, there are usually openings in the septa that make the cytoplasm of adjacent "cells" continuous; therefore, these fungi are actually coenocytic organisms, too. The hyphae of a thallus grow by elongating at the tips (Figure 12.1c). Each part of a hypha is capable of growth, and, when a fragment breaks off, it can elongate to form a new hypha. In the laboratory, fungi are usually grown from fragments obtained from a fungal thallus.

When environmental conditions are suitable, the hyphae grow, intertwine, and form a filamentous mass called a **mycelium**, which is visible to the unaided eye (Figure 12.2a). The portion of the mycelium concerned with obtaining nutrients is called the *vegetative mycelium;* the portion concerned with reproduction is the *reproductive* or *aerial mycelium,* so called because it projects above the surface of the medium on which the fungus is growing. The aerial mycelium often bears reproductive spores (Figure 12.2b), which we will discuss later.

Yeasts. Yeasts are nonfilamentous, unicellular fungi that are typically spherical or oval. Like molds, yeasts are widely distributed in nature; they are frequently found as a white powdery coating on fruits and leaves. Because most yeasts are colonies of unicellular organisms, they do not reproduce as a unit. Instead, the colony grows as the number of yeast cells increases by mitosis. Yeasts that undergo *fission,* such as *Schizosaccharomyces,* divide evenly to produce two new cells. Yeasts that undergo *budding,* such as *Saccharomyces,* divide unevenly.

In budding (Figure 12.3), the parent cell forms a protuberance (bud) on its outer surface. As the bud elongates,

FIGURE 12.1 Characteristics of fungal hyphae. (a) Septate hyphae have crosswalls dividing the hyphae into cell-like units. (b) Coenocytic hyphae lack crosswalls. (c) Hyphae grow by elongating at the tips.

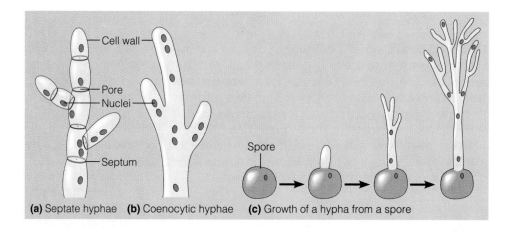

(a) Septate hyphae   **(b)** Coenocytic hyphae   **(c)** Growth of a hypha from a spore

TABLE 12.1 Major Differences Among Eucaryotic Microorganisms: Fungi, Algae, Protozoa, and Helminths				
	FUNGI	**ALGAE**	**PROTOZOA**	**HELMINTHS**
Kingdom	Fungi	Protista and Plantae	Protista	Animalia
Nutritional type	Chemoheterotroph	Photoautotroph	Chemoheterotroph	Chemoheterotroph
Multicellularity	All, except yeasts	Some	None	All
Cellular arrangement	Unicellular, filamentous, fleshy (such as mushrooms)	Unicellular, colonial, filamentous; tissues	Unicellular	Tissues and organs
Food acquisition method	Absorptive	Absorptive	Absorptive; ingestive (cytostome)	Ingestive (mouth); absorptive
Characteristic features	Sexual and asexual spores	Pigments	Motility; some form cysts	Many have elaborate life cycles, including egg, larva, and adult
Embryo formation	None	Some	None	All

Aerial mycelium

Vegetative mycelium

(a) *Aspergillus niger* on plate

(b) *A. niger* aerial mycelium   **LM**   100 μm

FIGURE 12.2 The fungal mycelium. (a) *Aspergillus niger* colony grown on glucose agar plate, showing both vegetative and aerial mycelia. **(b)** Micrograph of part of an aerial mycelium showing reproductive spores.

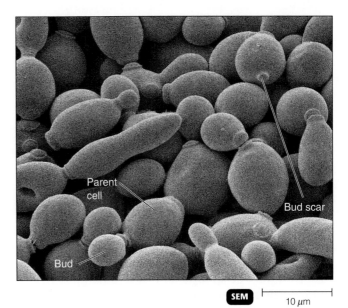

Parent cell

Bud scar

Bud

SEM   10 μm

FIGURE 12.3 A budding yeast. Micrograph of *Saccharomyces cerevisiae* in various stages of budding.

the parent cell's nucleus divides, and one nucleus migrates into the bud. Cell wall material is then laid down between the bud and parent cell, and the bud eventually breaks away.

One yeast cell can in time produce up to 24 daughter cells by budding. Some yeasts produce buds that fail to detach themselves; these buds form a short chain of cells called a **pseudohypha** (see Figure 21.15a).

During fission, the parent cell elongates, its nucleus divides, and two daughter cells are produced. Increases in the number of yeast cells on a solid medium produce a colony similar to a bacterial colony.

Yeasts are capable of facultative anaerobic growth. As you may recall from Chapter 5, yeasts can use oxygen or an organic compound as the final electron acceptor; this is a valuable attribute because it allows these fungi to

TABLE 12.2 Comparison of Selected Features of Fungi and Eubacteria

	FUNGI	EUBACTERIA
Cell type	Eucaryotic with well-defined nuclear membrane	Procaryotic
Cell membrane	Sterols present	Sterols absent, except in *Mycoplasma*
Cell wall	Glucans; mannans; chitin (no peptidoglycan)	Peptidoglycan
Spores	Produce a wide variety of sexual and asexual reproductive spores	Endospores (not for reproduction); some asexual reproductive spores
Metabolism	Limited to heterotrophic; aerobic, facultatively anaerobic	Heterotrophic, chemoautotrophic, photoautotrophic; aerobic, facultatively anaerobic, anaerobic
Sensitivity to antibiotics	Often sensitive to polyenes, imidazoles, and griseofulvin	Often sensitive to penicillins, tetracyclines, and aminoglycosides

SOURCE: After B. D. Davis et al. *Microbiology,* 4th ed. Philadelphia: J. B. Lippincott, 1990, p. 746.

survive in different environments. If given access to oxygen, yeasts perform aerobic respiration to metabolize carbohydrates to carbon dioxide and water; denied oxygen, they ferment carbohydrates and produce ethanol and carbon dioxide. This fermentation is the basis of the brewing, wine-making, and baking industries. Species of *Saccharomyces* produce ethanol in brewed beverages and carbon dioxide for leavening bread dough.

Dimorphic Fungi. Some fungi, most notably the pathogenic species, exhibit **dimorphism**—that is, two forms of growth. Such fungi can grow either as a mold or as a yeast. The moldlike forms produce vegetative and aerial mycelia; the yeastlike forms reproduce by budding. Dimorphism in pathogenic fungi is temperature-dependent: at 37°C the fungus is yeastlike, and at 25°C it is moldlike. However, the appearance of the dimorphic (in this instance, nonpathogenic) fungus shown in Figure 12.4 changes with CO_2 concentration. Recently, researchers at the Massachusetts Institute of Technology showed that a filamentous form of the baker's yeast *Saccharomyces cerevisiae* is induced by starvation.

Life Cycle

As mentioned previously, fungi increase their cell numbers by mitotic division. Filamentous fungi can reproduce asexually by fragmentation of their hyphae. In addition, both sexual and asexual reproduction in fungi occurs by the formation of **spores**. In fact, fungi are classified and identified by spore type; thus, if a fungus is to be identified, its spores must be examined.

Fungal spores, however, are quite different from bacterial endospores. Bacterial endospores (see Chapter 4) are formed so the bacterial cell can survive adverse environmental conditions. A single vegetative bacterial cell forms one endospore, which eventually germinates to produce a single vegetative bacterial cell. This process is not reproduction because it does not increase the total number of bacterial cells. But after a mold forms a spore, the spore detaches from the parent and germinates into a new mold (see Figure 12.1c). Unlike the bacterial endospore, this is a true reproductive spore; a second organism grows from the spore. Although fungal spores can survive for extended periods in dry or hot environments, they do not exhibit the extreme tolerance and longevity of bacterial endospores.

Spores are formed from the aerial mycelium in a great variety of ways, depending on the species. Fungal spores can be either asexual or sexual. **Asexual spores** are formed by the aerial mycelium of one organism. When these spores germinate, they become organisms that are genetically identical to the parent. **Sexual spores** result from the fusion of nuclei from two opposite mating strains of the same species of fungus. Fungi produce sexual spores less frequently than asexual spores. Organisms that grow from sexual spores will have genetic characteristics of both parental strains. Because spores are of considerable

Yeastlike growth   Moldlike growth

LM |—————| 10 μm

FIGURE 12.4 Fungal dimorphism. Dimorphism in the fungus *Mucor rouxii* depends on CO_2 concentration. On the agar surface *Mucor* exhibits yeastlike growth, but in the agar it is moldlike.

importance in the identification of fungi, we will next look at some of the various types of asexual and sexual spores.

Asexual Spores. Asexual spores are produced by an individual fungus through mitosis and subsequent cell division; there is no fusion of the nuclei of cells. Fungi are identified by examining their asexual spores. Several types of asexual spores are produced by fungi. One type, an **arthrospore**, is formed by the fragmentation of a septate hypha into single, slightly thickened cells (see Figures 12.5a and 24.16b). One species that produces such spores is *Coccidioides immitis* (kok-sid-ē-oi'dēz im'mi-tis). Another type of asexual spore is a **chlamydospore**, a thick-walled spore formed by rounding and enlargement within a hyphal segment (Figure 12.5b). A fungus that produces chlamydospores is the yeast *Candida albicans* (kan'did-ä al'bi-kanz). A **sporangiospore** is an asexual spore formed within a sac (**sporangium**) at the end of an aerial hypha called a **sporangiophore**. The sporangium can contain hundreds of sporangiospores (Figure 12.5c). Such spores are produced by *Rhizopus*. A fourth principal type of asexual spore is a **conidiospore**, which is a unicellular or multicellular spore that is not enclosed in a sac (Figure 12.5d). Conidiospores are produced in a chain at the end of a **conidiophore**. Such spores are produced by *Aspergillus*. A fifth type of asexual spore, a **blastospore**, consists of a bud coming off the parent cell (Figure 12.5e). Such spores are found in some yeasts, such as *C. albicans* and *Cryptococcus*.

FIGURE 12.5 Representative asexual spores. (a) Fragmentation of hyphae results in formation of arthrospores in this *Coccidioides immitis*. **(b)** Chlamydospores are thick-walled cells within hyphae of this *Candida albicans*. **(c)** Sporangiospores are formed within a sporangium (spore sac) of this *Rhizopus*. **(d)** Conidiospores are arranged in chains at the end of a conidiophore on this *Aspergillus flavus*. **(e)** Blastospores are formed from buds of a parent cell of *C. albicans*.

Sexual Spores. A fungal sexual spore results from sexual reproduction, consisting of three phases:

1. A haploid nucleus of a donor cell (+) penetrates the cytoplasm of a recipient cell (−).

2. The (+) and (−) nuclei fuse to form a diploid zygote nucleus.

3. By meiosis, the diploid nucleus gives rise to haploid nuclei (sexual spores), some of which may be genetic recombinants.

Often, sexual spores are produced only under special circumstances. For example, the plant-pathogenic ascomycetes produce ascospores only when host plants are at the end of a growing season. This fact suggests that the ascospores are initiated by changes in moisture, temperature, or nutrient availability. The ascospores survive until the plant hosts are growing again. The sexual spores produced by fungi are the criterion used to classify the fungi into several phyla.

Nutritional Adaptations

Fungi are generally adapted to environments that would be hostile to bacteria; thus, they can survive in places where microbial growth is not expected. Fungi are chemoheterotrophs, and, like bacteria, they absorb nutrients rather than ingesting them, as animals do. However, fungi differ from bacteria in certain environmental requirements and in the following nutritional characteristics:

1. Fungi usually grow better in an environment with a pH of 5.0, which is too acidic for the growth of most common bacteria.

2. Most molds are aerobic, so they grow on surfaces rather than throughout a substrate. Yeasts are facultative anaerobes.

3. Most fungi are more resistant to osmotic pressure than bacteria are; most fungi are therefore able to grow in high sugar or salt concentrations.

4. Fungi can grow on substances with a very low mois-

ture content, generally too low to support the growth of bacteria.

5. Fungi require somewhat less nitrogen than bacteria for an equivalent amount of growth.

6. Fungi are capable of metabolizing complex carbohydrates, such as lignin (wood), that most bacteria cannot use for nutrients.

These characteristics allow fungi to grow on such unlikely substrates as painted walls, shoe leather, or discarded newspapers.

Medically Important Phyla of Fungi

This section provides an overview of medically important phyla of fungi. The actual diseases they cause will be studied in Part Four, Chapters 21–26. Note that not all fungi cause disease. Other phyla of fungi are not discussed here because they contain no pathogens.

The genera named in the following phyla include many that are readily found as contaminants in foods and in laboratory bacterial cultures. Although these genera are not all of primary medical importance, they are typical examples of their respective groups.

Deuteromycota

Fungi are placed in the category Deuteromycota because they have not yet been found to produce sexual spores. Members of this phylum produce the asexual chlamydospores, arthrospores, and conidiospores; budding also occurs. Deuteromycota have septate hyphae.

Most of the pathogenic fungi are, or once were, classified as Deuteromycota. This phylum might be described as a "holding category" in which fungi are placed until sexual spores are observed and the fungus can be properly classified. Table 12.3 lists some well-known deuteromycetes that have recently been classified. Note that the generic names are changed with reclassification. When different species of a deuteromycete genus are observed to have morphologically different sexual spores, the species might be reclassified into two or more different genera, as was done with *Penicillium*. However, the asexual stage of the fungus is often still called *Penicillium*.

A Case of Mistaken Identity? *Pneumocystis* (nü-mō-sis′tis) pneumonia is the most common opportunistic infection in immunosuppressed individuals and is a significant cause of death in AIDS patients. The causative agent of this infection, *Pneumocystis carinii* (kär-i′-nē-ē), was not considered a human pathogen until the 1970s.

Pneumocystis lacks structures that can be easily used for identification (see Figure 24.18), and its taxonomic position has been uncertain since its discovery in 1908. Although it was provisionally classified as a protozoan, recent studies comparing its ribosomal RNA sequence with

TABLE 12.3 Several Genera of Reclassified Deuteromycota

DEUTEROMYCETE NAME	CURRENT PHYLUM	RECLASSIFIED NAME
Aspergillus	Ascomycota	*Sartorya, Eurotium, Emericella*
Blastomyces	Ascomycota	*Ajellomyces*
Candida	Ascomycota	*Pichia*
Cryptococcus	Basidiomycota	*Filobasidiella*
Histoplasma	Ascomycota	*Ajellomyces*
Microsporum	Ascomycota	*Nannizia*
Penicillium	Ascomycota	*Eupenicillium, Talaromyces*
Trichophyton	Ascomycota	*Arthroderma*

those of other protozoa, *Euglena*, cellular slime molds, plants, mammals, and fungi have shown that it might actually be a member of the Kingdom Fungi. Researchers have not been able to culture *Pneumocystis*, but they have developed some useful treatments for *Pneumocystis* pneumonia. Perhaps as researchers take into account that this organism may be a fungus, appropriate culture methods and treatments will result. If enough evidence is gathered to reclassify *Pneumocystis* as a fungus, it will be placed in the Deuteromycota until sexual spores are observed.

Zygomycota

The Zygomycota, or conjugation fungi, are saprophytic molds that have coenocytic hyphae. An example is *Rhizopus nigricans*, the common black bread mold. The asexual spores of *Rhizopus* are sporangiospores (Figure 12.6, upper right). The dark sporangiospores inside the sporangium give *Rhizopus* its descriptive common name. When the sporangium breaks open, the sporangiospores are dispersed. If they fall on a suitable medium, they will germinate into a new mold thallus.

The sexual spores are zygospores. A **zygospore** is a large spore enclosed in a thick wall (Figure 12.6, lower left). This type of spore results from the fusion of the nuclei of two cells that are morphologically similar to each other.

Ascomycota

The Ascomycota, or sac fungi, include molds with septate hyphae and yeasts. Their asexual spores are usually conidiospores produced in long chains from the conidiophore. The arrangement of conidiospores in *Eupenicillium* is shown in Figure 12.7, right (p. 302). The term *conidia* means dust, and these spores freely detach from the chain at the slightest disturbance and float in the air like dust.

An **ascospore** results from the fusion of the nuclei of two cells that can be either morphologically similar or dissimilar. These spores are produced in a saclike structure called an **ascus** (Figure 12.7, left). The members of this phylum are called sac fungi because of the ascus.

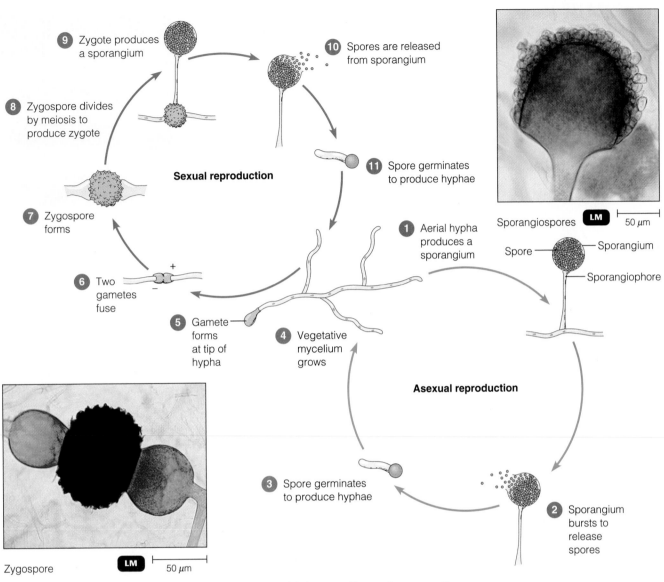

9 Zygote produces a sporangium

10 Spores are released from sporangium

8 Zygospore divides by meiosis to produce zygote

Sexual reproduction

7 Zygospore forms

Sporangiospores LM |—— 50 μm ——|

Spore — Sporangium

Sporangiophore

1 Aerial hypha produces a sporangium

6 Two gametes fuse

5 Gamete forms at tip of hypha

4 Vegetative mycelium grows

11 Spore germinates to produce hyphae

Asexual reproduction

3 Spore germinates to produce hyphae

2 Sporangium bursts to release spores

Zygospore LM |—— 50 μm ——|

FIGURE 12.6 Life cycle of *Rhizopus*, a zygomycete. This fungus will reproduce asexually most of the time. Two opposite mating strains (designated as + and –) are necessary for sexual reproduction.

Basidiomycota

The Basidiomycota, or club fungi, also possess septate hyphae. This phylum includes fungi that produce mushrooms. **Basidiospores** are formed externally on a base pedestal called a **basidium** (Figure 12.8, p. 303). (The common name of the fungus is derived from the shape of the basidium.) There are usually four basidiospores per basidium. Some of the basidiomycota produce asexual conidiospores. Representative basidiomycetes are shown in Figure 12.9 (p. 304).

Fungal Diseases

Any fungal infection is called a **mycosis**. Mycoses are generally chronic (long-lasting) infections because fungi grow slowly. Mycoses are divided into five groups according to the degree of tissue involvement and mode of entry into the host. They are classified as systemic, subcutaneous, cutaneous, superficial, or opportunistic. Characteristics of the medically important fungi are summarized in Table 12.4 on pages 306–307.

Systemic mycoses are fungal infections deep within the body. They are not restricted to any particular region of the body but can affect a number of tissues and organs. Systemic, or deep, mycoses are usually caused by saprophytic fungi that live in the soil. Inhalation of spores is the route of transmission; these infections typically begin in the lungs and then spread to other body tissues. They are not contagious from animal to human or from human to human. Two systemic mycoses, histoplasmosis and coccidioidomycosis, are discussed in Chapter 24.

FIGURE 12.7 Life cycle of _Eupenicillium,_ an ascomycete. Occasionally, when two opposite mating cells from two different strains (+ and –) fuse, sexual reproduction occurs.

Conidiospores

Conidiophore

SEM   5 μm

Conidiospores

Conidiophore

7 Ascus opens to release ascospores

8 Ascospore germinates to produce hyphae

1 Hypha produces conidiophore

Sexual reproduction

Asexual reproduction

4 Vegetative mycelium grows

2 Conidiospores are released from conidiophore

6 Ascospores form from fused cells by meiosis

– +

5 Two hyphal cells fuse

3 Conidiospore germinates to produce hyphae

Subcutaneous mycoses are fungal infections beneath the skin and are caused by saprophytic fungi that live in soil and on vegetation. Infection occurs by direct implantation of spores or mycelial fragments into a puncture wound in the skin.

Fungi that infect only the epidermis, hair, and nails are called **dermatophytes,** and their infections are called dermatomycoses or **cutaneous mycoses** (see Figure 21.14). Dermatophytes secrete _keratinase,_ an enzyme that degrades keratin. Keratin is a protein found in hair, skin, and nails. Infection is transmitted from human to human or from animal to human by direct contact or by contact with infected hairs and epidermal cells (as from barber shop clippers or shower room floors).

The fungi that cause **superficial mycoses** are localized along hair shafts and in superficial (surface) epidermal

cells. These infections are prevalent in tropical climates.

An **opportunistic pathogen** is generally harmless in its normal habitat but can become pathogenic in a host who is seriously debilitated or traumatized, who is under treatment with broad-spectrum antibiotics, or whose immune system is suppressed by drugs or by an immune disorder. AIDS patients are quite susceptible to opportunistic pathogens. A person's normal microbiota can become opportunistic pathogens under these conditions.

Mucormycosis is an opportunistic mycosis caused by _Rhizopus_ and _Mucor_ (mū-kôr); the infection occurs mostly in patients with ketoacidosis resulting from diabetes mellitus, leukemia, or treatment with immunosuppressive drugs. **Aspergillosis** is also an opportunistic mycosis; it is caused by _Aspergillus_ (see Figure 12.2). This disease occurs in people who have debilitating lung diseases or cancer

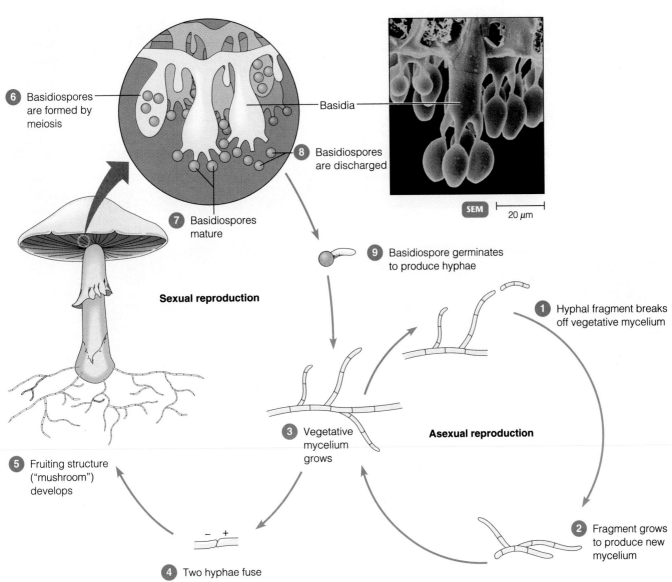

6 Basidiospores are formed by meiosis

Basidia

8 Basidiospores are discharged

SEM ⊢ 20 μm

7 Basidiospores mature

9 Basidiospore germinates to produce hyphae

Sexual reproduction

1 Hyphal fragment breaks off vegetative mycelium

3 Vegetative mycelium grows

Asexual reproduction

5 Fruiting structure ("mushroom") develops

2 Fragment grows to produce new mycelium

– +

4 Two hyphae fuse

FIGURE 12.8 Generalized life cycle of a basidiomycete. Mushrooms appear after cells from two mating strains (+ and –) have fused.

and have inhaled *Aspergillus* spores. The mycosis **candidiasis** (see Figure 21.15b) is most frequently caused by *Candida albicans* and may occur as vulvovaginal candidiasis during pregnancy. **Thrush**, a mucocutaneous candidiasis, is an inflammation of the mouth and throat; it frequently occurs in newborns and in people with AIDS.

Some fungi cause disease by producing toxins. These toxins will be discussed in Chapter 15.

Economic Effects of Fungi

Fungi have been used in biotechnology for many years. The yeast *Saccharomyces cerevisiae* is used to make bread and wine. *Saccharomyces* and another yeast, *Torulopsis,* are used as protein supplements in foods. *Trichoderma* is used commercially to produce the enzyme cellulase, which is used to remove plants' cell walls to produce a clear fruit juice. When the anticancer drug taxol, which is produced by yew trees, was discovered, there was concern that the yew forests of the northwestern coast of the U.S.A. would be decimated to harvest the drug. However, in 1993, Andrea and Donald Stierle saved the yews by discovering that the fungus *Taxomyces* also produces taxol. Fungi are also being used to produce herbicides.

In contrast to these beneficial effects, fungi can have undesirable effects for industry and agriculture because of their nutritional adaptations. As most of us have observed, mold spoilage of fruits, grains, and vegetables is relatively common, but bacterial spoilage of such foods is not. There is little moisture on the unbroken surfaces of such foods, and the interiors of fruits are too acidic for many bacteria

(a) *Boletus edulis*

(b) *Crucibulum vulgare*

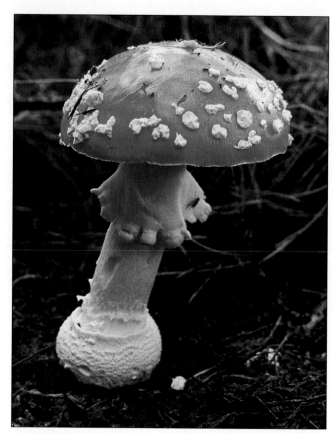

(c) *Amanita muscaria*

FIGURE 12.9 **Representative basidiomycetes.** **(a)** Two fruiting bodies of the edible basidiomycete *Boletus edulis*. **(b)** Bird's nest fungus (*Crucibulum vulgare*) growing on a twig. The basidiospores seen in one of the cups will pop out when the cup is hit by a raindrop. **(c)** *Amanita muscaria* produces amanitin neurotoxin.

to grow there. Jams and jellies also tend to be acidic, and they have a high osmotic pressure from the sugars they contain. These factors all discourage bacterial growth but readily support the growth of molds. A paraffin layer on top of a jar of homemade jelly helps deter mold growth because molds are aerobic and the paraffin layer keeps out the oxygen. However, fresh meats and certain other foods are such good substrates for bacterial growth that bacteria not only will outgrow molds but also will actively suppress mold growth in these foods.

The ability of fungi to grow at low moisture levels is of particular importance in their role as plant pathogens. Bacterial plant pathogens are far less common than fungal pathogens. In fact, the fungus that caused the great potato blight in Ireland during the early 1800s, *Phytophthora infestans* (fī-tof'thô-rä in-fes'tans), was one of the first microorganisms to be associated with a disease.

The spreading chestnut tree, of which Longfellow wrote, no longer grows in the U.S.A. except in a few widely isolated sites; a fungal blight killed virtually all of them. This blight was caused by the ascomycete *Cryphonectria parasitica* (kri-fō-nek'trē-ä par-ä-si'ti-kä), which was introduced into the U.S.A. from China about 1904. The fungus

allows the tree roots to live and put forth shoots regularly, but then it kills the shoots just as regularly. Another imported fungal plant disease is Dutch elm disease, caused by *Ceratocystis ulmi* (sē-rä-tō-sis'tis ul'mē). Carried from tree to tree by a bark beetle, the fungus blocks the afflicted tree's circulation. The disease has devastated the U.S.A.'s elm population.

ALGAE

LEARNING OBJECTIVES

- *List the defining characteristics of algae.*
- *List the outstanding characteristics of the six phyla of algae discussed in this chapter.*

Algae are familiar as the large brown kelp in coastal waters, the green scum in a puddle, and the green stains on soil or on rocks. Some algae are unicellular, others form chains of cells (are *filamentous*), and a few have plantlike bodies called *thalli* (singular: thallus). Algae are mostly

FIGURE 12.10 **Algae and their habitats.** **(a)** Although unicellular and filamentous algae can be found on land, they frequently exist in marine and freshwater environments as plankton. Multicellular green, brown, and red algae require a suitable attachment site, adequate water for support, and light of the appropriate wavelengths. **(b)** *Macrocystis pyrifera,* a brown alga. The hollow stipe and gas-filled pneumatocyst hold the thallus upright to ensure that sufficient sunlight is received for growth. **(c)** *Microcladia,* a red alga. The delicately branched red algae get their color from phycobiliprotein accessory pigments.

aquatic, although some are found in soil or on trees when sufficient moisture is available there. Unusual algal habitats include the hair of both the sedentary South American sloth and the polar bear. Water is necessary for physical support, reproduction, and diffusion of nutrients. Generally, algae are found in cool temperate waters, although the large floating mats of the brown alga *Sargassum* (sär-gas′sum) are found in the subtropical Sargasso Sea. Some species of brown algae grow in antarctic waters.

Characteristics of Algae

All algae have eucaryotic cells and have had flagella at some time during their lives, usually on sperm. Identification of unicellular and filamentous algae requires micro-

scopic examination. Most algae are found in the ocean. Their locations depend on the availability of appropriate nutrients, wavelengths of light, and surfaces for them to grow on. Red, brown, and multicellular green algae require rocks, snails, or even other algae for attachment. Probable locations for representative algae are shown in Figure 12.10a.

Vegetative Structures

The body of a multicellular alga is called a **thallus.** Thalli of the larger multicellular algae, those commonly called seaweeds, consist of branched **holdfasts** (which anchor the alga to a rock), stemlike and often hollow **stipes,** and leaflike **blades** (Figure 12.10b). The cells covering the thallus can carry on photosynthesis. The thallus lacks the con-

TABLE 12.4 Characteristics of Some Parasitic Fungi

PHYLUM	GROWTH CHARACTERISTICS	ASEXUAL SPORE TYPES	HUMAN PATHOGENS
Zygomycota	Nonseptate hyphae	Sporangiospores	*Rhizopus*
			Mucor
Ascomycota	Septate hyphae	Conidiospores	*Allescheria boydii*
	Dimorphic	Conidiospores	*Aspergillus**
			Blastomyces dermatitidis*
			Histoplasma capsulatum*
	Septate hyphae, strong affinity for keratin	Conidiospores	*Microsporum**
		Arthrospores	*Trichophyton**
Basidiomycota	Septate hyphae; includes the rusts and smuts, and plant pathogens; yeastlike encapsulated cells	Conidiospores	*Cryptococcus neoformans**
Deuteromycota	Septate hyphae	Conidiospores	*Epidermophyton*
		Chlamydospores	*Cladosporium werneckii*
	Dimorphic	Conidiospores	*Sporothrix schenckii*
		Arthrospores	*Coccidioides immitis*
	Yeastlike, pseudohyphae	Chlamydospores	*Candida albicans**
	Dimorphic	Chlamydospores Arthrospores Conidiospores	*Paracoccidioides brasiliensis*
	Septate hyphae, produce melaninlike pigments	Conidiospores	*Phialophora Fonsecaea*
	Yeastlike pseudohyphae	Arthrospores	*Trichosporon*

*Deuteromycete name. Refer to Table 12.3 for current classification and names.

ductive tissue (xylem and phloem) characteristic of vascular plants; algae absorb nutrients from the water over their entire surface. The stipe is not lignified or woody, so it does not offer the support of a plant's stem; instead, the pressure of surrounding water supports the algal thallus; some algae are also buoyed by a floating, gas-filled bladder.

Life Cycle

All algae can reproduce asexually. Multicellular algae with thalli and filamentous forms can fragment; each piece is capable of forming a new thallus or filament. When a unicellular alga divides, its nucleus divides (mitosis) and then the two nuclei move to opposite parts of the cell. The cell then divides into two complete cells (cytokinesis).

Sexual reproduction occurs in some algae (see Figure 12.11b). In some species, asexual reproduction may occur for several generations and then, under different conditions, the same species reproduce sexually. Other species alternate generations so that the offspring resulting from sexual reproduction reproduce asexually, and the next generation then reproduces sexually.

Nutrition

Algae are photoautotrophs; hence they are found throughout the photic (light) zone of bodies of water. Chlorophyll *a* (light-trapping pigment) and accessory pigments involved in photosynthesis are responsible for the distinctive colors of many algae. Some algae exist as single cells; others exist as colonies of hundreds or thousands of cells.

Algae are classified as protists or plants according to their structures, pigments, and other qualities (Table 12.5). Following are descriptions of some phyla of algae.

Selected Phyla of Algae

The *brown algae* or **kelp** (see Figure 12.10b) are macroscopic; some reach lengths of 50 meters. Most brown algae are found in coastal waters. Brown algae have a phenomenal growth rate. Some grow at rates exceeding 20 cm per day and therefore can be harvested regularly. *Algin,* a thickener used in many foods (such as ice cream and cake decorations), is extracted from their cell walls. Algin is

HABITAT	TYPE OF MYCOSIS	CLINICAL NOTES	REFER TO CHAPTER
Ubiquitous	Systemic	Opportunistic pathogen	24
Ubiquitous	Systemic	Opportunistic pathogen	24
Soil	Subcutaneous	Primary cause of maduromycosis	—
Ubiquitous	Systemic	Opportunistic pathogen	24
Unknown	Systemic	Inhalation	24
Soil	Systemic	Inhalation	24
Soil, animals	Cutaneous	Tinea capitis (ringworm)	21
Soil, animals	Cutaneous	Tinea pedis (athlete's foot)	21
Soil, bird feces	Systemic	Inhalation	22
Soil, humans	Cutaneous	Tinea cruris (jock itch), tinea unguium (of fingernails or toenails)	21
Ubiquitous	Superficial	Tinea nigra	—
Soil	Subcutaneous	Puncture wound	21
Soil	Systemic	Inhalation	24
Human normal microbiota	Cutaneous, systemic, mucocutaneous	Opportunistic pathogen	21, 26
Presumably soil	Systemic	Inhalation	—
Soil, plant debris	Subcutaneous	Chromomycosis	—
Soil, plant debris	Subcutaneous	Chromomycosis	—
Soil (?), humans	Superficial	White piedra	—

TABLE 12.5 Characteristics of Selected Phyla of Algae

	BROWN ALGAE	RED ALGAE	GREEN ALGAE	DIATOMS	DINOFLAGELLATES	EUGLENOIDS
Kingdom	Protista	Protista	Often Plantae	Protista	Protista	Protista
Phylum	Phaeophyta	Rhodophyta	Chlorophyta	Bacillariophyta	Dinoflagellata	Euglenophyta
Color	Brownish	Reddish	Green	Brownish	Brownish	Green
Cell wall	Cellulose and alginic acid	Cellulose	Cellulose	Pectin and silica	Cellulose and silica	None; protein pellicle
Flagella	In gametes	No	In unicellular forms and in gametes	In gametes	2	1
Cell arrangement	Multicellular	Most are multicellular	Unicellular and multicellular	Unicellular	Unicellular	Unicellular
Nutrition	Autotrophic	Autotrophic	Autotrophic	Autotrophic	Autotrophic; some heterotrophic	Autotrophic; some heterotrophic
Pigments	Chlorophyll *a* and *c*, xanthophylls	Chlorophyll *a* and *d*, phycobiliproteins	Chlorophyll *a* and *b*	Chlorophyll *a* and *c*, carotene, xanthophylls	Chlorophyll *a* and *c*, carotene, xanthins	Chlorophyll *a* and *b*, carotene
Sexual reproduction	Yes	Yes	Yes	Yes	In a few (?)	No
Storage materia	Carbohydrate (laminarin)	Starchlike	Starchl	Oil	Starch	Glucose polymer

(a) Multicellular green alga (*Ulva*)　　**(b)** Life cycle of a unicellular green alga (*Chlamydomonas*)

FIGURE 12.11　Green algae. (a) The multicellular green alga *Ulva*. **(b)** Life cycle of the unicellular green alga *Chlamydomonas*. Two whiplike flagella propel this cell.

also used in the production of a wide variety of nonfood goods, including rubber tires and hand lotion. Most *red algae* (Figure 12.10c) have delicately branched thalli and live at greater ocean depths than other algae. The thalli of a few red algae form crustlike coatings on rocks and shells. The red pigments allow red algae to absorb the blue light that penetrates deepest into the ocean. Agar is extracted from the red alga *Gelidium* (jel-id'ē-um). Another gelatinous material, called carrageenan, comes from a species of red algae called Irish moss. Carrageenan can be a thickening ingredient in evaporated milk, ice cream, and pharmaceuticals.

Green algae (Figure 12.11) can be classified as plants because they have cellulose cell walls, contain chlorophyll *a* and *b*, and store starch, as plants do. Green algae are believed to have given rise to terrestrial plants. Most green algae are microscopic, although they may be either unicellular or multicellular. Some filamentous kinds form the grass-green scum in ponds.

Diatoms (Figure 12.12) are unicellular or filamentous algae with complex cell walls that consist of pectin and a layer of silica. The two parts of the wall fit together like the halves of a Petri plate. The distinctive patterns of the walls are a useful tool in diatom identification. Diatoms store energy captured through photosynthesis in the form of oil.

(a) Diatoms (*Achinanthes*)　**SEM** ├────┤ 10 μm

(b) Asexual reproduction of a diatom

FIGURE 12.12　Diatoms. (a) In this micrograph of *Achinanthes minutissima,* note how the two parts of the cell wall fit together. **(b)** Asexual reproduction of a diatom. During mitosis, each daughter cell retains one-half of the cell wall from the parent (green) and must synthesize the remaining half (blue).

FIGURE 12.13 Some types of unicellular algae. (a) *Peridinium,* a dinoflagellate. Like some other dinoflagellates, *Peridinium* has two flagella in perpendicular, opposing grooves. When the two flagella beat simultaneously, they cause the cell to spin. **(b)** *Euglena.* Semirigid rings supporting the pellicle allow *Euglena* to change shape.

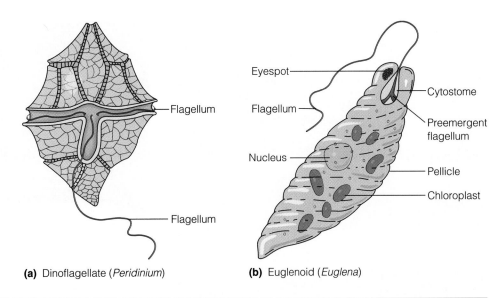

(a) Dinoflagellate (*Peridinium*)   **(b)** Euglenoid (*Euglena*)

Dinoflagellates (Figure 12.13a) are unicellular algae that are **plankton**—free-floating organisms. Their cell walls consist of many individual plates made of cellulose and silica. Marine dinoflagellates in the genus *Gonyaulax* (gō-nē-ō′laks) produce neurotoxins that cause paralytic shellfish poisoning. This disease spreads to humans when large numbers of dinoflagellates are eaten by mollusks—such as mussels or clams—that in turn are eaten by humans. Large concentrations of *Gonyaulax* give the ocean a deep red color, from which the name *red tide* originates. Mollusks should not be harvested for human consumption during a red tide. A similar disease called *ciguatera* occurs when the dinoflagellate *Gambierdiscus toxicus* (gam′bē-ėr-dis-kus toks′i-kus) passes up the food chain and is concentrated in large fish. Ciguatera is endemic (constantly present) in the South Pacific Ocean and the Caribbean Sea.

Euglenoids (Figure 12.13b) are unicellular, flagellated algae. Euglenoids have a semirigid plasma membrane called a *pellicle,* and they move by means of an anterior flagellum. All euglenoids except one have a red *eyespot* at the anterior end. This carotenoid-containing organelle senses light and directs the cell in the appropriate direction by using the *preemergent flagellum.* Some euglenoids are facultative chemoheterotrophs. In the dark, they ingest organic matter through the *cytostome.* Euglenoids are frequently studied with protozoa because they lack a cell wall and can move and feed using the flagellum.

Roles of Algae in Nature

Algae are an important part of any aquatic food chain because they fix carbon dioxide into organic molecules that can be consumed by chemoheterotrophs. Using the energy produced in photophosphorylation, algae convert carbon dioxide in the atmosphere into carbohydrates. Molecular oxygen (O_2) is a by-product of their photosynthesis. The top few meters of any body of water contain planktonic algae. As three-fourths of the Earth is covered with water, it is estimated that 80% of the Earth's O_2 is produced by planktonic algae.

Seasonal changes in nutrients, light, and temperature cause fluctuations in algal populations; periodic increases in numbers of planktonic algae are called **blooms.** Blooms of dinoflagellates are responsible for seasonal red tides. Blooms of a certain few species indicate that the water in which they grow is polluted because these algae thrive in high concentrations of organic materials that exist in sewage or industrial wastes. When algae die, the decomposition of the large numbers of cells associated with an algal bloom depletes the level of dissolved oxygen in the water. (This phenomenon is discussed in Chapter 27.)

Much of the world's petroleum was formed from diatoms that lived over 300 million years ago. When such diatoms and other planktonic organisms died and were buried by sediments, the organic molecules they contained did not decompose to be returned to the carbon cycle as CO_2. Heat and pressure resulting from the Earth's geologic movements altered the oil stored in the cells as well as the cell membranes. Oxygen and other elements were eliminated, leaving a residue of hydrocarbons in the form of petroleum and natural gas deposits.

Many unicellular algae are symbionts in animals. The giant clam, *Tridacna* (trī-dak′nä), has evolved special organs that host dinoflagellates. As the clam sits in shallow water, the algae proliferate in these organs when they are exposed to the sun. The algae release glycerol into the clam's bloodstream, thus supplying the clam's carbohydrate requirement. In addition, evidence suggests that the clam gets essential proteins by phagocytizing old algae.

(a) Crustose lichen

(b) Foliose lichen

(c) Fruticose lichen

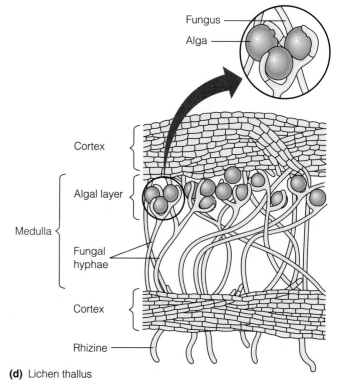

(d) Lichen thallus

FIGURE 12.14 Lichens. (a) Crustose. **(b)** Foliose. **(c)** Fruticose. **(d)** Cross section of a lichen thallus. The medulla is composed of fungal hyphae surrounding the algal layer. The protective cortex is a layer of fungal hyphae that covers the surface and sometimes the bottom of the lichen.

LICHENS

LEARNING OBJECTIVES

- *List the distinguishing characteristics of lichens, and describe their nutritional needs.*
- *Describe the roles of the fungus and the alga in a lichen.*

A **lichen** is a combination of a green alga (or a cyanobacterium) and a fungus. The two organisms exist in a *mutualistic* relationship, in which each partner benefits. The lichen is very different from either the alga or fungus growing alone, and if the partners are separated, the lichen no longer exists. Approximately 20,000 species of lichens occupy quite diverse habitats. Because lichens can inhabit areas in which neither fungi nor algae could sur-

vive alone, lichens are often the first life forms to colonize newly exposed soil or rock. Lichens secrete organic acids that chemically weather rock, and they accumulate nutrients needed for plant growth. They are some of the slowest-growing organisms on Earth.

Lichens can be grouped into three morphologic categories. *Crustose lichens* (Figure 12.14a) grow flush or encrusting onto the substratum; *foliose lichens* (Figure 12.14b) are more leaflike; and *fruticose lichens* (Figure 12.14c) have fingerlike projections. The lichen's thallus, or body, forms when fungal hyphae grow around algal cells to become the **medulla** (Figure 12.14d). Fungal hyphae project below the lichen body to form **rhizines,** or holdfasts. Fungal hyphae also form a **cortex,** or protective covering, over the algal layer and sometimes under it as well. After incorporation into a lichen thallus, the alga continues to grow by

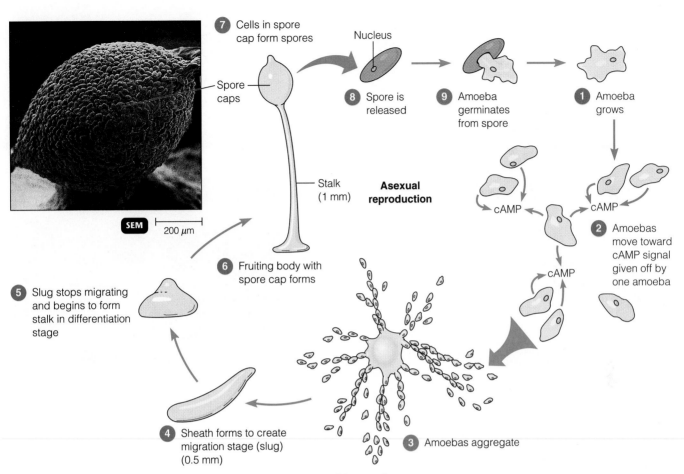

FIGURE 12.15 **Generalized life cycle of a cellular slime mold.** Inset shows a spore cap of *Dictyostelium* slime mold.

mitosis in the algal layer of the lichen, and the growing hyphae can incorporate new algal cells. In a lichen, the fungus reproduces sexually.

When the algal partner is cultured separately in vitro, about 1% of the carbohydrates produced during photosynthesis are released into the culture medium; however, when the alga is associated with a fungus, the algal plasma membrane is more permeable, and up to 60% of the products of photosynthesis are released to the fungus or are found as end-products of fungus metabolism. The fungus clearly benefits from this association. The alga, while giving up valuable nutrients, is in turn compensated; it receives from the fungus both protection from desiccation (cortex) and attachment (holdfast).

Populations of lichens readily incorporate cations (positively charged ions) into their thalli. Therefore, the concentrations and types of cations in the atmosphere can be determined by chemical analyses of lichen thalli. In addition, the presence or absence of species that are quite sensitive to pollutants can be used to ascertain air quality. A 1985 study in the Cuyahoga Valley (Ohio) revealed that 81% of the 172 lichen species that were present in 1917 were gone. Because this area is severely affected by air pol-

lution, the inference is that air pollutants, primarily sulfur dioxide (the major contributor to acid rain), caused the death of sensitive species.

SLIME MOLDS

LEARNING OBJECTIVE

- *Compare and contrast cellular slime molds and plasmodial slime molds.*

Slime molds have both fungal and animal characteristics and have membrane-bounded nuclei; thus, they are classified as protists. Cellular slime molds are typical eucaryotic cells that resemble amoebas. In the life cycle of cellular slime molds (Figure 12.15), ① the amoeboid cells live and grow by ingesting fungi and bacteria by phagocytosis. Cellular slime molds are of interest to biologists who study cellular migration and aggregation, for when conditions are unfavorable, ② and ③ large numbers of amoeboid cells aggregate to form a single structure. This aggre-

10 Zygote develops by nuclear division and cell growth

1 Multinucleate plasmodium forms

Asexual reproduction

2 Plasmodium grows, distributing nutrients by cytoplasmic streaming

9 Haploid gametes fuse, producing zygote

Channels of cytoplasmic streaming

LM |—— 5 mm

8 Gamete germinates from spore

Sexual reproduction

3 Plasmodium separates into groups of protoplasm

7 Spores are released

4 Each group forms sporangia on stalks

Sporangium

5 Spores develop in sporangia

Stalk

LM |—— 5 mm

6 Nuclei in spores go through meiosis, forming gametes

FIGURE 12.16 Life cycle of a plasmodial (acellular) slime mold. Pictured in photographs is *Physarum*.

gation occurs because some individual amoebas produce the chemical cyclic AMP (cAMP), toward which the other amoebas migrate. ④ The aggregated amoebas are enclosed in a slimy sheath called a *slug*. The slug migrates as a unit toward light. ⑤ After a period of hours, the slug ceases to migrate and begins to form differentiated structures. ⑥ Some of the amoeboid cells form a stalk; others swarm up the stalk to form a spore cap, and ⑦ most of these differentiate into spores. ⑧ When spores are released under favorable conditions, ⑨ they germinate to form single amoebas.

In 1973, a Dallas resident discovered a pulsating red blob in his backyard. The news media claimed that a "new

life form" had been found. For some people, the "creature" evoked a spine-chilling recollection of an old science fiction movie. Before imaginations got carried away too far, biologists calmed everyone's worst fears (or highest hopes). The amorphous mass was merely a plasmodial slime mold, they explained. Its unusually large size (46 cm in diameter), however, startled even scientists.

Plasmodial (acellular) slime molds were first scientifically reported in 1729. They belong to a separate phylum. A plasmodial slime mold exists as a mass of protoplasm with many nuclei (it is multinucleated). This mass of protoplasm is called a **plasmodium** (Figure 12.16). ① The entire plasmodium moves as a giant amoeba; it engulfs organic debris and bacteria. Biologists have found that musclelike proteins forming structures called microfilaments account for the movement of the plasmodium. ② When plasmodial slime molds are grown in laboratories, a phenomenon called **cytoplasmic streaming** is observed. During cytoplasmic streaming, the protoplasm within the plasmodium moves and changes both its speed and direction so that the oxygen and nutrients are evenly distributed.

The plasmodium grows as long as there is enough food and moisture. ③ When either is in short supply, the plasmodium separates into many groups of protoplasm; ④ each of these groups forms a stalked sporangium, ⑤ in which spores (a resistant, resting form of the slime mold) develop. ⑥ Nuclei within these spores undergo meiosis and form uninucleate haploid cells. ⑦ The spores are then released. ⑧ When conditions improve, these spores germinate, ⑨ fuse to form diploid cells, and ⑩ develop into a multinucleate plasmodium.

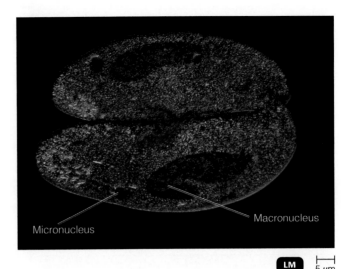

FIGURE 12.17 Conjugation in the ciliate protozoan ***Paramecium.*** Sexual reproduction in ciliates is by conjugation. Each cell has two nuclei, a micronucleus and a macronucleus. The micronucleus is haploid and is specialized for conjugation. One micronucleus from each cell will migrate to the other cell during conjugation. Both cells will then go on to produce two daughter cells with recombined DNA.

PROTOZOA

LEARNING OBJECTIVES

- *List the defining characteristics of protozoa.*
- *Describe the outstanding characteristics of the four phyla of protozoa discussed in this chapter, and give an example of each.*
- *Differentiate between an intermediate host and a definitive host.*

Protozoa are one-celled, eucaryotic organisms that belong to the Kingdom Protista. Among the protozoa are many variations on this cell structure, as we shall see. Protozoa inhabit water and soil and feed upon bacteria and small particulate nutrients. Some protozoa are part of the normal microbiota of animals. Of the nearly 20,000 species of protozoa, relatively few cause disease.

Characteristics of Protozoa

The term *protozoan* means "first animal," which generally describes its animal-like nutrition. In addition to getting food, a protozoan must reproduce, and parasitic species must be able to get from one host to another.

Life Cycle

Protozoa reproduce asexually by fission, budding, or schizogony. **Schizogony** is *multiple fission;* the nucleus undergoes multiple divisions before the cell divides. After many nuclei are formed, a small portion of cytoplasm concentrates around each nucleus, and then the single cell separates into daughter cells.

Sexual reproduction has been observed in some protozoa. The ciliates, such as *Paramecium*, reproduce sexually by **conjugation** (Figure 12.17), which is very different from the bacterial process of the same name. During protozoan conjugation, two cells fuse, and a haploid nucleus (the micronucleus) from each cell migrates to the other cell. This haploid micronucleus fuses with the haploid macronucleus within the cell. The parent cells separate, each now a fertilized cell. When the cells later divide, they produce daughter cells with recombined DNA. Some protozoa produce **gametes** or **gametocytes**, haploid sex cells. In reproduction, two gametes fuse to form a diploid zygote.

Encystment. Under certain adverse conditions, some protozoa produce a protective capsule called a **cyst.** A cyst permits the organism to survive when food, moisture, or oxygen are lacking, when temperatures are not suitable, or when toxic chemicals are present. A cyst also enables a parasitic species to survive outside a host. This is important because parasitic protozoa can have life cycles that involve more than one host.

Nutrition

Protozoa are mostly aerobic heterotrophs, although many intestinal protozoa are capable of anaerobic growth. One chlorophyll-containing flagellate, *Euglena* (ū-glē′nä) (see Figure 12.13b), can also grow in the dark as a heterotroph and is often included along with the protozoa.

(a) *Amoeba proteus*

(b) *Entamoeba histolytica*   LM   ⊢──⊣ 10 µm

FIGURE 12.18   Amoebas. (a) To move and to engulf food, amoebas (such as this *Amoeba proteus*) extend cytoplasmic structures called pseudopods. Food vacuoles are created when pseudopods surround food and bring it into the cell. **(b)** *Entamoeba histolytica*. Arrows point to red blood cells ingested by an amoeba.

All protozoa live in areas with a large supply of water. Some protozoa transport food across the plasma membrane. However, some have a protective covering called the **pellicle** and thus require specialized structures to take in food. Ciliates take in food by waving their cilia toward a mouthlike opening called the **cytostome**. Amoebas engulf food by surrounding it with pseudopods and phagocytizing it. In all protozoa, digestion takes place in membrane-bounded **vacuoles**, and waste may be eliminated through the plasma membrane or through a specialized **anal pore.**

Medically Important Phyla of Protozoa

Protozoa are a large and diverse group. Current schemes of classifying protozoan species into phyla and subphyla are based on motility, cell surface, feeding structures, nuclear structure, and even the presence of symbiotic bacteria. No

widespread agreement on taxonomic arrangement currently exists. Therefore, we will use one common taxonomic scheme that is based primarily on motility, and we will restrict our study to the groups that include disease-causing species.

Amoeboflagellates

The phylum Sarcomastigophora consists of protozoa that move using pseudopods or flagella. (A few species, such as *Naegleria*, use both types of motility.)

Amoebas. The amoebas (subphylum Sarcodina) move by extending blunt, lobelike projections of the cytoplasm called **pseudopods** (Figure 12.18a). Any number of pseudopods can flow from one side of the amoeba, and the rest of the cell will flow toward the pseudopods.

Entamoeba histolytica (en-tä-mē′bä his-tō-lī′ti-kä), the causative agent of amoebic dysentery, is the only pathogenic amoeba found in the human intestine (Figure 12.18b). The primary food of *E. histolytica* is red blood cells. *E. histolytica* is transmitted between humans through ingestion of the cysts that are excreted in the feces of the infected person. The amoebic form of *Naegleria* (nī-gle′rē-ä), as well as the amoeba *Acanthamoeba*, can infect the blood, brain, and spinal cord of humans, who become infected while swimming when waterborne cysts pass across mucous membranes.

Flagellates. The flagella of flagellates (subphylum Mastigophora) move in a whiplike manner that pulls the flagellated cells through their environment. Most flagellates have two or more flagella, but some parasitic species have as many as eight. Flagellates are typically spindle-shaped, with flagella projecting from the front end (Figure 12.19a). Their outer membrane is a tough, flexible pellicle; food is ingested through an oral groove or a cytostome.

An example of a flagellate that is a human parasite is *Trichomonas vaginalis* (trik-ō-mōn′as va-jin-al′is), shown in Figure 12.19b. Like some other flagellates, *T. vaginalis* has an **undulating membrane**, which appears to consist of highly modified flagella. *T. vaginalis* does not have a cyst stage and must be transferred from host to host quickly before desiccation occurs. *T. vaginalis* is found in the vagina and in the male urinary tract. It is usually transmitted by sexual intercourse but can be transmitted by toilet facilities or towels.

Another parasitic flagellate is *Giardia lamblia* (jē-är′dē-ä lam′lē-ä) (see Figures 12.19c and 25.16 for the vegetative form, called the **trophozoite**, and Figure 12.19d for the cyst stage). The parasite is found in the small intestine of humans and other mammals. It is excreted in the feces as a cyst and survives in the environment before being ingested by the next host. Diagnosis of giardiasis, the disease caused by *G. lamblia*, is often based on identification of cysts in feces.

The hemoflagellates are transmitted by the bites of blood-feeding insects and are found in the circulatory sys-

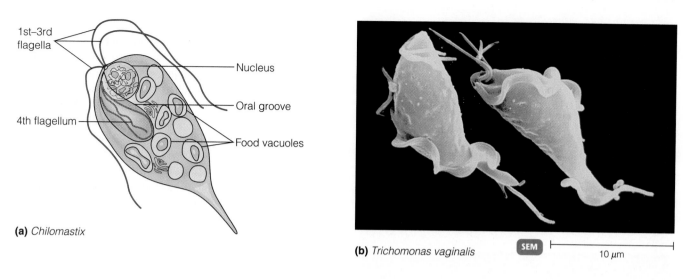

(a) *Chilomastix*

(b) *Trichomonas vaginalis*   SEM   10 μm

FIGURE 12.19 Flagellates. (a) *Chilomastix.* This flagellate, found in the human intestine, may be mildly pathogenic. The cysts survive for months outside a human host. The fourth flagellum is used to move food into the oral groove, where food vacuoles are formed. **(b)** *Trichomonas vaginalis.* This flagellate causes urinary and genital tract infections. Note the small undulating membrane. This flagellate does not have a cyst stage. **(c)** *Giardia lamblia.* The trophozoite of this intestinal parasite has eight flagella and two prominent nuclei, giving it a distinctive appearance. **(d)** The *G. lamblia* cyst provides protection from the environment before it is ingested by a new host.

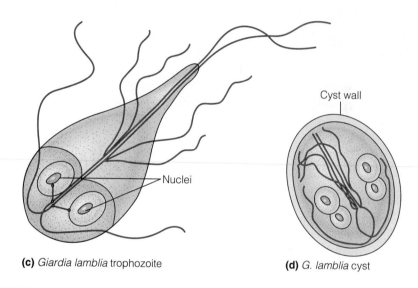

(c) *Giardia lamblia* trophozoite

(d) *G. lamblia* cyst

tem of the bitten host. To survive in this viscous fluid, hemoflagellates have long, slender bodies and an undulating membrane. The genus *Trypanosoma* (tri-pa-nō-sō'mä) includes the species that causes African sleeping sickness, *T. brucei gambiense* (brüs'ē gam-bē-ens'), which is transmitted by the tsetse fly. *T. cruzi* (kruz'ē; see Figure 23.21), the causative agent of Chagas' disease, is transmitted by the "kissing bug," so called because it bites on the face. After entering the insect, the trypanosome rapidly multiplies by fission. If the insect then defecates while biting a human, it can release trypanosomes that can contaminate the bite wound.

Ciliates

Species in the phylum Ciliophora have projections called cilia that are similar to but shorter than flagella. The cilia are arranged in precise rows on the cell (Figure 12.20). They are moved in unison to propel the cell through its environment and to bring food particles to the mouth. Ciliates possess complex cells. Almost all have a cytostome through

which food enters. The food then passes through the **cytopharynx**, at the end of which food vacuoles are formed. Waste particles leave the cell through the anal pore.

The only ciliate that is a human parasite is *Balantidium coli* (bal-an-tid'ē-um kō'lī), the causative agent of a severe, though rare, type of dysentery. When cysts are ingested by the host, they enter the colon, into which the trophozoites are released. The trophozoites feed on bacteria and fecal debris as they multiply, and cysts are passed out with feces.

Apicomplexans

The species in the third phylum we will study, the Apicomplexa, are not motile in their mature forms and are obligate intracellular parasites. Apicomplexans are characterized by the presence of a complex of special organelles at the apexes (tips) of their cells. The organelles in these apical complexes (hence the phylum name) contain enzymes used to penetrate the host's tissues. Apicomplexans have complex life cycles that involve trans-

(a) *Paramecium*

(b) *Vorticella*

LM | 50 μm

FIGURE 12.20 Ciliates. (a) Like all ciliates, *Paramecium* is covered with rows of cilia. *Paramecium* has specialized structures for ingestion (mouth), elimination of wastes (anal pore), and regulation of osmotic pressure (contractile vacuoles). The macronucleus is involved with protein synthesis and other ongoing cellular activities. The micronucleus functions in sexual reproduction. **(b)** *Vorticella* attaches to objects in water by the base of its stalk. The springlike stalk can expand so *Vorticella* can feed in different areas. Cilia surround its cytostome.

mission between several hosts. An example of an apicomplexan is *Plasmodium* (plaz-mō′dē-um), the causative agent of malaria. A vaccine against malaria is now being tested on humans (see the box in Chapter 18).

Plasmodium grows by sexual reproduction in the *Anopheles* (an-of′el-ēz) mosquito (Figure 12.21). ① When an *Anopheles* carrying the infective stage of *Plasmodium* (called a **sporozoite**) bites a human, sporozoites can be injected into the human. The sporozoites are carried by the blood to the liver. ② They undergo schizogony in liver cells and produce thousands of progeny called **merozoites**. ③ Merozoites enter the bloodstream and infect red blood cells. ④ The young trophozoite looks like a ring in which the nucleus and cytoplasm are visible. This is called a **ring stage**. ⑤ The ring stage enlarges and divides re-

peatedly, and ⑥ the red blood cells eventually rupture and release more merozoites. Upon release of the merozoites, their waste products, which cause fever and chills, are also released. Most of the merozoites infect new red blood cells and perpetuate their cycle of asexual reproduction. However, some develop into male and female sexual forms (gametocytes). Even though the gametocytes themselves cause no further damage, ⑦ they can be picked up by the bite of another *Anopheles* mosquito; they then enter the mosquito's intestine and begin their sexual cycle. ⑧ Here the male and female gametocytes unite to form a zygote. The zygote forms an oocyst, in which cell division occurs, and asexual sporozoites are formed. ⑨ When the oocyst ruptures, the sporozoites migrate to the salivary glands of the mosquito. They can then be injected into a new human host by the biting mosquito. The mosquito is the **definitive host** because it harbors the sexually reproducing stage of *Plasmodium*. The host in which the parasite undergoes asexual reproduction (in this case, the human) is the **intermediate host**.

Malaria is diagnosed in the laboratory by microscopic observation of thick blood smears for the presence of *Plasmodium* (see Figure 23.24). A peculiar characteristic of malaria is that the interval between periods of fever caused by release of merozoites is always the same for a given species of *Plasmodium* and is always a multiple of 24 hours. The reason and mechanism for such precision has intrigued scientists. After all, why should a parasite need a biological clock? Frank Hawking and his coworkers demonstrated that *Plasmodium*'s development is regulated by the host's body temperature, which normally fluctuates over a 24-hour period. The parasite's careful timing ensures that gametocytes are mature at night, when mosquitoes are feeding, to enable transmission of the parasite to a new host.

Another apicomplexan parasite of red blood cells is *Babesia microti* (ba-bē′sē-ä mī-krō′tē). *Babesia* causes fever and anemia in immunosuppressed individuals. In the U.S.A. it is transmitted by the tick *Ixodes scapularis*.

Toxoplasma gondii (toks-ō-plaz′mä gon′dē-ē) is another apicomplexan intracellular parasite of humans. The life cycle of this parasite involves domestic cats. The trophozoites, called *tachyzoites*, reproduce sexually and asexually in an infected cat, and oocysts, each containing eight sporozoites, are passed out with feces. If the oocysts are ingested by humans or other animals, the sporozoites emerge as trophozoites, which can reproduce in the tissues of the new host (see Figure 23.20). *T. gondii* is dangerous to pregnant women, as it can cause congenital infections in utero. Tissue examination and observation of *T. gondii* is used for diagnosis. Antibodies may be detected by ELISA and by indirect fluorescent-antibody tests (see Chapter 18).

Cryptosporidium (krip-tō-spô-ri′dē-um), an apicomplexan that is an intracellular parasite of mammals, was first reported as a cause of human disease in 1976. *Cryptosporidium* may be responsible for up to 30% of the di-

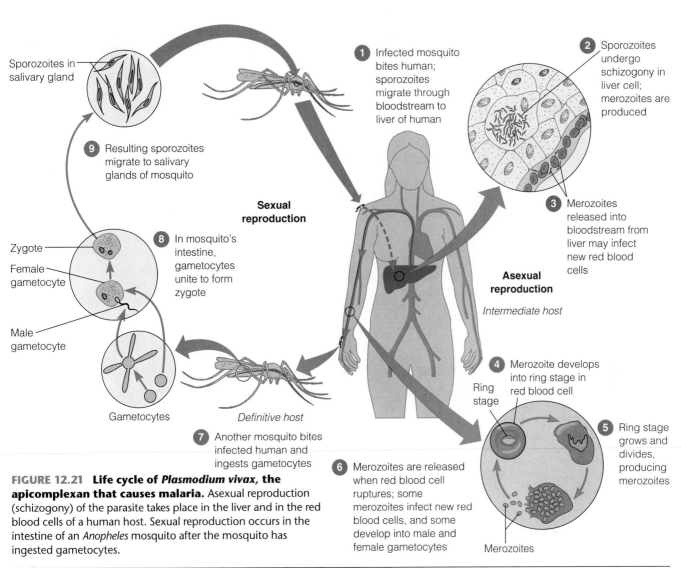

FIGURE 12.21 **Life cycle of *Plasmodium vivax,* the apicomplexan that causes malaria.** Asexual reproduction (schizogony) of the parasite takes place in the liver and in the red blood cells of a human host. Sexual reproduction occurs in the intestine of an *Anopheles* mosquito after the mosquito has ingested gametocytes.

Sexual reproduction

1 Infected mosquito bites human; sporozoites migrate through bloodstream to liver of human

Sporozoites in salivary gland

9 Resulting sporozoites migrate to salivary glands of mosquito

Zygote

Female gametocyte

Male gametocyte

8 In mosquito's intestine, gametocytes unite to form zygote

Gametocytes

7 Another mosquito bites infected human and ingests gametocytes

Definitive host

2 Sporozoites undergo schizogony in liver cell; merozoites are produced

3 Merozoites released into bloodstream from liver may infect new red blood cells

Asexual reproduction

Intermediate host

4 Merozoite develops into ring stage in red blood cell

Ring stage

5 Ring stage grows and divides, producing merozoites

6 Merozoites are released when red blood cell ruptures; some merozoites infect new red blood cells, and some develop into male and female gametocytes

Merozoites

---

arrheal illness in underdeveloped countries. In AIDS patients and other immunosuppressed people, *Cryptosporidium* can cause respiratory and gallbladder infections and may be a major cause of death. The organism, which lives inside the cells lining the small intestine, can be transmitted to humans through the feces of cows, rodents, dogs, and cats. Waterborne and hospital-acquired infections have also been reported. Inside the host cell, each *Cryptosporidium* organism forms four oocysts (see Figure 25.18), each containing four sporozoites. When the oocyst ruptures, sporozoites may infect new cells in the host or be released with the feces. The disease is diagnosed by acid-fast staining or fluorescent-antibody tests.

Microsporans

Microsporans—recently assigned to the phylum Microspora—are obligate intracellular parasites that lack mitochondria and microtubules (see Chapter 4). Microsporidial protozoa have been reported since 1984 to be the cause of

a number of human diseases, including chronic diarrhea and keratoconjunctivitis (inflammation of the conjunctiva near the cornea), most notably in AIDS patients.

Table 12.6 lists some typical parasitic protozoa and the diseases they cause.

HELMINTHS

LEARNING OBJECTIVES

- *List the distinguishing characteristics of parasitic helminths.*

- *Provide a rationale for the elaborate life cycles of parasitic worms.*

- *List the characteristics of the three groups of parasitic helminths, and give an example of each.*

- *Describe a parasitic infection in which humans serve as a definitive host, as an intermediate host, and as both.*

TABLE 12.6 Some Representative Parasitic Protozoa

PHYLUM SUBPHYLUM	HUMAN PATHOGENS	DISTINGUISHING FEATURES	DISEASE	SOURCE OF HUMAN INFECTIONS	SEE FIGURE	REFER TO CHAPTER
Sarcomastigophora Sarcodina (amoebas)	Acanthamoeba	Pseudopods	Keratitis	Water	——	21
	Entamoeba histolytica	Pseudopods	Amoebic dysentery	Fecal contamination of drinking water	12.18b	25
	Naegleria fowleri	Some flagellated forms	Microencephalitis	Water in which people swim	22.16	22
Mastigophora (flagellates)	Giardia lamblia	Two nuclei, eight flagella	Giardial enteritis	Fecal contamination of drinking water	12.19c 25.16	25
	Trichomonas vaginalis	No encysting stage	Urethritis; vaginitis	Contact with vaginal-urethral discharge	12.19b Table 26.1	26
	Trypanosoma cruzi		Chagas' disease	Bite of Triatoma (kissing bug)	23.21	23
	T. brucei gambiense, T.b. rhodesiense	Undulating membrane	African trypanosomiasis	Bite of tsetse fly	——	22
Ciliophora	Balantidium coli	Only parasitic ciliate of humans	Balantidial dysentery	Fecal contamination of drinking water	12.20b	——
Apicomplexa	Babesia microti		Babesiosis	Domestic animals, ticks	——	——
	Cryptosporidium	Complex life cycles may require more than one host	Diarrhea	Humans, other animals, water	25.18	25
	Isospora		Coccidiosis	Domestic animals	——	Table 19.7
	Plasmodium		Malaria	Bite of Anopheles mosquito	12.21 23.24	23
	Pneumocystis carinii*		Pneumonia	Humans	24.18	24
	Toxoplasma gondii		Toxoplasmosis	Cats, other animals; congenital	23.20	23
Microspora	Nosema	Unknown	Diarrhea, kerato-conjunctivitis	Other animals	——	——

*Classification is uncertain.

There are a number of parasitic animals that spend part or all of their lives in humans. Most of these animals belong to two phyla: Platyhelminthes (flatworms) and Nematoda (roundworms). These worms are commonly called **helminths**. There are also free-living species in these phyla, but we will limit our discussion to the parasitic species.

Characteristics of Helminths

Helminths are multicellular eucaryotic animals that generally possess digestive, circulatory, nervous, excretory, and reproductive systems. Parasitic helminths must be highly specialized to live inside their hosts. The following generalizations distinguish parasitic helminths from their free-living relatives:

1. They either *lack* a digestive system or have a greatly simplified one. They can absorb nutrients from the host's food, body fluids, and tissues.

2. Their nervous system is *reduced*. They do not need an extensive nervous system because they do not have to search for food or respond much to their environment. The environment within a host is fairly constant.

3. Their means of locomotion is either *reduced* or *completely lacking*. Because they are transferred from host to host, they do not need to search actively for a suitable habitat.

4. Their reproductive system is often complex; an individual produces large numbers of fertilized eggs, by which a suitable host is infected.

FIGURE 12.22 Flukes.
(a) General anatomy of an adult fluke, shown in cutaway view. The oral and ventral suckers attach the fluke to the host. The mouth is located in the center of the oral sucker. Flukes are hermaphroditic; each animal contains both testes and ovaries. (b) The Asian liver fluke *Clonorchis sinensis.* Heavy infestations may block bile ducts from the liver. Note the incomplete digestive system.

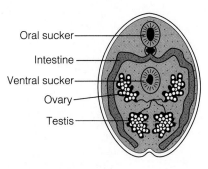

Oral sucker
Intestine
Ventral sucker
Ovary
Testis

(a) Fluke anatomy

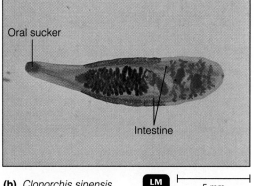

Oral sucker

Intestine

(b) *Clonorchis sinensis*   LM   5 mm

Life Cycle

The life cycle of parasitic helminths can be extremely complex, involving a succession of *intermediate hosts* for completion of each *larval* (developmental) stage of the parasite and a *definitive host* for the adult parasite.

Adult helminths may be **dioecious;** male reproductive organs are in one individual, and female reproductive organs are in another. In those species, reproduction occurs only when two adults of the opposite sex are in the same host.

Adult helminths may also be **hermaphroditic**—one animal has both male and female reproductive organs. Two hermaphrodites may copulate and simultaneously fertilize each other. A few types of hermaphrodites fertilize themselves.

Platyhelminths

Members of the phylum Platyhelminthes (**flatworms**) are flattened from back to front. They have what is called an *incomplete digestive system.* This type of digestive system has only one opening (mouth), through which food enters and wastes leave. The classes of parasitic flatworms include the trematodes and cestodes.

Trematodes

Trematodes, or **flukes,** have flat, leaf-shaped bodies with a ventral sucker and an oral sucker (Figure 12.22). The suckers hold the organism in place and suck fluids from the host. Flukes can also obtain food by absorbing it through their nonliving outer covering, called the **cuticle.** Flukes are given common names according to the tissue of the definitive host in which the adults live (for example, lung fluke, liver fluke, blood fluke). The Asian liver fluke *Clonorchis sinensis* (klo-nôr′kis si-nen′sis) is occasionally seen in immigrants in the U.S.A., but it cannot be transmitted because its intermediate hosts are not in the U.S.A.

To exemplify a fluke's life cycle, let us look at the lung fluke, *Paragonimus westermani* (pär-ä-gōn′e-mus we-ster-ma′nē), shown in Figure 12.23. The intermediate hosts for this fluke, and so the fluke itself, occur throughout the world, including the U.S.A. and Canada. The adult lung fluke lives in the bronchioles of humans and other mammals and is approximately 6 mm wide by 12 mm long. ① The hermaphroditic adults liberate eggs into the bronchi. Because sputum that contains eggs is frequently swallowed, the eggs are usually excreted in feces of the definitive host.

If the life cycle is to continue, ② the eggs must reach a body of water. ③ Inside the egg, a miracidial larva, or **miracidium,** then develops. When the egg hatches, ④ the larva enters a suitable snail. Only certain species of aquatic snails can be this first intermediate host. ⑤ Inside the snail, the miracidium develops into a redia, which then undergoes asexual reproduction and produces **rediae.** Each redia develops into a **cercaria** that ⑥ bores out of the snail and penetrates the cuticle of a crayfish. ⑦ The parasite encysts as a **metacercaria** in the muscles and other tissues of the crayfish. ⑧ When the crayfish is eaten by a human, the metacercaria is freed in the human's small intestine. It bores out and wanders around until it penetrates the lungs, enters the bronchioles, and develops into an adult lung fluke.

In a laboratory diagnosis, sputum and feces are examined microscopically for fluke eggs. Infection results from eating undercooked crayfish, and the disease can be prevented by thoroughly cooking crayfish.

The cercariae of the blood fluke *Schistosoma* (shis-tō-sō′mä) are not ingested. Instead, they burrow through the skin of the human host and enter the circulatory system. The adults are found in certain abdominal and pelvic veins. The disease schistosomiasis is a major world health problem; it will be discussed further in Chapter 23.

Cestodes

Cestodes, or **tapeworms,** are intestinal parasites. Their structure is shown in Figure 12.24. The head, or **scolex,** has suckers for attaching to the intestinal mucosa of the definitive host; some species also have small hooks for attachment. Tapeworms do not ingest the tissues of their hosts; in fact, they completely lack a digestive system. To obtain nutrients from the small intestine, they absorb food through their cuticle. The body consists of segments called **proglottids.** Proglottids are continually produced

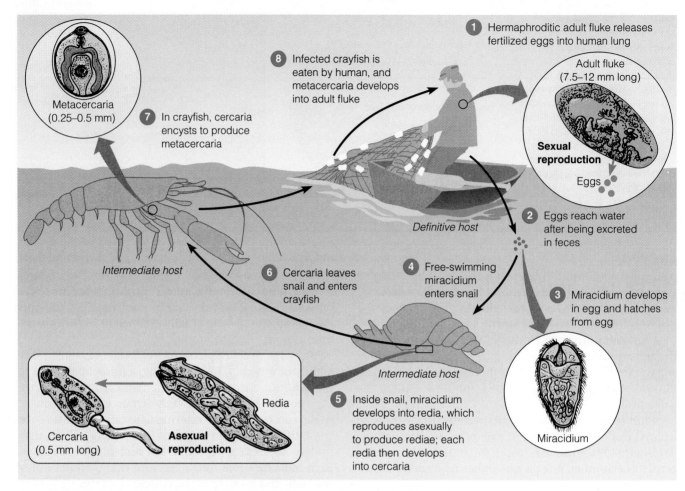

FIGURE 12.23 Life cycle of the lung fluke *Paragonimus westermani*. The trematode reproduces sexually in a human and asexually in a snail, its first intermediate host. Larvae encysted in the second intermediate host, a crayfish, infect humans when ingested.

by the neck region of the scolex, as long as the scolex is attached and alive. Each proglottid contains both male and female reproductive organs. The proglottids farthest away from the scolex are the mature ones containing fertilized eggs. Mature proglottids are essentially bags of eggs, each of which is infective to the proper intermediate host.

Humans as Definitive Hosts. The adults of *Taenia saginata* (te′nē-ä sa-ji-nä′tä), the beef tapeworm, live in humans and can reach a length of 6 meters. The scolex is about 2 mm in length and is followed by a thousand or more proglottids. The feces of an infected human contain mature proglottids, and each proglottid contains thousands of eggs. As the proglottids wriggle away from the fecal material, they increase their chances of being ingested by an animal that is grazing. Upon ingestion by cattle, the larvae hatch from the eggs and bore through the intestinal wall. The larvae migrate to muscle (meat), in which they encyst as **cysticerci**. When the cysticerci are ingested by humans, all but the scolex is digested. The

scolex anchors itself in the small intestine and begins producing proglottids.

Diagnosis of tapeworm infestation in humans is based on the presence of mature proglottids and eggs in feces. Cysticerci can be seen macroscopically in meat; their presence is referred to as "measly beef." Inspecting beef that is intended for human consumption for "measly" appearance is one way to prevent infections by beef tapeworm. Another method of prevention is to avoid the use of untreated human sewage as fertilizer in grazing pastures.

Humans are the only known definitive host of the pork tapeworm, *Taenia solium*. Adult worms living in the human intestine produce eggs, which are passed out in feces. When eggs are eaten by pigs, the larval helminth encysts in the pig's muscles; humans become infected when they eat undercooked pork. The human-pig-human cycle of *T. solium* is common in Latin America, Asia, and Africa. In the U.S.A., however, *T. solium* is virtually nonexistent in pigs; the parasite is transmitted from human to human. Eggs shed by one person and ingested by another person

Scolex

Sucker

Hooks

Neck

SEM ⊢ 200 µm

Testis

Genital pore

Ovary

Mature proglottid will disintegrate and release eggs

FIGURE 12.24 General anatomy of an adult tapeworm.
The scolex (see inset) consists of suckers and hooks that attach to the host's tissues. The body lengthens as new proglottids form at the neck. Each proglottid contains both testes and ovaries.

hatch, and the larvae encyst in the brain, causing **cysticercosis**. The human hosting *T. solium*'s larvae is serving as an intermediate host. Approximately 7 percent of the few hundred cases reported in recent years were acquired by people who had never been outside of the U.S.A. They may have become infected through household contact with people who were born in or had traveled in other countries.

Humans as Intermediate Hosts. Humans are the intermediate hosts for *Echinococcus granulosus* (ē-kīn-ō-kok′kus gra-nū-lō′sus), shown in Figure 12.25. Dogs and coyotes are the definitive hosts for this minute (2 to 8 mm) tapeworm. ① Eggs are excreted with feces and ② are ingested by deer, sheep, or humans. Humans can also become infected by

contaminating their hands with dog feces or saliva from a dog that has licked itself. ③ The eggs hatch in the human's small intestine, and the larvae migrate to the liver or lungs. ④ The larva develops into a **hydatid cyst.** The cyst contains "brood capsules," from which thousands of scoleces (singular: *scolex*) might be produced. ⑤ In the wild, the cysts might be in a deer that is eaten by a wolf. ⑥ The scoleces would be able to attach themselves in the wolf's intestine and produce proglottids. Diagnosis of hydatid cysts is frequently made only on autopsy, although X rays can detect the cysts (see Figure 25.21).

Nematodes

Members of the Phylum Nematoda, the roundworms, are cylindrical and tapered at each end. Roundworms have a *complete* digestive system, consisting of a mouth, an intestine, and an anus. Most species are dioecious. Males are smaller than females and have one or two hardened *spicules* on their posterior ends. Spicules are used to guide sperm to the female's genital pore.

Some species of nematodes are free-living in soil and water, and others are parasites on plants and animals. Parasitic nematodes do not have the succession of larval stages exhibited by flatworms. Some nematodes pass their entire life cycle, from egg to mature adult, in a single host.

Nematode infections of humans can be divided into two categories: those in which the egg is infective and those in which the larva is infective.

Eggs Infective for Humans

The pinworm *Enterobius vermicularis* (en-te-rō′bē-us vermi-kū-lar′is) spends its entire life in a human host (Figure 12.26). Adult pinworms are found in the large intestine. From there, the female pinworm migrates to the anus to deposit her eggs on the perianal skin. The eggs can be ingested by the host or by another person exposed through contaminated clothing or bedding. Pinworm infections are diagnosed by the Graham sticky-tape method. A piece of transparent tape is placed on the perianal skin in such a way that the sticky side picks up eggs that were deposited earlier. The tape is then microscopically examined for the presence of eggs.

Ascaris lumbricoides (as′kar-is lum-bri-koi′dēz) is a large nematode (30 cm in length). It is dioecious with **sexual dimorphism**; that is, the male and female worms look distinctly different, the male being smaller with a curled tail. The adult *Ascaris* lives in the small intestines of humans and domestic animals (such as pigs and horses); it feeds primarily on semidigested food. Eggs, excreted with feces, can survive in the soil for long periods until accidentally ingested by another host. The eggs hatch in the small intestine of the host, mature in the lungs, and from there migrate to the intestines.

Diagnosis is frequently made when the adult worms are excreted with feces. Prevention of infection in humans is

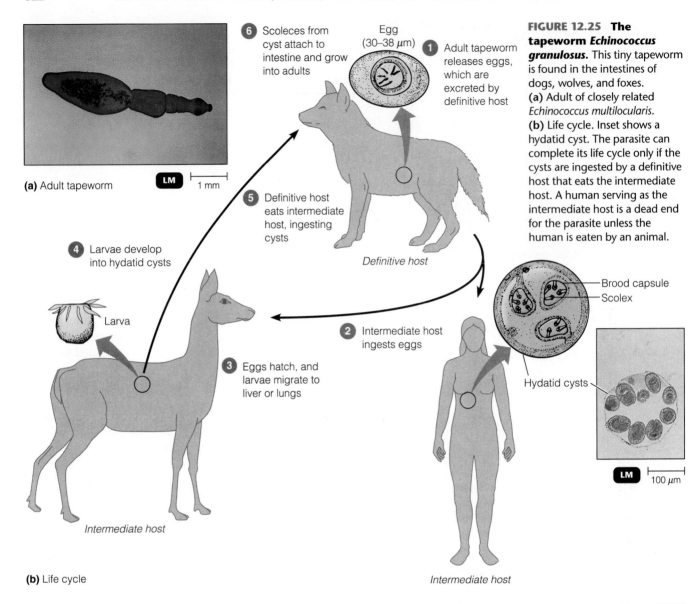

(a) Adult tapeworm

LM · 1 mm

❻ Scoleces from cyst attach to intestine and grow into adults

Egg (30–38 μm)

❶ Adult tapeworm releases eggs, which are excreted by definitive host

FIGURE 12.25 The tapeworm *Echinococcus granulosus*. This tiny tapeworm is found in the intestines of dogs, wolves, and foxes. **(a)** Adult of closely related *Echinococcus multilocularis*. **(b)** Life cycle. Inset shows a hydatid cyst. The parasite can complete its life cycle only if the cysts are ingested by a definitive host that eats the intermediate host. A human serving as the intermediate host is a dead end for the parasite unless the human is eaten by an animal.

❺ Definitive host eats intermediate host, ingesting cysts

Definitive host

❹ Larvae develop into hydatid cysts

Larva

❸ Eggs hatch, and larvae migrate to liver or lungs

❷ Intermediate host ingests eggs

Brood capsule
Scolex

Hydatid cysts

LM · 100 μm

Intermediate host

(b) Life cycle

Intermediate host

managed by proper sanitary habits. The life cycle of *Ascaris* in pigs can be interrupted by keeping pigs in areas free of fecal material.

Larvae Infective for Humans

Adult hookworms, *Necator americanus* (ne-kā′tôr ä-me-ri-ka′nus), live in the small intestine of humans (Figure 12.27); the eggs are excreted in feces. The larvae hatch in the soil and feed on bacteria there. A larva enters its host by penetrating the host's skin. It then enters a blood or lymph vessel, which carries it to the lungs. It is coughed up in sputum, swallowed, and finally carried to the small intestine. Diagnosis is based on the presence of eggs in feces. People can avoid hookworm infections by wearing shoes.

Trichinella spiralis (trik-in-el′lä spī-ra′lis) infections, called *trichinosis,* are usually acquired by eating encysted larvae in poorly cooked pork or bear meat. In the human

digestive tract, the larvae are freed from the cysts. They mature into adults in the small intestine and sexually reproduce there. Eggs develop in the female, and she gives birth to live nematodes. The larvae enter lymph and blood vessels in the intestines and migrate from there throughout the body. They encyst in muscles and other tissues and remain there until ingested by another host (see Figure 25.24).

Diagnosis of trichinosis is made by microscopic examination for larvae in a muscle biopsy. Trichinosis can be prevented by thoroughly cooking meat prior to consumption.

Four genera of roundworms called anisakine or wriggly worms can be transmitted to humans from infected fish and squid. The anisakine larvae are in the fish's intestine and migrate to the muscle during refrigerated storage. These roundworms are not transmitted in fresh fish. Freezing or thorough cooking will kill the larvae.

Table 12.7 lists representative parasitic helminths of each phylum and class and the diseases they cause.

FIGURE 12.26 The pinworm *Enterobius vermicularis.* (a) Adult pinworms live in the large intestine of humans. Like all roundworms, pinworms have a complete digestive system with a mouth, intestine, and anus. Most roundworms are dioecious, and the female (left and inset) is often distinctly larger than the male (right). **(b)** Pinworm eggs, which the female deposits on the perianal skin at night.

(a) Adult pinworms

Female
(8–13 mm long)

Male
(2–5 mm long)

(b) Pinworm egg

ARTHROPODS AS VECTORS

LEARNING OBJECTIVES

- *Define arthropod vector.*
- *Differentiate between a tick and a mosquito, and name a disease transmitted by each.*

Arthropods are animals characterized by segmented bodies, hard external skeletons, and jointed legs. With nearly 1 million species, this is the largest phylum in the animal kingdom. We will briefly describe arthropods here because a few suck the blood of humans and other animals and can transmit microbial diseases while doing so. Arthropods that carry disease-causing microorganisms are called **vectors.** Scabies, a disease caused by an arthropod, is described in Chapter 21.

Representative classes of arthropods include the following:

1. Arachnida (eight legs): spiders, mites, ticks

2. Crustacea (four antennae): crabs, crayfish

3. Chilopoda (two legs per segment): centipedes

4. Diplopoda (four legs per segment): millipedes

5. Insecta (six legs): bees, flies

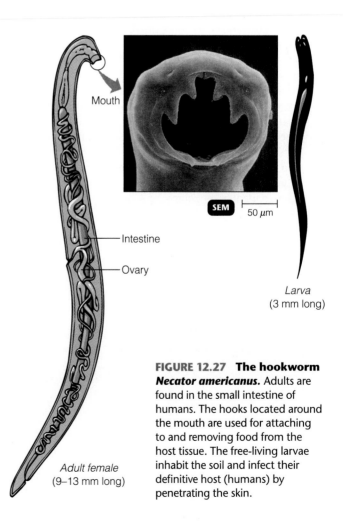

FIGURE 12.27 The hookworm *Necator americanus.* Adults are found in the small intestine of humans. The hooks located around the mouth are used for attaching to and removing food from the host tissue. The free-living larvae inhabit the soil and infect their definitive host (humans) by penetrating the skin.

Larva
(3 mm long)

Adult female
(9–13 mm long)

Morbidity and Mortality Weekly Report

AEDES ALBOPICTUS— WATCHING THE PROGRESS OF A NEW VECTOR

AN INFESTATION OF *Aedes albopictus* (Asian tiger mosquito), a mosquito known to transmit epidemic dengue fever in its native Asia, was discovered in 1985 in Harris County, Texas.

A. albopictus commonly breeds in standing water in tires stored outdoors. The mosquito was probably introduced into the U.S.A. in used-tire casings imported from Asia. Since January 1, 1988, used-tire casings must be clean and dry and be treated by an approved fumigation procedure. However, as shown in Figure (a), *A. albopictus* is now entrenched in the U.S.A., and, in many areas, it has replaced the native mosquito, *A. aegypti*. *A. albopictus* has not yet been proved to cause the spread of any viral disease in the Americas, but it represents a public health concern because of its aggressiveness and ability to feed continuously throughout the day.

In 1991, as part of an ongoing study to identify the sources of *A. al-bopictus* bloodmeals, researchers isolated eastern equine encephalitis (EEE) virus from the mosquito. By July 1991, Florida had reported early, widespread EEE virus activity, with 70 equine and 5 human cases. The CDC plans to improve surveillance for EEE and to more clearly define the role of *A. albopictus* in EEE transmission.

Laboratory studies have shown *A. albopictus* to be a more efficient vector of dengue virus than *A. aegypti*. Although endemic transmission of dengue has not occurred in the U.S.A. since 1986, introduction of the virus by international travelers could result in local transmission.

Worldwide there are 100 million cases of dengue per year. Dengue is now one of the leading causes of morbidity and mortality among Southeast Asian children. Dengue has been endemic in the Americas for over 200 years. Presently, the disease appears to be following a pattern similar to that in Southeast Asia 30 years ago. When dengue fever first breaks out in a geographical area where there have been no recent occurrences, it causes only a mild flulike illness. However, subsequent outbreaks occur as gastrointestinal hemorrhage and shock, called dengue hemorrhagic fever (DHF). For example, in 1978, Mexico reported its first cases of dengue in many years, and, in 1984 the first DHF cases associated with epidemic dengue were reported.

Since 1981, all four dengue virus serotypes have been imported into the U.S.A. Figure (b) shows the progression of dengue in the Americas over the past 13 years.

Scientists have investigated how dengue fever outbreaks progress from a mild illness to DHF. A person's first infection sensitizes the immune system, so that it reacts more rapidly and vigorously to the second infection. However, during the second infection, antibodies and live viruses may form immune complexes that attach to macrophages and other cells, actually increasing the number of viruses that infect a cell. This large number of viruses per cell results in massive tissue destruction. Because immunity increases the severity of the disease, immunization is impractical as a preventive measure. Instead, the most effective methods of preventing the spread of the disease appear to be general mosquito control measures.

SOURCES: *MMWR* 41:115;121 (2/21/92); *MMWR* 41:731–732 (10/2/92)

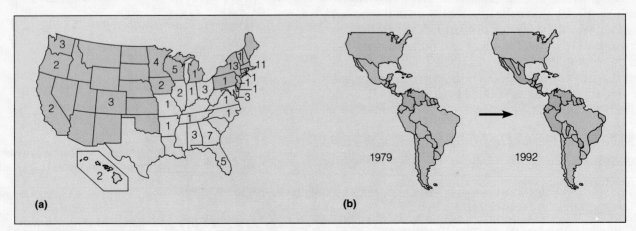

(a) **(b)**

***Aedes albopictus* and dengue fever. (a)** Distribution of *A. albopictus* in the U.S.A. and confirmed cases of dengue fever (1991). States with confirmed *A. albopictus* infestation are shown in yellow. Numbers indicate the number of confirmed dengue fever cases by state (total = 82). **(b)** Progression of dengue fever in the Americas, 1979–1992.

Table 12.8 lists those arthropods that are important vectors, and Figures 12.28, 12.29, and 12.30 illustrate some of them. These insects and ticks reside on an animal only when they are feeding. An exception to this is the louse, which spends its entire life on its host and cannot survive long away from a host.

Some vectors are just a mechanical means of transport for a pathogen. For example, houseflies lay their eggs on

TABLE 12.7 Representative Parasitic Helminths

PHYLUM	CLASS	HUMAN PARASITES	INTERMEDIATE HOST	DEFINITIVE HOST AND SITE	STAGE PASSED TO HUMANS; METHOD	DISEASE	LOCATION IN HUMANS	SEE FIGURE
Platyhelminthes	Trematodes	*Paragonimus westermani*	Freshwater snails and crayfish	Humans; lungs	Metacercaria in crayfish ingested	Paragonimiasis (lung fluke)	Lungs	12.23
		Schistosoma	Freshwater snails	Humans	Cercariae through skin	Schisto-somiasis	Veins	23.25 23.26
	Cestodes	*Taenia saginata*	Cattle	Humans; small intestine	Cysticerci in beef ingested	Tapeworm	Small intestine	——
		Taenia solium	Humans; pigs	Humans	Eggs ingested	Neurocysti-cercosis	Brain; any tissue	25.20
		Echinococcus granulosus	Humans	Dogs and other animals; intestines	Eggs from other animals ingested	Hydatidosis	Lungs, liver, brain	12.25 25.21
Nematoda		*Ascaris lumbricoides*	—	Humans; small intestine	Eggs ingested	Ascariasis	Small intestine	25.23
		Enterobius vermicularis	—	Humans; large intestine	Eggs ingested	Pinworm	Large intestine	12.26
		Necator americanus	—	Humans; small intestine	Larvae through skin	Hookworm	Small intestine	12.27
		Trichinella spiralis	—	Humans, swine, and other mammals; small intestine	Larvae ingested, especially in pork	Trichinosis	Muscles	25.24
		Anisakids	Marine fish and squid	Marine mammals	Larvae in fish ingested	Anisakiasis ("wriggly worms")	Gastro-intestinal tract	——

TABLE 12.8 Important Arthropod Vectors of Human Diseases

CLASS	ORDER	VECTOR	SEE FIGURE	DISEASE	REFER TO CHAPTER
Arachnida	Mites and ticks	*Dermacentor* (tick)	——	Rocky Mountain spotted fever	23
		Ixodes (tick)	12.29	Lyme disease, babesiosis	23
		Ornithodorus (tick)	——	Relapsing fever	23
Insecta	Sucking lice	*Pediculus* (human louse)	12.30a	Epidemic typhus	23
	Fleas	*Xenopsylla* (rat flea)	12.30b	Endemic murine typhus, plague	23
	True flies	*Chrysops* (deer fly)	12.30c	Tularemia	23
		Aedes (mosquito)	——	Dengue fever, yellow fever	23
		Anopheles (mosquito)	12.28b	Malaria	23
		Culex (mosquito)	——	Arboviral encephalitis	22
		Glossina (tsetse fly)	——	African trypanosomiasis	22
	True bugs	*Triatoma* (kissing bug)	12.30d	Chagas' disease	23

(a) Female mosquito

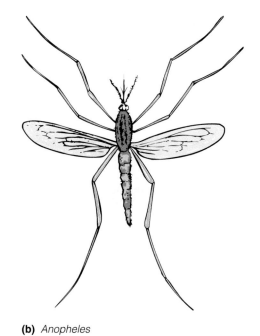

(b) *Anopheles*

FIGURE 12.28 **Mosquitoes.** **(a)** Female mosquito sucking blood from human skin. Mosquitoes transmit several diseases from person to person, including the viral diseases yellow fever and dengue fever. **(b)** *Anopheles* transmits malaria.

(a) *Ixodes scapularis* 1 mm

(b) *I. pacifus* 1 mm

FIGURE 12.29 **Ticks.** **(a)** *Ixodes scapularis* transmits Lyme disease on the East Coast of the U.S.A. **(b)** *I. pacificus* is the Lyme disease vector on the West Coast.

decaying organic matter such as feces. While doing so, a housefly can pick up a pathogen on its feet or body and transport the pathogen to our food.

Some parasites multiply in their vectors. When this happens, the parasites can accumulate in the vector's feces or saliva. Large numbers of parasites can then be deposited on or in the host while the vector is feeding there. The spirochete that causes Lyme disease is transmitted by ticks in this manner, and the dengue fever virus is transmitted in the same way by mosquitoes (see the box on p. 324).

Plasmodium is an example of a parasite that requires that its vector also be the definitive host. *Plasmodium* can sexually reproduce only in the gut of an *Anopheles* mosquito. *Plasmodium* is introduced into a human host with the mosquito's saliva, which acts as an anticoagulant to keep blood flowing.

To eliminate vector-borne diseases (such as African sleeping sickness), health workers focus on eradicating the vectors.

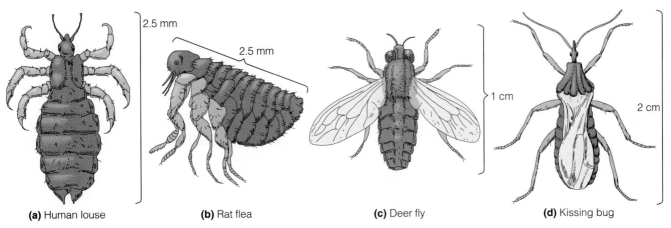

(a) Human louse **(b)** Rat flea **(c)** Deer fly **(d)** Kissing bug

FIGURE 12.30 Arthropod vectors. (a) The human louse, *Pediculus.* **(b)** The rat flea, *Xenopsylla.* **(c)** The deer fly, *Chrysops.* **(d)** The kissing bug, *Triatoma.* See Table 12.8 for the diseases these vectors carry.

STUDY OUTLINE

FUNGI (pp. 295–304)

1. The Kingdom Fungi includes yeasts, molds, and fleshy fungi (mushrooms).
2. Fungi are aerobic or facultatively anaerobic chemoheterotrophs.
3. Most fungi are decomposers, and a few are parasites of plants and animals.

Characteristics of Fungi (pp. 295–300)

1. A fungal thallus consists of filaments of cells called hyphae; a mass of hyphae is called a mycelium.
2. Yeasts are unicellular fungi. To reproduce, fission yeasts divide symmetrically, whereas budding yeasts divide asymmetrically.
3. Buds that do not separate from the mother cell form pseudohyphae.
4. Pathogenic dimorphic fungi are yeastlike at 37°C and moldlike at 25°C.
5. The following spores can be produced asexually: arthrospores, chlamydospores, sporangiospores, conidospores, and blastospores.
6. Fungi are classified according to the type of sexual spore that they form.
7. Sexual spores are usually produced in response to special circumstances, often changes in the environment.
8. Fungi can grow in acidic, low-moisture, aerobic environments.
9. They are able to metabolize complex carbohydrates.

Medically Important Phyla of Fungi (pp. 300–301)

1. Deuteromycota reproduce asexually by one or more of the following: chlamydospores, arthrospores, conidiospores, or budding.

2. Fungi are placed in the Deuteromycota because sexual spores have not yet been seen.
3. The classification of *Pneumocystis* as a fungus or protozoan is currently being studied.
4. The Zygomycota have coenocytic hyphae and produce sporangiospores and zygospores.
5. The Ascomycota have septate hyphae and produce ascospores and frequently conidiospores.
6. Basidiomycota have septate hyphae and produce basidiospores; some produce conidiospores.

Fungal Diseases (pp. 301–303)

1. Systemic mycoses are fungal infections deep within the body and affect many tissues and organs.
2. Subcutaneous mycoses are fungal infections beneath the skin.
3. Cutaneous mycoses affect keratin-containing tissues such as hair, nails, and skin.
4. Superficial mycoses are localized on hair shafts and superficial skin cells.
5. Opportunistic mycoses are caused by normal microbiota or fungi that are not usually pathogenic.
6. Opportunistic mycoses include mucormycosis, caused by some zygomycetes; aspergillosis, caused by *Aspergillus;* and candidiasis, caused by *Candida.*
7. Opportunistic mycoses can infect any tissues. However, they are usually systemic.

Economic Effects of Fungi (pp. 303–304)

1. *Saccharomyces* and *Trichoderma* are used in the production of foods.
2. Mold spoilage of fruits, grains, and vegetables is more common than bacterial spoilage of these products.

3. Many fungi cause diseases in plants (for example, in potatoes, chestnuts, and elms).

ALGAE (pp. 304–309)

1. Algae are unicellular, filamentous, or multicellular (thallic).
2. Most algae live in aquatic environments.

Characteristics of Algae (pp. 305–306)

1. All algae have eucaryotic cells and possess flagella at some time during their lives.
2. The thallus (or body) of multicellular algae usually consists of a stipe, a holdfast, and blades.
3. Algae reproduce asexually by cell division and fragmentation.
4. Many algae reproduce sexually.
5. Algae are photoautotrophs that produce oxygen.
6. Algae are classified as plants or protists according to their structures and pigments.

Selected Phyla of Algae (pp. 306–309)

1. Brown algae (kelp) are harvested for algin.
2. Red algae grow deeper in the ocean than other algae because their red pigments can absorb the blue light that penetrates to deeper levels.
3. Green algae have cellulose and chlorophyll *a* and *b* and store starch.
4. Diatoms are unicellular and have pectin and silica cell walls.
5. Multicellular algae include the brown algae, red algae, and green algae.
6. Unicellular dinoflagellates have cellulose and silica cell walls; some produce a neurotoxin that causes paralytic shellfish poisoning and ciguatera.
7. Euglenoids have a semirigid cell membrane and one flagellum; they are unicellular.

Roles of Algae in Nature (p. 309)

1. Algae are the primary producers in aquatic food chains.
2. Planktonic algae produce most of the molecular oxygen in the Earth's atmosphere.
3. Petroleum is the fossil remains of planktonic algae.
4. Unicellular algae are symbionts in such animals as *Tridacna*.

LICHENS (pp. 310–311)

1. A lichen is a symbiotic combination of an alga (or a cyanobacterium) and a fungus.
2. The alga photosynthesizes, providing carbohydrates for the lichen; the fungus provides a holdfast.
3. Lichens colonize habitats that are unsuitable for either the alga or fungus alone.
4. Lichens may be classified on the basis of morphology as crustose, foliose, or fruticose.

5. In the lichen thallus, the alga reproduces by cell division, and the fungus forms sexual spores.

SLIME MOLDS (pp. 311–313)

1. Cellular slime molds resemble amoebas and ingest bacteria by phagocytosis.
2. A plasmodial (acellular) slime mold is a multinucleated mass of protoplasm that engulfs organic debris and bacteria as it moves.

PROTOZOA (pp. 313–317)

1. Protozoa are unicellular eucaryotes in the Kingdom Protista.
2. Protozoa are found in soil and water and as normal microbiota in animals.

Characteristics of Protozoa (pp. 313–314)

1. Asexual reproduction is by fission, budding, or schizogony.
2. Sexual reproduction is by conjugation.
3. During ciliate conjugation, two haploid nuclei fuse to produce a zygote.
4. Some protozoa can produce a cyst for protection during adverse environmental conditions.
5. Most protozoa are heterotrophs and feed on bacteria and particulate organic material.
6. Protozoa are complex cells that have a pellicle, a cytostome, and an anal pore.

Medically Important Phyla of Protozoa (pp. 314–317)

1. *Entamoeba* is a parasitic amoeboflagellate that uses pseudopods for motility; it causes amoebic dysentery.
2. Parasitic amoeboflagellates that use flagella include the following: *Giardia lamblia*, causing an intestinal infection called giardiasis; *Trichomonas vaginalis*, causing genitourinary infections that may be transmitted by coitus; and trypanosomes, which are found in the blood of human hosts and are transmitted by blood-sucking insects.
3. The only ciliate that is a parasite of humans is *Balantidium coli*, the cause of one form of dysentery.
4. *Plasmodium* is an apicomplexan that causes malaria.
5. Asexual reproduction of *Plasmodium* occurs in red blood cells and the liver of humans.
6. Sexual reproduction of *Plasmodium* takes place in the intestine of the female *Anopheles* mosquito.
7. *Toxoplasma gondii* is an apicomplexan that infects humans; it can be transmitted to a fetus in utero.
8. *Cryptosporidium* causes respiratory and diarrheal diseases in immunosuppressed patients.
9. Microsporidia cause diarrhea and keratoconjunctivitis in AIDS patients.

HELMINTHS (pp. 317–323)

1. Parasitic flatworms belong to the Phylum Platyhelminthes.
2. Parasitic roundworms belong to the Phylum Nematoda.

Characteristics of Helminths (pp. 318–319)

1. Helminths are multicellular animals; a few are parasites of humans.
2. The anatomy and life cycles of parasitic helminths are modified for parasitism.
3. The adult stage of a parasitic helminth is found in the definitive host.
4. Each larval stage of a parasitic helminth requires an intermediate host.
5. Helminths can be hermaphroditic or dioecious.

Platyhelminths (pp. 319–321)

1. Flatworms are dorsoventrally flattened animals that have an incomplete digestive system.
2. Adult trematodes or flukes have an oral and ventral sucker with which they attach to and feed on host tissue.
3. Eggs of trematodes hatch into free-swimming miracidia that enter the first intermediate host; two generations of rediae develop in the first intermediate host; the rediae become cercariae that bore out of the first intermediate host and penetrate the second intermediate host; cercariae encyst as metacercariae in the second intermediate host; after they are ingested by the definitive host, the metacercariae develop into adults.
4. A cestode, or tapeworm, consists of a scolex (head) and proglottids.
5. Humans serve as the definitive host for the beef tapeworm, and cattle are the intermediate host.
6. Humans serve as the definitive host and can be an intermediate host for the pork tapeworm.
7. Humans serve as the intermediate host for *Echinococcus granulosus;* the definitive hosts are dogs, wolves, and foxes.

Nematodes (pp. 321–323)

1. Members of the Phylum Nematoda are roundworms with a complete digestive system.
2. The nematodes that infect humans with their eggs are *Enterobius vermicularis* (pinworm) and *Ascaris lumbricoides.*
3. The nematodes that infect humans with their larvae are *Necator americanus* (hookworm), *Trichinella spiralis,* and anisakine worms.

ARTHROPODS AS VECTORS (pp. 323–327)

1. Jointed-legged animals, including ticks and insects, belong to the Phylum Arthropoda.
2. Arthropods that carry diseases are called vectors.
3. Elimination of vector-borne diseases is best done by the control or eradication of the vectors.

STUDY QUESTIONS

Review

1. Contrast the mechanisms of conidiospore and ascospore formation by *Eupenicillium.*

2. Fill in the following table.

PHYLUM	SPORE TYPE(S)	
	SEXUAL	ASEXUAL
Zygomycota		
Ascomycota		
Basidiomycota		
Deuteromycota		

3. Fungi are classified into phyla on the basis of _____ .

4. The following is a list of fungi, their methods of entry into the body, and sites of infections they cause. Categorize each type of mycosis as cutaneous, opportunistic, subcutaneous, superficial, or systemic.

GENUS	METHOD OF ENTRY	SITE OF INFECTION	MYCOSIS
Blastomyces	Inhalation	Lungs	
Sporothrix	Puncture	Ulcerative lesions	
Microsporum	Contact	Fingernails	
Trichosporon	Contact	Hair shafts	
Aspergillus	Inhalation	Lungs	

5. A mixed culture of *Escherichia coli* and *Penicillium chrysogenum* is inoculated onto the following culture media. On which media would you expect each to grow? Why?
   (a) 0.5% peptone in tap water
   (b) 10% glucose in tap water

6. What is the role of the alga in a lichen? What is the role of the fungus?

7. Briefly discuss the importance of lichens in nature. Briefly discuss the importance of algae.

8. Complete the following table.

PHYLUM	CELL WALL COMPOSITION	SPECIAL FEATURES/ IMPORTANCE
Dinoflagellates		
Euglenoids		
Diatoms		
Red algae		
Brown algae		
Green algae		

Indicate which phyla consist primarily of unicellular forms. Which phylum would you include in the plant kingdom? Why? Into which kingdom would you place the others?

9. Differentiate between cellular and plasmodial slime molds. How does each survive adverse environmental conditions?

10. Transmission of helminthic parasites to humans usually occurs by _____ .

11. Why is it significant that *Trichomonas* does not have a cyst stage? Name a protozoan parasite that does have a cyst stage.

12. Protozoa are classified on the basis of _____ .

13. Recall the life cycle of *Plasmodium*. Where does asexual reproduction occur? Where does sexual reproduction occur? Identify the definitive host. Identify the vector.

14. To what phylum and class does this animal belong?

List two characteristics that put it in this phylum. Name the body parts. What is the name of the encysted larva of this animal?

15. Most nematodes are dioecious. What does this term mean? To what phylum do nematodes belong?

16. Vectors can be divided into three major types, according to the roles they play for the parasite. List the three types of vectors and a disease transmitted by each.

Critical Thinking

1. A generalized life cycle of the liver fluke *Clonorchis sinensis* is shown below. Identify the intermediate host(s). Identify the definitive host(s). To what phylum and class does this animal belong?

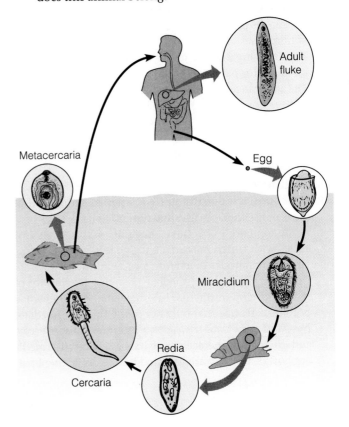

2. The size of a cell is limited by its surface-to-volume ratio; that is, if the volume becomes too great, internal heat cannot be dissipated and nutrients and wastes cannot be efficiently transported. How do plasmodial slime molds manage to circumvent the surface-to-volume rule?

3. The life cycle of the fish tapeworm *Diphyllobothrium* is similar to that of *Taenia saginata,* except that the intermediate host is fish. Describe the life cycle and method of transmission to humans. Why are freshwater fish more likely to be a source of tapeworm infestation than marine fish?

4. Microorganisms don't care what they are called. Why, then, might it be important to classify *Pneumocystis?*

5. *Trypanosoma brucei gambiense* (part a) is the causative agent of African sleeping sickness. To what phylum and class does it belong? Part b shows a simplified life cycle for *T. b. gambiense*. Identify the host and vector of this parasite.

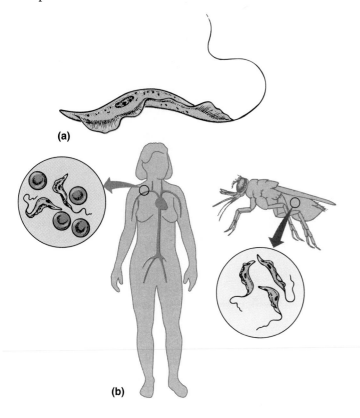

(a)

(b)

Clinical Applications

1. A girl developed generalized seizures. A CAT scan revealed a single brain lesion consistent with a tumor. Biopsy of the lesion showed a cysticercus. The patient lived in South Carolina and had never traveled outside of the state. What parasite caused her disease? How is this disease transmitted? How might it be prevented?

2. A California farmer developed a low-grade fever, myalgia, and cough. A chest X ray revealed an infiltrate in the middle lobe. Microscopic examination of the sputum revealed round, budding cells. A sputum culture grew mycelia and arthrospores. What organism is most likely the cause of the symptoms? What is causing the man's disease? How is this disease transmitted? How might it be prevented?

3. A teenage male in California complained of remittent fever, chills, and headaches. A blood smear revealed ring-shaped cells in his white blood cells. He was successfully treated with primaquine and chloroquine. The patient lives near the San Luis Rey River and has no history of foreign travel, blood transfusion, or IV drug use. What is the disease? How was it acquired?

Viruses

IN 1892, IN AN EXPERIMENT to isolate the cause of tobacco mosaic disease, Dmitri Iwanowski filtered the sap of diseased tobacco plants through a porcelain filter that was designed to retain bacteria. He expected to find the microbe trapped in the filter; instead, he found that the infectious agent had passed through the minute pores of the filter. When he injected healthy plants with the filtered fluid, they contracted tobacco mosaic disease.

Iwanowski still believed the infectious agent to be a bacterium that was small enough to pass through the filter. But later, other scientists, led by Martinus Beijerinck, observed that the behavior of this agent was different from that of bacteria. In the early 1900s, a distinction was finally made between bacteria and viruses, the filterable agents that cause tobacco mosaic disease and many other diseases. (*Virus* is the Latin word for poison.) In 1935, an American chemist, Wendell M. Stanley, isolated the tobacco mosaic virus (genus *Tobamovirus*), making it possible for the first time to carry out chemical and structural studies on a purified virus. At about the same time, the invention of the electron microscope made it possible to see viruses for the first time.

We now know that viruses are found as parasites in all types of organisms. They can reproduce only within cells. Many diseases of humans are known to be caused by viruses; some diseases of agriculturally important animals and plants are also known to be caused by viruses. Viruses infect fungi, bacteria, and protists as well.

Advances in molecular biology techniques in the 1980s and 1990s have led to the discovery of new human viruses. Hepatitis C virus and *Pestivirus*, which causes severe pediatric diarrhea, were discovered in 1989. Public health officials are concerned about health risks posed by these new viruses, and they are also concerned that the ease of world travel and changing environments will spread viruses into new areas. The 1993 outbreak of a *Hantavirus* infection in the southwestern U.S.A. (see the box in Chapter 24) illustrates the potential for new infections to emerge at any time. *Hantavirus* disease is an acute, often fatal, respiratory disease. Although the virus had previously been isolated from rodents, the disease had not been reported in humans in the U.S.A. The potential of epidemics caused by the dengue fever virus in the U.S.A. because of the introduction of a suitable vector is discussed in the box in Chapter 12, page 324.

GENERAL CHARACTERISTICS OF VIRUSES

LEARNING OBJECTIVE

- *Differentiate between a virus and a bacterium.*

The question of whether viruses are living organisms has an ambiguous answer. Life can be defined as a complex set of processes resulting from the actions of proteins specified by nucleic acids. The nucleic acids of living cells are

TABLE 13.1   A Comparison of Viruses and Bacteria

	BACTERIA		VIRUSES
	TYPICAL BACTERIA	RICKETTSIAS/ CHLAMYDIAS	
Intracellular parasite	−	+	+
Plasma membrane	+	+	−
Binary fission	+	+	−
Filterable through bacteriological filters	−	−/+	+
Possess both DNA and RNA	+	+	−
ATP-generating metabolism	+	+/−	−
Ribosomes	+	+	−
Sensitive to antibiotics	+	+	−
Sensitive to interferon	−	−	+

4. Cause the synthesis of specialized structures that can transfer the viral nucleic acid to other cells.

Viruses have few or no enzymes of their own for metabolism; for example, they lack enzymes for protein synthesis and ATP generation. To multiply, viruses must take over the metabolic machinery of the host cell. This fact has considerable medical significance for the development of antiviral drugs because most drugs that would interfere with viral multiplication would also interfere with the functioning of the host cell and hence are too toxic for clinical use. (Antiviral drugs are discussed in Chapter 20.) However, the presence of lipids in the coverings of some viruses makes these viruses sensitive to disinfection (outside of a host cell). For example, a 1:10 dilution of household chlorine bleach can inactivate HIV (the AIDS virus, which is in the genus *Lentivirus*) outside the body. Viral lipids may also be damaged by lipid solvents, such as ether, and by emulsifying agents, such as bile salts and detergents.

in action all the time. Because viruses are inert outside of living host cells, in this sense they are not considered to be living organisms. However, once viruses enter a host cell, the viral nucleic acids become active, and viral multiplication results. In this sense, viruses are alive when they multiply in the host cells they infect. From a clinical point of view, viruses can be considered alive because they cause infection and disease, just as pathogenic bacteria, fungi, and protozoa do. Depending on one's viewpoint, a virus may be regarded as an exceptionally complex aggregation of nonliving chemicals or as an exceptionally simple living microorganism.

How, then, do we define a virus? Viruses were originally distinguished from other infectious agents because they are especially small (filterable) and because they are **obligatory intracellular parasites**—that is, they absolutely require living host cells in order to multiply. However, both of these properties are shared by certain small bacteria, such as some rickettsias. Viruses and bacteria are compared in Table 13.1.

The truly distinctive features of viruses are now known to relate to their simple structural organization and composition and their mechanism of multiplication. Accordingly, **viruses** are entities that:

1. Contain a single type of nucleic acid, either DNA or RNA.

2. Contain a protein coat (sometimes itself enclosed by an envelope of lipids, proteins, and carbohydrates) that surrounds the nucleic acid.

3. Multiply inside living cells by using the synthesizing machinery of the cell.

Host Range

The **host range** of a virus is the spectrum of host cells that the virus can infect. There are viruses that infect invertebrates, vertebrates, plants, protists, fungi, and bacteria. However, most viruses are able to infect specific types of cells of only one host species. In this chapter we are concerned mainly with viruses that infect either humans or bacteria. Viruses that infect bacteria are called **bacteriophages** or **phages.**

The particular host range of a virus is determined by the virus's requirements for its specific attachment to the host cell and the availability within the potential host of cellular factors required for viral multiplication. For the virus to infect the host cell, the outer surface of the virus must chemically interact with specific receptor sites on the surface of the cell. The two complementary components are held together by weak bonds, such as hydrogen bonds. For some bacteriophages, the receptor site is part of the cell wall of the host; in other cases, it is part of the fimbriae or flagella. For animal viruses, the receptor sites are on the plasma membranes of the host cells.

Size

Viral sizes are ascertained with the aid of electron microscopy. Viruses vary considerably in size. Although most are quite a bit smaller than bacteria, some of the larger viruses (such as the vaccinia virus) are about the same size as some very small bacteria (such as the mycoplasmas, rickettsias, and chlamydias). Viruses range from 20 to 14,000 nm in length. The comparative sizes of several viruses and bacteria are shown in Figure 13.1.

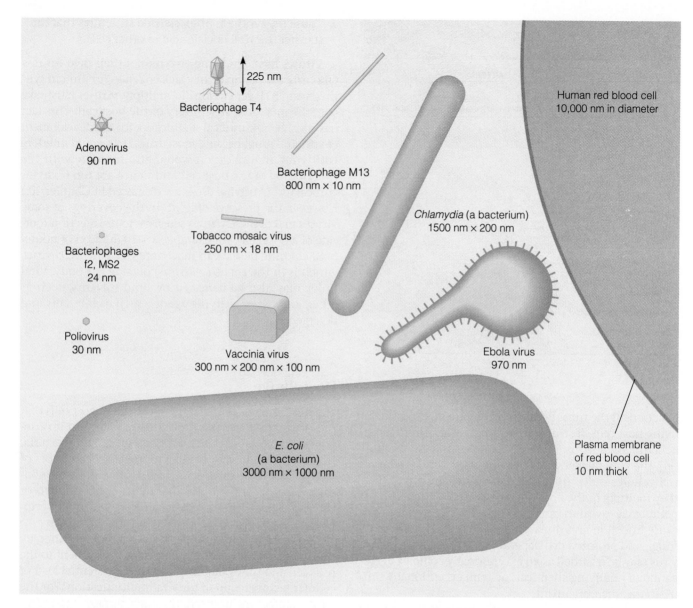

FIGURE 13.1 Virus sizes. Sizes of several viruses (blue) and bacteria (tan) are compared with a human red blood cell, shown to the right of the microbes. Dimensions are given in nanometers (nm) and are either diameters or length by width.

VIRAL STRUCTURE

LEARNING OBJECTIVE

- *Describe the chemical composition and physical structure of both an enveloped and a nonenveloped virus.*

A **virion** is a complete, fully developed viral particle composed of nucleic acid and surrounded by a protein coat. This coat protects it from the environment and serves as a vehicle of transmission from one host cell to another. Viruses are classified by differences in the structures of these coats.

Nucleic Acid

As we noted earlier, the core of a virus contains a single kind of nucleic acid, either DNA or RNA, which is the genetic material. The percentage of nucleic acid in relation to protein is about 1% for the influenza virus and about 50% for certain bacteriophages. The total amount of nucleic acid varies from a few thousand nucleotides (or pairs) to as many as 250,000 nucleotides. (*E. coli*'s chromosome consists of approximately four million nucleotide pairs.)

In contrast to procaryotic and eucaryotic cells, in which DNA is always the primary genetic material (and RNA plays an auxiliary role), a virus can have either DNA or RNA but never both. The nucleic acid of a virus can be single-stranded or double-stranded. Thus, there are viruses with the familiar double-stranded DNA, with single-stranded DNA, with double-stranded RNA, and with single-stranded RNA. Depending on the virus, the nucleic acid can be linear or circular. In some viruses (such as the influenza virus), the nucleic acid is in several separate segments.

Capsid and Envelope

The nucleic acid of a virus is surrounded by a protein coat called the **capsid** (Figure 13.2a). The structure of the capsid is ultimately determined by the viral nucleic acid and accounts for most of the mass of a virus, especially of small ones. Each capsid is composed of protein subunits called **capsomeres.** In some viruses, the proteins composing the capsomeres are of a single type; in other viruses, several types of protein may be present. Individual capsomeres are often visible in electron micrographs (see Figure 13.8a for an example). The arrangement of capsomeres is characteristic of a particular type of virus.

In some viruses, the capsid is covered by an **envelope** (Figure 13.2b), which usually consists of some combination of lipids, proteins, and carbohydrates. Some animal viruses are released from the host cell by an extrusion process that coats the virus with a layer of the host cell's plasma membrane; that layer becomes the viral envelope. In many cases, the envelope contains proteins determined by the viral nucleic acid and materials derived from normal host cell components.

Depending on the virus, envelopes may or may not be covered by **spikes,** which are carbohydrate–protein complexes that project from the surface of the envelope. Some viruses attach to host cells by means of spikes. Spikes are such a reliable characteristic of some viruses that they can be used as a means of identification. The ability of certain viruses, such as the influenza virus, to clump red blood cells is associated with spikes. Such viruses bind to red blood cells and form bridges between them. The resulting clumping is called **hemagglutination** and is the basis for several useful laboratory tests.

Viruses whose capsids are not covered by an envelope are known as **nonenveloped viruses** (see Figure 13.2a). The capsid of a nonenveloped virus protects the nucleic acid from nuclease enzymes in biological fluids and promotes the virus's attachment to susceptible host cells.

When the host has been infected by a virus, the host immune system is stimulated to produce antibodies (proteins that react with the surface proteins of the virus). This interaction between host antibodies and virus proteins inactivates the virus and stops the infection. However, some viruses can escape antibodies because regions of the genes that code for these viruses' surface proteins are susceptible to mutations. The progeny of mutant viruses have altered surface proteins, such that the antibodies are not able to react with them. Influenza virus frequently undergoes

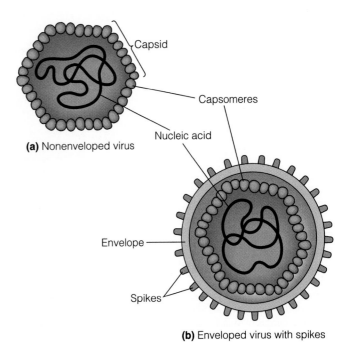

(a) Nonenveloped virus

(b) Enveloped virus with spikes

FIGURE 13.2 General structure of two types of viruses.

such changes in its spikes. This is the reason why you can get influenza more than once. Although you may have produced antibodies to one influenza virus, the virus can mutate and infect you again.

General Morphology

Viruses may be classified into several morphological types on the basis of their capsid architecture. The structure of these capsids has been revealed by electron microscopy and a technique called X-ray crystallography.

Helical Viruses

Helical viruses resemble long rods that may be rigid or flexible. The viral nucleic acid is found within a hollow, cylindrical capsid that has a helical structure. An example of a helical virus that is a rigid rod is the tobacco mosaic virus (Figure 13.3). Another is bacteriophage M13.

Polyhedral Viruses

Many animal, plant, and bacterial viruses are polyhedral viruses; that is, they are many-sided. The capsid of most polyhedral viruses is in the shape of an *icosahedron*, a regular polyhedron with 20 triangular faces and 12 corners (Figure 13.4a). The capsomeres of each face form an equilateral triangle. An example of a polyhedral virus in the shape of an isocahedron is the adenovirus (genus *Mastadenovirus*) (Figure 13.4b). Another icosahedral virus is the poliovirus.

FIGURE 13.3 Morphology of a helical virus. (a) Diagram of a portion of a helical virus. Several rows of capsomeres have been removed to reveal the nucleic acid. **(b)** Micrograph of a tobacco mosaic virus (*Tobamovirus*) showing helical rods.

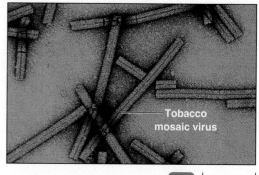

(a) A helical virus

(b) *Tobamovirus*   TEM   0.1 μm

FIGURE 13.4 Morphology of a nonenveloped polyhedral virus in the shape of an icosahedron. (a) Diagram of a polyhedral (icosahedral) virus. **(b)** Micrograph of an adenovirus (genus *Mastadenovirus*). Individual capsomeres in the protein coat are visible.

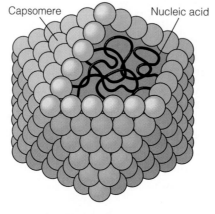

(a) A polyhedral virus

(b) A *Mastadenovirus*   TEM   0.1 μm

FIGURE 13.5 Morphology of an enveloped helical virus. (a) Diagram of an enveloped helical virus. **(b)** Micrograph of influenza viruses (genus *Influenzavirus*). Note the halo of spikes projecting from the outer surface of each envelope (see Chapter 24).

(a) An enveloped helical virus

(b) An *Influenzavirus*   TEM   0.1 μm

Enveloped Viruses

As we noted earlier, the capsid of some viruses is covered by an envelope. Enveloped viruses are roughly spherical. When helical or polyhedral viruses are enclosed by envelopes, they are called *enveloped helical* or *enveloped polyhedral viruses*. An example of an enveloped helical virus is the influenza virus (genus *Influenzavirus*) (Figure 13.5). An example of an enveloped polyhedral (icosahedral) virus is the herpes simplex virus (genus *Simplexvirus*) (Figure 13.6).

Complex Viruses

Some viruses, particularly bacterial viruses, have very complicated structures and are called **complex viruses.** One example of a complex virus is a bacteriophage. Certain

FIGURE 13.6 Morphology of an enveloped polyhedral (icosahedral) virus. (a) Diagram of an enveloped polyhedral virus. **(b)** Micrograph of a group of herpes simplex viruses (genus *Simplexvirus*). To the lower right, a virus particle is acquiring its envelope as it buds out through the nuclear membrane of a host cell.

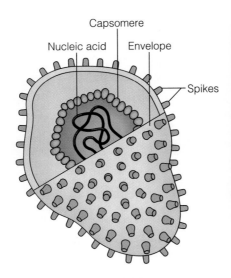

Capsomere
Nucleic acid
Envelope
Spikes

(a) An enveloped polyhedral virus

Virus acquiring envelope
Nuclear membrane of host cell
Enveloped virus

(b) A *Simplexvirus*

TEM 0.1 μm

65 nm
Capsid (head)
DNA
Sheath
Tail fiber
Pin
Baseplate

TEM 0.1 μm

(a) T-even bacteriophage

bacteriophages even have capsids to which additional structures are attached (Figure 13.7a). If you take a close look at the bacteriophages shown in this figure, you will note that the capsid (head) is polyhedral and the tail sheath is helical. The head contains the nucleic acid. Later in this chapter we will discuss the functions of the other structures, such as the tail sheath, tail fibers, plate, and pin. Another example of complex viruses is poxviruses, which do not contain clearly identifiable capsids but have several coats around the nucleic acid (Figure 13.7b).

TAXONOMY OF VIRUSES

LEARNING OBJECTIVES

- *Define viral species.*
- *Give an example of a family, genus, and common name for a virus.*

Virologists began addressing the problem of viral taxonomy in 1966 at the International Congress of Microbiology in Moscow. We need viral taxonomy, just as we need taxonomic groupings of plants, animals, or bacteria, to help us communicate and to understand newly discovered organisms. In general, the International Committee on Taxonomy of Viruses (ICTV) has not yet established higher taxa

(b) An *Orthopoxvirus*

TEM 0.1 μm

FIGURE 13.7 Morphology of complex viruses. (a) Diagram and micrograph of a T-even bacteriophage. **(b)** Micrograph of variola virus, a species in the genus *Orthopoxvirus* that causes smallpox.

FIGURE 13.8 DNA-containing animal viruses. (a) Negatively stained capsid of a herpesvirus, a member of the Herpesviridae. The individual capsomeres are clearly visible. **(b)** Negatively stained viruses that have been concentrated in a centrifuge gradient. The larger capsids are an adenovirus, and the smaller viruses are in the genus *Dependovirus*, which is in the family Parvoviridae.

(a) A herpesvirus   TEM   ⊢ 25 nm

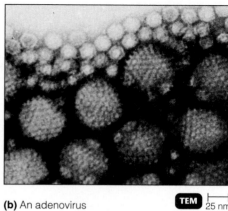

(b) An adenovirus   TEM   ⊢ 25 nm

(a) An *Influenzavirus*   TEM ⊢ 1 nm

(b) *Vesiculovirus*   TEM ⊢ 0.1 μm

(c) MMTV   TEM ⊢ 0.1 μm

FIGURE 13.9   RNA-containing animal viruses. (a) Envelopes of A2 influenza virus The spikes projecting from each envelope are visible. **(b)** Particles of vesicular stomatitis virus (*Vesiculovirus*), a member of the family Rhabdoviridae. **(c)** Mouse mammary tumor virus (MMTV), which is in the family Retroviridae.

(order through kingdom) for viruses. The two exceptions are that the complex bacteriophages (Figure 13.7a) will probably be placed in their own order, and enveloped viruses with one negative strand of RNA have been classified in the newly established order Mononegavirales.

Viruses are grouped by nucleic acid type, morphology, and the presence or absence of an envelope. Figure 13.8 shows electron micrographs of DNA-containing animal viruses, and Figure 13.9 presents electron micrographs of RNA-containing animal viruses. The single-stranded RNA viruses are further divided by their method of replication. The ICTV has organized the 2000 known species of viruses into 73 families, each of which shares morphologic characteristics and reproductive strategies. Family names end in *-viridae*. Likewise, genera of viruses share

certain characteristics; the suffix *-virus* is used for genus names.

A **viral species** is defined as a group of viruses sharing the same genetic information and ecological niche. Specific epithets for viruses have not yet been established, so viral species are designated by descriptive common names, such as human immunodeficiency virus (HIV), with subspecies (if any) designated by a number (HIV-1). Viral nomenclature is in the process of being standardized; this text introduces the new viral family and genus names and continues to use the common species names. Common names are written in regular type (herpes simplex virus), whereas genus names are capitalized and italicized (*Simplexvirus*). Table 13.2 presents a summary of the classification of viruses that infect humans.

TABLE 13.2 Classification of Human Viruses

CHARACTERISTICS	VIRAL FAMILY	VIRAL GENUS (WITH REPRESENTATIVE SPECIES) AND UNCLASSIFIED MEMBERS*	DIMENSIONS OF VIRION (DIAMETER IN NM)	CLINICAL OR SPECIAL FEATURES
Single-stranded DNA, nonenveloped	Parvoviridae	*Dependovirus*	18–25	Depend on coinfection with adenoviruses; cause fetal death, gastroenteritis
Double-stranded DNA, nonenveloped	Adenoviridae	*Mastadenovirus* (adenovirus)	70–90	Medium-sized viruses that cause various respiratory infections in humans; some cause tumors in animals.
	Papovaviridae	*Papillomavirus* (human wart virus) *Polyomavirus*	40–57	Small viruses that induce tumors; the human wart virus (papilloma) and certain viruses that produce cancer in animals (polyoma and simian) belong to this family. Refer to Chapters 21 and 26.
Double-stranded DNA, enveloped	Poxviridae	*Orthopoxvirus* (vaccinia and smallpox viruses) *Molluscipoxvirus*	200–350	Very large, complex, brick-shaped viruses that cause diseases such as smallpox (variola), molluscum contagiosum (wartlike skin lesion), cowpox, and vaccinia; vaccinia virus gives immunity to smallpox. Refer to Chapter 21.
	Herpesviridae	*Simplexvirus* (herpes simplex viruses 1 and 2) *Varicellavirus* (varicella-zoster virus) *Cytomegalovirus* *Lymphocryptovirus* (Epstein-Barr virus) Human herpes virus 6	150–200	Medium-sized viruses that cause various human diseases, such as fever blisters, chickenpox, shingles, and infectious mononucleosis; implicated in a type of human cancer called Burkitt's lymphoma. Refer to Chapters 21, 23, and 26.
	Hepadnaviridae	*Hepadnavirus* (hepatitis B virus)	42	After protein synthesis, hepatitis B virus uses reverse transcriptase to produce its DNA from mRNA; causes hepatitis B and liver tumors. Refer to Chapter 25.
Single-stranded RNA, nonenveloped + strand	Picornaviridae	*Enterovirus* *Rhinovirus* (common cold virus) Hepatitis A virus	28–30	At least 70 human enteroviruses are known, including the polio-, coxsackie-, and echoviruses; more than 100 rhinoviruses exist and are the most common cause of colds. Refer to Chapters 22, 23, 24, and 25.
Single-stranded RNA, enveloped + strand	Togaviridae	*Alphavirus* *Rubivirus* (rubella virus)	60–70	Included are many viruses transmitted by arthropods (*Alphavirus*); diseases include eastern equine encephalitis (EEE) and western equine encephalitis (WEE). Rubella virus is transmitted by the respiratory route. Refer to Chapters 21, 22, 23, and 26.
	Flaviviridae	*Flavivirus* *Pestivirus* Hepatitis C virus	40–50	Can replicate in arthropods that transmit them; diseases include yellow fever, dengue, St. Louis encephalitis, and Japanese encephalitis. The unclassified hepatitis C virus is most likely in this family. Refer to Chapters 22 and 23.
	Coronaviridae	*Coronavirus*	80–160	Associated with upper respiratory tract infections and the common cold. Refer to Chapter 24.

*Unclassified viruses have not been assigned to genera; therefore, only their common names are listed.

continued

TABLE 13.2 Continued

CHARACTERISTICS	VIRAL FAMILY	VIRAL GENUS (WITH REPRESENTATIVE SPECIES) AND UNCLASSIFIED MEMBERS*	DIMENSIONS OF VIRION (DIAMETER IN NM)	CLINICAL OR SPECIAL FEATURES
— strand, one strand of RNA	Rhabdoviridae	*Vesiculovirus* (vesicular stomatitis virus) *Lyssavirus* (rabies virus)	70–180	Bullet-shaped viruses with a spiked envelope; cause rabies and numerous animal diseases. Refer to Chapter 22.
	Filoviridae	*Filovirus*	80–14,000	Enveloped, helical viruses; Ebola and Marburg viruses are filoviruses.
	Paramyxoviridae	*Paramyxovirus* *Morbillivirus* (measles virus)	150–300	Paramyxoviruses cause parainfluenza, mumps, and Newcastle disease in chickens. Refer to Chapters 21 and 25.
— strand, multiple strands of RNA	Orthomyxoviridae	*Influenzavirus* (influenza viruses A and B) Influenza C virus	80–200	Envelope spikes can agglutinate red blood cells. Refer to Chapter 24.
	Bunyaviridae	*Bunyavirus* (California encephalitis virus) *Hantavirus*	90–120	Hantaviruses cause hemorrhagic fevers such as Korean hemorrhagic fever and *Hantavirus* pulmonary syndrome; associated with rodents. Refer to Chapter 24.
	Arenaviridae	*Arenavirus*	50–300	Helical capsids contain RNA-containing granules; cause lymphocytic choriomeningitis and hemorrhagic fevers. Refer to Chapter 24.
Produce DNA	Retroviridae	Oncoviruses *Lentivirus* (HIV)	100–120	Includes all RNA tumor viruses and double-stranded RNA viruses. Oncoviruses cause leukemia and tumors in animals; the *Lentivirus* HIV causes AIDS. Refer to Chapter 19.
Double-stranded RNA, nonenveloped	Reoviridae	*Reovirus* Colorado tick fever virus	60–80	Involved in mild respiratory infections and infantile gastroenteritis; an unclassified species causes Colorado tick fever. Refer to Chapter 23.

*Unclassified viruses have not been assigned to genera; therefore, only their common names are listed.

ISOLATION, CULTIVATION, AND IDENTIFICATION OF VIRUSES

LEARNING OBJECTIVES

- *Describe how bacteriophages are cultured.*
- *Describe how animal viruses are cultured.*
- *List three techniques used to identify viruses.*

The fact that viruses cannot multiply outside a living host cell complicates their detection, enumeration, and identification. Viruses must be provided with living cells instead of a fairly simple chemical medium. Living plants and animals are difficult and expensive to maintain, and disease-causing viruses that grow only in higher primates and human hosts cause additional complications. However, viruses that use bacterial cells as a host (bacteriophages) are rather easily grown on bacterial cultures. This is one reason why so much of our understanding of viral multiplication has come from bacteriophages.

Growth of Bacteriophages in the Laboratory

Bacteriophages can be grown either in suspensions of bacteria in liquid media or in bacterial cultures on solid media. The use of solid media makes possible the *plaque method* for detecting and counting viruses. A sample of bacteriophage is mixed with host bacteria and melted agar. The agar containing the bacteriophages and host bacteria is then poured into a Petri plate containing a hardened layer of agar growth medium. The virus–bacteria mixture solidifies into a thin top layer that contains a layer of bacteria approximately one cell thick. Each virus infects a bacterium, multiplies, and releases several hundred new viruses. These newly produced viruses infect other bacteria in the immediate vicinity, and more new viruses are produced. Following several virus multiplication cycles, all the bacteria in the area surrounding the original virus are destroyed. This produces a number of clearings, or **plaques**, visible against a "lawn" of bacterial growth on the surface of the agar (Figure 13.10). While the plaques form, uninfected bacteria elsewhere in the Petri plate multiply rapidly and produce a turbid background.

Each plaque theoretically corresponds to a single virus in the initial suspension. Therefore, the concentrations of viral suspensions measured by the number of plaques are usually given in terms of **plaque-forming units (pfu).**

Growth of Animal Viruses in the Laboratory

In the laboratory, three methods are commonly used for culturing animal viruses. These methods involve using living animals, embryonated eggs, or cell cultures.

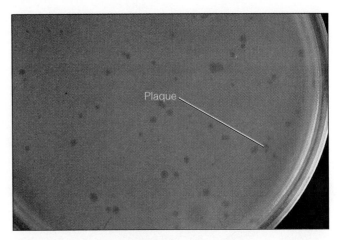

FIGURE 13.10 Viral plaques formed by bacteriophages. Clear viral plaques of various sizes have been formed by bacteriophage λ on a lawn of *E. coli.*

In Living Animals

Some animal viruses can be cultured only in *living animals,* such as mice, rabbits, and guinea pigs. Most experiments to study the immune system's response to viral infections must also be performed in virally infected live animals. Animal inoculation may be used as a diagnostic procedure for identifying and isolating a virus from a clinical specimen. After the animal is inoculated with the specimen, the animal is observed for signs of disease or is killed so that infected tissues can be examined for the virus.

Some human viruses cannot be grown in animals, or can be grown but do not cause disease. The lack of animal models for AIDS has slowed our understanding of its disease process and prevented experimentation with drugs that inhibit growth of the virus in vivo. Chimpanzees can be infected with one subspecies of human immunodeficiency virus (HIV-1, genus *Lentivirus*), but because they do not show symptoms of the disease, they cannot be used to study effects of viral growth and disease treatments. AIDS vaccines are presently being tested in humans, but the disease progresses so slowly in humans that it can take years to determine the effectiveness of these vaccines. In 1986 simian AIDS, an immune deficiency disease of green monkeys, was reported, followed in 1987 by feline AIDS, an immune deficiency disease of domestic cats. These diseases are caused by lentiviruses, which are closely related to HIV, and the diseases develop within a few months, thus providing a model for study of viral growth in different tissues. In 1990 a way to infect mice with human AIDS was found when mice were genetically engineered to produce human T cells and human gamma globulin.

In Embryonated Eggs

If the virus will grow in an *embryonated egg,* this can be a fairly convenient and inexpensive form of host for many

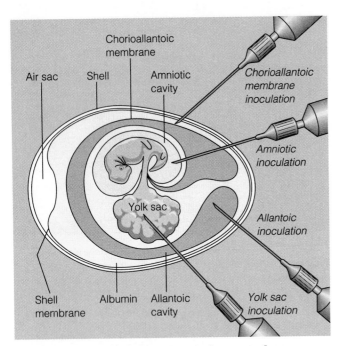

FIGURE 13.11 Inoculation of an embryonated egg.
The injection site determines the membrane on which the viruses will grow.

animal viruses. A hole is drilled in the shell of the embryonated egg, and a viral suspension or suspected virus-containing tissue is injected into the fluid of the egg. There are several membranes in an egg, and the virus is injected near the one most appropriate for its growth (Figure 13.11). Viral growth is signaled by the death of the embryo, by embryo cell damage, or by the formation of typical pocks or lesions on the egg membranes. This method was once the most widely used method of viral isolation and growth and is still used to grow viruses for some vaccines. You may be asked if you are allergic to eggs before receiving a vaccination because egg proteins may be present in the viral vaccine preparations. (Allergy is discussed in Chapter 19.)

In Cell Cultures

Cell cultures have replaced embryonated eggs as the preferred type of growth medium for many viruses. Cell cultures consist of cells grown in culture media in the laboratory. Because these cultures are generally rather homogeneous collections of cells and can be propagated and handled much like bacterial cultures, they are more convenient to work with than whole animals or embryonated eggs.

Cell culture lines are started by treating a slice of animal tissue with enzymes that separate the individual cells (Figure 13.12). These cells are suspended in a solution that provides the osmotic pressure, nutrients, and growth factors needed for the cells to grow. Normal cells tend to adhere to the glass or plastic container and reproduce to form a monolayer. Viruses infecting such a monolayer sometimes cause the cells of the monolayer to deteriorate as they multiply. This cell deterioration is called **cytopathic effect (CPE)** and is illustrated in Figure 13.13. CPE can be detected and counted in much the same way as plaques caused by bacteriophages on a lawn of bacteria.

Primary cell lines, derived from tissue slices, tend to die out after only a few generations. Certain cell lines, called **diploid cell lines**, developed from human embryos can be maintained for about 100 generations and are widely used for culturing viruses that require a human host. Cell lines developed from embryonic human cells are used to culture rabies virus (genus *Lyssavirus*) for a rabies vaccine called human diploid culture vaccine (see Chapter 22).

When viruses are routinely grown in a laboratory, **continuous cell lines** are used. These are transformed (or cancerous) cells that can be maintained through an indefinite number of generations, and they are sometimes called "immortal" cell lines (see the discussion of transformation on p. 356). One of these, the HeLa cell line, was isolated from the cancer of a woman who died in 1951. After years of laboratory cultivation, many such cell lines have lost almost all the original characteristics of the cell, but these changes have not interfered with the use of the cells for viral propagation. In spite of the success of cell culture in viral isolation and growth, there are still some viruses that have never been successfully cultivated in cell culture.

The idea of cell culture dates back to the end of the nineteenth century, but it was not a practical laboratory technique until the development of antibiotics in the years following World War II. A major problem with cell culture is that the cell lines must be kept free of microbial contamination. The maintenance of cell culture lines requires trained technicians with considerable experience working on a full-time basis. Because of these difficulties, most hospital laboratories and many state health laboratories do not isolate and identify viruses in clinical work. Instead, the tissue or serum samples are sent to central laboratories that specialize in such work.

Viral Identification

The identification of viral isolates is not an easy task. For one thing, viruses cannot be seen at all without the use of an electron microscope. Serological methods, such as Western blotting (see Chapter 10), are the most commonly used means of identification. In these tests, the virus is detected and identified by its reaction with antibodies. Antibodies, which are proteins produced by animals in response to exposure to certain substances, including viruses, are highly specific for the virus that causes their formation. We will discuss antibodies in detail in Chapter 17 and a number of immunological tests for identifying viruses in Chapter 18. Observation of the effects of viruses on host cells, described in Chapter 15, is also useful for identification of a virus.

1 A tissue is treated with enzymes to separate the cells.

2 Cells are suspended in culture medium.

3 Normal cells or primary cells grow in a monolayer across the glass or plastic container. Transformed cells or continuous cell cultures do not grow in a monolayer.

FIGURE 13.12 Cell cultures.

FIGURE 13.13 Cytopathic effect of viruses. (a) A monolayer of uninfected mouse L cells. **(b)** The same cells 24 hours after infection with VSV (see Figure 13.9b). Note the separation and "rounding up" of the cells.

(a) Uninfected mouse cells   LM   10 μm

(b) Transformed mouse cells after viral infection   LM   10 μm

Virologists can identify and characterize viruses by using such modern molecular methods as restriction fragment length polymorphisms, DNA fingerprints, or PCR. Restriction enzymes are used to cut DNA into specific nitrogenous base sequences, or fragments (see Chapter 9). Fragments of closely related viruses have similar electrophoretic patterns. PCR was used to amplify RNA from autopsy specimens to identify the 1993 *Hantavirus* outbreak in the U.S.A.

VIRAL MULTIPLICATION

LEARNING OBJECTIVES

- *Describe the lytic cycle of T-even bacteriophages.*
- *Describe the lysogenic cycle of T-even bacteriophages.*
- *Compare and contrast the multiplication cycle of DNA- and RNA-containing animal viruses.*

The nucleic acid in a virion contains only a few of the genes needed for the synthesis of new viruses. These include genes for the virion's structural components, such as the capsid proteins, and genes for a few of the enzymes

used in the viral life cycle. These enzymes are synthesized and functional only when the virus is within the host cell. Viral enzymes are almost entirely concerned with replicating or processing viral nucleic acid. Enzymes needed for protein synthesis, ribosomes, tRNA, and energy production are supplied by the host cell and are used for synthesis of virus proteins, including viral enzymes. Although the smallest nonenveloped virions do not contain any preformed enzymes, the larger virions may contain one or a few enzymes, which usually function in helping the virus to penetrate the host cell or to replicate its own nucleic acid.

Thus, for a virus to multiply, it must invade a host cell and take over the host's metabolic machinery. A single virus can give rise to several or even thousands of similar viruses in a single host cell. This process can drastically change the host cell and can even cause its death.

Multiplication of Bacteriophages

Although the means by which a virus enters and exits a host cell may vary, the basic mechanism of viral multiplication is similar for all viruses. The best-understood viral life cycles are those of the bacteriophages (phages). Phages can multiply by two alternative mechanisms: the lytic

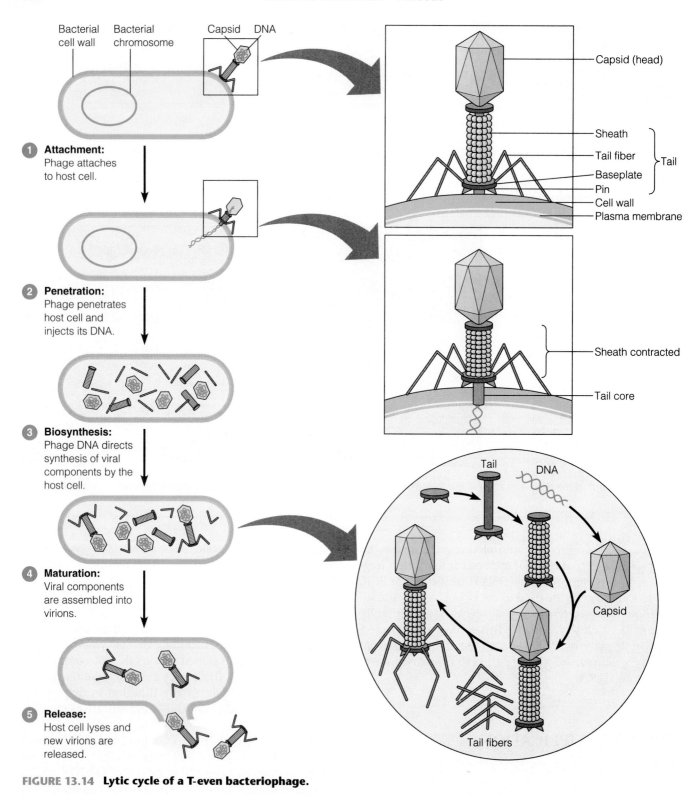

FIGURE 13.14 Lytic cycle of a T-even bacteriophage.

cycle or the lysogenic cycle. The **lytic cycle** ends with the lysis and death of the host cell, whereas the host cell remains alive in the **lysogenic cycle.** Because the so-called *T-even bacteriophages* (T2, T4, and T6) have been studied most extensively, we will describe the multiplication of T-even bacteriophages in their host, *E. coli,* as an example of the lytic cycle.

T-Even Bacteriophages: The Lytic Cycle
The virions of T-even bacteriophages are large, complex, and nonenveloped, with a characteristic head-and-tail structure shown in Figures 13.7a and 13.14. The length of DNA contained in these bacteriophages is only about 6% of that contained in *E. coli;* the bacteriophage has enough DNA for over 100 genes. The multiplication cycle of these

FIGURE 13.15 **A bacterio-phage one-step growth curve.** No new infective phage particles are found in a culture until after biosynthesis and maturation have taken place.

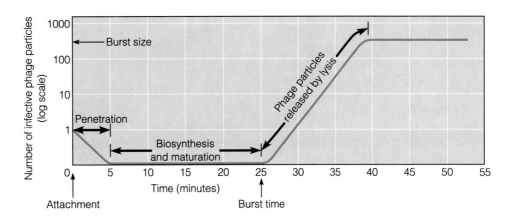

phages, like that of all viruses, can be divided into five distinct stages: attachment, penetration, biosynthesis, maturation, and release.

Attachment. ① After a chance collision between phage particles and bacteria, *attachment,* or *adsorption,* occurs. During this process, an attachment site on the virus attaches to a complementary receptor site on the bacterial cell. This attachment is a chemical interaction in which weak bonds are formed between the attachment and receptor sites. T-even bacteriophages use fibers at the end of the tail as attachment sites. The complementary receptor sites are on the bacterial cell wall.

Penetration. ② After attachment, the T-even bacteriophage injects its DNA (nucleic acid) into the bacterium. To do this, the bacteriophage's tail releases an enzyme, *phage lysozyme,* which breaks down a portion of the bacterial cell wall. During the process of *penetration,* the tail sheath of the phage contracts, and the tail core is driven through the cell wall. When the tip of the core reaches the plasma membrane, the DNA from the bacteriophage's head passes through the tail core and through the plasma membrane and enters the bacterial cell. The capsid remains outside the bacterial cell. Therefore, the phage particle functions like a hypodermic syringe to inject its DNA into the bacterial cell.

Biosynthesis. ③ Once the bacteriophage DNA has reached the cytoplasm of the host cell, biosynthesis of viral nucleic acid and protein occurs. Host protein synthesis is stopped by (a) virus-induced degradation of host DNA, (b) virus proteins that interfere with transcription, or (c) repression of translation.

Initially, the phage uses the host cell's nucleotides and several of its enzymes to synthesize many copies of phage DNA. Soon after, biosynthesis of viral proteins begins. Any RNA transcribed in the cell is mRNA transcribed from phage DNA for biosynthesis of phage enzymes and capsid protein. The host cell's ribosomes, enzymes, and amino acids are used for translation. There are genetic controls that regulate when different regions of phage DNA are transcribed into mRNA during the multiplication cycle. For example, there are early messages that are translated into early phage proteins, the enzymes used in the synthesis of phage DNA. Also, there are late messages that are translated into late phage proteins for the synthesis of capsid proteins.

For several minutes following infection, complete phages cannot be found in the host cell. Only separate components—DNA and protein—can be detected. The period during viral multiplication when complete, infective virions are not yet present is called the **eclipse period.**

Maturation. ④ In the next sequence of events, *maturation* occurs. In this process, bacteriophage DNA and capsids are assembled into complete virions. The viral components essentially assemble into a virus particle spontaneously, thus eliminating the need for many nonstructural genes and gene products. The phage heads and tails are separately assembled from protein subunits, and the head is filled with phage DNA and attached to the tail.

Release. ⑤ The final stage of viral multiplication is the *release* of virions from the host cell. The term **lysis** is generally used for this stage in the multiplication of T-even phages because in this case the plasma membrane actually breaks open (lyses). Lysozyme, whose code is provided by a phage gene, is synthesized within the cell. This enzyme causes the bacterial cell wall to break down, and the newly produced bacteriophages are released from the host cell. The released bacteriophages infect other susceptible cells in the vicinity, and the viral multiplication cycle is repeated within those cells.

The time that elapses from phage attachment to release is known as **burst time** and averages from 20 to 40 minutes. The number of newly synthesized phage particles released from a single cell is referred to as **burst size** and usually ranges from about 50 to 200.

The various stages involved in the multiplication of phages can be demonstrated experimentally in what is known as a *one-step growth experiment* (Figure 13.15). In

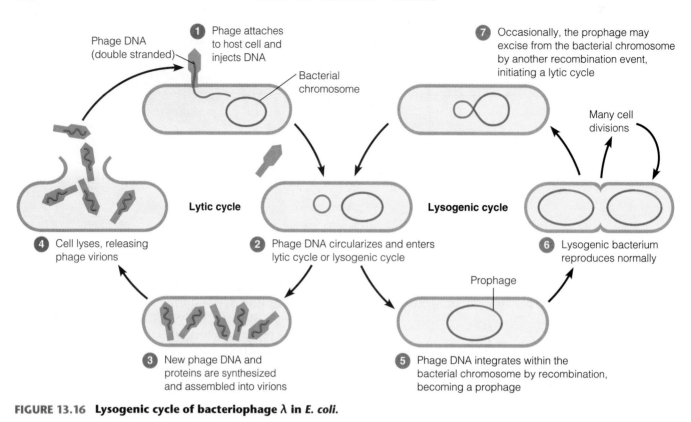

1 Phage attaches to host cell and injects DNA

Phage DNA (double stranded)

Bacterial chromosome

7 Occasionally, the prophage may excise from the bacterial chromosome by another recombination event, initiating a lytic cycle

Many cell divisions

Lytic cycle

Lysogenic cycle

4 Cell lyses, releasing phage virions

2 Phage DNA circularizes and enters lytic cycle or lysogenic cycle

6 Lysogenic bacterium reproduces normally

Prophage

3 New phage DNA and proteins are synthesized and assembled into virions

5 Phage DNA integrates within the bacterial chromosome by recombination, becoming a prophage

FIGURE 13.16   Lysogenic cycle of bacteriophage λ in *E. coli*.

this procedure, a phage suspension is diluted until a sample containing only a few phage particles is obtained. These particles are then introduced into a culture of host cells. Periodically, samples of phage particles are removed from the culture and inoculated onto a plate culture of susceptible host cells; the plaque method is used to determine the number of infective phage particles on this culture (see Figure 13.10). A few minutes after attachment, no infective particles are present. However, phage nucleic acid is found inside the infected cells, and capsid proteins will be synthesized. The time required for maturation is the interval between the appearance of phage nucleic acid and the synthesis of mature phages. After a few minutes, the number of infective phage particles found in subcultures begins to rise. The burst size is determined once the number of infective phage particles remains constant, indicating that no further phage multiplication will occur.

Bacteriophage Lambda: The Lysogenic Cycle

In contrast to T-even bacteriophages, some viruses do not cause lysis and death of the host cell when they multiply. Some phages can either proceed through a lytic cycle or begin a lysogenic cycle by incorporating their DNA into the host cell's DNA. In the latter state, called **lysogeny,** the phage remains latent. Phages that can multiply by either mechanism are called *lysogenic phages* or *temperate phages.* The participating bacterial host cells are known as *lysogenic cells.*

We will use the bacteriophage λ (lambda), a well-studied

temperate phage, as an example of the lysogenic cycle (Figure 13.16). ① Upon penetration into an *E. coli* cell, ② the linear phage DNA forms a circle. ③ This circle can multiply and be transcribed, ④ leading to the production of new phage and to cell lysis (lytic cycle). ⑤ Alternatively, the circle can recombine with and become part of the circular bacterial DNA (lysogenic cycle). The inserted phage DNA is now called a **prophage.** Most of the prophage genes are repressed by two repressor proteins that are the products of phage genes. These repressors stop transcription of all the other phage genes by binding to operators. Thus, the phage genes that would otherwise direct the synthesis and release of new virions are turned off, in much the same way that the genes of the *E. coli lac* operon are turned off by the *lac* repressor (see p. 208).

Every time the host cell's machinery replicates the bacterial chromosome, ⑥ it also replicates the prophage DNA. The prophage remains latent within the progeny cells. ⑦ However, a rare spontaneous event or the action of ultraviolet light or certain chemicals can lead to the excision (popping-out) of the phage DNA—and initiation of the lytic cycle.

There are three important results of lysogeny. First, the lysogenic cells are immune to reinfection by the same phage. (However, the host cell is not immune to infection by other phage types.) The second result of lysogeny is that the host cell may exhibit new properties. For example, the bacterium *Corynebacterium diphtheriae,* which causes diphtheria, is a pathogen whose disease-producing properties are related to the synthesis of a toxin. The organism can produce toxin only when it carries a temperate phage, because the prophage carries the gene coding for the toxin. As another example, only streptococci carrying a temperate phage are capable of producing the toxin associated with scarlet fever. The toxin produced by *Clostridium botulinum,* which causes botulism, is encoded by a prophage gene.

The third result of lysogeny is that it makes **specialized transduction** possible. Recall from Chapter 8 that bacterial genes can be picked up in a phage coat and transferred to another bacterium in a process called generalized transduction (see Figure 8.27). Any bacterial genes can be transferred by generalized transduction because the host chromosome is broken down into fragments, any of which can be packaged into a phage coat. In specialized transduction, however, only certain bacterial genes can be transferred.

Specialized transduction is mediated by a lysogenic phage, which packages bacterial DNA *along with* its own DNA in the same capsid. When a prophage is excised from the host chromosome, adjacent genes from either side may remain attached to the phage DNA. For example, the genes that may remain attached to bacteriophage λ are genes for galactose utilization and biotin synthesis. A phage particle newly synthesized from this DNA may thus contain one or more of the bacterial genes along with phage DNA. Upon lysogenizing a new host cell, the phage can confer new characteristics to the cell; a cell infected by bacteriophage λ, for example, could gain the ability to metabolize galactose or synthesize biotin. In Figure 13.17, bacteriophage λ has picked up the *gal* gene for galactose fermentation from its galactose-positive host. The phage carries this gene to a galactose-negative cell, which then becomes galactose-positive.

Certain animal viruses can undergo processes very similar to lysogeny. Animal viruses that can remain latent in cells for long periods without multiplying or causing disease may become inserted in a host chromosome or remain separate from host DNA in a repressed state (as some lysogenic phages). Cancer-causing viruses may also be latent, as will be discussed later in this chapter.

1. Prophage exists in galactose-using host (containing the *gal* gene).

2. Phage genome excises, carrying with it the adjacent *gal* gene from the host.

3. Phage matures and cell lyses, releasing phage carrying *gal* gene.

4. Phage infects a cell that cannot utilize galactose (lacking *gal* gene).

5. Along with the prophage, the bacterial *gal* gene becomes integrated into the new host's DNA.

6. Lysogenic cell can now metabolize galactose.

FIGURE 13.17 Specialized transduction. When a prophage is excised from its host chromosome, it can take with it a bit of the adjacent DNA from the bacterial chromosome.

Multiplication of Animal Viruses

The multiplication of animal viruses follows the basic pattern of bacteriophage multiplication but has several notable differences. Animal viruses differ from phages in their mechanism of entering the host cell. Also, once the virus is inside, the synthesis and assembly of the new viral components are somewhat different, partly because of the differences between procaryotic cells and eucaryotic cells. Animal viruses may have certain types of enzymes not found in phages. Finally, the mechanisms of maturation and release, and the effects on the host cell, differ in animal viruses and phages.

In the following discussion of the multiplication of animal viruses, we will first consider the processes that are shared by both DNA- and RNA-containing animal viruses. These processes are attachment, penetration, and uncoating. Then we will examine how DNA- and RNA-containing viruses differ with respect to their processes of biosynthesis and release from host cells.

Attachment

Like bacteriophages, animal viruses have attachment sites that attach to complementary receptor sites on the host

(a) Attachment   **(b)** Endocytosis   **(c)** Penetration   **(d)** Uncoating

FIGURE 13.18 **Entry of herpes simplex virus (*Simplexvirus*) into an animal cell.**
(a) Attachment of the viral envelope to the plasma membrane. **(b)** The cell's plasma membrane
folds inward, forming a vesicle around the virus; this results in loss of the envelope.
(c) Nonenveloped capsid penetrates the cytoplasm of the cell from the vesicle. **(d)** The
nucleic acid core is uncoated by digestion of the capsid.

cell's surface. However, the receptor sites of animal cells are proteins and glycoproteins of the plasma membrane (Figure 13.18a). Moreover, animal viruses do not possess appendages like the tail fibers of some bacteriophages. The attachment sites of animal viruses are distributed over the surface of the virus. The sites themselves vary from one group of viruses to another. In adenoviruses, which are icosahedral viruses, the attachment sites are small fibers at the corners of the icosahedron (see Figure 13.4b). In many of the enveloped viruses, such as *Influenzavirus,* the attachment sites are spikes located on the surface of the envelope (see Figure 13.5b). As soon as one spike attaches to a host receptor, additional receptor sites on the same cell migrate to the virus. Attachment is completed when many sites are bound.

Receptor sites are inherited characteristics of the host. Consequently, the receptor for a particular virus can vary from person to person. This could account for the individual differences in susceptibility to a particular virus. Understanding the nature of attachment can lead to the development of drugs that prevent viral infections. Monoclonal antibodies (discussed in Chapter 17) that combine with a virus's attachment site or the cell's receptor site may soon be used to treat some viral infections. One experimental treatment for AIDS involves injections of receptor sites. If the virus attaches to these, it can't attach to natural receptor sites on cells.

Penetration

Following attachment, penetration occurs. Penetration of enveloped animal viruses occurs by **endocytosis,** an active cellular process by which nutrients and other molecules are brought into a cell. A cell's plasma membrane continuously folds inward to form vesicles. These vesicles contain elements that originate outside the cell and are brought into the interior of the cell to be digested. If a

virion attaches to a small outfolding (called a microvillus) on the plasma membrane of a potential host cell, the host cell will enfold the virion into a fold of plasma membrane, forming a vesicle (Figure 13.18b). Once the virion is enclosed within the vesicle, its viral envelope is destroyed. The capsid is digested when the cell attempts to digest the vesicle's contents, or the nonenveloped capsid may be released into the cytoplasm of the host cell (Figure 13.18c).

Penetration sometimes occurs by an alternative method called **fusion,** in which the viral envelope fuses with the plasma membrane and releases the capsid into the cell's cytoplasm. For example, HIV appears to penetrate cells by this method.

Uncoating

Uncoating is the separation of the viral nucleic acid from its protein coat. This poorly understood process apparently varies with the type of virus. Uncoating occurs only in animal viruses; recall that phage DNA is inserted directly into the host cell. Some animal viruses accomplish uncoating by the action of lysosomal enzymes contained inside phagocytic vacuoles and coated vesicles of the host cell. These enzymes degrade the proteins of the viral capsid. The uncoating of poxviruses is completed by a specific enzyme encoded by the viral DNA and synthesized soon after infection. For other viruses, uncoating appears to be exclusively caused by enzymes in the host cell cytoplasm (Figure 13.18d). For at least one virus, the poliovirus, uncoating seems to begin while the virus is still attached to the host cell's plasma membrane.

Biosynthesis of DNA Viruses

Generally, DNA-containing viruses replicate their DNA in the nucleus of the host cell by using viral enzymes, and they synthesize their capsid and other proteins in the cytoplasm by using host cell enzymes. Then the proteins

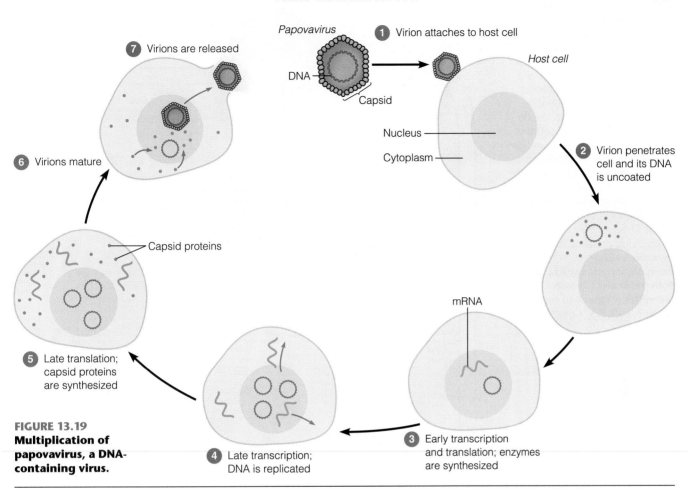

FIGURE 13.19
Multiplication of papovavirus, a DNA-containing virus.

migrate into the nucleus and are joined with the newly synthesized DNA to form virions. These virions are transported along the endoplasmic reticulum into the host cell's membrane for release. Herpesviruses, papovaviruses, adenoviruses, and hepadnaviruses (see Table 13.2) all follow this pattern of biosynthesis. Poxviruses are an exception because all of their components are synthesized in the cytoplasm.

Adenoviridae. Adenoviruses cause acute respiratory diseases ("common colds"). They are named after adenoids, from which they were first isolated.

Poxviridae. All diseases caused by poxviruses, including smallpox and cowpox, include skin lesions. *Pox* refers to pus-filled lesions. Viral multiplication is started by viral transcriptase; the viral components are synthesized and assembled in the cytoplasm of the host cell.

Herpesviridae. Nearly 100 herpesviruses are known. They are named after the spreading (*herpetic*) appearance of cold sores. Species of human herpesviruses (HHV) include those that cause cold sores (HHV1 and HHV2, both in the genus *Simplexvirus*), chickenpox (HHV3, genus *Varicellavirus*), and infectious mononucleosis (HHV4 or Epstein-Barr virus, genus *Lymphocryptovirus*).

Papovaviridae. Papovaviruses are named for *pa*pillomas (warts), *po*lyomas (tumors), and *va*cuolation (cytoplasmic vacuoles produced by some of these viruses). Warts are caused by the genus *Papillomavirus*. Some *Papillomavirus* species are capable of transforming cells and causing cancer. Viral DNA is replicated in the host cell's nucleus along with host cell chromosomes. Host cells may proliferate, resulting in a tumor.

To show an example of the multiplication of a DNA virus, the sequence of events in papovavirus is shown in Figure 13.19. After ①attachment, ②penetration, and uncoating, the viral DNA is released into the nucleus of the host cell. ③ and ④ Next occurs the transcription of a portion of the viral DNA—the "early" genes. Translation follows. The products of these genes are enzymes required for the multiplication of viral DNA. In most DNA viruses, early transcription is carried out with the host's transcriptase (RNA polymerase); poxviruses, however, contain their own transcriptase. Sometime after the initiation of DNA replication, transcription and translation of the remaining "late" viral genes occur. ⑤ This leads to the synthesis of capsid proteins, which occurs in the cytoplasm of the host cell. ⑥After the capsid proteins migrate into the nucleus of the host cell, maturation occurs; the viral DNA and capsid proteins assemble to form complete viruses, which ⑦ are released from the host cell.

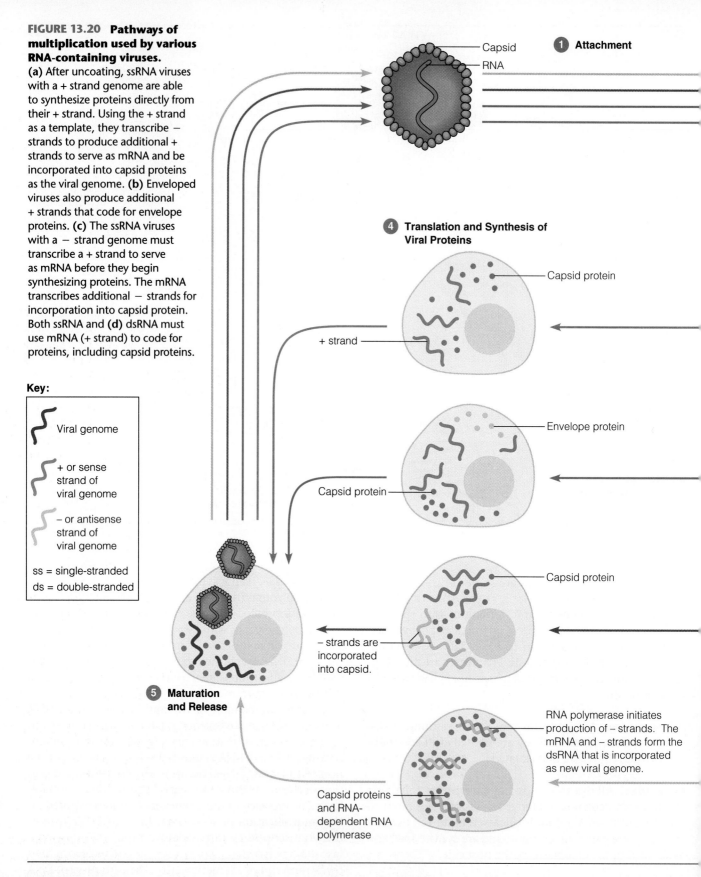

FIGURE 13.20 Pathways of multiplication used by various RNA-containing viruses.
(a) After uncoating, ssRNA viruses with a + strand genome are able to synthesize proteins directly from their + strand. Using the + strand as a template, they transcribe − strands to produce additional + strands to serve as mRNA and be incorporated into capsid proteins as the viral genome. (b) Enveloped viruses also produce additional + strands that code for envelope proteins. (c) The ssRNA viruses with a − strand genome must transcribe a + strand to serve as mRNA before they begin synthesizing proteins. The mRNA transcribes additional − strands for incorporation into capsid protein. Both ssRNA and (d) dsRNA must use mRNA (+ strand) to code for proteins, including capsid proteins.

Key:

Viral genome

+ or sense strand of viral genome

− or antisense strand of viral genome

ss = single-stranded
ds = double-stranded

1 Attachment

Capsid
RNA

4 Translation and Synthesis of Viral Proteins

Capsid protein

+ strand

Envelope protein

Capsid protein

Capsid protein

− strands are incorporated into capsid.

RNA polymerase initiates production of − strands. The mRNA and − strands form the dsRNA that is incorporated as new viral genome.

Capsid proteins and RNA-dependent RNA polymerase

5 Maturation and Release

Hepadnaviridae. These viruses are so named because they cause *hepa*titis and contain *DNA*. The only genus in this family causes hepatitis B. (Hepatitis A and C viruses, although not related to each other, are both RNA viruses.) Hepatitis is discussed in Chapter 25. Hepadnaviruses differ from other DNA viruses because they synthesize DNA by copying RNA, using viral reverse transcriptase. This en-

zyme is discussed on page 353 with the retroviruses, the only other family with reverse transcriptase.

Biosynthesis of RNA Viruses

The multiplication of RNA viruses is essentially the same as that of DNA viruses, except that several different mechanisms of mRNA formation occur among different groups

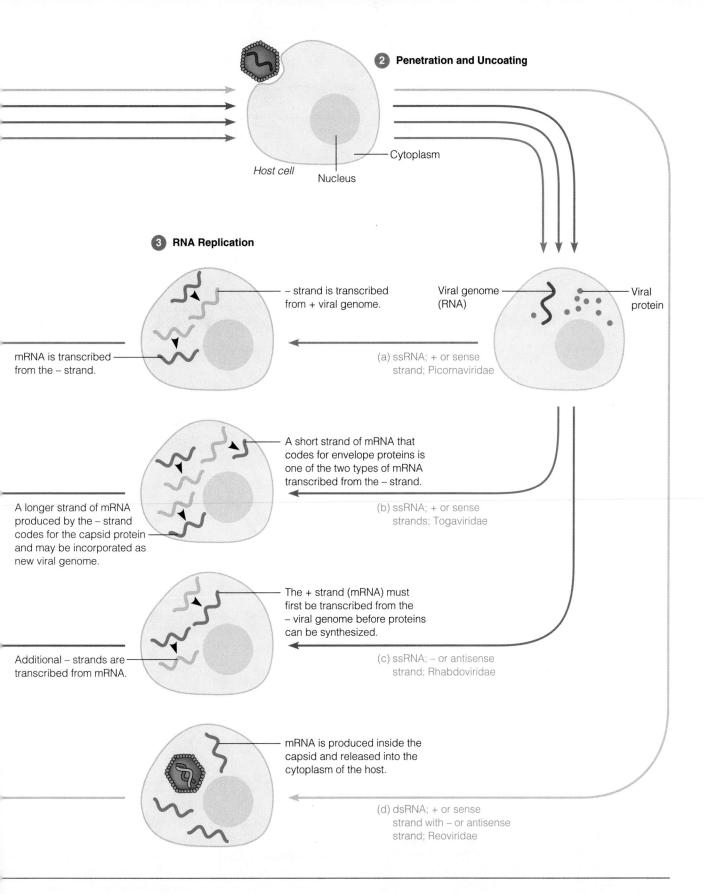

2 **Penetration and Uncoating**

Cytoplasm

Host cell   Nucleus

3 **RNA Replication**

– strand is transcribed from + viral genome.

Viral genome (RNA)

Viral protein

mRNA is transcribed from the – strand.

(a) ssRNA; + or sense strand; Picornaviridae

A short strand of mRNA that codes for envelope proteins is one of the two types of mRNA transcribed from the – strand.

A longer strand of mRNA produced by the – strand codes for the capsid protein and may be incorporated as new viral genome.

(b) ssRNA; + or sense strands; Togaviridae

The + strand (mRNA) must first be transcribed from the – viral genome before proteins can be synthesized.

Additional – strands are transcribed from mRNA.

(c) ssRNA; – or antisense strand; Rhabdoviridae

mRNA is produced inside the capsid and released into the cytoplasm of the host.

(d) dsRNA; + or sense strand with – or antisense strand; Reoviridae

of RNA viruses. Although the details of these mechanisms are beyond the scope of this text, for comparative purposes we will trace the multiplication cycles of the five nucleic acid types of RNA viruses (four of which are shown in Figure 13.20). RNA viruses multiply in the host cell's cytoplasm. The major differences among the multiplication processes of these viruses lie in how mRNA and viral RNA

are produced. Once viral RNA and viral proteins are synthesized, maturation occurs by similar means among all animal viruses, as will be discussed shortly.

Picornaviridae. Picornaviruses, such as poliovirus, are single-stranded RNA viruses. They are the smallest viruses;

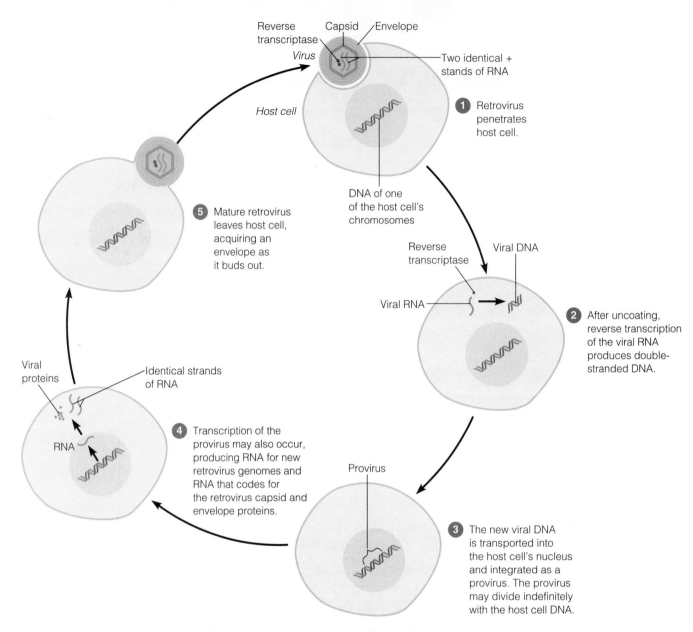

FIGURE 13.21   Multiplication and inheritance processes of the Retroviridae. A retrovirus may become a provirus that replicates in a latent state, and it may produce new retroviruses.

hence, the prefix *pico* (small) plus *RNA* gives these viruses their name. The RNA within the virion is identified as a + or **sense strand** because it can act as mRNA. After attachment, penetration, and uncoating are completed, the single-stranded viral RNA (Figure 13.20a) is translated into two principal proteins, which inhibit the host cell's synthesis of RNA and protein and which form an enzyme called *RNA-dependent RNA polymerase.* This enzyme catalyzes the synthesis of another strand of RNA, which is complementary in base sequence to the original infecting strand. This new strand, called a **−** or **antisense strand**, serves as a template to produce additional + strands. The + strands may serve as mRNA for the translation of capsid protein, may become incorporated into capsid protein to

form a new virus, or may serve as a template for continued RNA multiplication. Once viral RNA and viral protein are synthesized, maturation occurs.

Togaviridae. Togaviruses, which include arthropod-borne alphaviruses (see Chapter 22), also contain a single + strand of RNA. Togaviruses are enveloped viruses; their name is from the Latin word for covering, *toga.* Keep in mind that these are not the only enveloped viruses. After a − strand is made from the + strand, two types of mRNA are transcribed from the − strand (Figure 13.20b). One type of mRNA is a short strand that codes for envelope proteins; the other, longer strand serves as mRNA for capsid proteins and can become incorporated into a capsid.

Rhabdoviridae. Rhabdoviruses, such as rabies virus (genus *Lyssavirus*), are usually bullet-shaped. *Rhabdo* is from the Greek word for rod, which is not really an accurate description of their morphology. They contain a single − strand of RNA (Figure 13.20c). They also contain an RNA-dependent RNA polymerase that uses the − strand as a template from which to produce a + strand. The + strand serves as mRNA and as a template for synthesis of new viral RNA.

Reoviridae. Reoviruses are found in the respiratory and enteric (digestive) systems of humans. They were not associated with any diseases when first discovered; they were considered orphan viruses. Their name comes from the first letters of *respiratory, enteric,* and *orphan.*

After the capsid containing the double-stranded RNA enters a host cell, mRNA is produced inside the capsid and is released into the cytoplasm (Figure 13.20d). There it is used to synthesize more viral proteins. One of the newly synthesized viral proteins acts as RNA-dependent RNA polymerase to produce more − strands of RNA. The mRNA + and − strands form the double-stranded RNA that is then surrounded by capsid proteins.

Retroviridae. Many retroviruses infect vertebrates. One genus of retrovirus, *Lentivirus,* includes the subspecies HIV-1 and HIV-2, which cause AIDS (see the boxes on pp. 354–355 and in Chapter 19). The retroviruses that cause cancer will be discussed later in this chapter.

The formation of mRNA and RNA for new retrovirus virions is shown in Figure 13.21. ① These viruses carry their own polymerase, which is an RNA-dependent DNA polymerase called **reverse transcriptase,** so called because it carries out a reaction (RNA → DNA) that is exactly the reverse of the familiar transcription of DNA → RNA. The name *retrovirus* is derived from the first letters of *reverse transcriptase.* ② Reverse transcriptase uses the RNA of the virus to synthesize a complementary strand of DNA, which in turn is replicated to form double-stranded DNA. This enzyme also degrades the original viral RNA.

The formation of complete viruses requires that DNA be transcribed back into the RNA that will serve as mRNA for viral-protein synthesis and be incorporated into new virions. ③ However, before transcription can take place, the viral DNA must be integrated into the DNA of a host cell chromosome. In this integrated state, the viral DNA is called a **provirus.** Unlike a prophage, the provirus never comes out of the chromosome.

Once the provirus is integrated into the host cell's DNA, several things can happen. Sometimes the provirus simply remains in a latent state and replicates when the DNA of the host cell replicates. ④ In other cases, the provirus becomes transcribed and ⑤ produces new viruses, which may infect adjacent cells. The provirus can also convert the host cell into a tumor cell; possible mechanisms for this phenomenon will be discussed later.

(a) Release by budding

(b) *Alphavirus*   TEM   0.5 μm

FIGURE 13.22   Budding of an enveloped virus. (a) Diagram of the budding process. **(b)** The small "bumps" seen on this freeze-fractured plasma membrane are Sindbis virus (*Alphavirus*) particles caught in the act of budding out from an infected cell.

Maturation and Release

The first step in viral maturation is the assembly of the protein capsid; this assembly is usually a spontaneous process. The capsids of many of the RNA-containing animal viruses are enclosed by an envelope consisting of protein, lipid, and carbohydrate, as noted earlier. Examples of such viruses include orthomyxoviruses and paramyxoviruses. The envelope protein is synthesized by the virus and is incorporated into the plasma membrane of the host cell. The envelope lipid and carbohydrate are synthesized by host cell enzymes and are present in the plasma membrane. The envelope actually develops around the capsid by a process called **budding** (Figure 13.22).

M M W R

Morbidity and Mortality Weekly Report

AIDS: THE RISK TO HEALTH CARE WORKERS

WITH THE ADVENT of the AIDS epidemic, health care workers are understandably concerned about the risk of contracting AIDS after exposure to body fluids of infected patients. However, when precautions are observed, the risk to workers is very small, even for those treating AIDS patients.

UNDERSTANDING THE RISK

The first protection for health care workers is a clear understanding of how AIDS can (and cannot) be transmitted in the course of their work. To date, direct inoculation of infected material is the only proven method of transmission in the health care environment. Infected materials that can transmit AIDS are blood, semen, vaginal secretions, and breast milk. The most common route of transmission is through accidental needle sticks. However, inoculation is also possible if infected material contacts mucous membranes or a break in the health care worker's skin. There is no evidence of HIV transmission by aerosols, the fecal-oral route, mouth-to-mouth or casual contact, or contact with environmental surfaces such as floors, walls, chairs, and toilets. Although HIV has been detected in fluids such as saliva, tears, cerebrospinal fluid, amniotic fluid, and urine, HIV is unlikely to be transmitted through exposure to these fluids.

The number of health care workers who have become infected and the sources of their infections are being carefully monitored by the Centers for Disease Control and Prevention (CDC). In the U.S.A., by the end of 1990, 37 health care workers who denied other risk factors and who were exposed to blood from patients with HIV had contracted HIV infections, 29 of these following needlestick injuries. Six cases had extensive contact with blood of infected patients and did not observe routine barrier precautions. Additionally, two workers in industrial laboratories producing large quantities of highly concentrated HIV have been reported to have laboratory-acquired HIV infections.

In addition to monitoring all health care workers who have actually contracted HIV, the CDC and others also study workers who have been exposed to infected materials, to determine the magnitude of risk. The CDC estimates that the probability of infection following a needlestick injury with blood containing HIV is 3 out of 1,000 exposures, or 0.3%. The probability of infection from mucous membrane exposure to blood is 10 times lower. The probability of acquiring hepatitis B (HBV) from a needlestick injury with blood containing HBV is from 6 to 30%.

PRECAUTIONS

In 1985, the CDC developed the strategy of "universal precautions," which

---

After the sequence of attachment, penetration, uncoating, and biosynthesis of viral nucleic acid and protein, the assembled capsid containing nucleic acid pushes through the plasma membrane. As a result, a portion of the plasma membrane, now the envelope, adheres to the virus. This extrusion of a virus from a host cell is one method of release. Budding does not immediately kill the host cell, and in some cases the host cell survives.

Nonenveloped viruses are released through ruptures in the host cell plasma membrane. In contrast to budding, this type of release usually results in the death of the host cell.

VIRUSES AND CANCER

LEARNING OBJECTIVES

- *Define oncogene and transformed cell.*
- *Explain four methods of activating oncogenes.*
- *Discuss the relationship of DNA- and RNA-containing viruses to cancer.*

Several types of cancer are now known to be caused by viruses. Research on cancer-causing viruses has advanced our general understanding of cancer. Molecular biological should be followed in *all* health care settings. These are:

GLOVES. Disposable gloves should be used for direct exposure to infected blood, other body fluids, and tissues. Double-gloving is recommended during invasive surgical procedures. Personnel should not work when they have open skin lesions, weeping dermatitis, or cutaneous wounds.

GOWNS, MASKS, AND GOGGLES. Masks and protective eyewear are recommended when splashes are expected, such as during airway manipulation, endoscopy, and dental procedures, and in the laboratory.

NEEDLES. To minimize the risk of needle sticks, needles should not be resheathed and should be put in a puncture-proof container for sterilization and disposal.

DISINFECTION. Routine housekeeping in health care settings should include washing floors, walls, and other areas not normally associated with disease transmissions with 1:100 dilution of household bleach. A 1:10 dilution is recommended for disinfecting a spill.

PREVENTATIVE TREATMENT AFTER EXPOSURE

Avoidance of exposure is the health care worker's first line of defense. Admittedly, however, accidental exposure cannot always be prevented. Promising new evidence points to the possibility that exposed individuals may be able to reduce their risk by prophylactic (preventative) use of zidovudine (AZT). The National Institutes of Health and the University of California at San Francisco are developing recommendations for use of AZT.

THE RISK TO PATIENTS

Transmission of HBV from health care workers to patients during invasive dental procedures (tooth extractions) has been documented. The only documented instance of HIV transmission from a health care worker to patients involved a dental practice in Florida. The risk of HIV transmission from an infected health care worker to patients is small and can be reduced with universal precautions.

Restrictions on the procedures that can be performed by health care workers with HIV infection have also been considered by the American Medical Association, the American Dental Association, the CDC, and other organizations. Their recommendations state that the risk of HIV transmission from health care workers to patients is greatest during invasive procedures and that decisions regarding restriction of patient care by infected workers who perform such procedures should be made on an individual basis.

SOURCE: Adapted from *MMWR* 38(S-2) (5/12/89); *MMWR* 39(S-6) (6/23/89); *MMWR* 42:329–331; 337 (5/7/93).

research shows that the mechanisms of the diseases are similar, even when a virus does not cause the cancer.

Cancer is not a single disease but is a name given to many diseases. The human body contains more than 100 different kinds of cells, each of which can malfunction in its own distinctive way to cause cancer. When cells multiply in an uncontrolled way, the excess tissue that develops is called a tumor. A cancerous tumor is known as a malignant tumor, whereas a noncancerous tumor is called a benign tumor. (Tumors are generally named by attachment of the suffix *-oma* to the name of the tissue from which the tumor arises.)

Not all cancerous cells form solid tumors. When the malignant cells are those that give rise to white blood cells, the resulting cancer, characterized by the production of large numbers of white blood cells, is called leukemia.

The relationship between cancers and viruses was first demonstrated in 1908, when Wilhelm Ellerman and Olaf Bang were trying to isolate the causative agent of chicken leukemia. They found that leukemia could be transferred to healthy chickens by cell-free filtrates that contained viruses. Three years later, Francis Rous found that a chicken **sarcoma** (cancer of connective tissue) can be similarly transmitted. Virus-induced **adenocarcinomas** (cancers of glandular epithelial tissue) in mice were discovered in 1936. At that time, it was clearly shown that mouse mammary gland tumors are transmitted from mother to offspring through the mother's milk. In 1972, Sarah Stewart isolated the first oncovirus (a virus that can cause cancer) from a human sarcoma.

The viral cause of cancer can often go unrecognized for several reasons. First, most of the particles of some viruses infect cells but do not induce cancer. Second, cancer might not develop until long after viral infection. Third, cancers do not seem to be contagious, as viral diseases usually are.

Transformation of Normal Cells into Tumor Cells

Almost anything that can alter the genetic material of a cell has the potential to make a normal cell cancerous. These cancer-causing alterations to cellular DNA affect parts of the genome called **oncogenes.** Oncogenes were first identified in cancer-causing viruses and were thought to be a part of the normal viral genome. However, J. Michael Bishop and Harold E. Varmus of the U.S.A. received the 1989 Nobel Prize in Medicine for proving that the cancer-inducing genes carried by viruses are actually derived from animal cells. In 1976, Bishop and Varmus showed that the cancer-causing *src* gene in avian sarcoma viruses is derived from a normal part of chicken genes. At that time, they suggested that the *src* gene was involved in cell growth.

All nucleated cells have oncogenes, which are capable of transforming normal cells into cancerous cells. Under normal conditions, these genes probably code for proteins that are necessary to the cell's growth, but mutations can interfere with the normal regulation of oncogenes or cause oncogenes to code for abnormal proteins. Oncogenes that function normally are called **proto-oncogenes** to differentiate them from their abnormally activated counterparts. The wide occurrence of oncogenes suggests that they are not only normal but vital for cell functions.

Essentially the same oncogenes have been isolated from a wide variety of eucaryotes, from yeasts to humans, and from oncogenic viruses as well. More than 50 oncogenes have been isolated from animal cells; some of these are associated with human cancers. Many cancers are leukemias; different types of white blood cells proliferate in the different forms of leukemia. Because the same oncogenes are isolated from different types of cancer, the total number of different oncogenes is estimated to be less than 100. Furthermore, oncogenes can be grouped by nucleotide sequence and protein product to form a much smaller number of classes.

The most common types of oncogene products isolated so far are protein kinases. Normal **protein kinases** remove a phosphate group from ATP and add it to the amino acids *serine* or *threonine* within a protein. Phosphorylation modifies a protein's function. The protein kinases isolated from cancerous cells attach a phosphate group to the amino acid *tyrosine.* The exact role of the phosphorylated proteins in transforming normal cells into cancerous cells is unknown. However, hypotheses being investigated relate their activity to loss of normal controls on cell shape and cell growth.

Oncogenes can be activated by a variety of agents, including mutagenic chemicals, high-energy radiation, and viruses. Viruses capable of inducing tumors in animals are called **oncogenic viruses,** or **oncoviruses.** Approximately 10% of cancers are known to be virus-induced.

Both DNA- and RNA-containing viruses are capable of inducing tumors in animals. When this occurs, the tumor cells undergo **transformation;** that is, they acquire properties that are distinct from properties of uninfected cells or from infected cells that do not form tumors. An outstanding feature of all oncogenic viruses is that their genetic material integrates into the host cell's DNA and replicates along with the host cell's chromosome. This mechanism is similar to the phenomenon of lysogeny in bacteria, and it can alter the host cell's characteristics in the same way.

Normal animal cells in tissue culture move about randomly by amoeboid movement and divide repeatedly until they come into contact with each other. Then, both movement and cell division stop. This phenomenon is known as **contact inhibition.** Transformed cells in cell culture do not exhibit contact inhibition but instead form tumorlike cell masses. Transformed cells sometimes produce tumors when injected into susceptible animals.

After being transformed by viruses, many tumor cells contain a virus-specific antigen on their cell surface, called **tumor-specific transplantation antigen (TSTA),** or an antigen in their nucleus, called the **T antigen.** Transformed cells tend to be less round than normal cells and tend to exhibit certain chromosomal abnormalities, such as unusual numbers of chromosomes and fragmented chromosomes.

Activation of Oncogenes

Cancer appears to be a *multistep* process. A number of abnormalities must accumulate in the cell, and at least two oncogenes must be activated to abnormal functioning for malignancy to result. At present, four mechanisms for the activation of oncogenes are proposed.

1. A single **mutation** such as a nitrogenous base-pair substitution resulting from exposure to a chemical mutagen or radiation could alter the amino acid sequence of a gene product. For example, the *ras* oncogene in malignant cells differs from the normal *ras* proto-oncogene in only one nucleotide. The mutated

ras gene might increase a cell's response to growth-regulating hormones.

A new protein resulting from a mutation may contribute directly to transformation. Alternatively, the mutation might affect a gene whose product normally regulates an oncogene. If that oncogene codes for a cell growth factor and is no longer properly regulated, the growth factor might be produced in an uncontrolled way and lead to cancer.

2. **Transduction** of oncogenes by a virus can activate those oncogenes. For example, a cancer of red blood cells in chickens is caused when a particular type of retrovirus inserts the *erb* gene it carries. The viral *erb* gene product is slightly different from that of the normal chicken *erb* gene, and so it functions as an oncogene. Oncoviruses that lack oncogenes are also able to activate one of the host's oncogenes by inserting their proviral DNA next to the corresponding proto-oncogene.

3. **Translocation** of oncogenes from one chromosomal locus to another can occur during normal cellular activities and could place the oncogenes into a site in which normal controls are not active. In Burkitt's lymphoma, the *myc* oncogene is moved from its normal position on chromosome 8 to chromosome 14, and its product is increased because it is now regulated by an antibody-regulating gene. The continual presence of the *myc* gene product might allow cells to replicate indefinitely.

4. **Gene amplification** could cause the cell to produce unusually large amounts of oncogenic products. A gene is amplified when an unknown mechanism causes the gene to be duplicated several times. Excessive copies of the *neu* oncogene are seen in breast cancer.

Of these four major mechanisms of oncogene activation, viruses are or may be involved in the last three.

DNA Oncogenic Viruses

Oncogenic viruses are found within several families of DNA-containing viruses. These groups include the Adenoviridae, Herpesviridae, Poxviridae, Papovaviridae, and Hepadnaviridae. Among the papovaviruses, papillomaviruses cause uterine (cervical) cancer.

A genus of the Herpesviridae, *Lymphocryptovirus,* includes the Epstein-Barr virus (EB virus), which causes infectious mononucleosis, and two human cancers, Burkitt's lymphoma and nasopharyngeal carcinoma. Burkitt's lymphoma is a rare cancer of the lymphatic system; it affects mostly children in certain parts of Africa. Nasopharyngeal carcinoma, a cancer of the nose and throat, is found worldwide. Some researchers have also suggested that EB

virus is involved in Hodgkin's disease, a cancer of the lymphatic system.

About 90% of the U.S.A.'s population probably carry the latent stage of the EB virus in their lymphocytes but have no signs of disease. This latent stage is indicated by the presence of antibodies to the virus in blood serum. Although infection with this virus leads to mild symptoms in healthy children, it can cause infectious mononucleosis, mostly in teenagers.

EB virus was isolated from Burkitt's lymphoma cells in 1964 by Michael Epstein and Yvonne Barr. The proof that EB virus can cause cancer was accidentally demonstrated in 1985 when a 12-year-old boy known only as David received a bone marrow transplant. David was born without a functioning immune system and was kept from all microbes inside a sterile "bubble" his entire life; hence he was known as "the bubble boy." (The lymphocytes that provide immunity are produced in bone marrow.) Several months after the transplant, he developed signs of infectious mononucleosis, and a few months after that he died of cancer. An autopsy revealed that the virus had been unwittingly introduced into the boy with the bone marrow transplant. The case confirmed suspicions that EB virus can induce cancer in immunosuppressed individuals.

Another virus implicated in cancer is hepatitis B virus (HBV, genus *Hepadnavirus*). Many animal studies have clearly indicated the causal role of HBV in liver cancer. In one human study, virtually all people with liver cancer had previous HBV infections. Moreover, results of this study indicated that the risk of getting liver cancer is over 98 times higher in people with prior HBV infection than it is in those with no prior infection.

RNA Oncogenic Viruses

Among the RNA viruses, only the oncoviruses in the family Retroviridae cause cancer. Human T cell leukemia viruses (HTLV 1 and HTLV 2) cause adult T cell leukemia and lymphoma in humans. (T cells are a type of white blood cell involved in immune responses to disease.)

Sarcoma viruses of cats, chickens, and rodents and the mammary tumor viruses of mice are also retroviruses. Another retrovirus, feline leukemia virus (FeLV), causes leukemia in cats and is transmissible among cats. There is a test to detect the virus in cats' serum.

The ability of retroviruses to induce tumors is related to their production of a reverse transcriptase by the mechanism described earlier (see Figure 13.21). The provirus, which is the double-stranded DNA molecule synthesized from the viral RNA, becomes integrated into the host cell's DNA; new genetic material is thereby introduced into the host's genome, and this is the key reason why retroviruses can contribute to cancer. Some retroviruses contain oncogenes; others contain promoters that turn on oncogenes or other cancer-causing factors.

LATENT VIRAL INFECTIONS

LEARNING OBJECTIVE

- *Provide an example of a latent viral infection.*

A virus can remain in equilibrium with the host and not actually produce disease for a long period, often many years. The classic example of such a **latent infection** in viruses is the infection of the skin by herpes simplex virus, which produces cold sores. This virus inhabits the host's nerve cells but causes no damage until it is activated by such stimuli as fever and sunburn—hence the term *fever blister*.

In some individuals, viruses are produced but symptoms never appear. Even though a large percentage of the human population carries the herpes simplex virus, only 10% to 15% of people carrying the virus exhibit the disease. The virus of some latent infections can exist in a lysogenic state within host cells.

The chickenpox virus (genus *Varicellavirus*) can also exist in a latent state. Chickenpox (varicella) is a skin disease that is usually acquired in childhood. The virus gains access to the skin via the blood. From the blood, some viruses may enter nerves, where they remain latent. Later, changes in the immune (T cell) response can activate these latent viruses, causing shingles (zoster). The shingles rash appears on the skin along the nerve in which the virus was latent. Shingles occurs in 10% to 20% of people who have had chickenpox.

SLOW VIRAL INFECTIONS

LEARNING OBJECTIVE

- *Differentiate between slow viral infections and latent viral infections.*

The term **slow viral infection** was coined in the 1950s to refer to a disease process that occurs gradually over a long period and was thought to be caused by a virus. (The term does not imply that viral multiplication is unusually slow.) Typically, slow viral infections are fatal.

A number of diseases called slow viral infections have in fact been shown to be caused by conventional viruses. For example, several years after causing measles, the measles virus can be responsible for a rare form of encephalitis called *subacute sclerosing panencephalitis (SSPE)*. Slow viral infection is apparently different from latent viral infection in that, in most slow viral infections, detectable infectious virus gradually builds up over a long period, rather than appearing suddenly.

Several examples of slow viral infections caused by conventional viruses are listed in Table 13.3.

TABLE 13.3   Examples of Slow Viral Infections in Humans

DISEASE	PRIMARY EFFECT	CAUSATIVE VIRUS
Subacute sclerosing panencephalitis	Mental deterioration	Measles virus (*Morbillivirus*)
Progressive encephalitis	Rapid mental deterioration	Rubella virus (*Rubivirus*)
Progressive multifocal leukoencephalopathy	Brain degeneration	Papovavirus (family Papovaviridae)
Encephalopathy	Brain degeneration	HIV (*Lentivirus*)

PRIONS

LEARNING OBJECTIVE

- *Discuss three possible explanations for infectious diseases with no apparent cause.*

Other infectious diseases that have not been found to have a viral cause might be caused by prions. Stanley Prusiner proposed the name prion, for *pro*teinaceous *in*fectious particle. (The name would more logically be "proin," but Prusiner thought that "prion" sounded better.) Eight animal diseases now fall into this category; all eight are neurologic diseases. The human diseases are kuru, Creutzfeldt-Jakob disease (CJD), Gerstmann-Sträussler-Scheinker syndrome, and fatal familial insomnia. (Neurologic diseases are discussed in Chapter 22.) These diseases run in families, which indicates a possible genetic cause. However, CJD cannot be purely inherited because it has been transmitted with transplanted nerve (dura mater) tissue; perhaps the predisposition to the disease is inherited. The prototype of these diseases is scrapie, a neurologic disease of sheep, for which the infectious agent appears to be a pure protein. The mystery of how an infectious agent can lack nucleic acid has prompted several hypotheses, including the following:

1. The major portion of the infectious agent is a protein (PrP, for prion protein) whose gene is found in normal host DNA. The PrP gene is located on chromosome 20 in humans. An abnormal form of PrP, designated PrP^{Sc}, is found in the brains of animals with scrapie. Injection of PrP^{Sc} into experimental animals causes the disease in those animals. PrP^{Sc} may cause a post-translational change in PrP; however, a mechanism for this is not known. Other questions remain, including whether other genes are involved.

2. The prion protein might contain an undetectably small nucleic acid that directs synthesis of the abnormal PrP^{Sc}.

TABLE 13.4 Classification of Some Major Plant Viruses According to Morphological, Chemical, and Physical Properties

VIRUS	MORPHOLOGICAL CLASS	NUCLEIC ACID*	DIMENSIONS OF CAPSID (NM)	METHOD OF TRANSMISSION	FAMILY
Tobacco mosaic virus	Helical	ssRNA (+)	300 (length)	Wounds	Tetraviridae
Bean mosaic virus	Polyhedral	ssRNA (+)	23 (diameter)	Pollen	Picornaviridae
Potato yellow dwarf virus	Bullet-shaped, enveloped	ssRNA (−)	380 (length)	Leafhoppers and aphids	Rhabdoviridae
Wound tumor virus	Polyhedral	dsRNA	70 (diameter)	Leafhoppers	Reoviridae
Cauliflower mosaic virus	Polyhedral	dsDNA	50 (diameter)	Aphids	Papovaviridae

*ss = single-stranded; ds = double-stranded.

3. The infectious agent might actually be a bacterium, but one that is difficult to culture and can pass through filters designed to separate bacteria from viruses. A *Spiroplasma*-like bacterium has recently been seen in brain tissue from CJD patients.

PLANT VIRUSES AND VIROIDS

LEARNING OBJECTIVES

- *Differentiate among virus, viroid, and prion.*
- *Name a virus that causes a plant disease.*

Plant viruses resemble animal viruses in many respects: plant viruses are morphologically similar to animal viruses, and they have similar types of nucleic acid (Table 13.4). In fact, some plant viruses can multiply inside insect cells. Plant viruses cause many diseases of economically important crops, including beans (bean mosaic virus), corn and sugarcane (wound tumor virus), and potatoes (potato yellow dwarf virus). Viruses can cause color change, deformed growth, wilting, and stunted growth in their plant hosts. Some hosts, however, remain symptomless and only serve as reservoirs of infection.

Plant cells are generally protected from disease by an impermeable cell wall. Viruses must enter through wounds or be assisted by other plant parasites, including nematodes, fungi, and, most often, insects that suck the plant's sap. Once one plant is infected, it can spread infection to other plants in its pollen and seeds.

In laboratories, plant viruses are cultured in protoplasts (plant cells with the cell walls removed) and in insect cell cultures.

Some plant diseases are caused by **viroids**, short pieces of naked RNA, only 300 to 400 nucleotides long, with no protein coat. The nucleotides are often internally paired,

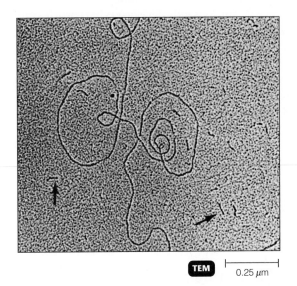

TEM   0.25 μm

FIGURE 13.23 Potato spindle tuber viroid (PSTV). The arrows point to the viroids, which are shown adjacent to much longer viral nucleic acid from bacteriophage T7 for comparison.

so the molecule has a closed, folded, three-dimensional structure that presumably helps protect it from attack by cellular enzymes. The RNA does not code for any proteins. Thus far, viroids have been conclusively identified as pathogens only of plants. Annually, infections by viroids, such as potato spindle tuber viroid (Figure 13.23), result in losses of millions of dollars due to crop damage.

Current research on viroids has discovered similarities between the base sequences of viroids and introns. Recall from Chapter 8 that introns are sequences of genetic material that do not code for polypeptides. This observation has led to the hypothesis that viroids evolved from introns—which leads to speculation that future work may discover animal viroids.

STUDY OUTLINE

GENERAL CHARACTERISTICS OF VIRUSES
(pp. 332–333)

1. Depending on one's viewpoint, viruses may be regarded as exceptionally complex aggregations of nonliving chemicals or as exceptionally simple living microbes.
2. Viruses contain a single type of nucleic acid (DNA or RNA) and a protein coat, sometimes enclosed by an envelope composed of lipids, proteins, and carbohydrates.
3. Viruses are obligatory intracellular parasites. They multiply by using the host cell's synthesizing machinery to cause the synthesis of specialized elements that can transfer the viral nucleic acid to other cells.
4. A virion is a complete, fully developed viral particle composed of nucleic acid surrounded by a coat.

Host Range (p. 333)
1. Host range refers to the spectrum of host cells in which a virus can multiply.
2. Most viruses infect only specific types of cells in one host species.
3. Host range is determined by the specific attachment site on the host cell's surface and the availability of host cellular factors.

Size (p. 333)
1. Viral size is ascertained by electron microscopy.
2. Viruses range from 20 to 14,000 nm in length.

VIRAL STRUCTURE (pp. 334–337)

Nucleic Acid (pp. 334–335)
1. The proportion of nucleic acid in relation to protein in viruses ranges from about 1% to about 50%.
2. Viruses contain either DNA or RNA, never both, and the nucleic acid may be single- or double-stranded, linear or circular, or divided into several separate molecules.

Capsid and Envelope (p. 335)
1. The protein coat surrounding the nucleic acid of a virus is called the capsid.
2. The capsid is composed of subunits, capsomeres, which can be a single type of protein or several types.
3. The capsid of some viruses is enclosed by an envelope consisting of lipids, proteins, and carbohydrates.
4. Some envelopes are covered with carbohydrate–protein complexes called spikes.

General Morphology (pp. 335–337)
1. Helical viruses (for example, tobacco mosaic virus) resemble long rods, and their capsids are hollow cylinders surrounding the nucleic acid.

2. Polyhedral viruses (for example, adenovirus) are many-sided. Usually the capsid is an icosahedron.
3. Enveloped viruses are covered by an envelope and are roughly spherical but highly pleomorphic. There are also enveloped helical viruses (for example, influenza virus) and enveloped polyhedral viruses (for example, herpes simplex virus).
4. Complex viruses have complex structures. For example, many bacteriophages have a polyhedral capsid with a helical tail attached.

TAXONOMY OF VIRUSES (pp. 337–340)

1. Classification of viruses is based on type of nucleic acid, morphological class, and presence or absence of an envelope.
2. Virus family names end in -viridae; genus names end in -virus; specific epithets have not been assigned.
3. A viral species is a group of viruses sharing the same genetic information and ecological niche.

ISOLATION, CULTIVATION, AND IDENTIFICATION OF VIRUSES (pp. 341–343)

1. Viruses must be grown in living cells.
2. The easiest viruses to grow are bacteriophages.

Growth of Bacteriophages in the Laboratory (p. 341)
1. The plaque method mixes bacteriophages with host bacteria and nutrient agar.
2. After several viral multiplication cycles, the bacteria in the area surrounding the original virus are destroyed; the area of lysis is called a plaque.
3. Each plaque originates with a single viral particle; the concentration of viruses is given as plaque-forming units.

Growth of Animal Viruses in the Laboratory (pp. 341–342)
1. Cultivation of some animal viruses requires whole animals.
2. Simian AIDS and feline AIDS provide models for study of human AIDS.
3. Some animal viruses can be cultivated in embryonated eggs.
4. Cell cultures are cells growing in culture media in the laboratory.
5. Primary cell lines and embryonic diploid cell lines grow for a short time in vitro.

6. Continuous cell lines can be maintained in vitro indefinitely.
7. Viral growth can cause cytopathic effects in the cell culture.

Viral Identification (pp. 342–343)

1. Serological tests are used most often to identify viruses.
2. Viruses may be identified by restriction enzyme fragments and appearance of host cells following infection.

VIRAL MULTIPLICATION (pp. 343–354)

1. Viruses do not contain enzymes for energy production or protein synthesis.
2. For a virus to multiply, it must invade a host cell and direct the host's metabolic machinery to produce viral enzymes and components.

Multiplication of Bacteriophages (pp. 343–347)

1. During a lytic cycle, a phage causes the lysis and death of a host cell.
2. Some viruses can either cause lysis or have their DNA incorporated as a prophage into the DNA of the host cell. The latter situation is called lysogeny.
3. The T-even bacteriophages that infect *E. coli* have been studied extensively.
4. During the attachment phase of the lytic cycle, sites on the phage's tail fibers attach to complementary receptor sites on the bacterial cell.
5. In penetration, phage lysozyme opens a portion of the bacterial cell wall, the tail sheath contracts to force the tail core through the cell wall, and phage DNA enters the bacterial cell. The capsid remains outside.
6. In biosynthesis, transcription of phage DNA produces mRNA coding for proteins necessary for phage multiplication. Phage DNA is replicated, and capsid proteins are produced. During the eclipse period, separate phage DNA and protein can be found.
7. During maturation, phage DNA and capsids are assembled into complete viruses.
8. During release, phage lysozyme breaks down the bacterial cell wall, and the multiplied phages are released.
9. The time from phage adsorption to release is called burst time (20 to 40 minutes). Burst size, the number of newly synthesized phages produced from a single infected cell, ranges from 50 to 200.
10. During the lysogenic cycle, prophage genes are regulated by a repressor coded for by the prophage. The prophage is replicated each time the cell divides.
11. Exposure to certain mutagens can lead to excision of the prophage and initiation of the lytic cycle.
12. Because of lysogeny, lysogenic cells become immune to reinfection with the same phage, and the host cell can exhibit new properties.

13. A lysogenic phage can transfer bacterial genes from one cell to another through transduction. Any genes can be transferred in generalized transduction, and specific genes can be transferred in specialized transduction.

Multiplication of Animal Viruses (pp. 347–354)

1. Animal viruses attach to the plasma membrane of the host cell.
2. Penetration of enveloped viruses occurs by endocytosis.
3. Animal viruses are uncoated by viral or host cell enzymes.
4. The DNA of most DNA viruses is released into the nucleus of the host cell. Transcription of viral DNA and translation produce viral DNA and, later, capsid protein. Capsid protein is synthesized in the cytoplasm of the host cell.
5. DNA viruses include members of the families Adenoviridae, Poxviridae, Herpesviridae, Papovaviridae, and Hepadnaviridae.
6. Multiplication of RNA viruses occurs in the cytoplasm of the host cell. RNA-dependent RNA polymerase synthesizes a double-stranded RNA.
7. Picornaviridae + strand RNA acts as mRNA and directs the synthesis of RNA-dependent RNA polymerase.
8. Togaviridae + strand RNA acts as a template for RNA-dependent RNA polymerase, and mRNA is transcribed from a new − RNA strand.
9. Rhabdoviridae − strand RNA is a template for viral RNA-dependent RNA polymerase, which transcribes mRNA.
10. mRNA is produced inside the capsid of the Reoviridae.
11. Retroviridae reverse transcriptase (RNA-dependent DNA polymerase) transcribes DNA from RNA.
12. After maturation, viruses are released. One method of release (and envelope formation) is budding. Naked viruses are released through ruptures in the host cell membrane.

VIRUSES AND CANCER (pp. 354–357)

1. An excess of tissue due to unusually rapid cell multiplication is called a tumor. Tumors are malignant (cancerous) or benign (noncancerous).
2. Tumors are usually named by attachment of the suffix *-oma* to the name of the tissue from which the tumor arises.
3. The earliest relationship between cancer and viruses was demonstrated in the early 1900s, when chicken leukemia and chicken sarcoma were transferred to healthy animals by cell-free filtrates.

Transformation of Normal Cells into Tumor Cells (p. 356)

1. Eucaryotic cells have proto-oncogenes that code for proteins necessary for the cells' normal growth. When activated to oncogenes, these genes transform normal cells into cancerous cells.

2. Viruses capable of producing tumors are called oncogenic viruses.
3. Several DNA viruses and retroviruses are oncogenic.
4. The genetic material of oncogenic viruses becomes integrated into the host cell's DNA.
5. Transformed cells lose contact inhibition, contain virus-specific antigens (TSTA and T antigen), exhibit chromosomal abnormalities, and can produce tumors when injected into susceptible animals.

Activation of Oncogenes (pp. 356–357)

1. A single mutation can result in the production of a protein required for transformation.
2. Transduction of oncogenes could result in oncogene products being made in abnormal amounts or at the wrong time.
3. Translocation of oncogenes could remove normal controls.
4. Gene amplification causes unusually large amounts of oncogene products.

DNA Oncogenic Viruses (p. 357)

1. Oncogenic viruses are found among the Adenoviridae, Herpesviridae, Poxviridae, and Papovaviridae.
2. The EB virus, a *Herpesvirus,* causes Burkitt's lymphoma and nasopharyngeal carcinoma. *Hepadnavirus* causes liver cancer.

RNA Oncogenic Viruses (p. 357)

1. Among the RNA viruses, only retroviruses seem to be oncogenic.
2. HTLV 1 and HTLV 2 have been associated with human leukemia and lymphoma.
3. The virus's ability to produce tumors is related to the production of reverse transcriptase. The DNA synthesized from the viral RNA becomes incorporated as a provirus into the host cell's DNA.
4. A provirus can remain latent, can produce viruses, or can transform the host cell.

LATENT VIRAL INFECTIONS (p. 358)

1. A latent viral infection is one in which the virus remains in the host cell for long periods without producing an infection.
2. Examples are cold sores and shingles.

SLOW VIRAL INFECTIONS (p. 358)

1. Slow viral infections are disease processes that occur over a long period and are generally fatal.
2. Some of these diseases are caused by conventional viruses; others apparently are not.

PRIONS (pp. 358–359)

1. The neurological diseases not caused by conventional viruses—such as scrapie, Creutzfeldt–Jakob disease, and certain other neurologic diseases—may be caused by unconventional infectious agents that are pure protein, without nucleic acid (prions).
2. Other hypotheses for the etiology of such diseases are a small bacterium and that prions may contain nucleic acid.

PLANT VIRUSES AND VIROIDS (p. 359)

1. Plant viruses must enter plant hosts through wounds or with invasive parasites, such as insects.
2. Some plant viruses also multiply in insect (vector) cells.
3. Viroids are infectious pieces of RNA that cause some plant diseases, such as potato spindle tuber viroid disease.

STUDY QUESTIONS

Review

1. Viruses were first detected because they are filterable. What do we mean by the term *filterable,* and how could this property have helped their detection before invention of the electron microscope?

2. Why do we classify viruses as obligatory intracellular parasites?

3. List the four properties that define a virus. What is a virion?

4. Describe the four morphological classes of viruses, then diagram and give an example of each.

5. On what bases are viruses assigned to families? Into what family would you place a + strand, enveloped RNA virus measuring 45 nm in diameter?

6. Describe how bacteriophages are detected and enumerated by the plaque method.

7. Explain how animal viruses are cultured in each of the following:
   (a) Whole animals
   (b) Embryonated eggs
   (c) Cell cultures

8. Why are continuous cell lines of more practical use than primary cell lines for culturing viruses? What is unique about continuous cell lines?

9. Describe the multiplication of a T-even bacteriophage. Be sure to include the essential features of attachment, penetration, biosynthesis, maturation, and release.

10. *Streptococcus pyogenes* produces erythrogenic toxin and is capable of causing scarlet fever only when it is lysogenic. What does this mean?

11. Describe the principal events of attachment, penetration, uncoating, biosynthesis, maturation, and release of an enveloped DNA-containing virus.

12. Recall from Chapter 1 that Koch's postulates are used to determine the etiology of a disease. Why is it difficult to determine the etiology of
    (a) a viral infection such as influenza?
    (b) cancer?

13. Assume that the following strand of RNA is the nucleic acid for an RNA-containing animal virus: UAGUCAAGGU.
    (a) Describe the steps of RNA replication for a virus that contains a + strand of RNA.
    (b) Describe the steps of RNA replication for a virus that contains a − strand of RNA.
    (c) Describe the steps of RNA replication for a virus that contains double-stranded RNA.
    (d) Describe the steps of RNA replication for a virus that contains reverse transcriptase.

14. Provide an explanation for the chronic recurrence of cold sores in some people. (*Note:* The cold sores almost always return at the same site.)

15. Slow viral infections such as _____ might be caused by _____ that are _____ .

16. Which viruses have been proven to cause cancer? How was this proven?

17. The DNA of DNA-containing oncogenic viruses can become integrated into the host DNA. When integrated, the DNA is called a _____ . How does this process result in transformation of the cell? Describe the changes of transformation. How can an RNA-containing virus be oncogenic?

18. Why was the discovery of simian AIDS and feline AIDS important?

19. Contrast viroids and prions. Name a disease thought to be caused by each.

20. Plant viruses cannot penetrate intact plant cells because _____ ; therefore, they enter cells by _____ . Plant viruses can be cultured in _____ .

Critical Thinking

1. Discuss the arguments for and against the classification of viruses as living organisms.

2. In some viruses, capsomeres function as enzymes as well as structural supports. Of what advantage is this to the virus?

3. Discuss four mechanisms of oncogene activation, and discuss how viruses may be implicated.

4. Prophages and proviruses have been described as being similar to bacterial plasmids. What similar properties do they exhibit? How are they different?

Clinical Applications

1. A 40-year-old male who was seropositive for HIV experienced abdominal pain, fatigue, and low-grade fever (38°C) for 2 weeks. Chest X ray revealed lung infiltrates. Gram and acid-fast stains were negative. A viral culture revealed the cause of his symptoms: a large, enveloped polyhedral virus with double-stranded DNA. What is the disease? Which virus causes it? Why was a viral culture done after the Gram and acid-fast stain results were obtained?

2. A newborn female developed extensive vesicular and ulcerative lesions over her face and chest. What is the most likely cause of her symptoms? How would you determine the viral cause of this disease without doing a viral culture?

3. Thirty-two people in Peters Creek reported to their physicians with fever (40°C), jaundice, and tender abdomen. All 32 had eaten an ice-slush beverage purchased from a local convenience store. Liver function tests were abnormal. Over the next several months, the symptoms subsided and liver function returned to normal. What is the disease? This disease could be caused by a member of the Picornaviridae, Hepadnaviridae, or Flaviviridae. Differentiate among these families by method of transmission, morphology, nucleic acid, and type of replication.

PART THREE

Interaction Between Microbe and Host

A technician checks the progress of protein separation by gel electrophoresis. The bands of proteins, which migrate at different rates in response to an electric current, are made visible with ultraviolet light. Electrophoresis is used in many different processes, including making the serum used to immunize people against diseases.

Principles of Disease and Epidemiology

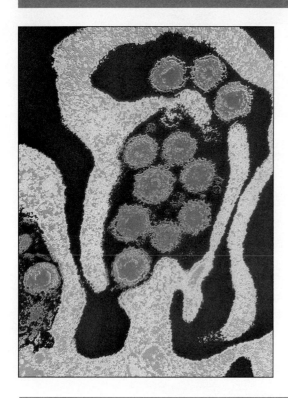

N OW THAT YOU HAVE a basic understanding of the structures and functions of microorganisms and some idea of the variety of microorganisms that exist, we can consider how the human body and various microorganisms interact in terms of health and disease.

We all have our defense mechanisms to keep us healthy. For instance, unbroken skin and mucous membranes are effective barriers against microbial invasion. Within the body, certain cells and certain specialized proteins called antibodies can work together to destroy microbes. In spite of our defenses, however, we are still susceptible to **pathogens** (disease-causing microorganisms). Some bacteria can invade our tissues and resist our defenses by producing capsules or enzymes; other bacteria release toxins that can seriously affect our health. A rather delicate balance exists between our defenses and the disease-producing mechanisms of microorganisms. When our defenses resist these disease-producing capabilities, we maintain our health. But when the disease-producing capability overcomes our defenses, disease results. After the disease has become established, an infected person may recover completely, suffer temporary or permanent damage, or die. The outcome depends on many factors.

In Part Three we examine some of the principles of infection and disease, the mechanisms of pathogenicity, the body's defenses against disease, and the ways that microbial diseases can be prevented by immunization and controlled by drugs. This first chapter discusses the general principles of disease, starting with a discussion of the meaning and scope of pathology. In the last section of this

chapter, Epidemiology, you will learn how these principles are useful in studying and controlling disease.

PATHOLOGY, INFECTION, AND DISEASE

LEARNING OBJECTIVE

• *Define pathology, etiology, infection, host, and disease.*

Pathology is the scientific study of disease (*pathos* means suffering; *logos* means science). Pathology is first concerned with the cause, or **etiology**, of disease. Second, it deals with **pathogenesis,** the manner in which a disease develops. Third, pathology is concerned with the structural and functional changes brought about by disease and with its final effects on the body.

Although the terms *infection* and *disease* are sometimes used interchangeably, they differ somewhat in meaning. **Infection** is the invasion or colonization of the body by pathogenic microorganisms; **disease** occurs when an infection results in any change from a state of health. Disease is an abnormal state in which part or all of the body is not properly adjusted or is not capable of carrying on its normal functions. An infection may exist in the absence of detectable disease. For example, the body may be infected with the virus that causes AIDS, but there may be no symptoms of the disease.

(a)   SEM   10 μm   **(b)**   SEM   10 μm   **(c)**   SEM   10 μm

FIGURE 14.1 Representative normal microbiota for different regions of the body.
(a) Bacteria on the surface of the skin. **(b)** Plaque on enamel near the gums. The bacteria that
cause dental plaque, although part of the normal microbiota, will cause gum disease and caries
unless removed frequently. **(c)** Bacteria of the large intestine.

The presence of a particular type of microorganism in a part of the body where it is not normally found is also called an infection—and may lead to disease. For example, although large numbers of *E. coli* are normally present in the healthy intestine, their infection of the urinary tract usually results in disease.

Few microorganisms are pathogenic. In fact, the presence of some microorganisms can even benefit the host. Therefore, before we discuss the role of microorganisms in causing disease, let us examine the relationship of the normal microbiota to the human body.

NORMAL MICROBIOTA

LEARNING OBJECTIVES

- *Define normal microbiota.*
- *Compare and contrast normal, transient, and opportunistic microbes.*
- *Compare commensalism, mutualism, and parasitism, and give an example of each.*

Animals, including humans, are generally germ-free in utero. (Viruses, some bacteria, and some parasitic protozoa and helminths may infect a fetus in the uterus.) At birth, however, normal and characteristic microbial populations begin to establish themselves. Just before a woman gives birth, lactobacilli in her vagina multiply rapidly. The newborn's first contact with microorganisms is usually with these lactobacilli, and they become the predominant organisms in the newborn's intestine. More microorganisms

are introduced to the newborn's body from the environment when breathing begins and feeding starts. After birth, *E. coli* and other bacteria acquired from foods begin to inhabit the large intestine. These microorganisms remain there throughout life and, in response to altered environmental conditions, may increase or decrease in number and contribute to disease.

Many other normally harmless microorganisms establish themselves inside other parts of the normal adult body and on its surface. A typical human body contains 1×10^{13} body cells, yet harbors an estimated 1×10^{14} bacterial cells. This gives you an idea of the abundance of microorganisms that normally reside in the human body. The microorganisms that establish more or less permanent residence (colonize) but that do not produce disease under normal conditions are members of the **normal microbiota** or **normal flora** (Figure 14.1). Others, called the **transient microbiota**, may be present for several days, weeks, or months and then disappear. Microorganisms are not found throughout the entire human body but are localized in certain regions, as shown in Figure 14.2.

The principal normal microbiotas in different regions of the body and some distinctive features of each region are listed in Table 14.1. The normal microbiota is also discussed more specifically in Part Four.

Relationships Between the Normal Microbiota and the Host

Once established, the normal microbiota can benefit the host by preventing the overgrowth of harmful microorganisms, a phenomenon called **microbial antagonism.**

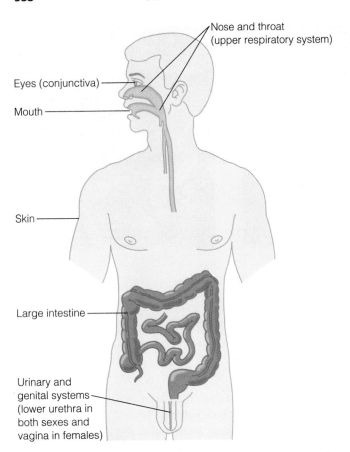

Nose and throat (upper respiratory system)

Eyes (conjunctiva)

Mouth

Skin

Large intestine

Urinary and genital systems (lower urethra in both sexes and vagina in females)

FIGURE 14.2 Locations of the normal microbiota on and in the human body. Various species of microbes are found in all the named regions.

Microbial antagonism involves competition among microbes; one consequence of this competition is that the normal microbiota protects the host against colonization by potentially pathogenic microbes by competing for nutrients, producing substances harmful to the invading microbes, and affecting conditions such as pH and available oxygen. When this balance between the normal microbiota and pathogenic microbes is upset, disease can result. For example, the normal bacterial microbiota of the adult human vagina maintains a local pH of from 3.5 to 4.5. The presence of the normal microbiota inhibits overgrowth of the yeast *Candida albicans,* which cannot grow under these conditions and is normally present in small numbers in the vagina. If the bacterial population is eliminated by antibiotics, excessive douching, or deodorants, the pH of the vagina reverts to nearly neutral, and *C. albicans* can flourish and become the dominant microorganism there. This condition can lead to a form of vaginitis (vaginal infection).

Another example of microbial antagonism is seen in the mouth, where streptococci produce compounds that prevent the growth of most gram-positive and gram-negative cocci. And in the large intestine, *E. coli* cells produce **bacteriocins,** proteins that inhibit the growth of

other bacteria, such as pathogenic *Salmonella* and *Shigella.* A final example involves another bacterium, *Clostridium difficile,* also in the large intestine. This microbe is responsible for nearly all gastrointestinal infections, ranging from mild diarrhea to severe or even fatal colitis, that follow antibiotic therapy. The normal microbiota of the large intestine effectively inhibits *C. difficile,* possibly by making host receptors unavailable, competing for available nutrients, or producing bacteriocins.

The relationship between the normal microbiota and the host is called **symbiosis** (living together). In the symbiotic relationship called **commensalism,** one of the organisms is benefited and the other is unaffected. Many of the microorganisms that make up our normal microbiota are commensals; these include the corynebacteria that inhabit the surface of the eye and certain saprophytic mycobacteria that inhabit the ear and external genitals. These bacteria live on secretions and sloughed-off cells, and they bring no apparent benefit or harm to the host.

Mutualism is a type of symbiosis that benefits both organisms. For example, the large intestine contains bacteria, such as *E. coli,* that synthesize vitamin K and some B vitamins. These vitamins are absorbed into the bloodstream and distributed for use by body cells. In exchange, the large intestine provides nutrients for the bacteria so that they can survive. In still another kind of symbiosis, one organism is benefited at the expense of the other; this relationship is called **parasitism.** Many disease-causing bacteria are parasites.

Opportunistic Organisms

Although categorizing symbiotic relationships by type is convenient, keep in mind that under certain conditions the relationship can change. For example, given the proper circumstances, a mutualistic organism, such as *E. coli,* can become harmful. *E. coli* is generally harmless as long as it remains in the large intestine; but if it gains access to other body sites, such as the urinary bladder, lungs, spinal cord, or wounds, it may cause urinary tract infections, pulmonary infections, meningitis, or abscesses, respectively.

Opportunists are potentially pathogenic organisms that ordinarily do not cause disease in their normal habitat in a healthy person. For example, organisms that gain access through broken skin or mucous membranes can cause opportunistic infections. Tooth decay and gum disease are caused by bacteria that are considered members of the normal microbiota of the mouth. Or, if the host is already weakened or compromised by infection, microbes that are usually harmless can cause disease. AIDS is often accompanied by a common opportunistic infection, *Pneumocystis* pneumonia, caused by the opportunistic organism *Pneumocystis carinii* (see Figure 24.18). This secondary infection can develop in AIDS sufferers because their immune systems are suppressed. Before the AIDS epidemic, this type of pneumonia was rare.

TABLE 14.1 Representative Members of the Normal Microbiota by Body Region*

REGION	PRINCIPAL COMPONENTS	COMMENTS
Skin	*Propionibacterium acnes, Staphylococcus epidermidis, Staphylococcus aureus, Corynebacterium xerosis, Pityrosporum* spp. (fungus), *Candida* spp. (fungus)	Most of the microbes in direct contact with skin do not become residents because secretions from sweat and oil glands have antimicrobial properties.
Eyes (conjunctiva)	*S. epidermidis, S. aureus,* diphtheroids	The conjunctiva, a continuation of the skin or mucous membrane, contains basically the microbiota found on the skin.
Nose and throat (upper respiratory system)	*S. aureus, S. epidermidis,* and aerobic diphtheroids in the nose; *S. epidermidis, S. aureus,* diphtheroids, *Streptococcus pneumoniae, Hemophilus,* and *Neisseria* in the throat	Although some members of the normal microbiota are potential pathogens, their ability to cause disease is reduced by microbial antagonism.
Mouth	Various species of *Streptococcus, Lactobacillus, Actinomyces, Bacteroides, Fusobacterium, Treponema, Corynebacterium,* and *Candida* (fungus)	Abundant moisture, warmth, and constant presence of food make the mouth an ideal environment that supports very large and diverse microbial populations on the tongue, cheeks, teeth, and gums.
Large intestine	*Bacteroides, Fusobacterium, Lactobacillus, Enterococcus, Bifidobacterium, Escherichia coli, Enterobacter, Citrobacter, Proteus, Klebsiella, Shigella, Candida* (fungus)	The large intestine contains the largest numbers of resident microbiota in the body because of its available moisture and nutrients.
Urinary and genital systems	*Staphylococcus epidermidis,* aerobic micrococci, *Enterococcus, Lactobacillus,* aerobic diphtheroids, *Pseudomonas, Klebsiella,* and *Proteus* in urethra; lactobacilli, aerobic diphtheroids, *Streptococcus, Staphylococcus, Bacteroides, Clostridium, Candida albicans* (fungus), and *Trichomonas vaginalis* (protozoan) in vagina	The lower urethra in both sexes has a resident population; the vagina has its acid-tolerant population of microbes because of the nature of its secretions.

*This table lists some members of the microbiota that are not discussed in this chapter; they will be discussed in Part Four. Unless indicated, the organisms are bacteria.

In addition to the usual symbionts, many people carry other microorganisms that are generally regarded as pathogenic but that may not cause disease in those people. Among the pathogens that are frequently carried in healthy individuals are echoviruses ("echo" comes from *e*nteric *c*ytopathogenic *h*uman *o*rphan), which can cause intestinal diseases, and adenoviruses, which can cause respiratory diseases. *Neisseria meningitidis,* which often resides benignly in the respiratory tract, can cause meningitis, a disease that inflames the coverings of the brain and spinal cord. *Streptococcus pneumoniae,* a normal resident of the nose and throat, can cause a type of pneumonia.

Cooperation Among Microorganisms

In addition to competition among microbes in causing disease, there are situations in which cooperation among microbes is a factor in causing disease.

For instance, mycoplasmas may play a role in the development of AIDS. You may recall from Chapter 4 that mycoplasmas are the smallest known bacteria that can grow and reproduce outside living host cells. Cultured cells infected with *Mycoplasma fermentans* and HIV, the virus that causes AIDS, have been shown to die more readily than cells infected with HIV alone. In this regard, mycoplasmas function as cofactors—that is, independent organisms (or molecules) that synergistically enhance or

cause disease. *Synergism* means that the effect of the two microbes acting together is greater than the effect of either acting alone.

Although the role of mycoplasmas in the development of AIDS is still unclear, antibodies against a specific segment of the mycoplasma's outer membrane have been shown to block HIV infection in vitro. In addition, cells infected with both mycoplasma and HIV do not clump together to form abnormal masses called syncytia, as do those infected with HIV alone.

Another example of cooperation among microbes in the development of disease involves oral streptococci that colonize the teeth. Pathogens that cause periodontal disease and gingivitis have been found to have receptors for the streptococci, rather than for the teeth.

ETIOLOGY OF INFECTIOUS DISEASES

LEARNING OBJECTIVE

• *List Koch's postulates.*

Some diseases, such as polio, have a well-known etiology. Some have an etiology that is not completely understood—for example, the relationship between chronic fatigue syndrome and EB virus. For still others, such as Creutzfeldt–Jakob disease, the etiology is unknown. Of course, not all diseases are caused by microorganisms. For example, the disease hemophilia is an *inherited (genetic) disease;* osteoarthritis and cirrhosis are considered *degenera-* *tive diseases.* There are several other categories of disease, but here we will discuss only *infectious diseases,* those caused by microorganisms. To see how microbiologists determine the etiology of an infectious disease, we will discuss in greater detail the work of Robert Koch that was introduced in Chapter 1.

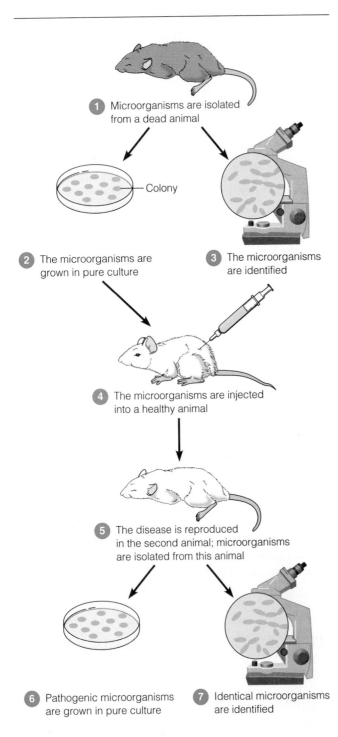

1 Microorganisms are isolated from a dead animal

Colony

2 The microorganisms are grown in pure culture

3 The microorganisms are identified

4 The microorganisms are injected into a healthy animal

5 The disease is reproduced in the second animal; microorganisms are isolated from this animal

6 Pathogenic microorganisms are grown in pure culture

7 Identical microorganisms are identified

FIGURE 14.3   Koch's postulates.

Koch's Postulates

In the historical overview of microbiology presented in Chapter 1, we briefly discussed Koch's famous postulates. Koch, you might recall, was a German physician who played a major role in establishing that microorganisms cause specific diseases. In 1877, Koch published some early papers on anthrax, a disease of cattle that can also occur in humans. Koch demonstrated that certain bacteria, today known as *Bacillus anthracis,* were always present in the blood of animals that had the disease and were not present in healthy animals. He knew that the mere presence of the bacteria did not prove that they had caused the disease; the bacteria could have been there as a result of the disease. Thus he experimented further. He took a sample of blood from a diseased animal and injected it into a healthy one. The second animal developed the same disease and died. He repeated this procedure many times, always with the same results. (A key criterion in the validity of any scientific proof is that experimental results be repeatable.) Koch also cultivated the microorganism in fluids outside the animal's body, and he demonstrated that the bacterium would cause anthrax even after many culture transfers. In short, he showed that a specific infectious disease (anthrax) is caused by a specific microorganism (*B. anthracis*) that can be isolated and cultured on artificial media.

Koch later used the same methods to show that the bacterium *Mycobacterium tuberculosis* is the causative agent of tuberculosis.

Koch's research provides a framework for the study of the etiology of any infectious disease. Today, we refer to Koch's experimental requirements as **Koch's postulates** (Figure 14.3). They are summarized as follows:

1. The same pathogen must be present in every case of the disease.

2. The pathogen must be isolated from the diseased host and grown in pure culture.

3. The pathogen from the pure culture must cause the disease when it is inoculated into a healthy, susceptible laboratory animal. In exclusively human diseases, this step is not morally acceptable, but it sometimes happens accidentally in a laboratory or through other circumstances. The case of the "bubble boy" (see Chapter 13, p. 357) is an example in which a pathogen accidentally introduced into a human confirmed the cause of a disease.

4. The pathogen must be isolated from the inoculated animal and must be shown to be the original organism.

Exceptions to Koch's Postulates

Although Koch's postulates are useful in determining the causative agent of most bacterial diseases, there are some exceptions. For instance, the bacterium *Treponema pallidum* is known to cause syphilis, but virulent strains have never been cultured on artificial media. The causative agent of leprosy, *Mycobacterium leprae,* has also never been grown on artificial media. Moreover, many rickettsial and viral pathogens cannot be cultured on artificial media because they multiply only within cells.

The discovery of microorganisms that cannot grow on artificial media has made necessary some modifications of Koch's postulates and the use of alternative methods of culturing and detecting particular microbes. For example, when investigators searching for the microbial cause of legionellosis (Legionnaires' disease) were unable to isolate the microbe directly from a victim, they took the alternative step of inoculating a victim's lung tissue into guinea pigs. These guinea pigs developed the disease's pneumonialike symptoms, whereas guinea pigs inoculated with tissue from an unafflicted person did not. Then tissue samples from the diseased guinea pigs were cultured in yolk sacs of chick embryos—a method, discussed in Chapter 13, that reveals the growth of extremely small microbes. After the embryos were incubated, electron microscopy revealed rod-shaped bacteria in the chick embryos. Finally, modern immunologic techniques (which will be discussed in Chapter 18) were used to show that the bacteria in the chick embryos were the same bacteria as those in the guinea pigs and in afflicted humans.

In a number of situations, a human host exhibits certain signs and symptoms that are associated only with a certain pathogen and its disease. For example, the pathogens responsible for diphtheria and tetanus give rise to distinguishing signs and symptoms that can be produced by no other microbe. They are unequivocally the only organisms that produce their respective diseases. But some infectious diseases are not so clear-cut. For example, nephritis (inflammation of the kidneys) can involve any of several different pathogens, all of which give rise to the same signs and symptoms. Thus, it is often difficult to know which particular microorganism is causing a disease. Other infectious diseases that sometimes have poorly defined etiologies are pneumonia, meningitis, and peritonitis (inflammation of the peritoneum, the membrane that lines the abdomen and pelvis and covers the organs within them).

Some pathogens can cause several disease conditions. *Mycobacterium tuberculosis,* for example, is implicated in diseases of the lungs, skin, bones, and internal organs. *Streptococcus pyogenes* can cause sore throat, scarlet fever, skin infections (erysipelas), puerperal fever, and osteomyelitis (inflammation of bone), among other diseases. When clinical signs and symptoms are used together with laboratory methods, these infections can usually be distinguished from infections of the same organs by other pathogens.

CLASSIFYING INFECTIOUS DISEASES

LEARNING OBJECTIVES

- *Differentiate between a communicable and a noncommunicable disease.*
- *Categorize diseases according to incidence and according to severity.*

Each disease affecting the body alters body structures and functions in particular ways, and these alterations are usually indicated by several kinds of evidence. For example, the patient may experience certain **symptoms**—changes in body function—such as pain and malaise (a vague feeling of body discomfort). These *subjective* changes are not apparent to an observer. The patient can also exhibit **signs,** which are *objective* changes that the physician can observe and measure. Frequently evaluated signs include lesions (changes produced in tissues by disease), swelling, fever, and paralysis. Sometimes a specific group of symptoms or signs always accompanies a particular disease; such a group is called a **syndrome.** The diagnosis of a disease is made by evaluation of the signs and symptoms together with the results of laboratory tests.

Diseases are often classified in terms of how they behave within a host and within a given population. Any disease that spreads from one host to another, either directly or indirectly, is said to be a **communicable disease.** Chickenpox, measles, genital herpes, typhoid fever, and tuberculosis are examples. Chickenpox and measles are also examples of **contagious diseases** that are *easily* spread from one person to another. A **noncommunicable disease** is not spread from one host to another. These diseases are caused by microorganisms that normally inhabit the body and only occasionally produce disease or by microorganisms that reside outside the body and produce disease only when introduced into the body. An example is tetanus: *Clostridium tetani* produces disease only when it is introduced into the body via abrasions or wounds.

Occurrence of Disease

To understand the full scope of a disease, we should know something about its occurrence. The **incidence** of a disease is the fraction of a population that contracts it during a particular period; the **prevalence** of a disease is the frac-

FIGURE 14.4 Reported AIDS cases in the U.S.A. Notice that the first 100,000 cases occurred over a 10-year period, whereas the second and third 100,000 cases in this epidemic occurred in just two years each. However, much of the increase shown for 1993 is due to an expanded definition of AIDS cases adopted in that year. [*Sources: MMWR 40(53), 1991; MMWR 41(52 & 53) (1/3/93); MMWR 42(52) (1/7/94.)*]

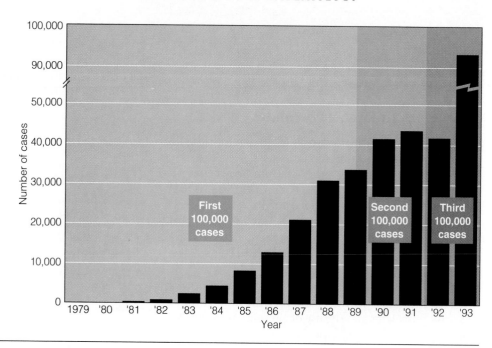

tion of the population having the disease at a specified time. Knowing the incidence and prevalence of a disease in different populations (for example, in populations representing different geographical regions or different racial groups) permits scientists to estimate the range of the disease's occurrence and its tendency to affect some groups of people more than others.

Frequency of occurrence is another criterion that is used in the classification of diseases. If a particular disease occurs only occasionally, it is called a **sporadic disease;** typhoid fever in the U.S.A. is such a disease. A disease constantly present in a population is called an **endemic disease;** an example of such a disease is the common cold. If many people in a given area acquire a certain disease in a relatively short period, it is called an **epidemic disease;** influenza is an example of a disease that often achieves epidemic status. Public health authorities currently consider AIDS near or at epidemic status (Figure 14.4). Some authorities consider gonorrhea and certain other sexually transmitted diseases to be epidemic at this time as well (see Figures 26.5 and 26.6). An epidemic disease that occurs worldwide is called a **pandemic disease.** We experience pandemics of influenza from time to time. Some authorities also consider AIDS to be pandemic.

Emerging Infectious Diseases

Despite predictions earlier in the twentieth century that infectious diseases would soon be eliminated as a public health problem, they still remain the major cause of death worldwide and a leading cause of illness and death in the U.S.A. Since the early 1970s, public health officials have been challenged by a variety of newly identified pathogens such as *E. coli* O157:H7, hepatitis C virus, and HIV.

New diseases arise when a pathogen moves from one host species to another, perhaps due to changes in the pathogen or changes in the susceptibility of a new host. For example, the seal plague that killed 300 seals in 1988 was caused by a paramyxovirus transferred from another species of seal in which it caused no symptoms. HIV is thought to have arisen from a virus infecting wild African monkeys. The incidences of many diseases thought to be under control, such as cholera, malaria, and tuberculosis, have increased in many areas. Increasing human population, ecological changes, antibiotic resistance, and air travel are believed to be factors in the changing patterns of some diseases. These new or changing diseases are called **emerging infectious diseases.** Some examples of these new diseases are described in the boxes in Chapters 12, 14, 23, 24, and 25.

Severity or Duration of Disease

Another useful way of defining the scope of a disease is in terms of its severity or duration. An **acute disease** is one that develops rapidly but lasts only a short time; a good example is influenza. A **chronic disease** develops more slowly, and the body's reactions may be less severe, but the disease is likely to be continual or recurrent for long periods. Infectious mononucleosis, tuberculosis, syphilis, and leprosy fall into this category. A disease that is intermediate between acute and chronic is described as a **subacute disease;** an example is subacute sclerosing panencephalitis, a rare brain disease characterized by diminished intellectual function and loss of nervous function. A **latent disease** is one in which the causative agent remains inactive for a time but then becomes active to produce symptoms of the disease; an example is shingles, one of the diseases caused by the varicella-zoster virus.

Extent of Host Involvement

Infections can also be classified according to the extent to which the host's body is affected. A **local infection** is one in which the invading microorganisms are limited to a relatively small area of the body. Examples of local infections are boils and abscesses. In a **systemic (generalized) infection**, microorganisms or their products are spread throughout the body by the blood or lymphatic system. Measles is an example of a systemic infection. Very frequently, agents of a local infection enter a blood or lymph vessel and spread to other specific parts of the body; we refer to this condition as a **focal infection**. Focal infections can arise from infections in the teeth, tonsils, or sinuses. The presence of bacteria in the blood is known as **bacteremia**, and if the bacteria actually multiply in the blood, the condition is called **septicemia. Toxemia** refers to the presence of toxins in blood (as occurs in tetanus), and **viremia** refers to the presence of viruses in blood.

The state of host resistance also determines the extent of infections. A **primary infection** is an acute infection that causes the initial illness. A **secondary infection** is one caused by an opportunist microbe after the primary infection has weakened the body's defenses. Secondary infections of the skin and respiratory tract are common and are sometimes more dangerous than the primary infections. *Pneumocystis* pneumonia as a consequence of AIDS is an example of a secondary infection; streptococcal bronchopneumonia following influenza is an example of a secondary infection that is more serious than the primary infection. An **inapparent (subclinical) infection** is one that does not cause any noticeable illness. Poliovirus and hepatitis A virus, for example, can be carried by people who never develop the illness.

SPREAD OF INFECTION

LEARNING OBJECTIVES

- *Define reservoir of infection; contrast human, animal, and nonliving reservoirs, and give one example of each.*
- *Explain four methods of disease transmission.*

Now that you have an understanding of the concept of the normal microbiota, the etiology of infectious diseases, and the types of infectious diseases, we will examine the source of pathogens and how diseases are transmitted.

Reservoirs of Infection

For a disease to perpetuate itself, there must be a continual source of the disease organisms. This source can be either a living organism or an inanimate object that provides a pathogen with adequate conditions for survival and multiplication and an opportunity for transmission. Such a source is called a **reservoir of infection.** These reservoirs may be human, animal, or nonliving (Figure 14.5).

Human Reservoirs

The principal living reservoir of human disease is the human body itself (Figure 14.5a). Many people harbor pathogens and transmit them directly or indirectly to others. People with signs and symptoms of a disease may transmit the disease; in addition, some people can harbor pathogens and transmit them to others without exhibiting any signs of illness. These people, called **carriers**, are important living reservoirs of infection. Some carriers have inapparent infections for which no signs or symptoms are ever exhibited. Other people, for example, those with latent diseases, carry a disease during its symptom-free stages—during the incubation period (before symptoms appear) or during the convalescent period (recovery). Human carriers play an important role in the spread of such diseases as AIDS, diphtheria, typhoid fever, hepatitis, gonorrhea, amoebic dysentery, and streptococcal infections.

Animal Reservoirs

Both wild (sylvatic) and domestic animals are living reservoirs of microorganisms that can cause human diseases (Figure 14.5b). Diseases that occur primarily in wild and domestic animals and can be transmitted to humans are called **zoonoses.** Some types of influenza (transmitted by wild birds), rabies (transmitted by bats, skunks, foxes, dogs, and cats), and Lyme disease (transmitted by ticks) are examples of zoonoses. Other representative zoonoses are presented in Table 14.2.

About 150 zoonoses are known. The transmission of zoonoses to humans can occur via one of many routes: by direct contact with infected animals; by direct contact with domestic pet waste (such as cleaning a litter box or bird cage); by contamination of food and water; by contact with contaminated hides, fur, or feathers; by consumption of infected animal products; or by insect vectors (insects that transmit pathogens).

Nonliving Reservoirs

The two major nonliving reservoirs of infectious disease are soil and water (Figure 14.5c). Soil harbors such pathogens as fungi, which cause mycoses, and *Clostridium botulinum*, the bacterial agent that causes botulism. Water that has been contaminated by the feces of humans and other animals is a reservoir for several pathogens, most notably those responsible for gastrointestinal diseases.

Transmission of Disease

The causative agents of disease can be transmitted from the reservoir of infection to a susceptible host by three principal routes: (1) contact, (2) vehicles, and (3) vectors (Figure 14.6).

(a) Human reservoir

(b) Animal reservoir

(c) Nonliving reservoir

FIGURE 14.5 Reservoirs of infection.

(a) Contact transmission (direct)

(b) Vehicle transmission (foodborne)

(c) Vector transmission (mechanical)

FIGURE 14.6 Means of disease transmission.

TABLE 14.2 Selected Zoonoses

DISEASE	CAUSATIVE AGENT	RESERVOIR	METHOD OF TRANSMISSION	REFER TO CHAPTER
Viral				
Influenza (some types)	*Influenzavirus* sp.	Swine, waterfowl	Direct contact	24
Rabies	*Lyssavirus* sp.	Bats, skunks, foxes, dogs, cats	Direct contact (bite)	22
Western equine encephalitis	*Alphavirus* spp.	Horses, birds	*Culex* mosquito bite	22
Yellow fever	*Alphavirus* spp.	Monkeys	*Aedes* mosquito bite	23
Bacterial				
Anthrax	*Bacillus anthracis*	Domestic livestock	Direct contact with contaminated hides or animals; air; food	23
Brucellosis	*Brucella* spp.	Domestic livestock	Direct contact with contaminated milk, meat, or animals	23
Bubonic plague	*Yersinia pestis*	Rodents	Flea bites	23
Cat-scratch disease	*Afipia felis* or *Rochalimaea henselae?*	Domestic cats	Direct contact	23
Leptospirosis	*Leptospira*	Wild mammals, domestic dogs and cats	Direct contact with urine, soil, water	26
Lyme disease	*Borrelia burgdorferi*	Field mice, deer	Tick bites	23
Pneumonic plague	*Yersinia pestis*	Rodents	Direct contact	23
Psittacosis (ornithosis)	*Chlamydia psittaci*	Birds, especially parrots	Direct contact	24
Q fever	*Coxiella burnetii*	Domestic livestock	Inhalation, tick bites (between animals), direct contact with contaminated milk	24
Rocky Mountain spotted fever	*Rickettsia rickettsii*	Rodents	Tick bites	23
Salmonellosis	*Salmonella* spp.	Poultry, rats, turtles	Ingestion of contaminated food, water	25
Tularemia	*Francisella tularensis*	Wild and domestic mammals, especially wild rabbits	Direct contact with infected animals; deerfly bites	23
Typhus fever	*Rickettsia typhi*	Rodents	Flea bites	23
Fungal				
Ringworms	*Trichophyton Microsporum Epidermophyton*	Domestic mammals	Direct contact; fomites (nonliving objects)	21
Protozoan				
Chagas' disease	*Trypanosoma cruzi*	Wild mammals	If bug defecates while feeding, bitten human or animal rubs feces into bite wound by scratching or rubbing	23
Malaria	*Plasmodium* spp.	Monkeys	*Anopheles* mosquito bite	23
Toxoplasmosis	*Toxoplasma gondii*	Cats and other mammals	Ingestion of contaminated meat or by direct contact with infected tissues or fecal material	23
Helminthic				
Hydatid cyst	*Echinococcus granulosus*	Dogs	Direct contact with fecal material	12
Tapeworm (beef)	*Taenia saginata*	Cattle	Ingestion of undercooked contaminated beef	25
Trichinosis	*Trichinella spiralis*	Pigs, bears	Ingestion of undercooked contaminated meat	25

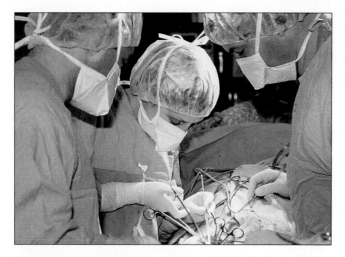

FIGURE 14.7 Prevention of contact transmission. Health care workers use masks, gloves, and other protective measures to avoid transmission of pathogens.

FIGURE 14.8 Droplet transmission. This high-speed photograph shows the spray of small droplets that come from the mouth during a sneeze.

Contact Transmission

Contact transmission (Figure 14.6a) is the spread of an agent of disease by direct contact, indirect contact, or droplet transmission. *Direct contact transmission* refers to the direct transmission of an agent by physical contact between its source and a susceptible host; no intermediate object is involved. This is also known as *person-to-person transmission*. The most common forms of direct contact transmission are touching, kissing, and sexual intercourse. Among the diseases that can be transmitted by direct contact are viral respiratory tract diseases (common cold and influenza), staphylococcal infections, hepatitis A, measles, scarlet fever, smallpox, and sexually transmitted diseases (syphilis, gonorrhea, and genital herpes). Direct contact is also one way of spreading AIDS and infectious mononucleosis. To guard against person-to-person transmission, health care personnel use gloves and other protective measures (Figure 14.7). Potential pathogens can also be transmitted by direct contact from animals (or animal products) to humans. Examples are the pathogens causing rabies and anthrax.

Indirect contact transmission occurs when the agent of disease is transmitted from its reservoir to a susceptible host by means of a nonliving object. The general term for any nonliving object involved in the spread of an infection is a **fomite.** Examples of fomites are tissues, handkerchiefs, towels, bedding, diapers, drinking cups, eating utensils, toys, money, and thermometers. Contaminated syringes serve as fomites in the transmission of AIDS and hepatitis B. Other fomites may transmit diseases such as tetanus.

Droplet transmission is a third type of contact transmission in which microbes are spread in *droplet nuclei* (mucus droplets) that travel only short distances (Figure 14.8). These droplets are discharged into the air by coughing, sneezing, laughing, or talking and travel less than one meter from the reservoir to the host. Disease agents that travel such short distances are not regarded as airborne (discussed shortly). Examples of diseases spread by droplet transmission are influenza, pneumonia, and whooping cough.

Vehicle Transmission

Vehicle transmission refers to the transmission of disease agents by a medium, such as water, food, and air (see Figure 14.6b). Other media include blood and other body fluids, drugs, and intravenous fluids. An outbreak of *Salmonella* infection caused by vehicle transmission is described in the box in Chapter 25. Here, we will discuss water, food, and air as vehicles of transmission.

In *waterborne transmission,* pathogens usually are spread by water contaminated with untreated or poorly treated sewage. Diseases transmitted via this route include cholera, waterborne shigellosis, and leptospirosis. In *foodborne transmission,* pathogens are generally transmitted in foods that are incompletely cooked, poorly refrigerated, or prepared under unsanitary conditions. Foodborne pathogens cause diseases such as food poisoning and tapeworm infection. *Airborne transmission* refers to the spread of agents of infection by droplet nuclei in dust that travel more than one meter from the reservoir to the host. In one type of airborne transmission, microbes are spread by droplets, which may be discharged in a fine spray from the mouth and nose during coughing and sneezing (see Figure 14.8) and which are small enough to remain airborne for prolonged periods. The virus that causes measles and the bacterium that causes tuberculosis can be transmitted via airborne droplets. Dust particles can harbor various pathogens. Staphylococci and streptococci can survive on dust and be transmitted by the airborne route. Spores produced

TABLE 14.3 Representative Arthropod Vectors and the Diseases They Transmit

DISEASE	CAUSATIVE AGENT	ARTHROPOD VECTOR	REFER TO CHAPTER
Malaria	*Plasmodium* spp.	*Anopheles* (mosquito)	23
African trypanosomiasis	*Trypanosoma brucei gambiense* and *Trypanosoma brucei rhodesiense*	*Glossina* sp. (tsetse fly)	22
American trypanosomiasis	*Trypanosoma cruzi*	*Triatoma* sp. (kissing bug)	23
Yellow fever	Arbovirus (yellow fever virus)	*Aedes* (mosquito)	23
Dengue	Arbovirus (dengue fever virus)	*Aedes aegypti* (mosquito)	12 (Box, p. 324), 23
Arthropod-borne encephalitis	Arbovirus (encephalitis virus)	*Culex* (mosquito)	22
Epidemic typhus	*Rickettsia prowazekii*	*Pediculus humanus* (louse)	23
Endemic murine typhus	*Rickettsia typhi*	*Xenopsylla cheopis* (rat flea)	23
Rocky Mountain spotted fever	*Rickettsia rickettsii*	*Dermacentor andersoni* and other species (tick)	23
Plague	*Yersinia pestis*	*Xenopsylla cheopis* (rat flea)	23
Relapsing fever	*Borrelia* sp.	*Ornithodorus* spp. (soft tick)	23
Lyme disease	*Borrelia burgdorferi*	*Ixodes* spp. (tick)	23

by certain fungi are also transmitted by the airborne route and can cause such diseases as histoplasmosis, coccidioidomycosis, and blastomycosis (see Chapter 24).

Vectors

Arthropods are the most important group of disease **vectors**—animals that carry pathogens from one host to another. (Insects and other arthropod vectors are discussed in Chapter 12.) Arthropod vectors transmit disease by two general methods. *Mechanical transmission* is the passive transport of the pathogens on the insect's feet or other body parts (Figure 14.6c). If the insect makes contact with a host's food, pathogens can be transferred to the food and later swallowed by the host. Houseflies, for instance, can transfer the pathogens of typhoid fever and bacillary dysentery (shigellosis) from the feces of infected people to food.

Biological transmission is an active process and is more complex. The arthropod bites an infected person or animal and ingests some of the infected blood. The pathogens then reproduce in the vector, and the increase in the number of pathogens increases the possibility that they will be transmitted to another host. Some parasites reproduce in the gut of the arthropod; these can be passed with feces. If the arthropod defecates or vomits while biting a potential host, the parasite can enter the wound. Other parasites reproduce in the vector's gut and migrate to the salivary gland; these are directly injected into a bite. Some protozoan and helminthic parasites use the vector as a host for a developmental stage in their life cycle.

Table 14.3 lists a few important arthropod vectors and the diseases they transmit.

Portals of Exit

In the next chapter you will learn how microorganisms enter the body through a preferred route or portal of entry, such as mucous membrane or skin. For now, because we are concerned mainly with the spread of disease throughout a population, it is important to know that pathogens also have definite routes of exit, called **portals of exit.** In general, portals of exit are related to the part of the body that has been infected.

The most common portals of exit are the respiratory and gastrointestinal tracts. For example, many pathogens living in the respiratory tract exit in discharges from the mouth and nose; these discharges are expelled during coughing or sneezing. These microorganisms are found in droplets formed from mucus. Pathogens that cause tuberculosis, whooping cough, pneumonia, scarlet fever, meningococcal meningitis, measles, mumps, smallpox, and influenza are discharged through the respiratory route. Other pathogens exit from the gastrointestinal tract in feces or saliva. Feces may be contaminated with pathogens associated with salmonellosis, cholera, typhoid fever, shigellosis, amoebic dysentery, and poliomyelitis. Saliva can also contain pathogens, such as the rabies virus.

Another important route of exit is the urogenital tract. Microbes responsible for sexually transmitted diseases are found in secretions from the penis and vagina. Urine can also contain the pathogens responsible for typhoid fever and brucellosis, which can exit via the urinary tract. Skin or wound infections are other portals of exit. Drainage from wounds can spread infections to another person directly or by contact with a contaminated fomite. Infected blood can be removed and reinjected by biting insects and contaminated needles and syringes to spread infection within a population. Examples of diseases transmitted by biting insects are yellow fever, Rocky Mountain spotted fever, tularemia, and malaria. AIDS and hepatitis B may be transmitted by contaminated needles and syringes.

NOSOCOMIAL (HOSPITAL-ACQUIRED) INFECTIONS

LEARNING OBJECTIVES

* *Define nosocomial infections and explain their importance.*
* *Provide an example of a compromised host.*

Infections acquired as a result of a hospital stay are called **nosocomial infections.** (The word *nosocomial* is derived from the Greek word for hospital; the term also includes infections acquired in nursing homes and other health-related facilities.) The Centers for Disease Control and Prevention (a public health service) estimates that from 5% to 15% of all hospital patients acquire some type of nosocomial infection. About 20,000 people die of nosocomial infections each year.

Nosocomial infections result from the interaction of several factors: (1) microorganisms in the hospital environment, (2) the compromised (or weakened) status of the host, and (3) the chain of transmission in the hospital. Figure 14.9 illustrates that the presence of any one of these factors alone is generally not enough to cause infection; it is the interaction of all three factors that poses a significant risk of nosocomial infection.

Microorganisms in the Hospital

Although every effort is made to kill or check the growth of microorganisms in the hospital, the hospital environment is a major reservoir for a variety of pathogens. One reason is that the bacteria of the normal microbiota of the human body are opportunists and present a particularly strong danger to hospital patients. In fact, most of the bacteria that cause nosocomial infections do not cause disease in healthy people but are pathogenic only for individuals whose defenses have been weakened by illness or therapy (see the box on p. 380).

In the past, most nosocomial infections were caused by gram-positive microbes. At one time, the gram-positive *Staphylococcus aureus* was the primary cause of nosocomial infections. Although antibiotic-resistant strains of that organism are still significant factors, the major causes today are gram-negative bacteria, such as *E. coli* and *Pseudomonas aeruginosa,* especially the antibiotic-resistant strains. *P. aeruginosa* has the ability to cause opportunistic skin infections, especially in surgical and burn patients. The

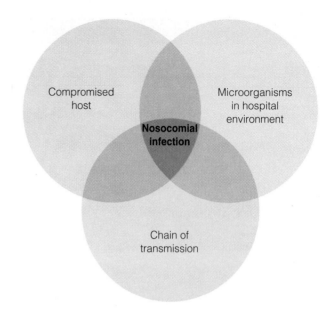

FIGURE 14.9 Nosocomial infections. The interaction of these three principal factors contributes to nosocomial infections.

major bacteria involved in nosocomial infections are summarized in Table 14.4.

In addition to being opportunists, some microorganisms in the hospital become resistant to antimicrobial drugs, which are commonly used there. For example, *P. aeruginosa* and other such gram-negative bacteria tend to be difficult to control with antibiotics because of their R factors, which carry genes that determine resistance to antibiotics (see Chapter 8). As the R factors recombine, new and multiple resistance factors are produced. These strains become part of the microbiota of patients and hospital personnel and become progressively more resistant to antibiotic therapy. In this way, people become part of the reservoir (and chain of transmission) for antibiotic-resistant strains of bacteria. Usually, if the host's resistance is high, the new strains are not much of a problem. But if disease, surgery, or trauma has weakened the host's defenses, secondary infections may be difficult to treat.

Compromised Host

A **compromised host** is one whose resistance to infection is impaired by disease, therapy, or burns. Two principal conditions can compromise the host: (1) broken skin or mucous membranes and (2) a suppressed immune system.

As long as the skin and mucous membranes remain intact, they provide formidable physical barriers against most pathogens. Burns, surgical wounds, trauma (such as accidental wounds), injections, invasive diagnostic procedures, intravenous therapy, and urinary catheters (used to drain urine) can all break the first line of defense and make an individual more susceptible to disease in hospitals. Burn patients are especially susceptible to nosocomial

TABLE 14.4 Bacteria Involved in Most Nosocomial Infections

BACTERIA	PERCENTAGE OF TOTAL INFECTIONS	INFECTIONS CAUSED
Enterobacteria (*E. coli, Klebsiella* species, *Proteus* species, *Enterobacter* species, and *Serratia marcescens*)	Over 40	Urinary tract infections, peritonitis, bacteremia, pneumonia, septicemia, gastrointestinal inflammations, and neonatal (in newborns) meningitis
Staphylococcus aureus	11	Urinary and respiratory tract infections and endocarditis (inflammation of the lining of the heart)
Enterococcus	10	Urinary tract surgical wound infections and endocarditis
*Pseudomonas aeruginosa**	9	Burn and surgical wound infections, septicemia, and pneumonia

*See the box in Chapter 7, page 170.

infections because their skin is no longer an effective barrier to microorganisms.

The risk of infection is also related to other invasive procedures, such as administering anesthesia, which may alter respiration and contribute to pneumonia, and tracheotomy, in which an incision is made into the trachea to assist breathing.

In healthy individuals, white blood cells called T lymphocytes provide resistance to disease by killing pathogens directly, mobilizing phagocytes and other lymphocytes, and secreting chemicals that kill pathogens. White blood cells called B lymphocytes, which develop into antibody-producing cells, also protect against infection. Antibodies provide immunity by such actions as neutralizing toxins, inhibiting attachment of a pathogen to host cells, and helping to lyse pathogens. Drugs, radiation therapy, steroid therapy, burns, diabetes, leukemia, kidney disease, stress, and malnutrition can all adversely affect the actions of T and B lymphocytes and compromise the host. In addition, the AIDS virus destroys certain T cells.

A summary of the principal kinds of nosocomial infections is presented in Table 14.5.

Chain of Transmission

Given the variety of pathogens (and potential pathogens) in the hospital and the compromised state of the host, routes of transmission are a constant concern. The principal routes of transmission of nosocomial infections are direct contact transmission from hospital staff to patient and from patient to patient, and indirect contact transmission through fomites and the hospital's ventilation system (airborne transmission).

Because hospital personnel are often in direct contact with patients, they can often transmit disease. For example, a physician or nurse may transmit her or his microbiota to a patient when changing a dressing, or a kitchen worker who carries *Salmonella* can contaminate a food supply.

Certain areas of a hospital are reserved for specialized

TABLE 14.5 The Principal Kinds of Nosocomial Infections

TYPE OF INFECTION	COMMENT
Urinary tract infection	Most common, usually accounting for about 50% of all nosocomial infections. Typically related to urinary catheterization.
Surgical wound infection	Ranks second in incidence (about 25%). An estimated 5% to 12% of all surgical patients develop postoperative infections; the percentage can reach 30% for certain surgeries, such as colon surgery and amputations.
Lower respiratory infection	Nosocomial pneumonias rank third in incidence (about 12%) and have high mortality rates. Most of these pneumonias are related to respiratory devices that aid breathing or administer medications.
Bacteremia	Bacteremias account for about 6% of hospital infections. Intravenous catheterization is implicated in nosocomial infections of the bloodstream, particularly infections caused by bacteria and fungi.
Cutaneous infection	Among the least common of all nosocomial infections. However, newborns have a high rate of susceptibility to skin and eye infections.

care; these include the burn, hemodialysis, recovery, intensive care, and oncology units. Unfortunately, these units also group patients together and provide environments for the epidemic spread of nosocomial infections from patient to patient.

Many diagnostic and therapeutic hospital procedures provide a fomite route of transmission. The urinary catheter used to drain urine from the bladder is a fomite in many nosocomial infections. Intravenous catheters, which pass through the skin and into a vein to provide fluids, nutrients, or medication, can also transmit nosocomial infections. Respiratory aids can introduce contaminated fluids into the lungs. Needles may introduce pathogens into muscle or blood, and surgical dressings can become contaminated and promote disease.

MMWR

Morbidity and Mortality Weekly Report

A CLUSTER OF *RHODOCOCCUS BRONCHIALIS* STERNAL WOUND INFECTIONS AFTER CORONARY ARTERY BYPASS SURGERY

BETWEEN MAY AND December 1988, seven patients who had undergone open-heart surgery developed *Rhodococcus bronchialis* infections in their surgical wounds. Each case was confirmed by biochemical and Southern blotting analysis. *R. bronchialis* had been isolated at the hospital in the five previous years; during those years the rate of sternal wound infections following open-heart surgery was approximately 1.5% per year. In 1988, the rate jumped to 6.3%.

All of the patients were men ages 51 to 68 who underwent coronary artery bypass surgery with multiple grafts. No early postoperative complications occurred, and all the patients were discharged from the hospital with healed incisions 6 days after surgery. The first clinical manifestation of infection was a blister or localized inflammation at the sternal wound. All seven patients were initially treated with antimicrobial agents. Only three patients improved with antimicrobial therapy alone; four subsequently required surgical debridement of the partially destroyed sternum and flap grafts.

Comparison of the seven patients with randomly selected patients revealed that the only factor shared by the seven patients was the presence of Nurse A during surgery. Nurse A's techniques in surgery were compared to those of another nurse, Nurse B. In the operating room, the nurse performs blood clotting-time tests on the patient's blood at 30-minute intervals during surgery. The test requires that a tube of blood be incubated in a water bath for several minutes. The only difference observed was that Nurse A held the clotting tube with both hands and got water from the water bath on both hands. Nurse B held the tube by the cap with one hand and did not wet her hands.

Finger imprints from Nurse A and Nurse B were cultured on blood-heart-infusion agar. Water from the water bath used to incubate clotting-time tests was also cultured before and after surgery, as were nasal swabs from operating room personnel. *R. bronchialis* was cultured from the hands and nasal swab of Nurse A and from the water bath. Plates left open in the operating rooms during surgery grew *R. bronchialis* only when Nurse A was present.

Nurse A was reassigned to noncardiac procedures while she was carrying *R. bronchialis*. Attempts to eliminate *R. bronchialis* from Nurse A included twice-daily hand scrubbing and daily shampooing with chlorhexidine, daily topical application of bacitracin, and daily oral rifampin. After 18 months, *R. bronchialis* had not been eliminated from Nurse A. However, because of the change in staffing, there have been no subsequent nosocomial infections caused by *R. bronchialis*.

Editorial Note: *Rhodococcus* is a gram-positive, nocardia-like bacterium found in soil, sewage, water, and mammal feces. Agar colonies produce aerial hyphae, and the cells are partially acid-fast. The only previous report of *R. bronchialis* infection in humans was published in 1971. Rhodococcal infections have been increasing over the past ten years, perhaps because of improved diagnostic methods, greater awareness of the nocardia-like bacteria, or a real increase in incidence. Opportunistic rhodococci have been reported in lobar pneumonias of AIDS patients.

SOURCE: *New England Journal of Medicine* 324:101–109 (1/10/91).

Control of Nosocomial Infections

Control measures aimed at preventing nosocomial infections vary from one institution to another, but certain procedures are generally implemented. It is important to reduce the number of pathogens to which patients are exposed by using aseptic techniques, handling contaminated materials carefully, insisting on frequent and conscientious handwashing, educating personnel about basic infection-control measures, and using isolation rooms and wards. Tubs used to bathe patients should be disinfected between uses so that bacteria from the previous patient will not contaminate the next one. Respirators and humidifiers provide both a suitable growth environment for some bacteria and a method of airborne transmission. These sources of nosocomial infections must be kept scrupulously clean and disinfected, and materials used for bandages and intubation (insertion of tubes into organs, such as the trachea) should be single-use disposable or sterilized before use. Packaging used to maintain sterility should be removed aseptically. Physicians can help improve patients' resistance to infection by prescribing antibiotics only when necessary, avoiding invasive procedures if possible, and minimizing the use of immunosuppressive drugs.

Accredited hospitals should have an infection-control committee. Most hospitals have at least an infection-control nurse or epidemiologist (an individual who studies disease in populations). The role of these personnel is to identify problem sources, such as antibiotic-resistant strains of bacteria and improper sterilization techniques. The infection-control officer should make periodic examinations of equipment to determine the extent of microbial contamination. Samples should be taken from tubing, catheters, respirator reservoirs, and other equipment.

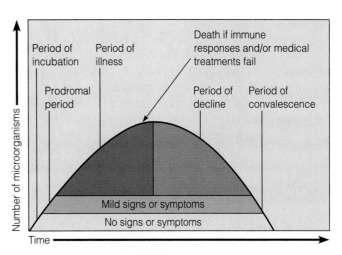

FIGURE 14.10 **Stages of a disease.**

PATTERNS OF DISEASE

LEARNING OBJECTIVES

- *Identify four predisposing factors for disease.*
- *Put the following terms in proper sequence in terms of the pattern of disease: period of decline, period of convalescence, period of illness, prodromal period, period of incubation.*

A definite sequence of events usually occurs during infection and disease. As you already know, there must be a reservoir of infection as a source of pathogens for an infectious disease to occur. Next, the pathogen must be transmitted to a susceptible host by direct contact, by indirect contact, or by vectors. Transmission is followed by invasion, in which the microorganism enters the host and multiplies. Following invasion, the microorganism injures the host through a process called pathogenesis (which you will learn more about in the next chapter). The extent of injury depends on the extent to which host cells are damaged, either directly or by toxins. Despite the effects of all these factors, the occurrence of disease ultimately depends on the resistance of the host to the activities of the pathogen.

Predisposing Factors

Certain predisposing factors also affect the occurrence of disease. A **predisposing factor** is one that makes the body more susceptible to a disease and may alter the course of the disease. Gender is sometimes a predisposing factor; for example, females have a higher incidence of urinary tract infections than males, whereas males have higher rates of pneumonia and meningitis. Other aspects of genetic background may play a role as well. Individuals with sickle-cell

anemia, for instance, are actually more resistant to malaria than are individuals who did not inherit this condition.

Climate and weather seem to have some effect on the incidence of infectious diseases. In temperate regions, the incidence of respiratory diseases increases during the winter. This increase may be related to the fact that when people stay indoors and have closer contact with each other, the spread of respiratory pathogens is facilitated.

Other predisposing factors include inadequate nutrition, fatigue, age, environment, habits, life-style or occupation, preexisting illness, chemotherapy, and emotional disturbances. It is often difficult to know the exact relative importance of the various predisposing factors.

Development of Disease

Once a microorganism overcomes the defenses of the host, development of the disease follows a certain sequence of steps that tends to be similar whether the disease is acute or chronic (Figure 14.10).

Period of Incubation
The period of incubation is the time interval between the initial infection and the first appearance of any signs or symptoms. In some diseases, the incubation period is always the same; in others, it is quite variable. The time of incubation depends on the specific microorganism involved, its virulence (degree of pathogenicity), the number of infecting microorganisms, and the resistance of the host. See Table 15.1 for a list of the incubation periods of a number of microbial diseases.

Prodromal Period
The prodromal period is a relatively short period that follows the period of incubation in some diseases. The prodromal period is characterized by early, mild symptoms of disease, such as general aches and malaise, and certain very specific symptoms, such as Koplik's spots in measles.

Period of Illness

During the period of illness, the disease is most acute. The person exhibits overt signs and symptoms of disease, such as fever, chills, muscle pain (myalgia), sensitivity to light (photophobia), sore throat (pharyngitis), lymph node enlargement (lymphadenopathy), and gastrointestinal disturbances. During the period of illness, the number of white blood cells may increase or decrease. Generally, the patient's immune response and other defense mechanisms overcome the pathogen, and the period of illness ends. When the disease is not successfully overcome (or successfully treated), the patient dies during this period.

Period of Decline

During the period of decline, the signs and symptoms subside. The fever decreases, and the feeling of malaise diminishes. During this phase, which may take from less than 24 hours to several days, the patient is vulnerable to secondary infections.

Period of Convalescence

During the period of convalescence, the person regains strength and the body returns to its prediseased state. Recovery has occurred.

We all know that during the period of illness, people can serve as reservoirs of disease and can easily spread infections to other people. However, you should also know that people can spread infection during incubation and convalescence. This is especially true of diseases such as typhoid fever and cholera, in which the convalescing person carries the pathogenic microorganism for months or even years.

EPIDEMIOLOGY

LEARNING OBJECTIVE

- *Define epidemiology, and describe three types of epidemiologic investigations.*

In today's crowded, overpopulated world, in which frequent travel and mass production and distribution of food and other goods are a way of life, diseases can spread rapidly. A contaminated food or water supply, for example, can affect many thousands of people very quickly. Identifying the causative agent is desirable so that a disease can be effectively controlled and treated. It is also desirable to understand the mode of transmission and geographical distribution of the disease. The science that deals with when and where diseases occur and how they are transmitted in populations is called **epidemiology.**

Modern epidemiology began in 1854 when John Snow, a British physician, reasoned that cholera was transmitted by contaminated water. He questioned some victims of a cholera epidemic in London and found that most of them obtained drinking water from the Broad Street pump. He also discovered that the incidence of cholera was higher among people who drank from the London section of the Thames than among the population using cleaner water upstream. Even before Pasteur's work, Snow used epidemiologic investigation to improve public health. Today, epidemiologists study specified groups of individuals to discover the causes and prevent the spread of diseases.

An epidemiologist not only determines the etiology of a disease but also identifies other possibly important factors and patterns concerning the people affected. An important part of the epidemiologist's work is assembling and analyzing such data as age, sex, occupation, personal habits, socioeconomic status, history of immunization, presence of any other diseases, and the common history of affected individuals (such as eating the same food or visiting the same doctor's office). Also important for prevention of future outbreaks is knowledge of the site at which a susceptible host came into contact with the agent of infection. In addition, the epidemiologist considers the period during which the disease occurs, either on a seasonal basis (to indicate whether the disease is prevalent during the summer or winter) or on a yearly basis (to indicate the effects of hygiene or immunization).

Figure 14.11 contains graphs indicating the incidence of selected diseases. Such graphs provide information about whether disease outbreaks are sporadic or epidemic and, if epidemic, how the disease might have spread. By establishing the frequency of a disease in a population and identifying the factors responsible for its transmission, an epidemiologist provides physicians with information that is important in determining the prognosis and treatment of a disease. Epidemiologists also evaluate how effectively a disease is being controlled in a community—by a vaccination program, for example. Finally, epidemiologists can provide data to help in evaluating and planning overall health care for a community.

Epidemiologists use three basic types of investigations when analyzing the occurrence of a disease: descriptive, analytical, and experimental.

Descriptive Epidemiology

Descriptive epidemiology entails collecting all data that describe the occurrence of the disease under study. Relevant information usually includes information about the affected individuals and the place and period in which the disease occurred.

Such a study is generally *retrospective* (looking backward after the episode has ended). In other words, the epidemiologist backtracks to the cause and source of the disease (see the box in Chapter 22). The search for the cause of toxic shock syndrome is an example of a fairly recent retrospective study. In the initial phase of an epidemiological study, retrospective studies are more common than

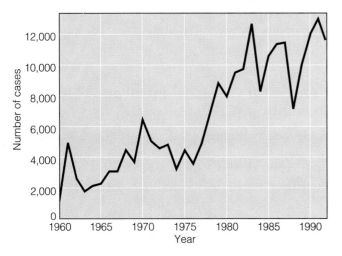

(a) Total viral meningitis cases by year, 1960–1992

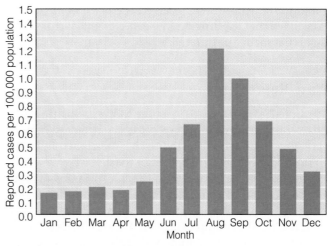

(b) Viral meningitis cases per 100,000 people by month for 1991

FIGURE 14.11 Epidemiological graphs. Epidemiologists use various types of graphs to describe patterns of disease. **(a)** This graph of viral meningitis cases shows the annual occurrence of the disease during the covered period. **(b)** This bar graph looks at viral meningitis from a different perspective, allowing epidemiologists to begin drawing some conclusions about the disease. Over a period of years, researchers saw that the incidence was consistently highest during the hot months. These data helped them realize that the disease was often contracted by waterborne transmission during swimming. Note that this graph records the number of cases per 100,000 people, rather than the total number of cases. **(c)** The recent annual incidence of toxic shock syndrome among menstruating women. This graph shows how public education can help in disease control. Once women learned how tampons were involved in the disease and how to use them more safely, the incidence of the disease fell sharply. (*Source:* CDC.)

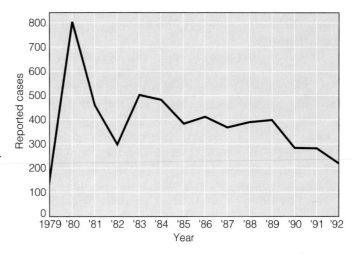

(c) Total toxic shock syndrome cases by year, 1979–1992

prospective (looking forward) studies, in which an epidemiologist chooses a group of people who are free of a particular disease to study. The group's subsequent disease experiences are then recorded for a given period. Prospective studies were used to test the Salk polio vaccine in 1954 and 1955.

Analytical Epidemiology

Analytical epidemiology analyzes a particular disease to determine its probable cause. This study can be done in two ways. With the *case control method,* the epidemiologist looks for factors that might have preceded the disease. A group of people who have the disease is compared with another group of people who are free of the disease. For example, one group with meningitis and one without the disease might be matched by age, sex, socioeconomic status, and location. These statistics are compared to determine which of all the possible factors—genetic, environmental, nutritional, and so forth—might be responsible

for the meningitis. With the *cohort method,* the epidemiologist studies two populations, one that has had contact with the agent causing a disease and another that has not (both groups are called *cohort groups*). For example, a comparison of one group composed of people who have received blood transfusions and one composed of people who have not could reveal an association between blood transfusions and the incidence of hepatitis B virus.

Experimental Epidemiology

Experimental epidemiology begins with a hypothesis about a particular disease; experiments to test the hypothesis are then conducted with a group of people. One such hypothesis could be the assumed effectiveness of a drug. A group of infected individuals is selected and divided randomly so that some receive the drug and others receive a placebo, a substance that has no effect. If all other factors are kept constant between the two groups, and if those people who received the drug recover more rapidly than those who received the placebo, it can be concluded that the drug was the experimental factor (variable) that made the difference.

Case Reporting

We noted earlier in this chapter that establishing the chain of transmission for a disease is extremely important. Once known, the chain can be interrupted in order to slow down or stop the spread of the disease.

An effective way to establish the chain of transmission is case reporting, a procedure that requires health personnel to report specified diseases to local, state, and national health officials. Examples of such diseases include AIDS, measles, gonorrhea, tetanus, and typhoid fever. Case reporting provides epidemiologists with an approximation of the incidence and prevalence of a disease. This information helps officials decide whether or not to investigate a given disease.

Case reporting provided epidemiologists with valuable leads regarding the origin and spread of AIDS. In fact, one of the first clues about AIDS came from reports of young males with Kaposi's sarcoma, formerly a disease of older males. Using these reports, epidemiologists began various studies of the patients. If an epidemiological study shows that a large enough segment of the population is affected by a disease, an attempt is then made to isolate and identify its causative agent. Identification is accomplished by a number of different microbiological methods. Very frequently, identification of the causative agent provides valuable information regarding the reservoir for the disease.

Once the chain of transmission is discovered, it is possible to apply control measures to stop the disease from spreading. These might include elimination of the source of infection, isolation and segregation of infected people, development of vaccines, and, as in the case of AIDS, education.

Centers for Disease Control and Prevention (CDC)

Epidemiology is a major concern of state and federal public health departments. The **Centers for Disease Control and Prevention (CDC)**, a U.S. Public Health Service branch located in Atlanta, Georgia, is a central source of epidemiological information in the U.S.A.

The CDC issues a publication called *Morbidity and Mortality Weekly Report (MMWR),* which is read by microbiologists, physicians, and other hospital and public health personnel. The *MMWR* contains data on the incidence of specific notifiable diseases (morbidity) and on the deaths from these diseases (mortality); these data are usually organized by state. (Notifiable diseases are those for which physicians are required by law to report cases to the Public Health Service.) Publication articles include reports of disease outbreaks, case histories of special interest, and summaries of the status of particular diseases during a recent period. These articles often include recommendations for procedures for diagnosis, immunization, and treatment. A number of graphs and other data in this textbook are from *MMWR,* and some of the boxes are direct excerpts from this publication.

• • •

In the next chapter we consider the mechanisms of pathogenicity. We will discuss in more detail the methods by which microorganisms enter the body and cause disease, the effects of disease on the body, and the means by which pathogens leave the body.

STUDY OUTLINE

INTRODUCTION (p. 366)

1. Disease-causing microorganisms are called pathogens.
2. Pathogenic microorganisms have special properties that allow them to invade the human body or produce toxins.
3. When a microorganism overcomes the body's defenses, a state of disease results.

PATHOLOGY, INFECTION, AND DISEASE
(pp. 366–367)

1. Pathology is the scientific study of disease.

2. Pathology is concerned with the etiology (cause) of disease, pathogenesis (development), and effects of disease.
3. Infection is the invasion and growth of pathogens in the body.
4. A host is an organism that shelters and supports the growth of pathogens.
5. Disease is an abnormal state in which part or all of the body is not properly adjusted or is not capable of carrying on normal functions.

NORMAL MICROBIOTA (pp. 367–369)

1. Animals, including humans, are usually germ-free in utero.

2. Microorganisms begin colonization in and on the surface of the body soon after birth.
3. Microorganisms that establish permanent colonies inside or on the body without producing disease make up the normal microbiota.
4. The transient microbiota is composed of microbes that are present for various periods and then disappear.

Relationships Between the Normal Microbiota and the Host (pp. 367–368)
1. The normal microbiota can prevent pathogens from causing an infection; this phenomenon is known as microbial antagonism.
2. The normal microbiota and the host exist in symbiosis (living together).
3. The three types of symbiosis are commensalism (one organism benefits and the other is unaffected), mutualism (both organisms benefit), and parasitism (one organism benefits and one is harmed).

Opportunistic Organisms (pp. 368–369)
1. Opportunists (opportunistic pathogens) do not cause disease under normal conditions but cause disease under special conditions.

Cooperation Among Microorganisms (p. 369)
1. The situation in which one microorganism makes it possible for another to cause a disease or to produce more severe symptoms is called cooperation.

ETIOLOGY OF INFECTIOUS DISEASES
(pp. 369–371)

Koch's Postulates (pp. 370–371)
1. Koch's postulates are a method for establishing that specific microbes cause specific diseases.
2. Koch's postulates have the following requirements: (1) the same pathogen must be present in every case of the disease; (2) the pathogen must be isolated in pure culture; (3) the pathogen isolated from pure culture must cause the same disease in a healthy, susceptible laboratory animal; and (4) the pathogen must be reisolated from the inoculated laboratory animal.

Exceptions to Koch's Postulates (p. 371)
1. Koch's postulates are modified to establish etiologies of diseases caused by viruses and some bacteria, which cannot be grown on artificial media.
2. Some diseases, such as tetanus, have unequivocal signs and symptoms.
3. Some diseases, such as pneumonia and nephritis, may be caused by a variety of microbes.
4. Some pathogens, such as *S. pyogenes,* cause several different diseases.

CLASSIFYING INFECTIOUS DISEASES
(pp. 371–373)

1. A patient may exhibit symptoms (subjective changes in body functions) and signs (measurable changes), which a physician uses to make a diagnosis (identification of the disease).
2. A specific group of symptoms or signs that always accompanies a specific disease is called a syndrome.
3. Communicable diseases are transmitted directly or indirectly from one host to another.
4. A contagious disease is one that is easily spread from one person to another.
5. Noncommunicable diseases are caused by microorganisms that normally grow outside the human body and are not transmitted from one host to another.

Occurrence of Disease (pp. 371–372)
1. Disease occurrence is reported by incidence (number of people contracting the disease) and prevalence (number of cases at a particular time).
2. Diseases are classified by frequency of occurrence: sporadic, endemic, epidemic, and pandemic.

Emerging Infectious Diseases (p. 372)
1. New diseases and diseases with changing incidences are called emerging infectious diseases.
2. Emerging diseases may result from increasing population, ecological changes, or increased air travel.

Severity or Duration of Disease (p. 372)
1. The scope of a disease may be defined as acute, chronic, subacute, or latent.

Extent of Host Involvement (p. 373)
1. A local infection affects a small area of the body; a systemic infection is spread throughout the body via the circulatory system.
2. A secondary infection can occur after the host is weakened from a primary infection.
3. An inapparent, or subclinical, infection does not cause any signs of disease in the host.

SPREAD OF INFECTION (pp. 373–378)

Reservoirs of Infection (p. 373)
1. A continual source of infection is called a reservoir of infection.
2. People who have a disease or are carriers of pathogenic microorganisms are human reservoirs of infection.
3. Zoonoses are diseases that affect wild and domestic animals and can be transmitted to humans.
4. Some pathogenic microorganisms grow in nonliving reservoirs, such as soil and water.

Transmission of Disease (pp. 373–377)

1. Transmission by direct contact involves close physical contact between the source of the disease and a susceptible host.
2. Transmission by fomites (inanimate objects) constitutes indirect contact.
3. Transmission via saliva or mucus in coughing or sneezing is called droplet transmission.
4. Transmission by a medium such as water, food, or air is called vehicle transmission.
5. Airborne transmission refers to pathogens carried on water droplets or dust for a distance greater than one meter.
6. Arthropod vectors carry pathogens from one host to another by both mechanical and biological transmission.

Portals of Exit (pp. 377–378)

1. Just as pathogens have preferred portals of entry, they also have definite portals of exit.
2. Three common portals of exit are the respiratory tract via coughing or sneezing, the gastrointestinal tract via saliva or feces, and the genital tract via secretions from the vagina or penis.
3. Arthropods and syringes provide a portal of exit for microbes in blood.

NOSOCOMIAL (HOSPITAL-ACQUIRED) INFECTIONS (pp. 378–381)

1. Nosocomial infections are defined as any infections that are acquired during the course of stay in a hospital, nursing home, or other health care facility.
2. From 5% to 15% of all hospitalized patients acquire nosocomial infections.

Microorganisms in the Hospital (p. 378)

1. Members of normal microbiota are often responsible for nosocomial infections when they are introduced into the body through such medical procedures as surgery and catheterization.
2. Opportunistic, drug-resistant gram-negative bacteria are the most frequent causes of nosocomial infections.

Compromised Host (pp. 378–379)

1. Patients with burns, surgical wounds, and suppressed immune systems are the most susceptible to nosocomial infections.

Chain of Transmission (p. 379)

1. Nosocomial infections are transmitted by direct contact between staff and patient and between patients.
2. Fomites such as catheters, syringes, and respiratory aids can transmit nosocomial infections.

Control of Nosocomial Infections (pp. 380–381)

1. Aseptic techniques can prevent nosocomial infections.
2. Hospital infection-control personnel are responsible for overseeing proper cleaning, storage, and handling of equipment and supplies.

PATTERNS OF DISEASE (pp. 381–382)

Predisposing Factors (p. 381)

1. A predisposing factor is one that makes the body more susceptible to disease or alters the course of a disease.
2. Examples include gender, climate, age, fatigue, and inadequate nutrition.

Development of Disease (pp. 381–382)

1. The period of incubation is the time interval between the initial infection and the first appearance of signs and symptoms.
2. The prodromal period is characterized by the appearance of the first mild signs and symptoms.
3. During the period of illness, the disease is at its height and all disease signs and symptoms are apparent.
4. During the period of decline, the signs and symptoms subside.
5. During the period of convalescence, the body returns to its prediseased state, and health is restored.

EPIDEMIOLOGY (pp. 382–384)

1. The science of epidemiology is the study of the transmission, incidence, and frequency of disease.
2. Data about infected people are collected and analyzed in descriptive epidemiology.
3. In analytical epidemiology, a group of infected people is compared with an uninfected group.
4. Controlled experiments designed to test hypotheses are performed in experimental epidemiology.
5. Case reporting provides data on incidence and prevalence to local, state, and national health officials.
6. The Centers for Disease Control and Prevention (CDC) is the main source of epidemiologic information in the U.S.A.
7. The CDC publishes *Morbidity and Mortality Weekly Report* to provide information on morbidity (incidence) and mortality (deaths).

STUDY QUESTIONS

Review

1. Differentiate between the following pairs of terms:
   (a) Etiology and pathogenesis
   (b) Infection and disease
   (c) Communicable disease and noncommunicable disease

2. What is meant by the normal microbiota? How does it differ from the transient microbiota?

3. Define symbiosis. Differentiate among commensalism, mutualism, and parasitism, and give an example of each.

4. What is a reservoir of infection? Match the following diseases with their reservoirs.

____ Influenza	(a)	Nonliving
____ Rabies	(b)	Human
____ Botulism	(c)	Animal

5. Describe how Koch's postulates establish the etiology of many infectious diseases. Why don't Koch's postulates apply to all infectious diseases?

6. Describe the various ways diseases can be transmitted in each of the following categories. Name one disease transmitted by each method.

(a) Transmission by direct contact	(d)	Droplet transmission
(b) Transmission by indirect contact	(e)	Vehicle transmission
(c) Transmission by arthropod vectors	(f)	Airborne transmission

7. List four predisposing factors to disease.

8. Indicate whether each of the following conditions is typical of subacute, chronic, or acute infections.
   (a) Patient experiences rapid onset of malaise; symptoms last five days.
   (b) Patient experiences cough and breathing difficulty for months.
   (c) Patient has no apparent symptoms and is a known carrier.

9. Of all the hospital patients with infections, one-third do not enter the hospital with an infection. How do they acquire these infections? What is the method of transmission of these infections? What is the reservoir of infection?

10. Differentiate between endemic and epidemic states of infectious disease.

11. What is epidemiology? What is the role of the Centers for Disease Control and Prevention (CDC)?

12. Distinguish between symptoms and signs as signals of disease.

13. How can a local infection become a systemic infection?

14. Why are some organisms that constitute the normal microbiota described as commensals, whereas others are described as mutualistic?

15. Put the following in the correct order to describe the pattern of disease: period of convalescence, prodromal period, period of decline, period of incubation, period of illness.

Critical Thinking

1. Ten years before Robert Koch published his work on anthrax, Anton De Bary showed that potato blight was caused by the fungus *Phytophthora infestans*. Why do you suppose we use Koch's postulates instead of something called "De Bary's postulates"?

2. Florence Nightingale gathered the following data in 1855.

POPULATION SAMPLED	DEATHS FROM CONTAGIOUS DISEASES
Englishmen	0.2%
English soldiers (in England)	18.7%
English soldiers (in Crimean War)	42.7%
English soldiers (in Crimean War) after Nightingale's sanitary reforms	2.2%

   Discuss how Nightingale used the three basic types of epidemiologic investigation. The contagious diseases were primarily cholera and typhus; how are these diseases transmitted and prevented?

3. Name the method of transmission of each of the following diseases:

(a) Malaria	(f)	Mononucleosis
(b) Tuberculosis	(g)	Measles
(c) Nosocomial infections	(h)	Hepatitis A
(d) Salmonellosis	(i)	Tetanus
(e) Streptococcal pharyngitis	(j)	Hepatitis B
	(k)	Chlamydial urethritis

4. The following graph shows the incidence of typhoid fever in the U.S.A. from 1955 to 1993. Mark the graph to show when this disease occurred sporadically and epidemically. What appears to be the endemic level? What would have to be shown to indicate a pandemic of this disease? How is typhoid fever transmitted?

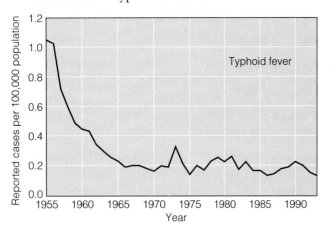

5. In 1991, a cholera epidemic caused by *Vibrio cholerae* O1 Inaba was discovered in Peru. During 1991 and 1992, 731,312 cholera cases were reported in 20 countries in the Western Hemisphere. The following map shows the location of cholera cases in the continental U.S.A. caused by *V. cholerae* O1 Inaba (gold), as well as where this strain has been found in coastal waters (purple).

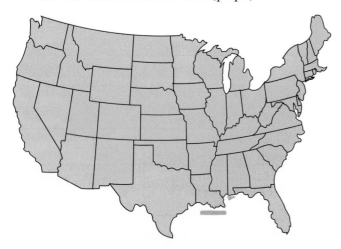

How is cholera transmitted? How did *V. cholerae* O1 Inaba get into the U.S.A.? Explain how eight people who did not travel outside the U.S.A. contracted *V. cholerae* O1 Inaba.

Clinical Applications

1. Three days before a nurse developed meningococcemia, she assisted with intubation of a patient with a *Neisseria meningitidis* infection. Of the 24 medical personnel involved, only this nurse became ill. The nurse recalled that she was exposed to nasopharyngeal secretions and did not receive antibiotic prophylaxis. What two mistakes did the nurse make? How is meningitis transmitted?

2. Three patients in a large hospital acquired infections of *Pseudomonas cepacia* during their stay. All three patients received cryoprecipitate, which is prepared from blood that has been frozen in a standard plastic blood transfer pack. The transfer pack is then placed in a water bath to thaw. What is the probable origin of the nosocomial infections? What characteristics of *Pseudomonas* would allow it to be involved in this type of infection?

3. The following is a case history of a 49-year-old man. Identify each period in the pattern of disease that he experienced. On February 7, he handled a parakeet with a respiratory illness. On March 9, he experienced intense pain in his legs, followed by severe chills and headaches. On March 16, he had chest pains, cough, and diarrhea, and his temperature was 40°C. Appropriate antibiotics were administered on March 17, and his fever subsided within 12 hours. He continued taking antibiotics for 14 days. (*Note:* The disease is psittacosis. Can you find the etiology?)

4. During 1989, 21% of the patients in a large hospital acquired *Clostridium difficile* diarrhea and colitis during their hospital stay. These patients required longer hospital stays than uninfected patients. Epidemiological studies provided the following information. What is the most likely mode of transmission of this bacterium in hospitals? How can transmission be prevented?

Rate of infection for patients

Single room	7%
Double room	17%
Triple room	26%

Rate of environmental isolations of *C. difficile*

Bedrail	10%
Commode	1%
Floor	18%
Call button	6%
Toilet	3%

Hands of hospital personnel after contact with patients that were culture-positive for *C. difficile*

Used gloves	0%
Did not use gloves	59%
Had *C. difficile* before patient contact	3%
Washed with nondisinfectant soap	40%
Washed with disinfectant soap	3%
Did not wash hands	20%

5. *Mycobacterium avium* is prevalent in AIDS patients. In an effort to determine the source of this infection, hospital water systems were sampled. The water contained chlorine.

Percentage of samples with *M. avium*

Hot water		Cold water	
February	88%	February	22%
June	50%	June	11%

What is the usual method of transmission for *Mycobacterium?* What is a probable source of infection in hospitals? How can such nosocomial infections be prevented?

Microbial Mechanisms of Pathogenicity

N OW THAT YOU HAVE a basic understanding of how microorganisms cause disease, we will take a look at some of the specific properties of microorganisms that contribute to **pathogenicity**, the ability to cause disease in a host, and **virulence**, the degree of pathogenicity. (As discussed throughout, the term *host* usually refers to humans.) Keep in mind that many of the properties contributing to microbial pathogenicity and virulence are unclear or unknown. We do know, however, that if the microbes' attack overpowers the host defenses, disease results.

ENTRY OF A MICROORGANISM INTO THE HOST

LEARNING OBJECTIVES

- *Define portal of entry, pathogenicity, virulence, and LD$_{50}$.*
- *Identify a microbe that enters the human body via (1) the mucous membranes, (2) the skin, and (3) the parenteral route.*
- *Using examples, explain how microbes adhere to host cells.*

Most pathogens must gain access to the host, adhere to host tissues, penetrate or evade host defenses, and damage the host tissue to cause disease. However, some microbes do not cause disease by directly damaging host tissue. Instead, disease is due to the accumulation of mi-

crobial waste products. Some microbes, such as those that cause dental caries and acne, can cause disease without penetrating the body. Pathogens can gain entrance to the human body and other hosts through several avenues, which are called **portals of entry.**

Portals of Entry

The portals of entry for pathogens are mucous membranes, skin, and direct deposition beneath the skin or membranes (parenteral).

Mucous Membranes

Many bacteria and viruses gain access to the body by penetrating mucous membranes lining the respiratory tract, gastrointestinal tract, genitourinary tract, and conjunctiva, a delicate membrane that covers the eyeball and lines the eyelids. Most pathogens enter through the mucous membranes of the gastrointestinal and respiratory tracts.

The respiratory tract is the easiest and most frequently traveled portal of entry for infectious microorganisms. Microbes are inhaled into the nose or mouth in drops of moisture and dust particles. Some diseases that are commonly contracted via the respiratory tract include the common cold, pneumonia, tuberculosis, influenza, measles, and smallpox.

Microorganisms can gain access to the gastrointestinal tract in food and water. Bringing contaminated fingers to

FIGURE 15.1 Adherence.
(a) Selective attachment of a pathogenic strain of *E. coli* to the intestinal tissue of a rabbit. **(b)** Bacteria adhering to the skin of a salamander.

Intestinal tissue

E. coli

Bacteria

Skin

(a)   TEM   1 μm   (b)   SEM   10 μm

the mouth is another common mode of entry. (See the box on p. 394 concerning how these and other human behaviors can, over time, influence the virulence of microorganisms.) Most microbes that enter the body in these ways are destroyed by hydrochloric acid (HCl) and enzymes in the stomach or by bile and enzymes in the small intestine. Those that survive can cause disease. Microbes in the gastrointestinal tract can cause poliomyelitis, hepatitis A, typhoid fever, amoebic dysentery, giardiasis, bacillary dysentery (shigellosis), and cholera. These pathogens are then eliminated with feces and can be transmitted to other hosts via water, food, or contaminated fingers.

An important pathogen capable of penetrating mucous membranes of the genitourinary tract is *Treponema pallidum,* the causative agent of syphilis.

Skin

The skin is one of the largest organs of the body and is a significant factor in defense against disease. When unbroken, the skin is impenetrable by most microorganisms, although even then it does not provide complete protection. Some microbes gain access to the body through openings in the skin, such as hair follicles and sweat gland ducts. In addition, larvae of the hookworm *Necator americanus* actually bore through intact skin, and some fungi grow on the keratin in skin or infect the skin itself.

Parenteral Route

Other microorganisms gain access to the body when they are deposited directly into the tissues beneath the skin or into mucous membranes when these barriers are penetrated or injured. This route is called the **parenteral route.** Punctures, injections, bites, cuts, wounds, surgery, and splitting due to swelling or drying can all establish parenteral routes.

Preferred Portal of Entry

Even after microorganisms have entered the body, they do not necessarily cause disease. The occurrence of disease depends on several factors, only one of which is the portal of entry. Many pathogens have a preferred portal of entry that is a prerequisite to their being able to cause disease. If they gain access to the body by another portal, disease might not occur. For example, the bacteria of typhoid fever, *Salmonella typhi,* produce all the signs and symptoms of the disease when swallowed (preferred route), but if the same bacteria are rubbed on the skin, no reaction (or only a slight inflammation) occurs. Streptococci that are inhaled (preferred route) can cause pneumonia; those that are swallowed generally do not produce signs or symptoms. Some pathogens, such as *Yersinia pestis,* the microorganism that causes plague, can initiate disease from more than one portal of entry. The preferred portals of entry for some common pathogens are given in Table 15.1.

Numbers of Invading Microbes

If only a few microbes enter the body, they will probably be overcome by the host's defenses. However, if large numbers of microbes gain entry, the stage is probably set for disease. Thus, the likelihood of disease increases as the number of pathogens increases.

The virulence of a microbe or the potency of its toxin is often expressed as the LD_{50} (lethal dose for 50% of hosts), the number of microbes in a dose that will kill 50% of inoculated test animals under normal conditions. The dose required to produce a demonstrable infection in 50% of the test animals is called the ID_{50} (infectious dose for 50% of hosts).

Adherence

Once pathogens gain entry to a host, almost all of them have some means of attaching themselves to host tissues. For most pathogens, this attachment, called **adherence,** is a necessary step in pathogenicity. (Of course, nonpathogens also have structures for attachment.) The attachment between pathogen and host is accomplished by means of surface molecules on the pathogen called **adhesins** or **ligands** that bind specifically to complementary surface receptors on the cells of certain host tissues (Figure 15.1). Adhesins may be located on a microbe's glyco-

TABLE 15.1 Portals of Entry for Causative Agents of Some Common Diseases

PORTAL OF ENTRY	CAUSATIVE AGENT*	DISEASE	INCUBATION PERIOD
Mucous membranes			
Respiratory tract	*Corynebacterium diphtheriae*	Diphtheria	2–5 days
	Neisseria meningitidis	Meningococcal meningitis	1–7 days
	Streptococcus pneumoniae	Pneumococcal pneumonia	Variable
	Mycobacterium tuberculosis†	Tuberculosis	Variable
	Bordetella pertussis	Whooping cough (pertussis)	12–20 days
	Influenzavirus	Influenza	18–36 hours
	Measles virus (*Morbillivirus*)	Measles (rubeola)	11–14 days
	Rubella virus (*Rubivirus*)	German measles (rubella)	2–3 weeks
	Epstein–Barr virus (*Lymphocryptovirus*)	Infectious mononucleosis	2–6 weeks
	Varicella-zoster virus (*Varicellavirus*)	Chickenpox (varicella)	14–16 days
	Smallpox virus (*Orthopoxvirus*)	Smallpox (variola)	12 days
	Coccidioides immitis (fungus)	Coccidioidomycosis (primary infection)	1–3 weeks
	Histoplasma capsulatum (fungus)	Histoplasmosis	5–18 days
Gastrointestinal tract	*Shigella* species	Bacillary dysentery (shigellosis)	1–2 days
	Brucella species	Brucellosis (undulant fever)	6–14 days
	Vibrio cholerae	Cholera	1–3 days
	Salmonella enteritidis, Salmonella typhimurium, Salmonella choleraesuis	Salmonellosis	7–22 hours
	Salmonella typhi	Typhoid fever	14 days
	Hepatitis A virus	Hepatitis A	15–50 days
	Paramyxovirus	Mumps	2–3 weeks
	Poliovirus (*Enterovirus*)	Poliomyelitis	4–7 days
	Trichinella spiralis (helminth)	Trichinosis	2–28 days
Genitourinary tract	*Neisseria gonorrhoeae*	Gonorrhea	3–8 days
	Treponema pallidum	Syphilis	9–90 days
Skin or parenteral route	*Clostridium perfringens*	Gas gangrene	1–5 days
	Clostridium tetani	Tetanus	3–21 days
	Leptospira interrogans	Leptospirosis	2–20 days
	Yersinia pestis	Plague	2–6 days
	Rickettsia rickettsii	Rocky Mountain spotted fever	3–12 days
	Hepatitis B virus† (*Hepadnavirus*)	Hepatitis B	6 weeks–6 months
	Rabies virus (*Lyssavirus*)	Rabies	10 days–1 year
	Flavivirus	Yellow fever	3–6 days
	Plasmodium species (protozoan)	Malaria	2 weeks

*All causative agents are bacteria, unless indicated otherwise. For viruses, the viral species and/or genus name is given.
†These pathogens can also cause disease after entering the body via the gastrointestinal tract.

calyx or on other microbial surface structures, such as fimbriae (see Chapter 4).

The majority of adhesins on the microorganisms we have studied so far are glycoproteins or lipoproteins. The receptors on host cells are typically sugars, such as mannose. Adhesins on different strains of the same species of pathogen can vary in structure. Different cells of the same host can also have different receptors that vary in structure. If adhesins, receptors, or both can be altered to interfere with adherence, infection can often be prevented (or at least controlled).

The following examples illustrate the diversity of adhesins. *Streptococcus mutans,* a bacterium that plays a key role in tooth decay, attaches to the surface of teeth by its glycocalyx. An enzyme produced by *S. mutans,* called *glucosyltransferase,* converts glucose (derived from sucrose or table sugar) into a sticky polysaccharide called dextran, which forms the glycocalyx. *Actinomyces* cells have fimbriae that adhere to the glycocalyx of *S. mutans,* further contributing to plaque. Enteropathogenic strains of *E. coli* (those responsible for gastrointestinal disease) have adhesins on fimbriae that adhere only to specific kinds of cells in certain regions of the small intestine. After adhering, *Shigella* and *E. coli* induce endocytosis as a vehicle to enter host cells and then multiply within them (see Figure 25.9). *Neisseria gonorrhoeae,* the causative agent of gonorrhea, also has fimbriae containing adhesins, which in this case permit attachment to cells with appropriate receptors in the genitourinary tract, eyes, and pharynx. *Staphylococcus aureus,* which can cause skin infections, binds to skin by a mechanism of adherence that resembles viral attachment (discussed in Chapter 13).

HOW BACTERIAL PATHOGENS PENETRATE HOST DEFENSES

LEARNING OBJECTIVES

- *Explain how capsules, cell wall components, and enzymes contribute to pathogenicity.*
- *Compare the effects of hemolysins, leukocidins, coagulase, kinases, hyaluronidase, and collagenase.*

Although some pathogens can cause damage on the surface of tissues, most must penetrate tissues to cause disease. Here we will consider several factors that contribute to the ability of bacteria to invade a host.

Capsules

We noted in Chapter 4 that some bacteria make glycocalyx material that forms capsules around their cell walls;

this property increases the virulence of the species. The capsule resists the host's defenses by impairing phagocytosis, a process by which certain cells of the body engulf and destroy microbes (to be discussed in Chapter 16). The chemical nature of the capsule appears to prevent the phagocytic cell from adhering to the bacterium. However, the human body can produce antibodies against the capsule, and when these antibodies are present on the capsule surface, the encapsulated bacteria are easily destroyed by phagocytosis.

One bacterium that owes its virulence to the presence of a polysaccharide capsule is *Streptococcus pneumoniae,* the causative agent of pneumococcal pneumonia (see Figure 24.10). Some strains of this organism have capsules, and others do not. Strains with capsules are virulent, but strains without capsules are avirulent because they are susceptible to phagocytosis. Other bacteria that produce capsules related to virulence are *Klebsiella pneumoniae,* a causative agent of bacterial pneumonia; *Hemophilus influenzae,* a cause of pneumonia and meningitis in children; *Bacillus anthracis,* the cause of anthrax; and *Yersinia pestis,* the causative agent of bubonic plague. Keep in mind that capsules are not the only cause of virulence. Many nonpathogenic bacteria produce capsules, and the virulence of some pathogens is not related to the presence of a capsule.

Components of the Cell Wall

The cell walls of certain bacteria contain chemical substances that contribute to virulence. For example, *Streptococcus pyogenes* produces a heat-resistant and acid-resistant protein called **M protein** (see Figure 21.5). This protein is found on both the cell surface and fimbriae. The M protein mediates attachment of the bacterium to epithelial cells of the host and helps the bacterium resist phagocytosis by white blood cells. The protein thereby increases the virulence of the microorganism. Immunity to *S. pyogenes* depends on the body's production of an antibody specific to M protein.

The waxes that make up the cell wall of *Mycobacterium tuberculosis* also increase virulence by resisting digestion by phagocytes. In fact, *M. tuberculosis* can even multiply inside phagocytes.

Enzymes

The virulence of some bacteria is thought to be aided by the production of extracellular enzymes (exoenzymes) and related substances. These chemicals can break cells open, dissolve materials between cells, and form or dissolve blood clots, among other functions.

Substances called **leukocidins,** produced by some bacteria, can destroy neutrophils, white blood cells (leukocytes) that are very active in phagocytosis. Leukocidins are

Microbiology in the News

HOW HUMAN BEHAVIOR INFLUENCES THE EVOLUTION OF VIRULENCE IN MICROORGANISMS

···

ACCORDING TO HUMAN logic, a parasite that kills its host is harming itself. Thus it seems reasonable that some parasites have evolved to a sort of commensalism with their human hosts. In this respect, tapeworms might be considered the "perfect parasite." Tapeworms cause no obvious symptoms in most patients, thus ensuring that their hosts will be able to move around and shed tapeworm eggs that will be ingested by intermediate hosts.

However, other diseases such as cholera, antibiotic-resistant tuberculosis, viral cancers, and syphilis sometimes do kill their hosts. Does it make sense for a parasite to kill its host? Remember that nature does not have a plan for evolution—the genetic variations that give rise to evolution are due to random mutations, not logic. However, according to natural selection, organisms best adapted to their environments will survive and reproduce. Coevolution between a parasite and its host seems to occur: the behavior of one influences that of the other.

Some parasites, such as the cholera bacterium, get transmitted before their host dies. *Vibrio cholerae* quickly induces diarrhea, threatening the

host's life due to loss of fluids and salts, but providing a way to transmit the parasite to another person through ingestion of contaminated water. This cycle continues in countries where war and poverty have prevented implementation of water purification systems. However, the virulent form of cholera has been selected against in countries where water-treatment systems prevent the bacteria from getting to a new host. After water purification was instituted in India, the milder agent, *V. cholerae* eltor, replaced the deadly *V. cholerae*.

When we interfere with a pathogen's survival by using antibiotics, we inadvertently select for antibiotic-resistant mutants. The rapid rise of multidrug-resistant *Mycobacterium tuberculosis,* documented in the popular and scientific media, demonstrates this. In an antibiotic-laden environment, the "fittest" bacterium is one that is resistant to antibiotics. Antibiotic resistance is selected for when chemotherapy is not completed and when antibiotics are not used correctly. For example, tuberculosis (TB)-chemotherapy protocols are very long (up to 2 years). In the U.S.A., only 75% of TB patients complete the prescribed chemotherapy. When a person doesn't take the prescribed antibiotic, the level of the drug in the body decreases, and some resistant bacteria survive. These bacteria reproduce, and some of the new population are able to survive in even higher levels of antibiotics.

Comparisons of human T cell leukemia virus type 1 (HTLV-1) infections in Japan and Jamaica also sug-

gest that human sexual behavior influences the virulence of sexually transmitted diseases. HTLV-1 is a sexually transmitted retrovirus that causes a type of adult leukemia. In Japan, the average age of onset of HTLV-1–caused cancers is 60 years, whereas in Jamaica onset occurs at about 45 years of age. In Japan, barrier contraceptives are more widely used than birth control pills, so the virus needs a healthy host for a longer period of time. In Jamaica, barrier contraceptives are not widely used and HTLV-1 is more virulent, perhaps because it has a greater potential for sexual transmission.

Human behavior may also have influenced the evolution of syphilis. *Treponema pallidum* has a long incubation period and goes into noninfectious latent periods between the primary, secondary, and tertiary stages of the disease. Although it can't be transmitted during the incubation and latent periods, these periods allow the bacteria to stay alive until the infected host changes sexual partners. Partner changes might take several years in a monogamous society.

Paul Ewald of Amherst College reminds us that commensalism is not the inevitable end point of evolution. In order to control diseases, we must modify our behavior to make commensalism the most favorable outcome for the pathogen. Some examples of these modifications are ensuring completion of the prescribed antibiotic protocol for tuberculosis, and the use of condoms to prevent transmission of sexually transmitted diseases.

also active against macrophages (phagocytic cells) present in tissues. Among the bacteria that secrete leukocidins are staphylococci and streptococci. Leukocidins produced by streptococci degrade lysosomes within leukocytes and thereby cause the death of the white blood cell (also see the box in Chapter 16). Enzymes released from the lysosomes can damage other cellular structures and thus intensify streptococcal lesions. This type of damage to white blood cells decreases host resistance.

Hemolysins are bacterial enzymes that cause the lysis of red blood cells. Bacteria produce hemolysins that differ in their ability to lyse different kinds of red blood cells (humans, sheep, and rabbits, for example) and the type of lysis they cause. Important producers of hemolysins are staphylococci; *Clostridium perfringens,* the most common causative agent of gas gangrene; and streptococci. *Streptolysins* are hemolysins produced by streptococci. One kind, called *streptolysin O (SLO),* is so named because it is inactivated by atmospheric oxygen. Another kind is called *streptolysin S (SLS)* because it has an affinity for a protein (albumin) in blood serum. Both streptolysins can cause lysis of not only red blood cells, but also white blood cells (whose function is to kill the streptococci) and other body cells.

Coagulases are bacterial enzymes that coagulate (clot) the fibrinogen in blood. Fibrinogen, a plasma protein produced by the liver, is converted by coagulases into fibrin, the threads that form a blood clot. The fibrin clot may protect the bacterium from phagocytosis and isolate it from other defenses of the host. Coagulases are produced by some members of the genus *Staphylococcus;* they may be involved in the walling-off process in boils produced by staphylococci. However, some staphylococci that do not produce coagulases are still virulent. (Capsules may be more important to their virulence.)

Bacterial kinases are bacterial enzymes that break down fibrin and thus dissolve clots formed by the body to isolate the infection. One of the better-known kinases is **streptokinase (fibrinolysin),** which is produced by such streptococci as *Streptococcus pyogenes.* Another kinase, **staphylokinase,** is produced by such staphylococci as *Staphylococcus aureus.* Injected directly into the blood, streptokinase has been used successfully to dissolve some types of blood clots in cases of heart attacks due to obstructed coronary arteries.

Hyaluronidase is another enzyme secreted by certain bacteria, such as streptococci. It hydrolyzes hyaluronic acid, a type of polysaccharide that holds together certain cells of the body, particularly cells in connective tissue. This digesting action is thought to be involved in the tissue blackening of infected wounds and to help the microorganism spread from its initial site of infection. Hyaluronidase is also produced by some clostridia that cause gas gangrene. For therapeutic use, hyaluronidase may be mixed with a drug to promote the spread of the drug through a body tissue.

Another enzyme, **collagenase,** produced by several species of *Clostridium,* facilitates the spread of gas gangrene. Collagenase breaks down the protein collagen, which forms the connective tissue of muscles and other body organs and tissues.

Other bacterial substances believed to contribute to virulence are *necrotizing factors,* which cause the death of body cells; *hypothermic factors,* which decrease body temperature; *lecithinase,* which destroys the plasma membrane, especially around red blood cells; *protease,* which breaks down proteins, especially in muscle tissue; and *siderophores,* which scavenge iron from the host's body fluids.

HOW BACTERIAL PATHOGENS DAMAGE HOST CELLS

LEARNING OBJECTIVES

- *Contrast the nature and effects of exotoxins and endotoxins.*
- *Outline the mechanisms of action of diphtheria toxin, botulinum toxin, tetanus toxin, cholera toxin, and lipid A.*
- *Identify the importance of the* Limulus *amoebocyte lysate assay.*

When a microorganism invades a body tissue, it initially encounters phagocytes of the host. If the phagocytes are successful in destroying the invader, no further damage is done to the host. But if the pathogen overcomes the host defense, then the microorganism can damage host cells in three basic ways: (1) by causing direct damage in the *immediate vicinity* of the invasion; (2) by producing toxins, transported by blood and lymph, that damage sites *far removed* from the original site of invasion; and (3) by inducing hypersensitivity reactions. This latter mechanism is considered in detail in Chapter 19. For now, we will discuss only the first two mechanisms.

Direct Damage

Once a pathogen is attached to host cells, it can pass through them to invade other tissues. During this invasion, the pathogens metabolize and multiply to kill host cells. Some bacteria, such as *E. coli, Shigella, Salmonella,* and *Neisseria gonorrhoeae,* can induce host epithelial cells to engulf them by a process that resembles phagocytosis. These pathogens can also be extruded from the host cells by a reverse phagocytosis process in order to enter other host cells. Some bacteria can also penetrate host cells by excreting enzymes and by their own motility; such penetration can itself damage the host cell. Most damage by bacteria, however, is done by toxins.

FIGURE 15.2 Exotoxins and endotoxins.

(a) **Exotoxins** are produced inside mostly gram-positive bacteria as part of their growth and metabolism. They are then released into the surrounding medium.

(b) **Endotoxins** are part of the outer portion of the cell wall (lipid A; see Figure 4.12c) of gram-negative bacteria. They are liberated when the bacteria die and the cell wall breaks apart.

Production of Toxins

Toxins are poisonous substances that are produced by certain microorganisms. They are frequently the primary factor that contributes to the pathogenic properties of those microbes. The capacity of microorganisms to produce toxins is called **toxigenicity.** Toxins transported by the blood or lymph can cause serious, and sometimes fatal, effects. Some toxins produce fever, cardiovascular disturbances, diarrhea, and shock. Toxins can also inhibit protein synthesis, destroy blood cells and blood vessels, and disrupt the nervous system by causing spasms. Of the 220 or so known bacterial toxins, nearly 40% cause disease by damaging eucaryotic cell membranes. The term **toxemia** refers to symptoms caused by toxins in the blood. Toxins are of two types: exotoxins and endotoxins.

Exotoxins

Exotoxins are produced inside some bacteria as part of their growth and metabolism and are released into the surrounding medium (Figure 15.2a). Exotoxins are proteins, and many are enzymes that catalyze only certain biochemical reactions. Most bacteria that produce exotoxins are gram-positive. The genes for most (perhaps all) exotoxins are carried on bacterial plasmids or phages. Because exotoxins are soluble in body fluids, they can easily diffuse into the blood and are rapidly transported throughout the body.

Exotoxins work by destroying particular parts of the host's cells or by inhibiting certain metabolic functions. Exotoxins may be grouped into three principal types, based on their mode of action: (1) **cytotoxins,** which kill host cells or affect their functions; (2) **neurotoxins,** which interfere with normal nerve impulse transmission; and (3) **enterotoxins,** which affect cells lining the gastrointestinal tract. Exotoxins are among the most lethal substances known. Only 1 mg of the botulinum exotoxin is enough to kill 1 million guinea pigs. Fortunately, only a few bacterial species produce such potent exotoxins.

Diseases caused by bacteria that produce exotoxins are often caused by minute amounts of exotoxins, not by the bacteria themselves. It is the exotoxins that produce the specific signs and symptoms of the disease. Thus, exotoxins are disease-specific. For example, the infection of a wound by *Clostridium tetani* need be no larger or more painful than a pin prick, yet the organisms in a wound that size can produce enough tetanus toxin to kill an unvaccinated human.

The body produces antibodies called **antitoxins** that provide immunity to exotoxins. When exotoxins are inactivated by heat or by formaldehyde, iodine, or other chemicals, they no longer cause the disease but are still able to stimulate the body to produce antitoxins. Such altered exotoxins are called **toxoids.** When toxoids are injected into the body as a vaccine, they stimulate antitoxin production so that immunity is produced to disease. Diphtheria and tetanus can be prevented by toxoid vaccination.

Next we will briefly describe a few of the more notable exotoxins (antitoxins will be discussed further in Chapter 18).

Diphtheria Toxin. *Corynebacterium diphtheriae* produces the diphtheria toxin only when it is infected by a lysogenic phage carrying the *tox* gene. This cytotoxin inhibits protein synthesis in eucaryotic cells. The mechanism by which it does this is an excellent example of how an exotoxin interacts with host cells (Figure 15.3). Diphtheria toxin is a protein consisting of two different polypeptides, designated A (active) and B (binding). Although only polypeptide A causes symptoms in the host, polypeptide B is required by polypeptide A to be active. Polypeptide B binds to surface receptors on the host cell and causes the transport of the entire protein across the plasma membrane into the cell. Once this is accomplished, polypeptide A inhibits protein synthesis within the target cell. (This model also applies to the cholera exotoxin, which is described shortly.)

Erythrogenic Toxins. *Streptococcus pyogenes* has the genetic material to synthesize three types of cytotoxins, designated A, B, and C. These erythrogenic toxins damage blood capillaries under the skin and produce a red skin rash. Scarlet fever, caused by *S. pyogenes* exotoxins, is named for this characteristic rash.

Botulinum Toxin. Botulinum toxin is produced by *Clostridium botulinum*. Although toxin production is associated with germination of endospores and with growth of vegetative cells, little of the toxin appears in the medium until it is released by lysis late in growth. Botulinum toxin is a neurotoxin; it acts at the neuromuscular junction (the junction between nerve cell and muscle cell) and prevents the transmission of impulses from the nerve cell to muscle. The toxin accomplishes this by binding to the nerve cell and inhibiting the release of a neurotransmitter called acetylcholine. As a result, botulinum toxin causes paralysis in which muscle tone is lacking (flaccid paralysis). *C. botulinum* produces several different types of botulinum toxin, and each possesses a different potency.

Tetanus Toxin. *Clostridium tetani* produces tetanus neurotoxin, also known as tetanospasmin. This toxin reaches the central nervous system and binds to nerve cells that control contraction of various skeletal muscles. These nerve cells normally send inhibiting impulses that prevent random contractions and terminate completed contractions. The binding of tetanospasmin blocks this relaxation pathway (see Chapter 22). The result is uncontrollable muscle contraction, producing the convulsive symptoms (spasmodic contractions) of tetanus, or "lockjaw."

Vibrio Enterotoxin. *Vibrio cholerae* produces an enterotoxin called cholera toxin. Like diphtheria toxin, cholera toxin consists of two polypeptides, A (active) and B (binding). The B component binds to plasma membranes of epithelial cells lining the small intestine, and the A component induces the formation of cyclic AMP from ATP in the cytoplasm (see Chapter 8). As a result, epithelial cells discharge large amounts of fluids and electrolytes (ions). Normal muscular contractions are disturbed, leading to severe diarrhea that may be accompanied by vomiting. *Heat-labile enterotoxin* (so called because it is more sensitive to heat than are most toxins) produced by some strains of *E. coli* has an action identical to that of cholera toxin.

Staphylococcal Enterotoxin. *Staphylococcus aureus* produces an enterotoxin that affects the intestines in the

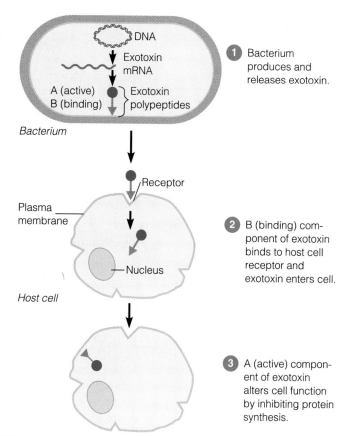

FIGURE 15.3 Action of an exotoxin. Proposed model for mechanism of action of diphtheria toxin.

same way as cholera toxin. A strain of *S. aureus* also produces enterotoxins that result in the symptoms associated with toxic shock syndrome, discussed in Chapter 1. A summary of diseases produced by exotoxins is presented in Table 15.2.

Endotoxins

Endotoxins differ from exotoxins in several ways. Endotoxins are part of the outer portion of the cell wall of gram-negative bacteria (Figure 15.2b). Recall from Chapter 4 that gram-negative bacteria have an outer membrane surrounding the peptidoglycan layer of the cell wall. This outer membrane consists of lipoproteins, phospholipids, and lipopolysaccharides (LPS) (see Figure 4.13c). The lipid portion of LPS, called **lipid A**, is the endotoxin. Thus, endotoxins are lipopolysaccharides, whereas exotoxins are proteins.

Endotoxins exert their effects when gram-negative bacteria die and their cell walls undergo lysis, thus liberating the endotoxin. Antibiotics used to treat diseases caused by gram-negative bacteria can lyse the bacterial cells; this reaction releases endotoxin and may lead to an immediate worsening of the symptoms, but the condition usually improves as the endotoxin breaks down. All endotoxins produce the same signs and symptoms, regardless of the

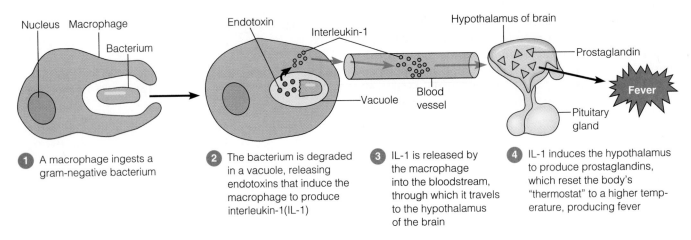

① A macrophage ingests a gram-negative bacterium

② The bacterium is degraded in a vacuole, releasing endotoxins that induce the macrophage to produce interleukin-1(IL-1)

③ IL-1 is released by the macrophage into the bloodstream, through which it travels to the hypothalamus of the brain

④ IL-1 induces the hypothalamus to produce prostaglandins, which reset the body's "thermostat" to a higher temp-erature, producing fever

FIGURE 15.4   Endotoxins and the pyrogenic response. The postulated mechanism by which endotoxins cause fever.

species of microorganism, although not to the same de-gree. Responses by the host include chills, fever, weakness, generalized aches, and, in some cases, shock and even death. Endotoxins can also induce miscarriage and pre-vent blood from clotting.

The fever (pyrogenic response) caused by endotoxins is believed to occur as depicted in Figure 15.4. ① When gram-negative bacteria are ingested by phagocytes and ② degraded in vacuoles, the lipopolysaccharides (LPS) of the bacterial cell wall are released. These endotoxins cause macrophages to produce small protein molecules called **interleukin-1 (IL-1)**, formerly called *endogenous pyrogen*, ③ which are carried via the blood to the hypo-thalamus, a temperature control center in the brain. ④ IL-1 induces the hypothalamus to release lipids called prostaglandins, which reset the thermostat in the hypo-thalamus at a higher temperature. The result is a fever. Bacterial cell death caused by lysis or antibiotics can also produce fever by this mechanism. Both aspirin and acet-aminophen reduce fever by inhibiting the synthesis of prostaglandins. The function of fever in the body is dis-cussed in Chapter 16.

Shock caused by gram-negative bacteria is called **septic (or endotoxic) shock.** The shock produced by endotoxins, like fever, is related to secretion of a substance by macro-phages. Phagocytosis of gram-negative bacteria causes the phagocytes to secrete a polypeptide called **tumor necro-sis factor (TNF),** or **cachectin.** This substance binds to many tissues in the body and alters their metabolism in a number of ways. One effect of TNF is damage to blood capillaries; their permeability is increased, and they lose large amounts of fluid. The result is a drop in blood pres-sure that results in shock. Low blood pressure has serious effects on the kidneys, lungs, and gastrointestinal tract. In addition, the presence of gram-negative bacteria such as *Hemophilus influenzae* type b in cerebrospinal fluid causes the release of IL-1 and TNF. These, in turn, cause a weak-ening of the blood-brain barrier that normally protects the central nervous system from infection. The barrier is weakened to let phagocytes in, but this also lets more bac-teria enter from the bloodstream.

Endotoxins do not promote the formation of effective antitoxins. Antibodies are produced, but they tend not to counter the effect of the toxin; sometimes, in fact, they ac-tually enhance its effect.

Representative microorganisms that produce endotox-ins are *Salmonella typhi* (the causative agent of typhoid fever), *Proteus* species (frequently the causative agents of urinary tract infections), and *Neisseria meningitidis* (the causative agents of meningococcal meningitis).

Because of the many diseases associated with endotox-ins, it is important to have a test to identify their presence in drugs, medical devices, and body fluids. One such lab-oratory test is called the **Limulus amoebocyte lysate (LAL) assay,** which can detect even minute amounts of endotoxin. The hemolymph (blood) of the Atlantic coast horseshoe crab, *Limulus polyphemus,* contains white blood cells called amoebocytes, which have large amounts of a protein (lysate) that causes clotting. In the presence of endotoxin, amoebocytes in the crab hemolymph lyse and liberate their clotting protein. The resulting gel-clot is a positive test for the presence of endotoxin. The degree of the reaction is measured by a spectrophotometer, an in-strument that measures degree of turbidity.

A comparison of exotoxins and endotoxins appears in Table 15.3.

PLASMIDS, LYSOGENY, AND PATHOGENICITY

LEARNING OBJECTIVE

- *Using examples, describe the roles of plasmids and lysogeny in pathogenicity.*

TABLE 15.2 Diseases Caused by Exotoxins

DISEASE	BACTERIUM	MECHANISM
Botulism	*Clostridium botulinum*	Neurotoxin prevents transmission of nerve impulses; flaccid paralysis results
Tetanus	*Clostridium tetani*	Neurotoxin blocks nerve impulses to muscle relaxation pathway
Gas gangrene and food poisoning	*Clostridium perfringens* and other species of *Clostridium*	One exotoxin (cytotoxin) causes massive red blood cell destruction (hemolysis); another exotoxin (enterotoxin) is related to food poisoning and causes diarrhea
Diphtheria	*Corynebacterium diphtheriae*	Cytotoxin inhibits protein synthesis, especially in nerve, heart, and kidney cells
Scalded skin syndrome, food poisoning, and toxic shock syndrome	*Staphylococcus aureus*	One exotoxin causes skin layers to separate and slough off (scalded skin); another exotoxin (enterotoxin) produces diarrhea and vomiting; still another exotoxin produces symptoms associated with toxic shock syndrome (TSS)
Cholera	*Vibrio cholerae*	Enterotoxin induces diarrhea
Scarlet fever	*Streptococcus pyogenes*	Cytotoxins cause vasodilation that results in the characteristic rash
Traveler's diarrhea	Enterotoxigenic *Escherichia coli* and *Shigella* spp.	Enterotoxin causes excessive secretion of ions and water; diarrhea results

TABLE 15.3 Comparison of Exotoxins and Endotoxins

PROPERTY	EXOTOXIN	ENDOTOXIN
Bacterial source	Mostly from gram-positive bacteria	Gram-negative bacteria
Relation to microorganism	Metabolic product of growing cell	Present in LPS of outer membrane of cell wall and released only with destruction of cell
Chemistry	Protein or short peptide	Lipid portion (lipid A) of LPS of outer membrane
Heat stability	Unstable; can usually be destroyed at 60°–80°C (except staphylococcal enterotoxin)	Stable; can withstand autoclaving (121°C for one hour)
Toxicity (ability to cause disease)	High	Low
Immunology (relation to antibodies)	Can be converted to toxoids; neutralized by antitoxin	Not easily neutralized by antitoxin; therefore, effective toxoids cannot be made
Pharmacology (effect on body)	Specific for a particular cell structure or function in the host (mainly affects cell functions, nerves, and gastrointestinal tract)	General, such as fever, weaknesses, aches, and shock; all produce the same effects
Lethal dose	Small	Considerably larger
Representative diseases	Gas gangrene, tetanus, botulism, diphtheria, scarlet fever	Typhoid fever, urinary tract infections, and meningococcal meningitis

Recall from Chapters 4 and 8 that plasmids are small circular DNA molecules that are not connected to the main bacterial chromosome and are capable of independent replication. One group of plasmids, called resistance (R) factors, is responsible for the resistance of some microorganisms to antibiotics. In addition, a plasmid may carry the information that determines a microbe's pathogenicity. Some examples of virulence factors that are encoded by plasmid genes include tetanospasmin, heat-labile enterotoxin, and staphylococcal enterotoxin. Other examples include dextransucrase, an enzyme produced by *Streptococcus mutans* that is involved in tooth decay; adhesins and coagulase produced by *Staphylococcus aureus;* and a type of fimbria specific to enteropathogenic strains of *E. coli.*

In Chapter 13, we noted that some bacteriophages (viruses that infect bacteria) can incorporate their DNA into the bacterial chromosome, becoming a prophage, and thus remain latent and not cause lysis of the bacterium. Such a state is called *lysogeny.* One outcome of lysogeny is that the host bacterial cell and its progeny may exhibit new properties coded for by the bacteriophage DNA. Among the bacteriophage genes that contribute to pathogenicity are the genes for diphtheria toxin, erythrogenic toxins, staphylococcal enterotoxin and pyrogenic toxin, botulinum neurotoxin, and the capsule produced by *Streptococcus pneumoniae*.

PATHOGENIC PROPERTIES OF NONBACTERIAL MICROORGANISMS

LEARNING OBJECTIVES

- *List nine cytopathic effects of viral infections.*
- *Discuss the causes of symptoms in fungal, protozoan, helminthic, and algal diseases.*

Viruses

The pathogenic properties of viruses depend on their gaining access to a host, evading the host's defenses, and then causing damage to or death of the host cell (cytopathic effects) while reproducing themselves.

Viral Mechanisms for Evasion of Host Defenses

Viruses have a variety of mechanisms that allow them to evade destruction by the host's immune response (see Chapter 17). For example, viruses can penetrate and grow inside of host cells, where components of the immune system cannot reach them. Viruses gain access to cells because they have attachment sites for receptors on their target cells. When such an attachment site is brought together with an appropriate receptor, the virus can bind to and penetrate the cell. Some viruses gain access to host cells because their attachment sites mimic substances useful to those cells. For example, the attachment sites of rabies virus can mimic the neurotransmitter acetylcholine. As a result, the virus can enter the host cell along with the neurotransmitter.

The AIDS virus (HIV) goes further by hiding its attachment sites from the immune response and by attacking components of the immune system directly. Like some viruses, HIV is cell-specific; that is, it attacks only particular body cells. HIV attacks only cells that have a surface marker called the CD4 protein, most of which are cells of the immune system called T lymphocytes. Binding sites on HIV are complementary to the CD4 protein. The surface of the virus is folded to form ridges and valleys, and the HIV binding sites are located on the floors of the valleys. CD4 proteins are long enough and slender enough to reach these binding sites, whereas antibody molecules made against HIV are too large to make contact with the sites. As a result, it is difficult for these antibodies to destroy HIV.

Cytopathic Effects of Viruses

Infection of a host cell by an animal virus usually kills the host cell (see Chapter 12). Death can be caused by the accumulation of large numbers of multiplying viruses, by the effects of viral proteins on the permeability of the host cell's plasma membrane, or by inhibition of host DNA, RNA, or protein synthesis. The different abnormalities that can lead to damage or death of a host cell are known as **cytopathic effects (CPE)**. These effects are used to diagnose many viral infections.

Cytopathic effects vary with the virus. One difference is the point in the viral infection cycle at which the effects occur. Some viral infections result in early changes in the host cell; in other infections, changes are not seen until a much later stage. Moreover, some viruses cause *cytocidal effects* (changes resulting in cell death), whereas others are noncytocidal. A virus can produce one or more of the following cytopathic effects:

1. At some stage in their multiplication, cytocidal viruses cause the macromolecular synthesis within the host cell to stop. Some viruses, such as herpes simplex virus, irreversibly stop mitosis.

2. When a cytocidal virus infects a cell, it causes the cell's lysosomes to release their enzymes.

3. **Inclusion bodies** (Figure 15.5a) are granules found in the cytoplasm or nucleus of some infected cells. These granules are sometimes viral parts—nucleic acids or proteins in the process of being assembled into virions. The granules vary in size, shape, and staining properties, according to the virus. Inclusion bodies are characterized by their ability to stain with an acidic stain (acidophilic) or with a basic stain (basophilic). Other inclusion bodies arise at sites of earlier viral synthesis but do not contain assembled viruses or their components. Inclusion bodies are important because their presence can help to identify the virus causing an infection. For example, rabies virus produces inclusion bodies (Negri bodies) in the cytoplasm of nerve cells, and their presence in the brain tissue of animals suspected of being rabid has been used as a diagnostic tool for rabies. Diagnostic inclusion bodies are also associated with measles virus, vaccinia virus, smallpox virus, herpesvirus, and adenoviruses.

4. At times, several adjacent infected cells fuse to form very large cells called **syncytia** or **giant cells** (Figure 15.5b). These are produced from infections by paramyxoviruses.

5. Some viral infections result in changes in the host cell's functions with no visible changes in the infected cells. For example, if a virus changes the production level of a hormone, the cells normally affected by that hormone will not function properly.

6. Some virus-infected cells produce a substance called **interferon.** Viral infection induces cells to produce

(a) An inclusion body   **TEM** ⊢——⊣ 1 μm

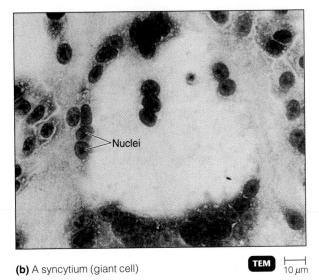

>Nuclei

(b) A syncytium (giant cell)   **TEM** ⊢——⊣ 10 μm

FIGURE 15.5 Some cytopathic effects of viruses.
(a) Cytoplasmic inclusion body (arrow) in a brain cell from a person who died of rabies. **(b)** The relatively clear area in the center is a syncytium (giant cell) formed in a culture of cells infected with measles virus. The numerous dark, oval bodies around the cell's edges are its multiple nuclei.

LM ⊢——⊣ 100 μm

FIGURE 15.6 Transformed cells in culture. In the center of this micrograph is a cluster of chick embryo cells transformed by Rous sarcoma virus. Such a cluster results from multiplication of a single cell infected with a transforming virus. Note how the transformed cells appear dark, in contrast to the monolayer of light, flat, normal cells around them. This appearance is caused by their spindle shapes and their uninhibited growth due to lack of contact inhibition.

TABLE 15.4 Cytopathic Effects of Selected Viruses	
VIRUS (GENUS)	**CYTOPATHIC EFFECT**
Poliovirus (*Enterovirus*)	Cell death
Papovavirus (family Papovaviridae)	Acidophilic inclusion bodies in nucleus
Adenovirus (*Mastadenovirus*)	Basophilic inclusion bodies in nucleus
Rhabdovirus (family Rhabdoviridae)	Acidophilic inclusion bodies in cytoplasm
Cytomegalovirus	Acidophilic inclusion bodies in nucleus and cytoplasm
Measles virus (*Morbillivirus*)	Cell fusion
Polyomavirus	Transformation
HIV (*Lentivirus*)	Destruction of T lymphocytes

interferon, but the host cell's DNA actually codes for the interferon. This substance protects neighboring uninfected cells from viral infection. More will be said about interferon in Chapter 16.

7. Many viral infections induce antigenic changes on the surface of the infected cells. These antigenic changes elicit a host antibody response against the infected cell, and thus they target the cell for destruction by the host's immune system.

8. Some viruses induce chromosomal changes in the host cell. Some viral infections result in chromosomal damage to the host cell, most often chromosomal breakage. Frequently oncogenes (cancer-causing genes) may be contributed or activated by a virus.

9. Most normal cells cease growing in vitro when they come close to another cell, a phenomenon known as *contact inhibition*. Viruses capable of causing cancer *transform* host cells, as discussed in Chapter 13. Transformation results in an abnormal, spindle-shaped cell that does not recognize contact inhibition (Figure 15.6). Loss of contact inhibition results in unregulated cell growth.

Some representative viruses that cause cytopathic effects are presented in Table 15.4. In subsequent chapters we will discuss the pathological properties of viruses in more detail.

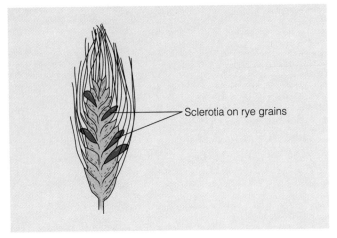

FIGURE 15.7 *Claviceps purpurea,* **the cause of ergotism.**
C. purpurea is a fungus that produces structures called sclerotia on grains such as rye. The sclerotia contain a toxin called ergot, which causes hallucinations and is a natural source of LSD.

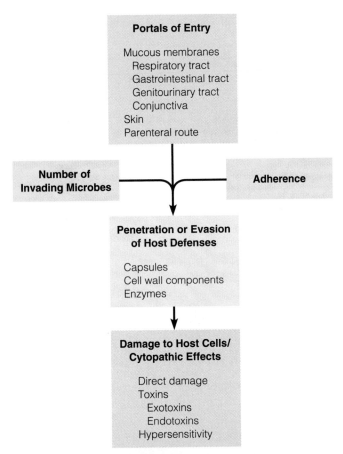

FIGURE 15.8 **Microbial mechanisms of pathogenicity.**
A summary of how microorganisms cause disease.

Fungi, Protozoa, Helminths, and Algae

This section describes some general pathological effects of fungi, protozoa, helminths, and algae that cause human disease. Most specific diseases caused by fungi, protozoa, and helminths, along with the pathological properties of these organisms, are discussed in detail in later chapters.

Fungi

Although fungi cause disease (see Chapter 12), they do not have a well-defined set of virulence factors. For example, *Cryptococcus neoformans* (krip-tō-kok′kus nē-ō-fôr′manz) is a fungus that causes a type of meningitis; it produces a capsule that helps it resist phagocytosis. Other fungi have metabolic products that are toxic to human hosts. In such cases, however, the toxin is only an indirect cause of disease, as the fungus is already growing in or on the host. Fungal infection can also provoke an allergic response in the host.

The disease called *ergotism,* which was common in Europe during the Middle Ages, is caused by a toxin produced by an ascomycete plant pathogen, *Claviceps purpurea* (kla′vi-seps pŭr-pŭ-rē′ä), that grows on grains (Figure 15.7). The toxin is contained in **sclerotia**, highly resistant portions of the fungus's mycelia that can detach. The toxin itself, **ergot,** is an alkaloid that can cause hallucinations resembling those produced by LSD; in fact, ergot is a natural source of LSD. Ergot also constricts capillaries and can cause gangrene of the extremities through prevention of proper blood circulation. Although *C. purpurea* still occasionally occurs on grains, modern milling usually removes the sclerotia.

A number of other toxins are produced by fungi that grow on grains or other plants. For example, peanut butter is occasionally recalled because of excessive amounts of **aflatoxin,** a toxin that has carcinogenic properties. Aflatoxin is produced by the growth of the mold *Aspergillus flavus.* When ingested, the toxin might be altered in a human's body to a mutagenic compound.

A few mushrooms produce toxins called **mycotoxins** (toxins produced by fungi). Examples are **phalloidin** and **amanitin,** produced by *Amanita phalloides* (am-an-ī′ta fal-loi′dēz), commonly known as the death angel. These neurotoxins are so potent that ingestion of the *Amanita* mushroom may result in death.

Protozoa

The presence of protozoa and their waste products often produces disease symptoms in the host (see Table 12.6, p. 318). Some protozoa can evade host defenses and cause disease for very long periods of time. For example, *Giardia,* which causes diarrhea, and *Trypanosoma,* which causes African trypanosomiasis (sleeping sickness), both have a mechanism that allows them to stay one step ahead of the host's immune system. The immune system is alerted to recognize foreign substances called *antigens;* the presence of antigens causes the immune system to produce antibodies designed to destroy them. When *Trypanosoma* is introduced into the bloodstream by a tsetse fly, it produces and displays a specific antigen. In response, the body produces antibodies against that antigen. However, within two weeks, the microbe stops displaying the original antigen and instead produces and displays a different one. Thus, the original antibodies are no longer effective. Because the microbe can make up to a thousand different antigens, such an infection can last for decades.

Helminths

The presence of helminths also often produces disease symptoms in a host (see Table 12.7, p. 325). Some of these organisms actually use host tissues for their own growth or produce large parasitic masses; the resulting cellular damage evokes the symptoms. An example is the roundworm *Wuchereria bancrofti* (vū-kėr-ār′ē-ä ban-krof′tē), the causative agent of elephantiasis. This parasite blocks lymphatic circulation, leading to the accumulation of lymph and eventually causing grotesque swelling of the legs and other body parts. Waste products of the metabolism of these parasites can also contribute to the symptoms of a disease.

Algae

A few species of algae produce neurotoxins. For example, some genera of dinoflagellates, such as *Gonyaulax,* are important medically because they produce a neurotoxin called **saxitoxin.** Although mollusks that feed on the dinoflagellates that produce saxitoxin show no symptoms of disease, people who eat the mollusks develop symptoms similar to botulism. Recall from Chapter 12 that extensive dinoflagellate growth is responsible for a situation called red tide. Public health agencies frequently prohibit human consumption of mollusks during red tides.

•   •   •

In the next chapter we will examine a group of nonspecific defenses of the host against disease. But before proceeding, examine Figure 15.8 very carefully. It summarizes some key concepts of the microbial mechanisms of pathogenicity we have discussed in this chapter.

STUDY OUTLINE

INTRODUCTION (p. 390)

1. Pathogenicity is the ability of a pathogen to produce a disease by overcoming the defenses of the host.
2. Virulence is the degree of pathogenicity.

ENTRY OF A MICROORGANISM INTO THE HOST (pp. 390–393)

1. The specific route by which a particular pathogen gains access to the body is called its portal of entry.

Portals of Entry (pp. 390–391)

1. Many microorganisms can penetrate mucous membranes of the conjunctiva and of the respiratory, gastrointestinal, and genitourinary tracts.
2. Microorganisms that are inhaled with droplets of moisture and dust particles gain access to the respiratory tract.
3. The respiratory tract is the most frequently used portal of entry.
4. Microorganisms enter the gastrointestinal tract via food, water, and contaminated fingers.
5. Most microorganisms cannot penetrate intact skin; they enter hair follicles and sweat ducts.
6. Some fungi infect the skin itself.
7. Some microorganisms can gain access to tissues by inoculation through the skin and mucous membranes in bites, injections, and other wounds.
8. This route of penetration is called the parenteral route.

Preferred Portal of Entry (p. 391)

1. Many microorganisms can cause infections only when they gain access through their specific portal of entry.

Numbers of Invading Microbes (p. 391)

1. Virulence can be expressed as LD_{50} (lethal dose for 50% of the inoculated hosts) or ID_{50} (infectious dose for 50% of the inoculated hosts).

Adherence (pp. 391–393)

1. Surface projections on a pathogen called adhesins (ligands) adhere to complementary receptors on the host cells.
2. Ligands can be glycoproteins or lipoproteins and are frequently associated with fimbriae.
3. Mannose is the most common receptor.

HOW BACTERIAL PATHOGENS PENETRATE HOST DEFENSES (pp. 393–395)

Capsules (p. 393)

1. Some pathogens have capsules that prevent them from being phagocytized.

Components of the Cell Wall (p. 393)

1. Proteins in the cell wall can facilitate adherence or prevent a pathogen from being phagocytized.
2. Some microbes can reproduce inside phagocytes.

Enzymes (pp. 393–395)

1. Leukocidins destroy neutrophils and macrophages.
2. Hemolysins lyse red blood cells.
3. Local infections can be protected in a fibrin clot caused by the bacterial enzyme coagulase.
4. Bacteria can spread from a focal infection by means of kinases (which destroy blood clots), hyaluronidase (which destroys a mucopolysaccharide that holds cells together), and collagenase (which hydrolyzes connective tissue collagen).

HOW BACTERIAL PATHOGENS DAMAGE HOST CELLS (pp. 395–398)

Direct Damage (p. 395)

1. Host cells can be destroyed when pathogens metabolize and multiply inside the host cells.

Production of Toxins (pp. 396–398)

1. Poisonous substances produced by microorganisms are called toxins; toxemia refers to symptoms caused by toxins in the blood.
2. The ability to produce toxins is called toxigenicity.
3. Exotoxins are produced by bacteria and released into the surrounding medium.
4. Exotoxins, not the bacteria, produce the disease symptoms.
5. Antibodies produced against exotoxins are called antitoxins.
6. Cytotoxins include diphtheria toxin (which inhibits protein synthesis) and erythrogenic toxins (which damage capillaries).
7. Neurotoxins include botulinum toxin (which prevents nerve transmission) and tetanus toxin (which prevents inhibitory nerve transmission).
8. *Vibrio cholerae* toxin and staphylococcal enterotoxin are enterotoxins, which induce fluid and electrolyte loss from host cells.
9. Endotoxins are lipopolysaccharides—the lipid A component of the cell wall of gram-negative bacteria.
10. Bacterial cell death, antibiotics, and antibodies may cause release of endotoxins.
11. Endotoxins cause fever (by inducing release of interleukin-1) and shock (because of TNF-induced decrease in blood pressure).
12. Endotoxins allow bacteria to cross the blood-brain barrier.
13. The *Limulus* amoebocyte lysate test is used to detect endotoxin in drugs and on medical devices.

PLASMIDS, LYSOGENY, AND PATHOGENICITY (pp. 398–400)

1. Plasmids may carry genes for antibiotic resistance, toxins, capsules, and fimbriae.
2. Lysogeny can result in bacteria with virulence factors, such as toxins or capsules.

PATHOGENIC PROPERTIES OF NONBACTERIAL MICROORGANISMS (pp. 400–403)

Viruses (pp. 400–401)

1. Viruses avoid the host's immune response by growing inside cells.
2. Viruses gain access to host cells because they have attachment sites for receptors on the host cell.
3. Signs of viral infections are called cytopathic effects (CPE).
4. Some viruses cause cytocidal effects (cell death), and others cause noncytocidal effects.
5. Cytopathic effects include the stopping of mitosis, lysis, formation of inclusion bodies, cell fusion, antigenic changes, chromosomal changes, and transformation.

Fungi, Protozoa, Helminths, and Algae (pp. 402–403)

1. Symptoms of fungal infections can be caused by capsules, toxins, and allergic responses.
2. Symptoms of protozoan and helminthic diseases can be caused by damage to host tissue or by the metabolic waste products of the parasite.
3. Some protozoa change their surface antigens while growing in a host so that the host's antibodies don't kill the protozoa.
4. Some algae produce neurotoxins that cause paralysis when ingested by humans.

STUDY QUESTIONS

Review

1. List three portals of entry, and describe how microorganisms gain access through each.

2. Compare pathogenicity with virulence.

3. Explain how drugs that bind each of the following would affect pathogenicity:
   (a) Mannose on human cell membranes
   (b) *Neisseria gonorrhoeae* fimbriae
   (c) *Streptococcus pyogenes* M protein

4. Define cytopathic effects, and give five examples.

5. Compare and contrast the following aspects of endotoxins and exotoxins: bacterial source, chemistry, toxicity, and pharmacology. Give an example of each toxin.

6. How are capsules and cell wall components related to pathogenicity? Give specific examples.

7. Describe how hemolysins, leukocidins, coagulase, kinases, and hyaluronidase might contribute to pathogenicity.

8. Describe the factors contributing to the pathogenicity of fungi, protozoa, and helminths.

9. Which of the following genera is the most infectious?

GENUS	ID_{50}
Escherichia	10^8 cells
Salmonella	10^5 cells
Shigella	200 cells
Treponema	52 cells

10. The LD_{50} of botulinum toxin is 0.000025 μg. The LD_{50} of Salmonella toxin is 200 μg. Which of these is the more potent toxin? How can you tell from the LD_{50} values?

11. Food poisoning can be divided into two categories: food infection and food intoxication. On the basis of toxin production by bacteria, explain the difference between these two categories.

12. How can viruses and protozoa avoid being killed by the host's immune response?

Critical Thinking

1. How can plasmids and lysogeny turn the normally harmless *E. coli* into a pathogen? Using the following graph, which shows confirmed cases of enteropathogenic *E. coli,* what is the usual portal of entry for this bacterium?

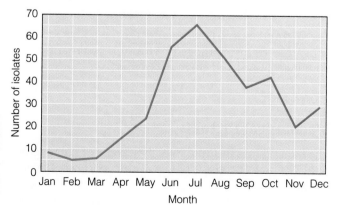

2. The cyanobacterium *Microcystis aeruginosa* produces a peptide that is toxic to humans. According to the following graph, when is this bacterium most toxic?

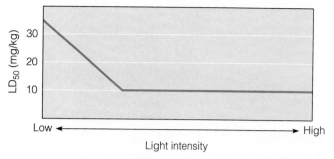

3. How do each of the following strategies contribute to the virulence of the pathogen? What disease does each organism cause?

STRATEGY	PATHOGEN
Changes its cell wall after entry into host	Yersinia pestis
Uses urea to produce ammonia	Helicobacter pylori
Causes host to make more receptors	Rhinovirus
Binds to site for epidermal growth factor on human cells	Salmonella

Clinical Applications

1. On July 8, a woman was given an antibiotic for presumptive sinusitis. However, her condition worsened, and she was unable to eat for 4 days because of severe pain and tightness of the jaw. On July 12 she was admitted to a hospital with severe facial spasms. She reported that on July 5 she incurred a puncture wound at the base of her great toe; she cleaned the wound but did not seek medical attention. What caused her symptoms? Was her condition due to an infection or an intoxication? Can she transmit this condition to another person?

2. Explain whether each of the following examples is a food infection or intoxication. What is the probable etiological agent in each case?
   (a) Eighty-two people who ate shrimp at a dinner in Port Allen, Louisiana, developed diarrhea, cramps, weakness, nausea, chills, headache, and fever from 4 hours to 2 days after eating.
   (b) Two people in Vermont who ate barracuda caught in Florida developed malaise, nausea, blurred vision, breathing difficulty, and numbness from 3 to 6 hours after eating.

3. Washwater containing *Pseudomonas* was sterilized and used to wash cardiac catheters. Three patients undergoing cardiac catheterization developed fever, chills, and hypotension. The water and catheters were sterile. Why did the patients show these reactions?

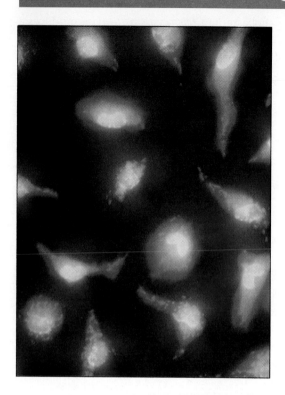

Nonspecific Defenses of the Host

FROM OUR DISCUSSION to this point, you can see that pathogenic microorganisms are endowed with special properties that enable them to cause disease if given the right opportunity. If microorganisms never encountered resistance from the host, we would constantly be ill and would eventually die of various diseases, but in most cases our body defenses prevent this from happening. Some of these defenses are designed to keep out microorganisms altogether; other defenses remove the microorganisms if they do get in; and still others combat them if they remain inside. Our ability to ward off disease through our defenses is called **resistance**. Vulnerability or lack of resistance is known as **susceptibility**.

In discussing resistance, we will divide our body defenses into two general kinds: nonspecific and specific (Figure 16.1). *Nonspecific resistance* refers to defenses that protect us from *any* pathogen, regardless of species. These include a first line of defense (skin and mucous membranes) and a second line of defense (phagocytes, inflammation, fever, and antimicrobial substances). *Specific resistance,* or *immune response,* is a third line of defense that the body offers against a particular pathogen. Specific defenses are based on specialized cells of the immune system called lymphocytes and the production of specific proteins called antibodies. Chapter 17 deals with specific resistance. In this chapter we consider nonspecific resistance.

SKIN AND MUCOUS MEMBRANES

LEARNING OBJECTIVES

- *Define resistance and susceptibility.*
- *Define nonspecific resistance.*
- *Describe the role of the skin and mucous membranes in nonspecific resistance, and differentiate between mechanical and chemical factors.*

The skin and mucous membranes are the body's first line of defense against pathogens. This function results from both mechanical and chemical factors.

Mechanical Factors

The intact skin is the human body's largest organ in terms of surface area. It consists of two distinct portions: the dermis and the epidermis (Figure 16.2). The *dermis,* the skin's inner, thicker portion, is composed of connective tissue. The *epidermis,* the outer, thinner portion, is in direct contact with the external environment. The epidermis consists of layers of continuous sheets of tightly packed epithelial cells with little or no material between the cells.

FIGURE 16.1 An overview of the body's defenses.

Nonspecific Defense Mechanisms		Specific Defense Mechanisms (Responses of the Immune System)
First line of defense	**Second line of defense**	**Third line of defense**
• Intact skin • Mucous membranes and their secretions	• Phagocytic white blood cells • Inflammation and fever • Antimicrobial substances	• Specialized lymphocytes: B cells and T cells • Antibodies

Two types of cells in the epidermis—Langerhans and Granstein cells—participate in immunity by assisting helper T (T_H) cells and suppressor T (T_S) cells, respectively. (These cells are discussed in Chapter 17.) The top layer of epidermal cells is dead and contains a waterproofing protein called **keratin.**

If we consider the closely packed cells, continuous layering, and presence of keratin, we can see why the intact skin provides such a formidable physical barrier to the entrance of microorganisms. The intact surface of healthy epidermis is rarely, if ever, penetrated by microorganisms. But when the epithelial surface is broken, a subcutaneous (below-the-skin) infection often develops. The bacteria most likely to cause such an infection are the staphylococci that normally inhabit the epidermis, hair follicles, and sweat and oil glands of the skin. Infections of the skin and underlying tissues frequently occur as a result of burns, cuts, stab wounds, or other conditions that break the skin. Moreover, when the skin is moist, as in hot, humid climates, skin infections are quite common, especially fungal infections such as athlete's foot.

Mucous membranes also consist of an epithelial layer and an underlying connective tissue layer. Mucous membranes line the entire gastrointestinal, respiratory, urinary, and reproductive tracts. The epithelial layer of a mucous membrane secretes a fluid called *mucus*, which prevents the tracts from drying out. Some pathogens that can thrive on the moist secretions of a mucous membrane are able to penetrate the membrane if the microorganism is present in sufficient numbers. *Treponema pallidum* is such a pathogen. This penetration may be facilitated by toxic substances produced by the microorganism, prior injury by viral infection, or mucosal irritation. Although mucous membranes inhibit the entrance of many microorganisms, they offer less protection than the skin.

Besides the physical barrier presented by the skin and mucous membranes, several other mechanical factors help protect certain epithelial surfaces. One such mechanism that protects the eyes is the *lacrimal apparatus* (Figure 16.3), a group of structures that manufactures and drains away tears. The lacrimal glands, located toward the upper, outermost portion of each eye socket, produce the

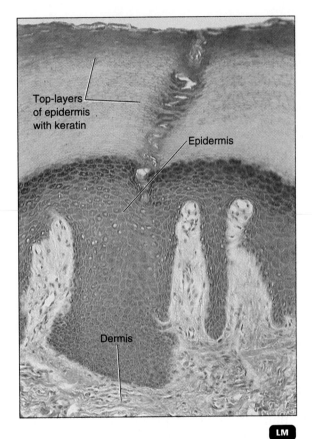

FIGURE 16.2 Section through human skin. The thin layers at the top of the photo contain keratin. These layers and the darker purple cells beneath them make up the epidermis. The lighter purple material near the bottom part of the photo is the dermis.

tears and pass them under the upper eyelid. From here, tears pass toward the corner of the eye near the nose and into two small holes that lead through tubes (lacrimal canals) to the nose. The tears are spread over the surface of the eyeball by blinking. Normally, the tears evaporate or pass into the nose as fast as they are produced. This continual washing action helps to keep microorganisms from settling on the surface of the eye. If an irritating substance

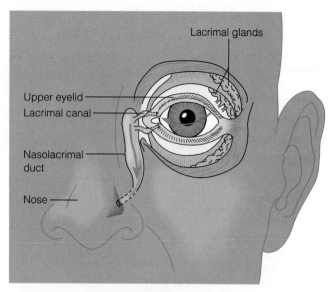

FIGURE 16.3   The lacrimal apparatus. The washing action of the tears, shown by the arrow passing over the surface of the eyeball, prevents microorganisms from settling on the surface of the eyeball. Tears produced by the lacrimal glands pass across the surface of the eyeball into two small holes that convey the tears into the lacrimal canals and the nasolacrimal duct.

or large numbers of microorganisms come in contact with the eye, the lacrimal glands start to secrete heavily, and the tears accumulate more rapidly than they can be carried away. This excessive production is a protective mechanism because the excess tears dilute and wash away the irritating substance or microorganisms.

In a cleansing action very similar to that of tears, saliva is produced by the salivary glands. Saliva helps to dilute the numbers of microorganisms and wash them from both the surface of the teeth and the mucous membrane of the mouth. This helps prevent colonization by microbes.

The respiratory and gastrointestinal tracts have many mechanical forms of defense. Mucus is slightly viscous (thick); thus, it traps many of the microorganisms that enter the respiratory and gastrointestinal tracts. The mucous membrane of the nose also has mucus-coated hairs that filter inhaled air and trap microorganisms, dust, and pollutants. The cells of the mucous membrane of the lower respiratory tract are covered with cilia. By moving synchronously, these cilia propel inhaled dust and microorganisms that have become trapped in mucus upward toward the throat. This so-called **ciliary escalator** (Figure 16.4) keeps the mucus blanket moving toward the throat at a rate of 1–3 cm per hour; coughing and sneezing speed up the escalator. (Some substances in cigarette smoke are toxic to cilia and can seriously impair the functioning of the ciliary escalator.) Microorganisms are also prevented from entering the lower respiratory tract by a small lid of cartilage called the epiglottis, which covers the larynx (voicebox) during swallowing.

The cleansing of the urethra by the flow of urine is an-

other mechanical factor that prevents microbial colonization in the urinary system. Vaginal secretions likewise move microorganisms out of the female body.

Chemical Factors

Mechanical factors alone do not account for the high degree of resistance of skin and mucous membranes to microbial invasion. Certain chemical factors also play important roles.

Sebaceous (oil) glands of the skin produce an oily substance called **sebum** that prevents hair from drying and becoming brittle. Sebum also forms a protective film over the surface of the skin. One of the components of sebum is unsaturated fatty acids, which inhibit the growth of certain pathogenic bacteria and fungi. The low pH of the skin, between pH 3 and 5, is caused in part by the secretion of fatty acids and lactic acid. The skin's acidity probably discourages the growth of many other microorganisms.

Bacteria that live as commensals on the skin decompose sloughed-off skin cells, and the resultant organic molecules and the end-products of their metabolism produce body odor. As you will see in Chapter 21, certain bacteria commonly found on the skin metabolize sebum, and this metabolism forms free fatty acids that cause the inflammatory response associated with acne. Isotretinoin (Accutane™), a derivative of vitamin A that prevents sebum formation, is a treatment for a very severe type of acne called cystic acne.

The sweat glands of the skin produce perspiration, which helps to maintain body temperature, eliminate certain wastes, and flush microorganisms from the surface of the skin. Perspiration also contains **lysozyme,** an enzyme capable of breaking down cell walls of gram-positive bacteria and, to a lesser extent, gram-negative bacteria (see Figure 4.12). Lysozyme is also found in tears, saliva, nasal secretions, and tissue fluids, where it exhibits its antimicrobial activity. Alexander Fleming was actually studying lysozyme in 1929 when he accidentally discovered the antimicrobial effects of penicillin (see Figure 1.5).

Gastric juice is produced by the glands of the stomach. It is a mixture of hydrochloric acid, enzymes, and mucus. The very high acidity of gastric juice (pH 1.2 to 3.0) is sufficient to preserve the usual sterility of the stomach. This acidity destroys bacteria and most bacterial toxins, except those of *Clostridium botulinum* and *Staphylococcus aureus*. However, many enteric pathogens are protected by food particles and are able to enter the intestines via the gastrointestinal tract. The bacterium *Helicobacter pylori* neutralizes stomach acid so it can grow in the stomach. Its growth results in gastritis and ulcers.

Blood also contains antimicrobial chemicals. For example, iron-binding proteins, called **transferrins,** inhibit bacterial growth by reducing the amount of available iron.

Next we will discuss phagocytosis, commonly regarded as the body's second line of defense against infection.

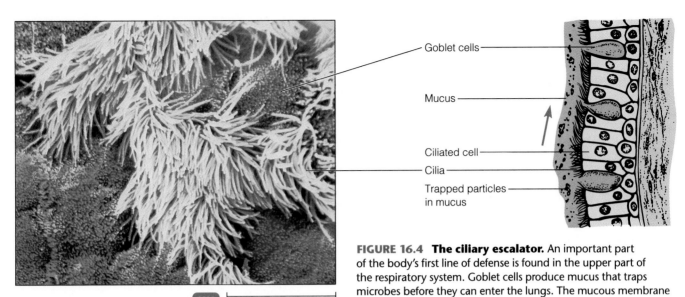

FIGURE 16.4 **The ciliary escalator.** An important part of the body's first line of defense is found in the upper part of the respiratory system. Goblet cells produce mucus that traps microbes before they can enter the lungs. The mucous membrane is also equipped with ciliated cells. Synchronized beating of the cilia expels mucus and the trapped microbes.

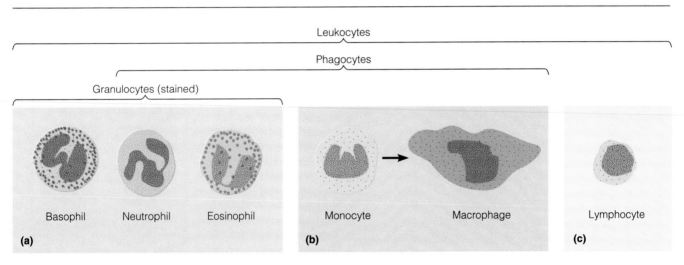

FIGURE 16.5 **Principal kinds of leukocytes.** (a) Granulocytes can be distinguished by staining. (b) Monocytes mature into actively phagocytic macrophages. Neutrophils, eosinophils, and macrophages are the phagocytic cells of the nonspecific defense system. (c) Lymphocytes are important in specific immunity.

PHAGOCYTOSIS

LEARNING OBJECTIVES

- *Define phagocytosis, and include the stages of adherence and ingestion.*
- *Classify phagocytic cells and describe the roles of granulocytes and monocytes.*

Phagocytosis (from the Greek words for *eat* and *cell*) is the ingestion of a microorganism or any particulate matter by a cell. We have previously mentioned phagocytosis as the method of nutrition of certain protozoa. In this chapter, phagocytosis is discussed as a means by which cells in the human body counter infection. The human cells that perform this function are collectively called **phagocytes.** All are types of white blood cells or derivatives of white blood cells. Before we look at phagocytes specifically, let us consider the various components of the blood more generally.

Formed Elements in Blood

Blood consists of a fluid called **plasma,** which contains **formed elements**—that is, cells and cell fragments (Table 16.1). Of the cells listed in Table 16.1, those that concern us at present are the **leukocytes,** or white blood cells (Figure 16.5).

TABLE 16.1 Formed Elements in Blood

TYPE OF CELL	ILLUSTRATION	NUMBERS PER CUBIC MILLIMETER (mm³)	FUNCTION
Erythrocytes (red blood cells)		4.8 to 5.4 million	Transport of O_2 and CO_2
Leukocytes (white blood cells)		5000 to 9000	
A. Granulocytes (stained) 1. Neutrophils (PMNs) (60% to 70% of leukocytes)			Phagocytosis
2. Basophils (0.5% to 1%)			Production of heparin and histamine
3. Eosinophils (2% to 4%)			Production of toxic proteins against certain parasites; some phagocytosis
B. Monocytes (3% to 8%)			Phagocytosis (when they mature into macrophages)
C. Lymphocytes (20% to 25%)			Antibody production (B lymphocytes); cell-mediated immunity (T lymphocytes)*
Platelets		250,000 to 400,000	Blood clotting

*Discussed in Chapter 17.

During many kinds of infections, especially bacterial infections, the total number of white blood cells increases; this increase is called *leukocytosis*. During the active stage of infection, the leukocyte count might double, triple, or quadruple, depending on the severity of the infection. Diseases that might cause such an elevation in the leukocyte count include meningitis, infectious mononucleosis, appendicitis, pneumococcal pneumonia, and gonorrhea. Other diseases, such as salmonellosis, brucellosis, pertussis, and some viral and rickettsial infections, may cause a *decrease* in the leukocyte count (called *leukopenia*). Leukopenia may be related to either impaired white blood cell production or the effect of increased sensitivity of white blood cell membranes to damage by complement, antimicrobial plasma proteins discussed later in the chapter. Leukocyte increase or decrease can be detected by a **differential white blood cell count**, which is a calculation of the percentage of each kind of white cell in a sample of 100 white blood cells. The percentages in a normal differential white blood cell count are shown in parentheses in the first column of Table 16.1.

Leukocytes are divided into three categories: granulocytes, monocytes, and lymphocytes. **Granulocytes** include some phagocytic types. They owe their name to the presence of large granules in their cytoplasm that can be seen under a light microscope. They are differentiated into three types on the basis of how the granules stain. The granules of **neutrophils** stain red and blue with a mixture of acidic and basic dyes, respectively; those of **basophils** stain blue with the basic dye methylene blue; and those of **eosinophils** stain red with the acidic dye eosin.

Neutrophils are also commonly called *polymorphonuclear leukocytes (PMNs)*, or *polymorphs*. The term *polymorphonuclear* refers to the fact that the nuclei of neutrophils contain from two to five lobes. Neutrophils, which are highly phagocytic and motile, are active in the initial stages of an infection. They have the ability to leave the blood, enter an infected tissue, and destroy microbes and foreign particles.

The role of basophils is not clear. However, they release substances, such as histamine, that are important in in-

FIGURE 16.6 Components of the lymphatic system. Fluid circulating between tissue cells (interstitial fluid) is picked up by lymphatic vessels. The fluid is then called lymph. Microorganisms can enter the lymphatic vessels. Macrophages in the lymph nodes remove microorganisms from the lymph. Lymph is transported into larger lymphatic vessels, to be collected in the right lymphatic duct and thoracic duct. Eventually, the lymph is returned to the blood at the subclavian veins, just before the blood enters the heart.

Tonsil

Right lymphatic duct

Right subclavian vein

Thymus gland

Lymphatic vessel

Left subclavian vein

Thoracic duct

Spleen

Small intestine

Peyer's patch

Lymph node

Bone marrow

Lymphatic vessel

flammation and allergic responses. They also release heparin, an anticoagulant.

Eosinophils are somewhat phagocytic and also have the ability to leave the blood. Their major function is to produce toxic proteins against certain parasites, such as helminths. Although eosinophils are physically too small to ingest and destroy helminths, they can attach to the outer surface of the parasites and discharge peroxide ions that destroy them. Their number increases significantly during certain parasitic worm infections and hypersensitivity (allergy) reactions.

Monocytes lack granules in their cytoplasm and are not actively phagocytic until they leave circulating blood, enter body tissues, and mature into **macrophages.** In fact, the maturation and proliferation of macrophages (along with lymphocytes) is one factor responsible for the swelling of lymph nodes during an infection.

Lymphocytes are not phagocytic but play a key role in specific immunity (see Chapter 17). They occur in lymphoid tissues of the lymphatic system: in the tonsils, lymph nodes, spleen, thymus gland, thoracic duct, bone marrow, appendix, Peyer's patches of the small intestine, and lymph nodes in the respiratory, gastrointestinal, and reproductive tracts (Figure 16.6). They also circulate in the blood.

Actions of Phagocytic Cells

When an infection occurs, both granulocytes (especially neutrophils) and monocytes migrate to the infected area. During this migration, monocytes enlarge and develop into actively phagocytic macrophages (Figure 16.7). Because these cells leave the blood and migrate through tissue to infected areas, they are called *wandering macrophages.* Some macrophages, called *fixed macrophages* or

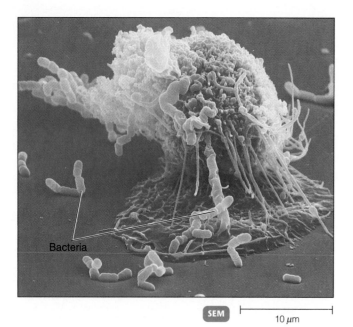

Bacteria

SEM | 10 μm

FIGURE 16.7   A macrophage engulfing rod-shaped bacteria.

TABLE 16.2 Classification and Functions of Phagocytes		
TYPE OF PHAGOCYTE	**FUNCTIONAL CELLS**	**FUNCTIONS**
Granulocytes	Neutrophils and eosinophils	Phagocytic against microbes during the initial phase of infection
Monocytes	Mononuclear phagocytic system: wandering macrophages and fixed macrophages (histiocytes)	Phagocytic against microbes as infection progresses and against worn-out blood cells as infection subsides; also involved in cell-mediated immunity (see Chapter 17)

histiocytes, are located in certain tissues and organs of the body. Fixed macrophages are found in the liver (Kupffer's cells), lungs (alveolar macrophages), nervous system (microglial cells), bronchial tubes, spleen, lymph nodes, bone marrow, and the peritoneal cavity surrounding abdominal organs. The various macrophages of the body constitute the **mononuclear phagocytic (reticuloendothelial) system.**

During the course of an infection, a shift occurs in the type of white blood cell that predominates in the bloodstream. Granulocytes, especially neutrophils, predominate during the initial phase of infection, at which time they are actively phagocytic; this dominance is indicated by their number in a differential white blood cell count. However, as the infection progresses, the macrophages predominate; they scavenge and phagocytize remaining living bacteria and dead or dying bacteria. The increased number of monocytes (which develop into macrophages) is reflected in a differential white blood cell count. As blood and lymph that contain microorganisms pass through organs with fixed macrophages, cells of the mononuclear phagocytic system remove the microorganisms by phagocytosis. The mononuclear phagocytic system also disposes of worn-out blood cells.

Table 16.2 summarizes the different types of phagocytic cells and their functions.

Mechanism of Phagocytosis

How does phagocytosis occur? For convenience of study, we will divide phagocytosis (Figure 16.8) into four main phases: chemotaxis, adherence, ingestion, and digestion.

Chemotaxis

① **Chemotaxis** is the chemical attraction of phagocytes to microorganisms. (The mechanism of chemotaxis is discussed in Chapter 4.) Among the chemotactic chemicals that attract phagocytes are microbial products, components of white blood cells and damaged tissue cells, and peptides derived from complement, a system of host defense discussed later in this chapter.

Adherence

As it pertains to phagocytosis, **adherence** is the attachment of the phagocyte's plasma membrane to the surface of the microorganism or other foreign material. In some instances, adherence occurs easily, and the microorganism is readily phagocytized. However, adherence can be hampered by the presence of large capsules or M protein. As mentioned in Chapter 15, the M protein of *Streptococcus pyogenes* inhibits the attachment of phagocytes to their surfaces and makes adherence more difficult. Organisms with large capsules include *Streptococcus pneumoniae* and *Hemophilus influenzae* type b. Heavily encapsulated microorganisms like these can be phagocytized only if the phagocyte traps the microorganism against a rough surface, such as a blood vessel, blood clot, or connective tissue fiber, from which the microbe cannot slide away.

Microorganisms can be more readily phagocytized if they are first coated with certain plasma proteins that promote the attachment of the microorganisms to the phagocyte. This coating process is called **opsonization.** The proteins that act as *opsonins* include some components of the complement system and antibody molecules (described later in this chapter and in Chapter 17).

Ingestion

② Following adherence, **ingestion** occurs. During the process of ingestion, the plasma membrane of the phagocyte extends projections called pseudopods that engulf the microorganism. ③ Once the microorganism is surrounded, the pseudopods meet and fuse, surrounding the

Microbe or other particle

Plasma membrane

Pseudopod

Cytoplasm

Lysosome

Digestive enzymes

Phagocyte

Phagosome (phagocytic vesicle)

Phagolysosomes

Partially digested microbe

Residual body

Indigestible material

1 Chemotaxis and adherence of microbe to phagocyte.

2 Ingestion of microbe by phagocyte.

3 Formation of a phagosome.

4 Fusion of the phagosome with a lysosome to form a phagolysosome.

5 Digestion of ingested microbe by enzymes.

6 Formation of residual body containing indigestible material.

7 Discharge of waste materials.

FIGURE 16.8 The mechanism of phagocytosis in a phagocyte.

microorganism with a sac called a *phagosome* or *phagocytic vesicle*.

Digestion

In this phase of phagocytosis, the phagosome pinches off from the plasma membrane and enters the cytoplasm. Within the cytoplasm, it contacts lysosomes that contain digestive enzymes and bactericidal substances (see Chapter 4). ④ Upon contact, the phagosome and lysosome membranes fuse to form a single, larger structure called a *phagolysosome*. ⑤ The contents of the phagolysosome take only 10–30 minutes to kill most types of bacteria. Microbial destruction probably occurs because of the contents of the lysosomes.

Lysosomal enzymes that attack microbial cells directly include lysozyme, which hydrolyzes peptidoglycan in bacterial cell walls. A variety of other enzymes, such as lipases, proteases, ribonuclease, and deoxyribonuclease, hydrolyze other macromolecular components of microorganisms. The hydrolytic enzymes are most active at about pH 4, which is the phagolysosome's usual pH because of lactic acid produced by the phagocyte. Lysosomes also contain enzymes that can produce toxic oxygen products such as superoxide radical ($O_2^-\cdot$), hydrogen peroxide (H_2O_2), singlet oxygen ($^1O_2$), and hydroxyl radical ($OH\cdot$) (see Chapter 6). Other enzymes can make use of these toxic oxygen products in killing ingested microorgan-

isms. For example, the enzyme myeloperoxidase converts chloride (Cl^-) ions and hydrogen peroxide into highly toxic hypochlorous acid (HOCl). The acid contains hypochlorous ions, which are found in household bleach and account for its antimicrobial activity.

⑥ After enzymes have digested the contents of the phagolysosome brought into the cell by ingestion, the phagolysosome contains indigestible material and is called a *residual body*. ⑦ This residual body then moves toward the cell boundary and discharges its wastes outside the cell.

Not all phagocytized microorganisms are killed by lysosomal enzymes. The toxins of some microbes, such as toxin-producing staphylococci and the dental plaque bacterium *Actinobacillus*, can actually kill phagocytes. Other microorganisms such as *Chlamydia, Shigella, Mycobacterium tuberculosis, Leishmania* (a protozoan), and malarial parasites can evade components of the immune system by entering phagocytes. In addition, these microbes can prevent both the fusion of a phagosome with a lysosome and the proper acidification of digestive enzymes. The microbes then multiply within the phagocyte, almost completely filling it. In most cases the phagocyte dies and the microbes are released by autolysis to infect other cells. Still other microbes, such as the causative agents of tularemia and brucellosis, can remain dormant for months or years at a time.

Phagocytosis, in addition to providing a nonspecific defense of the host, plays a role in immunity. Macrophages help T and B lymphocytes perform vital functions in immunity. In Chapter 17 we will discuss in more detail how phagocytosis supports immunity.

In the next section we will see how phagocytosis often occurs as part of another nonspecific mechanism of resistance: inflammation.

INFLAMMATION

LEARNING OBJECTIVES

- *Describe the stages of inflammation and their relation to nonspecific resistance.*
- *Describe phagocyte migration and phagocytosis.*

Damage to the body's tissues triggers a defensive response called **inflammation**. The damage can be caused by microbial infection, physical agents (such as heat, radiant energy, electricity, or sharp objects), or chemical agents (acids, bases, and gases). Inflammation is usually characterized by four symptoms: redness, pain, heat, and swelling. Sometimes a fifth symptom, loss of function, is present; its occurrence depends on the site and extent of damage. In apparent contradiction to the symptoms observed, however, the inflammatory response is beneficial. Inflammation has the following functions: (1) to destroy the injurious agent, if possible, and to remove it and its by-products from the body; (2) if destruction is not possible, to limit the effects on the body by confining or walling off the injurious agent and its by-products; and (3) to repair or replace tissue damaged by the injurious agent or its by-products.

For purposes of our discussion, we will divide the process of inflammation into three stages: vasodilation and increased permeability of blood vessels, phagocyte migration and phagocytosis, and tissue repair.

Vasodilation and Increased Permeability of Blood Vessels

Immediately following tissue damage, blood vessels dilate in the area of damage, and their permeability increases (Figure 16.9a and b). **Vasodilation** is an increase in the diameter of blood vessels. Vasodilation increases blood flow to the damaged area and is responsible for the redness (erythema) and heat associated with inflammation.

Increased permeability permits defensive substances normally retained in the blood to pass through the walls of the blood vessels and enter the injured area. The increase in permeability, which permits fluid to move from the blood into tissue spaces, is responsible for the **edema** (swelling) of inflammation. The pain of inflammation can be caused by nerve damage, irritation by toxins, or the pressure of edema.

① Vasodilation and the increase in permeability of blood vessels are caused by chemicals released by damaged cells in response to injury. One such substance is **histamine**, a chemical present in many cells of the body, especially in mast cells in connective tissue, circulating basophils, and blood platelets. Histamine is released in direct response to the injury of cells that contain it; it is also released in response to stimulation by certain components of the complement system (to be discussed later). Phagocytic granulocytes attracted to the site of injury can also produce chemicals that cause the release of histamine.

Kinins are another group of substances that cause vasodilation and increase the permeability of blood vessels. These chemicals are present in blood plasma and, once activated, also attract phagocytic granulocytes—chiefly neutrophils—to the injured area.

Prostaglandins (substances released by damaged cells) and **leukotrienes** (substances related to prostaglandins), produced by mast cells and basophils, are other chemicals that cause vasodilation.

Vasodilation and the increase in permeability of blood vessels also help to deliver clotting elements of blood into the injured area. ② The blood clots that form around the site of activity prevent the microorganism (or its toxins) from spreading to other parts of the body. ③ As a result, there may be a localized collection of pus in a cavity formed by the breakdown of body tissues. This focus of infection is called an **abscess**. Common abscesses include pustules and boils.

The next stage in inflammation involves the migration of phagocytes to the injured area.

Phagocyte Migration and Phagocytosis

Generally, within an hour after the process of inflammation is initiated, phagocytes appear on the scene (Figure 16.9c). ① As the flow of blood gradually decreases, phagocytes (both neutrophils and monocytes) begin to stick to the inner surface of the endothelium (lining) of blood vessels. This sticking process is called **margination**. ② Then the collected phagocytes begin to squeeze between the endothelial cells of the blood vessel to reach the damaged area. This migration, which resembles amoeboid movement, is called **diapedesis**, or cell walking; the migratory process can take as little as two minutes. ③ The phagocytes then begin to destroy invading microorganisms by phagocytosis.

As mentioned earlier, certain chemicals attract neutrophils to the site of injury (chemotaxis). These include chemicals produced by microorganisms and even other neutrophils; other chemicals are kinins and components

(a) Tissue damage

Bacteria entering on knife

Epidermis

Dermis

Subcutaneous tissue

Blood vessel

Nerve

1 Chemicals such as histamine, kinins, prostaglandins, and leukotrienes (represented as blue dots) are released by damaged cells

2 Blood clot forms

3 Abscess starts to form (yellow area)

(b) Vasodilation and increased permeability of blood vessels

Blood vessel endothelium

Neutrophil

Monocyte

Bacterium

Erythrocyte

1 Margination— phagocytes stick to endothelium

2 Diapedesis— phagocytes squeeze between endothelial cells

3 Phagocytosis of invading bacteria

Bacterium

Neutrophil   Macrophage

(c) Phagocyte migration and phagocytosis

Scab

Blood clot

Regenerated epidermis (parenchyma)

Regenerated dermis (stroma)

(d) Tissue repair

of the complement system. The availability of a steady stream of neutrophils is ensured by the production and release of additional granulocytes from bone marrow.

As the inflammatory response continues, monocytes follow the granulocytes into the infected area. Once the monocytes are contained in the tissue, they undergo changes in biological properties and become wandering macrophages. The granulocytes predominate in the early stages of infection but tend to die off rapidly. Macrophages enter the picture during a later stage of the infection, once granulocytes have accomplished their function. They are several times more phagocytic than granulocytes and are large enough to phagocytize tissue that has been destroyed, granulocytes that have been destroyed, and invading microorganisms.

After granulocytes or macrophages engulf large numbers of microorganisms and damaged tissue, they themselves eventually die. After a few days an area containing dead phagocytes and damaged tissue forms in the inflamed tissue. This collection of dead cells and various body fluids is called **pus.** Pus formation usually continues until the infection subsides. At times, the pus pushes to the surface of the body or into an internal cavity for dispersal. On other occasions the pus remains even after the infection is terminated. In this case, the pus is gradually destroyed over a period of days and is absorbed by the body.

FIGURE 16.9 The process of inflammation. (a) Damage to otherwise healthy tissue, in this case skin. **(b)** Vasodilation and increased permeability. **(c)** Phagocyte migration and phagocytosis of bacteria and cellular debris by macrophages and neutrophils. Macrophages develop from monocytes. **(d)** Repair of damaged tissue.

Tissue Repair

The final stage of inflammation is tissue repair, the process by which tissues replace dead or damaged cells (Figure 16.9d). Repair begins during the active phase of inflammation, but it cannot be completed until all harmful substances have been removed or neutralized at the site of injury. The ability of a tissue to regenerate, or repair itself, depends on the type of tissue. For example, skin has a high capacity for regeneration, whereas cardiac muscle tissue does not regenerate at all.

A tissue is repaired when its stroma or parenchyma produces new cells. The *stroma* is the supporting connective tissue, and the *parenchyma* is the functioning part of the tissue. For example, the capsule around the liver that encloses and protects it is part of the stroma because it is not involved in the functions of the liver; liver cells (hepatocytes) that perform the functions of the liver are part of the parenchyma. If only parenchymal cells are active in repair, a perfect or near-perfect reconstruction of the tissue occurs. A familiar example of perfect reconstruction is a minor skin cut, in which parenchymal cells are more active in repair. However, if repair cells of the stroma of the skin are more active, scar tissue is formed.

FEVER

LEARNING OBJECTIVE

- *Discuss the role of fever in nonspecific resistance.*

Inflammation is a local response of the body to injury. There are also systemic, or overall, responses; one of the most important is fever, an abnormally high body temperature. The most frequent cause of fever is infection from bacteria (and their toxins) or viruses.

Body temperature is controlled by a part of the brain called the hypothalamus. The hypothalamus is sometimes called the body's thermostat, and it is normally set at 37°C (98.6°F). It is believed that certain substances affect the hypothalamus by setting it at a higher temperature. Recall from Chapter 15 that when phagocytes ingest gram-negative bacteria, the lipopolysaccharides (LPS) of the cell wall (endotoxins) are released, causing the phagocytes to release interleukin-1 (endogenous pyrogen). Interleukin-1 causes the hypothalamus to release prostaglandins that reset the hypothalamic thermostat at a higher temperature, thereby causing fever (see Figure 15.4).

Assume that the body is invaded by pathogens and that the thermostat setting is increased to 39°C (102.2°F). To adjust to the new thermostat setting, the body responds with blood vessel constriction, increased rate of metabolism, and shivering, all of which raise body temperature. Even though body temperature is climbing higher than normal, the skin remains cold, and shivering occurs. This condition, called a *chill,* is a definite sign that body temperature is rising. When body temperature reaches the setting of the thermostat, the chill disappears. But the body will continue to maintain its temperature at 39°C until the interleukin-1 is eliminated. The thermostat is then reset at 37°C. As the infection subsides, heat-losing mechanisms such as vasodilation and sweating go into operation. The skin becomes warm and the person begins to sweat. This phase of the fever, called the *crisis,* indicates that body temperature is falling.

Up to a certain point, fever is considered a defense against disease. Interleukin-1 helps step up the production of T lymphocytes. High body temperature intensifies the effect of interferon (an antiviral protein that we will discuss shortly). It is believed to inhibit the growth of some microorganisms by decreasing the amount of iron available to them. Also, because the high temperature speeds up the body's reactions, it may help body tissues to repair themselves more quickly.

ANTIMICROBIAL SUBSTANCES

LEARNING OBJECTIVES

- *Discuss the function of complement.*
- *Discuss the role of interferon in nonspecific defenses.*

The body produces certain antimicrobial substances in addition to the chemical factors mentioned earlier. Among the most important of these are the proteins of the complement system and the interferons.

The Complement System

Complement is a defensive system consisting of serum proteins that participate in lysis of foreign cells, inflammation, and phagocytosis. The system can be activated by an immune reaction in the classical pathway or by direct interaction with a bacterium in the alternative pathway (Figure 16.10). Complement is nonspecific: the same proteins can be activated in response to any foreign cell. However, in the classical pathway, complement assists, or complements, specific immunity.

Components

The complement system consists of a group of at least 20 different interacting proteins found in normal serum. Proteins of the complement system make up about 5% of the serum proteins in vertebrates. The major components of the classical pathway are designated by a complex numbering system ranging from C1 through C9 (C stands for complement). The C1 protein also has three subcomponents: C1q, C1r, and C1s. The alternative pathway con-

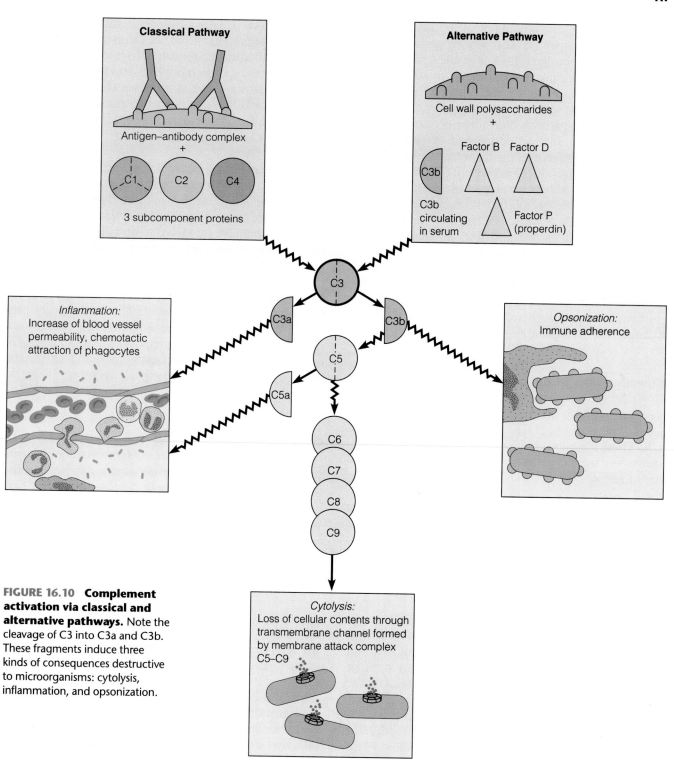

FIGURE 16.10 Complement activation via classical and alternative pathways. Note the cleavage of C3 into C3a and C3b. These fragments induce three kinds of consequences destructive to microorganisms: cytolysis, inflammation, and opsonization.

sists of proteins called factor B, factor D, and factor P (properdin), along with the C3 and C5 through C9 proteins active in the classical pathway.

Pathways of Activation

The proteins of the classical and alternative pathways act in an ordered sequence, or *cascade,* which, except for C4, follows the numerical designations. In a series of steps, each protein activates the next one in the series, usually by cleaving (splitting) it. The fragments of the cleaved proteins have new physiological or enzymatic functions. For example, one fragment might cause blood vessel dilation, whereas another fragment serves as part of the enzyme that cleaves the next protein in the series.

C3 plays a central role in the complement system. Its activation triggers several mechanisms that contribute to microbial destruction. As you can see in Figure 16.10, either of two pathways can activate C3. The classical pathway is initiated by the binding of antibodies to antigen (described in Chapter 17). The antigen could consist of bacteria or other foreign cells. The classical pathway also involves the activation of C1, C2, and C4 proteins. The details of complement activation by this pathway are described in the next section on consequences.

The alternative pathway does not involve antibodies; it is initiated by the interaction between certain polysaccharides and the proteins of the pathway. Factors B, D, and P react with C3 to produce low levels of C3b in the serum. The alternative pathway is initiated when C3b, factor B, factor D, and factor P combine with certain polysaccharides. Most of these polysaccharides are contained in the cell walls of certain bacteria and fungi, although they also include molecules on the surface of some foreign mammalian red blood cells. The alternative pathway is of particular importance in combating enteric gram-negative bacteria. The outer membrane of the bacterial cell wall contains a lipopolysaccharide that is an endotoxin (lipid A), triggering the alternative pathway. Note that this pathway does not involve C1, C2, or C4.

Consequences of Complement Activation

How does the complement system contribute to microbial destruction? Both the classical and alternative pathways lead to the cleavage of C3 into two fragments, C3a and C3b. These fragments induce three processes that are destructive to microorganisms: cytolysis, inflammation, and opsonization.

Cytolysis. The main function of the complement system is to destroy foreign cells by damaging their plasma membranes, causing the cellular contents to leak out. This is called **cytolysis** and is accomplished as follows (Figure 16.11a). ① Once a pair of antibodies recognizes and attaches to the antigen, the complement protein C1 (which actually consists of three subcomponents) binds to two or more adjacent antibodies, thus activating C1. ② Next, activated C1 in turn activates C2 and C4. It does this by functioning as a proteolytic (protein-splitting) enzyme and splitting the C2 and C4 proteins. C2 is split into fragments called C2a and C2b, and C4 into fragments called C4a and C4b. Then C2b and C4b combine to form another proteolytic enzyme, which in turn activates C3 by splitting it into two fragments, C3a and C3b. ③ C3b initiates a sequence of reactions involving C5, C6, C7, C8, and C9, which are known collectively as the **membrane attack complex**. The activated components of these proteins, with C9 proteins possibly playing a key role, attack the invading cell's membrane and ④ produce circular lesions, called **transmembrane channels**, that lead to loss of ions and eventual cytolysis. The utilization of the complement components in this process is called **complement fixation**; it forms the basis of an important clinical laboratory test that will be explained in Chapter 18. Figure 16.11b illustrates the effects of complement on a microbial cell.

Inflammation. C3a, a cleavage product from C3, and C5a, a cleavage product from C5, can contribute to the development of acute *inflammation* (Figure 16.12). C3a and C5a bind to mast cells, basophils, and blood platelets to trigger the release of histamine, which increases blood vessel permeability. C5a also functions as a powerful chemotactic factor that attracts phagocytes to the site of complement fixation.

Opsonization. When bound to the surface of a microorganism, C3b can interact with special receptors on phagocytes to promote phagocytosis (see Figure 16.10). This phenomenon is called *opsonization* or immune adherence. In the process, C3b functions as an opsonin by coating the microorganism and promoting attachment of the phagocyte to the microbe.

Inactivation of Complement

Once complement is activated, its destructive capabilities usually cease very quickly in order to minimize destruction of host cells. This is accomplished by various regulatory proteins found in the host's blood and on certain cells, such as blood cells. The proteins bring about the breakdown of activated complement and function as inhibitors and destructive enzymes.

Complement and Disease

In addition to its importance in defense, the complement system assumes a role in causing disease as a result of inherited deficiencies. For example, deficiencies of C1, C2, or C4 cause collagen vascular disorders that result in hypersensitivity (anaphylaxis); deficiency of C3, though rare, results in increased susceptibility to bacterial infections; and C5 through C9 defects result in increased susceptibility to *Neisseria meningitidis* and *N. gonorrhoeae* infections.

Interferons (IFNs)

Because viruses depend on their host cells to provide many functions of viral multiplication, it is difficult to inhibit viral multiplication without at the same time affecting the host cell itself. One way the infected host counters viral infections is with interferons. **Interferons** are a class of similar antiviral proteins produced by certain animal cells after viral stimulation. One of the principal functions of interferons is to interfere with viral multiplication.

One of the most interesting features of interferons is that they are host-cell-specific but not virus-specific. This means that interferon produced by human cells protects human cells but produces little antiviral activity for cells of other species, such as mice or chickens. However, the interferon of a species is active against a number of different viruses.

Not only do different animal species produce different interferons, but also different types of cells in an animal produce different interferons. Human interferons are of

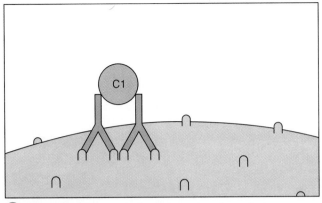

1 Once antibodies recognize and attach to the antigen, complement protein C1 binds to two adjacent antibodies.

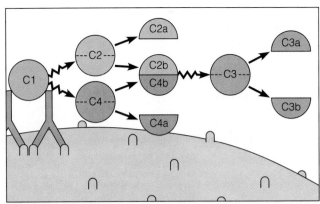

2 C1 acts as an enzyme that splits the C2 and C4 proteins into fragments. Fragments C2b and C4b combine to form another enzyme, which splits C3 into two fragments. The active fragment is called C3b.

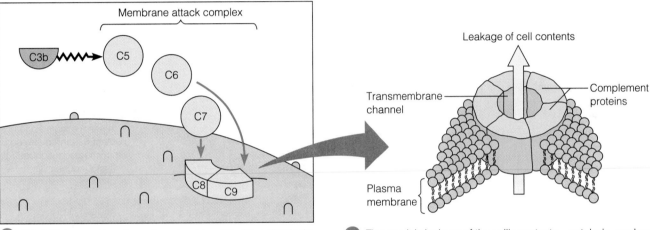

3 C3b initiates a series of reactions involving C5–C9, collectively called the membrane attack complex. This complex forms circular transmembrane channels (lesions) in the antigenic cell's membrane, with C9 proteins possibly playing a key role.

4 The result is leakage of the cell's contents—cytolysis, as shown in this enlarged view of a transmembrane channel.

(a)

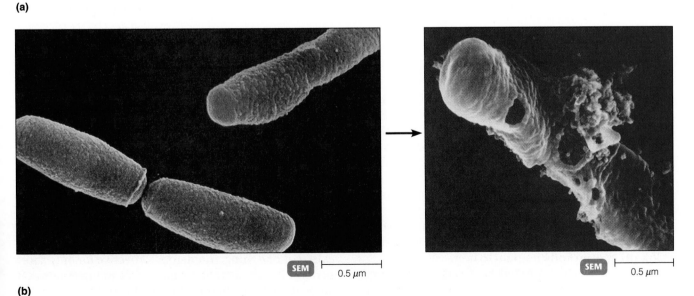

(b)

FIGURE 16.11 Cytolysis caused by complement via the classical pathway.
(a) Illustrations of a few key steps in the complex sequence that leads to the membrane attack complex forming holes in a plasma membrane. (b) Micrographs of a rod-shaped bacterium before cytolysis (left) and after cytolysis (right).

FIGURE 16.12 Inflammation stimulated by complement. (a) C3a and C5a bind to mast cells, basophils, and platelets to trigger the release of histamine, which increases blood vessel permeability. (b) C5a functions as a chemotactic factor that attracts phagocytes to the site of complement activation.

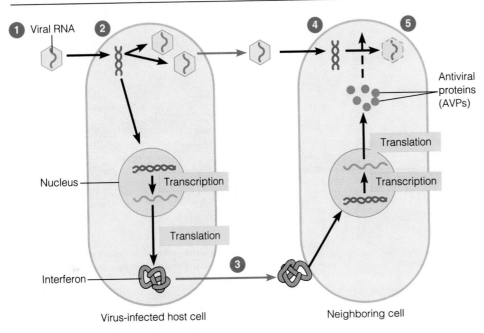

1. Viral RNA stimulates host cell to synthesize interferon.

2. New viruses are produced by multiplication.

3. Meanwhile, interferon reacts with plasma membrane or nuclear membrane receptors on uninfected neighboring cell and induces synthesis of antiviral proteins (AVPs).

4. New viruses are released and infect neighboring cell.

5. AVPs block viral protein synthesis and thus interfere with viral multiplication.

FIGURE 16.13 Antiviral action of interferon.

three principal types: (1) *alpha interferon* (α-IFN), (2) *beta interferon* (β-IFN), and (3) *gamma interferon* (γ-IFN). There are also various subtypes of interferon within each of the principal groups. In the human body, interferon is produced by fibroblasts in connective tissue, by lymphocytes, and by other leukocytes. Each of the three types of interferon produced by these cells can have a slightly different effect on the body.

All interferons are small proteins, with molecular weights between 15,000 and 30,000. They are quite stable at low pH and are fairly resistant to heat.

Produced by virus-infected host cells only in very small quantities, interferon diffuses to uninfected neighboring cells (Figure 16.13). It reacts with plasma or nuclear membrane receptors, inducing the uninfected cells to manufacture mRNA for the synthesis of **antiviral proteins (AVPs)**. These proteins are enzymes that disrupt vari-

ous stages of viral multiplication. For example, one AVP inhibits translation of viral mRNA by blocking initiation of protein synthesis. Another inhibits polypeptide elongation. Still another is involved in destroying viral mRNA before translation. The box on the facing page describes the role of γ-IFN in activating phagocytes.

The low concentrations at which interferon inhibits viral multiplication are nontoxic to uninfected cells. Because of its beneficial properties, interferon would seem to be an ideal antiviral substance. But certain problems do exist. For one thing, interferon is effective for only short periods. It typically plays a major role in infections that are acute and short-term, such as colds and influenza. Another problem is that it has no effect on viral multiplication in cells already infected.

The importance of interferon in protecting the body against viruses, as well as its potential as an anticancer

Microbiology in the News

GAMMA INTERFERON HELPS DEFECTIVE NEUTROPHILS

WHAT HAPPENS WHEN the body produces phagocytes such as neutrophils that do not function properly? Because these cells are a key part of the body's nonspecific defenses, it is clear that such malfunctioning cells would leave us more vulnerable to disease. This is exactly what happens to individuals with a rare, inherited condition called *chronic granulomatous disease (CGD)*—they suffer frequent, life-threatening bacterial and fungal infections because their neutrophils cannot kill their target cells.

Neutrophils generally kill cells with a process called a *respiratory burst* (see the figure). The adherence of a targeted cell such as a bacterium to the neutrophil's plasma membrane stimulates the neutrophil to produce NADPH by the pentose phosphate pathway. This NADPH is then reoxidized to NADP+ by a membrane complex called NADPH oxidase, which uses the electron to produce a superoxide free radical ($O_2^-\cdot$) from O_2. Finally, the enzyme superoxide dismutase (see Chapter 6) then converts this superoxide to hydrogen peroxide (H_2O_2). The resulting "burst" of hydrogen peroxide kills the bacterium.

In most individuals with CGD, the neutrophil plasma membrane lacks a cytochrome in the electron transport chain used to transport electrons to O_2 via NADPH oxidase. As a result, their phagocytes show normal chemotaxis and migration toward a target cell, but they cannot deliver this respiratory burst to kill their targets.

In looking for a way to treat this disease, researchers thought of gamma interferon (γ-IFN), which is normally produced by lymphocytes in order to activate macrophages to kill bacteria. A clinical trial was conducted in 10 hospitals and universities in the U.S.A. In this trial, 63 patients were given γ-IFN, and 65 patients in a control group were given a placebo. After 12 months of treatment, 81% of those receiving γ-IFN remained free of serious infection, compared to 20% of those who received the placebo.

Unfortunately, in order to receive this benefit, individuals with CGD must take γ-IFN for life, because it does not cure the disease. In addition, we do not know how γ-IFN causes these results. In the future it may be possible to cure CGD by replacing the defective cytochrome gene. In the meantime, however, γ-IFN allows individuals with CGD to lead healthier lives.

A correctly functioning neutrophil releases a respiratory burst of hydrogen peroxide to kill a bacterium.

agent, has made its production in large quantities a top public health priority. Several groups of scientists have successfully applied recombinant DNA technology in inducing certain species of bacteria to produce interferon. (This technique is described in Chapter 9.) The interferons produced with recombinant DNA techniques, called recombinant interferons (rINFs), are important for two reasons: they are pure, and they are plentiful.

Clinical trials to determine the anticancer effects of interferon were begun in 1981. At present, it appears that large doses of rINFs have only limited effects against some tumors and no effect on others. The clinical trials have also revealed that high doses of interferon can have side-effects that range from minor to quite serious, including fatigue, malaise, loss of appetite, fever, chills, pains in the joints, mental confusion, seizures, and cardiac complications. Several studies are also focusing on the use of rIFNs to treat AIDS, hepatitis, genital herpes, influenza, and the common cold.

Although glowing predictions about interferon's anticancer properties have not been justified by the results of clinical trials to date, research is continuing, for several reasons. First, there are many subtypes of interferon, and it might be possible to find one with definitive anticancer properties. Second, some laboratory studies indicate that interferon might work together with other chemotherapeutic agents so that the effects of interferon are enhanced. Encouraging results have been obtained by use of interferon in combination with either doxorubicin (used to treat a variety of blood cancers and solid tumors) or cimetidine (used to treat ulcers), for example. Third, it has been observed that patients who did not respond well to either specific chemotherapy or subsequent treatment with interferon showed improvement when returned to the original chemotherapy.

• • •

In Chapter 17 we will discuss specific immune responses. We will also look at the principal factors that contribute to immunity.

STUDY OUTLINE

INTRODUCTION (p. 406)

1. The ability to ward off disease through body defenses is called resistance.
2. Lack of resistance is called susceptibility.
3. Nonspecific resistance refers to all body defenses that protect the body from any kind of pathogen.
4. Specific resistance refers to defenses (antibodies) against specific microorganisms.

SKIN AND MUCOUS MEMBRANES (pp. 406–408)

Mechanical Factors (pp. 406–408)

1. The structure of intact skin and the waterproof protein keratin provide resistance to microbial invasion.
2. Some pathogens, if present in large numbers, can penetrate mucous membranes.
3. The lacrimal apparatus protects the eyes from irritating substances and microorganisms.
4. Saliva washes microorganisms from teeth and gums.
5. Mucus traps many microorganisms that enter the respiratory and gastrointestinal tracts; in the lower respiratory tract, the ciliary escalator moves mucus up and out.
6. The flow of urine moves microorganisms out of the urinary tract, and vaginal secretions move microorganisms out of the vagina.

Chemical Factors (p. 408)

1. Sebum contains unsaturated fatty acids, which inhibit the growth of pathogenic bacteria. Some bacteria commonly found on the skin can metabolize sebum and cause the inflammatory response associated with acne.
2. Perspiration washes microorganisms off the skin.
3. Lysozyme is found in tears, saliva, nasal secretions, and perspiration.
4. The high acidity (pH 1.2 to 3.0) of gastric juice prevents microbial growth in the stomach.
5. Normal microbiota prevents the growth of many pathogens.

PHAGOCYTOSIS (pp. 409–414)

1. Phagocytosis is the ingestion of microorganisms or particulate matter by a cell.
2. Phagocytosis is performed by phagocytes—certain types of white blood cells or derivatives of them.

Formed Elements in Blood (pp. 409–411)

1. Blood consists of plasma (fluid) and formed elements (cells and cell fragments).
2. Leukocytes (white blood cells) are divided into three categories: granulocytes (neutrophils, basophils, and eosinophils), lymphocytes, and monocytes.
3. During many infections, the number of leukocytes increases (leukocytosis); some infections are characterized by leukopenia (decrease in leukocytes).

Actions of Phagocytic Cells (pp. 411–412)

1. Among the granulocytes, neutrophils are the most important phagocytes.
2. Enlarged monocytes become wandering macrophages and fixed macrophages.
3. Fixed macrophages are located in selected tissues and are part of the mononuclear phagocytic system.
4. Granulocytes predominate during the early stages of infection, whereas monocytes predominate as the infection subsides.

Mechanism of Phagocytosis (pp. 412–414)

1. Chemotaxis is the process by which phagocytes are attracted to microorganisms.
2. The phagocyte then adheres to the microbial cells; adherence may be facilitated by opsonization—coating the microbe with plasma proteins.
3. Pseudopods of phagocytes engulf the microorganism and enclose it in a phagocytic vesicle to complete ingestion.
4. Many phagocytized microorganisms are killed by lysosomal enzymes and oxidizing agents.

5. Some microbes are not killed by phagocytes and can even reproduce in phagocytes.

INFLAMMATION (pp. 414–416)

1. Inflammation is a bodily response to cell damage; it is characterized by redness, pain, heat, swelling, and sometimes loss of function.

Vasodilation and Increased Permeability of Blood Vessels (p. 414)

1. The release of histamine, kinins, and prostaglandins causes vasodilation and increased permeability of blood vessels.
2. Blood clots can form around an abscess to prevent dissemination of the infection.

Phagocyte Migration and Phagocytosis (pp. 414–415)

1. Phagocytes have the ability to stick to the lining of the blood vessels (margination).
2. They also have the ability to squeeze through blood vessels (diapedesis).
3. Pus is the accumulation of damaged tissue and dead microbes, granulocytes, and macrophages.

Tissue Repair (p. 416)

1. A tissue is repaired when the stroma (supporting tissue) or parenchyma (functioning tissue) produces new cells.
2. Stromal repair by fibroblasts produces scar tissue.

FEVER (p. 416)

1. Fever is an abnormally high body temperature produced in response to a bacterial or viral infection.

2. Bacterial endotoxins and interleukin-1 can induce fever.
3. A chill indicates a rising body temperature; crisis (sweating) indicates that the body's temperature is falling.

ANTIMICROBIAL SUBSTANCES (pp. 416–422)

The Complement System (pp. 416–418)

1. The complement system consists of a group of serum proteins that activate one another to destroy invading microorganisms.
2. C1 binds to antigen–antibody complexes to eventually activate C3 protein. Factor B, factor D, factor P, and C3 bind to certain cell wall polysaccharides to activate C3b.
3. C3 activation can result in cell lysis, inflammation, and opsonization.
4. Complement is deactivated by host-regulatory proteins.
5. Complement deficiencies can result in increased susceptibility to disease.

Interferons (IFNs) (pp. 418–422)

1. Interferons are antiviral proteins produced in response to viral infection.
2. There are three types of human interferon: α-IFN, β-IFN, and γ-IFN. Recombinant interferon has also been produced.
3. Interferon's mode of action is to induce uninfected cells to produce antiviral proteins (AVPs) that prevent viral replication.
4. Interferons are host-cell-specific but not virus-specific.

STUDY QUESTIONS

Review

1. Define the following terms:
   (a) Resistance
   (b) Susceptibility
   (c) Nonspecific resistance

2. Describe the mechanical factors of the skin and mucous membranes that assume a role in nonspecific resistance.

3. Describe the chemical factors of the skin and mucous membranes that assume a role in nonspecific resistance.

4. Define phagocytosis.

5. Compare the structures and functions of granulocytes and monocytes in phagocytosis.

6. How do fixed and wandering macrophages differ?

7. Diagram the following processes that result in phagocytosis: margination, diapedesis, adherence, and ingestion.

8. Define inflammation, and list its characteristics.

9. Why is inflammation beneficial to the body?

10. How is fever related to nonspecific resistance?

11. What is the importance of chill and crisis during fever?

12. What is interferon? Discuss its role in nonspecific resistance.

13. What is complement?

14. Summarize the major outcomes of complement activation.

15. How does complement function in the alternative pathway?

Critical Thinking

1. Why do serum levels of transferrin increase during an infection? What can a bacterium do to respond to high levels of transferrin?

2. How is the complement system activated by bacterial endotoxin in the bloodstream? How does endotoxic shock result in massive host cell destruction?

3. A variety of drugs with the ability to reduce inflammation are available. Comment on the danger of misuse of these anti-inflammatory drugs.

Clinical Applications

1. To be a successful parasite, a microbe must avoid destruction by complement. The following list provides examples of complement-evading techniques. For each microbe, identify the disease it causes and describe how its strategy allows it to avoid destruction by complement.

PATHOGEN	STRATEGY
Group A streptococci	C3 does not bind to M protein
Hemophilus influenzae type b	Has a capsule
Pseudomonas aeruginosa	Sheds cell wall polysaccharides
Trypanosoma cruzi	Degrades C1

2. People with *Rhinovirus* infections of the nose and throat have an 80-fold increase in kinins and no increase in histamine. What do you expect for rhinoviral symptoms? What disease is caused by rhinoviruses?

3. A hematologist often performs a differential white blood cell count on a blood sample. Such a count determines the relative numbers of white blood cells. Why are these numbers important? What do you think a hematologist would find in a differential white blood cell count of a patient with mononucleosis? With neutropenia? With eosinophilia?

Specific Defenses of the Host: The Immune Response

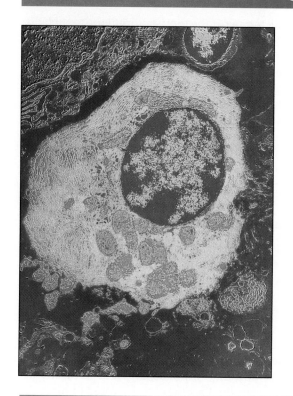

I N THE PREVIOUS CHAPTER we discussed nonspecific host defenses, including the skin and mucous membranes, phagocytosis, inflammation, and fever. In addition to these nonspecific defenses, we have **innate resistance** to certain illnesses. For example, all humans are resistant to many infectious diseases of other animals, including such diseases as canine distemper and chicken and hog cholera. However, resistance to human diseases, such as measles, can vary from person to person. Although the effect of measles is often relatively mild for individuals of European ancestry, the disease devastated the populations of Pacific Islanders when they were first exposed to measles by European explorers. Natural selection from many generations having been exposed to the measles virus presumably led to the more frequent inheritance of genes that conferred some resistance to the virus. An individual's resistance to disease also depends on other genetic factors, age, nutritional status, and general health.

IMMUNITY

LEARNING OBJECTIVE

- *Define specific defenses, innate resistance, and immunity.*

In this chapter we will study immunity, another aspect of the body's defenses. (See Figure 16.1 for a review of the body's defense systems.) In contrast to resistance, **immunity** involves a *specific* defensive response when a host is invaded by foreign organisms or other foreign substances.

The immune system of the body recognizes these substances as not belonging to itself, and it develops an immune response against them. Organisms or substances that provoke such a response are called **antigens.** This immune response involves the production of specialized lymphocytes and special proteins called **antibodies.** The specialized lymphocytes, in particular *B cells* and *T cells*, and the antibodies are specifically targeted at the antigens that caused their formation and can destroy or inactivate them if they encounter them again.

Invading organisms might be pathogenic bacteria, viruses, fungi, protozoa, or helminths; foreign materials include pollen, insect venom, and transplanted tissue. Body cells that become cancerous are also recognized as foreign and may be eliminated. (However, if they survive to become established as solid tumors, the immune system is no longer effective against them.) We will see in this and the next two chapters that the immune system is essential to our survival, but that it can also cause us harm when its efforts are misdirected.

TYPES OF ACQUIRED IMMUNITY

LEARNING OBJECTIVE

- *Contrast the four types of acquired immunity.*

Acquired immunity refers to the protection an animal develops against certain types of microbes or foreign substances. Acquired immunity is developed during an indi-

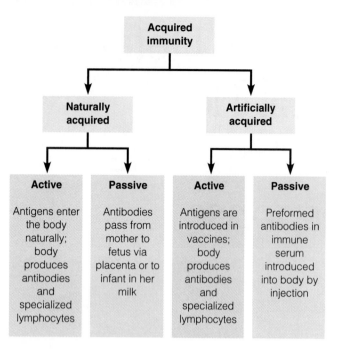

FIGURE 17.1   Types of acquired immunity.

vidual's lifetime. The various types of acquired immunity are summarized in Figure 17.1.

Immunity can be acquired either actively or passively. Immunity is acquired *actively* when a person is exposed to microorganisms or foreign substances and the immune system responds. Immunity is acquired *passively* when antibodies are transferred from one person to another. This type of immunity is considered passive because the recipient does not actively synthesize the antibodies. Passive immunity in the recipient lasts only as long as the antibodies are present—in most cases, for several weeks or months. Both actively acquired immunity and passively acquired immunity can be obtained by natural or artificial means.

Naturally Acquired Immunity

Naturally acquired active immunity is obtained when a person is exposed to antigens in the course of daily life. Once acquired, immunity is lifelong for some diseases, such as measles, chickenpox, and yellow fever. For other diseases, especially intestinal diseases, the immunity usually lasts for only a few years. Sometimes *subclinical infections* (those that produce no evident symptoms) can also confer immunity.

Naturally acquired passive immunity involves the natural transfer of antibodies from a mother to her infant. A pregnant woman's body passes some of her antibodies to her fetus across the placenta, a mechanism called *transplacental transfer*. If the mother is immune to such diseases as diphtheria, rubella, or polio, the newborn will be temporarily immune to these diseases as well. Certain antibodies are also passed from the mother to her nursing infant in breast milk, especially in the first secretions called *colostrum*. In the infant, immunity generally lasts only as long as the transmitted antibodies are present—usually a few weeks or months. These maternal antibodies are essential for providing immunity to the infant until his or her own immune system matures.

Artificially Acquired Immunity

Artificially acquired active immunity results from vaccination. **Vaccination**, also called **immunization**, introduces specially prepared antigens called **vaccines** into the body. Vaccines consist of inactivated bacterial toxins (toxoids), killed microorganisms, living but attenuated (weakened) microorganisms, or parts of microorganisms such as capsules. These substances are no longer able to cause disease, but they are still able to stimulate an immune response, much as naturally acquired pathogens do. Vaccines are discussed in more detail in Chapter 18.

Unlike artificially acquired active immunity, which occurs in response to the injection of antigens, **artificially acquired passive immunity** involves the introduction of antibodies into the body. These antibodies come from an animal or person who is already immune to the disease. For example, a person who is bitten by a poisonous snake might be injected with antibodies from a horse that has been immunized against snake venom.

The antibodies are found in the serum of the immune animal or individual. **Serum** (plural, *sera*) is the fluid remaining after blood has clotted and the blood cells and clotted matter have been removed. Because most of the antibodies remain in the serum, the name **antiserum** has become a generic term for blood-derived fluids containing antibodies. Hence, the study of antibodies and antigens is called **serology**. When serum is subjected to an electric current in the laboratory during electrophoresis (discussed in Chapter 10), the proteins within it move at different rates, as shown in Figure 17.2. One of the components that is separated out by this procedure contains most of the antibodies that were present in the original sample. This antibody-rich serum component is called **immune serum globulin** or **gamma globulin.**

When immune serum globulin from an individual who is immune to a disease is injected into the body, it confers immediate protection against the disease. Immune serum can be gathered from immune individuals and used to protect others. However, although artificially acquired passive immunity is immediate, it is short-lived because the body produces no new antibodies. The half-life of an injected antibody (time required for half of the antibodies to disappear) is typically about three weeks.

FIGURE 17.2 Separation of serum proteins by electrophoresis. In this procedure, serum is placed in a trough cut into a gel. In response to an electric current, the negatively charged proteins of the serum migrate through the gel from the negatively charged end (cathode) to the positively charged end (anode). Antibodies are concentrated mainly in the gamma (γ) fraction of globulins, hence the original term *gamma globulin*.

THE DUALITY OF THE IMMUNE SYSTEM

LEARNING OBJECTIVE

- *Differentiate between humoral (antibody-mediated) and cell-mediated immunity.*

Early in the history of immunological experimentation, biologists learned that some forms of immunity could be transferred between animals by transferring serum from immunized to nonimmunized animals. However, other forms of immunity were not transferable via blood serum. Much later it was learned that these forms of immunity were transferred only when certain lymphocytes were transferred.

These results are now known to reflect the two main components of the vertebrate immune system. The immunity associated with serum transfer reflects the actions of the humoral (antibody-mediated) immune system, whereas the immunity associated with the transfer of lymphocytes reflects the actions of the cell-mediated immune system. These components work both separately and together to protect us from disease.

The Humoral (Antibody-Mediated) Immune System

Humoral (antibody-mediated) immunity involves the production of antibodies that act against foreign organisms and substances. These antibodies are found in extracellular fluids, such as blood plasma, lymph, and mucus secretions (fluids once known as "humors"). Cells called **B cells** (or B lymphocytes) are responsible for the production of antibodies (Figure 17.3a). The humoral immune response defends mostly against bacteria, bacterial toxins, and viruses circulating freely in the body's fluids.

The Cell-Mediated Immune System

Cell-mediated immunity involves the production of specialized lymphocytes called **T cells** (T lymphocytes) that act against foreign organisms or tissues (Figure 17.3b). The cell-mediated immune response is most effective against bacteria and viruses located within phagocytic or infected host cells and against fungi, protozoa, and helminths. This system may also respond to transplanted tissue (such as a foreign skin graft) by mounting an immune response to reject it. The cell-mediated immune system is also an important defense against cancer.

ANTIGENS AND ANTIBODIES

LEARNING OBJECTIVES

- *Explain what an antigen is and what a hapten is.*
- *Explain what an antibody is, and describe the structural and chemical characteristics of antibodies.*
- *Describe at least one function for each of the five classes of antibodies.*

Antigens and antibodies play key roles in the response of the immune system. Antigens (sometimes called **immunogens**) provoke a highly specific immune response in an organism.

The Nature of Antigens

Normally, the immune system recognizes components of the body it protects as "self" and foreign matter as "nonself." This recognition is the reason why people's defense systems do not usually produce antibodies against their own body tissues (although we will see in Chapter 19 that this sometimes occurs).

Most antigens are proteins or large polysaccharides. Lipids and nucleic acids usually are antigenic only when combined with proteins and polysaccharides. Antigenic compounds are often components of invading microbes, such as the capsules, cell walls, flagella, fimbriae, and toxins of bacteria; the coats of viruses; or the surfaces of other types of microbes. Nonmicrobial antigens include pollen, egg white, blood cell surface molecules, serum proteins from other individuals or species, and surface molecules of transplanted tissues and organs.

Generally, antibodies recognize and interact with specific regions on antigens called **antigenic determinants** or **epitopes** (Figure 17.4). The nature of this interaction depends on the size, shape, and chemical nature of the antigenic determinant, as well as the chemical structure of the binding site on the antibody molecule. Most antigens have many different types of determinants on their

FIGURE 17.3 The duality of the immune system.

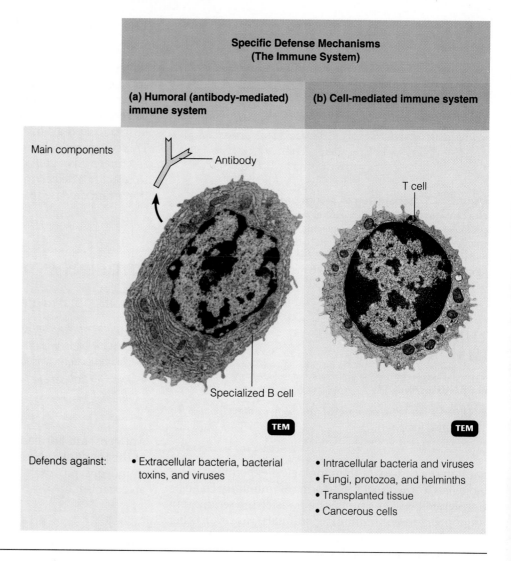

**Specific Defense Mechanisms
(The Immune System)**

	(a) Humoral (antibody-mediated) immune system	**(b) Cell-mediated immune system**
Main components	Antibody — Specialized B cell	T cell
	TEM	TEM
Defends against:	• Extracellular bacteria, bacterial toxins, and viruses	• Intracellular bacteria and viruses • Fungi, protozoa, and helminths • Transplanted tissue • Cancerous cells

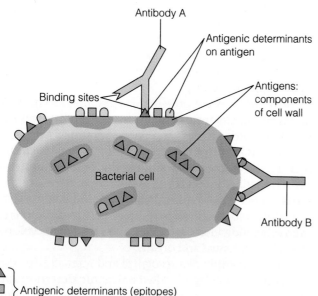

Antibody A

Antigenic determinants on antigen

Antigens: components of cell wall

Binding sites

Bacterial cell

Antibody B

△ ▢ ⌂ } Antigenic determinants (epitopes)

FIGURE 17.4 Antigenic determinants (epitopes). In this diagram, the antigenic molecules are components of the bacterial cell wall. Each antigen carries more than one antigenic determinant. Each antibody molecule has at least two binding sites that can attach to a specific kind of determinant site on an antigen. An antibody can also bind to identical determinants on two different cells at the same time, which can cause neighboring cells to aggregate.

surface. Because different determinants are recognized by different antibodies, the immune system may produce several distinct antibodies against a single antigen.

Most antigens have molecular weights of 10,000 or more. A foreign substance that has a low molecular weight is often not antigenic unless it is attached to a carrier molecule. These small compounds are called **haptens** (from *haptein*, to grasp; Figure 17.5). Once an antibody against the hapten has been formed, the hapten will react with the antibody independent of the carrier molecule. Penicillin is a good example of a hapten. This drug is not antigenic by itself, but some people develop allergic reactions to it. (Allergic reactions are a type of immune response.)

In these people, when penicillin combines with serum proteins, the resulting molecule initiates an immune response.

The Nature of Antibodies

Antibodies or **immunoglobulins (Igs)** are proteins that are made in response to an antigen and can recognize and bind to that antigen. Antibodies can therefore help cause the destruction of antigens. Antibodies are highly specific; that is, they can interact with only one type of antigenic determinant on an antigen.

Each antibody has at least two identical sites that bind to antigenic determinants. These sites are known as **antigen-binding sites.** The number of antigen-binding sites on an antibody is called the **valence** of that antibody. For example, most human antibodies have two binding sites; therefore, they are bivalent.

Antibody Structure

Because a bivalent antibody has the simplest molecular structure, it is called a **monomer.** A typical antibody monomer has four protein chains: two identical *light (L) chains* and two identical *heavy (H) chains*. (*Light* and *heavy* refer to the relative molecular weights.) The chains are joined by disulfide links (see Chapter 2) and other bonds to form a Y-shaped molecule (Figure 17.6). The Y-shaped molecule is flexible and can assume a T shape (note the hinge region indicated on the antibody molecule in Figure 17.6).

Both the H and L chains have sections located at the ends of the Y's arms called *variable (V) regions.* Together, the V regions from one L chain and one H chain form one *antigen-binding site;* hence, each antibody has two such sites. The amino-acid sequences of these V regions vary from one antibody to another. The resulting variations in chemical and three-dimensional structure account for the ability of different antibodies to recognize and bind to different antigens. Each antibody molecule has only one type of antigen-binding site and will bind to only one type of antigenic determinant. Later in this chapter we will describe the mechanism by which the body is able to make different antibodies against the huge array of antigens we may encounter.

The stem of the antibody monomer and the lower parts of the Y's arms are called *constant (C) regions* (see Figure 17.6a). The term *constant* refers to the fact that C regions of the L and H chains of different antibody monomers are relatively invariant in their amino-acid sequence. There are five major sequences found for C regions in heavy chains, which accounts for the five major classes of immunoglobulins discussed in the next section.

The stem of the Y-shaped antibody monomer is called the *Fc region.* The antibody molecule can attach to a host cell at the Fc region; complement (see Chapter 16) can also bind there. The two antigen-combining sites are left free to bind to compatible antigenic determinants.

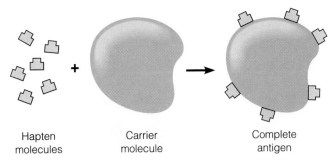

| Hapten molecules | Carrier molecule | Complete antigen |

FIGURE 17.5   Haptens. A hapten is a molecule too small to stimulate antibody formation by itself. However, when the hapten is combined with a larger carrier molecule, usually a serum protein, the hapten and its carrier together function as an antigen and can stimulate an immune response.

Immunoglobulin Classes

The five classes of immunoglobulins (Igs) are designated IgG, IgM, IgA, IgD, and IgE, according to the properties of their heavy chains. Each class plays a different role in the immune response. The structures of IgG, IgD, and IgE molecules resemble the structure shown in Figure 17.6a; molecules of IgM and IgA are usually two or more monomers joined together by disulfide bonds. The structures and characteristics of the immunoglobulin classes are summarized in Table 17.1.

IgG. IgG antibodies account for about 80% of all antibodies in serum. These monomer antibodies readily cross the walls of blood vessels and enter tissue fluids. Maternal IgG antibodies, for example, can cross the placenta and confer passive immunity to a fetus.

IgG antibodies protect against circulating bacteria and viruses, neutralize bacterial toxins, trigger the complement system, and, when bound to antigens, enhance the effectiveness of phagocytic cells.

IgM. Antibodies of the IgM (from *macro,* reflecting their large size) class make up 5% to 10% of the antibodies in serum. IgM has a pentamer structure consisting of five monomers held together by a polypeptide called a *J (joining) chain* (see Table 17.1). The large size of the molecule may prevent IgM from moving about as freely as IgG does. IgM antibodies generally remain in blood vessels and do not enter the surrounding tissues.

IgM is the predominant type of antibody involved in the response to the ABO blood group antigens on the surface of red blood cells (discussed in Chapter 19). It is also especially effective in aggregating antigens and in reactions involving complement. Finally, it can enhance the ingestion of target cells by phagocytic cells, as does IgG.

IgM antibodies are the first antibodies to appear in response to initial exposure to an antigen. A second exposure to antigen results mostly in increased production of IgG.

(a) Antibody molecule

(b) Enlarged antigen-binding site bound to an antigenic determinant

(c) Antibody symbol

(d) Model of an antibody molecule

FIGURE 17.6 Structure of a typical antibody molecule. (a) The Y-shaped molecule is composed of two *light chains* and two *heavy chains* linked by disulfide bridges (S—S). Most of the molecule is made up of *constant regions* (C), which are the same for all antibodies of the same class. The amino acid sequences of the *variable regions* (V), which form the two *antigen-binding sites*, differ from molecule to molecule. **(b)** One antigen-binding site is shown enlarged and bound to an antigenic determinant. **(c)** The symbol used to represent antibodies throughout the text. **(d)** A computer graphic image of an antibody molecule.

The fact that IgM appears first in response to a primary infection and is relatively short-lived makes it uniquely valuable in the diagnosis of disease. If high concentrations of IgM against a pathogen are detected in a patient, then it is likely that the disease observed is caused by that pathogen. Detection of IgG, which is relatively long-lived, may indicate only that immunity against a particular pathogen was acquired in the more distant past.

IgA. IgA accounts for only about 10% to 15% of the antibodies in serum, but it is by far the most common form in mucous membranes and in body secretions such as mucus, saliva, tears, and breast milk. Taking this into consideration, IgA is the most abundant immunoglobulin in the body. (IgG is the most abundant in serum.)

So-called *serum IgA* circulates in the serum mostly as a monomer. The most effective form of IgA is a dimer, called *secretory IgA,* consisting of two monomers connected by a J chain. It is produced in this form by plasma cells in the mucous membranes. Each dimer then enters and passes through a mucosal cell, where it acquires a polypeptide called a *secretory component* that protects it from enzymatic degradation. The main function of secretory IgA is probably to prevent the attachment of pathogens, particularly viruses and certain bacteria, to mucosal surfaces. Its presence in colostrum probably helps to protect infants from gastrointestinal infections especially.

IgD. IgD antibodies make up only about 0.2% of the total serum antibodies. Their structure resembles that of IgG molecules. IgD antibodies are found in blood and lymph and on the surfaces of B cells. These antibodies do not fix complement and cannot cross the placenta. IgD molecules have no known function in serum; they act as antigen receptors on the surfaces of B cells.

IgE. Antibodies of the IgE class are slightly larger than IgG molecules, but they constitute only 0.002% of the total serum antibodies. IgE molecules bind tightly by their Fc (stem) ends to receptors on mast cells and basophils,

TABLE 17.1 Summary of Immunoglobulin Classes

CHARACTERISTICS	IgG	IgM	IgA	IgD	IgE
Structure	Monomer	Pentamer	Dimer (with secretory component)	Monomer	Monomer
Percent of total serum antibody	80	5–10	10–15*	0.2	0.002
Location	Blood, lymph, intestine	Blood, lymph, B-cell surface (as monomer)	Secretions (tears, saliva, mucus, intestine, milk), blood, lymph	B-cell surface, blood, lymph	Bound to mast and basophil cells throughout body, blood
Molecular weight	150,000	970,000	405,000	175,000	190,000
Half-life (days) in serum	23	5	6	3	2
Complement fixation	Yes	Yes	No†	No	No
Placental transfer	Yes	No	No	No	No
Known functions	Enhances phagocytosis, neutralizes toxins and viruses, and protects fetus and newborn	Especially effective against microorganisms and agglutinating antigens; first antibodies produced in response to initial infection	Localized protection on mucosal surfaces	Not known, but presence on B cells may indicate function in initiation of immune response	Allergic reactions; possibly expulsion or lysis of protozoan parasites

*Percent in serum only; if mucous membranes and body secretions are included, percentage is much higher.
†May be "yes" via alternate pathway.

which are specialized cells that participate in allergic reactions (see Chapter 19). When an antigen such as pollen reacts with the IgE antibodies attached to a mast cell or basophil, that cell releases histamine and other chemical mediators. These chemicals provoke a response—for example, an allergic reaction such as hay fever. However, the response can be protective as well, for it causes the attraction of IgG, complement, and phagocytic cells. This is especially useful against parasitic worms.

B CELLS AND HUMORAL (ANTIBODY-MEDIATED) IMMUNITY

LEARNING OBJECTIVES

- *Describe the clonal selection theory.*
- *Explain how an antibody reacts with an antigen; identify the consequences of the reaction.*
- *Distinguish between a primary and a secondary response.*
- *Describe the genetic basis and molecular mechanisms that account for antibody diversity.*
- *Outline the procedure for making monoclonal antibodies.*

As mentioned earlier, the humoral (antibody-mediated) response is carried out by antibodies, which are produced by a special group of activated B cells. The process that leads to the production of antibodies starts when B cells are exposed to free, or extracellular, antigens. In response, the B cell becomes activated, and divides and differentiates into a clone of many effector cells called *plasma cells*. These plasma cells then produce antibodies that are directed against the specific antigen that activated the original B cell.

B cells develop from stem cells located in the bone marrow in adults and in the liver of fetuses (Figure 17.7). The stem cells destined to become B cells probably stay in the bone marrow and mature there. After maturation, the mature B cells migrate to lymphoid organs, such as the lymph nodes or the spleen (see Figure 16.6). Once in these organs, B cells recognize antigens by means of the **antigen receptors** found on their cell surfaces.

The Activation of Antibody-Producing Cells by Clonal Selection

Although the total population of B cells in a person can produce antibodies against an enormous assortment of

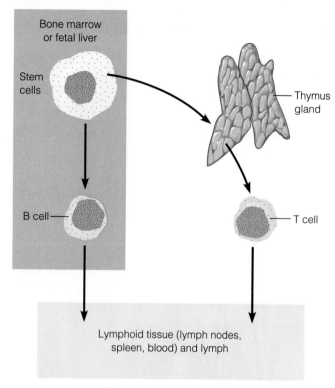

FIGURE 17.7 Differentiation of T cells and B cells. Both types of lymphocytes originate from stem cells in adult bone marrow or in the fetal liver. (Red blood cells, macrophages, neutrophils, and other white blood cells also originate from these same stem cells.) Some cells pass through the thymus and emerge as mature T cells. Other cells probably remain in the bone marrow and become B cells. Both types of cells then migrate to lymphoid tissues, such as the lymph nodes or spleen.

antigens, each B cell can produce antibodies against only one specific antigen. How then do B cells recognize their complementary antigens? In fact, IgM and IgD antibodies bound to the surface of B cells act as very specific *antigen receptors*. By the time a B cell is mature, it may carry as many as 100,000 antibody molecules embedded in its surface membrane.

The basic principle that allows a B cell to recognize an appropriate antigen is that the antibodies embedded in its surface (the antigen receptors) will bind only one specific antigen. However, the actual mechanism varies somewhat with the type of antigen. We will focus here on the simplest case of B-cell activation, in which antigens bind directly to antigen receptors on B cells. When the appropriate antigen binds to the antigen receptors on a B cell, the B cell proliferates into a large clone of cells (Figure 17.8), a phenomenon called **clonal selection.** Thus, an antigen *selects* the lymphocyte that will multiply to form a *clone* of cells with the same immunological specificity. (The same process occurs with T cells.)

As mentioned previously, some of these activated B cells differentiate into **plasma cells,** which secrete antibodies against the antigen. Each plasma cell lives only for

a few days but can produce about 2000 antibody molecules per second.

Stimulation of a B cell by an antigen also results in the production of a population of cells called **memory cells,** which function in long-term immunity. These cells will be discussed shortly.

An important question raised by the clonal selection theory is why the immune system does not react to the body's own cells and macromolecules, a phenomenon called *self-tolerance*. The exact answer is still a mystery, but apparently the B and T cells that interact with self antigens are somehow destroyed during fetal development. This mechanism is called **clonal deletion.**

Antigen–Antibody Binding and Its Results

When an antibody encounters an antigen for which it is specific, an **antigen–antibody complex** rapidly forms. An antibody binds to an antigen at the *antigen-binding site* formed by the variable regions of the light and heavy chains (see Figure 17.6a). Sometimes—especially when large amounts of antigen are present—some antigens attach to antigen-binding sites for which they are less than a perfect fit. Because of this, some antibodies are poorer matches for the antigen; these antibodies are said to have less *affinity*.

The binding of an antibody to an antigen basically protects the host by tagging foreign cells and molecules for destruction by phagocytes and complement (Figure 17.9). The foreign cells and substances are rendered harmless by four basic mechanisms: neutralization, agglutination and precipitation, and activation of complement. In *neutralization*, IgG antibodies can inactivate viruses by blocking their attachment to host cells, and can neutralize bacterial toxins by blocking their active sites. In *agglutination*, antibodies cause antigens to clump together. For example, the two antigen-binding sites of an IgG antibody can combine with antigenic determinants on two different foreign cells, aggregating the cells into clumps that are more easily ingested by phagocytes. Because of its numerous binding sites, however, the IgM molecule is even more effective at cross-linking foreign cells or other particulate antigens and causing their aggregation (see Figure 18.6). Aggregation of soluble antigenic molecules can cause them to *precipitate* (come out of solution), allowing their destruction by phagocytosis. (In Chapter 18 we will see how agglutination and precipitation are important in the diagnosis of some diseases.) Finally, both IgG and IgM antibodies trigger the *complement system* (see Chapter 16), which can lyse bacteria at the site of antibody attachment to the microbe. Ultimately, these mechanisms lead to destruction of the foreign cells and substances by phagocytosis, inflammation, and cell lysis.

However, as we will see in Chapter 19, the action of antibodies can also damage the host. For example, immune complexes of antibody, antigen, and complement can

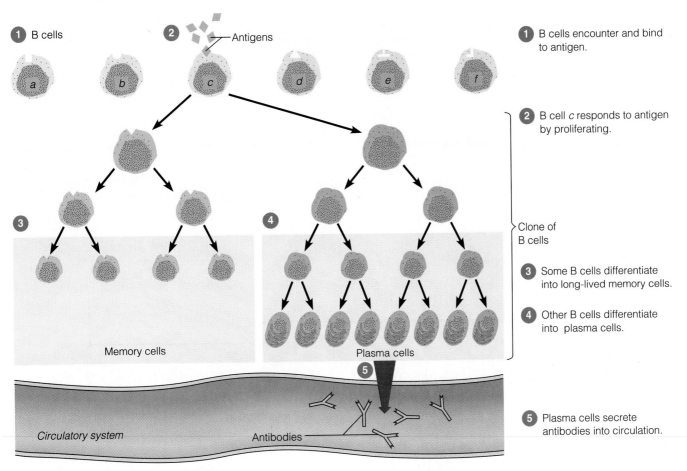

FIGURE 17.8 **Clonal selection and differentiation of B cells.** B cells can recognize an almost infinite number of antigens, but each particular cell recognizes only one type of antigen. An encounter with a particular antigen triggers the proliferation of a cell that is specific for that antigen (here B cell *c*) into a clone of cells with the same specificity—hence the term *clonal selection.*

damage host tissue. Antigens combining with IgE on mast cells can initiate allergic reactions, and antibodies can react with host cells and cause autoimmune disorders.

Immunological Memory

The intensity of the humoral response is reflected by the **antibody titer,** which is the amount of antibody in the serum. After the initial contact with an antigen, the exposed person's serum contains no detectable antibodies for several days. Then there is a slow rise in antibody titer; first IgM antibodies are produced, followed by IgG (Figure 17.10). Finally, a gradual decline in antibody titer occurs. This pattern is characteristic of a **primary response** to an antigen. The immune responses of the host intensify after a second exposure to an antigen. This **secondary response** is also called **memory** or an **anamnestic response.** As we saw in Figure 17.8, some activated B lymphocytes do not become antibody-producing plasma cells

but persist as long-lived **memory cells.** Years (perhaps decades) later, if these cells are stimulated by the same antigen, they rapidly differentiate into antibody-producing plasma cells. This is the basis of the secondary immune response seen in Figure 17.10. A similar response occurs with T cells, which, as we will see in Chapter 19, is necessary to establish the lifelong memory for distinguishing between self and nonself.

The Molecular Basis of Antibody Diversity

A problem that long troubled immunologists is how the body produces such an immense variety of antibodies— even antibodies against synthesized compounds that probably do not occur in nature. Fortunately, nature has evolved a mechanism by which the immune system can respond to almost any conceivable antigen. According to some estimates, an individual may be able to respond to as many as 100 million different antigens.

FIGURE 17.9 **The results of antigen–antibody binding.** The binding of antibodies to antigens to form antigen–antibody complexes tags foreign cells and molecules for destruction by phagocytes and complement.

The mechanism by which so many different antibodies can be made is analogous to the generation of huge numbers of words from a limited alphabet. As shown in Figure 17.6a, each antibody molecule is composed of two light chains and two heavy chains. Both types of chains are divided into constant regions and variable regions. The constant regions are not directly involved in antigen binding, but they differ over the different immunoglobulin classes (IgG, IgM, and so on). The variable regions of the antibody's light and heavy chains make up the antigen-binding sites (Figure 17.6a). It is in these regions that diversity is most required.

Because the *light chains* are produced in two major classes, kappa and lambda, three sets of genes are required to encode the information controlling the formation of

the antigen-binding sites (one set for heavy chains, one for kappa light, and one for lambda light). The genes that encode kappa light chains contain several hundred V segments, which code for different variable regions, and a single C segment, which codes for the constant region. In addition, between the V segments and the C segment, each gene has four J segments, which code for the end of the variable region that joins the constant region of the chain. (There is no relationship between the J segments of the antibody gene and the J chains that help hold together the antibody monomers in IgM and IgA molecules [see Table 17.1].) As immature lymphocytes develop in the bone marrow, the light chains are assembled by combining a V DNA segment from the collection of hundreds and one of the four J segments to connect it to the single C seg-

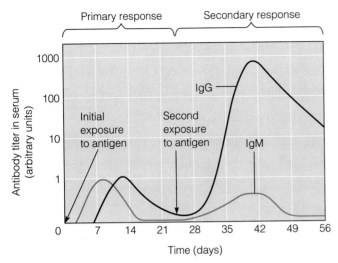

FIGURE 17.10 **The primary and secondary immune responses to an antigen.** IgM appears first in response to the initial exposure. IgG follows and provides longer-term immunity. The second exposure to the same antigen stimulates the memory cells formed at the time of initial exposure to rapidly produce a large amount of antibody. The antibodies produced in response to this second exposure are mostly IgG.

ment (Figure 17.11). As a result, we do not need extraordinarily large numbers of genes to generate huge numbers of different combinations. As an example, if there are 200 V segments and 4 J segments, then 800 (200 times 4) combinations are possible for the light chain alone.

The arrangement of the *heavy-chain* genes is similar in principle to that of the light-chain genes, and still more variation is available. Not only are there well over 100 V segments and 4 J segments, but also there are a dozen D (diversity) segments that can contribute to the entire variable region of the heavy chain. The number of possible variants becomes very large. For example, 100 V segments, 12 D segments, and 4 J segments would provide for 4800 variations of the heavy chains. If this were multiplied by the 800 variations of the light chains calculated earlier, the total would be nearly 4 million combinations—all derived from only a few hundred different genes. This example is conservative; a larger library of genes is actually available. Furthermore, we have not yet considered the heavy-chain *constant region* genes, which code for the information differentiating among antibody classes.

In addition to these gene-sorting mechanisms, antibody variability is generated by the mechanism of *somatic point mutations*—random mutations that arise as the B cell, after stimulation by a suitable antigen, begins proliferating and making antibodies. Many of the new B cells will synthesize antibodies that differ slightly, by one or only a few amino acids. The cells that make antibodies that are a better match to the antigen are more likely to be stimulated, with the result that the antigen-binding power of the antibodies produced tends to improve.

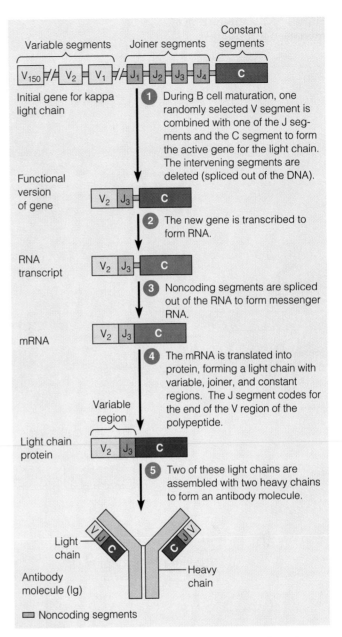

FIGURE 17.11 **The genetic basis of antibody diversity: formation of a kappa light antibody chain.** There is only one type of C region for all kappa light chains in animals.

Monoclonal Antibodies and Their Uses

Scientists have long observed that antibody-secreting B cells may become cancerous. When their proliferation is unchecked, they form tumors called *myelomas*. These cancer cells can be isolated and propagated indefinitely in cell culture. Cancer cells, in this sense, are "immortal," whereas normal B cells can be propagated in culture for only a few generations.

In 1975, Georges Köhler and César Milstein discovered how to fuse an "immortal" cancerous B cell (a mutant that

FIGURE 17.12 Production of monoclonal antibodies.

— Antigen

1 A mouse is injected with a specific antigen that will produce antibodies against that antigen.

Spleen

2 The spleen of the mouse is removed, and a suspension is made. The suspension includes B cells that produce antibodies against the injected antigen.

Suspension of spleen cells

Cultured myeloma cells (cancerous B cells)

Suspension of myeloma cells

Unfused spleen cells

Unfused myeloma cells

Fused hybrid cells

3 The spleen cells are then mixed with *myeloma cells* that are capable of continuous growth in culture but have lost the ability to produce antibodies. Some of the antibody-producing spleen cells and myeloma cells fuse to form hybrid cells. These hybrid cells are now capable of growing continuously in culture while producing antibodies.

Hybrid cell

4 The mixture of cells is placed in a selective medium that allows only hybrid cells to grow.

Hybridoma

5 Hybrid cells proliferate into clones called *hybridomas*. The hybridomas are screened for production of the desired antibody.

Desired monoclonal antibodies

6 The selected hybridomas are then cultured to produce large amounts of *monoclonal antibodies*.

had lost the ability to form antibodies) with a noncancerous antibody-secreting plasma (B) cell taken from a mouse that had been immunized with a particular antigen. The resulting hybrid cell could multiply to form a clone of cells called a **hybridoma.**

When a hybridoma is grown in culture, its genetically identical cells continue to produce the type of antibody characteristic of the ancestral B cell. The importance of the technique is that clones of the antibody-secreting cells

now can be maintained indefinitely in cell culture and can produce immense amounts of identical antibody molecules. Because all of these antibody molecules are produced by a single hybridoma clone, they are called **monoclonal antibodies** (Figure 17.12).

Monoclonal antibodies are useful for three reasons: they are uniform, they are highly specific, and they can be readily produced in large quantities. Because of these qualities, monoclonal antibodies have assumed enormous importance as diagnostic tools. For instance, commercial kits use monoclonal antibodies to recognize chlamydial and streptococcal bacteria, for example, and some non-prescription pregnancy tests use monoclonal antibodies to indicate the presence of a hormone excreted only in the urine of a pregnant woman (see Figure 18.16).

Monoclonal antibodies also have therapeutic applications. For example, when antibiotics are used to treat blood infections caused by gram-negative bacteria, endotoxins are released into the blood and cause septic shock (see Chapter 15). Monoclonal antibodies have been produced that are intended to neutralize these endotoxins. The box on page 438 describes this new treatment for septic shock.

Monoclonal antibodies are also being used to overcome unwanted effects of the immune system, such as rejection of transplanted organs (see Chapter 19). In this case, antibodies are prepared that react with the T cells involved in rejection of the transplanted tissue; the antibodies suppress the T-cell activity.

The prospect of using monoclonal antibodies to treat cancer is also of great interest. One approach is to combine monoclonal antibodies specifically targeted against cancer cells with a toxin to make an *immunotoxin*. The idea is that such immunotoxins could be carried throughout the body by the cardiovascular system to specifically attach to and kill cancer cells wherever they might be.

The therapeutic use of monoclonal antibodies has been limited because these antibodies are currently produced by mouse cells. The immune systems of some people have reacted against the foreign mouse proteins. Monoclonal antibodies derived from human cells would probably provoke fewer reactions. Several approaches are being taken to solve this problem. One is to construct, by recombinant-DNA methods, an antibody with variable regions derived from mouse cells and constant regions derived from human sources. This would still make use of the large number of hybridomas available from mouse cells and would also make the molecule more compatible with the human immune system. These antibodies are called *chimeric monoclonal antibodies* (named after the mythological monster with a lion's head, a goat's body, and a serpent's tail). Genetic engineering has also been proposed as a method for altering mouse antibodies so that they have characteristics that are "more human."

New techniques promise to expand greatly the scale and variety of monoclonal antibody production. For example, genetic engineering can be used to produce monoclonal antibodies without using hybridomas made from mammalian cells. The antibody-producing genes from animals are transplanted into bacteria such as *E. coli*. By such means, essentially the entire array of antibody chains that an animal might produce could be cloned into bacteria that can produce them in large amounts. This method could be used to make human monoclonal antibodies.

T CELLS AND CELL-MEDIATED IMMUNITY

LEARNING OBJECTIVES

- *Describe at least one function for each of the four types of T cells, memory cells, and natural killer cells.*
- *Identify at least one function of each of the following in cell-mediated immunity: cytokine, APC, MHC.*
- *Compare and contrast T-dependent antigens and T-independent antigens.*
- *Describe the role of antibody and natural killer cells in antibody-dependent cell-mediated cytotoxicity.*
- *Compare and contrast humoral (antibody-mediated) and cell-mediated immunity.*

In the cell-mediated immune response, T cells, like B cells, use antigen receptors to recognize and react to an antigen; each T cell reacts specifically with a single antigen. As in humoral immunity, clonal selection is the mechanism by which T cells with particular antigen receptors are stimulated to proliferate and differentiate into the T cells that carry out cell-mediated immunity. As with B cells, some of the activated (antigen-stimulated) T cells become memory cells, which provide the basis for a secondary immune response if they encounter the same antigen later.

However, T cells do not respond to free antigens, but only to intracellular antigens—those that have been processed and presented in specific ways by infected or phagocytic cells. In response, the activated T cell proliferates and differentiates into specialized effector cells. Before we discuss the cell-mediated immune response in more detail, we will first go over the components that make this response possible.

The Components of Cell-Mediated Immunity

T cells are the key component of cell-mediated immunity. However, other lymphocytes, as well as chemical messengers, are also integral to the cell-mediated response. T cells—like B cells and other cells involved in the immune response—develop from stem cells in the bone marrow (see Figure 17.7). However, unlike B cells, the precursors of T cells move to the thymus, where they differentiate into mature cells. After maturation, T cells migrate to lymphoid organs, where they encounter antigens.

When T cells recognize foreign antigens, they can differentiate into any of several different types of effector cells called **specialized T cells**. The different types of T cells are responsible for attacking most types of foreign antigens and regulating the immune response.

Types of T Cells

There seem to be four main functional types of T cells: helper T (T_H) cells, cytotoxic T (T_C) cells, delayed T hyper-

Microbiology in the News

DESIGNER ANTIBODIES MAY ELIMINATE SEPTIC SHOCK

·····································

THE CENTERS FOR Disease Control and Prevention (CDC) estimates that 100,000 Americans die every year from septic shock, making it the 13th leading cause of death in the U.S.A. Septic shock is caused by the endotoxin lipid A, which is found in the cell walls of all gram-negative bacteria, including *Salmonella, Neisseria, Pseudomonas,* and *Escherichia.*

A patient with a systemic gram-negative infection will feel much better after antibiotics kill the infecting bacteria. However, the body's reaction to the bacterial cell walls that remain can cause the fever and marked decrease in blood pressure that characterize septic shock.

Upon encountering the endotoxin, T cells release tumor necrosis factor (TNF), which stimulates the immune system by enhancing phagocytosis and causing macrophages to release the cytokine called interleukin-1 (IL-1). IL-1 promotes the response of B and T cells to an infection, and both IL-1 and TNF cause fever and increase

capillary permeability. This increased permeability allows fluid to escape from the bloodstream, and the resulting decrease in circulating blood volume leads to decreased blood pressure. Consequently, tissues receive less oxygen than they require, and death may follow.

In an effort to impede this deadly process, research teams at Centocor in Malvern, Pennsylvania, and Xoma in Berkeley, California, have produced monoclonal antibodies against lipid A. Human clinical trials of these designer antibodies are under way, and lives have already been saved.

In one dramatic instance, Mary Henderson of Los Angeles was on the brink of death from septic shock when one of her physicians learned of the experimental drug produced by Xoma. Within hours, the drug was obtained and was given to Henderson under a "compassionate use" exemption from the Food and Drug Administration (FDA). Henderson recovered very quickly, and her physicians believe she would have died without the monoclonal antibodies the drug contained.

Researchers at Celltech of Slough, England, and another team at Centocor are taking a different approach: producing monoclonal antibodies against TNF. The anti-TNF they produced has so far protected six pas-

sively immunized baboons against an injection of 40 billion gram-negative rods—an injection that would normally be lethal. Human clinical trials are planned for the near future.

In addition, researchers at Genentech in South San Francisco, California, and at Immunex of Seattle, Washington, are using genetic engineering in yet another approach to developing a treatment. Normally, when an antibody binds to its antigen, the Fc (stem) region becomes exposed and subsequently binds to a specific receptor on macrophages and monocytes, stimulating them to destroy the antigen–antibody complex by phagocytosis. These researchers have replaced the variable (antigen-binding) region of the monoclonal antibodies with the cell-surface receptor for TNF, leaving the Fc region intact. These engineered antibodies then bind to TNF, making it unavailable to react with its normal target cells and cause septic shock. In addition, the Fc region of the antibodies targets the bound TNF for destruction. In mouse studies, an injection of these engineered antibodies protected mice from an injection of endotoxin, if the injections were administered within three hours of each other.

---

sensitivity T (T_D) cells, and suppressor T (T_S) cells. Mature T cells of these four types can be identified by their characteristic cell-surface molecules.

Another classification of T cells that you may encounter is based on type of cell surface receptor called *CD,* for cluster of differentiation. Two types of these receptors are CD4 and CD8. **CD4 cells** are primarily helper T cells, whereas **CD8 cells** include both cytotoxic and suppressor T cells. The cause of the pathogenicity of the AIDS virus was identified by using monoclonal antibodies to show that a population of CD4 cells was decimated by this infection. These helper T cells are vital to a vigorous immune response.

Both of these systems of classification are currently used, and they overlap; for example, as mentioned, most helper T cells carry CD4 receptors. The following discussion uses the functional classification of these cells, but as we continue to learn more about these cells we may come to question its ultimate usefulness. Therefore, we have grouped the delayed hypersensitivity and suppressor T cells together in our discussion.

Helper T Cells. Helper T (T_H) **cells** play a central role in the immune response. Their major function is to stimulate other cells of the immune system to fight off intruders. For example, they activate cytotoxic T cells and other

TABLE 17.2 The Principal Cells That Function in Cell-Mediated Immunity

CELL	FUNCTION
Helper T (T_H) cell	Activates cytotoxic T cells and other helper T cells; necessary for B-cell activation by T-dependent antigens
Cytotoxic T (T_C) cell	Destroys target cells on contact
Delayed hypersensitivity T (T_D) cell	Provides protection against infectious agents; causes inflammation associated with allergic reactions and tissue transplant rejection
Suppressor T (T_S) cell	Regulates immune response and helps maintain tolerance
Natural killer (NK) cell	Attacks and destroys target cells; participates in antibody-dependent cell-mediated cytotoxicity

helper T cells. Some other T_H cells also help B cells respond to antigens, as we will see later.

Cytotoxic T Cells. Activated **cytotoxic T (T_C) cells** destroy target cells on contact. Protection against viral infections and intracellular bacterial infections is the most important function of T_C cells. Because viruses (and some bacteria) reproduce within host cells, they cannot be attacked there by antibodies. T_C cells recognize viral antigens on the surface of the virus-producing host cell and cause the destruction of that cell. T_C cells also play at least some role in the defense against a few diseases caused by protozoa and helminths. T_C cells may also kill cancer cells and cells of transplanted tissue. After it contacts and kills a target cell, the T_C cell continues to live and can kill repeatedly.

Other Functional Types of T Cells. Delayed hypersensitivity T (T_D) cells are associated with certain allergic reactions, such as that to poison ivy, and with rejection of transplanted tissues. In response to antigenic stimulation, T_D cells produce a variety of chemical substances that recruit defense cells such as macrophages. T_D cells are also important in the body's defense against cancer. We will discuss these topics more fully in Chapter 19.

T_D cells are the cells for which cell-mediated immunity was originally named, for the transfer of these cells between animals was found to transfer immunity to tuberculosis. However, T_D cells are probably not a separate population, but mostly T_H cells and a few T_C cells that are involved in these types of immune functions.

Suppressor T (T_S) cells are not well understood. They are generally thought to be T cells that regulate the immune response by turning it off when an antigen is no longer present. T_S cells are also important in the maintenance of self-tolerance, preventing self-directed T cells that avoid clonal deletion in the thymus from mounting

immune responses. Some immunologists consider it likely that T_S cells are not a distinct population but represent suppressive activity by the T_H and T_C cell populations; however, T_S cells usually carry CD8 receptors, whereas most T_H cells carry CD4 receptors.

Natural Killer Cells

The body's cell-mediated defense system also uses cells that are not T cells. Certain lymphocytes called **natural killer (NK) cells** are capable of destroying other cells, especially virus-infected cells and tumor cells. In contrast to cytotoxic T cells, NK cells do not seem to be immunologically specific; that is, they do not need to be stimulated by an antigen. They are not phagocytic but must contact the target cell to lyse it.

The functions of natural killer cells and the other principal cells involved in cell-mediated immunity are briefly summarized in Table 17.2.

Chemical Messengers of Immune Cells: Cytokines (Lymphokines)

As immunologists began working with cultures of immune system cells, they found that the fluids supporting these cultures contained a confusing array of soluble "factors"—chemical messengers—that regulated many other cells of the immune system. These factors were assigned names such as B-cell activating factor or B-cell differentiation factor. Only after decades of work beginning in the 1950s did the picture begin to clear.

Many of these factors were ultimately found to be the same chemical messenger, which was being produced by numerous different cell types. Such factors were at first called *lymphokines,* but as it became clear that they were also produced by cells other than lymphocytes, such as macrophages, the name **cytokines** was adopted. The convention today is to name cytokines for their biological activity until their amino-acid sequence is known. They may then be assigned a number as an *interleukin*—a name chosen to describe their function of communication between white cells. Many chemical messengers that were originally given different names are now known to be identical and are simply called *interleukin-1 (IL-1).* Another cytokine, important because of its role in T-cell proliferation, is *interleukin-2 (IL-2).*

Cytokines, in summary, are soluble chemical messengers by which cells of the immune system communicate with each other. Actually, we can only touch on this complex topic here; Table 17.3 lists a few of the better-known cytokines.

The Cell-Mediated Immune Response

As previously mentioned, individual T cells interact specifically with only a single antigen, which indicates the presence of T-cell receptors to recognize the antigen. These T-cell receptors are not the IgM and IgD molecules that serve

TABLE 17.3   Summary of Some Important Cytokines

CYTOKINE	REPRESENTATIVE ACTIVITY
Interleukin 1 (IL-1)	Stimulates T_H cells in presence of antigens; chemically attracts phagocytes in inflammatory response
Interleukin 2 (IL-2)	Involved in proliferation of antigen-stimulated T_H cells, proliferation and differentiation of B cells, and activation of T_C cells and NK cells
Gamma-interferon (γ-IFN)	Inhibits intracellular viral replication; increases activity of macrophages against microbes and tumor cells
Tumor necrosis factor-beta (TNF-β)	Cytotoxic to tumor cells; enhances activity of phagocytic cells
Granulocyte-macrophage colony-stimulating factor (GM-CSF)	Stimulates the formation of red and white blood cells from stem cells

SEM

FIGURE 17.13   An antigen-presenting cell. Antigen-presenting cells are usually macrophages or dendritic cells that present antigen to helper T cells. This micrograph shows a dendritic cell exhibiting the branching morphology that suggested their name. Dendritic cells are a type of leukocyte found in lymphoid tissue. (For a micrograph of a macrophage, see Figure 16.7.) Any antigen molecules being presented on the surface of the cells are too small to be seen here.

this purpose on B cells, but recently their structures have been identified, and they share many properties with immunoglobulins.

However, T cells do not respond to antigens in quite the same way that B cells do. For one thing, T cells cannot be stimulated by soluble antigens, only by antigens displayed on cell surfaces. Before a T cell can respond to an antigen, the antigen must be processed by macrophages or dendritic cells (leukocytes with a highly branched appearance), which are known as *antigen-presenting cells (APCs)* (Figure 17.13). Portions of the antigen, usually polypeptide fragments, are taken up by an antigen-presenting cell and are then positioned on the cell's surface.

In addition, unlike a B cell responding to an antigen, a T cell will recognize such an antigenic fragment only if it is in close association with a self molecule on the surface of the APC. This type of recognition is known as *associative recognition*. The T cell will bind to the antigen–self molecule complex. (If the T cell's activities were triggered by the self molecule alone, it would mount an immune attack on the body's own tissues.)

These self molecules are cell-surface proteins that are components of the **major histocompatibility complex (MHC)**. The particular collection of MHC proteins a person carries is unique to that individual. Thus, MHC proteins are well suited to serve as self molecules—signals by which the immune system distinguishes self from nonself. The MHC in humans will be discussed further in Chapter 19.

The first antigen encounter stimulates T cell clonal selection and differentiation into one of the functional types of T cells. However, further exposure to antigen is required to fully activate these specialized cells. In this discussion we will focus on the roles of cytotoxic and helper T cells in the cell-mediated immune response.

Cell-Mediated Cytotoxicity

Once they have been activated by exposure to the processed antigen on a target cell, T_C cells cause the lysis of other such target cells (Figure 17.14). T_C cells react only with cells carrying both the target antigen and a class of MHC molecule that are found on most body cells. ① The T_C cell binds to this complex on the cell's surface and ② then releases a protein called *perforin*. The protein forms a pore in the target cell membrane by a mechanism that is similar to the complement mechanism, ③ causing the lysis of the target cell. See Chapter 19 for a discussion of how cytotoxic T cells recognize and lyse cancerous cells.

The Role of Helper T Cells

In keeping with their "director" role in the immune system, T_H cells are especially prolific producers of cytokines. Once activated by an antigen, they can use these chemical messengers to influence the activity of other cells of the immune system.

In helper T cell activation (Figure 17.15), ① an antigen-presenting cell first encounters and processes an antigen. ② Binding of a T_H cell to an antigen-MHC complex on the APC stimulates the APC to secrete the cytokine IL-1. ③ The IL-1 in turn activates the T_H cell, which begins to synthesize interleukin-2 (IL-2). The IL-2 secreted by the T_H cell returns to receptors on the surface of the same cell, which then begins to proliferate and differentiate into mature T_H cells. Only T_H cells that have been stimulated by an antigen have receptors for IL-2, so although the cytokine IL-2 is nonspecific, the resulting T_H cells are specific to that antigen.

These activated T_H cells then begin to produce IL-2 and

FIGURE 17.14 **Cell-mediated cytotoxicity.** A cytotoxic T (T_C) cell binds to the MHC-antigen complex on the surface of an infected cell. The T_C cell then discharges a protein called perforin, which lyses the infected cell.

① Cytotoxic T (T_C) cell binds to infected cell.

② T_C cell releases perforin, which makes lesions in infected cell's membrane.

③ Infected cell lyses.

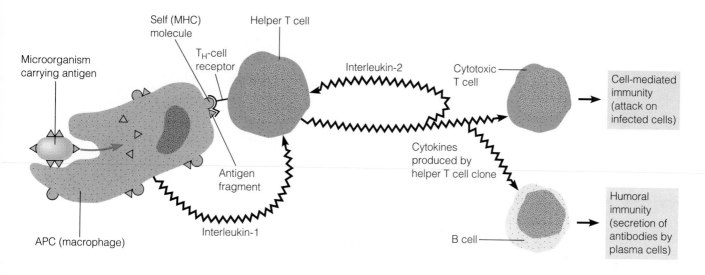

① An antigen-presenting cell (APC) encounters and processes an antigen, forming MHC-antigen complexes on its surface.

② A helper T (T_H) cell receptor binds to the complex, stimulating the APC to secrete interleukin-1.

③ This interleukin-1 stimulates the helper T cell to produce interleukin-2, which then stimulates that helper T cell to form a clone of helper T cells.

④ The cells of this clone in turn produce cytokines, stimulating cells of both immune systems.

FIGURE 17.15 **The central role of helper T cells.**

other cytokines. This IL-2 in turn stimulates other T_H cells specific to that antigen to proliferate and mature. In addition, the cytokines help activate macrophages, cytotoxic T cells, and B cells, which function in humoral immunity.

Some Interactions of the Cell-Mediated Immune System with the Humoral Immune System

In addition to the separate functions of the humoral and cell-mediated immune systems, the two systems work together to increase the immune response against foreign antigens. We will focus on two examples of such interactions.

Helper T Cells' Role in the Activation of B Cells

The production of antibodies by a B cell often requires the cooperation of other cells, particularly helper T cells. For example, with the help of certain macrophages, helper T cells stimulate B cells to produce antibodies against a type of antigen known as a **T-dependent antigen.** T-dependent antigens are mainly proteins such as those found on viruses, bacteria, foreign red blood cells, and hapten-carrier combinations.

FIGURE 17.16 How helper T cells may activate B cells to make antibodies against T-dependent antigens. B cells require the help of antigen-presenting cells (APCs) and specialized helper T cells to produce antibodies against T-dependent antigens.

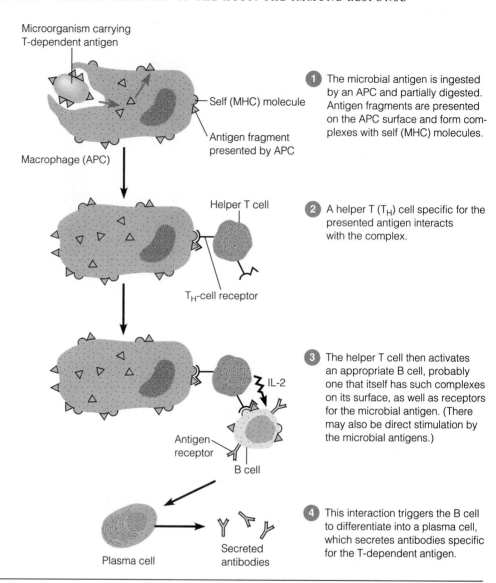

Microorganism carrying T-dependent antigen

Self (MHC) molecule

Antigen fragment presented by APC

Macrophage (APC)

Helper T cell

T_H-cell receptor

IL-2

Antigen receptor

B cell

Plasma cell

Secreted antibodies

1 The microbial antigen is ingested by an APC and partially digested. Antigen fragments are presented on the APC surface and form complexes with self (MHC) molecules.

2 A helper T (T_H) cell specific for the presented antigen interacts with the complex.

3 The helper T cell then activates an appropriate B cell, probably one that itself has such complexes on its surface, as well as receptors for the microbial antigen. (There may also be direct stimulation by the microbial antigens.)

4 This interaction triggers the B cell to differentiate into a plasma cell, which secretes antibodies specific for the T-dependent antigen.

When a T_H cell encounters a processed piece of T-dependent antigen combined with an MHC molecule on an antigen-presenting cell (Figure 17.16), the T_H cell is then primed to activate the appropriate B cell. The B cell also must carry self molecules; indeed, the T_H-cell receptor probably recognizes a single complex of the antigen and the self molecules on the B-cell surface.

This interaction with a helper T cell stimulates the B cell to differentiate into a plasma cell, which then produces antibodies against the T-dependent antigen. Helper T cells are involved in humoral immune responses that produce IgG, IgA, and IgE antibodies.

The antigens discussed earlier that can stimulate B cells directly, without the help of T cells, are called **T-independent antigens.** These T-independent antigens, usually composed of polysaccharides or lipopolysaccharides with repeating subunits, are able to bind to multiple B-cell receptors (Figure 17.17), which is probably why they do not seem to need T cell assistance. T-independent antigens generally provoke a weaker immune response than do T-dependent antigens, however.

T-independent antigen

FIGURE 17.17 T-independent antigens. T-independent antigens have repeating units that are able to cross-link several antigen receptors on the same B cell. These antigens stimulate the B cell to make antibodies without the aid of helper T cells.

HUMORAL (ANTIBODY-MEDIATED) IMMUNE SYSTEM **CELL-MEDIATED IMMUNE SYSTEM**

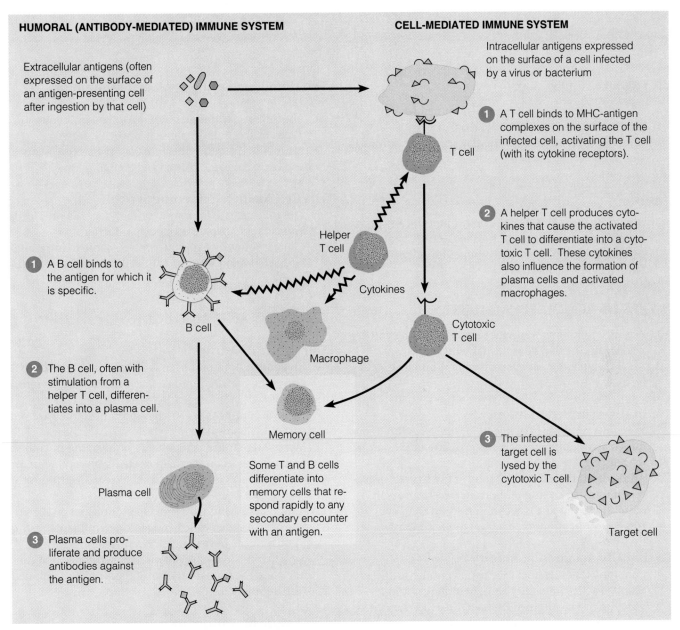

Extracellular antigens (often expressed on the surface of an antigen-presenting cell after ingestion by that cell)

Intracellular antigens expressed on the surface of a cell infected by a virus or bacterium

T cell

1 A T cell binds to MHC-antigen complexes on the surface of the infected cell, activating the T cell (with its cytokine receptors).

Helper T cell

2 A helper T cell produces cytokines that cause the activated T cell to differentiate into a cytotoxic T cell. These cytokines also influence the formation of plasma cells and activated macrophages.

Cytokines

1 A B cell binds to the antigen for which it is specific.

B cell

Cytotoxic T cell

Macrophage

2 The B cell, often with stimulation from a helper T cell, differentiates into a plasma cell.

Memory cell

3 The infected target cell is lysed by the cytotoxic T cell.

Some T and B cells differentiate into memory cells that respond rapidly to any secondary encounter with an antigen.

Plasma cell

Target cell

3 Plasma cells proliferate and produce antibodies against the antigen.

FIGURE 17.18 **Summary of the duality of the immune system.**

Antibody-Dependent Cell-Mediated Cytotoxicity

With the help of antibodies produced by the humoral immune system, the cell-mediated immune system can stimulate NK cells and cells of the nonspecific defense system to kill targeted cells. This process is referred to as **antibody-dependent cell-mediated cytotoxicity (ADCC).** (At one time the cells functioning in this manner were thought to be a special class of cells, called *killer cells,* but they now appear to be mostly NK cells.)

In this process, the target cell must first be coated with antibodies, leaving the antibodies' Fc (stem) regions pointing outward. A variety of cells in addition to NK cells, such as macrophages, neutrophils, and eosinophils, have re-

ceptors that will bind to these protruding Fc regions, and thus to the target cell. The target cell is then lysed by various substances secreted by the attacking cells. By this process the immune systems can successfully attack many relatively large organisms such as the various stages of parasitic helminths.

• • •

Figure 17.18 summarizes the duality of the immune system, with its interconnected humoral (antibody-mediated) and cell-mediated responses. In the next chapter we will discuss immunization and how it stimulates the immune system to prevent disease, as well as some common tests used to diagnose diseases.

STUDY OUTLINE

INTRODUCTION (p. 425)

1. An individual's genetically predetermined resistance to certain diseases is called innate resistance.
2. Individual resistance is affected by sex, age, nutritional status, and general health.

IMMUNITY (p. 425)

1. Immunity is the ability of the body to specifically counteract foreign organisms or substances called antigens.
2. Immunity results from the production of specialized lymphocytes and antibodies.

TYPES OF ACQUIRED IMMUNITY (pp. 425–427)

1. Acquired immunity is specific resistance to infection developed during the life of the individual.
2. An individual may develop or acquire immunity after birth.

Naturally Acquired Immunity (p. 426)

1. Immunity resulting from infection is called naturally acquired active immunity; this type of immunity may be long lasting.
2. Antibodies transferred from a mother to a fetus (transplacental transfer) or to a newborn in colostrum results in naturally acquired passive immunity in the newborn; this type of immunity can last up to a few months.

Artificially Acquired Immunity (p. 426)

1. Immunity resulting from vaccination is called artificially acquired active immunity and can be long lasting.
2. Vaccines can be prepared from attenuated, inactivated, or killed microorganisms and toxoids.
3. Artificially acquired passive immunity refers to humoral antibodies acquired by injection; this type of immunity can last for a few weeks.
4. Antibodies made by a human or other mammal may be injected into a susceptible individual.
5. Serum containing antibodies is often called antiserum.
6. When serum is separated by electrophoresis, antibodies are found in the gamma fraction of the serum and are termed immune serum globulin or gamma globulin.

THE DUALITY OF THE IMMUNE SYSTEM (p. 427)

1. B lymphocytes and antibodies are responsible for humoral (antibody-mediated) immunity and T lymphocytes for cell-mediated immunity.

The Humoral (Antibody-Mediated) Immune System (p. 427)

1. The humoral immune system involves antibodies produced by B cells in response to a specific antigen.
2. Antibodies primarily defend against bacteria, viruses, and toxins in blood plasma and lymph.

The Cell-Mediated Immune System (p. 427)

1. The cell-mediated immune system depends on T cells and does not involve antibody production.
2. Cellular immunity is primarily a response to intracellular bacteria and viruses, multicellular parasites, transplanted tissue, and cancer cells.

ANTIGENS AND ANTIBODIES (pp. 427–431)

The Nature of Antigens (pp. 427–429)

1. An antigen or immunogen is a chemical substance that causes the body to produce specific antibodies or sensitized T cells.
2. As a rule, antigens are foreign substances; they are not part of the body's chemistry.
3. Most antigens are components of invading microbes: proteins, nucleoproteins, lipoproteins, glycoproteins, or large polysaccharides with a molecular weight greater than 10,000.
4. Antibodies are formed against specific regions on the surface of an antigen called antigenic determinants.
5. Most antigens have many different determinants.
6. A hapten is a low-molecular-weight substance that cannot cause the formation of antibodies unless combined with another molecule.

The Nature of Antibodies (pp. 429–431)

1. An antibody or immunoglobulin is a protein produced by B lymphocytes in response to the presence of an antigen and is capable of combining specifically with the antigen.
2. An antibody has at least two identical antigen-binding (valence) sites.
3. A single bivalent antibody unit is a monomer.
4. Most antibody monomers consist of four polypeptide chains. Two are heavy chains, and two are light chains.
5. Within each chain is a variable portion (where antigen binding occurs) and a constant portion (which serves as a basis for distinguishing the classes of antibodies).
6. An antibody monomer is Y- or T-shaped; the variable regions form the tips, and the constant regions form the base and Fc (stem) region.
7. The Fc region can attach to a host cell or complement.
8. IgG antibodies are the most prevalent in serum; they provide naturally acquired passive immunity, neutralize

bacterial toxins, participate in complement fixation, and enhance phagocytosis.

9. IgM antibodies consist of five monomers held by a joining chain; they are involved in agglutination and complement fixation.
10. Serum IgA antibodies are monomers; secretory IgA antibodies are dimers that protect mucosal surfaces from invasion by pathogens.
11. IgD antibodies are antigen receptors on B cells.
12. IgE antibodies bind to mast cells and basophils and are involved in allergic reactions.

B CELLS AND HUMORAL (ANTIBODY-MEDIATED) IMMUNITY (pp. 431–437)

1. Humoral immunity involves antibodies that are produced by B cells.
2. Bone marrow stem cells give rise to B cells.
3. Mature B cells migrate to lymphoid organs.
4. A mature B cell recognizes an antigen with antigen receptors.

The Activation of Antibody-Producing Cells by Clonal Selection (pp. 431–432)

1. According to the clonal selection theory, a B cell becomes activated when an antigen reacts with antigen receptors on its surface.
2. The activated B cell produces a clone of plasma cells and memory cells.
3. Plasma cells secrete antibodies. Memory cells recognize pathogens from previous encounters.
4. T cells and B cells that react with self antigens are destroyed during fetal development; this is called clonal deletion.

Antigen–Antibody Binding and Its Results (pp. 432–433)

1. An antigen binds to the antigen-binding site (variable regions) of an antibody to form an antigen–antibody complex.
2. IgG antibodies inactivate viruses and neutralize bacterial toxins.
3. Agglutination of cellular antigens occurs when an IgG or IgM antibody combines with two cells.
4. Soluble antigens are precipitated when an IgG or IgM antibody combines with two antigen molecules.
5. Antigen–antibody complexes involving IgG and IgM antibodies can fix complement, resulting in lysis of a bacterial (antigenic) cell.

Immunological Memory (p. 433)

1. The amount of antibody in serum is called the antibody titer.
2. The response of the body to the first contact with an antigen is called the primary response. It is characterized by the appearance of IgM followed by IgG.
3. Subsequent contact with the same antigen results in a very high antibody titer and is called the secondary, anamnestic, or memory response. The antibodies are primarily IgG.
4. T cells also have primary and secondary responses.

The Molecular Basis of Antibody Diversity (pp. 433–435)

1. The constant region of the heavy chain of each immunoglobulin class (IgG, IgM, etc.) is identical in a person. The light chains are either a kappa or lambda class peptide.
2. The constant region of an antibody is coded for by the C segment; the variable region is coded for by hundreds of V segments and 4 J segments (and 12 D segments in heavy chains).
3. When lymphocytes develop, the V segments, a J segment (D segments in heavy chains), and the C segment recombine to form an antibody gene.
4. Somatic point mutations can occur in a formed antibody gene.
5. Recombination and mutations result in the ability to produce huge numbers of different antibody molecules.

Monoclonal Antibodies and Their Uses (pp. 435–437)

1. Hybridomas are produced in the laboratory by fusing a cancerous cell with an antibody-secreting plasma cell.
2. A hybridoma cell culture produces large quantities of the plasma cell's antibody, called monoclonal antibodies.
3. Monoclonal antibodies are used in serologic identification tests, to prevent tissue rejections, and to treat septic shock.
4. Immunotoxins can be made by combining a monoclonal antibody and a toxin; the toxin will then kill a specific antigen.

T CELLS AND CELL-MEDIATED IMMUNITY (pp. 437–443)

1. Cell-mediated immunity is associated with lymphocytes called T cells.
2. T cells respond to intracellular antigens.

The Components of Cell-Mediated Immunity (pp. 437–439)

1. T cells and other lymphocytes are responsible for cell-mediated immunity.
2. After maturation in the thymus gland, T cells migrate to lymphoid tissue.
3. T cells differentiate into several types of specialized T cells when they are stimulated by an antigen.

Types of T Cells (pp. 437–439)

1. Helper T (T_H) or CD4 cells activate other cells of the immune system.
2. Cytotoxic T (T_C) or CD8 cells lyse target cells, such as virus-infected cells and cancer cells.

3. Delayed hypersensitivity T (T_D) cells produce cytokines.

4. Suppressor T (T_S) cells (CD8 cells) appear to regulate the immune response.

5. Memory cells are responsible for a secondary immune response.

6. Natural killer (NK) cells lyse viral-infected and tumor cells. They are not T cells and are not antigenically specific.

Chemical Messengers of Immune Cells: Cytokines (Lymphokines) (p. 439)

1. Cells of the immune system communicate with each other by means of chemicals called cytokines.

2. Interleukins (IL) are cytokines.

The Cell-Mediated Immune Response (pp. 439–441)

1. T cells are specific for a single antigen but recognize the antigen only when it is associated with self proteins (associative recognition).

2. The antigen must be attacked by an antigen-presenting cell (APC) and positioned on the surface of the APC.

3. The major histocompatibility complex (MHC) consists of cell surface proteins that are unique to each individual and provide self molecules.

4. T_C cells release perforin to lyse cells carrying the target antigen and self molecule.

5. An APC secretes IL-1 after binding with a T_H cell; the T_H cell then secretes IL-2.

6. IL-2 causes proliferation of T_H cells specific to that antigen.

7. Activated T_H cells stimulate macrophages, T_C cells, and B cells.

Some Interactions of the Cell-Mediated Immune System with the Humoral Immune System (pp. 441–443)

1. T_H cells activate B cells to produce antibodies against T-dependent antigens.

2. Antigens that directly activate B cells are called T-independent antigens.

3. In antibody-dependent cell-mediated cytotoxicity (ADCC), NK cells, macrophages, and other white blood cells lyse antibody-coated cells.

STUDY QUESTIONS

Review

1. Define immunity.

2. Distinguish between the terms in the following pairs of terms.
   (a) Nonspecific resistance and immunity
   (b) Humoral and cell-mediated immunity
   (c) Active and passive immunity
   (d) Innate resistance and acquired immunity
   (e) Natural and artificial immunity
   (f) T-dependent and T-independent antigens
   (g) CD4 and CD8

3. Classify the following examples of immunity as naturally acquired active immunity, naturally acquired passive immunity, artificially acquired active immunity, or artificially acquired passive immunity.
   (a) Immunity following injection of diphtheria toxoid
   (b) Immunity following an infection
   (c) Newborn's immunity to yellow fever
   (d) Immunity following an injection of antirabies serum

4. Explain what an antigen is. Distinguish between immunogen and hapten.

5. Explain what an antibody is by describing the characteristics of antibodies. Diagram the structure of a typical antibody; label the heavy chain, light chain, and constant, variable, and Fc regions.

6. Discuss the clonal selection mechanism.

7. Define each of the following on the basis of its role in the immune response: IgG, IgM, IgA, IgD, IgE.

8. By means of a diagram, explain the role of T cells and B cells in immunity.

9. Explain a function for the following types of cells: T_C, T_D, T_H, and T_S. What is a cytokine?

10. (a) At time A, in the following graph, the host was injected with tetanus toxoid. At time B, the host was given a booster dose. Explain the meaning of the areas of the curve that are marked a and b.
    (b) Identify in the graph the antibody response of this same individual to exposure to a new antigen indicated at time B.

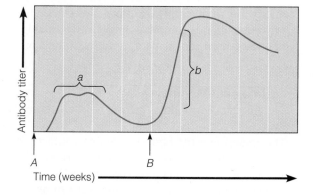

11. What effect does an antibody have on an antigen?

12. How does a T cell recognize an antigen?

13. What are natural killer cells?

14. How would each of the following prevent infection:
    (a) antibodies against *Neisseria gonorrhoeae* fimbriae?
    (b) antibodies against host cell mannose?

15. How are monoclonal antibodies produced?

16. Explain why a person who recovers from a disease can attend others with the disease without fear of contracting the disease.

17. Pooled human immune serum globulin is sometimes administered to a patient after exposure to hepatitis A. What is human immune serum globulin? What type of immunity might this confer on the patient?

Critical Thinking

1. Assume that the gene pool for heavy chains consists of five V, four D, and four J segments. How many different heavy chains are possible from these segments? Diagram the events leading to the production of a heavy chain consisting of $V_3D_2J_3C$.

2. Provide an explanation for the following:
   (a) IL-2 has been used to treat pancreatic cancer.
   (b) IL-2 would exacerbate autoimmune diseases.

3. A positive tuberculin skin test shows cell-mediated immunity to *Mycobacterium tuberculosis*. How could a person acquire this immunity?

Clinical Applications

1. A woman had life-threatening salmonellosis that was successfully treated with monoclonal antibodies. Why did this treatment work, when antibiotics and her own immune system failed?

2. A patient with acquired immune deficiency syndrome (AIDS) has a low T_H/T_S cell ratio. What clinical manifestations would this cause?

3. A patient with chronic diarrhea was found to lack IgA in his secretions, although he had a normal level of serum IgA. What was this patient found to be unable to produce?

4. Newborns (under 1 year) who contract dengue fever have a higher chance of dying from it if their mothers had dengue fever prior to pregnancy. Explain why.

Practical Applications of Immunology

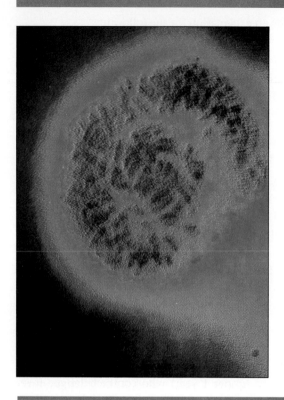

THE SCIENCE OF IMMUNOLOGY has its roots in observations about humans and disease. In this chapter we will look at two tools that resulted from such observations: vaccines, which protect people from disease by inducing immunity, and diagnostic tests, which tell us what microorganism is causing a disease. As our understanding of the immune system at the cellular and molecular levels increases, you can expect to see rapid advances in these areas, especially diagnostic tests.

VACCINES

LEARNING OBJECTIVES

• *Define vaccination, herd immunity, and attenuated, toxoid, and acellular vaccines.*

• *Explain why vaccination works.*

• *Differentiate between whole-agent vaccines and subunit vaccines. Provide an example of each.*

A **vaccine** is a suspension of microorganisms that will induce immunity in a host. Long before the invention of vaccines, it was known that people who recovered from certain diseases, such as smallpox, were immune to the disease thereafter. As discussed in Chapter 1, Chinese physicians may have been the first to try to exploit this phenomenon to prevent disease when they had children inhale dried smallpox scabs. We can only speculate as to whether the results were successful or not.

In the 1700s in Turkey, Lady Montague, the wife of the British ambassador, reported that it was the practice there that an "... old woman comes with a nutshell full of the matter of the best sort of smallpox and asks what veins you please to have opened, ... and puts into the vein as much venom as can lie upon the head of her needle. ..." This practice apparently often led to only a couple of days of mild illness. Success probably depended on two factors: first, the choice of a different portal of entry for the pathogen, which normally infects the body by entering through the respiratory system, and second, the use of a less virulent form of the virus. This process, which is called **variolation**, can be a useful form of protection, but the virulence of the pathogen is uncertain, and on occasion, the procedure may backfire and kill the recipient.

One person who received this treatment, at the age of eight, was Edward Jenner. Later in life, as a physician, Jenner was intrigued by a dairymaid's assertion that she had no fear of smallpox because she had already had cowpox. (Cowpox is a mild disease that causes lesions on cows' udders; dairymaids' hands often became infected during milking.) Motivated by his childhood memory of variolation, Jenner began a series of experiments in 1798 in which he deliberately inoculated people with cowpox in an attempt to prevent smallpox (Figure 18.1).

FIGURE 18.1 Jenner and the principle of vaccination.
Edward Jenner's inoculations with cowpox virus, which conferred immunity against smallpox, are shown here in a historical painting.

Principles and Effects of Vaccination

We now know that Jenner's inoculations worked because the cowpox virus provoked a primary immune response in the recipients, leading to the formation of antibodies and memory cells. When the recipient of such a vaccination then encountered the smallpox virus, these two viruses are so similar that the memory cells were stimulated, producing the faster and more intense secondary immune response (see Figure 17.10). This strategy, called **vaccination** or **immunization**, therefore confers artificially acquired active immunity (see Figure 17.1) on the recipient. A campaign of such vaccinations led (in 1977) to the worldwide eradication of smallpox, the first infectious disease for which deliberate eradication has been accomplished. The development of conventional vaccines based on the model of the smallpox vaccine is the single most important application of immunology.

Many communicable diseases can be controlled by behavioral and environmental methods. For example, proper sanitation can prevent the spread of cholera, and use of condoms can prevent the spread of sexually transmitted diseases. If prevention fails, bacterial diseases can often be treated with antibiotics. Viral diseases, however, are not readily treated once contracted. Therefore, vaccination is often the only feasible method of controlling viral disease. *Control* of a disease does not necessarily require that everyone be immune to it. If most of the population is immune, outbreaks are limited to sporadic cases because there are not enough susceptible individuals to support the spread of epidemics. This is known as **herd immunity.**

The principal vaccines used to prevent bacterial and viral diseases in the U.S.A. are listed in Tables 18.1 and 18.2. Recommendations for childhood immunizations against some of these diseases are given in Table 18.3. U.S. travelers who might be exposed to cholera, yellow fever, or other diseases not endemic in the U.S.A., can obtain current inoculation recommendations from the U.S. Public Health Service and local public health agencies.

Experience has shown that vaccines against enteric bacterial pathogens, such as those causing cholera and typhoid, are not nearly as effective or long-lived as those against viral diseases such as measles and smallpox. Vaccines that are effective against bacteria (including rickettsias and mycoplasmas) and against viruses have been produced, but to date no useful vaccines against chlamydias, fungi, protozoa, or helminthic parasites are in use in humans. However, researchers are currently working to develop a vaccine against malaria, which is caused by a protozoan parasite (see the box on p. 462).

Types of Vaccines and Their Characteristics

There are two basic types of vaccines: whole-agent and subunit vaccines. As the name implies, whole-agent vaccines contain whole, nonvirulent microorganisms. Subunit vaccines contain some part or product of microorganisms that can induce an immune response in the recipient.

Whole-Agent Vaccines

In whole-agent vaccines, the microorganisms may be either *inactivated* (killed) or alive but *attenuated* (weakened). Attenuated virus vaccines tend to mimic an actual infection and usually provide better immunity than that provided by inactivated viruses.

Viruses for vaccines may be inactivated by treatment with formalin or other chemicals. Heat is not used for this treatment because it is likely to alter the surface components of the virus enough to interfere with its ability to provoke an effective immune response. Inactivated virus vaccines commonly used in humans include those against rabies (animals sometimes receive a live vaccine considered too hazardous for humans), influenza, and polio (the Salk polio vaccine).

Many attenuated virus vaccines provide recipients life-long immunity without booster immunizations, and an effectiveness of 95% is not unusual. This long-term effectiveness probably occurs because the attenuated viruses tend to replicate in the body, and the original dose thus increases considerably over time. Examples of attenuated vaccines are the Sabin polio vaccines and those used against yellow fever, measles, rubella, and mumps. The at-

TABLE 18.1 Principal Vaccines Used in the U.S.A. to Prevent Bacterial Diseases in Humans

DISEASE	VACCINE	RECOMMENDATION	BOOSTER
Cholera	Crude fraction of *Vibrio cholerae*	For people who work and live in endemic areas	Every 6 months as needed
Diphtheria	Purified diphtheria toxoid	See Table 18.3	Every 10 years for adults
Meningococcal meningitis	Purified polysaccharide from *Neisseria meningitidis*	For people with substantial risk of infection	Need not established
Pertussis (whooping cough)	Killed whole or fragments of *Bordetella pertussis*	Children prior to school age; see Table 18.3	For high-risk adults
Plague	Crude fraction of *Yersinia pestis*	For people who come in regular contact with wild rodents in endemic areas	Every 6 to 12 months for continuing exposure
Pneumococcal pneumonia	Purified polysaccharide from *Streptococcus pneumoniae*	For adults with certain chronic diseases; people over 65	Normally not recommended
Tetanus	Purified tetanus toxoid	See Table 18.3	Every 10 years for adults
Typhoid fever	Killed or attenuated *Salmonella typhi*	For people in endemic areas or areas having outbreak	Every 3 years with killed, every 5 years with attenuated
Hemophilus influenzae b meningitis	Polysaccharide from *Hemophilus influenzae* b conjugated with protein to enhance effectiveness	See Table 18.3	None recommended

tenuated viruses are usually derived from mutations accumulated during long-term cell culture. One danger of such vaccines is that the live viruses can back-mutate to a virulent form, although this very rarely happens.

Subunit Vaccines

A subunit vaccine uses only those antigenic fragments of a microorganism that are best suited to stimulating an immune response. When subunit vaccines are produced by genetic engineering techniques—when other microbes are programmed to produce the desired antigenic fraction—the vaccines are called **recombinant vaccines.** For example, the vaccine against the hepatitis B virus consists of a portion of the viral protein coat that is produced by a genetically engineered yeast. Subunit vaccines are inherently safer because they cannot reproduce in the recipient. They also contain little or no extraneous material and therefore produce fewer adverse effects. However, this type of vaccine is usually less effective in stimulating antibody formation.

Toxoids (inactivated bacterial toxins) are a form of subunit vaccine that have been in use for many years. When injected, the toxoid stimulates an immune response that neutralizes any circulating toxin. Tetanus and diphtheria toxoids are familiar examples.

Protection against some pathogens depends on an immune response to their polysaccharide capsules, which are often poorly antigenic, especially in very young children. (See the discussion of T-cell–independent antigens in Chapter 17.) Vaccines against *Hemophilus influenzae* type b and *Streptococcus pneumoniae* contain only polysaccharide fractions. To enhance their effectiveness, they are com-

bined with proteins such as the toxoids of tetanus or diphtheria. Such vaccines are known as **conjugated vaccines.**

Another approach to making a subunit vaccine is to fragment a conventional whole-agent vaccine and collect only those portions that contain the desired antigens. Because the complete cells are not used, these vaccines are called **acellular vaccines.** A new vaccine against pertussis (whooping cough) is of this type (see Table 18.1).

Development of Vaccines

Interest in vaccine development declined with the introduction of antibiotics but has intensified in recent years. This is especially true for diseases affecting the health of populations in less developed parts of the world. Ideally, a vaccine gives lifelong immunity with a single, inexpensive administration as close to birth as possible. This seems improbable with our present technology, but it is a worthwhile goal.

Animal Cultivation and Cell Culture Methods

The development of a vaccine has historically required a method of growing the pathogen in usefully large amounts. The early successful viral vaccines were developed by animal cultivation. The vaccinia virus (of mysterious origins but closely enough related to the smallpox virus to induce immunity against it) was grown on the shaved bellies of calves. The rabies virus used by Pasteur in his vaccine over a hundred years ago was grown in the central nervous system of rabbits. The introduction of vac-

TABLE 18.2 Principal Vaccines Used in the U.S.A. to Prevent Viral Diseases in Humans

DISEASE	VACCINE	RECOMMENDATION	BOOSTER
Influenza	Killed virus	For chronically ill people, especially with respiratory diseases, or for healthy people over 65 years old	Annual
Measles	Attenuated virus	For infants 15 months old	See Table 18.3
Mumps	Attenuated virus	For infants 15 months old	(Duration of immunity not known)
Rubella	Attenuated virus	For infants 15 months old; for females of childbearing age who are not pregnant	(Duration of immunity not known)
Poliomyelitis	Attenuated or killed virus (enhanced potency type)	For children, see Table 18.3; for adults, as risk to exposure warrants	(Duration of immunity not known)
Rabies	Killed virus	For field biologists in contact with wildlife in endemic areas; for veterinarians	Every 2 years
Yellow fever	Attenuated virus	For people traveling to endemic areas; for military personnel	Every 10 years
Hepatitis B	Antigenic fragments of virus	For children, see Table 18.3; for adults, especially health care workers, homosexual males, injecting drug users, heterosexual persons with multiple partners, and household contacts of hepatitis B carriers	Duration of protection at least 7 years, need for boosters uncertain

TABLE 18.3 Schedule of Childhood Immunizations

VACCINE	BIRTH	2 MO.	4 MO.	6 MO.	15 MO.	4–6 YR. (SCHOOL ENTRY)	14–16 YR. (JR. HIGH ENTRY)
DTP (Diphtheria, Tetanus, Pertussis)		+	+	+	+*	+	
Td† (Tetanus, diphtheria toxoids)							+
OPV (Oral Polio Vaccine)		+	+		+	+	
MMR (Measles, Mumps, Rubella)					+‡	+	
Hib-conjugate vaccine§		+ (approx.)	+ (approx.)	+‖			
HBV#	+	+		+**			

*15–18 mo.
†Note DTP above
‡Or at 12 mo., if at high risk of measles
§Hib = *Hemophilus influenzae* type b; a combination vaccine including DTP is proposed
‖6–12 mo., depending on vaccine
#Hepatitis B; a second option is at 1–2 mo., 4 mo., and 6–18 mo., simultaneously with Hib, OPV, and MMR
**6–18 mo.

cines against polio, measles, mumps, and a number of other viral diseases that would not grow in anything but a living human awaited the development of cell culture techniques. Cell cultures from human sources, or more often from animals such as monkeys that are closely related to humans, enabled growth of these viruses on a large scale. Interestingly, the first vaccine against hepatitis B virus used viral antigens extracted from the blood of chronically infected humans because no other source was available.

Although in many cases cell culture techniques have been successful in allowing the development of vaccines against human viruses, they have not eliminated the need for a suitable test animal. For example, HIV, the AIDS virus, can be grown in cell cultures, but presently not even apes or monkeys develop AIDS. (Chimpanzees, which are an endangered species, become infected but do not develop the disease. Although the pigtail macaque monkey [not an endangered species] has been reported to support HIV infection, to date, the infection has not yet progressed to full-blown AIDS in any of them.) It is unlikely that humans would be administered even a killed version of such a lethal agent as HIV.

Scientists continue to remember the manufacturing defect in making the first polio vaccine in the 1950s: not all viruses were killed, and some vaccine batches contained infective viruses. Therefore, most believe that a successful AIDS vaccine will probably not contain any genetic information at all but will be genetically engineered to contain only antigenic subunits of the virus (see Chapter 19).

Recombinant Methods

As previously mentioned, some vaccines have been developed with the genetic engineering techniques described in Chapter 9. Scientists have been able to insert genes for the surface proteins of pathogenic microbes such as the smallpox virus into the generally harmless vaccinia virus. The result is a sort of "sheep in wolf's clothing," a recombinant virus that can function as a whole-agent vaccine because it has the external proteins of a pathogen but does not cause disease. In addition, because the vaccinia viral genome is unusually large and has room for several extra genes, a genetically engineered vaccinia virus might theoretically be used as a vaccine for several diseases simultaneously.

The genes for proteins that make up subunit vaccines can also be introduced into the genome of a bacterium or yeast by genetic engineering techniques. These subunits are produced in quantity by the bacterium or yeast and are then harvested in pure form for use as a subunit vaccine. For example, the latest hepatitis B vaccine is a subunit vaccine produced in a yeast (see Table 18.2).

DIAGNOSTIC IMMUNOLOGY

LEARNING OBJECTIVES

- *Differentiate between direct and indirect diagnostic tests.*

- *Explain how precipitation, immunodiffusion, agglutination, neutralization, viral hemagglutination, immunofluorescence, and ELISA can be used to diagnose diseases.*

- *Explain how countercurrent immunoelectrophoresis, hemagglutination, viral hemagglutination inhibition, complement fixation, and radioimmunoassay can be used to diagnose a disease or detect a compound.*

More than 100 years ago, when Robert Koch was trying to develop a vaccine against tuberculosis, he observed that when tubercular guinea pigs were injected with a suspension of *Mycobacterium tuberculosis*, the site of the injection became red and slightly swollen a day or two later. You may recognize this symptom as essentially the same test we use today to determine whether a person has been infected by the tuberculosis pathogen. Koch, of course, had no idea of the mechanism of cell-mediated immunity that caused this phenomenon, and he did not know of the existence of antibodies.

Since then, immunology has given us many other invaluable diagnostic tools, most of which are based on interactions of humoral antibodies with antigens. In this section we will discuss several of the most important techniques used to detect antigens and antibodies.

One problem our diagnostic tools must overcome is that antibodies cannot be seen directly. Even at magnifications of well over 100,000×, they appear only as fuzzy,

ill-defined particles. All we can do is infer their presence indirectly through a variety of reactions. In most instances the presence of antibodies is of primary interest, but the procedures can be reversed so that known antibodies can be used to detect antigens.

Precipitation Reactions

Precipitation reactions involve the reaction of soluble antigens with IgG or IgM antibodies (described in Chapter 17) to form large, interlocking molecular aggregates called *lattices*.

Precipitation reactions occur in two distinct stages. First, the antigens and antibodies rapidly form small antigen–antibody complexes (see Chapter 17). This interaction occurs within seconds and is followed by a slower reaction, which may take minutes to hours, in which the antigen–antibody complexes form lattices that precipitate from solution. Precipitation reactions normally occur only when the ratio of antigen to antibody is optimal. Figure 18.2 shows that no visible precipitate forms when either component is in excess. The optimal ratio is produced when separate solutions of antigen and antibody are placed adjacent to each other and allowed to diffuse together. In a **precipitin ring test** (Figure 18.3), a cloudy line of precipitation (ring) appears in the area in which the optimal ratio has been reached (the *zone of equivalence*).

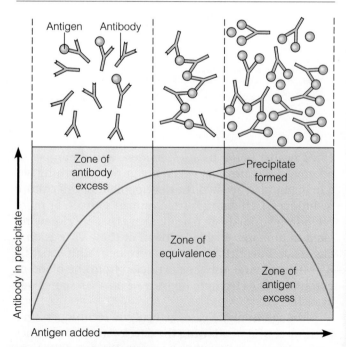

FIGURE 18.2 Precipitation curve. The curve is based on the ratio of antigen to antibody. The maximum amount of precipitate forms in the zone of equivalence, where the ratio is optimal.

Immunodiffusion tests are precipitation reactions carried out in an agar gel medium, on either a Petri plate or a microscope slide. In one such test, the **Ouchterlony test**, wells are cut into a purified agar gel in a Petri plate. A serum containing antibodies (antiserum) is added to one well, usually centrally located, and soluble test antigens are added to each surrounding well. A line of visible precipitate develops between the wells at the point where the optimal antigen–antibody ratio is reached (Figure 18.4). The Ouchterlony test is most useful in determining whether the antigens are identical, partially identical, or totally different.

Other precipitation tests use electrophoresis to speed up the movement of antigen and antibody in a gel. Protein mixtures can be separated rapidly, sometimes in less than an hour, with this method. A modification of the precipitation reaction combines the techniques of immunodiffusion and electrophoresis in a procedure called **immunoelectrophoresis.** The procedure is used in research to separate proteins in human serum and is the basis of certain diagnostic tests. For example, the **countercurrent immunoelectrophoresis (CIE) test**, also called *counter-immunoelectrophoresis*, can be used in the diagnosis of bacterial meningitis and other diseases. This test uses serum containing known antibodies to identify an unknown antigen in some body fluid (Figure 18.5). For suspected cases of meningitis, the antigens associated with the pathogen are detected in cerebrospinal fluid. CIE is based on the fact that some antigens and antibodies have opposite charges when they are placed in buffers of correct ionic strength and pH. When an electrical current is applied, the antigens and antibodies move toward the pole with the electrical charge opposite to the one they carry. They pass through each other to do so, and a precipitation line appears within an hour if a reaction occurs. Once an important type of test, its use is now declining.

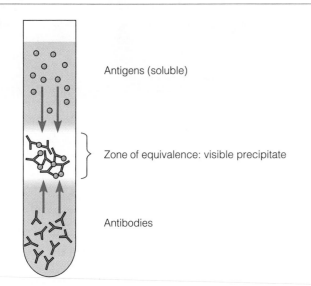

FIGURE 18.3   Precipitin ring test. This diagrammatic representation shows the diffusion of antigens and antibodies toward each other in a small-diameter test tube. Where they reach equal proportions, in the zone of equivalence, a visible line or ring of precipitate is formed.

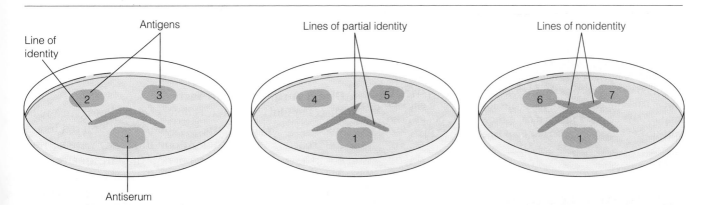

(a) Line of identity. Neither antigens (in wells 2 and 3) nor antibodies (in well 1) are able to diffuse past each other because they react and precipitate to form the line (shown in purple). Therefore, antigens 2 and 3 are identical.

(b) Lines of partial identity. The antigens in wells 4 and 5 are not identical, but they share many antigenic determinant sites and are therefore related. The spur shape of the line indicates that some of the antigens did not react with antibodies and diffused through the precipitation zone.

(c) Lines of nonidentity. The antiserum contains antibodies against antigens in wells 6 and 7. The antigens are unrelated because they diffused across each other's zones of precipitation.

FIGURE 18.4   The Ouchterlony immunodiffusion precipitation test. This figure shows three possible outcomes of this test, which is generally used to determine the degree of relatedness among antigens.

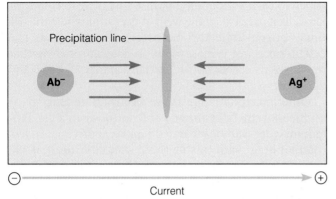

FIGURE 18.5 Countercurrent immunoelectrophoresis.
Antigens (Ag⁺) and antibodies (Ab⁻) are placed in opposite wells, and the pH of the surrounding medium is adjusted so that the antigen and antibody have opposite charges. When an electrical current is applied, antigens and antibodies move in opposite directions. A line of precipitate forms where antigen and antibody meet.

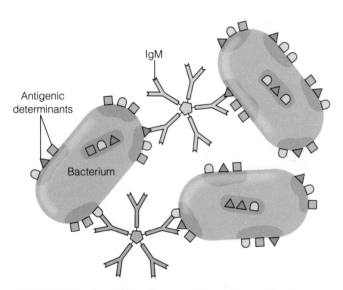

FIGURE 18.6 Agglutination reaction. When antibodies react with antigenic determinant sites on antigens carried on neighboring cells, such as these bacteria (or red blood cells), the particulate antigens (cells) agglutinate. IgM, the most efficient immunoglobulin for agglutination, is shown here, but IgG also participates in agglutination reactions (see Figure 17.9).

FIGURE 18.7 Measuring antibody titer with the direct agglutination test. (a) Each well in this microtiter plate contains, from left to right, only half the concentration of serum that is contained in the preceding well. Each well contains the same concentration of particulate antigens, in this instance red blood cells. (b) In a positive (agglutinated) reaction, sufficient antibodies are present in the serum to link the antigens together, forming a mat of antigen–antibody complexes on the bottom of the well. (c) In a negative (nonagglutinated) reaction, not enough antibodies are present to cause the linking of antigens. The particulate antigens roll down the sloping sides of the well, forming a pellet at the bottom. In this example, the antibody titer is 160 because the well with a 1:160 concentration is the most dilute concentration that produces a positive reaction.

Agglutination Reactions

Whereas precipitation reactions involve *soluble* antigens, agglutination reactions involve either *particulate* antigens (particles such as cells that carry antigenic molecules) or soluble antigens adhering to particles. These antigens can be linked together by antibodies to form visible aggregates, a reaction called **agglutination** (Figure 18.6). Agglutination reactions are very sensitive, relatively easy to read (see Figure 10.9), and available in great variety. Agglutination tests are classified as either *direct* or *indirect*.

Direct Agglutination Tests

Direct agglutination tests detect antibodies against relatively large cellular antigens, such as those on red blood cells, bacteria, and fungi. At one time they were carried out in a series of test tubes, but now they are usually done in plastic *microtiter plates,* which have many shallow wells that take the place of the individual test tubes. The amount of particulate antigen in each well is the same, but the amount of serum containing antibodies is diluted so that each successive well has half the antibodies of the previous well. These tests are used, for example, to test for brucellosis and to separate *Salmonella* isolates into serovars, types defined by serological means.

Clearly, the more antibody we start with, the more dilutions it will take to lower the amount to the point at which agglutination does not occur. This is the measure of **titer,** or concentration of serum antibody (Figure 18.7).

(a) Reaction in a positive indirect test for antibodies

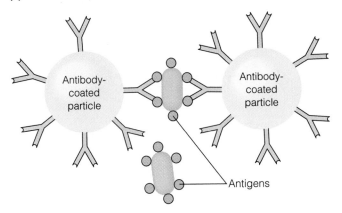

(b) Reaction in a positive indirect test for antigens

FIGURE 18.8 Reactions in indirect agglutination tests.
These tests are performed using antigens or antibodies coated onto particles such as minute latex spheres. **(a)** When particles are coated with antigens, agglutination indicates the presence of antibodies such as the IgM shown here. **(b)** When particles are coated with monoclonal antibodies, agglutination indicates the presence of antigens.

For infectious diseases in general, the higher the serum antibody titer, the greater the immunity to the disease. However, the titer alone is of limited use in diagnosing an existing illness because there is no way to know whether the measured antibodies were generated in response to the immediate infection or to an earlier illness. For diagnostic purposes, a *rise in titer* is significant; that is, the titer is higher later in the disease than at its onset. If it can be demonstrated that the person's blood had no antibody titer before the illness but has a significant titer while the disease is progressing, this change, called **seroconversion**, is also diagnostic. This is a situation frequently encountered with HIV infections.

Indirect (Passive) Agglutination Tests

Antibodies against soluble antigens can be detected by agglutination tests if the antigens are adsorbed onto particles such as red blood cells, bentonite clay, or most often, minute latex spheres, each about a tenth of the diameter of a bacterium. Such tests, known as *latex agglutination tests,* are commonly used for rapid detection of serum antibodies against many bacterial and viral diseases. In such **indirect (passive) agglutination tests** (Figure 18.8), the antibody reacts with the soluble antigen adhering to the

particles. The particles then agglutinate with one another, much as particles do in the direct agglutination tests. The same principle can be applied in reverse by using particles coated with antibodies to detect the antigens against which they are specific. This approach is especially common in tests for the streptococci that cause sore throats. A diagnosis can be completed in about 10 minutes.

Hemagglutination

When agglutination reactions involve the clumping of red blood cells, the reaction is called **hemagglutination.** These reactions, which involve red blood cell surface antigens and their complementary antibodies, are used routinely in blood typing (see Figure 19.4) and in the diagnosis of infectious mononucleosis.

Certain viruses, such as those causing mumps, measles, and influenza, have the ability to agglutinate red blood cells without an antigen–antibody reaction; this process is called **viral hemagglutination** (Figure 18.9). This type of hemagglutination can be inhibited by antibodies that neutralize the agglutinating virus. Diagnostic tests based on such neutralization reactions are covered later in the chapter.

Neutralization Reactions

As previously mentioned, **neutralization reactions** are antigen–antibody reactions in which the harmful effects of a bacterial exotoxin or a virus are blocked by specific antibodies. Neutralization reactions are used both in disease treatment and in diagnostic tests, especially for viruses. These reactions were first described in 1890, when investigators observed that immune serum could neutralize the toxic substances produced by *Corynebacterium diphtheriae.* Such a neutralizing substance, which is called an **antitoxin,** is a specific antibody produced by a host as it responds to a bacterial exotoxin or its corresponding toxoid (inactivated toxin). The antitoxin combines with the exotoxin to neutralize it (Figure 18.10a). Antitoxins produced in an animal can be injected into humans to provide passive immunity against a toxin. Antitoxins from horses are routinely used for prevention or treatment of diphtheria and botulism; tetanus antitoxin is usually of human origin.

Neutralization tests are used in the diagnosis of viral infections. The body responds to such infections by producing specific antibodies that bind to receptor sites on the viral surface. The binding of these antibodies prevents the virus from attaching to a host cell and thus destroys the virus's infectivity.

Viruses that exhibit their cytopathic (cell-damaging) effects (see Chapter 15) in cell culture or embryonated eggs can be used to detect the presence of neutralizing viral antibodies. If the serum to be tested contains antibodies against the particular virus, the antibodies will prevent that virus from infecting cells in the cell culture or

FIGURE 18.9 Viral hemagglutination. Viral hemagglutination is not an antigen–antibody reaction. Certain viruses simply bind to red blood cells and cause them to agglutinate.

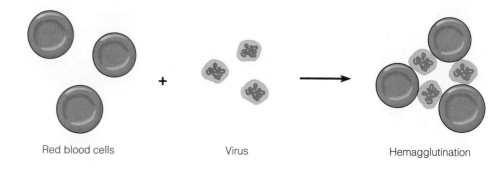

Red blood cells   Virus   Hemagglutination

FIGURE 18.10 Reactions in neutralization tests. (a) Effects of a toxin on a susceptible cell and neutralization of the toxin by antitoxin. **(b)** In the viral hemagglutination inhibition test, serum that may contain antibodies to a virus is mixed with red blood cells and the virus. If antibodies are present, as shown here, they neutralize the virus and inhibit hemagglutination.

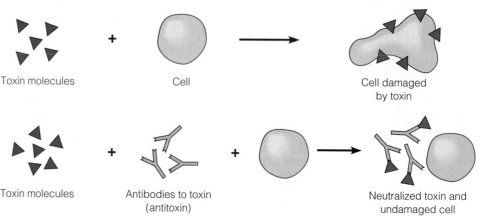

Toxin molecules   Cell   Cell damaged by toxin

Toxin molecules   Antibodies to toxin (antitoxin)   Neutralized toxin and undamaged cell

(a) Effects and neutralization of a toxin

Red blood cells   Antiviral antibodies from serum   Virus   Viruses neutralized and hemagglutination inhibited

(b) Neutralization of viruses in positive hemagglutination inhibition test

eggs, and no cytopathic effects will be seen. Such tests can thus be used both to identify a virus and to ascertain the viral antibody titer. In vitro neutralization tests are comparatively complex to carry out and are becoming less common in modern clinical laboratories.

A neutralization test that is more frequently used is the **viral hemagglutination inhibition test**. This test is used in the diagnosis of influenza, measles, mumps, and a number of other infections caused by viruses that can agglutinate red blood cells. If a person's serum contains antibodies against these viruses, these antibodies will react with the viruses and neutralize them (Figure 18.10b). For example, if hemagglutination occurs in a mixture of measles

virus and red blood cells but does not occur when the patient's serum is added to the mixture, this result indicates that the serum contains antibodies that have bound to and neutralized the measles virus.

Complement-Fixation Reactions

In Chapter 16 we discussed a group of serum proteins collectively called complement. During most antigen–antibody reactions, complement binds to the antigen–antibody complex and is used up, or fixed. This process of **complement fixation** can be used to detect very small

FIGURE 18.11 Complement-fixation test. This test is used to indicate the presence of antibodies to a known antigen. Complement will combine (be fixed) with an antibody that is reacting with an antigen. If all the complement is fixed in the complement-fixation stage, then none will remain to cause hemolysis of the red blood cells in the indicator stage. **(a)** All available complement is fixed by the antigen–antibody reaction; no hemolysis occurs, so the test is positive for the presence of antibodies. **(b)** No antigen–antibody reaction occurs. The complement remains, and the red blood cells are lysed in the indicator stage, so the test is negative.

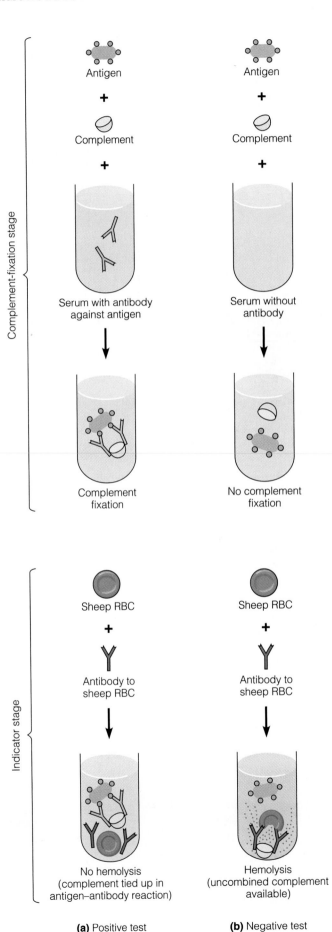

amounts of antibody. Antibodies that do not produce a visible reaction, such as precipitation or agglutination, can be demonstrated by the fixing of complement during the antigen–antibody reaction. Complement fixation was once used in the diagnosis of syphilis (Wasserman test) and is still used in the diagnosis of certain viral, fungal, and rickettsial diseases. Execution of the complement-fixation test requires great care and good controls. The test is performed in two stages, the complement-fixation and indicator stages (Figure 18.11).

Radioimmunoassay

Radioimmunoassay techniques aid in the detection of several compounds, including hormones, drugs, and immunoglobulins. The technique is especially useful for diagnosis of several types of viral hepatitis. Suppose, for example, we want to determine the amount of hepatitis B virus in a serum sample (Figure 18.12). The serum sample is combined with radioactively labeled virus and with antibodies to the virus. The radioactive virus and the virus in the sample will compete for antibody. Antigen–antibody complexes can then be isolated and analyzed for their radioactivity. If the antigen–antibody complexes have a high level of radioactive antigen, the serum sample did not have much antigen (virus) to combine with the antibody.

Immunofluorescence and Fluorescent-Antibody Techniques

Fluorescent-antibody (FA) techniques (Figure 18.13) can identify microorganisms in clinical specimens and can detect the presence of a specific antibody in serum. These techniques combine fluorescent dyes such as *fluorescein isothiocyanate (FITC)* with antibodies to make them fluoresce when exposed to ultraviolet light (see Chapter 3). These procedures are quick, sensitive, and very specific; the fluorescent-antibody test for rabies can be done in a few hours and has an accuracy rate close to 100%.

Fluorescent-antibody tests are of two types, direct and indirect. **Direct FA tests** are usually used to identify a microorganism in a clinical specimen. During this proce-

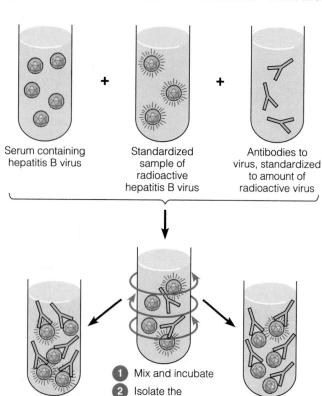

Serum containing hepatitis B virus

+

Standardized sample of radioactive hepatitis B virus

+

Antibodies to virus, standardized to amount of radioactive virus

1 Mix and incubate
2 Isolate the antigen–antibody complexes
3 Determine level of radioactivity

High level of radioactivity indicates relatively few viruses in serum

Low level of radioactivity indicates many viruses in serum

FIGURE 18.12 Radioimmunoassay. Radioimmunoassay techniques determine the concentration of material in a clinical sample. Here the procedure is used to detect hepatitis B virus.

dure, the specimen containing the antigen to be identified is fixed onto a slide. Fluorescein-labeled antibodies are then added, and the slide is incubated briefly. Next the slide is washed to remove any antibody not bound to antigen and is then examined under the fluorescence microscope for yellow-green fluorescence (Figure 18.13).

Indirect FA tests (Figure 18.13b) are used to detect the presence of a specific antibody in serum following exposure to a microorganism. They are often more sensitive than direct tests. During this procedure, a known antigen is fixed onto a slide. The test serum is then added, and, if

FIGURE 18.13 Fluorescent-antibody (FA) techniques. (a) A direct FA test to identify group A streptococci. (b) In an indirect FA test such as that used in the diagnosis of syphilis, the fluorescent dye is attached to anti-human immune serum globulin, which reacts with any human immunoglobulin (such as the *Treponema pallidum*–specific antibody) that has previously reacted with the antigen. The reaction is viewed through a fluorescence microscope (photo), and the antigen with which the dye-tagged antibody has reacted—in this case, *T. pallidum*—fluoresces (glows) in the ultraviolet illumination.

Group A streptococci from patient's throat

+

Fluorescent dye–labeled antibodies to group A streptococci

Fluorescent streptococci

(a) Reactions in a positive direct fluorescent-antibody test

LM   1 μm

T. pallidum from laboratory stock

+

Specific antibodies in serum of patient

Antibodies bind to *T. pallidum*

+

Fluorescent dye–labeled anti-human immune serum globulin

Fluorescent spirochetes

(b) Reactions in a positive indirect fluorescent-antibody test antibody that is specific to that microbe is present, it reacts with the antigen to form a bound complex. So that the antigen–antibody complex can be seen, fluorescein-labeled anti-human immune serum globulin (anti-HISG), an antibody that reacts specifically with human antibody, is added to the slide. After the slide has been incubated and washed (to remove unbound antibody), it is examined under a fluorescence microscope. If the known antigen fixed to the slide appears fluorescent, the antibody specific to the test antigen is present.

An especially interesting adaptation of fluorescent antibodies is the **fluorescence-activated cell sorter (FACS).** In Chapter 17, we learned that T cells carry antigenically specific receptors such as CD4 and CD8 on their surface, and these are characteristic of certain groups of T cells. The FACS is a modification of a *flow cytometer,* in which a suspension of cells leaves a nozzle as droplets containing no more than one cell each. Rapid vibration of the nozzle encourages formation of such droplets. After a laser beam strikes the cell-containing droplet, the beam is then received by a detector (Figure 18.14) that determines certain characteristics such as size. If the cells carry fluorescent-antibody markers to identify them as CD4 or CD8 T cells, the detector can measure this fluorescence. Equipped with such a detector, the flow cytometer becomes a FACS. As the laser beam detects a cell of a preselected size or fluorescence, an electric charge—either positive or negative—can be imparted to it. As the charged droplet falls between electrically charged plates, it is made to fall into one receiving tube or another, effectively separating cells of different types. Millions of cells can be separated in an hour with this process, all under sterile conditions, which allows them to be used in experimental work.

Enzyme-Linked Immunosorbent Assay (ELISA)

The **enzyme-linked immunosorbent assay (ELISA),** also known as the *enzyme immunoassay (EIA),* has become a widely used serological technique. There are two basic methods. The *direct ELISA* detects antigens, and the *indirect ELISA* detects antibodies. A microtiter plate with numerous shallow wells is used in both procedures. Variations of the test exist; for example, the reagents can be bound to tiny latex particles rather than to the surfaces of the microtiter plates. ELISA procedures are popular in large part because they require little interpretive skill to read; they tend to be either clearly positive or clearly negative.

Direct ELISA

In the first step of this method, antibody against the antigen to be detected is adsorbed to the surface of the wells of the microtiter plate (Figure 18.15a). A patient's sample containing unidentified antigen is then added to each well. If the antigen reacts specifically with the antibodies

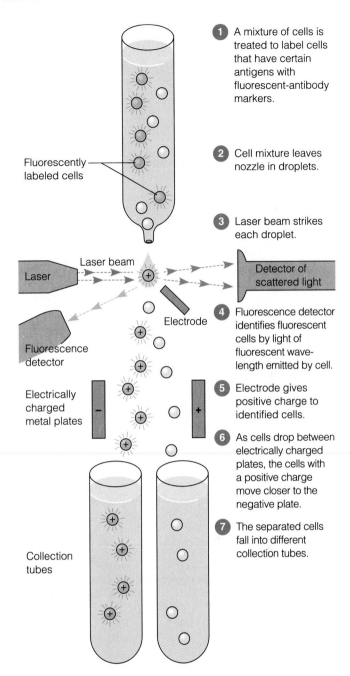

1. A mixture of cells is treated to label cells that have certain antigens with fluorescent-antibody markers.

2. Cell mixture leaves nozzle in droplets.

3. Laser beam strikes each droplet.

4. Fluorescence detector identifies fluorescent cells by light of fluorescent wavelength emitted by cell.

5. Electrode gives positive charge to identified cells.

6. As cells drop between electrically charged plates, the cells with a positive charge move closer to the negative plate.

7. The separated cells fall into different collection tubes.

Fluorescently labeled cells

Laser beam

Laser

Detector of scattered light

Electrode

Fluorescence detector

Electrically charged metal plates

Collection tubes

FIGURE 18.14 Fluorescence-activated cell sorter. This technique can be used to separate different classes of T cells. A fluorescence-labeled antibody reacts with, for example, the CD4 receptor on a T cell.

adsorbed to the well, the antigen will be retained there when the well is washed free of unbound antigen. A second antibody specific for the antigen is then added. If both the antibody adsorbed to the wall of the well and the antibody known to be specific for the antigen have reacted with the antigen, a "sandwich" will be formed, with the antigen between two antibody molecules. This reaction is visible only because the second added antibody is linked

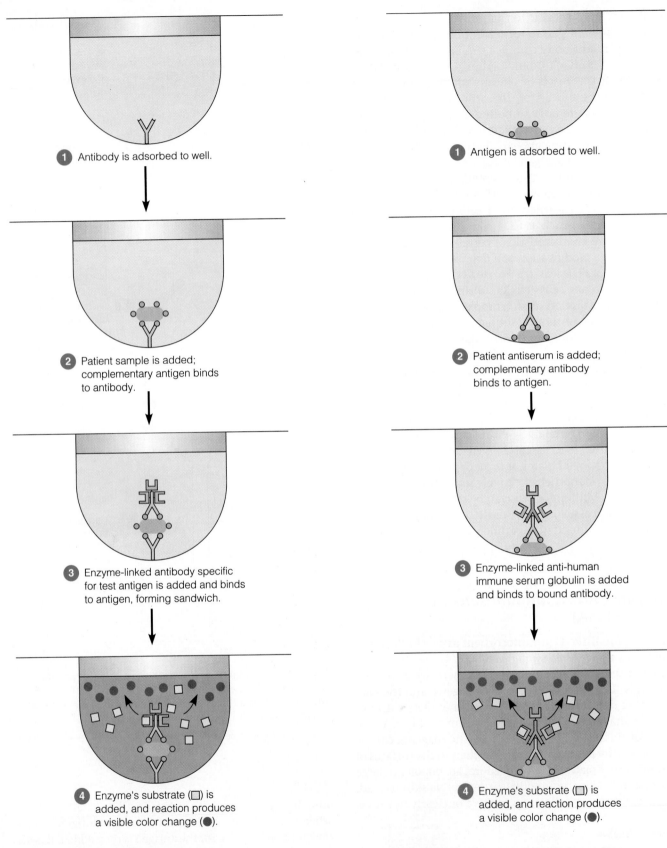

1 Antibody is adsorbed to well.

2 Patient sample is added; complementary antigen binds to antibody.

3 Enzyme-linked antibody specific for test antigen is added and binds to antigen, forming sandwich.

4 Enzyme's substrate (□) is added, and reaction produces a visible color change (●).

(a) A positive direct ELISA to detect antigens

1 Antigen is adsorbed to well.

2 Patient antiserum is added; complementary antibody binds to antigen.

3 Enzyme-linked anti-human immune serum globulin is added and binds to bound antibody.

4 Enzyme's substrate (□) is added, and reaction produces a visible color change (●).

(b) A positive indirect ELISA to detect antibodies

FIGURE 18.15 The ELISA, or EIA, method. The components are usually contained in small wells of a microtiter plate.

FIGURE 18.16 Use of monoclonal antibodies in a home pregnancy test. These home pregnancy tests detect a hormone called human chorionic gonadotropin (hCG) that is excreted only in the urine of a pregnant woman. The test area of the dipstick contains two types of monoclonal antibodies: free monoclonal antibodies, and capture monoclonal antibodies that are bound to the substrate in the test window. The free antibodies are specific for hCG and are color-labeled. The capture antibodies are specific for the hCG-free antibody complex. When the dipstick comes in contact with urine that contains hCG, the free antibodies bind the hCG and are then bound in turn by the capture antibodies, forming a "sandwich." The color-labeled antibodies then create a visible color change in the test window. (Other antibodies create the color change in the control window.)

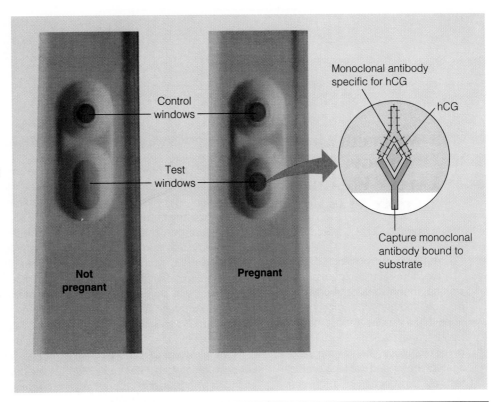

to an enzyme, such as horseradish peroxidase or alkaline phosphatase. Unbound enzyme-linked antibody is washed from the well, and then the enzyme's substrate is added to it. Enzymatic activity is indicated by a color change that can be visually detected or read with a special spectrophotometer. The test will be positive if the antigen has reacted with adsorbed antibodies in the first step. If the test antigen was not specific for the antibody adsorbed to the wall of the well, the test will be negative because the unbound antigen will have been washed away.

Indirect ELISA

In the first step of this method, a known antigen from the laboratory, rather than the antibody, is adsorbed to the walls of the shallow wells on the plate (Figure 18.15b). To see whether a serum sample contains antibodies against this antigen, the antiserum is added to the well. If the serum contains antibody specific to the antigen, the antibody will bind to the adsorbed antigen. All unreacted antiserum is washed from the well. Anti-HISG (an antibody that reacts with any human immunoglobulin) is then allowed to react with the antigen–antibody complex. The anti-HISG, which has been linked with an enzyme, reacts with the antibodies that are bound to the antigens in the well. Finally, all unbound anti-HISG is rinsed away, and the correct substrate for the enzyme is added. A colored enzymatic reaction occurs in the wells in which

bound antigen has reacted with antibody in the serum sample. This procedure resembles that of the indirect fluorescent-antibody test, except that the anti-HISG of the indirect assay is linked with an enzyme rather than with a fluorescing dye. This assay is used to test for antibodies against HIV.

Many ELISA-type tests are available for clinical use in the form of commercially prepared kits, and the procedures are often highly automated, with the results printed out directly by computer. Some tests based on this principle are also available for use by the public; one example is a commonly available home pregnancy test (Figure 18.16).

THE FUTURE OF VACCINES AND DIAGNOSTIC IMMUNOLOGY

LEARNING OBJECTIVE

- *Explain how genetic engineering has been revolutionizing vaccine development and production.*

Vaccines and diagnostic immunology are changing rapidly in response to new knowledge and new techniques. For example, genetic engineering has already begun to revolutionize current vaccine production. Most

Microbiology in the News

MOLECULAR BIOLOGY: THE LATEST WEAPON AGAINST MALARIA

THE LATEST microbiological and biochemical techniques of gene cloning, monoclonal-antibody production, and genetic engineering give hope that a vaccine against malaria may become available soon. Traditional techniques for creating vaccines against viruses and bacteria are ineffective against larger and more complex organisms, such as protozoa.

The need for an antimalarial vaccine has become desperate. In the early 1960s we thought that malaria would soon disappear completely from the U.S.A. and other temperate-zone countries and would become relatively rare even in tropical countries. The World Health Organization planned to reduce the number of *Anopheles* mosquitoes with massive sprayings of DDT and planned to treat already infected people with a variety of drugs related to quinine.

But that battle has gone against us. The mosquitoes became resistant to DDT, and *Plasmodium* has begun to develop resistance to antimalarial drugs. About 300 million new cases of malaria occur every year throughout the world. Moreover, the number of malaria cases in the U.S.A. has increased dramatically because of increased immigration from tropical countries and increased travel to tropical countries.

Researchers headed by Ruth and Victor Nussenzweig at New York University Medical Center identified a surface antigen of the sporozoite stage of all *Plasmodium* species.

The discovery of this antigen, called circumsporozoite (CS), has led to the production of monoclonal antibodies against it, which can be used in immunologic experiments. In 1983 the CS gene was cloned, making it possible to produce large quantities of the CS protein for analysis and antibody preparation.

Human tests of CS vaccines were conducted between 1987 and 1989, and their results were disappointing. Only one-third of the volunteers produced antibodies, and the vaccines did not initiate the production of memory cells.

Research leading to a generally effective malaria vaccine is hindered by the fact that the life cycle of *Plasmodium* is so complex (see Figure 12.21). The CS vaccines were directed against the sporozoite, which the mosquito injects into the victim. The immune response that is triggered by the vaccine must kill all the sporozoites within 30 minutes after the mosquito bites its victim, before the sporozoites enter the protection of the liver cells. A separate vaccine would be needed to provoke an immune response against the merozoites, which destroy red blood cells and cause the typical symptoms of malaria. This is proving to be exceptionally difficult because the antigen the merozoite causes to be displayed on the surface of red blood cells is not constant. *P. falciparum* can mutate these antigens at a

rate of 2% each generation. Yet a third vaccine would be needed to destroy the gametocytes, which continue the life cycle within the mosquito's body.

Research on vaccines is moving ahead at research centers worldwide. In 1994, Manuel Patarroyo's research team at the National University of Colombia reported the results of human trials of a vaccine consisting of protein fragments from several stages of the parasite's life cycle. This vaccine is called a "cocktail" because it consists of a variety of antigens: three merozoite peptides and a peptide sequence from the CS protein. Children suffering from *falciparum* malaria showed a 77% decrease in recurrences after receiving the vaccine and adults, a 40% decrease. However, the vaccine did not offer protection against *vivax* malaria.

An alternate approach may be to provide passive immunity to prevent the often-fatal anemia caused by malaria. The anemia is caused by the host's tumor necrosis factor (TNF). Kathleen Miller and her colleagues at Lawrence Berkeley Laboratory in California found that the number of red blood cells increased after malaria-infected mice were injected with monoclonal antibodies against TNF.

If experiments in developing a malaria vaccine are successful, it will be the first antiparasitic vaccine for humans and will provide a model for the development of vaccines against other parasitic diseases, such as trypanosomiasis and schistosomiasis.

of our present vaccines will soon be considered primitive. We may eventually develop new vaccines for such diseases as AIDS, malaria (see the box on the facing page), African trypanosomiasis, and leprosy, to name just a few. There will also be safer and even more effective vaccines for such diseases as cholera, pertussis, and many others.

New diagnostic tests that are more sensitive, specific, rapid, and simpler than those currently used will also be developed. For example, the development of monoclonal antibodies has already made it possible, for the first time, to test for chlamydial sexually transmitted diseases.

One of the most interesting speculations on the future of vaccines is the idea of *genetic immunization.* Rather than injecting protein antigen into an animal, it may prove possible to inject the DNA *coding for* a protein into the animal. Experiments have shown that the animal's body responds by producing the protein, which is then recognized as a foreign antigen, provoking the production of antibodies. The DNA is injected by the "gun" method described in Chapter 9 for injecting DNA into plant cells.

The techniques of molecular genetics such as DNA probes and PCR (see Chapter 9) are also powerful identification tools. However, it is virtually certain that rapid new developments in the practical applications of immunology can be expected in the near future, promising dramatic improvements in human health.

STUDY OUTLINE

VACCINES (pp. 448–452)

1. Edward Jenner developed the modern practice of vaccination when he inoculated people with cowpox virus to protect them against smallpox.

Principles and Effects of Vaccination (p. 449)

1. Vaccination, or immunization, confers artificially acquired active immunity.
2. Herd immunity results when most of a population is immune to a disease.

Types of Vaccines and Their Characteristics
(pp. 449–450)

1. Whole-agent vaccines consist of inactivated or attenuated microorganisms.
2. Attenuated virus vaccines generally provide lifelong immunity.
3. Subunit vaccines consist of antigenic fragments of a microorganism.
4. Subunit vaccines include recombinant vaccines (e.g., hepatitis B), toxoids (e.g., tetanus), bacterial capsules (e.g., *Hemophilus influenzae* type b), and parts of bacterial cells in acellular vaccines (e.g., whooping cough).

Development of Vaccines (pp. 450–452)

1. Viruses for vaccines may be grown in animals in cell cultures.
2. Recombinant whole-agent vaccines can be made by genetically engineering a harmless virus to produce a pathogen's antigen.
3. Recombinant subunit vaccines can be made by engineering a bacterium or yeast to produce an antigenic protein.

DIAGNOSTIC IMMUNOLOGY (pp. 452–461)

1. Many tests based on the interactions of antibodies and antigens have been developed to determine the presence of antibodies or antigens in a patient.

Precipitation Reactions (pp. 452–453)

1. The interaction of soluble antigens with IgG or IgM antibodies (precipitins) leads to precipitation reactions.
2. Precipitation reactions depend on the formation of lattices and occur best when antigen and antibody are present in optimal proportions. Excesses of either component decrease lattice formation and subsequent precipitation.
3. The precipitin ring test is performed in a small tube.
4. Immunodiffusion procedures such as the Ouchterlony test involve precipitation reactions carried out in an agar gel medium.
5. Such tests detect the presence of multiple antigens and provide information about their relatedness.
6. Immunoelectrophoresis combines electrophoresis with immunodiffusion for the analysis of serum proteins.
7. Countercurrent immunoelectrophoresis can be used to detect antigens in serum.

Agglutination Reactions (pp. 454–455)

1. The interaction of particulate antigens (cells that carry antigens) with antibodies leads to agglutination reactions.
2. Diseases may be diagnosed by combining the patient's serum with a known antigen.
3. Direct agglutination reactions can be used to determine antibody titer.
4. Diseases can be diagnosed by a rising titer or seroconversion (from no antibodies to the presence of antibodies).

5. Antibodies cause visible agglutination of soluble antigens affixed to latex spheres in indirect or passive agglutination tests.

6. Hemagglutination reactions involve agglutination reactions using red blood cells. Hemagglutination reactions are used in blood typing, diagnosis of certain diseases, and identification of viruses.

Neutralization Reactions (pp. 455–456)

1. In these reactions, the harmful effects of a bacterial exotoxin or virus are eliminated by a specific antibody.

2. An antitoxin is an antibody produced in response to a bacterial exotoxin or a toxoid that neutralizes the exotoxin.

3. In a virus neutralization test, the presence of antibodies against a virus can be detected by the antibodies' ability to prevent cytopathic effects of viruses in cell cultures.

4. Antibodies against certain viruses can be detected by their ability to interfere with viral hemagglutination in viral hemagglutination inhibition tests.

Complement-Fixation Reactions (pp. 456–457)

1. Complement-fixation reactions are serological tests based on the depletion of a fixed amount of complement in the presence of an antigen–antibody reaction.

Radioimmunoassay (p. 457)

1. Radioimmunoassay is used to detect the presence of a desired compound.

2. Antibodies against the compound are combined with a radioactively labeled antigen and a sample containing an unknown amount of antigen.

3. Analysis of radioactivity in the resulting antigen–antibody complexes indicates the amount of antigen in the sample.

Immunofluorescence and Fluorescent-Antibody Techniques (pp. 457–459)

1. Immunofluorescence techniques use antibodies labeled with fluorescent dyes.

2. Direct fluorescent-antibody tests are used to identify specific microorganisms.

3. Indirect fluorescent-antibody tests are used to demonstrate the presence of antibody in serum.

4. A fluorescence-activated cell sorter can be used to detect and count cells labeled with fluorescent antibodies.

Enzyme-Linked Immunosorbent Assay (ELISA) (pp. 459–461)

1. ELISA techniques use antibodies linked to an enzyme such as horseradish peroxidase or alkaline phosphatase.

2. Antigen–antibody reactions are detected by enzyme activity. If the indicator enzyme is present in the test well, an antigen–antibody reaction has occurred.

3. The direct ELISA is used to detect antigens against a specific antibody bound in a test well.

4. The indirect ELISA is used to detect antibodies against an antigen bound in a test well.

THE FUTURE OF VACCINES AND DIAGNOSTIC IMMUNOLOGY (pp. 461–463)

1. New and safer vaccines will be developed by genetic engineering.

2. Subunit, acellular, and conjugated vaccines will be developed.

3. The use of monoclonal antibodies will continue to make new diagnostic tests possible.

4. Genetic immunization may be possible; that is, an animal might be genetically engineered to make an antigen against which the animal's body will produce antibodies, thus conferring immunity against that antigen.

STUDY QUESTIONS

Review

1. Classify by type the following vaccines. Which could cause the disease it is supposed to prevent?
   (a) Attenuated poliomyelitis virus
   (b) Dead *Rickettsia prowazekii*
   (c) *Vibrio cholerae* toxoid

2. An antigen of the hepatitis B virus is used in the HBV vaccine. Can this vaccine cause hepatitis?

3. Explain the effects of excess antigen and antibody on the precipitation reaction. How is the precipitin ring test different from an immunodiffusion test?

4. How does the antigen in an agglutination reaction differ from that in a precipitation reaction?

5. Define the following terms, and give an example of how each reaction is used diagnostically.
   (a) Viral hemagglutination
   (b) Hemagglutination inhibition
   (c) Passive agglutination

6. Influenza virus, patient's serum, and red blood cells are mixed in a tube. If the patient has antibodies against influenza, what should be observed?

7. Measles virus, patient's serum, complement, red blood cells, and anti-red blood cells are mixed in a tube. If the patient has antibodies against measles, what should be observed?

8. Explain the fluorescent-antibody techniques.

9. Describe the direct and indirect fluorescent-antibody tests in the following situations. Sketch how the reactions might appear in these tests if the antigens and antibodies could be seen.
   (a) Rabies can be diagnosed postmortem by mixing fluorescent-labeled antibodies with brain tissue.
   (b) Syphilis can be diagnosed by adding the patient's serum to a slide fixed with *Treponema pallidum*. Anti-human immune serum globulin tagged with a fluorescent dye is added.

10. Explain the ELISA techniques.

11. Describe the direct and indirect ELISA in the following situations:
    (a) Respiratory secretions to detect respiratory syncytial virus.
    (b) Blood to detect human immunodeficiency virus antibodies.
    Which of these tests provides definitive proof of disease?

12. Match the following serological tests to the descriptions.

    ____ Precipitation
    ____ Immuno-electrophoresis
    ____ Agglutination
    ____ Radioimmuno-assay
    ____ Complement fixation
    ____ Neutralization
    ____ ELISA

    (a) Occurs with particulate antigens
    (b) Peroxidase activity indicates a positive test
    (c) Hemolysis is used as an indicator
    (d) Uses anti-human immune serum globulin
    (e) Occurs with a free soluble antigen
    (f) Used to determine the presence of antitoxin
    (g) Uses radioactive isotopes

Critical Thinking

1. What problems are associated with the use of live vaccines?

2. The World Health Organization has announced the complete eradication of smallpox and is working toward eradication of measles. Why would vaccination be more likely to eradicate a viral disease than a bacterial disease?

3. Many of the serological tests require a supply of antibodies against pathogens. For example, to test for *Salmonella*, anti-*Salmonella* antibodies are mixed with the unknown bacterium. How are these antibodies obtained?

4. A test for antibodies against *Treponema pallidum* uses an antigen called cardiolipin and the patient's serum (suspected of having antibodies). Why do the antibodies react with cardiolipin? What is the disease?

Clinical Applications

1. Which of the following is proof of a disease state? Why doesn't the other situation confirm a disease state? What is the disease?
   (a) *Mycobacterium tuberculosis* is isolated from a patient.
   (b) Antibodies against *M. tuberculosis* are found in a patient.

2. Streptococcal erythrogenic toxin is injected into a person's skin in the Dick test. What results are expected if a person has antibodies against this toxin? What type of immunological reaction is this? What is the disease?

3. The following data were obtained from fluorescent-antibody tests for anti-*Legionella* in four people. What conclusions can you draw? What is the disease?

	ANTIBODY TITER			
	DAY 1	DAY 7	DAY 14	DAY 21
Patient A	1:128	1:256	1:512	1:1024
Patient B	0	0	0	0
Patient C	1:256	1:256	1:256	1:256
Patient D	0	0	1:128	1:512

Disorders Associated with the Immune System

I N THIS CHAPTER we will see that not all immune system responses produce a desirable result, such as immunity to disease. Occasionally the reactions of the immune system are harmful. A familiar example is hay fever, which results from repeated exposure to plant pollen. Most of us also know that a blood transfusion will be rejected if the blood of the donor and the blood of the recipient are not compatible, and that rejection is also a potential problem with transplanted organs. Rejection of transfusions or transplants occurs because the immune system identifies the donated cells as nonself and attacks them. One's own tissue may also be mistakenly attacked by the immune system, causing diseases we classify as autoimmune.

We may also be harmed if the immune system fails to function properly. The obvious example is that we may be the victim of some pathogen if a vaccine, or recovery from a previous illness, failed to provide immunity to it. Cancer can be another, less apparent, indication of a failure of the immune system. When properly functioning, the immune system perceives new cancer cells to be nonself and eliminates them. Some people are born with defective immune systems, and in all of us our immune systems decline in effectiveness with age. In addition, many individuals have their immune systems deliberately crippled (immunosuppressed) to minimize rejection of transplanted organs, and others suffer from the effects of the AIDS virus, which attacks the immune system. Individuals with defective immune systems are very susceptible to otherwise rare opportunistic infections and to cancer.

Certain antigens provoke such a drastic immune response that they are termed **superantigens.** Prominent among them are the staphylococcal enterotoxins that cause food poisoning and the toxin that causes toxic shock syndrome (see Chapter 15). These toxins are nonspecific antigens and indiscriminately activate many T cell receptors at once. The result is the production of large amounts of cytokines from other stimulated cells that in turn stimulate production of a flood of T cells, perhaps 2,000 times the normal number. Most of these T cells do not match the antigens closely enough to be useful in fighting the infection, and they may even be damaging. In fact, infusion of cytokines such as interleukin-2 from T cells or macrophages activated by the T cells can produce the same symptoms as staphylococcal enterotoxin.

HYPERSENSITIVITY

LEARNING OBJECTIVE

- *Define hypersensitivity.*

The term **hypersensitivity** refers to an antigenic response beyond that considered normal; the term **allergy** is probably more familiar and is essentially synonymous. Hypersensitivity responses occur in individuals who have been *sensitized* by previous exposure to an antigen, which in this context is sometimes called an **allergen.** Once sensitized, when an individual is exposed to that antigen again, his or her immune system reacts to it in a damaging manner. The four principal types of hypersensitivity reactions, summarized in Table 19.1, are anaphylactic, cytotoxic, immune complex, and cell-mediated reactions.

TYPE I (ANAPHYLACTIC) REACTIONS

LEARNING OBJECTIVES

- *Define shocking dose, sensitizing dose, and desensitization.*
- *Describe the mechanism of anaphylaxis.*

Type I or anaphylactic reactions often occur within a few minutes after a person sensitized to an antigen is reexposed to that antigen. Anaphylaxis comes from the Greek word *phylaxis* (protection), plus the prefix *ana* (against), which makes the word mean "the opposite of protected." **Anaphylaxis** is an inclusive term for the reactions caused when certain antigens combine with IgE antibodies. Anaphylactic responses can be *systemic reactions*, which produce shock and breathing difficulties and are sometimes fatal, or *localized reactions*, which include common allergic conditions such as hay fever, asthma, and hives (slightly raised, often itchy and reddened areas of the skin).

The IgE antibodies produced in response to an antigen such as insect venom or plant pollen bind to the surfaces of cells such as mast cells and basophils. These two cell types are similar in morphology and in their contribution to allergic reactions. **Mast cells** are especially prevalent in the connective tissue of the skin and respiratory tract and in surrounding blood vessels. **Basophils** circulate in the bloodstream, where they constitute less than 1% of the leukocytes. Both are packed with granules containing a variety of chemicals called *mediators.*

Mast cells and basophils can have as many as 500,000 sites for IgE attachment. The Fc (stem) region of an IgE antibody (see Figure 17.6) can attach to one of these specific receptor sites on such a cell, leaving two antigen-combining sites free. Of course, the attached IgE monomers will not all be specific for the same antigen. But when an antigen such as plant pollen encounters two adjacent antibodies of the same appropriate specificity, it can bind to one antigen-combining site on each antibody, bridging the space between them. This bridge triggers the mast cell or basophil to undergo *degranulation,* which releases the granules inside these cells and also the mediators they contain (Figure 19.1).

These mediators cause the unpleasant and damaging effects of an allergic reaction. The best-known mediator is **histamine.** The release of histamine increases the permeability and dilation of blood capillaries, resulting in edema (swelling) and erythema (redness). Other effects include increased mucus secretion (a runny nose, for example) and smooth-muscle contraction, which in the respiratory bronchi results in breathing difficulty.

Other mediators include **leukotrienes** of various types and **prostaglandins.** These mediators are not preformed and stored in the granules but instead are synthesized by the antigen-triggered cell. Because leukotrienes tend to cause prolonged contractions of certain smooth muscles, their action contributes to the spasms of the bronchial tubes that occur during asthmatic attacks. Prostaglandins affect smooth muscles of the respiratory system and cause increased mucus secretion. A number of other mediators, including various cytokines, are also released with the degranulation of mast cells and basophils.

Collectively, all these mediators serve as chemotactic agents that, in a few hours, attract neutrophils and eosinophils to the site of the degranulated cell. They then activate various factors that cause inflammatory symptoms, such as dilation of the capillaries, swelling, increased secretion of mucus, and involuntary contractions of smooth muscles.

TABLE 19.1 Types of Hypersensitivity

TYPE OF REACTION	TIME BEFORE CLINICAL SIGNS	CHARACTERISTICS	EXAMPLES
Type I (anaphylactic)	<30 min	IgE binds to mast cells or basophils. Causes degranulation of mast cell or basophil and release of reactive substances such as histamine	Anaphylactic shock from drug injections and insect venom; common allergic conditions, such as hay fever, asthma
Type II (cytotoxic)	5–12 hours	Antigen causes formation of antibodies that bind to target cell. When combined with action of complement, destroys target cell	Transfusion reactions, Rh incompatibility
Type III (immune complex)	3–8 hours	Antibodies and antigens form complexes that cause damaging inflammation	Arthus reactions, serum sickness
Type IV (cell-mediated)	24–48 hours	No antibodies involved. Antigens cause formation of cytotoxic T (T_C) cells that kill target cells	Rejection of transplanted tissues; contact dermatitis such as poison ivy

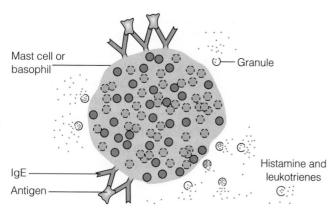

(a) IgE antibodies, produced in response to an antigen, coat mast cells and basophils. When an antigen bridges the gap between two adjacent antibody molecules of the same specificity, the cell undergoes degranulation and releases mediators such as histamine and leukotrienes.

SEM ⊢———————⊣
1 μm

(b) A degranulated mast cell that has reacted with an antigen and released granules of histamine and other reactive mediators.

FIGURE 19.1 Mechanism of anaphylaxis.

Systemic Anaphylaxis

At the turn of the century, two French biologists studied the responses of dogs to the venom of stinging jellyfish. Large doses of venom usually killed the dogs, but sometimes a few survived the injections. These surviving dogs were used for repeat experiments with the venom, and the results were surprising. Even a very tiny dose of the venom, one that should have been almost harmless, quickly killed the dogs. They suffered difficulty in respiration, entered shock as their cardiovascular systems collapsed, and quickly died. This phenomenon was named **anaphylactic shock.**

Systemic anaphylaxis (anaphylactic shock) can result when an individual sensitized to an antigen is exposed to it again. Injected antigens are more likely to cause a dramatic response than antigens introduced via other portals of entry. The release of mediators causes peripheral blood vessels throughout the body to dilate, resulting in a drop in blood pressure (shock). This reaction can be fatal within a few minutes. The drug epinephrine (adrenaline) counteracts these effects, and kits for epinephrine self-administration are available to people who are allergic to insect stings. However, there is very little time to act once someone develops systemic anaphylaxis. In the U.S.A., 50 to 60 people die each year from anaphylactic shock caused by insect stings.

You may be acquainted with someone who reacts to penicillin in this way. In these individuals, the penicillin, which is a nonimmunogenic hapten, combines with a carrier serum protein and in that form is able to induce antibody formation. Penicillin allergy probably occurs in about 2% of the population. Anyone who has had an adverse reaction to penicillin that included generalized hives, swelling of throat tissues, or chest constriction should not receive the drug again. A childhood rash without hives is probably not a significant reaction. However, there is no completely reliable skin test for sensitivity to penicillin.

Localized Anaphylaxis

Whereas sensitization to injected antigens is a common cause of systemic anaphylaxis, **localized anaphylaxis** is usually associated with antigens that are ingested (foods) or inhaled (pollen). The symptoms depend primarily on the route by which the antigen enters the body.

In allergies involving the upper respiratory system, such as hay fever (allergic rhinitis), sensitization usually involves mast cells in the mucous membranes of the upper respiratory tract. Reexposure to the airborne antigen—which might be a common environmental material such as plant pollen, fungal spores, animal dander (scaly, dried skin), or the feces of house dust mites (Figure 19.2)—results in a rapid release of large amounts of histamine, leukotrienes, and other mediators from mast cells in contact with the antigen. The typical symptoms are itchy and teary eyes, congested nasal passages, coughing, and sneezing. Antihistamine drugs, which compete for histamine receptor sites, are often used to treat these symptoms.

Asthma is an allergic reaction that mainly affects the lower respiratory system. Symptoms such as wheezing and shortness of breath are caused by the constriction of smooth muscles in the bronchial tubes. Mental or emotional stress can also be a contributing factor in precipitating an attack. Antihistamines are not an effective

FIGURE 19.2 **House dust mite.** Many people are allergic to these tiny creatures or, more specifically, to their airborne fecal pellets. House dust mites are found in bedding and house dust, where they feed on sloughed-off skin cells and similar materials. They are barely visible to the naked eye, about the size of the period printed at the end of this sentence.

FIGURE 19.3 **Skin test to identify allergens.** Drops of fluid containing test substances are placed on the skin. A light scratch is made with a needle to allow the substances to penetrate the skin. Reddening and swelling at the site identify the substance as a probable cause of an allergic reaction.

treatment for asthma because other mediators, such as leukotrienes and prostaglandins, are more important than histamine in this reaction. Treatment usually involves administration of epinephrine, which constricts the blood vessels and raises the blood pressure.

Antigens that enter the body via the gastrointestinal tract can also sensitize an individual. Many of us may know someone who is allergic to a particular food. Children tend to outgrow these problems, and in fact many food allergies may not be related to hypersensitivity at all and are more accurately described as food intolerances. For example, many people are unable to digest the lactose in milk because they lack the enzyme that breaks down this milk sugar. The lactose enters the intestine, where bacteria convert it into compounds that act as natural laxatives, causing diarrhea.

Gastrointestinal upset is a common symptom of food allergies, but it can also result from many other factors. Hives are more characteristic of a true food allergy, and ingestion of the antigen may result in systemic anaphylaxis. Death has even resulted when a person sensitive to fish ate french fries that had been prepared in oil previously used to fry fish. Skin tests are not reliable indicators for the diagnosis of food-related allergies, and completely controlled tests for hypersensitivity to ingested foods are very difficult to perform. Only eight foods are responsible for 97% of food-related allergies: eggs, peanuts, other nuts, milk, soy, fish, wheat, and peas.

Prevention of Anaphylactic Reactions

Avoiding contact with the sensitizing antigen is the most obvious way to prevent allergic reactions. Unfortunately, avoidance is not always possible. Some allergic individuals may never know exactly what antigen they are sensitive to. In other cases, skin tests might be of use in diagnosis (Figure 19.3). These tests involve inoculating small amounts of the suspected antigen just beneath the epidermal skin layer. Sensitivity to the antigen is indicated by a rapid inflammatory reaction that produces redness, swelling, and itching at the inoculation site. This small affected area is called a *wheal*.

Once the responsible antigen has been identified, the person can either try to avoid contact with it or undergo **desensitization.** This procedure consists of a series of dosages of the antigen carefully injected beneath the skin. The objective is to cause the production of IgG rather than IgE antibodies in the hope that the circulating IgG antibodies will act as *blocking antibodies* to intercept and neutralize the antigens before they can react with cell-bound IgE. Desensitization is not a routinely successful procedure, but it is effective in 65% to 75% of individuals whose allergies are induced by inhaled antigens, and in a reported 97% for insect venom.

TYPE II (CYTOTOXIC) REACTIONS

LEARNING OBJECTIVES

- *Describe the mechanism of cytotoxic reactions and how drugs can induce them.*
- *Describe the basis of the human blood group systems and their relationship to blood transfusions and hemolytic disease of the newborn.*

TABLE 19.2 The ABO Blood Group System

BLOOD GROUP	FREQUENCY (% U.S.) POPULATION			RBC ANTIGENS	ILLUSTRATION	PLASMA ANTIBODIES	BLOOD THAT CAN BE RECEIVED
	WHITE	BLACK	ASIAN				
AB	3	4	5	A B		None	A, B, AB, O Universal recipient
B	9	20	27	B		Anti-A	B, O
A	41	27	28	A		Anti-B	A, O
O	47	49	40	None		Anti-A Anti-B	O Universal donor

Type II (cytotoxic) reactions generally involve the activation of complement by the combination of IgG or IgM antibodies with an antigenic cell. This activation stimulates complement to lyse the affected cell, which might be either a foreign cell or a host cell that carries a foreign antigenic determinant (such as a drug) on its surface. Additional cellular damage may be caused by the action of macrophages and other cells that attack antibody-coated cells.

The most familiar cytotoxic hypersensitivity reactions are *transfusion reactions,* in which red blood cells are destroyed as a result of reaction with circulating antibodies. These involve blood group systems that include the ABO and Rh antigens.

The ABO Blood Group System

In the early 1900s, Karl Landsteiner discovered that human blood could be grouped into four principal types, which were designated A, B, AB, and O. This method of classification is called the **ABO blood group system.** Since then, at least 15 other blood group systems have been discovered, but our discussion will be limited to two of the best known, the ABO and the Rh systems. The main features of the ABO blood group system are summarized in Table 19.2.

A person's ABO blood type depends on the presence or absence of carbohydrate antigens located on the cell membranes of red blood cells (RBCs). Cells of blood type O lack both A and B antigens. Table 19.2 shows that the plasma of individuals with a given blood type, such as A, have antibodies against the alternative blood type (anti-B antibody). Individuals with type AB cells have plasma with no antibodies to either A or B antigens and so can receive transfusions with either A or B blood without a reaction. Type O individuals have antibodies against both A and B antigens. Lacking antigens, their blood can be transfused without difficulty to others, but they can receive transfusions of type O blood only.

When a transfusion is incompatible, as when type B blood is transfused into a person with type A blood, the antigens on the type B blood cells will react with anti-B antibodies in the recipient's serum. This antigen–antibody reaction activates complement, which in turn causes lysis of the donor's red blood cells as they enter the recipient's system.

In about 80% of the population (individuals called **secretors**), soluble antigens of the ABO type appear in saliva and other bodily fluids. It has been possible in criminal investigations to type saliva residues from a cigarette or to type semen in cases of rape.

The Rh Blood Group System

In the 1930s, the presence of a different surface antigen on human red blood cells was discovered. Soon after rab-

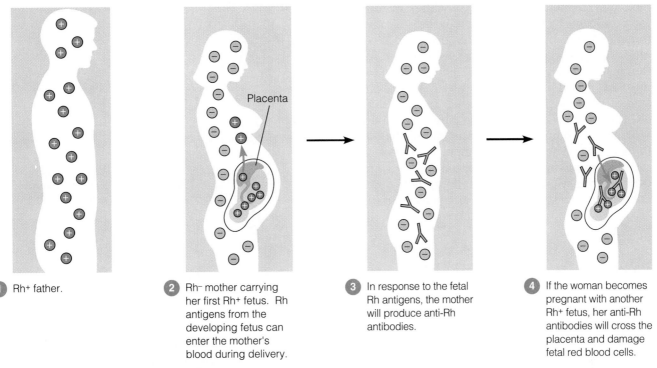

1. Rh+ father.

2. Rh− mother carrying her first Rh+ fetus. Rh antigens from the developing fetus can enter the mother's blood during delivery.

3. In response to the fetal Rh antigens, the mother will produce anti-Rh antibodies.

4. If the woman becomes pregnant with another Rh+ fetus, her anti-Rh antibodies will cross the placenta and damage fetal red blood cells.

FIGURE 19.4 Hemolytic disease of the newborn.

bits were injected with red blood cells from Rhesus monkeys, their serum was found to contain antibodies that were directed against the monkey blood cells but that would also agglutinate some human red blood cells. This indicated that a common antigen was present on both human and monkey red blood cells. The antigen was named the **Rh factor** (*Rh* for *Rhesus*). The roughly 85% of the population whose cells possess this antigen are called Rh+; those lacking this RBC antigen (about 15%) are Rh−. Antibodies that react with the Rh antigen do not occur naturally in the serum of Rh− individuals, but exposure to this antigen can sensitize their immune systems to produce anti-Rh antibodies.

Blood Transfusions and Rh Incompatibility

If blood from an Rh+ donor is given to an Rh− recipient, the donor's red blood cells stimulate the production of anti-Rh antibodies in the recipient. If the recipient then receives Rh+ red blood cells in a subsequent transfusion, a rapid, serious hemolytic reaction will develop.

Hemolytic Disease of the Newborn

Blood transfusions are not the only way in which an Rh− person can become sensitized to Rh+ blood. When an Rh− woman and an Rh+ man produce a child, there is a 50% chance that the child will be Rh+ (Figure 19.4). If the child is Rh+, the Rh− mother can become sensitized to this antigen during birth, when the placental membranes tear and the fetus's Rh+ red blood cells enter the maternal circula-

tion, causing the mother's body to produce anti-Rh antibodies of the IgG type. If the fetus in a later pregnancy is Rh+, her anti-Rh antibodies will cross the placenta and destroy the fetal red blood cells. The fetus's body responds to this immune attack by producing large numbers of immature red blood cells called erythroblasts; hence the name *erythroblastosis fetalis* was once used to describe what is now called **hemolytic disease of the newborn (HDNB)**. Before the birth of a fetus with this condition, the maternal circulation removes most of the toxic by-products of fetal red cell disintegration. After birth, however, the fetal blood is no longer purified by the mother, and the newborn develops jaundice and severe anemia.

HDNB is usually prevented today by immediate passive immunization of the Rh− mother with anti-Rh antibodies, which are available commercially. These anti-Rh antibodies combine with any fetal Rh+ red blood cells that have entered the mother's circulation, so it is much less likely that she will become sensitized to the Rh antigen. If the disease is not prevented, the newborn's Rh+ blood, contaminated with maternal antibodies, may have to be replaced by transfusion of uncontaminated blood.

Drug-Induced Cytotoxic Reactions

Blood platelets (thrombocytes) are minute cell-like bodies that are destroyed by drug-induced cytotoxic reactions in the disease called **thrombocytopenic purpura**. The drug molecules are usually haptens because they are too small to be antigenic by themselves, but in the situation illustrated in Figure 19.5, a platelet has become coated with molecules of a drug (quinine is a familiar example), and the combination is antigenic. Both antibody and complement are needed for lysis of the platelet. Because platelets are necessary for blood clotting, their loss results in hemorrhages that appear on the skin as purple spots (purpura).

Drugs may bind similarly to white or red blood cells. Immune-caused destruction of granulocytic white cells is called **agranulocytosis**, and it affects the body's phagocytic defenses. When red blood cells are destroyed in the same manner, the condition is termed **hemolytic anemia**.

TYPE III (IMMUNE COMPLEX) REACTIONS

LEARNING OBJECTIVE

• *Describe the mechanism of immune complex reactions.*

Type III reactions involve antibodies against soluble antigens circulating in the serum. (In contrast, type II immune reactions are directed against antigens located on cell or tissue surfaces.) The antigen–antibody complexes are deposited in organs and cause inflammatory damage.

Immune complexes form only when certain ratios of antigen and antibody occur. The antibodies involved are usually IgG. A significant excess of antibody leads to the formation of complement-fixing complexes that are rapidly removed from the body by phagocytosis (see Figure 17.9). When there is a significant excess of antigen, soluble complexes form that do not fix complement and do not cause inflammation. However, when a certain antigen–antibody ratio exists, usually with a slight excess of antigen, the soluble complexes that form are small and escape phagocytosis. These complexes circulate in the blood, pass between endothelial cells of the blood vessels, and become trapped in the basement membrane beneath the cells (Figure 19.6). In this location they may activate complement and cause a transient inflammatory reaction. Repeated introduction of the same antigen can lead to more serious inflammatory reactions.

Such damage from formation of immune complexes was first observed when rabbits were immunized repeatedly by injections of horse serum. In the area of the injections, edema (swelling) and reddening increased with each injection, and eventually the area became necrotic (that is, the tissue died). This response was named the **Arthus reaction**, after the researcher.

In the days before antibiotics, and when fewer vaccines were available, diseases caused by bacterial toxins were sometimes treated by antiserum from horses immunized with these toxins. The antibodies in the horse serum were

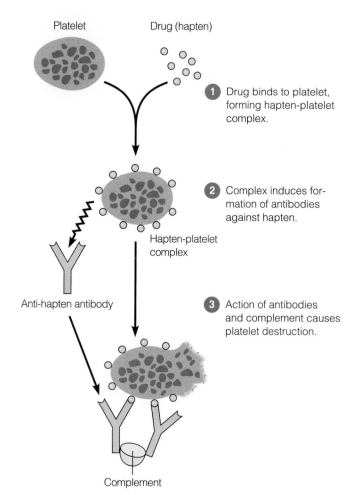

FIGURE 19.5 Drug-induced thrombocytopenic purpura. Molecules of a drug such as quinine accumulate on the surface of a platelet and stimulate an immune response that destroys the platelet.

intended to neutralize the toxins, but repeated injections often led to fever, itchy rashes, and swollen, painful joints in the recipient. This phenomenon became known as **serum sickness.**

Glomerulonephritis is an immune complex condition that causes inflammatory damage to the kidney glomeruli, which are sites of blood filtration. Antibodies generated in response to the M-protein of streptococci are believed to be one cause of this disease (see Chapter 15). Most affected individuals can be treated successfully with anti-inflammatory drugs, but in some the condition progresses to fatal kidney failure.

TYPE IV (CELL-MEDIATED) REACTIONS

LEARNING OBJECTIVE

• *Describe the mechanism of cell-mediated reactions.*

FIGURE 19.6 Immune complex–mediated hypersensitivity. ① Immune complexes are deposited on the basement membrane of the wall of a blood vessel, ② where they activate complement and attract inflammatory cells such as neutrophils to the site. ③ The neutrophils discharge enzymes as they react with the immune complexes, resulting in damage to tissue cells.

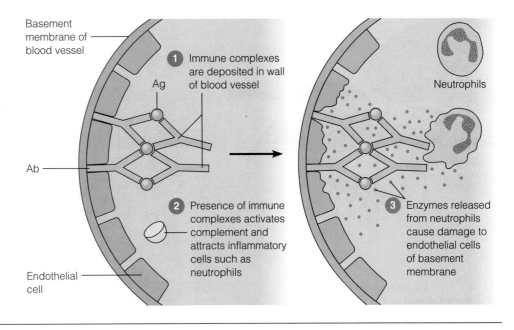

Basement membrane of blood vessel

Ag

Ab

Endothelial cell

1 Immune complexes are deposited in wall of blood vessel

2 Presence of immune complexes activates complement and attracts inflammatory cells such as neutrophils

Neutrophils

3 Enzymes released from neutrophils cause damage to endothelial cells of basement membrane

Up to this point we have discussed humoral types of immune reactions involving IgE, IgG, or IgM. Type IV reactions involve cell-mediated immune responses and are caused mainly by T cells. Instead of occurring within a few minutes or hours after a sensitized individual is again exposed to an antigen, these delayed reactions are not apparent for a day or more. A major factor in the delay is the time required for the participating T cells and macrophages to migrate to and accumulate near the foreign antigens. The T cells involved in delayed-type hypersensitivity reactions are primarily T_D cells. In some types of hypersensitivities resulting in tissue damage, T_C cells may also participate.

Causes of Type IV Reactions

Sensitization for type IV hypersensitivity reactions occurs when certain foreign antigens, particularly those that bind to tissue cells, are phagocytized by macrophages and then presented to receptors on the T cell surface. Contact between the antigenic determinant sites and the appropriate T cell causes the T cell to proliferate into mature differentiated T cells and memory cells.

When a person sensitized in this way is reexposed to the same antigen, a cell-mediated hypersensitivity reaction might result. Memory cells from the initial exposure activate T cells, and these T cells release cytokines in their interaction with the target antigen. The cytokines are a principal factor in this type of reaction. In addition, some cytokines, as we saw in Chapter 17, contribute to the inflammatory reaction to the foreign antigen by attracting macrophages to the site and activating them.

Cell-Mediated Hypersensitivity Reactions of the Skin

We have seen that hypersensitivity symptoms are frequently displayed on the skin. One cell-mediated hypersensitivity reaction that involves the skin is the familiar skin test for tuberculosis. Because *Mycobacterium tuberculosis* is often located within macrophages, this organism can stimulate a cell-mediated immune response. As a screening test, protein components of the bacteria are injected into the skin. If the recipient has (or has had) a previous infection by tuberculosis bacteria, an inflammatory reaction to the injection of these antigens will appear on the skin in one or two days; this interval is typical of delayed hypersensitivity reactions.

Allergic contact dermatitis, another common manifestation of type IV hypersensitivity, is usually caused by haptens that combine with proteins in the skin of some individuals to produce an immune response. Reactions to poison ivy (Figure 19.7), cosmetics, and the metals in jewelry (especially nickel) are familiar examples of these allergies.

The increasing exposure to latex in condoms, in certain catheters, and in gloves used by medical personnel has led to greater awareness of hypersensitivity to latex. In physicians and nurses, 3% or more have reported this type of hypersensitivity to the use of surgical gloves. Nonlatex substitutes are often less satisfactory to many users. Delayed-type hypersensitivity reactions are not to latex itself, but to chemicals used in its manufacture. The result is a contact dermatitis such as that shown in Figure 19.8. (Protein components of the latex itself cause an immedi-

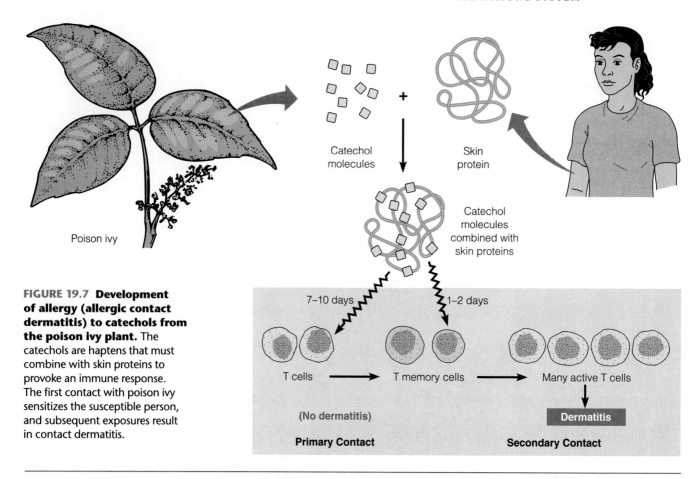

FIGURE 19.7 Development of allergy (allergic contact dermatitis) to catechols from the poison ivy plant. The catechols are haptens that must combine with skin proteins to provoke an immune response. The first contact with poison ivy sensitizes the susceptible person, and subsequent exposures result in contact dermatitis.

FIGURE 19.8 Allergic contact dermatitis. This person's hand exhibits a severe case of delayed-type contact dermatitis from wearing latex surgical gloves.

ate-type hypersensitivity characterized by hives on the affected area.)

The identity of the environmental factor causing the dermatitis can usually be determined by a *patch test*. Samples of suspected materials are taped to the skin; after 48 hours, the area is examined for inflammation.

AUTOIMMUNE DISEASES

LEARNING OBJECTIVES

- *Describe a mechanism for self-tolerance.*
- *Give an example of a type II, type III, and type IV auto-immune diseases.*

When the action of the immune system is in response to self antigens and causes damage to one's own organs, the result is an **autoimmune disease.**

Loss of Immunological Self-Tolerance

Autoimmune diseases occur when there is a loss of **self-tolerance**, the immune system's ability to discriminate

TABLE 19.3 Autoimmune Diseases

DISEASE	HYPERSENSITIVITY CLASS	DESCRIPTION
Hemolytic anemia	II	Destruction of red blood cells
Goodpasture's syndrome	II	Damage to basement membranes, but most symptoms result from attack on glomeruli of kidney
Graves' disease	II	Binding of antibodies to receptors for thyroid-stimulating hormone leads to overstimulation of thyroid activity
Myasthenia gravis	II	Blockage of acetylcholinesterase receptors prevents transmission of nerve signals
Thrombocytopenia purpura	II	Destruction of blood platelets
Rheumatoid arthritis	III	Inflammation of joints due to deposition of complexes of IgG and anti-IgG, and IgM and anti-IgM
Systemic lupus erythematosus	III	Antigen–antibody complexes (including antibodies against RNA, IgG, red blood cells, platelets, and chromosomes) are deposited in many locations, triggering inflammation; complexes containing antibodies to DNA are often deposited in kidneys
Addison's disease	IV	Destruction of adrenal gland cells that produce cortical hormones
Hashimoto's thyroiditis	IV	Destruction of cells of the thyroid gland
Insulin-dependent diabetes mellitus	IV	Destruction of pancreatic islet cells that produce insulin

between self and nonself. Apparently tolerance arises during fetal development or shortly after birth. If a fetus is exposed to an antigen, exposure to the same antigen after birth does not stimulate the production of antibodies or sensitized T cells.

In the generally accepted model by which T cells become able to distinguish between self and nonself, the cells acquire this ability during their passage through the thymus. During this period, any T cells that will target host cells are either eliminated (**clonal deletion**) or inactivated. This requirement makes it unlikely that the T cell will attack its own tissue cells. In autoimmune diseases, loss of self-tolerance leads to production of antibodies or a response by sensitized T cells against a person's own tissue antigens. Some of the major autoimmune diseases are summarized in Table 19.3. Autoimmune reactions, and the diseases they cause, can be cytotoxic, immune complex, or cell-mediated in nature.

Type II (Cytotoxic) Autoimmune Reactions

Graves' disease and myasthenia gravis are two examples of disorders caused by type II autoimmune reactions. Both diseases involve antibody reactions to cell surface antigens, although there is no cytotoxic destruction of the cells.

Graves' disease is caused by antibodies called long-acting thyroid stimulators. These antibodies attach to receptors on thyroid gland cells that are the normal target cells of the thyroid-stimulating hormone produced by the pituitary gland. The result is that the thyroid gland is stimulated to produce increased amounts of thyroid hormones and becomes greatly enlarged. The most striking signs of the disease are goiter, which results from the enlarged thyroid, and markedly bulging, staring eyes.

Myasthenia gravis is a disease in which muscles become progressively weaker. It is caused by antibodies that coat the acetylcholine receptors at the junctions at which nerve impulses reach the muscles. Eventually, the muscles controlling the diaphragm and the rib cage fail to receive the necessary nerve signals, and respiratory arrest and death result.

Type III (Immune Complex) Autoimmune Reactions

Systemic lupus erythematosus is a systemic autoimmune disease, involving immune complex reactions, that mainly affects women. The etiology of the disease is not completely understood, but afflicted individuals produce antibodies directed at components of their own cells, including DNA, which is probably released during the normal breakdown of tissues, especially the skin. The most damaging effects of the disease result from deposits of immune complexes in the kidney glomeruli.

Crippling **rheumatoid arthritis** is a disease in which immune complexes of IgM, IgG, and complement are deposited in the joints. In fact, immune complexes called rheumatoid factors may be formed by IgM binding to the Fc (stem) region of normal IgG. These factors are found in 70% of individuals suffering from rheumatoid arthritis. The chronic inflammation caused by this deposition eventually leads to severe damage to the cartilage and bone of the joint.

TABLE 19.4 Diseases Related to Specific HLAs

DISEASE	INCREASED RISK OF OCCURRENCE WITH SPECIFIC HLA*	DESCRIPTION
Inflammatory Diseases		
Multiple sclerosis	5 times	Progressive inflammatory disease affecting nervous system
Rheumatic fever	4–5 times	Cross-reaction with antibodies against streptococcal antigen
Endocrine Diseases		
Addison's disease	4–10 times	Deficiency in production of hormones by adrenal gland
Graves' disease	10–12 times	Antibodies attached to certain receptors in the thyroid gland cause it to enlarge and produce excessive hormones
Malignant Disease		
Hodgkin's disease	1.4–1.8 times	Cancer of lymph nodes

*As compared with general population.

Type IV (Cell-Mediated) Autoimmune Reactions

Hashimoto's thyroiditis is a result of the destruction of the thyroid gland, primarily by T cells of the cell-mediated immune system. It is a fairly common disorder and is often found in related family members. **Insulin-dependent diabetes mellitus** is a familiar condition caused by immunological destruction of insulin-secreting cells of the pancreas. T cells are clearly implicated in this disease; animals that are genetically likely to develop diabetes fail to do so when their thymus is removed in infancy.

DISEASES AND REACTIONS RELATED TO THE HUMAN LEUKOCYTE ANTIGEN (HLA) COMPLEX

LEARNING OBJECTIVES

- *Define HLA complex and explain its importance in disease susceptibility and tissue transplants.*
- *Explain how rejection of a transplant occurs and how rejection is prevented.*

The inherited genetic characteristics of individuals are expressed not only in the color of their eyes and the curl of their hair but also in the composition of the self molecules on their cell surfaces. Some of these are called **histocompatibility antigens.** The genes controlling the production of the most important of these self molecules are known as the **major histocompatibility complex (MHC)**; in humans these genes are called the **human leukocyte antigen (HLA) complex.** We encountered these self molecules in Chapter 17, where we saw that most antigens can stimulate an immune reaction only if they are associated with an MHC molecule.

A process called *HLA typing* is used to identify and compare HLAs. Certain HLAs are related to increased susceptibility to specific diseases; one medical application of HLA typing is to identify such susceptibility. A few of these relationships are summarized in Table 19.4.

Another important medical application of HLA typing is in transplant surgery, in which the donor and the recipient must be matched by *tissue typing*. The serological technique shown in Figure 19.9 is the one most often used. In serological tissue typing, the laboratory uses standardized antisera or monoclonal antibodies that are specific for particular HLAs. ① Lymphocytes from the person being tested are incubated with a selected specific antiserum. ② Complement and a dye, such as trypan blue, are then added. ③ If antibodies in the antiserum have reacted specifically with the lymphocyte, the cell is damaged. The damaged cell will take up the dye (undamaged cells will not), thus indicating that the lymphocyte possesses a certain antigen. This method is simple and rapid.

There are two main classes of HLAs that are important in such matches: class I (HLA-A, -B, and -C) and class II (HLA-DR, -DP, and -DQ). The genes that control production of these self molecules are located close together on human chromosome 6, as shown in Figure 19.10. Matching donors and recipients for class I self molecules has long been a standard procedure. These self molecules, which are found on the surface of almost all nucleated cells, stimulate a strong immune response by the antibodies and T_C cells mediating transplant rejection. However, matching for class II self molecules might well be more important, especially for tissue from an unrelated person. These HLAs are present, along with the processed foreign antigens, on the surface of the antigen-presenting cells (see Figure 17.15) and are simultaneously recognized by T_H cells. The

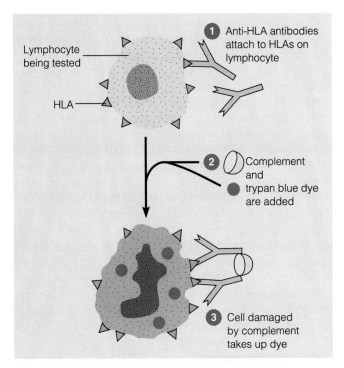

FIGURE 19.9 Tissue typing, serological method.
Lymphocytes from the person being tested are incubated with laboratory test stocks of anti-HLA antibodies specific for a particular HLA. If the antibodies react with the antigens on a lymphocyte, then complement damages the lymphocyte and dye can enter the cell. Such a positive test result indicates that the person has the particular HLA being tested for.

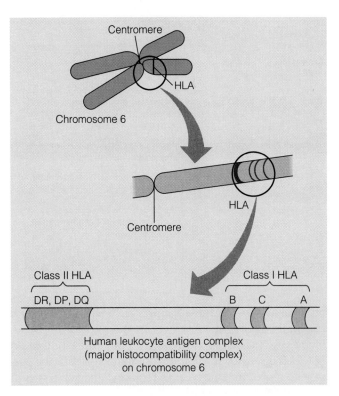

FIGURE 19.10 Genes controlling the human leukocyte antigen (HLA) complex. The genes that control HLA are located on chromosome 6. Note how the genes for class I self molecules (A, B, and C) and for class II self molecules (DR, DP, and DQ) are all located within a very short length of chromosome.

DR self molecules are especially important; if the donor and the recipient do not share these self molecules, the graft is unlikely to be accepted. The donor and the recipient must also be of the same ABO blood type.

A promising new technique for analysis of HLA, especially for class II self molecules, is the use of the polymerase chain reaction (see Figure 9.13) to amplify the cell's DNA. If this is done for both donor and recipient, a match between donor DNA and recipient DNA can then be made. Such a match should result in a much higher success rate in transplant surgery.

Transplantation Reactions

In sixteenth-century Italy, crimes were often punished by cutting off the offender's nose. A surgeon of the time, in his attempts to repair this mutilation, observed that if skin was taken from the patient, it healed properly, but if it was taken from another person, it did not. He called this a manifestation of "the force and power of individuality." Four hundred years later, we know the principles behind this phenomenon. Transplants recognized as nonself are *rejected*—attacked by T cells that directly lyse the grafted cells, by macrophages activated by T cells, and also by antibodies, which activate the complement system and injure blood vessels supplying the graft. However, transplants that are not rejected can add many healthy years to the life of an individual.

Since the first kidney transplant was performed in 1954, this particular type of transplant has become a nearly routine medical procedure. Other types of transplants that are now feasible include bone marrow, thymus, heart, liver, and cornea. Tissues and organs for transplant are usually taken from recently deceased individuals, although one of a pair of organs, such as a kidney, occasionally comes from a living donor.

Types of Transplants and Their Chances of Rejection

Some transplants or grafts do not stimulate an immune response. A transplanted cornea, for example, is rarely rejected, mainly because antibodies usually do not circulate into that portion of the eye, which is therefore considered an immunologically **privileged site.** (However, rejections do occur, especially when the cornea has developed many blood vessels from corneal infections or damage.) The brain is also an immunologically privileged site, probably

TABLE 19.5 Types of Grafts*

GRAFT	DESCRIPTION
Autograft (*autos,* "self")	A graft transplanted from one site to another site on the same person—for example, the skin from the thigh grafted over a burned area on the arm
Isograft (*isos,* "equal")	A graft between genetically identical individuals—identical twins
Allograft (*allos,* "other")	A graft between genetically different members of the same species
Xenograft (*xenos,* "strange")	A graft between different species—for example, transplant of a baboon heart to a human

*May include transplantation of an organ's cells rather than the complete organ.

because it does not have lymphatic vessels and because the walls of the blood vessels in the brain differ from blood vessel walls elsewhere in the body (the blood-brain barrier is discussed in Chapter 22). Someday it may be possible to graft foreign nerves to replace damaged nerves in the brain and spinal cord.

It is now possible to transplant **privileged tissue** that does not stimulate an immune rejection. An example is replacement of a person's damaged heart valve with a valve from a pig's heart. However, privileged sites and tissues are more the exception than the rule.

How animals tolerate pregnancy without rejecting the fetus is only partially understood. The uterus is not a privileged site, and yet in pregnancy the tissues of two genetically different individuals are in direct contact. Current theories are that there is a barrier that masks the fetal antigens and that the maternal immune system is suppressed.

When one's own tissue is grafted to another part of the body, as is done in burn treatment or in plastic surgery, the graft is not rejected. Recent technology has made it possible to use a few cells of a burn patient's uninjured skin to culture extensive sheets of new skin. The graft of this new skin is an example of an **autograft.** Identical twins have the same genetic makeup; therefore, skin or organs such as kidneys may be transplanted between them without provoking an immune response. These transplants are called **isografts.**

Most transplants, however, are made between people who are not identical twins; these transplants do trigger an immune response. Attempts are made to match the HLAs of the donor and recipient as closely as possible so that the chances of rejection are reduced. Because HLAs of close relatives are most likely to match, blood relatives, especially siblings, are the preferred donors. Grafts between people who are not identical twins are called **allografts.**

Because of the shortage of available organs, it would be helpful if **xenografts** (transplanted tissue or organs from

animals) could be more successfully transplanted to humans. However, the body tends to mount an especially severe immune assault on such transplants. Unsatisfactory attempts have been made to use organs from baboons and other nonhuman primates. Research interest is high in genetically engineering pigs—an animal in plentiful supply, of the right size, and inciting relatively little public sympathy—to make acceptable donors of organs. The various types of grafts are summarized in Table 19.5.

When bone marrow is transplanted into individuals with a defective immune system, the transplanted tissue may attack the host. The purpose of the transplant is to provide such people with the B-cell and T-cell manufacturing capability they lack. However, the result can be **graft-versus-host (GVH) disease.** The transplanted bone marrow contains immunocompetent cells that mount primarily a cell-mediated immune response against the tissue into which they have been transplanted. Because the recipients lack effective immunity, GVH disease is a serious complication and can even be fatal.

Immunosuppression

To keep the problem of rejection of transplanted tissue in perspective, it is useful to remember that the immune system is simply doing its job and has no way of recognizing that its attack against the transplant is not helpful. In an attempt to prevent rejection, the recipient of an allograft usually receives treatment to suppress this normal immune response against the graft.

The drug **cyclosporine,** isolated from a mold in 1976, revolutionized transplant medicine. Cyclosporine selectively inhibits the synthesis of certain cytokines without interfering with the synthesis of other proteins. The drug prevents the formation of the T_C cells that are largely responsible for transplant rejection, inhibits the release by macrophages of cytokines that cause inflammation, and prevents the differentiation of B cells into antibody-secreting plasma cells. It does this without suppressing the ability of bone marrow to produce red blood cells and platelets.

As a result of this discovery, the success rate of organ transplants has risen dramatically. Before cyclosporine came into general use, only about 50% of kidneys from donors who were not closely related to the patient were accepted; now as many as 90% of these transplants succeed. Liver transplants became practical only after cyclosporine became available, and heart transplants have achieved unprecedented success rates.

The greatest drawback to the use of the drug appears to be an increased risk of cancer, for that disease is combated by the same cell-mediated immune system that rejects transplanted organs. Problems with liver and kidney toxicity have also necessitated careful attention to dose levels. The newer immunosuppressant drugs tacrolimus (FK506) and rapamycin have a mode of action similar to that of cyclosporine but cause fewer undesirable effects.

TABLE 19.6 Immune Deficiency Diseases

DISEASE	CELLS AFFECTED	COMMENTS
Acquired immunodeficiency syndrome (AIDS)	T cells (virus destroys T_H [CD4] cells)	Allows cancer and bacterial, viral, fungal, and protozoan diseases
Selective IgA immunodeficiency	B, T cells	Affects about 1 in 700, causing frequent mucosal infections; specific cause uncertain
Common variable hypogammaglobulinema	B, T cells (decreased immunoglobulins)	Frequent viral and bacterial infections; second most common immune deficiency disease, about 1 in 70,000
Reticular dysgenesis	B, T, and stem cells (a combined immunodeficiency; deficiencies in B and T cells and neutrophils)	Usually fatal in early infancy; very rare; bone marrow transplant a possible treatment
Severe combined immunodeficiency	B, T, and stem cells (deficiency of both B and T cells)	Occurs in several forms to about 1 in 100,000; allows severe viral, fungal infections; treated with transplantation of bone marrow, fetal thymus
Thymic hypoplasia (DiGeorge syndrome)	T cells (defective thymus causes deficiency of T cells)	Lack of cell-mediated immunity; usually fatal in infancy from *Pneumocystis* pneumonia or viral or fungal infections
Wiskott-Aldrich syndrome	B, T cells (few platelets in blood, abnormal T cells)	Frequent infections by viruses, fungi, protozoa; eczema, defective blood clotting; usually causes death in childhood
X-linked infantile (Bruton's) agammaglobulinemia	B cells (decreased immunoglobulins)	Frequent extracellular bacterial infections; occurs in about 1 in 200,000; the first immunodeficiency disorder recognized (1952); affects males

IMMUNE DEFICIENCIES

LEARNING OBJECTIVE

- *Discuss the causes and effects of immune deficiencies.*

The condition of lacking sufficient immune response is called an immune deficiency. Immune deficiencies can be either congenital or acquired.

Congenital Immune Deficiencies

Occasionally, people are born with defective immune systems. Defects in or absence of a number of inherited genes can result in such **congenital immune deficiencies.** For example, individuals with a certain recessive trait may not have a thymus gland and therefore lack cell-mediated immunity. A different recessive trait causes lower numbers of B cells, affecting humoral immunity.

Acquired Immune Deficiencies

A wide variety of drugs, cancers, or infectious agents can result in **acquired immune deficiencies.** For example, Hodgkin's disease (cancer) lowers the cell-mediated response. Many viruses are capable of infecting and killing lymphocytes, hence lowering the immune response. Removal of the spleen decreases humoral immunity. Table 19.6 summarizes several of the better-known acquired immune deficiency conditions, including AIDS.

ACQUIRED IMMUNODEFICIENCY SYNDROME (AIDS)

LEARNING OBJECTIVES

- *Describe the effects of HIV on the immune system and the stages of HIV infection.*
- *Describe how HIV can be transmitted.*
- *Discuss current and possible future treatment and prevention of HIV infection.*

In 1981, a cluster of cases of *Pneumocystis* pneumonia (see Chapter 24) appeared in the Los Angeles area. This extremely rare disease was usually seen only in immunosuppressed individuals. Investigators soon correlated the appearance of this disease with an unusual incidence of a rare form of cancer of the skin and blood vessels called Kaposi's sarcoma. The people affected were all young homosexual men, and all showed loss of immune function. By 1983, the pathogen causing the loss of immune function had been identified as a retrovirus (see Chapter 13) that selectively infects helper T cells. This virus is now known as human immunodeficiency virus (HIV).

The Origin of AIDS

There have been several hypotheses about the origin of HIV. It is now believed to have arisen by mutation of a virus that had been endemic in some areas of central Africa for many years. The virus has been found in blood samples preserved from as early as 1959 in several African nations and in England.

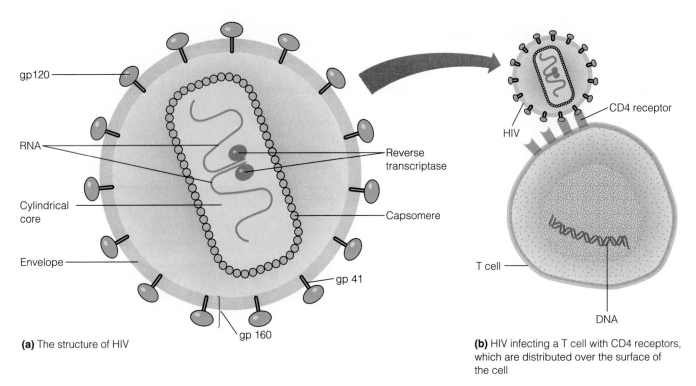

(a) The structure of HIV

(b) HIV infecting a T cell with CD4 receptors, which are distributed over the surface of the cell

FIGURE 19.11 HIV and its infective ability.

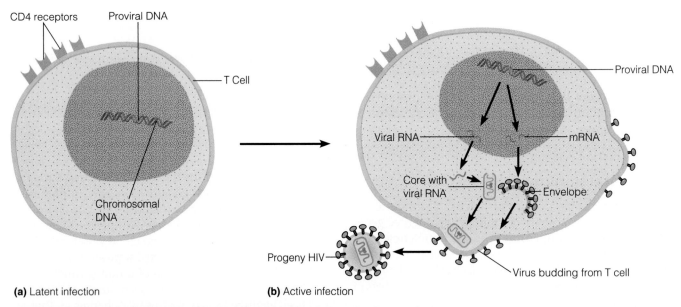

(a) Latent infection

(b) Active infection

FIGURE 19.12 Latent and active HIV infection in CD4 T cells. (a) In latent infections, no viral genes are being transcribed and no new viruses are being formed. **(b)** In active infections, proviral DNA controls the synthesis of new viruses, which bud from the host cell.

The HIV Infection

HIV is a retrovirus (see Figure 13.21). It has an RNA genome, the enzyme reverse transcriptase, and an envelope of lipoprotein (Figure 19.11a). The envelope has spikes termed gp120 (a glycoprotein with a molecular weight of 120,000). These spikes allow HIV to attach to the CD4 receptor on a host cell (Figure 19.11b). These receptors are found on T cells and macrophages, which are the main target cells of HIV. CD4 receptors are also found on a number of other lymphocytic cells and some epithelial cells. (Certain cells without CD4 receptors may also be infected, but less efficiently.) Attachment is followed by entry into the host cell. The viral RNA is transcribed by the enzyme reverse transcriptase into DNA. This viral DNA then becomes integrated into the chromosomal DNA of the host cell. There it may control the production of new HIVs that bud from the host cell, as shown in Figure 19.12 (see also Figure 1.1a).

Alternatively, this integrated DNA may remain latent in the cell's chromosome as a provirus. In this form it is not detected by the immune system. This ability of the virus to remain latent is one reason why the anti-HIV antibodies developed by infected individuals fail to prevent the disease. The virus also evades immune defenses by undergoing rapid antigenic changes. The mutation rate for the formation of DNA from the HIV RNA is so high that there is almost always at least one error in the replication of each HIV genome; each virus probably differs from its parent in at least some small way. HIV also evades the immune system by staying in vacuoles within macrophages (Figure 19.13). Moreover, infected cells can fuse to uninfected cells and spread the virus directly, often without exposure to circulating antibodies.

AIDS is actually only the end stage of an HIV infection. Shortly after the initial infection, the person undergoes seroconversion—that is, tests positive for antibodies to HIV. This interval is almost always less than six months. The Centers for Disease Control and Prevention classification system divides the progress of HIV infection into three clinical categories (Figure 19.14).

- *Category A.* The infection may be asymptomatic or cause persistent lymphadenopathy (swollen lymph nodes). The blood concentration of CD4 T cells is 500 per mm^3 or above.

- *Category B.* Certain symptoms appear as the CD4 T cell counts fall into the 200–499/mm^3 range and the immune system weakens. Especially common are persistent infections by the yeastlike fungus *Candida albicans,* which can appear in the mouth, throat, or vagina. Other conditions may include shingles, persistent diarrhea and fever, whitish patches on the oral mucosa (hairy leukoplakia), and certain cancerous or precancerous conditions such as cervical (of the cervix) growths.

- *Category C.* This stage is clinical AIDS. The CD4 T cell count is 200/mm^3 or lower. Less than 14% of the T cells present are CD4 T cells. Many AIDS indicator conditions now appear. Important among them are *Candida albicans* infections of the esophagus, bronchi, and lungs; cytomegalovirus eye infections; tuberculosis; *Pneumocystis* pneumonia; toxoplasmosis of the brain; and Kaposi's sarcoma.

A number of the disease conditions associated with HIV infection and AIDS are summarized in Table 19.7.

The average time from infection to the development of AIDS is about 10 years. Because of this long incubation period, it is not yet known what percentage of individuals infected with HIV will develop full-blown AIDS. Current indications are that almost all will eventually, but there are

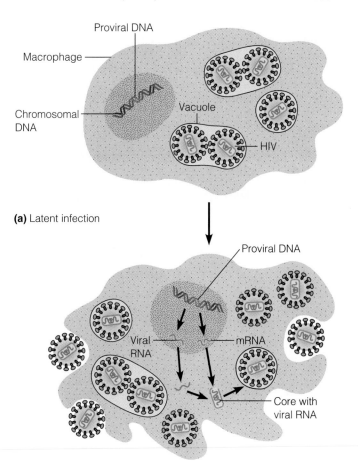

(a) Latent infection

(b) Active infection

FIGURE 19.13 Latent and active HIV infection in macrophages. (a) In latent infections, the newly made viruses remain in vacuoles. If the macrophage (not the infecting virus) is activated, these viruses will be released. **(b)** In active infections, new viruses are being produced; some of these remain in vacuoles while others are released.

now a relative handful of very long-term survivors. However, infants who acquire HIV infections from their mothers lack the well-established immune systems of adults and usually die within about a year.

Modes of HIV Transmission

Transmission of HIV requires transfer of or exposure to infected bodily fluids. The most important of these are blood, which contains an estimated 10–1000 infectious particles (IP) per milliliter, and semen (about 10–50 IP/ml). The viruses are often located within cells in these fluids, especially macrophages.

Routes of HIV transmission include intimate sexual contact, breast milk, blood-contaminated needles, organ transplants, artificial insemination, and blood transfusions. Probably the most dangerous form of sexual contact is anal receptive intercourse. Vaginal intercourse is much

FIGURE 19.14 The stages of an HIV infection. Note how the HIV population increases rapidly upon initial infection and drops rapidly as the immune system responds. Then the HIV population slowly increases, and the CD4 T cell population slowly decreases. The CD4 T cell population represented by the line primarily includes those in circulation. Many CD4 T cells evidently remain sequestered in the lymph nodes; only late in the HIV infection is this population affected. AIDS is the final stage of the infection process.

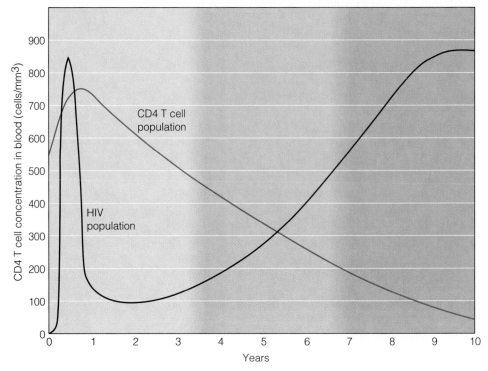

Category A: CD4 T cell count 500/mm³ or above. Asymptomatic or chronic lymphadenopathy.

Category B: CD4 T cell count from 200 to 499/mm³. Symptomatic. Early indications of immune failure.

Category C: CD4 T cell count below 200/mm³. AIDS indicator conditions.

more likely to transmit HIV from male to female than vice versa, and transmission either way is much greater when genital lesions are present. Although rare, transmissions have also occurred by oral-genital contact.

In sub-Saharan Africa, heterosexual transmission is so common that the male:female ratio among AIDS patients is about 1:1, as compared with about 13:1 in the U.S.A. Heterosexual transmission is also predominant in parts of Asia. This mode of transmission is increasing in the U.S.A., especially among women who are injecting drug users (IDUs) or sexual partners of men who are IDUs. Approximately 30% of children born to HIV-positive mothers are infected with the virus. A very few people in high-risk groups such as homosexuals, spouses of HIV-positive hemophiliacs, and African prostitutes seem to be completely resistant to infection by HIV. Preliminary research indicates that their HLAs have distinctive antigenic markers, and this may account for their apparent natural immunity. Table 19.8 summarizes the patterns of HIV transmission.

HIV is not transmitted by insects or casual contact such as hugging or sharing household facilities, drinking glasses, or towels. Saliva generally contains less than 1 IP/ml, and kissing is not known to spread the infection. In developed countries, transmission by blood transfusion is unlikely because blood is tested for HIV antibodies. However, there will always be a slight risk because blood might be donated during the interval between infection and ap-

pearance of detectable antibodies. Tests for the virus itself are also available, but they have not proved to be superior in screening blood. There is no risk at all in donating blood.

AIDS Research: Treatments and Vaccines

There are two basic areas of AIDS research. Some researchers are looking for drugs to treat AIDS, and others are working on vaccines to prevent the disease.

Chemotherapy

Most anti-HIV drugs such as zidovudine (AZT), ddI, and ddC are inhibitors of the enzyme reverse transcriptase. HIV is a retrovirus, you will recall, that copies RNA into DNA. The drugs—analogues of nucleic acids—trick the enzyme into terminating the synthesis of viral DNA. Such drugs have slowed the progress of the disease but have not led to a cure. Other than the reverse transcriptase step, there are at least 13 other points at which the production of HIV could be selectively interrupted by drugs. For example, because it has no equivalent in human cells, an attractive target is the viral enzyme protease. This enzyme cuts proteins into pieces that are then reassembled into new HIV particles. Inhibiting it would prevent viral synthesis. Numerous other approaches are being inten-

TABLE 19.7 Some Common Diseases Associated with AIDS

AGENT OR DISEASE	DISEASE DESCRIPTION
Protozoa	
*Pneumocystis carinii**	Life-threatening pneumonia
Cryptosporidium parvum	Persistent diarrhea
Toxoplasma gondii	Encephalitis
Isospora belli	Gastroenteritis
Viruses	
Cytomegalovirus	Fever, encephalitis, blindness
Herpes simplex virus	Vesicles of skin and mucous membranes
Varicella-zoster virus	Shingles
Bacteria	
Mycobacterium tuberculosis	Tuberculosis
Mycobacterium avium-intracellulare	May infect many organs; gastroenteritis and other highly variable symptoms
Fungi	
Histoplasma capsulatum	Pulmonary infection
Cryptococcus neoformans	Life-threatening meningitis
Candida albicans	Overgrowth on oral and vaginal mucous membranes (Category B)
Candida albicans	Overgrowth in esophagus, lungs (Category C)
Cancers or Precancerous Conditions	
Kaposi's sarcoma	Cancer of skin and blood vessels
Leukoplakia	Whitish patches on mucous membranes; commonly considered precancerous

*Possibly a fungus

TABLE 19.8 Global Patterns of HIV Transmission

PATTERN	DISTRIBUTION
I. Initially, homosexual men and IDUs primarily; then increasing heterosexual and pediatric transmission	North America, western Europe, Australia, New Zealand
II. Primarily heterosexual and pediatric transmission; male: female ratio approximately 1:1	Sub-Saharan Africa, southern Asia, parts of the Caribbean
III. HIV infections uncommon but increasing among homosexual men, prostitutes, and IDUs	Eastern Europe, North Africa, some parts of Asia and Pacific Rim nations

sively studied (see the box on pp. 484–485), and the most successful solutions might well be some that have not yet been anticipated.

The Possibility of Preventive Vaccines

There are great obstacles to production of an AIDS vaccine, among them the lack of a suitable animal host for the virus. However, researchers are now optimistic that a vaccine, once produced, could be effective. Because of the extreme virulence of the virus, many think it unlikely that any whole-virus vaccine, either killed or attenuated, would be acceptable for use in uninfected individuals. However, such a vaccine would be acceptable for use in attempting to clear the virus after infection. Most efforts are directed at subunit vaccines (see Chapter 18) based on surface envelope antigens of the virus. However, such vaccines must overcome the problem of many antigenic variants, which we mentioned earlier. In addition, an effective vaccine would have to stimulate cell-mediated as well as humoral immunity to deal with HIV contained within macrophages or other cells.

The Future Course of AIDS

The World Health Organization (WHO) has estimated that the HIV pandemic will infect 30 million adults by the year 2000, and that as many as 10 million children also will be born with the infection. More than 80% of the transmissions will probably be by heterosexual contact. New HIV infections among homosexual men and IDUs, and from contaminated blood supplies, are expected to be a declining percentage of the total number of infections.

The greatest concern of WHO now is the prospect of an explosive spread of the infection in southern Asia. The epidemic in sub-Saharan Africa, predominantly transmitted by heterosexual contact, is stabilizing in some countries but not in others. In the U.S.A. there are currently an estimated 1 million infected people. In the U.S.A. and Canada, AIDS cases are increasingly acquired heterosexually and among IDUs. Cases also occur at a disproportionately higher rate among urban minority populations; at present, AIDS is the leading cause of death among women ages 20 to 40 in New York City. The number of infected females is also increasing. This trend will probably continue, at least in the near future.

There is speculation that in societies in which people have many different sexual contacts, natural selection favors a more virulent form of the virus. An earlier death for the host is not likely to minimize spread of the infection. In societies tending to practice monogamous or protected sex, natural selection may favor viruses that are not as virulent. The host must live longer to transmit the virus.

The AIDS epidemic gives clear evidence of the value of basic scientific research. Without the advances of the past few decades in molecular biology, we would have been unable even to identify the agent of AIDS, to develop the tests used to screen donated blood, or to monitor the course of the infection.

Microbiology in the News

NEW WEAPONS AGAINST AIDS

HIV'S ABILITY to destroy, evade, and even hide inside cells of the human immune system makes it extremely difficult to prevent by vaccination and to treat. Because the virus is so unusual and so deadly, it has stimulated intense research into both conventional and extremely unconventional approaches to vaccination and chemotherapy.

VACCINES

Experimental vaccines are being made with gp120 and other envelope glycoproteins. It is thought that antibodies against gp120 would not only attack free viruses, but also block the gp120 on the surfaces of infected T cells, thus preventing cell fusion and cell-to-cell transmission. Human trials of these and other vaccines have already begun. Initially, these human tests can determine only whether the vaccines produce an immune response and whether they are safe. Determining whether they actually prevent AIDS is much more difficult. Even if an uninfected volunteer became infected with HIV early in these tests, researchers don't expect to know whether the vaccines actually prevent AIDS for three to five years because AIDS develops so slowly.

During September of 1987, Micro-GeneSys, Inc., of West Haven, Connecticut, began human trials of a vaccine that uses a different envelope glycoprotein, gp160. The gp160 gene from a genetically engineered HIV-1 (supplied by the National Institute of Allergy and Infectious Diseases, NIAID) is inserted into baculovirus, which infects moths and butterflies. The virus is grown in cultures, and the viral protein is harvested from the cultures and used in the vaccine.

NIAID scientists are studying the effect of the vaccine in male volunteers who are healthy and free of HIV. By 1990, one-third of the volunteers had developed anti-HIV antibodies. Cell-mediated immunity and the effectiveness of these antibodies in vivo have not been evaluated.

In 1990, researchers at the University of Washington reported that a recombinant vaccine called HIVAC-1e successfully protected mice against HIV infection. HIVAC-1e, made by Oncogen of Seattle, Washington, is a vaccinia virus into which the gene for gp120 has been inserted. The vaccinia virus causes host cells to produce gp120 proteins, which bind to host cell membranes. This stimulates the production of antibodies against the viral proteins. Currently, researchers at the University of Washington are testing the effects of the two vaccines, HIVAC-1e and MicroGeneSys's gp160, together.

Interestingly, Jonas Salk and Albert Sabin, the researchers responsible for different polio vaccines, have also been involved in developing an HIV vaccine (Sabin died in 1993). Another vaccine consisting of a synthetic protein (gp41) carried by the Sabin vaccine attenuated poliovirus was developed by Albert Sabin. Twenty-four infected patients have been given this vaccine.

Although these envelope glycoprotein–based vaccines hold promise, research on other types of vaccines is also under way. Jonas Salk has developed a killed-HIV vaccine in which the virus is killed with gamma radiation and freeze dried. Results of human trials of the Salk vaccine were discouraging, however. The study included 103 infected people, two-thirds of whom had been given the Salk AIDS vaccine. The findings showed no difference in the number of viruses in patients given the vaccine compared to the control (unvaccinated) patients.

Finally, NIAID scientists are suggesting that passive immunization of infants born to AIDS-infected women may prevent infection of newborns.

CHEMOTHERAPY

The viral enzymes reverse transcriptase and protease may provide potential targets for chemotherapeutic agents. For example, zidovudine (AZT) was approved for treating AIDS in 1987. AZT works by competitive inhibition. Reverse transcriptase inserts it into a growing molecule of viral DNA where thymidine would normally go. The DNA molecule is terminated because the next nucleoside cannot be added to the AZT.

The observed effects of AZT are a decrease in the incidence of opportunistic infections, Kaposi's sarcoma, and conversion to category C (AIDS). However, AZT helps only certain AIDS patients. It also has adverse effects that include immunosuppression and anemia.

Based on the success of AZT, other nucleoside analogues are currently being tested (see the table). In 1991, ddI (dideoxyinosine, Videx®) was approved for patients who do not respond to AZT. A third nucleoside analogue, ddC (dideoxycytidine, Hivid®), was approved for use in mid-1992. d4T (Stauvidine®) is an experimental drug available free of charge to patients who do not respond to AZT or ddI.

The use of immunotoxins—an antibody bound to a toxin—has been suggested. The antibody would bind to glycoproteins on the surface of infected cells, and the toxin would kill the targeted cell. Two immunotoxins are being produced, using a cytotoxic *Pseudomonas* exotoxin and ricin, a protein synthesis inhibitor.

Biochemists at the National Cancer Institute are using an antisense RNA that binds to the mRNA transcribed from HIV's *rev* gene. Products of the *rev* gene initiate viral synthesis. If antisense RNA binds to the *rev* mRNA, the proteins would not be made.

In 1993, Flossie Wong-Staal's group at the University of California, San Diego, received permission to test gene therapy against HIV in human patients. Infected T cells will be removed from each patient. The cells from some patients will receive DNA containing the gene for a ribozyme. The control patients' cells will get DNA without the gene for ribozyme. The cells will be cloned and then returned to their donors. In laboratory tests, the ribozyme cut the virus's genome and so reduced viral reproduction significantly.

Another plan for gene therapy is being explored at the Dana-Farber Cancer Institute in Boston. Researchers have engineered cells to allow antibodies access to the interior of viral-infected cells so the antibodies can interfere with viral synthesis. They did this by introducing into the cells the gene for anti-gp120 antibodies spliced with a gene that causes the protein to be synthesized in the endoplasmic reticulum and kept inside the cell.

In 1993, a number of hospitals began testing cell therapy using a device called CELLector™. Applied Immune Sciences of California developed the instrument, which uses monoclonal antibodies to remove selected cells from a patient's sample. Researchers plan to remove CD8 T cells from AIDS patients. The T_C cells that recognize HIV will be cloned, and an extra billion or so of these cells will be returned to the patients. These cells are able to kill HIV, so perhaps more of them will be able to eliminate HIV in a patient.

In a novel approach, Genentech of California and MicroGeneSys are testing gp120 and gp160 vaccines, respectively, on people infected with HIV. Curative rather than preventative vaccines are an unusual concept; however, the researchers hope that vaccination will stimulate a stronger immune response in the patients.

Research into new treatments and vaccines against HIV continues to be intense. However, it will take years of continued research and testing to discover effective drugs and vaccines.

AIDS DRUGS CURRENTLY BEING TESTED	MECHANISM OF ACTION

Nucleoside analogues
    Dideoxyadenosine (ddA)
    Ribavirin
    Phosphorothioate
    deoxynucleotides
	Stop DNA synthesis by competitive inhibition

Castanospermine	Inhibits action of glucosidase necessary for HIV biosynthesis
Dextran sulfate	Blocks attachment of viruses
Rifabutin	Inhibits reverse transcriptase
Alpha-interferon	Prevents virus replication
Ampligen	Induces interferon
Phosphonoformate	Inhibits reverse transcriptase
Interleukin-2	Causes proliferation of T cells
Peptide T	Blocks virus receptor sites
Soluble CD4	Virus binds to this instead of to real CD4 sites on lymphocytes
DP-107	Blocks virus receptor sites
A-747404	Inhibits protease

(a) After binding to the tumor-associated antigens of a cancerous cell, a T$_C$ cell (lower right) releases the cell after causing lesions in its surface (see also Figure 17.14).

(b) The cancerous cell has disintegrated, leaving only its cytoskeleton beneath the T$_C$ cell.

FIGURE 19.15 Interaction of a cytotoxic T (T$_C$) cell and a cancer cell.

IMMUNE RESPONSES TO CANCER

LEARNING OBJECTIVE

- *Describe the immune responses to cancer and immunotherapy.*

Cancer, like an infectious disease, represents a failure of the body's defenses, including the immune system. One of the most promising avenues for effective therapy of cancer makes use of immunological techniques.

Immunological Surveillance

The immune system's responsibility for patrolling the body for cancer cells is called **immunological surveillance.** Individuals who are immunosuppressed by either natural or artificial means are markedly more susceptible to cancers than the rest of the population. The conventional explanation for this susceptibility is that the immune system normally recognizes and destroys cancer cells before they become established in the body. The tendency of cancer to occur most often in the elderly, whose immune systems are becoming less efficient, or in the very young, whose immune systems may not have developed fully or properly, supports this view.

A cell becomes cancerous when it undergoes transformation (see Chapter 13) and begins to proliferate without control. Because of transformation, the surfaces of tumor cells may acquire tumor-associated antigens that mark them as nonself to the immune system. Activated T$_C$ cells respond to these nonself antigens by attaching to and lysing the cancerous cells that carry them (Figure 19.15).

Immunological Escape

Cancer sometimes occurs even in people with presumably healthy immune systems. Once established, the cancer seems to become resistant to immune rejection, even though cell-mediated and humoral immune responses are directed against the tumor-associated surface antigens. This resistance to rejection is called **immunological escape,** and several mechanisms have been proposed to explain it. One possibility is *antigen modulation,* in which tumor cells are believed to shed their associated antigens, thus evading recognition by the immune system. Another proposed mechanism is that some cancers, such as Hodgkin's disease, release factors that eventually suppress the entire cell-mediated immune system.

Immunotherapy

At the turn of the twentieth century, Dr. William B. Coley at a New York City hospital observed that if cancer patients contracted typhoid fever, their cancers often diminished noticeably. Following this lead, Coley made vaccines of killed gram-negative bacteria, called Coley's toxins, to simulate a bacterial infection. Some of this work was very promising, but its results were inconsistent, and advances in surgery and radiation treatment caused it to be nearly forgotten. We now know that endotoxins from such bacteria are powerful stimulants for the production of **tumor necrosis factor** (TNF) by macrophages. Tumor necrosis factor is a small protein that interferes with the blood supply of cancers in animals. There is now also much interest in cancer therapy with TNF and other cytokines (see Chapter 17) such as interleukin-2 and interferons.

The treatment of cancer by immunological means is termed **immunotherapy.** It seems likely to be an effective future approach to cancer treatment. One promising technique involves immunotoxins. In Chapter 17 we defined an immunotoxin as a combination of a monoclonal antibody and a toxic agent, such as the drug ricin or a radioactive compound. The monoclonal antibody is directed against a particular type of tumor-associated antigen. The antibody selectively locates the cancer cell, and the attached toxic agent destroys the cell but causes little or no damage to healthy tissue. This approach has the potential to become analogous to the use of antibiotics in treating infectious diseases. The principal types of cancers are summarized in Table 19.9.

During the final years of the twentieth century we can expect to see more advances in the battle against cancer than have occurred in all of previous history. Almost certainly, immunology will be one of the major weapons in that fight.

TABLE 19.9 Principal Types of Cancers

TYPE OF CANCER	DESCRIPTION
Carcinoma	The most common form of cancer; arises from the cells forming the skin, the glands (such as the mammary and prostate), the uterus, and the membranes of the respiratory and gastrointestinal tracts; metastasizes (spreads to other parts of the body) mainly through the lymph vessels
Leukemia	Cancer of the tissues that form blood; characterized by uncontrolled multiplication and accumulation of abnormal white blood cells
Lymphoma	Malignant tumor of the lymphoid tissue
Sarcoma	Another form of malignant tumor; arises from connective tissues, such as muscle, bone, cartilage, and membranes covering muscles and fat; metastasizes through the blood vessels

STUDY OUTLINE

INTRODUCTION (p. 466)

1. Hay fever, transplant rejection, and autoimmunity are examples of harmful immune reactions.
2. Infection, immunosuppression, and the carrier state are examples of failure of the immune system.
3. Superantigens activate many T-cell receptors, resulting in the release of excessive cytokines that can cause adverse host responses.

HYPERSENSITIVITY (pp. 466–467)

1. Hypersensitivity reactions represent immunologic responses to an antigen (allergen) that lead to tissue damage rather than immunity.
2. Hypersensitivity reactions occur when a person has been sensitized to an antigen.
3. Hypersensitivity reactions can be divided into four classes: types I, II, and III are immediate reactions based on humoral immunity, and type IV is a delayed reaction based on cell-mediated immunity.

TYPE I (ANAPHYLACTIC) REACTIONS (pp. 467–469)

1. Anaphylactic reactions involve the production of IgE antibodies that bind to mast cells and basophils to sensitize the host.
2. Binding of two adjacent IgE antibodies to an antigen causes the target cell to release chemical mediators, such as histamine, leukotrienes, and prostaglandins, which cause the observed allergic reactions.

3. Systemic anaphylaxis may develop in minutes after injection or ingestion of the antigen; this may result in circulatory collapse and death.
4. Localized anaphylaxis is exemplified by hives, hay fever, and asthma.
5. Skin testing is useful in determining sensitivity to an antigen.
6. Desensitization to an antigen can be achieved by repeated injections of the antigen, which leads to the formation of blocking (IgG) antibodies.

TYPE II (CYTOTOXIC) REACTIONS (pp. 469–472)

1. Type II reactions are mediated by IgG or IgM antibodies and complement.
2. The antibodies are directed toward foreign cells or host cells. Complement fixation may result in cell lysis. Macrophages and other cells may also damage the antibody-coated cells.

The ABO Blood Group System (p. 470)

1. Human blood may be grouped into four principal types, designated A, B, AB, and O.
2. The presence or absence of two carbohydrate antigens designated A and B on the surface of the red blood cell determines a person's blood type.
3. Naturally occurring antibodies are present or absent in serum against the opposite AB antigen.
4. Incompatible blood transfusions lead to the complement-mediated lysis of the donor red blood cells.

The Rh Blood Group System   (pp. 470–471)

1. Approximately 85% of the human population possesses another blood group antigen, designated the Rh antigen; these individuals are designated Rh$^+$.
2. The absence of this antigen in certain individuals (Rh$^-$) can lead to sensitization upon exposure to it.
3. An Rh$^+$ person can receive Rh$^+$ or Rh$^-$ blood transfusions.
4. When an Rh$^-$ person receives Rh$^+$ blood, that person will produce anti-Rh antibodies.
5. Subsequent exposure to Rh$^+$ cells will result in a rapid, serious hemolytic reaction.
6. An Rh$^-$ mother carrying an Rh$^+$ fetus will produce anti-Rh antibodies.
7. Subsequent pregnancies involving Rh incompatibility may result in hemolytic disease of the newborn.
8. The disease may be prevented by passive immunization of the mother with anti-Rh antibodies.

Drug-Induced Cytotoxic Reactions   (pp. 471–472)

1. In the disease thrombocytopenic purpura, platelets are destroyed by antibodies and complement.
2. Agranulocytosis and hemolytic anemia result from antibodies against one's own blood cells coated with drug molecules.

TYPE III (IMMUNE COMPLEX) REACTIONS
(p. 472)

1. Immune complex diseases occur when IgG antibodies and soluble antigen form small complexes that lodge in the basement membranes of cells.
2. Subsequent complement fixation results in inflammation.
3. Serum sickness and glomerulonephritis are immune complex diseases.

TYPE IV (CELL-MEDIATED) REACTIONS
(pp. 472–474)

1. Delayed hypersensitivity responses are due primarily to T$_D$ cell proliferation.
2. Sensitized T cells secrete cytokines in response to the appropriate antigen.
3. Cytokines attract and activate macrophages and initiate tissue damage.
4. The tuberculin skin test and allergic contact dermatitis are examples of delayed hypersensitivities.

AUTOIMMUNE DISEASES   (pp. 474–476)

1. Autoimmunity results from loss of self-tolerance.
2. Self-tolerance occurs during fetal development; T cells that will target host cells are eliminated (clonal deletion) or inactivated.

3. Graves' disease and myasthenia gravis are type II autoimmune reactions in which antibodies react to cell-surface antigens.
4. Systemic lupus erythematosus and rheumatoid arthritis are type III autoimmune reactions in which deposition of immune complexes results in tissue damage.
5. Hashimoto's disease and insulin-dependent diabetes mellitus are type IV autoimmune reactions mediated by T cells.

DISEASES AND REACTIONS RELATED TO THE HUMAN LEUKOCYTE ANTIGEN (HLA) COMPLEX   (pp. 476–478)

1. Histocompatibility self molecules located on cell surfaces express genetic differences among individuals; these antigens are coded for by MHC or HLA gene complexes.
2. To prevent rejection of transplants, HLA and ABO blood group antigens of the donor and recipient are matched as closely as possible.

Transplantation Reactions   (pp. 477–478)

1. Transplants recognized as foreign antigens may be lysed by T cells and attacked by macrophages and complement-fixing antibodies.
2. Transplantation to a privileged site (such as the cornea) or of a privileged tissue (such as pig heart valves) does not cause an immune response.
3. Four types of transplants have been defined on the basis of genetic relationships between the donor and the recipient: autografts, isografts, allografts, and xenografts.
4. Bone marrow (with immunocompetent cells) can cause graft-versus-host disease.
5. Successful transplant surgery often requires immunosuppressive treatment, which increases the host's susceptibility to infections and cancer.
6. Immunosuppressant drugs inhibit the synthesis of certain cytokines, the formation of T$_C$ cells, and the differentiation of B cells.

IMMUNE DEFICIENCIES   (p. 479)

1. Immune deficiencies can be congenital or acquired.
2. Congenital immune deficiencies are due to defective or absent genes.
3. A variety of drugs, cancers, and infectious diseases can cause acquired immune deficiencies.

ACQUIRED IMMUNODEFICIENCY SYNDROME (AIDS)   (pp. 479–485)

The Origin of AIDS   (p. 479)

1. HIV is thought to have originated in central Africa and was brought to other countries by immigration, vacationing, and imported blood products.

The HIV Infection (pp. 480–481)

1. AIDS is caused by the retrovirus HIV. HIV is an enveloped RNA virus.
2. Helper T cells are the primary host for HIV. gp120 attachment spikes on the viral envelope attach to CD4 receptors on host cells such as T_H cells.
3. The virus evades host defenses by latency, antigenic changes, and being spread by cell-to-cell contact rather than lysis.
4. Seroconversion usually occurs within six months after infection with HIV.
5. Category A HIV disease may be asymptomatic or have persistent lymphadenopathy. The CD4 T cell count is 500/mm^3 or greater.
6. Category B HIV disease is characterized by persistent *Candida* infections, shingles, diarrhea, and hairy leukoplakia. The CD4 T cell count is 200–499/mm^3.
7. Category C is called clinical AIDS. Indicator infections or diseases are *Candida*, cytomegalovirus, tuberculosis, *Pneumocystis*, toxoplasmosis, and Kaposi's sarcoma. The CD4 T cell count is 200/mm^3 or lower.

Modes of HIV Transmission (pp. 481–482)

1. HIV is transmitted in blood, semen, vaginal secretions, and breast milk and across the placenta.
2. In developed countries, blood transfusions are not a likely source of infection because blood is tested for HIV antibodies.
3. In the U.S.A., HIV is transmitted primarily by homosexual activity and intravenous drug use; heterosexual transmission is increasing.
4. In Africa and Asia, HIV is transmitted primarily by heterosexual activity.

AIDS Research: Treatments and Vaccines
(pp. 482–483)

1. The nucleotide analogues zidovudine, ddI, and ddC inhibit reverse transcriptase.

2. Another unique viral enzyme, protease, could be a target for selective chemotherapy.
3. Subunit vaccines are being tested. Vaccines must overcome the antigenic variation in HIV and its ability to stay inside host cells.

The Future Course of AIDS (p. 483)

1. The number of AIDS cases is expected to increase during the next few years. In addition, more than 80% of transmission will be by heterosexual contact, with females making up an increasing percentage of new infections.
2. It appears that less virulent viruses are selected for in cultures that favor monogamous or protected sex.

IMMUNE RESPONSES TO CANCER (pp. 486–487)

Immunological Surveillance (p. 486)

1. Cancer cells are normal cells that have undergone transformation, divide uncontrollably, and possess tumor-associated antigens. Metastasis refers to the spread of such cancer cells to other parts of the body.
2. The response of the immune system to cancer is called immunological surveillance.
3. T_C cells recognize and lyse cancerous cells.

Immunological Escape (p. 486)

1. The ability of tumor cells to escape immune responses against them is called immunological escape.
2. Antigen modulation and immunological enhancement represent two ways tumor cells escape detection and destruction by the immune system.

Immunotherapy (pp. 486–487)

1. Tumor necrosis factor, interleukin-2, and interferons may be useful in treating some cancers.
2. Immunotoxins are chemical poisons linked to a monoclonal antibody; the antibody selectively locates the cancer cell for release of the poison.

STUDY QUESTIONS

Review

1. Define hypersensitivity.

2. Compare and contrast the characteristics of the four types of hypersensitivity reactions. Give an example of each type.

3. List three mediators released in anaphylactic hypersensitivities, and explain their effects.

4. Contrast systemic and localized anaphylactic reactions. Which type is more serious? Cite an example of each.

5. Explain what happens when a person develops a contact sensitivity to poison oak.
   (a) What causes the observed symptoms?
   (b) How did the sensitivity develop?
   (c) How might this person be desensitized to poison oak?

6. Discuss the roles of antibodies and antigens in an incompatible tissue transplant.

7. What happens to the recipient of an incompatible blood type?

8. Explain how hemolytic disease of the newborn develops and how this disease might be prevented.

9. Which type of graft (autograft, isograft, allograft, or xenograft) is most compatible? Least compatible?

10. Which of the following blood transfusions is compatible? Explain your answers.

	DONOR	RECIPIENT
(a)	AB, Rh$^-$	AB, Rh$^+$
(b)	B, Rh$^+$	B, Rh$^-$
(c)	A, Rh$^+$	O, Rh$^+$

11. Define immunosuppression. Why is it used, and what problems does it cause?

12. Define autoimmunity. Present a hypothesis that could explain autoimmune responses.

13. Differentiate among type II, type III, and type IV autoimmune diseases. Name an example of each type.

14. Summarize the causes of immune deficiencies. What is the effect of an immune deficiency?

15. In what ways do tumor cells differ antigenically from normal cells? Explain how tumor cells may be destroyed by the immune system.

16. If tumor cells can be destroyed by the immune system, how does cancer develop? What does immunotherapy involve?

17. Differentiate between HIV infection and AIDS. How is HIV transmitted? How can it be prevented?

Critical Thinking

1. When and how does our immune system discriminate between self and nonself antigens?

2. The first preparations used for artificially acquired passive immunity were antibodies in horse serum. A complication that resulted from the therapeutic use of horse serum was immune complex disease. Why did this occur?

3. Do people with AIDS make antibodies? If so, why are they said to have an immune deficiency?

4. What is the method of action of the anti-AIDS drugs?

Clinical Applications

1. Fungal infections such as athlete's foot are chronic. These fungi degrade keratin but are not invasive and do not produce toxins. Why do you suppose that many of the symptoms of a fungal infection are due to hypersensitivity to the fungus?

2. After working in a mushroom farm for several months, a worker develops these symptoms: hives, edema, and swelling of lymph nodes.
   (a) What do these symptoms indicate?
   (b) What mediators cause these symptoms?
   (c) How may sensitivity to a particular antigen be determined?
   (d) Other employees do not appear to have any immunological reactions. What could explain this?
   (*Note:* The allergen is conidiospores from molds growing in the mushroom farm.)

3. Physicians administering live, attenuated mumps and measles vaccines prepared in chick embryos are instructed to have epinephrine available. Because epinephrine will not treat these viral infections, what is the purpose of keeping this drug on hand?

Antimicrobial Drugs

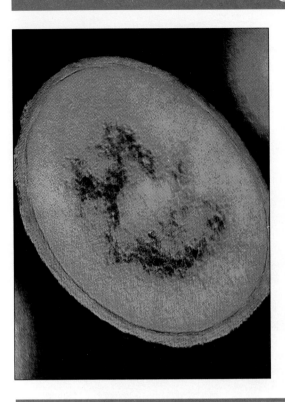

WHEN THE BODY'S normal defenses cannot prevent or overcome a disease, it is often treated with **chemotherapy** (drugs).

In this chapter, we focus on **antimicrobial drugs**, the class of **chemotherapeutic agents** used to treat infectious diseases. Like the disinfectants discussed in Chapter 7, these chemicals act by interfering with the growth of microorganisms. Unlike disinfectants, however, they must often act *within* the host. Therefore, their effects on the cells and tissues of the host are important. The ideal antimicrobial drug kills the harmful microorganism without damaging the host, the principle of **selective toxicity**.

Drugs used in the chemotherapy of infectious disease are classified into two groups. Drugs that have been synthesized by chemical procedures in the laboratory are called **synthetic drugs**. Drugs produced by bacteria and fungi are called **antibiotics**. The distinction is often ignored in practice.

HISTORY OF CHEMOTHERAPY

LEARNING OBJECTIVES

- *Define a chemotherapeutic agent, and distinguish between a synthetic drug and an antibiotic.*
- *Identify the contributions of Paul Ehrlich, Alexander Fleming, René Dubos, and Selman Waksman to chemotherapy.*

The birth of modern chemotherapy is credited to the efforts of Paul Ehrlich in Germany during the early part of the twentieth century. While attempting to stain bacteria without staining the surrounding tissue, he speculated about some "magic bullet" that would selectively find and destroy pathogens but not harm the host. This idea provided the basis for chemotherapy, a term he coined. Recall from Chapter 1 that Ehrlich eventually discovered a chemotherapeutic agent called salvarsan, an arsenic derivative that was effective against syphilis (by the standards of the day). Prior to Ehrlich's discovery, there had been only one effective chemotherapeutic agent in the medical arsenal: quinine, a drug used for the treatment of malaria.

In the late 1930s, the discovery of sulfa drugs, a group of synthetic drugs, touched off a new interest in chemotherapy. This discovery resulted from a systematic survey of chemicals, including many synthesized derivatives of aniline dyes. The chemical prontosil, which had been first synthesized and used as a dye many years before, was found to be an effective antimicrobial agent. Oddly enough, it worked only within living animals and was completely ineffective when the same bacteria were exposed to it in a test tube. The active ingredient in prontosil proved to be *sulfanilamide*, which forms as the animal metabolizes prontosil. This discovery led to the rapid development of a number of related drugs: the *sulfonamides*, or *sulfa drugs*, which are still used today.

We also mentioned in Chapter 1 that in 1928, Alexander Fleming observed that the growth of the bac-

FIGURE 20.1 Discovery of penicillin. Alexander Fleming took this photograph in 1928. The colony of *Penicillium* mold accidentally contaminated the plate and is inhibiting nearby bacterial growth.

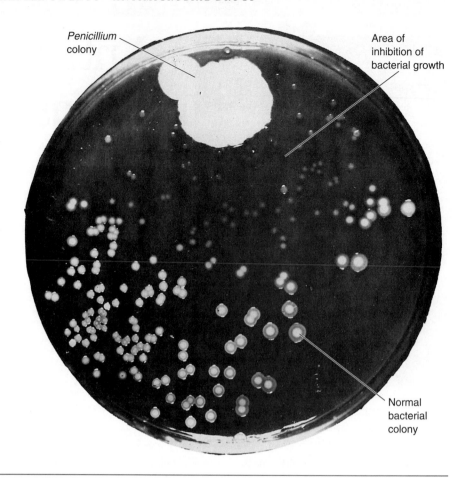

Penicillium colony

Area of inhibition of bacterial growth

Normal bacterial colony

terium *Staphylococcus aureus* was inhibited in the area surrounding the colony of a mold that had contaminated a Petri plate (Figure 20.1). The mold was identified as *Penicillium notatum,* and its active compound, which was isolated a short time later, was named *penicillin.* Similar inhibitory reactions between colonies on solid media are commonly observed in microbiology, and the mechanism of inhibition is called *antibiosis.* From this term comes the term *antibiotic,* meaning a substance produced by microorganisms that in small amounts inhibits another microorganism. Therefore, the wholly synthetic sulfa drugs technically are not antibiotics.

In 1940, a group of scientists at Oxford University headed by Howard Florey and Ernst Chain succeeded in the first clinical trials of penicillin. Intensive research in the U.S.A. then led to the isolation of especially productive *Penicillium* strains for use in mass production of the antibiotic. (The most famous of these high-producing strains was originally isolated from a cantaloupe bought at a market in Peoria, Illinois.) Penicillin is still one of our most effective antibiotics, and its enormous success led to the search for others, a search that has revolutionized medicine. It is an interesting fact that penicillin has a singular toxicity for guinea pigs, and it is entirely possible that this drug would never have left the experimental laboratory under the more demanding safety requirements that must be met today in the development of drugs.

Antibiotics are actually rather easy to discover, but few are of medical or commercial value. Some are used commercially other than for treating disease—for example, as a supplement in animal feed (see the box on p. 505). Many antibiotics are toxic to humans or lack any advantage over antibiotics already in use.

In 1939, René Dubos isolated a substance from a *Bacillus* bacterium that he had cultured from soil. This early antibiotic, gramicidin, did not prove especially useful, but it stimulated interest in soil as a source of antibiotic-producing microbes. A few years later, Selman Waksman, a soil microbiologist, isolated a species of *Streptomyces* that produced an antibiotic effective against many gram-negative bacteria and also the *Mycobacterium tuberculosis* bacteria that cause tuberculosis. This antibiotic, which he called streptomycin, was especially useful at the time because penicillin's activity had little effect on gram-negative organisms. Perhaps more important, it focused interest on the streptomycetes in soil as sources of antibiotics.

More than half of our antibiotics are produced by species of *Streptomyces,* filamentous bacteria that commonly inhabit soil. A few antibiotics are produced by bacteria of the genus *Bacillus,* and others are produced by molds, mostly of the genera *Penicillium* and *Cephalosporium* (sef-ä-lō-spô′rē-um). See Table 20.1 for the sources of many antibiotics in use today—a surprisingly limited group of organisms.

SPECTRUM OF ANTIMICROBIAL ACTIVITY

LEARNING OBJECTIVES

- *Define the following terms: spectrum of activity, super-infection, broad spectrum.*
- *Describe the problems of chemotherapy for viral, fungal, protozoan, and helminthic infections.*

It is comparatively easy to find or develop drugs that are effective against procaryotic cells (bacteria) and do not affect the eucaryotic cells of humans. These two cell types differ substantially in many ways, such as in the presence or absence of cell walls, the fine structure of their ribosomes, and details of their metabolism. Thus, selective toxicity has numerous targets. The problem is more difficult when the pathogen is a eucaryotic cell, such as a fungus, protozoan, or helminth. At the cellular level, these organisms resemble the human cell much more closely than a bacterial cell does. We will see that our arsenal against these types of pathogens is much more limited than our arsenal of antibacterial drugs. Viral infections are particularly difficult to treat because the pathogen is within the human host's cells, and the genetic information of the virus is directing the human cell to make viruses rather than to synthesize normal cellular materials. The more we discover about the reproduction, metabolism, and structure of pathogens, the better equipped we are to discover fresh targets for antimicrobials. No doubt the search will never end.

Some antimicrobial drugs have a narrow **spectrum of activity** (range of different microbial types they affect); penicillin, for instance, affects gram-positive bacteria but very few gram-negative bacteria. Other drugs affect a large number of gram-positive or gram-negative bacteria and

are therefore called **broad-spectrum drugs.** Table 20.2 summarizes the spectrum of activity of a number of chemotherapeutic drugs. Because the identity of the pathogen is not always immediately known, a broad-spectrum drug would seem to have an advantage in the treatment of disease by saving valuable time. The disadvantage is that many members of the normal microbiota of the host are destroyed by broad-spectrum drugs. The normal microbiota ordinarily competes with and checks the growth of

TABLE 20.1 Representative Sources of Antibiotics

MICROORGANISM	ANTIBIOTIC
Gram-Positive Rods	
Bacillus subtilis	Bacitracin
Bacillus polymyxa	Polymyxin
Actinomycetes	
Streptomyces nodosus	Amphotericin B
Streptomyces venezuelae	Chloramphenicol
Streptomyces aureofaciens	Chlortetracycline and tetracycline
Streptomyces erythraeus	Erythromycin
Streptomyces fradiae	Neomycin
Streptomyces noursei	Nystatin
Streptomyces griseus	Streptomycin
Micromonospora purpurea	Gentamicin
Fungi	
Cephalosporium	Cephalothin
Penicillium griseofulvum	Griseofulvin
Penicillium notatum	Penicillin

TABLE 20.2 Spectrum of Activity of Antibiotics and Other Antimicrobials

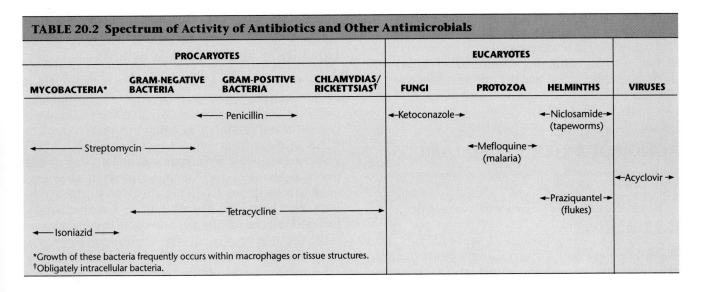

PROCARYOTES				EUCARYOTES			
MYCOBACTERIA*	GRAM-NEGATIVE BACTERIA	GRAM-POSITIVE BACTERIA	CHLAMYDIAS/ RICKETTSIAS†	FUNGI	PROTOZOA	HELMINTHS	VIRUSES
		◄— Penicillin —►		◄Ketoconazole►		◄Niclosamide► (tapeworms)	
◄— Streptomycin —►					◄Mefloquine► (malaria)		
							◄Acyclovir►
	◄———————— Tetracycline ————————►					◄Praziquantel► (flukes)	
◄Isoniazid►							

*Growth of these bacteria frequently occurs within macrophages or tissue structures.
†Obligately intracellular bacteria.

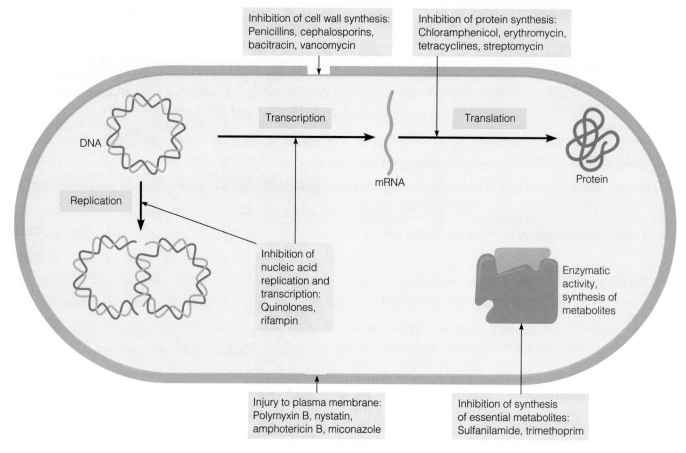

FIGURE 20.2 Summary of the major modes of action of antimicrobial drugs. This illustration shows these actions in a highly diagrammatic composite of a microbial cell.

pathogens or other microbes. If certain organisms in the normal microbiota are not destroyed by the antibiotic and their competitors are destroyed, the survivors may flourish and become opportunistic pathogens. An example that sometimes occurs is overgrowth by the yeastlike fungus *Candida albicans,* which is not sensitive to bacterial antibiotics. This overgrowth is called a **superinfection,** a term that is also applied to growth of a target pathogen that has developed resistance to the antibiotic. In this situation, such an antibiotic-resistant strain replaces the original sensitive strain, and the infection continues.

ACTION OF ANTIMICROBIAL DRUGS

LEARNING OBJECTIVE

- *Identify five methods of action of antimicrobial drugs.*

We will next examine the various ways in which antibiotics exert their antimicrobial activity. Antimicrobial

drugs either kill microorganisms directly (**bactericidal**) or simply prevent them from growing (**bacteriostatic**). In *bacteriostasis,* the host's own defenses, such as phagocytosis and antibody production, usually destroy the microorganisms. The major modes of action are summarized in Figure 20.2.

Inhibition of Cell Wall Synthesis

Recall from Chapter 4 that the cell wall of a bacterium consists of a macromolecular network called peptidoglycan. Peptidoglycan is found only in bacterial cell walls. Penicillin and certain other antibiotics prevent the synthesis of intact peptidoglycan; consequently, the cell wall is greatly weakened and the cell undergoes lysis (Figure 20.3). Because penicillin targets the synthesis process, only actively growing cells are affected by these antibiotics. And, because human cells do not have peptidoglycan cell walls, penicillin has very little toxicity for host cells. You may also recall that gram-negative bacteria have an outer membrane of lipopolysaccharides and other materials over a comparatively thin layer of pepti-

FIGURE 20.3 Inhibition of bacterial cell wall synthesis by an antibiotic. (a) *Staphylococcus aureus* before exposure to the antibiotic ampicillin. (b) *S. aureus* after exposure to ampicillin, which—like other penicillins—inhibits proper formation of cell walls.

(a) Before antibiotic   SEM ⊢—⊣ 1 μm

(b) After antibiotic   SEM ⊢—⊣ 1 μm

doglycan. Gram-negative bacteria are relatively resistant to penicillin because this outer membrane apparently prevents penicillin from reaching the site of peptidoglycan synthesis.

Antibiotics can affect cell wall synthesis in several ways. Two commonly used antibiotics, bacitracin and vancomycin, interfere with the synthesis of the linear strands of peptidoglycan. Penicillin and cephalosporins prevent the final cross-linking of the peptidoglycans, which interferes with the construction of the macromolecular cell wall.

Inhibition of Protein Synthesis

Because protein synthesis is a common feature of all cells, whether procaryotic or eucaryotic, it would seem an unlikely target for selective toxicity. One notable difference between procaryotes and eucaryotes, however, is the structure of their ribosomes. As we discussed in Chapter 4, eucaryotic cells have 80S ribosomes; procaryotic cells have 70S ribosomes. (The 70S ribosome is made up of a 50S and a 30S unit.) The difference in ribosomal structure accounts for the selective toxicity of antibiotics that affect protein synthesis. However, mitochondria (important eucaryotic organelles) also contain 70S ribosomes similar to those of bacteria. Antibiotics targeting the 70S ribosomes can therefore have adverse effects on the cells of the host. Among the antibiotics that interfere with protein synthesis are chloramphenicol, erythromycin, streptomycin, and the tetracyclines (Figure 20.4).

Reacting with the 50S portion of the 70S procaryotic ribosome, chloramphenicol inhibits the formation of peptide bonds in the growing polypeptide chain. Erythromycin also reacts with the 50S portion of the 70S procaryotic ribosome. Most drugs that inhibit protein synthesis have a broad spectrum of activity; erythromycin is an exception. Because it does not penetrate the gram-negative cell wall, it affects mostly gram-positive bacteria.

Some other antibiotics react with the 30S portion of the 70S procaryotic ribosome. The tetracyclines interfere with the attachment of the tRNA carrying the amino acids to the ribosome, preventing the addition of amino acids to the growing polypeptide chain. Tetracyclines do not interfere with mammalian ribosomes because they do not penetrate very well into intact mammalian cells. However, at least small amounts are able to enter the host cell, as is apparent from the fact that the intracellular pathogenic rickettsias and chlamydias are sensitive to tetracyclines. The selective toxicity of the drug in this case is due to a greater sensitivity of the bacteria at the ribosomal level.

Aminoglycoside antibiotics, such as streptomycin and gentamicin, interfere with the initial steps of protein synthesis by changing the shape of the 30S portion of the 70S procaryotic ribosome. This interference causes the genetic code on the mRNA to be read incorrectly.

Injury to the Plasma Membrane

Certain antibiotics, especially polypeptide antibiotics, bring about changes in the permeability of the plasma membrane; these changes result in the loss of important metabolites from the microbial cell. One such antibiotic is polymyxin B, which causes disruption of the plasma membrane by attaching to the phospholipids of the membrane.

Some antifungal drugs, such as nystatin, amphotericin B, miconazole, and ketoconazole, are effective against many fungal diseases. Such drugs combine with sterols in the fungal plasma membrane to disrupt the membrane (Figure 20.5). Because bacterial plasma membranes generally lack sterols, these antibiotics do not act on bacteria. However, the plasma membranes of animal cells do contain sterols, and nystatin and amphotericin B can be toxic to the host. Fortunately, animal cell membranes have mostly *cholesterol,* and fungal cells have mostly *ergosterol,* against which the drug is most effective, so that the balance of the toxicity is tilted toward the fungus.

FIGURE 20.4 Inhibition of protein synthesis by antibiotics. The black arrows indicate the specific points at which chloramphenicol, erythromycin, the tetracyclines, and streptomycin exert their activities.

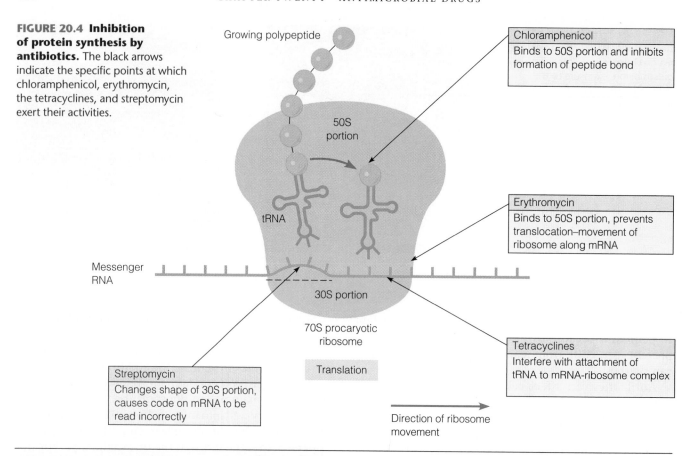

Growing polypeptide

50S portion

Chloramphenicol
Binds to 50S portion and inhibits formation of peptide bond

Erythromycin
Binds to 50S portion, prevents translocation–movement of ribosome along mRNA

tRNA

Messenger RNA

30S portion

70S procaryotic ribosome

Translation

Streptomycin
Changes shape of 30S portion, causes code on mRNA to be read incorrectly

Tetracyclines
Interfere with attachment of tRNA to mRNA-ribosome complex

Direction of ribosome movement

SEM

FIGURE 20.5 Injury to the plasma membrane of a yeast cell caused by an antifungal drug. The cell releases its cytoplasmic contents as the plasma membrane is disrupted by the antifungal drug miconazole.

Inhibition of Nucleic Acid Replication and Transcription

A number of antibiotics interfere with the processes of DNA replication and transcription in microorganisms. Some drugs with this mode of action, such as the antiviral idoxuridine, have an extremely limited usefulness because they interfere with mammalian DNA and RNA as well. Others, such as rifampin and the quinolones, are more widely used in chemotherapy because they are more selectively toxic.

Inhibition of Synthesis of Essential Metabolites

In Chapter 5, we mentioned that a particular enzymatic activity of a microorganism can be *competitively inhibited* by a substance (antimetabolite) that closely resembles the normal substrate for the enzyme (see Figure 5.6). An example of competitive inhibition is the relationship between the antimetabolite sulfanilamide (a sulfa drug) and para-aminobenzoic acid (PABA) (Figure 20.6). In many microorganisms, PABA is the substrate for an enzymatic reaction leading to the synthesis of folic acid, a vitamin that functions as a coenzyme for the synthesis of the purine and pyrimidine bases of nucleic acids and many amino acids. In the presence of sulfanilamide, the enzyme that normally converts PABA to folic acid combines with the drug instead of with PABA. This combination prevents folic acid synthesis and stops the growth of the microorganism. Because humans do not produce folic acid from PABA (they obtain it as a vitamin in ingested foods), sulfanilamide exhibits selective toxicity—it affects microorganisms that synthesize their own folic acid but does not harm the human host. Other chemotherapeutic agents that act as antimetabolites are the sulfones and trimethoprim.

FIGURE 20.6 Comparison of the structures of PABA and sulfanilamide, a sulfa drug that inhibits synthesis of essential metabolites. Sulfanilamide's structural resemblance to PABA allows it to act as a competitive inhibitor, stopping the synthesis of folic acid, a compound necessary for microorganisms to grow.

SURVEY OF COMMONLY USED ANTIMICROBIAL DRUGS

LEARNING OBJECTIVES

- *Describe the methods of action of each of the commonly used antibacterial drugs.*
- *Explain the actions of currently used antiviral drugs.*
- *Explain the actions of currently used antifungal, antiprotozoan, and antihelminthic drugs.*

We will next examine the properties and activities of some commonly used antimicrobial drugs. Commonly used drugs are listed in Table 20.3.

Antibacterial Synthetics

Isoniazid (INH)

Isoniazid (INH) is a very effective synthetic antimicrobial drug against *Mycobacterium tuberculosis*. It has little effect on nonmycobacteria. The primary effect of INH is to inhibit synthesis of mycolic acids, which are components of cell walls only of the mycobacteria. When used to treat tuberculosis, INH is usually administered simultaneously with other drugs, such as rifampin or ethambutol. Because the tubercle bacillus is usually found only within macrophages or walled off in tissue, any antitubercular drug must be able to penetrate into such sites.

Ethambutol

Ethambutol is effective only against mycobacteria. The drug apparently inhibits incorporation of mycolic acid into the cell wall. It is a comparatively weak antitubercular drug; its principal use is as the secondary drug to avoid resistance problems (see p. 510).

Sulfonamides

As we noted earlier, **sulfonamides (sulfa drugs)** were among the first synthetic antimicrobial drugs used to treat microbial diseases. Unfortunately, many pathogens have developed resistance to them. Moreover, sulfonamides cause allergic reactions in many people. For these reasons, antibiotics have tended to diminish the importance of sulfonamides in chemotherapy. The sulfonamides continue to be used to treat certain urinary tract infections and have other specialized uses, as in the combination drug **silver sulfadiazine**, used in the control of infections in burn patients. Sulfonamides are bacteriostatic; as we mentioned, their action is due to their structural similarity to para-aminobenzoic acid (PABA) (see Figure 20.6).

Probably the most widely used sulfa today is a combination of **trimethoprim** and **sulfamethoxazole** (TMP–SMZ). This combination is an excellent example of drug *synergism* (see p. 510). When the drugs are used in combination, only about one-tenth of the concentration is needed as compared to when each is used alone. The combination also has a broader spectrum of action and greatly reduces the emergence of resistant strains.

Figure 20.7 (p. 500) illustrates how the two drugs interfere with different steps of a metabolic sequence leading to the synthesis of precursors of proteins, DNA, and RNA. Sulfamethoxazole inhibits the synthesis of dihydrofolic acid from PABA, so that the amount of dihydrofolic acid made is much less than usual. Then trimethoprim (which is not classified as a sulfa drug) inhibits the conversion of any dihydrofolic acid present into tetrahydrofolic acid. TMP–SMZ is especially useful against many gram-negative pathogens of the urinary tract and intestinal tract.

Quinolones and Fluoroquinolones

In the early 1960s the synthetic drug *nalidixic acid* was developed—the first of the quinolone group of antimicrobials. It exerted a unique bactericidal effect by selectively inhibiting an enzyme (DNA gyrase) needed for the replication of DNA. Although nalidixic acid found only limited use (its only application being for urinary tract infections), it led to the development in the 1980s of a prolific group of new antibiotics, the **fluoroquinolones.** *Norfloxacin* and *ciprofloxacin* are used most. Fluoroquinolones have a broad spectrum of activity, penetrate tissues well, and are generally safe for adults, especially for treatment of urinary tract infections. However, they may adversely affect the development of cartilage, which limits their use in children, adolescents, and pregnant women.

Antifungal Synthetics

The antifungal drug *flucytosine,* an antimetabolite of the base cytosine, interferes with protein synthesis. It is preferentially taken up by fungal cells. Its spectrum of activity is limited to only a few systemic fungal infections, and it is almost always used in combination with other antifungal drugs.

TABLE 20.3 Commonly Used Antimicrobial Drugs

ANTIMICROBIAL DRUG	EFFECT ON MICROORGANISMS	MODE OF ACTION	CLINICAL USE/TOXICITY
Antibacterial Synthetics			
Isoniazid (INH)	Bacteriostatic	Competitive inhibition	With streptomycin and rifampin to treat tuberculosis
Ethambutol	Bacteriostatic	Competitive inhibition	With INH to treat tuberculosis
Sulfonamides	Bacteriostatic	Competitive inhibition	*Neisseria meningitidis* meningitis and some urinary tract infections
Quinolones (fluoroquinolones)	Bactericidal	Inhibit DNA synthesis	Urinary tract infections
Antibacterial Antibiotics			
Penicillin, natural penicillins G, V	Bactericidal	Inhibition of cell wall synthesis	Gram-positive bacteria
Penicillin, semisynthetic ampicillin, methicillin, oxacillin, and others	Bactericidal	Inhibition of cell wall synthesis	Broad spectrum (ampicillin); resistant to penicillinase and stomach acid (methicillin, oxacillin)
Cephalosporins	Bactericidal	Inhibition of cell wall synthesis	Similar to penicillins
Carbapenems	Bactericidal	Inhibition of cell wall synthesis	Very broad spectrum
Aminoglycosides	Bactericidal	Inhibition of protein synthesis	
Streptomycin			Gram-negative bacteria
Neomycin			Topical ointment; gram-negative bacteria
Gentamicin			Gram-negative bacteria
Tetracyclines	Bacteriostatic	Inhibition of protein synthesis	Very broad spectrum; some toxic effects include tooth discoloration and liver and kidney damage
Chloramphenicol	Bacteriostatic	Inhibition of protein synthesis	*Salmonella*/aplastic anemia is an occasional toxic effect
Macrolides Erythromycin	Bacteriostatic	Inhibition of protein synthesis	Diphtheria; legionellosis; various infections in individuals allergic to penicillin
Polypeptides Bacitracin	Bactericidal	Inhibition of cell wall synthesis	Gram-positive bacteria, primarily; usually used as topical ointment/toxic to kidneys
Polymyxin B	Bactericidal	Injury to plasma membranes	Gram-negative bacteria, including *Pseudomonas*/toxic to brain and kidneys
Vancomycin	Bactericidal	Inhibition of cell wall synthesis	Staphylococci/very toxic
Rifamycins Rifampin	Bactericidal	Inhibition of RNA synthesis	Gram-positive bacteria, some gram negatives, chlamydias, and poxviruses; often used with INH and ethambutol for tuberculosis

Antibacterial Antibiotics

Penicillin

The term **penicillin** refers to a group of over 50 chemically related antibiotics (Figure 20.8). All penicillins have a common core structure containing a β-lactam ring called the *nucleus*. Penicillin molecules are differentiated by the chemical side chains attached to their nuclei. Penicillins can be produced either naturally or semisynthetically.

Natural Penicillins. Penicillin extracted from cultures of the mold *Penicillium* exists in several closely related forms.

These are the so-called **natural penicillins** (Figure 20.8a). The prototype compound of all the penicillins is *penicillin G*. It has a narrow but useful spectrum of activity and is often the drug of choice against some staphylococci, streptococci, and several spirochetes. When injected intramuscularly, penicillin G is rapidly excreted from the body in three to six hours (Figure 20.9); when the drug is taken orally, the acidity of the digestive fluids in the stomach diminishes its concentration. *Procaine penicillin,* a combination of the drugs procaine and penicillin G, is retained at detectable concentrations for up to 24 hours; concentration peaks at about four hours. Still longer retention times

TABLE 20.3 Continued

ANTIMICROBIAL DRUG	EFFECT ON MICROORGANISMS	MODE OF ACTION	CLINICAL USE/TOXICITY
Antifungal Drugs			
Polyenes	Fungicidal	Injury to plasma membrane	
Nystatin			*Candida* infections of skin and vagina
Amphotericin B			Histoplasmosis, coccidioidomycosis, and blastomycosis/ kidney toxicity and anemia
Imidazoles and triazoles	Fungicidal	Injury to plasma membrane	
Clotrimazole, miconazole			Topical use for fungal infections
Ketoconazole			Systemic fungal infections/gastrointestinal upset, rash
Griseofulvin	Fungistatic	Inhibition of mitotic spindle function	Fungal infections of skin, primarily ringworm
Antiviral Drugs			
Amantadine	Antiviral	Blocks entry or uncoating	Influenza A prophylaxis and treatment
Ribavirin			Viral pneumonia/occasional anemia at high doses
Acyclovir			Herpes infections/some kidney toxicity; neurological effects at high doses
Thymine analogs	Antiviral	Inhibition of DNA synthesis	Herpes keratitis
Idoxuridine, zidovudine (AZT)			AIDS virus/occasional anemia and blood disorders
Adenine analog Vidarabine	Antiviral	Inhibition of DNA and RNA synthesis	Herpes encephalitis
Antiprotozoan Drugs			
Chloroquine	Antiprotozoan	Inhibition of DNA synthesis	Malaria
Pentamidine isethionate	Antiprotozoan	Reacts with DNA	African trypanosomiasis, *Pneumocystis** pneumonia
Diiodohydroxyquin	Antiprotozoan	Unknown	Amoebic infections/optic nerve damage
Metronidazole	Antiprotozoan	Anaerobic metabolism products damage DNA	Amoebic infections, *Trichomonas*, anaerobic bacteria
Antihelminthic Drugs			
Praziquantel	Antihelminthic	Unknown	Schistosomiasis and other fluke infestations
Niclosamide	Antihelminthic	Prevents generation of ATP	Tapeworm infestations

*May be a fungus.

can be achieved with *benzathine penicillin,* a combination of benzathine and penicillin G. Although retention times of as long as four months can be obtained, the concentration of the drug is so low that the organisms must be very sensitive to the drug. Penicillin V, which is acid-stable and can be best taken orally, and penicillin G are the natural penicillins most often used.

Natural penicillins have some disadvantages, however. Chief among them are their narrow spectrum of activity and their susceptibility to penicillinases. **Penicillinases** are enzymes produced by many bacteria, especially *Staphylococcus* spp., that cleave the β-lactam ring of the penicillin molecule (Figure 20.10). Because of this characteristic, penicillinases are sometimes called *β-lactamases.*

Semisynthetic Penicillins. A large number of **semisynthetic penicillins** (Figure 20.8b) have been developed in attempts to overcome the disadvantages of natural

FIGURE 20.7 Actions of the antibacterial synthetics trimethoprim and sulfamethoxazole. TMP–SMZ works by inhibiting different stages in the synthesis of precursors of DNA, RNA, and proteins. Together the drugs are synergistic. ① Sulfamethoxazole, a sulfonamide that is a structural analog of PABA, competitively inhibits the synthesis of dihydrofolic acid from PABA. ② Then trimethoprim, a structural analog of a portion of dihydrofolic acid, competitively inhibits synthesis of tetrahydrofolic acid.

FIGURE 20.8 Structure of penicillins, antibacterial antibiotics. The portion that all penicillins have in common—their nucleus, which contains the β-lactam ring—is shaded in purple. The unshaded portions represent the side chains that distinguish one penicillin from another.

Penicillin G

Penicillin V

β-lactam ring

(a) Natural (antibiotic) penicillins

Common nucleus

Ampicillin

Methicillin

Carbenicillin

Oxacillin

β-lactam ring

(b) Semisynthetic penicillins

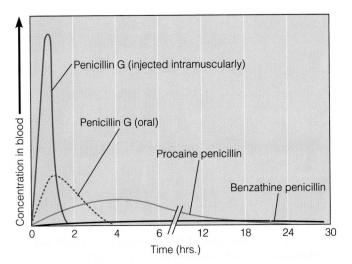

FIGURE 20.9 Retention of different forms of penicillin G in the blood. This drug reaches different concentrations in the blood when administered by different routes and can be combined with compounds that prolong its retention in the body. Notice that the time scale has been expanded by increasing the number of hours in the intervals to the right of the double-line break.

penicillins. Scientists develop these penicillins in either of two ways. First, they can stop synthesis of the molecule by *Penicillium* and obtain only the common penicillin nucleus for use. Second, they can remove the side chains from the completed natural molecules and then chemically add other side chains that, among other things, make them more resistant to penicillinase. Thus the term *semisynthetic* is applied to these penicillins; part of the penicillin is produced by the mold, and part is added synthetically. *Methicillin* was the first semisynthetic penicillin used widely, but many bacteria developed resistance to it, so others such as *oxacillin* have largely replaced it.

To overcome the problem of the narrow spectrum of activity of natural penicillins, extended-spectrum semisynthetic penicillins have been developed. These new penicillins are effective against many gram-negative bacteria as well as gram-positive ones, although they are not resistant to penicillinases. The first such penicillins were the aminopenicillins such as *ampicillin* and *amoxicillin*. When

bacterial resistance to these became more common, the carboxypenicillins were developed. Members of this group, such as *carbenicillin* and *ticarcillin,* have even greater activity against gram-negative bacteria and have the special advantage of activity against *Pseudomonas aeruginosa.* Among the more recent additions to the penicillin family are the ureidopenicillins such as *mezlocillin* and *azlocillin.* These extended-spectrum penicillins are modifications of the structure of ampicillin. The search for even more effective modifications of penicillin continues.

A different approach to the proliferation of penicillinase is to combine such penicillins with **potassium clavulanate** (*clavulanic acid*), a product of a streptomycete. Potassium clavulanate is a noncompetitive inhibitor of penicillinase with essentially no antimicrobial activity of its own. It has been combined with some new extended-spectrum penicillins such as amoxicillin (producing *Augmentin*®) and ticarcillin (producing *Timentin*®).

Monobactams

Another method of avoiding the effects of penicillinase is shown by *aztreonam,* a synthetic antibiotic that has only a single ring (and is therefore known as a *monobactam*) rather than the conventional β-lactam double ring. Its spectrum of activity is remarkable for a penicillin-related compound—it affects only gram-negative bacteria, including pseudomonads. Of special interest is the unusually low toxicity of this monobactam, the first available of the type.

Cephalosporins

In structure, the nuclei of **cephalosporins** resemble those of penicillin (Figure 20.11). Cephalosporins are also similar in action to penicillin; they inhibit the synthesis of cell walls. However, cephalosporins are sufficiently different from penicillin that they are resistant to penicillinases and are effective against more gram-negative organisms than the natural penicillins. However, the cephalosporins are susceptible to a separate group of β-lactamases.

The number of second- and third-generation cephalosporins has proliferated in recent years; there are now more than 70 versions. Each generation tends to be more effective against gram-negatives and has a broader spectrum of activity than the previous generation. Some char-

FIGURE 20.10 The effect of penicillinase on penicillins. Bacterial production of this enzyme, which is shown breaking the β-lactam ring, is by far the most common form of resistance to penicillins. R is an abbreviation for the chemical side groups that differentiate similar or otherwise-identical compounds.

β-lactam ring

Penicillin

Penicillinase →

Penicilloic acid

FIGURE 20.11 Comparison of the nuclear structure of cephalosporin with that of penicillin.

FIGURE 20.12 Structure of the antibacterial antibiotic gentamicin, a representative aminoglycoside. Glycosidic bonds (magenta) are bonds between amino sugars.

acteristic cephalosporins are *cephalothin, cefamandole,* and *cefotaxime.* Most cephalosporins are injected, but a few may be taken orally. Cephalosporins generally are more expensive than penicillins.

Carbapenems

A new class of antibiotics, the **carbapenems,** is remarkable for their extremely broad spectrum of activity. They represent another modification of the β-lactam structure. The first of this group is *Primaxin®,* a combination of a β-lactam antibiotic (*imipenem*) and *cilastatin sodium.* The cilastatin sodium prevents degradation of the combination in the kidneys. Tests have demonstrated that Primaxin is active against 98% of all organisms isolated from hospital patients. Carbapenems work by inhibiting cell wall synthesis.

Aminoglycosides

Aminoglycosides are a group of antibiotics in which amino sugars are linked by glycoside bonds. Probably the best-known aminoglycoside is *streptomycin,* which was discovered in 1944. Streptomycin is still used as an alternative drug in the treatment of tuberculosis, but rapid development of resistance and serious toxic effects have diminished its usefulness.

Aminoglycosides are bactericidal and inhibit protein synthesis. They can affect hearing by causing permanent damage to the auditory nerve, and damage to the kidneys has also been reported. Because of this, their use has been declining. *Neomycin* is present in many topical preparations. *Gentamicin* (Figure 20.12) plays a role in treatment of many enteric gram-negative infections and *Pseudomonas aeruginosa* infections.

Tetracyclines

Tetracyclines are a group of closely related broad spectrum antibiotics, produced by *Streptomyces* spp., that inhibit protein synthesis. Not only are they effective against

FIGURE 20.13 Structure of the antibacterial antibiotic tetracycline. Other tetracycline-type antibiotics share the four-cyclic-ring structure of tetracycline and closely resemble it.

gram-positive and gram-negative bacteria, but they penetrate body tissues well and are especially valuable against the intracellular rickettsias and chlamydias. Three of the more commonly used tetracyclines are *oxytetracycline* (*Terramycin®*), *chlortetracycline* (*Aureomycin®*), and tetracycline itself (Figure 20.13). Some newer semisynthetic tetracyclines, such as *doxycycline* and *minocycline,* are now available. They have the advantage of being retained longer in the body.

Tetracyclines are used for many urinary tract infections and are useful in treatment of mycoplasmal pneumonia and especially in chlamydial and rickettsial infections. They are also frequently used as alternative drugs for such diseases as syphilis and gonorrhea. Tetracyclines often suppress the normal intestinal microbiota because of their broad spectrum, causing gastrointestinal upsets and often leading to superinfections, particularly by *Candida albicans.* They are not advised for administration to children, who might experience a brownish discoloration of the teeth, or to pregnant women, in whom they might cause liver damage. Tetracyclines are among the most common antibiotics added to animal feeds, where their use results in significantly faster weight gains; however, some human health problems can also result (see the box on p. 505).

FIGURE 20.14 Structure of the antibacterial antibiotic chloramphenicol. Note the simple structure that makes synthesizing this drug less expensive than isolating it from *Streptomyces*.

FIGURE 20.15 Structure of the antibacterial antibiotic erythromycin, a representative macrolide. All macrolides have the macrocyclic lactone ring shown here.

Chloramphenicol

Chloramphenicol is a broad spectrum bacteriostatic antibiotic that interferes with protein synthesis on the 50S ribosome (Figure 20.14). Because of its relatively simple structure, it is less expensive for the pharmaceutical industry to synthesize it chemically than to isolate it from *Streptomyces*. Its relatively small molecular size promotes its diffusion into areas of the body that are normally inaccessible to many other drugs. However, chloramphenicol has serious adverse effects; most important is the suppression of bone marrow activity. This suppression affects the formation of blood cells. In about 1 in 40,000 users, the drug appears to cause aplastic anemia, a potentially fatal condition; the normal rate for this condition is only about 1 in 500,000 individuals. Physicians have been advised not to use the drug for trivial conditions or ones for which suitable alternatives are available. However, it is still a drug of choice in the treatment of certain types of meningitis and typhoid fever.

Macrolides

Macrolides are a group of antibiotics named for the presence of a macrocyclic lactone ring. The best-known macrolide in clinical use is *erythromycin* (Figure 20.15). Its

mode of action is the inhibition of protein synthesis. However, erythromycin is not able to penetrate the cell walls of most gram-negative bacilli. Its spectrum of activity is therefore similar to that of penicillin G, and it is a frequent alternative drug to penicillin. Erythromycin is, however, effective against *Legionella* and some *Neisseria* spp.—important gram-negative pathogenic genera. Because it can be administered orally, an orange-flavored preparation of erythromycin is a frequent penicillin substitute for the treatment of streptococcal and staphylococcal infections in children. Erythromycin is the drug of choice for the treatment of legionellosis and mycoplasmal pneumonia.

Polypeptides

A number of antibiotics are **polypeptides,** chains of amino acids linked by peptide bonds. Almost all these drugs have been isolated from the genus *Bacillus*. Two well-known examples are bacitracin and polymyxin B (Figure 20.16). Both drugs are relatively toxic, and their use is mostly topical.

Bacitracin (the name is derived from its source, a *Bacillus* isolated from a wound on a girl named Tracy) is effective primarily against gram-positive bacteria, such as staphylococci and streptococci. It is also used against some important gram-negative pathogens, such as *Neisseria*. It inhibits the synthesis of cell walls. Its use is restricted to topical application for superficial infections.

Polymyxin B is a bactericidal antibiotic effective against gram-negative bacteria. For many years it was one of very few drugs used against infections by gram-negative *Pseudomonas*. The mode of action of polymyxin B is to injure plasma membranes. Polymyxin B is seldom used today except in topical treatment of superficial infections because of toxicity when it is administered by injection.

Both bacitracin and polymyxin B are available in nonprescription antiseptic ointments, in which the polypeptides are usually combined with neomycin, a broad spectrum aminoglycoside. These are rare exceptions to the requirement for a prescription for antibiotics.

Vancomycin

Vancomycin is a member of a small group of glycopeptide antibiotics. It is a very toxic drug, is difficult to administer, and has a very narrow spectrum of activity that is based on inhibition of cell wall synthesis (peptidoglycans). The drug is important because it is probably the most effective drug in clinical use against penicillinase-producing staphylococci. Vancomycin is also used in special situations, such as treatment of streptococcal endocarditis and staphylococcus infections of such devices as prosthetic heart valves.

Rifamycins

The best-known derivative of the **rifamycin** family of antibiotics is *rifampin*. These drugs inhibit the synthesis of mRNA. By far the most important use of rifampin is

FIGURE 20.16 Structure of polymyxin B, a representative polypeptide antibacterial antibiotic. The amino acid subunits (names in magenta) and peptide bonds (blue) are shown.

Polymyxin B

Amphotericin B

FIGURE 20.17 Structure of the antifungal drug amphotericin B, representative of the polyenes.

Miconazole

FIGURE 20.18 Structure of the antifungal drug miconazole, representative of the imidazoles.

against mycobacteria in the treatment of tuberculosis and leprosy. A valuable characteristic of rifampin is its ability to penetrate tissues and reach therapeutic levels in cerebrospinal fluid and abscesses. This characteristic is probably an important factor in its antitubercular activity, because the tuberculosis pathogen is usually located inside tissues or macrophages. An unusual side effect of rifampin is the appearance of orange-red urine, feces, saliva, sweat, and even tears.

Other Antifungal Drugs

Fungal infections have become a more frequently encountered problem because of their role as opportunistic infections in immunosuppressed individuals, especially those with AIDS. This increased frequency of infection has intensified the search for more effective antifungals, especially for those that can treat systemic infections.

Polyenes

Two of the more commonly used **polyene antibiotics**, both of which are products of *Streptomyces* spp., are *nys-*

tatin and *amphotericin B* (Figure 20.17). Both drugs are fungicidal. The drugs combine with the sterols in fungal plasma membranes, making the membranes excessively permeable and killing the cell. Bacterial plasma membranes (except those of mycoplasmas) do not contain sterols, and the drugs therefore do not affect bacteria. Nystatin is used to treat local infections of the vagina and skin and can be taken orally for *Candida* infections of the intestinal tract. It is so poorly soluble that it does not cross the intestinal lining, so it does not enter the body tissue in toxic amounts. For many years amphotericin B has been a mainstay of clinical treatment for systemic fungal diseases, such as histoplasmosis, coccidioidomycosis, and blastomycosis. The drug's toxicity, particularly to the kidneys, is a strongly limiting factor in these uses.

Imidazoles and Triazoles

Imidazole antifungals, such as *clotrimazole, miconazole,* and *ketoconazole,* primarily interfere with sterol synthesis in fungi (see Figure 20.5), although they probably have other antimetabolic effects. Clotrimazole and miconazole (Figure 20.18) are generally used topically, most often in treatment of cutaneous mycoses such as athlete's foot.

ANTIBIOTICS IN ANIMAL FEED LINKED TO HUMAN DISEASE

OVER FORTY YEARS AGO, livestock growers began using antibiotics in the feed of closely penned animals that were being fattened for market. The drugs helped reduce the number of bacterial infections and controlled their spread in such ripe conditions. They also unexpectedly accelerated the animals' growth, an effect that led to wider and wider use. The growth-promoting effects are thought to be due to suppression of intestinal *Clostridium perfringens,* whose toxins may retard the animals' growth. Today, nearly half the antimicrobial drugs produced in the U.S.A. are given to farm animals. The largest percentage is used to promote weight gain in dairy cows.

Meat and milk that reach the consumer's table are not heavily ridden with antibiotics because the Food and Drug Administration (FDA) has established limits for antibiotic residues in edible tissues. The issue is that the constant presence of antibiotics in these animals puts selective pressure on their normal microbiota as well as on their pathogens to develop drug resistance.

THE RISK TO HUMANS

Bacteria such as *Salmonella* and *Listeria* can be transferred from animals to humans in meat or milk. The first known case of salmonellosis occurred in Germany in 1888, when 50 people became ill after eating ground beef from one cow. The use of antibiotics in animal feed preferentially allows the growth of strains of bacteria that are resistant to drugs commonly used to treat human infections. In the 1980s, researchers proved that antibiotic-

resistant bacteria are being transferred from animals to humans directly in meat or milk. In 1984, Scott Holmberg and his associates at the Centers for Disease Control and Prevention (CDC) traced an antibiotic-resistant *Salmonella newport* infection from South Dakota beef cattle to 18 people in four states. Antibiotic-resistant *S. typhimurium* originating from one Illinois dairy infected 16,000 people in seven states in 1985. Reports from other countries have shown similar patterns of resistance. Before 1984 in Europe, *S. dublin* was sensitive to antibiotics. Since 1984, isolates from cattle and from children with salmonellosis are resistant to chloramphenicol and streptomycin.

People taking antibiotics may have a higher risk of infection by resistant organisms. In an outbreak of *S. typhimurium* linked to milk consumption in the Midwest, people taking antimicrobials were more rapidly infected by the bacteria. Apparently, antimicrobial therapy gave the *S. typhimurium* from milk a selective advantage over other bacteria, many of which were killed by the antimicrobial drugs. Before 1980, only one case of salmonellosis that was exacerbated by antibiotic therapy had been reported.

The number of reported cases of *Salmonella* infections of humans is increasing, and 20% to 25% of these are due to antibiotic-resistant strains of bacteria. CDC is urging public health laboratories to identify strains of *Salmonella* by studying antibiotic resistance patterns and by using serotype and molecular biology techniques to trace the origins of infection.

MOLECULAR METHODS USED TO TRACE SOURCES

Molecular methods have been invaluable for tracing the sources of infections. In 1985, nearly 1000 cases of infection by *S. newport* resistant to multiple antibiotics were reported in Los Angeles County, California. A ra-

dioactive DNA probe was used to identify an identical plasmid carrying multiple antibiotic resistance genes in 99% of the *S. newport* isolates. A plasmid profile analysis was used to track the bacterium from the sick people back to a slaughterhouse, to a meat-deboning plant, and finally to three farms.

Nonpathogenic, commensal bacteria such as *E. coli* also develop resistance to these antimicrobials. This can be viewed as positive because the normal microbiota is not usually the targets of chemotherapy. However, *E. coli* and other microbiota are reservoirs of resistance genes for pathogens. For example, an Arizona woman was treated for five years with TMP-SMZ for a urinary tract infection. She then acquired *Shigella* dysentery that was resistant to TMP-SMZ, ampicillin, streptomycin, sulfasoxazole, and tetracycline. This multiply resistant strain had not been seen before. Within three weeks the same *Shigella* was isolated from six other local residents. Robert Tauxe and his colleagues at the CDC found identical resistance plasmids in all the *Shigella* and in the woman's *E. coli*. Although this case is not linked to livestock, antibiotic-resistant *E. coli* in cattle can be transferred to humans and then transfer their resistance plasmids to pathogens in the human intestine.

• • •

In spite of this evidence, cattle growers, feed manufacturers, and the drug industry still argue that the link between antibiotics in feed and human disease is sketchy, and claim that the benefits derived from the use of antibiotics outweigh the potential harmful effects. However, some farmers have reported that their sick animals are not responding to penicillin and tetracycline therapy. Perhaps substituting narrow-spectrum drugs for the broad-spectrum antimicrobials now in use may actually benefit the farmer.

When taken orally, ketoconazole is effective and is a less toxic alternative to amphotericin B for many systemic fungal infections, although occasional liver damage has been reported. Ketoconazole also has an unusually wide spectrum of activity.

Systemic fungal infections are often effectively treated by *fluconazole* and *itraconazole,* members of a new group of triazole antifungals. They tend to be less toxic than other systemic antifungals.

Griseofulvin

Griseofulvin is an antibiotic produced by a species of *Penicillium.* It has the interesting property of being active against superficial dermatophytic fungal infections of the hair (tinea capitis, or ringworm) and nails, even though its route of administration is oral. The drug apparently binds selectively to the keratin found in the skin, hair follicles, and nails. Its mode of action is to interfere with mitosis and thereby inhibit fungal reproduction.

Additional Antifungal Drugs

Tolnaftate is a common alternative to miconazole as a topical agent for the treatment of athlete's foot. Its mechanism of action is not known. *Undecylenic acid* is a fatty acid that has antifungal activity against athlete's foot, although it is not as effective as tolnaftate or the imidazoles.

Antiviral Drugs

In developed parts of the world it is estimated that at least 60% of infective illnesses are caused by viruses, and only about 15% by bacteria. Every year, at least 90% of the population of the U.S.A. suffers from some viral disease. Yet, only a handful of antiviral drugs have been approved in the U.S.A., and these are effective against only an extremely limited group of diseases.

The first antiviral drug licensed for systemic use in the U.S.A. was *amantadine.* Although how amantadine works is not precisely known, it prevents the virus either from entering the cell or, in at least one case, from uncoating after it enters. The drug has limited usefulness in the prevention of influenza A. Amantadine has little effect on the course of the disease once it has been contracted, but in confined situations, such as institutions or nursing homes, it can limit the spread of an outbreak. Another drug with activity against influenza is *ribavirin,* an analog of guanine, a purine in DNA and RNA. We will see that most antiviral drugs are synthetic nucleotides that interfere with DNA or RNA synthesis. Ribavirin has also been tested on infants suffering from a viral pneumonia. However, this drug must be administered by aerosol apparatus, is too expensive for general use, and is very toxic. Nevertheless, interest in ribavirin remains high because it is effective in vitro against a wide range of viruses.

The newest antiviral drugs, such as acyclovir, ganciclovir, and zidovudine (AZT), are not active until acted upon by enzymes that are synthesized by cells infected by a virus. This factor gives them more antiviral specificity.

Acyclovir is an analog of a guanine-containing nucleoside (Figure 20.19a). Its activity is limited to herpesviruses. It is available for administration topically as an ointment, orally, or by injection. Its best known use is in the treatment of genital herpes, but it is generally useful against infections by many herpesviruses, especially in immunosuppressed individuals. These diseases include encephalitis or eye infections caused by herpes simplex viruses and shingles or chickenpox. The mode of action of acyclovir is illustrated in Figures 20.19b and c. The drug terminates the synthesis of viral DNA but has little effect on host cell DNA synthesis. However, it has no effect on latent herpesviruses (see Chapter 13).

Ganciclovir is a newer antiviral that, like acyclovir, is an analog of guanine. However, it has a wider spectrum of activity that includes all the herpesviruses, most important the cytomegalovirus affecting AIDS or transplant patients.

Several other analogs of DNA nucleosides containing purine and pyrimidine have been synthesized and have limited use against herpesviruses. *Idoxuridine* and *trifluridine* are both analogs of thymidine (the thymine-containing nucleoside), and *vidarabine* is an analog of the purine adenine. All these drugs are available as topical ointments and are useful in treating herpes simplex virus infections of the eye (herpes keratitis). Vidarabine has also been approved for intravenous use in the treatment of encephalitis caused by herpesviruses and for treatment of immunosuppressed individuals infected with the varicella-zoster virus (chickenpox, shingles). Zidovudine (AZT) is an analog of thymidine that is used in chemotherapy of AIDS patients. Its mode of action is to interfere with the synthesis of DNA from RNA by the enzyme reverse transcriptase (see the box in Chapter 19, p. 484). Didanosine (dideoxyinosine; ddI) and dideoxycytosine (ddC) are other nucleoside analogs used to block this enzyme in HIV.

A new approach to viral chemotherapy that may bear good results involves interferon, a protein secreted by certain host cells responding to viral infections (see Chapter 16). Interferon inhibits cell infection by many kinds of viruses. Genetic-engineering techniques using bacteria have made sufficient amounts of interferon available for clinical studies of its possible effectiveness as a chemotherapeutic agent against cancer as well as against viruses.

Antiprotozoan and Antihelminthic Drugs

For hundreds of years, quinine from the Peruvian cinchona tree was the only drug known to be effective for the treatment of a parasitic infection (malaria). It was first introduced into Europe in the early 1600s and was known as "Jesuit's powder." There are now many antiprotozoan and antihelminthic drugs, although many of them are still considered experimental. This does not preclude their use,

FIGURE 20.19 Structure and function of the antiviral drug acyclovir. (a) Structural resemblance between acyclovir and the nucleoside 2'-deoxyguanosine. (b) The viral enzyme thymidine kinase combines phosphates with such nucleosides to form nucleotides. The nucleotides are then incorporated into DNA. (c) Because acyclovir is a structural analog of the nucleoside, the enzyme uses it to assemble a false nucleotide that cannot be incorporated into DNA.

(a) Structural resemblance of acyclovir to guanine-containing nucleoside

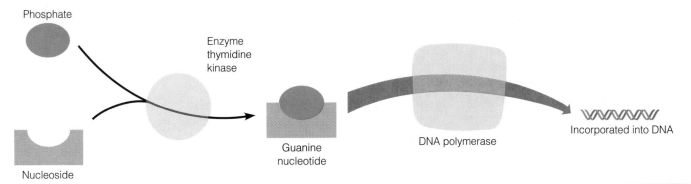

(b) Synthesis of normal viral DNA guanine nucleotide

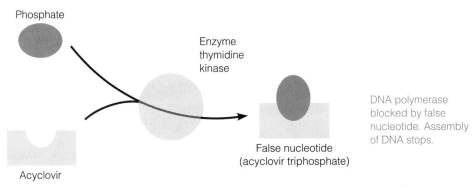

(c) Synthesis of false viral DNA nucleotide with acyclovir

however, by qualified physicians. The Centers for Disease Control and Prevention (CDC) provide several of them on request when they are not available commercially.

Antiprotozoan Drugs

Quinine still has very limited use as an alternative in the control of the protozoan disease malaria, but synthetic derivatives, such as *chloroquine,* have largely replaced it. For prevention of malaria in areas where resistance to chloroquine is a problem, the new drug *mefloquine* is recommended. *Quinacrine* is the drug of choice for treating the protozoan disease giardiasis. *Diiodohydroxyquin (iodoquinol)*

is an important drug prescribed for several intestinal amoebic diseases, but its dosage must be carefully controlled to avoid optic nerve damage. Its mode of action is unknown.

Metronidazole (Flagyl®) is one of the most widely used antiprotozoan drugs. It is unique in having activity not only against parasitic protozoa but also against obligate anaerobic bacteria. For example, as an antiprotozoan agent, it is the drug of choice for vaginitis caused by *Trichomonas vaginalis*. It is also used in the treatment of giardiasis and amoebic dysentery. The mode of action is damage to DNA by products of anaerobic metabolism.

Pentamidine isethionate is used in the early stages of African trypanosomiasis, but in the U.S.A. its most familiar clinical use is in the treatment of *Pneumocystis* pneumonia, a common complication in immunocompromised individuals, particularly victims of AIDS. The drug's mode of action is unknown, but it appears to bind to DNA. *Nifurtimox* is a member of the synthetic nitrofuran drugs and is effective against the trypanosome-caused Chagas' disease. Treatment may produce adverse effects of nausea and even convulsions.

Antihelminthic Drugs

With the increased popularity of sushi, a Japanese specialty often made with raw fish, the CDC began to notice an increased incidence of tapeworm infestations. To estimate the incidence, the CDC notes requests for *niclosamide,* which is the usual first choice in treatment. The drug is effective because it inhibits ATP production under anaerobic conditions. *Praziquantel* is about equally effective for treatment of tapeworms, although its mode of action is not known. Praziquantel has a broad spectrum and is highly recommended for treatment of several fluke-caused diseases, especially schistosomiasis. It causes the helminths to undergo muscular spasms and apparently makes them susceptible to attack by the immune system. *Mebendazole* is an antihelminthic drug frequently used in the treatment of several of the most common intestinal helminths: ascariasis, *Ascaris lumbricoides;* pinworms, *Enterobius vermicularis;* and whipworms, *Trichuris trichiura* (trik′ĕr-is trik-ē-yĕr′a). The mode of action is disruption of the microtubules in the cytoplasm, which indirectly results in an inability to generate ATP from carbohydrates. *Piperazine* is used in the treatment of pinworms and ascariasis. It paralyzes the organisms, which soon pass out of the victim's body.

TESTS TO GUIDE CHEMOTHERAPY

LEARNING OBJECTIVE

• *Describe two tests for microbial susceptibility to chemotherapeutic agents.*

Different microbial species and strains have different degrees of susceptibility to different chemotherapeutic agents. Moreover, the susceptibility of a microorganism can change with time, even during therapy with a specific chemotherapeutic agent. Thus a physician must know the sensitivities of the disease-causing microorganism before treatment can be started.

Several tests can be used to indicate which chemotherapeutic agent is most likely to combat a specific pathogen. These tests, however, are not always necessary. If the organisms have been identified—for example, as *Pseudomonas aeruginosa*, β-hemolytic streptococci, or gonococci—certain drugs can be selected without specific testing for

FIGURE 20.20 Disk-diffusion method for determining the activity of antimicrobials. Each disk contains a different chemotherapeutic agent, which diffuses into the surrounding agar. The clear zones indicate inhibition of growth of the microorganism swabbed onto the plate surface.

susceptibility. These tests are necessary only when susceptibility is not predictable, or when antibiotic resistance problems develop.

Disk-Diffusion Method

Probably the most widely used—but not necessarily the best—method of testing is the **disk-diffusion method** (Figure 20.20), also known as the **Kirby–Bauer test.** A Petri plate containing an agar medium is inoculated ("seeded") uniformly over its entire surface with a standardized amount of a test organism. Next, filter-paper disks impregnated with known concentrations of chemotherapeutic agents are placed on the solidified agar surface. During incubation, the chemotherapeutic agents diffuse from the disks into the agar. The farther the agent diffuses from the disk, the lower its concentration. If the chemotherapeutic agent is effective, a *zone of inhibition* forms immediately around the disk. The diameter of the zone of inhibition can be measured. Because the size of the zone of inhibition is affected by the diffusion rate of the chemotherapeutic agent, a wider zone does not always indicate greater antimicrobial activity. The zone diameter is compared to a standard table for that drug and concentration, and the results report that the organism is *sensitive, intermediate,* or *resistant.* This result is often inadequate for many clinical purposes. However, the test is simple and inexpensive and is most often used when more sophisticated laboratory facilities are not available.

Broth Dilution Tests

A weakness of the disk-diffusion method is that it does not determine if a drug is bactericidal and not just bacteriostatic. Some form of a **broth dilution test** is needed to de-

Broth only (negative control)

Doxycycline

Sulfameth-
oxazole

Ethambutol

Streptomycin

Kanamycin

FIGURE 20.21 A microdilution minimal inhibitory concentration (MIC) plate. Growth is shown by a white dot in the bottom of the well. In this test, doxycycline had no effect at all. Kanamycin and ethambutol were about equally effective. Streptomycin was effective at all concentrations. A trailing endpoint, as observed here with sulfamethoxazole, is read where there is an 80% reduction in growth. The negative control well contained broth only (no antimicrobials and no inoculum); the positive control well (not shown) contained no antimicrobials but was inoculated.

termine the *minimal inhibitory concentration (MIC)* and *minimal bactericidal concentration (MBC)* of an antimicrobial. The MIC is determined by making a sequence of decreasing concentrations of the drug in a broth, which is then inoculated with the test bacteria (Figure 20.21). The lowest concentration that prevents visible growth is the MIC. The wells that do not show growth can be cultured in broth free of the antimicrobial. If growth occurs in this broth, the drug was not bactericidal, and the MBC can be determined.

Dilution tests are often highly automated. The drugs are purchased already diluted into broth in wells formed in a plastic tray. A suspension of the test organism is prepared and inoculated into all the wells simultaneously by a special inoculating device. After incubation, the turbidity may be read visually, although clinical laboratories with high work loads may read the trays with special scanners that enter the data into a computer that provides a printout of the MIC.

Other tests are also useful for the clinician; a determination of the microbe's ability to produce β-lactamase is one example. One popular, rapid method makes use of a cephalosporin that changes color when its β-lactam ring is opened. In addition, a measurement of the *serum concentration* of an antimicrobial is especially important when toxic drugs are used. These assays tend to vary with the drug and may not always be suitable for smaller laboratories.

EFFECTIVENESS OF CHEMOTHERAPEUTIC AGENTS

LEARNING OBJECTIVES

- *Describe the mechanisms of drug resistance.*
- *Compare and contrast synergism and antagonism.*

Drug Resistance

Bacteria become resistant to antimicrobials by three major mechanisms: (1) destruction or inactivation of the drug (by β-lactamase, for example), (2) prevention of penetration to the target site within the microbe (a frequent mechanism for tetracycline resistance), and (3) alteration of the antimicrobial's target sites (for example, a single amino acid change in the ribosome may be enough to make the microbe resistant to certain macrolide antibiotics). Variations on these mechanisms also occur. For example, a microbe could become resistant to trimethoprim by synthesizing very large amounts of the enzyme against which the drug is targeted. Conversely, polyene antibiotics can become less effective when resistant organisms produce smaller amounts of the sterols against which the drug is effective. Of particular concern is the possibility that such *resistant mutants* will increasingly replace the susceptible normal populations (see the box on p. 505). Figure 20.22 shows how rapidly bacterial numbers increase during infection as resistance develops. The increasing appearance of antibiotic-resistant gonorrhea is discussed in the box in Chapter 26.

Hereditary drug resistance is often carried by extrachromosomal genetic elements called plasmids. Some plasmids, including those called **resistance (R) factors**, can be transferred between bacterial cells in a population and between different but closely related bacterial populations (see Chapter 8). R factors often contain genes for resistance to several antibiotics.

Strains of bacteria that are resistant to antibiotics are particularly common among people who work in hospitals, where antibiotics are in constant use. *Staphylococcus aureus,* a common opportunistic pathogen that is carried in the nasal passages, develops resistance to antibiotics very frequently. Every day in hospitals, surprisingly large volumes of antibiotics are dispensed as aerosols by the universal practice of clearing syringes of air before an injection is made. As these aerosols are generally inhaled in small amounts, the staphylococci harmlessly residing in a nurse's nose, for example, can quickly develop resistance. If these resistant microbes are transferred to surgical wounds and other exposed sites, they can cause nosocomial infections (see Chapter 14).

Drugs should not be used indiscriminately, and the concentrations administered should always be of optimum strength so that the chances of survival of resistant

FIGURE 20.22 Development of an antibiotic-resistant mutant during antibiotic therapy. The patient, suffering from a chronic kidney infection caused by a gram-negative bacterium, was treated with streptomycin. Antibiotic sensitivity is expressed as the minimum inhibitory concentration of the antibiotic in micrograms per milliliter.

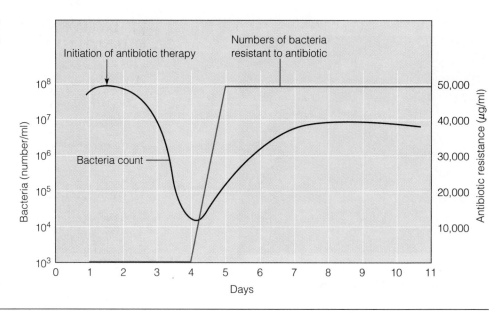

mutants are minimized. A few years ago, one medical journal reported on a survey of antibiotic use in one hospital. The results indicated that more than 60% of the antibiotics prescribed there were either given in the wrong dosage or used in applications for which they would not be effective. Therefore, many hospitals now have special monitoring committees who review the use of antibiotics for effectiveness and cost.

Another way the development of resistant strains can be reduced is through the administration of two or more drugs simultaneously. If a strain is resistant to one of the drugs, the other might be effective against it. The probability that the organism will acquire resistance to both drugs is roughly the product of the two individual probabilities. However, multidrug-resistant tuberculosis is an emerging problem (see the box in Chapter 15, p. 394).

Effects of Combinations of Drugs

The chemotherapeutic effect of two drugs given simultaneously is sometimes greater than the effect of either given alone, a phenomenon called **synergism.** For example, in the treatment of bacterial endocarditis, penicillin and streptomycin are much more effective when taken together than when either drug is taken alone. Damage to bacterial cell walls by penicillin makes it easier for streptomycin to enter.

Other combinations of drugs can show **antagonism.** For example, the simultaneous use of penicillin and tetracycline is often less effective than when either drug is used alone. By stopping the growth of the bacteria, the bacteriostatic drug tetracycline interferes with the action of penicillin, which requires bacterial growth.

The Future of Chemotherapeutic Agents

Antibiotics are clearly one of the greatest triumphs of medical science, although they are not without their problems and disadvantages. Our most pressing concern is with the misuse of antibiotics, which increases the likelihood of the emergence of resistant strains of pathogens. Unfortunately, the problem of antibiotic resistance has now reached a crisis point. Dealing with this crisis will require both intensified educational campaigns to promote the wise use of existing antibiotics and redoubled efforts toward the design of new ones.

Most current drugs have been discovered by screening natural sources and then manipulating the identified chemicals in a largely trial-and-error manner to increase their effectiveness. In the future, as our knowledge of the mechanisms of drug action becomes more detailed, computer-driven design at the molecular level will become more effective and thus more important. At the same time there will also be special interest in new antiviral, antifungal, and antiparasitic drugs because our arsenal in these categories is so limited.

•   •   •

The past few chapters have shown how profoundly science has changed the effects of infectious disease on human mortality and life span. At the turn of the twentieth century, the most common causes of death were infectious diseases such as tuberculosis, typhoid fever, or diphtheria. In the closing decade of this century, only influenza and the pneumonias remain among the 10 leading causes of death in the U.S.A. These figures bear testimony to the effectiveness of sanitation, vaccines, and (as discussed in this chapter) the discovery and use of antibiotics. In the forthcoming chapters on diseases we will see that the struggle against infectious disease continues.

INTRODUCTION (p. 491)

1. A chemotherapeutic agent is a chemical that combats disease in the body.
2. An antimicrobial drug is a chemical substance that destroys disease-causing microorganisms with minimal damage to host tissues.
3. Antimicrobial drugs may be synthetic drugs (prepared in the laboratory) or antibiotics (produced by bacteria or fungi).

HISTORY OF CHEMOTHERAPY (pp. 491–492)

1. The first chemotherapeutic agent, discovered by Paul Ehrlich, was salvarsan, used to treat syphilis.
2. Sulfa drugs came into prominence in the late 1930s.
3. Alexander Fleming discovered the first antibiotic, penicillin, in 1929; its first clinical trials were done in 1940.
4. In 1939, René Dubos isolated the antibiotic gramicidin from a soil *Bacillus*.
5. During the 1940s, Selman Waksman isolated streptomycin-producing *Streptomyces* spp.

SPECTRUM OF ANTIMICROBIAL ACTIVITY (pp. 493–494)

1. Antibacterial drugs affect many targets in a procaryotic cell.
2. Fungal, protozoan, and helminthic infections are more difficult to treat because these organisms have eucaryotic cells.
3. Narrow spectrum drugs affect only a select group of microbes—gram-positive cells, for example.
4. Broad spectrum drugs affect a large number of microbes.
5. Antimicrobial agents should not cause excessive harm to normal microbiota.
6. Superinfections occur when a pathogen develops resistance to the drug being used or when normally resistant microbiota multiplies excessively.

ACTION OF ANTIMICROBIAL DRUGS (pp. 494–496)

1. General action is either by directly killing microorganisms (bactericidal) or by inhibition of growth (bacteriostatic).
2. Some agents, such as penicillin, inhibit cell wall synthesis in bacteria.
3. Other agents, such as chloramphenicol, erythromycin, tetracyclines, and streptomycin, inhibit protein synthesis by acting on 70S ribosomes.
4. Agents such as polymyxin B cause injury to plasma membranes.

5. Rifampin and the quinolones inhibit nucleic acid synthesis.
6. Agents such as sulfanilamide act as antimetabolites by competitively inhibiting enzyme activity.

SURVEY OF COMMONLY USED ANTIMICROBIAL DRUGS (pp. 497–508)

Antibacterial Synthetics (p. 497)

1. Isoniazid (INH) inhibits mycolic acid synthesis in mycobacteria. INH is administered with rifampin or ethambutol to treat tuberculosis.
2. The antimetabolite ethambutol is used with other drugs to treat tuberculosis.
3. Sulfonamides competitively inhibit folic acid synthesis. Sulfonamides are bacteriostatic. They are used to treat urinary tract infections and can cause allergies and often develop drug resistance.
4. Trimethoprim and sulfamethoxazole competitively inhibit dihydrofolic acid synthesis for treatment of urinary tract and intestinal infections.
5. Quinolones and fluoroquinolones inhibit DNA gyrase for treatment of urinary tract infections.

Antifungal Synthetics (p. 497)

1. The antifungal agent flucytosine is an antimetabolite of cytosine.

Antibacterial Antibiotics (pp. 498–504)

1. All penicillins contain a β-lactam ring. Penicillins inhibit peptidoglycan synthesis.
2. Natural penicillins produced by *Penicillium* are effective against gram-positive cocci and spirochetes.
3. Penicillinases (β-lactamases) are bacterial enzymes that destroy natural penicillins.
4. Semisynthetic penicillins are made in the laboratory by adding different side chains onto the β-lactam ring made by the fungus.
5. Semisynthetic penicillins are resistant to penicillinases and have a broader spectrum of activity than natural penicillins.
6. The monobactam aztreonam affects only gram-negative bacteria.
7. Cephalosporins inhibit cell wall synthesis and are used against penicillin-resistant strains.
8. Carbapenems are broad spectrum antibiotics that inhibit cell wall synthesis.
9. Aminoglycosides include streptomycin, neomycin, spectinomycin, and gentamicin; all inhibit protein synthesis and are bactericidal.
10. Tetracyclines inhibit protein synthesis and are bacteriostatic toward many bacteria, including rickettsias and chlamydias.

11. Chloramphenicol inhibits protein synthesis and is bacteriostatic; a side effect of prolonged use is aplastic anemia.

12. Macrolides, such as erythromycin, inhibit protein synthesis and are bacteriostatic; erythromycin is used to treat legionellosis and *Neisseria* infections.

13. Polypeptides include bacitracin and polymyxin B. They are applied topically to treat superficial infections.

14. Bacitracin inhibits cell wall synthesis primarily in gram-positive bacteria.

15. Polymyxin B damages plasma membranes and is effective against gram-negative bacteria.

16. Vancomycin inhibits cell wall synthesis and may be used to kill penicillinase-producing staphylococci.

17. Rifamycins are bactericidal and are especially effective against gram-positive bacteria, including mycobacteria.

Other Antifungal Drugs (pp. 504–506)

1. Polyenes, such as nystatin and amphotericin B, combine with plasma membrane sterols and are fungicidal.

2. Imidazoles and triazoles interfere with sterol synthesis and are used to treat cutaneous and systemic mycoses.

3. Griseofulvin interferes with eucaryotic cell division and is used primarily to treat skin infections caused by fungi.

Antiviral Drugs (p. 506)

1. Amantadine blocks penetration or uncoating of influenza A virus.

2. Purine and pyrimidine analogs include ribavirin, acyclovir, ganciclovir, zidovudine (AZT), idoxuridine, trifluridine, vidarabine, ddI, and ddC.

3. Interferon inhibits cell infection by viruses.

Antiprotozoan and Antihelminthic Drugs (pp. 506–508)

1. Chloroquine, quinacrine, diiodohydroxyquin, pentamidine, and metronidazole are used to treat protozoan infections.

2. Antihelminthic drugs include niclosamide, mebendazole, praziquantel, and piperazine.

3. Mebendazole disrupts microtubules; piperazine paralyzes intestinal roundworms.

TESTS TO GUIDE CHEMOTHERAPY (pp. 508–509)

1. These tests are used to determine the degree of susceptibility of different microorganisms to chemotherapeutic agents.

2. They help to determine which chemotherapeutic agent is most likely to combat a specific pathogen.

3. These tests are used when susceptibility cannot be predicted or when drug resistance arises.

Disk-Diffusion Method (p. 508)

1. In this test, also known as the Kirby–Bauer test, a bacterial culture is inoculated on an agar medium, and filter-paper disks impregnated with chemotherapeutic agents are overlayed on the culture.

2. After incubation, the absence of microbial growth around a disk is called a zone of inhibition.

3. The diameter of the zone of inhibition, when compared with a standardized reference table, is used to determine whether the organism is sensitive, intermediate, or resistant to the drug.

Broth Dilution Tests (pp. 508–509)

1. In the broth dilution test, the microorganism is grown in liquid media containing different concentrations of a chemotherapeutic agent.

2. The minimum inhibitory concentration (MIC) is the lowest concentration of chemotherapeutic agent capable of preventing microbial growth.

3. The lowest concentration of chemotherapeutic agent that kills bacteria is called the minimum bactericidal concentration (MBC).

EFFECTIVENESS OF CHEMOTHERAPEUTIC AGENTS (pp. 509–510)

Drug Resistance (pp. 509–510)

1. Resistance may be due to enzymatic destruction of a drug, prevention of penetration of the drug to its target site, or cellular or metabolic changes at target areas.

2. Hereditary drug resistance is carried by plasmids called resistance (R) factors.

3. Resistance can be minimized by the discriminating use of drugs in appropriate concentrations and dosages.

Effects of Combinations of Drugs (p. 510)

1. Some combinations of drugs are synergistic; they are more effective when taken together.

2. Some combinations of drugs are antagonistic; when taken together, both drugs become less effective than when taken alone.

The Future of Chemotherapeutic Agents (p. 510)

1. Many bacterial diseases, previously treatable with antibiotics, have become resistant to antibiotics.

2. New antibacterial, antiviral, antifungal, and antiparasitic drugs are needed.

Review

1. Fill in the following table.

ANTIMICROBIAL AGENT	SYNTHETIC OR ANTIBIOTIC	METHOD OF ACTION	PRINCIPAL USE
Isoniazid			
Sulfonamides			
Ethambutol			
Trimethoprim			
Fluoroquinolones			
Penicillin, natural			
Penicillin, semisynthetic			
Cephalosporins			
Carbapenems			
Aminoglycosides			
Tetracyclines			
Chloramphenicol			
Macrolides			
Polypeptides			
Vancomycin			
Rifamycins			
Polyenes			
Griseofulvin			
Amantadine			
Zidovudine			
Niclosamide			

2. Define a chemotherapeutic agent. Distinguish between a synthetic chemotherapeutic agent and an antibiotic.

3. Discuss the contributions to chemotherapy made by each of the following people: Ehrlich; Fleming; Dubos; Waksman.

4. List and explain five criteria used to identify an effective antimicrobial agent.

5. What similar problems are encountered with antiviral, antifungal, antiprotozoan, and antihelminthic drugs?

6. Identify three methods of action of antiviral drugs. Give an example of a currently used antiviral drug for each method of action.

7. Compare and contrast the broth dilution and disk-diffusion tests. Identify at least one advantage of each.

8. Describe the disk-diffusion test for microbial susceptibility. What information can you obtain from this test?

9. Define drug resistance. How is it produced? What measures can be taken to minimize drug resistance?

10. List the advantages of using two chemotherapeutic agents simultaneously to treat a disease. What problem can be encountered using two drugs?

11. Why does a cell die from the following antimicrobial actions?
    (a) Colistimethate binds to phospholipids.
    (b) Kanamycin binds to 70S ribosomes.

12. How is translation inhibited by each of the following?
    (a) Chloroamphenicol
    (b) Erythromycin
    (c) Tetracycline
    (d) Streptomycin

13. Dideoxyinosine (ddI) is an antimetabolite of guanine. The 3'-OH is missing from ddI. How does ddI inhibit DNA synthesis?

Critical Thinking

1. Which of the following can affect human cells? Explain why or why not.
   (a) Penicillin
   (b) Nystatin
   (c) Erythromycin
   (d) Polymyxin

2. Why is idoxuridine effective if host cells also contain DNA?

3. Some bacteria become resistant to tetracycline because they don't make porins. Why can a porin-deficient mutant be detected by its inability to grow on a medium containing a single carbon source such as succinic acid?

4. Why do you suppose that *Streptomyces griseus* produces an enzyme that inactivates streptomycin? Why is this enzyme produced early in metabolism?

5. The following data were obtained from a disk-diffusion test.

ANTIBIOTIC	ZONE OF INHIBITION
A	15 mm
B	0 mm
C	7 mm
D	15 mm

(a) Which antibiotic was most effective against the bacteria being tested?
(b) Which antibiotic would you recommend for treatment of a disease caused by this bacterium?
(c) Was antibiotic A bactericidal or bacteriostatic? How can you tell?

6. The following results were obtained from a broth dilution test for microbial susceptibility.

TUBE NUMBER	ANTIBIOTIC CONCENTRATION	GROWTH	GROWTH IN SUBCULTURE
1	200 μg	−	−
2	100 μg	−	−
3	50 μg	−	+
4	25 μg	+	+

(a) The MIC of this antibiotic is _____.
(b) The MBC of this antibiotic is _____.

Clinical Applications

1. Penicillin and streptomycin can be used together under certain circumstances, but penicillin and tetracycline cannot be used together. Offer an explanation for this.

2. A patient with a urinary bladder infection took nalidixic acid, but her condition did not improve. Explain why her infection disappeared when she switched to a sulfonamide.

3. A patient with streptococcal sore throat takes penicillin for two days of a prescribed 10-day regime. Because he feels better, he then saves the remaining penicillin for some other time. After three more days, he suffers a relapse of the sore throat. Discuss the probable cause of the relapse.

PART FOUR

Microorganisms and Human Disease

In Guinea, West Africa, workers examine blackflies, carriers of the nematode that causes the disease river blindness, to determine if a bacterial insecticide has been effective against them. This biodegradable insecticide, which is a bacterial toxin that targets the insect larvae, is now sprayed on infested areas in an attempt to eradicate the disease.

Microbial Diseases of the Skin and Eyes

IN CHAPTER 16 we saw that the human body possesses a number of defenses that contribute to *nonspecific resistance* against many types of pathogens. These defenses include antimicrobial substances, phagocytosis, the inflammatory response, fever, and the skin and mucous membranes.

The skin, which covers and protects the body, is the body's first line of defense against pathogens. Basically, the skin is an inhospitable place for most microorganisms because the secretions of the skin are acidic and most of the skin contains little moisture. Moreover, much of the skin is exposed to radiation, which discourages microbial life. Some parts of the body, however, such as the axilla (armpit), have enough moisture to support microbial growth. The axillary region can support relatively large bacterial populations, whereas other regions, such as the scalp, support rather small numbers of microorganisms. The skin is a physical as well as an ecological barrier, and it is almost impossible for pathogens to penetrate it. However, some can enter through openings that are not readily apparent, and the larval forms of a few parasites can penetrate intact skin.

STRUCTURE AND FUNCTION OF THE SKIN

LEARNING OBJECTIVE

- *Describe the structure of the skin and mucous membranes and the ways pathogens can invade the skin.*

The skin of an average adult occupies a surface area of about 1.9 m² and varies in thickness from 0.05 to 3.0 mm. As we mentioned in Chapter 16, skin consists of two principal parts, the epidermis and the dermis (Figure 21.1). The **epidermis** is the thin, outer portion, composed of several layers of epithelial cells. The outermost layer of the epidermis, the **stratum corneum**, consists of dead cells that contain a waterproofing protein called **keratin**. The epidermis, when unbroken, is an effective physical barrier against microorganisms.

The **dermis** is the inner, relatively thick portion of skin that is composed mainly of connective tissue. The hair follicles, sweat-gland ducts, and oil-gland ducts contained in the dermis provide passageways through which microorganisms can enter the skin and penetrate deeper tissues.

Perspiration provides moisture and some nutrients for microbial growth. However, it contains salt, which inhibits many microorganisms, and lysozyme, which is capable of breaking down the cell walls of certain bacteria.

Sebum, secreted by oil glands, is a mixture of lipids (unsaturated fatty acids), proteins, and salts that prevents skin and hair from drying out. Although the fatty acids inhibit the growth of certain pathogens, sebum, like perspiration, is also nutritive for many microorganisms, most notably the corynebacteria.

Mucous Membranes

In the linings of body cavities, such as those associated with the gastrointestinal, respiratory, urinary, and genital tracts, the outer protective barrier differs from the skin. It consists of sheets of tightly packed *epithelial cells.* These

FIGURE 21.1 Structure of human skin. Note the passageways between the hair follicle and hair shaft through which microbes can penetrate the deeper tissues. Microbes can also enter the skin through sweat pores.

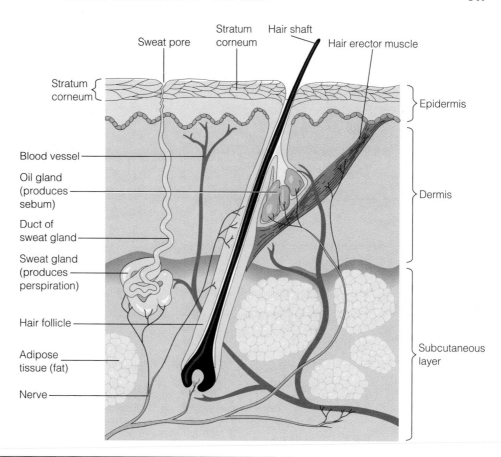

cells are attached at their bases to a layer of extracellular material called the *basement membrane.* Many of these cells secrete mucus—hence the name **mucous membrane (mucosa).** Other mucosal cells have cilia, and in the respiratory system the mucous layer traps particles, including microorganisms, which the cilia sweep upward out of the body (see Figure 16.4). Mucous membranes are often acidic, which tends to limit their microbial populations. Also, the membranes of the eyes are mechanically washed by tears, and the enzyme lysozyme in tears destroys the cell walls of certain bacteria. Mucous membranes are often folded to maximize surface area; the total surface area in an average human is about 400 m², much more than the surface area of the skin.

NORMAL MICROBIOTA OF THE SKIN

LEARNING OBJECTIVE

- *Provide examples of normal skin microbiota, and state the general locations and ecological roles of its members.*

Although the skin is generally inhospitable to most microorganisms, it supports the growth of certain microbes, which are established as part of the normal microbiota. On superficial skin surfaces, certain aerobic bacteria produce fatty acids from sebum. These acids tend to inhibit many microbes and allow better-adapted bacteria to flourish.

Microorganisms that find the skin a satisfactory environment tend to be resistant to drying and to relatively high salt concentrations. The skin's normal microbiota contains large numbers of gram-positive bacteria such as staphylococci and micrococci. Some of these are capable of growth at sodium chloride concentrations of 7.5% or more. Scanning electron micrographs (see Figure 14.1a) show that bacteria on the skin tend to be grouped into small clumps. Vigorous washing can reduce their numbers but will not eliminate them. Microorganisms remaining in hair follicles and sweat glands after washing will soon reestablish the normal populations. Areas of the body with more moisture, such as the armpits and between the legs, have higher populations of microbes. These metabolize the secretions from the sweat glands and are the main contributors to what the advertising world advises us is offensive body odor.

Also part of the skin's normal microbiota are grampositive pleomorphic rods called diphtheroids. Some diphtheroids, such as *Propionibacterium acnes,* are typically anaerobic and inhabit hair follicles. Their growth is supported by secretions from the oil glands (sebum), which we will see makes them a factor in acne. These bacteria produce propionic acid, which helps maintain the low pH of skin,

(a) Vesicle

(b) Bulla

(c) Macule

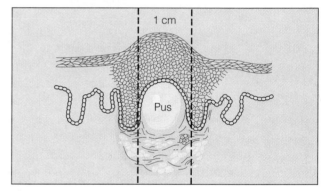

(d) Pustule (papule)

FIGURE 21.2 Skin lesions. (a) Vesicles are small, fluid-filled lesions. **(b)** Bullae are larger fluid-filled lesions. **(c)** Macules are flat lesions that are often reddish. **(d)** Papules are raised lesions; when they contain pus, as shown here, they are called pustules.

generally between 3 and 5. This acidity has a bacteriostatic effect on many potentially harmful microbes. Other diphtheroids, such as *Corynebacterium xerosis* (ze-rō′sis), are aerobic and occupy the skin surface. A yeast belonging to the genus *Pityrosporum* (pit-i-ros′pô-rum) is capable of growing on oily skin secretions and is a frequent inhabitant of the skin. Some consider it to be responsible for the scaling skin condition known as dandruff.

MICROBIAL DISEASES OF THE SKIN

LEARNING OBJECTIVES

- *Differentiate between staphylococci and streptococci, and list several skin infections caused by each.*
- *List the etiologic agent, method of transmission, and clinical symptoms of the following skin infections: otitis externa, acne, warts, smallpox, chickenpox, measles, rubella, cold sores, and scabies.*
- *Differentiate between cutaneous and subcutaneous mycoses, and provide an example of each.*

Many diseases manifest themselves by rashes and other lesions that mainly affect the skin. Variations in these lesions are often useful in describing the symptoms of the disease. For example, small, fluid-filled lesions are *vesicles* (Figure 21.2a). Vesicles larger than about 1 cm in diameter are termed *bullae* (Figure 21.2b). Flat, reddened lesions are known as *macules* (Figure 21.2c). Raised lesions are called *papules,* or *pustules* when they contain pus (Figure 21.2d). Often the focus of infection is elsewhere in the body, but it is convenient to classify these diseases by the organ most obviously affected, the skin. Table 21.1 summarizes the most important diseases associated with the skin. The following discussion covers these and other relatively common bacterial, fungal, and viral skin diseases.

Bacterial Diseases of the Skin

Two genera of bacteria, *Staphylococcus* and *Streptococcus* (commonly called staphylococci and streptococci), are frequent causes of skin-related diseases and merit special discussion. We will also discuss these bacteria in subsequent chapters in relation to other organs and conditions. Su-

TABLE 21.1 Microbial Diseases Associated with the Skin

	CAUSATIVE AGENT	CHARACTERISTICS	TREATMENT
Bacterial Diseases			
Impetigo	*Staphylococcus aureus;* occasionally, *Streptococcus pyogenes*	Superficial skin infection; isolated pustules	Penicillin (for *Streptococcus* infections only); erythromycin
Erysipelas	*Streptococcus pyogenes*	Reddish patches on skin; often with high fever	Penicillin, erythromycin
Pseudomonas dermatitis	*Pseudomonas aeruginosa*	Superficial rash	Usually self-limiting
Otitis externa	*Pseudomonas aeruginosa*	Superficial infection of external ear canal	Gentamicin, carbenicillin
Acne	*Propionibacterium acnes*	Inflammatory lesions originating with accumulations of sebum that rupture a hair follicle	Benzoyl peroxide, tetracyclines, isotretinoin
Viral Diseases			
Warts	*Papillomavirus* spp.	A horny projection of the skin formed by proliferation of cells	May be removed by liquid nitrogen cryotherapy, electrodesiccation, acids, or lasers
Smallpox (variola)	Smallpox virus	Pustules that may be nearly confluent on skin. Systemic viral infection affects many internal organs	None
Chickenpox (varicella)	Varicella-zoster virus	Vesicles in most cases confined to face, throat, and lower back	Acyclovir for immunocompromised patients
Shingles (zoster)	Varicella-zoster virus	Vesicles similar to chickenpox; typically on one side of waist, face and scalp, or upper chest	Acyclovir for immunocompromised patients
Herpes simplex	Herpes simplex virus type 1	Most commonly as cold sores—vesicles around mouth; can also affect other areas of skin and mucous membranes	Acyclovir may modify symptoms
Measles (rubeola)	Measles virus	Skin rash of reddish macules first appearing on face and spreading to trunk and extremities	None
Rubella (German measles)	Rubella virus	Mild disease with a rash resembling measles, but less extensive and disappears in 3 days or less	None
Fungal Diseases			
Ringworm (tinea)	*Microsporum, Trichophyton, Epidermophyton* species	Skin lesions of highly varied appearance; on scalp may cause local loss of hair	Griseofulvin (orally); miconazole, clotrimazole (topically)
Sporotrichosis	*Sporothrix schenckii*	Ulcer at site of infection spreading into nearby lymphatic vessels	Potassium iodide solution (orally)
Candidiasis	*Candida albicans*	Symptoms vary with infection site; usually affects mucous membranes or moist areas of skin	Miconazole, clotrimazole, nystatin (topically)
Parasitic Infestation			
Scabies	*Sarcoptes scabiei* (mite)	Papules due to hypersensitivity reaction to mites	Gamma benzenehexachloride, permethrin (topically)

perficial staphylococcal and streptococcal infections of the skin are very common. The bacteria frequently come into contact with the skin and have adapted fairly well to the physiological conditions there. Both genera also produce invasive enzymes and damaging toxins that contribute to the disease process. (See the discussion of superantigens in Chapter 19.)

Staphylococcal Skin Infections

Staphylococci are spherical gram-positive bacteria about 0.5 to 1.5 μm in diameter. They tend to form in irregular clusters like grapes because the cells divide at random points about their circumference and the daughter cells do not completely separate from each other (see Figures 4.1d and 11.11). For almost all clinical purposes these bac-

SEM

FIGURE 21.3 Coagulase-negative staphylococci. These slime-producing bacteria are shown here adhering to the surface of a plastic catheter.

teria are divided into those that produce **coagulase,** an enzyme that coagulates (clots) fibrin in blood, and those that do not (coagulase-positive and coagulase-negative strains). The coagulase-negative strains are very common on the skin, where they may represent 90% of the normal microbiota. They are generally pathogenic only when the skin barrier is broken or is invaded by medical procedures such as the insertion and removal of catheters (Figure 21.3). At one time the coagulase-negative staphylococci were considered to be all one species, *Staphylococcus epidermidis.* They have since been subdivided into several species, and the name *S. epidermidis* is applied now only to the predominant species from human skin.

Staphylococcus aureus is the most pathogenic of the staphylococci. Typically, it forms golden-yellow colonies (see Figure 11.11). Almost all pathogenic strains of *S. aureus* are coagulase-positive. Fibrin clots may protect the microorganisms from phagocytosis and isolate them from other defenses of the host. There is a high correlation between the bacterium's ability to form coagulase and its production of damaging toxins, several of which may injure tissues. Staphylococci may also produce *leukocidin,* which destroys phagocytic leukocytes, and *exfoliative toxin,* which is responsible for scalded skin syndrome, to be discussed shortly. Some staphylococcal toxins, *enterotoxins,* affect the gastrointestinal tract and will be discussed in Chapter 25 with diseases of the digestive system.

S. aureus is often a problem in the hospital environment. Because *S. aureus* is carried by patients, hospital personnel, and hospital visitors, the danger of infection of surgical wounds and other breaks in the skin is great. Moreover, such infections are difficult to treat because *S. aureus* is exposed to so many antibiotics that it quickly becomes resistant to them. At one time, this organism was

almost uniformly extremely susceptible to penicillin, but now only about 10% of *S. aureus* strains are sensitive. Most of the microorganism's resistance derives from the production of penicillinases.

The nasal passages provide an especially favorable environment for *S. aureus,* which is often present there in very large numbers. Its presence on unbroken skin is usually the result of transport from the nasal passages. *S. aureus* often enters the body through a natural opening in the skin barrier, the hair follicle's passage through the epidermal layer. Infections of hair follicles, or **folliculitis,** often occur as **pimples.** The infected follicle of an eyelash is called a **sty.** A more serious hair follicle infection is the **furuncle (boil),** which is a type of **abscess,** a localized region of pus surrounded by inflamed tissue. Antibiotics do not penetrate well into abscesses, which are difficult to treat. Draining pus from the abscess is frequently a preliminary step to successful treatment.

When the body fails to wall off a bacterial infection, neighboring tissue can be progressively invaded. The extensive damage is called a **carbuncle,** a hard, round deep inflammation of tissue under the skin. At this stage of infection, the patient usually exhibits the symptoms of generalized illness with fever.

Staphylococci are the primary cause of a very troublesome problem in hospital nurseries, **impetigo of the newborn.** Symptoms of this disease are thin-walled vesicles on the skin that rupture and later crust over. To prevent outbreaks, which can reach epidemic proportions, hexachlorophene-containing skin lotions are commonly prescribed (see Chapter 7).

Staphylococcal infections always carry the risk that the underlying tissue will become infected or that the infection will enter the bloodstream. The circulation of toxins, such as those produced by staphylococci, is called **toxemia.** One such toxin, which is produced by staphylococci lysogenized by certain phage types, causes **scalded skin syndrome.** (See Chapters 13 and 15 for a discussion of lysogeny and phage conversion.) This condition is first characterized by a lesion around the nose and mouth. The lesion develops rapidly into a bright red area and spreads; within 48 hours the skin of the affected areas peels off in sheets when it is touched (Figure 21.4). The scalded skin syndrome is frequently observed in children under the age of two, especially in newborns, as a complication of staphylococcal infections. These patients are seriously ill, and vigorous antibiotic therapy is required.

The scalded skin syndrome is also characteristic of the late stages of **toxic shock syndrome (TSS).** In this potentially life-threatening condition, fever, vomiting, and a sunburnlike rash are followed by shock (a sudden drop in blood pressure). TSS originally became known as a staphylococcal growth associated with the use of a new type of highly absorbent vaginal tampon; the correlation is especially high for cases in which the tampons remain in place too long. Staphylococcal toxin (mainly toxic shock syndrome toxin-1, or TSST-1) enters the bloodstream from the

FIGURE 21.4 Lesions of scalded skin syndrome. Some staphylococci produce a toxin that causes the skin to peel off in sheets, as on the hand of this infant. It is especially likely to occur in children under two years of age.

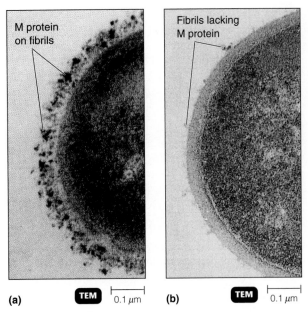

FIGURE 21.5 Group A *β*-hemolytic streptococci. (a) Part of a cell that carries M protein on its hairlike surface fibrils. (b) Part of a cell that lacks M protein.

bacterial growth site in and around the tampon, and its circulation causes the symptoms.

Today only little more than about half of the cases of TSS are associated with menstruation. Nonmenstrual TSS occurs from staphylococcal infections that follow nasal surgery in which absorbent packing is used, after surgical incisions, and in women who have just given birth.

Streptococcal Skin Infections

Like staphylococci, **streptococci** are gram-positive spherical bacteria. Unlike staphylococci, streptococcal cells grow in chains. Prior to division, the individual cocci elongate on the axis of the chain, and then the cells divide into pairs. When the dividing pairs do not separate, chaining occurs (see Figures 4.1a and 11.12).

Many nonpathogenic streptococci are found inhabiting the mouth, gastrointestinal tract, and upper respiratory system, but some streptococci are responsible for skin infections and even severe tissue damage. As streptococci grow, they secrete several toxins and enzymes. Among these toxins are *hemolysins,* which damage red blood cells. Depending on the type of destruction caused by the hemolysins, streptococci can be divided into *α*-hemolytic, *β*-hemolytic, and *γ*-hemolytic streptococci (see Chapter 11). *β*-hemolytic streptococci have remained highly susceptible to penicillin. Unlike the staphylococci, they have no ability to produce penicillinases.

β-hemolytic streptococci are further differentiated into a number of groups, designated A through O, according to

the antigenic carbohydrates in their cell walls. As causes of human disease, *β*-hemolytic group A streptococci are the most important. *β*-hemolytic group A streptococci can be further subdivided into over 55 immunologic types according to the antigenic properties of the *M protein* found in some strains (Figure 21.5). This protein is found external to the cell wall on fibrils. The M protein has antiphagocytic properties that contribute to the strain's pathogenicity. It also appears to aid the bacteria in adhering to and colonizing mucous membranes.

In addition to hemolysins and M proteins, a number of other substances contribute to the pathogenicity of *β*-hemolytic group A streptococci. These substances include *erythrogenic toxin* (responsible for the scarlet fever rash), *deoxyribonucleases* (enzymes that degrade DNA), *NADase* (an enzyme that breaks down NAD), *streptokinases* (enzymes that dissolve blood clots), *hyaluronidase* (an enzyme that dissolves hyaluronic acid, the cementing substance of connective tissue), and *leukocidins* (enzymes that kill white blood cells).

β-hemolytic (or invasive) group A streptococci cause a wide variety of diseases, a number of which will be discussed in subsequent chapters. When such streptococcal infections affect the dermis, they cause **erysipelas.** In this disease, the skin erupts into reddish patches that enlarge and thicken and swell at the margins (Figure 21.6). High fever is common. The reddening is caused by toxins produced by the streptococci as they invade new areas. Usually, the skin outbreak is preceded elsewhere in the body

FIGURE 21.6 Lesions of erysipelas, caused by group A β-hemolytic streptococcal toxins. The affected patch of skin has been reddened by streptococcal toxins.

FIGURE 21.7 Lesions of impetigo. This disease is characterized by isolated pustules that become crusted. Impetigo in older children, as depicted here, is usually caused by streptococci.

by a streptococcal infection, such as a streptococcal sore throat. It is not certain whether the skin is infected from an external invasion or via the lymphatic system. Erysipelas is most likely to occur in the very young and the very old. The drugs of choice for the treatment of erysipelas are penicillin and erythromycin.

The most common species of the β-hemolytic group A streptococci, *Streptococcus pyogenes,* may be implicated in some cases of impetigo. Although impetigo of the newborn is caused by staphylococci, **impetigo** is also common in children of toddler and grade-school age and is likely in these cases to be caused by streptococci. A superficial skin infection, impetigo is characterized by isolated pustules that become crusted and rupture (Figure 21.7). The disease is spread largely by contact, and the bacteria penetrate the skin through some minor abrasion or wound whose scab has been prematurely removed. Staphylococci can also be found in these lesions but are probably only secondary invaders. Fortunately, the condition is seldom serious. The drugs of choice are penicillin and erythromycin.

Infections by Pseudomonads

Pseudomonads are aerobic gram-negative rods that are widespread in soil and water. Able to survive in any moist environment, they can grow on traces of even unusual organic matter, such as soap films or cap liner adhesives, and are resistant to many antibiotics and disinfectants. The most prominent species is *Pseudomonas aeruginosa.*

Pseudomonads frequently cause outbreaks of ***Pseudomonas* dermatitis.** This is a self-limiting rash of about two weeks' duration often associated with swimming pools and pool-type saunas and hot tubs. When large numbers of people use these facilities, the alkalinity rises and the chlorines become less effective; at the same time, the concentration of nutrients that support growth of pseudomonads increases. Hot water facilitates the entry of

bacteria into hair follicles by causing the follicles to dilate. Competition swimmers are often troubled with **otitis externa,** or swimmer's ear, a pseudomonad infection of the external ear canal leading to the eardrum.

P. aeruginosa produces several exotoxins that account for much of its pathogenicity. It also has an endotoxin. Except for superficial skin infections and otitis externa, infection by *P. aeruginosa* is rare in healthy individuals. However, it often causes respiratory infections in people already compromised by immunologic deficiencies (natural or drug induced) or by chronic pulmonary disease, especially cystic fibrosis. (Respiratory infections are discussed in Chapter 24.)

P. aeruginosa is also a very common and serious opportunistic pathogen in burn patients, particularly those with second- and third-degree burns. In such cases, infection may produce blue-green pus, whose color is caused by the bacterial pigment **pyocyanin** (sometimes spelled *pyocyanine*). Of major concern in many hospitals is the ease with which *P. aeruginosa* is carried by flowers or plants sent by well-wishers.

Pseudomonads are relatively resistant to many of the commonly used disinfectants (see Chapter 7) and antibiotics (see Chapter 20). Antibiotic resistance by pseudomonads is still a problem, but in recent years a number of new antibiotics have been developed, so the number of antimicrobials available for chemotherapy of these infections is not as restricted as it once was. Two antibiotics used against pseudomonad infections are gentamicin and carbenicillin, often used in combination. Silver sulfadiazine is very useful in the treatment of burn infections by *P. aeruginosa.*

Acne

Acne is probably the most common skin disease of humans, affecting an estimated 17 million people in the

U.S.A. More than 85% of all teenagers have the problem to some degree. In the U.S.A., about 350,000 people each year are estimated to suffer from unusually severe acne, called **cystic acne**, which produces inflamed cysts and subsequent severe scarring of the face and upper body. Although the occurrence of acne decreases after the teen years, the scarring from these severe cases often remains.

Acne begins when channels for passage of sebum to the skin surface are blocked. The sebum accumulation leads to the formation of the familiar whiteheads; if the opening remains open, the lesions known as blackheads form. The accumulation of sebum often ruptures the lining of the hair follicle. Bacteria, especially *Propionibacterium acnes,* a diphtheroid commonly found on the skin, become involved at this stage. By metabolizing the sebum, *P. acnes* forms free fatty acids that cause an inflammatory response by the body. It is this inflammation that leads to the appearance of pustules and subsequent acne scars. Picking or scratching the lesions—or even having tight collars or other clothing in contact with lesions—increases the incidence of scar formation. Cosmetics frequently aggravate the condition, but diet—including consumption of chocolate—has been demonstrated to have no significant effect on the disease.

Topical applications of preparations containing benzoyl peroxide are often useful, and antibiotics such as tetracycline have also been used. Tretinoin (Retin-A) is effective in the prevention of acne lesions but it inactivates benzoyl peroxide and so must be applied at different times of the day. Severe cases should receive a dermatologist's treatment rather than household remedies. The most important recent development in treatment of severe cystic acne is isotretinoin (Accutane®). This drug, taken orally, inhibits sebum formation, and dramatic improvement follows. Isotretinoin is not recommended for use against the usual mild cases of acne because of its adverse effects; for one thing, it is *teratogenic* (causes damage to the fetus) if taken by a pregnant woman for even a few days.

Viral Diseases of the Skin

Many viral diseases, although systemic in nature, are most apparent by their effects on the skin.

Warts

Warts (benign tumors, or papillomas) are caused by viruses that stimulate an uncontrolled but benign growth of skin cells (see Figure 26.15). It was long known that warts can be transmitted from one person to another by contact, even sexually, but it was not until 1949 that viruses were identified in wart tissues. More than 50 different types of papillomavirus are now known to cause different types of warts, often of greatly differing appearance. After infection, there is an incubation period of several weeks before the warts appear.

The most common medical treatments for warts are ap-

FIGURE 21.8 Lesions of smallpox. In some severe cases, the lesions nearly run together.

plying extremely cold liquid nitrogen to them (cryotherapy), drying them with an electric current (electrodesiccation), or burning them with acids. Burning with acids is also the basis of some home remedies. Warts that do not respond to any other treatments, especially genital warts, can now be treated with injected interferon or lasers. The use of lasers results in a virus-laden aerosol. Physicians using lasers to remove warts have contracted warts themselves, especially in their nostrils. Although warts are not a form of cancer, some skin and cervical cancers seem to be associated with papillomaviruses.

Smallpox (Variola)

During the Middle Ages, an estimated 80% of the population of Europe contracted **smallpox** at some time during their lives. Those who recovered from the disease carried disfiguring scars. The disease was even more devastating to Native Americans, who had had no previous exposure and thus little resistance.

Smallpox is caused by a poxvirus known as the smallpox (variola) virus. There are two basic forms of this disease: **variola major**, which has a mortality rate of 20% or higher, and **variola minor**, which has a mortality rate of less than 1%. Recovery from one form of the disease produces effective immunity against both forms.

Transmitted first by the respiratory route, the viruses infect many internal organs before their eventual movement into the bloodstream (**viremia**) leads to infection of the skin and the production of more recognizable symptoms. The growth of the virus in the epidermal layers of the skin causes lesions (Figure 21.8).

Smallpox was the first disease to which immunity was artificially induced (see Chapters 1 and 18) and the first to be eradicated from the human population. The last victim of a natural case of smallpox is believed to be an individ-

(a) Initial infection: chickenpox (varicella)

(b) Recurrence of infection: shingles (herpes zoster)

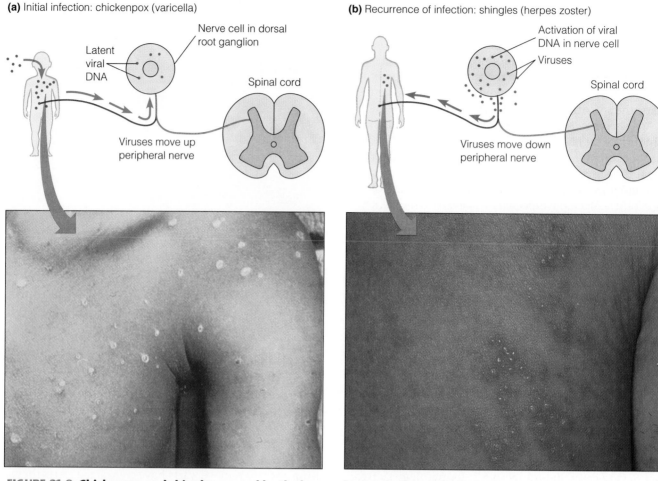

FIGURE 21.9 Chickenpox and shingles, caused by the herpesvirus varicella-zoster.
(a) Initial infection with the virus, usually during childhood, causes chickenpox, which is characterized by the formation of the lesions shown on the left. The virus then moves to a dorsal root ganglion, near the spine, where it remains latent indefinitely. **(b)** Later, usually in adulthood, immune system depression or stress can trigger the reactivation of the virus, causing shingles. The typical lesions shown on the right are on the back of a patient.

ual who recovered from variola minor in 1977 in Somalia, Africa.

The eradication of smallpox was possible because there are no animal host reservoirs for the disease. Once an effective vaccine became available, eradication was accomplished by a concerted vaccination effort coordinated by the World Health Organization.

In recent years, the smallpox virus collections in laboratories have been the most likely sources of new infections. The risk of such infection is not merely a hypothetical concern, as there have already been several laboratory-associated infections, one of which caused death. Today, only two sites maintain the smallpox virus, one in the U.S.A. and one in Russia.

Chickenpox (Varicella) and Shingles (Herpes Zoster)

Chickenpox (varicella) is a relatively mild childhood disease. After gonorrhea, it is the most common reportable

infectious disease in the U.S.A. It is probably greatly underreported, and estimates are that more than 2 million cases occur each year in the U.S.A. Disease summaries of the CDC show that about 100 deaths per year, usually from encephalitis (infection of the brain) or viral pneumonia, are attributed to chickenpox.

The initial infection with the herpesvirus varicella-zoster virus causes chickenpox (Figure 21.9a). The disease is acquired by entry of the virus into the respiratory system, and the infection localizes in skin cells after about two weeks. The infected skin is vesicular for three to four days. During that time, the vesicles fill with pus, rupture, and form a scab before healing. Lesions are mostly confined to the face, throat, and lower back but also can occur on the chest and shoulders. When chickenpox occurs in adults—which is not frequent because the high incidence in childhood grants immunity to most individuals—it is a more severe disease with a significant mortality rate.

Reye's syndrome is an occasional severe complication

of chickenpox, influenza, and some other viral diseases. A few days after the initial infection has receded, the patient persistently vomits and exhibits signs of brain dysfunction. Coma, fatty degeneration of the liver, and death can follow. Death or brain damage in survivors is from brain swelling, which prevents blood circulation. At one time, the death rate of reported cases approached 90%, but this rate has been declining with improved care and is now 30% or lower when the disease is recognized and treated in time. Reye's syndrome affects children and teenagers almost exclusively. The use of aspirin to lower fevers in chickenpox and influenza increases the chances of acquiring Reye's syndrome.

Like all herpesviruses, a characteristic of varicella-zoster virus is its ability to remain latent within the body. Following a primary infection, the virus enters the peripheral nerves and moves to a central nerve ganglion (a group of nerve cells in bodies lying outside the central nervous system), where it persists as viral DNA (see Figure 21.9a). The DNA may or may not be transcribed, but in any event it persists indefinitely. Humoral antibodies cannot penetrate into the nerve cell, and because no viral antigens are expressed on the surface of the nerve cell, cytotoxic T cells are not activated. Therefore, neither arm of the specific immune system disturbs the latent virus.

Latent varicella-zoster virus is located in the dorsal root ganglion near the spine. Later, perhaps as long as decades later, the virus may be reactivated (Figure 21.9b). The trigger can be stress or simply the lower immune competence associated with aging. The virions produced by the reactivated DNA move along the peripheral nerves to the cutaneous sensory nerves of the skin, where they cause a new outbreak of the virus in the form of **shingles** (herpes zoster).

In shingles, vesicles similar to those of chickenpox occur but are localized in distinctive areas. Typically, they are distributed about the waist, although facial shingles and infections of the upper chest and back also occur (see Figure 21.9b). The infection follows the distribution of the affected cutaneous sensory nerves and is usually limited to one side of the body at a time because these nerves are unilateral. Occasionally, such nerve infections can result in nerve damage that impairs vision or even causes paralysis. Severe pain is also frequently reported.

Shingles is simply a different expression of the virus that causes chickenpox; it expresses differently because the patient, having had chickenpox, now has partial immunity to the virus. Exposing children to shingles has led to their contracting chickenpox. Shingles seldom occurs in people under age 20, and by far the highest incidence is in the elderly population.

Immunocompromised patients are in serious danger from infection by varicella-zoster virus; multiple organs become infected, and a mortality rate of 17% is common. In such cases, the antiviral drug *acyclovir* has proven helpful. A live, attenuated vaccine is expected to be approved soon.

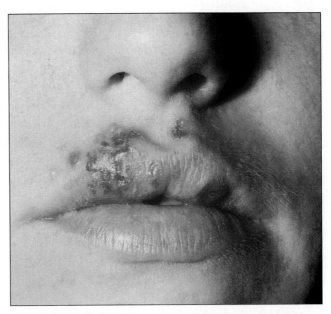

FIGURE 21.10 Cold sores, or fever blisters, caused by herpes simplex virus.

Herpes Simplex

Serologic surveys show that about 90% of the population of the U.S.A. has been infected with the herpes simplex virus. The initial infection usually occurs in infancy. Frequently, this infection is subclinical, but perhaps as many as 15% of the cases develop lesions known as **cold sores** or **fever blisters** (Figure 21.10). Usually occurring in the oral mucous membrane, these lesions heal as the infection subsides. Recurrences are usually associated with some trauma; excessive exposure to UV radiation from the sun is a frequent cause, as are emotional upsets or the hormonal changes associated with menstrual periods. These infections are caused by *herpes simplex virus type 1* (HSV-1), which is transmitted primarily by oral or respiratory routes. An important complication is herpetic keratitis (discussed later in this chapter), in which the cornea of the eye becomes infected. Between recurrences, HSV-1 is latent in the trigeminal nerve ganglia communicating between the face and the central nervous system. Transmission of HSV-1 infections occurs by skin contact among high school and college wrestlers (**herpes gladiatorum**). An incidence as high as 3% has been reported in high school wrestlers; see the box on page 528 for more information.

A different infection by a very similar virus, *herpes simplex virus type 2* (HSV-2), is transmitted primarily by sexual contact. HSV-2 is differentiated from HSV-1 by its antigens and by its effect on cells in tissue cultures. We will discuss this infection in Chapter 26 in our discussion of diseases of the urinary and genital systems. HSV-2 causing genital herpes is latent in the sacral ganglia near the base of the spine.

Very rarely, either type of the herpes simplex virus may spread to the brain, causing **herpes encephalitis**. Infec-

FIGURE 21.11 Number of reported measles cases in the U.S.A., 1960 to 1993. Note the sharp decline in cases after introduction of the measles vaccine in 1963. Inset: Measles is at its lowest reported level in the U.S.A. However, previous periods of low activity have been followed by resurgences. [*Sources:* CDC, *Summary of Notifiable Diseases 1990, MMWR 39(53); MMWR 41(53) (1/8/93); MMWR 42(50) (12/24/93).*]

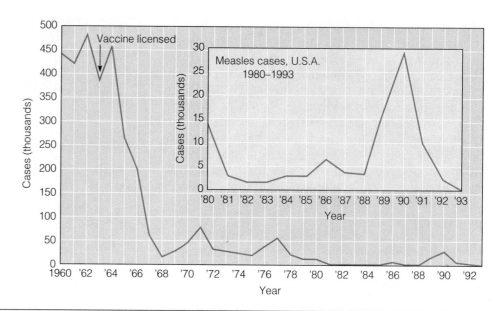

tions by HSV-2 are more serious, with a fatality rate as high as 70% if untreated. Only about 10% of survivors can expect to lead healthy lives. However, this is one of the rare viral diseases for which a chemotherapeutic treatment is reasonably effective. When administered promptly, acyclovir often cures such encephalitis. Even so, the mortality in tests was still 28%, and only 38% of the survivors escaped serious neurological damage.

Measles (Rubeola)

Measles is an extremely contagious viral disease that is spread by the respiratory route. Because a person with measles is infectious before symptoms appear, quarantine is not an effective measure of prevention.

Humans are the only reservoir for measles, although monkeys are also susceptible. Therefore, measles can potentially be eradicated, much as smallpox was. Since the licensing of a vaccine in 1963, the number of measles cases has declined from more than 400,000 cases per year to a record low of 1497 cases in 1983 (Figure 21.11). Unfortunately, the number of cases has been slowly creeping upward since then. Despite laws requiring immunization for school entry, the immunization rate is low in some densely populated inner-city groups, and about 40% of measles cases now occur in preschool children in these groups. In addition, although the measles vaccine is about 95% effective, cases continue to occur among those who do not develop or retain good immunity. Some of these infections are caused by contact with infected people who come from outside the U.S.A.

An unexpected result of the measles vaccine is that many cases of measles today occur in children less than one year of age. Measles is especially hazardous to infants—they are more likely to have serious complications. In prevaccination days, measles was rare at this age because infants were protected by maternal antibodies de-

rived from their mothers' recovery from the disease. Unfortunately, antibodies made in response to the vaccine are not as effective in providing protection as are antibodies made in response to the disease. Therefore, because the vaccine is effective only when administered at the age of about 15 months, the child is vulnerable for a considerable time.

The development of rubeola is similar to that of smallpox and chickenpox. Infection begins in the upper respiratory system. After an incubation period of 10 to 12 days, symptoms develop resembling those of a common cold—sore throat, headache, and cough. Shortly thereafter, a macular rash appears on the skin, beginning on the face and spreading to the trunk and extremities (Figure 21.12). Lesions of the oral cavity include the diagnostically useful *Koplik spots* (tiny red patches with central white specks) on the oral mucosa opposite the molars.

Measles is an extremely dangerous disease, especially among very young and elderly individuals. It is frequently complicated by middle ear infection or pneumonia caused by the virus itself or by secondary bacterial infection. Encephalitis strikes approximately 1 in 1000 measles victims; its survivors are often left with permanent brain damage. As many as 1 in 3000 cases is fatal, mostly in infants. The virulence of the virus seems to vary with different epidemic outbreaks. Complications such as encephalitis occur, if at all, about a week after the rash appears.

Rubella

Rubella, or German measles, is a much milder viral disease than rubeola (measles) and often goes undetected. A macular rash of small red spots and a light fever are the usual symptoms (Figure 21.13). Complications are rare, especially in children, but encephalitis occurs in about 1 case in 6000, mostly in adults. Transmission is by the respiratory route, and an incubation time of two to three

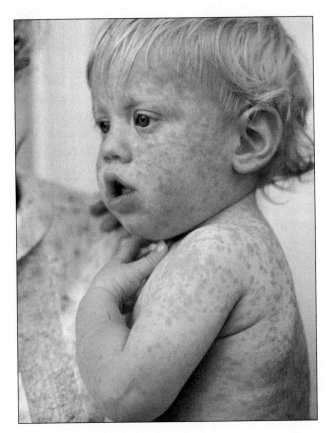

FIGURE 21.12 The rash of small raised spots typical of measles (rubeola). The rash typically begins on the face and spreads to the trunk and extremities.

FIGURE 21.13 The rash of red spots characteristic of rubella. The spots are not raised above the surrounding skin.

weeks is the norm. Recovery from clinical or subclinical cases appears to give a firm immunity.

The seriousness of rubella was not appreciated until 1941, when the association was made between serious birth defects and maternal infection during the first trimester (three months) of pregnancy—the **congenital rubella syndrome.** If a pregnant woman contracts the disease during this time, there is about a 35% incidence of serious fetal damage, including deafness, eye cataracts, heart defects, mental retardation, and death. Fifteen percent of babies with congenital rubella syndrome die during their first year. The last major epidemic of rubella in the U.S.A. was during 1964 and 1965. About 20,000 severely impaired children were born during this epidemic.

It is therefore important to identify women of childbearing age who are not immune to rubella. In some states, the blood test required for a marriage license includes a test for rubella antibodies. Serum antibody can be assayed by a number of commercially available laboratory tests. Accurate diagnosis of immune status always requires such tests; histories alone are unreliable.

In addition to this surveillance, a rubella vaccine was introduced in 1969; in 1979, a more effective version, which produces fewer adverse effects, replaced it. It is still unclear, however, whether these vaccines confer an immunity that lasts from childhood through the childbearing years. Follow-up studies indicate that more than 90% of vaccinated individuals are protected for at least 15 years. Due to these preventive measures, fewer than 10 cases of congenital rubella syndrome are now reported every year.

For pregnant women who are found not to be immune to rubella, there has been concern about the possibility that administration of the vaccine could cause fetal damage. However, in hundreds of cases in which women were vaccinated three months before or three months after their presumed dates of conception, no case of congenital rubella syndrome defects has occurred. Nevertheless, the vaccine is still not recommended for pregnant women.

Fungal Diseases of the Skin

The skin is most susceptible to microorganisms that are able to resist high osmotic pressure and low moisture. It is not surprising, therefore, that fungi cause a number of skin disorders. Any fungal infection of the body is called a **mycosis.**

Cutaneous Mycoses

Fungi that colonize the hair, nails, and the outer layer (stratum corneum) of the epidermis (see Figure 21.1) are called **dermatophytes,** and their infections are called **dermatomycoses.** Dermatophytes grow on the keratin present in those locations, causing infections called **tineas,** or **ringworms.** The name *ringworm* arose from the ancient Greek belief that the infections, which tend to expand circularly, were caused by a worm. The Romans incorrectly associated ringworm with lice, and the term *tinea* is

Morbidity and Mortality Weekly Report

HERPES GLADIATORUM AT A HIGH SCHOOL WRESTLING CAMP IN MINNESOTA

THE MINNESOTA DEPARTMENT of Health investigated an outbreak of herpes simplex virus type 1 (HSV-1) dermatitis (herpes gladiatorum) in participants at a Minnesota wrestling camp. The camp was held July 2 through July 28 and attended by 175 male high school wrestlers from throughout the U.S.A. The outbreak was detected during the final week of camp, and wrestling contact was subsequently discontinued for the final two days.

A case was defined as isolation of HSV-1 from involved skin or eye or the presence of cutaneous vesicles. To identify cases, a clinic was held at the camp to obtain viral cultures and examine skin lesions. Additional clinical data were obtained from review of emergency department records at the facility where all affected wrestlers were referred for medical care. A questionnaire was administered to wrestlers by telephone following the conclusion of camp.

Clinical and questionnaire data were available for 171 persons (98%). The mean age of these participants was 16 years (range: 14–18 years); 153 (89%) were white; 137 (80%) were high school juniors or seniors. The median length of time in competitive wrestling was four years.

Sixty persons (35%) met the case definition (see the figure), including 21 (12%) who had HSV-1 isolated from the skin or eye. All affected wrestlers had onset during the camp session or within one week after leaving camp. Two wrestlers had a proba-

ble recurrence of HSV, one oral and one cutaneous, during the first week of camp. Lesions were located on the head or neck in 44 persons (73%), the extremities in 25 (42%), and the trunk in 17 (28%). Herpetic conjunctivitis occurred in five persons; none developed keratitis. Associated signs and symptoms included lymphadenopathy (60%), fever and/or chills (25%), sore throat (40%), and headache (22%). Forty-four) persons (73%) were treated with acyclovir.

Thirty-eight wrestlers interviewed reported a past history of oral HSV-1 infection. The attack rate was 24% for wrestlers who reported a past history of oral herpes and 38% for wrestlers without a history of oral herpes. Twenty-three percent of affected wrestlers continued to wrestle for at least two days after rash onset. Athletes who reported wrestling with a participant with a rash were more likely to have the infection.

Editorial Note: Herpes gladiatorum was first described in the mid-1960s. In 1988, an outbreak of herpes gladiatorum was reported among three Wisconsin high school wrestling teams. Lesions occur most often on the head and neck. Primary infection may cause constitutional symptoms with fever, malaise, weight loss, and regional lymphadenopathy. Ocular involvement includes keratitis and conjunctivitis.

Transmission occurs primarily through skin-to-skin contact. Autoinoculation may lead to involvement of multiple sites. Previous infection with HSV-1 may reduce the risk of acquiring herpes gladiatorum. However, the prevalence of antibody to HSV-1 is low among white adolescents, and many adolescents are susceptible when they enter competitive wrestling. Control methods should include education of athletes and trainers regarding herpes gladiatorum, routine skin examinations before wrestling contact, and exclusion of wrestlers with suspicious skin lesions. The outbreak in the Minnesota camp might have been prevented if athletes with such lesions had been promptly excluded from contact competition.

SOURCE: *MMWR* 39:69–70 (2/9/90).

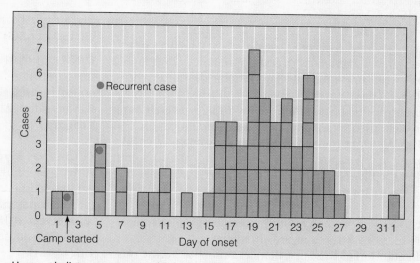

Herpes gladiatorum cases at a high school wrestling camp, by day of onset.

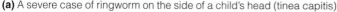

(a) A severe case of ringworm on the side of a child's head (tinea capitis).

(b) Ringworm of the foot, or athlete's foot (tinea pedis).

FIGURE 21.14 Dermatomycoses.

derived from the Latin for worms or insect larvae. **Tinea capitis,** or ringworm of the scalp, is fairly common among elementary school children and can result in bald patches (Figure 21.14a). It is usually transmitted by contact with fomites. Dogs and cats are also frequently infected with fungi that cause ringworm in children. Ringworm of the groin, or jock itch, is known as **tinea cruris,** and ringworm of the feet, or athlete's foot, is known as **tinea pedis** (Figure 21.14b). The moisture in such areas favors fungal infections.

Three genera of fungi are involved in cutaneous mycoses. *Trichophyton* (trik-ō-fī'ton) can infect hair, skin, or nails; *Microsporum* (mī-krō-spô'rum) usually involves only the hair or skin; *Epidermophyton* (ep-i-dėr-mō-fī'ton) affects only the skin and nails. The topical drugs available without prescription for tinea infections are miconazole or clotrimazole. An oral antibiotic, griseofulvin, is often useful in these infections because it can localize in keratinized tissue. If the infection is unusually serious, oral ketoconazole may be prescribed.

Diagnosis of cutaneous mycoses is usually made by microscopic examination of scrapings of the affected areas. Some species growing on hairs will fluoresce when illuminated by ultraviolet light (Wood's light). The dermatophytes can be readily cultured. Physicians with little specialized knowledge of fungal morphology often use a selective medium, Dermatophyte Test Medium. If, after a few days of growth, the white, fluffy fungal colonies turn red, the culture is recorded as positive for dermatophytes.

Subcutaneous Mycoses

Subcutaneous mycoses are more serious than cutaneous mycoses. Even when the skin is broken, cutaneous fungi do not seem to be able to penetrate past the stratum corneum, perhaps because they cannot obtain sufficient iron for growth in the epidermis and the dermis. Usually these diseases are caused by fungi that inhabit the soil, especially decaying vegetation, and penetrate the skin through a small wound that allows entry into subcutaneous tissues. In this country, the disease of this type most commonly encountered is **sporotrichosis,** caused by the dimorphic fungus *Sporothrix schenckii* (spô-rō'thriks shen'kē-ē). Most cases occur among gardeners or others working with soil; the infection frequently forms a small ulcer on the hands. The fungus often enters the lymphatic system in the area and there forms similar lesions. The condition is seldom fatal and is effectively treated by ingesting a dilute solution of potassium iodide, even though the organism is not affected in vitro by even a 10% solution of potassium iodide.

Candidiasis

The bacterial microbiota of the mucous membranes in the genitourinary tract and mouth usually suppresses the growth of such fungi as *Candida albicans* (Figure 21.15a). Because the fungus is not affected by antibacterial drugs, it sometimes overgrows mucosal tissue when the antibiotics suppress the normal bacterial microbiota. Changes in the normal mucosal pH may have a similar effect. Such

(a) *Candida albicans*

LM

(b) Oral candidiasis, or thrush

FIGURE 21.15 Candidiasis. (a) *Candida albicans*: Note the spherical chlamydospores (asexual spores formed from hyphal cells) and the smaller blastospores (asexual spores produced by budding) (see Chapter 12). The pseudohyphae (long cells that resemble hyphae) are resistant to phagocytosis and so may be a factor in pathogenicity. **(b)** This case of oral candidiasis, or thrush, produced a thick, creamy coating on the tongue.

overgrowths by the yeastlike *C. albicans* are called **candidiasis.** Newborn infants, whose normal microbiota has not become established, often suffer from a whitish overgrowth of the oral cavity, called **thrush** (Figure 21.15b). *C. albicans* is also a very common cause of vaginitis (see Chapter 26). Immunosuppressed individuals are unusually prone to candida infections of the skin and mucous membranes. On people who are obese or diabetic, the areas of the skin with more moisture tend to become infected with this fungus. The infected areas become bright red, with lesions on the borders. Skin and mucosal infections by *C. albicans* are usually treated with topical applications of miconazole, clotrimazole, or nystatin. If candidiasis becomes systemic, as can happen in immunosuppressed individuals, fulminating disease and death can result. Oral ketoconazole is the usual treatment for systemic candidiasis.

Parasitic Infestation of the Skin

Parasitic organisms such as some protozoa, helminths, and microscopic arthropods can infest the skin and cause disease conditions. We will cover only one example of a common arthropod infestation, scabies.

SEM ⊢—⊣ 10 μm

FIGURE 21.16 *Sarcoptes scabiei,* **the cause of scabies.** Note the two eggs below the mite in this burrow in a patient's skin.

TABLE 21.2 Microbial Diseases Associated with the Eye

	CAUSATIVE AGENT	CHARACTERISTICS	TREATMENT
Bacterial Diseases			
Neonatal gonorrheal ophthalmia	*Neisseria gonorrhoeae*	Acute infection with much pus formation; if treatment is delayed, ulcers form on cornea	Silver nitrate, tetracycline, or erythromycin for prevention; penicillin for treatment
Inclusion conjunctivitis	*Chlamydia trachomatis*	Swelling of eyelid; mucus and pus formation	Tetracycline
Trachoma	*Chlamydia trachomatis*	Conjunctivitis causes scarring of eyelid that mechanically damages cornea, often causing secondary infections	Tetracycline
Viral Disease			
Herpetic keratitis	Herpes simplex type 1 virus	Initially a mild conjunctivitis, but may progress to corneal ulcers and severe damage	Trifluridine may be effective
Protozoan Disease			
Acanthamoeba keratitis	*Acanthamoeba* spp.	Frequently results in severe eye damage	Topical propamidine isethionate, neosporin, miconazole; corneal transplant or eye removal surgery may be required

Scabies

Probably the first documented connection between a microscopic organism and a disease in humans was scabies, which was described by an Italian physician in 1687. The disease involves intense local itching and is caused by the tiny mite *Sarcoptes scabiei* burrowing under the skin to lay its eggs (Figure 21.16). The burrows are often visible as slightly elevated, serpentine lines about a millimeter in width. However, scabies may appear as a variety of inflammatory skin lesions, many of them secondary infections from scratching. The mite is transmitted by intimate contact, even sexual, and is most often seen in family members, nursing home residents, and teenagers infected by children they baby-sit. The mite lives about 25 days, but by that time eggs have hatched and produced a dozen or so progeny. The recommended treatment is a solution of permethrin insecticide or gamma benzene hexachloride (Kwell®) applied topically.

MICROBIAL DISEASES OF THE EYE

LEARNING OBJECTIVES

- *List the etiologic agent, method of transmission, and clinical symptoms of the following eye infections: conjunctivitis, trachoma, herpetic keratitis, and* Acanthamoeba *keratitis.*
- *Describe the epidemiologies of neonatal gonorrheal ophthalmia and inclusion conjunctivitis.*

The epithelial cells covering the eye can be considered a continuation of the skin or mucosa. Many microbes can infect the eye, largely through the conjunctiva, the mucous membrane that lines the eyelids and covers the outer surface of the eyeball. A summary of diseases associated with the eye is presented in Table 21.2.

Inflammation of the Eye Membranes: Conjunctivitis

Conjunctivitis is an inflammation of the conjunctiva. The popularity of contact lenses has been accompanied by problems with microbial infections of the eye. This is especially true of the soft-lens designs often worn for extended periods. The most common bacterial pathogens are pseudomonads, which can cause serious eye damage. Homemade saline solutions, which are a frequent source of infection, should not be used, and the manufacturer's recommendations for disinfection should be followed scrupulously. The most effective methods are thermal; lenses that cannot be heated can be disinfected with hydrogen peroxide, which is then neutralized.

Bacterial Diseases of the Eye

The bacterial microorganisms most commonly associated with the eye are *Staphylococcus epidermidis* (e-pi-dèr'mi-dis), *S. aureus*, and diphtheroids. These microbes usually originate from the skin and upper respiratory tract.

Neonatal Gonorrheal Ophthalmia
Neonatal gonorrheal ophthalmia is a serious form of conjunctivitis caused by *Neisseria gonorrhoeae* (the cause

FIGURE 21.17 Lesions of trachoma. The eyelids have been pulled back to show the inflammatory nodules on the conjunctival membrane in contact with the cornea, which is damaged by the abrasion and is susceptible to secondary infections.

FIGURE 21.18 The trophozoite of an *Acanthamoeba*, the cause of *Acanthamoeba* keratitis.

of gonorrhea). Large amounts of pus are formed; if treatment is delayed it will frequently lead to ulceration of the cornea. It is acquired as the infant passes through the birth canal, and infection carries a high risk of blindness. Early in this century, legislation required that the eyes of all newborn infants be treated with a 1% solution of silver nitrate, which proved to be a very effective treatment in preventing this eye infection. Between 1906 and 1959, the percentage of admissions to schools for the blind that could be attributed to neonatal ophthalmia declined from 24% to only 0.3%. Silver nitrate is being replaced with antibiotics because there are frequent coinfections by the gonococcus and sexually transmitted chlamydias, and silver nitrate is not effective against chlamydias.

Inclusion Conjunctivitis

Chlamydial conjunctivitis, or **inclusion conjunctivitis**, is quite common today. It is caused by *Chlamydia trachomatis*, a bacterium that grows only as an obligate intracellular parasite. In infants, who acquire it in the birth canal, the condition tends to resolve spontaneously in a few weeks or months, but in rare cases it can lead to scarring of the cornea, much as in trachoma, discussed shortly. The current recommendation is to treat the eyes of newborn infants within an hour of birth (the timing is very important) with erythromycin or tetracyclines, which are usually effective against both the gonococcus and chlamydial organisms. Chlamydial conjunctivitis also appears to spread in the unchlorinated waters of swimming pools and in this context is called swimming pool conjunctivitis. Tetracycline applied as an ophthalmic ointment is an effective treatment.

Trachoma

A much more serious infection caused by *Chlamydia trachomatis* is **trachoma**, the greatest single infectious cause of blindness in the world today. Millions now have the disease, and many millions have been blinded by the infection. In the arid parts of Africa and Asia, almost all children are infected early in their lives. Trachoma also occurs occasionally in the southwestern U.S.A., especially among Native Americans. The disease is transmitted largely by hand contact or by such objects as towels. Flies may also carry the bacteria.

The disease is a conjunctivitis that eventually leads to permanent scarring. Blindness is a result of long-term mechanical abrasion of the cornea by these scars and by turned-in eyelashes (Figure 21.17). Secondary infections by other bacterial pathogens are also a factor. Antibiotic ointments, especially tetracycline, are useful in treatment. Partial immunity is generated by recovery. The disease can be controlled through sanitary practices and health education.

Other Infectious Diseases of the Eye

Microorganisms such as viruses or protozoa can also cause eye diseases. The diseases discussed here are characterized by inflammation of the cornea, which is called *keratitis*.

Herpetic Keratitis

Herpetic keratitis is usually caused by herpes simplex type 1 virus. This is a localized cornea infection that can recur with an epidemiology similar to that of cold sores. It is characterized by inflammation and corneal ulcers that can be quite deep. Herpetic keratitis is one of the few viral diseases for which there is an effective chemotherapeutic treatment; trifluridine is the drug of choice.

Acanthamoeba Keratitis

The first case of *Acanthamoeba* (a-kan-thä-mē′bä) **keratitis** was reported in 1973 in a Texas rancher. Since then, well over 100 cases have been diagnosed. This amoeba has been found in fresh water, hot tubs, soil, and even homemade contact lens saline solutions. Most recent cases have been associated with the wearing of contact lenses. Contributing factors are faulty disinfecting procedures (only heat will reliably kill the cysts), homemade saline solutions, and wearing the lenses while swimming. The infection in its early stages is only a mild inflammation, but later stages are often accompanied by severe pain. Damage is often so severe as to require a corneal transplant, or even removal of the eye. Diagnosis is confirmed by the presence of trophozoites (Figure 21.18) and cysts in stained scrapings of the cornea.

STUDY OUTLINE

INTRODUCTION (p. 516)

1. The skin is a physical and chemical barrier against microorganisms.
2. Moist areas of the skin (such as the axilla) support larger populations of bacteria than dry areas (such as the scalp).

STRUCTURE AND FUNCTION OF THE SKIN (pp. 516–517)

1. The outer portion of the skin, called the epidermis, contains keratin, a waterproof coating.
2. The inner portion of the skin, the dermis, contains hair follicles, sweat ducts, and oil glands that provide passageways for microorganisms.
3. Sebum and perspiration are secretions of the skin that can inhibit growth of microorganisms.
4. Sebum and perspiration provide nutrients for some microorganisms.
5. Body cavities are lined with epithelial cells. When these cells secrete mucus, they constitute the mucous membrane.

NORMAL MICROBIOTA OF THE SKIN (pp. 517–518)

1. Microorganisms that live on skin are resistant to desiccation and high concentrations of salt.
2. Gram-positive cocci predominate on the skin.
3. The normal skin microbiota is not completely removed by washing.
4. Members of the genus *Propionibacterium* metabolize oil from the oil glands and colonize hair follicles.

MICROBIAL DISEASES OF THE SKIN (pp. 518–531)

1. Vesicles are small fluid-filled lesions; bullae are vesicles larger than 1 cm; macules are flat, reddened lesions; papules are raised lesions; and pustules are raised lesions containing pus.

Bacterial Diseases of the Skin (pp. 518–523)

Staphylococcal Skin Infections (pp. 519–521)

1. Staphylococci are gram-positive bacteria that often grow in clusters.
2. The majority of skin microbiota consists of coagulase-negative *S. epidermidis*.
3. Almost all pathogenic strains of *S. aureus* produce coagulase.
4. Pathogenic *S. aureus* can produce enterotoxins, leukocidins, and exfoliative toxin.
5. Localized infections (sties, pimples, and carbuncles) result from *S. aureus* entering openings in the skin.
6. Many strains of *S. aureus* produce penicillinase.
7. Impetigo of the newborn is a highly contagious superficial skin infection caused by *S. aureus*.
8. Toxemia occurs when toxins enter the bloodstream; staphylococcal toxemias include scalded skin syndrome and toxic shock syndrome.

Streptococcal Skin Infections (pp. 521–522)

1. Streptococci are gram-positive cocci that often reproduce in chains.
2. Streptococci are classified according to their hemolytic enzymes and cell-wall antigens.
3. Group A β-hemolytic streptococci (including *S. pyogenes*) are the pathogens most important to humans.
4. Group A β-streptococci produce a number of virulence factors: M protein, erythrogenic toxin, deoxyribonuclease, NADase, streptokinases, and hyaluronidase.
5. Streptococci are susceptible to penicillin.
6. Erysipelas (reddish patches) and impetigo (isolated pustules) are skin infections caused by *S. pyogenes*.

Infections by Pseudomonads (p. 522)

1. Pseudomonads are gram-negative rods. They are aerobes found primarily in soil and water. They are resistant to many disinfectants and antibiotics.
2. *Pseudomonas aeruginosa* is the most prominent species.
3. *P. aeruginosa* produces an endotoxin and several exotoxins.

4. Diseases caused by *P. aeruginosa* include otitis externa, respiratory infections, burn infections, and dermatitis.
5. Infections have a characteristic blue-green pus caused by the pigment pyocyanin.
6. Carbenicillin, gentamicin, and silver sulfadiazine are useful in treating *P. aeruginosa* infections.

Acne (pp. 522–523)

1. *Propionibacterium acnes* can metabolize sebum trapped in hair follicles.
2. Metabolic end products (fatty acids) cause an inflammatory response known as acne.
3. Tretinoin, benzoyl peroxide, tetracycline, and Accutane® are used to treat acne.

Viral Diseases of the Skin (pp. 523–527)

Warts (p. 523)

1. Papillomaviruses cause skin cells to proliferate and produce a benign growth called a wart or papilloma.
2. Warts are spread by direct contact.
3. Warts may regress spontaneously or be removed chemically or physically.

Smallpox (Variola) (pp. 523–524)

1. Variola virus causes two types of skin infections, variola major and variola minor.
2. Smallpox is transmitted by the respiratory route, and the virus is moved to the skin via the bloodstream.
3. The only host for smallpox is humans.
4. Smallpox has been eradicated as a result of a vaccination effort by the WHO.

Chickenpox (Varicella) and Shingles (Herpes Zoster) (pp. 524–525)

1. Chickenpox (varicella) is caused by varicella-zoster virus.
2. Varicella-zoster virus is transmitted by the respiratory route and is localized in skin cells, causing a vesicular rash.
3. Complications of chickenpox include encephalitis and Reye's syndrome.
4. After chickenpox, the virus can remain latent in nerve cells and subsequently activate as shingles.
5. Shingles (herpes-zoster) is characterized by a vesicular rash along the affected cutaneous sensory nerves.
6. The virus can be treated with acyclovir. An attenuated live vaccine should soon be approved for use.

Herpes Simplex (pp. 525–526)

1. Herpes simplex infection of mucosal cells results in cold sores and occasionally encephalitis.
2. The virus remains latent in nerve cells, and cold sores can recur when the virus is activated.
3. Herpes simplex type 1 virus is transmitted primarily by oral and respiratory routes.
4. Herpes encephalitis occurs when herpes simplex viruses infect the brain.

5. Acyclovir has proven successful in treating herpes encephalitis.

Measles (Rubeola) (p. 526)

1. Measles is caused by measles virus and transmitted by the respiratory route.
2. Vaccination provides effective long-term immunity.
3. After the virus has incubated in the upper respiratory tract, macular lesions appear on the skin, and Koplik spots appear on the oral mucosa.
4. Complications of measles include middle ear infections, pneumonia, encephalitis, and secondary bacterial infections.

Rubella (pp. 526–527)

1. The rubella virus is transmitted by the respiratory route.
2. A red rash and light fever might occur in an infected individual; the disease can be asymptomatic.
3. Congenital rubella syndrome can affect a fetus when a woman contracts rubella during the first trimester of her pregnancy.
4. Damage from congenital rubella syndrome includes stillbirth, deafness, eye cataracts, heart defects, and mental retardation.
5. Vaccination with live rubella virus provides immunity of unknown duration.

Fungal Diseases of the Skin (pp. 527–530)

Cutaneous Mycoses (pp. 527–529)

1. Fungi that colonize the outer layer of the epidermis cause dermatomycoses.
2. *Microsporum*, *Trichophyton*, and *Epidermophyton* cause dermatomycoses called ringworm, or tinea.
3. These fungi grow on keratin-containing epidermis, such as hair, skin, and nails.
4. Ringworm and athlete's foot are usually treated with topical application of antifungal chemicals.
5. Diagnosis is based on microscopic examination of skin scrapings or fungal culture.

Subcutaneous Mycoses (p. 529)

1. Sporotrichosis results from a soil fungus that penetrates the skin through a wound.
2. The fungi grow and produce subcutaneous nodules along the lymphatic vessels.

Candidiasis (pp. 529–530)

1. *Candida albicans* causes infections of mucous membranes and is a common cause of thrush (in oral mucosa) and vaginitis.
2. *C. albicans* is an opportunistic pathogen that may proliferate when the normal bacterial microbiota is suppressed.
3. Topical antifungal chemicals may be used to treat candidiasis.

Parasitic Infestation of the Skin (pp. 530–531)

1. Scabies is caused by a mite burrowing and laying eggs in the skin.
2. Topical application of gamma benzene hexachloride is used to treat scabies.

MICROBIAL DISEASES OF THE EYE (pp. 531–533)

1. The mucous membrane lining the eyelid and covering the eyeball is the conjunctiva.

Inflammation of the Eye Membranes:
Conjunctivitis (p. 531)

1. Conjunctivitis is caused by a number of bacteria.
2. Conjunctivitis can be transmitted by improperly disinfected contact lenses.

Bacterial Diseases of the Eye (pp. 531–532)

1. *Staphylococcus epidermidis*, *S. aureus*, and diphtheroids are members of the normal microbiota of the eye.
2. These microorganisms usually originate from the skin and upper respiratory tract.

3. Neonatal gonorrheal ophthalmia is caused by the transmission of *Neisseria gonorrhoeae* from an infected mother to an infant during its passage through the birth canal.
4. At one time, all newborn infants were treated with 1% silver nitrate or an antibiotic to prevent the growth of *Neisseria*. Antibiotics that also prevent *Chlamydia* infection are now used.
5. Inclusion conjunctivitis is an infection of the conjunctiva caused by *Chlamydia trachomatis*. It is transmitted to infants during birth and is transmitted in unchlorinated swimming water.
6. In trachoma, which is caused by *C. trachomatis*, scar tissue forms on the cornea.
7. Trachoma is transmitted by hands, fomites, and perhaps flies.

Other Infectious Diseases of the Eye (pp. 532–533)

1. Inflammation of the cornea is called keratitis.
2. Herpetic keratitis causes corneal ulcers. The etiology is herpes simplex virus type 1 that invades the central nervous system and can recur.
3. Trifluridine is an effective treatment for herpes keratitis.
4. *Acanthamoeba*, transmitted via water, can cause a serious form of keratitis.

STUDY QUESTIONS

Review

1. Discuss the usual mode of entry of bacteria into the skin. Compare bacterial skin infections with those caused by fungi and viruses with respect to method of entry.

2. What bacteria are identified by a positive coagulase test? What bacteria are characterized as group A β-hemolytic?

3. Compare and contrast impetigo and erysipelas.

4. Complete the table of epidemiology below.

DISEASE	ETIOLOGIC AGENT	CLINICAL SYMPTOMS	METHOD OF TRANSMISSION
Acne			
Pimples			
Warts			
Chickenpox			
Fever blisters			
Measles			
Rubella			

5. How do sporotrichosis and athlete's foot differ? In what ways are they similar? How is each disease treated?

6. (a) Differentiate between conjunctivitis and keratitis.
   (b) Select a bacterial and viral eye infection, and discuss the epidemiology of each.

7. Describe the symptoms, etiologic agent, and treatment of candidiasis. How is candidiasis contracted?

8. Why does the blood test required for a marriage license include a test for antibodies against rubella for women?

9. Identify the following diseases based on the symptoms in the following chart.

SYMPTOMS	DISEASE
Koplik spots	
Macular rash	
Vesicular rash	
Small, spotted rash	
Recurrent "blisters" on oral mucosa	
Corneal ulcer and swelling of lymph nodes	

10. What complications can occur from herpes simplex virus type 1 infections?

11. What is in the MMR vaccine? Why are children not vaccinated against smallpox?

12. Explain the relationship between shingles and chicken-pox.

13. Why are the eyes of all newborn infants washed with an antiseptic or antibiotic?

14. What is the leading infectious cause of blindness in the world?

15. What is scabies? How is it treated?

Critical Thinking

1. A laboratory test used to determine the identity of *Staphylococcus aureus* is its growth on mannitol salt agar. The medium contains 7.5% sodium chloride (NaCl). Why is it considered a selective medium for *S. aureus?*

2. Is it necessary to treat a patient for warts? Explain briefly.

3. Analyses of nine conjunctivitis cases provided the data in the following chart. How were these infections transmitted? How could they be prevented?

NO.	ETIOLOGY	ISOLATED FROM EYE COSMETICS OR CONTACT LENSES
5	*S. epidermidis*	+
1	*Acanthamoeba*	+
1	*Candida*	+
1	*P. aeruginosa*	+
1	*S. aureus*	+

Clinical Applications

1. A hospitalized patient recovering from surgery develops an infection that has blue-green pus and a grapelike odor. What is the probable etiology? How might the patient have acquired this infection?

2. A 12-year-old diabetic girl using continuous subcutaneous insulin infusion to manage her diabetes developed a fever (39.4°C), low blood pressure, abdominal pain, and erythroderma. She was supposed to change the needle insertion site every three days after cleaning the skin with an iodine solution. Frequently she did not change the insertion site more often than every ten days. Blood culture was negative, and abscesses at insertion sites were not cultured. What is the probable cause of her symptoms?

3. A teenage male with confirmed influenza was hospitalized when he developed respiratory distress. He had a fever, rash, and low blood pressure. *S. aureus* was isolated from his respiratory secretions. Discuss the relationship between his symptoms and the etiological agent.

Microbial Diseases of the Nervous System

S OME OF THE MOST devastating infectious diseases affect the nervous system—especially the brain and spinal cord. Because of the crucial importance of these control systems, they are strongly protected from accident and infection by bone and other structures. Pathogens capable of causing diseases of the nervous system often have virulence characteristics of a special nature to penetrate these defenses.

STRUCTURE AND FUNCTION OF THE NERVOUS SYSTEM

LEARNING OBJECTIVES

- *Define the following terms: central nervous system and blood–brain barrier.*
- *Differentiate between meningitis and encephalitis.*

The human nervous system is organized into two divisions: the central nervous system and the peripheral nervous system (Figure 22.1). The **central nervous system** consists of the brain and the spinal cord. As the control center for the entire body, it picks up sensory information from the environment, interprets the information, and sends impulses that coordinate the body's activities. The **peripheral nervous system** consists of all the nerves that branch off from the brain and the spinal cord. These nerves are the lines of communication between the cen-

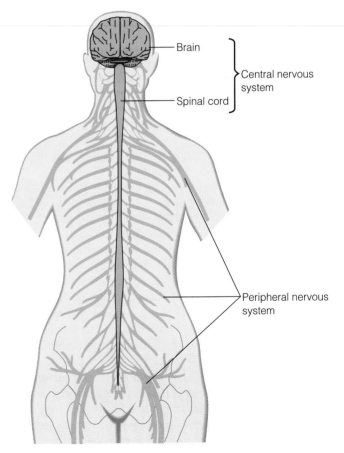

FIGURE 22.1 Organization of the human nervous system. This frontal view shows the central and peripheral nervous system divisions.

FIGURE 22.2 The meninges and cerebrospinal fluid. The meninges, whether cranial or spinal, consist of three layers: dura mater, arachnoid, and pia mater. Between the arachnoid and the pia mater is the subarachnoid space, in which cerebrospinal fluid circulates. Note how the cerebrospinal fluid is vulnerable to contamination by microbes carried in the blood that are able to penetrate the blood–brain barrier at the walls of the blood vessels.

tral nervous system, the various parts of the body, and the external environment.

Both the brain and the spinal cord are covered and protected by three continuous membranes called *meninges* (Figure 22.2). These are the outermost dura mater, the middle arachnoid, and the innermost pia mater. Between the pia mater and arachnoid membranes is a space called the *subarachnoid space,* in which an adult has 100 to 160 ml of *cerebrospinal fluid* circulating. Because cerebrospinal fluid has low levels of complement and circulating antibodies and few phagocytic cells, bacteria can multiply in it with few checks.

An important feature of the brain is the **blood–brain barrier.** Certain capillaries permit some substances to pass from the blood into the brain but restrict others. These capillaries are less permeable than others within the body and are therefore more selective in passing materials.

Drugs cannot cross the blood–brain barrier unless they are lipid-soluble. (Glucose and many amino acids are not lipid-soluble, but they can cross the barrier because special

transport systems exist for them.) The lipid-soluble antibiotic chloramphenicol enters the brain readily. Penicillin is only slightly lipid-soluble, but if it is given in very large doses, enough may cross the barrier to be effective. Inflammations of the brain tend to alter the blood–brain barrier in such a way as to allow antibiotics to cross that would not be able to cross if there were no infection.

Even though the central nervous system has considerable protection, it can still be invaded by microorganisms in several ways. For example, microorganisms can gain access through trauma, such as a skull or backbone fracture, or through a medical procedure such as a spinal tap, in which a needle is inserted into the subarachnoid space of the spinal meninges to obtain a sample of fluid for diagnosis. Some microorganisms can also move along peripheral nerves. But probably the most common routes of central nervous system invasion are the bloodstream and lymphatic system (see Chapter 23), when inflammation alters permeability of the blood–brain barrier.

An infection of the meninges is called **meningitis.** An

infection of the brain itself is called **encephalitis.** The following sections cover bacterial, viral, fungal, and protozoan infections of the nervous system, as well as those currently thought to be caused by prions (see Chapter 13).

BACTERIAL DISEASES OF THE NERVOUS SYSTEM

LEARNING OBJECTIVES

- *Discuss the epidemiology of meningitis caused by* Hemophilus influenzae, Streptococcus pneumoniae, Neisseria meningitidis, *and* Listeria monocytogenes.
- *Discuss the epidemiology of tetanus and leprosy, including method of transmission, etiology, disease symptoms, and preventive measures.*
- *Provide the etiologic agent, symptoms, suspect foods, and treatment for botulism.*

Microbial infections of the central nervous system are infrequent but often have serious consequences.

Bacterial Meningitis

Most patients suffering from any type of meningitis complain of a headache and have symptoms of nausea and vomiting. Convulsions and coma accompany the infection in many cases. The mortality rate varies with the pathogen but is generally high for an infectious disease today. Many people who survive an attack suffer some sort of neurological damage.

Meningitis can be caused by different types of pathogens, including viruses, bacteria, fungi, and protozoa. Viral meningitis is probably more common than bacterial meningitis but tends to be a mild disease. In this section we will discuss bacterial meningitis. Table 22.1 presents data about important types of bacterial meningitis.

There are three major types of bacterial meningitis: *Hemophilus influenzae* meningitis, pneumococcal meningitis caused by *Streptococcus pneumoniae*, and meningococcal meningitis caused by *Neisseria meningitidis*. These three types make up the great majority of meningitis cases. *H. influenzae* is the most common cause of bacterial meningitis; it produces an estimated 20,000 cases each year, mostly in children (Figure 22.3). However, only meningococcal meningitis is a notifiable disease (see Chapter 14).

Nearly 50 other species of bacteria have been reported to be opportunistic pathogens that occasionally cause meningitis. These include such common organisms as *Escherichia coli*, *Pseudomonas aeruginosa*, and *Klebsiella pneumoniae*. Although the number of cases of meningitis caused by these gram-negative organisms is low, the fatality rate is in excess of 30%.

TABLE 22.1 Types of Bacterial Meningitis

CAUSATIVE AGENT	PERCENTAGE OF REPORTED MENINGITIS CASES	COMMENTS
Hemophilus influenzae	45%	Fatality rate of about 6%
Streptococcus pneumoniae	18%	Fatality rate of about 26%
Neisseria meningitidis	14%	Can occur as epidemics; fatality rate of about 10%
Listeria monocytogenes	Uncommon	Causes spontaneous abortions and fetal damage; meningitis in immuno-suppressed individuals
Mycobacterium tuberculosis	Uncommon	Complication of tuberculosis; fatality rate of about 20%
Leptospira interrogans	Uncommon	Common symptom of leptospirosis
Nocardia asteroides	Uncommon	Common complication of pulmonary infections by *Nocardia*

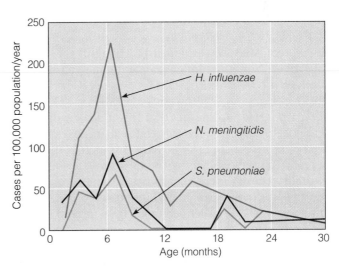

FIGURE 22.3 Yearly cases of the most common types of bacterial meningitis in children. The graph shows the incidence by age of bacterial meningitis caused by *Hemophilus influenzae, Neisseria meningitidis,* and *Streptococcus pneumoniae*. (*Source: J. Infect. Dis.* 127:271, 1973.)

Hemophilus influenzae Meningitis

Hemophilus influenzae is a nonmotile, aerobic, gram-negative, pleomorphic bacterium (Figure 22.4). It is a member of the normal throat microbiota, in addition to causing meningitis. The bacterium, which is encapsulated, is further divided into six types on the basis of its antigenic capsular carbohydrates. One type has special importance to medical microbiology, type b. The virulence of *H. influenzae* type b is related to its capsule.

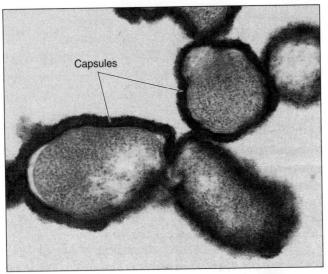

Capsules

TEM ⊢————⊣ 100 μm

FIGURE 22.4 *Hemophilus influenzae* **type b, the cause of** *H. influenzae* **meningitis.** Note the thick layer of capsular material that makes the bacterium resistant to phagocytosis.

The name *Hemophilus influenzae* was given because the microorganism was erroneously thought to be the causative agent of the influenza pandemics of 1890 and World War I. *H. influenzae* was probably only a secondary invader during those virus-caused pandemics. The term *Hemophilus* refers to the fact that the microorganism requires factors in blood for growth (*hemo* means blood).

H. influenzae meningitis accounts for most of the cases of reported bacterial meningitis (45%) but has the lowest mortality rate of the common types of bacterial meningitis (6%). It occurs mostly in children under 4 years of age and is especially prevalent at about age 6 months. Residual damage to the nervous system occurs in many cases. The antibodies acquired from the mother provide protection for about 2 months. The prevalence of the carrier state in the throats of children is quite high, and transmission in day-care centers is now an increasing problem.

Invasion of the meninges by *H. influenzae* is usually preceded by a viral infection of the respiratory tract, which opens the way for the bacteria to enter the bloodstream and be transported to the meninges. Conjugated vaccines directed at the antigenic polysaccharide capsule have been developed. Current recommendations are to begin the vaccination series at 2 months. In most cases, good immunity is achieved at 6 or 7 months.

Streptococcus pneumoniae Meningitis (Pneumococcal Meningitis)

Streptococcus pneumoniae, like *H. influenzae*, is a common inhabitant of the nasopharyngeal region. The capsule of *S. pneumoniae* makes it very resistant to phagocytosis and is the most important element in its pathogenicity. About

half of the cases of pneumococcal meningitis occur among children ages 1 month to 4 years. However, the hospitalized elderly are another susceptible group. For this group, some protection may be available from the vaccine for pneumococcal pneumonia, which is caused by the same organism.

Although pneumococcal meningitis accounts for only about 18% of the cases of reported bacterial meningitis, the mortality rate of about 26% is very high.

Neisseria Meningitis (Meningococcal Meningitis)

Meningococcal meningitis is caused by the bacterium *Neisseria meningitidis*. The neisseriae are aerobic, nonmotile, gram-negative cocci.

N. meningitidis can inhabit the portion of the throat behind the nose (nasopharynx) without causing any symptoms. This carrier state can last from several days to several months and provides the reservoir for the organism. If an individual without adequate immunity acquires the microorganism, usually through contact with a healthy carrier, the resulting throat infection can lead to bacteremia, which is followed by meningitis.

The bacteria observed in extracellular or cerebrospinal fluid are found in leukocytes that have entered the fluid in response to the infection. Most of the symptoms associated with meningitis are thought to be caused by an endotoxin produced by the bacteria.

Meningococcal meningitis usually strikes children younger than 2 years of age, with the highest incidence occurring during the first year of life. Children are usually born with maternal immunity and become susceptible as the immunity weakens, at about 6 months of age. The death rate for meningococcal meningitis has been about 10% in recent years. A vaccine against this type of meningitis is used routinely by the U.S.A.'s military, but it is not effective in very young children.

Diagnosis and Treatment of the Most Common Types of Bacterial Meningitis

Prompt treatment of any bacterial meningitis is essential, so chemotherapy of suspected cases is usually initiated before identification of the pathogen is complete. The broad spectrum third-generation cephalosporins are usually the first choice of antibiotics. As soon as identification is confirmed, or perhaps when the antibiotic sensitivity has been determined from cultures, the antibiotic treatment may be changed. If *H. influenzae* meningitis outbreaks occur in a day-care center, other children in the center can be given antibiotics as a preventive measure. This also has the effect of lowering the carrier rate in the group.

Diagnosis of bacterial meningitis requires a sample of cerebrospinal fluid obtained by a spinal tap. A simple Gram stain is often useful; it will frequently determine the identity of the pathogen with considerable reliability. Cultures are also made from the fluids. For this purpose, prompt and careful handling is required because many of

Listeria monocytogenes

Pseudopod

TEM

FIGURE 22.5 Cell-to-cell spread of *Listeria monocytogenes*, the cause of listeriosis. Note that the bacterium has caused the macrophage on the right, in which it resided, to form a pseudopod that is now engulfed by the macrophage on the left. The pseudopod will soon be pinched off and the microbe transferred to the macrophage on the left.

the likely pathogens are very sensitive and will not survive much storage time or even changes in temperature. The cultures are usually inoculated into an enrichment medium that helps protect them from oxygen, and they are then streaked onto media such as blood agar and incubated in an atmosphere of 5 to 10% CO_2. The most frequently used type of serological tests performed on cerebrospinal fluid are the latex agglutination tests.

Listeriosis

Listeria monocytogenes is a gram-positive rod known to cause stillbirths and neurological disease in animals long before it was recognized as causing human disease. It is excreted in animal feces and hence is widely distributed in soil and water. The name is derived from the proliferation of monocytes (a type of leukocyte) found in some animals infected by it. The disease **listeriosis** usually is a mild, often symptomless disease in healthy adult humans, but recovering or apparently healthy individuals often shed the pathogen indefinitely in their feces. *L. monocytogenes* is ingested by phagocytic cells but is not destroyed; it even proliferates within them. It also has the unusual capability of moving directly from one macrophage to an adjacent one (Figure 22.5).

As a pathogen, *L. monocytogenes* has two principal characteristics—it mainly affects adults who are immunosuppressed, pregnant, or have cancer, and it has a special affinity for growth in the central nervous system and the placenta supplying the fetus with nutrients. Growth in the central nervous system is usually expressed as meningitis. When it infects a pregnant woman, the growth on the placenta leads to a high rate of spontaneous abortion or still-

birth. A surviving newborn may be acutely ill with septicemia and meningitis. The infant mortality related to this type of infection is about 60%.

In human outbreaks, the organism is mostly foodborne. It is frequently isolated from a wide variety of foods; dairy products have been involved in several outbreaks. *L. monocytogenes* is one of the few pathogens that is capable of growth at refrigerator temperatures, which can lead to an increase in its numbers during a food's shelf life.

Considerable effort is being made to develop methods by which *L. monocytogenes* can be better detected in foods. Considerable improvement has been made with selective growth media and rapid biochemical tests. However, eventually DNA probes (see Chapter 10) and serological tests using monoclonal antibodies are expected to be the most satisfactory. Diagnosis in humans depends on isolation and culturing of the organism, usually from blood or cerebrospinal fluid. Penicillin and ampicillin are the antibiotics of choice for treatment.

Tetanus

The causative agent of **tetanus**, *Clostridium tetani*, is an obligately anaerobic, endospore-forming, gram-positive rod. It is especially common in soil contaminated with animal fecal wastes.

The symptoms of tetanus are caused by an extremely potent neurotoxin (*tetanospasmin*) that is released upon death and lysis of the growing bacteria (see Chapter 15). It enters the central nervous system via peripheral nerves or the blood. An amount of tetanospasmin weighing as much as the ink in one period on this page could kill 30 people. The bacteria themselves do not spread from the infection site, and there is no inflammation.

In a muscle's normal operation, a nerve impulse initiates the contraction of the muscle. At the same time, an opposing muscle receives a signal to relax so as not to oppose the contraction. The tetanus neurotoxin blocks the relaxation pathway so that both sets of muscles contract, resulting in the characteristic muscle spasms called *spastic paralysis* (Figure 22.6). The muscles of the jaw are affected early in the disease, which prevents opening of the mouth (lockjaw). Gradually, other skeletal muscles become affected, including those involved in swallowing. Death results from spasms of the respiratory muscles.

Because the organism is an obligate anaerobe, the wound through which it enters the body must provide anaerobic growth conditions. Improperly cleaned deep puncture wounds, especially those with little or no bleeding, serve very well. Of the approximately 100 tetanus cases that have occurred each year in the U.S.A. in recent years (about 21% of which were fatal), many arise from trivial but fairly deep injuries that are thought to be too minor to bring to a physician.

Most people in the U.S.A. have received the DPT (diphtheria, pertussis, tetanus) immunization, which includes

FIGURE 22.6 An advanced case of tetanus. Drawing of a British soldier during the Napoleanic wars. Called opisthotonos, these spasms can actually result in a fractured spine. (Drawing by Charles Bell of the Royal College of Surgeons, Edinburgh.)

the tetanus toxoid that stimulates the formation of antibodies to fight tetanus toxin. Immunization is nearly 100% effective, but recovery from the clinical disease does not confer immunity because the amount of toxin usually produced is too small to be immunogenic.

When a wound is severe enough to be brought to the attention of a physician, the decision is usually made to provide protection against tetanus, and a booster of toxoid may be given. Table 22.2 summarizes the questions involved in this decision. A patient who has had the full sequence of three doses of toxoid and a booster every ten years should be immune. However, people often do not get boosters on schedule, and serological surveys show that at least 50% of the U.S.A.'s population does not have adequate protection. Almost 70% of tetanus cases in the U.S.A. occur in individuals over 50 years of age.

If toxoid has been given before, the secondary immune response (see Figure 17.10) to a booster given after the injury will be rapid enough to be protective. Tetanus immune globulin (TIG), which is prepared from the serum of immunized humans, is often given simultaneously to provide temporary immunity. It is injected at a different site, however, to avoid neutralizing the toxoid. To minimize production of more toxin, damaged tissue should be removed (debridement) and antibiotics should be administered. However, once the toxin has attached to the nerves, such therapy is of little use.

In less developed areas of the world, tetanus arising from the severed umbilical cords of infants is a major cause of death. In some cultures, the practice of dressing the cut umbilical cord with such materials as soil, clay, and cow dung is a major contributor to the development of tetanus. Worldwide, there are probably several hundred thousand cases of tetanus from all causes each year.

Botulism

Botulism, a form of food poisoning, is caused by *Clostridium botulinum,* an obligately anaerobic, endospore-forming,

gram-positive rod that is found in soil and in many freshwater sediments. Ingesting the endospores usually does no harm, as will be explained shortly. However, in anaerobic environments, such as sealed cans, the microorganism produces an exotoxin that animal assays show to be the most potent of all natural toxins. This neurotoxin is highly specific for the synaptic end of the nerve, where it blocks the release of acetylcholine, a chemical necessary for transmission of nerve impulses across synapses. Individuals suffering from botulism undergo a progressive *flaccid paralysis* for one to ten days and may die from respiratory and cardiac failure. Nausea, but no fever, may precede the neurological symptoms. The initial neurological symptoms vary, but nearly all sufferers have double or blurred vision. Other symptoms include difficulty in swallowing and general weakness. Incubation time varies, but symptoms typically appear within a day or two. As with tetanus, recovery from the disease does not confer immunity because the toxin is usually not present in amounts large enough to be immunogenic. A vaccine is available for exposed laboratory workers and is used by veterinarians to protect valuable horse-breeding stock in some states.

Botulism was first described as a clinical disease in the early 1800s, when it was known as the sausage disease (*botulus* is the Latin word for sausage). Blood sausage was made by filling a pig stomach with blood and ground meats, tying shut all the openings, boiling it for a short time, and smoking it over a wood fire. The sausage was then stored at room temperature. This attempt at food preservation included most of the requirements for an outbreak of botulism; it killed competing bacteria but allowed the more heat-stable *C. botulinum* endospores to survive, and it provided anaerobic conditions and an incubation period for toxin production.

Most botulism results from attempts at food preservation that fail to eliminate the *C. botulinum* endospores. The botulinal toxin is very heat labile and will be destroyed by most ordinary cooking methods that bring the food to a boil. Sausage and bacon rarely cause botulism

TABLE 22.2 Wound Management for Prevention of Tetanus

| HISTORY OF TETANUS IMMUNIZATION | MINOR WOUNDS | | EXTENSIVE WOUNDS | |
	ADMINISTER TOXOID	ADMINISTER TETANUS IMMUNE GLOBULIN	ADMINISTER TOXOID	ADMINISTER TETANUS IMMUNE GLOBULIN
Unknown or 0 to 2 doses	Yes	No	Yes	Yes
3 or more doses	No*	No	No†	No

*Yes, if more than 10 years since last dose.
†Yes, if more than 5 years since last dose.

today, largely because of the addition of nitrites. Nitrites prevent *C. botulinum* growth following germination of the endospores.

Botulinal toxin is not formed in acidic foods (below pH 4.7). Such foods as tomatoes therefore can be safely preserved without the use of a pressure cooker. There have been cases of botulism from acidic foods that normally would not have supported the growth of the botulism organisms. Most of these episodes are related to mold growth, which metabolized enough acid to allow the initiation of growth of the botulism organisms.

There are several serological types of the botulinal toxin produced by different strains of the pathogen. These differ considerably in their virulence and other factors.

Type A toxin is probably the most virulent. There have been deaths from type A toxin when the food was only tasted but not swallowed. It is even possible to absorb lethal doses through skin breaks while handling laboratory samples. In untreated cases, there is a 60% to 70% mortality rate. The type A endospore is the most heat resistant of all *C. botulinum* strains. In the U.S.A., it is found mainly in California, Washington, Colorado, Oregon, and New Mexico. Some eastern states have never had a type A outbreak. The type A organism is usually proteolytic (the breakdown of proteins by clostridia releases amines with unpleasant odors), but obvious spoilage odor is not always apparent in low-protein foods, such as corn or beans.

Type B toxin is responsible for most European outbreaks of botulism and is the most common type in the eastern U.S.A. The mortality rate in cases without treatment is perhaps 25%. Type B botulism organisms occur in both proteolytic and nonproteolytic strains. The endospores of the proteolytic strains are more heat resistant and more likely to survive in food.

Type E toxin is produced by botulism organisms that are often found in marine or lake sediments. Therefore, outbreaks often involve seafood and are especially common in the Pacific Northwest, Alaska, and the Great Lakes area of the U.S.A. The endospore of type E botulism is less heat resistant than that of other strains and is usually destroyed by boiling. Types A and B will survive much higher temperatures. Type E is nonproteolytic, so the

chance of detecting spoilage in high-protein foods such as fish is minimal. It is also capable of producing toxin at refrigerator temperatures and requires less strictly anaerobic conditions for growth.

Botulism is not a common disease. Only a few cases are reported each year, but outbreaks from restaurants occasionally involve 20 or 30 cases. Native Americans in Alaska probably have the highest rate of botulism in the world, mostly of type E. The problem arises from food preparation methods that reflect a cultural need to avoid use of scarce fuels for heating or cooking. One of the more common foods involved in Alaskan outbreaks of botulism is *muktuk*. This is prepared by slicing the flippers of seals or whales into strips and then drying them for a few days. To tenderize them, they are stored anaerobically in a container of seal oil for several weeks until they approach putrefaction. The 40% mortality rate for type E botulism observed in recent years among Alaskan natives reflects the difficulty in getting prompt treatment for isolated ethnic groups.

Botulism organisms do not seem to be able to compete successfully with the normal intestinal microbiota, so production of toxin by ingested bacteria almost never causes botulism in adults. However, the intestinal microbiota of infants is not well established, and they may suffer from **infant botulism.** This disease was first reported in 1976; an estimated 250 cases occur in the U.S.A. every year, many more than any other form of botulism. Infants have many opportunities to ingest soil and other materials contaminated with the endospores of the organism, but 30% of reported cases have been associated with honey. Endospores of *C. botulinum* are recovered with some frequency from honey, and a lethal dose may be as few as 2000 bacteria. The recommendation is not to feed honey to infants under one year of age; there is no problem with older children or adults who have normal intestinal microbiota.

The treatment of botulism relies heavily on supportive care. Antibiotics are of almost no use because the toxin is preformed. Antitoxins aimed at the neutralization of A, B, and E toxins are available and are usually administered together. This trivalent antitoxin will not affect the toxin al-

FIGURE 22.7 Diagnosis of botulism by identification of botulinal toxin type. To determine whether botulinum toxin is present, mice are injected with the liquid portion of food extracts or cell-free cultures. If the mice die within 72 hours, this is evidence of the presence of toxin. To determine the specific type of the toxin, groups of mice are passively immunized with antisera specific for *C. botulinum* types A, B, and E. If one group of mice receiving a specific antitoxin lives and the other mice die, the type of toxin in the food or culture has been identified.

ready attached to the nerve ends and is probably more effective on type E than on types A and B. The toxin is quite firmly bound. Recovery requires that the nerve endings regenerate; it therefore proceeds slowly. Extended respiratory assistance may be needed, and some neurological impairment may persist for months.

Botulism is diagnosed by the inoculation of mice with samples from patient serum, stool, or vomitus specimens (Figure 22.7). Different sets of mice are immunized with type A, B, or E antitoxin. All the mice are then inoculated with the test toxin; if, for example, those protected with type A antitoxin are the only survivors, then the toxin is type A. The toxin in food can similarly be identified by mouse inoculation.

The botulism pathogen can also grow in wounds in a manner similar to that of clostridia, causing tetanus or gas gangrene (see Chapter 23). Such episodes of **wound botulism** occur occasionally.

Leprosy

Mycobacterium leprae is probably the only bacterium that grows in the peripheral nervous system, although it can also grow in skin cells. It is an acid-fast rod closely related to the tuberculosis pathogen, *Mycobacterium tuberculosis*. The organism was first isolated and identified about 1870 by Gerhard A. Hansen of Norway; his discovery was one

of the first links ever made between a specific bacterium and a disease. **Leprosy** is sometimes called **Hansen's disease** in an effort to avoid using the dreaded name *leprosy*.

The organism has an optimum growth temperature of 30°C and shows a preference for the outer, cooler portions of the human body. A very long generation time of about 12 days has been estimated. The microorganism has never been grown on artificial media, but in 1969 armadillos, which have a body temperature of only 30° to 35°C, were experimentally infected with *M. leprae*. We now know that armadillos can contract a leprosylike disease in the wild, although they are not considered a source of human infection. Armadillos are used in studies of the disease itself and in evaluations of the effectiveness of chemotherapeutic agents.

Leprosy occurs in two main forms (although borderline forms are also recognized) that apparently reflect the effectiveness of the host's cell-mediated immune system. The *tuberculoid (neural) form* is characterized by regions of the skin that have lost sensation and are surrounded by a border of nodules (Figure 22.8a). This form of the disease occurs in people with effective immune reactions. Recovery sometimes occurs spontaneously. The disease can be diagnosed by detecting acid-fast bacilli in the fluids from a slit cut in a cool site, such as an earlobe. The **lepromin test** uses an extract of lepromatous tissue injected into the skin. A visible skin reaction that develops at the injection site indicates that the body has developed an immune response to the leprosy bacillus. This test is negative during the later lepromatous stage of the disease.

In the *lepromatous (progressive) form* of leprosy, skin cells are infected and disfiguring nodules form all over the body. Patients with this type of leprosy have the least effective cell-mediated immune response, and the disease has progressed from the tuberculoid stage. Mucous membranes of the nose tend to become affected, and a lion-faced appearance is associated with this type of leprosy. Deformation of the hand into a clawed form and considerable necrosis of tissue can also occur (Figure 22.8b). The progression of the disease is unpredictable, and remissions may alternate with rapid deterioration.

The exact means of transfer of the leprosy bacillus is uncertain, but patients with lepromatous leprosy shed large numbers in their nasal secretions and in exudates (oozing matter) of their lesions. Most people probably acquire the infection when secretions containing the pathogen contact their nasal mucosa. Leprosy, however, is not very contagious, and transmission usually occurs only between people in fairly intimate and prolonged contact. The time from infection to the appearance of symptoms is usually measured in years, although children can have a much shorter incubation period. Death is not usually a result of the leprosy itself but of complications, such as tuberculosis.

Much of the public's fear of leprosy can probably be attributed to biblical and historical references to the disease. In the Middle Ages, patients with leprosy were rigidly ex-

(a) Tuberculoid leprosy

(b) Lepromatous leprosy

FIGURE 22.8 Lesions of leprosy. (a) Tuberculoid (neural) leprosy, showing the depigmented area of skin surrounded by a border of nodules that is typical of this earlier stage. **(b)** If the immune system fails to control the disease, the result is lepromatous (progressive) leprosy. This severely deformed hand shows the progressive tissue damage to the cooler parts of the body typical of this later stage.

cluded from normal European society and sometimes were even given bells to wear so that people could avoid contact with them. This isolation might have contributed to the near disappearance of the disease in Europe. But patients with leprosy are no longer kept in isolation, because patients can be made noncommunicable within a few days by the administration of sulfone drugs. The National Leprosy Hospital in Carville, Louisiana, once housed several hundred patients but now houses only a few elderly ones and will eventually be closed. Most patients today are treated at several centers on an outpatient basis.

The number of leprosy cases in the U.S.A. has been gradually increasing. Currently, over 200 cases are reported each year, and the total number of cases is estimated at about 2600. Many of these cases are imported, for the disease is usually found in tropical climates. Millions of people, most of them in Asia and Africa, suffer from leprosy today.

Although the disease is still difficult to cure, prolonged treatment with sulfone drugs (such as dapsone) has been effective in arresting its progress. Dapsone has been the mainstay of treatment, but resistance to the drug is a problem. Rifampin and a fat-soluble dye, clofazimine, are the other main drugs that are used, often along with dapsone.

A vaccine for leprosy is undergoing field trials. However, because of the extremely long incubation period for leprosy, it will be many years before the vaccine's effectiveness is established. An encouraging sign that a vaccine might be effective is the fact that the BCG vaccine used to provide immunity to tuberculosis (also caused by a *Mycobacterium* species) is somewhat protective against lep-

rosy. A BCG program in Africa to control tuberculosis showed that the vaccine gave at least 50% protection against leprosy at the same time.

VIRAL DISEASES OF THE NERVOUS SYSTEM

LEARNING OBJECTIVES

- *Discuss the epidemiology of poliomyelitis, rabies, and arbovirus encephalitis, including method of transmission, etiology, disease symptoms, and preventive measures.*

- *Compare the Salk and Sabin polio vaccines.*

- *Compare the pre-exposure and postexposure treatments for rabies.*

Most viruses affecting the nervous system enter it by circulation in the blood or lymph. However, some viruses can enter peripheral nerve axons and move up them.

Poliomyelitis

Poliomyelitis (polio) is best known as a cause of paralysis. However, only about 10% of infected people develop any identifiable symptoms, and the paralytic form of poliomyelitis probably affects less than 1% of those infected with the poliovirus. The great majority of cases are asymp-

FIGURE 22.9 Iron lungs at a March of Dimes Respiratory Center in Los Angeles in the 1950s. Breathing was possible for many only with these mechanical aids. A few survivors from these polio epidemics still use these machines, at least part-time. Others are able to use portable respiratory aids.

tomatic or cause mild symptoms, such as headache, sore throat, fever, and nausea, which are often interpreted as mild meningitis or influenza. Asymptomatic or mild cases of polio are most common in the very young.

In some parts of the world, because of poor sanitary conditions, most of the population contract asymptomatic poliomyelitis as infants (while they are still protected by maternal antibodies) and thus develop immunity. Nonetheless, the incidence of the disease in the tropical and subtropical zones is so high that there are an estimated 250,000 cases of paralytic polio each year. As sanitation has improved in the more developed regions of the world, the chance of developing the disease in infancy has decreased, particularly for those in the upper and middle socioeconomic groups. When infection occurs in adolescence or early adulthood, the paralytic form of the disease occurs more frequently.

Humans are the only known natural host for polioviruses. Polioviruses are more stable than most other viruses and can remain infectious for relatively long periods in water and food. The primary mode of transmission is ingestion of water contaminated with feces containing the virus.

Because the infection is initiated by ingestion of the virus, its primary areas of multiplication are the throat and small intestine. This accounts for the initial sore throat and nausea. Next, the virus invades the tonsils and the lymph nodes of the neck and ileum (terminal portion of the small intestine). From the lymph nodes, the virus enters the blood *(viremia).* In most cases the viremia is only transient, the infection does not progress past the lymphatic stage, and clinical disease does not result. If the viremia is persistent, however, the virus penetrates the

capillary walls and enters the central nervous system. Once in the central nervous system, the virus displays a high affinity for nerve cells, particularly motor nerve cells in the upper spinal cord (anterior horn cells). The virus does not infect the peripheral nerves or the muscles. As the virus multiplies within the cytoplasm of the motor nerve cells, the cells die and paralysis results. Death can result from respiratory failure (Figure 22.9).

Diagnosis of polio is usually based on isolation of the virus from feces and throat secretions. Cell cultures can be inoculated, and cytopathic effects on the cells can be observed (see Table 15.4, p. 401).

The incidence of polio in the U.S.A. has decreased markedly since the availability of the polio vaccines. The last cases acquired in the U.S.A. that were attributed to a wild-type virus were reported in 1979 (Figure 22.10). The development of the first polio vaccine was made possible by the introduction of practical techniques for cell culture, for the virus does not grow in any common laboratory animal. The polio vaccine was the prototype for the vaccines for mumps, measles, and rubella.

There are three different serotypes of the poliovirus, and immunity must be provided for all three. Two vaccines are available. The *Salk vaccine,* which was developed in 1954, uses viruses that have been inactivated by treatment with formalin. Vaccines of this type are inactivated polio vaccines (IPV). They require a series of injections. Their effectiveness is high, perhaps 90% against paralytic polio. The antibody levels decline with time, and booster shots are needed every few years to maintain full immunity. Using only IPV, several European countries have almost eliminated polio from their populations. A newer IPV has been introduced that is produced on human

FIGURE 22.10 Reported cases of poliomyelitis. Number of cases of poliomyelitis in the U.S.A. from 1950 to 1993. Note the decrease after introduction of the Salk and Sabin vaccines. [*Sources: CDC, Summary of Notifiable Diseases 1991, MMWR* 40(53) (10/2/92); *MMWR* 41(53) (1/8/93); *MMWR* 42(50) (12/24/93).]

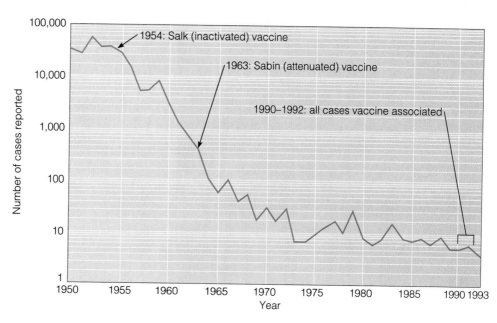

diploid cells. It is called *enhanced inactivated polio vaccine* (E-IPV). An important use of E-IPV is in the immunization of immunosuppressed individuals.

The *Sabin vaccine,* introduced in 1963, contains three living, attenuated strains of the virus and is more popular in the U.S.A. than the Salk vaccine. It is less expensive to administer, and most people prefer taking a sip of orange-flavored drink containing the virus (oral polio vaccine, or OPV) to having a series of injections. The immunity achieved with the OPV resembles that acquired by natural infection. One disadvantage is that, on rare occasions—one in 560,000 first doses, one in about 2,000,000 on subsequent doses—one of the attenuated strains of the virus (type 3) seems to revert to virulence and cause the disease. These cases often occur in secondary contacts, not in the person who received the vaccine. This illustrates that recipients of the vaccine can infect contacts, leading—in most cases—to immunization. Some medical scientists have suggested that a return to the IPV might be desirable despite its disadvantages.

The World Health Organization has targeted the year 2000 for elimination of polio in the world. This goal has apparently been met in North and South America, where cases due to wild-type virus have not been reported for several years. A 1991 case in Peru may have been the last in the Western Hemisphere. However, eliminating polio will be more difficult than the eradication of smallpox was. Whereas smallpox expressed obvious symptoms, many cases of polio are asymptomatic, and the fecal-oral route of transmission makes containment more difficult.

The OPV has been invaluable in suppressing wild-type virus, but, as we have seen in the U.S.A., it is also now the main source of new polio cases. The most likely next step

will be to combine E-IPV with the initial doses of DPT, which can then be safely reinforced with easily administered OPV boosters.

Rabies

Rabies is an acute infectious disease that usually results in fatal encephalitis. The causative agent is *rabies virus* (a rhabdovirus; *rhabdos*—rod) that has a characteristic bullet shape (Figure 22.11). Humans usually acquire the rabies virus from the bite of an infected animal. Even the lick of such an animal is dangerous because rabies virus in its saliva can enter minute scratches. The virus can cross the intact mucous membrane lining the eyes or body openings. Scientists investigating bat behavior in caves have contracted rabies, presumably from rabies virus present in aerosols of bat secretions in the cave atmosphere. Humans have also acquired rabies from aerosols of the virus when macerating infected tissue in a laboratory blender.

Rabies is unique, as the incubation period is usually long enough to allow development of immunity from post-exposure vaccination. (The amount of virus introduced into the wound is usually too small to provoke adequate natural immunity in time.) Initially, the virus multiplies in skeletal muscle and connective tissue, where it remains localized for periods ranging from days to months. Then it enters and travels along the peripheral nerves to the central nervous system, where it causes encephalitis. Bites in areas rich in nerve fibers, such as the hands and face, are especially dangerous, and the resulting incubation period tends to be short. Bites on the hands have an average incubation period of eight weeks; bites on the face, five

TEM |————| 0.1 μm

FIGURE 22.11 Rabies virus. Note the unusual bullet shape.

weeks. Once the virus enters the peripheral nerves, it is no longer accessible to the immune system.

The preliminary symptoms at the onset are mild and varied, resembling several common infections. When the central nervous system becomes involved, the patient tends to alternate between periods of agitation and intervals of calm. At this time, a frequent symptom is spasms of the muscles of the mouth and pharynx that occur when the patient feels air drafts or swallows liquids. In fact, even the mere sight or thought of water can set off the spasms—thus the common name *hydrophobia* (fear of water). The final stages of the disease result from extensive damage to the nerve cells of the brain and the spinal cord. Death in a few days is almost inevitable.

Animals with **furious rabies** are at first restless, then become highly excitable and snap at anything within reach. The biting behavior is essential to maintain the virus in the animal population. When paralysis sets in, the flow of saliva increases as swallowing becomes difficult, and nervous control is progressively lost. The disease is almost always fatal within a few days. Some animals suffer from **paralytic rabies**, in which there is only minimal excitability. This form is especially common in cats. The animal remains relatively quiet and even unaware of its surroundings, but it might snap irritably if handled. A similar manifestation of rabies occurs in humans and is often misdiagnosed as Guillain–Barré syndrome (a form of paralysis that is usually transient but is sometimes fatal) or other conditions. There is some speculation that the

two forms of the disease may be caused by slightly different forms of the virus.

Laboratory diagnosis of rabies in humans and animals is based on several findings. When the patient or animal is alive, a diagnosis can sometimes be confirmed by immunofluorescence studies, in which viral antigens are detected in saliva, serum, or cerebrospinal fluid. After death, diagnosis is confirmed by a fluorescent-antibody test performed on brain tissue.

Any person bitten by an animal that is positive for rabies must take antirabies treatment. Another indication for antirabies treatment is any unprovoked bite by a skunk, bat, fox, coyote, bobcat, or raccoon not available for examination. Treatment after a dog or cat bite, if the animal cannot be found, is determined by the prevalence of rabies in the area.

The original Pasteur treatment, in which the virus was attenuated by drying in the dissected spinal cords of rabies-infected rabbits, has been replaced in the U.S.A. by human diploid cell vaccine (HDCV). The HDCV is administered in a series of 5 or 6 injections at intervals during a 28-day period. Passive immunization is provided simultaneously by injecting human rabies immune globulin (RIG) that has been harvested from people who are immunized against rabies, such as laboratory workers.

Some problems have arisen with the HDCV. Veterinarians and laboratory workers, who must be immunized periodically, have developed allergic reactions. There have been cases of rabies in HDCV-immunized individuals in Africa, causing concern that some African rabies may be caused by viruses that are antigenically different from the standard strains used to make the vaccines. Furthermore, the HDCV is much too expensive for many parts of the world; in these areas, cheaper but less effective vaccines are prepared from animal nerve tissue. However, new vaccines are being developed.

Rabies occurs all over the world. In the U.S.A., 20,000 to 30,000 people are given the rabies vaccine each year. Nearly twice that many *die* of rabies each year in India, where the vaccine is given to about 3 million people. Australia, Great Britain, New Zealand, and Hawaii are free of the disease, a condition maintained by rigid quarantine. In North America, rabies is widespread among wildlife, predominantly skunks, foxes, and raccoons, although it is also found in domesticated animals (Figure 22.12). In fact, worldwide, dogs are the most common carriers of rabies. Rabies is seldom found in squirrels, rabbits, rats, or mice. Rabies has long been endemic in vampire bats of South America. In Europe and North America there are ongoing experiments to immunize wild animals with live rabies vaccine produced in genetically engineered vaccinia viruses, in food dropped for the animals to find. These experiments have been highly successful in Europe. In the U.S.A., up to 6000 cases of rabies are diagnosed in animals each year, but in recent years, only one to three cases per year have been diagnosed in humans (see the box on pp. 550–551).

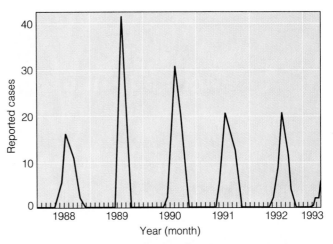

FIGURE 22.13 Arboviral infections of the central nervous system. Cases per month due to California encephalitis viruses in the U.S.A., 1988–1993. Note the seasonal occurrence of cases during the mosquito season. [*Sources: MMWR* 41(55) (9/24/93) and CDC Division of Vector-Borne Diseases.]

(a) Predominant wildlife species affected by rabies in the U.S.A. Several antigenically different strains of the virus are associated with different animals. This is often useful in tracing the possible origin of cases when it is otherwise unknown, as described in the box on pages 550–551. Almost all states report cases of rabies in bats.

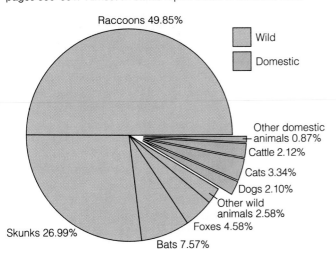

(b) Rabies cases in various wild and domestic animals in the U.S.A. in 1992.

FIGURE 22.12 Reported cases of rabies in animals. (*Source: J. Amer. Vet. Med. Assoc.*, Dec. 15, 1992.)

Arthropod-Borne Encephalitis

Encephalitis caused by mosquito-borne arboviruses (encephalitis viruses) is rather common in the U.S.A. Figure 22.13 shows the incidence of encephalitis in the U.S.A. from 1988 to 1993. The increase in the summer months coincides with the proliferation of adult mosquitoes during these months. Sentinel animals, such as caged rabbits or chickens, are tested periodically for antibodies to arboviruses. This gives health officials information on the incidence and types of virus in their area.

A number of clinical types of arthropod-borne (sometimes shortened to arbo) encephalitis have been identified; all can cause symptoms ranging from the subclinical to rapid death. Active cases of these diseases are characterized by chills, headache, and fever. As the disease progresses, mental confusion and coma are observed. Survivors may suffer from permanent neurological problems.

Horses as well as humans are frequently affected by these viruses; thus there are strains causing Eastern equine encephalitis (EEE) and Western equine encephalitis (WEE). These two viruses are the most likely to cause severe disease in humans. EEE is the least prevalent arboencephalitis (its main mosquito vector prefers to feed on birds), but it is the most severe. The mortality rate is about 25% or more, and a high incidence of brain damage, deafness, and similar neurological problems occurs in survivors. St. Louis encephalitis (SLE) acquired its name from the location of an early major outbreak (in which the fact that mosquitoes were involved in transmission of these diseases was discovered). SLE is the most common of the arboviral encephalitis diseases, even though less than 1% of infections exhibit symptoms. California encephalitis (CE) was first identified in that state, but most cases occur elsewhere. The La Crosse strain of CE is the most commonly encountered arbovirus. Some important characteristics of arboencephalitis types are summarized in Table 22.3.

The Far East also has endemic encephalitis. **Japanese B encephalitis** is the best known; it is a serious public health problem, especially in Japan, Korea, and China. Vaccines are used to control the disease in these countries and are often recommended for visitors.

Diagnosis of arthropod-borne encephalitis is usually made by serological tests, such as complement fixation. The most effective control measure is local control of the mosquitoes.

MMWR

Morbidity and Mortality Weekly Report

HUMAN RABIES— NEW YORK, 1993

IN AUGUST 1993, A FATAL case of human rabies in an 11-year-old girl was reported to the New York State Department of Health; this was the first indigenously acquired fatal case diagnosed in New York in 39 years. This report summarizes the investigation of this case.

On July 5, the girl complained of pain in the knuckles on her left hand. During July 6–7, she had increasing pain that extended up to the left shoulder. On July 8, a throat culture was obtained and amoxacillin was prescribed.

On July 9, the patient developed fever, severe muscle spasms of the left arm, difficulty walking, and hallucinations. On evaluation in an emergency room on July 10, she had fever (38.6°C [101.1°F]), otitis media, pharyngitis, and a maculopapular rash on the chest. The throat culture

obtained July 8 was positive for presumed *Streptococcus* group A, and recurrent streptococcal pharyngitis and otitis media were diagnosed. She was treated with ceftriaxone.

She subsequently would not drink, withdrew when offered a drink, and had difficulty swallowing oral secretions. On evaluation in an emergency room on July 11, she had a temperature of 40.7°C (105.3°F) and a white blood cell count elevated to 13,300. Viral meningoencephalitis or meningococcal infection was diagnosed. She was transported by helicopter ambulance to a tertiary-care medical center.

On admission to the pediatric intensive-care unit, she was alert, oriented, and cooperative but agitated. Acyclovir was added to her treatment regimen. The patient developed respiratory distress, hypertension, and tachycardia and was placed on mechanical ventilation; cardiac arrhythmias (irregular heartbeats) subsequently occurred, and she died from cardiac arrest.

During August 2–3, examination of routine histopathologic slides of brain tissue revealed possible Negri bodies. Specimens tested by the rabies fluorescent-antibody (FA) technique indicated fluorescent inclusions in the brain stem, midbrain, and cerebellum. Rabies diagnosis was confirmed at CDC by FA testing and histologic examination of formalin-fixed and paraffin-embedded tissue. The RNA extracted from formalin-fixed brain tissue was reverse transcribed and amplified by PCR. The nucleotide sequence identified a viral variant associated with rabies in insectivorous bats.

The patient lived in a heavily wooded area of the Catskill Mountains and had no history of foreign travel. She had no known history of contact with a bat, and examination of her home and outbuildings on the property revealed no evidence of bat infestation. She had been active outdoors, and her family kept horses, dogs, cats, rabbits, hamsters, and gerbils as pets; none of these pets had died with clinical signs consistent with rabies or dis-

TABLE 22.3   Types of Arthropod-Borne Encephalitis

DISEASE	MOSQUITO VECTOR	HOST ANIMALS	GEOGRAPHIC DISTRIBUTION IN U.S.A.	COMMENT
St. Louis encephalitis	*Culex*	Birds	Throughout country	Mostly urban outbreaks; affects mainly adults over 40.
California encephalitis	*Aedes*	Small mammals	North Central states, New York State	Affects mostly 5–18 age group in rural or suburban areas; La Crosse strain medically most important. Mainly in upper Midwest.
Western equine encephalitis	*Culex*	Birds, horses	Throughout country	Severe disease; frequent neurological damage, especially in infants.
Eastern equine encephalitis	*Aedes, Culiseta*	Birds, horses	East coast	More severe even than WEE; affects mostly young children and younger adults; relatively uncommon in humans.

appeared. A survey of all neighbors indicated the same about their pets.

As a result of close contact with the patient and/or her secretions, rabies postexposure prophylaxis was administered to 55 people, including family members and 35 health-care workers.

Editorial Note: Human rabies in the U.S.A. is uncommon, primarily because of canine rabies-control programs and access to improved human rabies antibodies and vaccines. Since 1980, 16 human rabies cases have been reported in the U.S.A. Of these, seven were acquired from exposure outside the U.S.A.; for nine of the 16 cases, no definitive history of exposure was identified. Potential reasons for the failure of public health authorities to establish definitive exposures include unrecognized exposure, communication (i.e., language) barriers, and memory loss and impaired speech because of encephalitis at presentation.

Rabies is not usually diagnosed when patients initially receive medical evaluation. Since 1980, of the 16 people with rabies diagnosed in the U.S.A., rabies was diagnosed postmortem in nine. In addition, six cases of human-to-human transmission were diagnosed postmortem among recipients of transplanted corneas, whose donors died of an illness unrecognized as rabies.

The case in this report is the sixth since 1980 in which insectivorous bats were implicated. A definite history of exposure through a bat's bite was identified for only one of the six cases, whereas contact with a bat was associated with two additional cases; for three cases, the nature of exposure was not determined, but bat rabies variants were identified by molecular typing.

Bat rabies is endemic in the U.S.A., and cases have been reported from all of the 48 contiguous states. The rabies virus variant identified in this case, and in three of the other five occurring since 1980, is associated with the silver-haired bat (*Lasionycteris noc-* *tivagans*), a solitary, migratory species with a preferred habitat of old-growth forest.

Exposure to potentially rabid animals (for example, paralyzed bats) should be avoided. Postexposure prophylaxis is recommended for all individuals bitten or scratched by such animals. Bat bites may be more difficult to recognize than those inflicted by terrestrial animals. Treatment should be considered for any physical contact with bats when bite or mucous membrane contact cannot be excluded. Because reduction of bat populations is neither feasible nor desirable as a means for controlling rabies in bats, efforts to prevent this problem should be directed toward the exclusion of bats from human dwellings to minimize direct contact with humans and companion animals. In addition, all dogs and cats in the 48 contiguous states and Alaska should have a current rabies vaccination.

SOURCE: *MMWR* 42:799, 805–806 (10/22/93).

FUNGAL DISEASE OF THE NERVOUS SYSTEM

LEARNING OBJECTIVE

- *Identify the causative agent, vector, symptoms, and treatment for cryptococcosis.*

The central nervous system is seldom invaded by fungi. However, one pathogenic fungus in the genus *Cryptococcus* is well adapted to growth in central nervous system fluids.

Cryptococcus neoformans Meningitis (Cryptococcosis)

Fungi of the genus *Cryptococcus* are spherical cells resembling yeasts; they reproduce by budding and produce polysaccharide capsules, some much thicker than the cells themselves (Figure 22.14). Only one species, *Cryptococcus neoformans*, is pathogenic for humans, causing the disease called **cryptococcosis**. The organism is widely distributed in soil, especially soil contaminated with pigeon droppings. It is also found in pigeon roosts and nests on the window ledges of urban buildings. Most cases of cryptococcosis occur in urban areas. It is thought to be transmitted by the inhalation of dried infected pigeon droppings.

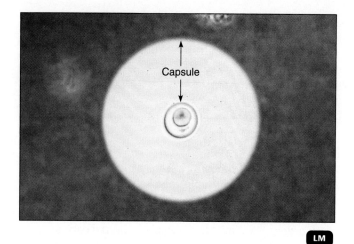

FIGURE 22.14 *Cryptococcus neoformans,* **the cause of cryptococcosis.** This yeastlike fungus has an unusually thick capsule. In this micrograph, the capsule is made visible by suspending the cells in dilute India ink.

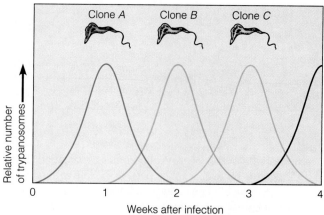

FIGURE 22.15 **How trypanosomes evade the immune system.** The population of each trypanosome clone drops nearly to zero as the immune system supresses its members, but a new clone with a different antigenic surface then replaces the previous clone. The black line represents the population of clone *D*.

Inhalation of *C. neoformans* initially causes infection of the lungs, frequently subclinical, and often the disease does not proceed beyond this stage. However, it can spread through the bloodstream to other parts of the body, including the brain and meninges, especially in immunosuppressed individuals and those receiving steroid treatments for major illnesses. The disease is usually expressed as chronic meningitis, which is usually progressive and fatal if untreated.

The best serological diagnostic test is a latex agglutination test to detect cryptococcal antigens in patients' serum or cerebrospinal fluid. The drugs of choice for treatment are amphotericin B and flucytosine.

PROTOZOAN DISEASES OF THE NERVOUS SYSTEM

LEARNING OBJECTIVE

- *Identify the causative agent, vector, symptoms, and treatment for African trypanosomiasis and* Naegleria *microencephalitis.*

Protozoa capable of invading the central nervous system are rare. However, those that can reach it cause devastating effects.

African Trypanosomiasis

African trypanosomiasis, or sleeping sickness, is a protozoan disease that affects the nervous system. In 1907, Winston Churchill described Uganda during an epidemic of sleeping sickness as a "beautiful garden of death." Even today, the disease affects about a million people in Central and East Africa, and about 20,000 new cases are reported each year.

The disease is caused by *Trypanosoma brucei gambiense* and *Trypanosoma brucei rhodesiense,* flagellates that are injected by the bite of a tsetse fly (*Glossina* [gläs-sē'nä]). Animal reservoirs for the two trypanosomes are similar, but they occur in different habitats and are spread by different species of tsetse fly vector. Human-to-human transmission by bites of the insect vector is more likely with infections by *T. b. gambiense,* which circulates in the patient's blood during the two-to four-year course of the disease. *T. b. rhodesiense* causes a disease that is more acute, running its course in only a few months.

During the early stages of either form of the disease, a few trypanosomes can be found in the blood. During later stages, the trypanosomes move into the cerebrospinal fluid. Symptoms of the disease include decreases in physical activity and mental acuity. Untreated, the host enters a coma, and death is almost inevitable.

There are some moderately effective chemotherapeutic agents, such as suramin and pentamidine isethionate. However, the drugs produce toxic effects, and, because they do not cross the blood–brain barrier, they are effective only when the central nervous system has not become affected. A toxic arsenical called melarsoprol is usually chosen when the central nervous system is affected. Recently, the new drug eflornithine, which blocks an enzyme required for proliferation of the parasite, has come into use. It is so dramatically effective against *T. b. gambiense* that it has been called the resurrection drug. However, its effectiveness against *T. b. rhodesiense* is variable.

A vaccine is being developed, but a major obstacle is that the trypanosome is able to change protein coats at least 100 times and can thus evade antibodies aimed at only one or a few of the proteins. Each time the body's im-

FIGURE 22.16 *Naegleria fowleri,* **the cause of** *Naegleria* **microencephalitis.** This micrograph shows the numerous pseudopods of this deadly parasite.

mune system is successful in suppressing the trypanosome, a new clone of parasites appears with a different antigenic coat (Figure 22.15).

Naegleria Microencephalitis

Naegleria fowleri (nī-gle′rē-ä fou′lėr-ē) is a protozoan (amoeba) that causes the neurological disease *Naegleria* **microencephalitis.** Although cases of it are reported in most parts of the world, only a few cases per year are reported in the U.S.A. The most common victims are children who swim in ponds or streams. The organism (Figure 22.16) initially infects the nasal mucosa and later proliferates in the brain. The fatality rate is nearly 100%—diagnosis is typically made at autopsy.

NERVOUS SYSTEM DISEASES THOUGHT TO BE CAUSED BY PRIONS

LEARNING OBJECTIVE

- *List the characteristics of diseases thought to be caused by prions.*

A number of fatal diseases of the human central nervous system are thought to be caused by what have been called prions (see Chapter 13). These diseases have long incuba-

tion times and progress slowly. The damage to the central nervous system is insidious and progressive, without the fever and inflammation seen in encephalitis. Autopsies show a characteristic spongiform (porous, like a sponge) degeneration of the brain. In recent years, the study of these diseases has been one of the most interesting areas of microbiology in medical science.

A typical prion disease is **sheep scrapie.** The infected animal rubs against fences and walls until areas on its body are raw. During a period of several weeks or months, the animal gradually loses motor control and dies. The prion can be passed to other animals, such as mice, by the injection of brain tissue from one animal to the next.

Humans suffer from diseases similar to scrapie. **Creutzfeldt–Jakob disease (CJD)** is an example. This disease is rare (about 200 cases per year in the U.S.A.) but often occurs in families. There is no doubt that an infective agent is involved because transmission via corneal transplants and accidental scalpel nicks of a surgeon during autopsy has been reported. Several cases have been traced to injection of a growth hormone derived from human tissue.

Some tribes in New Guinea have suffered from a prion disease called **kuru.** Kuru is apparently related to the practice of smearing brain tissue onto the body during cannibalistic rituals. The infection results when the agent is introduced into open sores or cuts. The disease is disappearing as cannibalism dies out.

A current international public health concern is **bovine spongiform encephalopathy.** Since 1986 thousands of cattle in the British Isles have come down with this condition, known locally as mad cow disease. It causes the animals to become unmanageable, and they must be destroyed. The current theory is that the disease is probably due to prions transmitted in feed supplements derived from animals such as sheep. There is great concern in the international medical community that the condition might be transmitted to humans via the ingestion of beef.

As we discussed in Chapter 13, prions seem to be pure protein. If nucleic acid is present, it has not been detected to date. The exact nature of this agent and its mode of replication are still mysteries. In fact, some researchers dispute the existence of prions as an infectious agent in transmission of these diseases. Among other proposals are that the infectious agent is a spiroplasma (a tightly coiled microbe lacking a cell wall, which is usually associated with plant and insect diseases), that it is a metabolic disorder in which a toxic prion protein accumulates and is mistaken for an infectious agent, or that it is a genetic disorder and prions are "phantoms of virology." Certainly, the exact nature of these diseases and the responsible agents will continue to be the subject of intense research—and contention.

• • •

Table 22.4 summarizes the main microbial diseases associated with the nervous system.

TABLE 22.4   Microbial Diseases Associated with the Nervous System

DISEASE	CAUSATIVE AGENT	MODE OF TRANSMISSION	TREATMENT
Bacterial Diseases			
Meningococcal meningitis	*Neisseria meningitidis*	Contact with a healthy carrier via respiratory tract	Third-generation cephalosporins
Hemophilus influenzae meningitis	*H. influenzae*	Via respiratory tract	Third-generation cephalosporins
Pneumococcal meningitis	*Streptococcus pneumoniae*	Via respiratory tract	Third-generation cephalosporins
Botulism	*Clostridium botulinum*	Ingestion of toxin	Antitoxins and respiratory support
Listeriosis	*Listeria monocytogenes*	Ingestion of contaminated food	Penicillins
Leprosy	*Mycobacterium leprae*	Transfer of exudates from lesions	Sulfone drugs (dapsone), rifampin, clofazimine
Tetanus	*Clostridium tetani*	Deep anaerobic wound contamination	Immune globulin and respiratory support
Fungal Disease			
Cryptococcosis	*Cryptococcus neoformans*	Inhalation of dried infected pigeon droppings	Amphotericin B, flucytosine
Viral Diseases			
Poliomyelitis	Poliovirus	Ingestion of virus	None
Rabies	Rabies virus	Bite of a rabid animal	Immune globulin and postexposure vaccination
Arthropod-borne encephalitis	Arboviruses	Mosquitoes	None
Protozoan Diseases			
African trypanosomiasis	*Trypanosoma brucei rhodesiense, T. b. gambiense*	Tsetse flies	Suramin, pentamidine isethiouate, melarsoprol, eflornithine
Naegleria microencephalitis	*Naegleria fowleri*	From fresh water, probably nasal passages	None
Prion Diseases			
Creutzfeldt–Jakob disease	Prion	Ingestion, wound contamination, or transplant	None
Kuru	Prion	Ingestion or wound contamination	None

STUDY OUTLINE

STRUCTURE AND FUNCTION OF THE NERVOUS SYSTEM (pp. 537–539)

1. The central nervous system consists of the brain, which is protected by the skull bones, and the spinal cord, which is protected by the backbone.
2. The peripheral nervous system consists of the nerves that branch from the central nervous system.
3. The central nervous system is covered by three layers of membranes called meninges; they are called the dura mater, arachnoid, and pia mater. Cerebrospinal fluid circulates between the arachnoid and the pia mater in the subarachnoid space.
4. The blood–brain barrier normally prevents many substances, including antibiotics, from entering the brain.

5. Microorganisms can enter the central nervous system through trauma, along peripheral nerves, and through the bloodstream and lymphatic system.
6. An infection of the meninges is called meningitis. An infection of the brain is called encephalitis.

BACTERIAL DISEASES OF THE NERVOUS SYSTEM (pp. 539–545)

Bacterial Meningitis (pp. 539–541)

1. Meningitis can be caused by viruses, bacteria, fungi, and protozoa.
2. The three major causes of bacterial meningitis are *Hemophilus influenzae*, *Streptococcus pneumoniae*, and *Neisseria meningitidis*.

3. Nearly 50 species of opportunistic bacteria can cause meningitis.

Hemophilus influenzae Meningitis (pp. 539–540)

1. *H. influenzae* is part of the normal throat microbiota.
2. *H. influenzae* requires blood factors for growth; there are six types of *H. influenzae* based on capsule differences.
3. *H. influenzae* type b is the most common cause of meningitis in children under 4 years old.
4. The disease often occurs as a secondary infection following a viral respiratory infection such as influenza.
5. A conjugated vaccine directed against the capsular polysaccharide antigen is available.

Streptococcus pneumoniae Meningitis (Pneumococcal Meningitis) (p. 540)

1. *S. pneumoniae* is commonly found in the nasopharynx.
2. Hospitalized patients and young children are most susceptible to *S. pneumoniae* meningitis. It is rare but has a high mortality rate.
3. The vaccine for pneumococcal pneumonia may provide some protection against pneumococcal meningitis.

Neisseria Meningitis (Meningococcal Meningitis) (p. 540)

1. *N. meningitidis* causes meningococcal meningitis. This bacterium is found in the throats of healthy carriers.
2. The bacteria probably gain access to the meninges through the bloodstream. The bacteria may be found in leukocytes in cerebrospinal fluid.
3. Symptoms are due to endotoxin. The disease occurs most often in young children.
4. Military recruits are vaccinated with purified capsular polysaccharide to prevent epidemics in training camps.

Diagnosis and Treatment of the Most Common Types of Bacterial Meningitis (pp. 540–541)

1. Cephalosporins may be administered before identification of the pathogen.
2. Diagnosis is based on isolation and identification of the bacteria in cerebrospinal fluid.
3. Cultures are usually made on blood agar and incubated in an atmosphere containing reduced oxygen levels.

Listeriosis (p. 541)

1. *Listeria monocytogenes* causes meningitis in newborns, the immunosuppressed, pregnant women, and cancer patients.
2. It is acquired by ingestion of contaminated food.
3. Listeriosis may be asymptomatic in healthy adults.
4. *L. monocytogenes* can cross the placenta and cause spontaneous abortion and stillbirth.

Tetanus (pp. 541–542)

1. Tetanus is caused by a localized infection of a wound by *Clostridium tetani*.
2. *C. tetani* produces the neurotoxin tetanospasmin, which causes the symptoms of tetanus: spasms, contraction of muscles controlling the jaw, and death resulting from spasms of respiratory muscles.
3. *C. tetani* is an anaerobe that will grow in deep, unclean wounds and wounds with little bleeding.
4. Acquired immunity results from DPT immunization that includes tetanus toxoid.
5. Following an injury, an immunized person may receive a booster of tetanus toxoid. An unimmunized person may receive (human) tetanus immune globulin.
6. Debridement (removal of tissue) and antibiotics may be used to control the infection.

Botulism (pp. 542–544)

1. Botulism is caused by an exotoxin produced by *C. botulinum* growing in foods.
2. Serological types of botulinum toxin vary in virulence, with type A being the most virulent.
3. The toxin is a neurotoxin that inhibits transmission of nerve impulses.
4. Blurred vision occurs in one to two days; progressive flaccid paralysis follows for one to ten days, possibly resulting in death from respiratory and cardiac failure.
5. *C. botulinum* will not grow in acidic foods or in an aerobic environment.
6. Endospores are killed by proper canning. Addition of nitrites to foods inhibits growth after endospore germination.
7. The toxin is heat labile and is destroyed by boiling (100°C) for five minutes.
8. Infant botulism results from the growth of *C. botulinum* in an infant's intestines.
9. Wound botulism occurs when *C. botulinum* grows in anaerobic wounds.
10. For diagnosis, mice protected with antitoxin are inoculated with toxin from the patient or foods.

Leprosy (pp. 544–545)

1. *Mycobacterium leprae* causes leprosy, or Hansen's disease.
2. *M. leprae* has never been cultured on artificial media. It can be cultured in armadillos.
3. The tuberculoid form of the disease is characterized by loss of sensation in the skin surrounded by nodules. The lepromin test is positive.
4. Laboratory diagnosis is based on observations of acid-fast bacilli (rods) in lesions or fluids and the lepromin test.
5. In the lepromatous form, disseminated nodules and tissue necrosis occur. The lepromin test is negative.
6. Leprosy is not highly contagious and is spread by prolonged contact with exudates.
7. Untreated individuals often die of secondary bacterial complications, such as tuberculosis.
8. Patients with leprosy are made noncommunicable within four or five days with sulfone drugs and then treated as outpatients.
9. Leprosy occurs primarily in the tropics.

VIRAL DISEASES OF THE NERVOUS SYSTEM
(pp. 545–550)

Poliomyelitis (pp. 545–547)

1. The symptoms of poliomyelitis are usually headache, sore throat, fever, stiffness of the back and neck, and occasionally paralysis (less than 1% of the cases).
2. Poliovirus is found only in humans and is transmitted by ingestion of water contaminated with feces.
3. Poliovirus first invades lymph nodes of the neck and small intestine. Viremia and spinal cord involvement may follow.
4. At present, outbreaks of polio in the U.S.A. are uncommon because of the use of vaccines.
5. Diagnosis is based on isolation of the virus from feces and throat secretions.
6. The Salk vaccine (an inactivated polio vaccine [IPV]) involves injection of formalin-inactivated viruses and boosters every few years. The Sabin vaccine (an oral polio vaccine [OPV]) contains three live, attenuated strains of poliovirus and is administered orally.
7. Through vaccination, the WHO plans to eliminate polio by the year 2000.

Rabies (pp. 547–548)

1. Rabies virus (a rhabdovirus) causes an acute, usually fatal, encephalitis called rabies.
2. Rabies may be contracted through the bite of a rabid animal, by inhalation of aerosols, or invasion through minute skin abrasions. The virus multiplies in skeletal muscle and connective tissue.
3. Encephalitis occurs when the virus moves along peripheral nerves to the central nervous system.
4. Symptoms of rabies include spasms of mouth and throat muscles followed by extensive brain and spinal cord damage and death.
5. Laboratory diagnosis may be made by direct immunofluorescent tests of saliva, serum, and cerebrospinal fluid or brain smears.
6. Reservoirs for rabies in the U.S.A. include skunks, bats, foxes, and raccoons. Domestic cattle, dogs, and cats may get rabies. Rodents and rabbits seldom get rabies.
7. The Pasteur treatment for rabies involved multiple subcutaneous injections of rabies virus grown in rabbit spinal cord tissue.
8. Current postexposure treatment includes administration of human rabies immune globulin (RIG) along with multiple intramuscular injections of human diploid cell vaccine (HDCV).
9. Pre-exposure immunization consists of injections of HDCV.

Arthropod-Borne Encephalitis (pp. 549–550)

1. Symptoms of encephalitis are chills, headache, fever, and eventually coma.
2. Many types of arboviruses transmitted by mosquitoes cause encephalitis.
3. The incidence of arthropod-borne encephalitis increases in the summer months when mosquitoes are most numerous.
4. Horses are frequently infected by EEE and WEE viruses.
5. Diagnosis is based on serological tests.
6. Control of the mosquito vector is the most effective way to control encephalitis.

FUNGAL DISEASE OF THE NERVOUS SYSTEM
(pp. 551–552)

Cryptococcus neoformans Meningitis (Cryptococcosis) (pp. 551–552)

1. *Cryptococcus neoformans* is an encapsulated yeastlike fungus that causes *Cryptococcus neoformans* meningitis.
2. The disease may be contracted by inhalation of dried infected pigeon droppings.
3. The disease begins as a lung infection and may spread to the brain and meninges.
4. Immunosuppressed individuals are most susceptible to *Cryptococcus neoformans* meningitis.
5. Diagnosis is based on latex agglutination for cryptococcal antigens in patient's serum or cerebrospinal fluid.

PROTOZOAN DISEASES OF THE NERVOUS SYSTEM (pp. 552–553)

African Trypanosomiasis (pp. 552–553)

1. African trypanosomiasis is caused by the protozoa *Trypanosoma brucei gambiense* and *T. b. rhodesiense* and transmitted by the bite of the tsetse fly (*Glossina*).
2. The disease affects the nervous system of the human host, causing lethargy and eventually coma. It is commonly called sleeping sickness.
3. Vaccine development is hindered by the protozoan's ability to change its surface antigens.

Naegleria Microencephalitis (p. 553)

1. Encephalitis caused by the protozoan *N. fowleri* is almost always fatal.
2. The protozoan invades the brain from the nasal mucosa.

NERVOUS SYSTEM DISEASES THOUGHT TO BE CAUSED BY PRIONS (pp. 553–554)

1. Diseases of the central nervous system that progress slowly and cause spongiform degeneration are thought to be caused by prions.
2. Sheep scrapie and bovine spongiform encephalopathy are examples of diseases thought to be caused by prions that are transferable from one animal to another.
3. Creutzfeldt–Jakob disease and kuru are human diseases similar to scrapie. Kuru occurs in isolated groups of cannibals who eat brains.
4. Prions may be pure protein or protein associated with undetectable nucleic acid.
5. Other agents may be responsible for causing these diseases.

STUDY QUESTIONS

Review

1. Differentiate between meningitis and encephalitis.

2. Fill in the following table.

CAUSATIVE AGENT OF MENINGITIS	SUSCEPTIBLE POPULATION	METHOD OF TRANS-MISSION	TREATMENT
N. meningitidis			
H. influenzae			
S. pneumoniae			
L. monocytogenes			
C. neoformans			

3. Briefly explain the derivation of the name *Hemophilus influenzae*.

4. If *Clostridium tetani* is relatively sensitive to penicillin, then why doesn't penicillin cure tetanus?

5. Compare and contrast the Salk and Sabin vaccines with respect to composition, advantages, and disadvantages.

6. What treatment is used against tetanus under the following conditions?
   (a) Before a person suffers a deep puncture wound
   (b) After a person suffers a deep puncture wound

7. Why is the following description used for wounds that are susceptible to *C. tetani* infection: ". . . Improperly cleaned deep puncture wounds . . . ones with little or no bleeding . . ."?

8. List the following information for botulism: etiologic agent, suspect foods, symptoms, treatment, conditions necessary for microbial growth, basis for diagnosis, prevention.

9. Provide the following information on leprosy: etiology, method of transmission, symptoms, treatment, prevention, and susceptible population.

10. Provide the following information on poliomyelitis: etiology, method of transmission, symptoms, prevention. Why aren't the Salk and Sabin vaccines considered treatments for poliomyelitis?

11. Provide the etiology, method of transmission, reservoirs, and symptoms for rabies.

12. Outline the procedures for treating rabies after exposure. Outline the procedures for preventing rabies prior to exposure. What is the reason for the differences in the procedures?

13. Why is the incidence of arbovirus encephalitis in the U.S.A. higher in the summer months?

14. Describe the symptoms and laboratory test results that would lead to a diagnosis of possible prion disease.

15. Fill in the following table.

DISEASE	ETIOLOGY	VECTOR	SYMPTOMS	TREAT-MENT
Arthropod-borne en-cephalitis				
African trypano-somiasis				

16. Why are meningitis and encephalitis generally difficult to treat?

17. Provide evidence that Creutzfeldt–Jakob disease is caused by a transmissible agent that resembles viruses.

Critical Thinking

1. Most of us have been told that a "rusty nail" causes tetanus. What do you suppose is the origin of this adage?

2. A BCG vaccination will result in positive lepromin and tuberculin tests. What is the relationship between leprosy and tuberculosis?

3. Some health-care professionals feel that the OPV should no longer be used. Provide an argument to support this view.

Clinical Applications

1. A one-year-old infant was lethargic and had a fever. When he was admitted to the hospital, he had multiple brain abscesses with gram-negative coccobacillary rods. Identify the disease, etiology, and treatment.

2. A 40-year-old bird handler was admitted to the hospital with soreness over his upper jaw, progressive vision loss, and bladder dysfunction. He had been well two months earlier. Within weeks he lost reflexes in his lower extremities and subsequently died. Examination of cerebrospinal fluid showed lymphocytes. What etiology do you suspect? What further information do you need?

3. One week after bathing in a hot spring, a nine-year-old girl was hospitalized after a three-day history of progressive, severe headaches, nausea, lethargy, and stupor. Examination of cerebrospinal fluid revealed amoeboid organisms. Identify the disease, etiology, and treatment.

Microbial Diseases of the Cardiovascular and Lymphatic Systems

THE CARDIOVASCULAR SYSTEM consists of the heart, blood, and blood vessels. The **lymphatic system** consists of the lymph, lymph vessels, lymph nodes, and the lymphoid organs (tonsils, appendix, spleen, and thymus gland—see Figure 16.6). Because both systems circulate various substances throughout the body, they can serve as vehicles for the spread of infection. A summary of diseases associated with the cardiovascular and lymphatic systems is presented in Table 23.1.

STRUCTURE AND FUNCTION OF THE CARDIOVASCULAR SYSTEM

LEARNING OBJECTIVE

- *Identify the role of the cardiovascular system in spreading infections and in eliminating infections.*

The center of the cardiovascular system (Figure 23.1) is the *heart.* The function of the heart is to circulate the blood through the body's tissues so it can deliver certain substances to cells and remove other substances from them. *Blood vessels* are the tubes that carry circulating blood throughout the body. An *artery* is a blood vessel that carries blood from the heart to different tissues of the body. The body's main artery is the aorta. When the arteries reach their destinations, they branch into smaller vessels called *arterioles,* which in turn branch into even smaller

vessels called *capillaries.* Capillary walls are only one cell thick, and it is through these walls that blood and tissue cells exchange materials. The capillaries that take blood away from the tissue cells converge to form small vessels called *venules.* Venules unite to form *veins,* and veins return the blood to the heart. The body's main veins are the *superior vena cava* and the *inferior vena cava.*

The *blood* itself is a mixture of formed elements and a liquid called plasma (see Chapter 16). The *plasma* transports dissolved nutrients to body cells and carries wastes away from the cells. The formed elements in blood include red blood cells, white blood cells, and platelets. Red blood cells (*erythrocytes*) carry oxygen and some carbon dioxide, although most of the carbon dioxide in blood is dissolved in the plasma. White blood cells (*leukocytes*) play several roles in defending the body against infection, as was discussed in Part Three. Some white cells, in particular the neutrophils, are phagocytes. B cells and T cells, two types of lymphocytes, play key roles in immunity.

STRUCTURE AND FUNCTION OF THE LYMPHATIC SYSTEM

LEARNING OBJECTIVE

- *Identify the role of the lymphatic system in spreading infections and in eliminating infections.*

TABLE 23.1 Diseases of the Cardiovascular and Lymphatic Systems

DISEASE	CAUSATIVE AGENT	MODE OF TRANSMISSION	TREATMENT
Bacterial Diseases			
Puerperal sepsis and infections related to abortions	Primarily *Streptococcus pyogenes*; *Clostridium* and *Bacteroides* spp. often cause postabortion infections	Unsanitary conditions during childbirth or abortion	Penicillin
Endocarditis			
Subacute bacterial	Many organisms, especially α-hemolytic streptococci	Bacteremia from some focus of infection localizing in the heart	Varies with agent; penicillin for α-hemolytic streptococci
Acute bacterial	*Staphylococcus aureus*, *Streptococcus pyogenes*	Bacteremia from some focus of infection localizing in the heart	Penicillin
Pericarditis	*Streptococcus pyogenes*	Bacteremia from some focus of infection localizing in the heart	Penicillin
Rheumatic fever	Group A β-hemolytic streptococci	Inhalation leading to streptococcal sore throat	Penicillin (for sore throat)
Tularemia	*Francisella tularensis*	Animal reservoir (rabbits); skin abrasions, ingestion, inhalation, bites	Streptomycin or gentamicin
Brucellosis	*Brucella* spp.	Animal reservoir (cows); ingestion in milk, direct contact with skin abrasions	Streptomycin, tetracycline
Anthrax	*Bacillus anthracis*	Reservoir is soil or animals; inhalation or ingestion of heavy spore concentrations	Penicillin
Gangrene	*Clostridium perfringens*	Contamination of open wound by clostridial endospores	Debridement, amputation, hyperbaric chamber, penicillin
Rat bite fever	*Spirillum* or *Streptobacillus* spp.	Rat bite	Penicillin or tetracycline
Cat-scratch disease	Uncertain; bacteria *Afipia felis* or *Rochalimaea henselae*	Cat scratches	Rifampin
Plague	*Yersinia pestis*	*Xenopsylla cheopis* (rat flea)	Tetracycline and streptomycin
Relapsing fever	*Borrelia* spp.	Soft ticks	Tetracycline
Lyme disease	*Borrelia burgdorferi*	*Ixodes* spp. (tick)	Penicillin, tetracycline
Epidemic typhus	*Rickettsia prowazekii*	*Pediculus humanus corporis* (body louse)	Tetracycline and chloramphenicol
Endemic murine typhus	*Rickettsia typhi*	*Xenopsylla cheopis* (rat flea)	Tetracycline and chloramphenicol
Rocky Mountain spotted fever	*Rickettsia rickettsii*	*Dermacentor andersoni* and other species (tick)	Tetracycline and chloramphenicol
Viral Diseases			
Myocarditis	Several agents, especially coxsackievirus (enterovirus)	Inhalation and ingestion	None
Infectious mononucleosis	Epstein–Barr virus	Oral secretions, kissing	None
Burkitt's lymphoma	Epstein–Barr virus	Uncertain	None
Yellow fever	Arbovirus (yellow fever virus)	*Aedes aegypti* (mosquito)	None
Dengue	Arbovirus (dengue fever virus)	*Aedes aegypti* (mosquito)	None
Protozoan Diseases			
Toxoplasmosis	*Toxoplasma gondii*	Contact with cat feces or contaminated, undercooked meats	Pyrimethamine in combination with sulfadiazine
Chagas' disease	*Trypanosoma cruzi*	Contamination of reduviid bug bite wound with bug feces	Nifurtimox in early stages
Malaria	*Plasmodium* spp.	*Anopheles* mosquito	Chloroquine, mefloquine
Helminthic Disease			
Schistosomiasis	*Schistosoma* spp.	Contaminated water; cercariae enter the body through intact skin	Praziquantel

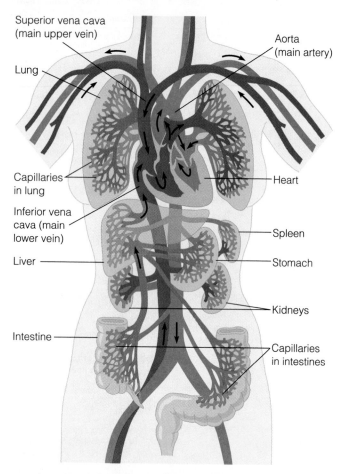

FIGURE 23.1 The human cardiovascular system. Details of circulation to the head and extremities are not shown in this simplified diagram. The blood circulates from the heart through the arterial system (red) to the capillaries (purple) in the lungs and other parts of the body. From these capillaries, the blood returns through the venous system (blue) to the heart.

As part of the overall pattern of circulation, some plasma filters out of the blood capillaries and into spaces between tissue cells. These spaces are called *interstitial spaces.* The fluid circulating around and between tissue cells is called *interstitial fluid.* Microscopic lymphatic vessels that surround tissue cells are called *lymph capillaries;* they are larger and more permeable than blood capillaries. As the interstitial fluid moves around the tissue cells, it is picked up by the lymph capillaries; the fluid is then called *lymph* (Figure 23.2).

Because lymph capillaries are very permeable, they readily pick up microorganisms or their products. From lymph capillaries, lymph is transported into larger lymph vessels called *lymphatics,* which contain valves that keep the lymph moving toward the heart. Eventually, all the lymph is returned to the blood just before the blood enters the right atrium of the heart. As a result of this circulation, proteins and fluid that have filtered from the plasma are returned to the blood.

At various points along the lymphatic system are oval structures called *lymph nodes,* through which lymph flows. Within the lymph nodes are fixed macrophages that help to clear the lymph of microorganisms. At times, the number of microorganisms circulating through lymph nodes is so great that the nodes themselves become infected and hence become enlarged and tender. The lymph nodes are also important to the body's immune reaction against invading microorganisms. Microorganisms entering the lymph nodes and other lymphoid tissues encounter two types of lymphocytes: B cells that are stimulated to produce humoral antibodies, and T cells that differentiate into mature T cells that carry out important defensive functions (see Figure 17.3).

BACTERIAL DISEASES OF THE CARDIOVASCULAR AND LYMPHATIC SYSTEMS

LEARNING OBJECTIVES

- *List the signs and symptoms of septicemia, and explain the importance of infections that develop into septicemia.*
- *Describe the epidemiologies of puerperal sepsis, bacterial endocarditis, and myocarditis.*
- *List the cause of and treatment and preventive measures for rheumatic fever.*
- *Discuss the epidemiologies of tularemia, brucellosis, anthrax, and gas gangrene.*
- *List four diseases acquired by animal bites.*
- *Compare and contrast the causative agents, vectors, reservoirs, symptoms, treatments, and preventive measures for plague, relapsing fever, Lyme disease, and typhus.*

Once bacteria gain access to the bloodstream, they may become widely disseminated and sometimes are also able to reproduce rapidly.

Septicemia

Although blood is normally sterile, moderate numbers of microorganisms are usually not harmful. However, if the defenses of the blood and lymphatic systems fail, the microorganisms can undergo uncontrolled proliferation in the blood. That condition is called **septicemia**, or blood poisoning. One sign of septicemia is the appearance of **lymphangitis** (Figure 23.3), inflamed lymph vessels visible as red streaks under the skin, running along the arm or leg from the infection site. Sometimes the streaks end at a lymph node, where the lymphocytes attempt to stop the invading microorganisms. Clinically, a person suffering from septicemia exhibits the symptoms of **septic shock.** The small arteries and veins are constricted, pe-

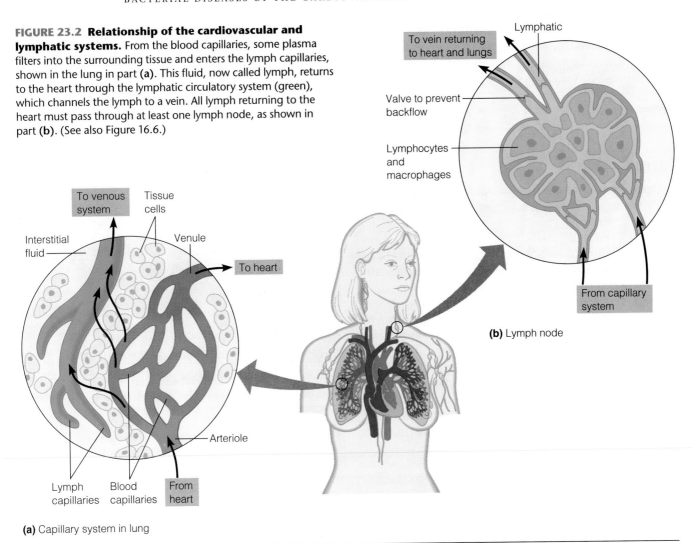

FIGURE 23.2 Relationship of the cardiovascular and lymphatic systems. From the blood capillaries, some plasma filters into the surrounding tissue and enters the lymph capillaries, shown in the lung in part **(a)**. This fluid, now called lymph, returns to the heart through the lymphatic circulatory system (green), which channels the lymph to a vein. All lymph returning to the heart must pass through at least one lymph node, as shown in part **(b)**. (See also Figure 16.6.)

(b) Lymph node

(a) Capillary system in lung

ripheral blood circulation decreases, and tissues served by the peripheral circulation become oxygen-starved. The patient develops rapid breathing and heartbeat, a drop in blood pressure, and mental confusion. The extremities become cool and may turn blue. Septic shock may progress to coma and death.

The organisms most frequently associated with septicemia are gram-negative rods, although a few gram-positive bacteria and fungi are also implicated. Among the gram-negative rods are *Escherichia coli, Serratia marcescens, Proteus mirabilis, Enterobacter aerogenes, Pseudomonas aeruginosa,* and *Bacteroides* species. These bacteria enter the blood from a focus of infection in the body. As you may recall, the cell walls of many gram-negative bacteria contain endotoxins that are released upon the lysis of the cell. It is the endotoxin that actually causes the symptoms. Once released, the endotoxin damages blood vessels; this damage causes the low blood pressure and subsequent shock. In some cases, antibiotics aggravate the condition by causing the lysis of large numbers of bacteria that then release damaging endotoxins. Products that will neutral-

FIGURE 23.3 Lymphangitis, one sign of septicemia. As the infection spreads from its original site along the lymph vessels, the inflamed walls of the vessels become visible as red streaks.

ize these endotoxins with monoclonal antibodies are currently being investigated. Many of these gram-negative rods are of nosocomial origin and are introduced into the bloodstream by medical procedures, such as catheters or feeding tubes, that bypass the normal barriers between the environment and the blood.

Puerperal Sepsis

Puerperal sepsis, also called **puerperal fever** and **childbirth fever,** is a nosocomial infection that frequently leads to septicemia. This begins as an infection of the uterus as a result of childbirth or abortion. *Streptococcus pyogenes,* a group A β-hemolytic streptococcus, is the most frequent cause, although other organisms may cause infections of this type. Recall from Chapter 21 that group A β-hemolytic streptococci have an M protein on their surface that increases virulence by improving adherence to the mucous membranes and by increasing resistance to phagocytosis.

Puerperal sepsis progresses from an infection of the uterus to an infection of the abdominal cavity (peritonitis) and in many cases to septicemia. At one Paris hospital between 1861 and 1864, of the 9886 women who gave birth, 1226 (12%) died of such infections. At that time, the death rate from puerperal sepsis outside such hospitals was frequently twice that. These deaths were largely unnecessary. Some 20 years before, Oliver Wendell Holmes in the U.S.A. and Ignaz Semmelweiss in Austria had clearly demonstrated that the disease was transmitted by the hands and instruments of the attending midwives or physicians and that disinfection of hands and instruments could prevent such transmission. Yet, in 1879, Louis Pasteur still found it necessary to lecture physicians about the cause of the disease.

Antibiotics, especially penicillin, and modern hygienic practices have now made *S. pyogenes* puerperal sepsis an uncommon complication of childbirth. Infections related to improperly performed abortions—not strictly puerperal sepsis—are often of mixed bacterial types, many of them caused by anaerobic bacteria of the *Bacteroides* or *Clostridium* genera. Penicillin is also effective in these cases.

Bacterial Endocarditis

The wall of the heart consists of three layers. The outer layer, the *pericardium,* is a sac that encloses the heart. The middle layer, the *myocardium,* is the thickest layer and consists of cardiac muscle tissue. The inner layer is a lining of epithelium and is called the *endocardium.* This layer lines the heart muscle itself and covers the valves in the heart. An inflammation of the endocardium is called **endocarditis.**

One type of bacterial endocarditis, **subacute bacterial endocarditis,** is characterized by fever, anemia, general

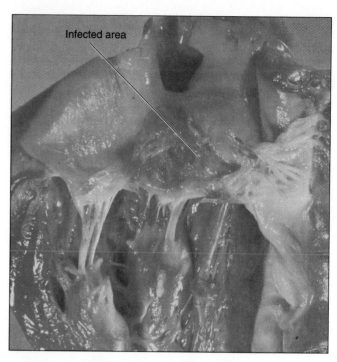

FIGURE 23.4 Acute bacterial endocarditis. This heart has been dissected to expose the bicuspid (mitral) valve. The cordlike structures connect the heart valve to the operating muscles. In a healthy heart, the top region would be completely smooth and light colored. The bacterial infection is responsible for the rough, reddened area; the inflammation resulted in a narrowing of the valve opening.

weakness, and heart murmur. It is usually caused by α-hemolytic streptococci such as are common in the oral cavity, although enterococci or staphylococci are often involved. The condition probably arises from a focus of infection elsewhere in the body, such as in the teeth or tonsils. Microorganisms are released by tooth extractions or tonsillectomies, enter the blood, and find their way to the heart. Normally, such bacteria would be quickly cleared from the blood by the body's defensive mechanisms. However, in people whose heart valves are abnormal, because of either congenital heart defects or such diseases as rheumatic fever and syphilis, the bacteria lodge in the pre-existing lesions. Within the lesions, the bacteria multiply and become entrapped in blood clots that protect them from phagocytes and antibodies. As multiplication progresses and the clot gets larger, pieces of the clot break off and can block blood vessels or lodge in the kidneys. In time, the function of the heart valves is impaired. Left untreated, subacute bacterial endocarditis is fatal within a few months.

A more rapidly progressive type of bacterial endocarditis is **acute bacterial endocarditis** (Figure 23.4), which is usually caused by *Staphylococcus aureus.* The organisms find their way from the initial site of infection to normal or abnormal heart valves; the rapid destruction of the heart valves is frequently fatal within a few days or weeks if untreated.

FIGURE 23.5 Nodule caused by rheumatic fever. Rheumatic fever was named, in part, because of the characteristic subcutaneous nodules that appear at the joints, as shown in this patient's elbow. Infection with group A β-hemolytic streptococci sometimes leads to this autoimmune complication.

Penicillin is sometimes used prophylactically to prevent endocarditis during procedures such as tooth extractions and tonsillectomies. Streptococci can also cause **pericarditis,** inflammation of the sac around the heart (the pericardium). Infections of the heart muscle tissue, or myocardium, are called **myocarditis** and can involve a widely varied array of microbes, including bacteria, viruses, fungi, protozoa, and helminths. In the U.S.A., the most common cause is enteroviruses of the coxsackie B type. In Central and South America, *Trypanosoma cruzi*, the protozoan causing Chagas' disease (see p. 577), is the most prevalent agent.

Rheumatic Fever

Group A β-hemolytic streptococcal infections, such as those caused by *Streptococcus pyogenes*, sometimes lead to **rheumatic fever,** which is generally considered an autoimmune type of complication (see Chapter 19). It occurs mainly among people between the ages of 4 and 18. The disease is usually expressed as an arthritis, but another frequent manifestation is an inflammation of the heart, which damages the valves. Subcutaneous nodules at joints often accompany this stage (Figure 23.5). Other symptoms include fever and malaise. Although rheumatic fever usually does no permanent damage to the joints, there can be permanent heart damage.

In addition, as many as 10% of people with rheumatic fever develop **Sydenham's chorea,** an unusual complication known in the Middle Ages as St. Vitus' dance. Several months following an episode of rheumatic fever, the patient (much more likely to be a girl than a boy), exhibits purposeless, involuntary movements during waking hours.

Occasionally, sedation is required to prevent self-injury from flailing arms and legs. The condition disappears in a few months.

Rheumatic fever is usually precipitated by a streptococcal sore throat. One to five weeks later, probably after the original infection has disappeared, evidence of heart abnormalities appears. Reinfection with streptococci (for example, another sore throat) renews the attack and causes further damage to the heart. Perhaps as many as 3% of children with untreated β-hemolytic streptococcal infections contract rheumatic fever. However, many of these cases are essentially subclinical.

The exact mechanism by which streptococci produce rheumatic fever is still obscure, but some data point to an immunological reaction that is localized in the heart and joints. For example, infection with group A β-hemolytic streptococci may result in deposition of streptococcal antigens such as M proteins in the joints and heart. When the body produces antibodies against the antigens, the antigen–antibody reaction might cause the damage. It is also possible that a streptococcal antigen is cross-reactive with components of heart muscle and produces antibodies that subsequently damage the heart. Individuals with symptoms of damage from rheumatic fever do have relatively high antibody titers against streptococcal antigens. Genetic factors might also play a role in this disease. People of certain HLA types (see Chapter 19) are about 15 times more at risk than are the rest of the population.

Early in this century, rheumatic fever killed more school-age children in the U.S.A. than all other diseases combined. For reasons never clearly established, the incidence declined steadily even before the introduction of effective antimicrobials in the 1930s and 1940s. In the 1940s, especially among military personnel, there were still about 250,000 cases annually; by the 1980s, the disease had become rare in the U.S.A. However, beginning about 1984, clusters of cases have occurred in various parts of the country, primarily in children of middle-class families. The long decline of the disease in this country is thought to be due to some loss of virulence in the streptococci. The recent outbreaks have usually been associated with certain serological types of M protein and a heavily mucoid growth habit on blood agar. Because streptococci have remained sensitive to penicillin, the disease is not likely to again cause as many deaths as it did earlier in the century. However, group A streptococcal infections have recently reemerged as a public health problem in the U.S.A. (see the box on pp. 564–565). In addition, rheumatic fever remains a major health problem in Africa, South America, and much of Asia. (In India alone, an estimated 6 million children are affected.)

Penicillin, usually as long-lasting benzathine penicillin, is administered to patients with rheumatic fever as a prophylactic. The penicillin will not alleviate the symptoms (which are treated with anti-inflammatory drugs), but it is intended to prevent subsequent infections that could cause a recurrence of rheumatic fever.

MMWR

Morbidity and Mortality Weekly Report

INVASIVE GROUP A STREPTOCOCCAL BACTEREMIA— COLORADO

FROM JANUARY through August 1989, invasive group A *Streptococcus* (IGAS) was isolated from blood cultures obtained from 19 patients who had been admitted to a hospital in metropolitan Denver, Colorado. In comparison, this organism was cultured from blood from eight patients in 1988 and three in 1987 at this hospital (see the graph). The Colorado Department of Health was notified of this increase, and in August the department reviewed these patients' medical charts to describe IGAS bacteremia cases and to determine whether they represented community-acquired disease, nosocomial acquisition, or laboratory artifact.

Seven patients had no identified source of bacteremia, four had a possible cutaneous source of infection (cellulitis, impetigo, open sores, or an abscess), seven had pneumonia, and one had both cutaneous and respira-

tory infections. Two of the seven without identified sources of infection developed IGAS bacteremia following major trauma in motor vehicle crashes. A third patient developed IGAS bacteremia following elective laser hemorrhoidectomy (removal of hemorrhoids). Two patients developed acute respiratory distress syndrome, two developed acute renal failure, and nine (47%) died after developing IGAS bacteremia.

Retrospective surveys of metropolitan Denver hospital microbiology laboratories identified a rate of seven

IGAS cases per 100,000 per year in the general population of the Denver metropolitan area. Ongoing case investigation is examining potential explanations for the increasing incidence of group A streptococcal disease and possible risk factors, such as age, underlying illnesses, socioeconomic status, and race/ethnicity.

Editorial Note: During the last half century, the incidence of severe infections with group A *Streptococcus* (including bacteremia, puerperal sepsis, and scarlet fever) and sequelae (such

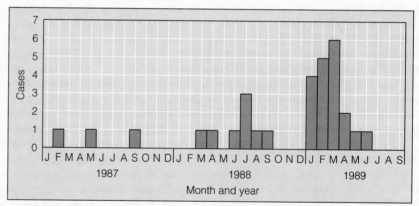

Invasive group A streptococcal bacteremia cases in a hospital—Denver, Colorado, January 1987–September 1989.

Tularemia

Tularemia is a disease caused by a small, gram-negative, facultatively anaerobic, pleomorphic, rod-shaped bacterium, *Francisella tularensis*. The microorganism was named for Tulare County, California, where it was originally observed in ground squirrels in 1911.

Tularemia exhibits one of several forms, depending on whether the infection is acquired by inhalation, ingestion, bites, or, most commonly, contact through minor skin breaks. The first sign of infection is usually local inflammation and a small ulcer at the site of infection. About a week after infection, the regional lymph nodes

enlarge; many will contain pockets filled with pus. If the disease is not contained by the lymphatic system, the microorganisms can produce septicemia, pneumonia, and abscesses throughout the body. Ingestion of infected, inadequately cooked meat leads to a focus of infection in the mouth or throat. The highest mortality occurs with pneumonic tularemia.

Humans most frequently acquire the infection through minor skin abrasions and by rubbing the eyes after handling small wild mammals. Probably 90% of the cases in this country are contracted from rabbits; it is estimated that 1% of all American rabbits are infected. The disease as rheumatic fever) has decreased markedly in the U.S.A. Potential explanation for this decline may relate to antibiotic therapy, improved living conditions, and decreased virulence of the infecting organism. Recently, however, group A streptococcal infections have re-emerged as a public health problem. In 1985 and 1986, clusters of patients with rheumatic fever were reported from several areas, and since 1987 a severe syndrome associated with group A streptococcal infection (streptococcal toxic-shock–like syndrome) has been recognized.

The findings in Denver are consistent with a trend toward increasing rates of IGAS bacteremia. Although a direct comparison of the IGAS bacteremia incidence rate in Denver for 1989 and for previous years is not possible because of incomplete laboratory records, the 1989 rate is more than double the most recent population-based estimate of three cases per 100,000 individuals per year from Charleston County, South Carolina (1985–1987). In addition, during 1989, CDC received an increased number of reports of IGAS bacteremia from other areas in the U.S.A., several

Scandinavian countries, England, and Wales. These reports suggest widespread changes in the epidemiology of group A streptococcal disease.

One form of IGAS, necrotizing fasciitis, is characterized by destruction of muscle and fat tissue. Development of IGAS appears to be facilitated by the presence of specific virulent strains and predisposing host factors. Protease activity is significantly associated with necrotizing fasciitis. Penicillin is the treatment of choice and surgical intervention is usually needed in cases of necrotizing fasciitis.

IGAS bacteremia occurs primarily in the elderly. Underlying medical conditions in affected people may include chronic debilitating disease, immunosuppressive illness or medication, and intravenous drug use. In addition, surgery, traumatic injury, or other disruption of the cutaneous barrier may predispose people to invasive infection. Despite appropriate therapy, illness can progress to shock, disseminated intravascular coagulation, and death; reported mortality rates range from 5% to 45%.

Factors contributing to the apparent recent increase in IGAS bacteremia are unclear. No single serotype,

to suggest a common source, was found in the isolates from Denver. Serologic typing of group A streptococcal strains based on antigenic differences in the M-protein suggests that an increase in the proportion of virulent organisms may play a role. M-types 1 and 3, comprising 75% of the Denver isolates, may be more invasive and more likely to cause clusters of infection than most other streptococcal M-types. Analysis of serotyping data from isolates submitted to CDC since 1972 shows an increase in the proportion of these two M-types in the 1980s.

Efforts to prevent and control invasive group A streptococcal disease should be directed at surveillance to establish the incidence of this problem in different geographic areas; further epidemiologic studies of toxic-shock–like syndrome; and improved understanding of the roles of immune responses, exotoxin, and other virulence factors. Physicians should continue to diagnose and treat mild streptococcal infections to prevent their progression to severe invasive disease.

SOURCES: *MMWR* 39(1): 3–6 (1/12/90); *MMWR* 43(21): 401–402 (6/3/94).

can also be spread by the bites of arthropods, such as deer flies (see Figure 12.30c), ticks, or rabbit lice, particularly in the western states; contact with small animals is a more likely cause in the central and eastern states (Figure 23.6).

Although no well-defined toxins have been identified to account for its virulence, *F. tularensis* survives for long periods within body cells, including phagocytes. In fact, resistance to phagocytosis might account for the low infective dose required. Naturally acquired immunity is usually permanent, but recurrences have been reported. An attenuated live vaccine is available for high-risk laboratory workers.

Serological diagnosis can be done with an agglutination test that becomes positive after about ten days of infection. A rising titer is diagnostic. The organism is considered so dangerous to handle (because of the ease with which aerosol infections are transmitted) that such diagnosis should not be attempted without special isolation hoods.

Streptomycin and gentamicin are the antibiotics of choice, but prolonged administration is necessary to prevent relapses. Before the days of antibiotics, mortality rates were as high as 15%. The intracellular location of the bacterium is a problem in chemotherapy.

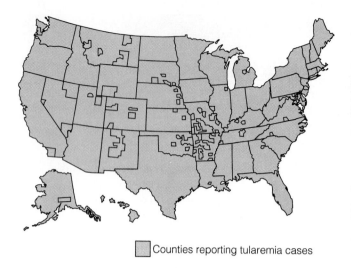

Counties reporting tularemia cases

FIGURE 23.6 Tularemia distribution. The geographical distribution, by county, of 163 cases of tularemia reported in the U.S.A. during 1992. (*Source:* CDC.)

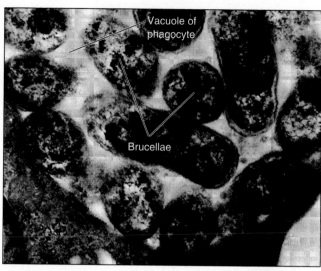

TEM ⊢—————⊣ 0.25 μm

FIGURE 23.7 *Brucella abortus*, the cause of brucellosis. Phagocyte containing *B. abortus* in a vacuole. The organisms are able to survive and reproduce in phagocytic cells, where chemical and immunological defenses have difficulty reaching them.

Brucellosis (Undulant Fever)

Like the bacteria that cause tularemia, the *Brucella* that cause **brucellosis** favor intracellular growth, move through the lymphatic system, and travel to organs via the bloodstream. These bacteria are very small, gram-negative, aerobic rods. They are difficult to grow on artificial media because they require that carbon dioxide be added to the atmosphere of the incubator, and they also require specially enriched media.

The most common *Brucella* species in the U.S.A. is *Brucella abortus* (ä-bôr′tus; Figure 23.7). It is found in cattle and is transmitted by unpasteurized milk products. The disease caused by this species is usually mild and often self-limited. The most common form in the rest of the world is *Brucella melitensis* (me-li-ten′sis), which has a reservoir in goats, sheep, and camels and is also transmitted by unpasteurized milk. The disease it causes is severe and often results in disability or death. *Brucella suis* (sü′is) is transmitted by contact with swine. Infections by *B. suis* are distinguished by the formation of destructive abscesses. All species are closely related genetically and are sometimes considered biovars of *B. melitensis*.

Brucella cells tend to multiply in the uterus of a susceptible animal, where their growth is favored by the presence of a carbohydrate, *mesoerythritol,* that is produced in the fetus and the membranes surrounding it. When the animal is not pregnant, it excretes the bacteria in milk. About 100 or fewer cases of brucellosis are reported in the U.S.A. each year, but there are few if any deaths. Because of pasteurization, dairy products are involved in only about 10% of brucellosis cases today; most cases occur as a result of contact with diseased animal tissue by farmers, veterinarians, and meat packers.

The organisms apparently enter the human body by passing through minute abrasions in the skin or mucous membrane of the mouth, throat, or gastrointestinal tract. *B. melitensis* and *B. suis* are the most easily transmitted, and some think that they may even be able to penetrate intact skin. Once in the body, the microorganisms are ingested by macrophages, in which they multiply and travel via the lymphatic system to the lymph nodes. From there, the microorganisms can be transported to the liver, spleen, or bone marrow. The ability of the microorganisms to grow inside the phagocytes partly accounts for their virulence and resistance to antibiotic therapy and antibodies. If any immunity is achieved, it is not reliable. A vaccine is used for cattle but not for humans.

Brucellosis is highly variable in its signs and symptoms because different organisms infect different organs. Many cases are subclinical; others are acute with fever and chills. Chronic brucellosis may last more than a year and may have very low-grade symptoms. The fever, particularly with *B. melitensis,* typically spikes to about 40°C (104°F) each evening; that is why the disease is sometimes called *undulant fever.*

Because the symptoms are difficult to interpret, serological testing is important in the diagnosis of brucellosis. Several tests are in use, the most common being an agglutination test. An ELISA test is considered more specific and sensitive, however. A simple agglutination test used on serum or milk to identify brucellosis-positive cattle has been an important factor in establishing a number of brucellosis-free areas.

FIGURE 23.8 A cutaneous pustule of anthrax on a human arm. If this localized infection is not contained by the body's defenses, septicemia might be the next stage. Anthrax is caused by *Bacillus anthracis.*

FIGURE 23.9 Toes of a patient with gangrene. This disease is caused by *Clostridium perfringens* and other clostridia. The black, necrotic tissue, resulting from poor circulation or injury, furnishes anaerobic growth conditions for the bacteria, which then progressively destroy adjoining tissue.

Treatment must be prolonged because of the intracellular growth habit of the bacteria. Frequently, combinations of tetracycline plus streptomycin are used for several weeks.

Anthrax

In 1877, Robert Koch isolated *Bacillus anthracis,* the bacterium that causes **anthrax** in animals. The endospore-forming bacillus is a large, aerobic, gram-positive microorganism that is apparently able to grow slowly in soil types having specific moisture conditions. The endospores have survived in soil tests for up to 60 years. The disease strikes primarily grazing animals, such as cattle and sheep. The *B. anthracis* endospores are ingested along with grasses, causing a fulminating, fatal septicemia.

There is often less than one case of human anthrax on average in the U.S.A. each year. People at risk are those who handle animals, hides, wool, and other animal products from certain foreign countries. Goat hair and handicrafts containing animal hides from the Middle East have been a repeated source of infection.

If contact is made with materials containing anthrax endospores, the organisms can enter the skin through a cut or abrasion and cause a pustule there (Figure 23.8). This pustular infection is sometimes kept localized by the defenses of the body, but there is always danger of septicemia. Probably the most dangerous form of anthrax is pulmonary anthrax, contracted when the endospores are inhaled. *Woolsorters' disease,* a dangerous form of pneumonia, results. It begins abruptly with high fever, difficulty in breathing, and chest pain. This disease eventually results in septicemia, and the mortality rate is high.

In humans, anthrax is best diagnosed by isolation of the bacterium. A number of morphological and biological tests can be used for precise identification. Penicillin is the drug of choice in human treatment. However, once septicemia is well advanced, antibiotic therapy may prove useless, probably because the exotoxins remain despite the death of the bacteria.

Gangrene

If a wound causes the blood supply to be interrupted (a condition known as *ischemia*), the wound becomes anaerobic. Ischemia leads to *necrosis,* or death of the tissue. The death of soft tissue from loss of blood supply is called **gangrene** (Figure 23.9). These conditions can also occur as a complication of diabetes.

Substances released from dying and dead cells provide nutrients for many bacteria. Various species of the genus *Clostridium,* which are gram-positive, endospore-forming anaerobes widely found in soil and in the intestinal tracts of humans and domesticated animals, grow readily in such conditions. *Clostridium perfringens* is the species most commonly involved in gangrene, but other clostridia and several other bacteria can also grow in such wounds.

Once ischemia and the subsequent necrosis have developed, **gas gangrene** can develop, especially in muscle tissue. As the *C. perfringens* microorganisms grow, they ferment carbohydrates in the tissue and produce gases (carbon dioxide and hydrogen) that swell the tissue. The bacteria produce toxins that move along muscle bundles, killing cells and producing necrotic tissue that is favorable for further growth. Eventually, these toxins and bacteria enter the bloodstream and cause systemic illness. Enzymes produced by the bacteria degrade collagen and proteinaceous tissue, facilitating spread of the disease. Without treatment, the condition is uniformly fatal.

FIGURE 23.10 Hyperbaric chambers used to treat gas gangrene. A multiplace hyperbaric chamber that can accommodate up to 12 patients is shown in the foreground. Such chambers are usually available at major medical centers. In the left background is a smaller, monoplace chamber designed to accommodate a single patient.

One complication of improperly performed abortions is the invasion of the uterine wall by *C. perfringens,* which resides in the genital tract of about 5% of all women. This infection can lead to gas gangrene and result in a life-threatening invasion of the bloodstream.

The surgical removal of necrotic tissue is called debridement; it and amputation are the most common medical treatments for gas gangrene. When gas gangrene occurs in such regions as the abdominal cavity, the patient can be treated in a *hyperbaric chamber,* which contains a pressurized oxygen-rich atmosphere (Figure 23.10). The oxygen saturates the infected tissues and thereby prevents the growth of the obligately anaerobic clostridia. Small chambers are available that can accommodate a gangrenous limb. Prompt cleaning of serious wounds and precautionary antibiotic treatment are the most effective steps in the prevention of gas gangrene. Penicillin is effective against *C. perfringens.*

Systemic Diseases Caused by Bites and Scratches

Animal bites can result in serious infections. Most serious bites are by domestic animals, such as dogs and cats, because they live in close contact with humans. It has been reported that 69% of dog bites become infected unless effectively cleansed. Domestic animals often harbor *Pasteurella multocida* (pas-tyér-el'lä mul-tō'si-dä), a nonmotile, gram-negative rod similar to the *Yersinia* bacterium that causes plague. *P. multocida* is primarily a pathogen of animals; it causes septicemia, hence the name *multocida,* meaning many killing. Humans infected with *P. multocida* may show varied responses. For example, local infections

with severe swelling and pain can develop at the site of the wound. Development of forms of pneumonia and septicemia is possible and life-threatening. Penicillin and tetracycline are usually effective in treatment of these infections.

Clostridium species and other anaerobes, such as species of *Bacteroides* and *Fusobacterium,* can also infect deep animal bites.

Rat Bite Fever

Rat bites are typically infected by organisms of the genus *Streptobacillus* (strep-tō-bä-sil'lus) or *Spirillum.* These infections occur more commonly in Asiatic countries than in the U.S.A. The disease, often called **rat bite fever,** is characterized by recurring fever, arthritislike symptoms (inflammation, pain, stiffness), and infections of the lymphatic vessels.

Bites by Humans

In one typical recent year, there were 210 rat bites of humans reported in New York City—and 760 human bites of humans! Bites by humans are even more prone to infection and complications such as septicemia and tissue destruction than are animal bites. Bites can be self-inflicted, such as a bitten lip or one's clenched fist striking another's unoffending tooth. The infections are often mixed; most frequently isolated are *Staphylococcus aureus,* hemolytic streptococci, and anaerobes such as *Bacteroides* and *Fusobacterium.* The wounds should be thoroughly washed and debrided and, to further avoid anaerobic conditions, perhaps not closed for a few days. Penicillin is the usual initial choice for antibiotic treatment.

Cat-Scratch Disease

Encounters with animals that result in minor scratches can cause serious illness. Because cats are the most frequent animal involved, the condition is known as **cat-scratch disease;** the cats themselves do not appear ill. Symptoms begin with a papule at the infection site that progresses to a rash, swelling of lymph nodes, and prolonged fever. Of an estimated 24,000 cases that occur each year in the U.S.A., about 2000 require hospitalization, and deaths occasionally occur. The causative organism is of uncertain identity. Some believe that a gram-negative rod, *Afipia felis* (ä-fi'pē-ä fe fē'lis), is the pathogen. Others propose that a rickettsialike bacterium, *Rochalimaea henselae* (rō-chä-lē'mä-ä hen'sel-ī) causes the disease. A number of antibiotics, especially rifampin, are effective in treatment.

Plague

Few diseases have affected human history more dramatically than **plague,** known in the Middle Ages as the Black Death. This term comes from one of its characteristics, the dark blue areas of skin caused by hemorrhages. In the fourteenth century this disease destroyed perhaps one-fourth

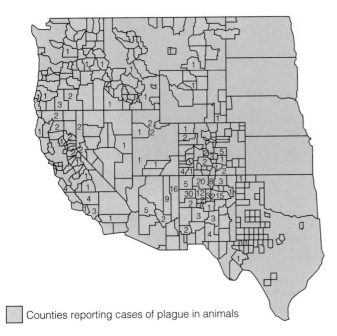

☐ Counties reporting cases of plague in animals

FIGURE 23.11 Plague distribution. Geographical distribution of human and animal plague in the U.S.A., 1970–1992. Counties reporting plague in animals are shown in gold; human cases are indicated by numerals. (*Source:* CDC.)

Bubo

FIGURE 23.12 A case of bubonic plague. Bubonic plague is caused by infection with *Yersinia pestis*. This photograph shows a bubo (swollen lymph node) under the arm of a patient. Swollen lymph nodes are a common indication of systemic infection.

of the total population of Europe. Near the turn of the twentieth century, as many as 10 million people are reported to have died from plague in India in a 20-year period. The number of plague cases reported in the U.S.A. has been increasing in recent years, and there are now about 10 to 15 cases annually (Figure 23.11). A mortality rate of 25% is not unusual.

The disease is caused by a gram-negative, rod-shaped bacterium, *Yersinia pestis*. Normally a disease of rats, plague is transmitted from one rat to another by the rat flea, *Xenopsylla cheopis* (ze-nop-sil′lä kē-ō′pis) (see Figure 12.30b). European rats, introduced into the U.S.A. many years ago, are the primary reservoirs of plague. In the U.S.A., particularly in the far West and Southwest, the disease is endemic in wild rodents, especially ground squirrels, prairie dogs, and chipmunks.

If its host dies, the rat flea seeks a replacement host, which may be another rodent or a human. A plague-infected flea is hungry for a meal because the growth of the bacteria blocks the flea's digestive tract, and the blood the flea ingests is quickly regurgitated. An arthropod vector is not always necessary for plague transmission. Contact from the skinning of infected animals, scratches of domestic cats, and similar incidents have been reported to cause infection.

In parts of the world where human proximity to rats is common, infection from this source still prevails. From the flea bite, bacteria enter the human's bloodstream and proliferate in the lymph and blood. One factor in the virulence of the plague bacterium is its ability to survive and

proliferate inside phagocytic cells rather than being destroyed by them. An increased number of highly virulent organisms eventually emerges, and an overwhelming infection results. The lymph nodes in the groin and armpit become enlarged, and fever develops as the body's defenses react to the infection. Such swellings are called *buboes,* which accounts for the name **bubonic plague** (Figure 23.12). The mortality rate of untreated bubonic plague ranges from 50% to 75%. Death from bubonic plague, if it occurs, is usually within less than a week after the appearance of symptoms.

A particularly dangerous condition arises when the bacteria are carried by the blood to the lungs and a form of the disease called **pneumonic plague** results. The mortality rate for this type of plague is nearly 100%. Even today, this disease can rarely be controlled if it is not recognized within 12 to 15 hours of the onset of fever. People with pneumonic plague usually die within three days. Like influenza, pneumonic plague is easily spread by airborne droplets from human or animal cases. Great care must be taken to prevent airborne infection of people in contact with patients. Contacts of pneumonic plague cases should receive prophylactic antibiotics.

Diagnosis of plague is most commonly done by isolating the bacterium and then identifying it by fluorescent-antibody and phage tests. People exposed to infection can be given prophylactic antibiotic protection. A number of antibiotics, including streptomycin and tetracycline, are effective. Recovery from the disease gives a reliable immunity. No vaccines are available, except for people likely to come into contact with infected fleas during field operations or for laboratory personnel exposed to the organism. The long-term effectiveness of the plague vaccine is not great.

Relapsing Fever

Except for the species that causes Lyme disease (discussed in the next section), all members of the spirochete genus *Borrelia* cause **relapsing fever.** In the U.S.A., the disease is transmitted by soft ticks that feed on rodents. The incidence of relapsing fever increases during the summer months, when the activity of rodents and arthropods increases. The disease is characterized by a fever, sometimes in excess of 40.5°C (105°F), jaundice, and rose-colored spots. After 3 to 5 days, the fever subsides. Three or four relapses may occur, each shorter and less severe than the initial fever. Each recurrence is caused by a different antigenic type of the spirochete. Diagnosis is made by observation of bacteria in the patient's blood, unusual for a spirochete disease.

Lyme Disease (Lyme Borreliosis)

In 1975, a cluster of disease cases in young people that was first diagnosed as rheumatoid arthritis was reported near the city of Lyme, Connecticut. The seasonal occurrence (summer months), lack of contagiousness among family members, and descriptions of an unusual skin rash that appeared several weeks before the first symptoms suggested a tick-borne disease. The fact that penicillin alleviated the progression of symptoms suggested a bacterial pathogen. In 1983, a spirochete that was later named *Borrelia burgdorferi* was identified as the cause (Figure 23.13a). **Lyme disease** may now be the most common tick-borne disease in this country. It is also found in Europe, China, Japan, and Australia. In the U.S.A., the disease is most prevalent on the Atlantic Coast (Figure 23.13b). Deer and field mice are the most important animal reservoirs that maintain the spirochete in the tick population.

The tick (one of two *Ixodes* spp.) feeds three times during its life cycle (Figure 23.14a). The first and second feedings, as a larva and then as a nymph, are usually on a field mouse. The third feeding, as an adult, is usually on a deer. These feedings are separated by several months, and the ability of the spirochete to remain viable in the disease-tolerant field mice is crucial to maintaining the disease in the wild. Other animals also serve as reservoirs for the pathogen. None of these wild animals shows symptoms of being affected by the disease, although dogs and horses sometimes do.

On the Pacific Coast, the tick is the western blacklegged tick *Ixodes pacificus* (iks-ō'dēs pas-i'fi-kus). In the rest of the country, it is mostly *Ixodes scapularis* (scap-ū-lăr'is). This latter tick is so small that it is often missed on the body

(a) The cause of Lyme disease, *Borrelia burgdorferi*

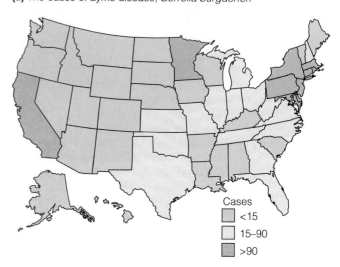

Cases
■ <15
□ 15–90
■ >90

(b) Distribution of Lyme disease in the U.S.A. in 1993

FIGURE 23.13 The causative agent and distribution of Lyme disease. [*Source:* Distribution data from *MMWR* 42(52) (1/7/94).]

(Figures 23.14b; see also Figure 12.29a). On the Atlantic Coast, almost all *Ixodes* ticks carry the spirochete; on the Pacific Coast few are infected, because that tick feeds on lizards that do not carry the spirochete effectively. The spirochete is maintained in the wild by a higher infection rate in a second tick that does not bite humans. However, it does infect a wood rat upon which both ticks feed.

The first symptom of Lyme disease is usually a rash that appears at the bite site. It is a red area that clears in the center as it expands to a final diameter of about 15 cm (Figure 23.15). This distinctive rash is seen in about 75% of the cases. Flulike symptoms appear in a couple of weeks as the rash fades. Antibiotics given during this interval are very effective in limiting the disease. During a second phase (when it occurs) there is often evidence that the heart is affected. Heartbeats may become so irregular that a pacemaker is required. Neurological symptoms such as facial paralysis, meningitis, and encephalitis may be seen. Months or years later, some patients develop arthritis that may affect them for years. Immune responses to the presence of the bacteria are probably the cause of this joint

FIGURE 23.14 **The tick vector (*Ixodes* spp.) of Lyme disease.**

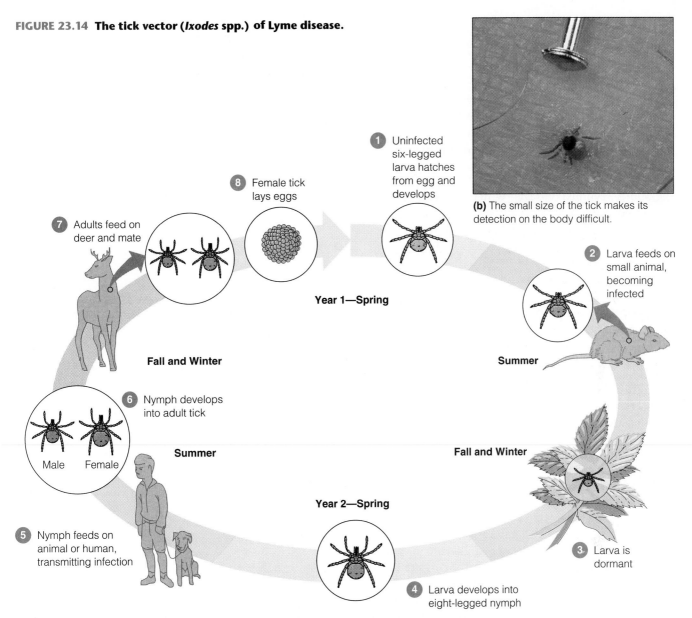

(1) Uninfected six-legged larva hatches from egg and develops

(b) The small size of the tick makes its detection on the body difficult.

(8) Female tick lays eggs

(7) Adults feed on deer and mate

Year 1—Spring

(2) Larva feeds on small animal, becoming infected

Fall and Winter

Summer

(6) Nymph develops into adult tick

Male   Female

Summer

Fall and Winter

Year 2—Spring

(3) Larva is dormant

(5) Nymph feeds on animal or human, transmitting infection

(4) Larva develops into eight-legged nymph

(a) The tick, *Ixodes scapularis*, has a two-year life cycle in which it requires three blood meals. The tick is infected by its first blood meal, and can pass on the infection to a human in its second.

damage. Many of the symptoms of long-term Lyme disease resemble those of the later stages of syphilis, also caused by a spirochete.

A number of antibiotics are effective in treatment of the disease, although in the later stages large amounts may be needed. Diagnosis of Lyme disease depends partly on the symptoms and an index of suspicion based on the prevalence in the geographical area. For the diagnosis of Lyme disease, an almost bewildering array of serological tests have been developed, but at this time it is unclear which give the best results. False-negative and false-positive results are quite common. Physicians are cautioned that serological tests must be interpreted in conjunction with clinical symptoms and the likelihood of exposure to infection.

FIGURE 23.15 **The classic "bull's-eye" rash of Lyme disease.**

Typhus

The various typhus diseases are caused by rickettsias (bacteria that are obligate intracellular parasites of eucaryotes). Rickettsias, which are spread by arthropod vectors, infect mostly the endothelial cells of the vascular system and multiply within them. The resulting inflammation causes local blockage of the small blood vessels. There are several related rickettsial diseases that differ mainly in their severity and in their arthropod vectors, including epidemic typhus, endemic murine typhus, and the spotted fevers (tick-borne typhus). Q fever, which is discussed in the next chapter (on respiratory diseases), is an atypical rickettsial disease caused by an atypical rickettsia.

Epidemic Typhus

Epidemic typhus (louse-borne typhus) is caused by *Rickettsia prowazekii* and carried by the human body louse *Pediculus humanus corporis* (ped-ik´ū-lus hü´ma-nus kôr´pô-ris) (see Figure 12.30a). The pathogen grows in the gastrointestinal tract of the louse and is excreted by it. The pathogen is not transmitted directly by the bite of an infected louse but rather is transmitted when the feces of the louse are rubbed into the wound when the bitten host scratches the bite. The disease can flourish only in crowded and unsanitary conditions, when lice can transfer readily from an infected host to a new host. In recent years, a number of cases of a mild form of epidemic typhus have been traced to contact with a reservoir in flying squirrels in the eastern U.S.A. There have been no deaths. Although the vector in these cases is unknown, it is probably something other than a louse.

Epidemic typhus disease produces a high and prolonged fever for two or more weeks. Stupor and a rash of small red spots caused by subcutaneous hemorrhaging are characteristic as the rickettsias invade blood vessel linings. Mortality rates are very high when the disease is untreated.

Tetracycline and chloramphenicol are usually effective against epidemic typhus, but the elimination of conditions in which the disease can flourish is more important in its control. The most sensitive diagnostic test is an indirect fluorescent-antibody test. The microbe is considered especially hazardous, and attempts to culture it require extreme care. Vaccines are available for military populations, which historically have been highly susceptible to the disease. Recovery from the disease gives a solid immunity and also renders a person immune to the related endemic murine typhus.

Endemic Murine Typhus

Endemic murine typhus occurs sporadically rather than in epidemics. The term *murine* (derived from Latin for mouse) refers to the fact that rodents, such as rats and squirrels, are the common hosts for this typhus. Endemic murine typhus is transmitted by the rat flea *Xenopsylla cheopis* (see Figure 12.30b), and the pathogen responsible for the disease is *Rickettsia typhi,* a common inhabitant of rats. Texas has had a number of outbreaks of murine ty-

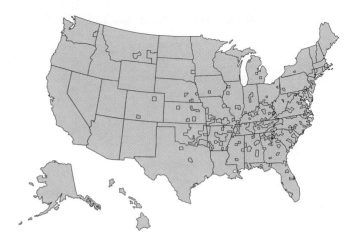

Counties reporting cases of Rocky Mountain spotted fever

FIGURE 23.16 Distribution of Rocky Mountain spotted fever (tick-borne typhus). Geographical distribution of the 628 cases reported in the U.S.A. in 1991. [*Source: MMWR* 40(53) (10/2/92).]

phus in recent years, often associated with campaigns to eliminate rodents, which caused the rat fleas to seek new hosts. With a mortality rate of less than 5%, the disease is considerably less severe than the epidemic form. Except for the reduced severity of the disease, endemic murine typhus is clinically indistinguishable from epidemic typhus.

Tetracycline and chloramphenicol are effective treatments for endemic murine typhus, and the diagnostic procedure for the disease is the same as for epidemic typhus. Rat control is the best preventive measure for the disease.

Spotted Fevers

Tick-borne typhus, or **Rocky Mountain spotted fever,** is probably the best-known rickettsial disease in the U.S.A. It is caused by *Rickettsia rickettsii.* Despite its name (it was first recognized in the Rocky Mountain area), it is most common in the southeastern U.S.A. and Appalachia (Figure 23.16). This rickettsia is a parasite of ticks and is usually passed from one generation of ticks to another through their eggs, a mechanism called *transovarian passage* (Figure 23.17). Surveys show that in endemic areas perhaps 1 out of every 1000 ticks is infected. The rickettsias do not kill the tick. In different parts of the country, different ticks are involved—in the West, the wood tick *Dermacentor andersoni* (dėr-mä-sen´tôr an-dėr-sōn´ē), in the East the dog tick *Dermacentor variabilis* (vār-ē-a´bil-is).

About a week after the tick bites, a rash develops that is sometimes mistaken for measles (Figure 23.18); however, it often appears on palms and soles, which does not occur with viral rashes. The rash is evidence of the vascular damage, resulting in leakage of blood into surrounding tissues, that is characteristic of many rickettsial infections. The rash is accompanied by fever and headache. Death, which occurs in about 3% of the approximately 1000 cases re-

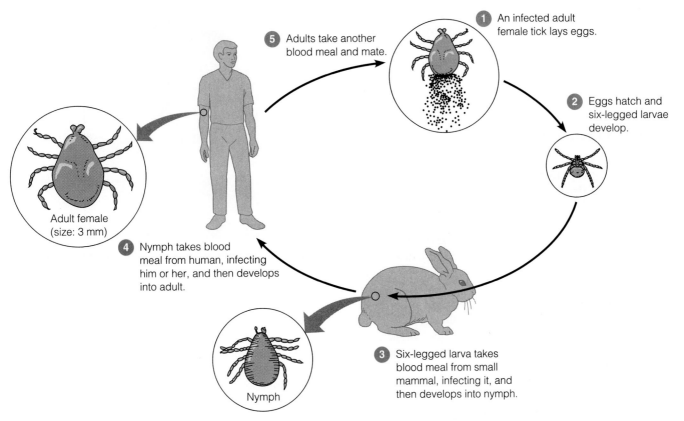

FIGURE 23.17 Life cycle of the tick vector (*Dermacentor* spp.) of Rocky Mountain spotted fever. Mammals are not essential to survival of the pathogen, *Rickettsia rickettsii,* in the tick population; the bacteria may be passed by transovarian passage, so new ticks are infected upon hatching. A blood meal is required for ticks to advance to the next stage in the life cycle.

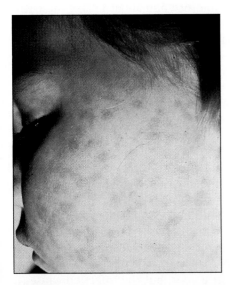

FIGURE 23.18 The rash caused by Rocky Mountain spotted fever. This rash is often mistaken for measles. People with dark skin have a higher mortality rate; often the rash is not recognized early enough for effective treatment.

ported each year, is usually caused by kidney and heart failure. Antibiotics such as tetracycline and chloramphenicol are very effective if administered early enough. No vaccine is available.

Serological tests are available but do not become positive until late in the illness. Diagnosis before the typical rash appears is difficult; symptoms vary widely. A misdiagnosis can be costly—if treatment is not prompt and correct, the mortality rate is about 20%.

In addition, there are a number of other tick-borne rickettsial diseases known by local names in various regions of the world.

VIRAL DISEASES OF THE CARDIOVASCULAR AND LYMPHATIC SYSTEMS

LEARNING OBJECTIVES

- *Describe the epidemiologies of Burkitt's lymphoma and infectious mononucleosis.*
- *Compare and contrast the causative agents, vectors, reservoirs, symptoms, and treatments for yellow fever and dengue.*

Viruses are the cause of a number of cardiovascular and lymphatic diseases, mostly prevalent in tropical areas. However, one viral disease of this type, infectious mononucleosis, is an especially familiar infectious disease among college-age individuals in the U.S.A.

Burkitt's Lymphoma

In the 1950s, Denis Burkitt, a physician working in eastern Africa, noticed the frequent occurrence in young children of a fast-growing tumor of the jaw (Figure 23.19). Now well known as **Burkitt's lymphoma,** this is the most common childhood cancer in Africa. The disease has a limited geographical distribution similar to that of malaria in central Africa. Burkitt suspected a viral cause of the tumor and a mosquito vector. At that time there was no known virus that caused human cancer, although several viruses were clearly associated with animal cancers. Intrigued by this possibility, the British virologist Tony Epstein and his student Yvonne Barr performed biopsies on the tumors. A virus was cultured from this material, and the electron microscope showed a herpeslike virus in the culture cells—the Epstein–Barr (EB) virus. EB virus is clearly associated with Burkitt's lymphoma, but the mechanism by which the virus causes the tumor is not understood. The cancer is known to occur in people who lack EB antibodies. Research eventually showed, however, that mosquitoes do not transmit the virus or the disease. Instead, mosquito-borne malarial infections apparently foster the development of Burkitt's lymphoma by impairing the immune response to the EB virus, which is almost universally present in human adults. In areas without endemic malaria, such as the U.S.A., Burkitt's lymphoma is rare. The appearance of the lymphoma in AIDS patients is an indication of the importance of immune surveillance in preventing expression of the disease. Another medical problem that was later associated with the EB virus is **nasopharyngeal carcinoma.** This disease is a major cause of death in Southeast Asia, where it has an incidence of 10–20 cases per 100,000 population.

Infectious Mononucleosis

The EB virus was the subject of one of the accidental discoveries that often advance science. A technician in a laboratory investigating the EB virus was being used as the negative control in serological tests for the EB virus. While on vacation she contracted an infection, characterized by fever, sore throat, swollen lymph nodes in the neck, and general weakness—the symptoms familiar to many young people as **infectious mononucleosis,** or mono. The most interesting facet of the technician's disease was that she became serologically positive for the EB virus. It was soon confirmed that the same virus that is associated with Burkitt's lymphoma also causes infectious mononucleosis.

FIGURE 23.19 A child with Burkitt's lymphoma. Cancerous tumors of the jaw closely associated with Epstein–Barr virus are seen mainly in children. This child was successfully treated.

In less developed parts of the world, infection with the EB virus is acquired in early childhood and 90% of the children over four years of age have acquired antibodies. Childhood infections by the EB virus are usually asymptomatic, but if infection is delayed until young adulthood, as is often the case in the U.S.A., the result is infectious mononucleosis. The peak incidence of the disease in this country occurs at about 15–25 years of age. College populations, particularly those from the upper socioeconomic strata, have a high incidence of the disease. Most college students have no immunity, and about 15% of these students can expect to contract the disease. The disease is generally self-limiting and seldom fatal. A principal cause of the rare deaths is rupture of the enlarged spleen (a common response to a systemic infection) during vigorous activity. Recovery is usually complete in a few weeks, and immunity is permanent.

The usual route of infection is by transfer of saliva by kissing or, for example, by sharing drinking vessels. Reproduction of the virus appears to occur mainly in the parotid glands, which accounts for its presence in saliva. It does not spread among casual household contacts, so aerosol transmission is unlikely. The incubation period before appearance of symptoms is from four to seven weeks.

Although reproduction is considered to occur in epithelial cells of the parotid glands, the infection is almost exclusively selective for B cells. The virus does not reproduce significantly in B cells, but the infection causes B cells to transform into plasma cells, which reproduce rapidly. These are then attacked by cytotoxic T cells of the cell-mediated immune system. The symptoms of infectious mononucleosis are associated with B-cell proliferation and the immune response. After recovery, the virus remains latent in a small number of B cells.

The disease name *mononucleosis* refers to lymphocytes with unusual lobed nuclei that proliferate in the blood during the acute infection. The infected B cells produce nonspecific antibodies called heterophil antibodies. Older diagnostic tests depended heavily on blood counts of the distinctive lymphocytes and on detection of the heterophil antibodies. A fluorescent-antibody test that detects IgM antibodies against the EB virus is the most specific diagnostic method.

Yellow Fever

Arboviruses, or arthropod-borne viruses, can reproduce either in arthropods, such as mosquitoes, or in humans. **Yellow fever** is caused by an arbovirus (*yellow fever virus*) and is historically important because it was the first such virus discovered and provided the first confirmation that an insect could transmit a virus.

The yellow fever virus is injected into the skin by the mosquito. The virus then spreads to local lymph nodes, where it multiplies; from the lymph nodes, it advances to the liver, spleen, kidney, and heart, where it can persist for days. In the early stages of the disease, the person experiences fever, chills, headache, and backache, followed by nausea and vomiting. This stage is followed by jaundice, a yellowing of the skin that gave the disease its name. The jaundice is caused by the deposition of bile pigments in the skin and mucous membranes as a result of liver damage. The mortality rate for yellow fever is high, about 20%.

Yellow fever is still endemic in many tropical areas, such as Central America, tropical South America, and Africa. Monkeys are a natural reservoir for the virus. Control of the disease depends on local control of *Aedes aegypti* (ä-e′dēz ē-jip′tē), the usual mosquito vector. Immunization of the exposed population is also an effective control.

There is no specific treatment for yellow fever. Diagnosis is usually by clinical signs, but it can be confirmed by a rise in antibody titer or isolation of the virus from the blood. The vaccine used is an attenuated live viral strain and yields a very effective immunity with few adverse effects.

Dengue

Dengue (den′ghē) is a similar but milder viral disease also borne by the *Aedes aegypti* mosquito. This disease, endemic in the Caribbean and other tropical environments around the world, is caused by *dengue fever virus,* an arbovirus. It is characterized by fever, severe muscle and joint pain, and rash. Except for the painful symptoms, which have led to the name **breakbone fever,** classic dengue fever is a relatively mild disease and is rarely fatal.

The countries surrounding the Caribbean are reporting an increasing number of cases of dengue. In most years, more than 100 cases are imported into the U.S.A., mostly by travelers from the Caribbean and South America. The mosquito vector for dengue is common in the Gulf states,

and there is some concern that the virus will sooner or later be introduced into this region and become endemic. A special apprehension of health officials is the introduction into the U.S.A. of an Asian mosquito, *Aedes albopictus* (al-bō-pik′tus), that is an efficient carrier of the virus and an aggressive biter. It transmits the virus by transovarian passage and from person to person. The range of this mosquito can potentially cover much of the country (see the box in Chapter 12, p. 324). Control measures are directed at eliminating the *Aedes* mosquitoes. No specific treatment is effective against the virus.

PROTOZOAN DISEASES OF THE CARDIOVASCULAR AND LYMPHATIC SYSTEMS

LEARNING OBJECTIVES

- *Compare and contrast the causative agents, methods of transmission, reservoirs, symptoms, and treatments for toxoplasmosis, American trypanosomiasis, and malaria.*
- *Discuss the worldwide effects of these diseases on human health.*

Protozoa that cause diseases of the cardiovascular and lymphatic systems often have complex life cycles, and their presence may affect human hosts seriously. Malaria, a devastating disease in much of the world, is included in this group, along with Chagas' disease and toxoplasmosis.

Toxoplasmosis

Toxoplasmosis, a disease of blood and lymphatic vessels, is caused by the small, crescent-shaped protozoan *Toxoplasma gondii. T. gondii* is a spore-forming protozoan (sporozoan), like the malarial parasite.

Cats are an essential part of the life cycle of *T. gondii* (Figure 23.20). Random tests on urban cats have shown that a large number of them are infected with the organism, which causes no apparent illness in the cat. The organism undergoes its only sexual phase in the intestinal tract of the cat. Oocysts are then shed in the cat's feces and contaminate food or water that can be ingested by other animals. The *oocysts* contain *sporozoites* that invade host cells and form trophozoites called *tachyzoites* (about the size of large bacteria, $2 \times 7 \mu m$). The intracellular parasite reproduces rapidly (*tachys* means rapid), and the increased numbers cause the rupture of the host cell, with release of more tachyzoites. As the immune system becomes increasingly effective, the disease enters a chronic phase in which the infected host cell develops a wall to form a *tissue cyst.* The numerous parasites within such a cyst (in this stage called *bradyzoites; bradys* means slow) reproduce very

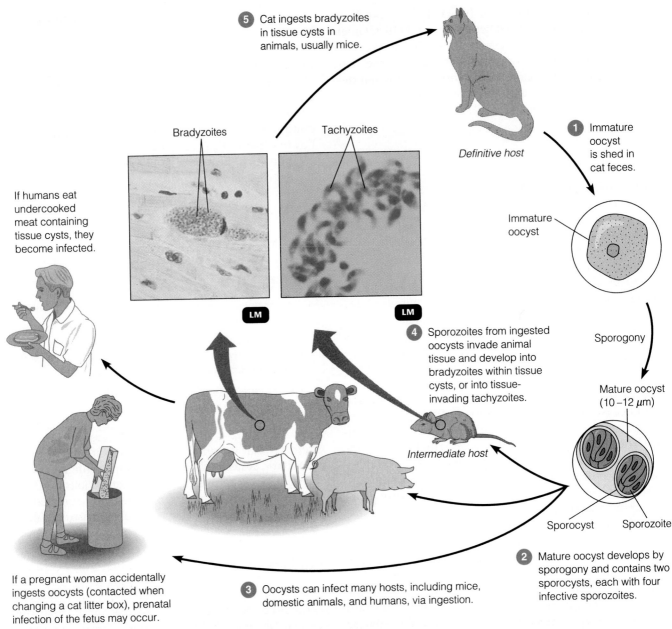

5 Cat ingests bradyzoites in tissue cysts in animals, usually mice.

Bradyzoites

Tachyzoites

Definitive host

1 Immature oocyst is shed in cat feces.

If humans eat undercooked meat containing tissue cysts, they become infected.

Immature oocyst

LM

LM

Sporogony

4 Sporozoites from ingested oocysts invade animal tissue and develop into bradyzoites within tissue cysts, or into tissue-invading tachyzoites.

Mature oocyst (10–12 μm)

Intermediate host

Sporocyst   Sporozoite

2 Mature oocyst develops by sporogony and contains two sporocysts, each with four infective sporozoites.

If a pregnant woman accidentally ingests oocysts (contacted when changing a cat litter box), prenatal infection of the fetus may occur.

3 Oocysts can infect many hosts, including mice, domestic animals, and humans, via ingestion.

FIGURE 23.20 Life cycle of *Toxoplasma gondii*, the cause of toxoplasmosis. The domestic cat is the definitive host, in which the protozoa reproduce sexually.

slowly, if at all, and persist for years, especially in the brain. Loss of immune function, AIDS being the best example, allows the infection to be reactivated from such cysts. Humans generally acquire the infection by ingestion of undercooked meats containing tachyzoites or tissue cysts, although there is a possibility of contracting the disease more directly by contact with cat feces.

The disease is a rather undefined mild illness, and the primary danger is in the congenital infection of a fetus. The effects on the fetus can be drastic, including convulsions, severe brain damage, blindness, and death. Fetal damage occurs only when the primary infection is acquired during pregnancy. Some surveys have shown that approximately 30% of the population carries antibodies for this organism; this percentage indicates the high rate of subclinical, unrecognized infections.

T. gondii can be isolated and grown in cell cultures. The preferred serological tests use an indirect fluorescent-antibody method or an ELISA test. Interpretation of tests to justify extreme action such as a therapeutic abortion is difficult. Toxoplasmosis is treated with pyrimethamine in combination with sulfadiazine.

FIGURE 23.21 *Trypanosoma cruzi,* **the cause of American trypanosomiasis (Chagas' disease).** The flagellum follows the outer margin of the undulating membrane and then extends beyond the trypanosome.

FIGURE 23.22 Distribution of American trypanosomiasis. The disease is found mostly in Mexico, Central America, and South America; a very small area of the U.S.A. (in southern Texas) is affected.

American Trypanosomiasis (Chagas' Disease)

American trypanosomiasis (Chagas' disease) is a protozoan disease of the cardiovascular system. The causative agent is *Trypanosoma cruzi* (tri-pa-nō-sō′mä kruz′ē), a flagellated protozoan (Figure 23.21). The disease occurs in the extreme southern U.S.A., Mexico, Central America, and into South America (Figure 23.22). Only a few cases have been reported in the U.S.A., but the disease infects 40% to 50% of the population in some rural areas of South America.

The reservoir for *T. cruzi* includes a wide variety of wild animals, including rodents, opossums, and armadillos. The arthropod vector is the reduviid bug, called the "kissing bug" because it often bites people near the lips (see Figure 12.30d). The insects live in the cracks and crevices of mud or stone huts with thatched roofs. The trypanosomes, which grow in the gut of the bug, are passed on if the bug defecates while feeding. The bitten human or animal often rubs the feces into the bite wound or other skin abrasions by scratching, or into the eye by rubbing. An unusual means of transmission occasionally occurs in remote areas of Mexico, where reduviid bugs are eaten as an aphrodisiac.

At the site of inoculation, the trypanosomes reproduce and pass through various stages of their life cycle. A swollen lesion develops at the inoculation site (often near the eye). The parasites are carried by the blood to many organs of the body, where they multiply in host cells and cause destructive inflammation of the affected tissues. The disease is most dangerous to children, whose death rate may be as high as 10%. Although many organs may be affected, the principal cause of death is damage to the heart. Growth in the brain is also a common development. If the parasites damage the nerves controlling the peristaltic action of the esophagus or colon, these organs no longer transport food and become grossly enlarged, conditions known as megaesophagus and megacolon, respectively.

Laboratory procedures for the diagnosis of American trypanosomiasis include observing the rather sparse population of trypanosomes in blood or growing the microorganisms from blood cultures. An extraordinary form of diagnosis, which is actually done with some frequency, is *xenodiagnosis*. Reduviid bugs, reared in parasite-free conditions, are allowed to feed on the arm of a suspected patient. The trypanosome is then identified in the intestinal tract of the insect 10–20 days later. Serological tests are not very satisfactory.

Treatment of this disease is very difficult when chronic, progressive stages have been reached. The trypanosome multiplies intracellularly and is difficult to reach chemotherapeutically. Nifurtimox, the drug of choice in treatment, decreases the number of parasites but is not a cure.

Malaria

Malaria is characterized by chills and fever and often by vomiting and severe headache. These symptoms typically appear at two- or three-day intervals that alternate with asymptomatic periods. Malaria is found wherever the mosquito vector *Anopheles* is found (see Figure 12.28b) and there are human hosts for the protozoan parasite *Plasmodium.* The disease was once common in the U.S.A., but effective mosquito control and a reduction in the number

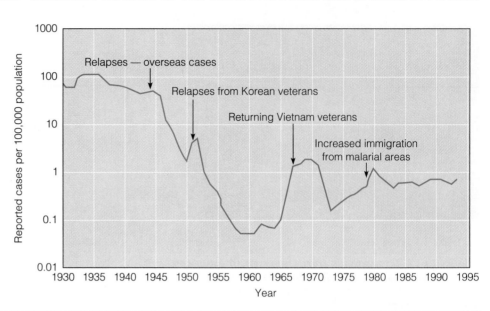

FIGURE 23.23 Reported cases of malaria in the U.S.A., 1930 to 1993. Malaria was a common disease as recently as 1935. Its decline coincided with government control programs such as the Tennessee Valley Authority (TVA) flood control program, the Work Projects Administration (WPA) malarial control program, and war area control programs during World War II. Note the recent increases in incidence as veterans returned from Vietnam and immigration from malarial areas increased. [*Sources:* CDC, *Summary of Notifiable Diseases 1992; MMWR 40(53) (10/2/92); MMWR 41(53) (1/8/93); MMWR 42(52) (1/7/94).*]

of human carriers caused the reported cases to drop below 100 by 1960 (Figure 23.23). In recent years, however, there has been an upward trend in the number of cases in the U.S.A., reflecting a worldwide resurgence of malaria, increased travel to malarial areas, and an increase in immigration from malarial areas. Occasionally, malaria has been transmitted by unsterilized syringes used by drug addicts, and blood transfusions from people who have been in an endemic area are also a potential risk. In tropical Asia, Africa, and Central and South America, malaria is still a serious problem. In fact, as the protozoa develop resistance to antimalarial drugs and as political instability continues to interfere with control measures, malaria is more often being described as a resurgent disease. Current estimates are that it affects some 300 million people worldwide each year and causes from 2 to 4 million deaths.

The causative organisms of malaria are the sporozoans of the genus *Plasmodium*. Four pathogenic species are recognized, each of which causes a distinctive form of the disease. The most dangerous species is *Plasmodium falciparum* (fal-sip'är-um), a species that is also widespread geographically. Also widely distributed is *P. vivax* (vī'vaks). *P. malariae* (mä-lā'rē-ī) and *P. ovale* (ō-vä'lē) cause geographically restricted, relatively milder diseases.

The mosquito carries the *sporozoite* form of the *Plasmodium* protozoa in its saliva (see Figure 12.21). The sporozoite enters the bloodstream of the bitten human and within about 30 minutes enters the liver cells. The sporozoites in the liver cells undergo reproductive schizogony by a series of stages that finally results in the release of large numbers of *merozoite* forms into the bloodstream. These merozoites infect red blood cells and undergo reproductive schizogony again. Laboratory diagnosis of malaria is made by examining a blood smear for infected red blood cells (Figure 23.24). Eventually, the red blood cells rupture and release large numbers of merozoites. This rupturing of the red blood cells is fairly simultaneous, and the release of toxic compounds is the cause of the paroxysms (recurrent intensifications of symptoms) of chills and fever that are the characteristic signs of malaria. The fever reaches 40°C (104°F), and a sweating stage begins as the fever subsides. Between paroxysms, the patient feels normal. Anemia results from the loss of red blood cells, and excessive enlargement of the liver and spleen is an added complication.

When the red blood cells rupture, many of the released merozoites infect other red blood cells within a few seconds to renew the cycle in the bloodstream. If only 1% of the red blood cells contain parasites, there are an estimated 100,000,000,000 parasites in circulation at one time in a typical malaria patient! Other merozoites develop into male or female *gametocytes*. When these enter the digestive tract of a feeding mosquito, they pass through a sexual cycle that produces new infective sporozoites. It took the combined labors of several generations of scientists to discover this complex life cycle of the malaria parasite.

The highest mortality from malaria is in young children. People who survive malaria acquire a limited immunity. Although they can be reinfected, they tend to have a less severe form of the disease. This relative immunity almost disappears if the person leaves an endemic area with its periodic reinfections. People who have the genetic sickle-cell trait—common in many areas where malaria is endemic—are relatively resistant to malaria.

Considerable effort is currently being expended on the search for an effective vaccine (see the box in Chapter 18, p. 462). The sporozoite stage is of primary interest as a target for a vaccine because neutralizing it would prevent the initial infection from becoming well established. However, some believe that any effective vaccine would have to protect against all three stages.

Malaria was once fairly effectively treated with quinine, but newer derivatives of quinine, such as primaquine and especially chloroquine, have been the mainstays for prevention and treatment of malaria for many years. Resis-

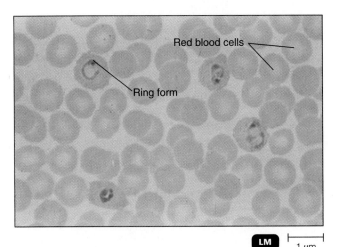

Red blood cells

Ring form

LM | 1 μm

FIGURE 23.24 Blood smear from a malaria patient. The protozoa grow within infected red blood cells (RBCs). The ring form is an early stage. The dark spot of the ring is the nucleus, and the light central area within the circular ring is the food vacuole of the parasite. The food in the vacuole is the cytoplasm of the RBC. In later stages, just before lysis of the RBC, the cell becomes crowded with developing merozoites (not shown here).

tance to these drugs is spreading rapidly, and current chemical alternatives are more expensive, which prohibits widespread use in the parts of the world most troubled by malaria. The latest drug of choice for many applications is mefloquine, a quinine derivative. Another approach to chemotherapy has been to use drug combinations such as Fansidar®, a combination of pyrimethamine and sulfadoxine. Both drugs are inhibitors of folic acid synthesis (see the discussion of sulfa drugs in Chapter 20); together, they have a synergistic effect. Resistance against such combinations is also emerging.

Effective control of malaria is not in sight. It will probably require a combination of chemotherapeutic and immunological approaches. The expense and the need for an effective political organization in malarial areas are probably going to be as important in controlling the disease as advances in medical research.

HELMINTHIC DISEASES OF THE CARDIOVASCULAR AND LYMPHATIC SYSTEMS

LEARNING OBJECTIVE

- *Diagram the life cycle of* Schistosoma. *Show where the cycle can be interrupted to prevent human disease.*

Many helminths find the cardiovascular system a convenient highway to travel as part of their life cycle. Schisto-

somes find a home there, shedding eggs that are distributed in the bloodstream.

Schistosomiasis

Schistosomiasis is a debilitating disease caused by a small fluke. Eggs shed by the fluke lodge in body organs and initiate a damaging inflammatory reaction. The disease is not contracted in the U.S.A., but more than 400,000 immigrants to the U.S.A. have the disease, and about 200 million people in Asia, Africa, South America, and the Caribbean are also affected. Worms of the genus *Schistosoma* (shis-tō-sō'mä) cause the disease; they vary with geographical location, but the characteristics of the disease are similar.

When waters become contaminated with *Schistosoma* ova excreted in human wastes (Figure 23.25), a motile, ciliated larval form of *Schistosoma* called a *miracidium* is released from the ova and enters certain species of snails. (The lack of a suitable host snail is a primary reason why schistosomiasis is not transmitted in the U.S.A.). Eventually, the pathogen emerges from the snail in an infective form called the *cercaria*. These cercariae have forked swimming tails. When they contact the skin of a person wading or swimming in the water, they discard the tail and enzymatically penetrate the skin. They are then carried by the bloodstream to the veins of the liver or urinary bladder, depending partly on the species of schistosome. There they mature into an adult form, 15 to 20 mm in length, in which the slender female lives in a groove in the body of the male. The union produces a supply of new ova, some of which cause local tissue damage (*granulomas*) from defensive body reactions (Figure 23.26). Other ova are excreted and enter the water to continue the cycle.

The adult worms appear to be unaffected by the host's immune system. Apparently, they quickly coat themselves with a layer that mimics the host's tissues. However, a previous infection tends to be protective to some degree against subsequent infections, which encourages work toward development of a vaccine.

Laboratory diagnosis consists of microscopic identification of the flukes or their ova in fecal and urine specimens, intradermal tests, and serological tests such as complement-fixation and precipitin tests.

Praziquantel and oxamniquine are approved for use against schistosomes in the U.S.A. Sanitation and elimination of the host snail are also useful forms of control.

Swimmer's Itch

Swimmers in lakes in the northern U.S.A. are sometimes troubled by **swimmer's itch**. This is a cutaneous allergic reaction to cercariae similar to that of schistosomiasis. However, these parasites mature only in wildfowl and not in humans, so infection does not progress beyond penetration of the skin and a local inflammatory response.

Male

Mouth

Sucker

Female
Mouth

SEM ⊢————⊣
10 μm

(a) Male and female schistosomes. The female lives in a groove on the ventral surface of the male schistosome ("split-body"), is continuously fertilized, and continuously lays eggs. The sucker is used by the male to attach to the host.

1 Adult flukes lay eggs

Adults
Female (size:15–20 mm)
Eggs
Male

8 Cercariae travel through circulatory system to intestinal blood vessels, where they mature into adults

Definitive host

2 Eggs reach body of water after being excreted in feces or urine

7 Free-swimming cercariae penetrate human skin, losing tail

Cercaria (0.33 mm)

Cercaria (0.13 mm)

Egg (0.15 mm)

6 Cercariae are released from snail

Intermediate host

3 Eggs hatch into free-swimming larva (miracidia)

5 Miracidium reproduces in snail, forming several cercariae

4 Miracidium penetrates snail

Miracidium (0.2 mm)

(b) Life cycle of *Schistosoma*, cause of schistosomiasis.

FIGURE 23.25 Schistosomiasis.

Egg

Granuloma

FIGURE 23.26 Granuloma from a patient with schistosomiasis. Some of the eggs laid by the adult schistosomes lodge in the tissue, and the body responds to the irritant by surrounding it with scarlike tissue, forming a granuloma.

LM

STUDY OUTLINE

INTRODUCTION (p. 558)

1. The heart, blood, and blood vessels make up the cardiovascular system.
2. Lymph, lymph vessels, lymph nodes, and lymphoid organs constitute the lymphatic system.

STRUCTURE AND FUNCTION OF THE CARDIOVASCULAR SYSTEM (p. 558)

1. The heart circulates substances to and from tissue cells.
2. Arteries and arterioles transport blood away from the heart. Venules and veins bring blood to the heart.
3. Materials are exchanged between blood and tissue cells at capillaries. Capillaries connect arterioles to venules.
4. Blood is a mixture of plasma and cells.
5. Plasma transports dissolved substances. Red blood cells carry oxygen. White blood cells are involved in the body's defense against infection.

STRUCTURE AND FUNCTION OF THE LYMPHATIC SYSTEM (pp. 558–560)

1. Fluid that filters out of capillaries into spaces between tissue cells is called interstitial fluid.
2. Interstitial fluid enters lymph capillaries and is called lymph.
3. Lymphatics return lymph to the blood.
4. Lymph nodes contain fixed macrophages, B cells, and T cells.

BACTERIAL DISEASES OF THE CARDIOVASCULAR AND LYMPHATIC SYSTEMS (pp. 560–573)

Septicemia (pp. 560–562)

1. The growth of microorganisms in blood is called septicemia.
2. Signs include lymphangitis (inflamed lymph vessels). Septicemia can lead to septic shock, characterized by decreased blood pressure.
3. Septicemia usually results from a focus of infection in the body.
4. Gram-negative rods are usually implicated. Endotoxin causes the symptoms.

Puerperal Sepsis (p. 562)

1. Puerperal sepsis begins as a uterine infection following childbirth or abortion; it can progress to peritonitis or septicemia.
2. *Streptococcus pyogenes* is the most frequent cause.

3. Oliver Wendell Holmes and Ignaz Semmelweiss demonstrated that puerperal sepsis was transmitted by the hands and instruments of midwives and physicians.
4. Puerperal sepsis is now uncommon because of modern hygienic techniques and antibiotics.
5. Anaerobic bacteria, including *Bacteroides* and *Clostridium,* can cause infections after improperly performed abortions.

Bacterial Endocarditis (pp. 562–563)

1. The outer layer of the heart is the pericardium. The inner layer is the endocardium.
2. Subacute bacterial endocarditis is usually caused by α-hemolytic streptococci, staphylococci, or enterococci.
3. The infection arises from a focus of infection, such as a tooth extraction.
4. Preexisting heart abnormalities are predisposing factors.
5. Signs include fever, anemia, and heart murmur.
6. Acute bacterial endocarditis is usually caused by *Staphylococcus aureus.*
7. The bacteria cause rapid destruction of heart valves.
8. *Streptococcus* can also cause pericarditis.
9. Inflammation of the heart muscle (myocardium) is called myocarditis.
10. Coxsackievirus (enterovirus) and *T. cruzi* are common causes of myocarditis.

Rheumatic Fever (p. 563)

1. Rheumatic fever is an autoimmune complication of group A β-hemolytic streptococcal infections.
2. Rheumatic fever is expressed as arthritis, inflammation of the heart, or Sydenham's chorea and can result in permanent heart damage.
3. Antibodies against group A β-hemolytic streptococci react with streptococcal antigens deposited in joints or heart valves or cross-react with the heart muscle.
4. Rheumatic fever can follow a streptococcal infection, such as streptococcal sore throat. Streptococci might not be present at the time of rheumatic fever.
5. Prompt treatment of streptococcal infections can reduce the incidence of rheumatic fever.
6. Rheumatic fever is treated with anti-inflammatory drugs.

Tularemia (pp. 564–565)

1. Tularemia is caused by *Francisella tularensis.* The reservoir is small wild mammals, especially rabbits.
2. Signs include ulceration at the site of entry, followed by septicemia and pneumonia.
3. Humans contract tularemia by handling diseased carcasses, eating undercooked meat of diseased animals, and being bitten by certain vectors (such as deer flies).

4. *F. tularensis* is resistant to phagocytosis.
5. Laboratory diagnosis is based on an agglutination test on isolated bacteria.

Brucellosis (Undulant Fever) (pp. 566–567)

1. Brucellosis can be caused by *Brucella abortus, B. melitensis,* and *B. suis.*
2. Domesticated animals (cattle, pigs, goats, and camels) constitute the reservoir.
3. The bacteria enter through minute breaks in the mucosa or skin, reproduce in macrophages, and spread via lymphatics to liver, spleen, or bone marrow.
4. Signs include malaise and fever that spikes each evening (undulant fever).
5. A vaccine for cattle is available.
6. Diagnosis is based on serological tests.

Anthrax (p. 567)

1. *Bacillus anthracis* causes anthrax. In soil, endospores can survive for up to 60 years.
2. Grazing animals acquire an infection after ingesting the endospores.
3. Humans contract anthrax by handling hides from infected animals. The bacteria enter through cuts in the skin or through the respiratory tract.
4. Entry through the skin results in a pustule that can progress to septicemia. Entry through the respiratory tract can result in pneumonia.
5. Diagnosis is based on isolation and identification of the bacteria.

Gangrene (pp. 567–568)

1. Soft tissue death from ischemia (loss of blood supply) is called gangrene.
2. Microorganisms grow on nutrients released from gangrenous cells.
3. Gangrene is especially susceptible to the growth of anaerobic bacteria such as *Clostridium perfringens,* the causative agent of gas gangrene.
4. *C. perfringens* can invade the uterine wall during improperly performed abortions.
5. Debridement, hyperbaric chambers, and amputation are used to treat gas gangrene.

Systemic Diseases Caused by Bites and Scratches

(p. 568)
1. *Pasteurella multocida,* introduced by the bite of a dog or cat, can cause septicemia.
2. Anaerobic bacteria such as *Clostridium, Bacteroides,* and *Fusobacterium* infect deep animal bites.
3. Rat bite fever is caused by *Streptobacillus* or *Spirillum.*
4. *S. aureus, Bacteroides,* α-hemolytic streptococci, and *Fusobacterium* can infect human bites.
5. Cat-scratch disease may be caused by *Afipia felis* or *Rochalimaea henselae.*

Plague (pp. 568–570)

1. Plague is caused by *Yersinia pestis.* The vector is usually the rat flea (*Xenopsylla cheopis*).
2. Reservoirs for plague include European rats and North American rodents.
3. Signs of bubonic plague include bruises on the skin and enlarged lymph nodes (buboes).
4. The bacteria can enter the lungs and cause pneumonic plague.
5. Laboratory diagnosis is based on isolation and identification of the bacteria.
6. Antibiotics are effective to treat plague, but they must be administered promptly after exposure to the disease.

Relapsing Fever (p. 570)

1. Relapsing fever is caused by *Borrelia* species and transmitted by soft ticks.
2. The reservoir for the disease is rodents.
3. Signs include fever, jaundice, and rose-colored spots. Signs recur three or four times after apparent recovery.
4. Laboratory diagnosis is based on the presence of spirochetes in the patient's blood.

Lyme Disease (Lyme Borreliosis) (pp. 570–571)

1. Lyme disease is caused by *Borrelia burgdorferi* and is transmitted by a tick (*Ixodes*).
2. Lyme disease is prevalent on the Atlantic Coast of the U.S.A.
3. Deer and field mice are the animal reservoirs.
4. Diagnosis is based on serological tests and clinical symptoms.

Typhus (pp. 572–573)

1. Rickettsias are obligate intracellular parasites of eucaryotic cells.

Epidemic Typhus (p. 572)

1. The human body louse *Pediculus humanus corporis* transmits *Rickettsia prowazekii* in its feces, which are deposited while the louse is feeding.
2. Epidemic typhus is prevalent in crowded and unsanitary living conditions that allow the proliferation of lice.
3. The signs of typhus are rash, prolonged high fever, and stupor.
4. Tetracyclines and chloramphenicol are used to treat typhus.

Endemic Murine Typhus (p. 572)

1. Endemic murine typhus is a less severe disease caused by *Rickettsia typhi* and transmitted from rodents to humans by the rat flea.

Spotted Fevers (pp. 572–573)

1. *Rickettsia rickettsii* is a parasite of ticks (*Dermacentor* species) in the southeastern U.S.A., Appalachia, and the Rocky Mountain states.
2. The rickettsia may be transmitted to humans, in whom it causes tick-borne typhus fever.
3. Chloramphenicol and tetracyclines are effective to treat the disease.
4. Serological tests are used for laboratory diagnosis.

VIRAL DISEASES OF THE CARDIOVASCULAR AND LYMPHATIC SYSTEMS (pp. 573–575)

Burkitt's Lymphoma (p. 573)

1. Epstein–Barr (EB) virus causes Burkitt's lymphoma and nasopharyngeal cancer.
2. EB virus causes cancer in laboratory inoculated monkeys.
3. Burkitt's lymphoma tends to occur in patients whose immune system has been weakened, for example, by malaria or AIDS.

Infectious Mononucleosis (pp. 574–575)

1. Infectious mononucleosis is caused by EB virus.
2. The virus multiplies in the parotid glands and is present in saliva. It causes the proliferation of atypical lymphocytes.
3. The disease is transmitted by ingestion of saliva from infected individuals.
4. Diagnosis is made by an indirect fluorescent-antibody technique.

Yellow Fever (p. 575)

1. Yellow fever is caused by an arbovirus (yellow fever virus). The vector is the mosquito *Aedes aegypti*.
2. The virus multiplies in lymph nodes and then disseminates to liver, spleen, kidney, and heart.
3. Signs and symptoms include fever, chills, headache, nausea, and jaundice.
4. Diagnosis is based on the presence of virus-neutralizing antibodies in the host.
5. No treatment is available. An attenuated, live viral vaccine is available.

Dengue (p. 575)

1. Dengue is caused by an arbovirus (dengue fever virus) and is transmitted by the mosquito *Aedes aegypti*.
2. Signs are fever, muscle and joint pain, and rash.
3. Mosquito abatement is necessary to control the disease.

PROTOZOAN DISEASES OF THE CARDIOVASCULAR AND LYMPHATIC SYSTEMS (pp. 575–579)

Toxoplasmosis (pp. 575–576)

1. Toxoplasmosis is caused by the sporozoan *Toxoplasma gondii*.
2. *T. gondii* undergoes sexual reproduction in the intestinal tract of domestic cats, and oocysts are eliminated in cat feces.
3. Oocysts can be ingested by cattle and other animals.
4. Sporozoites hatch from an oocyst and invade host cells.
5. In the host cell, sporozoites reproduce to form either tissue-invading tachyzoites or bradyzoites.
6. Bradyzoites reproduce in tissue cysts.
7. Humans contract the infection by ingesting tachyzoites or tissue cysts in undercooked meat from an infected animal, or oocysts in dried cat feces.
8. Subclinical infections are probably common because the disease symptoms are rather mild.
9. Congenital infections can occur. Signs and symptoms include convulsions, brain damage, blindness, and death.
10. *T. gondii* can be isolated and grown in all cultures and identified by serological tests.

American Trypanosomiasis (Chagas' Disease) (p. 577)

1. *Trypanosoma cruzi* causes Chagas' disease. The reservoir includes many wild animals. The vector is the reduviid bug, the "kissing bug."
2. The infection begins at the site of inoculation and disseminates via the bloodstream to other organs. It causes destructive inflammation of infected organs.
3. Observation of the trypanosomes in blood confirms diagnosis.

Malaria (pp. 577–579)

1. The signs and symptoms of malaria are chills, fever, vomiting, and headache, which occur at two- or three-day cycles.
2. Malaria is transmitted by *Anopheles* mosquitoes. The causative agent is any one of four species of *Plasmodium*.
3. Sporozoites reproduce in the liver and release merozoites into the bloodstream, where they infect red blood cells and produce more merozoites.
4. Laboratory diagnosis is based on microscopic observation of merozoites in red blood cells.
5. A vaccine is being developed.
6. New drugs are being developed as the protozoa develop resistance to drugs such as chloroquine.

HELMINTHIC DISEASES OF THE CARDIOVASCULAR AND LYMPHATIC SYSTEMS (pp. 579–580)

Schistosomiasis (p. 579)

1. Species of the blood fluke *Schistosoma* cause schistosomiasis.
2. Eggs eliminated with feces hatch into larvae that infect the intermediate host—a snail. Free-swimming cercariae are released from the snail and penetrate the skin of a human.
3. The adult flukes live in the veins of the liver or urinary bladder in humans.
4. Adult flukes reproduce, and eggs are excreted or remain in the host.
5. Symptoms are from the host's defense to eggs that remain in the body.
6. Observation of eggs or flukes in feces, skin tests, or indirect serological tests may be used for diagnosis.
7. Chemotherapy is used to treat the disease; sanitation and snail eradication are used to prevent the disease.

Swimmer's Itch (pp. 579–580)

1. Swimmer's itch is a cutaneous allergic reaction to cercariae that penetrate the skin. The definitive host for this fluke is wildfowl.

STUDY QUESTIONS

Review

1. What are the signs of septicemia?
2. How can septicemia result from a single focus of infection, such as an abscess?
3. Complete the following table.

DISEASE	FREQUENT CAUSATIVE AGENT	PREDISPOSING CONDITION(S)
Puerperal sepsis		
Subacute bacterial endocarditis		
Acute bacterial endocarditis		
Myocarditis		
Pericarditis		

4. Describe the probable cause of rheumatic fever. How is rheumatic fever treated? How is it prevented?
5. Compare and contrast epidemic typhus, endemic murine typhus, and tick-borne typhus.
6. Plot the temperature of a patient with brucellosis for a one-week period.
7. Complete the following table.

DISEASE	CAUSATIVE AGENT	VECTOR	SIGNS/ SYMPTOMS	TREATMENT
Malaria				
Yellow fever				
Dengue				
Relapsing fever				

8. Complete the following table.

DISEASE	CAUSATIVE AGENT	METHOD OF TRANSMISSION	RESERVOIR	SIGNS/SYMPTOMS	PREVENTION
Tularemia					
Brucellosis					
Anthrax					
Lyme disease					

9. Provide the following information on plague: causative agent; vector; U.S.A. reservoir; control; treatment; prognosis (probable outcome).
10. List the causative agent, method of transmission, and reservoir for schistosomiasis, toxoplasmosis, and American trypanosomiasis. Which disease are you most likely to get in the U.S.A.? Where are the other diseases endemic?
11. Cite evidence implicating EB virus in Burkitt's lymphoma. Offer an explanation of why a large number of people who are infected with EB virus do not get Burkitt's lymphoma or nasopharyngeal cancer.
12. List four bacterial infections that might result from animal bites.
13. Why is *Clostridium perfringens* likely to grow in gangrenous wounds?
14. List the causative agents and methods of transmission of infectious mononucleosis.
15. Differentiate between the transmission and symptoms of bubonic plague and pneumonic plague.

Critical Thinking

1. Indirect fluorescent-antibody tests on the serum of three 25-year-old women, each of whom is considering pregnancy, provided the following information. Which of these women may have toxoplasmosis? What advice might be given to each woman with regard to toxoplasmosis?

PATIENT	ANTIBODY TITER		
	DAY 1	DAY 5	DAY 12
Patient A	1:1024	1:1024	1:1024
Patient B	1:1024	1:2048	1:3072
Patient C	0	0	0

2. What is the most effective way to control malaria and dengue?

3. In adults, the second dengue virus infection results in dengue hemorrhagic fever (DHF), which is characterized by bleeding from the skin and mucosa. DHF can be fatal. In infants under one year old, the first dengue virus infection results in DHF. Offer an explanation for this.

Clinical Applications

1. A 19-year-old man went deer hunting. While on the trail, he found a partially dismembered dead rabbit. The hunter picked up the front paws for good luck charms and gave them to another hunter in the party. The rabbit had been handled with bare hands that were bruised and scratched from the hunter's work as an automobile mechanic. Festering sores on his hands, legs, and knees were noted two days later. What infectious disease do you suspect the hunter has? How would you proceed to prove it?

2. On March 30, a 35-year-old veterinarian experienced fever, chills, and vomiting. On March 31, he was hospitalized with diarrhea, left armpit bubo, and secondary bilateral pneumonia. On March 27, he had treated a cat that had labored respiration; an X ray revealed pulmonary infiltrates. The cat died on March 28 and was disposed of. Chloramphenicol was administered to the veterinarian. On April 10, his temperature returned to normal, and on April 20, he was released from the hospital. Sixty human contacts were given tetracycline. Identify the incubation and prodromal periods for this case. Explain why the 60 contacts were treated. What was the etiologic agent? How would you identify the agent?

3. Three of five patients who underwent heart valve replacement surgery developed bacteremia. The causative agent was *Enterobacter cloacae*. What were the patients' signs and symptoms? How would you identify this bacterium? A manometer used in the operations was culture-positive for *E. cloacae*. What is the most likely source of this contaminant? Suggest a way of preventing such occurrences.

4. In August and September, six people who each at different times spent a night in the same cabin developed the symptoms shown in the following graph. Three recovered after tetracycline therapy, two recovered without therapy, and one was hospitalized with septic shock. What is the disease? What is the incubation period of this disease? How do you account for the periodic temperature changes? What caused septic shock in the sixth patient?

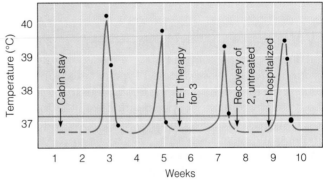
—— Average healthy human body temperature

5. A 67-year-old man worked in a textile mill that processed imported goat hair into fabrics. He noticed a painless, slightly swollen pimple on his chin. Two days later he developed a 1-cm ulcer at the pimple site and a temperature of 37.6°C. He was treated with tetracycline. What is the etiology of this disease? Suggest ways to prevent it.

Microbial Diseases of the Respiratory System

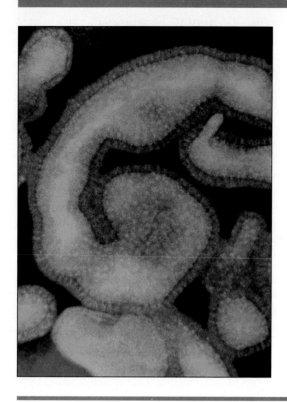

WITH EVERY BREATH we inhale several microorganisms, so the upper respiratory system is a major portal of entry for pathogens. In fact, respiratory system infections are the most common type of infection—and among the most damaging. Some pathogens that enter the body via the respiratory system can infect other parts of the body—for example, the pathogens that cause measles, mumps, and rubella. Another very serious aspect of respiratory infections is the ease with which they are spread, both by *direct contact* with droplets emitted during sneezing, coughing, and talking, and by *fomites* (contaminated objects). Transmission by droplets is especially common for highly communicable respiratory infections, such as pneumonic plague and influenza. Microbial diseases associated with the respiratory system are summarized in Table 24.1.

STRUCTURE AND FUNCTION OF THE RESPIRATORY SYSTEM

LEARNING OBJECTIVE

- *Describe how microorganisms are prevented from entering the respiratory system.*

It is convenient to think of the respiratory system as being composed of two divisions, the upper respiratory system and the lower respiratory system. The **upper respiratory system** consists of the nose, the pharynx (throat), and the

structures associated with them, including the middle ear and the auditory (eustachian) tubes (Figure 24.1). Ducts from the sinuses and the nasolacrimal ducts from the lacrimal (tear-forming) apparatus empty into the nasal cavity (see Figure 16.3). The auditory tubes from the middle ear empty into the upper portion of the throat.

The upper respiratory system has several anatomical defenses against airborne pathogens. Coarse hairs in the nose filter large dust particles from the air. The nose is lined with a mucous membrane that contains numerous mucus-secreting cells and cilia. The upper portion of the throat also contains a ciliated mucous membrane. The mucus moistens inhaled air and traps dust and microorganisms, especially particles larger than 4 to 5 μm. The cilia assist in removing these particles by moving them toward the mouth for elimination.

At the junction of the nose and throat are masses of lymphoid tissue, the tonsils and adenoids, that contribute to immunities to certain infections. Occasionally, however, these tissues become infected and help spread infection to the ears via the auditory tubes. Because the nose and throat are connected to the sinuses, nasolacrimal apparatus, and middle ear, infections commonly spread from one region to another.

The **lower respiratory system** consists of the larynx (voice box), trachea (windpipe), bronchial tubes, and alveoli (Figure 24.2). Alveoli are the air sacs that make up the lung tissue; within them, oxygen and carbon dioxide are exchanged between the lungs and blood. Our lungs contain millions of alveoli, with an absorptive area of about 90 square meters. The double-layered membrane covering

TABLE 24.1 Microbial Diseases Associated with the Respiratory System

DISEASE	CAUSATIVE AGENT	MODE OF TRANSMISSION	TREATMENT
Bacterial Diseases of the Upper Respiratory System			
Streptococcal pharyngitis (strep throat)	Streptococci, especially *Streptococcus pyogenes*	Respiratory secretions	Penicillin
Scarlet fever	Erythrogenic toxin-producing strains of *Streptococcus pyogenes*	Respiratory secretions	Penicillin and erythromycin
Diphtheria	*Corynebacterium diphtheriae*	Respiratory secretions, healthy carriers	Antitoxin and penicillin, tetracyclines, erythromycin
Cutaneous diphtheria	*Corynebacterium diphtheriae*	Respiratory secretions, healthy carriers	Antitoxin and penicillin, tetracyclines, erythromycin
Otitis media	Several agents, especially *Staphylococcus aureus*, *Streptococcus pneumoniae*, β-hemolytic streptococci, and *Hemophilus influenzae*	Complication of cold, or nose or throat infection	Penicillin and amoxicillin
Viral Disease of the Upper Respiratory System			
Common cold	Coronaviruses, rhinoviruses	Respiratory secretions	None
Bacterial Diseases of the Lower Respiratory System			
Pertussis (whooping cough)	*Bordetella pertussis*	Respiratory secretions	Erythromycin
Tuberculosis	*Mycobacterium tuberculosis*	Respiratory secretions	Isoniazid and rifampin
	Mycobacterium bovis	Usually milk-borne	Isoniazid and rifampin
Pneumococcal pneumonia	*Streptococcus pneumoniae*	Healthy carriers; from colonization of upper respiratory system	Penicillin
Hemophilus influenzae pneumonia	*Hemophilus influenzae*	Healthy carriers; from colonization of upper respiratory system	Cephalosporins
Mycoplasmal pneumonia	*Mycoplasma pneumoniae*	Respiratory secretions	Tetracycline, erythromycin
Legionellosis	*Legionella pneumophila*	Aerosols from contaminated water	Erythromycin, rifampin
Psittacosis (ornithosis)	*Chlamydia psittaci*	Animal reservoirs; aerosols of dried droppings and other exudates of birds; person-to-person transmission is rare	Tetracyclines
Chlamydial pneumonia	*Chlamydia pneumoniae*	Respiratory secretions	Tetracyclines
Q fever	*Coxiella burnetii*	Animal reservoirs; aerosols in dairy barns; unpasteurized milk	Tetracyclines
Viral Diseases of the Lower Respiratory System			
Viral pneumonia	Several viruses	Aerosols; complication of other viral diseases	Ribivirin diminishes symptoms
Influenza	Influenza viral strains; many serotypes	Respiratory secretions	Amantadine (prophylaxis)
Hantavirus pulmonary syndrome	*Hantavirus*	Rodent reservoir; airborne virus particles	None
Fungal Diseases of the Lower Respiratory System			
Histoplasmosis	*Histoplasma capsulatum*	Bird or bat reservoirs; airborne fungal spores from growth in soil	Amphotericin B, itraconazole
Coccidioidomycosis	*Coccidioides immitis*	Soil organisms; airborne arthrospores	Amphotericin B, ketoconazole
Blastomycosis	*Blastomyces dermatitidis*	Soil organisms; airborne spores	Amphotericin B
Protozoan Disease of the Lower Respiratory System			
Pneumocystis pneumonia	*Pneumocystis carinii*	Probably inhalation of respiratory secretions	Trimethoprim-sulfamethoxazole, pentamidine isethionate

*May be a fungus.

Sinus

Nasal cavity

Oral cavity

Tongue

Epiglottis

Larynx (voice box)

Trachea (windpipe)

Sinus

Middle ear

Auditory (eustachian) tube

Opening of auditory tube

Tonsils

Pharynx (throat)

Spine

Esophagus

FIGURE 24.1 Structures of the upper respiratory system.

the lungs is the *pleura*. A ciliated mucous membrane lines the lower respiratory system down to the smaller bronchial tubes and helps to prevent microorganisms from reaching the lungs. As we discussed in Chapter 16, particles trapped in the larynx, trachea, and larger bronchial tubes are moved up toward the throat by a ciliary action called the ciliary escalator (see Figure 16.4). If microorganisms actually reach the lungs, then certain phagocytic cells—the *alveolar macrophages*—usually locate, ingest, and destroy most of them. IgA antibodies in such secretions as respiratory mucus, saliva, and tears also help to protect mucosal surfaces of the respiratory system from many pathogens. Thus, the body has several mechanisms for removing the pathogens that cause airborne infections. However, if all these mechanisms fail, then the microorganism wins the host–parasite competition and a respiratory disease results.

NORMAL MICROBIOTA OF THE RESPIRATORY SYSTEM

LEARNING OBJECTIVE

- *Characterize the normal microbiota of the upper and lower respiratory systems.*

A number of potentially pathogenic microorganisms are part of the normal microbiota in the upper respiratory system. However, they usually do not cause illness because the predominant microorganisms of the normal microbiota suppress their growth by competing with them for nutrients and producing inhibitory substances.

By contrast, the lower respiratory tract is nearly sterile—although the trachea may contain a few bacteria—because of the normally efficient functioning of the ciliary escalator in the bronchial tubes.

MICROBIAL DISEASES OF THE UPPER RESPIRATORY SYSTEM

LEARNING OBJECTIVES

- *Differentiate among pharyngitis, laryngitis, tonsillitis, and sinusitis.*
- *List the etiologic agent, symptoms, prevention, preferred treatment, and laboratory identification tests for streptococcal pharyngitis, scarlet fever, diphtheria, cutaneous diphtheria, otitis media, and the common cold.*

As most of us know from personal experience, the respiratory system is the site of many common infections. We

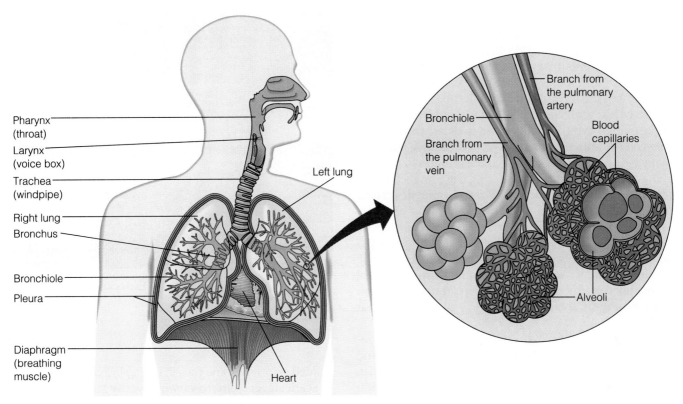

FIGURE 24.2 Structures of the lower respiratory system.

will soon discuss **pharyngitis**—inflammation of the mucous membranes of the throat, or sore throat. When the larynx is the site of infection, we suffer from **laryngitis**, which affects our ability to speak. This infection is caused by bacteria, such as *Streptococcus pneumoniae* or *S. pyogenes*, or viruses, often in combination. The organisms that cause pharyngitis also can cause inflamed tonsils, or **tonsillitis.**

The nasal sinuses are cavities in certain cranial bones that open into the nasal cavity. They have a mucous membrane lining that is continuous with that of the nasal cavity. When a sinus becomes infected with such organisms as *S. pneumoniae* or *Hemophilus influenzae*, the mucous membranes become inflamed, and there is a heavy nasal discharge of mucus. This condition is called **sinusitis.** If the opening by which the mucus leaves the sinus becomes blocked, internal pressure can cause pain or a sinus headache. These diseases are almost always *self-limiting*, meaning that recovery will usually occur even without medical intervention.

Probably the most threatening infectious disease of the upper respiratory system is **epiglottitis**, inflammation of the epiglottis. The epiglottis is a flaplike structure of cartilage that prevents ingested material from entering the larynx (see Figure 24.1). Inflammation of the epiglottis is a rapidly developing disease that can result in death within a few hours. It is caused by opportunistic pathogens, usually *H. influenzae* type b.

Bacterial Diseases of the Upper Respiratory System

Airborne pathogens make their first contact with the body's mucous membranes as they enter the upper respiratory system. Many respiratory or systemic diseases initiate infections here.

Streptococcal Pharyngitis (Strep Throat)

Streptococcal pharyngitis (strep throat) is an upper respiratory infection caused by group A β-hemolytic streptococci. This gram-positive bacterial group consists solely of *Streptococcus pyogenes*, the same bacterium responsible for many skin and soft tissue infections, such as impetigo, erysipelas, and acute bacterial endocarditis.

The pathogenicity of group A streptococci is enhanced by their resistance to phagocytosis. They are also able to produce special enzymes that lyse fibrin clots (*streptokinase*), discussed in Chapter 15. They also produce *streptolysins* that are cytotoxic to tissue cells, red blood cells, and protective leukocytes. There is some concern that group A streptococci of unusual virulence are now in circulation.

Without some analysis, streptococcal pharyngitis cannot be distinguished from pharyngitis of other causes (mostly viral). Probably no more than half of so-called strep throats are actually streptococcal in origin. At one time, the diagnosis of strep throat was based on culturing of bacteria from a throat swab. Results took overnight or longer, and this test has been almost universally replaced by new indirect agglutination diagnostic tests for streptococcal pharyngitis that use microscopic latex particles coated with antibodies against group A streptococci. A swab of the throat is still used to collect bacteria, and the streptococcal antigens are extracted from the swab. An agglutination-type reaction is used to detect the reaction between the antigen and antibody-coated latex particles. Many of these tests take as little as ten minutes to perform and are highly specific in detecting the presence or absence of group A streptococci. In fact, *S. pyogenes* can be detected in the throats of many people who are only asymptomatic carriers. Therefore, a positive result is not a certain indication that the symptoms are caused by the detected streptococci. A rise in IgM antibody titer is the best indication of a sore throat of genuinely streptococcal origin, but this is not a practical option for day-to-day use. A negative result should be checked by culturing a throat swab.

Streptococcal pharyngitis is characterized by local inflammation and a fever. Frequently, tonsillitis occurs and the lymph nodes in the neck become enlarged and tender. Another frequent complication is otitis media (infection of the middle ear).

Nearly all streptococci are sensitive to penicillin, the drug of choice. Failures with penicillin therapy are often due to β-lactamases produced by other members of the throat microbiota. Treatment of streptococcal pharyngitis is important to prevent such complications as rheumatic fever and glomerulonephritis, both of which are caused by an immune response to streptococcal infections.

More than 80 serological types of group A streptococci are responsible for at least a dozen different illnesses. Because resistance to streptococcal diseases is type-specific, a person who has recovered from infection by one type is not necessarily immune to infection by another type.

Strep throat is now most commonly transmitted by respiratory secretions, but epidemics spread by unpasteurized milk were once frequent.

Scarlet Fever

When the *S. pyogenes* strain causing streptococcal pharyngitis produces an *erythrogenic* (reddening) *toxin,* the resulting infection is called **scarlet fever.** When the strain produces this toxin, it has been lysogenized by a bacteriophage (see Chapter 13). As you may recall, this means that the genetic information of a bacteriophage (bacterial virus) has been incorporated into the chromosome of the bacterium, so the characteristics of the bacterium have been altered. The toxin causes a pinkish-red skin rash (Figure 24.3), which is probably the skin's hypersensitivity reaction to the circulating toxin, and a high fever. The tongue has a spotted, strawberrylike appearance and then, as it loses its upper membrane, becomes very red and enlarged. As the disease runs its course, the affected skin frequently peels off, as if sunburned, very similar to the

FIGURE 24.3 The typical red skin rash caused by scarlet fever.

scalded skin syndrome caused by *Staphylococcus aureus.*

Scarlet fever seems to vary in severity and frequency at different times and in different locations but generally has been declining in recent years. It is a communicable disease spread mainly by inhalation of infective droplets from an infected person. Classically, scarlet fever has been considered to be associated with streptococcal pharyngitis, but it might accompany a streptococcal skin infection.

Penicillin is effective in treating the original streptococcal infection.

Diphtheria

Another bacterial infection of the upper respiratory system is **diphtheria;** until 1935, it was the leading infectious killer of children in the U.S.A. The disease begins with a sore throat and fever, followed by general malaise and swelling of the neck. The pathogenesis of diphtheria is better understood than that of most other infectious diseases. The organism responsible is *Corynebacterium diphtheriae,* which is a gram-positive, non-endospore-forming rod. Its morphology is pleomorphic, frequently club-shaped (*corynes,* Gk. "clubbed"), and it stains unevenly (Figure 24.4).

Part of the normal immunization program for children in the U.S.A. is the *DPT vaccine* (see Chapter 18). The *D* stands for diphtheria toxoid, an inactivated toxin that causes the body to produce antibodies against the diphtheria toxin.

C. diphtheriae has adapted to a generally immunized population, and relatively nonvirulent strains are found in the throats of many symptomless carriers. The bacterium is well suited to airborne transmission and is very resistant to drying. Its cell walls are very similar to those of the tubercle bacillus, which also shows high resistance to environmental stress.

FIGURE 24.4 *Corynebacterium diphtheriae,* **the organism that causes diphtheria.** This Gram stain shows the club-shaped morphology; the dividing cells are often observed to fold together to form V- and Y-shaped figures.

Characteristic of diphtheria is a tough (*diphtheria,* Gk. "leather"), grayish membrane that forms in the throat in response to the infection. It contains fibrin, dead tissue, and bacterial cells and can totally block the passage of air to the lungs.

Although the bacteria do not invade tissues, those that have been lysogenized by a phage can produce a powerful exotoxin (see Chapter 15). Historically, it was the first disease for which a toxic cause was identified. Circulating in the bloodstream, the toxin interferes with protein synthesis. Only 0.01 mg of this highly virulent toxin is enough to kill a 91-kg (200-pound) person. Thus, if antitoxin therapy is to be effective, it must be administered before the toxin enters the tissue cells. When such organs as the heart or kidneys are affected by the toxin, the disease can rapidly be fatal. In other cases the nerves can be involved and partial paralysis results.

Laboratory diagnosis by bacterial identification is difficult, requiring several selective and differential media. Identification is complicated by the need to differentiate between toxin-forming isolates and strains that are not toxigenic; both may be found in the same patient.

Even though antibiotics such as penicillin and erythromycin control the growth of the bacteria, they do not neutralize the diphtheria toxin. Thus, to treat diphtheria, antibiotics should be used only in conjunction with antitoxin.

The number of diphtheria cases reported in the U.S.A. each year is presently five or fewer, but the death rate for respiratory cases is still from 5% to 10%, mostly among the elderly or the very young. In young children, the disease occurs mainly in groups that have not been immunized for religious or other reasons. When diphtheria was more common, repeated contacts with toxigenic strains reinforced the immunity, which otherwise weakens with time. Many adults now lack immunity because routine immunization was less available during their childhood. Some surveys indicate effective immune levels in as few as 25% of the adult population. Because of this, when any wound or trauma requires tetanus toxoid, it is usually combined with diphtheria toxoid (DT vaccine).

Cutaneous Diphtheria

Diphtheria is also expressed as **cutaneous diphtheria.** In this form of the disease, *C. diphtheriae* infects the skin, usually at a wound or similar skin lesion, and there is minimal systemic circulation of the toxin. In cutaneous infections, the bacteria cause slow-healing ulcerations covered by a gray membrane.

Cutaneous diphtheria is fairly common in tropical countries. In the U.S.A., it occurs mostly among Native Americans and among adults in low socioeconomic groups. It is responsible for most of the reported cases of diphtheria in people over 30 years old.

In the past, diphtheria was spread mainly to healthy carriers by droplet infection. Respiratory cases have been known to arise from contact with cutaneous diphtheria. With the advent of immunization, however, the carrier rate of toxigenic strains has decreased dramatically.

Otitis Media

One of the more uncomfortable complications of the common cold, or of any infection of the nose or throat, is infection of the middle ear, **otitis media,** leading to earache. The middle ear can be infected directly by contaminated water from swimming pools and by some severe trauma, such as injury to the eardrum and certain skull fractures. The infecting microorganisms cause the formation of pus, which builds up pressure against the eardrum and causes it to become inflamed and painful. The condition is most frequent in early childhood, probably because the auditory tube connecting the middle ear to the throat is small and is more easily blocked by infection (see Figure 24.1). Enlarged adenoids might be a contributing factor for children with an unusually high frequency of infection.

A number of bacteria can cause otitis media. The one most commonly isolated is *S. pneumoniae* (about 35% of cases). Other bacteria frequently involved include nonencapsulated *H. influenzae* (25% to 30%), *Moraxella (Branhamella) catarrhalis* (mô-raks-el′lä bran-ha-mel′lä ka-tär′al-is) (10% to 15%), *S. pyogenes* (8% to 10%), and *S. aureus* (1% to 2%). In about 3% to 5% of cases, no bacteria can be detected. Most of these cases are probably caused by obligate anaerobic bacteria. Viruses are possibly a cause, but treatment always assumes that bacteria are the causative agents. Broad spectrum penicillins, such as amoxicillin, are usually the first choice for therapy for children under eight. Penicillin is the recommended antibiotic in older patients.

Viral Disease of the Upper Respiratory System

Probably the most prevalent disease of humans, at least those living in the temperate zones, is a viral disease affecting the upper respiratory system—the common cold.

Common Cold

A number of different viruses are involved in the etiology of the **common cold.** About 50% of all colds are caused by *rhinoviruses. Coronaviruses* probably cause another 15% or 20%. About 10% of all colds are caused by one of an assortment of other viruses. In about 40% of colds, no agent can be identified.

We tend to accumulate immunities against cold viruses during our lifetimes, which may be a reason why older people tend to get fewer colds. Young children have three or four colds per year, but adults at age 60 average less than one cold per year. Immunity is based on the ratio of IgA antibodies to single serotypes and has a reasonably high short-term effectiveness. Isolated populations may develop a group immunity, and their colds disappear until a new set of viruses is introduced. Altogether, there are probably more than 200 agents that cause colds. There are at least 113 serotypes of rhinoviruses alone, so a vaccine effective against so many different agents does not seem practical.

However, a new approach has been suggested. Most cold viruses, although antigenically different, often use as few as two different receptor sites on the mucosa. Instead of trying to produce antibodies against many different virus configurations, this may make it possible to produce an antibody that will achieve the same results by blocking the relatively few receptor sites in the body.

The symptoms of the common cold are familiar to all of us. They include sneezing, excessive nasal secretion, and congestion. (In ancient times, one school of medical thought believed that the nasal discharges were waste products from the brain.) The infection can easily spread from the throat to the sinus cavities, the lower respiratory system, and the middle ear, leading to complications of laryngitis and otitis media. The uncomplicated cold usually is not accompanied by fever.

The rhinoviruses prefer a temperature slightly below that of normal body temperature—such as might be found in the upper respiratory system, which is open to the outside environment. No one knows exactly why the number of colds seems to increase with colder weather in temperate zones. It is not known whether closer indoor contact promotes epidemic-type transmission or whether physiological changes increase susceptibility.

A single rhinovirus deposited on the nasal mucosa is often sufficient to cause a cold. However, there is surprisingly little agreement on how the cold virus is transmitted to a site in the nose. One line of experimentation tends to show that cold sufferers deposit the viruses on doorknobs, telephones, and other surfaces, where they remain viable for hours. Healthy people can then transfer these viruses to their hands and thence to their nasal passages. This theory was supported by an experiment in which healthy people who used virucidal iodine solutions on their hands had a much reduced incidence of colds. However, another series of experiments involving a group of cardplayers, half with colds and half without colds (Figure 24.5) supported a different conclusion. Half of the healthy players were restrained so that they could not use their hands to transfer to their noses viruses picked up from the playing cards, and the other half were not restrained. The players who could not touch their noses came down with as many colds as those who could—an argument for airborne transmission. Even when healthy participants in card games were placed in a room separate from anyone suffering from a cold and isolated from their airborne secretions, but handled cards literally soaked in nasal secretions, none developed colds. In a perhaps less repellent series of experiments, researchers required healthy volunteers to kiss cold sufferers for 60 to 90 seconds; only 8% of the volunteers came down with colds.

Because colds are caused by viruses, antibiotics are of no use in treatment. The common cold will usually run its course to recovery in about a week. Recovery time is not affected by over-the-counter drugs presently available, although such drugs may lessen the severity of certain signs and symptoms. Research to develop chemotherapeutic agents that interfere with viral replication is ongoing.

MICROBIAL DISEASES OF THE LOWER RESPIRATORY SYSTEM

LEARNING OBJECTIVES

- *List the etiologic agent, symptoms, prevention, preferred treatment, and laboratory identification tests for pertussis and tuberculosis.*
- *Compare and contrast the eight bacterial pneumonias discussed in this chapter.*
- *List the etiologic agent, symptoms, prevention, and preferred treatment for viral pneumonia, influenza, and Hantavirus pulmonary syndrome.*
- *List the etiologic agent, method of transmission, preferred treatment, and laboratory identification tests for four fungal and protozoan diseases of the respiratory system.*

The lower respiratory system can be infected by many of the same bacteria and viruses that infect the upper respiratory system. As the bronchi (see Figure 24.2) become involved, **bronchitis** or **bronchiolitis** develops. A severe complication of bronchitis is **pneumonia**, in which the pulmonary alveoli become involved.

FIGURE 24.5 An experiment in the transmission of colds. An experiment with cardplayers that included cold sufferers and healthy people tended to support the theory of airborne transmission of cold-causing viruses.

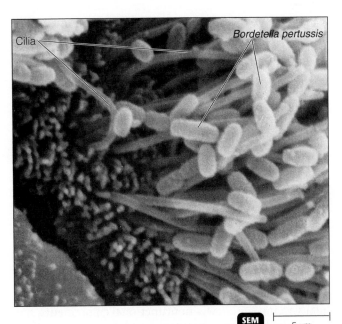

SEM | 5 μm

FIGURE 24.6 Ciliated cells of the respiratory system infected with *Bordetella pertussis*. Cells of *B. pertussis* can be seen growing on the cilia; they will eventually cause the loss of the ciliated cells.

Bacterial Diseases of the Lower Respiratory System

Bacterial diseases of the lower respiratory system include tuberculosis and the many types of pneumonia caused by bacteria. Less well-known diseases such as psittacosis and Q fever also fall under this category.

Pertussis (Whooping Cough)

Infection by the bacterium *Bordetella pertussis* results in pertussis (whooping cough). *B. pertussis* is a small, non-motile, gram-negative coccobacillus. The virulent strains possess a capsule. It is an obligate aerobe and produces an endotoxin as part of its cell wall. In the cytoplasm, the bacteria produce an exotoxin that is apparently released along with the endotoxin when the cell autolyses upon death. The precise role of the toxins in pathogenesis has yet to be determined; the organism does not invade tissues. However, it attaches to the cilia in the trachea and impedes their action, allowing mucus to accumulate. Eventually, a toxic fragment of the bacterial cell wall, *tracheal cytotoxin*, causes loss of the ciliated cells (Figure 24.6).

Pertussis, which is primarily a childhood disease, can be quite severe. The initial stage, which is called the *catarrhal stage,* resembles a common cold. Prolonged sieges of coughing characterize the second, *paroxysmal stage*. (The name *pertussis* is derived from the Latin *per,* "thoroughly," and *tussis,* "cough.") When ciliary action is compromised, mucus accumulates, and the infected person desperately attempts to cough up these mucus accumulations. The violence of the coughing in small children can actually result in broken ribs. Gasping for air between coughs causes a whooping sound, hence the informal name of the disease. Coughing episodes occur several times a day for one to six weeks. The *convalescence stage,* the third stage, may last for months. The disease is of unusually long duration for a respiratory infection. The disease is transmitted by inhaling pathogens expelled by the coughing of infected patients. It is highly infectious, with a transmission rate of 90% among nonimmune contacts.

Diagnosis of pertussis depends in large part on clinical signs and symptoms. The organism can be cultured from a throat swab inserted through the nose on a thin wire and held in the throat while the patient coughs. The pathogen is very sensitive to environmental stresses and must be transferred to the culture medium quickly. A direct fluorescent-antibody test is the diagnostic serological test most often used. It is not especially satisfactory, and there is considerable interest in developing improved serological tests or DNA probes for diagnosis of pertussis.

The disease is most severe in infants under one year of age, and nearly half the cases occur in this age group. Severe cases, especially in infants, are usually treated with erythromycin. Although antibiotics might not result in rapid improvement, the drugs do render the patient noninfectious.

After recovery, immunity is good—at least, second attacks tend to be very mild. A vaccine prepared from heat-inactivated whole bacteria is a part of the regular immunization schedule for children (the P in DPT vaccine stands for *pertussis*). Vaccination is usually recommended at about two months of age, when the titer of maternal antibodies in the baby's circulation is dropping and the child's own antibodies have not yet replaced them. Vaccination has lowered the annual number of cases in the

U.S.A. from more than 250,000 per year to fewer than 3000, generally with fewer than ten deaths. There has been considerable concern about the safety of the pertussis vaccine. Because it contains higher levels of endotoxins than any other vaccine, it frequently produces feverish reactions. The main question, however, is whether it causes permanent neurological damage in recipients. The consensus of the medical community is that it does not, but that concern about the possibility is leading to unacceptably low rates of vaccination in the general population. In Japan and England, adverse publicity has led to widespread avoidance of pertussis vaccination. The result was major epidemics.

Because infants are less capable of coping with the effort of coughing to maintain an airway, the mortality rate in this group is relatively high. Authorities point to these figures as indications that vaccination should continue while research on better vaccines is intensified. An acellular vaccine, with fewer nonfunctional components, has been tested in Japan, Sweden, and limited areas of the U.S.A. Very few adverse effects have been reported with this newer vaccine. In the U.S.A., the whole-cell vaccine is used for the initial three doses, and the subunit vaccine is now licensed for the fourth and fifth doses.

Tuberculosis

Tuberculosis is an infectious disease caused by the bacterium *Mycobacterium tuberculosis,* a slender rod and an obligate aerobe. The rods grow slowly (20-hour or longer generation time), sometimes form filaments (Figure 24.7), and tend to grow in clumps. On the surface of liquid media, their growth appears moldlike, which suggested the genus name *Mycobacterium,* from the Greek *mykes,* meaning fungus.

These bacteria are relatively resistant to normal staining procedures. When stained by the Ziehl–Neelson or Kinyoun technique that stains the cell with carbolfuchsin dye, they cannot be decolorized with a mixture of acid and alcohol and are therefore classified as *acid-fast* (see Chapter 3). This characteristic reflects the unusual composition of the cell wall, which contains large amounts of lipid materials. These lipids might also be responsible for the resistance of mycobacteria to environmental stresses, such as drying. In fact, these bacteria can survive for weeks in dried sputum and are very resistant to chemical antimicrobials used as antiseptics and disinfectants.

Tuberculosis is a good example of the importance of the ecological balance between host and parasite in infectious disease. Hosts are not usually aware of pathogens that invade the body and are defeated. If defenses fail, however, hosts become very much aware of the resulting disease. Several factors may affect host resistance levels: the presence of other illness and physiological and environmental factors such as malnutrition, overcrowding, and stress.

Tuberculosis is most commonly acquired by inhaling the tubercle bacillus. Only very fine particles containing one to three bacilli reach the lungs, where they are usually phagocytized by a macrophage in the alveoli (see

LM   5 μm

FIGURE 24.7 *Mycobacterium tuberculosis.* The filamentous, funguslike growth shown here often occurs in liquid growth media and is responsible for the organism's name (*myco* means fungus).

Figure 24.2). The macrophages of a healthy individual become activated by the presence of the bacilli and usually destroy them (Figure 24.8). ①If they do not, the macrophages protect the microbe from the defenses of the body, and many of the bacilli survive and multiply within the macrophages.

②This infection causes the chemotactic attraction of new macrophages from the bloodstream and the formation of an early *tubercle* (a small lump that is characteristic of tuberculosis and gives the disease its name). The newly attracted macrophages tend not to be activated, so the bacilli multiply rapidly within them. ③After a few weeks, hypersensitivity reactions against these bacillus-laden macrophages kill many of them and a caseous (meaning of cheeselike consistency) area forms in the center of the tubercle. This *caseous center* is surrounded by tightly packed layers of macrophages and lymphocytes; the highly aerobic bacteria fail to multiply in this location. However, a few may remain dormant for many years and serve as the basis for reactivation of the disease. As the caseous lesions slowly heal, they become calcified and show up clearly on X rays, on which they are called *Ghon complexes.*

④In some individuals, activated macrophages capable of effectively destroying the bacilli never develop, and the caseous area continues to enlarge in a process called *liquefaction.* An air-filled *tuberculous cavity* is formed. Conditions within the cavity begin to favor the proliferation of the tubercle bacilli, which then begin to multiply for the first time outside of macrophages. ⑤ The number of bacilli soon grows very large, and eventually the lesion ruptures, releasing the microorganisms into the airways and then to the blood and lymphatic systems. This dis-

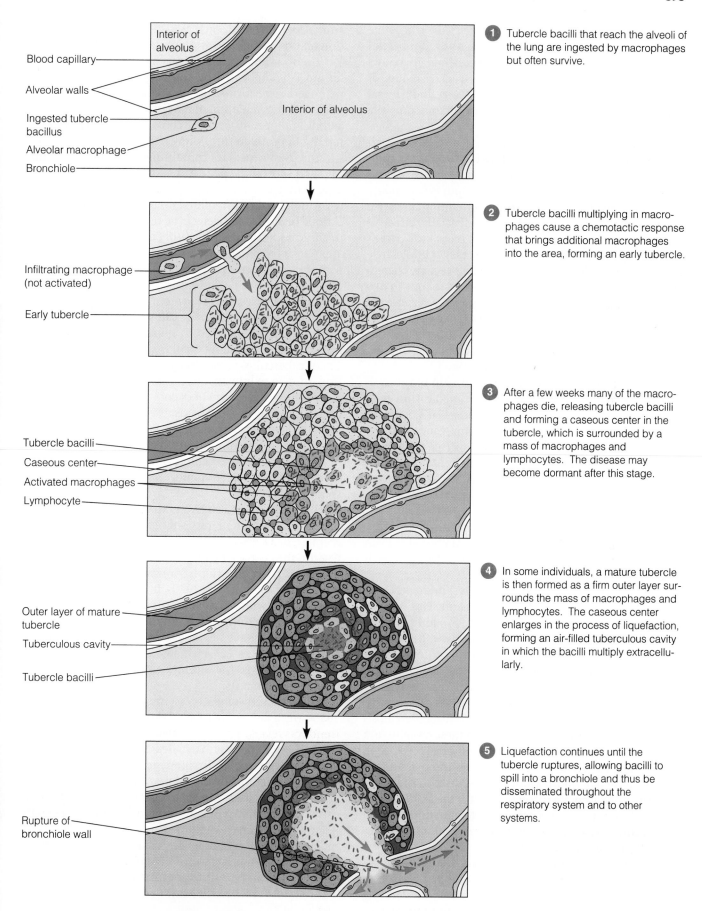

Interior of alveolus

Blood capillary

Alveolar walls

Ingested tubercle bacillus

Alveolar macrophage

Bronchiole

Interior of alveolus

1 Tubercle bacilli that reach the alveoli of the lung are ingested by macrophages but often survive.

Infiltrating macrophage (not activated)

Early tubercle

2 Tubercle bacilli multiplying in macrophages cause a chemotactic response that brings additional macrophages into the area, forming an early tubercle.

Tubercle bacilli

Caseous center

Activated macrophages

Lymphocyte

3 After a few weeks many of the macrophages die, releasing tubercle bacilli and forming a caseous center in the tubercle, which is surrounded by a mass of macrophages and lymphocytes. The disease may become dormant after this stage.

Outer layer of mature tubercle

Tuberculous cavity

Tubercle bacilli

4 In some individuals, a mature tubercle is then formed as a firm outer layer surrounds the mass of macrophages and lymphocytes. The caseous center enlarges in the process of liquefaction, forming an air-filled tuberculous cavity in which the bacilli multiply extracellularly.

Rupture of bronchiole wall

5 Liquefaction continues until the tubercle ruptures, allowing bacilli to spill into a bronchiole and thus be disseminated throughout the respiratory system and to other systems.

FIGURE 24.8 The pathogenesis of tuberculosis.

seminated infection is called *miliary tuberculosis* (the name is derived from the numerous millet seed–sized tubercles formed in the infected tissues). It overwhelms the body's remaining defenses and is characterized by a loss of weight, coughing (often bringing up blood), and general loss of vigor. At one time tuberculosis was known by the common name of *consumption.*

The first effective antibiotic in the treatment of tuberculosis was streptomycin. Probably the drug most commonly used today is the synthetic chemical isoniazid (INH). A combination of INH and rifampin is commonly used. Some treatments use a capsule containing these two drugs plus pyrazinamide, which is especially effective against bacteria within macrophages. Ethambutol, pyrazinamide, and streptomycin are used in cases of intolerance or resistance to the primary drugs. To be effective, chemotherapy for tuberculosis must be continued for one to two years. Prolonged treatment is necessary, partly because the tubercle bacillus grows very slowly and many antibiotics are effective only against growing cells. Also, the bacillus is often hidden for long periods in macrophages or other locations difficult to reach with antibiotics.

A major practical problem with treatment of tuberculosis is the patient's compliance with medication regimens during the long course of treatment. This behavior has been a major contributor to the emergence of drug-resistant strains of the tubercle bacillus (see the box in Chapter 15, p. 394). Indeed, some strains are now resistant to every antitubercular drug available. Medical science is little more capable in the treatment of such cases than it was a century ago.

For preventive prophylaxis for people exposed to tuberculosis, a 6- to 12-month course of INH is usually given alone because there are relatively few bacteria and development of resistance is not likely.

People infected with tuberculosis respond with cell-mediated immunity against the bacterium. This form of immune response, rather than humoral immunity, develops because the pathogen is located mostly within macrophages. This immunity, involving sensitized T cells, is the basis for the *tuberculin skin test.* In this test, a purified protein derivative of the tuberculosis bacterium, derived by precipitation from broth cultures, is injected cutaneously. If the injected person has been infected with tuberculosis in the past, sensitized T cells react with these proteins and a delayed hypersensitivity reaction appears in about 48 hours. This reaction appears as an induration (hardening) and reddening of the area around the injection site. Probably the most accurate tuberculin test is the *Mantoux test,* in which dilutions of 0.1 ml of antigen are injected and the reacting area of the skin is measured. A number of similar tests are also in common use.

A positive tuberculin test in the very young is a probable indication of an active case of tuberculosis. In older individuals it might indicate only hypersensitivity resulting from a previous infection or vaccination, not a current active case. Nonetheless, it is an indication that further examination is needed, such as a chest X ray for the detection of lung lesions and attempts to isolate the bacterium.

Confirmation of tuberculosis by isolation of the bacterium is complicated by the very slow growth of the pathogen. The formation of a colony might take 3–6 weeks, with completion of a reliable identification series adding another 3–6 weeks. Because the bacteria grow more rapidly in broth culture, many clinical laboratories grow specimens in a medium containing radiolabeled palmitic acid as a substrate. The release of CO_2 indicates the presence of mycobacteria. This procedure still takes 7–14 days, with additional time required for positive species identification. Tests that can detect the tubercle bacillus directly in clinical specimens are urgently needed. The most likely approaches are to amplify the bacterial DNA by the polymerase chain reaction and then use a DNA probe assay (see Chapter 18).

Another species, *Mycobacterium bovis* (bō′vis), is a pathogen mainly of cattle. It is the cause of **bovine tuberculosis**, which is transmitted to humans via contaminated milk or food. Bovine tuberculosis accounts for less than 1% of tuberculosis cases in the U.S.A. It seldom spreads from human to human, but before the days of pasteurized milk and the development of control methods such as tuberculin testing of cattle herds, this disease was a frequent form of tuberculosis in humans. *M. bovis* infections cause tuberculosis that primarily affects the bones or lymphatic system. At one time, a common manifestation of this type of tuberculosis was hunchbacked deformation of the spine.

The *BCG vaccine* is a live culture of *M. bovis* that has been made avirulent by long cultivation on artificial media. (BCG stands for the *b*acillus of *C*almette and *G*uerin, the people who originally isolated the strain.) The vaccine is fairly effective in preventing tuberculosis and has been in use since the 1920s. In the U.S.A., the vaccine has not been widely used, but this may be reconsidered. It is presently recommended only for certain children at high risk who have negative skin tests. Apparently, the vaccine enhances cell-mediated immunity, and people who have received the vaccine show a positive reaction to tuberculin skin tests.

After the introduction of effective antibiotics in the 1950s, the incidence of tuberculosis declined steadily. This trend was reversed in 1985, when the number of reported cases showed a perceptible increase. Currently, more than 20,000 new cases of tuberculosis are reported in the U.S.A. each year (Figure 24.9a). The mortality, currently nearly 2000 per year, has continued to decrease. Many of the new cases are found among immigrant groups, especially Asian-born refugees. Certain ethnic groups tend to have much higher rates of tuberculosis. For example, African-Americans, Native Americans, Asians, and Hispanics account for about two-thirds of tuberculosis cases (Figure 24.9b). Most cases in the white population occur in the very elderly.

Recently, however, there has been a marked increase of

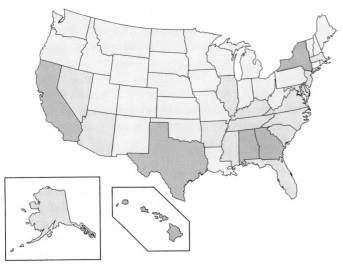

Cases per 100,000 population

☐ 0.0–5.6   ▨ 11.4–17.0

☐ 5.7–11.3   ▨ >17.1

(a) Geographic distribution of reported cases of tuberculosis in the U.S.A. [*Source: MMWR* 42(53) (1/7/94).]

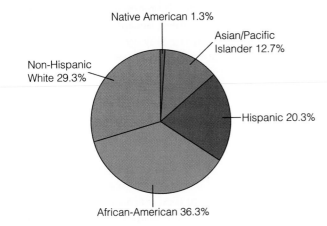

(b) Percentage of cases by race or ethnic origin in the U.S.A. [*Source: MMWR* 40(53) (10/2/92).]

FIGURE 24.9 Tuberculosis distribution.

tuberculosis in people about 25–44 years of age. The Centers for Disease Control and Prevention believe that the increase in this age group is caused primarily by increased susceptibility associated with HIV infection. Other mycobacterial diseases also affect people in the late stages of HIV infections. A majority of the isolates are of a related group of organisms known as the *M. avium-intracellulare* (ā′vē-um in′trä-cel-ū-lä-rē) complex. In the general population, infections by these organisms are uncommon.

In the U.S.A., 10–12 million are estimated to be infected by the tubercle bacillus, many with only latent infections. Worldwide, the number of active cases in developing countries is estimated in the tens of millions; at least 3 million die of the disease each year.

LM   0.5 µm

FIGURE 24.10 *Streptococcus pneumoniae*, the most common cause of pneumococcal pneumonia. Note the paired arrangement of the cells. The capsule, important in pathogenicity, has been made more apparent here by reaction with a specific pneumonococcal antiserum (quellung reaction).

Bacterial Pneumonias

The term *pneumonia* is applied to many pulmonary infections. Most pneumonias are caused by bacteria. Pneumonias caused by *Streptococcus pneumoniae* and *Hemophilus influenzae* are by far the most common and are therefore referred to as *typical pneumonias*. Many other bacteria are also frequent causes of pneumonia. In addition, fungi, protozoa, and viruses are responsible for numerous cases. Pneumonias caused by organisms that are often difficult to identify and cultivate under clinical conditions, such as mycoplasmas, chlamydias, and viruses, are known as *atypical pneumonias*. Actually, a third to a half of pneumonia cases that are serious enough to be hospitalized cannot be associated with a specific pathogen. The physician usually makes an initial "best-guess" diagnosis based on symptoms, history, and microscopic examination of a sputum specimen. An antibiotic treatment is then directed at the suspected organisms.

Pneumococcal Pneumonia. The most common cause of bacterial pneumonia in adults is *S. pneumoniae* (the pneumococcus); the disease it causes is **pneumococcal pneumonia**. *S. pneumoniae* is a gram-positive, ovoid bacterium (Figure 24.10). This organism is also a common cause of otitis media, meningitis, and septicemia. Because it usually forms cell pairs, it was formerly called *Diplococcus pneumoniae*. It produces a dense capsule that makes the pathogen resistant to phagocytosis. These capsules are also the basis of serological differentiation of pneumococci into some 83 serotypes. Before antibiotic therapy became available, antisera directed at these capsular antigens were used to treat the disease.

Pneumococcal pneumonia involves both the bronchi and the alveoli (see Figure 24.2). Symptoms include fever, difficult breathing, and chest pain. The lungs have a reddish appearance because blood vessels are dilated. In response to the infection, alveoli fill with some red blood cells, polymorphonuclear leukocytes (PMNs), and fluid from surrounding tissues. The sputum is often rust-colored from blood coughed up from the lungs. Pneumococci can invade the bloodstream, the pleural cavity surrounding the lung, and occasionally the meninges. No bacterial toxin has been clearly related to pathogenicity.

A presumptive diagnosis can be made by isolation of the pneumococci from the throat, sputum, and other fluids. Pneumococci can be distinguished from other α-hemolytic streptococci by observing the inhibition of growth next to a disk of optochin (ethylhydrocupreine hydrochloride) or by performing a bile solubility test. They can also be serologically typed. The bacteria are observed under a microscope; when they react with a positive antiserum, their capsules appear to swell. This swelling is called the *quellung reaction* (from the German word for swelling).

There are many healthy carriers of the pneumococcus. Virulence of the bacteria seems to be based mainly on the carrier's resistance, which can be lowered by stress. Many illnesses of the elderly terminate in pneumococcal pneumonia.

Recurrence of pneumococcal pneumonia is not uncommon, but the second serological type is usually different. Before chemotherapy was available, the fatality rate was as high as 25%. This has now been lowered to less than 1% for younger patients treated early in the course of their disease. Penicillin is the drug of choice, but a few strains of penicillin-resistant pneumococci have been reported. A vaccine has been developed from the purified capsular material of the 23 types of pneumococci that cause at least 80% of the pneumococcal pneumonias in the U.S.A. This vaccine is used for the groups most susceptible to infection, the elderly and debilitated individuals.

Hemophilus influenzae Pneumonia.
Clinically, pneumonias caused by *S. pneumoniae* and *H. influenzae* are similar. A Gram stain of sputum will often help in differentiation—*H. influenzae* is a gram-negative coccobacillus. Patients with conditions such as alcoholism, poor nutrition, cancer, or diabetes are especially susceptible to this type of pneumonia. Second-generation cephalosporins are resistant to the β-lactamases produced now by many *H. influenzae* strains and so are usually the drug of choice.

Mycoplasmal Pneumonia.
The mycoplasmas, which do not have cell walls, do not grow under the conditions normally used to recover most bacterial pathogens. Because of this characteristic, pneumonias caused by mycoplasmas are often confused with viral pneumonias.

The bacterium *Mycoplasma pneumoniae* is the causative agent of **mycoplasmal pneumonia (primary atypical pneumonia)**. This type of pneumonia was first discovered when such atypical infections responded to tetracyclines, indicating that the agent was nonviral.

The disease is endemic and is a fairly common cause of pneumonia in young adults and children. The pathogens are transmitted in airborne droplets. Mycoplasmas usually infect the upper respiratory tract, and relatively few cases develop into pneumonia. The mild symptoms, which persist for three weeks or longer, are low fever, cough, and headache. Death is very rare.

When isolates from throat swabs and sputum grow on a medium containing horse serum and yeast extract, they form distinctive colonies with a "fried-egg" appearance. The colonies are usually so small that they must be observed with a hand lens or microscope. *M. pneumoniae* is β-hemolytic for guinea pig red blood cells; this trait differentiates it from other mycoplasmas. Conclusive identification is made by fluorescent-antibody methods. The mycoplasmas are highly varied in appearance because they lack cell walls (see Figure 11.10). Their flexibility allows them to pass through filters with pores as small as 0.2 μm in diameter, which is small enough to retain most other bacteria.

Diagnosis based on the recovery of organisms might not be useful in treatment because as long as two weeks may be required for the slow-growing organisms to develop. Serological tests can be used in diagnosis. A complement fixation test has been the standard, and ELISA tests to detect IgG and IgM are also available.

Legionellosis.
Legionellosis (Legionnaires' disease) first received public attention in 1976, when a series of deaths occurred among members of the American Legion who had attended a meeting in Philadelphia. A total of 182 people contracted pulmonary disease, apparently at this meeting, 29 of whom died. Because no obvious bacterial cause could be found, the deaths were attributed to viral pneumonia. Close investigation, mostly with techniques directed at locating a suspected rickettsial agent, eventually identified an unknown bacterium, an aerobic gram-negative rod now known as *Legionella pneumophila*. Several species of *Legionella* have now been identified.

The disease is characterized by a high fever of 41°C (105°F), cough, and general symptoms of pneumonia. No person-to-person transmission seems to be involved. Recent studies have shown that the bacterium can be isolated with some frequency from natural waters. In addition, the organisms can grow in the water of air-conditioning cooling towers, which might mean that some epidemics in hotels, urban business districts, and hospitals were caused by airborne transmission. A recent outbreak was traced to an indoor artificial waterfall. The organism has also been found to inhabit the water lines of many hospitals. Most hospitals keep the temperature of hot water lines relatively low (43° to 55°C) as a safety measure, and in cooler parts of the system this inadvertently maintains a good growth temperature for this organism. The bacterium is considerably more resistant to chlorine than most other bacteria and can survive for long periods in water with a low level of chlorine. One possible explanation for the resistance of *Legionella* to chlorine and heat is an interrelationship with waterborne amoebae. The bacteria are ingested by the amoebae but continue to proliferate; they may even survive within encysted amoebae.

Although the disease has only recently been identified, it seems now to have always been fairly common. Over 1000 cases are reported each year, but the actual incidence is estimated at more than 25,000. *L. pneumophila* is often overlooked as a possible cause of disease. One hospital that had never recorded a case of nosocomial legionellosis found upon investigation that more than 14% of nosocomial pneumonias had this cause and that most of the sites in its water distribution system contained *L. pneumophila*. Males over 50 years of age are most likely to contract the disease, especially if they are heavy smokers or alcohol abusers or have a chronic illness.

L. pneumophila is also responsible for *Pontiac fever;* its symptoms include fever, muscle ache, and usually a cough. The condition is mild and self-limiting.

The best diagnostic method is culture on a selective charcoal–yeast extract medium. Examination of respiratory specimens by fluorescent-antibody methods (see Chapter 18) and ELISA tests may be used, and a DNA probe test is available (see Figure 10.18). Erythromycin and rifampin are the drugs of choice for treatment.

Psittacosis (Ornithosis). The term **psittacosis (ornithosis)** was derived from the disease's association with psittacine birds, such as parakeets and other parrots. It was later found that the disease can also be contracted from many other birds, such as pigeons, chickens, ducks, and turkeys. Therefore, the more general term *ornithosis* has come into use.

The causative organism is *Chlamydia psittaci* (sit'tä-sē), a gram-negative, obligate intracellular bacterium (see Figure 11.9a). One way chlamydias differ from the rickettsias, which are also obligate intracellular bacteria, is that chlamydias form tiny **elementary bodies** as one part of their life cycle (see Figure 11.9b). Unlike most rickettsias, elementary bodies are resistant to environmental stress; therefore, they can be transmitted through air and do not require a bite to transfer the infective agent directly from one host to another.

Psittacosis is a form of pneumonia that usually causes fever, headache, and chills, Subclinical infections are very common, and stress appears to enhance susceptibility to the disease. Disorientation and even delirium in some cases indicate that the nervous system can be involved.

The disease is seldom transmitted from one human to another but is usually spread by contact with the droppings and other exudates of fowl. One of the most common modes of transmission is inhalation of dried particles from droppings. The birds themselves usually have diarrhea, ruffled feathers, respiratory illness, and a generally droopy appearance. The parakeets and other parrots sold commercially are usually (but not always) free of the disease. Pet store employees and people involved in raising turkeys are at greatest risk of contracting the disease.

Diagnosis is made by isolation of the bacterium in embryonated eggs or mice, or by cell culture. A fluorescent-antibody staining technique can then be used to identify the organism by the presence of its specific antigens. If organisms are not isolated, a complement fixation test that detects rising serum antibody titer in the patient can be used. No vaccine is available, but tetracyclines are effective antibiotics in treating humans and animals. Effective immunity does not result from recovery, even when high titers of antibody are present in the person's serum.

Most years, fewer than 100 cases and very few deaths are reported in the U.S.A. The main danger is late diagnosis. Before antibiotic therapy was available, mortality was about 20%.

Chlamydial Pneumonia. In recent years there have been outbreaks of a respiratory illness, especially common in populations of college students, that were found to be caused by a chlamydial organism. Originally it was considered a strain of *C. psittaci* (known by the letters *TWAR*, which were derived from laboratory markers, *TW* for Taiwan and *AR* for acute respiratory, that were being used to identify the original isolates). The pathogen has now been assigned the species name *Chlamydia pneumoniae*, and the disease is known as chlamydial pneumonia. It is apparently transmitted from person to person, probably by the respiratory route, but not as readily as infections such as influenza. Nearly half of the population is serologically positive to antibodies against the organism, an indication that this is a fairly common illness. Clinically, it resembles mycoplasmal pneumonia. Several serological tests have been developed that are useful in diagnosis; the one most used at present is a fluorescent-antibody test. The most effective antibiotic has been tetracycline.

Q Fever. In Australia in the mid-1930s, a disease was identified that was characterized by a fever lasting one or two weeks, chills, chest pain, severe headache, and other evidence of a pneumonia-type infection. The disease was rarely fatal. In the absence of an obvious cause, the affliction was labeled **Q fever** (for *query*)—much as one might say "X fever." The causative agent was subsequently identified as the obligately parasitic, intracellular bacterium *Coxiella burnetii*, a rickettsia (Figure 24.11a). Most rickettsias are not resistant enough to survive airborne transmission, but this organism is an exception.

The most serious complication of Q fever is endocarditis, which occurs in about 10% of cases. Five to ten years might elapse between the initial infection and appearance of endocarditis. The organisms apparently reside in the liver during this interval.

Coxiella burnetii in vacuole

(a) Note the intracellular growth of this rickettsia in vacuoles of the host cell

TEM |—————| 5 μm

E

(b) This cell has just divided; note the endospore-like body (E), which is probably responsible for the relative resistance of the organism

TEM |—————| 100 nm

FIGURE 24.11 *Coxiella burnetii*, the cause of Q fever.

The organism is a parasite of a number of arthropods, especially cattle ticks, and is transmitted among animals by tick bites. The infection in animals is usually subclinical. The cattle ticks first spread the disease among dairy herds, and the organisms are shed in the feces, milk, and urine of infected cattle. Once the disease is established in a herd, it is maintained by aerosol transmission. The disease is spread to humans by ingestion of unpasteurized milk and inhalation of aerosols of microbes generated in dairy barns, especially from placental material at calving time. Inhalation of a single organism is enough to cause infection. Many dairy workers have acquired at least subclinical infections. Workers in meat- and hide-processing plants are also at risk. The pasteurization temperature of milk, which was originally aimed at eliminating tuberculosis organisms, was raised slightly in 1956 to ensure the killing of *C. burnetii*. In 1981, an endosporelike body was discovered, which may account for this heat resistance (Figure 24.11b).

The organism may be identified by isolation and growth in chick embryos in eggs or in cell culture. Laboratory personnel testing for *Coxiella*-specific antibodies in the patient's serum can use serological tests, mostly complement fixation and agglutination types.

Most cases of Q fever in the U.S.A. are reported from the western states. The disease is endemic to California, Arizona, Oregon, and Washington. A vaccine for laboratory workers and other high-risk personnel is available. Tetracyclines are very effective in treatment.

Other Bacterial Pneumonias. An expanding number of bacteria have been implicated as causes of pneumonia. Important among them are *Staphylococcus aureus, M. (B.) catarrhalis, Streptococcus pyogenes,* and certain anaerobes found in the oral cavity. Gram-negative bacilli such as *Klebsiella pneumoniae* and *Pseudomonas* spp. occasionally cause bacterial pneumonia.

Viral Diseases of the Lower Respiratory System

For a virus to reach the lower respiratory system and initiate disease, it must pass numerous host defenses designed to trap and destroy it.

Viral Pneumonia

Viral pneumonia can occur as a complication of influenza, measles, or even chickenpox. A number of enteric and other viruses have been shown to cause viral pneumonia, but viruses are isolated and identified in less than 1% of pneumonia-type infections because few laboratories are equipped to properly test clinical samples for viruses. In those cases of pneumonia for which no cause is determined, viral etiology is often assumed if mycoplasmal pneumonia has been ruled out.

Respiratory syncytial virus is probably the most common cause of viral respiratory disease among infants. The name is derived from its characteristic of causing cell fusion (syncytium formation) in cell culture. It is worldwide in occurrence. Epidemics in hospital nurseries are of particular concern. The symptoms are coughing and wheezing that last for more than one week. If an infant is ill enough to be hospitalized, the mortality rate is 1% or 2%. Aerosol administration of the antiviral drug ribavirin at least significantly diminishes the severity of symptoms.

Influenza (Flu)

The developed countries of the world are probably more aware of **influenza (flu)** than of any other disease, except for the common cold. The flu is characterized by chills, fever, headache, and general muscular aches. Recovery normally occurs in a few days, and coldlike symptoms appear as the fever subsides. Nonetheless, in the U.S.A. an estimated 10,000 to 20,000 die of the "flu," even in nonepidemic years. Incidentally, diarrhea is not a normal symptom of the disease, and the intestinal discomforts attributed to "stomach flu" are probably from some other cause.

Viruses in the genus *Influenzavirus* consist of eight weakly linked RNA segments of differing lengths enclosed by an inner layer of protein and an outer lipid bilayer (Fig-

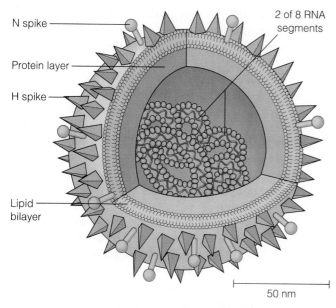

FIGURE 24.12 Detailed structure of the influenza virus.
The envelope is composed of a protein layer, a lipid bilayer, and two types of spikes. Inside the envelope are eight segments of RNA.

FIGURE 24.13 Antigenic shifts cause differences among influenza viruses. Because the eight RNA segments can be readily rearranged, the virus can evade previous immunity via major changes in the protein composition of the H and N spikes. Such antigenic shifts account for most of the major flu epidemics of this century.

ure 24.12). Embedded in the lipid bilayer are numerous projections that characterize the virus. There are two types of projections, *hemagglutinin (H) spikes* and *neuraminidase (N) spikes*.

The H spikes, of which there are about 500 on each virus, allow the virus to recognize and attach to body cells before infecting them. Antibodies against the influenza virus are directed mainly at these spikes. The term *hemagglutinin* refers to the agglutination of red blood cells (hemagglutination) that occurs when the viruses are mixed with them. This reaction is important in serological tests, such as the hemagglutination inhibition test often used to identify influenza and some other viruses.

The N spikes, of which there are about 100 per virus, differ from the H spikes in appearance and function. Apparently they help the virus separate from the infected cell as the virus exits after intracellular reproduction. N spikes also stimulate formation of antibodies, but these are less important in the body's resistance to the disease than those produced in response to the H spikes.

Viral strains are identified by variation in the H and N antigens. The different forms of the antigens are assigned numbers—for example, H_1, H_2, H_3, N_1, and N_2. Each number change represents a substantial alteration in the protein makeup of the spike. The first antigenic type was isolated in 1933 and was named H_0N_1. The Asiatic flu pandemic of 1957 was of type H_2N_2, and the Hong Kong pandemic of 1968 was designated H_3N_2. These changes are called *antigenic shifts* (Figure 24.13), and they are great enough to evade most of the immunity developed in the human population. These antigenic shifts are probably caused by a major genetic recombination. Because influenza virus RNA occurs as eight segments, recombina-

tion is likely in infections caused by more than one strain. Recombination between the RNA of animal viral strains (found in swine, horses, and birds, for example) and the RNA of human strains might be involved. Swine, ducks, and chickens in southern Chinese farming communities have come under suspicion as being the animals most likely to be involved in genetic shifts (and are thus called "mixing vessels"). Wild ducks and other migratory birds then become carriers that spread the virus over large geographical areas.

Between episodes of such major antigenic shifts, there are minor annual variations in the antigenic makeup called *antigenic drift*. The virus might still be designated as H_3N_2, for example, but viral strains arise reflecting minor antigenic changes within the antigenic group. These strains are sometimes assigned names related to the locality in which they were first identified. They usually reflect an alteration of only a single amino acid in the protein makeup of the H or N spike. Such a minor, one-step mutation is probably a response to selective pressure by antibodies (usually IgA in the mucous membranes) that neutralizes all viruses except for the new mutations. Such mutations can be expected about once in each million multiplications of the virus. High mutation rates are a characteristic of RNA viruses, which lack much of the "proofreading" ability of DNA viruses.

The usual result of antigenic drift is that a vaccine effective against H_3, for example, will be less effective against H_3 isolates circulating ten years after the event. There will have been enough drift in that time that the virus can largely evade the antibodies originally stimulated by the earlier strain.

The influenza viruses are also classified into major groups according to the antigens of their protein coats. These groups are A, B, and (rarely) C. The A-type viruses are the ones responsible for major pandemics. The B-type virus also circulates and mutates, but it is usually responsible for more geographically limited and milder infections.

To date it has not been possible to make a vaccine for influenza that gives long-term immunity to the general population. Although it is not difficult to make a vaccine for a particular antigenic strain of virus, each new strain of circulating virus must be identified in time for useful development and distribution of a new vaccine. Representatives from the CDC have trained medical workers in mainland China in the techniques needed to detect antigenic changes in the influenza virus. By combining information from China, Japan, and Taiwan, we now get earlier warning of new influenza virus types. This usually allows greater opportunity to develop annual vaccines directed at the likely antigenic types. Unless the strain appears to be unusually virulent, vaccine administration is usually limited to the elderly, to hospital personnel, and to similar high-risk groups. The vaccines are often *multivalent*—that is, directed at several strains in circulation at the time. At present, influenza viruses for manufacturing vaccines are grown in embryonated egg cultures. The vaccines are usually 70% to 90% effective, but the duration of protection is probably no more than three years.

One new approach to the problem of antigenic variation might be to use genetic-engineering techniques. A laboratory bacterium can theoretically be made to produce very quickly a large number of antigens of the influenza virus when genes for this purpose have been inserted into the bacterial DNA.

Almost every year, epidemics of the flu spread rapidly through large populations. The disease is so readily transmissible that epidemics are quickly propagated through populations susceptible to the newly changed strain of virus. The mortality rate from the disease is not high, usually less than 1%, and these deaths are mainly among the very young and the very old. However, so many people are infected in a major epidemic that the total number of deaths is often large. Very often, the cause of death is not the influenza virus but secondary bacterial infections. The bacterium *H. influenzae* was named under the mistaken belief that it was the principal pathogen causing influenza rather than being a secondary invader. *S. aureus* and *S. pneumoniae* are other prominent secondary bacterial invaders.

In any discussion of influenza, the great pandemic of 1918–1919 must be mentioned. Worldwide, more than 20 million people died. No one is sure why it was so unusually lethal. Usually, the very young and very old are the principal victims, but in 1918–1919, young adults had the highest mortality rate. Normally the infection is restricted to the upper respiratory system, but some change in virulence allowed the virus to invade the lungs and frequently cause viral pneumonia. Bacterial complications also frequently accompanied the infection and, in those preantibiotic days, were often fatal. The 1918 viral strain apparently became endemic in the swine population of the U.S.A. Occasionally, influenza is still spread to humans from this reservoir, but the disease has not propagated like the virulent disease of 1918 did.

An antiviral drug, amantadine, has been found to significantly reduce the symptoms of influenza if administered promptly. In prophylactic use it apparently reduces the rate of infection and illness by perhaps as much as 70%. The bacterial complications of influenza are amenable to treatment with antibiotics.

A complete serological identification of influenza virus isolates is generally conducted in a large, central laboratory. However, many hospital laboratories can now identify an isolate as an influenza virus and rapidly distinguish between A and B serotypes using immunofluorescence testing with monoclonal antibodies. Because only type A is susceptible to antivirals such as amantadine, this can be valuable information.

Hantavirus Pulmonary Syndrome

In 1993, several cases of severe respiratory illness, mostly fatal, were observed in New Mexico. The symptoms are mild at first, resembling a severe cold or influenza. Then the lungs fill with fluids leaking from the lung capillary system, literally drowning the victim. This flooding of the lungs with blood plasma is a type of internal hemorrhaging, for which another disease, Korean Hemorrhagic Fever (KHF), is named.

Within four weeks the cause was identified by PCR as a strain of *Hantavirus*, named for the Hantaan River in Korea. KHF was first noticed by the Western world during the Korean War. The KHF *Hantavirus*, identified in 1976, infects 150,000 people per year and has a 5% mortality rate.

At present, five strains of *Hantavirus* are recognized, and each strain is carried by a different rodent. Infections are usually associated with contact with rodents. There is no evidence of human-to-human transmission. *Hantavirus* has been found in rodents throughout the U.S.A.; however, the majority of human infections have occurred in the Four Corners region, where Colorado, Utah, Arizona, and New Mexico meet (see the box, p. 606). Treatment at this time is only supportive, consisting of treating symptoms and putting the patient on a respirator.

Fungal Diseases of the Lower Respiratory System

Fungi often produce spores that are disseminated through the air. It is therefore not surprising that several serious fungal diseases affect the lower respiratory system. The rate of fungal infections has been increasing over the past 15 years or so. Opportunistic fungi are able to grow in immunosuppressed patients, and AIDS, transplant drugs, and anticancer drugs have created more immunosuppressed people than ever before.

Histoplasmosis

Histoplasmosis superficially resembles tuberculosis. In fact, it was first recognized as a disease in the U.S.A. when X-ray surveys showed lung lesions in many people who were tuberculin-test negative. Although the lungs are the organs most likely to be initially infected, the organisms cause lesions in almost all organs of the body. Normally, symptoms of the disease are rather poorly defined and mostly subclinical, and the disease passes for a minor respiratory infection. In a few cases, perhaps fewer than 0.1%, histoplasmosis becomes progressive and is a severe, generalized disease.

The causative organism, *Histoplasma capsulatum* (his-tō-plaz'mä kap-su-lä'tum), is a dimorphic fungus; that is, it has a yeastlike morphology in tissue growth (Figure 24.14a), and, in soil or artificial media, it forms a filamentous mycelium carrying reproductive conidia (Figure 24.14b). In the body, the yeastlike form is found intracellularly in macrophages, where it survives and multiplies.

Although histoplasmosis is rather widespread throughout the world, it has a limited geographical range in the U.S.A. (Figure 24.15). For example, of 513 cases reported in a recent year, 403 were from Minnesota, Iowa, Missouri, and Alabama. In general, the disease is found in the states adjoining the Mississippi and Ohio rivers. More than 75% of the population in some of these states have antibodies against the infection. In other states—Maine, for example—a positive test is a rather rare event. Approximately 50 deaths are reported in the U.S.A. each year from histoplasmosis. The total number of infected people may run into the millions, so the death rate is low.

Humans acquire the disease from airborne conidia produced under conditions of particular moisture and pH lev-

(a) Yeastlike form typical of growth in tissue or at 37°C; note that one cell near the center is budding

(b) Filamentous, spore-forming phase found in soil or at temperatures below 35°C; the spores are usually the infectious particle

FIGURE 24.14 *Histoplasma capsulatum,* **a dimorphic fungus that causes histoplasmosis.**

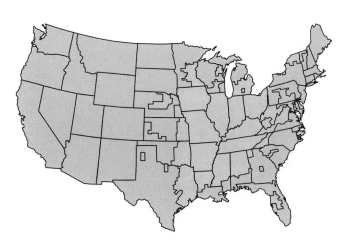

FIGURE 24.15 Histoplasmosis distribution. Geographical distribution of histoplasmosis (gold) in the U.S.A.

(a) Two spherical bodies found in mouse lung tissue; note the thick wall (SW) of the empty spherule and the mass of spores in the other spherule

SEM | 10 μm

(b) Arthrospores, the airborne infectious particles that spread coccidioidomycosis

SEM | 1 μm

FIGURE 24.16 *Coccidioides immitis*, the cause of coccidioidomycosis.

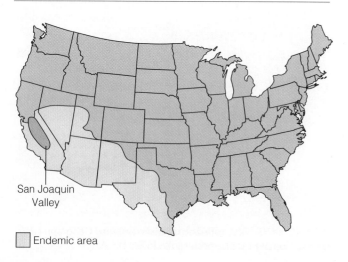

San Joaquin Valley

☐ Endemic area

els. These conditions occur particularly where droppings from birds and bats have accumulated. Birds themselves do not carry the disease, but their droppings provide nutrients, particularly a source of nitrogen, for the fungus. Bats, which have a lower body temperature than birds, carry the fungus, shed it in their feces, and probably infect new soil sites.

Clinical signs and history, complement fixation tests, and, most important, isolation of the organism or its identification in tissue specimens are necessary for proper diagnosis. Currently, the most effective chemotherapy is with amphotericin B or itraconazole.

Coccidioidomycosis

Another fungal pulmonary disease, also rather restricted geographically, is **coccidioidomycosis**. The causative organism is *Coccidioides immitis,* a dimorphic fungus. The spores are found in dry, highly alkaline soils of the American Southwest and in similar soils of South America and northern Mexico. In tissues, the organism forms a thick-walled body filled with spores (*spherules*) (Figure 24.16a). In soil, it forms filaments that reproduce by formation of arthrospores (Figure 24.16b). The wind carries the arthrospores to transmit the infection. Arthrospores are often so abundant that simply driving through an endemic area can result in infection, especially during a dust storm.

The symptoms of coccidioidomycosis include chest pain and perhaps fever, coughing, and loss of weight. Most infections are inapparent, and almost all patients recover in a few weeks, even without treatment. Sound immunity remains. However, in less than 1% of the infections, a progressive disease resembling tuberculosis disseminates throughout the body. The resemblance to tuberculosis is so close that isolation of the organism is necessary to properly diagnose coccidioidomycosis. Diagnosis is most reliably made by identification of the spherules in tissue or fluids. The organism is also readily cultured from fluids or lesions. Laboratory workers must handle it with unusual care because of the possibility of infectious aerosols. A tuberculinlike skin test is used in screening for cases. Of those tested in California, more than half had positive skin tests, although most had not been aware of the infection. Farm workers are most likely to be infected, and predisposing factors, such as fatigue and poor nutrition, can lead to more severe disease.

Approximately 50–100 deaths are reported each year from this disease in the U.S.A. Most cases are reported from California. In 1992, 92% of the reported cases were in the San Joaquin Valley, with only a scattering of cases

FIGURE 24.17 Endemic area for coccidioidomycosis in the U.S.A. High-altitude and nonarid regions within the indicated area are associated with only a slightly increased risk of exposure. The dark gold area in California is the San Joaquin Valley. Because of very high incidence there, the disease is also sometimes called Valley fever.

FIGURE 24.18 Life cycle of *Pneumocystis carinii*, the cause of *Pneumocystis* pneumonia.

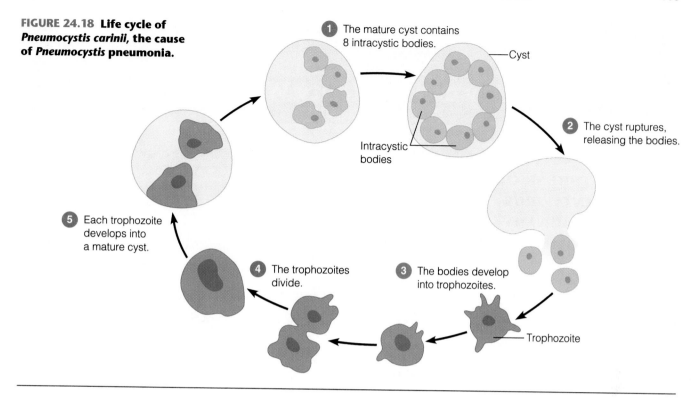

1. The mature cyst contains 8 intracystic bodies.

Cyst

Intracystic bodies

2. The cyst ruptures, releasing the bodies.

3. The bodies develop into trophozoites.

Trophozoite

4. The trophozoites divide.

5. Each trophozoite develops into a mature cyst.

reported from other states (Figure 24.17). Amphotericin B has been used to treat serious cases, but less toxic imidazole drugs, such as ketoconazole and itraconazole, are useful alternatives.

Blastomycosis (North American Blastomycosis)

Blastomycosis is usually called **North American blastomycosis** to differentiate it from a similar South American blastomycosis. It is caused by the fungus *Blastomyces dermatitidis* (blas-tō-mī′sēz dėr-mä-tit′i-dis), a dimorphic fungus found most often in the Mississippi Valley, where it probably grows in soil. Approximately 30–60 deaths are reported each year, although most infections are asymptomatic. The infection begins in the lungs and can spread rapidly. Cutaneous ulcers commonly appear, and there is extensive abscess formation and tissue destruction. The organism can be isolated from pus and biopsy specimens. Amphotericin B is usually an effective treatment.

Other Fungi Involved in Respiratory Disease

Many other opportunistic fungi may cause respiratory disease, particularly in immunosuppressed hosts or when there is exposure to massive numbers of spores. **Aspergillosis** is an important example; it is transmitted by the spores of *Aspergillus fumigatus* (fü-mi-gä′tus) and other species of *Aspergillus,* which are fairly widespread in decaying vegetation. Compost piles are ideal sites for growth, and farmers and gardeners are most often exposed to infective amounts of such spores. Similar pulmonary infections sometimes result when individuals are exposed to spores of mold genera, such as *Rhizopus* or *Mucor.* Such diseases can be very dangerous, particularly invasive infections of pulmonary aspergillosis. Predisposing factors include an impaired immune system, cancer, and diabetes. As with most systemic fungal infections, there is only a limited arsenal of antifungal agents available; amphotericin B has proved the most useful.

Protozoan Disease of the Lower Respiratory System

Protozoan diseases of the respiratory system are unusual; most protozoan diseases affect the cardiovascular or digestive system. Probably the most interesting aspect of the protozoan causing the respiratory disease listed here is that it may well not be a protozoan at all.

Pneumocystis Pneumonia

***Pneumocystis* pneumonia** is caused by *Pneumocystis carinii* (Figure 24.18). The taxonomic position of this organism has been uncertain ever since its discovery in 1909, when it was thought to be a developmental stage of a trypanosome. Since that time there has been no universal agreement about whether it is a protozoan or a fungus. It has some characteristics of both groups of organisms. Recent analysis of RNA shows that it is strongly related to certain yeasts.

The disease occurs throughout the world and can be endemic in hospitals. The organism is found in healthy human lungs but causes disease among immunosuppressed patients. This group includes people receiving im-

MMWR

Morbidity and Mortality Weekly Report

OUTBREAK OF *HANTAVIRUS* PULMONARY SYNDROME—1993

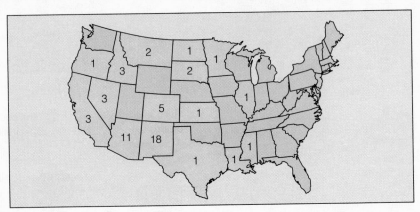

Reported cases of *Hantavirus* pulmonary syndrome in the U.S.A., 1993.

WEEK 1: On May 14, the New Mexico Department of Health (NMDOH) was notified that two people living in the same household had died within five days of each other. Their illnesses were characterized by a prodrome of fever, aches, headache, and cough, followed by the rapid development of respiratory failure. Tests for *Yersinia pestis* and other bacterial and viral pathogens were negative.

Week 2: Seven confirmed cases of *Hantavirus* Pulmonary Syndrome (HPS) have been identified by the CDC and NMDOH. A confirmed case is now defined as an unexplained adult respiratory distress syndrome (ARDS) or acute bilateral pulmonary interstitial infiltrates and/or prodromal symptoms in a person who had onset during 1993 and who has laboratory evidence of recent *Hantavirus* infection. *Hantavirus*-specific nucleotide sequences were amplified from autopsy specimens from the two case-patients using the PCR. Of the seven cases, four were from New Mexico, two from Arizona, and one from Colorado.

Rodents were collected around the domestic settings of several case-patients. Of the 52 rodents tested, 12 had serological evidence of *Hantavirus* infections; all 12 were *Peromyscus maniculatus* (deer mouse).

Week 3: Laboratory evidence of HPS infection has been confirmed in

12 patients. The high prevalence of *Hantavirus* antibodies in deer mice and the similarity of PCR products in the deer mice and human case-patients suggest that this genus may be involved in *Hantavirus* transmission to humans.

Week 4: Laboratory evidence of HPS has been confirmed in 15 patients who had onset of illness from January 1 through June 30. Each of these patients had one or more of the following: positive ELISA serology with elevated IgM titers indicating recent infection, seroconversion by ELISA, positive immunohistochemistry on formalin-fixed lung tissue, or amplification of *Hantavirus* nucleotide sequences from frozen tissue. Eleven patients died. Similar illnesses in an additional 23 people, 10 of whom died, are being investigated for possible *Hantavirus* infections.

A total of 668 rodents have been trapped in and around houses in 14 rural sites. Of the first 283 rodents tested, *Hantavirus* antibodies were detected in 23%.

Editorial Note: As of December 31, 1993, 55 HPS cases have been confirmed in the U.S.A. (see map). The deer mouse reservoir inhabits all areas of the country. People participating in agricultural activities near habitats of infected rodents are likely to be at greater risk for infection. Interim recommendations to reduce the risk of exposure to potentially infected rodents in the southwestern U.S.A. have been published; these recommendations are generally applicable to rural settings in other parts of the country.

The current outbreak appears to be caused by a newly recognized *Hantavirus* called the Muerto Canyon Virus. This is probably not a new virus but an emerging infection. Comparisons of *Hantavirus* genomes suggest that this virus probably came to North America with the first rats from the Old World.

SOURCES: *MMWR* 42(22):421–424 (6/11/93), *MMWR* 42(23):441–443 (6/18/93), *MMWR* 42(24):477–479 (6/25/93), *MMWR* 42(25):495–496 (7/1/93), *MMWR* 42(26):517–519 (7/9/93), *MMWR* 42(36):707 (12/17/93), *MMWR* 43(3):45–48 (1/28/94).

munosuppressive drugs to minimize rejection of transplanted tissue and those whose immunity is depressed because of cancer. People with acquired immunodeficiency syndrome (AIDS) are also very susceptible to this organism.

Before the AIDS epidemic, *Pneumocystis* pneumonia was an uncommon disease; perhaps 100 cases occurred each year. But a leading AIDS researcher observed in 1990 that, "during the past year, the number of deaths from *Pneumocystis carinii* pneumonitis {pneumonia} in the U.S.A. exceeded the combined number of deaths from meningococcal infections, all types of viral hepatitis and encephalitis, gonorrhea, syphilis, varicella, measles, mumps, rubella, diphtheria, tetanus, pertussis, polio, amebiasis, shigellosis, salmonellosis, typhoid fever, typhus fevers, cholera, rabies, brucellosis, anthrax, tularemia, botulism, and malaria. More patients died of *P. carinii* pneumonitis than of all types of tuberculosis."

In the human lung the organisms are found mostly in the lining of the alveoli. There they form a thick-walled cyst in which spherical intracystic bodies successively divide as part of a sexual cycle. The mature cyst contains eight such bodies (see Figure 24.18). Eventually it ruptures and releases them, each body developing into a vegetative cell. The vegetative cells can reproduce asexually by binary fission or budding, but may also enter the encysted sexual stage.

The drug of choice is currently trimethoprim-sulfamethoxazole, but pentamidine isethionate is a frequent alternative.

STUDY OUTLINE

INTRODUCTION (p. 586)

1. Infections of the upper respiratory system are the most common type of infection.
2. Pathogens that enter the respiratory system can infect other parts of the body.
3. Respiratory infections are transmitted by direct contact and fomites.

STRUCTURE AND FUNCTION OF THE RESPIRATORY SYSTEM (pp. 586–588)

1. The upper respiratory system consists of the nose, pharynx, and associated structures, such as the middle ear and auditory tubes.
2. Coarse hairs in the nose filter large particles from air entering the respiratory tract.
3. The ciliated mucous membranes of the nose and throat trap airborne particles and remove them from the body.
4. Lymphoid tissue, tonsils, and adenoids provide immunity to certain infections.
5. The lower respiratory system consists of the larynx, trachea, bronchial tubes, and alveoli.
6. The ciliary escalator of the lower respiratory system helps prevent microorganisms from reaching the lungs.
7. Microorganisms in the lungs can be phagocytized by alveolar macrophages.
8. Respiratory mucus contains IgA antibodies.

NORMAL MICROBIOTA OF THE RESPIRATORY SYSTEM (p. 588)

1. The normal microbiota of the nasal cavity and throat can include pathogenic microorganisms.
2. The lower respiratory system is usually sterile because of the action of the ciliary escalator.

MICROBIAL DISEASES OF THE UPPER RESPIRATORY SYSTEM (pp. 588–592)

1. Specific areas of the upper respiratory system can become infected to produce pharyngitis, laryngitis, tonsillitis, sinusitis, and epiglottitis.
2. These infections may be caused by several bacteria and viruses, often in combination.
3. *H. influenzae* type b can cause epiglottitis.
4. Most respiratory tract infections are self-limiting.

Bacterial Diseases of the Upper Respiratory System (pp. 589–591)

Streptococcal Pharyngitis (Strep Throat) (pp. 589–590)
1. This infection is caused by group A β-hemolytic streptococci, the group that consists of *Streptococcus pyogenes*.
2. Symptoms of this infection are inflammation of the mucous membrane and fever; tonsillitis and otitis media may also occur.
3. Preliminary rapid diagnosis is made by indirect agglutination tests. Definitive diagnosis is based on a rise in IgM antibodies.

4. Penicillin is used to treat streptococcal pharyngitis.
5. Immunity to streptococcal infections is type-specific.
6. Streptococcal sore throat is usually transmitted by droplets but at one time was commonly associated with unpasteurized milk.

Scarlet Fever (p. 590)

1. Streptococcal sore throat, caused by an erythrogenic toxin-producing *S. pyogenes,* results in scarlet fever.
2. *S. pyogenes* produces erythrogenic toxin when lysogenized by a phage.
3. Symptoms include a red rash, high fever, and a red, enlarged tongue.

Diphtheria (pp. 590–591)

1. Diphtheria is caused by exotoxin-producing *Corynebacterium diphtheriae.*
2. Exotoxin is produced when the bacteria are lysogenized by a phage.
3. A membrane, containing fibrin and dead human and bacterial cells, forms in the throat and can block the passage of air.
4. The exotoxin inhibits protein synthesis, and heart, kidney, or nerve damage may result.
5. Laboratory diagnosis is based on isolation of the bacteria and appearance of growth on differential media.
6. Antitoxin must be administered to neutralize the toxin, and antibiotics can stop growth of the bacteria.
7. Routine immunization in the U.S.A. includes diphtheria toxoid in the DPT vaccine.

Cutaneous Diphtheria (p. 591)

1. Slow-healing skin ulcerations are characteristic of cutaneous diphtheria.
2. There is minimal dissemination of the exotoxin in the bloodstream.

Otitis Media (p. 591)

1. Earache, or otitis media, can occur as a complication of nose and throat infections or through direct inoculation from an external source.
2. Pus accumulation causes pressure on the eardrum.
3. Bacterial causes include *Streptococcus pneumoniae,* nonencapsulated *Hemophilus influenzae, Moraxella (Branhamella) catarrhalis, Streptococcus pyogenes,* and *Staphylococcus aureus.*

Viral Disease of the Upper Respiratory System (p. 592)

Common Cold (p. 592)

1. Any one of approximately 200 different viruses can cause the common cold; rhinoviruses cause about 50% of all colds.
2. Symptoms include sneezing, nasal secretions, and congestion.

3. Sinus infections, lower respiratory tract infections, laryngitis, and otitis media can occur as complications of a cold.
4. Colds are most often transmitted by indirect contact.
5. Rhinoviruses prefer temperatures slightly lower than body temperature.
6. The incidence of colds increases during cold weather, possibly because of increased interpersonal indoor contact or physiological changes.
7. Antibodies are produced against the specific viruses.

MICROBIAL DISEASES OF THE LOWER RESPIRATORY SYSTEM (pp. 592–607)

1. Many of the same microorganisms that infect the upper respiratory system also infect the lower respiratory system.
2. Diseases of the lower respiratory system include bronchitis and pneumonia.

Bacterial Diseases of the Lower Respiratory System (pp. 593–600)

Pertussis (Whooping Cough) (pp. 593–594)

1. Pertussis is caused by *Bordetella pertussis.*
2. The initial stage of pertussis resembles a cold and is called the catarrhal stage.
3. The accumulation of mucus in the trachea and bronchi causes deep coughs characteristic of the paroxysmal (second) stage.
4. The convalescence (third) stage can last for months.
5. Laboratory diagnosis is based on isolation of the bacteria on enrichment and selective media followed by serological tests.
6. Regular immunization for children includes dead *B. pertussis* cells as part of the DPT vaccine.

Tuberculosis (pp. 594–597)

1. Tuberculosis is caused by *Mycobacterium tuberculosis.*
2. Large amounts of lipids in the cell wall account for the bacterium's acid-fast characteristic as well as its resistance to drying and disinfectants.
3. *M. tuberculosis* may be ingested by alveolar macrophages; if not killed, the bacteria reproduce in the macrophages.
4. Lesions formed by *M. tuberculosis* are called tubercles; dead macrophages and bacteria form the caseous center that might calcify and appear in an X ray as a Ghon complex.
5. Liquefaction of the caseous center results in a tuberculous cavity in which *M. tuberculosis* can grow.
6. New foci of infection can develop when a caseous lesion ruptures and releases bacteria into blood or lymph vessels; this is called miliary tuberculosis.
7. Miliary tuberculosis is characterized by weight loss, coughing, and loss of vigor.
8. Chemotherapy usually involves two drugs taken for one to two years; multiple–drug-resistant *M. tuberculosis* is becoming prevalent.

9. A positive tuberculin skin test can indicate either an active case of tuberculosis, or prior infection, or vaccination and immunity to the disease.
10. Laboratory diagnosis is based on isolation of the bacteria and requires incubation of up to eight weeks.
11. *Mycobacterium bovis* causes bovine tuberculosis and can be transmitted to humans by unpasteurized milk.
12. *M. bovis* infections usually affect the bones or lymphatic system.
13. BCG vaccine for tuberculosis consists of a live, avirulent culture of *M. bovis*.
14. *M. avium-intracellulare* complex infects patients in the late stages of HIV infection.

Bacterial Pneumonias (pp. 597–600)

1. Most pneumonias are caused by *S. pneumoniae* and *H. influenzae;* these are called typical pneumonias.
2. Atypical pneumonias are caused by organisms that are difficult to identify (e.g., viruses).

Pneumococcal Pneumonia (pp. 597–598)

1. Pneumococcal pneumonia is caused by encapsulated *Streptococcus pneumoniae.*
2. Symptoms of this disease are fever, difficult breathing, chest pain, and rust-colored sputum.
3. The bacteria can be identified by the production of α-hemolysins, inhibition by optochin, bile solubility, and the quellung reaction.
4. A vaccine consists of purified capsular material from 23 serotypes of *S. pneumoniae.*

Hemophilus influenzae Pneumonia (p. 598)

1. Alcoholism, poor nutrition, cancer, and diabetes are predisposing factors to *H. influenzae* pneumonia.
2. *H. influenzae* is a gram-negative coccobacillus.

Mycoplasmal Pneumonia (p. 598)

1. *Mycoplasma pneumoniae* causes mycoplasmal pneumonia; it is an endemic disease.
2. *M. pneumoniae* produces small "fried-egg" colonies after two weeks' incubation on enriched media containing horse serum and yeast extract.
3. A complement fixation test, used to diagnose the disease, is based on the rising of antibody titer.

Legionellosis (pp. 598–599)

1. The disease is caused by the aerobic gram-negative rod *Legionella pneumophila.*
2. The bacterium can grow in water, such as air-conditioning cooling towers, and then be disseminated in the air.
3. This pneumonia does not appear to be transmitted from person to person.
4. Bacterial culture, fluorescent-antibody tests, or DNA probes are used for laboratory diagnosis.

Psittacosis (Ornithosis) (p. 599)

1. *Chlamydia psittaci* is transmitted by contact with contaminated droppings and exudates of fowl.

2. Elementary bodies allow the bacteria to survive outside a host. Reticulate bodies develop in host cells.
3. The bacteria are isolated in embryonated eggs, mice, or cell culture; identification is based on fluorescent-antibody staining.
4. Commercial bird handlers are most susceptible to this disease.

Chlamydial Pneumonia (p. 599)

1. *Chlamydia pneumoniae* causes pneumonia; it is transmitted from person to person.
2. A fluorescent-antibody test is used for diagnosis.

Q Fever (pp. 599–600)

1. Obligately parasitic, intracellular *Coxiella burnetii* causes Q fever.
2. The disease is usually transmitted to humans through unpasteurized milk or inhalation of aerosols in dairy barns.
3. Laboratory diagnosis is made with the culture of bacteria in embryonated eggs or cell culture.

Other Bacterial Pneumonias (p. 600)

1. Gram-positive bacteria that cause pneumonia include *S. aureus* and *S. pyogenes.*
2. Gram-negative bacteria that cause pneumonia include *M.(B.) catarrhalis, K. pneumoniae,* and *Pseudomonas* spp.

Viral Diseases of the Lower Respiratory System (pp. 600–603)

Viral Pneumonia (p. 600)

1. A number of viruses can cause pneumonia as a complication of infections such as influenza.
2. The etiologies are not usually identified in a clinical laboratory because of the difficulty in isolating and identifying viruses.
3. Respiratory syncytial virus is a common cause of pneumonia in infants.

Influenza (Flu) (pp. 600–602)

1. Influenza is caused by influenza virus and is characterized by chills, fever, headache, and general muscular aches.
2. Hemagglutinin (H) and neuraminidase (N) spikes project from the outer lipid bilayer of the virus.
3. Viral strains are identified by antigenic differences in the H and N spikes; they are also divided by antigenic differences in their protein coats (A, B, and C).
4. Viral isolates are identified by hemagglutination-inhibition tests and immunofluorescence testing with monoclonal antibodies.
5. Antigenic shifts that alter the antigenic nature of the H and N spikes make natural immunity and vaccination of questionable value. Minor antigenic changes are caused by antigenic drift.
6. Deaths during an influenza epidemic are usually from secondary bacterial infections.

7. Multivalent vaccines are available for the elderly and other high-risk groups.
8. Amantadine is an effective prophylactic and curative drug against *Influenzavirus* type A.

Hantavirus Pulmonary Syndrome (pp. 602–603)

1. A strain of *Hantavirus* carried by deer mice causes this disease; another strain causes Korean Hemorrhagic fever.
2. The disease has a high fatality rate; treatment consists of managing patients' symptoms.

Fungal Diseases of the Lower Respiratory System (pp. 603–605)

1. Fungal spores are easily inhaled; they may germinate in the lower respiratory tract.
2. The incidence of fungal diseases has been increasing over the past 15 years or so.
3. The following mycoses can be treated with amphotericin B.

Histoplasmosis (pp. 603–604)

1. *Histoplasma capsulatum* causes a subclinical respiratory infection that only occasionally progresses to a severe, generalized disease.
2. The disease is acquired by inhalation of airborne conidia.
3. Isolation of the fungus or identification of the fungus in tissue samples is necessary for diagnosis.

Coccidioidomycosis (pp. 604–605)

1. Inhalation of the airborne arthrospores of *Coccidioides immitis* can result in coccidioidomycosis.
2. Most cases are subclinical, but when there are predisposing factors such as fatigue and poor nutrition, a progressive disease resembling tuberculosis can result.

Blastomycosis (North American Blastomycosis) (p. 605)

1. *Blastomyces dermatitidis* is the causative agent of blastomycosis.
2. The infection begins in the lungs and can spread to cause extensive abscesses.

Other Fungi Involved in Respiratory Disease (p. 605)

1. Opportunistic fungi can cause respiratory disease in immunosuppressed hosts, especially when large numbers of spores are inhaled.
2. Among these fungi are *Aspergillus*, *Rhizopus*, and *Mucor*.

Protozoan Disease of the Lower Respiratory System (pp. 605–607)

Pneumocystis Pneumonia (pp. 605–607)

1. Immunosuppressed patients and cancer patients are susceptible to an endemic parasite, *Pneumocystis carinii*.
2. *Pneumocystis* pneumonia is currently being treated with trimethoprim or pentamidine.
3. *P. carinii* may actually be a fungus.

STUDY QUESTIONS

Review

1. Describe how microorganisms are prevented from entering the upper respiratory system. How are they prevented from causing infections in the lower respiratory system?

2. Respiratory diseases are usually transmitted by _____. List other methods by which they can be transmitted.

3. How do the normal microbiota of the respiratory system illustrate microbial antagonism?

4. Compare and contrast mycoplasmal pneumonia and viral pneumonia.

5. How is otitis media contracted? What causes otitis media? Why was otitis media included in a chapter on diseases of the respiratory system?

6. List the causative agent, symptoms, and treatment for four viral diseases of the respiratory system. Separate the diseases according to whether they infect the upper or lower respiratory system.

7. Complete the following table.

DISEASE	CAUSATIVE AGENT	SYMPTOMS	TREATMENT
Streptococcal sore throat			
Scarlet fever			
Diphtheria			
Whooping cough			
Tuberculosis			
Pneumococcal pneumonia			
H. influenzae pneumonia			
Chlamydial pneumonia			
Legionellosis			
Psittacosis			
Q fever			
Epiglottitis			

8. Under what conditions can the saprophytes *Aspergillus* and *Rhizopus* cause infections?

9. Why is the DPT vaccine routinely given to children in the U.S.A.? What does it consist of?

10. A patient has been diagnosed as having pneumonia. Is this sufficient information to begin treatment with antimicrobial agents? Briefly discuss why or why not.

11. List the causative agent, method of transmission, and endemic area for the diseases histoplasmosis, coccidioidomycosis, blastomycosis, and *Pneumocystis* pneumonia.

12. Briefly describe the procedures and positive results of the tuberculin test, and indicate what is revealed by a positive test.

13. Discuss reasons for the increased incidences of colds and pneumonias during cold weather.

14. Match the bacteria involved in respiratory infections with the following laboratory test results.

Gram-positive cocci
   Catalase-positive _____
   Catalase-negative
      β-Hemolytic, bacitracin inhibition _____
      α-Hemolytic, optochin inhibition _____
Gram-positive rods
   Not acid-fast _____
   Acid-fast _____
Gram-negative cocci _____
Gram-negative rods
   Aerobes
      Coccobacilli _____
      Rods _____
   Facultative anaerobes
      Coccobacilli _____
      Rods _____
Intracellular parasites
   Form elementary bodies _____
   Do not form elementary bodies _____
Wall-less _____

15. Explain why it would be impractical to include cold and influenza vaccinations in the required child vaccinations.

Critical Thinking

1. Differentiate between the following.
  (a) *S. pyogenes* causing strep throat and *S. pyogenes* causing scarlet fever.
  (b) Diphtheroids and *C. diphtheriae*.

2. Why might vaccination against influenza be of questionable value?

3. Over the past 50 years, Western medicine has begun to think that infectious diseases were a thing of the past. Tuberculosis cases began to increase in 1979, *Pneumocystis* pneumonia was seen for the first time in 1981, and *Hantavirus* infections were seen for the first time in the U.S.A. in 1993. Provide some explanations for these data.

Clinical Applications

1. In August, a 24-year-old man from Virginia developed difficult breathing and bilateral lobe infiltrates two months after driving through California. During initial evaluation, typical pneumonia was suspected and he was treated with antibiotics. Efforts to diagnose the pneumonia were unsuccessful. In October, a laryngeal mass was detected and laryngeal cancer was suspected; treatment with steroids and bronchodilators did not produce improvement. Lung biopsy and laryngoscopy detected diffuse granular tissue. He was treated with amphotericin B and discharged after five days. What was the disease? What might have been done differently to decrease the patient's recovery time from three months to one week?

2. During a 6-month period, 72 clinic staff members became tuberculin-positive. A case–control study was undertaken to determine the most likely source of *M. tuberculosis* infection among the staff. A total of 16 cases and 34 tuberculin-negative controls were compared.

   Pentamidine isethionate is not used for TB treatment. What disease was probably being treated with this drug? What is the most likely source of infection?

	CASES	CONTROL
Works ≥ 40 hr/week	100	62
In room during aerosolized pentamidine isethionate therapy in TB patients	31	3
Patient contact	94	94
Lunch eaten in staff lounge	38	35
Resident of western Palm Beach	75	65
Female	81	77
Cigarette smoker	6	15
Contact with nurse diagnosed with TB	15	12
In unventilated room during collection of TB-positive sputum samples	13	8

Microbial Diseases of the Digestive System

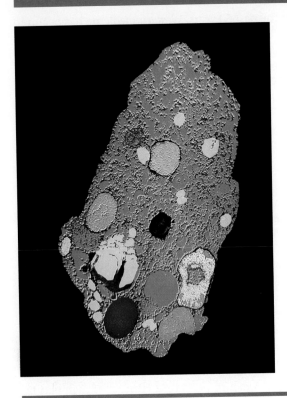

MICROBIAL DISEASES of the digestive system are second only to respiratory diseases as causes of illness in the U.S.A. Most such diseases result from the ingestion of food or water contaminated with pathogenic microorganisms or their toxins. These pathogens usually enter the food or water supply after being shed in the feces of people or animals infected with them. Therefore, microbial diseases of the digestive system are typically transmitted by a *fecal–oral* cycle. This cycle is interrupted by good sanitation practices in food handling as well as by modern methods of sewage treatment and disinfection of drinking water.

the accessory structures lie outside the tract and produce secretions that are conveyed by ducts into the tract.

The purpose of the digestive system is to digest foods—that is, to break them down into small molecules that can be taken up and used by body cells. In a process called *absorption,* these end-products of digestion then pass from the small intestine into the blood or lymph for distribution to body cells. Then the food moves through the large intestine where water, vitamins, and nutrients are absorbed from it and microbial cells are added to it. The resulting undigested solids, called *feces,* are eliminated from the body through the anus. Feces carry intestinal pathogens, and it is necessary to prevent them from contaminating food and water.

STRUCTURE AND FUNCTION OF THE DIGESTIVE SYSTEM

LEARNING OBJECTIVE

- List the structures of the digestive system that contact food.

The **digestive system** may be divided into two principal groups of organs (Figure 25.1). One group is the *gastrointestinal (GI) tract* or *alimentary canal,* essentially a tubelike structure that includes the mouth, pharynx (throat), esophagus (food tube), stomach, small intestine, and large intestine. The other group of organs, the *accessory structures,* consists of the teeth, tongue, salivary glands, liver, gallbladder, and pancreas. Except for the teeth and tongue,

NORMAL MICROBIOTA OF THE DIGESTIVE SYSTEM

LEARNING OBJECTIVE

- List examples of the microbiota for each part of the gastrointestinal tract.

Bacteria heavily populate most of the digestive system. In the mouth, each milliliter of saliva can contain millions of bacteria. The stomach and small intestine have relatively few microorganisms because of the hydrochloric acid produced by the stomach and the rapid movement of food through the small intestine. By contrast, the large intestine has enormous microbial populations, exceeding 100 billion bacteria per gram of feces. (Up to 40% of fecal mass is microbial cell material.) The population of the large intestine is composed mostly of anaerobes of the genera *Lactobacillus* and *Bacteroides* and facultative anaerobes, such as *E. coli, Enterobacter, Klebsiella,* and *Proteus.* Most of these bacteria assist in the enzymatic breakdown of foods, and some of them synthesize useful vitamins.

BACTERIAL DISEASES OF THE MOUTH

LEARNING OBJECTIVE

- *Describe the events that lead to dental caries and periodontal disease.*

The entrance to the digestive system provides an environment that supports a large and varied microbial population.

Dental Caries (Tooth Decay)

The teeth are unlike any other exterior surface of the body. They are hard, and they do not shed surface cells (Figure 25.2). This allows the accumulation of masses of microorganisms and their products. These accumulations are called **dental plaque,** and they are intimately involved in the formation of **dental caries,** or tooth decay.

Oral bacteria convert sucrose and other carbohydrates into lactic acid, which in turn attacks the tooth enamel. The microbial population on and around the teeth is very complex; more than 300 species have been identified. Probably the most important *cariogenic* (causing caries) bacterium is *Streptococcus mutans,* a gram-positive coccus. Some other species of streptococci are also cariogenic but play a lesser role in initiating caries.

The initiation of caries depends on the attachment of *S. mutans,* or other streptococci, to the tooth (Figure 25.3). These bacteria do not adhere to a clean tooth, but within minutes a freshly brushed tooth will become coated with a pellicle (thin film) of proteins from saliva. Within a couple of hours, cariogenic bacteria become established on this pellicle and begin to produce a gummy polysaccharide of glucose molecules called *dextran* (Figure 25.3b). In the production of dextran, the bacteria first hydrolyze sucrose into its component monosaccharides, fructose and glucose. The enzyme glucosyltransferase then assembles the glucose molecules into dextran. The residual fructose is the primary sugar fermented into lactic acid. Accumulations of bacteria and dextran adhering to the teeth make up dental plaque. The bacterial population of plaque is predominantly streptococci and filamentous members of the genus *Actinomyces.* (Older, calcified deposits of plaque are called *dental calculus* or *tartar.*) *S. mutans* especially fa-

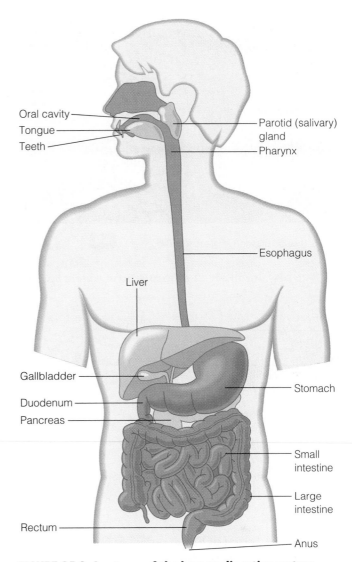

FIGURE 25.1 Anatomy of the human digestive system.

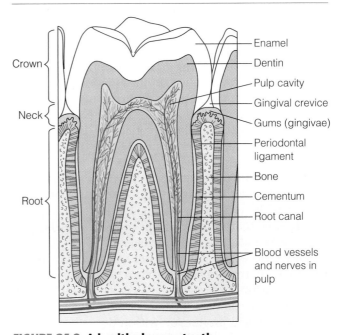

FIGURE 25.2 A healthy human tooth.

(a) *S. mutans* growing in glucose broth   SEM ⊢——⊣ 5 μm

(b) *S. mutans* growing in sucrose broth; note the large accumulation of dextran   SEM ⊢——⊣ 5 μm

FIGURE 25.3 Involvement of *Streptococcus mutans* and sucrose in dental caries.

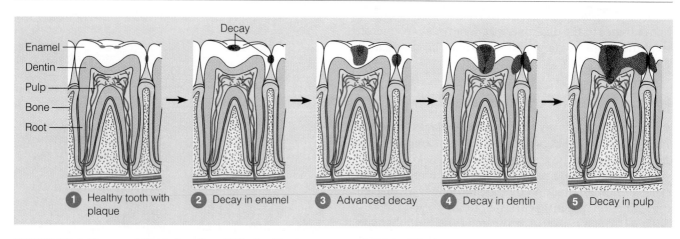

FIGURE 25.4 Stages of dental caries. ① Tooth with plaque accumulation in difficult-to-clean areas. ② Decay begins as enamel is attacked by acids formed by bacteria. ③ Decay advances through the enamel. ④ Decay advances into the dentin. ⑤ Decay enters the pulp and may form abscesses in the tissues surrounding the root.

vors crevices or other sites on the teeth protected from the shearing action of chewing, or the flushing action of the liter or so of saliva produced in the mouth each day. On protected areas of the teeth, plaque accumulations can be several hundred cells thick. Because plaque is not very permeable to saliva, the lactic acid produced by bacteria is not diluted or neutralized, and it breaks down the enamel of the teeth to which the plaque adheres.

Although saliva contains nutrients that encourage the growth of bacteria, it also contains antimicrobial substances, such as lysozyme (see Chapter 16), that help protect exposed tooth surfaces. Some protection is also provided by *crevicular fluid,* a tissue exudate that flows into the gingival crevice (see Figure 25.2) and is closer in composition to serum than to saliva. It protects teeth by virtue of both its flushing action and its phagocytic cell and immunoglobulin content.

Localized acid production within deposits of dental plaque results in a gradual softening of the external *enamel*

(Figure 25.4). Enamel low in fluoride is more susceptible to the effects of the acid. This is the reason for fluoridation of water and toothpastes, which has been an important factor in the recent decline in dental caries in the U.S.A.

If the initial penetration of the enamel by caries remains untreated, bacteria can penetrate into the interior of the tooth. The composition of the bacterial population involved in advancement of the decayed area from the enamel into the *dentin* is entirely different from that of the population initiating the decay. The dominant microorganisms are gram-positive rods and filamentous bacteria; *S. mutans* is present in small numbers only. Although once considered the cause of dental caries, *Lactobacillus* species actually play no role in initiating the process. However, these very prolific lactic acid producers are important in advancing the front of the decay once it is established.

The decayed area eventually advances to the *pulp* (see Figure 25.4), which connects with the tissues of the jaw and contains the blood supply and the nerve cells. Almost

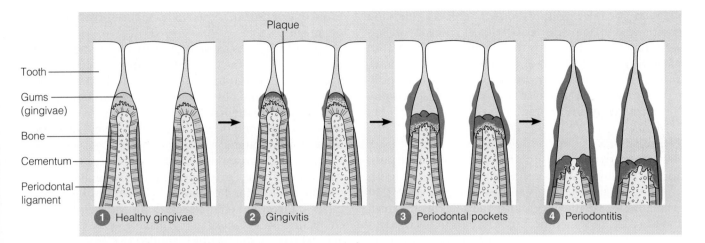

FIGURE 25.5 Stages of periodontal disease. ① Teeth firmly anchored by healthy bone and gum tissue (gingiva). ② Toxins in plaque irritate gums, causing gingivitis. ③ Periodontal pockets form as tooth separates from gingiva. ④ Gingivitis progresses to periodontitis. Toxins destroy the gingiva and bone that support the tooth and the cementum that protects the root.

any member of the normal microbiota of the mouth can be isolated from the infected pulp and roots. Once this stage is reached, root canal therapy is required to remove the infected and dead tissue and to provide access for antimicrobials that suppress renewed infection. If untreated, the infection may advance from the tooth to the soft tissues, producing dental abscesses caused by mixed bacterial populations that contain many anaerobes. These, and most other dental soft tissue infections, can be treated by penicillin and its derivatives.

Although dental caries are probably one of the more common infectious diseases in humans today, they were not much seen in the Western world until about the seventeenth century. In human remains from older times, only about 10% of the teeth contain caries. The introduction of table sugar, or sucrose, into the diet is highly correlated with our present level of caries in the Western world. Studies have shown that sucrose, a disaccharide composed of glucose and fructose, is much more cariogenic than either glucose or fructose individually (see Figure 25.3). People living on high-starch diets (starch is a polysaccharide of glucose) have a low incidence of caries unless sucrose is also a significant part of their diet. The contribution of bacteria to tooth decay has been shown by experiments with germ-free animals. Such animals do not develop caries even when fed a sucrose-rich diet designed to encourage their formation.

Sucrose is pervasive in the modern Western diet. However, if sucrose is ingested only at regular mealtimes, the protective and repair mechanisms of the body are usually not overwhelmed. It is the sucrose that is ingested between meals that is most damaging to the teeth. One set of experiments showed that 330 grams of sucrose per day at mealtimes did not cause caries, but only 120 grams per day at frequent intervals between meals caused an increase of five to seven times in the number of caries. Although

such sugars as mannitol, sorbitol, and xylitol are metabolized to acid by oral bacteria, they are not cariogenic. This is why they are used to sweeten "sugarless" candies and chewing gums.

The best strategies for the prevention of caries are minimal ingestion of sucrose; brushing, flossing, and professional cleaning to remove plaque; and the use of fluorides. As for the numerous mouthwashes that are said to prevent or reduce plaque, the Food and Drug Administration does not recognize any as being of proven effectiveness. The ancient Chinese and a number of other cultures made use of urine as a mouthwash to improve oral health. Indeed, it tends to lower acidity and may be useful, but this should not be construed as a recommendation. A vaccine based on IgA antibodies in the saliva is theoretically possible, and tests conducted on animals have given encouraging results.

Periodontal Disease

Even people who avoid tooth decay might, in later years, lose their teeth to **periodontal disease.** Periodontal ("surrounding the tooth") disease is a collective term for a number of conditions characterized by inflammation and degeneration of structures that support the teeth (Figure 25.5).

The roots of the tooth are protected by a covering of specialized connective tissue called cementum. As the gums recede with age, formation of caries on the cementum becomes more common.

Gingivitis

In many cases of periodontal disease, the infection is restricted to the gums, or gingivae. These infections, called **gingivitis,** are characterized by bleeding of the gums (see

Figure 25.5) while brushing the teeth. This is a condition experienced to some degree by almost everyone. It has been shown experimentally that gingivitis will appear in a few weeks if brushing is discontinued and plaque is allowed to accumulate. An assortment of streptococci, actinomycetes, and anaerobic gram-negative bacteria predominate in these infections.

Periodontitis

Gingivitis can progress to a chronic condition called **periodontitis**, which is responsible for nearly 10% of tooth loss in older adults. This is an insidious condition that generally causes little discomfort. The gums are inflamed and bleed easily. Sometimes pus is formed in pockets surrounding the teeth (periodontal pockets, see Figure 25.5). As the infections continue, they progress toward the root tips. The bone and tissue that support the teeth are destroyed, leading eventually to loosening and loss of the teeth. Numerous bacteria of many different types are involved in these infections; the damage to tissue is done by an inflammatory response to the presence of these bacteria. Periodontitis is treated either by surgically eliminating the periodontal pockets or by using specialized cleaning techniques on the tooth surfaces normally protected by the gums.

Acute necrotizing ulcerative gingivitis, also termed **Vincent's disease** or **trench mouth**, is one of the more common serious mouth infections. The disease causes enough pain that normal chewing is difficult. Foul breath (halitosis) also accompanies the infection. The bacteria usually associated with this condition are the anaerobic *Fusobacterium* and actinomycetes. Because these causative organisms are usually anaerobic, treatment with oxidizing agents, debridement, and administration of antibiotics may be temporarily effective.

BACTERIAL DISEASES OF THE LOWER DIGESTIVE SYSTEM

LEARNING OBJECTIVE

- *List the etiologic agents, suspect foods, signs and symptoms, and treatments for staphylococcal food poisoning, salmonellosis, typhoid fever, bacillary dysentery, cholera, peptic disease syndrome, and gastroenteritis.*

Diseases of the digestive system are essentially of two types, infections and intoxications.

An **infection** occurs when a pathogen enters the gastrointestinal (GI) tract and multiplies. Microorganisms can penetrate into the intestinal mucosa and grow there or can pass through to other systemic organs. Infections are characterized by a delay in the appearance of gastrointestinal disturbance while the pathogen increases in numbers or affects invaded tissue. There is also usually a

fever, one of the body's general responses to an infective organism.

Some pathogens cause disease by elaborating toxins that affect the GI tract. An **intoxication** is caused by the ingestion of such a preformed toxin. Most intoxications, such as that caused by *Staphylococcus aureus,* are characterized by a very sudden appearance (usually in only a few hours) of symptoms of a GI disturbance. Fever is less often one of the symptoms.

Both infections and intoxications often cause *diarrhea,* which most of us have experienced. Severe diarrhea accompanied by blood or mucus is called *dysentery.* Both types of digestive system diseases are also frequently accompanied by *abdominal cramps, nausea,* and *vomiting.* The general term *gastroenteritis* is applied to diseases causing inflammation of the stomach and intestinal mucosa. Botulism (see Chapter 22) is a special case of intoxication because the ingestion of the preformed toxin affects the nervous system rather than the GI tract.

In the developing countries, diarrhea is a major factor in infant mortality. Approximately one child in every ten dies of it before the age of five. It also affects the absorption of nutrients from their food and adversely affects the growth of the survivors. The cause of diarrhea may be any of several organisms. Most are not identified, but surveys in such countries as Bangladesh indicate that the three most common causes are enterotoxigenic *E. coli, Shigella* spp., and intestinal rotaviruses. It is estimated that mortality from childhood diarrhea could be halved by *oral rehydration therapy.* This is usually a solution of sodium chloride, potassium chloride, and sodium bicarbonate. Such solutions are sold in the infant supply department of many stores.

Staphylococcal Food Poisoning (Staphylococcal Enterotoxicosis)

A very common cause of gastroenteritis is **staphylococcal food poisoning.** This is an *intoxication* caused by ingestion of an *enterotoxin* produced by *S. aureus.* Staphylococci are comparatively resistant to environmental stresses, as we discussed in Chapter 11. Staphylococci have a fairly high resistance to heat; vegetative cells can tolerate 60°C for half an hour. They are also resistant to drying and radiation, which helps them survive on skin surfaces. Resistance to high osmotic pressures helps them grow in foods, such as cured ham, in which the high osmotic pressure of salts inhibits the growth of competitors.

S. aureus is often an inhabitant of the nasal passages, from which it contaminates the hands. It is also a frequent cause of skin lesions on the hands. From these sources, it can readily enter food. If the organisms are allowed to incubate in the food (a situation called *temperature abuse*), they reproduce and release enterotoxin into the food. These events, which lead to outbreaks of staphylococcal intoxication, are illustrated in Figure 25.6.

1. Food containing protein is cooked (bacteria usually killed).

2. Then food is contaminated by worker with staphylococci on hands (competing bacteria have been eliminated).

OR

Slow cooling in large masses

Room temperature cooling

3. Organisms incubate in food (temperature abuse) long enough to form and release toxins. (Reheating will eliminate staphylococci but not the toxin.)

4. Food containing toxins is eaten.

Staphylococcal intoxication

5. In one to six hours, intoxication occurs.

FIGURE 25.6 The sequence of events in a typical outbreak of staphylococcal food poisoning.

is useful in the tentative identification of types that are likely to be virulent.

Generally, a population of about 1 million bacteria per gram of food will produce enough enterotoxin to cause illness. The growth of the organism is facilitated if the competing organisms in the food have been eliminated—by cooking, for example. The organism is also more likely to grow if competing bacteria are inhibited by a higher-than-normal osmotic pressure or by a relatively low moisture level. *S. aureus* tends to outgrow most competing bacteria under these conditions.

Custards, cream pies, and ham are examples of high-risk foods. Competing organisms are restrained in custards by the high osmotic pressure of sugar and by cooking. In ham they are inhibited by curing agents such as salts and preservatives. Poultry products can also harbor staphylococci if they are handled and allowed to stand at room temperatures. Hamburger, however, is seldom associated with any type of bacterial food poisoning. (For an important exception, see the discussion of *E. coli* later in the chapter.) Hamburger contains microorganisms of the normal microbiota that are competitive with staphylococci, and it is usually cooked immediately before consumption. Any foods that are prepared in advance and are not kept chilled are a potential source of staphylococcal food poisoning. Because contamination of foods cannot be avoided completely, the most reliable method of preventing staphylococcal food poisoning is adequate refrigeration during storage to prevent toxin formation.

The toxin itself is heat stable and can survive up to 30 minutes of boiling. So, once the toxin is formed, it is not destroyed when the food is reheated, although the bacteria will be killed.

The toxin quickly triggers the brain's vomiting reflex center; abdominal cramps and usually diarrhea then ensue. This reaction is essentially immunological in character; the staphylococcal enterotoxin is a model example of a superantigen (discussed in Chapter 19). Recovery is usually complete within 24 hours.

The mortality rate of staphylococcal food poisoning is almost nil among otherwise healthy individuals, but it can be significant in weakened individuals, such as residents of nursing homes. No reliable immunity results from recovery. However, there is a great deal of variation in individual susceptibility to the toxin, and it is suspected that immunity acquired from a previous exposure might account for some of this variation.

The diagnosis of staphylococcal food poisoning is usually based on the symptoms, particularly the short incubation time characteristic of intoxication. If the food has not been reheated so that the bacteria are not killed, the organism can be recovered and grown. *S. aureus* isolates can be tested by the *phage-typing* method, which is useful in tracing the source of the contamination (see Figure 10.12). These bacteria grow well in 7.5% sodium chloride, so this concentration is often used in media for their selective isolation. Pathogenic staphylococci usually fer-

S. aureus produces several toxins that damage tissues or increase the organism's virulence. The enterotoxins causing food poisoning are classified as serological types A (which is responsible for most cases) through D. Production of this toxin is often correlated with production of an enzyme that coagulates blood plasma. Such bacteria are described as *coagulase-positive* (see Chapter 11). No direct pathogenic effect can be attributed to the enzyme, but it ment mannitol, produce blood hemolysins and coagulase, and form golden-yellow colonies. They cause no obvious spoilage when growing in foods.

No simple procedure will identify the toxin in foods. Identification requires a complex extraction and a method of serological testing against antisera; these procedures are not yet commercially available, but kits that make use of monoclonal antibodies to detect them are in limited use. Food can be easily tested, however, for the presence of *thermostable nuclease,* an enzyme produced by *S. aureus* that survives heating to temperatures that will kill the cells themselves. It can be detected by a simple test in which an extract of the food is allowed to react with DNA. The presence of detectable amounts of the enzyme usually indicates a microbial population large enough to have produced sufficient toxin to cause illness.

Salmonellosis (*Salmonella* Gastroenteritis)

In bacterial infections, disease is a result of microbial growth in body tissues rather than of the ingestion of food and drink already contaminated by preformed toxins resulting from microbial growth. Bacterial infections, such as salmonellosis, usually have longer incubation periods (12 hours to 2 weeks) than bacterial intoxications, reflecting the time needed for the organism to grow in the tissues of the host. Bacterial infections are also often characterized by some fever, indicating the host's response to the infection.

The *Salmonella* bacteria (named for the discoverer, Daniel Salmon) are gram-negative, facultatively anaerobic, nonendospore-forming, usually motile rods. Their normal habitat is the intestinal tracts of humans and many animals. All salmonellae are considered pathogenic to some degree, causing **salmonellosis,** or *Salmonella* **gastroenteritis.**

The nomenclature of the *Salmonella* organisms is confusing. According to *Bergey's Manual,* none of the present methods for naming salmonellae is satisfactory from a scientific viewpoint. Rather than recognized species, there are more than 2000 *serovars* (closely related organisms differentiated by serological testing). Only about 50 such serovars are isolated with any frequency in the U.S.A. Some of these are named much like species—for example, *Salmonella dublin* and *Salmonella eastbourne* (ēst'bôrn) were named after the sites where they were first isolated, and *Salmonella typhimurium* (tī-fi-mūr'ē-um) causes a typhoidlike disease in mice. Others are represented by the Kauffmann–White scheme (see the discussion of *Salmonella* in Chapter 11), which is generally used in clinical laboratories. This method assigns numbers and letters to different antigens: O (somatic or body), Vi (capsular), and H (flagellar).

Salmonellosis has an incubation time of about 12 to 36 hours. The salmonellae first invade the intestinal mucosa and multiply there. Sometimes they manage to pass

FIGURE 25.7 Salmonellosis. How the infection of endothelial cells in the intestinal wall spreads to other parts of the body.

through the intestinal mucosa to enter the lymphatic and cardiovascular systems, and from there they may spread to eventually affect many organs (Figure 25.7). The fever associated with *Salmonella* infections might be from endotoxins released by lysed cells, but this relationship is not certain. There is usually a moderate fever accompanied by nausea, abdominal pain and cramps, and diarrhea. As many as 1 billion salmonellae per gram can be found in an infected person's feces during the acute phase of the illness.

The mortality rate is overall very low, probably lower than 1%. However, the death rate is higher in infants and among the very old; death is usually from septicemia. The severity and incubation time can depend on the number of *Salmonella* ingested. Normally, recovery will be complete in a few days, but many patients will continue to shed the organisms in their feces for up to six months. Antibiotic therapy is not useful in treating salmonellosis or, indeed, many diarrheal diseases; treatment consists of oral rehydration therapy.

Salmonellosis is probably greatly underreported. The 45,000 to 50,000 cases reported each year are probably only from 1% to 10% of the actual total (Figure 25.8).

Meat products, especially poultry, are particularly susceptible to contamination by *Salmonella*. If these products are mishandled, the bacteria can grow to infective numbers rather quickly. The sources of the bacteria are the intestinal tracts of many animals, and meats can be contaminated readily in processing plants. Poultry, eggs, and egg products are often contaminated by *Salmonella*.

Prevention of salmonellosis depends on good sanitation practices to deter contamination and on proper refrigeration to prevent increases in bacteria. The organisms are generally destroyed by normal cooking that heats the food to an internal temperature of at least 60°C(140°F). However, contaminated food can contaminate a surface, such as a cutting board. Although the food first prepared on the board might later be cooked and its bacteria killed, another food subsequently prepared on the board might not be cooked.

Outbreaks of salmonellosis have been traced to contaminated whole eggs. Indications are that the bacteria are transmitted to the eggs before they are laid, although the chickens may be asymptomatic. People in the endemic areas, especially, have been cautioned to eat only well-cooked eggs. Eggs fried "sunny-side up" or boiled for only four minutes are not well enough cooked to reliably kill *Salmonella*. An often unsuspected factor is the presence of inadequately cooked or raw eggs in foods such as hollandaise sauce and Caesar salad dressing (see the box on p. 622).

Diagnosis of salmonellosis usually depends on isolating the organism from the patient's stool or from leftover food. Isolation requires specialized selective and differential media; these methods are relatively slow. Also, the small numbers of *Salmonella* generally present in foods present a special problem in detection. Because this disease is important, much effort has been applied to improving detection and identification methods. The most promising of these are DNA probes in kit form (see Figure 10.18) and ELISA-type kits that make use of monoclonal antibodies. These have made one-day detection and identification of *Salmonella* possible and are especially useful in screening food for small numbers of the organisms.

Typhoid Fever

A few serovars of *Salmonella* are much more virulent than others. The most virulent, *Salmonella typhi,* causes the bacterial infection **typhoid fever**. This pathogen is not found

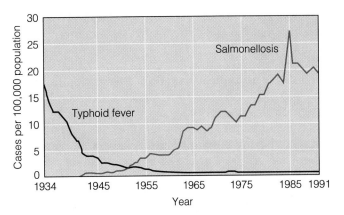

FIGURE 25.8 Incidence of salmonellosis and typhoid fever. The incidence of typhoid fever has been declining in the U.S.A. while that of salmonellosis (gastroenteritis) has been increasing. (*Source:* CDC, *Summary of Notifiable Diseases,* 1991.)

in animals; the disease is spread only in the feces of other humans.

The incubation period is much longer than that of salmonellosis, normally about two weeks. The patient first suffers from a high fever of 40°C (104°F) and continual headache. Diarrhea appears only during the second or third week, and the fever tends to decline. Unlike salmonellosis, the bacteria do not multiply in the intestinal epithelial cells. Also, they differ by continuing to multiply in phagocytic cells. The organism becomes disseminated in the body and can be isolated from the blood, urine, and feces. In severe cases, there can be perforation of the intestinal wall. The mortality rate is now about 1% to 2%; at one time, it was at least 10%. Before the days of proper sewage disposal, water treatment, and food sanitation, typhoid was an extremely common disease. Its incidence has been declining in the U.S.A., whereas that of salmonellosis has been increasing (see Figure 25.8). Typhoid fever is still a frequent cause of death in parts of the world with poor sanitation.

A substantial number of recovered patients become chronic carriers. They harbor the pathogen in the gallbladder and continue to shed bacteria for several months. A number of such carriers continue to shed the organism indefinitely. Most of us are familiar with the case of Typhoid Mary. Her name was Mary Mallon; she worked as a cook in New York state in the early part of the twentieth century and was responsible for several outbreaks of typhoid and three deaths. Her case became well known through the attempts of the state to restrain her from working at her chosen trade.

In each recent year there have been about 400–500 cases of typhoid fever in the U.S.A. Approximately 70% of these cases were acquired during foreign travel. Normally, there are fewer than three deaths each year.

The third-generation cephalosporins, such as ceftriaxone, are the usual drugs of choice for the treatment of severe typhoid fever. However, drug sensitivity testing is required because of the continuing appearance of resistant strains. Recovery from typhoid confers a lifelong immunity.

Immunization for typhoid is normally not done in more developed countries, except for high-risk laboratory and military personnel. The vaccine that has long been in use is a killed-organism type, which must be injected. Live, orally ingested vaccines have become available. Their effectiveness is not much greater than older, killed vaccines (conferring immunity lasting only a few years on 65% of recipients), but they have many fewer adverse effects.

Bacillary Dysentery (Shigellosis)

Bacillary dysentery (shigellosis) is a severe form of diarrhea caused by a group of facultatively anaerobic gram-negative rods of the genus *Shigella*. This genus was named for the Japanese microbiologist Kiyoshi Shiga.

There are four species of pathogenic *Shigella*: *S. sonnei* (sōn′ne-ē), *S. dysenteriae* (dis-en-te′rē-ī), *S. flexneri* (fleks′-nér-ē), and *S. boydii* (boi′dē-ē). These bacteria are residents only of the intestinal tract of humans, apes, and monkeys. They are closely related to the pathogenic *E. coli*. *S. sonnei*, the most common in the U.S.A., causes a relatively mild dysentery. At the other extreme, infection with *S. dysenteriae* (sometimes called the Shiga bacillus) often results in a severe dysentery and prostration. The toxin responsible is unusually virulent and is known as the Shiga toxin. This is the species found least frequently in the U.S.A.

Virulent shigellae produce exotoxins that inhibit protein synthesis, thereby killing cells (Figure 25.9). The bacteria proliferate to immense numbers in the small intestine, but the damage they cause is to the large intestine. There they produce tissue destruction in the intestinal mucosa, causing severe diarrhea with blood and mucus in the stools. Infected people may have as many as 20 bowel movements in a day. The ulcerations in the intestinal mucosa eventually heal but form scar tissue. Additional symptoms of infection are abdominal cramps and fever. Except for *S. dysenteriae*, *Shigella* rarely invades the bloodstream. Diagnosis is usually based on recovery of the organisms from rectal swabs.

In recent years, the number of cases reported in the U.S.A. has been about 20,000 to 25,000, with 5 to 15 deaths. *S. dysenteriae* has a significant mortality rate, however, and the death rate in tropical areas can be much higher, perhaps 20%. The disease probably occurs more often than the reported numbers indicate. Some cases of so-called traveler's diarrhea might be mild forms of bacillary dysentery. This dysentery is especially common in institutionalized patients and in people living on Native American reservations. Some immunity seems to result from recovery, but no satisfactory vaccine has yet been developed.

In severe cases of bacillary dysentery, antibiotic therapy and fluid and electrolyte replacement are indicated. At present, fluoroquinolones are the antibiotics of choice.

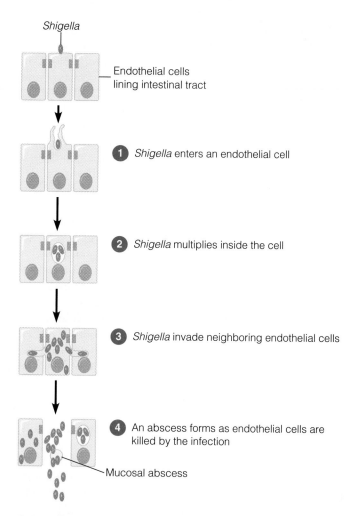

FIGURE 25.9 Shigellosis. How infection of endothelial cells in the intestinal wall can cause intestinal damage.

Shigella

Endothelial cells lining intestinal tract

1 *Shigella* enters an endothelial cell

2 *Shigella* multiplies inside the cell

3 *Shigella* invade neighboring endothelial cells

4 An abscess forms as endothelial cells are killed by the infection

Mucosal abscess

Cholera

During the 1800s, the bacterial infection called Asiatic cholera crossed Europe and North America in repeated epidemics. Today, **cholera** is endemic in Asia, particularly India, and has only occasional outbreaks in Western countries. These outbreaks are caused by temporary lapses in sanitation. The 1991–1992 epidemic in South America resulted in over 700,000 cases and 6000 deaths (Figure 25.10).

The causative organism is *Vibrio cholerae*, a slightly curved, gram-negative rod with a single polar flagellum (Figure 25.11). The serogroup O:1 of *V. cholerae* causes the classically recognized epidemic form of the disease. It can be subdivided into two *biotypes* (differentiated other than by serology) as classical or *eltor* (named for the original culture isolated at a quarantine camp, El Tor, for pilgrims to Mecca). At present, the most widespread cholera pathogen is serogroup O:1, biotype eltor. It grows in the small intestine and produces an enterotoxin that results in the secretion of chlorides, bicarbonates, and water. The excess water and mineral electrolytes are excreted, taking

Initial epidemics: January 1991

August 1991

February 1992

December 1992

FIGURE 25.10 Cholera epidemic in Latin America, 1991–1992. [*Source: MMWR 42(89) (2/12/93).*]

(a) *Vibrio cholerae* SEM | 5 μm

(b) Healthy intestine SEM | 5 μm

(c) Infected intestine SEM | 5 μm

FIGURE 25.11 Cholera. (a) *Vibrio cholerae,* the pathogen that causes cholera. Note the slightly curved morphology. **(b)** Healthy intestinal wall of an infant rabbit. **(c)** Rabbit intestine after infection with *V. cholerae.*

on the typical appearance of "rice water stools" from the masses of intestinal mucus, epithelial cells, and bacteria. The sudden loss of these fluids and electrolytes (from 12 to 20 liters of fluid might be lost in one day) causes shock, collapse, and often death. Because of the loss of fluid, the blood becomes so viscous that vital organs are unable to function properly. Violent vomiting sometimes occurs. The organisms are not invasive, and a fever is usually not present. The severity of the disease varies considerably, and the number of subclinical cases might be several times the number that are reported.

In the U.S.A. there have been occasional cases of cholera caused by the O:1 serogroup. These have all been in the Gulf coast area, and it is speculated that the organism may be endemic in coastal waters. Most gastroenteritis due to *V. cholerae* in this country has been caused by the non-O:1 serogroup, usually ingested in contaminated seafood. The waters of the Gulf and Pacific coasts support an indigenous population of these organisms. These bacteria differ from the O:1 serotypes in several ways, and they are more likely to invade the intestinal mucosa, causing bloody stools and fever.

The cholera bacteria can be readily isolated from the feces, partly because the organisms can grow in media that are alkaline enough to suppress the growth of many other organisms. The non-O:1 serogroup is occasionally isolated from blood and wounds.

Recovery from the disease results in an effective immunity based on the antigenic activity of both the cells and the enterotoxin. However, because of antigenic differences among bacterial strains, the same person can have cholera more than once. Most victims in endemic areas are children. The available vaccines are protective to some extent but provide immunity of relatively short duration as compared with the immunity from a natural infection.

Tetracycline is used for treatment, but chemotherapy is not as effective as replacement of the lost fluids and electrolytes. Untreated cases of cholera may have a 50% mortality rate, whereas the rate for cases having proper supportive care can be less than 1%.

OUTBREAK OF SALMONELLA ENTERITIDIS INFECTION ASSOCIATED WITH CONSUMPTION OF RAW SHELL EGGS, 1991

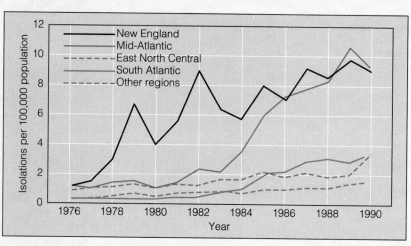

Isolation rate for *Salmonella enteritidis,* by region in the U.S.A., 1976–1990.

SALMONELLA ENTERITIDIS (SE) is the most frequently reported *Salmonella* serotype in the U.S.A. From January through December 1991, state health departments reported 66 outbreaks of SE in the U.S.A. to CDC. This report describes an SE outbreak associated with consumption of raw shell eggs.

During October 1991, 15 persons who ate at a restaurant during a 9-day period developed gastroenteritis. Predominant symptoms were diarrhea (100%), fever (92%), abdominal cramping (92%), nausea (83%), and chills (75%). The median incubation period was 24 hours; median duration of illness was 7 days. Eight ill patrons required intravenous rehydration, and six were hospitalized. *Salmonella* group D was isolated from the stool of all 13 ill patrons who submitted specimens; all of the eight isolates further typed were identified as SE. Fourteen of the 15 ill patrons and none of 11 well patrons interviewed had eaten Caesar salad.

During the outbreak, 23 of the restaurant's 78 employees had onset of gastroenteritis. Of the 66 employees for whom cultures were obtained, SE was isolated from stool samples of 15 of the 22 ill employees and six of 44 asymptomatic employees. Confirmed SE infection among employees was associated with eating the restaurant's raw egg dishes or handling raw eggs in the restaurant kitchen.

The Caesar salad dressing was pre-pared early in the morning by combining 36 yolks from hand-cracked eggs with olive oil, anchovies, garlic, and warm water. Batches of Caesar dressing were prepared daily except for one 3-day period when a single batch was used. The dressing was refrigerated until the restaurant opened, when it was placed in a chilled compartment in the salad preparation area for approximately 8–12 hours until the restaurant closed.

The restaurant obtained eggs from a single supplier twice weekly and stored them in a walk-in refrigerator until use. No eggs from the shipment implicated in the outbreak were available for testing, but three cases of eggs from a different shipment from the same supplier were available. Two pools of 10 eggs each from each of the three cases were sampled and submitted for culture. SE was isolated from one of the six pools. A traceback by the SE Task Force, U.S. Department of Agriculture (USDA), determined that the source flock for eggs used during the outbreak was the same flock from which the SE-positive eggs were obtained. The flock had been destroyed before recognition of the outbreak.

Editorial Note: From 1976 through 1990, isolation rates for SE increased in the U.S.A. (see the figure). In 1990, the 8591 SE isolates reported through CDC's *Salmonella* Surveillance System represented 21% of all reported *Salmonella* isolates, surpassing *S. typhimurium* to become the most frequently reported serotype. During 1985–1991, 375 SE outbreaks accounted for 12,784 cases of illness and 49 deaths.

An estimated 0.01% of all shell eggs contain SE. Because most serious illnesses or deaths associated with these infections occur among infants, the elderly, or immunocompromised persons, special attention should be directed to the diets of these persons to prevent the consumption of foods containing raw or undercooked eggs.

Most cases of SE infection occur as sporadic cases or in limited family outbreaks. The outbreak described in this report may be the first time SE infection has been documented as a potential occupational hazard for employees preparing raw egg dishes in restaurants. Commercial food-service establishments can reduce the risk for outbreaks and infections among employees by using pasteurized egg products or eliminating eggs in such recipes.

SOURCE: *MMWR* 41(21):369–372 (5/29/92).

Vibrio parahaemolyticus Gastroenteritis

Vibrio parahaemolyticus is found in salt water estuaries in many parts of the world. It is morphologically similar to *V. cholerae,* but it is halophilic and requires 2% or more sodium chloride for optimum growth. It is the most common cause of gastroenteritis in Japan, with thousands of cases reported annually. It is also a frequent cause of gastroenteritis in many nations of Southeast Asia. The organism is present in coastal waters of the continental U.S.A. and Hawaii. Crustaceans, such as shrimp and crabs, have been associated with several outbreaks caused by *V. parahaemolyticus* in the U.S.A. in recent years.

Symptoms include abdominal pain, vomiting, a burning sensation in the stomach, and watery stools, resembling those of cholera. The organism has also been known to cause cutaneous infections of cuts that have come in contact with contaminated clams and oysters. The incubation time is normally less than 24 hours. The generation time under optimum conditions is less than 10 minutes, and large numbers are reached quickly. Recovery usually follows in a few days.

Because *V. parahaemolyticus* has a requirement for sodium, as well as for a high osmotic pressure, isolation media containing 2% to 4% sodium chloride are used in diagnosis of the disease.

Escherichia coli Gastroenteritis (Traveler's Diarrhea)

One of the most prolific microorganisms in the human intestinal tract is *E. coli.* Because it is so common and so easily cultivated, microbiologists often regard it as something of a laboratory pet. *E. coli* is normally harmless, but certain strains can be pathogenic. All pathogenic strains have specialized fimbriae that allow them to bind to certain intestinal epithelial cells. They also produce toxins that cause gastrointestinal disturbances.

There are several distinct pathogenic groups of *E. coli.* The *enterotoxigenic* form is not invasive but forms an enterotoxin that produces a watery diarrhea that resembles a mild form of cholera. This group of organisms is a primary cause of **traveler's diarrhea** and, in developing countries, most infant diarrhea. *Enteroinvasive E. coli* strains produce an enterotoxin that has an activity much like that of the *Shigella* toxins. These organisms penetrate the lining of the intestinal tract, resulting in inflammation, fever, and sometimes a *Shigella*-like dysentery.

The causative agent is not always identified, but 50% to 65% of traveler's diarrhea is probably caused by enterotoxigenic *E. coli.* Of the remaining cases, from 10% to 20% are probably caused by *Shigella.* Other enteric bacteria, such as *Salmonella* and *Campylobacter,* as well as an assortment of unidentified bacterial pathogens, viruses, and protozoan parasites, are also occasionally involved.

Traveler's diarrhea is almost impossible to avoid in parts of the world with poor sanitation practices. Prophylactic antibiotic therapy is generally not recommended, although it can be effective. An alternative preventive regimen is to take two bismuth subsalicylate (Pepto-Bismol®) tablets four times a day. One trial showed that only 14% of those following this regimen developed diarrhea during a three-week trip, whereas 40% of the control group, who took a placebo, developed traveler's diarrhea.

There has been considerable research on a vaccine, but results have been disappointing. In adults the disease is usually self-limiting, and chemotherapy is not attempted. Once contracted, the best treatment is the usual oral rehydration recommended for all diarrhea. In severe cases, antimicrobials may be necessary.

In recent years, the *enterohemorrhagic E. coli* strains have become well known in the U.S.A. as the cause of several outbreaks of disease associated with ingestion of undercooked hamburger or raw milk. Cattle are suspected as the reservoir. These organisms produce verotoxins, which are responsible for hemorrhagic colitis (an inflammation of the colon with bleeding). The best known organism in this group is *E. coli* serotype O157:H7. Many patients, especially children, infected by this organism produce stools combined with copious amounts of blood, but without fever. This toxin is nearly identical to the Shiga toxin produced by the unusually virulent strains of *S. dysenteriae* previously mentioned. A dangerous complication is hemolytic uremic syndrome (blood in the urine, kidney failure), which occurs when the kidneys are affected by the toxin. Some 5% to 10% of small children progress to this stage. A significant mortality rate occurs among the children in whom the hemolytic uremic syndrome appears. Also, some of them may require kidney dialysis or even transplants.

Because of the attention this pathogen has attracted, there are now a number of biochemical and immunological tests to differentiate isolates from other enteric pathogens. The enterohemorrhagic *E. coli* cells are present in 1% or more meat samples; the best preventative measure is to cook meats thoroughly.

In Chapter 26, *E. coli* is discussed as one of the most common causes of urinary tract infections.

Campylobacter Gastroenteritis

Campylobacter strains are gram-negative, microaerophilic, spirally curved rods that were once considered only minor causes of human gastroenteritis. Like *Salmonella,* they are part of the intestinal microbiota of a number of animals, especially poultry and cattle, but their infective dose is much smaller than that of *Salmonella.* Cows with no evidence of illness may excrete them in their milk. After improved methods for their isolation and identification were developed in the late 1970s, it became clear that they often caused human disease. In fact, they have replaced *Shigella* spp. as the second most common cause of diarrhea in the U.S.A. In some states, they are even more frequently isolated in diarrhea cases than salmonellae.

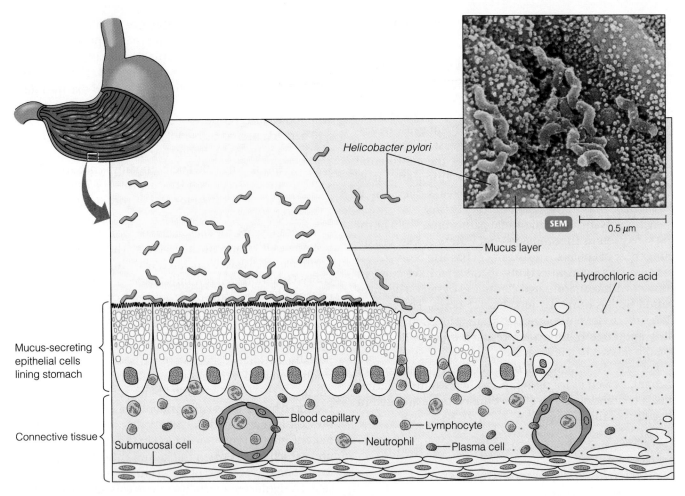

FIGURE 25.12 *Helicobacter pylori* **infection leading to ulceration of the stomach wall.** These spiral bacteria are able to grow and cause damage in the acidic environment of the stomach by producing urease, which neutralizes the acidity. Such destruction of the protective mucus layer leads to erosion of the wall.

Campylobacter **gastroenteritis** is usually caused by *C. jejuni.* Clinically, it is characterized by fever, cramping abdominal pain, and diarrhea with blood and mucus present in the stools. A billion organisms per gram of stool have been isolated at the peak of an infection. Normally, recovery follows within a week.

Helicobacter Peptic Disease Syndrome

In 1982, a physician in Australia cultured a spiral-shaped bacterium observed in the biopsied tissue of stomach ulcer patients. After some taxonomic confusion it was named *Helicobacter pylori.* Today, it is widely accepted that this organism is responsible for **peptic disease syndrome.** This syndrome includes gastric (stomach) or duodenal (first few inches of small intestine) ulcers.

The stomach mucosa contains cells that secrete gastric juice—proteolytic enzymes and hydrochloric acid that activates the enzymes. Other specialized cells produce a layer of mucus that protects the stomach itself from digestion. If this defense is disrupted, an inflammation of the stomach (gastritis) results. This inflammation can then progress to an ulcerated area (Figure 25.12). Acid has been thought to be the primary agent of this inflammation, and therapy has been directed at drugs that inhibit acid release. Recently, however, *H. pylori* has been shown to bind to type O blood group antigens on the gastric epithelial cells. (People of this blood group are twice as likely to develop gastric cancer and ulcers.) The immune system may react to its presence and inadvertently initiate the damage, which is mainly due to the corrosive effects of acid.

Through an interesting adaptation, *H. pylori* can grow in the highly acidic environment of the stomach, which is lethal for most microorganisms. *H. pylori* produces large amounts of an especially efficient urease, an enzyme that converts urea to the alkaline compound ammonia. This results in a locally high pH in the area of growth.

Eradication of *H. pylori* with antimicrobial drugs usually leads to disappearance of peptic ulcers, and their recurrence rate is low. A number of antibiotics have proven effective, including metronidazole. Bismuth subsalicylate (Pepto-Bismol®) is also an effective antimicrobial for this application and is usually part of the drug regimen.

A number of diagnostic tests have been developed for peptic disease syndrome, the simplest being the breath test. Radioactively labeled urea is swallowed and about 30 minutes later radioactively labeled CO_2 can be detected in the breath of an infected patient. Serological testing and biopsies are also available.

Yersinia Gastroenteritis

Other enteric pathogens that are being identified with increasing frequency are *Yersinia enterocolitica* (en-tėr-ō-kōl-it'ik-ä) and *Yersinia pseudotuberculosis* (sū-dō-tü-bėr-kū-lō'sis). These gram-negative organisms are intestinal inhabitants of many domestic animals and are often transmitted in meat and milk. Both organisms are distinctive in their ability to grow at refrigerator temperatures of 4°C (39°F). The organisms cause a gastroenteritis known as **yersiniosis.** The symptoms of yersiniosis are diarrhea, fever, headache, and abdominal pain. The pain is often severe enough to cause a misdiagnosis of appendicitis. Diagnosis requires culture of the organism, which can then be evaluated by serological tests.

Clostridium perfringens Gastroenteritis

Probably one of the more common, if underrecognized, forms of food poisoning in the U.S.A. is caused by *Clostridium perfringens,* a large, gram-positive, endospore-forming, obligately anaerobic rod. This organism is also responsible for human gas gangrene (see Chapter 23).

Most outbreaks are associated with meats or meat stews contaminated with intestinal contents of the animal during slaughter. The organism's nutritional requirement for amino acids is met by such foods, and when the meats are cooked, the oxygen level is lowered enough for clostridial growth. The endospores survive most routine heatings, and the generation time of the vegetative bacterium is less than 20 minutes under ideal conditions. Large populations can therefore build up rapidly when foods are being held for serving or when inadequate refrigeration leads to slow cooling.

The organism grows in the intestinal tract and produces an exotoxin that causes the typical symptoms of abdominal pain and diarrhea. Most cases are mild and self-limiting and probably are never clinically diagnosed. The symptoms usually appear from 8 to 12 hours after ingestion.

Diagnosis is usually based on isolation and identification of the organism in stool samples.

Bacillus cereus Gastroenteritis

Bacillus cereus (se'rē-us) is a large, gram-positive, endospore-forming organism that is very common in soil and vegetation and is usually considered harmless. It has, however, been identified as the cause of outbreaks of foodborne illness. Heating the food does not always kill the spores, which germinate as the food cools. Because competing organisms have been eliminated in the cooked food, *B. cereus* grows rapidly and produces toxins. Rice dishes served in oriental restaurants seem especially susceptible. Some cases resemble *C. perfringens* infections and are almost entirely diarrheal in nature. Other episodes involve nausea and vomiting. It is suspected that different toxins are involved in producing the differing symptoms. Both forms of the disease are self-limiting.

VIRAL DISEASES OF THE DIGESTIVE SYSTEM

LEARNING OBJECTIVES

- *List the etiologic agents, methods of transmission, sites of infection, and symptoms for mumps, CMV inclusion disease, and viral gastroenteritis.*

- *Differentiate among hepatitis A, hepatitis B, hepatitis C, hepatitis D, and hepatitis E.*

Although viruses do not reproduce within the contents of the digestive system like bacteria, they invade many organs associated with the system.

Mumps

The targets of the mumps virus, the parotid glands, are located just below and in front of the ears (see Figure 25.1). Because the parotids are one of the three pairs of salivary glands of the digestive system, it is appropriate to include a discussion of mumps in this chapter.

Mumps typically begins with painful swelling of one or both parotid glands 16 to 18 days after exposure to the virus (Figure 25.13). The virus is transmitted in saliva and respiratory secretions, and its portal of entry is the respiratory tract. An infected person is most infective to others during the first 48 hours before clinical symptoms appear. Once the viruses have begun to multiply in the respiratory tract and local lymph nodes in the neck, they reach the salivary glands via the blood. Viremia (the presence of virus in the blood) begins several days before the onset of mumps symptoms and before the virus appears in saliva. The virus is present in the blood and saliva for three to five days after the onset of the disease and in the urine after ten days or so.

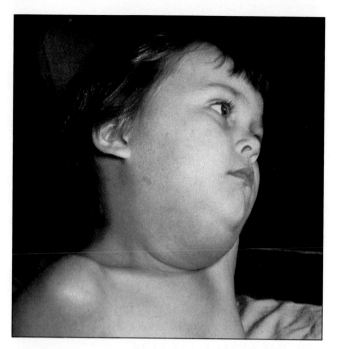

FIGURE 25.13 A case of mumps. This patient shows the typical swelling of mumps.

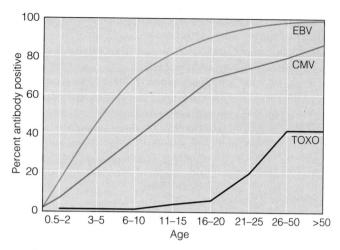

FIGURE 25.14 Typical prevalence of antibodies against Epstein–Barr (EB) virus, cytomegalovirus (CMV), and *Toxoplasma gondii* by age in the U.S.A.

Mumps is characterized by inflammation and swelling of the parotid glands, fever, and pain during swallowing. About four to seven days after the onset of symptoms, the testes can become inflamed (orchitis). This happens in about 20% to 35% of males past puberty. Sterility is a possible but rare consequence. Other complications of mumps include meningitis, inflammation of the ovaries, and pancreatitis. In children, mumps is less common than chickenpox or measles because the disease is less infectious.

An effective attenuated live vaccine is available and is often administered as part of the trivalent measles–mumps–rubella (MMR) vaccine. The number of cases of mumps has dropped sharply since the introduction of the vaccine in 1968. For example, in 1968, 152,209 cases of mumps were reported, whereas in 1993 there were 1602 cases. Second attacks are rare, and cases involving only one parotid gland, or subclinical cases (about 30% of those infected), are as effective as bilateral mumps in conferring immunity.

Serological diagnosis is not usually necessary. If laboratory confirmation of a diagnosis based only on symptoms is desired, the virus can be isolated by embryonated-egg or cell culture techniques. The virus is now usually identified by ELISA tests.

Cytomegalovirus (CMV) Inclusion Disease

The cytomegalovirus (CMV) is a herpesvirus that induces a cellular swelling characterized as *cytomegaly* (large cell).

These cells are known as "owl's eyes" because of their distinctive appearance. Once a person is infected with CMV, the infection persists for life. Blood macrophages and T lymphocytes are the probable sites of latency. The virus is shed at intervals in such body secretions as saliva, urine, semen, cervical secretions, and breast milk.

Some estimate that 80% of the population of the U.S.A. may carry the virus. It has been said that if CMV were accompanied by a skin rash, it would be one of the better known childhood diseases. Figure 25.14 contains plots of the prevalence of antibodies against CMV, Epstein–Barr (EB) virus, and *Toxoplasma gondii*. In less developed parts of the world, infection may approach 100%. It can be spread by kissing and other personal contacts, especially to day-care workers. There is a high incidence of the virus in the male homosexual population. In adults and older children, most infections are subclinical or, at worst, much like a mild case of infectious mononucleosis. However, immunosuppressed individuals may develop life-threatening pneumonia. Commercial products are now available that contain antibodies that passively neutralize CMV present in donated kidneys.

A primary CMV infection acquired during pregnancy by a nonimmune mother can seriously harm the fetus. At the time of her first pregnancy, a woman of affluent background has about a 50% chance of being naturally immune to CMV infection; a woman from a disadvantaged background, who has probably experienced poor sanitation and more crowding, has about an 80% chance. Virus transmission to the fetus by an immune mother will sometimes occur, but almost no damage ensues. Tests for immune status are readily available, and young women should realize that the main source of infection is young children. Any physician should determine the immune status of a woman of childbearing age and inform any nonimmune woman of the risks of pregnancy. Each year

TABLE 25.1 Characteristics of the Various Types of Viral Hepatitis

CHARACTERISTIC	HEPATITIS A	HEPATITIS B	HEPATITIS C	HEPATITIS D	HEPATITIS E
Transmission	Fecal–oral (ingestion of contaminated food and water)	Parenteral (injection of contaminated blood or other body fluids)	Parenteral	Parenteral (host must be coinfected with hepatitis B)	Fecal–oral
Agent	Hepatitis A virus (HAV); single-stranded RNA; no envelope	Hepatitis B virus (HBV); double-stranded DNA; envelope	Hepatitis C virus (HCV); single-stranded RNA; envelope	Hepatitis D virus (HDV); single-stranded RNA; envelope from HBV	Hepatitis E virus (HEV); single-stranded RNA; no envelope
Incubation period	2 to 6 weeks	4 to 26 weeks	2 to 22 weeks	Uncertain	2 to 6 weeks
Manifestations or symptoms	Mostly subclinical; severe cases: fever, headache, malaise, jaundice	Frequently subclinical; similar to HAV, but fever, headache absent, and more likely to progress to severe liver damage	Similar to HBV	Severe liver damage; high mortality rate	Similar to HAV, but pregnant women may have high mortality rate
Chronic liver disease	No	Yes	Yes	Yes	No
Vaccines	Available soon, but immunoglobulins give temporary protection	Genetically engineered	None	HBV vaccine is protective because coinfection required	None

nearly 4000 infants are born with symptomatic disease; many die, and the survivors are severely damaged. In addition, another 4500 to 6000 children, while showing no symptoms at birth, later develop handicaps such as deafness or neurological problems.

A reliable vaccine would be valuable. Some have been tested but are not presently available.

The virus is susceptible in vitro to the antiviral drug ganciclovir. At this time the drug has not been approved for infected infants; however, it has had beneficial effects when used on immunocompromised adult patients. One of the leading microbial complications for AIDS patients is reactivated cytomegalovirus infections, especially affecting the eyesight.

The most reliable diagnostic method is isolation of the virus from bodily fluids during the first two weeks of life; the sample is usually shipped to a central reference laboratory. Serological tests that detect IgG are indicative of a latent infection. The presence of IgM is evidence of a probable primary infection.

Hepatitis

Hepatitis is an inflammation of the liver. Viral hepatitis is now the second most frequently reported infectious disease in the U.S.A. There are at least five different viruses causing hepatitis, and probably more remain to be discovered or become better known. Hepatitis is also an occasional result of infections by other viruses such as EB virus or cytomegalovirus. Drug and chemical toxicity can also cause acute hepatitis that is clinically identical to viral hepatitis. The characteristics of the various forms of viral hepatitis are summarized in Table 25.1.

Hepatitis A

The *hepatitis A virus (HAV)* is the etiological agent of **hepatitis A.** The HAV contains single-stranded RNA and lacks an envelope. It can be grown in cell culture.

After a typical entrance via the oral route, HAV multiplies in the epithelial lining of the intestinal tract. Viremia eventually occurs, and the virus spreads to the liver, kidneys, and spleen. The virus is shed in the feces and can also be detected in the blood and urine. The amount of virus excreted is greatest before symptoms appear and then declines rapidly. Therefore, a food handler responsible for spreading the virus might not appear to be ill at the time. The virus is probably able to survive for several days on such surfaces as cutting boards. Contamination of food or drink by feces is aided by the resistance of HAV to chlorine disinfectants at concentrations ordinarily used in water. Mollusks, such as oysters, that live in contaminated waters are also a source of infection. After all, what other animals do we often eat raw, digestive system and all?

At least 50% of infections with HAV are subclinical, especially in children. In clinical cases, the initial symptoms are anorexia (loss of appetite), malaise, nausea, diarrhea, abdominal discomfort, fever, and chills. These symptoms are more likely to appear in adults; they last from 2 to 21 days, and the mortality is low. In some cases there is also jaundice, with yellowing of the skin and the whites of the eyes and the dark urine typical of liver infections. In such cases the liver becomes tender and enlarged.

There is no chronic form of HAV infection, and the virus is usually shed only during the acute stage of disease, although it is difficult to detect. The incubation time averages about four weeks and ranges from two to six weeks, which makes epidemiological studies for the source of infections difficult. There are no animal reservoirs.

FIGURE 25.15 Hepatitis B virus (HBV). Micrograph and illustrations depict the distinct types of HBV particles discussed in the text.

In the U.S.A., the percentage of the population with HAV is much higher among lower socioeconomic groups (72% to 88%) than among middle and upper socioeconomic groups (18% to 30%). Most cases are unreported, and the 20,000 or more cases reported in the U.S.A. each year represent only a fraction of the actual number.

Acute disease is diagnosed by detection of IgM anti-HAV, because these antibodies appear about four weeks after infection and disappear about three to four months after infection. Recovery results in lifelong immunity.

No specific treatment for the disease exists, but people at risk for exposure to hepatitis A can be given immune globulin, which provides protection for several months. This is sufficient for travelers to high-risk areas. A vaccine is available in some European countries and will probably be licensed soon in the U.S.A.

Hepatitis B

Hepatitis B is caused by the hepatitis B virus (**HBV**). HBV and HAV are completely different viruses: HBV is larger, its genome is double-stranded DNA, and it is enveloped. HBV is a unique form of DNA virus; instead of replicating its DNA directly, it passes through an intermediate RNA stage resembling a retrovirus. Because HBV has often been transmitted by blood transfusions, this virus has been actively studied in order to determine how to identify contaminated blood.

The serum from patients with hepatitis B contains three distinct particles. The largest, the *Dane particle,* is the complete virion; it is infectious and capable of replicating. There are also smaller *spherical particles,* about half the size of a Dane particle, and *filamentous particles,* which are tubular particles similar in diameter to the spherical particles but about ten times as long (Figure 25.15). The spherical and filamentous particles are unassembled components of Dane particles without nucleic acids; assembly is evidently not very efficient, and large numbers of these unassembled components accumulate. Fortunately, these numerous unassembled particles contain *hepatitis B surface antigen (HB$_s$Ag),* which can be detected with antibodies specific against them. Such antibody tests make large-scale screening of blood for HBV possible.

Physicians, nurses, dentists, medical technologists, and others who are in daily contact with blood have a considerably higher incidence of this disease than does the general population. It is estimated that as many as 10,000 health care workers become infected each year in the U.S.A. Regulations now require that employees exposed to blood be offered free vaccinations by their employers. There have also been instances of transmission to patients by surgeons and dentists. Intravenous drug users often share needles and fail to sterilize them properly; as a consequence, they also have a high incidence of hepatitis B.

In addition to transmission by virus-carrying blood, recent evidence indicates that hepatitis B can be transmitted by any secretion of bodily fluids, such as saliva, breast milk, and semen. Transmission by semen donated for artificial insemination has been documented, and semen has been implicated in transmission between heterosexuals with multiple partners and in male homosexuals. A mother who is positive for HB$_s$Ag, especially if she is a chronic carrier, may transmit the disease to her infant, usually at birth. In most cases, this type of transmission can be prevented by giving the newborn immune globulin (HBIG) immediately after birth. These babies should also be vaccinated.

The incubation period averages about 12 weeks but ranges from 4 to 26 weeks. Because this incubation period is lengthy and of uneven duration, determination of the origin of an infection can be difficult.

An estimated 300,000 people, mostly young adults, are infected with hepatitis B each year. Clinical signs of hepatitis B vary widely. Probably about one-half of the cases are entirely asymptomatic. Symptoms are highly variable and include, in the early stages, loss of appetite, low-grade fever, and joint pains. Later, jaundice usually appears. These clinical manifestations are more likely to appear in adults. It is difficult to distinguish between hepatitis A and B solely on clinical grounds.

Ninety percent of acute hepatitis B infections end in complete recovery. The mortality rate is higher than that of hepatitis A, but it is probably less than 1% of the infected population. It might be as high as 2% or 3% among hospitalized patients, of which there are about 10,000 cases in the U.S.A. each year. Overall, up to 10% of patients become chronic carriers. This is age related; as many as 90% of infected newborns become carriers, but only 3% to 5% of adults. These carriers are reservoirs for transmission of the virus, and they also have a high rate of liver disease. There may be as many as 1 million HBV carriers in the U.S.A. A special concern is the strong correlation between the occurrence of liver cancer and the incidence of chronic hepatitis B infections. Chronic carriers are about 200 times more likely to get liver cancer than is the general population. Liver cancer is the most prevalent form of cancer in sub-Saharan Africa and the Far East, areas where hepatitis B is extremely common.

Researchers have been unable to cultivate HBV in cell culture, a step that was necessary for the development of vaccines for polio, mumps, measles, and rubella. The HBV vaccines available use HB$_s$Ag produced by a genetically engineered yeast. Vaccination is recommended for high-risk groups; a partial listing would include medical personnel exposed to blood and blood products, people undergoing hemodialysis, patients and staff at institutions for the mentally retarded, and homosexually active males.

Taiwan, which has a high incidence of the disease, has instituted a program to prevent mother-to-infant transmission. All infants delivered to HB$_s$Ag-positive mothers are vaccinated, and infants of highly infectious carrier mothers also receive HBIG to provide passive immunity. Similar recommendations are in effect in the U.S.A.

A promising treatment for chronic HBV infection is alpha interferon. A significant number of recipients show improvement.

Hepatitis C (Non-A, Non-B Hepatitis)

Screening of donated blood for HB$_s$Ag became so effective that transmission of hepatitis B by blood transfusion declined dramatically. However, in the 1960s a previously unsuspected form of hepatitis appeared that could also be transmitted by transfusion. This form was named, by elimination, non-A, non-B (NANB) hepatitis. It is now named **hepatitis C (HCV)**. This virus has a single strand of RNA and is enveloped. Soon, almost all cases of transfusion-transmitted hepatitis were of this type. Hepatitis C has an incubation period of from 2 to 22 weeks and causes symptoms that are subclinical or relatively mild. However, almost 50% of cases progress to chronic hepatitis.

Serological tests have been developed to detect HCV antibodies. These tests have reduced transmission of HCV via transfused blood to very low levels. However, cases still occur. A basic problem with the test is that it detects antibodies and not the virus itself. It also lacks specificity, and false-positives are frequent. Because a delay occurs between infection and appearance of antibodies, there will always be some risk. A test that would detect the virus itself is still needed. It is also strongly suspected that other, as yet unidentified NANB viruses exist.

Many chronic cases of hepatitis C respond to therapy with alpha interferon, but relapses are frequent.

Hepatitis D (Delta Hepatitis)

In 1977, a new hepatitis virus, now known as **hepatitis D virus (HDV)**, was discovered in carriers of HBV in Italy. People who carried this so-called *delta antigen* and were also infected with hepatitis B had a much higher incidence of severe liver damage and a much higher mortality rate than people who had antibodies against HBV alone. With time, it became clearer that HDV can occur as either acute (*coinfection form*) or chronic (*superinfection form*) hepatitis. In people with a case of self-limiting acute HBV, the coinfection with HDV disappeared as the HBV was cleared from the system, and the condition resembled a typical case of acute HBV. However, if the HBV infection progressed to the chronic stage, the superinfection with HDV was often accompanied by progressive liver damage and a fatality rate several times that of people infected with HBV alone.

Epidemiologically, HDV is linked to the epidemiology of HBV. Most cases occur in the Middle East and Mediterranean area, where the infection appears to spread mostly by intimate contact. In the Far East, however, HBV is extremely common but HDV is not. Furthermore, in the Amazon Basin and northern South America the disease takes the form of sudden epidemics with a very high mortality rate. In the U.S.A. and northern Europe the disease occurs predominantly in high-risk groups such as injecting drug users.

Structurally, the HDV antigen is a single strand of RNA, shorter than any other animal-infecting virus, that is protected by a capsid. This particle is not capable of causing an infection. It becomes infectious when an external envelope of HB_SAg, whose formation is controlled by the genome of HBV, covers the HDV protein core (the delta antigen) (see Figure 25.15).

Hepatitis E (Infectious NANB Hepatitis)

Another form of NANB hepatitis is spread by fecal–oral transmission, much like hepatitis A, which it clinically resembles. The agent is now known as **hepatitis E virus (HEV)** and is endemic in areas of the world with poor sanitation, especially India and southeast Asia. It resembles HAV in being an unenveloped virus with a single strand of RNA but is not related serologically to it. Like hepatitis A virus, it does not cause chronic liver disease, but for some unexplained reason it is responsible for a very high mortality rate in pregnant women.

Other Types of Hepatitis

New techniques in molecular biology and serology have provided evidence of hepatitis F, transmitted by blood, and a severe disease, hepatitis G. Hepatitis F was discovered in 1990, and hepatitis G in 1991.

Viral Gastroenteritis

A number of viruses, such as polioviruses, echoviruses, and coxsackieviruses, are transmitted by the fecal–oral route. However, despite the name *enteroviruses,* these viruses generally do not directly affect the digestive system. About 90% of cases of viral gastroenteritis are caused by rotavirus or by the Norwalk agent.

Rotavirus (rota means wheel) is the most common cause of viral gastroenteritis, mostly causing disease in small children. Following an incubation period of two or three days, the patient suffers from low-grade fever, diarrhea, and vomiting, which persist for a mean of five to eight days. There may be more than one serotype of rotavirus causing disease, and immunity is only partially protective. An ELISA and several other serological tests can be used to detect the virus in feces.

Major epidemics of viral gastroenteritis have been caused by a virus known as the *Norwalk agent* (named after an outbreak in Norwalk, Ohio, in 1968). All age groups are affected by this virus. Following a two-day incubation period, infected people typically suffer nausea, abdominal cramps, diarrhea, and vomiting for one to three days. Symptoms are mild, as compared with those of rotavirus infections. About one-half of middle-aged Americans show evidence, by serum antibodies, that they have been infected. Susceptibility to the virus varies considerably; some people are not affected and do not even respond with an immune reaction. Others develop gastroenteritis and an immunity that might be only short term.

The only treatment for viral gastroenteritis is oral rehydration or, in exceptional cases, intravenous rehydration.

FUNGAL DISEASES OF THE DIGESTIVE SYSTEM

LEARNING OBJECTIVE

• *Identify the causes of ergot and aflatoxin poisoning.*

Some fungi produce toxins called **mycotoxins.** These toxins cause blood diseases, nervous system disorders, kidney damage, and liver damage.

Ergot Poisoning

Some mycotoxins are produced by *Claviceps purpurea,* a fungus causing smut infections on grain crops (see Figure 15.7). The mycotoxins produced by *C. purpurea* cause **ergot poisoning,** or *ergolism,* which results from the ingestion of rye or other cereal grains contaminated with the fungus. Ergot poisoning was very widespread during the Middle Ages. The toxin can restrict blood flow in the extremities, and gangrene results. It may also cause hallucinogenic symptoms, producing bizarre behavior similar to that caused by LSD.

Aflatoxin Poisoning

Aflatoxin is a mycotoxin produced by the fungus *Aspergillus flavus,* a common mold. Aflatoxin has been found in many foods but is particularly likely to be found on peanuts. Aflatoxin is highly toxic and can cause serious damage to livestock when their feed is contaminated with *A. flavus.* Although the risk to humans is unknown, there is strong evidence that aflatoxin contributes to cirrhosis of the liver and cancer of the liver in parts of the world, such as India and Africa, where food is subject to aflatoxin contamination.

PROTOZOAN DISEASES OF THE DIGESTIVE SYSTEM

LEARNING OBJECTIVE

• *List the etiologic agents, methods of transmission, symptoms, and treatments for giardiasis, amoebic dysentery, and cryptosporidiosis.*

A number of pathogenic protozoa complete their life cycles in the human digestive tract. Usually they are ingested as resistant, infective cysts and are shed in greatly increased numbers as newly produced cysts.

FIGURE 25.16 **The trophozoite form of *Giardia lamblia*, the flagellated protozoan that causes giardiasis.** Note the circular mark left on the intestinal wall by the ventral sucker disc the parasite uses to attach itself. The dorsal side is smoothly streamlined, and the intestinal contents move easily around the attached organism.

FIGURE 25.17 **A typical flask-shaped ulcer in an intestinal wall, caused by *Entamoeba histolytica*.**

Giardiasis

Giardia lamblia is a flagellated protozoan that is able to attach firmly to a human's intestinal wall (Figure 25.16). In 1681 Leeuwenhoek described them as having ". . . bodies . . . somewhat longer than broad and their belly, which was flatlike, furnished with sundry little paws." *G. lamblia* is the cause of **giardiasis**, a prolonged diarrheal disease of humans. The disease sometimes persists for weeks and is characterized by malaise, nausea, flatulence (intestinal gas), weakness, weight loss, and abdominal cramps. The protozoa sometimes occupy so much of the intestinal wall that they interfere with food absorption. There are frequent outbreaks of giardiasis in the U.S.A.; it is probably the most common cause of epidemic waterborne diarrheal disease. About 7% of the population are healthy carriers and shed the cysts in their feces.

The organism is also shed by a number of wild mammals, especially beavers, and the disease occurs in backpackers who drink from wilderness waters. Most outbreaks in the U.S.A. are transmitted by contaminated water supplies. In a recent national survey of surface waters serving as sources for municipalities, the organism was detected in 18% of the samples. Because the cyst stage of the protozoa is relatively insensitive to chlorine, filtration of water supplies is usually necessary to eliminate the cysts from water.

Giardia is not reliably found in stools, and the *string test* is recommended for diagnosis. In this test, a gelatin capsule packed with about 140 cm of fine string is swallowed by the patient. One end of the string is taped to the cheek. The gelatin capsule dissolves in the stomach, and an enclosed weighted rubber bag attached to the other end of

the string enters the upper bowel. After a few hours, the string is drawn up through the mouth and examined for parasites. A number of commercially available tests use monoclonal antibodies specific for *G. lamblia*; specimens for the tests are taken by rectal swab. These tests are especially useful for epidemiological screening.

Treatment with metronidazole or quinacrine hydrochloride is usually effective within a week.

Amoebic Dysentery (Amoebiasis)

Amoebic dysentery (amoebiasis) is found worldwide and is spread mostly by food or water contaminated by cysts of the protozoan amoeba *Entamoeba histolytica* (see Figure 12.18b). Although stomach acid (HCl) can destroy vegetative cells, it does not affect the cysts. In the intestinal tract the cyst wall is digested away and the vegetative forms are released. The vegetative forms then multiply in the epithelial cells of the wall of the large intestine. A severe dysentery results, and the feces characteristically contain blood and mucus. The vegetative forms feed on red blood cells and destroy tissue in the gastrointestinal tract (Figure 25.17).

Severe bacterial infections result if the intestinal wall is perforated. Abscesses might have to be treated surgically, and invasion of other organs, particularly the liver, is not uncommon. Perhaps 5% of the population in the U.S.A. are asymptomatic carriers of *E. histolytica*. Worldwide, one person in ten is estimated to be infected, mostly asymptomatically, and about a tenth of these infections progress to the more serious stages.

Diagnosis depends largely on recovery and identification of the organisms in feces. (Red blood cells observed within the trophozoite stage of an amoeba are considered an indication of *E. histolytica*.) There are several serological tests that can also be used for diagnosis, including latex-agglutination and fluorescent-antibody tests. Such tests are especially useful when the affected areas are outside the intestinal tract and the patient is not passing amoebae.

Metronidazole plus iodoquinol are the drugs of choice in treatment.

Cryptosporidiosis

Cryptosporidiosis is caused by the protozoan *Cryptosporidium parvum,* long recognized as a pathogen causing diarrhea in calves but not known to infect humans until 1976. Since then, the number of reported cases has been increasing. This is another example of a disease with particularly serious effects on immunosuppressed individuals, including those with AIDS. In otherwise healthy individuals, the organism causes intestinal distress of short duration. Immune-deficient individuals, however, suffer from severe life-threatening diarrhea that becomes progressively worse.

Transmission to humans is largely via water systems contaminated by animal wastes, especially cattle. Thousands may be affected in such outbreaks. The cysts (actually oocysts), like those of *G. lamblia,* are resistant to chlorination and must be removed from water by filtration. Occasional cases also occur by fecal–oral transmission from close contact with animals.

Diagnosis is based on microscopic examination of feces for *Cryptosporidium* oocysts (Figure 25.18). An ELISA test that can detect the parasite in stool specimens is also available. Widely used for water testing, especially during outbreaks of water-spread diarrhea, is a fluorescent-antibody test that can simultaneously detect both *G. lamblia* cysts and *C. parvum* oocysts.

HELMINTHIC DISEASES OF THE DIGESTIVE SYSTEM

LEARNING OBJECTIVE

- *List the etiologic agents, methods of transmission, symptoms, and treatments for tapeworms, hydatid disease, pinworms, hookworms, ascariasis, and trichinosis.*

Helminthic parasites are very common in the human intestinal tract, especially under conditions of poor sanitation. Figure 25.19 shows the worldwide incidence of infestation with some intestinal helminths. In spite of their size and formidable appearance, they often produce few symptoms. They have become so well adapted to their human hosts, and vice versa, that when their presence is revealed it is often a surprise.

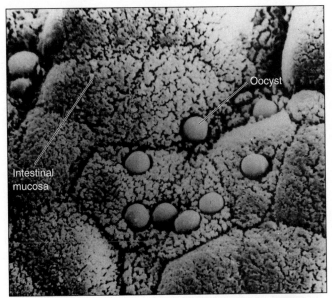

SEM ⊢⊣ 1 µm

FIGURE 25.18 Cryptosporidiosis. Oocysts of *Cryptosporidium parvum* are shown here embedded in the mucosal surface of the intestine.

Tapeworm Infestations

Taenia saginata is a typical tapeworm whose life cycle extends through three stages. The adult worm lives in the intestine of a human host, where it produces eggs that are excreted in the feces. The eggs are ingested with contaminated feed by animals, such as grazing cattle, and hatch into a larval form called a cysticercus that lodges in the animal's muscles. When a human ingests undercooked meat containing cysticerci, the cysticerci develop into adult worms. Infestations by tapeworms begin with the consumption of undercooked beef, pork, or fish containing larval forms of the tapeworm. The larvae develop into adults that attach to the intestinal wall by suckers on the scolex. The adult beef tapeworm, *T. saginata,* can live in the human intestine for 25 years and reaches a length of six meters (18 feet) or longer. Even a worm of this size seldom causes significant symptoms beyond a vague abdominal discomfort. There is, however, psychological distress when a meter or more of detached segments (proglottids) break loose and unexpectedly slip out of the anus, which happens occasionally.

Taenia solium (sō'lē-um), the pork tapeworm, has a life cycle similar to the beef tapeworm. An important difference is that *T. solium* may produce the larval stage in the human host, a disease condition called **neurocysticercosis.** In this disease, the infection can arise from the tapeworm egg rather than only from ingestion of undercooked

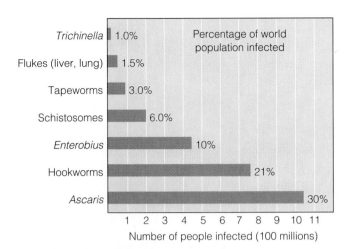

FIGURE 25.19 **Prevalence of infestation with selected intestinal helminths in the worldwide human population.**

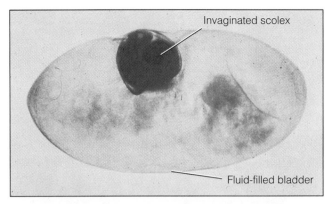

(a) A cysticercus of *Taenia solium;* note the scolex of the larval tapeworm invaginated in the fluid-filled bladder

(b) Cysticerci can occur in many tissues—this one has formed in an eye; note the scolex has evaginated from the bladder

FIGURE 25.20 **Neurocysticercosis in humans.**

meat containing cysticerci. The eggs can be ingested as a result of poor sanitary practices or possibly by autoinfection (when eggs produced by the adult tapeworm in the intestinal tract somehow enter the stomach).

Cysticerci can develop in many human organs (Figure 25.20). When this occurs in muscle tissue, the symptoms are seldom severe. Cysticerci that develop in the eyes and in the brain, however, may be more serious. Neurocysticercosis, which is endemic in Mexico and Central America, has become a fairly common condition in parts of the U.S.A. with substantial immigrant populations from those areas. The symptoms often mimic those of a brain tumor. The number of cases reported reflects, in part, the use of CT (computerized tomography) scanning in diagnosis. These expensive machines X ray the body in continuous "slices." In endemic areas, neurological patients can be screened with serological tests for antibodies to *T. solium.*

The fish tapeworm, *Diphyllobothrium latum* (dī-fil-lō-bo'thrē-um lā'tum), is found in pike, trout, perch, and salmon. The CDC has issued warnings about the risks of fish tapeworm infection from the increasingly popular sushi and sashimi (Japanese dishes prepared from raw fish). To relate a vivid example, about ten days after eating sushi, one person developed symptoms of abdominal distention, flatulence (intestinal gas), belching, intermittent abdominal cramping, and diarrhea. Eight days later, the patient passed a 1.2-meter (four-feet) long tapeworm identified as a species of *Diphyllobothrium.*

Laboratory diagnosis of human infestation consists of identification of the tapeworm eggs or segments in feces. The drug of choice for eliminating tapeworms is niclosamide.

Hydatid Disease

Not all tapeworms are large. One of the most dangerous is *Echinococcus granulosus,* which is only a few millimeters in

length (see Figure 12.25a). Humans are not the definitive hosts. The adult form lives in the intestinal tract of carnivorous animals, such as dogs and wolves. Typically, humans become infected from the feces of a dog that has become infected by eating the flesh of a sheep or deer containing the cyst form of the tapeworm. Unfortunately, humans can be an intermediate host, and cysts can develop in the body.

Once ingested by a human, the egg of *E. granulosus* may migrate to various tissues of the body. The liver and lungs are the most common sites, but the brain and numerous other sites also may be infected. Once in place, the egg develops into a **hydatid cyst** that can grow to a diameter of 10 mm in a few months (Figure 25.21). In some locations, cysts may not be apparent for many years. Some become enormous, containing up to 15 liters (4 gallons) of fluid. Damage may arise from the size of the cyst in such areas as the brain or the interior of bones. If the cyst ruptures in the host, it can lead to the development of a great many daughter cysts. Another factor in the pathogenicity of such cysts is that the fluid contains proteinaceous material, to which the host becomes sensitized. If the cyst suddenly ruptures, the result can be life-threatening anaphylactic shock. Treatment is usually surgical removal, but care must be taken to avoid release of the fluid and poten-

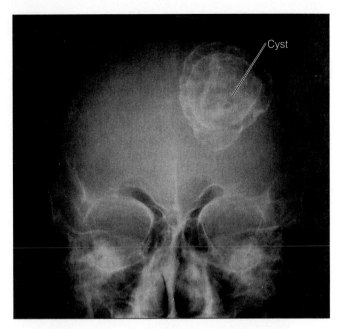

FIGURE 25.21 Hydatid cyst formed by *Echinococcus granulosus.* A large cyst can be seen in this X ray of the brain of an infected individual.

FIGURE 25.22 *Ancylostoma* sp. hookworm attached to intestinal mucosa. Note how the mouth of the worm is adapted to feeding on the tissue.

tial spread of infection or anaphylactic shock. If removal is not feasible, a new drug, albendazole, can kill the cysts. The disease is seen most frequently among people who raise sheep or hunt or trap wild animals.

Nematode Infestations

Pinworm Infestation

Most of us are familiar with the **pinworm**, *Enterobius vermicularis* (see Figure 12.26). This tiny worm migrates out of the anus of the human host to lay its eggs, causing local itching. Whole households may become infected. Such drugs as pyrantel pamoate or mebendazole are usually effective in treatment.

Hookworm Infestation

Hookworm infestations were once a very common parasitic disease in the southeastern states. In the U.S.A., the species most often seen is *Necator americanus* (see Figure 12.27). The hookworm attaches to the intestinal wall and feeds on blood and tissue (Figure 25.22) rather than on partially digested food, so the presence of large numbers of worms can lead to anemia and lethargic behavior. Large infestations can also lead to a bizarre symptom, *pica,* which is a craving for peculiar foods, such as laundry starch or soil containing a certain type of clay. Pica is a symptom of iron deficiency anemia. Because the life cycle of the hookworm requires human feces to enter the soil and bare skin to contact contaminated soil, the incidence

of the disease has declined greatly with improved sanitation and the practice of wearing shoes. Hookworms can be treated effectively with mebendazole.

Ascariasis

One of the most widespread helminthic infections is **ascariasis**, caused by *Ascaris lumbricoides.* This condition is familiar to many physicians in the U.S.A. As described in Chapter 12, diagnosis is often made when an adult worm emerges from the anus. These worms can be quite large, up to 30 cm (about 1 foot) in length (Figure 25.23). In the intestinal tract they live on partially digested food and cause few symptoms.

The worm's life cycle begins when eggs are shed in a person's feces and, under poor sanitary conditions, are ingested by another person. In the upper intestine the eggs hatch into small wormlike larvae that pass into the bloodstream and then into the lungs. There they migrate into the throat and are swallowed. The larvae develop into egg-laying adults in the intestines. (All this migration just to return to the place where they started!)

In the lungs the larvae may cause some pulmonary symptoms, and extremely large numbers cause pneumonia. Otherwise, the worms do not usually cause severe symptoms, but their presence can be manifested in distressing ways. The most dramatic consequences of infection with *A. lumbricoides* are from the migrations of adult worms. The worms are equipped with minute cutting teeth and can penetrate tissue. If they pass through the in-

FIGURE 25.23 *Ascaris lumbricoides,* **the cause of ascariasis.** This photograph shows a male (the smaller worm with the coiled end) and a female. These worms are up to 30 cm (1 foot) in length.

testinal wall, they can infect the abdominal cavity. Worms have been known to leave the body of small children through the umbilicus (navel) and to escape through the nostrils of a sleeping person. Once ascariasis is diagnosed, it can be effectively treated with mebendazole.

Trichinosis

Most infestations by the small roundworm *Trichinella spiralis,* called **trichinosis,** are insignificant. The larvae, in encysted form, are located in muscles of the host. In 1970, routine autopsies of human diaphragm muscles showed that about 4% of those tested carried this parasite. Severe cases of trichinosis can be fatal—sometimes in only a few days. The severity of the disease is generally proportional to the number of larvae ingested. Ingestion of undercooked pork is the most common vehicle of infestation (Figure 25.24), but ingestion of the flesh of animals that feed on garbage (bears, for example) also causes outbreaks.

1 Adult *Trichinella spiralis* develop, invade intestinal wall of pig, and produce larvae that invade muscles.

Garbage, including undercooked or raw pork

2 *T. spiralis* larvae encyst in pig's muscle tissue (capsule is 0.25 to 0.5 mm in length).

LM   0.5 mm

5 Meanwhile, other animals are infected by eating infected meat that has been dumped.

Undercooked pork

3 Human eats undercooked pork containing cysts.

4 In human intestine, cyst walls are removed, and *T. spiralis* adults develop. Adults produce larvae that encyst in muscles.

(a) Life cycle of *Trichinella spiralis,* the causative agent of trichinosis

(b) *T. spiralis* adult   LM   0.1 mm

FIGURE 25.24 Trichinosis.

TABLE 25.2 Microbial Diseases Associated with the Digestive System

DISEASE	CAUSATIVE AGENT	MODE OF TRANSMISSION	TREATMENT/VACCINE
Bacterial Diseases of the Mouth			
Dental caries	*Streptococcus mutans* is most prominent	Bacteria use sucrose to form plaque	Drilling and filling. Prevention includes restricting ingestion of sucrose; brushing, flossing, and fluoridation
Periodontal disease	Variety of organisms	Presence of bacterial plaque initiates an inflammatory response	Antibiotics such as penicillin and tetracycline; cleaning teeth may prevent occurrence
Bacterial Diseases of the Lower Digestive System			
Staphylococcal food poisoning	*Staphylococcus aureus*	Ingestion of exotoxin in food, usually improperly refrigerated	Replace lost water and electrolytes
Salmonellosis	*Salmonella* spp.	Ingestion of contaminated food and drink	Replace lost water and electrolytes
Typhoid fever	*Salmonella typhi*	Ingestion of contaminated food and drink	Third-generation cephalosporins
Bacillary dysentery (shigellosis)	*Shigella* spp.	Ingestion of contaminated food and drink	Replace lost water and electrolytes; fluoroquinolones
Cholera	*Vibrio cholerae*	Ingestion of contaminated food and drink	Replace lost water and electrolytes
Vibrio parahaemolyticus gastroenteritis	*Vibrio parahaemolyticus*	Often ingestion of contaminated shellfish	Replace lost water and electrolytes
Escherichia coli gastroenteritis	Enterotoxigenic, enteroinvasive, enteropathogenic, and enterohemorrhagic strains of *Escherichia coli*	Ingestion of contaminated food and drink	Replace lost water and electrolytes
Campylobacter gastroenteritis	*Campylobacter jejuni*	Ingestion of contaminated food and drink	Replace lost water and electrolytes
Helicobacter peptic disease syndrome	*Helicobacter pylori*	Presumably ingested	Bismuth-containing drugs, metronidazole
Yersinia gastroenteritis	*Yersinia enterocolitica*	Ingestion of contaminated food and drink	Replace lost water and electrolytes
Clostridium perfringens gastroenteritis	*Clostridium perfringens*	Ingestion of contaminated food (usually meat) and drink	Replace lost water and electrolytes
Bacillus cereus gastroenteritis	*Bacillus cereus*	Ingestion of contaminated food	None, self-limiting
Viral Diseases of the Digestive System			
Mumps	Mumps virus	Saliva and respiratory secretions	None; vaccine available (MMR)
Cytomegalovirus (CMV) inclusion disease	Cytomegalovirus (CMV)	Placental transfer and blood transfusion	Ganciclovir
Hepatitis A	Hepatitis A virus (HAV)	Ingestion of contaminated food or drink	Vaccine available soon; now passive immunization
Hepatitis B	Hepatitis B virus (HBV)	Blood or blood contamination by needles, transfusions; sexual contact, especially between male homosexuals	Vaccine available
Hepatitis C	Hepatitis C virus (HCV)	Blood or blood contamination by needles, transfusions	None
Hepatitis D	Hepatitis D virus (HDV)	Probably same as hepatitis B	Hepatitis B vaccine effective because coinfection required
Hepatitis E	Hepatitis E virus (HEV)	Ingestion of contaminated food and water	None
Viral gastroenteritis	Norwalk agent or rotavirus	Ingestion of contaminated food and water	Replace lost water and electrolytes

TABLE 25.2 (Continued)

DISEASE	CAUSATIVE AGENT	MODE OF TRANSMISSION	TREATMENT/VACCINE
Fungal Diseases of the Digestive System			
Ergot poisoning	Mycotoxin produced by *Claviceps purpurea*	Ingestion of contaminated food, usually cereal grain	Antispasmodic drugs
Aflatoxin poisoning	Mycotoxin produced by *Aspergillus flavus*	Ingestion of contaminated food, usually peanuts	None
Protozoan Diseases of the Digestive System			
Giardiasis	*Giardia lamblia*	Ingestion of contaminated water	Metronidazole and quinacrine hydrochloride
Amoebic dysentery (amoebiasis)	*Entamoeba histolytica*	Ingestion of contaminated food and water	Metronidazole plus iodoquinol
Cryptosporidiosis	*Cryptosporidium*	Fecal–oral, mostly in water supply	None
Helminthic Diseases of the Digestive System			
Tapeworm infestations	*Taenia saginata* (beef tapeworm); *T. solium* (pork tapeworm); *Diphyllobothrium latum* (fish tapeworm)	Ingestion of contaminated meat or fish	Niclosamide
Hydatid disease	*Echinococcus granulosus*	Ingestion of eggs	Surgical removal, albendazole
Pinworms	*Enterobius vermicularis*	Ingestion of eggs	Pyrantel pamoate, mebendazole
Hookworms	*Necator americanus*	Larvae in soil penetrate skin	Mebendazole
Ascariasis	*Ascaris lumbricoides*	Ingestion of eggs	Mebendazole
Trichinosis	*Trichinella spiralis*	Ingestion of larvae in meat	Mebendazole and corticosteroids

Any ground meat can be contaminated from machinery previously used to grind contaminated meats. Eating raw sausage or hamburger is a risky habit. One person acquired trichinosis by chewing the fingernails after handling infected pork. Prolonged freezing of meats containing *Trichinella* tends to eliminate the worms, but freezing should not be considered a substitute for thorough cooking.

In the muscles of hosts such as pigs, the *T. spiralis* larvae are encysted in the form of short worms about 1 mm in length. When the flesh of an infected animal is ingested by humans, the cyst wall is removed by digestive action in the intestine. The organism then matures into the adult form. The adult worms spend only about one week in the intestinal mucosa and produce larvae that invade tissue. Eventually, the encysted larvae localize in muscle (common sites include the diaphragm and eye muscles), where they are barely visible in biopsied specimens.

Symptoms of trichinosis include fever, swelling about the eyes, and gastrointestinal upset. Small hemorrhages under the fingernails are often observed. Biopsy specimens, as well as a number of serological tests, can be used in diagnosis. Recently, a serological ELISA test that detects the parasite in meats has been developed. Treatment consists of the administration of mebendazole to kill intestinal worms and corticosteroids to reduce inflammation.

In the past ten years, the number of cases reported annually in the U.S.A. has varied from 30 to 129. Deaths are rare, and in most years there are none.

•   •   •

Table 25.2 summarizes microbial diseases associated with the digestive system.

STUDY OUTLINE

INTRODUCTION (p. 612)

1. Diseases of the digestive system are the second most common illnesses in the U.S.A.
2. Diseases of the digestive system usually result from ingestion of microorganisms and their toxins in food and water.
3. The fecal–oral cycle of transmission can be broken by proper disposal of sewage, drinking-water disinfection, and proper preparation and storage of foods.

STRUCTURE AND FUNCTION OF THE DIGESTIVE SYSTEM (p. 612)

1. The gastrointestinal (GI) tract, or alimentary canal, consists of the mouth, pharynx, esophagus, stomach, small intestine, and large intestine.
2. The teeth, tongue, salivary glands, liver, gallbladder, and pancreas are accessory structures.
3. In the GI tract, with mechanical and chemical help from the accessory structures, large food molecules are broken down into smaller molecules that can be transported by blood or lymph to cells.
4. Feces, solids resulting from digestion, are eliminated through the anus.

NORMAL MICROBIOTA OF THE DIGESTIVE SYSTEM (pp. 612–613)

1. A wide variety of bacteria colonize the mouth.
2. The stomach and small intestine have few resident microorganisms.
3. The large intestine is the habitat for *Lactobacillus, Bacteroides, E. coli, Enterobacter, Klebsiella,* and *Proteus.*
4. Bacteria in the large intestine assist in degrading food and synthesizing vitamins.
5. Up to 40% of fecal mass is microbial cells.

BACTERIAL DISEASES OF THE MOUTH

(pp. 613–616)

Dental Caries (Tooth Decay) (pp. 613–615)
1. Dental caries begin when tooth enamel and dentin are eroded and the pulp is exposed to bacterial infection.
2. *Streptococcus mutans,* found in the mouth, uses sucrose to form dextran from glucose and lactic acid from fructose.
3. Bacteria adhere to teeth and produce sticky dextran, forming dental plaque.
4. Acid produced during carbohydrate fermentation destroys tooth enamel at the site of the plaque.
5. Gram-positive rods and filamentous bacteria can penetrate into dentin and pulp.

6. Carbohydrates such as starch, mannitol, and sorbitol are not used by cariogenic bacteria to produce dextran and do not promote tooth decay.
7. Caries are prevented by restricting ingestion of sucrose and by physical removal of plaque; a vaccine against *S. mutans* is theoretically possible.

Periodontal Disease (pp. 615–616)
1. Caries of the cementum and gingivitis are caused by streptococci, actinomycetes, and anaerobic gram-negative bacteria.
2. Chronic gum disease (periodontitis) can cause bone destruction and tooth loss; periodontitis is due to an inflammatory response to a variety of bacteria growing on the gums.
3. Acute necrotizing ulcerative gingivitis is caused by *Fusobacterium* and actinomycetes.

BACTERIAL DISEASES OF THE LOWER DIGESTIVE SYSTEM (pp. 616–625)

1. A gastrointestinal infection is caused by growth of a pathogen in the intestines.
2. Incubation times, the times required for bacterial cells to grow and their products to produce symptoms, range from 12 hours to 2 weeks. Symptoms of infection generally include a fever.
3. A bacterial intoxication results from ingestion of preformed bacterial toxins.
4. Symptoms appear from 1 to 48 hours after ingestion of the toxin. Fever is not usually a symptom of intoxication.
5. Infections and intoxications cause diarrhea, dysentery, or gastroenteritis.
6. These conditions are usually treated with fluid and electrolyte replacement.

Staphylococcal Food Poisoning (Staphylococcal Enterotoxicosis) (pp. 616–618)
1. Staphylococcal food poisoning is caused by ingestion of an enterotoxin produced in improperly stored foods.
2. *S. aureus* is inoculated into foods during preparation. The bacteria grow and produce enterotoxin in food stored at room temperature.
3. The exotoxin is not denatured by boiling for 30 minutes.
4. Foods with high osmotic pressure and those not cooked immediately before consumption are most often the source of staphylococcal enterotoxicosis.
5. Nausea, vomiting, and diarrhea begin between one and six hours after eating, and the symptoms last approximately 24 hours.
6. Laboratory identification of *S. aureus* isolated from foods or the presence of thermostable nuclease in foods can confirm diagnosis.

Salmonellosis (*Salmonella* Gastroenteritis)
(pp. 618–619)

1. Salmonellosis, or *Salmonella* gastroenteritis, is caused by many *Salmonella* species.
2. Symptoms include nausea, abdominal pain, and diarrhea and begin 12 to 36 hours after ingestion of large numbers of *Salmonella*. Septicemia can occur in infants and in the elderly.
3. Fever might be caused by endotoxin.
4. Mortality is lower than 1%, and recovery can result in a carrier state.
5. Heating food to 68°C will usually kill *Salmonella*.
6. Laboratory diagnosis is based on isolation and identification of *Salmonella* from feces and foods.

Typhoid Fever (pp. 619–620)

1. *Salmonella typhi* causes typhoid fever; the bacteria are transmitted by contact with human feces.
2. Fever and malaise occur after a two-week incubation. Symptoms last from two to three weeks.
3. *S. typhi* is harbored in the gallbladder of carriers.
4. Vaccines are available for high-risk people.

Bacillary Dysentery (Shigellosis) (p. 620)

1. Bacillary dysentery is caused by four species of *Shigella*.
2. Symptoms include blood and mucus in stools, abdominal cramps, and fever. Infections by *S. dysenteriae* result in ulceration of the intestinal mucosa.
3. Isolation and identification of the bacteria from rectal swabs are used for diagnosis.

Cholera (pp. 620–621)

1. *Vibrio cholerae* produces an exotoxin that alters membrane permeability of the intestinal mucosa; the resulting vomiting and diarrhea cause loss of body fluids.
2. The incubation period is approximately three days. The symptoms last for a few days. Untreated cholera has a 50% mortality rate.
3. Diagnosis is based on isolation of *Vibrio* from feces.
4. *Vibrio cholerae* non-O:1 causes gastroenteritis in the U.S.A. It is usually transmitted via contaminated seafood.

Vibrio parahaemolyticus Gastroenteritis (p. 623)

1. Gastroenteritis can be caused by the halophile *V. parahaemolyticus*.
2. Onset of symptoms begins within 24 hours after ingestion of contaminated foods. Recovery occurs within a few days.
3. The disease is contracted by ingestion of contaminated crustaceans or contaminated mollusks.

Escherichia coli Gastroenteritis (Traveler's Diarrhea) (p. 623)

1. Gastroenteritis may be caused by enterotoxigenic, enteroinvasive, enteropathogenic, or enterohemorrhagic strains of *E. coli*.
2. The disease occurs as epidemic diarrhea in nurseries, as traveler's diarrhea, as endemic diarrhea in less developed countries, and as hemorrhagic colitis.
3. In adults, the disease is usually self-limiting and does not require chemotherapy.
4. Enterohemorrhagic *E. coli*, such as *E. coli* O157:H7, produces verotoxins that cause inflammation and bleeding of the colon.
5. Verotoxins can affect the kidneys to cause hemolytic uremic syndrome.

Campylobacter Gastroenteritis (pp. 623–624)

1. *Campylobacter* is the second most common cause of diarrhea in the U.S.A.
2. *Campylobacter* is transmitted in cow's milk.

Helicobacter Peptic Disease Syndrome (pp. 624–625)

1. *Helicobacter pylori* produces ammonia, which neutralizes stomach acid; the bacteria colonize the stomach mucosa and cause gastritis.
2. Bismuth and several antibiotics may be useful in treating gastritis.

Yersinia Gastroenteritis (p. 625)

1. *Y. enterocolitica* and *Y. pseudotuberculosis* are transmitted in meat and milk.
2. *Yersinia* can grow at refrigeration temperatures.

Clostridium perfringens Gastroenteritis (p. 625)

1. A self-limiting gastroenteritis is caused by *C. perfringens*.
2. Endospores survive heating and germinate when foods (usually meats) are stored at room temperature.
3. Exotoxin produced when the bacteria grow in the intestines is responsible for the symptoms.
4. Diagnosis is based on isolation and identification of the bacteria in stool samples.

Bacillus cereus Gastroenteritis (p. 625)

1. Ingesting food contaminated with the soil saprophyte *Bacillus cereus* can result in diarrhea, nausea, and vomiting.

VIRAL DISEASES OF THE DIGESTIVE SYSTEM (pp. 625–630)

Mumps (pp. 625–626)

1. Mumps virus enters and exits the body through the respiratory tract.
2. About 16 to 18 days after exposure, the virus causes inflammation of the parotid glands, fever, and pain during swallowing. About 4 to 7 days later, orchitis may occur.
3. After onset of the symptoms, the virus is found in the blood, saliva, and urine.
4. A measles–mumps–rubella (MMR) vaccine is available.
5. Diagnosis is based on symptoms or an ELISA on viruses cultured in embryonated eggs or cell culture.

Cytomegalovirus (CMV) Inclusion Disease
(pp. 626–627)

1. CMV (a herpesvirus) causes intranuclear inclusion bodies and cytomegaly of host cells.
2. CMV is transmitted by saliva, urine, semen, cervical secretions, and human milk.
3. CMV inclusion disease can be asymptomatic, a mild disease, or progressive and fatal. Immunosuppressed patients may develop pneumonia.
4. If the virus crosses the placenta, it can cause congenital infection of the fetus resulting in impaired mental development, neurological damage, and stillbirth.
5. Diagnosis is based on isolation of the virus or detection of IgG and IgM antibodies.

Hepatitis (pp. 627–630)

1. Inflammation of the liver is called hepatitis. Symptoms include loss of appetite, malaise, fever, and jaundice.
2. Viral causes of hepatitis include hepatitis viruses, Epstein–Barr (EB) virus, and CMV.

Hepatitis A (pp. 627–628)

1. Hepatitis A virus causes hepatitis; at least 50% of all cases are subclinical.
2. HAV is ingested in contaminated food or water, grows in the cells of the intestinal mucosa, and spreads to the liver, kidneys, and spleen in the blood.
3. The virus is eliminated with feces.
4. The incubation period is 2 to 6 weeks; the period of disease is 2 to 21 days, and recovery is complete in 4 to 6 weeks.
5. Diagnosis is based on tests for IgM antibodies.
6. Passive immunization can provide temporary protection. A vaccine will be available soon.

Hepatitis B (pp. 628–629)

1. Hepatitis B virus causes hepatitis that is frequently serious.
2. HBV is transmitted by blood transfusions, contaminated syringes, saliva, sweat, breast milk, and semen.
3. Blood is tested for HB_sAg before being used in transfusions.
4. The average incubation period is three months; recovery is usually complete, but some patients develop a chronic infection or become carriers.
5. A vaccine against HB_sAg is available.

Hepatitis C (Non-A, Non-B Hepatitis) (p. 629)

1. Hepatitis C virus is transmitted via blood.
2. The incubation period is 2 to 22 weeks; the disease is usually mild, but some patients develop chronic hepatitis.
3. Blood is tested for HCV antibodies before being used in transfusions.

Hepatitis D (Delta Hepatitis) (pp. 629–630)

1. Hepatitis D virus agent has a circular strand of RNA and uses HB_sAg as a coat.

Hepatitis E (Infectious NANB Hepatitis) (p. 630)

1. Hepatitis E virus is spread by the fecal–oral route.

Other Types of Hepatitis (p. 630)

1. There is evidence of the existence of hepatitis types F and G.

Viral Gastroenteritis (p. 630)

1. Viral gastroenteritis is most often caused by rotavirus or the Norwalk agent.

FUNGAL DISEASES OF THE DIGESTIVE SYSTEM (p. 630)

1. Mycotoxins are toxins produced by some fungi.
2. Mycotoxins affect the blood, nervous system, kidneys, or liver.

Ergot Poisoning (p. 630)

1. Ergot poisoning, or ergotism, is caused by the mycotoxin produced by *Claviceps purpurea*.
2. Cereal grains are the crop most often contaminated with the *Claviceps* mycotoxin.

Aflatoxin Poisoning (p. 630)

1. Aflatoxin is a mycotoxin produced by *Aspergillus flavus*.
2. Peanuts are the crop most often contaminated with aflatoxin.

PROTOZOAN DISEASES OF THE DIGESTIVE SYSTEM (pp. 630–632)

Giardiasis (p. 631)

1. *Giardia lamblia* grows in the intestines of humans and wild animals and is transmitted in contaminated water.
2. Symptoms of giardiasis are malaise, nausea, flatulence, weakness, and abdominal cramps that persist for weeks.
3. Diagnosis is based on identification of the protozoa in the small intestine.

Amoebic Dysentery (Amoebiasis) (pp. 631–632)

1. Amoebic dysentery is caused by *Entamoeba histolytica* growing in the large intestine.
2. The amoeba feeds on red blood cells and GI tract tissues. Severe infections result in abscesses.
3. Diagnosis is confirmed by observation of trophozoites in feces and several serological tests.

Cryptosporidiosis (p. 637)

1. *Crytosporidium parvum* causes diarrhea; in immunosuppressed patients the disease is prolonged for months.
2. It is transmitted in contaminated water.
3. Diagnosis is based on identification of oocysts in feces.

HELMINTHIC DISEASES OF THE DIGESTIVE SYSTEM (pp. 632–637)

Tapeworm Infestations (pp. 632–634)

1. Tapeworms are contracted by the consumption of undercooked beef, pork, or fish containing encysted larvae (cysticerci).

2. The scolex attaches to the intestinal mucosa of humans (the definitive host) and matures into an adult tapeworm.
3. Eggs are shed in the feces and must be ingested by an intermediate host.
4. Adult tapeworms can be undiagnosed in a human.
5. Diagnosis is based on observation of proglottids and eggs in feces.
6. Neurocysticercosis in humans occurs when the pork tapeworm larvae encyst in humans.

Hydatid Disease (pp. 633–634)

1. Humans infested with the tapeworm *Echinococcus granulosus* might have hydatid cysts in their lungs or other organs.
2. Dogs and wolves are usually the definitive hosts and sheep or deer are the intermediate hosts for *E. granulosus*.

Nematode Infestations (pp. 634–637)

Pinworm Infestation (p. 634)

1. Humans are the definitive host for pinworms, *Enterobius vermicularis*.

2. The disease is acquired by ingesting *Enterobius* eggs.

Hookworm Infestation (p. 634)

1. Hookworm larvae bore through skin and migrate to the intestine to mature into adults.
2. In the soil, hookworm larvae hatch from eggs shed in feces.

Ascariasis (pp. 634–635)

1. *Ascaris lumbricoides* adults live in human intestines.
2. The disease is acquired by ingesting *Ascaris* eggs.

Trichinosis (pp. 635–637)

1. *Trichinella spiralis* larvae encyst in muscles of humans, swine, and other mammals to cause trichinosis.
2. The roundworm is contracted by ingesting undercooked meat containing larvae.
3. Adults mature in the intestine and lay eggs. The new larvae migrate to invade muscles.
4. Symptoms include fever, swelling around the eyes, and gastrointestinal upset.
5. Biopsy specimens and serological tests are used for diagnosis.

STUDY QUESTIONS

Review

1. State examples of the representative normal microbiota, if any, in the mouth, pharynx, esophagus, stomach, small intestine, large intestine, and rectum.

2. What properties of *S. mutans* implicate this bacterium in the formation of dental caries? Why is sucrose, more than any other carbohydrate, responsible for the formation of dental caries?

3. Complete the following table.

DISEASE	ETIOLOGIC AGENT	SUSPECT FOODS	SYMP-TOMS	TREAT-MENT
Staphylococcal food poisoning				
Salmonellosis				
Bacillary dysentery				
Cholera				
Gastroenteritis				
Traveler's diarrhea				

4. Differentiate between salmonellosis and typhoid fever.

5. You probably listed *E. coli* in answer to Questions 1 and 3. Explain why this one bacterial species is both beneficial and harmful.

6. What preventive treatments are currently used for hepatitis A? For hepatitis B? For hepatitis C?

7. How is blood that is to be used for transfusions tested for HBV? For HCV?

8. Define mycotoxin. Provide an example of a mycotoxin.

9. Explain how the following diseases differ and how they are similar: giardiasis, amoebic dysentery, and cryptosporidiosis.

10. Differentiate between amoebic dysentery and bacillary dysentery.

11. List the general symptoms of gastroenteritis. Because there are many etiologies, what is the laboratory diagnosis usually based on?

12. Differentiate among the following factors of bacterial intoxication and bacterial infection: prerequisite conditions, etiologic agents, onset, duration of symptoms, and treatment.

13. Peptic disease syndrome is caused by _____ . How can this bacterium tolerate stomach acid?

14. Complete the following table.

DISEASE	ETIOLOGIC AGENT	METHOD OF TRANS-MISSION	SITE OF INFEC-TION	SYMP-TOMS	PREVEN-TION
Mumps					
CMV inclusion disease					
Hepatitis A					
Hepatitis B					
Viral gastro-enteritis					

15. Diagram the life cycle of a human tapeworm.

16. Diagram the life cycle of *Trichinella,* and include humans in the cycle.

17. How can bacterial and protozoan infections of the GI tract be prevented?

18. Look at your diagrams for Questions 15 and 16. Indicate sequences in the life cycles that could be easily broken to prevent these diseases.

Critical Thinking

1. Why is a human infection of trichinosis considered a "dead end" for the parasite?

2. Complete the following table.

DISEASE	CONDITIONS NECESSARY FOR MICROBIAL GROWTH	BASIS FOR DIAGNOSIS	PREVENTION
Staphylococcal food poisoning			
Salmonellosis			

3. Match the following foods with the genus of microorganism most likely to contaminate each.

Beef _____   (a) *Vibrio*
Delicatessen meats _____   (b) *Campylobacter*
Eggs _____   (c) *E. coli* O157:H7
Milk _____   (d) *Listeria*
Oysters _____   (e) *Salmonella*
Pork _____   (f) *Trichinella*

What disease does each microbe cause? How can these diseases be prevented?

4. Which diseases of the gastrointestinal tract can be acquired by swimming in a pool or lake? Why are these diseases not likely to be acquired while swimming in the ocean?

Clinical Applications

1. In New York on April 26, patient A was hospitalized with a two-day history of diarrhea. An investigation revealed that patient B had onset of watery diarrhea on April 22. On April 24, three other people (patients C, D, and E) had onset of diarrhea. All three had vibriocidal antibody titers ≥1:640. In Ecuador on April 20, B bought crabs that were boiled and shelled. He shared crab meat with two people (F and G), then froze the remaining crab in a bag. Patient A returned to New York on April 21 with the frosted bag of crab meat in his suitcase. The bag was placed in a freezer overnight and thawed on April 22 in a double-boiler for 20 minutes. The crab was served 2 hours later in a crab salad. The crab was consumed during a 6-hour period by A, C, D, and E. F and G did not become ill. What is the etiology of this disease? How was it transmitted and how could it have been prevented?

2. 2130 students and employees of a public school system developed diarrheal illness on April 2. The cafeteria served chicken that day. On April 1, part of the chicken was placed in water-filled pans and cooked in an oven for two hours at a dial setting of 177°C. The oven was turned off, and the chicken was left overnight in the warm oven. The remainder of the chicken was cooked for two hours in a steam cooker and then left in the device overnight at the lowest possible setting (43°C). Two serotypes of a gram-negative, oxidase-negative, lactose-negative rod were isolated from 32 patients. What is the pathogen? How could this outbreak have been prevented?

3. Staff in one hospital ward noted an increase in the number of cases of HBV. Fifty cases occurred during a six-month period compared with four cases during the previous six months. Between January 1 and 15, all 50 patients had multiple invasive procedures as shown below:

Transfusion, fingerstick, IV catheter, heparin injection	78%
Transfusion, insulin injection, surgery, fingerstick	64%
Fingerstick, IV catheter, insulin injection, heparin injection	80%
Transfusion, heparin injection, surgery, IV catheter	2%
Heparin injection, IV catheter, insulin injection, surgery	0%

How did the patients acquire HBV? Provide an explanation for the 2% and 0%.

Microbial Diseases of the Urinary and Reproductive Systems

THE URINARY SYSTEM is composed of organs that regulate the chemical composition and volume of the blood and excrete the waste products of metabolism. The **reproductive system** is composed of organs that produce gametes to propagate the species and, in the female, organs that support and nourish the developing embryo. Because the urinary and reproductive systems are closely related anatomically, we will discuss them in the same chapter. In fact, some diseases that affect one system also affect the other, especially in the male. Both systems open to the external environment and thus share portals of entry for microorganisms that can cause disease. The normal microbiota of these and other bodily systems also cause opportunistic infections of the urinary and reproductive systems.

nary bladder, where it is stored prior to elimination from the body. Elimination occurs through the urethra. In the female, the urethra conveys only urine to the exterior. In the male, the urethra is a common tube for both urine and seminal fluid.

The urinary tract has certain characteristics that help prevent infection. Where the ureters enter the urinary bladder, there are valves that normally prevent the backflow of urine to the kidneys. This mechanism helps to shield the kidneys from lower urinary tract infections. In addition, the acidity of normal urine has some antimicrobial properties. The flushing action of urine to the exterior also aids in preventing microorganisms from setting up foci of infection.

STRUCTURE AND FUNCTION OF THE URINARY SYSTEM

LEARNING OBJECTIVE

- *List the antimicrobial features of the urinary system.*

The urinary system consists of two *kidneys*, two *ureters*, a single *urinary bladder*, and a single *urethra* (Figure 26.1). Certain wastes are removed from the blood as it circulates through the kidneys. The wastes are collectively called *urine*. The urine passes through the ureters into the uri-

STRUCTURE AND FUNCTION OF THE REPRODUCTIVE SYSTEM

LEARNING OBJECTIVE

- *Identify the portals of entry for microbes into the reproductive system.*

The **female reproductive system** consists of two *ovaries*, two *uterine (fallopian) tubes*, the *uterus*, the *vagina*, and *external genitals* (Figure 26.2). The ovaries produce female sex hormones and ova (eggs). When an ovum is released in the process called ovulation, it enters a uterine tube;

FIGURE 26.1 Organs of the human urinary system, shown here in the female.

there fertilization may occur if viable sperm are present. The fertilized ovum (zygote) descends the uterine tube and enters the uterus. It implants in the inner wall and remains there while it develops into an embryo and, later, a fetus. At birth, the infant is expelled from the uterus through the vagina. The vagina also serves as a copulatory canal. The external genitals (vulva) include the clitoris, labia, and glands that produce a lubricating secretion during copulation.

The **male reproductive system** consists of two *testes,* a system of *ducts, accessory glands,* and the *penis* (Figure 26.3). The testes produce male sex hormones and sperm. To exit from the body, sperm cells pass through a series of ducts: epididymis, ductus (vas) deferens, ejaculatory duct, and urethra.

NORMAL MICROBIOTA OF THE URINARY AND REPRODUCTIVE SYSTEMS

LEARNING OBJECTIVE

- *Describe the normal microbiota of the upper urinary tract, the male urethra, and the female urethra and vagina.*

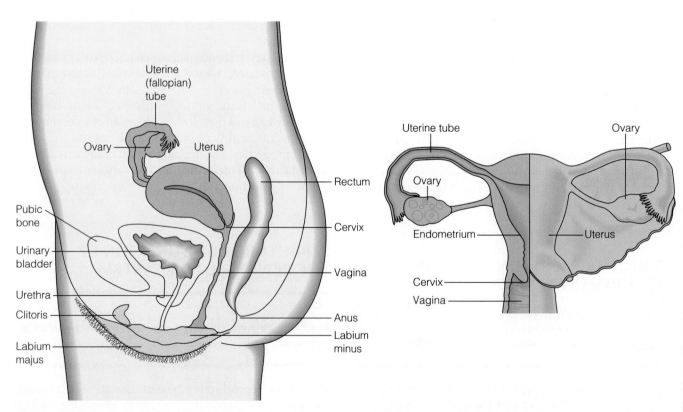

(a) Side view cross-section of female pelvis showing reproductive organs

(b) Front view of female reproductive organs, with left side of the structures shown sectioned

FIGURE 26.2 Female reproductive organs.

FIGURE 26.3 Male reproductive organs. This side view of a male pelvis in cross section shows the reproductive organs.

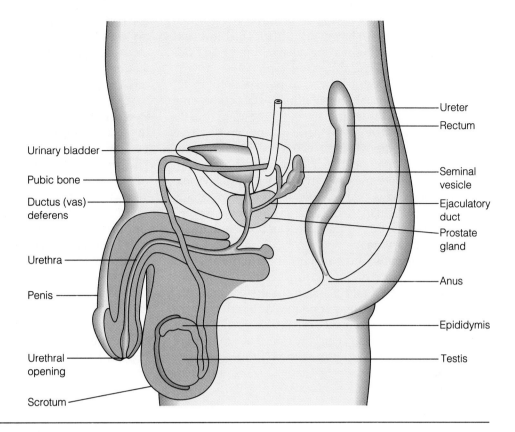

Urinary bladder

Pubic bone

Ductus (vas) deferens

Urethra

Penis

Urethral opening

Scrotum

Ureter

Rectum

Seminal vesicle

Ejaculatory duct

Prostate gland

Anus

Epididymis

Testis

Normal urine is sterile in the urinary bladder and the organs of the upper urinary tract. The female urethra, however, does contain a small resident microbiota. Urine also becomes contaminated with members of the skin microbiota during its passage through the urethra. Therefore, urine collected directly from the bladder has fewer microbial contaminants than voided urine.

In the female genital system, the normal microbiota of the vagina is greatly influenced by sex hormones. For example, within a few weeks after birth, the female infant's vagina is populated by lactobacilli. This population grows because estrogens are transferred from maternal to fetal blood and cause glycogen to accumulate in the cells lining the vagina. Lactobacilli convert the glycogen to lactic acid, and the pH of the vagina becomes acidic. This glycogen–lactic acid sequence provides the conditions under which the acid-tolerant normal microbiota grows in the vagina.

The physiological effects of estrogens diminish several weeks after birth, and other bacteria—including corynebacteria and a variety of cocci and bacilli—become established and dominate the microbiota. As a result, the pH of the vagina becomes more neutral until puberty. At puberty, estrogen levels increase, lactobacilli again dominate, and the vagina again becomes acidic. During the reproductive years, small numbers of other bacteria and fungi become part of the microbiota. In the adult, a disturbance of this ecosystem by an increase in glycogen (caused by oral contraceptives or pregnancy, for example) or elimination of the normal microbiota by antibiotics can lead to

vaginitis, an infection of the vagina. When the female reaches menopause, estrogen levels again decrease, the composition of the microbiota returns to that of childhood, and the pH again becomes neutral.

The male urethra is normally sterile, except for a few contaminating microbes near the external opening.

DISEASES OF THE URINARY SYSTEM

LEARNING OBJECTIVES

- *Describe methods of transmission for urinary and reproductive system infections.*
- *List the microorganisms that cause cystitis, pyelonephritis, and leptospirosis, and name the predisposing factors for these diseases.*
- *Describe the cause and treatment of glomerulonephritis.*

The urinary system normally contains relatively few microbes, but it is often subject to opportunistic infections that can be quite troublesome.

Bacterial Diseases of the Urinary System

Many infections of the urinary system appear to be opportunistic and related to a number of predisposing factors. These factors include disorders of the nervous system, toxemia associated with pregnancy, diabetes mellitus, and obstructions to the flow of urine, such as tumors and kidney stones. Because the anus, from which feces are excreted, is located close to the urethra, especially in females, it is not unusual for the urinary tract to become contaminated with intestinal bacteria. These are usually *Escherichia coli, Proteus* species, and other gram-negative enteric bacteria; staphylococci, enterococci, and pseudomonads are also common in such infections. These *Pseudomonas* infections are, as elsewhere, unusually troublesome to treat. The fungus *Candida albicans* is also an opportunistic agent of urinary tract infections. Moreover, a number of sexually transmitted diseases can cause inflammation in the urinary tract.

Infections usually cause inflammation of the affected tissue. Inflammation of the urethra is called *urethritis;* of the urinary bladder, *cystitis;* and of the ureters, *ureteritis.* The most significant danger from lower urinary tract infections is that they can affect the kidneys (causing *pyelonephritis*) and impair their function. The kidneys are also sometimes affected by systemic bacterial diseases, such as leptospirosis; when the kidneys are thus infected, the pathogens causing these diseases can be found in excreted urine.

Many infections of the urinary tract are of nosocomial origin. In fact, about 35% of all nosocomial infections occur in the urinary tract. Operations on the urinary bladder and prostate gland and catheterization for draining the urinary bladder are procedures that can introduce bacteria into the bladder and ureters. *E. coli* causes more than half of the nosocomial infections of the urinary tract; fecal streptococci, *Proteus, Klebsiella,* and *Pseudomonas* also commonly cause such infections.

Treatment of diseases of the urinary tract depends on diagnostic isolation of the causative organism and determination of its antibiotic sensitivity. The traditional rule of thumb—that urine containing more than 100,000 bacteria per milliliter indicates a symptomatic urinary tract infection—has been modified. Counts as low as 1000/ml of any single bacterial type, or as few as 100/ml of coliforms (intestinal bacteria such as *E. coli*), are now considered an indication of significant infection.

Cystitis

Cystitis is an inflammation of the urinary bladder and is very common, especially among females. Symptoms often include *dysuria* (difficult or painful urination) and *pyuria* (the presence of leukocytes in the urine). The female urethra has many microorganisms around its opening, and it is shorter than the male urethra, so microorganisms can traverse it more readily. Sexual intercourse or careless personal hygiene facilitates such transfer. Contributing factors in females include gastrointestinal system infections and preexisting infections of the vagina, uterus, or urethra. In males, cystitis might be associated with infections of the gastrointestinal system, kidneys, or urethra. Most cases are due to infection by *E. coli.* The second most com-

mon bacterial cause is the coagulase-negative *Staphylococcus saprophyticus* (sap-rō-fit′i-kus).

Trimethoprim-sulfamethoxazole usually clears the infection quickly. Quinolone antibiotics or ampicillin are often successful if resistance is encountered.

Pyelonephritis

In 25% of untreated cases, cystitis may progress to **pyelonephritis,** an inflammation of one or both kidneys. The disease is generally a complication of infection elsewhere in the body. In females, it is often a complication of lower urinary tract infections. The causative agent in about 75% of the cases is *E. coli.* Should pyelonephritis become chronic, scar tissue forms in the kidneys and severely impairs their function. Because pyelonephritis is a potentially life-threatening condition, treatment usually begins with intravenous administration of a broad-spectrum antibiotic such as a second- or third-generation cephalosporin. An extended course of 14 days is recommended.

Leptospirosis

Leptospirosis is primarily a disease of domestic or wild animals, but it can be passed to humans and sometimes causes severe kidney or liver disease. The causative organism is the spirochete *Leptospira interrogans* (in-tėr′rä-ganz), shown in Figure 26.4. *Leptospira* has a characteristic shape—an exceedingly fine spiral wound so tightly that it is barely discernible under the darkfield microscope. Like other spirochetes, *L. interrogans* (so named because the hooked ends suggest a question mark) stains poorly and is difficult to see under a normal light microscope. It is an obligate aerobe that can be readily grown in a variety of artificial media supplemented with rabbit serum.

Animals infected with the spirochete shed the bacteria in their urine for extended periods. Humans become infected by contact with urine-contaminated water or soil, or sometimes with animal tissue. People whose occupations expose them to animals or animal products are most at risk. Usually the pathogen enters through minor abrasions in the skin or mucous membranes. In the U.S.A., dogs and rats are the most common sources. Domestic dogs have a sizable rate of infection; even when immunized they may continue to shed leptospira.

After an incubation period of one or two weeks, headaches, muscle aches, chills, and fever abruptly appear. Several days later the acute symptoms disappear and the temperature returns to normal. A few days later, however, a second episode of fever may occur. In a small number of cases the kidneys and liver become seriously infected (**Weil's disease**); kidney failure is the most common cause of death. Recovery results in a solid immunity. There are usually about 50 cases reported each year in the U.S.A., but because the clinical symptoms are not distinctive, many cases are probably not diagnosed correctly.

Serological tests usually must be done by central reference laboratories, which use suspensions of either killed or living leptospira. If antibodies are present, they cause agglutination of the suspension. Most diagnoses are made

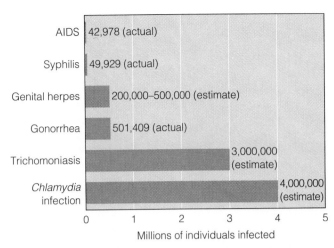

AIDS | 42,978 (actual)
Syphilis | 49,929 (actual)
Genital herpes | 200,000–500,000 (estimate)
Gonorrhea | 501,409 (actual)
Trichomoniasis | 3,000,000 (estimate)
Chlamydia infection | 4,000,000 (estimate)

Millions of individuals infected

FIGURE 26.5 Incidence of sexually transmitted diseases (STDs) in the U.S.A., 1993. The data include reported cases plus the CDC's estimates of unreported cases. [*Source: MMWR* 42(51 & 52) (1/7/94).]

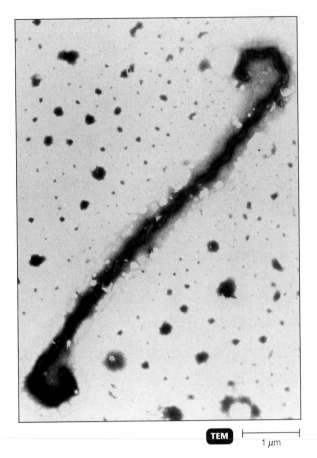

FIGURE 26.4 *Leptospira interrogans*, the causative agent of leptospirosis. The hooked ends are often seen in preparations of this spirochete. Remnants of the threadlike axial filaments are barely visible on the hooks.

TEM |— 1 μm —|

by isolation of the pathogen from the blood or cerebrospinal fluids. Treatment of leptospirosis with antibiotics is often unsatisfactory because the correct diagnosis is delayed, which minimizes the effectiveness of the treatment. The drug of choice is penicillin.

Glomerulonephritis

Glomerulonephritis (Bright's disease) is an inflammation of the glomeruli in the kidneys. The glomeruli are blood capillaries that assist in filtering blood as it passes through the kidneys. Glomerulonephritis is an immune-complex hypersensitivity disease. Most cases are a sequel to infection with β-hemolytic streptococci (*S. pyogenes*). Unlike rheumatic fever (see Chapter 23, p. 563), this includes cases of impetigo, as well as pharyngitis. Soluble streptococcal antigens combine with specific antibodies to form antigen–antibody complexes that interact with complement (see Chapter 16). The complexes are deposited in the glomeruli, where they cause inflammation and kidney damage.

The disease is characterized by fever, high blood pressure, and the presence of protein and red blood cells in the urine. The protein and red blood cells are in the urine because the permeability of the glomeruli increases as a result of inflammation. Although a few patients die of glomerulonephritis and some develop chronic conditions, most recover completely.

DISEASES OF THE REPRODUCTIVE SYSTEM

LEARNING OBJECTIVES

- *List the etiologic agents, symptoms, methods for diagnosis, and treatments for gonorrhea, PID, syphilis, NGU, vaginitis, lymphogranuloma venereum, chancroid, candidiasis, and trichomoniasis.*
- *Discuss the epidemiology of genital herpes and genital warts.*
- *List reproductive system diseases that can cause congenital and neonatal infections, and explain how these infections can be prevented.*

Microbes causing infections of the reproductive system are usually very sensitive to environmental stresses and require intimate contact for transmission.

Bacterial Diseases of the Reproductive System

Most diseases of the reproductive system are transmitted by sexual activity and are therefore called **sexually transmitted diseases (STDs)**. Most of these diseases can be cured with antibiotics, if treated early, and can largely be prevented by the use of condoms. Nevertheless, STDs are a major public health problem in the U.S.A. Figure 26.5 illustrates the relative frequency of the more common STDs in the U.S.A.

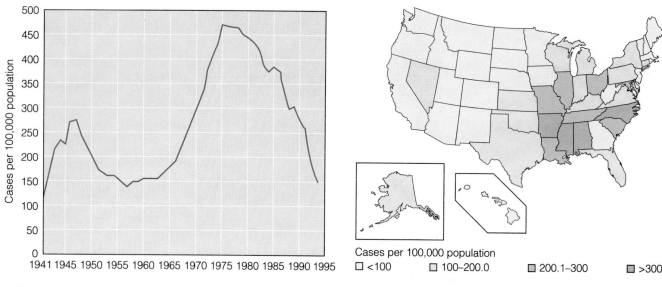

(a) Incidence of gonorrhea in the U.S.A. from 1941 to 1993

Cases per 100,000 population
☐ <100   ☐ 100–200.0   ▨ 200.1–300   ▦ >300

(b) Geographical distribution of cases in 1993

FIGURE 26.6 Incidence and distribution of gonorrhea in the U.S.A. [*Sources:* CDC, *Summary of Notifiable Diseases 1991. MMWR 42(52) (1/7/94).*]

Gonorrhea

By far the most common reportable communicable disease in the U.S.A. is **gonorrhea**, an STD caused by the gram-negative diplococcus *Neisseria gonorrhoeae*. (Health care providers must report "reportable" disease cases to local, state, or federal health agencies.) An ancient disease, gonorrhea was described and given its present name by the Greek physician Galen in A.D. 150. At present, the number of cases is decreasing after steadily and steeply increasing for many years (Figure 26.6). Somewhat more than 600,000 cases in the U.S.A. are reported to the Centers for Disease Control and Prevention (CDC) each year, and the true number of cases is probably much larger. More than 60% of the cases are in the 15- to 24-year-old age group.

To infect, the gonococcus must attach to the mucosal cells of the epithelial wall by means of fimbriae. The organism does not infect the layered squamous cells characteristic of the external skin but invades the spaces separating mucosal cells. Mucosal cells are found in the oral–pharyngeal area, the eyes, joints, and rectum, both male and female genitals, and external genitals of prepubertal females. The invasion sets up an inflammation, and, when leukocytes move into the inflamed area, the characteristic pus formation results. In men, a single unprotected exposure results in infection with gonorrhea from 20% to 35% of the time. Women become infected from 60% to 90% of the time from a single exposure.

Males become aware of a gonorrheal infection by painful urination and discharge of pus-containing material from the urethra (Figure 26.7). About 80% of infected males show these obvious symptoms after an incubation period of only a few days; most others show symptoms in less than a week. In the days before antibiotic therapy, symptoms persisted for weeks. In untreated cases, recovery may eventually occur without complication, but, when complications do occur, they can be serious. In some cases, the urethra is scarred and partially blocked. Sterility can result when the testes become infected or when the ductus (vas) deferens, the tube carrying sperm from the testes, becomes blocked by scar tissue.

In females, the disease is more insidious. Very few women are aware of the early stages of the infection. Later in the course of the disease there might be abdominal pain from complications such as *pelvic inflammatory disease,* which is discussed on page 651.

In both sexes, untreated gonorrhea can become a serious, systemic infection. Complications of gonorrhea can involve the joints, heart (**gonorrheal endocarditis**), meninges (**gonorrheal meningitis**), eyes, pharynx, or other parts of the body. **Gonorrheal arthritis**, which is caused by the growth of the gonococcus in fluids in joints, occurs in about 1% of gonorrhea cases. Joints commonly affected include the wrist, knee, and ankle.

Gonorrheal eye infections occur most often in newborns. If the mother is infected, then the eyes of the infant can become infected as it passes through the birth canal. This condition, **ophthalmia neonatorum**, can result in blindness. Because of the seriousness of this condition and the difficulty of being certain that the mother is free of gonorrhea, antibiotics are placed in the eyes of all newborn infants. If the mother is known to be infected, an intramuscular injection of antibiotic is also administered to the infant. Some sort of prophylaxis is required by law in

FIGURE 26.7 Pus-containing discharge from the urethra of a male with an acute case of gonorrhea.

FIGURE 26.8 A smear of pus from a patient with gonorrhea. The *Neisseria gonorrhoeae* bacteria are contained within phagocytic leukocytes. These gram-negative bacteria are visible here as pairs of cocci. The large stained bodies are the nuclei of the leukocytes.

most states. Gonorrheal infections can also be transferred by hand contact from infected sites to the eyes of adults.

Gonorrheal infections can be acquired at any point of sexual contact; pharyngeal and anal gonorrhea are not uncommon. The symptoms of **pharyngeal gonorrhea** often resemble those of the usual septic sore throat. **Anal gonorrhea** can be painful and accompanied by discharges of pus. In some cases, however, the symptoms are limited to itching.

The increase in sexual activity with a series of partners and the fact that the disease in the female may go unrecognized contributed considerably to the increased incidence of gonorrhea and other STDs during the 1960s and 1970s. The widespread use of oral contraceptives also contributed to the increase. Oral contraceptives tend to increase the moisture content and raise the pH of the vagina while increasing the susceptibility of mucosal cells. Oral contraceptives have often replaced condoms and spermicides, which helped to prevent disease transmission. All these factors predispose the female to infection and therefore have increased the overall incidence of disease. Immunity to reinfection does not result from recovery from gonorrhea.

Antibiotic resistance to the gonococcus has accelerated at an alarming rate in recent years. Penicillin has been an effective treatment for gonorrhea for many years, although the dosages have had to be much increased because of penicillin-resistant bacteria (see the box on p. 650). Ceftriaxone, a third-generation cephalosporin, is currently the drug of choice, but this is subject to change. Because of frequent coinfection with chlamydia, antibiotic treatment should include an effective antichlamydial antibiotic, usually of the tetracycline group.

Gonorrhea in men is diagnosed by finding gonococci in a stained smear of pus from the urethra. The typical gram-negative diplococci within the phagocytic leuko-

cytes are readily identified (Figure 26.8). Gram staining of exudates is not as reliable with women. Usually, a culture is taken from within the cervix (see Figure 26.2b) and grown on special media. Cultivation of the nutritionally fastidious bacterium requires an atmosphere containing carbon dioxide. The gonococcus is very sensitive to adverse environmental influences (desiccation and temperature) and survives poorly outside the body. It even requires special transporting media to keep it viable for short intervals before the cultivation is under way. Cultivation has the advantage of allowing determination of antibiotic sensitivity. Diagnosis of gonorrhea has been aided by the development of an ELISA that detects *N. gonorrhoeae* in urethral pus or on cervical swabs within about three hours with high accuracy. Other rapid tests now available use monoclonal antibodies against antigens on the surface of the gonococcus. Colorimetric results can be seen within 40 minutes. DNA probe tests are available now and are very accurate for identification of clinical isolates from suspected cases.

Nongonococcal Urethritis

Nongonococcal urethritis (NGU), also known as **nonspecific urethritis (NSU)**, refers to any inflammation of the urethra not caused by *Neisseria gonorrhoeae*. Symptoms include painful urination and a watery discharge.

Trauma (such as passage of a catheter) can cause this condition, but at least 40% of the cases of NGU are acquired sexually. In fact, NGU might be the most common sexually transmitted disease in the U.S.A. today. Although it is not a reportable disease and exact data are lacking, the CDC estimates that 3 to 5 million new cases of NGU occur each year.

ANTIBIOTIC-RESISTANT GONOCOCCI AND NATURAL SELECTION

STRAINS OF GONOCOCCI resistant to each of the agents recommended for gonorrhea therapy have emerged. Penicillinase-producing *Neisseria gonorrhoeae* (PPNG) was first reported in 1976. Although the number of reported cases of gonorrhea has declined during the last ten years, the incidence of PPNG infection has increased. In 1990, PPNG cases made up 11% of all gonorrhea cases reported (see the figure). This represents a 244% increase over the PPNG cases reported in 1988. PPNG has increased so rapidly that many health departments have discontinued testing for β-lactamase and penicillins are no longer recommended therapy.

IMPORTATION FROM OTHER COUNTRIES

PPNG accounts for about 30% of all gonococcal isolates in the Philippines and 16% in the Republic of Singapore. (Among the factors thought to contribute to the high prevalence of PPNG in the Philippines and Singapore is the preventive use of oral penicillins, especially by prostitutes.) CDC established the Gonococcal Isolate Surveillance Project (GISP) to monitor antibiotic resistance in *N. gonorrhoeae*. GISP tests isolates from "ports of import" (e.g., New Orleans, Boston, and San Francisco) to provide an early warning for changes in treatment recommendations.

MULTIPLE-DRUG RESISTANCE

Tetracycline-resistant *N. gonorrhoeae* (TRNG) emerged in 1986, and the incidence of TRNG infection in the U.S.A. has risen each year since. It appears that *N. gonorrhoeae* acquired the *tetM* plasmid from streptococci.

In May 1981, a PPNG strain resistant to spectinomycin was reported from Travis Air Force Base, California. The patient had acquired the infection in the Philippines. Spectinomycin was used for five years before resistance was seen; resistance to ceftriaxone was reported in 1988, and to ciprofloxacin in 1991, which raises the question of how long the life span of these and other drugs will be.

In 1983, 56 cases of penicillin-resistant gonococcal infection occurred in North Carolina. These cases represent the first reported outbreak of gonorrhea caused by bacteria that are resistant to penicillin but do not produce penicillinase, a phenomenon called chromosomally mediated resistance (CMRNG). The number of CMRNG cases has increased 200%. CMRNG cells make a penicillin-binding protein instead of penicillinase. This is significant because the isolates are negative in a test for PPNG but are not treatable with penicillin.

Combinations of plasmid and chromosomally mediated resistance to multiple antibiotics have been isolated.

Eradication of endemic antibiotic-resistant gonorrhea is extremely difficult. The once-effective single dose of penicillin or tetracycline is no longer an acceptable treatment. Resistance can result in delays in treatment until an effective drug is found, and patients with inadequately treated gonorrhea are at high risk for complications.

Asked to list the reasons for the emergence of drug-resistant bacterial strains, a spokesman for the National Institutes of Health blamed the unnecessary use of prophylaxis, the use of antibiotics in animal feeds, the availability of over-the-counter antibiotics in many countries, and misuse by health professionals. The indiscriminate use of antibiotics may result in a vast majority of infections being caused by antibiotic-resistant bacteria.

SOURCES: *MMWR* 39:167–169, 284–287, 293 (1990); *MMWR* 42(SS-3) (1993).

Trends in percentage of gonococcal isolates with plasmid-mediated resistance to penicillin and tetracycline, U.S.A., 1988–1991.

Because the symptoms are often mild in males, and females are usually asymptomatic, many cases go untreated. Complications are not common but can be serious. Males may develop inflammation of the epididymis. In females, inflammation of the uterine tubes may cause sterility by scarring the tubes. As many as 60% of such cases may be from chlamydial infection.

Probably the most common pathogen associated with NGU is *Chlamydia trachomatis*. A substantial number of people suffering from gonorrhea are coinfected with *C. trachomatis,* which is responsible for the sexually transmitted disease lymphogranuloma venereum (see p. 655) and trachoma (an eye infection). Like *N. gonorrhoeae, C. trachomatis* can infect the eyes of a newborn during birth.

At one time, culture of chlamydias required cell culture lines not available in all laboratories. This diagnostic picture has changed dramatically with the availability of several new rapid tests for chlamydial infections. Some test kits require no specialized equipment and give reliable results in 30 minutes.

Bacteria other than *N. gonorrhoeae* and *C. trachomatis* can also be implicated in NGU. The next most common cause of urethritis and infertility is probably *Ureaplasma urealyticum* (ū-rē-ä-lit'i-kum). This organism is a member of the mycoplasma group (bacteria without a cell wall). Another mycoplasma, *Mycoplasma hominis* (ho'mi-nis), commonly inhabits the normal vagina but can opportunistically cause uterine tube infection.

Both chlamydia and mycoplasma are sensitive to tetracycline-type antibiotics such as doxycycline. Erythromycin is an acceptable alternative drug.

Pelvic Inflammatory Disease

Pelvic inflammatory disease (PID) is a collective term for any extensive bacterial infection of the female pelvic organs, particularly the uterus, cervix, uterine tubes, or ovaries. Probably 1 million cases of PID occur each year, and 15,000 of them require surgery, such as removal of the infected uterine tube. The number of cases has been increasing in recent years, in part because more states require the reporting of chlamydial infections.

PID is often caused by *N. gonorrhoeae*. Approximately 20% to 30% of untreated gonorrhea cases progress to PID. However, coinfection with chlamydial bacteria is frequent, and they are also an important agent of this type of infection. In later stages of PID, anaerobes and various facultative anaerobes predominate.

The bacteria may attach to spermatozoa and be transported by them to the uterine tubes. Women who use barrier-type contraceptives, especially with spermicides, have a significantly lower rate of PID.

Infection of the uterine tubes **(salpingitis)** is the most serious of these infections. Salpingitis can result in scarring that blocks the passage of ova from the ovary to the uterus (Figure 26.9), possibly causing sterility. One episode of salpingitis causes infertility in 10% to 15% of women; 50% to 75% become infertile after three or more such infections. Tens of thousands of women have been rendered

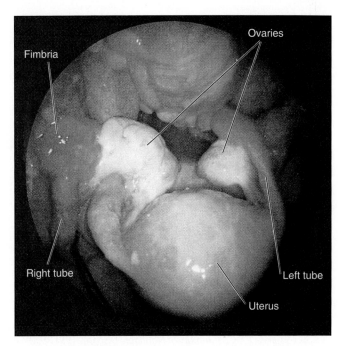

FIGURE 26.9 **Salpingitis.** This photo taken by laparoscope (a specialized endoscope) shows an inflamed left uterine tube and red, swollen fimbriae caused by salpingitis.

sterile by such infections in recent years. Only about 20% of such cases can be reversed surgically.

A blocked uterine tube may cause a fertilized ovum to be implanted in the uterine tube rather than the uterus. This *ectopic,* or *tubal, pregnancy* is a life-threatening condition because of the possibility of rupture of the tube and resulting hemorrhage. The reported cases of ectopic pregnancies have been increasing steadily, which corresponds to the increasing occurrence of PID.

The mortality rate of PID is under 1%; deaths usually occur because misdiagnosis prevented appropriate treatment. The recommended treatment is simultaneous administration of doxycycline and cefoxitin sodium (a cephalosporin). This combination is active against both the gonococcus and chlamydia. Such recommendations are constantly being reviewed.

Syphilis

The earliest reports of **syphilis** date back to the end of the fifteenth century in Europe. This was coincidental with the return of Columbus from the New World and gave rise to a hypothesis that syphilis was introduced to Europe by his men. Another hypothesis is that syphilis existed in Europe and Asia before the fifteenth century but became widespread as urban living became more common. One description of the "Morbus Gallicus" (French disease) seems to clearly describe syphilis as early as 1547 and ascribes its transmission in these terms: ". . . It is taken when one pocky person doth synne in lechery one with another."

In contrast to the change in the number of gonorrhea cases, the number of new syphilis cases in the U.S.A. has remained fairly stable (Fig 26.10). This relative stability as

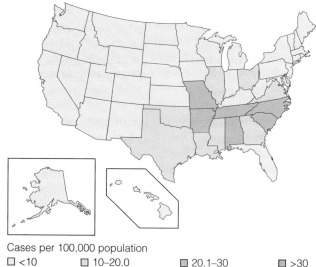

Cases per 100,000 population
☐ <10   ☐ 10–20.0   ▨ 20.1–30   ▨ >30

(a) Incidence of syphilis in the U.S.A. from 1941 to 1993

(b) Geographical distribution of cases in 1993

FIGURE 26.10 Incidence and distribution of primary and secondary syphilis in the U.S.A. [*Sources: CDC, Summary of Notifiable Diseases 1991. MMWR 42(52) (1/7/94).*]

compared with gonorrhea (see Figure 26.6) is remarkable because the epidemiology of the two diseases is quite similar and concurrent infections are not uncommon.

The highest incidence of syphilis is in the 20- to 39-year-old group, which is also the most sexually active. Many states discontinued the requirements for premarital syphilis tests because so few cases were detected. At present, the population most at risk is economically disadvantaged inner-city residents, especially drug-using prostitutes of both sexes.

The causative organism of syphilis is a weakly staining, gram-negative spirochete, *Treponema pallidum*. The spirochete is a thin, tightly coiled helix no more than 20 μm in length (Figure 26.11). The virulent strains of the spirochetes have been successfully cultured only in cell cultures, which is not very useful for routine clinical diagnosis. Separate strains of *T. pallidum* are responsible for certain tropically endemic skin diseases such as **yaws**. These diseases are not sexually transmitted.

Syphilis is transmitted by sexual contact of all kinds, via syphilitic infections of the genitals or other bodily parts. The incubation period averages about three weeks but can range from two weeks to several months. The disease progresses through several recognized stages.

In the *primary stage* of the disease, the initial sign is a small, hard-based **chancre**, or sore, which usually appears at the site of infection (Figure 26.12a). The chancre is painless, and a serous exudate is formed in the center. This fluid is highly infectious, and examination with a dark-field microscope shows many spirochetes. In a few weeks, this lesion disappears. None of these symptoms causes any distress. In fact, many women are entirely unaware of the chancre, which is often on the cervix. In males, the chancre sometimes forms in the urethra and is not visible. Sero-

FIGURE 26.11 *Treponema pallidum,* **the causative agent of syphilis.** (The organisms are made more visible in this micrograph by the use of special silver stain.)

logical diagnostic tests are positive in about 80% of patients in the primary stage. During this stage, bacteria enter the bloodstream and lymphatic system, which distribute them widely in the body.

Several weeks after the primary stage (the exact length of time varies), the disease enters the *secondary stage,* characterized mainly by skin rashes of varying appearance (Figure 26.12b). Other symptoms often observed are the loss of patches of hair, malaise, and mild fever. The rash is widely distributed on the skin and is also found in the mucous membranes of the mouth, throat, and cervix. At this stage, the lesions of the rash contain many spirochetes and are very infectious. Dentists or other medical workers coming into contact with fluid from these lesions can become infected by the spirochete entering through minute breaks in the skin. Such nonsexual transmission is possible, but the organisms do not survive long on environmental surfaces and are very unlikely to be transmitted via such objects as toilet seats. Serological tests for syphilis are almost always positive at this stage.

The symptoms of secondary syphilis usually subside after a few weeks, and the disease enters a *latent period*. During this period, there are no symptoms of the disease, although serological tests are positive. After two to four years of latency, the disease is not normally infectious, except for transmission from mother to fetus. The majority of cases do not progress beyond the latent stage, even without treatment. Some resistance to reinfection is observed during this period.

Because the symptoms of primary and secondary syphilis are not disabling, people may enter the latent period without having received medical attention. In less than half of these cases, the disease reappears in a *tertiary stage*. This stage occurs only after an interval of many years—usually at least ten—after the onset of the latent phase. *T. pallidum* has an outer layer of lipids that stimulates remarkably little effective immune response, especially from cell-destroying complement reactions. It has been described as a Teflon® pathogen. Nonetheless, most of the symptoms of tertiary syphilis are probably due to the body's immune reactions, of a cell-mediated nature, to surviving spirochetes. Inflammatory responses by phagocytic cells such as neutrophils and macrophages also contribute. This causes lesions called *gummas,* rubbery masses of tissue that appear in many organs and sometimes on the external skin (Figure 26.12c). Although many of these lesions are not very harmful, some can cause extensive tissue damage, such as perforation of the palate (roof of the mouth), which interferes with speech. The cardiovascular system can be affected, most seriously by a weakening of the aorta. If the central nervous system is damaged, a loss of motor control occurs (*tabes dorsalis*); the brain may be affected (*paresis*), with personality changes, blindness, and seizures. Few, if any, organisms are found in the lesions of the tertiary stage, and they are not considered very infectious. Today, cases of syphilis allowed to progress to this stage have become rare.

One of the most distressing and dangerous forms of syphilis, called **congenital syphilis**, is transmitted across the placenta to the unborn fetus. Damage to mental development and other neurological symptoms are among the more serious consequences. This type of infection is most common when pregnancy occurs during the latent stage of syphilis. A pregnancy during the primary or secondary stage is likely to produce a stillbirth.

Serological testing for primary syphilis is not completely reliable. Often, the best method for diagnosis is an examination of exudates removed from punctured lymph nodes in the affected area or from lesions. A darkfield microscope is necessary because the organisms do not stain well and are only about 0.2 μm in diameter, a size that approaches the limits of resolution for a brightfield microscope. Figure 26.11 shows the spirochete in a special silver-impregnated stain.

The **Venereal Disease Research Laboratory (VDRL)** slide flocculation test is probably the most widely used

(a) Chancre of primary stage on a male in the genital area

(b) Lesions of secondary syphilis rash on a palm; any surface area of the body may be afflicted with such lesions

(c) Gummas of tertiary stage on the back of an arm; gummas such as these are rarely seen today, in the era of antibiotics

FIGURE 26.12 Characteristic lesions associated with various stages of syphilis.

serological screening test. The **rapid plasma reagin (RPR)** card test, which is similar, is also in common use. Both of these tests are nonspecific, in that they do not detect antibodies produced against the spirochete itself but detect *reagin-type antibodies*. Reagin-type antibodies are apparently a response to lipid materials that the body forms as an indirect response to infection by the spirochete. The antigen used in such precipitation-type slide tests is thus not the syphilis spirochete but is an extract of beef heart

TABLE 26.1 Characteristics of the Most Common Types of Vaginitis and Vaginosis

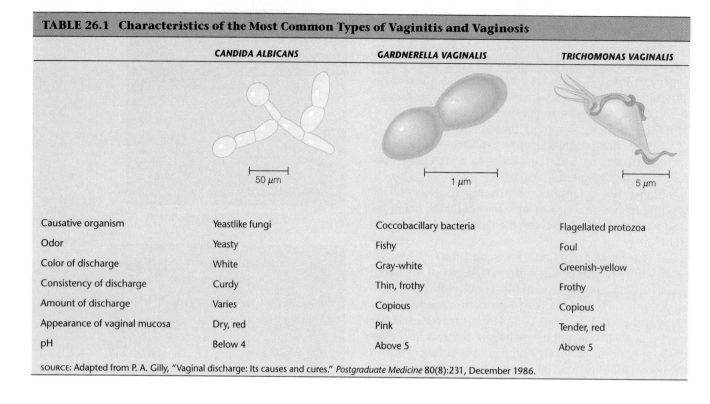

	CANDIDA ALBICANS	GARDNERELLA VAGINALIS	TRICHOMONAS VAGINALIS
Causative organism	Yeastlike fungi	Coccobacillary bacteria	Flagellated protozoa
Odor	Yeasty	Fishy	Foul
Color of discharge	White	Gray-white	Greenish-yellow
Consistency of discharge	Curdy	Thin, frothy	Frothy
Amount of discharge	Varies	Copious	Copious
Appearance of vaginal mucosa	Dry, red	Pink	Tender, red
pH	Below 4	Above 5	Above 5

SOURCE: Adapted from P. A. Gilly, "Vaginal discharge: Its causes and cures." *Postgraduate Medicine* 80(8):231, December 1986.

(cardiolipin) that seems to contain lipids similar to those that stimulated the reagin-type antibody production. These antibodies are formed from three to five weeks after infection, so the VDRL and RPR tests are of only limited usefulness in detecting primary syphilis. Each of these serological tests will detect only about 70% to 80% of primary syphilis cases, but they detect 99% of secondary syphilis cases.

Most slide tests produce a significant percentage of false-positive reactions for syphilis. (These are most likely to occur in people with immune disorders.) Therefore, positive reactions should be confirmed by a more exacting test based on spirochete-type antigens. These tests often have less than 1% false-positive reaction. One such test is the **fluorescent treponemal antibody absorption (FTA–ABS) test.** This is an indirect immunofluorescence test in which a preparation of an avirulent strain of *T. pallidum* (one that can be cultured) is allowed to react with a sample of a patient's serum on a microscope slide (see Figure 18.13b). The spirochetes are detected by their fluorescent glow when struck by ultraviolet light. A disadvantage of the FTA–ABS test is that it remains positive even after syphilis has been successfully treated. The nonspecific reagin-based tests tend to become negative upon successful treatment and are therefore more useful for determining the effect of therapy.

Benzathine penicillin, a long-acting formulation that remains effective in the body for about two weeks, is the usual antibiotic treatment of syphilis. The serum concentrations achieved by this formulation are low, but the spirochete has remained very sensitive to this treatment.

Penicillin therapy is especially effective during the primary stage, but antibiotics are generally not effective in tertiary syphilis, probably because the organisms are usually not present.

For penicillin-sensitive people, a number of other antibiotics, such as erythromycin and the tetracyclines, have also proved effective. Antibiotic therapy aimed at gonorrhea and other infections is not likely to eliminate syphilis as well because such therapy for those other diseases is usually administered for too short a period for the slow-growing spirochete. Similarly, penicillin directed at syphilis does not reach a high enough concentration to successfully treat gonorrheal infections.

Gardnerella Vaginosis

Inflammation of the vagina due to infection, **vaginitis,** is most commonly caused by one of three organisms: the fungus *Candida albicans* (kan'did-ä al'bi-kans), the protozoan *Trichomonas vaginalis* (trik-ō-mōn'as va-jin-al'is), or the bacterium *Gardnerella vaginalis,* which is a small, pleomorphic gram-variable rod (Table 26.1). Vaginal infections not attributed to *Trichomonas* or *Candida* were once termed *nonspecific vaginitis.* Now most of these cases are attributed to the presence of *G. vaginalis.* There is no sign of inflammation (and therefore the term *vaginosis* is preferred to vaginitis). The condition is something of an ecological mystery. The population of *G. vaginalis* is many times higher than in uninfected women, the population of *Lactobacillus* bacteria is correspondingly decreased, and there is an overgrowth by anaerobic bacteria such as *Bacteroides* spp. It is not exactly known whether the decrease in acid-producing lactobacilli, and the higher pH characteristic of the condition, is a cause or an effect of the overall microbial changes. The condition is probably sexually transmitted, although the organism is a common inhabitant of the vaginas of asymptomatic women. There is no corresponding disease condition in males, but *Gardnerella* is often present in their urethras.

Gardnerella vaginosis occurs when the vaginal pH is above 5 (normally it would be less than 4). The condition is characterized by a pronounced fishy odor of a frothy vaginal discharge, which is usually copious. Diagnosis is based on the fishy odor, the vaginal pH, and the microscopic observation of *clue cells* in the discharge. These clue cells are sloughed-off vaginal epithelial cells covered with bacteria, mostly *G. vaginalis* (Figure 26.13). Treatment is primarily by metronidazole, a drug that eradicates the anaerobes essential to continuation of the disease but allows the normal lactobacilli to repopulate the vagina.

Lymphogranuloma Venereum

There are a number of STDs that are uncommon in the U.S.A. but are seen frequently in the tropical areas of the world. For example, *Chlamydia trachomatis,* the cause of trachoma (see Chapter 21) and a major cause of NGU, is also responsible for **lymphogranuloma venereum (LGV)**, a disease found in tropical and near-tropical regions. It is apparently caused by a strain of *C. trachomatis* that is invasive and tends to infect lymphoid tissue. In the U.S.A. there are usually 200–400 cases per year, mostly in the southeastern states.

After a latent period from 7 to 12 days, a small lesion appears at the site of infection, usually on the genitals. The lesion ruptures and heals without scarring. From one week to two months later, the microorganisms invade the lymphatic system, and the regional lymph nodes become enlarged and tender. Suppuration (a discharge of pus) may also occur. The inflammation of the lymph nodes results in scarring, which occasionally obstructs the lymph vessels. This scarring leads to edema of the genital skin and massive enlargement of the external genitals in males. In females, rectal narrowing results from involvement of the lymph nodes in the rectal region. Enlarged nodes may be slow to subside, even after successful antibiotic therapy. These conditions can eventually require surgery.

For diagnosis, pus can be aspirated from infected lymph nodes. When infected cells are properly stained with an iodine preparation, the clumped, intracellular chlamydias can be seen as inclusions. The isolated organisms can also be grown in cell culture or in embryonated eggs. Several serological tests are available, but all are troubled with frequent false-positive reactions. The drug of choice for treatment is doxycycline.

Chancroid (Soft Chancre)

The STD known as chancroid occurs more frequently in tropical areas, where it is seen more often than syphilis. The number of reported cases in the U.S.A. has been in-

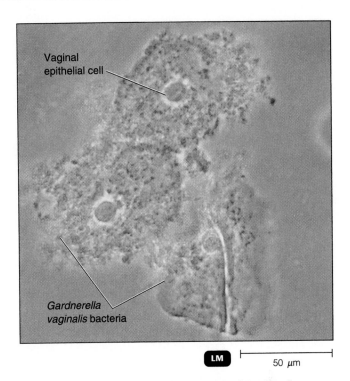

creasing steadily, from 840 to 5000 during a recent nine-year period. Almost all occur in the states of New York, Texas, California, Florida, and Georgia.

In **chancroid (soft chancre)**, a swollen, painful ulcer that forms on the genitals involves an infection of the adjacent lymph nodes. About a week elapses between infection and appearance of the symptoms. Infected lymph nodes in the groin area sometimes even break through and discharge pus to the surface. These ulcers are highly infective, as are the other genital lesions that occur. Lesions might also occur on such diverse areas as the tongue and lips. Because chancroid is so seldom seen by some physicians, they might confuse it with primary syphilis or genital herpes infections. The causative organism is *Hemophilus ducreyi* (dü-krā′ē), a small gram-negative rod that can be isolated from exudates of lesions. The recommended antibiotics are erythromycin or ceftriaxone.

Viral Diseases of the Reproductive System

Viral diseases of the reproductive system are difficult to treat effectively, and so they represent an increasing health problem.

Genital Herpes

A much publicized STD is genital herpes, usually caused by *herpes simplex virus type 2.* (The herpes simplex virus occurs as either type 1 or type 2.) Herpes simplex virus type 1

FIGURE 26.13 Clue cells. In the diagnosis of *Gardnerella vaginalis* infections, it is essential to find clue cells. These cells are vaginal epithelial cells coated with bacilli, mostly *G. vaginalis.*

Vaginal epithelial cell

Gardnerella vaginalis bacteria

LM   50 µm

FIGURE 26.14 Vesicles of genital herpes on a penis.

is primarily responsible for the common cold sore or fever blister (see Chapter 21). Herpes simplex virus type 2, and sometimes type 1, causes an STD called **genital herpes.** The lesions appear after an incubation period of about one week or less and cause a burning sensation. After this, vesicles appear (Figure 26.14). In both sexes, urination can be painful and walking is quite uncomfortable; the patient is even irritated by clothing. Usually, the vesicles heal in a couple of weeks.

The vesicles contain fluid that is infectious. Condoms provide uncertain protection because the vesicles on women are usually on the external genitals (seldom on the cervix or within the vagina), and the vesicles on men are often at the base of the penis.

One of the most distressing characteristics of genital herpes is the possibility of recurrences. As in other herpes infections, such as cold sores or chickenpox–shingles, the virus enters a latent state in nerve cells. The site of latency is in nerve ganglia near the base of the spine. There is considerable individual variation in latency. Some people have frequent attacks; for others, recurrence is rare. Reactivation appears to be triggered by a number of factors, including menstruation, emotional stress or illness (especially if accompanied by fever, a factor that is also involved in appearance of cold sores), and perhaps just scratching the affected area. About 88% of patients with type 2 virus and about 50% of those with type 1 virus will have recurrences. The average patient will have recurrences every three or four months. If a person is going to experience recurrences, the first will usually appear within about six months after the primary infection.

In the U.S.A., the incidence of genital herpes has increased so much that there are probably about 30 million infected individuals. Females are less likely to be aware that they have the disease. This lack of awareness can have serious effects because of *neonatal herpes.* The virus can cross the placental barrier and affect the fetus if the mother suffers from an acute infection early in pregnancy.

How frequently this happens is not known, but for practical purposes the fetus is considered infected if the virus can be grown from the amniotic fluid. Infection can result in spontaneous abortion or serious damage to the fetus, such as mental retardation and defective sight and hearing.

If the fetus remains free of the virus, it needs to be protected from infection in the birth canal. Women with a history of genital herpes are often asymptomatic during late pregnancy. Tests to detect asymptomatic shedding of viruses are not very reliable, so screening for them is of very limited use. It is hoped that DNA probes will change this picture. If there are obvious viral lesions at delivery time, a cesarean section is probably wise. The operation should be done before the fetal membrane ruptures and the viruses spread to the uterus. However, the risk of infection is low in instances of mothers with recurrent but presently asymptomatic infections. In these cases, vaginal deliveries are often possible. Maternal antibodies are apparently responsible.

Herpes infection of the newborn is most likely to have serious consequences when the mother acquires the primary herpes infection during the pregnancy. In this situation, the newborn is ten times more likely to be infected than in instances of recurrent asymptomatic herpes. Therefore, any pregnant woman without a history of genital herpes should avoid sexual contact with anyone who might be carrying the virus.

At present there is no cure for genital herpes, although research on its prevention and treatment is intensive. Acyclovir, administered orally, is proving to be reasonably effective in alleviating the symptoms of a primary outbreak. There is some relief of pain and other symptoms and slightly faster healing. Continuous administration for several months has been found to lower the chances of recurrence; however, it has no effect on the frequency of recurrence after it is discontinued.

Genital Warts

Warts are an infectious disease; since 1907 it has been known that they are caused by viruses known as papillomaviruses. It is probably less well known that warts can be transmitted sexually and that this is an increasing problem. An estimated half million new cases are reported each year. There are more than 60 serotypes of papillomaviruses, and certain serotypes tend to be linked with certain forms of genital warts. Morphologically some warts are extremely large and "warty" in appearance, with multiple fingerlike projections; others tend to be relatively smooth or flat (Figure 26.15). The incubation time is usually a matter of weeks or months. The greatest danger from genital warts is their connection to cancer. Some 90% of genital warts are caused by viruses of serotypes that do not cause cancer, but there are several serotypes that are frequently associated with a progression to cancer. In women this is usually cervical cancer, and in men it is usually can-

FIGURE 26.15 Genital warts on a vulva.

bicans is the cause of thrush (oral candidiasis). It is also responsible for occasional cases of NGU in males and for **vulvovaginal candidiasis,** which is the most common cause of vaginitis. About 75% of all women experience at least one episode.

The lesions of vulvovaginal candidiasis resemble those of thrush but produce more irritation; severe itching; a thick, yellow, cheesy discharge; and a yeasty odor. *C. albicans* is an opportunistic pathogen. Predisposing conditions include use of oral contraceptives and pregnancy, which cause an increase of glycogen in the vagina (see the discussion of the normal vaginal microbiota on p. 645). Diabetes and treatment with broad-spectrum antibiotics are also associated with the occurrence of *C. albicans* vaginitis.

Vulvovaginal candidiasis is diagnosed by microscopic identification of the fungus in scrapings of lesions and by isolation of the fungus in culture. Treatment consists of topical application of azole antifungals such as clotrimazole or miconazole.

cer of the penis. Genital warts in women are much more likely to be precancerous than those in men. It is hoped that serological typing of warts will eventually become a routine aid in determining which are the most dangerous, but at present this is expensive and limited to relatively few laboratories. Treatments of warts were discussed previously in Chapter 21, and none of them is completely satisfactory.

AIDS

In concluding this section on viral STDs, we should recall that AIDS is a viral disease that is frequently transmitted by sexual contact. However, its effect is on the immune system, so it was discussed in Chapter 19 (p. 482) and the boxes on pages 354–355 and 484–485. It is important to remember that the lesions resulting from many of the diseases of bacterial and viral origin facilitate transmission of HIV.

Fungal Disease of the Reproductive System

The fungal disease described under this grouping is now well known, primarily because of the advertising devoted to nonprescription treatments for its cure.

Candidiasis

Candida albicans is a yeastlike fungus that often grows on mucous membranes of the mouth, intestinal tract, and genitourinary tract (see Table 26.1 and Figure 21.15). Infections are usually a result of opportunistic overgrowth when the competing microbiota is suppressed by antibiotics or other factors. As we discussed in Chapter 21, *C. al-*

Protozoan Disease of the Reproductive System

The only STD caused by a protozoan affects only women. Although fairly common, it is not widely known.

Trichomoniasis

The protozoan *Trichomonas vaginalis* is frequently a normal inhabitant of the vagina in females and of the urethra in many males (see Table 26.1 and Figure 12.19b). If the normal acidity of the vagina is disturbed, the protozoan may overgrow the normal microbial population of the genital mucosa and cause **trichomoniasis.** (Males rarely have any symptoms as a result of the presence of the organism.) It is often accompanied by a coinfection with gonorrhea. In response to the protozoan infection, the body accumulates leukocytes at the infection site. The resulting purulent discharge is profuse, greenish-yellow, and characterized by a foul odor. This discharge is accompanied by irritation and itching. This type of vaginitis is usually sexually transmitted. It may be associated with premature births.

Diagnosis is easily made by microscopic examination and identification of the organisms in the discharge. The organism can be found in semen or urine of male carriers. Treatment is by oral metronidazole, administered to both sexual partners, which readily clears the infection. The organism can also be isolated and grown on laboratory media.

•   •   •

The major microbial diseases of the urinary and reproductive systems are summarized in Table 26.2.

TABLE 26.2 Microbial Diseases Associated with the Urinary and Reproductive Systems

DISEASE	CAUSATIVE AGENT	MODE OF TRANSMISSION	TREATMENT
Bacterial Diseases of the Urinary System			
Cystitis (bladder infection)	*Escherichia coli, Staphylococcus saprophyticus*	Opportunistic infections	Trimethoprim-sulfamethoxazole
Pyelonephritis (kidney infection)	Most common is *E. coli*	From systemic or lower urinary tract infections	Broad spectrum cephalosporins
Leptospirosis (kidney infection)	*Leptospira interrogans*	Direct contact with infected animals, urine of infected animals, or contaminated water	Penicillin
Glomerulonephritis (kidney infection)	Immune-complex hypersensitivity reaction to certain strains of *Streptococcus pyogenes*	Sequel to infection by particular strains of *S. pyogenes* in another part of the body	None, autoimmune condition
Bacterial Diseases of the Reproductive System			
Gonorrhea	*Neisseria gonorrhoeae*	Usually sexual contact	Third-generation cephalosporins
Nongonococcal urethritis (NGU)	*Chlamydia* or other bacteria, including *Mycoplasma hominis* and *Ureaplasma urealyticum*	Sexual contact or opportunistic infections	Tetracyclines or erythromycin
Pelvic inflammatory disease (PID)	*N. gonorrhoeae, Chlamydia trachomatis*	A complication of STDs	Doxycycline, cefoxitin sodium
Syphilis	*Treponema pallidum*	Usually sexual contact	Benzathine penicillin
Gardnerella vaginosis	*Gardnerella vaginalis*	Opportunistic pathogen, vaginal pH of 5 to 6	Metronidazole
Lymphogranuloma venereum (LGV)	*Chlamydia trachomatis*	Usually sexual contact	Doxycycline
Chancroid (soft chancre)	*Hemophilus ducreyi*	Usually sexual contact	Erythromycin or ceftriaxone
Viral Diseases of the Reproductive System			
Genital herpes	Herpes simplex virus type 2, and occasionally type 1	Usually sexual contact	Acyclovir may alleviate symptoms
Genital warts	Papillomavirus	Direct contact	Removal by various mechanical and chemical means; interferon injection
Fungal Disease of the Reproductive System			
Candidiasis	*Candida albicans*	Opportunistic pathogen, may be transmitted by sexual contact	Clotrimazole, miconazole
Protozoan Disease of the Reproductive System			
Trichomoniasis (usually vaginitis)	*Trichomonas vaginalis*	Usually sexual contact	Metronidazole

STUDY OUTLINE

INTRODUCTION (p. 643)

1. The urinary system regulates the chemical composition of the blood and excretes nitrogenous waste.
2. The reproductive system produces gametes for reproduction and, in the female, supports the growing embryo.
3. The urinary and reproductive systems are closely related anatomically, and both open to the external environment.
4. Microbial diseases of these systems can result from infection from an outside source or from opportunistic infection by members of the normal body microbiota.

STRUCTURE AND FUNCTION OF THE URINARY SYSTEM (p. 643)

1. Urine is transported from the kidneys through ureters to the urinary bladder and is eliminated through the urethra.
2. Valves prevent urine from flowing back to the urinary bladder and kidneys.
3. The flushing action of urine and the acidity of normal urine have some antimicrobial value.

STRUCTURE AND FUNCTION OF THE REPRODUCTIVE SYSTEM (pp. 643–644)

1. The female reproductive system consists of two ovaries, two uterine tubes, the uterus, the vagina, and external genitals.
2. The vagina functions as the birth canal and copulatory canal.
3. The male reproductive system consists of two testes, ducts, accessory glands, and the penis.
4. Seminal fluid leaves the male body through the urethra.

NORMAL MICROBIOTA OF THE URINARY AND REPRODUCTIVE SYSTEMS (pp. 644–645)

1. The urinary bladder and upper urinary tract are sterile under normal conditions.
2. The female urethra has a small resident microbiota.
3. Lactobacilli dominate the vaginal microbiota during reproductive years.
4. The male urethra is normally sterile.

DISEASES OF THE URINARY SYSTEM

(pp. 645–647)

Bacterial Diseases of the Urinary System

(pp. 645–647)

1. Opportunistic gram-negative bacteria from the intestines often cause urinary tract infections, especially in females.

2. Predisposing factors of urinary tract infections include nervous system disorders, toxemia, diabetes mellitus, and obstructions to the flow of urine.
3. Urethritis, cystitis, and ureteritis are terms describing inflammations of tissues of the lower urinary tract.
4. Pyelonephritis can result from lower urinary tract infections or from systemic bacterial infections.
5. Diagnosis and treatment of urinary tract infections depend on the isolation and antibiotic sensitivity testing of the etiologic agents.
6. About 35% of all nosocomial infections occur in the urinary system. *E. coli* causes more than half of these infections.
7. More than 1000 bacteria of one species per milliliter of urine or 100 coliforms per milliliter of urine indicates an infection.

Cystitis (p. 646)

1. Inflammation of the urinary bladder, or cystitis, is common in females.
2. Microorganisms at the opening of the urethra and along the length of the urethra, careless personal hygiene, and sexual intercourse contribute to the high incidence of cystitis in females.

Pyelonephritis (p. 646)

1. Inflammation of the kidneys, or pyelonephritis, is usually a complication of lower urinary tract infections.
2. About 75% of pyelonephritis cases are caused by *E. coli*.

Leptospirosis (pp. 646–647)

1. The spirochete *Leptospira interrogans* is the cause of leptospirosis.
2. Leptospirosis is characterized by chills, fever, headache, and muscle aches.
3. Serological identification is difficult because there are many serotypes of *L. interrogans*.

Glomerulonephritis (p. 647)

1. Glomerulonephritis, or Bright's disease, is an immune-complex disease occurring as a sequel to a β-hemolytic streptococcal infection.
2. Antigen–antibody complexes are deposited in the glomeruli and cause inflammation and damage.

DISEASES OF THE REPRODUCTIVE SYSTEM

(pp. 647–658)

Bacterial Diseases of the Reproductive System

(pp. 647–655)

1. Most diseases of the reproductive system are sexually transmitted diseases (STDs).

2. Most STDs can be prevented by the use of condoms and are treated with antibiotics.

Gonorrhea (pp. 648–649)

1. *Neisseria gonorrhoeae* causes gonorrhea.
2. Gonorrhea is the most common reportable communicable disease in the U.S.A.
3. *N. gonorrhoeae* attaches to mucosal cells of the oral–pharyngeal area, genitals, eyes, joints, and rectum by means of fimbriae.
4. Symptoms in males are painful urination and pus discharge. Blockage of the urethra and sterility are complications of untreated cases.
5. Females might be asymptomatic unless the infection spreads to the uterus and uterine tubes (see pelvic inflammatory disease).
6. Gonorrheal endocarditis, gonorrheal meningitis, and gonorrheal arthritis are complications that can affect both sexes if gonorrheal infections are untreated.
7. Ophthalmia neonatorum is an eye infection acquired by infants during passage through the birth canal of an infected mother.
8. Gonorrhea is diagnosed by Gram stain, ELISA, or DNA probe.

Nongonococcal Urethritis (pp. 649–651)

1. Nongonococcal urethritis (NGU), or nonspecific urethritis, is any inflammation of the urethra not caused by *N. gonorrhoeae.*
2. About 40% of NGU cases are transmitted sexually.
3. Most cases of NGU are caused by *Chlamydia trachomatis.*
4. Symptoms of NGU are often mild or lacking, although salpingitis and sterility may occur.
5. *C. trachomatis* can be transmitted to infants' eyes at birth.
6. *Ureaplasma urealyticum* and *Mycoplasma hominis* also cause NGU.

Pelvic Inflammatory Disease (p. 651)

1. Extensive bacterial infection of the female pelvic organs, especially of the reproductive system, is called pelvic inflammatory disease (PID).
2. PID is caused by *N. gonorrhoeae, Chlamydia trachomatis,* and other bacteria that gain access to the uterine tubes. Infection of the uterine tubes is called salpingitis.
3. PID can result in blockage of the uterine tubes and sterility.

Syphilis (pp. 651–654)

1. Syphilis is caused by *Treponema pallidum,* a spirochete that has not been cultured in vitro. Laboratory cultures are grown in cell cultures.
2. *T. pallidum* is transmitted by direct contact and can invade intact mucous membranes or penetrate through breaks in the skin.
3. The primary lesion is a small, hard-based chancre at the site of infection. The bacteria then invade the blood and lymphatic system, and the chancre spontaneously heals.
4. The appearance of a widely disseminated rash on the skin and mucous membranes marks the secondary stage. Spirochetes are present in the lesions of the rash.
5. The patient enters a latent period after the secondary lesions spontaneously heal.
6. At least ten years after the secondary lesion, tertiary lesions called gummas can appear on many organs.
7. Congenital syphilis, resulting from *T. pallidum* crossing the placenta during the latent period, can cause neurological damage in the newborn.
8. *T. pallidum* is identifiable through darkfield microscopy of fluid from primary and secondary lesions.
9. Many serological tests, such as VDRL, RPR, and FTA–ABS, can be used to detect the presence of antibodies against *T. pallidum* during any stage of the disease.

Gardnerella Vaginosis (pp. 654–655)

1. Vaginitis can be caused by *Candida albicans, Trichomonas vaginalis,* or *Gardnerella vaginalis.* Vaginosis is infection without inflammation.
2. Diagnosis of *G. vaginalis* is based on increased vaginal pH, fishy odor, and the presence of clue cells.

Lymphogranuloma Venereum (p. 655)

1. *C. trachomatis* causes lymphogranuloma venereum (LGV), which is primarily a disease of tropical and subtropical regions.
2. The initial lesion appears on the genitals and heals without scarring.
3. The bacteria are spread in the lymph system and cause enlargement of lymph nodes, obstruction of lymph vessels, and edema of the genital skin.
4. The bacteria are isolated and identified from pus taken from infected lymph nodes.

Chancroid (Soft Chancre) (p. 655)

1. Chancroid, a swollen, painful ulcer on the mucous membranes of the genitals or mouth, is caused by *Hemophilus ducreyi.*

Viral Diseases of the Reproductive System
(pp. 655–657)

Genital Herpes (pp. 655–656)

1. Herpes simplex type 2 virus causes an STD called genital herpes.
2. Symptoms of the infection are painful urination, genital irritation, and fluid-filled vesicles.
3. Neonatal herpes is contracted during fetal development or birth. It can result in neurological damage or infant fatalities.
4. The virus might enter a latent stage in nerve cells. Vesicles reappear following trauma and hormonal changes.
5. Genital herpes is associated with cervical cancer.
6. The drug acyclovir has proven effective in treating the

symptoms of genital herpes; however, it does not cure the disease.

Genital Warts (pp. 656–657)

1. Papillomaviruses cause warts.
2. The papillomaviruses that cause genital warts have been associated with cancer of the cervix or penis.

AIDS (p. 657)

1. AIDS is a sexually transmitted disease of the immune system (see Chapter 19, pp. 479–485).

Fungal Disease of the Reproductive System
(p. 657)

Candidiasis (p. 657)

1. *Candida albicans* causes NGU in males and vulvovaginal candidiasis in females.
2. Vulvovaginal candidiasis is characterized by lesions that produce itching and irritation.
3. Predisposing factors for candidiasis are pregnancy, diabetes, tumors, and broad-spectrum antibacterial chemotherapy.
4. Diagnosis is based on observation of the fungus and its isolation from lesions.

Protozoan Disease of the Reproductive System
(p. 657)

Trichomoniasis (p. 657)

1. *Trichomonas vaginalis* causes trichomoniasis when the pH of the vagina increases.
2. Diagnosis is based on observation of the protozoa in purulent discharges from the site of infection.

STUDY QUESTIONS

Review

1. List the members of the normal microbiota of the urinary system, and show their habitats in Figure 26.1.

2. List the normal microbiota of the genital system, and show their habitats in Figures 26.2 and 26.3.

3. How are urinary tract infections transmitted?

4. Explain why *E. coli* is frequently implicated in cystitis in females. List some predisposing factors for cystitis.

5. Name one organism that causes pyelonephritis. What are the portals of entry for organisms that cause pyelonephritis?

6. What is glomerulonephritis? How is it transmitted? How is it treated?

7. Complete the following table.

DISEASE	CAUSATIVE AGENT	SYMPTOMS	METHOD OF DIAGNOSIS	TREATMENT
Gardnerella vaginosis				
Gonorrhea				
Syphilis				
PID				
NGU				
Lymphogranuloma venereum				
Chancroid				

8. Leptospirosis is a kidney infection of humans and other animals. How is this disease transmitted? What types of activities would increase one's exposure to this disease? What is the etiology?

9. Describe the symptoms of genital herpes. What is the etiologic agent? When is this infection least likely to be transmitted?

10. Name one fungus and one protozoan that can cause genital system infections. What symptoms would lead you to suspect these infections?

11. List the genital infections that cause congenital and neonatal infections. How can transmission to a fetus or newborn be prevented?

Critical Thinking

1. The tropical skin disease called yaws is transmitted by direct contact. Its causative agent, *Treponema pallidum pertenue,* is indistinguishable from *T. pallidum.* The appearance of syphilis in Europe coincides with the return of Columbus from the New World. How might *T. pallidum pertenue* have evolved into *T. pallidum* in the temperate climate of Europe?

2. Why can frequent douching be a predisposing factor to bacterial vaginosis, vulvovaginal candidiasis, or trichomoniasis?

3. *Neisseria* is cultured on Thayer–Martin media, consisting of chocolate agar and nystatin, incubated in a 5% CO_2 environment. How is this selective for *Neisseria?*

4. The list below is a key to selected microorganisms that cause genitourinary infections. Complete this key by listing genera discussed in this chapter in the blanks that correspond to their respective characteristics.

Gram-negative bacteria
   Spirochete
      Aerobic
      Anaerobic   _____
   Coccus
      Oxidase-positive   _____
   Bacillus, nonmotile
      Requires X factor   _____
      Gram-positive wall   _____
   Obligate intracellular parasite   _____
   Lacking cell wall
      Urease-positive   _____
      Urease-negative   _____
Fungus
   Pseudohyphae   _____
Protozoa
   Flagella   _____
No organism observed/cultured from patient   _____

Clinical Applications

1. A previously well 19-year-old female was admitted to a hospital after two days of nausea, vomiting, headache, and neck stiffness. Cerebrospinal fluid and cervical cultures showed gram-negative diplococci in leukocytes; a blood culture was negative. What disease did she have? How was it probably acquired?

2. A 28-year-old woman was admitted to a Wisconsin hospital with a one-week history of arthritis of the left knee. Four days later, a 32-year-old man was examined for a two-week history of urethritis and a swollen, painful left wrist. A 20-year-old woman seen in a Philadelphia hospital had pain in the right knee, left ankle, and left wrist for three days. Pathogens cultured from synovial fluid or urethral culture were gram-negative diplococci that re-

quired proline to grow. Antibiotic sensitivity tests gave the following results:

ANTIBIOTIC	MIC TESTED (μg/ml)	SUSCEPTIBLE MIC (μg/ml)
Cefoxitin sodium	0.5	≤ 2
Penicillin	8	≤ 0.06
Spectinomycin	64	≤ 32
Tetracycline	4	≤ 0.25

What is the pathogen, and how is this disease transmitted? Which of the antibiotics should be used for treatment? What is the evidence that these cases are related?

3. Using the following information, determine what the disease is and how the infant's illness might have been prevented.

May 11: A 23-year-old woman has her first prenatal examination. She is 4½ months pregnant. Her VDRL results are negative.

June 6: The woman returns to her physician complaining of a labial lesion of a few days' duration. A biopsy is negative for malignancy, and herpes test results are negative.

July 1: The woman returns to her physician because the labial lesion continues to cause some discomfort.

Sept. 15: The baby's father has multiple penile lesions and a generalized body rash.

Sept. 25: The woman delivers her baby. Her RPR is 1:32 and the infant's is 1:128.

Oct. 1: The woman takes her infant to a pediatrician because the baby is lethargic. She is told the infant is well and not to worry.

Oct. 2: The baby's father has a persistent body rash and plantar and palmar rashes.

Nov. 8: The infant becomes acutely ill with pneumonia and is hospitalized. The admitting physician finds clinical and radiologic signs of osteochondritis.

Environmental and Applied Microbiology

A technician prepares bags of dried wheat bran coated with bacteria for use in bioremediation: cleaning up oil and toxic chemicals. The bran will absorb the chemicals and make them more accessible, so that the bacteria can then break them down into less harmful compounds. The technician's coverings protect the bacteria from contamination.

Soil and Water Microbiology

M ANY PEOPLE ASSOCIATE BACTERIA with disease or food spoilage and therefore regard microorganisms as harmful things to be avoided or controlled. Actually, harmful microorganisms are only a very small fraction of the total microbial population. Bacteria and other microorganisms are in fact essential to the maintenance of life on Earth. In previous chapters we focused mainly on disease-causing capabilities of microorganisms. In this chapter you will learn about many of the positive functions microbes perform in the environment.

We can find microbes, especially bacteria, in the most widely varied habitats on Earth. They are found in the frozen Antarctic and in boiling hot springs. Aircraft flying miles above the Earth can recover microbes from the thin atmosphere. Exploration of the deepest ocean sediments shows large numbers of bacteria living there, in eternal darkness and subjected to incredible pressures (see the box in Chapter 6, p. 149). Microorganisms are found in the clearest mountain streams flowing from a melting glacier and in waters nearly saturated with salts, such as those of the Dead Sea (see the box in Chapter 5, p. 131).

Most microorganisms in the environment are *saprophytes,* which obtain their energy from organic matter. Other microorganisms use photosynthesis directly. One group of microorganisms extracts energy from the oxidation of inorganic matter such as ammonia and sulfur— these are important contributors to many mineral cycles essential to life on earth. Some microbes are *parasites*— they derive their nutrients and their ability to reproduce from other microbes. *Bdellovibrio* bacteria, for example,

prey on other bacteria (see the box in Chapter 3, p. 57).

Mutualism, in which two organisms benefit, is exemplified by the association of a fungus and an alga in lichen (discussed later in this chapter). Some protozoa have even adapted to being hosts to single-celled algae that reside within their cytoplasm; the protozoa furnish protection, and the algae provide additional nutrients for the host.

Commensalism, in which one organism benefits from the relationship without affecting the other, is common among microbes. For example, the few bacteria that can degrade polymers (such as cellulose) release simpler compounds (such as glucose) that neighboring scavengers quickly utilize. Some organisms participate in *cometabolism.* These organisms might be incapable of directly metabolizing a highly resistant chemical, such as a pesticide, for example. However, if provided with some primary source of energy such as glucose, the organisms can degrade the pesticide.

Microorganisms must take advantage of any niches found in their environment. Situations only minute distances apart may be utterly different in their supplies of oxygen, sunlight, or nutrients, and may have a much different pH. However, some microorganisms are able to modify their environments. For example, some minimize their competition by creating an acidic environment that is inhospitable to competing organisms. Some microbial population or another fills almost every imaginable ecological niche. Many of these niches are in soil and water. The microorganisms that inhabit soil and water are the subject of this chapter.

SOIL MICROBIOLOGY AND CYCLES OF THE ELEMENTS

LEARNING OBJECTIVES

- *Explain how the components of soil affect soil microbiota.*
- *Outline the carbon and nitrogen cycles, and explain the roles of microorganisms in these cycles.*
- *Identify a role for microorganisms in removal of pollutants from soil.*

Soil microbiology, compared to the microbiology of industry or human disease, attracts little attention. But soil, which has been described as rock on its way to the seas, is pulsating with life. Billions of living organisms, some microscopic but including comparatively huge insects and earthworms, form a vibrant living community in the soil. We can think of soil as a "biologic fire." A leaf falling from a tree is consumed by it, and its organic matter is released as carbon, nitrogen, and other chemical compounds. These enter the cycles of the elements, without which life on Earth as we know it could not exist.

The Components of Soil

Soil is a complex mixture of solid inorganic matter (rocks and minerals), organic matter, water, air, and living organisms.

Minerals

The weathering of rocks adds such elements as silicon, aluminum, and iron to the soil. Calcium, magnesium, potassium, sodium, phosphorus, and small amounts of other elements that are essential for life are also present in the mineral component of soil.

Organic Matter

The organic matter in soil consists of carbohydrates, proteins, lipids, and other materials. Organic matter makes up from 2% to 10% of most agriculturally important soils. Swamps and bogs, by contrast, have a higher content of organic matter, up to 95% in some peat bog soils. The waterlogged soil of swamps and bogs is an anaerobic environment, and the microbial decomposition of organic matter occurs very slowly there.

All organic matter in soil is derived from the remains of microorganisms, plants, and animals, their waste products, and the biochemical activities of various microorganisms. A great portion of the organic matter is of plant origin—mostly dead roots, wood and bark, and fallen leaves. A second source of organic matter is the vast numbers of bacteria, fungi, algae, protozoa, and viruses, which can total billions per gram of fertile soil. These organisms break down organic substances, producing and maintaining a continuous supply of inorganic substances that

plants and other organisms require for growth. Much of the organic matter in soil is ultimately decomposed to inorganic substances, such as ammonia, water, carbon dioxide, and various compounds of nitrate, phosphate, and calcium.

A considerable part of the organic matter in soil occurs as **humus.** This dark material consists of partially decomposed organic matter, chiefly materials that are relatively resistant to decay.

The addition of organic matter, either completely or partially decomposed, is essential to the fertility of soil. Moreover, these spongy organic materials loosen the soil and thereby prevent the formation of heavy crusts and increase the pore spaces. This addition of pore spaces in turn increases aeration and water retention.

Water and Gases

Microbial life in soil depends on the availability of some water, which dissolves many inorganic and organic constituents in soil and makes them available to the living inhabitants. However, if the water content is too high, atmospheric oxygen cannot easily penetrate the soil, and this can be a factor in limiting growth efficiency and the types of organisms that flourish. Generally, because of respiration by soil organisms, soil gases contain a high proportion of carbon dioxide and a low proportion of oxygen. Most of these soil gases exist in pore spaces between soil particles or are dissolved in the water.

Organisms

Fertile soil contains a great many animals, ranging from nematodes (minute worms) to larger forms, including insects, millipedes, centipedes, spiders, slugs, snails, earthworms, mice, moles, gophers, and reptiles. Most of these animals are beneficial, in that they promote some mechanical movement of the soil and thereby help to keep the soil loose and open. All soil organisms also contribute to the organic matter of soils in the form of their waste products and eventual remains.

Soil also contains the root systems of higher plants and enormous numbers of microorganisms.

Soil Microbiota. The soil is one of the main reservoirs of microbial life (Table 27.1). A good agricultural soil the size of an American football field usually contains a microbial population in its upper few inches of a size approaching the weight of a cow eating grass on that field. The metabolic capabilities of these vast numbers of microorganisms are probably about 100,000 times those of the grazing cow. However, measurement of the carbon dioxide from soil and other evidence indicate that these microorganisms exist in near starvation and at low reproductive rates. When usable nutrients are added to soil, the microbial populations and their activity rapidly increase until the nutrients are depleted, and then the microbial activity returns to the lower, baseline levels.

TABLE 27.1 Microorganisms per Gram of Typical Garden Soil at Various Depths

DEPTH (CM)	BACTERIA	ACTINOMYCETES*	FUNGI	ALGAE
3–8	9,750,000	2,080,000	119,000	25,000
20–25	2,179,000	245,000	50,000	5000
35–40	570,000	49,000	14,000	500
65–75	11,000	5000	6000	100
135–145	1400	——	3000	——

*Filamentous bacteria
SOURCE: Adapted from M. Alexander, *Introduction to Soil Microbiology*, 2nd ed. New York: Wiley, 1977.

The most numerous organisms in soil are bacteria. Typical garden soil has millions of bacteria in each gram. As Table 27.1 shows, the population is highest in the top few centimeters of the soil and declines rapidly with depth. The populations are usually estimated via plate counts on nutrient media, and the actual numbers are probably greatly underestimated by this method. No single nutrient medium or growth condition can possibly meet all the myriad nutrient and other requirements of soil microorganisms.

Actinomycetes are bacteria, but they are usually considered separately in enumerations of soil populations because they differ so greatly from our conventional concept of bacteria. They typically form long filaments resembling those of molds, but with diameters of bacterial dimensions. This filamentous growth habit is a considerable advantage for a microorganism in soil. Under dry conditions, the filament can bridge the gap between one particle of soil and another. This type of growth also maximizes surface area for a given weight and may give the actinomycetes a nutritional advantage. These organisms are found in soil in very large numbers and produce a gaseous substance called *geosmin*, which gives the soil its characteristic musty odor. The **biomass** (total mass of living organisms in a given volume) of the actinomycetes is probably about equal to the biomass of all other bacteria in soil. Interest in actinomycetes was greatly stimulated by the discovery that some genera, particularly *Streptomyces*, produce valuable antibiotics.

Fungi are found in soil in much smaller numbers than actinomycetes and other bacteria. Because plate counts of fungal colonies are largely based on the germination of asexual spores, the actual fungal population is probably lower than such counts indicate. The biomass of fungi probably equals the biomass of all bacteria, including actinomycetes. This large biomass occurs because fungal mycelia are much larger than bacterial cells. Molds greatly outnumber yeasts in soil.

Algae and cyanobacteria are found even in dry desert soils. They sometimes form visible accumulations on the surfaces of moist soils. As photosynthetic organisms, they are found mainly on the soil surface, where sunlight is available. However, significant numbers of algae and cyanobacteria are also found more than 50 cm below the surface. The environmental contribution of these microorganisms is significant only in special cases. For example, fixation of atmospheric nitrogen (discussed shortly) by some species of cyanobacteria in grasslands and tundra regions and after rainfall in deserts adds significantly to the fertility of soils in those regions.

Protozoa are plentiful in soil. They become dormant as cysts under adverse conditions. When microbial numbers increase, many protozoa function as predators, and their numbers tend to rise and fall with the bacterial populations.

Microbial Pathogens in Soil. Human pathogens, which are mostly parasites, generally find the soil an alien, hostile environment. Even relatively resistant enteric pathogens, such as *Salmonella* species, have been observed to survive for only a few weeks or months when introduced into soil. Most human pathogens that can survive in soil are endospore-forming bacteria. For example, endospores of *Bacillus anthracis,* which causes anthrax in animals, can survive in certain soils for decades before finally germinating when ingested by grazing animals. Disposing of the body of an animal infected with anthrax requires care so that the soil is not seeded with the endospores from the dead animal.

Clostridium tetani (the causative agent of tetanus), *Clostridium botulinum* (the causative agent of botulism), and *Clostridium perfringens* (the causative agent of gas gangrene) are all endospore-forming pathogens whose normal habitat is the soil. From the soil they are introduced into foods or wounds, where they grow and produce toxins.

Certain helminthic pathogens, such as the hookworm, spend part of their life cycle as larval worms in the soil. These larvae penetrate the skin of humans who come into contact with them. Because the larval development takes place only in warm, moist soil, these diseases are limited to tropical or subtropical areas.

Pathogens of plants are much more likely to inhabit soil. Most plant pathogens are fungi, mainly because fungi are capable of growing at the low moisture level typical of plant surfaces. Many of the rusts, smuts, blights, and wilts affecting plants are caused by fungi that spend part of their life cycle in soil. Bacterial diseases of plants are less common, but bacterial pathogens in soil do cause plant diseases, especially rots, in which bacterial enzymes hydrolyze plant tissue.

Viruses also cause plant diseases. The mechanism by which they penetrate the thick plant cell walls involves a sap-feeding insect.

Some of the microorganisms found in soil are insect pathogens, which are potentially useful for pest control. *Bacillus thuringiensis,* for example, is a soil bacterium that is pathogenic to the larvae of many insects; it is now used

FIGURE 27.1 The carbon cycle.
On a global scale, the return of CO_2 to the atmosphere by respiration closely balances its removal by fixation. However, the burning of wood and fossil fuels adds more CO_2 to the atmosphere; as a result, the amount of atmospheric CO_2 is steadily increasing.

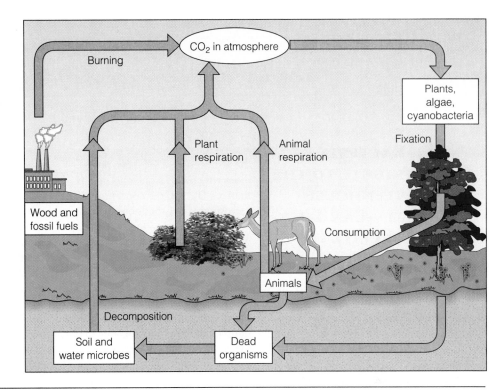

Microorganisms and Biogeochemical Cycles

Perhaps the most important role of soil microorganisms is their participation in **biogeochemical cycles**—the recycling of certain chemical elements so that they can be used over and over again. Among these elements are carbon, nitrogen, sulfur, and phosphorus. Were it not for the activities of microorganisms in the various biogeochemical cycles, essential elements would become depleted and all life would cease.

The Carbon Cycle

The first biogeochemical cycle we will consider is the carbon cycle (Figure 27.1). As you know, all organic compounds contain carbon. Most of the inorganic carbon used in the synthesis of organic compounds comes from the carbon dioxide in the atmosphere. Some carbon dioxide is also dissolved in water.

In photosynthesis—the first step in the carbon cycle—carbon dioxide is incorporated, or *fixed,* into organic compounds by such photoautotrophs as cyanobacteria, green plants, algae, and green and purple sulfur bacteria. In the next step in the cycle, chemoheterotrophs consume the organic compounds—animals eat photoautotrophs, espe-

cially green plants, and may in turn be eaten by other animals. Thus, as the organic compounds of the photoautotrophs are digested and resynthesized, the carbon atoms of carbon dioxide are transferred from organism to organism up the food chain.

Some of the organic molecules are used by chemoheterotrophs, including animals, to satisfy their energy requirements. When such energy is released through the process of respiration, carbon dioxide is released into the atmosphere. This carbon dioxide immediately becomes available to start the cycle over again. However, much of the carbon remains within the organisms until they excrete wastes or die. When the organisms die, the organic compounds of their bodies are deposited in the soil and are decomposed by microorganisms, principally by bacteria and fungi. During this decomposition, carbon dioxide is returned to the atmosphere. Although the carbon dioxide of the atmosphere makes up only about 0.03% of the atmospheric gases, it is essential for the growth of plants and algae.

Carbon is stored in rocks, such as limestone ($CaCO_3$), and is dissolved as carbonate ions (CO_3^{2-}) in oceans; it is also stored in fossil forms, such as coal and petroleum. Burning such fossil fuels releases carbon dioxide. The amount of carbon dioxide in the atmosphere has been increasing as a result of widespread use of fossil fuels. Some scientists believe that this is a problem, because the increase of atmospheric carbon dioxide may be causing a warming of the Earth known as the greenhouse effect (see the box on p. 668).

in their control (see Chapter 11, p. 284). A number of other insect pathogens, including viruses and fungi, are under investigation as biological control agents.

Microbiology in the News

BACTERIA CONTRIBUTE TO THE GREENHOUSE EFFECT—CAN THEY ALLEVIATE IT?

WATER, METHANE (CH_4) and carbon dioxide (CO_2) in the Earth's atmosphere absorb infrared radiation that would otherwise radiate out into space and reradiate it back to the Earth. This phenomenon is important for maintaining the temperature of the Earth, for without these gases, the average temperature of the Earth would be about $-18°C$.

However, scientists have been warning for years that the increased burning of fossil fuels is raising the CO_2 content of the atmosphere, causing a greenhouse effect that could eventually warm the atmosphere enough to melt the polar ice caps and flood major coastal cities. In addition, the more pessimistic scientists have warned that the warmer air could cause chronic droughts in major crop-growing areas, resulting in worldwide famine.

Other, more optimistic, scientists claim that nature takes care of the problem automatically. The extra CO_2 stimulates the growth of plants, which fix CO_2 into organic molecules, thus reestablishing a balance. Dr. Boyd Strain, a botanist at Duke University, suggests that much of the increase in agricultural productivity after the Industrial Revolution came not from the use of fertilizers and pesticides but from the increasing abundance of CO_2 in the atmosphere.

This increased plant growth may also be the cause of increased CH_4 in the atmosphere. CH_4 is five times more effective as a greenhouse gas than is CO_2, and the concentration of atmospheric CH_4 has been increasing at an alarming rate—more than doubling in the last 100 years. Additional plant growth also results in a corresponding increase in the amount of decaying plant matter. According to Paul Guthrie, an atmospheric scientist at the NASA Goddard Space Flight Center, the dangerous increase in the CH_4 content of the air may be largely a result of the decay of the extra plant matter whose growth was stimulated by the increased amount of CO_2. When dead plants decay in anaerobic pockets of organic debris in forests or grasslands, in the sediment of lakes and bays, or in rice paddies, bacterial fermentation breaks down the organic matter, releasing CO_2 and hydrogen (H_2). Anaerobic archaeobacteria known as *methanogens* then combine these into CH_4 and water (see the figure).

There are aerobic methane-oxidizing bacteria (methanotrophs) where the limnetic and profundal zones of lakes meet, and anaerobic methane oxidizers are present in the benthos. At present, little is known about these methanotrophic bacteria. In a natural ecosystem, the actions of the methanotrophs and methanogens are expected to be in balance. Why then is CH_4 increasing at such an alarming rate? Are the activities of humans, such as pollution and the use of fertilizers, inhibiting the growth of methanotrophs? While some scientists are measuring the increasing CH_4 in the atmosphere, other scientists must begin isolating methanotrophs and determining their numbers and optimum growth conditions. With more information, we may be able to use the methanotrophs for greenhouse-gas bioremediation.

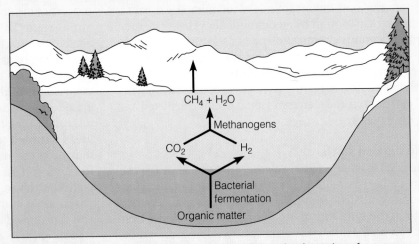

Methane is generated in the sediments of lakes and bays by the action of methanogens.

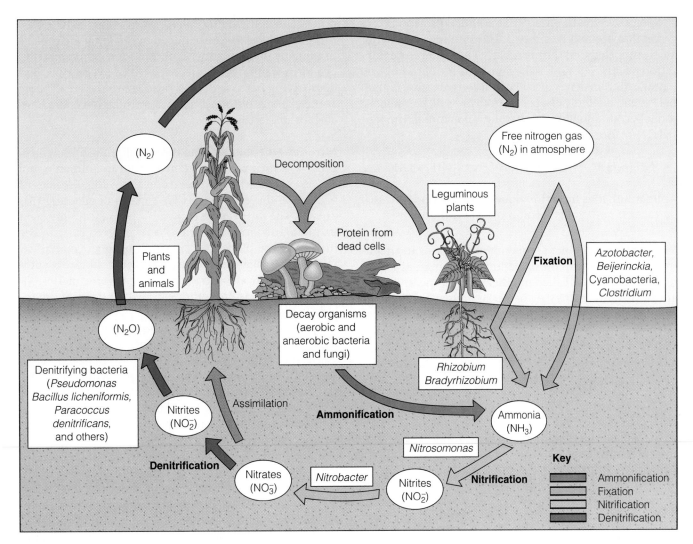

FIGURE 27.2 **The nitrogen cycle.** In general, nitrogen in the atmosphere goes through fixation, nitrification, and denitrification. Nitrates assimilated into plants and animals after nitrification go through decomposition, ammonification, and then nitrification again. All of these processes except assimilation are performed by bacteria.

The Nitrogen Cycle

The nitrogen cycle is shown in Figure 27.2. Nitrogen is needed by all organisms for their synthesis of protein, nucleic acids, and other nitrogen-containing compounds. Molecular nitrogen (N_2) makes up almost 80% of the Earth's atmosphere; the atmosphere over every acre (the area of an American football field from one goal line to the opposite 10-yard line) of fertile soil contains more than 30,000 tons of nitrogen. Despite the abundance of nitrogen as a molecular gas, however, no eucaryote is able to make direct use of it. Instead, the nitrogen must be fixed (combined) with other elements, such as oxygen and hydrogen. The resulting compounds, such as nitrate ion (NO_3^-) and ammonium ion (NH_4^+), are then used by autotrophic organisms. The chemical and physical forces operating in the soil, water, and air, together with the activities of specific microorganisms, are important to the conversion of nitrogen to usable forms.

Ammonification. Almost all the nitrogen in the soil exists in organic molecules, primarily in proteins. When an organism dies, the process of microbial decomposition results in the hydrolytic breakdown of proteins into amino acids. In a process called deamination, the amino groups of amino acids are removed and converted into ammonia (NH_3) (see Figure 27.2). This release of ammonia is called **ammonification**. Ammonification, brought about by numerous bacteria and fungi, can be represented as follows:

$$\text{Proteins from dead cells and waste products} \xrightarrow{\text{Microbial decomposition}} \text{Amino acids}$$

$$\text{Amino acids} \xrightarrow{\text{Microbial ammonification}} \text{Ammonia } (NH_3)$$

Microbial growth releases extracellular proteolytic enzymes that decompose proteins. The resulting amino acids are transported into the microbial cells, where ammonification occurs. The fate of the ammonia produced by ammonification depends on soil conditions. Because ammonia is a gas, it rapidly disappears from dry soil, but in moist soil it becomes solubilized in water, and ammonium ions (NH_4^+) are formed:

$$NH_3 + H_2O \longrightarrow NH_4^+OH \longrightarrow NH_4^+ + OH^-$$

Ammonium ions from this sequence of reactions are used by bacteria and plants for amino acid synthesis.

Nitrification. The next sequence of reactions in the nitrogen cycle involves the oxidation of the ammonium ion to nitrate, a process called **nitrification**. Living in the soil are autotrophic nitrifying bacteria, such as those of the genera *Nitrosomonas* and *Nitrobacter*. These organisms obtain energy by oxidizing ammonia or nitrite. In the first stage, *Nitrosomonas* oxidizes ammonium to nitrites:

$$\underset{\text{Ammonium ion}}{NH_4^+} \xrightarrow{\textit{Nitrosomonas}} \underset{\text{Nitrite ion}}{NO_2^-}$$

In the second stage, such organisms as *Nitrobacter* oxidize nitrites to nitrates:

$$\underset{\text{Nitrite ion}}{NO_2^-} \xrightarrow{\textit{Nitrobacter}} \underset{\text{Nitrate ion}}{NO_3^-}$$

Plants tend to use nitrate as their source of nitrogen for protein synthesis because nitrate is highly mobile in soil and is more likely to encounter a plant root than ammonium. Ammonium ions would actually make a more efficient source of nitrogen, as they require less energy to incorporate into protein, but these positively charged ions are usually bound to negatively charged clays in the soil, whereas the negatively charged nitrate ions are not bound.

Denitrification. The nitrate ions resulting from nitrification are fully oxidized and no longer contain any biologically usable energy. However, they can be used as electron acceptors by microbes metabolizing other organic energy sources in the absence of atmospheric oxygen (see *anaerobic respiration* in Chapter 5). This process, called **denitrification**, can lead to loss of nitrogen to the atmosphere, especially as nitrogen gas. Denitrification can be represented as follows:

$$\underset{\text{Nitrate ion}}{NO_3^-} \longrightarrow \underset{\text{Nitrite ion}}{NO_2^-} \longrightarrow \underset{\substack{\text{Nitrous}\\\text{oxide}}}{N_2O} \longrightarrow \underset{\substack{\text{Nitrogen}\\\text{gas}}}{N_2}$$

Pseudomonas species appear to be the most important group of bacteria in denitrification in soils. A number of

other genera, including *Paracoccus* (pär-ä-kok′kus), *Thiobacillus*, and *Bacillus*, also include species that are capable of carrying out denitrification reactions. Denitrification occurs in waterlogged soils depleted of oxygen. Because denitrifying bacteria deliver nitrogen to the atmosphere at the expense of soil nitrates, denitrification is unfavorable for soil fertility.

Nitrogen Fixation. During the initial phase of the nitrogen cycle, nitrogen gas is converted into ammonia in a process called **nitrogen fixation**. Only a few species of bacteria and cyanobacteria are capable of enacting this process. The nitrogenase enzyme responsible for nitrogen fixation is anaerobic, so it probably evolved early in the history of the planet, before the atmosphere contained molecular oxygen and before nitrogen-containing compounds were available from decaying organic matter. Nitrogen fixation is brought about by two types of organisms, nonsymbiotic and symbiotic.

Nonsymbiotic (free-living) *nitrogen-fixing bacteria* are found in particularly high concentrations in the *rhizosphere*, the region where the soil and roots make contact—especially in grasslands. Among the nonsymbiotic bacteria that can fix nitrogen are aerobic species such as *Azotobacter*. These aerobic organisms apparently shield the nitrogenase enzyme from oxygen by, among other things, having a very high rate of oxygen use that minimizes the diffusion of oxygen into the cell, where the enzyme is located.

Another nonsymbiotic obligate aerobe that fixes nitrogen is *Beijerinckia* (bī-yĕ-rink′ē-ä). Some anaerobic bacteria, such as certain species of *Clostridium*, also fix nitrogen. The bacterium *Clostridium pasteurianum* (pas-tyĕr-ē-a′num), an obligately anaerobic, nitrogen-fixing microorganism, is a prominent example. Other nonsymbiotic nitrogen-fixing bacteria include certain species of the facultatively anaerobic *Klebsiella*, *Enterobacter*, and *Bacillus* and the anaerobic photoautotrophic *Rhodospirillum* (rō-dō-spī-ril′um) and *Chlorobium*.

There are many species of aerobic, photosynthesizing cyanobacteria that fix nitrogen. Because their energy supply is independent of carbohydrates in soil or water, they are especially useful suppliers of nitrogen to the environment. Cyanobacteria usually carry their nitrogenase enzymes in specialized structures called *heterocysts* that provide anaerobic conditions for fixation.

Most of the nonsymbiotic nitrogen-fixing organisms are capable of fixing large amounts of nitrogen under laboratory conditions. However, in the soil there is usually a shortage of usable carbohydrates to supply the energy needed for the reduction of nitrogen to ammonia, which is then incorporated into protein. Nevertheless, these nitrogen-fixing bacteria make important contributions to the nitrogen economy of such areas as grasslands, forests, and the arctic tundra.

Symbiotic nitrogen-fixing bacteria serve an even more important role in plant growth for crop production. Members of the genera *Rhizobium*, *Bradyrhizobium*, and others

FIGURE 27.3 Formation of a root nodule. Members of the nitrogen-fixing genera *Rhizobium* and *Bradyrhizobium* form these nodules on legumes. This mutualistic association is beneficial to both the plant and the bacteria.

Pea plant

Root

Nodules

Root hairs

1 Rhizobia attach to root hair

Rhizobia

4 Enlarged root cells form a nodule

Infection thread

Bacteroids

3 Bacteria change into bacteroids; packed root cells enlarge

2 An infection thread is formed, through which bacteria enter root cells

infect the roots of leguminous plants, such as soybeans, beans, peas, peanuts, alfalfa, and clover. (These agriculturally important plants are only a few of the thousands of known leguminous species, many of which are bushy plants or small trees found in poor soils in many parts of the world.) Rhizobia are specially adapted to particular leguminous plant species, on which they form **root nodules** (Figure 27.3). ① The bacteria attach to the root of the host legume, usually at a root hair. ② In response to the bacterial infection, an indentation forms in the root hair, and an *infection thread* synthesized by the plant passes down the root hair into the root itself. The bacteria follow this infection thread and enter the cells in the root. ③ Inside these cells, the bacteria alter their morphology into larger forms called *bacteroids* that eventually pack the plant cell. ④ The root cells are stimulated by this infection to form a tumorlike nodule of bacteroid-packed cells.

Nitrogen is then fixed by a symbiotic process of the plant and the bacteria. The plant furnishes anaerobic conditions and growth nutrients for the bacteria, and the bacteria fix nitrogen to be incorporated into plant protein. Millions of tons of nitrogen are fixed in this way each year.

There are similar examples of symbiotic nitrogen fixation in nonleguminous plants, such as alder trees. These trees are among the first to appear in forests after fires or glaciation. The alder tree is symbiotically infected with an actinomycete (*Frankia*) and forms nitrogen-fixing root nodules. About 50 kg of nitrogen can be fixed each year by the growth of one acre of alder trees; the trees thus make a valuable addition to the forest economy.

Another important contribution to the nitrogen economy of forests is made by **lichens**, which are a symbiosis between a fungus and an alga or a cyanobacterium (see Figure 12.14). When one symbiont is a nitrogen-fixing

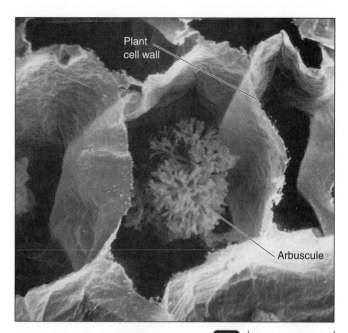

(a) Endomycorrhiza (vesicular-arbuscular mycorrhiza). A fully developed arbuscule of an endomycorrhiza in a plant cell. (The term arbuscule means little bush.) As the arbuscule decomposes, it releases nutrients for the plant.

FIGURE 27.4 **The *Azolla*–cyanobacteria symbiosis.**
A section through the floating sporangium of the freshwater fern *Azolla* shows the resident filamentous nitrogen-fixing cyanobacteria, *Anabaena azollae.*

(b) Ectomycorrhiza. The mycelial mantle of a typical ectomycorrhizal fungus surrounding a eucalyptus tree root.

FIGURE 27.5 **Mycorrhizae.**

cyanobacterium, the product is fixed nitrogen that eventually enriches the forest soil. Free-living cyanobacteria can fix significant amounts of nitrogen in desert soils after rains and on the surface of arctic tundra soils. Rice paddies can accumulate heavy growths of such nitrogen-fixing organisms. The cyanobacteria also form a symbiosis with a small floating fern, *Azolla,* which grows thickly in rice paddy waters (Figure 27.4). So much nitrogen is fixed by these organisms that other nitrogenous fertilizers are often unnecessary for rice cultivation.

The Role of Mycorrhizae. A very important contribution to plant growth is made by **mycorrhizae** (*myco* means fungus; *rhiza* means root). There are two primary types of these fungi: *endomycorrhizae,* also known as *vesicular-arbuscular mycorrhizae,* and *ectomycorrhizae.* Both types function as root hairs on plants; that is, they extend the surface area through which the plant can absorb nutrients, especially phosphorus, which is not very mobile in soil.

Vesicular-arbuscular mycorrhizae form large spores that can be isolated easily from soil by sieving. The hyphae from these germinating spores penetrate into the plant root and form two types of structures, vesicles and arbuscules. **Vesicles** are smooth oval bodies that are most prob-

ably storage structures. **Arbuscules** (Figure 27.5a) are formed inside plant cells. Nutrients travel from the soil through fungal hyphae to these arbuscules, which gradually break down and release the nutrients to the plants. Most grasses and other plants are surprisingly dependent on these fungi for proper growth, and their presence is nearly universal in the plant world.

Ectomycorrhizae mainly infect trees such as pine and oak. The fungus forms a mycelial *mantle* over the smaller roots of the tree (Figure 27.5b). Ectomycorrhizae do not form vesicles or arbuscules. Managers of commercial pine

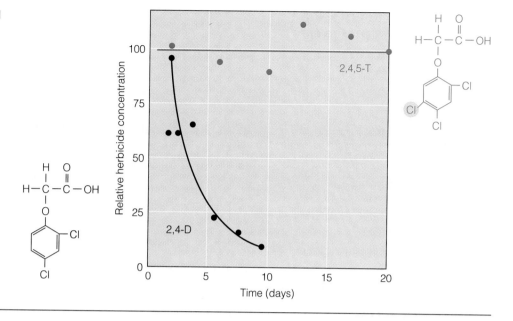

FIGURE 27.6 Slight structural differences in synthetic chemicals can result in great differences in their biodegradability. This graph shows the structures and rates of microbial decomposition of the herbicides 2,4-D (black) and 2,4,5-T (magenta) .

tree farms must take care to see that seedlings are inoculated with soil containing effective mycorrhizae. Truffles, known as a culinary delicacy, are ectomycorrhizae. In France, trained pigs are often used to find them by smell and root them up. To the female pig, the odor is that of a potential mate. In nature, the fungus is dependent upon ingestion by an animal, which distributes the undigested spores into new locations.

Other Biogeochemical Cycles

Microorganisms are also important in cycles involving other elements, such as sulfur. In addition, there are important microbial transformations of potassium, iron, manganese, mercury, selenium, zinc, and other minerals. The various chemical reactions in these cycles are often essential to make the minerals available in soluble form for plants' metabolism.

Degradation of Synthetic Chemicals in the Soil

We seem to take for granted that soil microorganisms will degrade materials entering the soil. Natural organic matter, such as falling leaves or animal residues, are in fact readily degraded. However, in this industrial age there are many chemicals that do not occur in nature—such as plastics—that enter the soil in large amounts. Some plastics used for garbage bags now contain starch incorporated into their chemical structure. The starch is readily biodegradable, but after it is consumed, plastic polymers remain.

Many synthetic chemicals such as pesticides are highly resistant to degradation by microbial attack. A well-known example is the insecticide DDT. When DDT was first introduced, its property of *recalcitrance* (resistance to degradation) was considered quite beneficial because one application remained effective in the soil for an extended time. However, it was soon found that the chemical tended to accumulate and concentrate in parts of the food chain because of its solubility in fat. Eagles and other predatory birds accumulated DDT from contaminated food and suffered impaired reproductive ability; that is, their eggs had soft shells and broke during incubation.

Not all synthetic chemicals are as recalcitrant as DDT. Some are made up of chemical bonds and subunits that are subject to attack by bacterial enzymes. Small differences in chemical structure can make large differences in biodegradability. The classic example is that of two herbicides: 2,4-D, the common chemical used to kill lawn weeds, and 2,4,5-T, which is used to kill shrubs. The addition of a single chlorine atom to the structure of 2,4-D extends its life in soil from a few days to an indefinite period (Figure 27.6).

A growing problem is the leaching into groundwaters of toxic materials that are not biodegradable or that degrade very slowly. The sources of these materials may include landfills, illegal industrial dumps, or pesticides applied to agricultural crops. Once groundwater becomes contaminated, the environmental and economic damage can be devastating. Researchers are developing processes and isolating bacteria that promote degradation or removal of these toxic materials. One example of current work is described in the box in Chapter 2 (p. 31).

AQUATIC MICROBIOLOGY AND SEWAGE TREATMENT

LEARNING OBJECTIVES

- *Describe the freshwater and seawater habitats of microorganisms.*

- *Discuss the causes and effects of eutrophication.*

- *Explain how water is tested for bacteriologic quality.*

- *Compare primary, secondary, and tertiary sewage treatments.*

- *List some of the biochemical activities that take place in an anaerobic sludge digester.*

- *Define BOD, septic tank, oxidation pond, activated sludge, and trickling filter.*

- *List three examples illustrating the use of bacteria to remove pollutants.*

Aquatic microbiology refers to the study of microorganisms and their activities in natural waters, such as lakes, ponds, streams, rivers, estuaries, and the sea. Sewage often enters lakes and streams, and its degradation and effects on the microbial life in these waters is an important factor in aquatic microbiology. We will also see that the treatment of sewage by municipalities mimics the natural action of microorganisms in streams.

Aquatic Microorganisms

Large numbers of microorganisms in a body of water generally indicate high nutrient levels in the water. Water contaminated by inflows from sewage systems or from biodegradable industrial organic wastes is relatively high in bacterial counts. Similarly, ocean estuaries (fed by rivers) have higher nutrient levels and hence higher microbial counts than other shoreline waters.

In water, particularly in water with low nutrient concentrations, microorganisms tend to grow on stationary surfaces and on particulate matter. In this way, a microorganism has contact with more nutrients than if it were randomly suspended and floating freely with the current. Many bacteria whose main habitat is water often have appendages and holdfasts that attach to various surfaces. One example is *Caulobacter* (see Figure 11.16a). Some bacteria also have gas vesicles that they can fill and empty to adjust buoyancy.

Freshwater Microbiota

Figure 27.7 shows a typical lake or pond that serves as an example to represent the various zones and the kinds of microbiota found in a body of fresh water. The **littoral zone** along the shore has considerable rooted vegetation, and light penetrates throughout it. The **limnetic zone** consists of the surface of the open water area away from the shore. The **profundal zone** is the deeper water under the limnetic zone. The **benthic zone** contains the sediment at the bottom.

Microbial populations of freshwater bodies tend to be affected mainly by the availability of oxygen and light. Light is, in many ways, the more important resource because photosynthetic algae are the main source of organic matter, and hence of energy, for the lake. These organisms are the *primary producers* of a lake that supports a population of bacteria, protozoa, fish, and other aquatic life. Photosynthetic algae are located in the limnetic zone.

Areas of the limnetic zone with sufficient oxygen contain pseudomonads and species of *Cytophaga* (sī-täf'äg-ä), *Caulobacter,* and *Hyphomicrobium.* Oxygen does not diffuse into water very well, as any aquarium owner knows. Microorganisms growing on nutrients in stagnant water quickly use up the dissolved oxygen in the water. In the oxygenless water, fish die, and odors (from hydrogen sulfide and organic acids, for example) are produced from anaerobic activity. Wave action in shallow layers, or water movement in rivers, tends to increase the amount of oxygen throughout the water and aid in the growth of aerobic populations of bacteria. Movement thus improves the quality of water and aids in the degradation of polluting nutrients.

Deeper waters of the profundal and benthic zones have low oxygen concentrations and less light. Algal growth near the surface often filters the light, and it is not unusual for photosynthetic microorganisms in deeper zones to use different wavelengths of light from those used by surface-layer photosynthesizers (see Figure 12.10).

In the profundal zone, the purple and green sulfur bacteria are found. These bacteria are anaerobic photosynthetic organisms that metabolize hydrogen sulfide to sulfur and sulfate in the bottom sediments of the benthic zone.

The sediment in the benthic zone includes bacteria such as *Desulfovibrio* that use sulfate (SO_4^{2-}) as an electron acceptor and reduce it to hydrogen sulfide (H_2S), which is responsible for the rotten-egg odor of many lake muds. Methane-producing bacteria are also part of these anaerobic benthic populations. In swamps, marshes, or bottom sediments, they produce methane gas. *Clostridium* species are common in bottom sediments and may include botulism organisms, particularly those causing outbreaks of botulism in waterfowl.

Seawater Microbiota

The open ocean is relatively high in osmotic pressure, low in nutrients, and quite cold at great depths. The pH also tends to be higher than is optimal for most microorganisms. Bacterial populations in such waters tend to be much smaller than in estuaries and in small freshwater bodies fed by rivers and streams. Much of the microscopic life of the ocean is composed of photosynthetic diatoms and other algae. Largely independent of preformed organic nutrient sources, these microbes use energy from photosynthesis and atmospheric carbon dioxide for carbon. These organisms constitute the marine **phytoplankton**

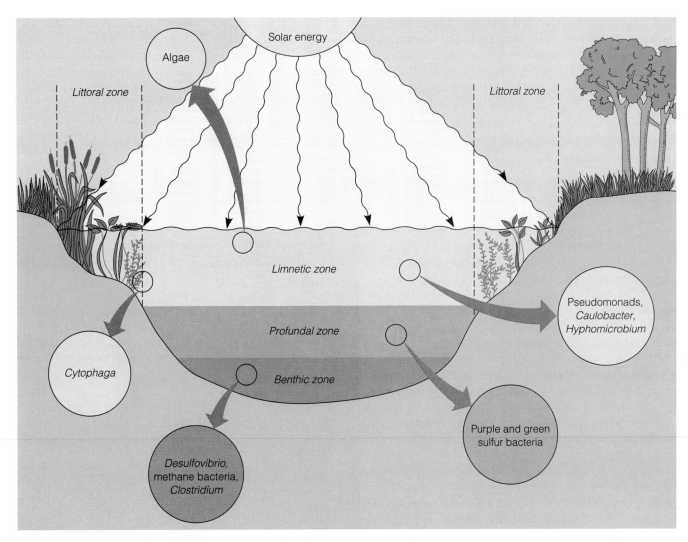

FIGURE 27.7 The zones of a typical lake or pond and some representative microorganisms of each zone. The microorganisms fill niches that vary in light, nutrients, and oxygen availability.

community, the basis of the oceanic food chain. Ocean bacteria benefit from the eventual death and decomposition of phytoplankton and also attach to their living bodies. Protozoa in turn feed on bacteria and the smaller phytoplankton. Krill—shrimplike crustaceans—feed on the phytoplankton and in turn are an important food supply of larger sea life. Many fish and whales are also able to feed directly on the phytoplankton.

Microbial luminescence is an interesting aspect of deep-sea life. Certain bacteria produce light flashes when they are agitated by wave action or by a boat's wake. Many bacteria are also luminescent, and some have established symbiotic relationships with benthic-dwelling fish. These fish sometimes use the glow of their resident bacteria as an aid in attracting and capturing prey in the complete darkness of the ocean depths (Figure 27.8). These biolumines-

FIGURE 27.8 Luminescent bacteria as light organs in fish. This is a flashlight fish (*Photoblepharon palpebratus*).

cent organisms have an enzyme called luciferase that picks up electrons from flavoproteins in the electron transport chain and then emits some of the electron's energy as a photon of light. An industrial application of bioluminescence is discussed in the box in Chapter 28.

The Role of Microorganisms in Water Quality

Water in nature is seldom totally pure. Even rainfall is contaminated as it falls to Earth. For example, combustion of fossil fuels puts sulfur compounds in the air. These encounter the falling rain and form sulfurous acid—the notorious *acid rain*. Water quality is measured by the degree of its contamination by soluble chemicals, suspended colloids such as clay particles (turbidity), and microbes.

Water Pollution

The form of water pollution that is our primary interest is microbial pollution, especially by pathogenic organisms.

Transmission of Infectious Diseases. Water that moves below the ground's surface undergoes a filtering that tends to remove microorganisms. For this reason, water from springs and deep wells is generally of good quality. The most dangerous form of water pollution occurs when human feces enter the water supply. Many diseases are perpetuated by the fecal–oral route of transmission, in which a pathogen shed in human, or sometimes animal, feces contaminates water or food and is later ingested (see Chapter 25). Typical of these diseases are typhoid fever and cholera, which are caused by bacteria, and hepatitis A, which is caused by a virus. An especially troublesome waterborne disease is diarrhea caused by ingestion of cysts of the protozoa *Giardia lamblia*. This protozoan is a problem not only in municipal water systems but also in mountain streams, which are contaminated by beavers and other animals. Shellfish that feed by filtering tend to concentrate waterborne viruses and bacteria in their tissues, so shellfish from polluted water may be dangerous to eat.

Not all pathogens must be ingested to cause disease. For example, the helminthic disease schistosomiasis is spread among people who swim or wade in waters contaminated by human wastes. The infective pathogens are not usually ingested but are free-swimming cercaria that bore through the skin.

As good sanitation practices have nearly eliminated a few diseases, such as typhoid fever and cholera, in the U.S.A., attention has focused on other waterborne diseases. Table 27.2 summarizes some recent data on illnesses associated both with ingestion of water intended for drinking and with recreational water use. Outbreaks caused by protozoa such as *G. lamblia* and *Cryptosporidium* are usually related to their cyst or spore forms that are resistant to chlorination. Fifty percent of the infections associated with recreational water occurred in inadequately chlori-

TABLE 27.2 Waterborne Disease Outbreaks in the U.S.A., 1991–1992

AGENT	OUTBREAKS	CASES
Associated with Water Intended for Drinking		
Giardia lamblia (protozoan)	4	123
Cryptosporidium parvum (protozoan)	3	3551
Hepatitis A virus	1	10
Shigella sonnei	1	150
Acute gastrointestinal illness of unknown etiology	23	13,367
Associated with Recreational Water Use		
Naegleria fowleri (protozoan)	6	6
G. lamblia	4	34
Legionella	3	16
C. parvum	2	526
Shigella sonnei	2	226
Adenovirus 3	1	595
E. coli O157:H7*	1	80
Pseudomonas	1	61
Leptospira	1	6
Acute gastrointestinal illness of unknown etiology	2	76

*Mixed outbreak of *E. coli* O157:H7 and *Shigella sonnei*
SOURCE: *MMWR* 42(SS-5):1–21 (11/19/93).

nated pools; the others were associated with natural lakes. There is some anxiety about pathogens such as *Legionella*, the cause of a pneumonia often spread by inhalation of water aerosols. *Mycobacterium* spp., *Pseudomonas* spp., and other opportunistic pathogens that are dangerous to immunosuppressed individuals are being found more and more often in tap water.

Chemical Pollution. Preventing chemical contamination of water is a difficult problem. Industrial and agricultural chemicals leached from the land enter water in great amounts and in forms that are resistant to biodegradation. Many of these chemicals become biologically concentrated in some of the organisms in the food chain.

A striking example of industrial water pollution involved mercury, used in the manufacture of paper. The metallic mercury was allowed to flow into waterways as waste. It was assumed that the mercury was inert and would remain segregated in the sediments. However, bacteria in the sediments converted the mercury into a soluble chemical compound, methyl mercury, which was then taken up by fish and invertebrates in the waters. When such seafood is a substantial part of the human diet, the mercury concentrations can accumulate with devastating effects on the nervous system.

① A rocky beach at Green Island, Alaska, contaminated with oil spilled from the *Exxon Valdez*

② The same beach after three applications of carbon-free nutrients (fertilizer) to encourage microbial activity

FIGURE 27.9 The use of bioremediation to clean up an oil spill.

Another example of chemical pollution is the synthetic detergents developed immediately after World War II. These rapidly replaced many of the soaps then in use. Because these new detergents were not biodegradable, they rapidly accumulated in the waterways. In some rivers, large rafts of detergent suds could be seen traveling downstream; in some cities a small head of bubbles might appear on the surface of a glass of water. These detergents were replaced in 1964 by new biodegradable formulations.

The new detergents, however, brought new problems. Substantial amounts of phosphates were added to many of these detergents to improve their effectiveness. Unfortunately, phosphates pass virtually unchanged through most sewage treatment systems and can cause **eutrophication** (*eu* means well; *troph* means nourish) of lakes and streams—an overabundance of nutrients, resulting in overgrowth of algae or cyanobacteria and the eventual death of other organisms.

To understand how phosphate pollution leads to eutrophication, we must first consider the nutritional requirements of algae and cyanobacteria. These organisms get their energy from sunlight and their carbon from carbon dioxide dissolved in the water. Most waters contain adequate amounts of the minerals required for algal growth, except for nitrogen and phosphorus. Both of these nutrients can enter water from domestic, farm, and industrial wastes when waste treatment is absent or inefficient. These additional nutrients cause dense aquatic growths called **blooms**. Because many cyanobacteria are able to fix nitrogen from the atmosphere, these photosynthesizing organisms require only traces of phosphorus to initiate blooms. Once eutrophication results in blooms of algae or cyanobacteria, the eventual effect is the same as the addition of biodegradable organic matter. Blooms generate additional organic matter by photosynthesis and nitrogen fixation from the atmosphere. In the short run,

these algae and cyanobacteria produce oxygen. However, they eventually die and are degraded by bacteria. During the degradation process the oxygen in the water is used up, which may kill the fish. Undegraded remnants of organic matter settle to the bottom and hasten the filling of the lake.

Municipal waste containing detergents is likely to be the main source of phosphates in lakes and streams because soil phosphorus is relatively insoluble and tends to be retained in soil. Because of this, phosphate-containing detergents are banned in some localities.

Oil spills on land and at sea represent some of the most dramatic examples of chemical pollution. Bacteria that are especially effective in degrading petroleum products can be selected and applied to such spills (see the box in Chapter 2, p. 31). Oil is rich in carbon and sulfur but low in the nitrogen and phosphorus that are essential for microbial growth. If oil-soluble "fertilizer" containing these nutrient factors is spread over oil spills along with the adapted bacteria (Figure 27.9), the oil is metabolized efficiently. Indigenous bacteria that can utilize the spilled oil are also stimulated into activity by the additional nutrients. This is termed **bioremediation.** Oil also requires oxygen for its metabolism; therefore, in cases when oil is spilled on land, vigorous tilling of the soil to maximize aeration is also necessary.

Coal-mining wastes, particularly in the eastern U.S.A., are very high in sulfur content, mostly iron sulfide (FeS_2). In the process of obtaining energy from the oxidation of the ferrous ion (Fe^{2+}), bacteria such as *Thiobacillus ferrooxidans* convert the sulfide into sulfate. The sulfate enters streams as sulfuric acid, which lowers the pH of the water and damages aquatic life. The low pH also promotes the formation of insoluble iron hydroxides, which form the yellow precipitates often seen clouding such polluted waters.

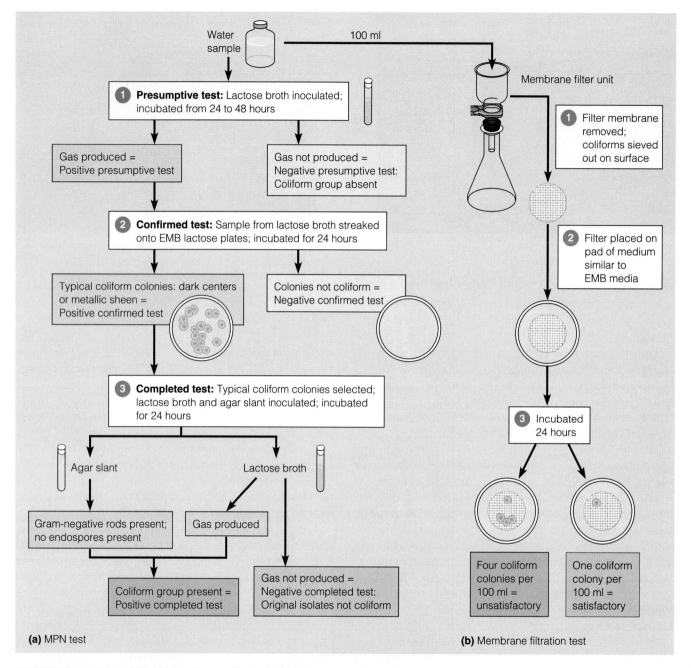

FIGURE 27.10 Two methods for the analysis of drinking water for coliforms.
(a) MPN test method. Note how the series of tests uses the characteristics of coliforms: the production of gas from lactose fermentation and gram-negative rod morphology. Coliform numbers are estimated by inoculating a number of tubes of lactose broth. The percentage of tubes positive for coliforms is a statistical indication of the number of coliforms in the sample. **(b)** Membrane filtration method, used by many laboratories to test for coliforms. Coliform colonies have a distinctive appearance (dark centers or metallic sheen).

Tests for Water Purity

Historically, most of our concern about water purity has been related to the transmission of disease. Therefore, tests have been developed to determine the safety of water; many of these tests are also applicable to foods.

It is not practical, however, to look only for pathogens in water supplies. For one thing, if we were to find the pathogen causing typhoid or cholera in the water system,

the discovery would already be too late to prevent an outbreak of the disease. Moreover, such pathogens would probably be present only in small numbers and might not be included in tested samples.

The tests for water safety in use today are aimed instead at detecting particular **indicator organisms**. There are several criteria for an indicator organism. The most important criterion is that the organism be consistently present in human feces in substantial numbers so that its detection is a good indication that human wastes are entering the water. The indicator organisms should also survive in the water at least as well as the pathogenic organisms would. The indicator organisms must also be detectable by simple tests that can be carried out by people with relatively little training in microbiology.

In the U.S.A., the usual indicator organisms are the *coliform bacteria*. **Coliforms** are defined as aerobic or facultatively anaerobic, gram-negative, non-endospore-forming, rod-shaped bacteria that ferment lactose to form gas within 48 hours of being placed in lactose broth at 35°C. Because some coliforms are not solely enteric bacteria but are more commonly found in plant and soil samples, many standards for food and water specify the identification of *fecal coliforms*. The predominant fecal coliform is *E. coli,* which constitutes a large proportion of the human intestinal population. There are specialized tests to distinguish between fecal coliforms and nonfecal coliforms. Note that coliforms are not themselves pathogenic under normal conditions, although certain strains can cause diarrhea and opportunistic urinary tract infections.

The United States Environmental Protection Agency has established guidelines for evaluating the bacteriologic quality of drinking water and fresh and marine recreational waters. The criteria include the minimum number of water samples to be examined each month and the maximum number of coliform organisms permitted in each 100 ml of water. Figure 6.16 shows the most probable number (MPN) method for determining the number of bacteria in a 100-ml sample; Appendix B provides the MPN table used for calculations. For detection and enumeration of coliforms, selective and differential media are required, and one of two specialized methods is used. These are summarized in Figure 27.10.

A more convenient method of detecting coliforms, specifically the coliform *E. coli,* makes use of media containing the two substrates *o*-nitrophenyl-β-D-galactopyranoside (ONPG) and 4-methylumbelliferyl-β-D-glucuronide (MUG). Coliforms produce the enzyme β-galactosidase, which acts on ONPG and forms a yellow color, indicating their presence in the sample. *E. coli* is unique among coliforms in almost always producing the enzyme β-glucuronidase, which acts on MUG to form a fluorescent compound that glows blue when illuminated by long-wave ultraviolet light (Figure 27.11). These simple tests or variants of them can detect the presence or absence of coliforms or *E. coli* and can be combined with the MPN test to enumerate them.

Coliforms have been very useful as indicator organisms in water sanitation, but they have limitations. One problem has been the growth of coliform bacteria in *biofilms* on the inner surfaces of water pipes. Their presence in tap water has led to a number of community orders to boil water even though these coliforms are not necessarily related to fecal contamination and there is no evidence that they have a significant effect on public health. A more serious problem is that some pathogens are

FIGURE 27.11 ONPG and MUG test. A yellow color (positive ONPG) indicates the presence of coliforms, and blue fluorescence (positive MUG) indicates the presence of the fecal coliform *E. coli.* The clear result indicates an uncontaminated sample.

even more resistant than coliforms to chemical disinfection—in their order of increasing resistance to chlorination, viruses and protozoan cysts. By using sophisticated methods of detecting viruses, it has been found that chemically disinfected water samples that are free of coliforms are often contaminated with enteric viruses. The cysts of *G. lamblia* and *Cryptosporidium* are so resistant to chlorination that eliminating them by this method is probably impractical; mechanical methods such as filtration are necessary.

Concern for these problems has led to recommendations that all water treatment plants producing drinking water should follow the stricter purification methods already in use at some plants. These methods are described in the next section.

Water Treatment

When water is obtained from uncontaminated reservoirs fed by clear mountain streams or from deep wells, it requires minimal treatment to make it safe to drink. Many cities, however, obtain their water from badly polluted sources, such as rivers that have received municipal and industrial wastes upstream. The steps used to purify this water are shown in Figure 27.12. Very turbid (cloudy) water is allowed to stand in a holding reservoir for a time to allow as much particulate suspended matter as possible to settle out. The water then undergoes **flocculation**—that is, removal of colloidal materials such as clay, which would otherwise remain in suspension indefinitely. A flocculant chemical, such as aluminum potassium sulfate (alum), forms aggregations of fine suspended particles called floc. As these aggregations slowly settle out, they entrap colloidal material and carry it to the bottom. Large

FIGURE 27.12 **Steps of water treatment in a typical municipal water purification plant.**

---

numbers of viruses and bacteria are also removed by this treatment. Alum was used to clear muddy river water during the first half of the nineteenth century in the military forts of the American West, long before the germ theory of disease was developed. There were also observations at the time in European cities that people who used water filtered through sand to remove turbidity had a lower incidence of cholera during outbreaks.

After flocculation treatment, water is treated by **filtration**—that is, passing it through beds of sand or diatomaceous earth. As we mentioned previously, some protozoan cysts, such as those of *G. lamblia,* are removed from water only by such filtration treatment. The microorganisms are trapped mostly by surface adsorption in the particulate beds. They do not penetrate the tortuous routing between the particles, even though the openings might be larger than the organisms that are filtered out. These filters are periodically backflushed to clear them of accumulations. Water systems of cities that have an exceptional concern for toxic chemicals supplement sand-type filtration with filters of activated charcoal (carbon). Charcoal removes not only particulate matter but also most dissolved organic chemical pollutants.

Before entering the municipal distribution system, the filtered water is chlorinated. Because organic matter neutralizes chlorine, the plant operators must pay constant attention to maintaining effective levels of chlorine. There has been some concern that chlorine itself might be a health hazard, that it might react with organic contaminants of the water to form carcinogenic compounds. At present, this possibility is considered an acceptable risk when compared with the proven usefulness of chlorination of water.

As we noted in Chapter 7, one substitute for chlorination is ozone treatment. Ozone (O_3) is a highly reactive

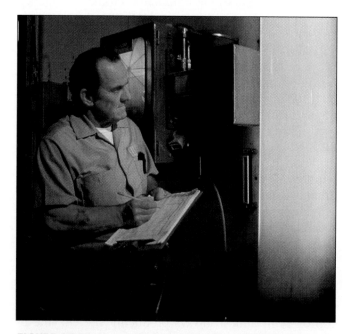

FIGURE 27.13 **An ozone generator.** The technician here is recording the operation of an ozone generator in a water treatment plant.

---

form of oxygen that is formed by electrical spark discharges and ultraviolet light. (The fresh odor of air following an electrical storm or around an ultraviolet light bulb is from ozone.) Ozone for water treatment is generated electrically at the site of treatment (Figure 27.13). Use of ultraviolet light is also a possible alternative to chemical disinfection. Arrays of ultraviolet tube lamps are arranged in quartz tubes so that water flows close to the lamps. This is necessary because of the low penetrating power of ultraviolet radiation.

FIGURE 27.14 Steps of typical sewage treatment. Microbial activity occurs aerobically in trickling filters or in activated sludge aeration tanks and anaerobically in the anaerobic sludge digester. A particular system would use either activated sludge aeration tanks or trickling filters, not both, as shown in this figure. Methane produced by sludge digestion is burned off or used to power heaters or pump motors.

Sewage Treatment

After water has been used, it becomes sewage. Sewage includes all the water from a household that is used for washing, as well as toilet wastes. Rainwater flowing into street drains and some industrial wastes enter the sewage systems in some cities. Sewage is mostly water and contains little particulate matter, perhaps only about 0.03%. Even so, in large cities this portion of sewage can total more than 1000 tons of solid material per day.

Until environmental awareness intensified, a surprising number of large cities in this country had only rudimentary sewage treatment systems or no system at all. Raw sewage, untreated or nearly so, was simply discharged into rivers or oceans. A flowing, well-aerated stream is capable of considerable self-purification. Therefore, until increases in populations and their wastes exceeded this capability, casual treatment of municipal

wastes caused little complaint. In the U.S.A., most cases of simple discharge have been improved. However, this is not the picture in much of the world. Seventy percent of the communities bordering the Mediterranean dump their unprocessed sewage into the sea. Even in a developed nation such as Great Britain, this 1990 description reported in the *British Medical Journal* of conditions near one ocean resort is still valid: "In summer it [the sewage] causes a visible brown buoyant stain extending from the outfall pipe and spreading its way along the bays as it is brought in by the incoming tide."

Primary Treatment. The usual first step in sewage treatment is called **primary treatment** (Figure 27.14a). In this process, incoming sewage receives preliminary treatment; large floating materials are screened out, the sewage is allowed to flow through settling chambers to remove sand

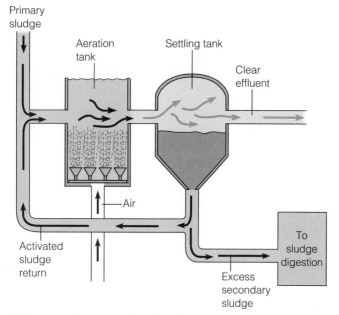

(a) Cross-section of an activated sludge system

(b) An aeration tank; note that the surface is frothing from aeration

LM   30 μm

(c) Particle of floc formed by a species of *Zooglea*, the organism that contributes to floc formation in activated sludge systems; note the cells embedded in gelatinous matter synthesized by the bacteria

FIGURE 27.15 Activated sludge system of secondary treatment.

and similar gritty material, skimmers remove floating oil and grease, and floating debris is shredded and ground. After this step, the sewage passes through sedimentation tanks, where more solid matter settles out. (The design of these primary settling tanks varies.) Sewage solids collecting on the bottom are called **sludge**, at this stage called *primary sludge*. From 40% to 60% of suspended solids are removed from sewage by this settling treatment, and flocculating chemicals that increase the removal of solids are sometimes added at this stage. Biological activity is not particularly important in primary treatment, although some digestion of sludge and dissolved organic matter can occur during long holding times. The sludge is removed on either a continuous or an intermittent basis, and the effluent (the liquid flowing out) then undergoes secondary treatment.

Biochemical Oxygen Demand. Primary treatment removes approximately 25% to 35% of the **biochemical oxygen demand (BOD)** of the sewage. An important concept in sewage treatment and in the general ecology of waste treatment, BOD is a measure of the biologically degradable organic matter in water. BOD is determined by the amount of oxygen required by bacteria to metabolize the organic matter. The classic method of measurement is to use special bottles with airtight stoppers. Each bottle is first filled with the test water or dilutions of the test water. The water is initially aerated to provide a relatively high level of dissolved oxygen and is seeded with bacteria if necessary. The filled bottles are then incubated in the dark for five days at 20°C, and the decrease in dissolved oxygen is determined by a chemical or electronic testing method. The more oxygen that is used up as the bacteria degrade the organic matter in the sample, the greater the BOD, which is usually expressed in milligrams of oxygen per liter of water. The amount of oxygen that normally can be dissolved in water is only about 10 mg/liter; typical BOD values of waste water may be twenty times this amount. If this waste water enters a lake, for example, bacteria in the lake begin to consume the organic matter responsible for the high BOD, rapidly depleting the oxygen in the lake water.

Secondary Treatment. After primary treatment, the greater part of the BOD remaining in the sewage is in the form of dissolved organic matter. **Secondary treatment**, which is predominantly biological, is designed to remove most of this organic matter and reduce the BOD (Figure 27.14b). In this process, the sewage undergoes strong aeration to encourage the growth of aerobic bacteria and other microorganisms that oxidize the dissolved organic matter to carbon dioxide and water. Two commonly used methods of secondary treatment are activated sludge systems and trickling filters.

In the aeration tanks of the **activated sludge system**, air or pure oxygen is added to the effluent from primary treatment (Figure 27.15a and b). The name is derived from

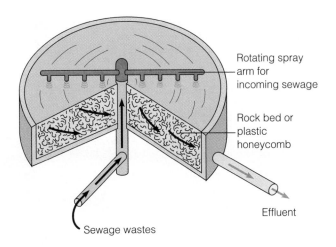

(a) Trickling filter system showing the rotating spray arm in action

(b) Cutaway of a trickling filter system

FIGURE 27.16 Trickling filter method of secondary treatment. The sewage is sprayed from the system of rotating pipes onto a bed of rocks or plastic honeycomb designed to have a maximum surface area and to allow oxygen to penetrate deeply into the bed. Microorganisms grow on the enormous surface area, forming a microbial slime layer that aerobically metabolizes the organic matter in the sewage trickling down through the bed.

the practice of adding to the incoming sewage some of the sludge from a previous batch. This sludge inoculum contains large numbers of metabolizing bacteria, together with yeasts, molds, and protozoa. An especially important ingredient of the sludge are species of *Zoogloea* bacteria, which form flocculant masses (floc) in the aeration tanks (Figure 27.15c). The activity of these aerobic microorganisms oxidizes much of the effluent's organic matter into carbon dioxide and water. When the aeration phase is completed, the floc (secondary sludge) is allowed to settle to the bottom, just as the primary sludge settles in primary treatment.

Soluble organic matter in the sewage is adsorbed onto the floc and is incorporated into microorganisms in the floc. As the floc settles out, this organic matter is removed with the floc and is subsequently treated in an anaerobic sludge digester. More organic matter is probably removed by this settling-out process than by the relatively short-term aerobic oxidation.

Most of the settled sludge is removed for treatment in an anaerobic sludge digester; a small amount is used to inoculate the next sewage batch. The effluent is sent on for final treatment. Occasionally, when aeration is stopped, the sludge will float rather than settle out; this phenomenon is called *bulking*. When this happens, the organic matter in the floc flows out with the discharged effluent and often causes serious problems of local pollution. A considerable amount of research has been devoted to the causes of bulking and its possible prevention. It is apparently caused by the growth of filamentous bacteria of various types; the sheathed bacterium *Sphaerotilus natans* is often mentioned as the primary offender. Activated sludge

systems are quite efficient: they remove from 75% to 95% of the BOD from sewage.

Trickling filters are the other commonly used method of secondary treatment. In this method, the sewage is sprayed over a bed of rocks or molded plastic (Figure 27.16). The rocks or other components of the bed must be large enough so that air penetrates to the bottom but small enough to maximize the surface area available for microbial activity. Actually, there is no filtering action. A slimy, gelatinous film of aerobic microorganisms grows on the rocks' surfaces. In many ways, this film is functionally similar to the organisms found in activated sludge systems. Because air circulates throughout the rock bed, these aerobic microorganisms in the slime layer are able to oxidize much of the organic matter trickling over the surfaces into carbon dioxide and water. Trickling filters remove from 80% to 85% of the BOD, so they are generally less efficient than activated sludge systems. However, trickling filters are usually less troublesome to operate and have fewer problems from overloads or toxic sewage.

Disinfection and Release. Treated sewage is disinfected, usually by chlorination, before being discharged (Figure 27.14c). After disinfection, the water is dechlorinated so chlorine will not kill aquatic organisms when the water is discharged. Chlorine is removed in an exchange reaction by the addition of sulfur dioxide. Not all microbes are necessarily killed by chlorination. The number of coliforms allowed in the effluent depends on where it is discharged. The discharge is usually into an ocean or into flowing streams, although spray irrigation fields are sometimes used to avoid phosphorus and bacterial contamination of

(a) An anaerobic sludge digester at a California treatment plant. Much or all of a typical digester is below ground level. Methane from such a digester is often used to run pumps or heaters in the treatment plant. Excess methane is being burned off in the flame shown here.

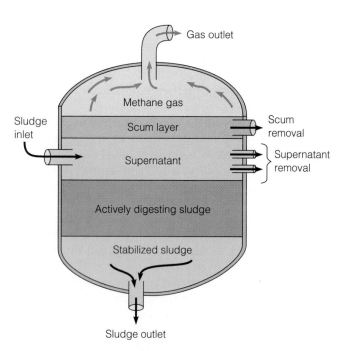

(b) Cross-section of a sludge digester. The scum and supernatant layers are low in solids and are recirculated through secondary treatment.

FIGURE 27.17 Sludge digestion.

Sludge Digestion. Primary sludge accumulates in primary sedimentation tanks; sludge also accumulates in activated sludge and in trickling filter secondary treatments. For further treatment, these sludges are often pumped to **anaerobic sludge digesters** (Figures 27.14d and 27.17). The process of sludge digestion is carried out in large tanks from which oxygen is almost completely excluded. In secondary treatment, emphasis is placed on the maintenance of aerobic conditions so that organic matter is converted to carbon dioxide, water, and solids that can settle out.

An anaerobic sludge digester, however, is designed to encourage the growth of anaerobic bacteria, especially methane-producing bacteria that decrease organic solids by degrading them to soluble substances and gases, mostly methane (60% to 70%) and carbon dioxide (20% to 30%). Methane and carbon dioxide are relatively innocuous end-products, comparable to the carbon dioxide and water from aerobic treatment. The methane is routinely used as a fuel for heating the digester and is also frequently used to run power equipment in the plant.

There are essentially three stages in the activity of an anaerobic sludge digester. The first stage is the production of carbon dioxide and organic acids from anaerobic fermentation of the sludge by various anaerobic and facultatively anaerobic microorganisms. In the second stage, the organic acids are metabolized to form considerable hydrogen and carbon dioxide as well as organic acids, such as acetic acid. These products are the raw materials for a third stage, in which the methane-producing bacteria produce methane (CH_4). Most of the methane is derived from the energy-yielding reduction of carbon dioxide by hydrogen gas:

$$CO_2 + 4H_2 \longrightarrow CH_4 + 2H_2O$$

Other methane-producing microbes split acetic acid (CH_3COOH) to yield methane and carbon dioxide:

$$CH_3COOH \longrightarrow CH_4 + CO_2$$

After anaerobic digestion is completed, considerable amounts of undigested sludge still remain, although it is relatively stable and inert. To reduce its volume, this sludge is pumped to shallow drying beds or water-extracting filters. It is then carried away for disposal. It can be used for landfill or as a soil conditioner. Sludge has about one-fifth of the value of normal commercial lawn fertilizers but has desirable soil-conditioning qualities, much as soil humus does.

Septic Tanks. Homes and businesses in areas of low population density that are not connected to municipal sewage systems often use a **septic tank** (Figure 27.18), a device whose operation is similar in principle to primary treatment. Sewage enters a holding tank, and suspended solids settle out. The sludge in the tank must be pumped out periodically and disposed of. The effluent flows through a

waterways. For ocean discharge, 240 coliforms per 100 ml of water may be allowed, but in a smaller body of water coliforms may have to be reduced to 23 per 100 ml.

In cities with limited freshwater supplies, there is already some recycling of sewage into drinking water. This is safe and practical and will probably become commonplace in the future.

FIGURE 27.18 Septic tank system. (a) Overall plan. Most soluble organic matter is disposed of by percolation into soil. (b) Cross-section of a septic tank. Solids settle out in the septic tank as sludge, which is pumped out periodically through the access manhole.

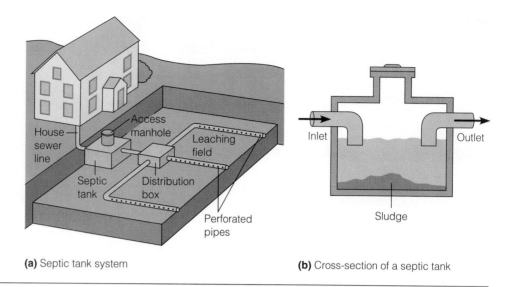

(a) Septic tank system

(b) Cross-section of a septic tank

system of perforated piping into a leaching (soil drainage) field. The effluent entering the soil is decomposed by soil microorganisms. These systems work well when not overloaded and when the drainage system is properly sized to the load and soil type. Heavy clay soils require extensive drainage systems because of the soil's poor permeability. The high porosity of sandy soils can allow chemical or bacterial pollution of nearby water supplies.

Oxidation Ponds. Many small communities and many industries use **oxidation ponds**, also called *lagoons* or *stabilization ponds,* for water treatment. These are inexpensive to build and operate but require considerable land. Designs vary, but most incorporate two stages. The first stage is analogous to primary treatment; the sewage pond is deep enough that conditions are almost entirely anaerobic. Sludge settles out in this stage. In the second stage, which roughly corresponds to secondary treatment, effluent is pumped into an adjoining pond or system of ponds that are shallow enough to be aerated by wave action. Because it is difficult to maintain aerobic conditions for bacterial growth in ponds with so much organic matter, the growth of algae is encouraged to produce oxygen. Bacterial action in decomposing the organic matter in the wastes generates carbon dioxide. Algae, which use carbon dioxide in their photosynthetic metabolism, grow and produce oxygen, which in turn encourages activity of aerobic microorganisms in the sewage. Considerable amounts of organic matter in the form of algae accumulate, but this is not a problem because the oxidation pond, unlike a lake, already has a large nutrient load.

Some small sewage-producing operations, such as isolated campgrounds or highway rest stop areas, use an *oxidation ditch* for sewage treatment. In this method, a small oval channel in the shape of a race track is filled with sewage water. A paddle wheel similar to that on a Mississippi steamboat propels the water in a self-contained flowing stream aerated enough to oxidize the wastes.

Tertiary Treatment. As we have seen, primary and secondary treatments of sewage do not remove all the biologically degradable organic matter. Amounts of organic matter that are not excessive can be released into a flowing stream without causing a serious problem. Eventually, however, the pressures of increased population might increase wastes beyond a body of water's carrying capacity, and additional treatments might be required. Even now, primary and secondary treatments are inadequate in certain situations, such as when the effluent is discharged into small streams or recreational lakes. Some communities have therefore developed **tertiary treatment** plants. Lake Tahoe in the Sierra Nevada, surrounded by extensive development, is the site of one of the best-known tertiary systems.

The effluent from secondary treatment plants contains some residual BOD. It also contains about 50% of the original nitrogen and 70% of the original phosphorus, which can greatly affect a lake's ecosystem. Tertiary treatment is designed to remove essentially all the BOD, nitrogen, and phosphorus. Tertiary treatment depends less on biological treatment than on physical and chemical treatments. Some systems encourage denitrifying bacteria to form volatile nitrogen gas. Nitrogen is converted to ammonia and evaporates into the air in stripping towers. Phosphorus is precipitated out by combining with such chemicals as lime, alum, and ferric chloride. Filters of fine sands and activated charcoal remove small particulate matter and dissolved chemicals. Finally, chlorine is added to the purified water to kill or inhibit any remaining microorganisms and to oxidize any remaining odor-producing substances.

Tertiary treatment provides water that is suitable for drinking, but the process is extremely costly. Secondary treatment is less costly, but water that has undergone only secondary treatment contains many substances that pollute water. Much work is being done to design secondary treatment plants in which the effluent can be used for irrigation. This design would eliminate a source of water

① Nonbiodegradable materials are removed as the wastes pass on a moving belt

② The biodegradable wastes are placed in a rotating digester that increases the efficiency of bacterial digestion and slowly moves the material through

③ The composted material leaves the digester, where microbial action caused temperatures to reach about 60°C (140°F)

④ The composted material has much less mass than the original wastes, thus lessening the demand on landfill; the compost is also useful as a soil conditioner

FIGURE 27.19 Composting municipal wastes.

pollution, provide nutrients for plant growth, and reduce the demand on already scarce water supplies. The soil would act as a trickling filter to remove chemicals and microorganisms before the water reaches groundwater and surface water supplies.

Solid Municipal Waste

Solid municipal waste (garbage) is most frequently placed into large compacted landfills. Conditions are largely anaerobic, and even presumably biodegradable materials such as paper are not very effectively attacked by microorganisms. In fact, recovering a 20-year-old newspaper in readable condition is not at all unusual. Such anaerobic conditions, however, do promote the activity of the same methanogens encountered in anaerobic sludge digesters in sewage treatment. The methane they produce can be tapped with drill holes and burned to generate electricity or purified and introduced into natural gas pipeline systems (see Figure 28.16). Such systems are part of the design of more than a hundred large landfills, some of which provide energy for several thousand homes.

The amount of organic matter entering landfills can be considerably lessened if it is first separated from material that is not biodegradable and *composted*. Composting is a process long used by gardeners to convert plant remains into the equivalent of natural humus. A pile of leaves or grass clippings will undergo microbial degradation. Under favorable conditions thermophilic bacteria will raise the temperature of the compost to 55 or 60°C in a couple of days. After the temperature declines, the pile can be turned to renew the oxygen supply and a second temperature rise will occur. Over time, the thermophilic microbial populations are replaced by mesophilic populations that slowly continue the conversion to a stable material similar to humus. Where space is available, municipal wastes are composted in windrows (long, low piles) that are distributed and periodically turned over by specialized machinery. Municipal waste disposal now also makes increasing use of composting methods. Continuous-flow digesters can be used to compost solid municipal wastes very efficiently (Figure 27.19).

• • •

It is hoped that this chapter on aquatic and soil microbiology has left you with a greater appreciation of the microbial influences on the environment around us. Without the natural and human-directed applications of microbes, life would be very different, and perhaps could not sustain itself at all.

STUDY OUTLINE

INTRODUCTION (p. 664)

1. Microorganisms live in a wide variety of habitats.
2. Most microorganisms do not cause disease.
3. Some parasites get their nutrients from other microbes.
4. Saprophytes get their nutrients from dead organic matter.
5. Both partners benefit in mutualism.
6. Commensals benefit from each other without causing harm.
7. In cometabolism, one organism can perform a biochemical process if provided with certain chemicals by another organism.

SOIL MICROBIOLOGY AND CYCLES OF THE ELEMENTS (pp. 665–673)

The Components of Soil (pp. 665–667)

1. Soil consists of solid inorganic matter, water, air, and living organisms and products of their decay.
2. Weathering of rocks adds minerals to soil.
3. Organic matter in soil comes from microorganisms, plants, animals and their waste products, and the biochemical activities of microorganisms.
4. Organic matter that is relatively resistant to decay is called humus.
5. Microbial life in soil depends on the availability of water and oxygen.
6. Microorganisms in the soil decompose organic matter and transform nitrogen- and sulfur-containing compounds into usable forms.
7. Bacteria are the most numerous organisms in the soil.
8. The characteristic musty odor of soil is from geosmin produced by actinomycetes.
9. Soil is not a reservoir for human pathogens, except for some endospore-forming bacteria and helminthic larvae; human pathogens generally find soil to be a hostile environment.
10. Insect and plant pathogens are found in soil.

Microorganisms and Biogeochemical Cycles (pp. 667–673)

1. In biogeochemical cycles, certain chemical elements are recycled.
2. Microorganisms are essential to the continuation of biogeochemical cycles.

The Carbon Cycle (pp. 667–668)

1. CO_2 is fixed into organic compounds by photoautotrophs.
2. These organic compounds provide nutrients for chemoheterotrophs.
3. Chemoheterotrophs release CO_2 that is then used by photoautotrophs.

The Nitrogen Cycle (pp. 669–673)

1. Microorganisms decompose proteins from dead cells and release amino acids.
2. Ammonia is liberated by microbial ammonification of the amino acids.
3. Ammonia is oxidized to nitrates for energy by nitrifying bacteria.
4. Denitrifying bacteria reduce nitrates to molecular nitrogen (N_2).
5. N_2 is converted into ammonia by nitrogen-fixing bacteria.
6. Nitrogen-fixing bacteria include free-living genera such as *Azotobacter*, cyanobacteria, and the symbiotic bacteria *Rhizobium* and *Frankia*.
7. Ammonium and nitrate are used by bacteria and plants to synthesize amino acids that are assembled into proteins.
8. Symbiotic fungi called mycorrhizae live in and on plant roots.
9. Mycorrhizae increase the surface area and nutrient absorption of the plant.

Other Biogeochemical Cycles (p. 673)

1. Microorganisms are also involved in the transformation of other elements, including sulfur, potassium, iron, manganese, mercury, zinc, and selenium.
2. These reactions make minerals available in soluble form for plants' metabolism.

Degradation of Synthetic Chemicals in the Soil (p. 673)

1. Many synthetic chemicals, such as pesticides and plastics, are recalcitrant (resistant to degradation).
2. Recalcitrance is based on the nature of the chemical bonding.

AQUATIC MICROBIOLOGY AND SEWAGE TREATMENT (pp. 674–686)

Aquatic Microorganisms (pp. 674–676)

1. The study of microorganisms and their activities in natural waters is called aquatic microbiology.
2. Natural waters include lakes, ponds, streams, rivers, estuaries, and the sea.
3. The concentration of bacteria in water is proportional to the amount of organic material in the water.
4. Most aquatic bacteria tend to grow on surfaces rather than in a free-floating state.

Freshwater Microbiota (p. 674)

1. The number and location of freshwater microbiota depend on the availability of oxygen and light.
2. Photosynthetic algae are the primary producers of a lake. They are found in the limnetic zone.

3. Pseudomonads, *Cytophaga, Caulobacter,* and *Hyphomicrobium* are found in the limnetic zone, where oxygen is abundant.

4. Microbial growth in stagnant water uses available oxygen and can cause odors and the death of fish.

5. The amount of dissolved oxygen is increased by wave action.

6. Purple and green sulfur bacteria are found in the profundal zone, which contains light and H_2S but no oxygen.

7. *Desulfovibrio* reduces SO_4^{2-} to H_2S in benthic mud.

8. Methane-producing bacteria are also found in the benthic zone.

Seawater Microbiota (pp. 674–676)

1. The open ocean is not a favorable environment for most microorganisms because of its high osmotic pressure, low nutrients, and high pH.

2. Phytoplankton, consisting mainly of diatoms, are the primary producers of the open ocean.

3. Some algae and bacteria are bioluminescent. They possess the enzyme luciferase, which can emit light.

The Role of Microorganisms in Water Quality
(pp. 676–686)

Water Pollution (pp. 676–677)

1. Microorganisms are filtered from water that percolates into groundwater supplies.

2. Some pathogenic microorganisms are transmitted to humans in drinking and recreational waters.

3. Recalcitrant chemical pollutants may be concentrated in animals in an aquatic food chain.

4. Mercury is metabolized by certain bacteria into a soluble compound that is concentrated in animals.

5. Nutrients such as phosphates cause algal blooms, which can lead to eutrophication of aquatic ecosystems.

6. Eutrophication means "well nourished." It is the result of the addition of pollutants or natural nutrients.

7. The use of microorganisms to remove pollutants is called bioremediation.

8. The growth of oil-degrading bacteria can be enhanced by the addition of nitrogen and phosphorous fertilizer.

9. *Thiobacillus ferrooxidans* produces sulfuric acid at coal-mining sites.

Tests for Water Purity (pp. 678–679)

1. Tests for the bacteriological quality of water are based on the presence of indicator organisms.

2. The most common indicator organisms are coliforms.

3. Coliforms are aerobic or facultatively anaerobic, gram-negative, non-endospore-forming rods that ferment lactose with the production of acid and gas within 48 hours of being placed in a medium at 35°C.

4. Fecal coliforms, predominantly *E. coli,* are used to indicate the presence of human feces.

Water Treatment (pp. 679–680)

1. Drinking water is held in a holding reservoir long enough that suspended matter settles.

2. Flocculation treatment uses a chemical such as alum to coalesce and then settle colloidal material.

3. Filtration removes protozoan cysts and other microorganisms.

4. Drinking water is disinfected with chlorine to kill remaining pathogenic bacteria.

Sewage Treatment (pp. 681–686)

1. Domestic wastewater is called sewage.

2. It includes household water, toilet wastes, industrial wastes, and rainwater.

Primary Treatment (pp. 681–682)

1. Primary sewage treatment is the removal of solid matter called sludge.

2. Biological activity is not very important in primary treatment.

Biochemical Oxygen Demand (p. 682)

1. Primary treatment removes approximately 25% to 35% of the biochemical oxygen demand (BOD) of the sewage.

2. BOD is a measure of the biologically degradable organic matter in water.

3. It is determined by measuring the amount of oxygen bacteria require to degrade the organic matter.

Secondary Treatment (pp. 682–683)

1. Secondary treatment is the biological degradation of organic matter in sewage after primary treatment.

2. Activated sludge and trickling filters are methods of secondary treatment.

3. Microorganisms degrade the organic matter aerobically.

4. Secondary treatment removes up to 95% of the BOD.

Disinfection and Release (pp. 683–684)

1. Treated sewage is disinfected before discharge onto land or into water.

Sludge Digestion (p. 684)

1. Sludge is placed in an anaerobic sludge digester; bacteria degrade organic matter and produce simpler organic compounds, methane, and CO_2.

2. The methane produced in the digester is used to heat the digester and to operate other equipment.

3. Excess sludge is periodically removed from the digester, dried, and disposed of (as landfill or as soil conditioner) or incinerated.

Septic Tanks (pp. 684–685)

1. Septic tanks can be used in rural areas to provide primary treatment of sewage.

2. They require a large leaching field for the effluent.

Oxidation Ponds (p. 685)

1. Small communities can use oxidation ponds for secondary treatment.

2. These require a large area in which to build an artificial lake.

Tertiary Treatment (pp. 685–686)

1. Tertiary treatment uses physical filtration and chemical precipitation to remove all the BOD, nitrogen, and phosphorus from water.
2. Tertiary treatment provides drinkable water, whereas secondary treatment provides water usable only for irrigation.

Solid Municipal Waste (p. 686)

1. Municipal landfills prevent decomposition of solid wastes because they are dry and anaerobic.
2. In some landfills, methane produced by methanogens can be recovered for an energy source.
3. Composting and continuous-flow digesters can be used to promote biodegradation of organic matter.

STUDY QUESTIONS

Review

1. Write a one-sentence description of each of the following soil constituents: solid inorganic matter, water, gases, organic matter.

2. The precursor to coal is peat found in bogs. Why does peat accumulate in bogs?

3. The metabolic activities of microorganisms often produce acids. Why would microbial growth increase the solid inorganic matter in soil?

4. Compare and contrast humus and recalcitrant chemicals.

5. Diagram the carbon cycle in the presence and absence of oxygen. Name at least one microorganism that is involved at each step.

6. Fill in the following table with the information provided below.

PROCESS	CHEMICAL REACTIONS	MICROORGANISMS
Ammonification		
Nitrification		
Denitrification		
Nitrogen fixation		

Choices:
$NO_3^- \longrightarrow N_2$ *Bacillus*
$N_2 \longrightarrow NH_3$ *Rhizobium*
$-NH_2 \longrightarrow NH_3$ *Nitrosomonas*
$NH_3 \longrightarrow NO_2^-$ *Azotobacter*
$NO_2^- \longrightarrow NO_3^-$ *Nitrobacter*
   Proteolytic bacteria

7. The following organisms have important roles as symbionts with plants and fungi. Describe the symbiotic relationship of each organism with its host: cyanobacteria, mycorrhizae, *Rhizobium*, *Frankia*.

8. Compare and contrast the physical conditions of the ocean with those of fresh water.

9. Matching.

____ $CO_2 + H_2S \xrightarrow{\text{light}} C_6H_{12}O_6 + S^0$
____ $SO_4^{2-} + H^+ \longrightarrow H_2S$
____ $CO_2 + H_2 \longrightarrow CH_4$

(a) Methane-producing bacteria
(b) *Desulfovibrio*
(c) Photosynthetic bacteria

10. Indicate which reactions in Question 9 require atmospheric oxygen (O_2).

11. Outline the treatment process for drinking water.

12. What is the purpose of a coliform count on water?

13. If coliforms are not normally pathogenic, why are they used as indicator organisms for bacteriological quality?

14. The following processes are used in wastewater treatment. Match the type of treatment with the processes. Each choice can be used once, more than once, or not at all.

PROCESSES
____ Leaching field
____ Removal of solids
____ Biological degradation
____ Activated sludge
____ Chemical precipitation of phosphorus
____ Trickling filter
____ Results in drinking water
____ Effluent can be used for irrigation
____ Produces methane

TYPES OF TREATMENT
(a) Primary
(b) Secondary
(c) Tertiary

15. Define BOD.

16. Why is activated sludge a more efficient means of removing BOD than a sludge digester?

17. Why are septic tanks and oxidation ponds not feasible for large municipalities?

18. Explain the effect of dumping untreated sewage into a pond on the eutrophication of the pond. The effect of sewage that has primary treatment? The effect of sewage that has secondary treatment? Contrast your previous answers with the effect of each type of sewage on a fast-moving river.

19. Bioconversion refers to the generation of energy sources by bacteria. Describe an example of bioconversion.

20. Bioremediation refers to the use of living organisms to remove pollutants. Describe three examples of bioremediation.

Critical Thinking

1. Here are the formulas of two detergents that have been manufactured:

Which of these would be recalcitrant, and which would be readily degraded by microorganisms? (*Hint:* Refer to the degradation of fatty acids in Chapter 5.)

2. A sewage treatment plant in Redwood City, California, is a tertiary treatment plant. Outline the flow of water and solids through this plant. What is the final quality of the water with respect to BOD, nitrogen, and phosphorus?

3. What is the MPN of a water sample with these multiple-tube fermentation test results: 10-ml portions: 4 positive for acid and gas; 1-ml portions: 3 positive for acid and gas; 0.1-ml portions: 1 positive? (See MPN table, Appendix B.)

Clinical Applications

1. Complete the following graph to show the effect of dumping phosphates into a body of water at time *x*.

Algae
Concentration of   O₂
Fish
x
Time

2. The vacuum-heap process (see photo) is used to remove benzene and other hydrocarbons from soil contaminated by gasoline and oil. The pipes are used to add nitrates, phosphates, oxygen, or water. Why are each of these added? Why is it not always necessary to add bacteria?

3. Flooding after two weeks of heavy rainfall in Tooele, Utah, preceded the high rate of diarrheal illness shown in the following graph. *G. lamblia* was isolated from 26% of the patients. A comparison study of a town 65 miles away revealed that there was diarrheal illness in 2.9% of the 103 people interviewed. Tooele has a municipal water system and a municipal sewage treatment plant. Explain the probable cause of this epidemic and method(s) of stopping it.

Distribution of Cases of Diarrheal Illness

Applied and Industrial Microbiology

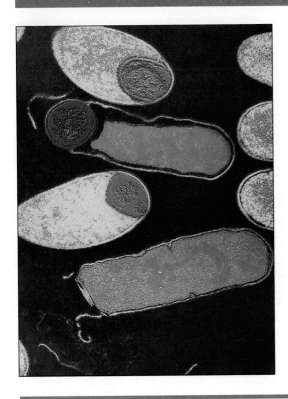

IN THE PREVIOUS CHAPTER on soil and aquatic microbiology we saw that microbes are an essential factor in many natural phenomena that make life possible on Earth. In this chapter we will look at how microorganisms are harnessed in such useful applications as the making of food and industrial products. Many of these processes have roots early in the progress of civilization but require today a thorough understanding of biology and chemistry. Earlier, in Chapter 9, we encountered industrial applications of microorganisms that are at the cutting edge of our knowledge of molecular biology.

FOOD MICROBIOLOGY

LEARNING OBJECTIVES

- *Provide a brief history of the development of food preservation.*
- *Describe thermophilic anaerobic spoilage and flat sour spoilage by mesophilic bacteria.*
- *Describe how canning, pasteurization, low temperature, aseptic packaging, and radiation are used to preserve foods.*
- *Provide examples of chemical food preservatives, and explain why they are used.*
- *Outline at least four beneficial activities of microorganisms in food production.*

Modern civilization and its large populations could not be supported without effective methods of food preservation. In fact, civilization arose only after agriculture produced a year-round stable food supply in a single site so that people were able to give up nomadic hunting and gathering cultures. Many of the methods of food preservation used today were probably discovered by chance in centuries past. People in these early cultures observed that dried meat and salted fish resisted decay. Nomads no doubt noticed that soured animal milk resisted further decomposition and was still palatable. Moreover, if the curd of the soured milk was pressed to remove moisture and allowed to ripen (in effect, cheese-making), it was even more effectively preserved and tasted better. Farmers soon learned that if grains were kept dry, they did not become moldy.

Food Preservation and Spoilage

The preservation and spoilage of food involve the factors that control the growth of microorganisms (see Chapters 6 and 7). Bacteria and fungi require a minimum amount of available moisture for growth. Drying food and adding salt or sugar minimizes the available moisture and thus prevents spoilage. Acidity created by the natural fermentation of milk and vegetable juices, as in yogurt and sauerkraut, also prevents the growth of many spoilage bacteria.

These methods of preservation are still in use today.

However, a tour of any supermarket will demonstrate that heat sterilization, pasteurization, refrigeration, and freezing are now much more popular methods for the control of food spoilage. These modern methods preserve food nearer its natural state and palatability than the methods that were used in the distant past.

Canning

Preservation by heat originated in the early 1800s, before microorganisms were known to cause human disease and food spoilage. The technology of canning was promoted by the military. When armies were small professional organizations, they could be largely supported off the land, with supplements of portable rations of wine, cheese, and dried grains. But with the rise of large armies of citizen soldiers, a development of the French Revolution, this strategy was no longer viable. Thus, the French government offered a prize to anyone who could devise a method of preserving food, particularly meat. In 1810, the prize was won by a confectioner, Nicholas Appert, who showed that food could be preserved if it was sealed in tightly stoppered containers and boiled for specified periods. He developed detailed tables of the boiling times required for different foods and container sizes. As we know today, many endospore-forming bacteria will survive hours of such boiling. Failures were manageably few, however, and were usually attributed to faulty sealing.

The concept of heat preservation was followed quickly by the invention of the metal can. Temperatures above the boiling point of water were introduced into processing by oil or salt baths. Achieving these temperatures with pressurized steam in closed containers, the method used today, was introduced late in the nineteenth century, when reliable pressure controls and safety valves were developed (Figure 28.1). By then, of course, there was a better theoretical understanding of microbe control.

Commercial Sterilization. Preserving foods by heating a properly sealed container is not difficult. The problem is to use the minimum amount of heat necessary to kill spoilage organisms and dangerous microorganisms, such as the endospore-forming *Clostridium botulinum*. Because heating often degrades the appearance and palatability of food, much research was needed to determine the exact heat treatment that would succeed in sterilizing the food but minimize these reductions in quality.

Canned goods today undergo what is called **commercial sterilization** (Figure 28.2), which is not as rigorous as true sterilization. Commercial sterilization is intended to destroy *C. botulinum* endospores. If this is accomplished, then any other significant spoilage or pathogenic bacteria will also be destroyed. To ensure complete sterilization, enough heat is applied for the **12D treatment** (12-decimal reductions), by which a theoretical population of *C. botulinum* endospores would be decreased by 12 logarithmic cycles. What this means is that if there were 10^{12} (1,000,000,000,000) endospores in a can, after treat-

FIGURE 28.1 A commercial canning retort.

ment there would be only one survivor. Because 10^{12} is an improbably large population, this treatment is considered quite safe. Certain thermophilic endospore-forming bacteria have endospores that are more resistant to heat treatment than are those of *C. botulinum*. However, these bacteria are obligate thermophiles and generally remain dormant at temperatures lower than about 45°C (112°F). Therefore, they are not a problem at normal storage temperatures.

Spoilage of Canned Food. If canned foods are incubated at high temperatures, such as exist in a truck in the hot sun or next to a steam radiator, the thermophilic bacteria that often survive commercial sterilization can germinate and grow. **Thermophilic anaerobic spoilage** is a fairly common cause of spoilage in low-acid canned foods. The can usually swells from gas, and the contents have a lowered pH and a sour odor. A number of thermophilic species of *Clostridium* can cause this type of spoilage. When thermophilic spoilage occurs but the can is not swollen by gas production, the spoilage is termed **flat sour spoilage.** This type of spoilage is caused by thermophilic organisms such as *Bacillus stearothermophilus* (ste-rō-thér-mä'fil-us), which is found in the starch and sugars used in food preparation. Many industries have standards for the numbers of such thermophilic organisms permitted in raw materials. Both types of spoilage occur only when the cans are stored at higher than normal temperatures, which permits the growth of bacteria whose endospores are not destroyed by normal processing.

Mesophilic bacteria can spoil canned foods if the food is underprocessed or if the can leaks. Normally, these bacteria are killed by proper processing procedures. Underprocessing is more likely to result in spoilage by endospore-

FIGURE 28.2 The commercial sterilization process in industrial canning. ① The food to be canned is washed, sorted, and blanched. Blanching is a treatment with hot water or live steam that softens the product so that the can will fill better. It also destroys enzymes that might alter the color, flavor, or texture of the product and lowers the microbial count. ② Cans are filled to capacity, leaving as little dead space as possible. ③ Cans are placed in a steam box, where the heat exhausts (drives out) dissolved air. ④ Cans are sealed. ⑤ Cans are sterilized in a large retort with steam under pressure. ⑥ Cans are cooled by submerging them in or spraying them with water. ⑦ Cans are then labeled, stored, and delivered.

(a) Formation of a side seam   **(b)** Formation of a double seam for top or bottom

FIGURE 28.3 Construction of a metal can. Note the seam construction. During cooling after sterilization (see Figure 28.2, step ⑥), the vacuum formed in the can may actually force contaminating organisms into the can along with water.

formers; the presence of non-endospore-formers strongly suggests that the can leaks. Leaking cans are often contaminated during the cooling of cans after processing by heat. The hot cans are sprayed with cooling water or passed through a trough filled with water. As the can cools, a vacuum is formed inside, and external water can be sucked through a leak past the heat-softened sealant in the crimped lid (Figure 28.3). Contaminating bacteria in the cooling water are drawn into the can with the water. Spoilage from underprocessing or can leakage is likely to produce odors of putrefaction, at least in high-protein foods, and occurs at normal storage temperatures. In such types of spoilage there is always the potential that botulinal bacteria will be present.

Some acidic foods, such as tomatoes or preserved fruits, are preserved by processing temperatures of 100°C or below. As a rule, the only important spoilage organisms in acidic foods are molds, yeasts, and occasional species of acid-tolerant, non-endospore-forming bacteria. These organisms are capable of growing at the normal pH of these

TABLE 28.1   Types of Canned Food Spoilage

	INDICATIONS OF SPOILAGE	
TYPE OF SPOILAGE	APPEARANCE OF CAN	CONTENTS OF CAN
Low- and Medium-Acid Foods (pH above 4.5)		
Flat sour (*Bacillus stearothermophilus*)	Possible loss of vacuum on storage	Appearance not usually altered; pH markedly lowered; sour; may have slightly abnormal odor; sometimes cloudy liquid
Thermophilic anaerobic (*Clostridium thermosaccharolyticum*)	Swollen	Fermented, sour, cheesy, or butyric acid odor
Sulfide (*Desulfotomaculum nigrificans*)	No swelling	Usually blackened; "rotten-egg" odor
Putrefactive anaerobic (*Clostridium sporogenes*)	Swollen	May be partially digested; pH slightly above normal; typical putrid odor
Aerobic endospore-formers (*Bacillus* spp.)	Usually no swelling, except in cured meats when nitrate and sugar are present	Coagulated evaporated milk; foods usually have sour taste
High-Acid Foods (pH below 4.5)		
Flat sour (*Bacillus coagulans*)	Flat, little change in vacuum	Slight pH change; off odor and flavor
Butyric anaerobic (*Clostridium butyricum*)	Swollen	Fermented, butyric acid odor
Non-endospore-formers (mostly lactic acid bacteria)	Swollen	Acid odor
Yeasts	Swollen, may burst	Fermented; yeasty odor
Molds	Flat	Surface growth; musty odor

SOURCE: Data from the National Canners Association, 1950 6th Street, Berkeley, CA 94710.

foods but are easily killed by temperatures lower than 100°C. Occasional problems in acidic foods develop from a few organisms that are both heat-resistant and acid-tolerant. Examples of heat-resistant fungi are the mold *Byssochlamys fulva* (bis-sō-klam′is fül′vä), which produces a *heat-resistant ascospore,* and a few molds, especially species of *Aspergillus,* that sometimes produce specialized resistant bodies called *sclerotia.* A spore-forming bacterium, *Bacillus coagulans* (kō-ag′ū-lanz), is unusual in that it is capable of growth at a pH of almost 4.0. A summary of the types of canned food spoilage is presented in Table 28.1.

Home canning is an important means of preserving food. Because of the possibility of botulism food poisoning resulting from improper canning methods, people involved in home canning should obtain reliable directions and follow them exactly.

Pasteurization

As we discussed in Chapter 1, Louis Pasteur began his career as a microbiologist when the French brewing industry commissioned him in 1871 to investigate the causes of spoilage in beer and wine. He concluded that spoilage was caused by the growth of microorganisms, and he worked out a method of eliminating them by mild heating. Too much heat would alter the beverage's characteristics to an unacceptable degree, but killing all the microorganisms was found to be unnecessary. The principle of the selective killing of microorganisms, now called pasteurization, was later applied to other foods, milk in particular (see Chapter 7). Other products, such as ice cream, yogurt, and beer, all have individual pasteurization times and temperatures, which often differ considerably. Reasons for variation include less efficient heating in more viscous foods and the protective effects of fats.

Aseptic Packaging

The use of **aseptic packaging** to preserve food has been increasing recently. Packages are usually made of some material that cannot tolerate conventional heat treatment, such as laminated paper or plastic. The packaging materials come in continuous rolls that are fed into a machine that sterilizes the material with a hot hydrogen peroxide solution, sometimes aided by ultraviolet light (Figure 28.4). Metal containers can be sterilized with superheated steam or other high-temperature methods. High-energy electron beams can also be used to sterilize the packaging materials. While still in the sterile environment, the material is formed into packages, which are then filled with liquid foods that have been conventionally sterilized by heat. The filled package is not sterilized after it is sealed.

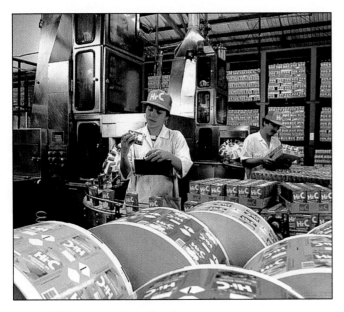

FIGURE 28.4 Aseptic packaging.

Low-Temperature Preservation

Low temperatures lengthen the reproduction time of microorganisms. Even so, some molds and bacteria can grow at a significant rate at temperatures below the freezing point of water, 0°C. (Food typically does not freeze solid until it is several degrees below this point.) Growth of microorganisms in foods a few degrees below 0°C is common, and growth at minus 18°C has been reported. A properly set refrigerator maintains a temperature range from 0° to 5°C, although some sections such as the door shelves may be higher. Many microorganisms grow slowly at these temperatures and will alter the taste and appearance of foods that are stored for too long.

Pathogenic bacteria, with a few exceptions, will not grow at these temperatures. Among the exceptions are the clostridia that cause type E botulism, *Yersinia enterocolitica*, and *Listeria monocytogenes*. Freezing does not immediately kill significant numbers of bacteria, but frozen bacterial populations are dormant and decline slowly with time. Many parasites, such as the roundworms that cause trichinosis, are killed by several days of freezing. Some important temperatures associated with microorganisms and food spoilage are shown in Figure 28.5.

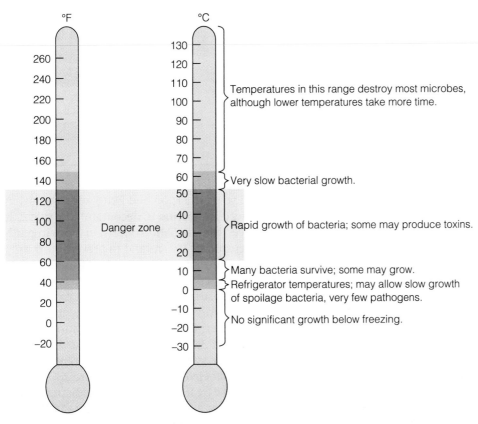

FIGURE 28.5 Spoilage temperatures.

FIGURE 28.6 The effect of a food's mass on its cooling rate and its chance of spoilage.
Note that in this example the pan of rice with a depth of 5 cm (2 in.) cooled through the incubation temperature of the *Bacillus cereus* in about an hour, whereas the pan of rice with a depth of 15 cm (6 in.) remained in this temperature range for about five hours.

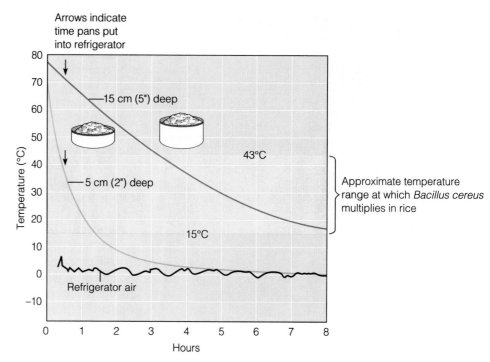

Arrows indicate time pans put into refrigerator

15 cm (5") deep

5 cm (2") deep

43°C

15°C

Refrigerator air

Approximate temperature range at which *Bacillus cereus* multiplies in rice

Hours

Anyone responsible for preparing food in large amounts should be aware of the spoilage that can arise from the slow cooling of large amounts of hot food. Hot food placed in a refrigerator cools more slowly than most people realize, and the food can spend a considerable time at the incubation temperatures of pathogenic or spoilage bacteria before its temperature drops to that of the refrigerator (Figure 28.6).

Radiation and Food Preservation

There has been considerable research into the use of ionizing radiation for food preservation, especially for military applications. It is possible to sterilize food by radiation, but this procedure is unlikely to replace conventional heat sterilization for most purposes in the near future. The taste and appearance of irradiated food have been much improved in recent years. This improvement has been accomplished mainly by freezing the food in liquid nitrogen and exhausting oxygen from the package before it is irradiated.

However, a major factor affecting the adoption of this type of preservation is that the public is apprehensive about foods exposed to radiation, even though there is no evidence that harmful effects result from the amount and types of radiation used. Irradiation is considered to be another form of food additive. As such, it must be reported on the label and must meet the stringent requirements for safety testing required of chemical food additives. These are expensive and can be justified only when a considerable market is expected.

Pasteurization-like radiation treatments that fall short of total sterilization may prove to be commercially practi-

FIGURE 28.7 Irradiation logo.
This logo indicates that a food has received irradiation treatment.

cal. For several years the Food and Drug Administration (FDA) has allowed irradiation to decontaminate spices. It has also approved the irradiation of certain foods in moderate doses to eliminate insects and to prevent sprouting or other undesirable maturation of certain fruits and vegetables. Pork may also be irradiated to eliminate the parasitic worms that cause trichinosis. One of the most potentially useful applications now approved is the irradiation of poultry to reduce the number of disease-causing bacteria. Presently, about 35% of retail poultry carries pathogenic bacteria such as *Salmonella*. Foods treated in this manner must be labeled with a logo (Figure 28.7) and a statement such as "Treated with gamma radiation to extend shelf life." After a few years to allow for familiarization, only the logo will be required.

Many other countries have allowed the use of irradiation on foods for some time. In the U.S.A., the greatest commercial interest is in irradiation treatment of spices and pet foods. The preferred method for irradiation when deep penetration is a requirement is gamma rays produced by cobalt-60. However, this type of treatment requires sev-

Irradiation sources lifted from storage pool for processing period

Shielding

Material to be irradiated

Shielding

Conveyors to move material in and out of processing position

(a) Irradiation facility

(b) Source emitting gamma rays

FIGURE 28.8 Gamma ray source used at an irradiation facility. The gamma ray source shown here is raised and lowered by remote control.

eral hours of exposure in isolation behind protective walls (Figure 28.8). High-energy electron accelerators are much faster and sterilize in a few seconds, but this treatment has low penetrating power and is suitable only for sliced meats, bacon, or similar thin products.

Microwaves, such as those used for cooking, kill bacteria only by heating foods and have little or no direct killing effect. Bacteria can survive (and are easily isolated) on the interior walls of microwave ovens.

Chemical Preservatives

Chemical preservatives are frequently added to foods to retard spoilage. Among the more common additives are sodium benzoate, sorbic acid, and calcium propionate. These chemicals are simple organic acids, or salts of organic acids, which the body readily metabolizes and which are generally judged to be safe in foods. *Sorbic acid,* or its more soluble salt *potassium sorbate,* and *sodium benzoate* prevent molds from growing in certain acidic foods, such as cheese and soft drinks. Such foods, usually with a pH of 5.5 or less, are most susceptible to mold-type spoilage. *Calcium propionate,* an effective fungistat used in bread, prevents the growth of surface molds and the *Bacillus* bacterium that causes ropy bread. These organic acids inhibit mold growth, not by affecting the pH but by interfering with the mold's metabolism or the integrity of the plasma membrane.

Sodium nitrate and *sodium nitrite* are added to many meat products, such as ham, bacon, hot dogs, and sausage. The active ingredient is sodium nitrite, which certain bacteria in the meats can also produce from sodium nitrate. These bacteria use nitrate as a substitute for oxygen under anaerobic conditions, much as it is used in denitrification in soil. The nitrite has two main functions: to preserve the pleasing red color of the meat by reacting with blood components in the meat and to prevent the germination and growth of any botulism endospores that might be present. There has been some concern that the reaction of nitrites with amino acids can form certain carcinogenic products known as **nitrosamines,** and the amount of nitrites added to foods has generally been reduced recently for this reason. However, the use of nitrites continues because of their established value in preventing botulism. Because nitrosamines are formed in the body from other sources, the added risk posed by limited use of nitrates and nitrites in meats might be lower than was once thought.

Antibiotics have only very limited use as food preservatives. A principal concern is development of resistant strains (see the box in Chapter 20, p. 505). Approved antibiotics generally do not have significant clinical uses. Nisin is probably the most widely used antibiotic in foods (see the box in Chapter 8, p. 220). Because of its poor solubility it has no clinical use; it is nontoxic and readily degraded when ingested. It is mainly active against endospore-forming bacteria. The polyene antifungal natamycin (pimaricin) is occasionally used in dairy products.

Foodborne Infections and Microbial Intoxications

We mentioned in Chapter 25 that illness resulting from microbial growth in food is associated with two principal

TABLE 28.2 Foodborne Microbial Intoxications

DISEASE	FOODS INVOLVED	PREVENTION	CLINICAL FEATURES	ONSET AND DURATION
Aflatoxin poisoning	Moldy grains, peanuts	Avoid eating contaminated grains, peanut products	Low doses may induce liver cancer; high doses cause general liver damage	Carcinogenic effects have indefinite onset; direct toxicity to animals usually sudden
Bacillus cereus intoxication	Custard, cereal, starchy foods	Refrigeration of foods	Cramps, diarrhea, nausea, vomiting	Onset: 8–16 hours Duration: Less than 1 day
Botulism	Canned foods	Proper canning procedures; boiling food before consumption	Difficulty swallowing, double vision, respiratory paralysis	Onset: 2 hours–6 days Duration: Weeks
Ciguatera	Carnivorous tropical fish in which dinoflagellate toxin is concentrated	Avoid eating large tropical fish	Tingling, rash, fever, breathing discomfort	Onset: 2–6 hours Duration: To 2 weeks
Ergotism	Moldy grains	Avoid eating contaminated grains	Burning abdominal pain, hallucinations	Onset: 1–2 hours Duration: Months
Methyl mercury poisoning	Freshwater or ocean fish	Stop dumping mercury into waters	Blurred vision, numbness, apathy, coma	Onset: 1 week Duration: May be chronic
Mushroom poisoning	*Amanita* species	Avoid eating poisonous mushrooms	Vomiting, liver necrosis, neurotoxic effects	Onset: Less than 1 day Duration: Less than 10 days
Paralytic mollusk poisoning	Bivalve mollusks during red tide (dinoflagellate blooms)	Avoiding eating mollusks during red tide	Tingling, rash, fever, respiratory paralysis	Onset: Less than 1 hour Duration: Less than 12 hours
Scombroid poisoning	Histaminelike substance produced by *Proteus* growing on certain fish, such as tuna	Refrigeration of fish	Headache, cramps, hives, shock (rare)	Onset: Several minutes–1 hour Duration: Less than 24 hours
Staphylococcal intoxication	High-osmotic-pressure foods not cooked before eating	Refrigeration of foods	Nausea, vomiting, diarrhea	Onset: Less than 1 day Duration: Less than 3 days

mechanisms. In the mechanism known as *foodborne infection,* the contaminating microorganism infects the person who ingests contaminated food. As the pathogen grows in this host, it produces damaging toxins. Diseases caused by this mechanism include gastroenteritis, typhoid fever, and dysentery. In the other mechanism, called *microbial intoxication,* the toxin is formed in the food by microbial growth and then is ingested with the food. Diseases associated with this mechanism include botulism, staphylococcal food poisoning, and mycotoxicoses (intoxications caused by fungal toxins, such as ergot or aflatoxin). Several microbial intoxication diseases are described in Table 28.2.

Dairy products, which are often consumed without having been cooked, are particularly likely to transmit food-related diseases. Standards for sanitation in the dairy industry are therefore very stringent. Because most dairy milk is drawn mechanically and is promptly put into cooled holding tanks, most of the bacteria in milk today are gram-negative psychrophiles. Grade A pasteurized cultured products, such as buttermilk and cultured sour cream, have high counts of lactic acid bacteria as a natural result of their method of manufacture. Therefore, the standards require only that they contain fewer than 10 coliforms per milliliter. Pasteurized Grade A milk is required to have a standard plate count of fewer than 20,000 bacteria and not more than 10 coliforms per milliliter. Grade

A dry milk products are required to have a standard plate count of fewer than 30,000 bacteria per gram or a coliform count of fewer than 10 per gram.

Some people prefer raw milk because of the lack of additives or because they believe it has not been altered chemically. In the U.S.A., unpasteurized (raw) dairy products are available commercially only in California, where they carry a warning. The standard for certified raw milk is a maximum bacterial count of 10,000 per milliliter and 10 coliforms per milliliter. The microbial standards guaranteed by certification do not prevent transmission of disease in raw milk; in fact, certified raw milk has been the source of several outbreaks of salmonellosis.

The Role of Microorganisms in Food Production

In the latter part of the nineteenth century, microbes used in food production were grown in pure culture for the first time. This development quickly led to improved understanding of the relationships between specific microbes and their products and activities. This period can be considered the beginning of industrial food microbiology. For example, once it was understood that a certain yeast grown under certain conditions produced beer and that certain bacteria could spoil the beer, brewers were better

① ② ③

FIGURE 28.9 Making cheddar cheese. ① The milk has been coagulated by the action of rennin (forming curd) and is inoculated with ripening bacteria for flavor and acidity. Here the workers are cutting the curd into slabs. ② The liquid whey is drained away from the chopped curd through a central trough. ③ The curd is milled to allow even more drainage of whey and is compressed into blocks for extended ripening. The longer the ripening period, the more acidic (sharper) the cheese.

able to control the quality of their products. Specific industries became active in microbiological research and selected certain microbes for their special qualities. For example, the brewing industry extensively investigated the isolation and identification of yeasts and selected those that were able to produce more alcohol.

As our knowledge of microbial genetics increases, we will continue to improve our ability to select and use desirable strains of microorganisms. Tremendous advances in molecular biology have permitted the alteration of microorganisms by techniques that are increasingly replacing the relatively hit-or-miss approaches of the past. The application of these new techniques, known as *genetic engineering,* is discussed in Chapter 9. In this section we will discuss the role of microorganisms in the production of several common foods.

Cheese

The U.S.A. leads the world in the manufacture of cheese, producing more than 1.5 million tons each year. Although there are many types of cheeses, all require the formation of a *curd,* which can then be separated from

the main liquid fraction, or *whey* (Figure 28.9). The curd is made up of a protein, *casein,* and is usually formed by the action of an enzyme, *rennin* (or chymosin), which is aided by acidic conditions provided by certain lactic acid-producing bacteria. These inoculated lactic acid bacteria also provide the characteristic flavors and aromas of fermented dairy products during the ripening process. The curd undergoes a microbial ripening process, except for a few unripened cheeses, such as ricotta and cottage cheese.

Cheeses are generally classified by their hardness, which is produced in the ripening process. The more moisture lost from the curd and the more the curd is compressed, the harder the cheese. Romano and Parmesan cheeses, for example, are classified as *very hard* cheeses; Cheddar and Swiss are *hard* cheeses. Limburger, blue, and Roquefort cheeses are classified as *semisoft;* Camembert is an example of a *soft* cheese.

The hard Cheddar and Swiss cheeses are ripened by lactic acid bacteria growing anaerobically in the interior. Such hard, interior-ripened cheeses can be quite large. The longer the incubation time, the higher the acidity and the sharper the taste of the cheese. A *Propionibacterium* (prō-pē-on-ē-bak-ti′rē-um) species produces carbon dioxide, which forms the holes in Swiss cheese. Semisoft cheeses, such as Limburger, are ripened by bacteria and other contaminating organisms growing on the surface. Blue and Roquefort cheeses are ripened by *Penicillium* molds inoculated into the cheese. The texture of the cheese is loose enough that adequate oxygen can reach the aerobic molds. The growth of the *Penicillium* molds is visible as blue-green clumps in the cheese. Camembert cheese is ripened in small packets so that the enzymes of *Penicillium* mold growing aerobically on the surface will diffuse into the cheese for ripening.

Other Dairy Products

Butter is made by churning cream until the fatty globules of butter separate from the liquid buttermilk. The typical flavor and aroma of butter and buttermilk are from diacetyls, a combination of two acetic acid molecules that is a metabolic end-product of fermentation by some lactic acid bacteria. Today, buttermilk is usually not a by-product of butter-making but is made by inoculating skim milk with bacteria that form lactic acid and the diacetyls. The inoculum is allowed to grow for 12 hours or more before the buttermilk is cooled and packaged. Sour cream is made from cream inoculated with organisms similar to those used to make buttermilk.

A wide variety of slightly acidic dairy products—probably a heritage of a nomadic past—are found around the world. Many of them are part of the daily diet in the Balkans, Eastern Europe, and Russia. One such product is yogurt, which is also popular in the U.S.A. Commercial yogurt is made from low-fat milk, from which much of the water has been evaporated in a vacuum pan. The resulting thickened milk is inoculated with a mixed culture of *Streptococcus thermophilus*, primarily for acid production, and *Lactobacillus bulgaricus* (bul-gã′ri-kus), to contribute flavor and aroma. The temperature of the fermentation is about 45°C for several hours, during which time *S. thermophilus* outgrows *L. bulgaricus*. Maintaining the proper balance between the flavor-producing and the acid-producing organisms is the secret of making yogurt.

Kefir and kumiss are popular beverages in Eastern Europe. The usual lactic acid-producing bacteria are supplemented with a lactose-fermenting yeast to give these drinks an alcohol content of 1% or 2%.

Nondairy Fermentations

Microorganisms are also used in baking, especially for bread. The sugars in bread dough are fermented by yeasts, much as they are in the fermentation of alcoholic beverages (discussed in the next section). Anaerobic conditions for production of ethanol by the yeasts are mandatory for production of alcoholic beverages. In baking, carbon dioxide forms the typical bubbles of leavened bread. Aerobic conditions favor carbon dioxide production and are encouraged as much as possible. This is the reason the bread dough is kneaded repeatedly. Whatever ethanol is pro-

duced evaporates during baking. In some breads, such as rye or sourdough, the growth of lactic acid bacteria produce the typical tart flavor (see the box in Chapter 1, p. 6).

Fermentation is also used in the production of such foods as sauerkraut, pickles, and olives. In Asia, extremely large amounts of soy sauce are produced by molds that form starch-degrading enzymes to produce fermentable sugars. This principle is used in making other Asian fermented foods, such as miso. In soy sauce production, molds such as *Aspergillus oryzae* (a-spėr-jil′lus ô′ri-zī) are grown on wheat bran and then are allowed to act with lactic acid bacteria on cooked soybean and crushed wheat mixtures. After this process has produced fermentable carbohydrates, a prolonged fermentation results in soy sauce. Table 28.3 lists many fermented foods.

Alcoholic Beverages and Vinegar

Microorganisms are involved in the production of almost all alcoholic beverages. Beer and ale are products of grain starches fermented by yeast (Table 28.4, p. 703). Because yeasts are unable to use starch directly, the starch from grain must be converted to glucose and maltose, which the yeasts can ferment into ethanol and carbon dioxide. In this conversion, called **malting**, starch-containing grains, such as malting barley, are allowed to sprout and then are dried and ground. This product, called **malt**, contains starch-degrading enzymes (amylases) that convert cereal starches into carbohydrates that can be fermented by yeasts. Light beers use amylases to convert more of the starch to fermentable glucose and maltose—which results in fewer carbohydrates and more alcohol. The beer is then diluted to arrive at an alcohol percentage in the usual range. Sake, the Japanese rice wine, is made from rice without malting because the mold *Aspergillus* is first used to convert the rice's starch to sugars that can be fermented. For **distilled spirits**, such as whiskey, vodka, and rum, carbohydrates from cereal grains, potatoes, and molasses are fermented to alcohol. The alcohol is then distilled to make a concentrated alcoholic beverage.

Wines are made from fruits, typically grapes, that contain sugars that can be used directly by yeasts for fermentation; malting is unnecessary in wine-making. Grapes usually need no additional sugars, but other fruits might be supplemented with sugars to ensure enough alcohol production. The steps of wine-making are shown in Figure 28.10. Lactic acid bacteria are important when wine is made from grapes that are especially acidic from high concentrations of malic acid. These bacteria convert the malic acid to the weaker lactic acid in a process called **malolactic fermentation**. The result is a less acidic, better-tasting wine than would otherwise be produced.

Wine producers who allowed wine to be exposed to air found that it soured from the growth of aerobic bacteria that converted the ethanol in the wine to acetic acid. The result was vinegar (*vin* means wine; *aigre* means sour). The process is now used deliberately to make vinegar. Ethanol is first produced by anaerobic fermentation of carbohy-

TABLE 28.3 Some Fermented Foods and Related Products

FOODS AND PRODUCTS	RAW INGREDIENTS	FERMENTING ORGANISMS	LOCATION PRODUCED
Dairy Products			
Cheeses (ripened)	Milk curd	*Streptococcus* spp., *Leuconostoc* spp.	Worldwide
Kefir	Milk	*Streptococcus lactis, Lactobacillus bulgaricus, Candida* spp.	Primarily southwestern Asia
Kumiss	Mare's milk	*Lactobacillus bulgaricus, L. leichmannii, Candida* spp.	Russia
Taette	Milk	*S. lactis* var. *taette*	Scandinavia
Yogurt	Milk, milk solids	*S. thermophilus, L. bulgaricus*	Worldwide
Meat and Fish Products			
Country-cured hams	Pork hams	*Aspergillus, Penicillium* spp.	Southern U.S.A.
Dry sausages	Pork, beef	*Pediococcus cerevisiae*	Europe, U.S.A.
Fish sauces	Small fish	Halophilic *Bacillus* spp.	Southeast Asia
Izushi	Fresh fish, rice, vegetables	*Lactobacillus* spp.	Japan
Nonbeverage Plant Products			
Cocoa beans	Cacao fruits (pods)	*Candida krusei, Geotrichum* spp.	Africa, South America
Coffee beans	Coffee cherries	*Erwinia dissolvens, Saccharomyces* spp.	Brazil, Congo, Hawaii, India
Kimchi	Cabbage and other vegetables	Lactic acid bacteria	Korea
Miso	Soybeans	*Aspergillus oryzae, Saccharomyces rouxii*	Primarily Japan
Olives	Green olives	*Leuconostoc mesenteroides, Lactobacillus plantarum*	Worldwide
Poi	Taro roots	Lactic acid bacteria	Hawaii
Sauerkraut	Cabbage	*Leuconostoc mesenteroides, Lactobacillus plantarum*	Worldwide
Soy sauce	Soybeans	*A. oryzae* or *A. soyae, S. rouxii, Lactobacillus delbrueckii*	Japan, China, U.S.A.
Tempeh	Soybeans	*Rhizopus oligosporus, R. oryzae*	Indonesia, New Guinea, Surinam
Breads			
Idli	Rice and bean flour	*Leuconostoc mesenteroides*	Southern India
Rolls, cakes, etc.	Wheat flours	*Saccharomyces cerevisiae*	Worldwide
San Francisco sourdough bread	Wheat flour	*S. exiguus, Lactobacillus sanfrancisco*	Northern California
Sour pumpernickel	Wheat flour	*Leuconostoc mesenteroides*	Switzerland, other areas

SOURCE: J. M. Jay, *Modern Food Microbiology,* 3rd ed. New York: Van Nostrand Reinhold, 1986.

drates by yeasts. The ethanol is then aerobically oxidized to acetic acid by acetic acid-producing bacteria of the genera *Acetobacter* and *Gluconobacter*.

Microorganisms as a Food Source

In the future—as human populations increase and if the amount of arable land does not correspondingly increase—microorganisms may become more important to the food requirements of the world. When used as a food source, microorganisms are referred to as **single-cell protein (SCP).** Microorganisms can use as substrates many carbonaceous materials, such as cellulose, methanol, and petroleum hydrocarbons.

Experimentation is under way to develop SCP that is palatable, economical, and without adverse effects for humans. So far, many SCPs have failed to meet one or more of these criteria. However, the use of genetic engineering techniques may increase our rate of success in this field.

FIGURE 28.10 **Basic steps in making red wine.** For white wines, the pressing precedes fermentation so that the color is not extracted from the solid matter.

One microorganism already sold as food in the U.S.A. and Mexico is the cyanobacterium *Spirulina* (spī-rü-lī′nä), which has long been grown as food in Lake Chad, Africa, and was also used by the Aztecs in Mexico. The yeast *Torulopsis* is used as a protein supplement in foods. In addition, bacterial proteins grown on methanol derived from methane gas have been commercially produced for animal feed, thus providing food for humans indirectly.

INDUSTRIAL MICROBIOLOGY

LEARNING OBJECTIVES

- *Differentiate between primary and secondary metabolites.*
- *Describe the role of microorganisms in the production of alternative energy sources, industrial chemicals, and pharmaceuticals.*
- *Define industrial fermentation, bioreactor, and biotechnology.*

Much of industrial microbiology—the use of microbes to make industrial products—has centered on the lactic acid and ethanol fermentations that are the basis of the major food industries we just discussed. Although the lactic acid and ethanol that are produced are incidental to food fermentation, they have industrial uses. Microbes are used to

make a variety of products, many of which are not related to foods. During the two World Wars, microbes were used to make chemical compounds such as glycerol and acetone. In the future, as the petroleum products from which many chemicals are synthesized become scarce, the world may come to depend on microbial fermentations of renewable substrates to produce these chemicals. The potential shortage of petroleum has also led to interest in microbes as alternative energy producers.

In recent years, industrial microbiology has been revolutionized by the application of genetically engineered organisms. An example of a genetically engineered biosensor to detect pollution is explained in the box on page 707. In Chapter 9 we discussed the methods for making these modified organisms and some of the products derived from them—what is now known as **biotechnology**.

Fermentation Technology

Industrial production of microbial products usually involves fermentation. In industry, this term refers to large-scale cultivation of microbes or other single cells to produce a commercially valuable substance (see the box in Chapter 5 on p. 124 for other definitions of fermentation). We have just discussed the most familiar examples: the anaerobic food fermentations used in the dairy, brewing, and wine-making industries. Much of the same technology,

TABLE 28.4 Production of Alcoholic Beverages by Yeasts

BEVERAGE	YEAST	METHOD OF PREPARATION	FUNCTION OF YEAST
Beer and Wine			
Root beer (pre-1905 and homemade)	*Saccharomyces cerevisiae*	Molasses, sassafras bark, wintergreen bark, and sarsaparilla root added for flavor; yeast added; incubated aerobically.	Converts sugar into carbon dioxide by aerobic metabolism; 0.03% alcohol.
Beer	*S. carlsbergensis* (bottom yeast)	Germinated barley releases starches and amylase enzymes (malting). Enzymes in malt hydrolyze starch to fermentable sugars (mashing). Liquid (wort) sterilized. Hops (flowers) added for flavor. Yeast added, incubated at 37°–49°C.	Converts sugar into alcohol and carbon dioxide; 4.0% alcohol. Yeast grows on bottom of fermenting vessel.
Ale	*S. cerevisiae* (top yeast)	As in beer; incubated at 50°–70°C.	Converts sugar into alcohol; 6% alcohol. Yeast grows at top of fermentation vessel.
Sake	*S. cerevisiae*	*Aspergillus oryzae* converts starch in steamed rice into sugar; yeast added; incubated at 20°C.	Converts sugar into alcohol; 14%–16% alcohol.
Wine, natural	*S. cerevisiae*	Strain of grape provides various flavors and sugar concentrations. Grapes crushed into must; sulfur dioxide added to inhibit wild yeast; yeast added. Red wines: incubated at 25°C. Aged in oak for 3–5 years and in bottle for 5–15 years. White wines: incubated at 10°–15°C. Aged 2–3 years in bottle.	Converts grape sugar into alcohol; 14% or less alcohol.
Wine, sherry	*S. cerevisiae* and *S. beticus* or *S. bayanus*	As natural wine, with additional surface growth (flor) at 27°C. Alcohol added to 18%–21%.	*S. beticus* grows as surface film, producing aldehydes from alcohol.
Wine, sparkling (champagne)	*S. cerevisiae*	As natural wine, with secondary fermentation in bottle. 2.5% sugar and yeast added to bottled wine; incubated at 15°C; bottle inverted to collect yeast in neck.	In secondary fermentation, produces carbon dioxide; yeast settles quickly.
Distilled Beverages			
Rum, Jamaica	Wild yeast	Cane molasses inoculated from previous fermentation. Oak aging adds color. Distilled to concentrate.	Converts sugar to alcohol; 50%–95% alcohol.
Brandy	*S. cerevisiae*	Fruits pressed; yeast added. Distilled to concentrate alcohol, blended with other brandies.	Converts sugar into alcohol; 40%–43% alcohol.
Whiskey	*S. cerevisiae*	Wort (see beer) is fermented by yeast. Distilled to concentrate alcohol; aged in charred oak barrels.	Converts sugar to alcohol; 50%–95% alcohol.

with the frequent addition of aeration, has been adapted to make other industrial products, such as insulin and human growth hormone, from genetically engineered microorganisms, including aerobic cells. Industrial fermentation is also used in biotechnology to obtain useful products from genetically engineered plant and animal cells. For example, animal cells are used to make monoclonal antibodies (see Chapter 17).

Vessels for industrial fermentations are called **bioreactors** and are designed with close attention to aeration, pH control, and temperature control. There are many different designs, but the most widely used bioreactors are of the continuously stirred type (Figure 28.11). The air is introduced through a diffuser at the bottom (which breaks up the incoming airstream to maximize aeration) and a series of impeller paddles and stationary wall baffles keep the microbial suspension agitated. Oxygen is not very soluble in water, and keeping the heavy microbial suspension well aerated is difficult. Highly sophisticated designs have been developed to achieve maximum efficiency in aeration and other growth requirements, including medium formulation. The high value of the products of genetically

engineered microorganisms and eucaryotic cells has stimulated development of newer types of bioreactors and computerized controls for them.

Generally speaking, the microbes in industrial fermentation produce either primary metabolites, such as ethanol, or secondary metabolites, such as penicillin. A **primary metabolite** is formed essentially at the same time as the cells, and the production curve follows the cell population curve almost in parallel, with only minimal lag (Figure 28.12a). **Secondary metabolites** are not produced until the microbe has largely completed its growth phase (known as the **trophophase**) and has entered the stationary phase of the growth cycle (Figure 28.12b). The following period, during which most of the secondary metabolite is produced, is known as the **idiophase.** The secondary metabolite may be a microbial conversion of a primary metabolite. Alternatively, it may be a product of metabolism of the original growth medium that the microbe makes only after considerable numbers of cells and a primary metabolite have accumulated.

Strain improvement is also an ongoing activity in industrial microbiology. Probably the best-known example

Motor

Acid/base for pH control

Steam for sterilization

Foam breaker

Liquid level

Flat-bladed impeller

Cooling jacket

Culture broth

Baffle

Diffuser

Sterile air

Harvesting drain

(a)

(b)

FIGURE 28.11 Bioreactors. (a) Cutaway view of a continuously stirred bioreactor. (b) A bioreactor tank.

FIGURE 28.12 Primary and secondary fermentation.
(a) Primary metabolite. A primary metabolite such as ethanol from yeasts has a production curve that lags only slightly behind the line showing cell growth.
(b) Secondary metabolite. A secondary metabolite such as penicillin from mold begins to be produced only after the main growth phase of the cell (trophophase) is completed. The main production of the secondary metabolite occurs during the stationary phase of cell growth (idiophase).

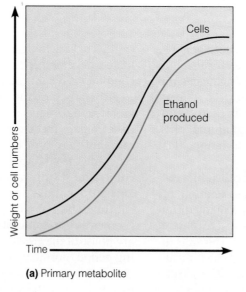

Cells

Ethanol produced

Weight or cell numbers

Time

(a) Primary metabolite

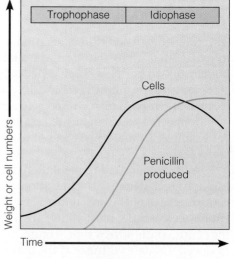

Trophophase | Idiophase

Cells

Penicillin produced

Weight or cell numbers

Time

(b) Secondary metabolite

is that of the mold used for penicillin production. The original culture of *Penicillium* did not produce penicillin in large enough quantities for commercial use. A more efficient culture was isolated from a cantaloupe from a Peoria, Illinois, supermarket. This strain was treated variously with ultraviolet light, X rays, and nitrogen mustard (a chemical mutagen). Selections of mutants, including some that arose spontaneously, quickly increased the production rates by a factor of more than 100. Improvements in fermentation techniques have nearly tripled even this yield.

Immobilized Enzymes and Organisms

Microbes are, in many ways, packages of enzymes. Industries have been making limited use of free enzymes to manufacture many products such as high-fructose syrups or semisynthetic antibiotics such as ampicillin. To do this, the enzyme must be immobilized on the surface of some solid support or otherwise manipulated so that it can convert a continuous flow of substrate to product without being lost. These techniques have also been adapted to live whole cells, and sometimes even to dead cells (Figure 28.13). Whole-cell systems are difficult to aerate, and they lack the single-enzyme specificity of immobilized enzymes. However, whole cells are advantageous if the process requires a series of steps that can be carried out by one microbe's enzymes. They also have the advantage of allowing continuously running processes with large cell populations operating at high reaction rates. Immobilized cells, which are usually anchored to microscopically small spheres or fibers, are presently used to make high-fructose syrups, aspartic acid, and numerous other products of biotechnology.

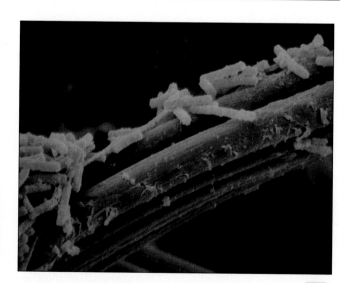

SEM

FIGURE 28.13 Immobilized cells. In some industrial processes, the cells are immobilized on surfaces such as the silk fibers shown here. The substrate flows past the immobilized cells.

Industrial Products

In this section we will discuss some of the more important commercial microbial products and the growing alternative energy industry.

Amino Acids

Amino acids have become a major industrial product from microorganisms. For example, over 165,000 tons of glutamic acid, used to make the flavor enhancer monosodium glutamate, are produced every year. Certain amino acids, such as lysine, cannot be synthesized by animals and are present only at low levels in the normal diet. Therefore, the commercial synthesis of lysine and some of the other essential amino acids as food supplements is an important industry. More than 35,000 tons of lysine are produced every year. In nature, microbes rarely produce amino acids in excess of their own needs because feedback inhibition (see Chapter 5) prevents wasteful production of primary metabolites. Commercial microbial production of amino acids depends on specially selected mutants and sometimes on ingenious manipulations of metabolic pathways.

Lysine is produced by a bacterium, *Corynebacterium glutamicum* (glü-tam′i-kum), that normally produces both threonine and lysine. If both accumulate, feedback inhibition prevents further production of both amino acids, which share a common enzyme in their synthetic pathway. A mutant bacterium that lacks the enzymes necessary to make threonine is used to produce lysine in large amounts. Enough threonine is added to the medium to allow this mutant organism to grow, but not enough to trigger feedback inhibition. The organism then continues to produce lysine because lysine alone does not cause feedback inhibition.

Glutamic acid is produced by the same bacterium that produces lysine. The bacterium is induced to excrete glutamic acid when it is provided with only a minimal amount of the vitamin biotin. The plasma membrane is weakened by the vitamin deficiency, and the glutamic acid leaks into the medium. This technique of inducing the cell to form a leaky membrane is also sometimes used in antibiotic production.

Two microbially synthesized amino acids, *phenylalanine* and *aspartic acid*, have become important as ingredients in the sugar-free sweetener aspartame (NutraSweet®).

In applications in which only the L-isomer of an amino acid is wanted, microbial production, which forms only the L-isomer, has an advantage over chemical production, which forms both the D- and L-isomers.

About the only amino acids that are not produced by microbial fermentation are methionine, alanine, glycine, and cysteine, which are chemically synthesized.

Citric Acid

Citric acid is a constituent of citrus fruits, such as oranges and lemons, and at one time these were its only industrial source. However, over a hundred years ago, citric acid was identified as a product of mold metabolism. It can be chemically synthesized, but not cheaply enough to replace production by mold fermentation. Citric acid has an extraordinary range of uses beyond the obvious ones of giving tartness and flavor to foods. It is an antioxidant and pH adjuster in many foods, and in dairy products it often serves as an emulsifier. Most citric acid is used in foods, and only about 30% of its applications are industrial and pharmaceutical. Well over 300,000 tons of citric acid are produced every year. Much of it is produced by a mold, *Aspergillus niger* (nī′jér), using molasses as a substrate. The mold excretes citric acid when provided with only a limited supply of iron and manganese.

Enzymes

Enzymes are widely used in different industries. For example, amylases are used in the production of syrups from corn starch, in the production of paper sizing (a coating for smoothness, as on this page), and in the production of glucose from starch. Glucose isomerase is an important enzyme; it converts the glucose that amylases form from starches into fructose, which is used in place of sucrose as a sweetener in many foods. Probably half of the bread baked in this country is made with the aid of proteases, which adjust the amounts of glutens (protein) in wheat so that baked goods are improved or made uniform. Other proteolytic enzymes are used as meat tenderizers or in detergents as an additive to remove proteinaceous stains. Rennin, which is an enzyme used to form curds in milk, is usually produced commercially by fungi. Several enzymes produced commercially by microorganisms are listed in Table 28.5.

Vitamins

Vitamins are sold in large quantities combined in tablet form and are used as individual food supplements. Microbes can provide an inexpensive source of some vitamins. *Vitamin B$_{12}$* is produced by *Pseudomonas* and *Propionibacterium* spp. Cobalt is a part of the molecule, and cobalt chloride must be provided in the medium. **Riboflavin** is another vitamin produced by fermentation, mostly by fungi such as *Ashbya gossypii* (ash′bē-ä gos-sip′ē-ē).

Pharmaceuticals

Modern pharmaceutical microbiology developed after World War II, with the production of antibiotics. To increase the number of random mutations, antibiotic-producing organisms were subjected to ultraviolet radiation in the hope that some mutations would be useful. These and other techniques soon led to strains that produce antibiotics in amounts thousands of times greater than the original isolates did. All antibiotics were originally the products of microbial metabolism. Many antibiotics are still produced by microbial fermentations, and work continues on selection of more productive mutants by nutritional and genetic manipulations. Antibiotics are

TABLE 28.5 Some Microbial Enzymes Produced Commercially

ENZYME	MICROORGANISM	USE OF ENZYME
α-Amylase	*Aspergillus niger, Aspergillus oryzae*	Baking: Flour supplement Starch: Cold-water laundry Textiles: Desizing agent Food: Converts starch to glucose Pharmaceuticals: Digestive aid used when eating beans (Beano®)
β-Amylase	*Bacillus subtilis*	Brewing: Converts starch to maltose
Cellulase	*Trichoderma viride*	Food: Liquid coffee concentrate
Chymosin (rennin)	*Escherichia coli* (recombinant)	Food: Making cheese curd
Dextransucrase	*Leuconostoc mesenteroides*	Pharmaceuticals: Dextran
Glucanase	*Saccharomyces* (recombinant)	Brewing: "Lite" beer
β-Glucanase	*B. subtilis, A. niger*	Brewing: Degradation of barley cell walls
Glucose oxidase	*A. niger*	Food: Glucose removal from egg solids Pharmaceuticals: Test papers
Invertase	*Saccharomyces cerevisiae*	Candy: Prevents granulation in soft center Food: Artificial honey
Lactase	*Saccharomyces fragilis*	Dairy: Prevents crystallization of lactose in ice cream and concentrated milk
	E. coli	Pharmaceuticals: Digestive aid used when eating dairy products (Lactaid®)
Lipase	*A. niger*	Dairy: Flavor production in cheese
Pectinase	*A. niger*	Wine and juice: Clarification
Penicillinase	*B. subtilis*	Pharmaceuticals: Diagnostic agent
Protease	*A. oryzae*	Brewing: Beer stabilizer Baking: Bread making Food: Meat tenderizer Pharmaceuticals: Digestive aid Textiles: Desizing agent
Streptokinase	Group C β-hemolytic *Streptococcus*	Pharmaceuticals: Lysis of embolisms and thrombolisms
Subtilisins	*Bacillus subtilis*	Laundry: Added to detergents

SOURCE: Adapted from J. R. Porter, "Microbiology and the food and energy crisis." *ASM News* 40(11):822, November 1974.

typically made industrially by inoculating a solution of growth medium with spores of the appropriate mold or streptomycete and vigorously aerating it. After the antibiotic reaches a satisfactory concentration, it is extracted by solution, precipitation, and other standard industrial procedures.

Applications of Microbiology

BIOSENSORS— BACTERIA THAT DETECT TOXIC POLLUTANTS

Each year in the U.S.A., industrial plants generate 265 million metric tons of hazardous waste, 80% of which make their way into landfills. Burying these chemicals doesn't remove them from the ecosystem, however; it just moves them to other places, where they may still find their way into bodies of water. Traditional chemical analyses to locate these chemicals are expensive and can't distinguish between chemicals that affect biological systems from those that lie inert in the environment.

In response to this problem, Olga Selifonova and her colleagues at the Center for Environmental Diagnostics and Bioremediation in Florida are among a group of scientists developing biosensors—bacteria that can locate biologically active pollutants. Biosensors do not require costly chemicals or equipment, and they work quickly—within minutes.

In order to work, these bacterial biosensors require both a receptor that is activated in the presence of pollutants and a reporter that will make such a change apparent. Biosensors use the *lux* operon from *Vibrio* or *Photobacterium* as a reporter. This operon contains inducer and structural genes for the enzyme luciferase. In the presence of a coenzyme called $FMNH_2$, luciferase reacts with the molecule in such a way that the enzyme–substrate complex emits blue-green light, which then oxidizes the $FMNH_2$ to produce FMN. Therefore, a bacterium containing the *lux* gene will emit visible light (see photos) when the receptor is activated.

The *lux* operon is readily transferred to many bacteria. It has already been transferred to the plant pathogen *Xanthomonas* in order to monitor the progression of its infection in plants. Because the altered bacteria emit light, their presence can be detected by photographic film. *Bradyrhizobium* has also been engineered to contain the *lux* genes in an effort to understand root nodule formation in soybean plants.

In another application, *Lactococcus* bacteria containing the *lux* operon are being used to detect the presence of antibiotics in milk that is to be used for cheese production. (If milk contains antibiotics, the cheese starter cultures will not grow.) Because the emission of light requires a living cell, the presence of antibiotics is measured as a decrease in light output by such recombinant *Lactococcus* bacteria.

The Microtox System developed in Great Britain uses the marine bacterium *Photobacterium* directly to detect toxic pollutants. These bacteria cannot emit light if they are killed by pollutants.

Other environmental biosensors use recombinant bacteria such as *Escherichia coli*. Some *E. coli* bacteria have been genetically engineered to contain both a *mer* resistance operon, which acts as a receptor, and the *lux* operon. The *mer* acts as the promotor for the *lux* structural genes. When such an *E. coli* cell is placed in a soil sample containing mercury, the mercury acts as an inducer, activating the production of luciferase in the bacteria.

Detecting harmful pollutants in soil and water is necessary to protect humans and animals. However, after detection, bioremediation processes are still required to remove the pollutants.

(a)

(b)

Vibrio fischeri emits light when energy is released by the transport of electrons to luciferase. These photos show colonies of *V. fischeri* photographed **(a)** in daylight and **(b)** in the dark, illuminated by their own light.

FIGURE 28.14 Production of steroids. Conversion of a precursor compound such as a sterol into a steroid by *Streptomyces*. The addition of a hydroxyl group to carbon number 11 (highlighted in purple on the steroid) might require more than 30 steps by chemical means, but the microorganism can add it in only one step.

FIGURE 28.15 Biological leaching of copper ores.
(a) The chemistry of the process is considerably more complicated than shown here. The *Thiobacillus* bacteria essentially derive their energy from the oxidation of sulfides to sulfates. (b) Aerial photograph of a copper ore leaching facility.

(a) Simplified copper ore leaching process

(b) Copper ore leaching facility

Vaccines are a product of industrial microbiology. Many antiviral vaccines are mass produced in chicken eggs or cell cultures. Production of vaccines against bacterial diseases usually requires the growth of large amounts of the bacteria. Genetic engineering techniques are increasingly important in the area of vaccine development and production (see Chapter 9).

Steroids (see Figure 2.14) are a very important group of chemicals that include cortisone, which is used as an anti-inflammatory drug, and estrogens and progesterone, which are used in birth control pills. Recovering steroids from animal sources is difficult, but microorganisms can synthesize steroids from sterols or from related compounds derived from animals and plants. Certain strains of molds and bacteria (particularly streptomycetes) can produce steroids by converting chemical groups on sterols and similar molecules (Figure 28.14). Genetic engineering has already begun to revolutionize the microbial production of medical pharmaceuticals.

Uranium and Copper

Thiobacillus ferrooxidans is used in the recovery of otherwise unprofitable grades of uranium and copper ores (Figure 28.15). For example, when solutions containing ferric ion (Fe^{3+}) are washed through deposits of insoluble copper compounds, the copper is oxidized to soluble compounds. In this process, Fe^{3+} is reduced to Fe^{2+}. The Fe^{2+} can be reoxidized by *T. ferrooxidans*. The soluble copper then moves out of the ores and is reclaimed.

Microorganisms

Microorganisms themselves sometimes constitute an industrial product. A simple example is the sale of yeast cultures for making wines at home. Bakers' yeast is produced in large aerated fermentation tanks. At the end of the fermentation, the contents of the tank are about 4% yeast solids. The concentration can be much higher, but only at a cost in efficiency. The cells are harvested by continuous centrifuges and are pressed into the familiar yeast cakes sold in the supermarket for home baking. Wholesale bakers purchase yeast in 50-lb boxes.

Other important microbes that are sold industrially are the symbiotic nitrogen-fixing bacteria *Rhizobium* and *Bradyrhizobium*. These organisms are usually mixed with peat moss to preserve moisture; the farmer mixes the peat moss and bacterial inoculum with the seeds of legumes to ensure infection of the plants with efficient nitrogen-fixing strains (see Chapter 27). Gardeners for many years have used the insect pathogen *Bacillus thuringiensis* (Bt) to control leaf-eating insect larvae. The Bt strain *israelensis* is active against mosquito larvae and is widely used in municipal control programs. Commercial preparations containing toxic crystals and endospores of this organism are available at almost any gardening supply store.

Alternative Energy Sources Using Microorganisms

Each year, the U.S.A. produces many hundreds of tons of organic wastes from crops, forests, and municipalities. This waste is called biomass, and so far it has been disposed of by burning or placing in a dump. Conversion of biomass into alternative fuel sources, called **bioconversion**, would help meet our energy needs as well as help dispose of troublesome accumulations of solid waste. The bioconversion of organic materials into alternative fuels by microorganisms is a developing industry.

Methane is probably one of the most convenient energy sources produced from biomass conversion. Methane was discussed in Chapter 27 in the box on page 668 as a

FIGURE 28.16 Methane production from solid waste. Methane begins to accumulate a few months after a landfill is completed and sealed. It remains productive for 5 to 10 years. This photo shows a methane flare from such a landfill.

product of the anaerobic treatment of sewage sludge. Many communities produce useful amounts of methane from wastes in landfill sites (Figure 28.16). Large cattle-feeding lots must dispose of immense amounts of animal manure, and considerable effort has been devoted to devising practical methods for producing methane from these wastes. A major problem with any scheme for large-scale methane production is the need to economically concentrate the widespread biomass material. If it could be concentrated, the animal and human wastes in the U.S.A. could supply much of our energy now supplied by fossil fuels and natural gas.

An interesting aspect of methane production from animal wastes is the inhibition of the process by certain antibiotics fed to cattle. The purpose of these antibiotics is to suppress methane production from the largely cellulosic foods in the animal's rumen. Converting as much as possible of the nutrients in these foods into meat or milk is desirable, but certain anaerobic bacteria convert some of these nutrients into methane gas. This gas tends to be lost to the atmosphere and represents an inefficient use of nutrients. The antibiotics used to inhibit these bacteria are excreted in the animal's wastes, and, unfortunately, they inhibit desired methane production from the wastes.

The agricultural industry has encouraged the production of ethanol from agricultural products. Gasoline-containing ethanol (90% gasoline + 10% ethanol) is available in many parts of the U.S.A. and is used in automobiles worldwide. Corn is currently the most frequently used substrate, but eventually agricultural waste products may be used. A by-product of ethanol production is a large quantity of ethanol-producing yeasts. The FDA has approved use of such yeasts as a food additive in human snack food. In the future we will see more efforts to recover the energy now lost in industrial, municipal, and rural wastes.

Industrial Microbiology and the Future

Microbes have been exceedingly useful to humankind, even when their existence was unknown. They will always remain an essential part of many basic food-processing technologies. The development of genetic engineering has further intensified interest in industrial microbiology by expanding the potential for new products and applications. As new biotechnology applications and products enter the marketplace, they will affect our lives and well-being in ways that we can only guess about today.

STUDY OUTLINE

FOOD MICROBIOLOGY (pp. 691–702)

1. The earliest methods of preserving foods were drying, the addition of salt or sugar, and fermentation.

Food Preservation and Spoilage (pp. 691–697)
1. Modern food preservation includes the use of heat, cold, and radiation.

Canning (pp. 692–694)
1. In 1810, Nicholas Appert devised a canning process that involved keeping air out of sealed cans and heating food in the cans.
2. In the nineteenth century, steam under pressure (pressure cooking) was used in canning.
3. Today, commercial sterilization heats canned foods to the minimum temperature necessary to destroy *Clostridium botulinum* endospores while minimizing alteration of the food.
4. The commercial sterilization process uses sufficient heat to reduce a population of *C. botulinum* by 12 logarithmic cycles (12D treatment).
5. Endospores of thermophiles can survive commercial sterilization.
6. Canned foods stored above 45°C can be spoiled by thermophilic anaerobes.
7. Thermophilic anaerobic spoilage is sometimes accompanied by gas production; if no gas is formed, the spoilage is called flat sour spoilage.
8. Spoilage by mesophilic bacteria is usually from improper heating procedures or leakage.
9. Acidic foods can be preserved by heat of 100°C because microorganisms that survive are not capable of growth in a low pH.
10. *Byssochlamys, Aspergillus,* and *Bacillus coagulans* are acid-tolerant and heat-resistant organisms that can spoil acidic foods.

Pasteurization (p. 694)
1. Louis Pasteur determined that, because spoilage was caused by only certain microorganisms, it was not necessary to kill all microorganisms to prevent spoilage.
2. Pasteurization is the application of heat that selectively kills microorganisms without greatly altering the quality of the food.

Aseptic Packaging (p. 694)
1. Presterilized materials are assembled into packages and aseptically filled with heat-sterilized liquid foods.

Low-Temperature Preservation (pp. 695–696)
1. Many microorganisms, although few pathogens, will grow slowly at refrigeration temperatures (0°C to 7°C).
2. *Clostridium botulinum* type E, *Yersinia enterocolitica,* and *Listeria monocytogenes* are pathogens that will grow at low temperatures.
3. Freezing might not kill bacteria, but it will prevent their growth.

Radiation and Food Preservation (pp. 696–697)
1. Gamma radiation can be used to sterilize food, kill insects and parasitic worms, and prevent sprouting of fruits and vegetables.
2. Microwaves kill bacteria indirectly by heating the food.

Chemical Preservatives (p. 697)
1. Sodium benzoate, sorbic acid or potassium sorbate, and calcium propionate are added to foods to inhibit the growth of molds.
2. These chemicals are metabolized by humans.
3. Sodium nitrate and sodium nitrite are added to meats to preserve the red color and to prevent the germination of *C. botulinum* endospores.
4. Nitrates are converted to nitrites by some bacteria during anaerobic respiration; nitrites are the active ingredient.

5. The antibiotics nisin and natamycin are used to preserve dairy products.

Foodborne Infections and Microbial Intoxications
(pp. 697–698)

1. In a foodborne infection, the pathogen grows in the infected host.
2. Gastroenteritis, typhoid fever, and dysentery are foodborne infections.
3. Microbial intoxication results from ingestion of a toxin produced by microbes growing in food.
4. Botulism, staphylococcal food poisoning, and mycotoxicoses are examples of microbial intoxications.
5. Pasteurized dairy milk might contain gram-negative psychrotrophs.
6. Pasteurized products of cultured milk (such as buttermilk and sour cream) contain large numbers of lactic acid bacteria.
7. Public health service standards for Grade A pasteurized milk are a standard plate count of fewer than 20,000 bacteria per milliliter and not more than 10 coliforms per milliliter.
8. Dry milk products must have a standard plate count of fewer than 30,000 bacteria per gram or a coliform count fewer than 10 per gram.
9. The species of bacteria in unpasteurized (raw) milk are not controlled, and raw milk can transmit disease.

The Role of Microorganisms in Food Production
(pp. 698–702)

Cheese (pp. 699–700)

1. The milk protein casein curdles because of the action by lactic acid bacteria or the enzyme rennin or chymosin.
2. Cheese is the curd separated from the liquid portion of milk, called whey.
3. Hard cheeses are produced by lactic acid bacteria growing in the interior of the curd.
4. The growth of microorganisms in cheeses is called ripening.
5. Semisoft cheeses are ripened by bacteria growing on the surface.
6. Soft cheeses are ripened by *Penicillium* growing on the surface.

Other Dairy Products (p. 700)

1. Old-fashioned buttermilk was produced by lactic acid bacteria growing during the butter-making process.
2. Commercial buttermilk is made by letting lactic acid bacteria grow in skim milk for 12 hours.
3. Sour cream, yogurt, kefir, and kumiss are produced by lactobacilli, streptococci, or yeasts growing in low-fat milk.

Nondairy Fermentations (p. 700)

1. Sugars in bread dough are fermented by yeast to ethanol and CO_2; the CO_2 causes the bread to rise.

2. Sauerkraut, pickles, olives, and soy sauce are the products of microbial fermentations.

Alcoholic Beverages and Vinegar (pp. 700–701)

1. Carbohydrates obtained from grains, potatoes, or molasses are fermented by yeasts to produce ethanol in the production of beer, ale, sake, and distilled spirits.
2. The sugars in fruits such as grapes are fermented by yeasts to produce wines.
3. In wine-making, lactic acid bacteria convert malic acid into lactic acid in malolactic fermentation.
4. *Acetobacter* and *Gluconobacter* oxidize ethanol in wine to acetic acid (vinegar).

Microorganisms as a Food Source (pp. 701–702)

1. Microorganisms can be used to produce single-cell protein (SCP) from otherwise unusable substrates such as cellulose.

INDUSTRIAL MICROBIOLOGY (pp. 702–710)

1. Microorganisms produce alcohol, lactic acid, glycerol, and antibiotics that are used in food manufacturing and other industrial processes.
2. Industrial microbiology has been revolutionized by the ability of genetically engineered cells to make many new products.
3. Biotechnology is a way of making commercial products by using living organisms.

Fermentation Technology (pp. 702–705)

1. In industry, the growth of cells on a large scale is called fermentation.
2. Industrial fermentations are carried on in bioreactors, which control aeration, pH, and temperature.
3. Primary metabolites such as ethanol are formed as the cells grow (during the trophophase).
4. Secondary metabolites such as penicillin are produced during the stationary phase (idiophase).
5. Mutant strains that produce a desired product can be selected.

Immobilized Enzymes and Organisms (p. 705)

1. Enzymes or whole cells can be bound to solid spheres or fibers. When substrate passes over the surface, enzymatic reactions change the substrate to the desired product.

Industrial Products (pp. 705–709)

1. Most amino acids used in foods and medicine are produced by bacteria.
2. Microbial production of amino acids can be used to produce L-isomers; chemical production results in both D- and L-isomers.
3. Lysine and glutamic acid are produced by *Corynebacterium glutamicum*.
4. Citric acid, used in foods, is produced by *Aspergillus niger*.

5. Enzymes used in manufacturing foods, medicines, and other goods are produced by microbes.

6. Some vitamins used as food supplements are made by microorganisms.

7. Vaccines, antibiotics, and steroids are products of microbial growth.

8. The metabolic activities of *Thiobacillus ferrooxidans* can be used to recover uranium and copper ores.

9. Yeasts are grown for wine- and bread-making; other microbes (*Rhizobium, Bradyrhizobium,* and *Bacillus thuringiensis*) are grown for agricultural use.

Alternative Energy Sources Using Microorganisms (pp. 709–710)

1. Organic waste, called biomass, can be converted by microorganisms into alternative fuels, a process called bioconversion.

2. Fuels produced by microbial fermentation are methane and ethanol.

Industrial Microbiology and the Future (p. 710)

1. Genetic engineering will enhance the ability of industrial microbiology to produce medicines and other useful products.

STUDY QUESTIONS

Review

1. Why are foods preserved?

2. List five methods used to preserve foods.

3. Matching.

____ Sorbic acid	(a)	Sterilization
____ 5°C	(b)	Pasteurization
____ 72°C for 15 seconds	(c)	Low-temperature preservation
____ 121°C for 15 minutes	(d)	Chemical preservation
____ 12D treatment		

4. Even though many mesophilic bacteria can grow at low temperatures, what value is it to store food in a refrigerator?

5. Define pasteurization and aseptic packaging.

6. Pasteurization does not kill all microorganisms, so food can still spoil. Why, then, are dairy products pasteurized?

7. State one advantage and one disadvantage of the addition of nitrites to foods.

8. Outline the steps in the process of wine production.

9. Outline the steps in the production of cheese, and compare the production of hard and soft cheeses.

10. List two examples of whole cells as the desired product of industrial fermentation.

11. How can microorganisms provide energy sources? What metabolic processes can result in fuels?

12. Nicholas Appert's invention was the forerunner of hermetically sealed cans. Why do hermetically sealed cans prevent food spoilage?

13. Why is a can of blackberries preserved by commercial sterilization typically heated to 100°C instead of at least 116°C?

14. Under what conditions would you expect to find thermophiles spoiling foods? Psychrotrophs?

15. Discuss the importance of industrial microbiology.

16. Label the trophophase and idiophase in this graph. Indicate when primary and secondary metabolites are formed.

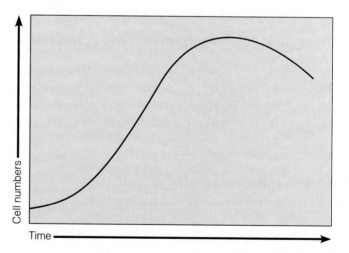

17. Why is a bioreactor better than a large flask for industrial production of an antibiotic?

Critical Thinking

1. Which bacteria seem to be most frequently used in the production of food? Can you guess why?

2. Why do the following processes preserve food?
   (a) Fermentation
   (b) Salting
   (c) Drying

3. *Methylophilus methylotrophus* can convert methane (CH_4) into proteins. Amino acids are represented by

$$H_2N-\underset{\underset{H}{|}}{\overset{\overset{R}{|}}{C}}-C\overset{\diagup O}{\underset{\diagdown OH}{}}$$

Diagram a pathway illustrating the production of at least one amino acid.

Clinical Applications

1. Three to five days after eating Thanksgiving dinner at a restaurant, 112 people developed fever and gastroenteritis. All the food had been consumed except for five "doggie" bags. Bacterial analysis of the mixed contents of the bags (containing roast turkey, giblet gravy, and mashed potatoes) showed the same bacterium that was isolated from the patients. The gravy had been prepared from giblets of 43 turkeys that had been refrigerated for three days prior to preparation. The uncooked giblets were ground in a blender and added to a thickened, hot stock mixture. The gravy was not reboiled and was stored at room temperature throughout Thanksgiving Day. What was the source of the illness? What was the most likely etiologic agent? Was this an infection or an intoxication? (*Hint:* Refer to Chapter 25.)

2. An outbreak of typhoid fever occurred following a family gathering of 293 people. Cultures from 17 of these people yielded *Salmonella typhi*. Nine foods prepared by three foodhandlers were available at the event.

FOODS CONSUMED	PERCENT ILL
Salad and roast beef	60
Noodles and baked beans	42
Noodles and egg salad	12
Egg salad and roast beef	0
Baked beans and fruit	0

What is the most likely source of *S. typhi?* Which tests would verify this?

3. Suppose you are culturing an organism that produces enough lactic acid to kill itself in a few days.
   (a) How can the use of a bioreactor help you maintain the culture for weeks or months? The following graph shows conditions in the bioreactor.

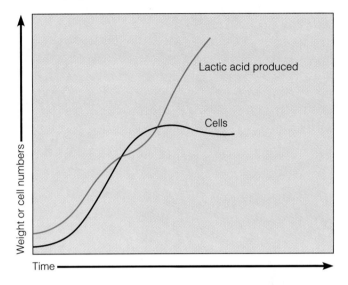

   (b) If your desired product is a secondary metabolite, when can you begin collecting it?
   (c) If your desired product is the cells themselves and you want to maintain a continuous culture, when can you begin harvesting?

Classification of Bacteria According to *Bergey's Manual**

KINGDOM PROCARYOTAE

Division I **Gracilicutes**
Procaryotes with thin cell walls, implying a gram-negative type of cell wall

Division II **Firmicutes**
Procaryotes with thick and strong skin, indicating a gram-positive type of cell wall

Division III **Tenericutes**
Procaryotes of a pliable and soft nature, indicating the lack of a rigid cell wall

Division IV **Mendosicutes**
Procaryotes that lack conventional peptidoglycan (archaeobacteria)

The Spirochetes

Spirochaetales
    Spirochaetaceae
        Borrelia
        Cristispira
        Spirochaeta

Treponema
Leptospiraceae
    Leptospira
Other Organisms: Hindgut Spirochetes of Termites and *Cryptocercus punctulatus* (wood-eating cockroach)

Aerobic/Microaerophilic, Motile, Helical/Vibrioid Gram-Negative Bacteria

Alteromonas
Aquaspirillum
Azospirillum
Bdellovibrio
Campylobacter
Cellvibrio
Halovibrio
Helicobacter
Herbaspirillum
Marinomonas
Micavibrio
Oceanospirillum
Spirillum
"Sporospirillum"
Vampirovibrio

Nonmotile (or Rarely Motile), Gram-Negative Curved Bacteria

Spirosomaceae
    Flectobacillus
    Runella
    Spirosoma

*Bergey's Manual of Systematic Bacteriology, 1st ed., 4 vols. (1984), is the reference for classification. *Bergey's Manual of Determinative Bacteriology,* 9th ed. (1994), should be used for identification of unknown bacteria.

Note: Names in quotation marks have not been validated by publication in the *International Journal of Systematic Bacteriology.*

Other Genera
  Ancyclobacter
  "Brachyarcus"
  Cyclobacterium
  Meniscus
  "Pelosigma"

Gram-Negative Aerobic Rods and Cocci

Pseudomonadaceae
  Frateuria
  Pseudomonas
  Xanthomonas
  Zooglea
Azotobacteriaceae
  Azomonas
  Azotobacter
Rhizobiaceae
  Agrobacterium
  Bradyrhizobium
  Phyllobacterium
  Rhizobium
Methylococcaceae
  Methylococcus
  Methylomonas
Acetobacteraceae
  Acetobacter
  Gluconobacter
Legionellaceae
  Legionella
Neisseriaceae
  Acinetobacter
  Kingella
  Moraxella
  Neisseria
Other Genera
  Acidiphilium
  Acidomonas
  Acidothermus
  Afipia
  Agromonas
  Alcaligenes
  Alteromonas
  Aminobacter
  Aquaspirillum
  Azorhizobium
  Beijerinckia
  Bordetella
  Brucella
  Chromohalobacter
  Chryseomonas
  Comoamonas
  Cupriavidas
  Deleya
  Derxia
  Ensifer
  Erythrobacter

  Flavimonas
  Flavobacterium
  Francisella
  Halomonas
  Hydrogenophaga
  Janthinobacterium
  Lampropedia
  Marinobacter
  Marinomonas
  Mesophilobacter
  Methylobacillus
  Methylobacterium
  Methylophaga
  Methylophilus
  Methylovorus
  Morococcus
  Oceanospirillum
  Oligella
  Paracoccus
  Phenylobacterium
  Psychrobacter
  Rhizobacter
  Roseobacter
  Rugamonas
  Sinorhizobium
  Sphingobacterium
  Thermoleophilum
  Thermomicrobium
  Thermus
  Variovorax
  Volcaniella
  Weeksella
  Xanthobacter
  Xylella
  Xylophilus
  Zoogloea

Facultatively Anaerobic Gram-Negative Rods

Enterobacteriaceae
  Arsenophonus
  Budvicia
  Buttiauxella
  Cedecea
  Citrobacter
  Edwardsiella
  Enterobacter
  Erwinia
  Escherichia
  Ewingella
  Hafnia
  Klebsiella
  Kluyvera
  Leclercia
  Leminorella
  Moellerella
  Morganella

Obesumbacterium
Pantoea
Pragia
Proteus
Providencia
Rahnella
Salmonella
Serratia
Shigella
Tatumella
Xenorhabdus
Yersinia
Yokenella
Vibrionaceae
  Aeromonas
  Enhydrobacter
  Photobacterium
  Plesiomonas
  Vibrio
Pasturellaceae
  Actinobacillus
  Haemophilus
  Pasteurella
Other Genera
  Calymmatobacterium
  Cardiobacterium
  Chromobacterium
  Eikenella
  Gardnerella[†]
  Streptobacillus
  Zymomonas

Fusobacterium
Haloanaerobium
Halobacteroides
Ilyobacter
Lachnospira[†]
Leptotrichia
Malonomonas
Megamonas
Mitsuokella
Oxalobacter
Pectinatus
Pelobacter
Porphyromonas
Prevotella
Propionigenium
Propionispira
Rikenella
Roseburia
Ruminobacter
Sebaldella
Selenomonas
Sporomusa
Succinimonas
Succinivibrio
Syntrophobacter
Syntrophosmonas
Thermobacteroides
Thermospipho
Thermotoga
Tissierella
Wolinella
Zymophilus

Gram-Negative Anaerobic, Straight, Curved, and Helical Bacteria

Bacteroidaceae
  Acetivibrio
  Acetoanaerobium
  Acetofilamentum
  Acetogenium
  Acetomicrobium
  Acetothermus
  Acidaminobacter
  Anaerobiospirillum
  Anaerorhabdus
  Anaerovibrio
  Bacteroides
  Butyrivibrio[†]
  Centipeda
  Fervidobacterium
  Fibrobacter

Dissimulatory Sulfate- or Sulfur-Reducing Bacteria

Desulfobacter
Desulfobacterium
Desulfobulbus
Desulfococcus
Desulfomicrobium
Desulfomonas
Desulfomonile
Desulfonema
Desulfosarcina
Desulfotomaculum[†]
Desulfovibrio
Desulfurella
Desulfuromonas
Thermodesulfobacterium

Anaerobic Gram-Negative Cocci

Veillonellaceae
  Acidaminococcus
  Megasphaera
  Syntrophococcus
  Veillonella

[†]Genera that have been placed in more than one volume. *Gardnerella, Lachnospira,* and *Butyrivibrio* have thin, gram-positive walls but stain as gram-negatives. *Desulfotomaculum* is an endospore-forming bacterium that is also included in the dissimilatory sulfate reducers. *Gallionella* is both iron-oxidizing and appendaged.

The Rickettsias and Chlamydias

Rickettsiales
  Rickettsiaceae
    Cowdria
    Coxiella
    Ehrlichia
    Neorickettsia
    Rickettsia
    Rickettsiella
    Rochalimaea
    Wolbachia
  Bartonellaceae
    Bartonella
    Grahamella
  Anaplasmataceae
    Aegyptianella
    Anaplasma
    Eperythrozoon
    Haemobartonella
Chlamydiales
  Chlamydiaceae
    Chlamydia

The Mycoplasmas

Mycoplasmataceae
  Mycoplasma
  Ureaplasma
Acholeplasmataceae
  Acholeplasma
Spiroplasmataceae
  Spiroplasma
Other Genera
  Anaeroplasma
  Asteroleplasma

Endosymbionts

Endosymbionts of Protozoa
  Caedibacter
  Holospora
  Lyticum
  Pseudocaedibacter
  Tectibacter
Endosymbionts of Insects
  Blattabacterium
Endosymbionts of Fungi and Invertebrates Other than Arthropods

Gram-Positive Cocci

Aerobic, Catalase-Positive Genera
  Deinobacter
  Deinococcus
  Marinococcus
  Micrococcus
  Planococcus
  Saccharococcus
  Staphylococcus
  Stomatococcus
Aerotolerant, Catalase-Negative Genera
  Aerococcus
  Enterococcus
  Gemella
  Lactococcus
  Leuconostoc
  Melissococcus
  Pediococcus
  Streptococcus
  Trichococcus
  Vagococcus
Anaerobic, Catalase-Negative Genera
  Coprococcus
  Peptococcus
  Peptostreptococcus
  Ruminococcus
  Sarcina

Endospore-Forming Gram-Positive Rods and Cocci

Amphibacillus
Bacillus
Clostridium
Desulfotomaculum[†]
Oscillospira
Sporohalobacter
Sporolactobacillus
Sporosarcina
Sulfidobacillus
Syntrophospora

Regular, Nonsporing Gram-Positive Rods

Brochothrix
Carnobacterium
Caryophanon
Erysipelothrix
Kurthia
Lactobacillus
Listeria
Renibacterium

Irregular, Nonsporing Gram-Positive Rods

Acetobacterium
Acetogenium
Actinomyces
Aeromicrobium
Agromyces
Arachnia
Arcanobacterium
Arthrobacter
Aureobacterium
Bifidobacterium
Brachybacterium
Brevibacterium

Butyrivibrio†
Caseobacter
Cellulomonas
Clavibacter
Coriobacterium
Curtobacterium
Dermabacter
Eubacterium
Exigouibacterium
Falcivibrio
Gardnerella†
Jonesia
Lachnospira†
Microbacterium
Mobiluncus
Pimelobacter
Propionibacterium
Rarobacter
Rothia
Rubrobacter
Sphaerobacter
Terrabacter
Thermoanaerobacter

Mycobacteria

Mycobacteriaceae
  Mycobacterium

Anoxygenic Phototrophic Bacteria

Purple Sulfur Bacteria; Internal Sulfur Granules
  Chromatiaceae
    Amoebobacter
    Chromatium
    Lamprobacter
    Lamprocystis
    Thiocapsa
    Thiocystis
    Thiodictyon
    Thiopedia
    Thiospirillum
Purple Sulfur Bacteria; External Sulfur Granules
  Ectothiorhodopsin
Purple Nonsulfur Bacteria
  Rhodobacter
  Rhodocyclus
  Rhodomicrobium
  Rhodopila
  Rhodopseudomonas
  Rhodospirillum
Green Nonsulfur Bacteria
  Heliobacillus
  Heliobacterium
Green Sulfur Bacteria
  Anacalochloris
  Chlorobium

Chloroherpeton
Pelodictyon
Prosthecochloris
Multicellular Filamentous Green Bacteria
  Chloroflexus
  Chloronema
  Heliothrix
  "Oscillochloris"
Anaerobic Chemotrophic Bacteria
  Erythrobacter

Oxygenic Phototrophic Bacteria

Cyanobacteria
Chroococcales
  Chamaesiphon
  Cyanothece
  Gloeobacter
  Gloeocapsa
  Gloethece
  Microcystis
  Myxobaktron
  Synechococcus
  Synechocystis
Pleurocapsales
  Chroococcidiopsis
  Dermocarpa
  Dermocarpella
  Myxosarcina
  Pleurocapsa
  Xenococcus
Oscillatoriales
  Arthrospira
  Crinalium
  Lyngbya
  Microcoleus
  Oscillatoria
  Pseudanabaena
  Spirulina
  Starria
  Trichodesmium
Nostocales
  Nostocaceae
    Anabaena
    Aphanizomenon
    Cylindrospermum
    Nodularia
    Nostoc
  Scytonemataceae
    Scytonema
  Rivulariaceae
    Calothrix
Stigonematales
  Chlorogloeopsis
  Fischerella
  Geitleria
  Stigonema

Prochlorales
  Prochloraceae
    Prochloron
    Prochlorothrix

Aerobic Chemolithotrophic Bacteria and Associated Organisms

Nitrifying Bacteria
  Nitrobacteraceae
  Nitrate-Oxidizing Bacteria
    Nitrobacter
    Nitrococcus
    Nitrospina
    Nitrospira
  Ammonia-Oxidizing Bacteria
    Nitrosococcus
    Nitrosolobus
    Nitrosomonas
    Nitrosospira
    "Nitrosovibrio"
Colorless Sulfur-Oxidizing Bacteria
    Acidiphilium
    Macromonas
    Thermothrix
    Thiobacillus
    Thiobacterium
    Thiodendron
    Thiomicrospira
    Thiosphaera
    Thiospira
    Thiovulum
Obligate Chemolithotrophic Hydrogen Bacteria
    Hydrogenobacter
Iron- and Manganese-Oxidizing and/or -Depositing Bacteria
    Aquaspirillum
    "Bilophococcus"
    Gallionella†
    "Leptospirillum"
    "Metallogenium"
    "Naumaniella"
    "Ochrobium"
    "Siderocapsa"
    "Siderococcus"
    Sulfobacillus

Budding and/or Appendaged Bacteria

Prosthecate Bacteria
    Ancalomicrobium
    Asticcacaulis
    Caulobacter
    Dichotomicrobium
    Filomicrobium
    Hirschia
    Hyphomicrobium
    Hyphomonas
    Labrys
    Pedomicrobium
    Prosthecobacter
    Prosthecomicrobium
    Stella
    Verrucomicrobium
Planctomycetales
    Gemmata
    "Isophaera"
    Pirelulla
    Planctomyces
Other Budding and/or Appendaged Bacteria
    Angulomicrobium
    Blastobacter
    Ensifer
    Gallionella†
    Gemmiger
    Nevskia
    Seliberia

Sheathed Bacteria

    "Clonothrix"
    Crenothrix
    Haliscominobacter
    Leptothrix
    "Lieskeella"
    "Phragmidiothrix"
    Sphaerotilus

Nonphotosynthetic, Nonfruiting Gliding Bacteria

Single-Celled, Rod-Shaped Gliding Bacteria
    Capnocytophaga
    Chitinophaga
    Cytophaga
    Flexibacter
    Flexithrix
    Lysobacter
    Microscilla
    Sporocytophaga
    Thermonema
Flattened, Filamentous Gliding Bacteria
    Alysiella
    Simonsiella
Sulfur-Oxidizing Gliding Bacteria
    Achromatium
    Beggiatoa
    Thioploca
    "Thiospirillopsis"
    Thiothrix
The Pelonemas
    "Achroonema"
    "Desmanthus"
    "Pelonema"
    "Peloploca"
Other Genera
    Agitococcus

Desulfonema
Herpetosiphon
"*Isosphaera*"
Leucothrix
Saprospira
Toxothrix
Vitreoscilla

Gliding, Fruiting Bacteria

Angiococcus
Archangium
Chondromyces
"*Corallococcus*"
Cystobacter
"*Haploangium*"
Melittangium
Myxococcus
Nannocystis
Polyangium
"*Sorangium*"
Stigmatella

ARCHAEOBACTERIA

The Methanogens

Cell Walls with Pseudomurein
Methanobacterium
Methanobrevibacter
Methanosphaera
Methanothermus
Cell Walls Without Pseudomurein
Methanococcus
Methanocorpusculum
Methanoculleus
Methanogenium
Methanolacinia
Methanomicrobium
Methanoplanus
Methanospirillum
Other Genera
Methanococcoides
Methanohalobium
Methanohalophilus
Methanolobus
Methanosarcina
Methanothrix

Archaeal Sulfate Reducers

Archaeoglobus

Extremely Halophilic, Aerobic Archaeobacteria

Haloarcula
Halobacterium
Halococcus

Haloferax
Natronobacterium
Natronococcus

Cell Wall–Less Archaeobacteria

Thermoplasma

Extremely Thermophilic and Hyperthermophilic S°-Metabolizers

Cocci, Optimum Growth Below pH 4
Acidianus
Desulfurolobus
Metallosphaera
Sulfolobus
Rods That Use H_2 as Energy Source
Pyrobaculum
Thermofilum
Thermoproteus
Cocci or Disc-Shaped That Oxidize H_2S
Desulfurococcus
Hyperthermus
Pyrococcus
Pyrodictium
Staphylothermus
Thermococcus
Thermodiscus

ACTINOMYCETES

Nocardioform Actinomycetes

Actinobispora
Actinokineospora
Actinopolyspora
Amycolata
Amycolatopsis
Gordona
Jonesia
Kibdelosporangium
Nocardia
Nocardioides
Oerskovia
Promicromonospora
Pseudoamycolata
Pseudonocardia
Rhodococcus
Saccharomonospora
Saccharopolyspora
Terrabacter
Tsukamurella

Actinomycetes with Multilocar Sporangia

Dermatophilus
Frankia
Geodermatophilus

Actinoplanetes

Actinoplanes
Ampullariella
Catelbatospora
Dactylosporangium
Micromonospora
Pilimelia

Streptomyces and Related Genera

Intrasporangium
Kineosporia
Sporichthya
Streptomyces
Streptoverticillium

Maduromycetes

Actinomadura
Microbispora
Microtetraspora

Planobispora
Planomonospora
Spirillospora
Streptosporangium

Thermomonospora and Related Genera

Actinosynnema
Nocardiopsis
Streptoalloteichus
Thermomonospora

Thermoactinomycetes

Thermoactinomyces

Other Actinomycete Genera

Glycomyces
Kitasatosporia
Saccharothrix

Most Probable Numbers (MPN) Table

MPN Index for Various Combinations of Positive and Negative Results When Five 10-ml Portions, Five 1-ml Portions, and Five 0.1-ml Portions Are Used (see pp. 159–161)

NUMBER OF TUBES GIVING POSITIVE REACTION OUT OF			MPN INDEX PER 100 ml	NUMBER OF TUBES GIVING POSITIVE REACTION OUT OF			MPN INDEX PER 100 ml
5 OF 10 ml EACH	5 OF 1 ml EACH	5 OF 0.1 ml EACH		5 OF 10 ml EACH	5 OF 1 ml EACH	5 OF 0.1 ml EACH	
0	0	0	<2	4	3	1	33
0	0	1	2	4	4	0	34
0	1	0	2	5	0	0	23
0	2	0	4	5	0	1	30
1	0	0	2	5	0	2	40
1	0	1	4	5	1	0	30
1	1	0	4	5	1	1	50
1	1	1	6	5	1	2	60
1	2	0	6	5	2	0	50
2	0	0	4	5	2	1	70
2	0	1	7	5	2	2	90
2	1	0	7	5	3	0	80
2	1	1	9	5	3	1	110
2	2	0	9	5	3	2	140
2	3	0	12	5	3	3	170
3	0	0	8	5	4	0	130
3	0	1	11	5	4	1	170
3	1	0	11	5	4	2	220
3	1	1	14	5	4	3	280
3	2	0	14	5	4	4	350
3	2	1	17	5	5	0	240
4	0	0	13	5	5	1	300
4	0	1	17	5	5	2	500
4	1	0	17	5	5	3	900
4	1	1	21	5	5	4	1600
4	1	2	26	5	5	5	≥1600
4	2	0	22				
4	2	1	26				
4	3	0	27				

SOURCE: *Standard Methods for the Examination of Water and Wastewater,* 18th ed. Washington, D.C.: American Public Health Association, 1992.

Methods for Taking Clinical Samples

To diagnose a disease, it is often necessary to obtain a sample of material that may contain the disease-causing organism. Samples must be taken aseptically. The sample container should be labeled with the patient's name, room number (if hospitalized), date, time, and medications being taken. Samples must be transported to the laboratory immediately for culture. Delay in transport may result in growth of some organisms, and their toxic products may kill other organisms. Pathogens tend to be fastidious and die if not kept in optimum environmental conditions.

In the laboratory, samples from infected tissues are cultured on differential and selective media in an attempt to isolate and identify any pathogens or organisms that are not normally found in association with that tissue.

UNIVERSAL PRECAUTIONS

The following procedures should be used by all health care workers, including students, whose activities involve contact with patients or with blood or other body fluids. These procedures were developed to minimize the risk of transmitting AIDS in a health care environment, but adherence to these guidelines will minimize transmission of *all* nosocomial infections.

1. Gloves should be worn when touching blood and body fluids, mucous membranes, and nonintact skin and when handling items or surfaces soiled with blood or body fluids. Gloves should be changed after contact with each patient.

2. Hands and other skin surfaces should be washed immediately and thoroughly if contaminated with blood or other body fluids. Hands should be washed immediately after gloves are removed.

3. Masks and protective eyewear or face shields should be worn during procedures that are likely to generate droplets of blood or other body fluids.

4. Gowns or aprons should be worn during procedures that are likely to generate splashes of blood or other body fluids.

5. To prevent needlestick injuries, needles should not be recapped, purposely bent or broken, or otherwise manipulated by hand. After disposable syringes and needles, scalpel blades, and other sharp items are used, they should be placed in puncture-resistant containers for disposal.

6. Although saliva has not been implicated in HIV transmission, mouthpieces, resuscitation bags, and other ventilation devices should be available for use in areas in which the need for resuscitation is predictable. Emergency mouth-to-mouth resuscitation should be minimized.

7. Health care workers who have exudative lesions or weeping dermatitis should refrain from all direct patient care and from handling patient-care equipment.

8. Pregnant health care workers are not known to be at a greater risk of contracting HIV infection than health care workers who are not pregnant; however, if a health care worker develops HIV infection during

Source: Centers for Disease Control and Prevention and National Institutes for Health. *Biosafety in Microbiological and Biomedical Laboratories,* 2nd ed. Government Printing Office, 1988.

pregnancy, the infant is at risk of infection. Because of this risk, pregnant health care workers should be especially familiar with and strictly adhere to precautions to minimize the risk of HIV transmission.

INSTRUCTIONS FOR SPECIFIC PROCEDURES

Wound or Abscess Culture

1. Cleanse the area with a sterile swab moistened in sterile saline.

2. Disinfect the area with 70% ethanol or iodine solution.

3. If the abscess has not ruptured spontaneously, a physician will open it with a sterile scalpel.

4. Wipe the first pus away.

5. Touch a sterile swab to the pus, taking care not to contaminate the surrounding tissue.

6. Replace the swab in its container, and properly label the container.

Ear Culture

1. Clean the skin and auditory canal with 1% tincture of iodine.

2. Touch the infected area with a sterile cotton swab.

3. Replace the swab in its container.

Eye Culture

This procedure is usually performed by an ophthalmologist.

1. Anesthetize the eye with topical application of a sterile anesthetic solution.

2. Wash the eye with sterile saline solution.

3. Collect material from the infected area with a sterile cotton swab. Return the swab to its container.

Blood Culture

1. Close room windows to avoid contamination.

2. Clean skin around selected vein with 2% tincture of iodine on a cotton swab.

3. Remove dried iodine with gauze moistened with 80% isopropyl alcohol.

4. Draw a few milliliters of venous blood.

5. Aseptically bandage the puncture.

Urine Culture

1. Provide the patient with a sterile container.

2. Instruct the patient to collect a midstream sample. This is obtained by voiding a small volume from the bladder before collection. This washes away extraneous bacteria of the skin microbiota.

3. A urine sample may be stored under refrigeration (4° to 6°C) for up to 24 hours.

Fecal Culture

For bacteriological examination, only a small sample is needed. This may be obtained by inserting a sterile swab into the rectum or feces. The swab is then placed in a tube of sterile enrichment broth for transport to the laboratory. For examination for parasites, a small sample may be taken from a morning stool. The sample is placed in a preservative (polyvinyl alcohol, buffered glycerol, saline, or formalin) for microscopic examination for eggs and adult parasites.

Sputum Culture

1. A morning sample is best as microorganisms will have accumulated while the patient is sleeping.

2. Patient should rinse his or her mouth thoroughly to remove food and normal microbiota.

3. Patient should cough deeply from the lungs and expectorate into a sterile glass wide-mouth jar.

4. Care should be taken to avoid contamination of personnel.

5. In cases such as tuberculosis in which there is little sputum, stomach aspiration may be necessary.

6. Infants and children tend to swallow sputum. A fecal sample may be of some value in these cases.

Biochemical Pathways

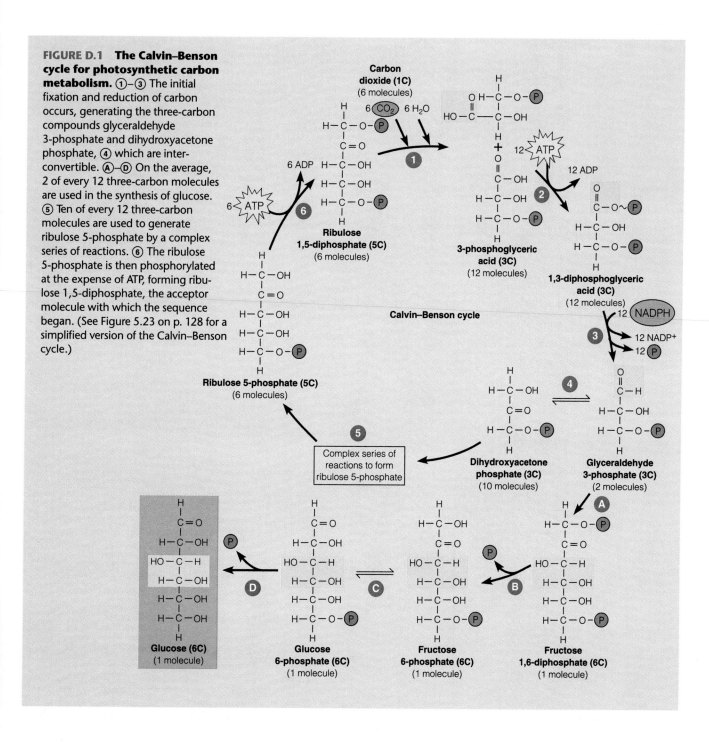

FIGURE D.1   The Calvin–Benson cycle for photosynthetic carbon metabolism. ①–③ The initial fixation and reduction of carbon occurs, generating the three-carbon compounds glyceraldehyde 3-phosphate and dihydroxyacetone phosphate, ④ which are interconvertible. Ⓐ–Ⓓ On the average, 2 of every 12 three-carbon molecules are used in the synthesis of glucose. ⑤ Ten of every 12 three-carbon molecules are used to generate ribulose 5-phosphate by a complex series of reactions. ⑥ The ribulose 5-phosphate is then phosphorylated at the expense of ATP, forming ribulose 1,5-diphosphate, the acceptor molecule with which the sequence began. (See Figure 5.23 on p. 128 for a simplified version of the Calvin–Benson cycle.)

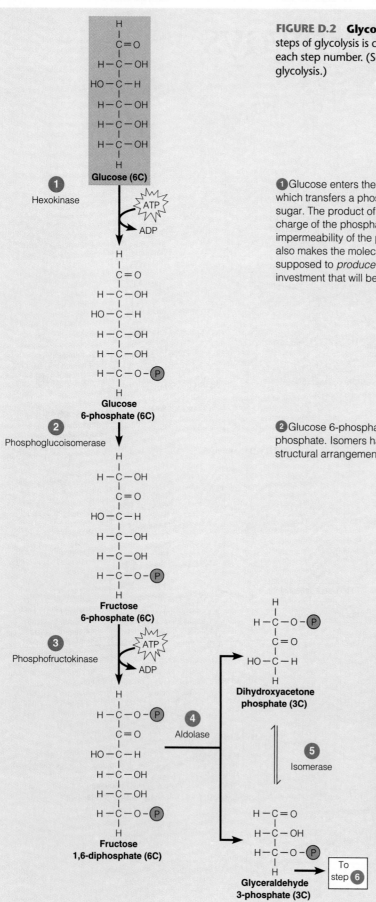

FIGURE D.2   Glycolysis (Embden–Meyerhof pathway). Each of the ten steps of glycolysis is catalyzed by a specific enzyme, which is named under each step number. (See Figure 5.12 on p. 115 for a simplified version of glycolysis.)

①Glucose enters the cell and is phosphorylated by the enzyme hexokinase, which transfers a phosphate group from ATP to the number 6 carbon of the sugar. The product of the reaction is glucose 6-phosphate. The electrical charge of the phosphate group traps the sugar in the cell because of the impermeability of the plasma membrane to ions. Phosphorylation of glucose also makes the molecule more chemically reactive. Although glycolysis is supposed to *produce* ATP, in step ①, ATP is actually consumed—an energy investment that will be repaid with dividends later in glycolysis.

②Glucose 6-phosphate is rearranged to convert it to its isomer, fructose 6-phosphate. Isomers have the same number and types of atoms but in different structural arrangements.

③In this step, still another molecule of ATP is invested in glycolysis. An enzyme transfers a phosphate group from ATP to the sugar, producing fructose 1,6-diphosphate.

④This is the reaction from which glycolysis gets its name ("sugar splitting"). An enzyme cleaves fructose 1,6-diphosphate into two different three-carbon sugars: glyceraldehyde 3-phosphate and dihydroxyacetone phosphate. These two sugars are isomers.

⑤The enzyme isomerase interconverts the three-carbon sugars. The next enzyme in glycolysis uses only glyceraldehyde 3-phosphate as its substrate. This pulls the equilibrium between the two three-carbon sugars in the direction of glyceraldehyde 3-phosphate, which is removed as fast as it forms.

6 Triose phosphate dehydrogenase

1,3-diphosphoglyceric acid (3C)
(2 molecules)

7 Phospho-glycerokinase

3-phosphoglyceric acid (3C)
(2 molecules)

8 Phospho-glyceromutase

2-phosphoglyceric acid (3C)
(2 molecules)

9 Enolase

Phosphoenolpyruvic acid (PEP) (3C)
(2 molecules)

10 Pyruvate kinase

Pyruvic acid (3C)
(2 molecules)

6 An enzyme now catalyzes two sequential reactions while it holds glyceraldehyde 3-phosphate in its active site. First, the sugar is oxidized at the number one carbon and NAD^+ is reduced, resulting in the formation of NADH + H^+. Second, the enzyme couples this reaction to the creation of a high-energy phosphate bond at the number one carbon of the oxidized substrate. The source of the phosphate is inorganic phosphate, which is always present in the cell. As products, the enzyme releases NADH + H^+ and 1,3-diphosphoglyceric acid. Notice in the figure that the new phosphate bond is symbolized with a high-energy bond (~), which indicates that the bond is at least as energetic as the phosphate bonds of ATP.

7 At this step, glycolysis produces ATP. The phosphate group, with its high-energy bond, is transferred from 1,3-diphosphoglyceric acid to ADP. For each glucose molecule that began glycolysis, step **7** produces two molecules of ATP, because every product after the sugar-splitting step (step **4**) is doubled. Of course, two ATPs were invested to get sugar ready for splitting. The ATP ledger now stands at zero. By the end of step **7**, glucose has been converted to two molecules of 3-phosphoglyceric acid.

8 Next, an enzyme relocates the remaining phosphate group of 3-phosphoglyceric acid to form 2-phosphoglyceric acid. This prepares the substrate for the next reaction.

9 An enzyme forms a double bond in the substrate by extracting a water molecule from 2-phosphoglyceric acid to form phosphoenolpyruvic acid. This results in the electrons of the substrate being arranged in such a way that the remaining phosphate bond becomes very unstable.

10 The last reaction of glycolysis produces another molecule of ATP by transferring the phosphate group from phosphoenolpyruvic acid to ADP. Because this step occurs twice for each glucose molecule, the ATP ledger now shows a net gain of two ATPs. Thus, the glycolysis of one molecule of glucose results in two molecules of pyruvic acid, two molecules of NADH + H^+, and two molecules of ATP. Each molecule of pyruvic acid can now undergo respiration or fermentation.

Acetyl CoA (2C)

To Krebs cycle

Aerobic conditions (respiration)

Anaerobic conditions (fermentation)

Ethanol (2C) and CO_2 (1C)

OR

Lactic acid (3C)

FIGURE D.3   The pentose phosphate pathway. This pathway, which operates simultaneously with glycolysis, provides an alternate route for the oxidation of glucose and plays a role in the synthesis of biological molecules, depending on the needs of the cell. Possible fates of the various intermediates are shown in blue. (See p. 114 for a discussion of the pentose phosphate pathway.)

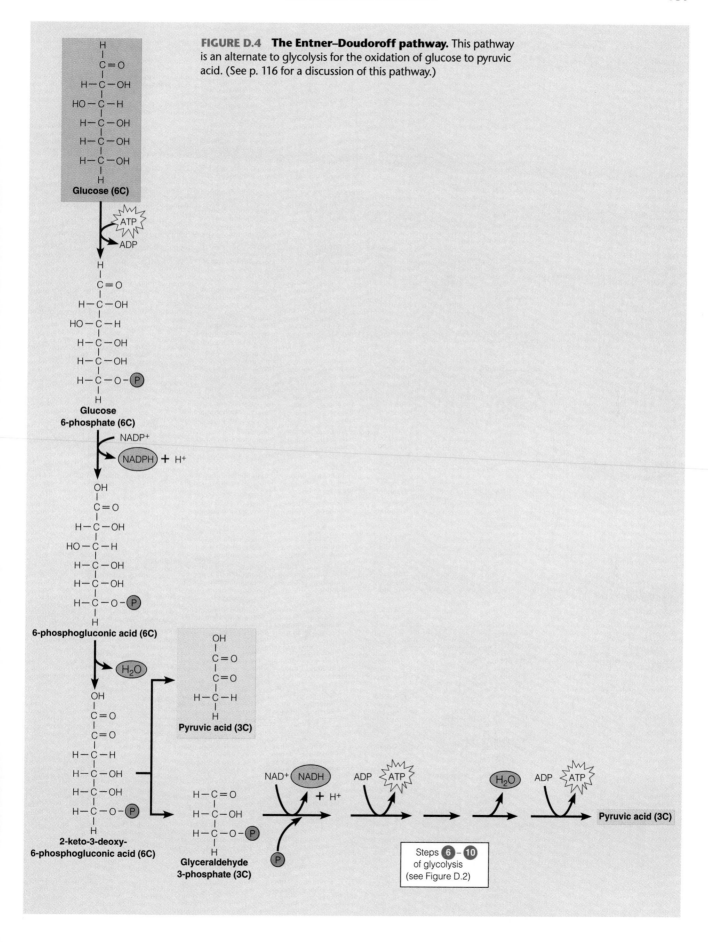

FIGURE D.4 **The Entner–Doudoroff pathway.** This pathway is an alternate to glycolysis for the oxidation of glucose to pyruvic acid. (See p. 116 for a discussion of this pathway.)

FIGURE D.5 **The Krebs cycle.**
(See Figure 5.13 on p. 117 for a simplified version.)

From glycolysis →

Pyruvic acid (3C)

NADH + H⁺

CoA

NAD⁺

CO₂

Acetyl ∼ CoA (2C)

1 Acetyl CoA adds its two-carbon acetyl fragment (*pink*) to oxaloacetic acid, a four-carbon compound. The unstable bond of acetyl CoA is broken as oxaloacetic acid displaces the coenzyme and attaches to the acetyl group. The product is the six-carbon citric acid. CoA is then free to prime another two-carbon fragment derived from pyruvic acid.

8 The last oxidative step produces another molecule of NADH + H⁺ and regenerates oxaloacetic acid, which accepts a two-carbon fragment from acetyl CoA for another turn of the cycle.

8

Oxaloacetic acid (4C)

CoA

1

NADH + H⁺

NAD⁺

Malic acid (4C)

Krebs cycle

Citric acid (6C)

2 A molecule of water is removed, and another is added back. The net result is the conversion of citric acid to its isomer, isocitric acid.

H₂O

7 Bonds in the substrate are rearranged in this step by the addition of a water molecule.

7

H₂O

Fumaric acid (4C)

To electron transport chain

Isocitric acid (6C)

FADH₂

FAD

NAD⁺

NADH + H⁺

3

CO₂

3 The substrate loses a CO₂ molecule (*gray*), and the remaining five-carbon compound is oxidized, reducing NAD⁺ to NADH + H⁺.

6 In another oxidative step, two hydrogens are transferred to FAD to form FADH₂. The function of this coenzyme is similar to that of NADH + H⁺, but FADH₂ stores less energy.

6

Succinic acid (4C)

CoA

NAD⁺

NADH + H⁺

5

GTP

GDP + P

P + ADP

ATP

Succinyl ∼ CoA (4C)

4

CO₂

α-Ketoglutaric acid

5 Substrate-level phosphorylation occurs in this step. CoA is displaced by a phosphate group, which is then transferred to GDP to form guanosine triphosphate (GTP). GTP is similar to ATP, which is formed when GTP donates a phosphate group to ADP.

4 CO₂ (*gray*) is lost; the remaining four-carbon compound is oxidized by the transfer of electrons to NAD⁺ to form NADH + H⁺ and is then attached to CoA by an unstable bond.

Exponents, Exponential Notation, Logarithms, and Generation Time

EXPONENTS AND EXPONENTIAL NOTATION

Very large and very small numbers—such as 4,650,000,000 and 0.00000032—are cumbersome to work with. It is more convenient to express such numbers in exponential notation—that is, as a power of 10. For example, 4.65×10^9 is in **standard exponential notation**, or **scientific notation**. 4.65 is the *coefficient*, and 9 is the power or *exponent*. In standard exponential notation, the coefficient is always a number between 1 and 10, and the exponent can be positive or negative.

To change a number into exponential notation, follow two steps. First, determine the coefficient by moving the decimal point so there is only one nonzero digit to the left of it. For example,

$$0.0000003\ 2$$

The coefficient is 3.2. Second, determine the exponent by counting the number of places you moved the decimal point. If you moved it to the left, the exponent is positive. If you moved it to the right, the exponent is negative. In the example, you moved the decimal point seven places to the right, so the exponent is -7. Thus

$$0.00000032 = 3.2 \times 10^{-7}$$

Now suppose you are working with a large number instead of a very small number. The same rules apply, but the exponential value will be positive rather than negative. For example,

$$4,650,000,000 = 4.65 \times 10^{+9}$$
$$= 4.65 \times 10^9$$

To multiply numbers written in exponential notation, multiply the coefficients and *add* the exponents. For example,

$$(3 \times 10^4) \times (2 \times 10^3) =$$
$$(3 \times 2) \times 10^{4+3} = 6 \times 10^7$$

To divide, divide the coefficient and *subtract* the exponents. For example,

$$\frac{3 \times 10^4}{2 \times 10^3} = \frac{3}{2} \times 10^{4-3} = 1.5 \times 10^1$$

Microbiologists use exponential notation in many situations. For instance, exponential notation is used to describe the number of microorganisms in a population. Such numbers are often very large (Chapter 6). Another application of exponential notation is to express concentrations of chemicals in a solution—chemicals such as media components (Chapter 6), disinfectants (Chapter 7), or antibiotics (Chapter 20). Such numbers are often very small. Converting from one unit of measurement to another in the metric system requires multiplying or dividing by a power of 10, which is easiest to carry out in exponential notation.

LOGARITHMS

A **logarithm (log)** is the power to which a base number is raised to produce a given number. Usually we work with logarithms to the base 10, abbreviated **log10**. The first step in finding the $\log_{10}$ of a number is to write the number in standard exponential notation. If the coefficient is exactly 1, the $\log_{10}$ is simply equal to the exponent. For example,

$$\log_{10} 0.00001 = \log_{10}(1 \times 10^{-5})$$
$$= -5$$

If the coefficient is not 1, as is often the case, the logarithm function on a calculator must be used to determine the logarithm.

Microbiologists use logs for calculating pH levels and for graphing the growths of microbial populations in culture (Chapter 6).

CALCULATING GENERATION TIME

As a cell divides, the population increases exponentially. Numerically this is equal to 2 (because one cell divides into two) raised to the number of times the cell divided (generations):

$$2^{\text{number of generations}}$$

To calculate the final concentration of cells:

$$\text{initial number of cells} \times 2^{\text{number of generations}} =$$
$$\text{number of cells}$$

For example: if 5 cells were allowed to divide 9 times, this would result in

$$5 \times 2^9 = 2560 \text{ cells}$$

To calculate the number of generations a culture has undergone, cell numbers must be converted to logarithms. Standard logarithm values are based on 10. 0.301 is the log of 2 that is used because one cell divides into two.

$$\text{no. of generations} = \frac{\text{log number of cells (end)} - \text{log number of cells (beginning)}}{0.301}$$

To calculate the generation time for a population:

$$\frac{60 \text{ min} \times \text{hours}}{\text{number of generations}} = \text{minutes/generation}$$

As an example we calculate the generation time if 100 bacterial cells growing for 5 hours produced 1,720,320 cells.

$$\frac{\log 1,720,320 - \log 100}{0.301} = 14 \text{ generations}$$

$$\frac{60 \text{ min} \times 5 \text{ hr}}{14 \text{ generations}} = 21 \text{ minutes/generation}$$

A practical application of the calculation is determining the effect of a newly developed food preservative on the culture. Suppose 900 of the same species were grown under the same conditions as the previous example, except that the preservative was added to the culture medium. After 15 hours, there were 3,276,800 cells. Calculate the generation time, and decide whether the preservative inhibited growth.

Answer: 75 min/generation. The preservative did inhibit growth.

Further Reading

PART ONE: FUNDAMENTALS OF MICROBIOLOGY

Chapter 1

Anderson, R.M., and R.M. May. "Understanding the AIDS pandemic." *Scientific American* 266(5):58–66, May 1992. A discussion of the evolution, spread, and control of epidemics.

Brock, T.D., ed. *Microorganisms from Smallpox to Lyme Disease.* New York: W.H. Freeman, 1990. A collection of *Scientific American* articles that describe diseases, their causes, and their effects on humans.

Brock, T.D., ed. *Milestones in Microbiology.* Washington, D.C.: American Society for Microbiology, 1971. The writings of Jenner, Lister, Fleming, and others, with instructive editorial comments.

De Kruif, P. *Microbe Hunters.* New York: Harcourt Brace Jovanovich, 1966. The stories of van Leeuwenhoek, Koch, Pasteur, and others, presented in an interesting narrative.

Jaret, P. "Our immune system: the war within." *National Geographic* 169(6):702–734, June 1986. An exciting look at AIDS, malaria, and conception through colorized electron micrographs.

Margulis, L., and K.V. Schwartz. *Five Kingdoms: An Illustrated Guide to the Phyla of Life on Earth,* 2nd ed. New York: W.H. Freeman, 1988. A discussion of the classification of living organisms and descriptions of phyla.

"Microbes for hire." *Science 85* 6(6):30–46, July/August 1985. Three articles that discuss and illustrate technological uses of microbes in large-scale production of foods, vaccines, and drugs.

Rosebury, T. *Life on Man.* New York: Viking Press, 1969. A humorous yet scientific account of the role of microbes on the human body.

Chapter 2

Hakomori, S. "Glycosphingolipids." *Scientific American* 254(5):44–53, May 1986. Discusses lipid structure and the role of lipids in cancer.

Harold, F.M. *The Vital Force: A Study of Bioenergetics.* New York: W.H. Freeman, 1987. Chapter 4 covers energy production in bacteria, including photosynthesis, respiration, and translocation. Chapter 5 describes energy utilization such as movement and transport. Chapter 6 uses metabolic examples to illustrate evolution of eucaryotic cells.

Holum, J.R. *Fundamentals of General, Organic, and Biological Chemistry,* 4th ed. New York: Wiley, 1990. An introductory chemistry textbook with chapters on cellular biochemistry.

Karplus, M., and J.A. McCammon. "The dynamics of proteins." *Scientific American* 254(4):42–51, April 1986. Relates the constant motion of proteins to their biological functions by using X-ray crystallography.

Mathews, C.K., and K.E. van Holde. *Biochemistry.* Redwood City, Calif.: Benjamin/Cummings, 1990. A biochemistry text with detailed descriptions of organic molecules and their functions in cells.

"Molecules of life." *Scientific American* 253(4), October 1985. Entire issue devoted to biochemistry; includes articles on RNA, DNA, and proteins.

Sharon, N., and H. Lis. "Carbohydrates in cell recognition." *Scientific American* 268(1):82–89, January 1993. Illustrates new information on the functions of carbohydrates other than as carbon and energy sources; shows the tremendous diversity of carbohydrates.

Watson, J.D. *The Double Helix,* ed. Gunther S. Stent. New York: W.W. Norton, 1980. Original papers and Watson's account of the struggle to decipher the structure of DNA give insight into creative scientific processes. Includes commentaries and reviews by other scientists.

Chapter 3

Balows, A., et al., eds. *Manual of Clinical Microbiology,* 5th ed. Washington, D.C.: American Society for Microbiology, 1991. Contains chapters on microscopy, specimen preparation, and staining.

Boatman, E.S., et al. "Today's microscopy." *BioScience* 37(6): 384–394, 1987. Provides a background in fundamentals and then describes recent advances in computer enhancement and laser and acoustic microscopy.

Isenberg, H.D., ed. *Clinical Microbiology Procedures Handbook.* Washington, D.C.: American Society for Microbiology, 1992. Stepwise descriptions of commonly used clinical procedures.

Lichtman, J.W. "Confocal microscopy." *Scientific American* 271(2):40–45, August 1994. Describes how a light microscope with computer enhancement can produce three-dimensional images; this microscope can also look deep inside living specimens.

Morrison, P., and P. Morrison. *Powers of Ten.* New York: Freeman, 1982. Beautiful photographs demonstrate the relative size of things; includes biological material between 10^{-1} and 10^{-6} m.

Taylor, D.L., M. Nederlof, F. Lanni, and A.S. Waggoner. "The new vision of light microscopy." *American Scientist* 90(4):322–335, July–August 1992. Discusses how cameras and computers make up the modern light microscopy workstation; includes a description of fluorescence microscopy.

Chapter 4

Adler, J. "The sensing of chemicals by bacteria." *Scientific American* 235(4):40–47, April 1976. Describes the chemotactic response of bacteria using molecules that detect the presence of chemicals in the environment.

Allen, M.M. "Cyanobacterial cell inclusions." *Annual Review of Microbiology* 38:1–25, 1984. Discusses the effects of starvation and feeding on carboxysomes, gas vacuoles, lipid inclusions, and other cell inclusions.

Benz, R. "Structure and function of porins from gram-negative bacteria." *Annual Review of Microbiology* 42:359–393, 1988. Describes the chemical structure of porins and their role in allowing materials into a bacterial cell.

Costerton, J.W., R.T. Irvin, and K.-J. Cheng. "The bacterial glycocalyx in nature and disease." *Annual Review of Microbiology* 35:299–324, 1981. A thorough discussion of the functions of the glycocalyx.

Ferris, F.G., and T.J. Beveridge. "Functions of bacterial cell surface structure." *BioScience* 35:172–177, 1985. Includes ultrastructures and functions of the cell wall, capsules, flagella, and pili.

Henrichsen, J. "Twitching motility." *Annual Review of Microbiology* 37:81–93, 1983. Describes a locomotion that is not related to flagella.

Moir, A., and D.A. Smith. "The genetics of bacterial spore germination." *Annual Review of Microbiology* 44:531–553, 1990. Provides a detailed description of the events of sporulation and germination.

Thomas, L. *The Lives of a Cell: Notes of a Biology Watcher.* New York: Viking Press, 1974. Thoughtful and entertaining essays by a physician and researcher, including essays on evolution of eucaryotic cells and the importance of microbes.

Chapter 5

Blakemore, R.P., and R.B. Frankel. "Magnetic navigation in bacteria." *Scientific American* 245(6):58–65, December 1981. Describes synthesis and function of magnets in bacterial cells.

Dawes, I.W., and I.W. Sutherland. *Microbial Physiology,* 2nd ed. Cambridge, Mass: Blackwell Scientific, 1992. A basic text on energy production and anabolism in microbes.

Devlin, T.M. *Textbook of Biochemistry,* 3rd ed. New York: Wiley-Liss, 1992. An advanced text that includes clinical applications of biochemistry.

Neidhardt, F.C., J.L. Ingraham, and M. Schaechter. *Physiology of the Bacterial Cell.* Sunderland, Mass.: Sinauer, 1990. Includes chapters on all aspects of microbial metabolism, with emphasis on the success of bacteria in a variety of environments.

Nicholls, D.G. *Bioenergetics: An Introduction to the Chemiosmotic Theory.* New York: Academic Press, 1982. An introductory text on this theory.

Oliver, D. "Protein secretion in *Escherichia coli.*" *Annual Review of Microbiology* 39:615–648, 1985. A summary of the processes by which membrane proteins and extracellular proteins are synthesized and excreted from cells.

"Proton pumping." *BioScience* 35:14–48, January 1985. Includes seven articles, with excellent illustrations, describing proton chemistry and photosynthetic, halobacterial, and eucaryotic membrane systems.

Stryer, L. *Biochemistry,* 3rd ed. New York: W.H. Freeman, 1988. The latest edition of a clearly written and widely used textbook.

Youvan, D.C., and B.L. Marrs. "Molecular mechanisms of photosynthesis." *Scientific American* 256(6):42–48, June 1987. Describes modern techniques from physics and molecular genetics used to identify the electron transport chain in bacterial (cyclic) photophosphorylation.

See also Further Reading for Chapter 2.

Chapter 6

Atlas, R.M., and R. Bartha. *Microbial Ecology: Fundamentals and Applications,* 3rd ed. Redwood City, Calif.: Benjamin/Cummings, 1993. Discusses the interrelationships of microbes with their environments; includes enumeration and isolation techniques.

Cooper, S. *Bacterial Growth and Division.* San Diego, Calif.: Academic Press, 1991. Describes the division cycle of selected bacteria.

Difco Manual: Dehydrated Culture Media and Reagents for Microbiology, 10th ed. Detroit: Difco Laboratories, 1984. Formulas and applications of media from the manufacturer.

Gerhardt, P., ed. *Methods for General and Molecular Bacteriology.* Washington, D.C.: American Society for Microbiology, 1993. A compendium on laboratory techniques, including culture media and bacterial growth curves.

Kirsop, B.E., and A. Doyle. *Maintenance of Microorganisms and Cultured Cells.* San Diego, Calif: Academic Press, 1991. Simple methods for maintaining microbial cultures are presented in this laboratory manual.

Krieg, N.R., and P.S. Hoffman. "Microaerophily and oxygen toxicity." *Annual Review of Microbiology* 40:107–130, 1986. Offers seven explanations for microaerophily.

Sutter, V., et al. *Wadsworth Anaerobic Bacteriology Manual,* 4th ed. Belmont, Calif.: Starr Publishing Co., 1986. Describes methods for isolation and cultivation of anaerobes.

Chapter 7

Balows, A., ed. *Manual of Clinical Microbiology,* 5th ed. Washington, D.C.: American Society for Microbiology, 1991. Includes chapters on preventing contamination in clinical laboratories.

Block, S.S., ed. *Disinfection, Sterilization and Preservation,* 4th ed. Philadelphia: Lea and Febiger, 1991. A comprehensive reference on principles and procedures for disinfection and sterilization.

Castle, M., and E. Ajemian. *Hospital Infection Control: Principles and Practice,* 2nd ed. New York: Delmar, 1987. An excellent reference for practical surveillance and maintenance of asepsis.

Collins, C.H., P.M. Lyne, and J.M. Grange, eds. *Microbiological Methods,* 6th ed. Stoneham, Mass.: Butterworths, 1989. This manual includes chapters on safety in microbiology and sterilization and disinfection.

Phillips, G.B., and W.S. Miller, eds. *Industrial Sterilization.* Durham, N.C.: Duke University Press, 1973. Methods and applications for the use of gas, formaldehyde, and radiation sterilization.

Richardson, J.H., ed. *Laboratory Safety: Principles and Practices,* 2nd ed. Washington, D.C.: American Society for Microbiology, 1994. Contains a chapter on sterilization and disinfection.

Russell, A. *The Destruction of Bacterial Spores.* New York: Academic Press, 1982. A discussion of physical and chemical control of endospores in industry and hospitals.

Chapter 8

Cech, T.R. "RNA as an enzyme." *Scientific American* 255(5):64–75, November 1986. Describes how eucaryotic RNA works as an enzyme to remove exons to produce an RNA molecule.

Devoret, R. "Bacterial tests for potential carcinogens." *Scientific American* 241(2):40–49, August 1979. Describes short tests using bacteria instead of animals to detect carcinogenic chemicals and the effects of carcinogens on DNA.

Grunstein, M. "Histone as regulators of genes." *Scientific American* 267(4):68–74B, October 1992. Provides a model for initiating and preventing transcription at histones.

Howard-Flanders, P. "Inducible repair of DNA." *Scientific American* 245(5):72–80, November 1981. Describes the roles of two enzymes that respond to an "SOS" for DNA damage.

Radman, M., and R. Wagner. "The high fidelity of DNA duplication." *Scientific American* 259(2):40–46, August 1988. A discussion of the roles of DNA polymerase, exonucleases, and repair enzymes in making perfect copies of DNA.

Rennie, J. "DNA's new twists." *Scientific American* 266(3):122–132, March 1993. Discussion of the role of mutations in evolution.

Steitz, J.A. "'Snurps.'" *Scientific American* 258(6):56–63, June 1988. Describes how introns are removed from eucaryotic RNA by small nuclear ribonucleoproteins.

Watson, J.D., N.H. Hopkins, J.W. Roberts, J.A. Steitz, and A.M. Weiner. *Molecular Biology of the Gene,* 4th ed. Redwood City, Calif.: Benjamin/Cummings, 1988. A recent, authoritative textbook that discusses topics of microbial genetics in great detail; very clear explanations.

Weinberg, R.A. "Finding the anti-oncogene." *Scientific American* 259(3):44–51, September 1988. Describes the first known growth-suppressing or anticancer gene.

Chapter 9

Antébi, E., and D. Fishlock. *Biotechnology: Strategies for Life.* Cambridge, Mass.: MIT Press, 1986. This oversized book includes hundreds of color photographs and illustrations concerning genetic engineering of cells.

Barton, J.H. "Patenting life." *Scientific American* 264(3):40–46, March 1991. This article discusses the legal issues of ownership of genetically engineered cells.

Binns, A.N., and M.F. Thomashow. "Cell biology of *Agrobacterium* infection and transformation of plants." *Annual Review of Microbiology* 42:575–606, 1988. This article describes T DNA integration and the host range of *Agrobacterium.*

Biotechnology. This journal is published monthly. Each issue contains articles on new products, methods, and current research.

Chilton, M.D. "A vector for introducing new genes into plants." *Scientific American* 248(6):50–59, June 1983. Describes the use of the *Agrobacterium* Ti plasmid for genetic engineering.

Erickson, D. "Hacking the genome." *Scientific American* 266(4):128–137, April 1992. Describes the goal of the Human Genome Project to develop a computer database containing the sequence of all three billion nucleotides in the human genome.

Gasser, C.S., and R.T. Fraley. "Transgenic crops." *Scientific American* 266(6):62–69, June 1992. Provides examples of the development of pest-resistant crops and spoilage-resistant produce.

Green, H. "Cultured cells for the treatment of disease." *Scientific American* 265(5):64–70, November 1991. Describes the use of cloned human cells to replace diseased and damaged cells.

Mullis, K.B. "The unusual origin of the polymerase chain reaction." *Scientific American* 262(4):56–65, April 1990. An interesting article by the discoverer of PCR.

Neufeld, P.J., and N. Colman. "When science takes the witness stand." *Scientific American* 262(5):46–53, May 1990. Illustrates separation of DNA fragments by electrophoresis and analysis by Southern blotting in describing use of DNA analysis in courts of law.

Newhouse, J.R. "Chestnut blight." *Scientific American* 263(1):106–111, July 1990. Describes the use of biotechnology to develop disease-resistant trees and a weakened fungus.

Präve, P., ed. *Basic Biotechnology.* New York: VCH Publishers, 1987. This text includes control of metabolic pathways, product recovery, and bioreactors.

Verma, I.M. "Gene therapy." *Scientific American* 266(3):122–132, March 1992. Describes the use of viral vectors to insert genes into human cells in attempts to cure inherited diseases.

PART TWO: SURVEY OF THE MICROBIAL WORLD

Chapter 10

deDuve, C. *Blueprint for a Cell.* Burlington, N.C.: Carolina Biological, 1991. A description of the origin of life; includes chapters on procaryotic and eucaryotic cell structure.

Holt, J.G., ed. *Bergey's Manual of Systematic Bacteriology,* 1st ed., 4 volumes. Baltimore: Williams and Wilkins, 1984. A discussion of bacterial classification that provides a perspective on the relative importance of classification and identification is in the front of each volume.

Horgan, J. "In the Beginning." *Scientific American* 264(2):116–125, February 1991. Describes how the first cells might have formed, the evolution of DNA from the first genetic material, and the earliest life known.

Isenberg, H.D., ed. *Clinical Microbiology Procedures Handbook,* 2 volumes. Washington, D.C.: American Society for Microbiology, 1992. Provides useful methods for standard bacteriological tests for identification and includes rapid methods and molecular biology techniques.

Knoll, A.H. "The distribution and evolution of microbial life in the late Proterozoic Era." *Annual Review of Microbiology* 39:391–417, 1985. Describes fossil and chemical evidence for early life and provides living models for early life.

Margulis, L. *Symbiosis in Cell Evolution,* 2nd ed. New York: Freeman. A good book that explains hypotheses and experiments pertaining to the evolution of procaryotes, eucaryotes, fermentation, and respiration.

Schneierson's Atlas of Diagnostic Microbiology, 9th ed. Abbott Park, Ill.: Abbott Laboratories, 1984. Color photographs of common laboratory tests with interpretive comments.

Woese, C.R. "Archaebacteria." *Scientific American* 244(6):98–122, June 1981. Gives detailed descriptions of the archaeobacteria and comparisons with eubacteria and eucaryotes.

Chapter 11

Balows, A., H.G. Trüper, M. Dworkin, W. Harder, and K.H. Schliefer, eds. *The Prokaryotes: A Handbook on the Biology of Bacteria,* 2nd ed., 3 volumes. New York: Springer-Verlag, 1991. Discusses characteristics and provides information on cultivation, isolation, and identification of bacteria.

Baron, E.J., and S.M. Finegold. *Diagnostic Microbiology,* 8th ed. St. Louis: Mosby, 1990. A good reference for procedures used in the clinical microbiology laboratory.

Dworkin, M., and D. Kaiser, eds. *Myxobacteria II.* Washington, D.C.: American Society for Microbiology, 1993. Descriptions of genetic and biochemical investigations of the biology, development, and social behavior of the gliding, fruiting bacteria.

Goodfellow, M., M. Mordarski, and S. Williams. *Biology of the Actinomycetes.* New York: Academic Press, 1984. A survey of the biology, ecology, and pathogenicity of this diverse group of bacteria.

Holt, J.G., ed. *Bergey's Manual of Determinative Bacteriology,* 9th ed. Baltimore: Williams & Wilkins, 1993. The standard reference for identification of bacteria.

Holt, J.G., ed. *Bergey's Manual of Systematic Bacteriology,* 1st ed., 4 volumes. Baltimore: Williams & Wilkins, 1984–1989. These four volumes are the standard reference for identification and classification of bacteria.

Maniloff, J., ed. *Mycoplasmas.* Washington, D.C.: American Society for Microbiology, 1992. A thorough description of the molecular biology and pathogenesis of these small, wall-less bacteria.

Mann, N., and N. Carr, eds. *Photosynthetic Prokaryotes.* New York: Plenum, 1992. A comprehensive reference on physiology, genetics, and commercial applications of cyanobacteria and anoxygenic phototrophs.

Shapiro, J.A. "Bacteria as multicellular organisms." *Scientific American* 258(6):82–89, June 1988. Describes bacterial colonies that function as multicellular organisms.

Starr, M.P., H. Stolp, H.G. Trüper, A. Balows, and H.G. Schlegel. *The Prokaryotes: A Handbook on Habitats, Isolation and Identification of Bacteria,* 2nd ed., 2 volumes. New York: Springer-Verlag, 1989. Discusses characteristics and provides information on cultivation, isolation, and identification of bacteria.

Winkler, H.H. "*Rickettsia* Species (as Organisms)." *Annual Review of Microbiology* 44:131–153, 1990. A survey of the biology and pathogenesis of these intracellular parasites.

Chapter 12

Anderson, D.M. "Red tides." *Scientific American* 271(2):62–68, August 1994. Describes algal toxins and the danger they pose to humans and marine mammals.

Barnett, H.L., and B.B. Hunter. *Illustrated Genera of Imperfect Fungi,* 4th ed. New York: Macmillan, 1987. Good line drawings facilitate identification of saprophytic molds.

Barnett, J.A., R.W. Payne, and D. Yarrow. *Yeasts: Characteristics and Identification,* 2nd ed. New York: Cambridge University Press, 1990. A complete key to yeasts based on morphologic and biochemical characteristics.

Bold, H.C., and M.J. Wynne. *Introduction to the Algae,* 2nd ed. Englewood Cliffs, N.J.: Prentice-Hall, 1985. Covers biology of freshwater and marine unicellular and multicellular algae.

Garcia, L.S., and D.A. Bruckner, eds. *Diagnostic Medical Parasitology,* 2nd ed. Washington, D.C.: American Society for Microbiology, 1991. Includes chapters on all groups of parasites and nosocomial and parasitic infections in AIDS patients.

Larone, D.H. *Medically Important Fungi,* 2nd ed. Washington, D.C.: American Society for Microbiology, 1993. A guide to identifying fungal pathogens based on microscopic examination and laboratory cultures; includes chapters on laboratory techniques.

Walsh, T.J., and P.A. Pizzo. "Nosocomial fungal infections: a classification for hospital-acquired fungal infections and mycoses arising from endogenous flora or reactivation." *Annual Review of Microbiology* 42:517–545, 1988. Discusses the fungal infections that are becoming more common in chronically ill and immunosuppressed patients.

Chapter 13

Diener, T.O. "Viroids and their interactions with host cells." *Annual Review of Microbiology* 36:239–258, 1982. Comprehensive coverage of the structure, pathogenesis, and possible origin of viroids by their discoverer.

Francki, R.I.B., C.M. Fauquet, O.L. Knudson, and F. Brown, eds. "Classification and nomenclature of viruses. Fifth report of the International Committee on Taxonomy of Viruses." *Archives of Virology:* Supplementum 2. New York: Springer-Verlag Wien, 1991. Comprehensive descriptions of each viral family, including viroids.

Hogle, J.M., M. Chow, and D.J. Filman. "The structure of poliovirus." *Scientific American* 256(3):42–49, March 1987. Describes virus assembly and provides a mechanism for protein coat changes that allow a virus to escape the host's immune response.

Hunter, T. "The proteins of oncogenes. "*Scientific American* 251 (2):70–79, August 1984. Explains what the products of oncogenes do.

Prusiner, S.B. "Scrapie prions." *Annual Review of Microbiology* 43:345–374, 1989. This work by the discoverer of prions includes descriptions of human prion diseases such as Creutzfeldt–Jakob disease.

Roizman, B., and A.E. Sears. "An inquiry into the mechanisms of herpes simplex virus latency." *Annual Review of Microbiology* 41:543–571, 1987. Describes host cell changes and provides models for establishment and maintenance of latency.

Spencer, D.C., and P.W. Price. "Human immunodeficiency virus and the central nervous system." *Annual Review of Microbiology* 46:655–693, 1992. Provides a model that accounts for how HIV-1 causes brain injury.

Sugimura, T. "Multistep carcinogenesis," *Science* 258:603–607, October 23, 1992. Evaluates the roles of diet, chemical carcinogens, and viruses in oncogene activation.

Varmus, H. "Reverse transcription." *Scientific American* 257(3): 56–64, September 1987. Presents the action of this enzyme and its role in retroviruses, hepatitis B virus, and evolution.

Watson, J.D., N.H. Hopkins, J.W. Roberts, J.A. Steitz, and A.M. Weiner. *Molecular Biology of the Gene,* 4th ed. Redwood City, Calif.: Benjamin/Cummings, 1988. Chapter 7 discusses phages and Chapter 24 gives an excellent overview of the molecular biology of eucaryotic viruses.

White, D.O., and F.J. Fenner. *Medical Virology*, 3rd ed. New York: Academic Press, 1986. A comprehensive textbook on the biology of medically important viruses.

Zur Hausen, R. "Viruses in human cancers." *Science* 254: 1167–1173, November 22, 1991. Describes the mechanisms by which papillomavirus, HBV, EB virus, and HTLV cause cancer.

PART THREE: INTERACTION BETWEEN MICROBE AND HOST

Chapter 14

Braun, M.M., W.L. Heyward, and J.W. Curran. "The global epidemiology of HIV infections and AIDS." *Annual Review of Microbiology* 44:555–577, 1990. This report covers the history of the AIDS epidemic and geographic differences in its epidemiology.

Epidemiology Issue. *Science* 234:921, 951–979, November 21, 1986. Three articles (on AIDS, salmonellosis, leishmaniasis, and malaria) concerning the tools and techniques used to identify and control an epidemic.

Jaret, P. "The disease detectives." *National Geographic* 179(1): 114–140, January 1991. Introduces scientists working on treatments and vaccines for cholera, Lassa fever, Lyme disease, and AIDS.

Mandell, G.L., R.G. Douglas, and J.E. Bennett, eds. "Nosocomial infections." In *Principles and Practice of Infectious Disease*, 3rd ed. New York: Wiley, 1989. Discusses specific nosocomial infections, such as pneumonia.

Mausner, J.S., and S. Kramer. *Mausner and Bahn Epidemiology*, 2nd ed. Philadelphia: Saunders, 1985. A comprehensive reference on the collection and interpretation of epidemiological data.

McEvedy, C. "The bubonic plague." *Scientific American* 258(2): 118–123, February 1988. A historical account of plague with hypotheses about the occurrence of epidemics.

Mills, J., and H. Masur. "AIDS-related infections." *Scientific American* 263(2):48–57, August 1990. A discussion of the causes and treatments of opportunistic infections affecting AIDS patients.

Roueche, B. *The Medical Detectives*, 2 vols. New York: Truman Talley Books. Vol. 1 by Times Books, Volume 2 by Washington Square Press, 1984. True accounts describing the scientific investigations of epidemic diseases. Written so that the reader identifies the etiology with the epidemiologic investigator.

Symposium Papers. *Respiratory Care* 32(2):81–124, February 1989. Five articles on the spread of infections in hospitals.

van der Waaij, D. "The ecology of the human intestine and its consequences for overgrowth by pathogens such as *Clostridium difficile*." *Annual Review of Microbiology* 43:69–87, 1989. Discusses the normal development of intestinal microbiota and conditions that allow infections.

Chapter 15

Carmichael, W.W. "The toxins of cyanobacteria." *Scientific American* 270(1):78–86, January 1994. Describes the mechanism of action of the neurotoxins and other toxins that poison water supplies.

Ewald, P.W. "The evolution of virulence." *Scientific American* 266(4):86–93, April 1993. Offers hypotheses on how human behavior influences whether a microbe evolves to cause a mild or deadly disease.

Fischetti, V.A. "Streptococcal M protein." *Scientific American* 264(6):58–65, June 1991. Describes the mechanism by which M protein repels phagocytes and allows antibodies to attach to a highly variable region.

Goodenough, U.W. "Deception by pathogens." *American Scientist* 79:344–355, July–August 1991. Discussion of bacteria and viruses that use camouflage, mimicry, and distraction.

Habicht, G.S., G. Beck, and J.L. Benach. "Lyme disease." *Scientific American* 251(1):78–83, July 1987. Describes how investigators discovered the cause of this new disease since the first reports in 1975. Also describes the actions of bacterial LPS and interleukin-1 in causing symptoms.

Hadley, T.J., F.W. Klotz, and L.H. Miller. "Invasion of erythrocytes by malaria parasites: a cellular and molecular overview." *Annual Review of Microbiology* 40:451–477, 1986. Describes the role of ligands in attachment to red blood cells.

Levins, R., et al. "The emergence of new diseases." *American Scientist* 82(1):52–60, January–February 1994. Discusses historical epidemics and new diseases and suggests how new diseases may evolve.

Middlebrook, J.L., and R. Dorland. "Bacterial toxins: cellular mechanisms of action." *Microbiology Reviews* 48:199–221, September 1984. A review of the methods of action of many microbial toxins.

Moberg, C.L., and Z.A. Cohn. "René Jules Dubos." *Scientific American* 264(5):66–74, May 1991. This biography recaps Dubos's work on the relationship between bacterial virulence and disease.

Rietschel, E.T., and H. Brade. "Bacterial endotoxins." *Scientific American* 267(2):54–61, August 1992. Explains how endotoxins exert their harmful and beneficial effects by stimulating the host's defenses.

Chapter 16

Beaman, L., and B.L. Beaman. "The role of oxygen and its derivatives in microbial pathogenesis and host defense." *Annual Review of Microbiology* 38:27–48, 1984. Describes the neutrophil's "respiratory burst" using myeloperoxidase and singlet oxygen to kill microbes.

Bretscher, M.S. "How animal cells move." *Scientific American* 257(6):72–90, December 1987. Describes the movement of phagocytes, fibroblasts, and cancer cells.

Edelson, R.L., and J.M. Fink. "The immunologic function of skin." *Scientific American* 256(6):46–53, June 1985. Discusses the role of T cells in skin.

Johnson, H.M., F.W. Bazer, B.E. Szente, and M.A. Jarpe. "How interferons fight disease." *Scientific American* 270(5):68–75, May 1994. Describes action of interferons and their use in treating diseases.

Joiner, K.A. "Complement evasion by bacteria and parasites." *Annual Review of Microbiology* 42:201–230, 1988. Identifies mechanisms of bacterial blocking of complement and bacterial use of complement to access cells.

Old, L.J. "Tumor necrosis factor." *Scientific American* 258(5):59–75, May 1988. Discusses how bacterial (LPS) endotoxin stimulates production of tumor-destroying proteins in animals.

Ross, G.D., ed. *Immunobiology of the Complement System: An Introduction for Research and Clinical Medicine*. New York: Academic Press, 1986. A compilation of information on complement.

Shafer, W.M., and R.F. Rest. "Interactions of gonococci with phagocytic cells." *Annual Review of Microbiology* 43:121–145, 1989. A discussion of why gonococci can be bound to or engulfed by PMNs and not be killed.

Smith, K.A. "Interleukin-2." *Scientific American* 262(3):50–57, March 1990. Describes how interleukin-2 causes T cells to proliferate.

Trager, W. "The biochemistry of resistance to malaria." *Scientific American* 244(3):154–164, March 1981. Describes the evolution of sickle-cell anemia and how this disease offers protection from malaria.

Chapter 17

Ada, G.L., and G. Nossal. "The clonal selection theory." *Scientific American* 257(2):62–69, August 1987. An excellent explanation of the development and current understanding of the clonal selection theory.

Engelhard, V.H. "How cells process antigens." *Scientific American* 271(2):54–61, August 1994. Describes the interaction between MHC and antigen receptors on T_C and T_H cells.

Grey, H.M., A. Sette, and S. Buus. "How T cells see antigen." *Scientific American* 261(5):56–64, November 1989. Describes how antigens are processed for presentation to T cells.

Lerner, R.A., and A. Tramontano. "Catalytic antibodies." *Scientific American* 258(3):58–70, March 1988. Illustrates how antigen-binding sites react with antigens and suggests new approaches to vaccines.

Tizard, I.R. *Immunology,* 3rd ed. San Diego, CA: Saunders College Publishing, 1992. A textbook covering all aspects of resistance and immunology.

Tonegawa, S. "The molecules of the immune system." *Scientific American* 253(4):122–131, October 1985. Describes how billions of antibodies can be formed from a few genes.

von Boehmer, H., and P. Kisielow. "How the immune system learns about self." *Scientific American* 265(4):74–81, October 1991. Describes a mechanism for deletion of anti-self clones.

Young, J.D.-E., and Z.A. Cohn. "How killer cells kill." *Scientific American* 258(1):38–44, January 1988. Shows how killer cells cause their target cancer and virus-infected cells to leak to death.

Chapter 18

Anderson, D.C., and E.R. Stiehm. "Immunization." *Journal of the American Medical Association* 268:2959–2963, November 25, 1992. A compendium of immunization schedules and available pre- and post-exposure treatments.

Boon, T. "Teaching the immune system to fight cancer." *Scientific American* 266(3):82–89, March 1993. Describes the use of genetic engineering to vaccinate patients with tumor cells.

Kemp, D.J., R.L. Coppel, and R.F. Anders. "Repetitive proteins and genes of malaria." *Annual Review of Microbiology* 41:181–208, 1987. Discusses how the malaria parasite evades the immune system and summarizes the possibility of development of a malaria vaccine.

Kennedy, R.C., J.L. Melnick, and G.R. Dreesman. "Anti-idiotypes and immunity." *Scientific American* 255(1):48–56, July 1986. Describes how these antibodies open the way to new manipulations of the immune system.

Langer, W.L. "Immunization against smallpox before Jenner." *Scientific American* 234(1):112–117, January 1976. An interesting history of the use of smallpox itself as a vaccination against smallpox.

Lerner, R.A. "Synthetic vaccines." *Scientific American* 248(2):66–74, February 1983. Discusses use of a synthetic viral protein to produce antibodies.

Matthews, T.J., and D.P. Bolognesi. "AIDS vaccines." *Scientific American* 259(4):120–127, October 1988. Describes the AIDS vaccines in human trials.

Sabin, A.B. "Evaluation of some currently available and prospective vaccines." *Journal of the American Medical Association* 246:236–241, 1981. A summary of the currently available vaccines against bacterial pneumonia, poliomyelitis, measles, rubella, and influenza, with a discussion of new vaccines under investigation.

Stites, D.P., and A.I. Terr. *Basic and Clinical Immunology,* 7th ed. East Norwalk, Conn.: Appleton & Lange, 1991. An immunology textbook with a comprehensive discussion of the serological and diagnostic aspects of immunology.

Winkler, W.G., and K. Bögel. "Control of rabies in wildlife." *Scientific American* 266(6):86–92, June 1992. A novel attempt to control rabies involving the use of vaccine-filled bait.

Chapter 19

Atkinson, M.A., and N.K. Maclaren. "What causes diabetes?" *Scientific American* 263(1):62–71, July 1990. Describes the development of insulin-dependent diabetes and the role of T cells in destroying insulin-producing cells.

Boon, T. "Teaching the immune system to fight cancer." *Scientific American* 268(3):82–88, March 1993. How genetically engineered tumor cells might be used to treat cancer.

Buisseret, P.D. "Allergy." *Scientific American* 247(2):86–95, August 1982. Discusses the roles of mast cells, IgE, and antigens in hypersensitivity responses.

Collier, J.R., and D.A. Kaplan. "Immunotoxins." *Scientific American* 251(1):56–64, July 1984. Illustrates how cancer cells might be targeted for destruction by a monoclonal antibody carrying a toxic agent.

deShazo, R.D., and D.L. Smith, eds. "Primer on allergic and immunological diseases, 3rd ed." *Journal of the American Medical Association* 268:2785–2993, November 25, 1992. This special issue contains 26 chapters on diseases associated with the immune system.

"HIV/AIDS Special Issue." *Journal of the American Medical Association* 269, June 9, 1993. This special issue contains 10 articles on AIDS.

Johnson, H.M., J.K. Russell, and C.H. Pontzer. "Superantigens in human disease." *Scientific American* 266(4):92–101, April 1992. Proteins that activate too many T cells can lead to autoimmunity or even kill the T cells.

Laurence, J. "The immune system in AIDS." *Scientific American* 253(6):84–93, December 1985. A well-illustrated article on the mechanism of AIDS pathogenicity.

"Life, death and the immune system." *Scientific American* 269(3), September 1993. Ten articles that cover a wide range of topics, including how the immune system works, autoimmune disease, and AIDS.

Rennie, J. "The body against itself." *Scientific American* 263(6):107–115, December 1990. A discussion of current trends in immunology, including treatments for autoimmune diseases.

Rodger, J.C., and B.L. Drake. "The enigma of the fetal graft." *American Scientist* 75:51–57, January–February 1987. Addresses the question of why the mother doesn't reject the fetus, which is genetically foreign tissue.

Rosenberg, S.A. "Adoptive immunotherapy for cancer." *Scientific American* 262(5):62–69, May 1990. Describes the use of cell-transfer therapy to treat cancer.

Schwartz, R.H. "T cell anergy." *Scientific American* 269(2):62–71. Describes how the failure of T cells to respond might be useful in preventing transplant rejections.

"What science knows about AIDS." *Scientific American* 259(4), October 1988. This issue is devoted entirely to AIDS. Ten articles cover a wide range of topics, including molecular biology, epidemiology, and treatments.

Chapter 20

Abraham, E.P. "The beta-lactam antibiotics." *Scientific American* 244(6):76–86, June 1981. A discussion of semisynthetic penicillins and cephalosporins.

Baldry, P. *The Battle Against Bacteria.* Cambridge: Cambridge University Press, 1976. The history of the fight against diseases and the development of antibacterial drugs.

Cundliffe, E. "How antibiotic-producing organisms avoid suicide." *Annual Review of Microbiology* 43:207–233, 1989. Discussion of mechanisms such as enzymatic degradation and target alternation.

Franklin, T.J., and G.A. Snow. *Biochemistry of Antimicrobial Action,* 4th ed. New York: Chapman & Hall, 1989. A thorough reference on antimicrobial drugs covering their development and methods of action.

Hirsch, M.S., and J.C. Kaplan. "Antiviral therapy." *Scientific American* 246(4):76–85, April 1987. A review of the mechanisms by which antiviral drugs work.

Hugo, W.B., and A.D. Russell. *Pharmaceutical Microbiology,* 5th ed. Oxford, England: Blackwell Scientific Publications, 1992. A comprehensive textbook covering the uses of antimicrobial agents and their production by microorganisms.

Moellering, R.C. *Frontiers in Antimicrobial Chemotherapy.* Washington, D.C.: American Society for Microbiology, 1993. Each chapter describes a current topic in research on antimicrobials, including antiretroviral agents and β-lactamases.

Schwarcz, S.K., et al. "National surveillance of antimicrobial resistance in *Neisseria gonorrhoeae.*" *Journal of the American Medical Association* 264:1413–1417, September 19, 1990. Shows drug resistance by drug and by city.

PART FOUR: MICROORGANISMS AND HUMAN DISEASE

The following medical microbiology textbooks discuss principles of microbiology with an emphasis on bacterial pathogens. Some texts also include viral, fungal, and parasitic diseases.

Baron, S., ed. *Medical Microbiology,* 3rd ed. New York: Churchill Livingstone, 1991.

Boyd, R.F., and B.G. Hoerl. *Basic Medical Microbiology,* 4th ed. Boston: Little, Brown, 1992.

Davis, B.D., R. Dulbecco, H.N. Eisen, and H.S. Ginsberg. *Microbiology,* 4th ed. New York: Harper & Row, 1990.

Joklik, W.K., H.P. Willett, D.B. Amos, and C.M. Wilfert, eds. *Zinsser Microbiology,* 20th ed. Norwalk, Conn.: Appleton–Century–Crofts, 1992.

Mandell, G.L., R.G. Douglas, and J.E. Bennett, eds. *Principles and Practice of Infectious Disease,* 3rd ed. New York: Wiley, 1989.

Murray, P.R., M.A. Pfaller, G.S. Kobayashi, and K.S. Rosenthal. *Medical Microbiology,* 2nd ed. St. Louis: C.V. Mosby, 1993.

Schaechter, M., G. Medoff, and D. Schlessinger. *Mechanisms of Microbial Disease.* Baltimore: Williams and Wilkins, 1989.

Sherris, J.C., ed. *Medical Microbiology,* 2nd ed. New York: Elsevier, 1990.

Wilson, G., A. Miles, and M.T. Parker. *Topley and Wilson's Principles of Bacteriology, Virology and Immunity,* 8th ed., 5 vols. Baltimore: Williams and Wilkins, 1990.

Chapter 21

Chesney, P.J., M.S. Bergdoll, J.P. Davis, and J.M. Vergeront. "The disease spectrum, epidemiology, and etiology of toxic shock syndrome." *Annual Review of Microbiology* 38:315–338, 1984. History and recent developments related to this disease.

Cliff, A., and P. Haggert. "Island epidemics." *Scientific American* 250(5):138–147, May 1984. An excellent review of epidemiology using measles as an example.

Fischetti, V.A. "Streptococcal M protein." *Scientific American* 264(6):58–65, June 1991. Compares the M protein fibrils on streptococcal bacteria with the defensive quills of the porcupine.

Henderson, D.A. "The eradication of smallpox." *Scientific American* 235(4):25–33, October 1976. A summary of the incidence of smallpox and investigation of the last endemic infection.

Iandolo, J.J. "Genetic analysis of extracellular toxins of *Staphylococcus aureus.*" *Annual Review of Microbiology* 43:375–402, 1989. Describes many of the 30+ toxins produced by *S. aureus.*

Langmuir, A.D., et al. "The Thucydides syndrome." *New England Journal of Medicine* 313:1027–1030, 1985. Descriptive epidemiology using translations of Thucydides' observations are used to understand an ancient epidemic.

Noble, W.C., ed. *The Skin Microflora and Microbial Skin Disease.* New York: Cambridge University Press, 1992. A thorough description of normal and pathogenic microbiota associated with the skin of humans and animals.

Sheagren, J.N. "*Staphylococcus aureus,* the persistent pathogen." *New England Journal of Medicine* 310:1368–1373, 1437–1442, 1984. Part One describes infections and immunity; Part Two discusses toxemias, treatment, and prevention.

Shepherd, M.G., R.T.M. Poulter, and P.A. Sullivan. "*Candida albicans:* biology, genetics, and pathogenicity." *Annual Review of Microbiology* 39:579–614, 1985. A thorough monograph on diseases caused by *C. albicans* and host defenses; includes an overview of antifungal drugs.

See also Further Reading for Part Four.

Chapter 22

Chasan, D.J. "The polio paradox." *Science 86* 7:37–39, April 1986. A discussion about one little-used vaccine and another that is the leading cause of polio.

Gaylord, H., and P.J. Brennan. "Leprosy and the leprosy bacillus: recent developments." *Annual Review of Microbiology* 41:645–675, 1987. A review of information about antigens and immunity to *M. leprae* from armadillo cultures.

Hall, S.S. "The La Crosse file." *Science 84* 5:54–62, July–August 1984. The epidemiology of California encephalitis virus.

John, D.T. "Primary amoebic meningoencephalitis and the biology of *Naegleria fowleri.*" *Annual Review of Microbiology* 36: 101–123, 1982. A good article on the natural history of the disease and the parasite.

Kristensson, K., and E. Norrby. "Persistence of RNA viruses in the central nervous system." *Annual Review of Microbiology* 40: 159–184, 1986. Descriptions of viruses such as measles virus that remain in the CNS for years and of the resultant host changes.

Rosen, L. "The natural history of Japanese encephalitis virus." *Annual Review of Microbiology* 40:395–414, 1986. A good lesson in natural history that describes alternate hosts and overwintering.

Smith, L., and H. Sugiyama. *Botulism: The Organism, Its Toxins, the Disease,* 2nd ed. Springfield, Ill.: Thomas, 1988. A summary of information on botulism.

Tuomanen, E. "Breaching the blood-brain barrier." *Scientific American* 268(2):80–84, February 1993. Explains how its gram-negative cell wall allows *Neisseria* to enter the central nervous system.

Winkler, W.G., and K. Bögel. "Control of rabies in wildlife." *Scientific American* 266(6):86–92, June 1992. This article describes successful use of vaccine-filled bait to control rabies.

See also Further Reading for Part Four.

Chapter 23

Barbour, A.G. "Antigenic variation of a relapsing fever *Borrelia* species." *Annual Review of Microbiology* 44:155–171, 1990. Antigenic variation may account for relapses of *Borrelia* infection.

Bisno, A.L. "Acute rheumatic fever: forgotten but not gone." *New England Journal of Medicine* 316:421–427, 476–478, 1987. A discussion of the 1985–1986 resurgence of rheumatic fever in the intermountain area.

Greg, C.T. *Plague: An Ancient Disease in the 20th Century.* Albuquerque: University of New Mexico Press, 1985. An interesting and factual account of plague in this century.

Habicht, G.S., G. Beck, and J.L. Benach. "Lyme disease." *Scientific American* 257(1):78–83, July 1987. Describes how investigators discovered the cause of this new disease since the first report in 1975. Also describes the actions of bacterial LPS on interleukin-1 in causing symptoms.

Henle, W., G. Henle, and E.T. Lennette. "The Epstein–Barr virus." *Scientific American* 241(1):48–59, July 1979. A summary of information on one of the most common viruses infecting humans.

McDade, J.E., and V.F. Newhouse. "Natural history of *Rickettsia rickettsii.*" *Annual Review of Microbiology* 40:287–309, 1986. Discusses the ecology of these bacteria and reasons for the increasing range and incidence of the disease.

McEvedy, C. "The bubonic plague." *Scientific American* 258(2): 118–123, February 1988. A history of plague illustrating its loss of virulence over time.

Miller, L.H., et al. "Research toward a malaria vaccine." *Science* 234:1349–1355, 1986. A survey of progress in the search for a vaccine against malaria.

See also Further Reading for Part Four.

Chapter 24

McKinlay, M.A., D.C. Pevear, and M.G. Rossmann. "Treatment of the picornavirus common cold by inhibitors of viral uncoating and attachment." *Annual Review of Microbiology* 46:635–654, 1992. Discusses the discovery, methods of action, and clinical trials for several cold drugs.

Miller, J.A. "Clinical opportunities for plant and soil fungi." *BioScience* 36:656–658, 1986. Describes the rise in fungal infections from species previously known only as plant pathogens and soil saprophytes and from new genera of fungi.

Mills, J., and H. Masur. "AIDS-related infections." *Scientific American* 263(2):50–57, August 1990. A summary of opportunistic infections with emphasis on *Pneumocystis* pneumonia.

Reichman, L.B. "Fear, embarrassment, and relief: the tuberculosis epidemic and public health." *American Journal of Public Health* 83:639–641, May 1993. This issue has three articles on tuberculosis in the United States.

Stuart-Harris, C. "The epidemiology and prevention of influenza." *American Scientist* 69:166–172, 1981. Discusses the relationship between antigenic shift in the virus and immunity to influenza.

See also Further Reading for Part Four.

Chapter 25

Betley, M.J., V.L. Miller, and J.J. Mekalanos. "Genetics of bacterial enterotoxins." *Annual Review of Microbiology* 40:577–605, 1986. Discusses the production of enterotoxins by *V. cholerae, E. coli, S. dysenteriae,* and *S. aureus.*

Dick, J.D. "*Helicobacter (Campylobacter) pylori:* a new twist to an old disease." *Annual Review of Microbiology* 44:249–269, 1990. An in-depth treatise on the general biology, diagnosis, and identification of *Helicobacter.*

Kolenbrander, P.E. "Intergeneric coaggregation among human oral bacteria and ecology of dental plaque." *Annual Review of Microbiology* 42:627–656, 1988. Describes how bacteria aggregate on tooth surfaces to form plaque.

Prince, A.M. "Non-A, non-B hepatitis viruses." *Annual Review of Microbiology* 37:217–232, 1983. Describes ongoing research seeking the causative agents of NANB hepatitis.

Reimann, H., and F.L. Bryan. *Food-borne Infections and Intoxications,* 2nd ed. New York: Academic Press, 1979. An authoritative discussion of nearly all foodborne diseases.

Riley, L.W. "The epidemiologic, clinical, and microbiologic features of hemorrhagic colitis." *Annual Review of Microbiology* 41:383–407, 1987. Discusses intestinal diseases caused by *E. coli* that are endemic in North America.

Taylor, J.M. "The structure and replication of hepatitis delta virus." *Annual Review of Microbiology* 46:253–276, 1992. Describes the life cycle of the delta virus and its need for hepadnavirus.

Tiollais, P., and M. Buendia. "Hepatitis B virus." *Scientific American* 264(4):116–123, April 1991. Describes the epidemiology, reproduction, and possible eradication of this virus.

See also Further Reading for Part Four.

Chapter 26

Aral, S.O., and K.K. Holmes. "Sexually transmitted diseases in the AIDS era." *Scientific American* 264(2):620–669, February 1991. Discusses the epidemiology and risk factors for STDs in the U.S.A.; includes suggestions for controlling these diseases.

Braun, M.M., W.L. Heyward, and J.W. Curran. "The global epidemiology of HIV infection and AIDS." *Annual Review of Microbiology* 44:555–577, 1990. Discusses geographic differences in the epidemiology of AIDS.

Corey, L., and P.G. Spear. "Infections with herpes simplex viruses." *New England Journal of Medicine* 314:686–691, 749–755, 1986. Part One describes the virus; Part Two describes clinical infections and treatment.

Holmes, K.K. "The *Chlamydia* epidemic." *Journal of the American Medical Association* 245:1718–1723, 1981. A readable summary of recent information on chlamydial bacteria and their sexually transmitted diseases.

Journal of the American Medical Association 264(11):1413–1417, 1432–1437, 1451–1452, September 19, 1990. This issue includes three articles on the rise in gonorrhea and syphilis in the U.S.A.

Shafer, W.M., and R.F. Rest. "Interactions of gonococci with phagocytic cells." *Annual Review of Microbiology* 43:121–145, 1989. Describes the conflicts between the antiphagocytic properties of *Neisseria* and PMNs.

Sheppard, H.W., and M.S. Ascher. "The natural history and pathogenesis of HIV infection." *Annual Review of Microbiology* 46:533–564, 1992. Complete description of the disease and its relation to the cellular immune system.

See also Further Reading for Part Four.

PART FIVE: ENVIRONMENTAL AND APPLIED MICROBIOLOGY

Chapter 27

Atlas, R.M., and R. Bartha. *Microbial Ecology: Fundamentals and Applications,* 3rd ed. Redwood City, Calif.: Benjamin/Cummings, 1993. The foremost textbook on the subject of microbial ecology.

AWWA Organisms in Water Committee. "Committee report: microbiological considerations for drinking water regulation revisions." *Journal of the American Water Works Association* 79:81–88, 1987. A discussion of the current status of the microbiology of drinking water and suggested changes for the future.

Brock, T.D., ed. *Thermophiles: General Molecular and Applied Microbiology.* New York: Wiley, 1986. Discusses the microorganisms, their metabolism, and growth requirements of the unique environment of the hot springs in Yellowstone National Park.

Childress, J.J., H. Felbeck, and G.N. Somero. "Symbiosis in the deep sea." *Scientific American* 256(5):114–120, May 1987. Describes the chemoautotrophic bacterial symbionts in worms living at hydrothermal vents.

Dowling, D.N., and W.J. Broughton. "Competition for nodulation of legumes." *Annual Review of Microbiology* 40:131–157, 1986. Describes the effects of the environment, other organisms, and host genes on root nodule formation.

Erlich, H.L. *Geomicrobiology,* 2nd ed. New York: Marcel Dekker, 1990. A textbook on the role of microorganisms in transformations of carbon, phosphorus, mercury, manganese, and other elements.

Margulis, L., D. Chase, and R. Guerero. "Microbial communities." *BioScience* 36:160–170, 1986. A discussion about bacteria in their natural environment. Photographs of living systems show relationships that are not apparent in vitro.

Padwa, D., ed. "Benefits to agriculturists from the studies of microorganisms." *Developments in Industrial Microbiology* 24:19–67, 1983. Papers on technological uses of nitrogen-fixing bacteria, mycorrhizae, yeasts, and agrobacteria.

Paul, E.A., and F.E. Clark. *Soil Microbiology and Biochemistry.* New York: Academic Press, 1989. This authoritative text includes chapters on microbial transformations of nitrogen and metals and methods for studying soil organisms.

Rogers, J.E., and W.B. Whitman, eds. *Microbial Production and Consumption of Greenhouse Gases.* Washington, D.C.: American Society for Microbiology, 1991. An excellent source for data on concentrations, environmental effects, and sources of greenhouse gases.

Shannon, M.J.R., and R. Unterman. "Evaluating bioremediation: distinguishing fact from fiction." *Annual Review of Microbiology* 47:715–738, 1993. Discusses methods and applications for bioremediation.

Chapter 28

Banwart, G.J. *Basic Food Microbiology,* 2nd ed. New York: Van Nostrand Reinhold, 1989. A comprehensive textbook including foodborne illnesses, control of microorganisms, and useful microorganisms.

Brierley, C.L. "Microbiological mining." *Scientific American* 247(2):44–53, August 1982. Discusses the role of *Thiobacillus* in leaching copper from ore.

Crueger, W., and A. Crueger. *Biotechnology: A Textbook of Industrial Microbiology,* 2nd ed. Sunderland, Mass.: Sinauer, 1990. This text covers the principles and processes for production of enzymes, antibiotics, single-cell protein, and fuels.

Hesseltine, C.W. "Microbiology of oriental fermented foods." *Annual Review of Microbiology* 37:575–601, 1983. An interesting monograph on the bacteria and fungi used to prepare traditional Asian foods.

Kosikowski, F.V. "Cheese." *Scientific American* 252(5):88–99, May 1985. Describes how types of milk, species of microbes, and ripening times are used to produce 2000 cheese varieties.

Reed, G. *Prescott and Dunn's Industrial Microbiology,* 4th ed. Westport, Conn.: AVI, 1983. A standard reference book in industrial microbiology.

Rose, A.H., ed. *Economic Microbiology,* 8 vols. New York: Academic Press. Published over several years, these volumes cover food, energy, and other aspects of applied microbiology.

Sawyer, C.A.D., and J.J. Pestka. "Foodservice systems: presence of injured bacteria in foods during food product flow." *Annual Review of Microbiology* 39:51–67, 1985. An analysis of microbiological quality of foods that are prepared and stored, both hot and cold, before serving.

Strobel, G.A. "Biological control of weeds." *Scientific American* 265(1):72–78, July 1991. Describes plant pathogens and fungal toxins that are proving useful as herbicides.

Wilson, M., and S.E. Lindow. "Release of recombinant microorganisms." *Annual Review of Microbiology* 47:913–944, 1993. Covers current applications of recombinant microbes and assessments of the release of natural and recombinant strains.

Zeikus, J.G. "Chemical and fuel production by anaerobic bacteria." *Annual Review of Microbiology* 34:423–464, 1980. A summary of the technology and practicality of producing hydrocarbon fuels and organic solvents from agricultural and municipal wastes.

Pronunciation of Scientific Names

RULES OF PRONUNCIATION

The easiest way to learn new material is to talk about it, and that requires saying scientific names. Scientific names may look difficult at first glance, but keep in mind that generally every *syllable* is pronounced and the primary requirement in saying a scientific name is to communicate it.

The rules for the pronunciation of scientific names depend, in part, on the derivation of the root word and its vowel sounds. We have provided some general guidelines here. Pronunciations frequently do not follow the rules because a common usage has become "accepted" or the derivation of the name cannot be determined. Furthermore, for many scientific names there are alternative correct pronunciations.

Vowels

Pronounce all the vowels in scientific names. Two vowels written together and pronounced as one sound are called a *diphthong* (for example, the *ou* in *sound*). A special comment is needed about the pronunciation of the vowel endings *-i* and *-ae:* There are two alternative ways to pronounce each of these. In this book, we usually give the pronunciation of a long *e* (*ē*) to the *-i* ending and a long *i* (*ī*) to the *-ae* ending. However, the reverse pronunciations are also correct and in some cases are preferred. For example, *coli* is usually pronounced kō′lī.

Consonants

When *c* or *g* is followed by *ae, e, oe, i,* or *y,* it has a soft sound. When *c* or *g* is followed by *a, o, oi,* or *u,* it has a hard sound. When a double *c* is followed by *e, i,* or *y,* it is pronounced as *ks* (e.g., cocci).

Accent

The accented syllable is usually the next-to-last or third-to-last syllable.

1. The accent is on the next-to-last syllable:
   a. When the name contains only two syllables, for example, pes′tis.
   b. When the next-to-last syllable is a diphthong, for example, a-kan-thä-mē′bä.
   c. When the vowel of the next-to-last syllable is long, for example, tre-pō-nē′mä. The vowel in the next-to-last syllable is long in words ending in the following suffixes:

SUFFIX	EXAMPLE
-ales	Orders such as Eubacteriales
-ina	*Sarcina*
-anus, -anum	*pasteurianum*
-uta	*diminuta*

   d. When the word ends in one of the following suffixes:

SUFFIX	EXAMPLE
-atus, -atum	*caudatum*
-ella	*Salmonella*

2. The accent is on the third-to-last syllable in family names. Families end in *-aceae,* which is always pronounced -ā′sē-ē.

PRONUNCIATION OF ORGANISMS IN THIS TEXT

Pronunciation key:

a hat	e see	o hot	th thin
ā age	ė term	ō go	u cup
ã care	g go	ô order	ů put
ä father	i sit	oi oil	ü rule
ch child	ī ice	ou out	ū use
e let	ng long	sh she	zh seizure

Acanthamoeba polyphaga   a-kan-thä-mē'bä pol-ē-fā'gä

Acetobacter   ä-sē-tō-bak'tėr

Achinanthes minutissima   ā'kin-an-thēs min-ü-tis'sē-mä

Acinetobacter   a-si-ne'tō-bak-tėr

Actinobacillus actinomycemcomitans   ak-tin-ō-bä-sil'lus ak-tin-ō-mī-sem-kō'mi-tanz

Actinomyces israelii   ak-tin-ō-mī'sēs is-rä'lē-ē

Aedes aegypti   ä-e'dēz ē-jip'tē

A. albopictus   al-bō-pik'tus

Aeromonas hydrophilia   ār-ō-mō'nas hī-dro'fil-ē-ä

Afipia felis   ä-fi'pē-ä fē'lis

Agrobacterium tumefaciens   ag-rō-bak-ti'rē-um tü-me-fā'shenz

Ajellomyces   ä-jel-lō-mī'sēs

Alcaligenes   al'kä-li-gen-ēs

Allescheria boydii   al-lesh-er'ē-ä boi'dē-ē

Amanita phalloides   am-an-ī'ta fal-loi'dēz

Anabaena   an-ä-bē'nä

Anopheles   an-of'e-l-ēz

Aquaspirillum bengal   ä-kwä-spī-ril'lum ben'gal

A. graniferum   gra-ni'fėr-um

A. magnetotacticum   mag-ne-tō-tak'ti-kum

Arthroderma   är-thrō-dėr'mä

Ascaris lumbricoides   as'kar-is lum-bri-koi'dēz

Ashbya gossypii   ash'bē-ä gos-sip'ē-ē

Aspergillus flavus   a-spėr-jil'lus flā'vus

A. fumigatus   fü-mi-gä'tus

A. niger   nī'jėr

A. oryzae   ô'ri-zī

A. soyae   soi'ī

Azolla   ä-zō'lä

Azomonas   ä-zō-mō'nas

Azospirillum   ä-zō-spī-ril'lum

Azotobacter   ä-zō-tō-bak'tėr

Babesia microti   ba-bē'sē-ä mī-krō'tē

Bacillus anthracis   bä-sil'lus an-thrā'sis

B. cereus   se'rē-us

B. coagulans   kō-ag'ū-lanz

B. licheniformis   lī-ken-i-fôr'mis

B. megaterium   meg-ä-tėr'ē-um

B. polymyxa   po-lē-miks'ä

B. sphaericus   sfe'ri-kus

B. stearothermophilus   ste-rō-thėr-mä'fil-us

B. subtilis   su'til-us

B. thuringiensis   thůr-in-jē-en'sis

Bacteroides hypermegas   bak-tė-roi'dēz hī'pėr-meg-äs

Balantidium coli   bal-an-tid'ē-um kō'lī (or kō'lē)

Bdellovibrio bacteriovorus   del-lō-vib'rē-ō bak-tė-rē-o'vô-rus

Beauveria   bō-vär'ē-a

Beggiatoa   bej-jē-ä-tō'ä

Beijerinckia   bī-yė-rink'ē-ä

Bifidobacterium globosum   bī-fi-dō-bak-ti'rē-um glob-ō'sum

B. pseudolongum   sū-dō-lông'um

Blastomyces dermatitidis   blas-tō-mī'sēz dėr-mä-tit'i-dis

Blattabacterium   blat-tä-bak-ti'rē-um

Boletus edulis   bō-lē'tus e'dū-lis

Bordetella pertussis   bòr-de-tel'lä pėr-tus'sis

Borrelia burgdorferi   bòr-rěl'ē-ä burg-dôr'fėr-ē

Bradyrhizobium   brad-ē-rī-zō'bē-um

Brucella abortus   brü-sel'lä ä-bôr'tus

B. melitensis   me-li-ten'sis

B. suis   sü'is

Byssochlamys fulva   bis-sō-klam'is fül'vä

Campylobacter fetus   kam-pī-lō-bak'tėr fē'tus

C. jejuni   jē-jū'nē

Candida albicans   kan'did-ä al'bi-kanz

C. krusei   krūs'ā-ē

C. utilis   ū'til-is

Canis familiaris   kānis fa-mil'ē-ār-is

Carpenteles   kär-pen'tel-ēz

Caulobacter   kô-lō-bak'tėr

Cephalosporium   sef-ä-lō-spô'rē-um

Ceratocystis ulmi   sē-rä-tō-sis'tis ul'mē

Chilomastix   kē'lō-ma-sticks

Chlamydia pneumoniae   kla-mi'dē-ä nü-mō'nē-ī

C. psittaci   sit'tä-sē

C. trachomatis   trä-kō'mä-tis

Chlamydomonas   klam-i-dō-mō'näs

Chlorobium   klô-rō'bē-um

Chloroflexus   klô-rō-flex'us

Chromatium vinosum   krō-mā'tē-um vi-nō'sum

Chroococcus turgidus   krō-ō-kok'kus tėr'ji-dus

Chrysops   krī'sops

Citrobacter   sit'rō-bak-tėr

Cladosporium werneckii   kla-dō-spô'rē-um wėr-ne'kē-ē

Claviceps purpurea   kla'vi-seps půr-pů-rē'ä

Clonorchis sinensis   klo-nôr'kis si-nen'sis

Clostridium acetobutylicum   klôs-tri'dē-um a-sē-tō-bū-til'li-kum

C. botulinum   bo-tū-lī'num

C. butyricum   bü-ti′ri-kum

C. difficile   dif′fi-sil-ē

C. pasteurianum   pas-tyėr-ē-ā′num

C. perfringens   pėr-frin′jens

C. sporogenes   spô-rä′jen-ēz

C. tetani   te′tan-ē

C. thermosaccharolyticum   thėr-mō-sak-kär-ō-li′ti-kum

Coccidioides immitis   kok-sid-ē-oi′dēz im′mi-tis

Corynebacterium diphtheriae   kô-rī-nē-bak-ti′rē-um dif-thi′rē-ī

C. glutamicum   glü-tam′i-kum

C. xerosis   ze-rō′sis

Coxiella burnetii   käks-ē-el′lä bėr-ne′tē-ē

Cristispira pectinis   kris-tē-spī′rä pek′tin-is

Crucibulum vulgare   krü-si-bū′lum vul′gär

Cryphonectria parasitica   kri-fō-nek′trē-ä par-ä-si′ti-kä

Cryptococcus neoformans   krip-tō-kok′kus nē-ō-fôr′manz

Cryptosporidium   krip-tō-spô-ri′dē-um

Culex   kū′leks

Culiseta   kū-li′se-tä

Cytophaga   sī-täf′äg-ä

Daptobacter   dap′to-bak-tėr

Dermacentor andersoni   dėr-mä-sen′tôr an-dėr-sōn′ē

D. variabilis   vär-ē-a′bil-is

Desulfotomaculum nigrificans   dē-sul-fō-to-ma′kū-lum nī′gri-fi-kans

Desulfovibrio desulfuricans   dē-sul-fō-vib′rē-ō dē-sul-fėr′i-kans

Didinium nasutum   dī-di′nē-um nä-sūt′um

Diphyllobothrium latum   dī-fil-lō-bo′thrē-um lā′tum

Echinococcus granulosus   ē-kīn-ō-kok′kus gra-nū-lō′sus

Ectothiorhodospira mobilis   ek′tō-thī-ō-rō-dō-spī-rä mō′bil-is

Emericella   em-ėr-ē-sel′lä

Emmonsiella   em-mon-sē-el′lä

Entamoeba histolytica   en-tä-mē′bä his-tō-li′ti-kä

Enterobacter aerogenes   en-te-rō-bak′tėr ã-rä′jen-ēz

E. cloacae   klō-ā′kī

Enterobius vermicularis   en-te-rō′bē-us ver-mi-kū-lar′is

Enterococcus faecalis   en-te-rō-kok′kus fē-kā′lis

Epidermophyton   ep-i-dėr-mō-fī′ton

Erwinia dissolvens   ėr-wi′nē-ä dis-solv′ens

Erysipelothrix rhusiopathiae   ār-i-si-pel′ō-thrix rus-ē-ō-path′ē-ī

Erysiphe grammis   ār′i-sīf gram′mis

Escherichia coli   esh-ėr-i′kē-ä kō′lī (or kō′lē)

Euglena   ū-glē′nä

Eurotium   yėr-ō′tē-um

Filobasidiella   fi-lō-ba-si-dē-el′lä

Fonsecaea pedrosoi   fon-se′kē-ä pe-drō′sō-ē

Francisella tularensis   fran-sis-el′lä tü-lä-ren′sis

Frankia   frank′ē-ä

Fusobacterium   fü-sō-bak-ti′rē-um

Gambierdiscus toxicus   gam′bē-ėr-dis-kus toks′i-kus

Gardnerella vaginalis   gärd-nė-rel′lä va-jin-al′is

Gelidium   jel-id′ē-um

Geotrichum   jē-ō-trik′um

Giardia lamblia   jē-är′dē-ä lam′lē-ä

Gloeocapsa   glē-ō-kap′sä

Glomus fasciculatus   glo′mus fa-sik-ū-lä′tus

Glossina   gläs-sē′nä

Gluconobacter   glü-kon-ō-bak′tėr

Gonyaulax   gō-nē-ō′laks

Gymnoascus   jim-nō-as′kus

Gymnodynium microadriaticus   jim-nō-din′ē-um mī-krō-ā′drē-at-i-kus

Haloarcula   hä′lō-är-kū-lä

Halobacterium halobium   ha-lō-bak-ti′rē-um hal-ō′bē-um

Halococcus   hä-lō-kok′kus

Helicobacter pylori   hē′lik-ō-bak-tėr pī′lô-rē

Hemophilus ducreyi   hē-mä′fi-lus dü-krā′ē

H. influenzae   in-flü-en′zī

Histoplasma capsulatum   his-tō-plaz′mä kap-su-lä′tum

Holospora   hō-lo′spô-rä

Homo sapiens   hō′mō sā′pē-ens

Hydrogenomonas   hī-drō-je-nō-mō′näs

Hyphomicrobium   hī-fō-mī-krō′bē-um

Isospora   ī-so′spô-rä

Ixodes scapularis   iks-ō′dēs scap-ū-lār′is

I. pacificus   pas-i′fi-kus

Klebsiella pneumoniae   kleb-sē-el′lä nü-mō′nē-ī

Lactobacillus bulgaricus   lak-tō-bä-sil′lus bul-gā′ri-kus

L. delbrueckii   del-brük′ē-ē

L. leichmannii   līk-man′nē-ē

L. plantarum   plan-tä′rum

L. sanfrancisco   sän-fran-sis′kō

Lactococcus lactis   lak-tō-kok′kus lak′tis

Legionella anisa   lē-jä-nel′lä a′nis-ä

L. feeleii   fē′lē-ē

L. micdadei   mik-dā′dē

L. pneumophila   nü-mō′fi-lä

Leptospira interrogans   lep-tō-spī′rä in-tėr′rä-ganz

Leuconostoc mesenteroides   lü-kō-nos′tok mes-en-ter-oi′dēz

Licmorphora   lik-môr′fôr-ä

Listeria monocytogenes   lis-te′rē-ä mo-nō-sī-tô′je-nēz

Meniscus   mē-nis′kus

Methanosarcina   meth-a-nō-sär′sī-nä

Methylophilus methylotrophus   meth-i-lo′fi-lus meth-i-lō-trōf′us

Microcladia   mī-krō-klād′ē-ä

Micrococcus luteus   mī-krō-kok′kus lū′tē-us

Micromonospora purpureae mī-krō-mo-nä'spô-rä pür-pü-rē'ī

Microsporum gypseum mī-krō-spô'rum jip'sē-um

Mixotricha mix-ō-trik'ä

Moraxella (Branhamella) mô-raks-el'lä bran-ha-mel'lä

M. B. catarrhalis ka-tär'al-is

M. lacunata la-kü-nä'tä

M. osloensis os'lō-en-sis

Mucor rouxii mū-kôr rō'ē-ē

Mycobacterium avium-intracellulare mī-kō-bak-ti'rē-um ā'vē-um-in'trä-cel-ū-lä-rē

M. bovis bō'vis

M. leprae lep'rī

M. smegmatis smeg-ma'tis

M. tuberculosis tü-bėr-kū-lō'sis

Mycoplasma hominis mī-kō-plaz'mä ho'mi-nis

M. pneumoniae nu-mō'nē-ī

Myxococcus fulvus micks-ō-kok'kus ful'vus

Naegleria fowleri nī-gle'rē-ä fou'lėr-ē

Nannizia nan-ni'zē-ä

Necator americanus ne-kā'tôr ä-me-ri-ka'nus

Neisseria gonorrhoeae nī-se'rē-ä go-nôr-rē'ī

N. meningitidis me-nin-ji'ti-dis

Nereocystis nē-rē-ō-sis'tis

Nitrobacter nī-trō-bak'tėr

Nitrosomonas nī-trō-sō-mō'näs

Nocardia asteroides nō-kär'dē-ä as-tėr-oi'dēz

Oocystis ō-ō-sis'tis

Ornithodorus ôr-nith-ō'dô-rus

Paracoccidioides brasiliensis pär-ä-kok-sid'ē-oi-dēz bra-sil-ē-en'sis

Paracoccus denitrificans pär-ä-kok'kus dē-nī-tri'fi-kanz

Paragonimus westermani pär-ä-gōn'e-mus we-stėr-ma'nē

Paramecium multimicronucleatum pär-ä-mē'sē-um mul-tē-mī-krō-nü-klē-ä'tum

Pasteurella haemolytica pas-tyėr-el'lä hēm-ō-lit'i-kä

P. multocida mul-tō'si-dä

Pediculus humanus corporis ped-ik'ū-lus hü'ma-nus kôr'pô-ris

Pediococcus cerevisiae pe-dē-ō-kok'kus se-ri-vis'ē-ī

Penicillium chrysogenum pen-i-sil'lē-um krī-so'gen-um

P. griseofulvum gri-sē-ō-fül'vum

P. notatum nō-tä'tum

Peridinium per-i-din'ē-um

Petriellidium pet-rē-el-li'dē-um

Phialophora verruscosa fē-ä-lō'fô-rä ver-rü-skō'sä

Phormidium luridum fôr-mi'dē-um ler'i-dum

Phytophthora infestans fī-tof'thô-rä in-fes'tans

Pichia pik'ē-ä

Pityrosporum pit-i-ros'pô-rum

Plasmodium falciparum plaz-mō'dē-um fal-sip'är-um

P. malariae mä-lā'rē-ī

P. ovale ō-vä'lē

P. vivax vī'vaks

Plesiomonas shigelloides ple-sē-ō-mō'nas shi-gel-loi'des

Pneumocystis carinii nü-mō-sis'tis kär-i'nē-ī (or kär-i'nē-ē)

Propionibacterium acnes prō-pē-on-ē-bak-ti'rē-um ak'nēz

P. freudenreichii froi-den-rīk'ē-ē

Proteus mirabilis prō'tē-us mi-ra'bi-lis

P. vulgaris vul-ga'ris

Prototheca prō-tō-thā'kä

Pseudomonas aeruginosa sū-dō-mō'nas ā-rü-ji-nō'sä

P. andropogonis ān'dro-po-gō-nis

P. cepacia se-pā'sē-ä

P. fluorescens flôr-es'ens

P. haemolytica hē-mō-lit'i-kä

P. multocida mul'tō-sid-ä

P. syringae sėr-in'gī

Quercus kwer'kus

Rhizobium japonicum rī-zō'bē-um jap-on'i-kum

R. meliloti mel-i-lot'ē

Rhizopus nigricans rī'zō-pus nī'gri-kans

R. oryzae ô'rī-zī

Rhodococcus bronchialis rō-dō-kok'kus bron-kē'al-is

Rhodopseudomonas rō-dō-su-dō-mō'nas

Rhodospirillum rubrum rō-dō-spī-ril'um rūb'rum

Rickettsia prowazekii ri-ket'sē-ä prou-wä-ze'kē-ē

R. rickettsii ri-ket'sē-ē

R. typhi tī'fē

Riftia rift'ē-ä

Rochalimaea henselae rō-chä-lē'mä-ä hen'sel-ī

R. quintana kin'tän-ä

Rosa multiflora rō-sä mul-ti-flô'rä

Saccharomyces beticus sak-ä-rō-mī'sēs bet'i-kus

S. bayanus bā'an-us

S. carlsbergensis kärls-bėr'gen-sis

S. cerevisiae se-ri-vis'ē-ī

S. ellipsoideus ē-lip-soi'dē-us

S. exiguus egz-ij'ū-us

S. fragilis fra'jil-is

S. rouxii rō-ē'ē

Salmonella choleraesuis sal-mōn-el'lä kol-ėr-ä-sü'is

S. dublin dub'lin

S. eastbourne ēst'bôrn

S. enteritidis en-tėr-ī'ti-dis

S. newport new'pôrt

S. typhi tī'fē

S. typhimurium tī-fi-mùr'ē-um

Sarcoptes sär-kop'tēs

Sargassum sär-gas'sum

Sartorya   sär-tô′rē-ä

Schistosoma   shis-tō-sō′mä or skis-tō-sō′mä

Serratia marcescens   ser-rä′tē-ä mär-ses′sens

Shigella boydii   shi-gel′lä boi′dē-ē

S. dysenteriae   dis-en-te′rē-ī

S. flexneri   fleks′nėr-ē

S. sonnei   sōn′ne-ē

Sphaerotilus natans   sfe-rä′ti-lus nā′tans

Spirillum ostreae   spī-ril′lum ost′rä-ī

S. volutans   vō′lū-tans

Spiroplasma   spī-ro-plaz′mä

Spirosoma   spī-rō-sō′mä

Spirulina   spī-rü-lī′nä

Spongomorphora   spon′jō-môr-fô-rä

Sporothrix schenkii   spô-rō′thriks shen′kē-ē

Staphylococcus aureus   staf-i-lō-kok′kus ô′rē-us

S. epidermidis   e-pi-dėr′mi-dis

S. saprophyticus   sap-rō-fit′i-kus

Stella   stel′lä

Stigmatella aurantiaca   stig-mä′tel-lä ô-rän-tē′ä-kä

Streptobacillus   strep-tō-bä-sil′lus

Streptococcus mutans   strep-tō-kok′kus mū′tans

S. pneumoniae   nü-mō′nē-ī

S. pyogenes   pī-äj′en-ēz

S. thermophilus   thėr-mo′fil-us

Streptomyces aureofaciens   strep-tō-mī′sēs ô-rē-ō-fa′si-ens

S. erythraeus   ā-rith′rē-us

S. fradiae   frā′dē-ī

S. griseus   gri′sē-us

S. nodosus   nō-dō′sus

S. noursei   nôr′sē-ī

S. olivoreticuli   ō-liv-ō-re-tik′ū-lē

S. venezuelae   ve-ne-zü-e′lī

Sulfolobus   sul-fō-lō′bus

Taenia saginata   te′nē-ä sa-ji-nä′tä

T. solium   sō′lē-um

Talaromyces   ta-lä-rō-mī′sēs

Tetrahymena   tet-rä-hī′me-nä

Thermoactinomyces vulgaris   thėr-mō-ak-tin-ō-mī′sēs vul-ga′ris

Thermoplasma   thėr-mō-plaz′mä

Thiobacillus ferrooxidans   thī-ō-bä-sil′lus fer-rō-oks′i-danz

T. thiooxidans   thī-ō-oks′i-danz

Thiocapsa floridana   thī-ō-kap′sä flôr-i′dä-nä

Thunnus   tün′nus

Toxoplasma gondii   toks-ō-plaz′mä gon′dē-ē

Trachelomonas   trä-kel-ō-mōn′as

Treponema pallidum   tre-pō-nē′mä pal′li-dum

Triatoma   trī-ä-tō′ma

Trichinella spiralis   trik-in-el′lä spī-ra′lis

Trichomonas vaginalis   trik-ō-mōn′as va-jin-al′is

Trichonympha sphaerica   trik-ō-nimf′ä sfe′ri-kä

Trichophyton   trik-ō-fī′ton

Trichosporon   trik-ō-spôr′on

Trichuris trichiura   trik′ėr-is trik-ē-yėr′a

Tridacna   trī-dak′nä

Tropheryma   trō-fer-ē′mä

Trypanosoma brucei gambiense   tri-pa-nō-sō′mä brüs′ē gam-bē-ens′

T. brucei rhodesiense   rō-dē-sē-ens′

T. cruzi   kruz′ē

Ureaplasma urealyticum   ū-rē-ä-plaz′mä ū-rē-ä-lit′i-kum

Veillonella   vi-lo-nel′lä

Vibrio cholerae   vib′rē-ō kol′ėr-ī

V. parahaemolyticus   pa-rä-hē-mō-li′ti-kus

Wuchereria bancrofti   vū-kėr-ār′ē-ä ban-krof′tē

Xenopsylla cheopis   ze-nop-sil′lä kē-ō′pis

Yersinia enterocolitica   yėr-sin′ē-ä en-tėr-ō-kōl-it′ik-ä

Y. pestis   pes′tis

Y. pseudotuberculosis   sū-dō-tü-bėr-kū-lō′sis

Zoogloea   zō-ō-glē′ä

Glossary

ABO blood group system Classification of red blood cells based on the presence or absence of A and B carbohydrate antigens.

Abscess A localized accumulation of pus.

Acellular vaccine Vaccine consisting of antigenic parts of cells.

Acetyl coenzyme A (acetyl CoA) A substance composed of an acetyl group and a carrier molecule called coenzyme A (CoA); provides the means for pyruvic acid to enter the Krebs cycle. *See* Krebs cycle.

Acetyl group
$$CH_3-\overset{\displaystyle O}{\overset{\displaystyle \|}{C}}-$$

Acid A substance that dissociates into one or more hydrogen ions (H^+) and one or more negative ions.

Acid-fast stain A differential stain used to identify bacteria that are not decolorized by acid-alcohol.

Acidic dye A salt in which the color is in the negative ion; used for negative staining.

Acidophile A bacterium that grows below pH 4.

Acquired immune deficiency The inability, obtained during the life of an individual, to produce specific antibodies or T cells. Due to drugs or disease.

Acquired immunity The ability, obtained during the life of the individual, to produce specific antibodies.

Actinomycetes Gram-positive bacteria that tend to form branching filaments; may form true mycelia; may produce conidiospores; see *Bergey's Manual,* Vols. 2 and 4.

Activated sludge Sludge being digested by aerobic organisms; used in a method of secondary sewage treatment that involves seeding of incoming sewage with sludge from previous batch.

Activation energy The minimum collision energy required for a chemical reaction to occur.

Actively acquired immunity Production of antibodies or specialized lymphocytes by an individual in response to an antigen.

Active site Region on an enzyme that interacts with the substrate.

Active transport Net movement of a substance across a membrane against a concentration gradient; requires energy.

Acute disease A disease in which symptoms develop rapidly but last for only a short time.

Adenine A purine nucleic acid base that pairs with thymine in DNA and uracil in RNA.

Adenocarcinoma Cancer of glandular tissue.

Adenosine diphosphate (ADP) The substance formed when ATP is split and energy is released.

Adenosine diphosphoglucose (ADPG) Glucose activated by ATP; precursor for glycogen synthesis.

Adenosine triphosphatase The enzyme that catalyzes the two reactions ADP + Ⓟ $\longrightarrow$ ATP and ATP $\longrightarrow$ ADP + Ⓟ.

Adenosine triphosphate (ATP) An important intracellular energy source.

Adherence Attachment of a microbe or phagocyte to another's plasma membrane or other surface.

Adhesin *See* Ligands.

Aerial mycelium A mycelium composed of fungal hyphae that project above the surface of the growth medium and produce asexual spores.

Aerobe An organism requiring oxygen (O_2) for growth.

Aerobic respiration Respiration in which the final electron acceptor in the electron transport chain is oxygen (O_2).

Aerotolerant anaerobe An organism that does not use oxygen (O_2) but is not affected by its presence.

Aflatoxin $C_{17}H_{10}O_6$, a carcinogenic toxin produced by *Aspergillus flavus.*

Agar A complex polysaccharide derived from a marine alga and used as a solidifying agent in culture media.

Agglutination A joining together or clumping of cells.

Airborne transmission The spread of pathogens farther than 1 meter in air from reservoir to susceptible host.

Alcohol An organic molecule with the functional group —OH.

Alcohol fermentation A catabolic process, beginning with glycolysis, that produces ethyl alcohol.

Aldehyde An organic molecule with the functional group

$$-C-H$$
$$\parallel$$
$$O$$

Algae A group of photosynthetic eucaryotes; some are included in the Kingdom Protista and some in the Kingdom Plantae.

Algin Sodium salt of mannuronic acid ($C_6H_8O_6$); found in brown algae.

Alkaline Having more hydroxide ions (OH^-) than hydrogen ions (H^+); pH is greater than 7.

Allergen An antigen that evokes a hypersensitivity response.

Allergy *See* Hypersensitivity.

Allograft A graft between persons who aren't identical twins.

Allosteric inhibition The process in which an enzyme's activity is changed because of binding on the allosteric site.

Allosteric site The site on an enzyme at which an inhibitor binds.

Alpha-amino acid An amino acid with —COOH and —NH_2 attached to the same carbon atom.

Alum Aluminum sulfate [$Al_2(SO_4)_3$].

Amanitin Polypeptide mushroom toxin that causes liver and nerve damage.

Ames test A procedure using bacteria to identify potential carcinogens.

Amination The addition of an amino group.

Amino acid An organic acid containing an amino group and a carboxyl group.

Amino acid activating enzyme Enzyme that attaches an amino acid to tRNA.

Aminoglycoside An antibiotic consisting of amino sugars and an aminocyclitol ring; for example, streptomycin.

Amino group —NH_2.

Ammonification Removal of amino groups from amino acids to form ammonia.

Amoeba An organism belonging to the Kingdom Protista that moves by means of pseudopods.

Amphibolic pathway A pathway that is anabolic and catabolic.

Amphitrichous Having tufts of flagella at both ends of a cell.

Anabolism All synthesis reactions in a living organism.

Anaerobe An organism that does not require oxygen (O_2) for growth.

Anaerobic respiration Respiration in which the final electron acceptor in the electron transport chain is an inorganic molecule other than oxygen (O_2); for example, a nitrate or sulfate.

Anaerobic sludge digester Anaerobic digestion used in secondary sewage treatment.

Anal pore A site in certain protozoa for elimination of waste.

Analytical epidemiology Comparison of a diseased group and a healthy group to determine the cause of the disease.

Anamnestic response *See* Memory response.

Anaphylaxis A hypersensitivity reaction involving IgE antibodies, mast cells, and basophils.

Angstrom (Å) A unit of measurement equal to 10^{-10} m, 10^{-4} μm, and 10^{-1} nm; no longer an official unit.

Animalia A kingdom composed of multicellular eucaryotes lacking cell walls.

Anion An ion with a negative charge.

Anoxygenic Not producing molecular oxygen; typical of cyclic photophosphorylation.

Antagonism Active opposition; for example, between two drugs or two microbes.

Antibiotic An antimicrobial agent produced naturally by a bacterium or fungus.

Antibody A protein produced by the body in response to an antigen and capable of combining specifically with that antigen, usually leading in vivo to its destruction or inactivation.

Antibody-dependent cell-mediated cytotoxicity (ADCC) Killing of antibody-coated cells by natural killer cells and phagocytes.

Antibody titer The amount of antibody in serum.

Anticodon The three nucleotides by which a tRNA recognizes an RNA codon.

Antigen Any substance that, when introduced into the body, causes antibody formation and reacts only with its specific antibody.

Antigen-binding sites Sites on an antibody that bind to antigenic determinants.

Antigenic determinant A specific region on the surface of an antigen against which antibodies are formed.

Antigenic drift Minor variations in the antigenic makeup of influenza viruses that occur with time.

Antigenic shift Major genetic changes in influenza viruses causing changes in H and N antigens.

Antigen modulation The process whereby tumor cells shed their specific antigens, thus avoiding the host's immune response.

Antigen-presenting cell A macrophage or dendritic cell that engulfs an antigen and presents fragments to T cells.

Antigen receptors Antibodylike molecules on B and T cells that allow them to recognize and bind to their specific antigens.

Antihuman immune serum globulin Antibodies that react specifically with human antibodies.

Antimetabolite Any substance that interferes with metabolism by competitive inhibition of an enzyme.

Antimicrobial drug A chemical that destroys pathogens without damaging body tissues.

Antiparallel DNA Double-stranded DNA molecule in which one strand is 5' ⟶ 3' and the complement is 3' ⟶ 5'.

Antisense strand (−strand) Viral RNA that cannot act as mRNA.

Antiseptic A chemical for disinfection of the skin, mucous membranes, or other living tissues.

Antiserum A blood-derived fluid containing antibodies.

Antitoxin A specific antibody produced by the body in response to a bacterial exotoxin or its toxoid.

Antiviral protein Protein made in response to interferon that blocks viral multiplication.

Apoenzyme The protein portion of an enzyme, which requires activation by a coenzyme.

Arbuscule Fungal mycelia in plant root cells.

Archaeobacteria Procaryotic organisms lacking peptidoglycan.

Arthropod   An animal phylum characterized by an exoskeleton and jointed legs; includes insects and ticks.

Arthrospore   An asexual fungal spore formed by fragmentation of a septate hypha.

Arthus reaction   Inflammation and necrosis at the site of injection of foreign serum, due to immune complex formation.

Artificially acquired active immunity   The production of antibodies by the body in response to a vaccination.

Artificially acquired passive immunity   The transfer of humoral antibodies formed by one individual to a susceptible individual, accomplished by injection of antiserum.

Ascospore   A sexual fungal spore produced in an ascus, formed by the Ascomycetes class of fungi.

Ascus   A saclike structure containing ascospores; found in the ascomycetes.

Asepsis   The absence of contamination by unwanted organisms.

Aseptic packaging   Commercial food preservation by filling sterile containers with sterile food.

Asexual reproduction   Reproduction without opposite mating strains.

Asexual spores   Reproductive cells produced by mitosis and cell division (eucaryotes) or binary fission (actinomycetes).

A site   Site on ribosome where new tRNA binds.

Associative recognition   T-cell response to a foreign antigen and MHC protein.

Atom   The smallest unit of matter that can enter into a chemical reaction.

Atomic number   The number of protons in the nucleus of an atom.

Atomic weight   The total number of protons and neutrons in the nucleus of an atom.

Attenuation   Lessening of virulence of a microorganism.

Autoclave   Equipment for sterilization by steam under pressure, usually operated at 15 psi and 121°C.

Autogenous hypothesis   Model for evolution of eucaryotes proposing that organelles arose from internal membranes in procaryotes.

Autograft   A tissue graft from one's self.

Autoimmune disease   Damage to one's own organs due to action of the immune system.

Autotroph   An organism that uses carbon dioxide (CO_2) as its principal carbon source.

Auxotroph   A mutant microorganism with a nutritional requirement not possessed by the parent.

Axial filament   The structure for motility found in spirochetes.

Bacillus   Any rod-shaped bacterium; when written as a genus, refers to rod-shaped, endospore-forming, facultatively anaerobic, gram-positive bacteria.

Bacteremia   A condition in which there are bacteria in the blood.

Bacteria   All living organisms with procaryotic cells.

Bacterial growth curve   A graph indicating the growth of a bacterial population over time.

Bactericidal   Capable of killing bacteria.

Bacteriochlorophyll   The light-absorbing pigment found in green and purple bacteria.

Bacteriocin   Toxic protein produced by bacteria that kills other bacteria.

Bacteriophage (phage)   A virus that multiplies in bacterial cells.

Bacteriorhodopsin   The light-absorbing purple pigment in the *Halobacterium* cell membrane.

Bacteriostatic   Capable of inhibiting bacterial growth.

Bacteroid   Enlarged *Rhizobium* cells found in root nodules.

Basal body   A structure that anchors flagella to the cell wall and plasma membrane.

Base   A substance that dissociates into one or more hydroxide ions (OH^-) and one or more positive ions.

Base analog   A chemical that is structurally similar to the normal nitrogenous bases in nucleic acids but with altered base-pairing properties.

Base pairs   The arrangement of nitrogenous bases in nucleic acids based on hydrogen bonding; in DNA, base pairs are A-T and G-C; in RNA, base pairs are A-U and G-C.

Base substitution   The replacement of a single base in DNA by another base, causing a mutation.

Basic dye   A salt in which the color is in the positive ion; used for bacterial stains.

Basidiospore   A sexual fungal spore produced in a basidium, characteristic of the basidiomycetes.

Basidium   A pedestal that produces basidiospores; found in the basidiomycetes.

Basophil   A granulocyte that readily takes up basic dye.

B cells   A type of lymphocyte that acts as one of the main components of the humoral immune system; differentiates into antibody-secreting plasma cells and memory cells.

BCG vaccine   A live, attenuated strain of *Mycobacterium bovis* used to provide immunity to tuberculosis.

Benign tumor   A noncancerous tumor.

Benthic zone   The sediment at the bottom of a body of water.

Bergey's Manual   The standard taxonomic reference on bacteria.

Beta oxidation   The removal of two carbon units from a fatty acid to form acetyl CoA.

Binary fission   Bacterial reproduction by division into two daughter cells.

Binomial nomenclature   The system of having two names (genus and specific epithet) for each organism.

Bioaugmentation   Addition of nutrients such as nitrogen and phosphorus to speed up bioremediation.

Biochemical oxygen demand (BOD)   A measure of the biologically degradable organic matter in water.

Biochemical pathway   A sequence of enzymatically catalyzed reactions occurring in a cell.

Biochemistry   The science of chemical processes in living organisms.

Bioconversion   Changes in organic matter brought about by the growth of microorganisms.

Biogenesis   The concept that living cells can arise only from preexisting cells.

Biogeochemical cycles   The recycling of chemical elements by microorganisms for use by other organisms.

Biological transmission   Transmission of a pathogen from one host to another when the pathogen reproduces in the vector.

Bioluminescence   The emission of light from the electron transport chain of certain living organisms.

Biomass   Organic matter produced by living organisms and measured by weight.

Bioreactor   Fermentation vessel with controls for environmental conditions, e.g., temperature and pH.

Bioremediation   The use of microbes to remove an environmental pollutant.

Biotechnology   The use of living cells in industry, usually to produce products such as pharmaceuticals.

Biovar   A subgroup of a *Salmonella* serovar based on biochemical testing.

Blade   Flat leaflike structure of multicellular algae.

Blastospore   An asexual fungal spore produced by budding from the parent cell.

Blocking antibody   An IgG antibody that reacts with an allergen to prevent a hypersensitivity reaction.

Blood–brain barrier   Cell membranes that allow some substances to pass from the blood to the brain but restrict others.

Bloom (algal)   Abundant growth of microscopic algae, producing visible colonies in nature.

Brightfield microscope   A microscope that uses visible light for illumination; the specimens are viewed against a white background.

Broad-spectrum drug   A chemical that has antimicrobial activity against many infectious microorganisms.

Broth dilution test   A method of determining the MIC by using serial dilutions of an antimicrobial drug.

Brownian movement   The movement of particles, including microorganisms, in a suspension owing to bombardment by the moving molecules in the suspension.

Bubo   An enlarged lymph node caused by inflammation.

Budding   (1) Asexual reproduction beginning as a protuberance from the parent cell that grows to become a daughter cell. (2) Release of an enveloped virus through the plasma membrane of an animal cell.

Buffer   A substance that tends to stabilize the pH of a solution.

Bulking   Condition arising when sludge floats rather than settles in secondary sewage treatment.

Burst size   The number of newly synthesized bacteriophage particles released from a single cell.

Burst time   The time required from bacteriophage adsorption to release.

Calvin–Benson cycle   Conversion of CO_2 into reduced organic compounds; used by autotrophs.

Cancer   A malignant, invasive cellular tumor that has the capability of spreading throughout the body or body parts.

Candle jar   A sealed jar containing a lighted candle, used to incubate bacterial cultures in high CO_2 atmosphere.

Capsid   The protein coat of a virus that surrounds the nucleic acid.

Capsomere   A protein subunit of a capsid.

Capsule   An outer, viscous covering on some bacteria composed of a polysaccharide or polypeptide.

Carbapenems   Antibiotics that contain a β-lactam antibiotic and cilastatin.

Carbohydrates   Organic compounds composed of carbon, hydrogen, and oxygen, with the hydrogen and oxygen present in a 2:1 ratio; include starches, sugars, and cellulose.

Carbon cycle   The series of processes that converts carbon dioxide (CO_2) to organic substances and back to carbon dioxide in nature.

Carbon fixation   *See* Calvin–Benson cycle.

Carbon skeleton   The basic chain or ring of carbon atoms in a molecule; for example,

$$-\overset{|}{\underset{|}{C}}-\overset{|}{\underset{|}{C}}-\overset{|}{\underset{|}{C}}-$$

Carboxyl group   —COOH.

Carboxysome   Procaryotic inclusion containing ribulose 1,5-diphosphate carboxylase.

Carcinogen   Any cancer-causing substance.

Cardiolipin   A beef heart extract used in the venereal disease research laboratory (VDRL) slide test to detect antibodies against syphilis.

Carrageenan   A galactose polymer in the cell walls of red algae.

Carrier   An individual who harbors a pathogen but exhibits no signs of illness.

Carrier proteins   Plasma membrane proteins that transport substances across the membrane.

Casein   Milk protein.

Catabolic repression   Inhibition of decomposition of alternate carbon sources by glucose.

Catabolism   All decomposition reactions in a living organism.

Catalase   An enzyme that catalyzes the breakdown of hydrogen peroxide to water and oxygen.

Catalyst   A substance that increases the rate of a chemical reaction but isn't changed in the reaction.

Cation   A positively charged ion.

CD   Cluster of differentiation. A cell surface receptor used to differentiate lymphocytes. **CD4** usually refers to helper T cells; **CD8** includes cytotoxic and suppressor T cells.

Cell   The basic microscopic unit of structure and function of all living organisms.

Cell culture   Animal or plant cells grown in vitro.

Cell-mediated immunity   An immune response that involves T cells binding to antigens presented on infected cells; T cells then differentiate into several types of effector T cells, including helper and cytotoxic.

Cell theory   The principle that all living things are composed of cells.

Cellulose   A glucose polysaccharide that is the main component of plant cell walls.

Cell wall   The outer covering of most bacterial, fungal, algal, and plant cells: in eubacteria, it consists of peptidoglycan.

Centrioles   A structure consisting of nine microtubule triplets, found in eucaryotic cells.

Cephalosporin   An antibiotic produced by the fungus *Cephalosporium* that inhibits the synthesis of gram-positive bacterial cell walls.

Cercaria   A free-swimming larva of trematodes.

Chancre   A hard sore, the center of which ulcerates.

Chemical bond Attractive force between atoms forming a molecule.

Chemical element A fundamental substance composed of atoms that have the same atomic number and behave the same way chemically.

Chemical energy The energy of a chemical reaction.

Chemically defined medium A culture medium in which the exact chemical composition is known.

Chemical reaction The process of making or breaking bonds between atoms.

Chemiosmosis A mechanism that uses a proton gradient across a cytoplasmic membrane to generate ATP.

Chemistry The science of the interactions of atoms and molecules.

Chemoautotroph An organism that uses an inorganic chemical as an energy source and carbon dioxide (CO_2) as a carbon source.

Chemoheterotroph An organism that uses organic molecules as a source of carbon and energy.

Chemostat An apparatus to keep a culture in log phase indefinitely.

Chemotaxis Motion in response to the presence of a chemical.

Chemotherapy Treatment of a disease with chemical substances.

Chemotroph An organism that uses oxidation–reduction reactions as its primary energy source.

Chimeric monoclonal antibodies Genetically engineered antibodies made of human constant regions and mouse variable regions.

Chitin A glucosamine polysaccharide that is the main component of fungal cell walls and arthropod skeletons.

Chlamydospore An asexual fungal spore formed within a hypha.

Chloramphenicol A broad spectrum bacteriostatic chemical.

Chlorobium vesicle *See* Chlorosome.

Chlorophyll *a* The light-absorbing pigment in cyanobacteria, algae, and plants.

Chloroplast The organelle that performs photosynthesis in photoautotrophic eucaryotes.

Chlorosome Plasma membrane folds in green sulfur bacteria containing bacteriochlorophylls.

Chromatin Threadlike, uncondensed DNA in an interphase eucaryotic cell.

Chromatophore An infolding in the plasma membrane where bacteriochlorophyll is located in photoautotrophic bacteria.

Chromosome The structure that carries hereditary information.

Chromosome number Diploid number of a eucaryote.

Chronic disease An illness that develops slowly and is likely to continue or recur for long periods.

Cilia Relatively short cellular projections that move in a wave-like manner.

Ciliate A member of the protozoan phylum Ciliata that uses cilia for locomotion.

Cisterns Stacked elements of the Golgi complex.

Class A taxonomic ranking between phylum and order.

Clonal deletion Elimination of B and T cells that react with self.

Clonal selection Development of clones of B and T cells against a specific antigen.

Clone A population of cells that are identical to the parent cell.

Clue cells Sloughed-off vaginal cells covered with *Gardnerella vaginalis*.

Coagulase A bacterial enzyme that causes blood plasma to clot.

Coccobacillus A bacterium that is an oval rod.

Coccus A spherical or ovoid bacterium.

Codon A group of three nucleotides in mRNA that specifies the insertion of an amino acid into a protein.

Coenocytic hyphae Fungal filaments that are not divided into uninucleate cell-like units because they lack septa.

Coenzyme A nonprotein substance that is associated with and that activates an enzyme.

Coenzyme A (CoA) A coenzyme that functions in decarboxylation.

Cofactor The nonprotein component of an enzyme.

Cohort method Comparison of two populations: control and experimental.

Coliforms Aerobic or facultatively anaerobic, gram-negative, nonspore-forming, rod-shaped bacteria that ferment lactose with acid and gas formation within 48 hours at 35°C; used as an indicator organism for tests of bacteriological purity.

Collagen The main structural protein of muscles.

Collision theory The principle that chemical reactions occur because energy is gained as particles collide.

Colony A clone of bacterial cells on a solid medium that is visible to the naked eye.

Commensalism A system of interaction in which two organisms live in association and one is benefited while the other is neither benefited nor harmed.

Commercial sterilization A process of treating canned goods aimed at destroying the endospores of *Clostridium botulinum*.

Communicable disease Any disease that can be spread from one host to another.

Competence The physiological state in which a recipient cell can take up and incorporate a large piece of donor DNA.

Competitive inhibitor A chemical that competes with the normal substrate for the active site of an enzyme.

Complement (C) A group of serum proteins involved in phagocytosis and lysis of bacteria.

Complementary DNA (cDNA) DNA made in vitro from an mRNA template.

Complement fixation The process in which complement combines with an antigen–antibody complex.

Complete digestive system A digestive system with a mouth and an anus.

Complex lipids Lipids containing phosphorus, nitrogen, and/or sulfur.

Complex medium A culture medium in which the exact chemical composition is not known.

Complex virus A virus with a complicated structure, such as a bacteriophage.

Compost Decomposing organic matter.

Compound A substance composed of two or more different chemical elements.

Compound light microscope An instrument with two sets of lenses that uses visible light as the source of illumination.

Compromised host A host whose resistance to infection is impaired.

Condensation reaction A chemical reaction in which a molecule of water is released.

Condenser A lens system located below the microscope stage that directs light rays through the specimen.

Congenital disease A disease present at birth as a result of genetics or some condition that occurred in utero.

Congenital immune deficiency The inability, due to an individual's genotype, to produce specific antibodies or T cells.

Conidiophore An aerial hypha bearing conidiospores.

Conidiospore An asexual spore produced in a chain from a conidiophore.

Conjugated protein Molecule consisting of amino acids and other organic or inorganic compounds.

Conjugated vaccine A vaccine consisting of the desired antigen and other proteins.

Conjugation The transfer of genetic material from one cell to another involving cell-to-cell contact.

Constitutive enzyme An enzyme that is produced regardless of how much substrate is present.

Contact inhibition The cessation of animal cell movement and division as a result of contact with other cells.

Contact transmission Spread of disease by direct or indirect contact or via droplets.

Contagious disease A disease that is easily spread from one person to another.

Continuous cell line Animal cells that can be maintained through an indefinite number of generations in vitro.

Convalescent period The period of recovery from a disease.

Corepressor A molecule that binds to a repressor, enabling the repressor to bind to the operator site.

Cortex (of a lichen thallus) Protective fungal covering of a lichen.

Countercurrent immunoelectrophoresis (CIE) The movement of antigen and antibody toward each other through an electric field. *See also* Electrophoresis.

Counterstain A stain used to give contrast in a differential stain.

Covalent bond A chemical bond in which the electrons of one atom are shared with another atom.

Cresol A mixture of isomers from petroleum.

Crisis The phase of a fever characterized by vasodilation and sweating.

Cristae Foldings of the inner membrane of a mitochondrion.

Crossing over Process by which a portion of one chromosome is exchanged with a portion of another chromosome.

Culture Microorganisms that grow and multiply in a container of culture medium.

Culture medium The nutrient material prepared for growth of microorganisms in a laboratory.

Curd The solid part of milk that separates from the liquid in the making of cheese, for example.

Cutaneous mycosis A fungal infection of the epidermis, nails, or hair.

Cuticle Nonliving outer covering of helminths.

Cyanobacteria Oxygen-producing photoautotrophic procaryotes; also called blue-green bacteria.

Cyclic AMP A molecule derived from ATP, in which the phosphate group has a cyclic structure; acts as a cell messenger.

Cyclic photophosphorylation Movement of an electron from chlorophyll through a series of electron acceptors and back to chlorophyll; purple and green bacterial photophosphorylation.

Cyclosporine A drug that suppresses the T-cell response.

Cyst A sac with a distinct wall containing fluid or other material; also, a protective capsule of some protozoa.

Cysticercus Encysted tapeworm larva.

Cystitis Inflammation of the urinary bladder.

Cytochrome oxidase An enzyme that oxidizes cytochrome *c*.

Cytochromes Proteins that function as electron carriers in respiration and photosynthesis.

Cytocidal Resulting in cell death.

Cytokines Small proteins released from human cells in response to bacterial infection; directly or indirectly may induce fever, pain, or T-cell proliferation.

Cytopathic effect (CPE) Tissue deterioration caused by viruses.

Cytoplasm In a procaryote, everything inside the plasma membrane; in a eucaryote, everything inside the plasma membrane and external to the nucleus.

Cytoplasmic streaming The flowing of cytoplasm in a eucaryotic cell.

Cytosine A pyrimidine nucleic acid base that pairs with guanine.

Cytoskeleton Microfilaments and microtubules that provide support and movement for eucaryotic cytoplasm.

Cytostome The mouthlike opening in some protozoa.

Cytotoxic T (T_C) cells Specialized T cells that destroy infected cells presenting antigens.

Cytotoxin Bacterial toxins that kill or alter the functions of host cells.

D- Prefix describing a stereoisomer.

Darkfield microscope A microscope that has a device to scatter light from the illuminator so that the specimen appears white against a black background.

Dark (light-independent) reaction *See* Calvin–Benson cycle.

Deamination Removal of an amino group.

Death phase Period of logarithmic decrease in a bacterial population.

Debridement Surgical removal of necrotic tissue.

Decarboxylation Removal of carbon dioxide (CO_2) from an amino acid.

Decimal reduction time (DRT) The time (in minutes) required to kill 90% of a bacterial population at a given temperature.

Decolorizing agent A solution used in the process of removing a stain.

Decomposition reaction A chemical reaction in which bonds are broken to produce smaller parts from a large molecule.

Deep freezing Preservation of bacterial cultures at $-50°C$ to $-95°C$.

Definitive host An organism that harbors the adult, sexually mature form of a parasite.

Degeneracy (of the genetic code) Redundancy of genetic code; that is, most amino acids are signaled by several codons.

Degranulation Release of contents of secretory granules from mast cells or basophils during anaphylaxis.

Dehydration The removal of water.

Dehydration synthesis *See* Condensation reaction.

Dehydrogenation The loss of hydrogen atoms from a substrate.

Delayed hypersensitivity Cell-mediated hypersensitivity.

Delayed hypersensitivity T (T_D) cells Specialized T cells that produce lymphokines in type IV hypersensitivities.

Denaturation A change in the molecular structure of a protein, usually making it nonfunctional.

Denitrification The reduction of nitrates to nitrites or nitrogen gas.

Dental plaque A combination of bacterial cells, dextran, and debris adhering to the teeth.

Deoxyribonucleic acid (DNA) The nucleic acid of genetic material.

Deoxyribose A five-carbon sugar contained in DNA nucleotides.

Dermatophyte A fungus that causes a cutaneous mycosis.

Dermis The inner portion of the skin.

Descriptive epidemiology Analysis of all data regarding the occurrence of a disease to determine the cause of the disease.

Desensitization The prevention of allergic inflammatory responses.

Desiccation The removal of water.

Detergent Any substance that reduces the surface tension of water.

Dextran A polymer of glucose.

Diacetyl $CH_3COCOCH_3$ produced from carbohydrate fermentation; chiefly responsible for the odor of dairy products.

Diagnosis Identification of a disease.

Diapedesis The process by which phagocytes move out of blood vessels.

Differential interference contrast (DIC) microscope An instrument that provides a three-dimensional, magnified image.

Differential medium A solid culture medium that makes it easier to distinguish colonies of the desired organism.

Differential stain A stain that distinguishes objects on the basis of reactions to the staining procedure.

Differential white blood cell count The number of each kind of leukocyte in a sample of 100 leukocytes.

Diffusion The net movement of molecules or ions from an area of higher concentration to an area of lower concentration.

Digestion The process of breaking down substances physically and chemically.

Digestive vacuole An organelle in which substrates are broken down enzymatically.

Dimorphism The property of having two growth forms.

Dioecious Referring to organisms in which organs of different sexes are located in different individuals.

Diphtheroid A gram-positive pleomorphic rod.

Dipicolinic acid Chemical substance found in bacterial endospores and not in vegetative cells.

Diplobacilli Rods that divide and remain attached in pairs.

Diplococci Cocci that divide and remain attached in pairs.

Diploid Having two sets of chromosomes; normal state of a eucaryotic cell.

Diploid cell lines Eucaryotic cells grown in vitro.

Direct agglutination Use of known antibodies to identify an unknown cell-bound antigen.

Direct contact A method of spreading infection from one host to another through some kind of close association of the hosts.

Direct FA test A fluorescent-antibody test to detect the presence of an antigen.

Direct microscopic count Enumeration of cells by observation through a microscope.

Direct test Identification of a pathogen.

Disaccharide A sugar consisting of two monosaccharides.

Disease Any change from a state of health.

Disinfectant Any substance used on inanimate objects to kill or inhibit the growth of microorganisms.

Disk-diffusion test *See* Kirby–Bauer test.

Dissimilation plasmids Plasmids containing genes coding for the production of enzymes that catalyze the catabolism of certain unusual sugars and hydrocarbons.

Dissociation Transformation of a compound into positive and negative ions in solution.

Distilled spirits Alcoholic beverages made by distillation of wine.

Disulfide bond Two atoms of sulfur held together by a covalent bond (S—S).

Division A phylum; used in botany and microbiology.

DNA fingerprinting Analysis of DNA by electrophoresis of restriction enzyme fragments of the DNA.

DNA ligase An enzyme that covalently bonds the phosphate of one nucleotide with the 3' carbon of another.

DNA sequencing Processes by which the nucleotide sequence of DNA is determined.

Donor cell A cell that gives DNA to a recipient cell in recombination.

DPT vaccine A combined vaccine used to provide active immunity, containing diphtheria and tetanus toxoids and killed *Bordetella pertussis* cells.

Droplet infection The transmission of infection by small liquid droplets carrying microorganisms.

D value *See* Decimal reduction time.

Dysentery A disease characterized by frequent, watery stools containing blood and mucus.

Eclipse period The time during viral multiplication when complete, infective virions are not present.

Edema An abnormal accumulation of fluid in body parts or tissues, causing swelling.

Electron A negatively charged particle in motion around the nucleus of an atom.

Electron acceptor   An ion that picks up an electron that has been lost from another atom.

Electron donor   An ion that gives up an electron to another atom.

Electronic configuration   The arrangement of electrons in shells or energy levels in an atom.

Electron microscope   A microscope that uses a flow of electrons instead of light to produce an image.

Electron shells   Regions of an atom corresponding to different energy levels.

Electron transport chain   A series of compounds that transfer electrons from one compound to another, generating ATP by oxidative phosphorylation.

Electrophoresis   The separation of substances (for example, serum proteins) by their rate of movement through an electric field.

Electroporation   Technique by which DNA is inserted into a cell using an electric current.

Elementary body   An infectious form of *Chlamydia*.

ELISA (enzyme-linked immunosorbent assay)   A group of serological tests that use enzyme reactions as indicators.

Embden–Meyerhof pathway   *See* Glycolysis.

Emulsification   The process of mixing two liquids that do not dissolve in each other.

Encephalitis   Inflammation of the brain.

Encystment   Formation of a cyst.

Endemic disease   A disease that is constantly present in a certain population.

Endergonic reaction   A chemical reaction that requires energy.

Endocarditis   Inflammation of the lining of the heart (endocardium).

Endocytosis   The process of moving material into a eucaryotic cell.

Endomembrane system   Organelles in eucaryotic cells; so called because of their interaction.

Endoplasmic reticulum (ER)   A membrane network in eucaryotic cells connecting the plasma membrane with the nuclear membrane.

Endospore   A resting structure formed inside some bacteria.

Endosymbiotic hypothesis   Model for evolution of eucaryotes proposing that organelles arose from procaryotic cells living inside a host procaryote.

Endotoxic shock   *See* Septic shock.

Endotoxin   Part of the outer portion of the cell wall (lipid A) of most gram-negative bacteria; released on destruction of the cell.

End-product inhibition   *See* Feedback inhibition.

Energy level   The energy of an electron and its position relative to the atomic nucleus.

Enrichment culture   A culture medium used for preliminary isolation that favors the growth of a particular microorganism.

Enterics   The common name for bacteria in the family Enterobacteriaceae.

Enterotoxin   An exotoxin that causes gastroenteritis, such as those produced by *Staphylococcus, Vibrio,* and *Escherichia.*

Entner–Doudoroff pathway   An alternate pathway for the oxidation of glucose to pyruvic acid.

Envelope   An outer covering surrounding the capsid of some viruses.

Enzyme   A protein that catalyzes chemical reactions in a living organism.

Enzyme induction   The process by which a substance can cause the synthesis of an enzyme.

Enzyme poison   Substance that can permanently inactivate enzymes, e.g., cyanide.

Enzyme-substrate complex   A temporary union of an enzyme and its substrate.

Eosinophil   A granulocyte whose granules take up the stain eosin.

Epidemic disease   A disease acquired by many people in a given area in a short time.

Epidemiology   The science dealing with when and where diseases occur and how they are transmitted.

Epidermis   The outer portion of the skin.

Epitope   *See* Antigenic determinant.

Equilibrium   The point of even distribution.

Equivalent treatments   Different methods that have the same effect on controlling microbial growth.

Ergot   A toxin produced in sclerotia by the fungus *Claviceps purpurea* that causes ergotism, a disease characterized by hallucinations, and contraction of arteries and uterine muscle.

Erythrogenic toxin   A substance, produced by some streptococci, that causes erythema.

Ester linkage   The bonding between two organic molecules (R) as R—C—O—R
$$\underset{\displaystyle O}{\overset{\displaystyle \|}{}}$$

Ethambutol   A synthetic antimicrobial agent that interferes with the synthesis of RNA.

Etiology   The study of the cause of a disease.

Eubacteria   Procaryotic organisms; they possess peptidoglycan cell walls.

Eucaryote   A cell with DNA enclosed within a distinct membrane-bounded nucleus.

Eugenics   Controlling the gene pool of a species by selective breeding (usually refers to humans).

Eutrophication   The addition of organic matter and subsequent removal of oxygen from a body of water.

Exchange reaction   A chemical reaction that has both synthesis and decomposition components.

Exergonic reaction   A chemical reaction that releases energy.

Exon   A region of a eucaryotic chromosome that codes for a protein.

Exonuclease   An enzyme that cuts DNA.

Exotoxin   Protein toxin released from living, mostly gram-positive bacterial cells.

Experimental epidemiology   The study of a disease using controlled experiments.

Exponential growth phase   *See* Log phase.

Extracellular enzyme   An enzyme released from a cell to break down large molecules.

Extreme halophile   An organism that requires a high salt concentration for growth.

Extreme thermophile An organism whose optimum growth temperature is at least 80°C.

Eyespot A pigmented area capable of detecting the presence of light.

Facilitated diffusion The transfer of a substance across a plasma membrane from an area of higher concentration to an area of lower concentration mediated by carrier proteins (permeases).

Facultative anaerobe An organism that can grow with or without oxygen (O_2).

Facultative halophile An organism capable of growth in, but not requiring, 1% to 2% salt.

Family A taxonomic group between order and genus.

Fat An organic compound consisting of glycerol and fatty acids.

Fatty acids Long hydrocarbon chains ending in a carboxyl group.

Feedback inhibition Inhibition of an enzyme in a particular pathway by the accumulation of the end-product of the pathway.

Fermentation The enzymatic degradation of carbohydrates in which the final electron acceptor is an organic molecule, ATP is synthesized by substrate-level phosphorylation, and oxygen (O_2) is not required.

Fever An abnormally high body temperature.

F factor Fertility factor; a plasmid found in the donor cell in bacterial conjugation.

Fibrinolysin A kinase produced by streptococci.

Filtration Passage of a liquid or gas through a screenlike material; a 0.45-μm filter removes bacteria.

Fimbriae Appendages on a bacterial cell used for attachment.

Five kingdom system Classification system in which organisms are assigned to one of five kingdoms: Animalia, Plantae, Fungi, Protista, or Monera.

Fixed macrophage A macrophage that is located in a certain organ or tissue—for example, in liver, lungs, spleen, or lymph nodes.

Fixing (chemical elements) Combining elements so that a critical element can enter the food chain. *See* Nitrogen fixation; Calvin–Benson cycle.

Fixing (in slide preparation) The process of attaching the specimen to the slide.

Flagella Thin appendages that arise from one or more locations on the surface of a cell and are used for cellular locomotion.

Flagellate A member of the protozoan phylum Mastigophora that uses flagella for locomotion.

Flaming The process of sterilizing an inoculating loop by holding it in an open flame.

Flat sour spoilage Thermophilic spoilage of canned goods not accompanied by gas production.

Flatworm An animal belonging to the phylum Platyhelminthes.

Flavin adenine dinucleotide (FAD) A coenzyme that functions in the removal and transfer of hydrogen ions and electrons from substrate molecules.

Flavin mononucleotide (FMN) A coenzyme that functions in the transfer of electrons in the electron transport chain.

Flavoprotein Protein with flavin coenzyme that functions as an electron carrier in respiration.

Flocculation Removal of colloidal material during water purification by addition of a chemical that causes the colloidal particles to coalesce.

Flow cytometer An instrument that separates and counts cells; it detects cells by the presence of a fluorescent tag on the cell surface.

Fluid mosaic model A way of describing the dynamic arrangement of phospholipids and proteins comprising the plasma membrane.

Fluke A flatworm belonging to the class Trematoda.

Fluorescence The ability to give off light of one color when exposed to light of another color.

Fluorescence-activated cell sorter (FACS) A modification of a flow cytometer that sorts cells in addition to separating and counting them.

Fluorescence microscope A microscope that uses an ultraviolet light source to illuminate specimens that will fluoresce.

Fluorescent-antibody technique A diagnostic tool using antibodies labeled with fluorochromes and viewed through a fluorescence microscope.

Fluorochromes Dyes used to stain bacteria that fluoresce when illuminated with ultraviolet light.

Fluoroquinolone Synthetic antibacterial agents that inhibit DNA synthesis.

Focal infection A systemic infection that began as an infection in one place.

Fomite A nonliving object that can spread infection.

Forespore A structure consisting of chromosome, cytoplasm, and endospore membrane inside a bacterial cell.

Frameshift mutation A mutation caused by the addition or deletion of one or more bases in DNA.

Free radical A highly reactive particle with an unpaired electron; designated X·.

Freeze-drying *See* Lyophilization.

FTA-ABS test An indirect fluorescent-antibody test used to detect syphilis.

Functional groups Arrangement of elements in organic molecules that are responsible for most of the chemical properties of those molecules.

Fungi Organisms that belong to the Kingdom Fungi; eucaryotic chemoheterotrophs.

Gamete A male or female reproductive cell.

Gametocyte A male or female protozoan cell.

Gamma globulin *See* Immune serum globulin.

Gangrene Tissue death from loss of blood supply.

Gastroenteritis Inflammation of the stomach and intestine.

Gas vacuole A procaryotic inclusion for buoyancy compensation.

Gene A segment of DNA or a sequence of nucleotides in DNA that codes for a functional product.

Gene amplification A mechanism that causes a gene to be replicated many times.

Gene library Storage of DNA by inserting restriction enzyme fragments in a bacterium, yeast, or phage.

Gene therapy   Treating a disease by replacing abnormal genes.

Generalized transduction   Transfer of bacterial chromosome fragments from one cell to another by a bacteriophage.

Generation time   The time required for a cell or population to double in number.

Genetic code   The mRNA codons and the amino acids for which they code.

Genetic engineering   Manufacturing and manipulating genetic material in vitro.

Genetic recombination   The process of joining pieces of DNA from different sources.

Genetics   The science of heredity.

Genetic screening   Techniques used to determine which genes are in a cell's genetic material.

Genotype   The genetic makeup of an organism.

Genus   The first name of the scientific name (binomial); the taxon between family and species.

Geosmin   An alcohol produced by actinomycetes that has an earthy odor.

Germ   Living substance capable of developing into an organ, part, or organism as a whole.

Germicidal   Capable of killing microorganisms.

Germicidal lamp   An ultraviolet light (wavelength = 260 nm) capable of killing bacteria.

Germination   The process of starting to grow from a spore or endospore.

Germ theory of disease   The principle that microorganisms cause disease.

Global regulation   Gene expression initiated by environmental stress.

Globulin   The protein type to which antibodies belong.

Glomerulonephritis   Inflammation of the glomeruli of the kidneys, but not a result of kidney infection.

Glucan   A polysaccharide component of yeast cell walls.

Glycerol   An alcohol; $C_3H_5(OH)_3$. *See* Fat.

Glycocalyx   A gelatinous polymer surrounding a cell.

Glycogen   A polysaccharide stored by some cells.

Glycolysis   The main pathway for the oxidation of glucose to pyruvic acid.

Golgi complex   An organelle involved in the secretion of certain proteins.

Graft-versus-host (GVH) disease   A condition that occurs when a transplanted tissue has an immune response to the tissue recipient.

Gram-negative cell wall   A peptidoglycan layer surrounded by a lipopolysaccharide outer membrane.

Gram-positive cell wall   Composed of peptidoglycan and teichoic acids.

Gram stain   A differential stain that divides bacteria into two groups, gram-positive and gram-negative.

Granulocyte   A leukocyte with visible granules in the cytoplasm; includes neutrophils, basophils, and eosinophils.

Griseofulvin   A fungistatic antibiotic.

Group translocation   In procaryotes, active transport in which a substance is chemically altered during transport across the plasma membrane.

Guanine   A purine nucleic acid base that pairs with cytosine.

Gumma   A rubbery mass of tissue characteristic of tertiary syphilis.

Halogen   One of the following elements: fluorine, chlorine, bromine, iodine, or astatine.

Halophile   An organism that grows in high concentrations of salt.

Haploid   A eucaryotic cell or organism with one of each type of chromosome.

Hapten   An antigen that has reactivity and not immunogenicity.

Heavy metals   Certain elements with specific gravity greater than 4, for example, silver (Ag), copper (Cu), and mercury (Hg); these elements can be used as antimicrobial agents.

Helminth   A parasitic roundworm or flatworm.

Helper T (T_H) cells   Specialized T cells that often interact with an antigen before B cells interact with the antigen.

Hemagglutination   Clumping of red blood cells.

Hemoflagellate   A parasitic flagellate found in the circulatory system of its host.

Hemolysins   Enzymes that lyse red blood cells.

Herd immunity   The presence of immunity in most of a population.

Hermaphroditic   Having both male and female reproductive capacities.

Heterocyst   A large cell in certain cyanobacteria; site for nitrogen fixation.

Heterolactic   Any organism producing lactic acid and other acids or alcohols as end-products of fermentation; e.g., *Escherichia*.

Heterotroph   An organism that requires an organic carbon source.

Hexachlorophene   A chlorinated phenol used as an antiseptic.

Hexose monophosphate shunt   *See* Pentose phosphate pathway.

Hfr   A bacterial cell in which the F factor has become integrated into the chromosome; stands for high frequency of recombination.

High-efficiency particulate air (HEPA) filter   A screenlike material that removes particles larger than 0.3 μm from air.

High-energy bond   A bond that can readily be broken to release usable energy; designated by ~.

High-temperature short-time (HTST) pasteurization   Pasteurizing at 72°C for 15 seconds.

Histamine   A substance released by tissue cells that causes vasodilation, capillary permeability, and smooth muscle contraction.

Histocompatibility antigens   Antigens on the surface of human cells.

Histones   Proteins associated with DNA in eucaryotic chromosomes.

Holdfast   The branched base of an algal stipe.

Holoenzyme   An enzyme consisting of an apoenzyme and a cofactor.

Homolactic   An organism producing only lactic acid from fermentation (for example, *Streptococcus*).

Homologous chromosome   A chromosome that has the same base sequence as another. In a diploid cell, one of a pair of chromosomes.

Host   An organism infected by a pathogen.

Host range   The species, strains, or cell types that a pathogen can infect.

Hot-air sterilization   Sterilization by the use of an oven at 170°C for approximately 2 hours.

H (hemagglutination) spikes   Antigenic projections from the outer lipid bilayer of influenza virus.

Human leukocyte antigen (HLA) complex   Human cell surface antigens. *See also* Major histocompatibility complex.

Humoral immunity   Immunity produced by antibodies dissolved in body fluids, mediated by B cells.

Humus   Organic matter that remains in soil following partial decomposition.

Hyaluronic acid   A mucopolysaccharide that holds together certain cells of the body.

Hybridoma   A cell made by fusing an antibody-producing B cell with a cancer cell.

Hydrogen bond   A bond between a hydrogen atom covalently bonded to oxygen or nitrogen and another covalently bonded oxygen or nitrogen.

Hydrogen ion (H$^+$)   A proton.

Hydrolysis   A decomposition reaction in which chemicals react with the H$^+$ and OH$^-$ of a water molecule.

Hydroxide   OH$^-$.

Hydroxyl   —OH.

Hyperbaric chamber   An apparatus to hold materials at pressures greater than 1 atmosphere; used to oxygenate gangrenous tissue.

Hypersensitivity   Altered, enhanced immune reactions leading to pathologic changes.

Hyperthermophile   *See* Extreme thermophile.

Hypertonic solution   A solution that has a higher concentration of solutes than an isotonic solution.

Hyphae   Long filaments of cells in fungi or actinomycetes.

Hypotonic solution   A solution that has a lower concentration of solutes than an isotonic solution.

Icosahedron   A polyhedron with 20 triangular faces and 12 corners.

ID$_{50}$ The bacterial concentration required to produce a demonstrable infection in 50% of the test host population.

Idiophase   Stationary growth phase.

IgA   The class of antibodies found in secretions.

IgD   The class of antibodies found on B cells.

IgE   The class of antibodies involved in hypersensitivities.

IgG   The most abundant class of antibodies in serum.

IgM   The first class of antibodies to appear after exposure to an antigen.

Imidazoles   Antifungal drugs that interfere with sterol synthesis.

Immune complex   A circulating antigen–antibody aggregate capable of fixing complement.

Immune serum globulin   The serum fraction containing immunoglobulins (antibodies).

Immunity   The body's defense against a particular microorganism.

Immunization   A process that produces immunity.

Immunodiffusion test   A test consisting of precipitation reactions carried out in an agar gel medium.

Immunoelectrophoresis   Identification of proteins by electrophoretic separation and then serological testing.

Immunofluorescence   Procedures using the fluorescent-antibody technique.

Immunogen   *See* Antigen.

Immunoglobulin (Ig)   An antibody.

Immunological escape   Resistance of cancer cells to immune rejection.

Immunological surveillance   The body's immune response to cancer.

Immunology   The study of a host's specific defenses to a pathogen.

Immunosuppression   Inhibition of the immune response.

Immunotherapy   Treatment using antibodies.

Immunotoxin   An immunotherapeutic agent consisting of a poison bound to a monoclonal antibody.

IMViC   Biochemical tests used to identify enterics, including indole, methyl red, Voges–Proskauer, and citrate.

Inapparent infection   *See* Subclinical infection.

Incidence   The fraction of the population that contracts a disease during a particular length of time.

Inclusion   Material held inside a cell, often consisting of reserve deposits.

Incomplete digestive system   A digestive system with one opening (mouth) for intake of food and elimination of waste.

Incubation period   The time interval between the actual infection and first appearance of any signs or symptoms of disease.

Indirect (passive) agglutination test   Indirect agglutination test using soluble antigens attached to latex or other small particles.

Indirect contact   Transmission of pathogens by fomites.

Indirect FA test   A fluorescent-antibody test to detect the presence of specific antibodies.

Indirect tests   Diagnostic tests based on the presence of antibodies against a pathogen.

Induction   *See* Enzyme induction.

Inert   Inactive.

Infection   Growth of microorganisms in the body.

Infection thread   An invagination in a root hair of a plant that allows *Rhizobium* to infect the root.

Infectious disease   A disease caused by pathogens.

Inflammation   A host response to tissue damage characterized by redness, pain, heat, and swelling.

Innate resistance   Resistance of an individual to diseases that affect other species and other individuals of the same species.

Inoculate   To introduce microorganisms into a culture medium or host.

Inoculating loop (needle)   An instrument used to transfer bacteria from one culture medium to another.

Inorganic compounds   Small molecules not usually containing carbon.

Insertion sequence   *See* Transposon.

Interferon   An antiviral protein produced by certain animal cells in response to a viral infection.

Interleukins   Chemicals that cause T-cell proliferation. *See* Cytokines.

Intermediate host   An organism that harbors the larval or asexual stage of a helminth or protozoan.

Intoxication   Poisoning.

Intron   A region in a eucaryotic gene that does not code for a protein or mRNA.

Invasiveness   The ability of microorganisms to establish residence in a host.

In vitro   "In glass"; not in a living organism.

In vivo   Within a living organism.

Iodophor   A complex of iodine and a detergent.

Ion   A negatively or positively charged atom or group of atoms.

Ionic bond   A chemical bond formed when atoms gain or lose electrons in the outer energy levels.

Ionization   Separation of a molecule into groups of atoms with electrical charges.

Ionizing radiation   High-energy radiation that causes ionization; for example, X rays and gamma rays; wavelengths less than 1 nm.

Ischemia   Local loss of blood supply.

Isoelectric focusing   Electrophoresis of a protein until it reaches its isoelectric point.

Isoelectric point   The pH at which a protein no longer migrates in electrophoresis.

Isograft   Tissue graft between identical twins.

Isomers   Two molecules with the same chemical formula but different structures.

Isoniazid (INH)   A bacteriostatic agent used to treat tuberculosis.

Isotonic solution   A solution in which, after immersion of a cell, osmotic pressure is equal across the cell's membrane.

Isotope   A form of a chemical element in which the number of neutrons in the nucleus is different from the other forms of that element.

Kelp   Multicellular brown algae.

Keratin   A protein found in epidermis, hair, and nails.

Kinases   (1) Bacterial enzymes that break down fibrin (blood clots). (2) Enzymes that remove a ⓟ from ATP and attach it to another molecule.

Kingdom   The highest category in the taxonomic hierarchy of classification.

Kinins   Substances released from tissue cells that cause vasodilation.

Kirby–Bauer test   An agar-diffusion test to determine microbial susceptibility to chemotherapeutic agents.

Koch's postulates   Criteria used to determine the causative agent of infectious diseases.

Krebs cycle   A pathway that converts two-carbon compounds to carbon dioxide (CO_2), transferring electrons to NAD^+ and other carriers.

L-   Prefix describing a stereoisomer; the amino acids found in proteins are always the L-isomers.

Lactic acid fermentation   A catabolic process, beginning with glycolysis, that produces lactic acid.

Lagging strand   During DNA replication, the daughter strand that is synthesized discontinuously.

Lag phase   The time interval in a bacterial growth curve during which there is no growth.

Larva   Sexually immature stage of a helminth or arthropod.

Latent disease   A disease characterized by a period of no symptoms when the pathogen is inactive.

Latent infection   A condition in which a pathogen remains in the host for long periods without producing disease.

LD_{50}   The lethal dose for 50% of the inoculated hosts within a given period.

Leading strand   During DNA replication, the daughter strand that is synthesized continuously.

Lepromin test   A skin test to determine the presence of antibodies to *Mycobacterium leprae*.

Leukemias   Cancers characterized by abnormally large numbers of leukocytes.

Leukocidins   Substances produced by some bacteria that can destroy neutrophils and macrophages.

Leukocyte   White blood cell.

Leukopenia   A condition in which the number of leukocytes is smaller than normal.

Leukotrienes   Mediators of anaphylaxis.

Lichen   A symbiosis between a fungus and an alga or a cyanobacterium.

Ligands   Carbohydrate-specific binding proteins projecting from procaryotic cells; used for adherence.

Light reaction   *See* Photophosphorylation.

Limnetic zone   The surface zone of an inland body of water away from the shore.

***Limulus* amoebocyte lysate (LAL) assay**   A test to detect the presence of bacterial endotoxins.

Lipase   An exoenzyme that breaks down fats into their component glycerol and fatty acids.

Lipid   A molecule composed of glycerol and fatty acids.

Lipid A   A component of gram-negative outer membrane; endotoxin.

Lipopolysaccharide (LPS)   A molecule consisting of a lipid and a polysaccharide, forming the outer layer of gram-negative cell walls.

Lipoprotein   A molecule consisting of a lipid and a protein.

Liposome   A fatty globule that may be used to administer chemotherapeutic agents.

Lithotroph   *See* Autotroph.

Littoral zone   The region along the shore of an inland body of water where there is considerable vegetation and where light penetrates to the bottom.

Local infection   An infection in which pathogens are limited to a small area of the body.

Localized anaphylaxis   An anaphylaxis-type reaction, such as hay fever, asthma, and hives.

Logarithm (log)   Power to which a base number is raised to produce a given number.

Log phase Period of bacterial growth or logarithmic increase in cell numbers.

Lophotrichous Having two or more flagella at one end of a cell.

Luciferase An enzyme that accepts electrons from flavoproteins and emits a photon of light.

Lymphangitis Inflammation of lymph vessels.

Lymphocyte A white blood cell involved in specific immune responses.

Lymphoma Cancer of lymphoid tissue.

Lyophilization Freeze-drying; freezing a substance and sublimating the ice in a vacuum.

Lysis (1) Disruption of the plasma membrane. (2) In disease, a gradual period of decline.

Lysogeny A state in which phage DNA is incorporated into the host cell without lysis.

Lysosome An organelle containing digestive enzymes.

Lysozyme An enzyme capable of lysing bacterial cell walls.

Lytic cycle A sequence for replication of phages that results in host cell lysis.

Macrolides Antibiotics that inhibit protein synthesis; for example, erythromycin.

Macromolecules Large organic molecules.

Macrophage A phagocytic cell; an enlarged monocyte.

Macrophage activation factor Increases macrophages' efficiency at destroying ingested cells.

Macrophage chemotactic factor Attracts macrophages to infection site.

Macrophage migration-inhibiting factor Prevents macrophages from leaving infection site.

Macular rash Small red spots appearing on the skin.

Major histocompatibility complex (MHC) The genes that code for the histocompatibility antigens; also known as human leukocyte antigen (HLA) complex.

Malolactic fermentation Conversion of malic acid to lactic acid by lactic acid bacteria.

Malt Germinated barley grains containing maltose and amylase.

Malting Germination of starchy grains resulting in glucose and maltose production.

Mannan A polysaccharide component of yeast cell walls.

Mantoux test A tuberculin skin test.

Margination The process by which phagocytes stick to the lining of blood vessels.

Mast cell A type of cell found throughout the body that contains histamine and other substances that stimulate vasodilation.

Maximum growth temperature The highest temperature at which a species can grow.

Mechanical transmission The process by which arthropods transmit infections by carrying pathogens on their feet and other body parts.

Medulla (of a lichen thallus) Lichen body consisting of algae (or cyanobacteria) and fungi.

Meiosis The process that leads to the formation of haploid gametes in a diploid organism.

Membrane attack complex Complement proteins 5 through 9, which together make lesions in cell membranes that lead to cell death.

Membrane filter A screenlike material with pores small enough to retain microorganisms.

Memory cells Long-lived B or T cells responsible for the memory or secondary response.

Memory response A rapid rise in antibody titer following exposure to an antigen after the primary response to that antigen.

Meningitis Inflammation of the meninges covering the central nervous system.

Merozoite A trophozoite of *Plasmodium* found in infected red blood cells or liver cells.

Mesophile An organism that grows between about 10°C and 50°C.

Mesosome An irregular fold in the plasma membrane of a procaryotic cell that is an artifact of preparation for microscopy.

Messenger RNA (mRNA) The type of RNA molecule that directs the incorporation of amino acids into proteins.

Metabolism The sum of all the chemical reactions that occur in a living cell.

Metacercaria The encysted stage of a fluke in its final intermediate host.

Metachromatic granule The intracellular volutin stored by some bacteria.

Meter (m) The standard unit of length in the metric system; one ten-millionth of the distance from the equator to the pole.

Methane Odorless, colorless, flammable gas; CH_4.

Microaerophile An organism that grows best in an environment with less oxygen (O_2) than is normally found in air.

Micrometer (μm) A unit of measure equal to 10^{-6} m.

Microorganism A living organism too small to be seen with the naked eye; includes bacteria, fungi, protozoa, and microscopic algae; also includes viruses.

Microtubule The structure of the proteins that make up eucaryotic flagella and cilia.

Microwave An electromagnetic wave with wavelength between 10^{-1} and 10^{-3} m.

Minimal bactericidal concentration (MBC) The lowest concentration of a chemotherapeutic agent that will kill test microorganisms.

Minimal inhibitory concentration (MIC) The lowest concentration of a chemotherapeutic agent that will prevent growth of the test microorganisms.

Minimum growth temperature The lowest temperature at which a species will grow.

Miracidium The free-swimming, ciliated larva of a fluke that hatches from the egg.

Missense mutation A mutation that results in substitution of an amino acid in a protein.

Mitochondria Organelles containing the respiratory ATP-synthesizing enzymes.

Mitosis A eucaryotic cell division process in which the chromosomes are duplicated, followed by division of the cytoplasm of the cell.

Mixed culture A culture containing more than one kind of microorganism.

MMR vaccine Attenuated measles, mumps, rubella viruses.

MMWR *Morbidity and Mortality Weekly Report;* weekly publication of the Centers for Disease Control and Prevention containing data on notifiable diseases and topics of special interest.

Molds Fungi that form mycelia and appear as cottony tufts.

Mole An amount of a chemical equal to the atomic weights of all the atoms in a molecule of the chemical.

Molecular biology The science dealing with DNA and protein synthesis of living organisms.

Molecular weight The sum of the atomic weights of all atoms making up a molecule.

Molecule A combination of atoms forming a specific chemical compound.

Monera The kingdom to which all procaryotic organisms belong.

Monoclonal antibodies Specific antibodies produced by in vitro clones of B cells hybridized with cancerous cells.

Monocyte A phagocyte that is the precursor of a macrophage.

Monolayer A single layer of cells in vitro due to the cessation of cell division by contact inhibition.

Monomers The units that combine to form polymers.

Mononuclear phagocytic system A system of fixed macrophages located in the spleen, liver, lymph nodes, and bone marrow.

Monotrichous Having a single flagellum.

Morbidity (1) The incidence of a specific disease. (2) The condition of being diseased.

Mordant A substance added to a staining solution to make it stain more intensely.

Morphology The external appearance.

Mortality The deaths from a specific notifiable disease.

Most probable number (MPN) method A statistical determination of the number of coliforms per 100 ml of water or 100 g of food.

Motility The ability of an organism to move by itself.

M protein A heat- and acid-resistant protein of streptococcal cell walls and fibrils.

Multiple-tube fermentation test A method of detecting the presence of coliforms. *See* Most probable number (MPN) method.

Murein *See* Peptidoglycan.

Mutagen An agent in the environment that brings about mutations.

Mutation Any change in the base sequence of DNA.

Mutation rate The probability of a gene mutating each time a cell divides.

Mutualism A symbiosis in which both organisms are benefited.

Mycelium A mass of long filaments of cells that branch and intertwine, typically found in molds.

Mycetoma A chronic infection caused by certain fungi and *Nocardia,* characterized by a tumorlike appearance.

Mycology The science dealing with fungi.

Mycorrhiza A fungus growing in symbiosis with plant roots.

Mycosis A fungal infection.

Mycotoxin A toxin produced by a fungus.

Myocarditis Inflammation of the heart muscle.

Nanometer (nm) A unit of measurement equal to 10^{-9} m, 10^{-3} μm, and 10 Å.

Natural killer (NK) cells Lymphoid cells that destroy tumor cells and virus-infected cells.

Naturally acquired active immunity Antibody production in response to an infectious disease.

Naturally acquired passive immunity The natural transfer of humoral antibodies; transplacental transfer.

Necrosis Tissue death.

Negative (indirect) selection *See* Replica plating.

Negative staining A procedure that results in colorless bacteria against a stained background.

Neurotoxin A chemical that is poisonous to the nervous system.

Neutralization reaction An antigen–antibody reaction that inactivates a bacterial exotoxin or virus.

Neutron An uncharged particle in the nucleus of an atom.

Neutrophil Also called polymorphonuclear leukocyte; a highly phagocytic granulocyte.

Nicotinamide adenine dinucleotide (NAD⁺) A coenzyme that functions in the removal and transfer of H⁺ and electrons from substrate molecules.

Nicotinamide adenine dinucleotide phosphate (NADP⁺) A coenzyme similar to NAD⁺.

Nitrification The oxidation of nitrogen from ammonia to nitrites and nitrates.

Nitrofuran A synthetic antimicrobial drug.

Nitrogen cycle The series of processes that converts nitrogen (N_2) to organic substances and back to nitrogen in nature.

Nitrogen fixation The conversion of nitrogen (N_2) into ammonia.

Nitrosamine A carcinogen formed by the combination of nitrite and amino acids; nitroso-: —N=O.

Nomenclature The system of naming things.

Noncommunicable disease A disease that is not transmitted from one person to another.

Noncompetitive inhibitor *See* Allosteric inhibition.

Noncyclic photophosphorylation Movement of an electron from chlorophyll to NAD⁺; plant and cyanobacterial photophosphorylation.

Nonionizing radiation Short wavelength radiation that does not cause ionization; ultraviolet radiation.

Nonsense codon A special terminator codon that does not code for any amino acid.

Nonsense mutation A base substitution in DNA that results in a nonsense codon.

Nonspecific resistance Host defenses that tend to afford protection from any kind of pathogen.

Normal microbiota (flora) Collection of microorganisms that colonize an animal without causing disease.

Nosocomial infection An infection that develops during the course of a hospital stay and was not present at the time the patient was admitted.

Notifiable disease A disease that physicians must report to the public health service.

N (neuraminidase) spikes Antigenic projections from the outer lipid bilayer of influenza virus.

Nuclear envelope   The double membrane that separates the nucleus from the cytoplasm in a eucaryotic cell.

Nuclear pores   Openings in the nuclear envelope through which materials enter and leave the nucleus.

Nucleic acid   A macromolecule consisting of nucleotides; for example, RNA and DNA.

Nucleic acid hybridization   The process of combining single complementary strands of DNA.

Nucleoid   The region in a bacterial cell containing the chromosome.

Nucleoli   Areas in a eucaryotic nucleus where rRNA is synthesized.

Nucleoplasm   The gel-like fluid within the nuclear envelope.

Nucleoprotein   A macromolecule consisting of protein and nucleic acid.

Nucleoside   A compound consisting of a purine or pyrimidine base and a pentose sugar.

Nucleotide   A compound consisting of a purine or pyrimidine base, a five-carbon sugar, and a phosphate.

Nucleus   (1) The part of a eucaryotic cell that contains the genetic material. (2) The part of an atom consisting of the protons and neutrons.

Numerical taxonomy   A method of comparing organisms on the basis of many characteristics.

Nutrient broth (agar)   A complex medium made of beef extracts.

Objective lenses   In a compound light microscope, the lenses closest to the specimen.

Obligate anaerobe   An organism that is unable to use oxygen.

Ocular lens   In a compound light microscope, the lens closest to the viewer.

Oligodynamic action   The ability of small amounts of a heavy metal compound to exert antimicrobial activity.

Oncogene   A gene that can bring about malignant transformation.

Oncogenic virus   A virus that is capable of producing tumors.

Oocyst   An encysted sporozoan zygote in which cell division occurs to form the next infectious stage.

Operator   The region of DNA adjacent to structural genes that controls their transcription.

Operon   The operator site and structural genes it controls.

Opportunist   An organism that does not ordinarily cause a disease but can become pathogenic under certain circumstances.

Opsonization   The enhancement of phagocytosis by coating microorganisms with certain serum proteins (opsonins).

Optimum growth temperature   The temperature at which a species grows best.

Oral groove   On some protozoa, the site at which nutrients are taken in.

Order   A taxonomic classification between class and family.

Organelles   Membrane-bounded structures within eucaryotic cells.

Organic compounds   Molecules that contain carbon and hydrogen.

Organic growth factor   An essential organic compound that an organism is unable to synthesize.

Osmosis   The net movement of solvent molecules across a selectively permeable membrane from an area of higher concentration to an area of lower concentration.

Osmotic pressure   The force with which a solvent moves from a solution of lower solute concentration to a solution of higher solute concentration.

Ouchterlony test   An immunodiffusion test.

Outer membrane   The outer layer of a gram-negative cell wall consisting of lipoproteins, lipopolysaccharides, and phospholipids.

Oxidase test   A diagnostic test for the presence of cytochrome c using p-aminodimethylaniline.

Oxidation   (1) The removal of electrons from a molecule. (2) The addition of oxygen to a molecule.

Oxidation pond   A method of secondary sewage treatment.

Oxidation–reduction (redox) reaction   A coupled reaction in which one substance is oxidized and one is reduced.

Oxidative phosphorylation   The synthesis of ATP coupled with electron transport.

Oxygen cycle   The processes that convert molecular oxygen (O_2) to oxides, water, and organic compounds and back to O_2.

Oxygenic   Producing oxygen, as in plant and cyanobacterial photosynthesis.

PABA   Para-aminobenzoic acid; a precursor for folic acid synthesis.

PAGE   Polyacrylamide gel electrophoresis. *See also* Electrophoresis.

Pandemic disease   An epidemic that occurs worldwide.

Papular rash   A skin rash characterized by raised spots.

Parasite   An organism that derives nutrients from a living host.

Parasitism   A symbiosis in which one organism (the parasite) exploits another (the host) without providing any benefit in return.

Parenteral route   Deposition directly into tissues beneath the skin and mucous membranes.

Passive agglutination   *See* Indirect agglutination test.

Passively acquired immunity   Immunity acquired when antibodies produced by another source are transferred to an individual.

Pasteurization   The process of mild heating to kill particular spoilage organisms or pathogens.

Pathogen   A disease-causing organism.

Pathology   Study of the causes and development of disease.

Pellicle   The flexible covering of some protozoa.

Penicillins   A group of antibiotics produced either by *Penicillium* (natural penicillins) or by adding side chains to the β-lactam ring (semisynthetic penicillins).

Pentose phosphate pathway   A metabolic pathway that can occur simultaneously with glycolysis to produce pentoses and NADH without ATP production.

Peptide   A chain of two (di-), three (tri-), or more (poly-) amino acids.

Peptide bond   A bond joining the amino group of one amino acid to the carboxyl group of a second amino acid with the loss of a water molecule.

Peptidoglycan (murein) The structural molecule of eubacterial cell walls consisting of the molecules N-acetylglucosamine, N-acetylmuramic acid, tetrapeptide side chain, and peptide side chain.

Peptone Short chains of amino acids produced by the action of acids or enzymes on proteins.

Perforin Protein that makes a pore in a target cell membrane, released by T_C cells.

Pericarditis Inflammation of the sac around the heart (pericardium).

Periplasmic space The region of a gram-negative cell wall between the outer membrane and the cytoplasmic membrane.

Peritrichous Having flagella distributed over the entire cell.

Permease A carrier protein in the plasma membrane.

Peroxidase An enzyme that breaks down hydrogen peroxide: $H_2O_2 + NADH + H^+ \longrightarrow 2H_2O + NAD^+$.

Peroxide An oxygen anion consisting of two atoms of oxygen; O_2^{2-}.

pH The symbol for hydrogen ion (H^+) concentration; a measure of the relative acidity or alkalinity of a solution.

Phage *See* Bacteriophage.

Phage typing A method of identifying bacteria using specific strains of bacteriophages.

Phagocyte A cell capable of engulfing and digesting particles that are harmful to the body.

Phagocytosis The ingestion of solids by cells.

Phagolysosome A digestive vacuole.

Phagosome Food vacuole of a phagocyte.

Phase-contrast microscope A compound light microscope that allows examination of structures inside cells through the use of a special condenser.

Phenol Carbolic acid.

Phenolic A synthetic derivative of phenol.

Phenotype The external manifestations of the genetic makeup of an organism.

Phosphate group A portion of a phosphoric acid molecule attached to some other molecule:

$$-O-\overset{\overset{\textstyle O}{\|}}{\underset{\underset{\textstyle OH}{|}}{P}}-O^-$$

Phospholipid A complex lipid composed of glycerol, two fatty acids, and a phosphate group.

Phosphorylation The addition of a phosphate group to an organic molecule.

Photoautotroph An organism that uses light as its energy source and carbon dioxide (CO_2) as its carbon source.

Photoheterotroph An organism that uses light as its energy source and an organic carbon source.

Photophosphorylation The production of ATP by photosynthesis.

Photosynthesis The light-fueled synthesis of carbohydrate from carbon dioxide (CO_2).

Phototaxis Movement in response to the presence of light.

Phototroph An organism that uses light as its primary energy source.

Phylogenetic Refers to phylogeny, the evolutionary history of a group of organisms.

Phylum A taxonomic classification between kingdom and class.

Phytoplankton Free-floating photoautotrophs.

Pilus An appendage on a bacterial cell used for transfer of genetic material during conjugation.

Pinocytosis The engulfing of fluid by infolding of the plasma membrane.

Plankton Free-floating aquatic organisms.

Plantae A kingdom composed of multicellular eucaryotes with cellulose cell walls.

Plaque A clearing in a confluent growth of bacteria due to lysis by phages. *See also* Dental plaque.

Plaque-forming units Visible plaques counted.

Plasma The liquid portion of blood in which the formed elements are suspended.

Plasma cells Cells that activated B lymphocytes differentiate into; plasma cells manufacture specific antibodies.

Plasma membrane The selectively permeable membrane enclosing the cytoplasm of a cell; the outer layer in animal cells, internal to the cell wall in other organisms.

Plasmid A small cyclic DNA molecule in bacteria replicating independently of the chromosome.

Plasmodium (1) A multinucleated mass of protoplasm, as in plasmodial slime molds. (2) When written as a genus, refers to the protozoan agent of malaria.

Plasmolysis Loss of water from a cell in a hypertonic environment.

Plate count A method of determining the number of bacteria in a sample by counting the number of colony-forming units on a solid culture medium.

Pleomorphic Having many shapes.

Pneumonia Inflammation of the lungs.

Point mutation *See* Base substitution.

Polar molecule A molecule with an unequal distribution of charges.

Poly-β-hydroxybutyric acid A fatty acid storage material unique to bacteria.

Polyene antibiotic An antimicrobial agent that alters sterols in eucaryotic plasma membranes and contains more than four carbon atoms and at least two double bonds.

Polyhedron A many-sided solid.

Polymer A molecule consisting of a sequence of similar units or monomers.

Polymerase An enzyme that synthesizes specific polymers.

Polymerase chain reaction (PCR) Use of DNA polymerase to make multiple copies of a DNA template in vitro.

Polymorphonuclear leukocyte (PMN) A neutrophil.

Polypeptide (1) A chain of amino acids. (2) A group of antibiotics.

Polyribosome An mRNA strand with several ribosomes attached to it, on which successive translation is occurring.

Porins Proteins in the outer membrane of gram-negative cell walls that allow passage of small molecules.

Portal of entry   The avenue by which a pathogen gains access to the body.

Portal of exit   The route by which a pathogen leaves the body.

Positive (direct) selection   A procedure for picking out mutant cells by growing them.

Potential energy   Energy that is stored.

Pour plate method   A method of inoculating a solid nutrient medium by mixing bacteria in the melted medium and pouring the medium into a Petri plate to solidify.

PPD (purified protein derivative)   Proteins extracted from a boiled culture of *Mycobacterium tuberculosis;* the antigen used in the tuberculin skin test.

Precipitation reaction   A reaction between soluble antigens and multivalent antibodies to form visible aggregates.

Precipitin ring test   A precipitation test performed in a capillary tube.

Predisposing factor   Anything that makes the body more susceptible to a disease or alters the course of a disease.

Prevalence   The fraction of a population having a specific disease at a given time.

Primary cell line   Human tissue cells that grow for only a few generations in vitro.

Primary infection   An acute infection that causes the initial illness.

Primary metabolite   A chemical formed during cell growth.

Primary producer   Photoautotroph that converts light energy into chemical energy.

Primary response   Antibody production to the first contact with an antigen. *See also* Memory response.

Primary treatment   Physical removal of solid matter from wastewater.

Prion   A proposed infectious protein particle.

Privileged site (tissue)   An area of the body (or a tissue) that does not elicit an immune response.

Probe   A short, labeled, single strand of DNA or RNA used to locate its complementary strand in a quantity of DNA.

Procaryote   A cell whose genetic material is not enclosed in a nuclear envelope. *See* Monera.

Prodromal period   The time following incubation when the first symptoms of illness appear.

Product   The substance formed in a chemical reaction.

Profundal zone   The deeper water under the limnetic zone in an inland body of water.

Proglottid   A body segment of a tapeworm containing both male and female organs.

Promoter   The starting point on DNA for transcription of RNA by RNA polymerase.

Prophage   Phage DNA inserted into the host cell's DNA.

Prostaglandins   Hormonelike substances that are synthesized in many tissues and circulate in the blood, the exact function of which is unknown.

Prostheca   A stalk or bud protruding from a procaryotic cell.

Prosthetic group   A coenzyme that is bound tightly to its apoenzyme.

Protein   A large molecule containing carbon, hydrogen, oxygen, and nitrogen (and sulfur); some proteins have a helical structure and others are pleated sheets.

Protein kinase   Enzyme that transfers a phosphate group from ATP to an amino acid.

Proteolytic   An enzyme capable of hydrolyzing a protein.

Protista   The kingdom to which protozoa belong; unicellular, eucaryotic organisms.

Proton   A positively charged particle in the nucleus of an atom.

Proto-oncogene   An inactive oncogene that is still functioning normally.

Protoplasmic streaming   Movement of protoplasm in a plasmodial slime mold.

Protoplast   A gram-positive bacterium or a plant cell treated to remove the cell wall.

Protoplast fusion   A method of joining two cells by first removing their cell walls; used in genetic engineering.

Protozoa   Unicellular eucaryotic organisms belonging to the kingdom Protista.

Provirus   Viral DNA that is integrated into the host cell's DNA.

Pseudohypha   A short chain of fungal cells that results from the lack of separation of daughter cells after budding.

Pseudomurein   The structural molecules of archaeobacterial cell walls consisting of N-acetyltalosaminuronic acid, N-acetyl-glucosamine, and L amino acids.

Pseudopods   Extensions of a cell that aid in locomotion and feeding.

P site   Site on ribosome where tRNA is attached to peptide chain.

Psychrophile   An organism that grows best at about 15°C and does not grow above 20°C.

Psychrotroph   An organism that is capable of growth between about 0°C and 30°C.

Pure culture   A population of one strain or species of bacteria.

Purines   The class of nucleic acid bases that includes adenine and guanine.

Pus   An accumulation of dead phagocytes, dead bacterial cells, and fluid.

Pyocin   A bacteriocin produced by *Pseudomonas aeruginosa.*

Pyocyanin   Blue-green pigment produced by *Pseudomonas aeruginosa.*

Pyogenic   Pus-forming.

Pyrimidines   The class of nucleic acid bases that includes uracil, thymine, and cytosine.

Quaternary ammonium compound (quat)   A cationic detergent with four organic groups attached to a central nitrogen atom; used as a disinfectant.

Quellung reaction   Apparent swelling of a bacterial capsule in the presence of a specific antibody.

Quinine   An antimalarial drug derived from the cinchona tree and effective against schizogomy stage in red blood cells.

Radioimmunoassay   A method of measuring the amount of a compound using radioactively labeled antibodies.

Rapid plasma reagin   A serologic test for syphilis.

r-determinant   Genes for antibiotic resistance carried on R factors.

Reactants   Substances that are combined in a chemical reaction.

Reactivity The ability of an antigen to combine with an antibody.

Reagin IgE antibodies made in response to a treponemal infection characterized by their ability to combine with lipids; reagin-lipid complex will fix complement.

Recalcitrance Being resistant to degradation.

Receptor An attachment for a pathogen on a host cell.

Recipient cell A cell that receives DNA from a donor cell in recombination.

Recombinant DNA A DNA molecule produced by recombination.

Recombinant DNA technology *See* Genetic engineering.

Recombinant vaccine A vaccine made by recombinant DNA techniques.

Redia A trematode larval stage that reproduces asexually to produce cercariae.

Redox reactions Paired oxidation–reduction reactions.

Red tide A bloom of planktonic dinoflagellates.

Reducing medium A culture medium containing ingredients that will remove dissolved oxygen from the medium to allow the growth of anaerobes.

Reduction (1) The addition of electrons to a molecule. (2) The gain of hydrogen atoms.

Refractive index The relative velocity with which light passes through a substance.

Rennin A proteolytic enzyme obtained from a calf's stomach that forms curd from milk.

Replica plating A method of inoculating a number of solid minimal culture media from an original plate of complete medium; mutant colonies that don't grow on the minimal media can be selected from the original plate.

Replication fork The point where DNA separates and new strands of DNA will be synthesized.

Replicative form A double-stranded RNA molecule produced during the multiplication of certain RNA viruses.

Reportable disease *See* Notifiable disease.

Repression The process by which a repressor can stop the synthesis of a protein.

Repressor A protein that binds to the operator site to prevent transcription.

Reservoir of infection A continual source of infection.

Resistance The ability to ward off diseases through nonspecific and specific defenses.

Resistance (R) factor A bacterial plasmid carrying genes that determine resistance to antibiotics.

Resistance transfer factor A group of genes for replication and conjugation on the R factor.

Resolution The ability to distinguish fine detail with a magnifying instrument.

Respiration (1) An ATP-generation process in which chemical compounds are oxidized and the final electron acceptor is usually an inorganic molecule. (2) The process by which living organisms produce carbon dioxide (CO_2).

Restriction enzymes Enzymes that cut double-stranded DNA, leaving staggered ends.

Reticulate body An intracellular stage of *Chlamydia*.

Reverse transcriptase RNA-dependent DNA polymerase; an enzyme that synthesizes a complementary DNA from an RNA template.

Reversible reaction A chemical reaction in which the end products can readily revert to the original molecules.

Rh factor Classification of red blood cells based on the presence or absence of Rh antigens.

Rhizine A rootlike hypha that anchors a fungus to a surface.

Rhizosphere The region in soil where the soil and roots make contact.

Riboflavin A B vitamin that functions as a flavoprotein.

Ribonucleic acid (RNA) The class of nucleic acids that comprises messenger RNA, ribosomal RNA, and transfer RNA.

Ribose A five-carbon sugar that is part of ribonucleotide molecules and RNA.

Ribosomal RNA (rRNA) The RNA molecules that form the ribosomes.

Ribosomes The sites of protein synthesis in a cell, composed of RNA and protein.

Ribozyme An enzyme consisting of RNA.

Ricin A toxic protein from castor oil beans.

Rifamycin An antibiotic that inhibits bacterial RNA synthesis.

Ring stage A young *Plasmodium* trophozoite that looks like a ring in a red blood cell.

RNA-dependent RNA polymerase An enzyme that synthesizes a complementary RNA from an RNA template.

RNA primer A short strand of RNA used to start synthesis of the lagging strand of DNA, and to start the polymerase chain reaction.

Roll tube Anaerobic culturing technique in which bacteria grow in agar medium that has solidified on the interior wall of a test tube.

Root nodule A tumorlike growth on the roots of certain plants containing symbiotic nitrogen-fixing bacteria.

Rough ER Endoplasmic reticulum with ribosomes on its surface.

Roundworms Animals belonging to the phylum Aschelminthes.

S (Svedberg unit) Notes the relative rate of sedimentation during ultra-high-speed centrifugation.

Sabin vaccine A preparation containing three attenuated strains of polio virus administered orally.

Saccharide Sugar: general formula $(CH_2O)_n$.

Salk vaccine A preparation of a formalin-inactivated polio virus that is injected.

Salt A substance that dissolves in water to cations and anions, neither of which is H^+ or OH^-.

Sanitation Removal of microbes from eating utensils and food preparation areas.

Saprophyte An organism that obtains its nutrients from dead organic matter.

Sarcina (1) A group of eight bacteria that remain in a packet after dividing. (2) When written as a genus, refers to gram-positive, anaerobic cocci.

Sarcoma A cancer of fleshy, nonepithelial tissue or connective tissue.

Saturation (1) The condition in which the active site on an enzyme is occupied by the substrate or product at all times. (2) In a hydrocarbon, having the maximum number of hydrogen atoms.

Saxitoxin Neurotoxin produced by some dinoflagellates.

Scanning electron microscope An electron microscope that provides three-dimensional views of the specimen magnified about 10,000 times.

Schaeffer–Fulton stain An endospore stain that uses malachite green to stain the endospores and safranin as a counterstain.

Schick test A skin test to detect the presence of antibodies to diphtheria.

Schizogony The process of multiple fission, in which one organism divides to produce many daughter cells.

Scientific notation Expression of numbers as powers of 10; e.g., $7,650 = 7.65 \times 10^3$ or $0.00032 = 3.2 \times 10^{-4}$.

Sclerotia The compact mass of hardened mycelia of the fungus *Claviceps purpurea* that fills infected rye flowers; produces the toxin ergot.

Scolex The head of a tapeworm, containing suckers and possibly hooks.

Secondary infection An infection caused by an opportunistic pathogen after a primary infection has weakened the host's defenses.

Secondary metabolite A chemical formed after cell growth has stopped.

Secondary response *See* Memory response.

Secondary treatment Biological degradation of the organic matter in wastewater, following primary treatment.

Secretion The production and release of fluid containing a variety of substances from a cell.

Secretory vesicle Membrane-bounded sac produced by the ER; transports synthesized material into cytoplasm.

Selective medium A culture medium designed to suppress the growth of unwanted microorganisms and encourage the growth of desired ones.

Selective permeability The property of a plasma membrane to allow certain molecules and ions to move through the membrane while restricting others.

Selective toxicity The property of some antimicrobial agents to be toxic for a microorganism and nontoxic for the host.

Self-tolerance The ability of an organism to recognize and not make antibodies against self.

Semiconservative replication The process of DNA replication in which each double-stranded molecule of DNA contains one original strand and one new strand.

Sense codon A codon that codes for an amino acid.

Sense strand (+strand) Viral RNA that can act as mRNA.

Sepsis The presence of unwanted bacteria.

Septate hypha A hypha consisting of uninucleate cell-like units.

Septicemia A condition characterized by the multiplication of bacteria in the blood.

Septic shock Sudden drop in blood pressure due to infection by gram-negative bacteria.

Septic tank A tank, built into the ground, in which waste water is treated by primary treatment.

Septum A crosswall dividing two parts.

Serial dilution The process of diluting a sample several times.

Seroconversion Change in a person's response to an antigen in a serological test.

Serology The branch of immunology concerned with the study of antigen–antibody reactions in vitro.

Serovar A subgroup of *Salmonella* based on the presence of antigens.

Serum The liquid remaining after blood plasma is clotted, which contains immunoglobulins.

Sexual dimorphism The distinctly different appearance of adult male and female organisms.

Sexual reproduction Reproduction that requires two opposite mating strains, usually designated male and female.

Sexual spore Spore formed by sexual reproduction.

Shuttle vector A plasmid that can exist in several different species; used in genetic engineering.

Signs Changes due to a disease that a physician can observe and measure.

Simple stain A method of staining microorganisms with a single basic dye.

Single-cell protein (SCP) A food substitute consisting of microbial cells.

Singlet oxygen Highly reactive O_2.

Skin test The intradermal injection of an antigen or antibody to determine susceptibility to an antigen.

Slide agglutination test A method of identifying an antigen by combining it with a specific antibody on a slide.

Slime mold Funguslike protists.

Slow viral infection A disease process that occurs gradually over a long period. *See also* Prion.

Sludge Solid matter obtained from sewage.

Smear A thin film of material on a slide.

Smooth ER Endoplasmic reticulum without ribosomes.

Soap A surface-active agent made from animal fats and lye (NaOH).

Solubility The ability to be dissolved, usually in water.

Solute A substance dissolved in another substance.

Solvent A dissolving medium.

Southern blot A technique that uses DNA probes to detect the presence of specific DNA in restriction fragments separated by electrophoresis.

Specialized T cell T cell with a specific function; differentiated after exposure to an antigen.

Specialized transduction The process of transferring a piece of cell DNA adjacent to a prophage to another cell.

Species (1) The most specific level in the taxonomic hierarchy. **Bacterial species:** A population of cells with similar characteristics. **Viral species:** A group of viruses sharing the same genetic information and ecological niche. (2) Specific epithet.

Specific epithet The second name in a scientific binomial.

Specific resistance *See* Immunity.

Spheroplast A gram-negative bacterium treated to damage the cell wall, resulting in a spherical cell.

Spike A carbohydrate-protein complex that projects from the surface of certain enveloped viruses.

Spiral *See* Spirillum.

Spirillum (1) A helical or corkscrew-shaped bacterium. (2) When written as a genus, refers to aerobic, helical bacteria with clumps of polar flagella.

Spirochete A corkscrew-shaped bacterium with axial filaments.

Spontaneous generation The idea that life could arise spontaneously from nonliving matter.

Spontaneous mutation A mutation that occurs without a mutagen.

Sporadic disease A disease that occurs occasionally in a population.

Sporangiophore Aerial hypha supporting a sporangium.

Sporangiospore An asexual fungal spore formed within a sporangium.

Sporangium A sac containing one or more spores.

Spore A reproductive structure formed by fungi and actinomycetes.

Sporogenesis *See* Sporulation.

Sporozoite A trophozoite of *Plasmodium* found in mosquitoes, infective for humans.

Sporulation The process of spore and endospore formation.

Stability The condition of resisting change.

Staphylococci A grapelike cluster of broad sheet of spherical cells.

Stationary phase The period in a bacterial growth curve when the number of cells dividing equals the number dying.

Stem cells Fetal cells that give rise to bone marrow, blood cells, and B and T cells.

Stereoisomers Two molecules consisting of the same atoms, arranged in the same manner but differing in their relative positions; mirror images.

Sterile Free of microorganisms.

Steroids A specific group of chemical substances, including cholesterol and hormones.

Sterol A lipid alcohol found in the plasma membranes of fungi and *Mycoplasma*.

Stipe A stemlike supporting structure of multicellular algae and basidiomycetes.

Strain A group of cells derived from a single cell.

Streak plate method A method of isolating a culture by spreading microorganisms over the surface of a solid culture medium.

Streptobacilli Rods that remain attached in chains after cell division.

Streptococci Cocci that remain attached in chains after cell division.

Structural gene A gene that codes for an enzyme.

Subacute disease A disease with symptoms between acute and chronic.

Subclinical infection An infection that does not cause a noticeable illness.

Subcutaneous mycosis A fungal infection of tissue beneath the skin.

Substrate Any compound with which an enzyme reacts.

Substrate-level phosphorylation The synthesis of ATP by direct transfer of a high-energy phosphate group from an intermediate metabolic compound to ADP.

Subunit vaccine A vaccine consisting of an antigenic fragment, often produced by genetic engineering.

Suicide gene A gene that codes for a product that kills the cell.

Sulfa drugs Any synthetic chemotherapeutic agent containing sulfur and nitrogen. *See also* Sulfonamides.

Sulfhydryl group —SH.

Sulfonamides Bacteriostatic compounds that interfere with folic acid synthesis by competitive inhibition.

Superantigen An antigen that activates many different T cells, thereby eliciting a large immune response.

Superficial mycosis A fungal infection localized in surface epidermal cells and along hair shafts.

Superinfection (1) Growth of the target pathogen that has developed resistance to the antimicrobial drug being used. (2) Growth of an opportunistic pathogen.

Superoxide dismutase Enzyme that destroys superoxide free radical $(O_2 \cdot)$; $O_2^- \cdot + O_2^- \cdot + 2H^+ \longrightarrow H_2O_2 + O_2$

Superoxide free radical $O_2^- \cdot$.

Suppressor T cells (T_S) T cells that are thought to end an immune response after an antigen is no longer present.

Surface-active agent (surfactant) Any compound that decreases the tension between molecules lying on the surface of a liquid.

Susceptibility The lack of resistance to a disease.

Symbiosis The living together of two different organisms.

Symptom A change in body function that is felt by a patient as a result of a disease.

Syncytium A multinucleated giant cell resulting from some viral infections.

Syndrome A specific group of signs or symptoms accompanying a particular disease.

Synergism The principle whereby the effectiveness of two drugs used simultaneously is greater than that of either drug used alone.

Synthesis reaction A chemical reaction in which two or more atoms combine to form a new, larger molecule.

Synthetic drug A chemotherapeutic agent that is prepared from chemicals in a laboratory.

Systemic anaphylaxis Hypersensitivity reaction causing vasodilation and resulting in shock.

Systemic (generalized) infection An infection throughout the body.

Systemic mycosis A fungal infection in deep tissues.

Tachyzoite Rapidly growing trophozoite form of a protozoan.

T antigen An antigen in the nucleus of a tumor cell.

Tapeworm A flatworm belonging to the class Cestoda.

Target cell An infected body cell to which defensive cells of the immune system bind.

Taxis Response to an environmental stimulus.

Taxon A taxonomic category.

Taxonomy The science of classification.

T cell A type of lymphocyte, which develops from a stem cell processed in the thymus gland, that is responsible for cell-mediated immunity.

T-dependent antigen B cells require T_H cells to respond to this type of antigen. *See also* T-independent antigen.

Teichoic acid A polysaccharide found in gram-positive cell walls.

Temperate phage A bacteriophage existing in lysogeny with a host cell.

Temperature abuse Improper food storage at a temperature that allows bacteria to grow.

Terminator The site on DNA at which transcription ends.

Tertiary treatment Physical and chemical treatment of wastewater to remove all biochemical oxygen demand, nitrogen, and phosphorus, following secondary treatment.

Tetracyclines Broad-spectrum antibiotics that interfere with protein synthesis.

Tetrad A group of four cocci.

Tetrahedron A four-sided solid structure.

T-even bacteriophage A complex virus with double-stranded DNA that infects *E. coli*; for example, T2, T4, T6.

Thallus The entire vegetative structure or body of a fungus, lichen, or alga.

Thermal death point (TDP) The temperature required to kill all the bacteria in a liquid culture in 10 minutes at pH 7.

Thermal death time (TDT) The length of time required to kill all bacteria in a liquid culture at a given temperature.

Thermoduric Heat resistant.

Thermophile An organism whose optimum growth temperature is between 50°C and 60°C.

Thermophilic anaerobic spoilage Spoilage of canned foods due to the growth of thermophilic bacteria.

Thermostable nuclease A heat-stable enzyme produced by *Staphylococcus aureus*.

Three kingdom system Classification system in which organisms are assigned to one of three kingdoms: Eucaryota, Eubacteria, or Archaeobacteria.

Thylakoid A chlorophyll-containing membrane in a chloroplast.

Thymine A pyrimidine nucleic acid base in DNA that pairs with adenine.

Tincture An alcoholic solution.

T-independent antigen B cells do not require T_H cells to respond to this type of antigen. *See also* T-dependent antigen.

Tinea A cutaneous fungal infection; ringworm.

Ti plasmid *Agrobacterium* plasmid carrying genes for tumor induction in plants.

Titer Reciprocal of a dilution.

Total magnification Magnification of a specimen, determined by multiplying the ocular lens magnification by the objective lens magnification.

Toxemia Symptoms due to toxins in the blood.

Toxigenicity The capacity of a microorganism to produce a toxin.

Toxin Any poisonous substance produced by a microorganism.

Toxoid An inactivated toxin.

Trace element A chemical element required in small amounts for growth.

Transamination The transfer of an amino group from an amino acid to an organic acid.

Transcription The process of synthesizing RNA from a DNA template.

Transduction Transfer of DNA from one cell to another by a bacteriophage. *See also* Generalized transduction, Specialized transduction.

Transferrins Human iron-binding proteins that reduce iron available for a pathogen.

Transfer RNA (tRNA) The class of molecules that brings amino acids to the ribosomal site where they are incorporated into proteins.

Transformation (1) The process in which genes are transferred from one bacterium to another as "naked" DNA in solution. (2) The changing of a normal cell into a cancerous cell.

Transient microbiota Collection of microorganisms that are present on an animal for a short time without causing a disease.

Translation The use of mRNA as a template in the synthesis of protein.

Translocation Movement of a gene from one chromosomal locus to another.

Transmembrane channels Pores in target cells' membranes produced by complement.

Transmission electron microscope An electron microscope that provides high magnifications of thin sections of a specimen.

Transposon A small piece of DNA that can move from one region of the DNA molecule to another.

Trickling filter A method of secondary sewage treatment.

Trophophase *See* Log phase.

Trophozoite The vegetative form of a protozoan.

Tuberculin test A skin test used to detect the presence of antibodies to *Mycobacterium tuberculosis*.

Tumor Excessive tissue caused by uncontrolled cell growth.

Tumor necrosis factor A polypeptide released by phagocytes in response to bacterial endotoxins; induces shock.

Tumor-specific transplantation antigen (TSTA) A viral antigen on the surface of a transformed cell.

Turbidity The cloudiness of a suspension.

Turnover number The number of substrate molecules metabolized per enzyme molecule per second.

12D treatment A sterilization process that would result in a decrease of the number of *Clostridium botulinum* endospores by 12 logarithmic cycles.

Ubiquinones Low molecular weight, nonprotein carriers in the electron transport chain.

UDP-N-acetylglucosamine (UDPNAc) A compound necessary for the biosynthesis of peptidoglycan.

Ultra-high-temperature (UHT) treatment Treatment of food with high temperatures (140–150°C) for very short times to make the food sterile so that it can be stored at room temperature.

Ultrastructure Fine detail not seen with a compound light microscope.

Ultraviolet (UV) radiation Radiation from 10 to 400 nm.

Uncoating The separation of viral nucleic acid from its protein coat.

Undulating membrane A highly modified flagellum on some protozoa.

Uracil   A pyrimidine nucleic acid base in RNA that pairs with adenine.

Uridine diphosphoglucose (UDPG)   Precursor for synthesis of glycogen.

Use-dilution test   A method of determining effectiveness of a disinfectant using serial dilutions.

Vaccination   The process of conferring immunity by administering a vaccine.

Vaccine   A preparation of killed, inactivated, or attenuated microorganisms or toxoids to induce artificially acquired active immunity.

Vacuole   An intracellular inclusion, in eucaryotic cells, surrounded by a plasma membrane; in procaryotic cells, surrounded by a proteinaceous membrane and containing gas.

Valence   The combining capacity of an atom or a molecule.

Vancomycin   An antibiotic that inhibits cell wall synthesis.

Variolation   Early method of vaccination using infected material from a patient.

Vasodilation   Dilation or enlargement of blood vessels.

VDRL test   A rapid screening test to detect the presence of antibodies against *Treponema pallidum*.

Vector   (1) An arthropod that carries disease-causing organisms from one host to another. (2) A plasmid or virus used in genetic engineering to insert genes into a cell.

Vegetative   Refers to cells involved with obtaining nutrients, as opposed to reproduction or resting.

Vehicle transmission   Transmission of a pathogen to a large number of people by an inanimate reservoir.

Vesicle   (1) The expanded, terminal area of the Golgi complex. (2) A fluid-filled blister. (3) In a procaryote, a protein-covered hollow cylinder in a gas vacuole. (4) In a mycorrhizal fungus, a storage structure.

V factor   NAD$^+$ or NADP$^+$.

Vibrio   (1) A curved or comma-shaped bacterium. (2) When written as a genus, refers to gram-negative, motile, facultatively anaerobic curved rods.

Viral hemagglutination   The ability of certain viruses to cause agglutination of red blood cells in vitro.

Viral hemagglutination inhibition   Antibodies against particular viruses prevent the viruses from agglutinating red blood cells in vitro.

Viremia   The presence of viruses in the blood.

Virion   A fully developed complete viral particle.

Viroid   An infectious piece of "naked" RNA.

Virulence   The degree of pathogenicity of a microorganism.

Virus   A submicroscopic, parasitic, filterable agent consisting of a nucleic acid surrounded by a protein coat.

Visible light   Radiation from 400 to 700 nm, which the human eye can see.

Volutin   Stored phosphate in a procaryotic cell.

Wandering macrophage   A macrophage that leaves the blood and migrates to infected tissue.

Wassermann test   A complement fixation test used to diagnose syphilis.

Western blot   A technique that uses antibodies to detect the presence of specific proteins separated by electrophoresis.

Wheal   A small, burning or itching swelling on the skin, such as that resulting from a skin test or a mosquito bite.

Whey   The fluid portion of milk that separates from curd.

Xenodiagnosis   A method of diagnosis based on exposing a parasite-free normal host to the parasite and then examining the host for parasites.

Xenograft   A tissue graft from another species.

X factor   A precursor necessary to synthesize cytochromes.

Yeast   A unicellular fungus belonging to the phylum Ascomycetes.

Zone of inhibition   The area of no bacterial growth around an antimicrobial agent in the disk-diffusion test.

Zoonosis   A disease that occurs primarily in wild and domestic animals but can be transmitted to humans.

Zygospore   A sexual fungal spore characteristic of the zygomycetes.

Zygote   A diploid cell produced by the fusion of two haploid gametes.

Credits

Chapter 5

Opener: © Institut Pasteur/CNRI/Phototake. 5.3b: Science Visuals Unlimited-IBMUKOU. Box photo: © Helen E. Carr/Biological Photo Service.

Chapter 6

Opener: © Alex Rakosy/Custom Medical Stock Photo. Box photo: © D. Foster/WHOI/Visuals Unlimited. 6.4: Forma Scientific, Marietta, Ohio. 6.6, 6.7a, b, d: Christine Case. 6.7c: © Yoav Levy/Phototake. 6.8b: © Janice Sheldon/Photo 20-20. 6.9b: © Lee D. Simon/Photo Researchers, Inc. 6.15a: H.W. Jannasch, Woods Hole Oceanographic Institute, Woods Hole, MA. 6.15b: Courtesy of Millipore Corporation.

Chapter 7

Opener: © David Scharf/Peter Arnold, Inc. 7.5: Christine Case. 7.8: CDC.

Chapter 8

Opener: © Dr. Gopal Murti/Photo Researchers, Inc. 8.1a: © Dr. Gopal Murti/Science Photo Library/Photo Researchers, Inc. 8.1b: Photo by H. Ris, University of Wisconsin, Madison. 8.2: E. Marieb, *Human Anatomy & Physiology,* 2nd Ed. (Redwood City, CA: Benjamin/Cummings, 1992), Figure 3.24, p. 89. © 1992 Benjamin/Cummings Publishing Company. 8.5a: J. Cairns, Imperial Cancer Research Fund Laboratory, Mill Hill, London. 8.5b, 8.16: Adapted from N. Campbell, *Biology,* 3rd Ed. (Redwood City, CA: Benjamin/Cummings, 1993), Figures 17.10, 16.22, pp. 356, 337. © 1993 Benjamin/Cummings Publishing Company. 8.7: P. Berg and M. Singer, *Dealing with Genes: The Language of Heredity* (Mill Valley, CA: University Science Books, 1992), Figure 3.1, p. 61. © 1992 University Science Books. 8.11: O.L. Miller Jr., B.A. Hamkalo, and C.A. Thomas, Jr., *Science* 169 (24 July 1970):392. Copyright 1970 by the American Association for the Advancement of Science. 8.25: © Dr. Dennis Kunkel/Phototake. 8.28a: R. Welch, University of Wisconsin Medical School.

Chapter 9

Opener: © SPL/Photo Researchers, Inc. 9.1, 9.9: Adapted from N. Campbell, *Biology,* 3rd Ed. (Redwood City, CA: Benjamin/Cummings, 1993), Figures 19.2, 19.7, pp. 392, 398. © 1993 Benjamin/Cummings Publishing Company. 9.3b: © Dr. Jeremy Burgess/SPL/Photo Researchers, Inc. 9.4: Reprinted with permission from *American Laboratory,* 8(5), 1990, p. 36. 9.7: © Matt Meadows/Peter Arnold, Inc. 9.10: © Secchi-Lecaque/Roussel-UCLAF/SPL/Photo Researchers, Inc. 9.11: Adapted from J.D. Watson, M. Gilman, J. Witkowski, M. Zoller, *Recombinant DNA,* 2nd Ed. (NY: W.H. Freeman, 1992), Figure 7.23, p. 129. © 1992 by James D. Watson, Richard Gilman, Jan Witkowski, and Mark Zoller. Reproduced by permission of W.H. Freeman and Company. Box photo: Kelco, A Division of Merck and Company. 9.12: Courtesy of Cellmark Diagnostics. 9.14: USDA.

Part Two

Opener: © Charles O'Rear/Westlight.

Chapter 10

Opener: © A.B. Dowsett/SPL/Photo Researchers, Inc. T10.2 (left): © Ralph Robinson/VU. T10.2 (center): © A.B. Dowsett/Science Source/Photo Researchers, Inc. T10.2 (right): © Eugene McArdle/Custom Medical Stock Photo. 10.3: J. Pickett/Photo Researchers, Inc. 10.5a: Courtesy of J.W. Schopf, University of California, Los Angeles. 10.5b: S.M. Awramik, University of California/Biological Photo Service. Box photo: Courtesy of Marine World/Africa USA, Vallejo, CA. 10.9, 10.16: Christine Case. 10.10a, b, 10.11: CDC. 10.12: Courtesy of the Microbial Diseases Laboratory, Berkeley, CA. 10.14: Biavatti et

al., *Int. J. Systematic Bacteriology,* 32:358–373, 1982. 10.15b, c: Courtesy of Steffen Kjelleberg. From Jouper-Jaan et al., *Applied and Environmental Microbiology,* 1986, 53:1419–1421.

Chapter 11

Opener, 11.22: © Charles O'Rear/Westlight. Intro text: *JAMA,* Nov. 18, 1988, Vol. 260, No. 19. 11.1: © Dr. David Chase/CNRI/Phototake. 11.2a: © Stanley Flegler/Visuals Unlimited. 11.2b: A.B. Dowsett/SPL/Photo Researchers, Inc. 11.3: Krieg, N.R., and Holt, J.G. (eds.), *Bergey's Manual of Systematic Bacteriology,* Vol. 1, The Williams and Williams Co., Baltimore. 11.4: Charles C. Brinton Jr. & Sam Chuen-Mo To, University of Pittsburgh. 11.5: © Institut Pasteur/CNRI/Phototake. 11.6: © SPL/Custom Medical Stock Photo. 11.7a: CDC. 11.7b: Sidney M. Finegold. 11.8a, b: Bergey's Manual Trust, Michigan State University. 11.9b: Courtesy of R.C. Cutlip, National Animal. 11.10: © Michael Gabridge/Visuals Unlimited. 11.11: Dr. Tony Brain/SPL/Custom Medical Stock Photo. 11.12, 11.15: © David M. Phillips/Visuals Unlimited. 11.13a: R.E. Strange and J.R. Hunter in G.W. Gould and A. Hurst, eds., *The Bacterial Spore* (New York: Academic Press, 1969), p. 461. 11.13b: C.L. Hannay and P. Fitz-James, *Canadian J. Microbiology,* 1:694, 1955. 11.13c: © T.J. Beveridge/University of Guelph/Biological Photo Service. 11.14: © Science VU/Visuals Unlimited. 11.16a: J. S. Poindexter, *Bacteriology Review,* 28:231–295, 1964. Courtesy of J.S. Poindexter. 11.16b: H. Rechenbach and M. Dworkin, 1981, "Introduction to the Gliding Bacteria", in *The Prokaryotes,* eds. M. Stars, H. Stolp. H.G. Trupu, A. Balows, and H.G. Schlegel, Springer-Verlag, pp. 315–327. 11.16c: K. Stephens, University of Washington/Biological Photo Service. 11.17a: © Biological Photo Service. 11.17b: © Michael Richard/Visuals Unlimited. 11.18: Courtesy of Ralph J. Pledger & John A. Baross. 11.19: © Paul Johnson/Biological Photo Service. 11.20a: © T.E. Adams/Visuals Unlimited. 11.20b: © Sherman Thomson/Visuals Unlimited. 11.20c: Susan M. Barns, Indiana University. 11.21b: © Dr. J. Burgess/Photo Researchers, Inc.

Chapter 12

Opener, 12.2b, 12.3, 12.15: © David Scharf/Peter Arnold, Inc. 12.2a, 12.4, 12.9a–c, 12.10c, 12.14a–c, 12.16 (both): Christine Case. 12.5a, b, e: Dr. G. Cole, University of Texas. 12.5c: © D. Kunkel/Fran Heyl & Associates. 12.5d: © Biological Photo Service. 12.6 (both): © Ed Reschke/Peter Arnold, Inc. 12.7, 12.18b: © Manfred Kage/Peter Arnold, Inc. 12.8: Biophoto Associates/Photo Researchers, Inc. 12.10b: © David Wright. 12.11a: © E.R. Degginger/Bruce Coleman. 12.12a: C.M. Pringle, University Of California, Berkeley/Biological Photo Service. 12.17: © CABISCO/Phototake. 12.19b: © David M. Phillips/Visuals Unlimited. 12.20b: © E.R. Degginger/Photo Researchers, Inc. 12.22b: © A.M. Siegelman/Visuals Unlimited. 12.24: © Stanley Flegler/Visuals Unlimited. 12.25a, b, 12.26a: © R. Calentine/Visuals Unlimited. 12.27: © Michael McClure/Phototake. 12.28a: © Hans Pfletschinger/Peter Arnold, Inc. 12.29a: CDC. 12.29b: © Jim Hendel, UC Berkeley.

Chapter 13

Opener: © CNRI/Phototake. 13.3b: © Oscar Bradfute/Peter Arnold, Inc. 13.4b: R.C. Valentine and H.G. Pereira, *J. Mol. Biol.* 13.5b, 13.6b, 13.18a–d: © Chris Bjornberg/Photo Researchers, Inc. 13.7a: Robley Williams. 13.7b: © Hans Gelderblom/Visuals Unlimited. 13.8a: © CDC/Science Source/Photo Researchers, Inc. 13.8b: C. Garon and J. Rose, National Institute of Allergy and Infectious Diseases, CDC. 13.9a: © Phototake/CNRI. 13.9b: J.J. Cardamone, Jr., University of Pittsburgh/Biological Photo Service. 13.9c: G. Smith, National Cancer Institute. T13.2: Art from R.I.B. Francki, C.M. Fauquet, D.L. Knudson, F. Brown, eds., "Classification and Nomenclature of Viruses. Fifth Report of the International Committee on Taxonomy of Viruses." *Archives of Virology: Supplementum 2* (Wien, New York: Springer-Verlag, 1991). Copyright © 1991 by Springer-Verlag. 13.10: Christine Case. 13.13a, b: © Gail W.T. Wertz, University of Alabama Medical

School/Biological Photo Service. 13.22b: D.R. Brown et al., *J. Virology*, 10:524–536, 1972. 13.23: T.O. Diener, USDA.

Part Three

Opener: © Peter Menzel.

Chapter 14

Opener: © A.B. Dowsett/SPL/Photo Researchers, Inc. 14.1a–c: © David M. Phillips/Visuals Unlimited. 14.5a: © Michael P. Gadomski 1988/Photo Researchers, Inc. 14.5b: © Erika Stone/Peter Arnold, Inc. 14.5c: © Tony Freeman/Photo Edit. 14.6a: © Bob Daemmrich/ Stock Boston. 14.6b: © Bonnie M. Freer/Peter Arnold, Inc. 14.6c: © Scott Camazine 1990/Photo Researchers, Inc. 14.7: © Mitch Kezar/ Phototake. 14.8: © Kent Wood/Photo Researchers, Inc.

Chapter 15

Opener: © Dennis Kunkel/Phototake. 15.1a: L.R. Inman/J.R. Cantey, *J. Clin. Investigation* 71:1–8, 1973. 15.1b: © Steve Murphree/Biological Photo Service. 15.5a, b: F.A. Murphy, Viral Pathology Branch, CDC. 15.6: © John P. Bader/Biological Photo Service.

Chapter 16

Opener: © Jan Hinsch/SPL/Photo Researchers, Inc. 16.1: Adapted from N. Campbell, *Biology*, 3rd Ed. (Redwood City, CA: Benjamin/ Cummings, 1993), Figure 39.2, p. 851. © 1993 Benjamin/Cummings Publishing Company. 16.2: © Ed Reschke/Peter Arnold Inc. 16.4: CNRI/SPL/Photo Researchers, Inc. 16.6, 16.8: G.J. Tortora, S.R. Grabowski, *Principles of Anatomy and Physiology,* 7th Ed. (New York: HarperCollins, 1993), Figure 22.1, p. 684. © 1993 HarperCollins College Publishers. Reprinted by permission of the publisher. 16.7: © Lennart Nilsson, Boehringer Ingelheim International, GmbH. 16.11b: © Lennart Nilsson/*The Body Victorious*, Delacorte Publ.

Chapter 17

Opener: © Dr. Brian Eyden/Science Photo Library/Photo Researchers, Inc. 17.1, 17.11: E. Marieb, *Human Anatomy & Physiology*, 2nd Ed. (Redwood City, CA: Benjamin/Cummings, 1992), Figures 22.11, 22.13, pp. 703, 706. © 1992 Benjamin/Cummings Publishing Company. 17.3 (left): © David M. Phillips/The Population Council/Photo Researchers, Inc. 17.3 (right): © Bloom & Fawcett/Photo Researchers, Inc. 17.6d: © A.J. Olson, Scripps Clinic and Research Foundation, 1986. 17.7, 17.9, 17.14, 17.15: Adapted from N. Campbell, *Biology*, 3rd Ed. (Redwood City, CA: Benjamin/Cummings, 1993), Figures 39.7, 39.13, 39.15a, 39.14, pp. 856, 862, 866, 865. © 1993 Benjamin/ Cummings Publishing Company. 17.13: A.K. Szakal et al., *Ann. Rev. Immunol.*, 1989, 7:91–109.

Chapter 18

Opener: © James Callagan/CNRI/Phototake. 18.1: © Science VU-NLM/Visuals Unlimited. 18.7: © Biological Photo Service. 18.13: CDC. 18.16: Courtesy of Parke-Davis.

Chapter 19

Opener: © Nancy Kedersha/Immunogen/SPL/Photo Researchers, Inc. 19.1b, 19.3, 19.15a, b: © Lennart Nilsson, Boehringer Ingelheim International GmbH. 19.2: © David Scharf. T19.2: E. Marieb, *Human Anatomy & Physiology*, 2nd Ed. (Redwood City, CA: Benjamin/ Cummings, 1992), T18.4, p. 599. © 1992 Benjamin/Cummings Publishing Company. 19.8: Harvard Medical School. 19.11: Hoth, Jr., Meyers, Stein, "Current Status of HIV Therapy," *Hospital Practice*, Vol. 27, No. 9, p. 154. Illustrations © Alan D. Iselin. Reprinted with permission. 19.12, 19.13: Greenberg, "Immunopathogenesis of HIV Infection," *Hospital Practice*, Vol. 27, No. 9, p. 154. Illustrations © Alan D. Iselin. Reprinted with permission.

Chapter 20

Opener: CNRI/Science Photo Library/Photo Researchers, Inc. 20.1: © The Bettmann Archive. 20.3a, b: D. Greenwood and F. O'Grady, *Science*, 163:1076, 1969. Copyright 1969 by the AAAS. 20.5: Madeline Bastide, Laboratorie Di Immunologie er Parasitologie, CNRI, Paris. 20.20: © Visuals Unlimited. 20.21: Courtesy of Donald Nash, from *J. Clinical Microbiology*, Dec. 1986, p. 977. Reprinted with permission of the American Society for Microbiology. 20.22: T.D. Brock and M.T. Madigan, *Biology of Microorganisms*, 6th Ed. (Englewood Cliffs, NJ: Prentice-Hall, 1991), p. 410. © 1991 Prentice-Hall. Adapted by permission of Prentice-Hall.

Part Four

Opener: © Charles O'Rear/Westlight.

Chapter 21

Opener: © E. Gueho/SPL/Custom Medical Stock Photo. 21.3: Lilly Research Center, Eli Lilly and Company. 21.4: © Charles Stoer, M.D. 1988/Camera M.D. Studios. 21.5a, b: P.P. Cleary, University of Minnesota School of Medicine/Biological Photo Service. 21.6: © Carroll H. Weiss, RBP, 1983/Camera M.D. Studios. 21.7: © Ken Greer/Visuals Unlimited. 21.8: © National Medical Slide Library/CMSP. 21.9a, 21.14a, 21.15a: CDC. 21.9b, 21.14b: Camera M.D. Studios. 21.10: © Carroll H. Weiss, RBP, 1991/Camera M.D. Studios. 21.12: © Lowell Georgia/ Science Source/Photo Researchers, Inc. 21.13: © Carroll H. Weiss, RBP, 1985/Camera M.D. Studios. 21.15b: © Science Photo Library/Photo Researchers, Inc. 21.16: Walter B. Shelley, M.D. 21.17: © AFIP/Visuals Unlimited. 21.18: David John, Oklahoma State University.

Chapter 22

Opener: © Alfred Pasieka/SPL/Photo Researchers, Inc. 22.4: Daniel Musher; from *J. Infectious Diseases*, 149:4, 4/84. 22.5: L.T. Tilney, P.S. Connelly, & D.H. Portnoy. 22.6: © The Bettmann Archive. 22.7: FDA. 22.8a: © Kenneth E. Greer/Visuals Unlimited. 22.8b: CDC. 22.9: Courtesy of Warren E. Collins. 22.11: A.K. Harrison/CDC. 22.14: Courtesy of Dr. Edward Bottone. 22.16: Courtesy of the University of Granada.

Chapter 23

Opener, 23.21: © James Dennis/CNRI/Phototake. 23.1: Adapted from N. Campbell, *Biology,* 3rd Ed. (Redwood City, CA: Benjamin/ Cummings, 1993), Figure 38.5a, p. 823. © 1993 Benjamin/Cummings Publishing Company. 23.3: © Ken Greer/Visuals Unlimited. 23.4: © Camera M.D. Studios. 23.5: Courtesy of Dr. Victor Marks. 23.7: Marilyn J. Tufte. 23.8: CDC/Biological Photo Service. 23.9: © SIU/ Visuals Unlimited. 23.10: © Giselle Hosgood. 23.12: Science VU/ Visuals Unlimited. 23.13a: © Scott Camazine/Photo Researchers, Inc. 23.14a: Steere, "Current Understanding of Lyme Disease," *Hospital Practice*, Vol. 28, No. 4, Figure 1, p. 39. Illustration © Hospital Practice. Reprinted by permission. 23.14b: © Bernard Furnival/Fran Heyl & Assoc. 23.15, 23.18, 23.24: CDC. 23.19: Science VU/CDC/Visuals Unlimited. 23.20 (left), 23.26: H. Zaiman, M.D. 23.20 (right): Courtesy AFIP. 23.25a: © NIH/Science Source/Photo Researchers, Inc.

Chapter 24

Opener: © Dennis Kunkel/Phototake. 24.2: Adapted from N. Campbell, *Biology*, 3rd Ed. (Redwood City, CA: Benjamin/Cummings, 1993), Figure 38.22, p. 840. © 1993 Benjamin/Cummings Publishing Company. 24.3: CDC. 24.4: © G.W. Willis M.D./Biological Photo Service. 24.5: Michael Kientz/*Discover* Magazine. 24.6: © Biological Photo Service. 24.7: Frederick S. Nolte. 24.8: Dannenberg, Jr., "Pulmonary Tuberculosis," *Hospital Practice*, Vol. 28, No. 1. Figure 2, p. 53 © Seward Hung; Figure 5, p. 57 © Laura Pardi Duprey. Reprinted by permission. 24.10: © Raymond B. Otero/Visuals Unlimited. 24.11a: © D. Paretsky, University of Kansas/Biological Photo Service. 24.11b:

Thomas F. McCaul and Jim C. Williams, "Developmental Cycle of *Coxiella Burnetti*: Structure and Morphogenesis of Vegetative and Sporogenic Differences," *J. Bacteriology,* Sept. 1981, p. 1068. 24.14a: © Science VU/Visuals Unlimited. 24.14b, 24.16b: © G.T. Cole/ University of Texas/Biological Photo Service. 24.16a: David Drutz and Milton Huppert, *J. Infectious Diseases,* 147:379. University of Chicago Press. Reprinted by permission. 24.17: Reprinted from the *American Family Physician,* Vol. 41, No. 5, May 1990, "Coccidioidomycosis: Office Diagnosis and Treatment," Hedges & Miller, Figure 1, p. 1499 (Kansas City, MO: American Academy of Family Physicians, 1990). Reprinted with permission. 24.18: B.J. Bogitsh and T.C. Cheng, *Human Parasitology* (Orlando, FL: Saunders College Publishing, 1990). © 1990 Saunders College Publishing. By permission of the publisher.

Chapter 25

Opener: © London School of Hygiene and Tropical Medicine/ SPL/CMSP. 25.3a, b: Hutton D. Slade, *Microbiol. Review,* 12 (2):331–384, June 1980. 25.7: From Schaechter et al., *Mechanisms of Microbial Disease,* 2nd Ed. (Baltimore, MD: Williams & Wilkins, 1993), Figure 18.1, p. 269. © 1993 Williams & Wilkins. 25.11a: G.T. Cole, University of Texas at Austin/Biological Photo Service. 25.11b, c: Reprinted from *Food Techonology,* 36(3):93–96, 1982. 25.12 (photo): © Veronika Burmeister/Visuals Unlimited. 25.12 (art): From "Helicobacterpylori Infection," *Hospital Practice,* Vol. 26, No.2, Figure 1, p. 47. © 1991 HP Publishing Company. Reprinted with permission. 25.13: CDC. 25.15: © Science VU/Visuals Unlimited. 25.16: Reprinted by permission of the publisher. From "Ultrastructural Observations on Giardiasis in a Murine Model," by Robert Owen, Paulina Nemanic. and David P. Stevens, *Gastroenterology,* 76:757–769. Copyright 1979 by the American Gastroenterology Association. 25.17: Air Force Institute of Pathology. 25.18: Marietta Voge, UCLA Medical Center. 25.20a, b: H. Zaiman, M.D. 25.21: Courtesy of AFIP. 25.22: © R. CalentineVisuals Unlimited. 25.23, 25.24b: © A.M. Siegelman/Visuals Unlimited. 25.24a: © L. Jensen/Visuals Unlimited.

Chapter 26

Opener, 26.8: © George J. Wilder/Visuals Unlimited. 26.4: D. Bromley. 26.7: © Science VU/Visuals Unlimited. 26.9: Courtesy of Dr.

David Soper/Medical College of VA. 26.11: © M. Abbey/Photo Researchers, Inc. 26.12a, b: © Carroll H. Weiss 1981/Camera M.D. Studios. 26.12c, 26.13: CDC. 26.14: © Mike Remmington, University of Washington Viral Disease Clinic. 26.15: © Kenneth Greer/Visuals Unlimited.

Part Five

Opener: © Charles O'Rear/Westlight.

Chapter 27

Opener: © CNRI/Science Photo Library/Photo Researchers, Inc. 27.1: Adapted from N. Campbell, L. Mitchell, J. Reece, *Biology: Concepts and Connections* (Redwood City, CA: Benjamin/Cummings, 1994), Figure 36.14, p. 710. © 1994 Benjamin/Cummings Publishing Company. 27.3: Hugo Spencer/Photo Researchers, Inc. 27.4: G.A. Peters, Virginia Commonwealth University. 27.5a: © M.F. Brown, University of Missouri/Biological Photo Service. 27.5b: © R.L. Peterson, University of Guelph/Biological Photo Service. 27.8: © Ken Lucas/Biological Photo Service. 27.9 (both): Courtesy of Exxon Company, USA. 27.11: Courtesy of Idexx Laboratories, Inc. 27.13: © N.R. Rowan/Custom Medical Stock Photo. 27.15b: © V. Paulson 1984/Biological Photo Service. 27.15c: R.F. Unz, *Int. J. Systematic Bacteriology,* 21:91–99. 1971. 27.16a: © Douglas Munnecke/Biological Photo Service. 27.17a: Christine Case. 27.19 (all): Steve Woit. p. 690: Randall Von Wedel, Cytoculture International.

Chapter 28

Opener: © A.B. Dowsett/Science Photo Library/Photo Researchers, Inc. 28.1: Courtesy of FMC Corporation. 28.4: Courtesy of International Paper. 28.8b: © Hank Morgan/Rainbow. 28.9 (all): © David R. Frazier/Photo Researchers, Inc. 28.11b: LSL Biolafitte, Inc. 28.13: © Manfred Kage/Peter Arnold, Inc. Box photo, 28.15b: Christine Case. 28.16: © Denis Finley/National Geographic.

Index

C